Endosymbiosis of Animals
with Plant Microorganisms

Endosymbiosis of Animals with Plant Microorganisms

Revised English Version

PAUL BUCHNER

Professor Emeritus, University of Munich

INTERSCIENCE PUBLISHERS

A DIVISION OF JOHN WILEY & SONS, INC.

NEW YORK · LONDON · SYDNEY

Preface to the English Edition

Ever since the first barrier-breaking researches of Pierantoni and Sulc in 1909 and 1910, the study of endosymbiosis occurring in animals and microorganic flora vitally essential to each other has held me enthralled. Not only have I dedicated the major part of my scientific labors to it, but also it has permitted me to enjoy productive collaboration with a large number of my students, some of whom subsequently followed in my footsteps and, in turn, inspired the efforts of students of their own. As a consequence, I have seen with my own eyes the great majority of the adaptations described in this book—some often bordering on the fantastic.

Recognizing the rapid development of research studies, I collated summaries in 1921, 1930, and 1953 and moved in 1939 and 1960 to make them available to a wider circle of readers. In retrospect, as I advance in years, I am especially pleased to have lived to see the present expanded version, in English, of the volume which first appeared in 1953. In the years elapsed since then, our knowledge of endosymbioses has broadened and deepened, and I venture to hope not only that this new edition will become more generally known, but also that it will attract new collaborators. The extent and morphological bases of endosymbioses, it is true, have become generally understood during the last fifty years, but the coordinated development of these branches of biology require, above all, more intensive teamwork between biochemistry and microbiology.

I thank all who have made this edition of my book possible. Professor Bertha Mueller, formerly of the University of Hawaii, undertook the onerous task of translation. Dr. Walter Carter of the Pineapple Research Institute of Hawaii kindly collaborated in checking it, and a portion of the expense of translation was borne by the National Institutes of Health, U.S. Department of Health, Education and Welfare. Inasmuch as Professor Mueller's labors extended over some years, Dr. Francis H. Foeckler, presently at the Biological Institute of the University of Regensburg, translated many insertions which became necessary. I must thank

v

him also for his help in correcting the proofs.   Last, but by no means least, I owe thanks to the publisher whose labors have resulted in the book in its present form.

PAUL BUCHNER

Porto d'Ischia, Naples
July, 1965

Contents

ENDOSYMBIOSIS; HISTORY OF ITS DISCOVERY; ITS DISTRIBUTION

mission by smearing of the eggs, 162. The numerous ways of localization in larvae and imagines in other snout beetles, 163. Transmission by egg infection, 172. Behavior during embryonic development, 173. Symbiosis in the apionines, 178.

GENERAL SECTION

The gut lumen as the oldest location of symbionts, 619. Acquisition of symbionts without participation of the gut, 619. Colonization of the gut without morphological accommodations, 619. Retention arrangements in the gut, 620. Confinement of the symbionts to gut evaginations, 622. Colonization of the gut epithelial cells, 622. Extra- and intracellular colonization of the Malpighian vessels, 624.

Sites between alimentary tract and hypodermis, 625. Rare transition stages, 626. Colonization of the fat tissue and derivatives of the same (pseudomycetomes), 627. Sites which are derived from indifferent embryonic cell material (genuine mycetomes), 629. Isolated mycetocytes and groups of mycetocytes in the region of the fat tissue, 629. Mycetomes with one kind of symbiont, 631. Syncytia and synsyncytia, 632. Mycetomes with several symbionts, 634. Additional symbionts in the epithelium of the mycetome, 637. Pseudomycetes derived from the gut or hypodermis in the mesodermal region, 638. Luminous organs of the tunicates derived from blood cells, 638.   Luminous organs arising as invaginations of the integument, 639.

The acquisition of symbionts is left each time to chance, 640.   Transmission by means of infected cocoon fluid, 641.   Soiling of the eggshell with symbionts from the anus, 641; without morphological accommodation (*Hylemyia* type), 641.   Bacterial reservoirs in the hindgut in trypetids, 642. Smearing arrangements of the Heteroptera, 642; acanthosomids, 643; *Coptosoma*, 643; *Brachypelta*, 643.   Intersegmental bacterial syringes and vaginal pockets in anobiids and cerambycids, 644.   Corresponding arrangements in the lagriids, 644.   Vaginal tubules in *Bromius* and *Cassida*, 645. Milk glands in the glossines, hippoboscids, and nycteribiids in the service of transmission, 645.

Transmission by means of egg infection, 646.   Entrance into the ovocyte from all sides, 647.   Bipolar egg infection, 647.   Infection only at the posterior pole of the egg between the follicle cells, 648; with temporary colonization of the follicle cells, 649.   Intact mycetocytes carry the symbionts into the egg cells, 649.   Ovarial ampulla in the service of egg infection, 651.   Egg infection at the anterior pole, 652.   Nutrient cords as a means of transport, 652.   Several kinds of symbionts infect together at the posterior or anterior pole, 653.   One portion of the same infect at the posterior pole, the other at the anterior, 654.   Depot mycetomes in the oviducts of *Fulgora* and *Bladina*, 654.   Transmission of symbionts in arachnoids, 655.   Spores serve transmission in Pyrosoma, 655.

Effective forces in the transport of symbionts, 656.   Shifting of mycetomes and mycetocytes, 656.   Active mobility or transport by the bloodstream? 656.   Participation of the host? 657.   Locally confined capability of the follicle for reception, 657.   The act of engulfment by the ovocyte, 658. Unknown directional forces? 658.

Early sealing off of the symbionts, 659.   Provision of the symbionts with cleavage, yolk, and blastoderm nuclei, 660.   Details of the process, 661. Provision of the symbionts with derivatives of the polar bodies, 662. Transitory mycetomes, 663.   They decompose into their building elements, 664.   Their cells die off, 664.   Primary mycetomes are retained, 665; are replaced by secondary mycetomes, 665.   Shifting of the symbionts in Anoplura, 666; in aleurodids, 666.   Enlargement of the *a*-organs through infection of the mycetome epithelium, 667.   Cell replacement by epithelial cells in *x*-organs, 667.

Relationships between egg infection and later embryonic processes, 668. Meaningful symbiont concentration before the beginning of development, 668.   Plasma streaming as a means of concentration, 669.   Movements of the symbionts, which have not yet been provided with nuclei, after development has begun, 669.   Differentiations of the egg plasma and the embryonal cells in the service of concentration and sealing off, 670. Plasma radiation at the top of the germ band, 671.   Regulated movements of still sterile or already infected elements, 672. Infection at the posterior pole and the relationships of the symbionts to the germ band yield true mycetomes, 673. Infection at the anterior pole, 673. Colonization of the lymph or scattered mycetocytes, 673. Relationships between transport of the symbionts through the germ band and final location of the mycetomes, 674.

The process of separation with several kinds of symbionts, 675; in the case of egg infection, 675; in the infection of aphid embryos, 677. Relationships to the age of the acquisition, 678. Postembryonic processes in the case of egg infection, 679; of infection of young larvae, 680. Early rudiments of symbiont habitations, 680. Measures which assure the preservation of the symbiosis in holometabolism, 680.

Control of the reproductive rate, 684. Different tempos of reproduction, 685. Sharply defined reproductive periods during embryonic development, 685. Symbiont reproduction during postembryonic development, 686. Bursts of reproduction in connection with egg infection, 688; in infection mounds, 690.

Control of the symbiont form, 692. Plasticity of the microorganisms, 692. Degeneration of original bacterial forms, 693. Repression of spore formation, 695. Cyclic changes in form, 695. Differences in form of the symbionts in both sexes, 697. Cultivation of specific transmission forms, 698.

Factors regulating reproduction, 703. Role of the intensity of the host's metabolism, 703. Age and death remove the controls, 704. Role of disposable space, 705; of the specific chemistry of the sexes, 706. Mycetome epithelium as site of cultivation of infecting stages, 707. Molding influences of active ingredients, 708. Significance of transmission forms, 709.

Control of the extension of the symbionts, 711. Temporary removal of local immunity, 712. Comparison of old established symbionts with young ones and companion forms, 713. Gradual perfection of transmission, 715. Regulation of mutual relationships in plurisymbioses, 715. Non-harmonious occurrences in the same, 716. Artificial transplantation of symbionts, 716. Regulated elimination of symbionts, 719; through the anus, 719; by means of bacteriolysins, 720; by phagocytic removal of mycetomes, 721; by suppression of egg infection, 721.

Constructive accomplishments of the hosts, 722. Comparison with the reactions to parasitic attack, 722. Cytological reactions of inhabited host cells, 723. Behavior of the nuclei, 726. Perfection rows, 728. Convergent formations, 728. Glands in the service of symbiosis, 729. New structures established in advance, 730. Early measures taken in egg infection, 731. Instincts released by symbiosis, 731.

Relationships of the symbiosis to the system of the hosts as a measure of its age, 732. The old age of blattid symbiosis, 732. Dissolution of intracellular symbiosis in termites, 733. Symbiont loss in ants and *Calandra*

16. The Significance of Endosymbiosis 764

Blattid symbionts as suppliers of vitamins and amino acids? 808. Symbionts producing cellulase, 815. Manifold accomplishments of one and the same symbionts, 816. Relationships of the symbiosis to the dietary mode of the host, 817.

Tasks of experimental symbiosis research, 821. Significance of symbiont accumulation, 821. Extent of the principle of symbiosis in the animal kingdom, 824. Role of intestinal bacteria in vertebrates and man, 824. The varied significance of symbiosis research for zoology and microbiology, and for the investigation of immunity and medical questions, 827.

Endosymbiosis:
History of its discovery;
its distribution

1

Algal symbiosis

By the term endosymbiosis we mean well-regulated and essentially undisturbed cooperative living between two differently constituted partners. It is usually a far more highly organized partner which shelters another within its body, and the mutual adaptation is so complete as to justify the assumption that the arrangement is useful to the host. Endosymbiosis may be established firstly between two plant organisms, as with lichen, mycorrhiza, the little tubers of legumes and alders, and the Rubiaceae; secondly, and far more rarely, between two animal organisms, as in the symbiosis of termites and flagellates or of ruminants and the ciliates in their rumen; and, thirdly, between animal hosts and plant guests.

In this book we shall discuss only the last-mentioned type. It includes the symbiotic relationship of animal hosts to algae and Cyanophyceae, a subject about which we have long been well informed and which is virtually a closed field today; it also includes the extraordinarily widespread and often fantastically complex bacterial and fungal symbiosis which has been set forth only within recent decades. It is this much more important branch of symbiosis research, together with numerous related problems of a general nature, which is the central concern of our observations. Algal symbiosis is included in this introductory section because of its historical interest, but later it is cited only occasionally by way of comparison.

It is not surprising that cases of algal symbiosis came to attention much earlier than did those of fungi and bacteria, for the unusual green or yellow of the algae is easily observable in the host animal, which as a rule is more or less transparent.

Nevertheless, many false trails were taken. From the very beginning the green of plants came to mind at the sight of the same green color in innumerable shelled and naked amoebae, infusorians, hydras, and turbellarians, and in the green branches and crusts of fresh-water sponges, but actual proof that genuine chlorophyll was involved was not presented until the mid-nineteenth century. Wöhler (1843) had made mention of algae-like bodies in infusorians, and Siebold (1849) had been among the first to assert that the color of *Chlorohydra viridissima* Pall. and other green animals should be attributed, if not to chlorophyll itself, at least to some

closely related substance.   Max Schultze (1851), on the basis of comparative tests with a series of chemicals, proved that the animal green color of
Stentor, *Chlorohydra*, and *Dalyellia* was indeed chlorophyll.   Somewhat later
in the same year Cohn, a botanist, arrived at a similar conclusion, and his
view was adopted by such authors as Stein (1854) and Claparède and
Lachmann (1858).

Our information on "animal chlorophyll" was broadened through spectroscopic investigations by Lankester (1868), Cohn and Schröder (1872),
and Sorby (1873, 1875), and through Geddes' demonstration (1878) that
the green marine turbellarian *Convoluta*, like plants, expires oxygen.   When
Engelmann investigated a green *Vorticella* in 1883, there was little that he
could do beyond confirming the conclusions of his predecessors.

All this information represented substantial progress but by no means
provided final clarification.   It was still necessary to decide whether the
chlorophyll represented an animal-specific product and thus whether the
distinction between plant and animal realms was being obliterated; or
whether the chlorophyll represented minute organisms which, like parasites, grow proliferously in the body of animals, usually in the cells.   A
number of authors like Geddes, Lankester, Kleinenberg, and McMunn
were convinced that chlorophyll was endogenous in its nature.   McMunn
added to the confusion in his belief that an "enterochlorophyll" could be
established in all kinds of animals.   He referred to it as animal chlorophyll,
although it differs markedly from the genuine.   Indeed, the authors of
this period were all too prone to suspect chlorophyll as soon as green or
greenish brown colors made their appearance, as in *Bonellia* and many
actinians.

On the other hand, arguments could be raised against interpretations
such as those of Geddes or Lankester.   Most of the green fresh-water animals mentioned occur frequently in the colorless state, but they by no
means turn green when exposed to light, as might be expected if their
coloring matter were a product of their own metabolism, as is the case in
plants.   Substantial evidence of the independent nature of chlorophyll
granules was presented by Hamann (1882), who proved that they do not
originate in the egg of *Chlorohydra*, as Kleinenberg had still assumed, but
migrate to the egg from the endoderm cells.   The transmission of a
symbiont by means of egg infection was thus established for the first time.
Later this type of transmission was found in hundred-fold form, especially
with fungal and bacterial endosymbioses.

Other serious doubts arose.   More accurate research on the doubtful
inclusions showed them to be colorless, protoplasmic bodies containing
nuclei and chromatophores, usually surrounded by a distinct, cellulose
membrane.   Moreover scientists were coming to the conclusion that

rhizopods and ciliates are unicellular organisms without room for other animal-specific cells. Finally it was shown that these "chlorophyll granules," unlike true chromatophores, may be forcibly removed from the animal and yet continue to live.

As a result a number of zoologists decisively rejected the hypothesis of animal chlorophyll and asserted that all these inclusions were plant parasites or commensals. Geza Entz (1876) was first to present this idea, but his report, written in Hungarian, went unnoticed in the beginning. Independently, Brandt and Geddes (1881, 1882) expressed similar views in reports so convincing that the validity of the new theory could no longer be seriously doubted, although Lankester in particular refused to accept it for a long time.

With these advances in the study of green inquilines, called Zoochlorella by Brandt, a second vast field was opening up in this intracellular inter-linking of animal and plant organisms. The green inclusions of fresh-water animals were long-familiar phenomena, and studies by scientists like Johannes Müller, Agassiz, Haeckel, Moseley, and Huxley had revealed that luminous yellow to brown formations occur regularly in all kinds of marine animals—protozoans, sponges, coelenterates, and turbellarians—and that these formations may likewise cause the animals to be colored. Although their cellular nature became evident with relative rapidity, they were considered for a long time to be integral parts of the animals, and this gave rise to various misleading ideas regarding their significance. Cien-kovsky (1871) was the first to give valid evidence of their parasitic nature. In the course of his research he had observed a radiolarian in which the so-called yellow cells, still termed hepatic cells by Haeckel, survived the animal host. They were transformed into flagellates and continued to reproduce for months. Richard Hertwig (1879), hesitant at first to accept this idea, came to realize that similar inclusions in actinians are independently vegetating algae. Geddes (1882) and particularly Brandt (1883) made substantial additions to the data on cooperative living between marine animals and *Zooxanthellae*, as the yellow cells were now being called in contradistinction to the *Zoochlorellae*.

Our present knowledge of the endosymbiosis of animals and algae developed step by step from many other isolated observations made in the course of systematic and anatomical investigations and from reports dealing specifically with algal symbiosis. The limits of this type of symbiosis are set by the dependence of the symbionts upon water and adequate light. Among the protozoans of fresh water, it is chiefly the thecamoebae which often live symbiotically with algae (Archer, 1869–1871, 1873; Leidy, 1879; Penard, 1890, 1902; and others); more rarely, naked amoebae (Gruber, 1904); heliozoans (Hertwig and Lesser, 1874; Leidy, 1879); and ciliates

(Stein, 1859–1867; Dangeard, 1900; W. I. Schmidt, 1921; and others). In general it is a very loose alliance, for species may occur without algae or may lose them at times, and various observations even indicate periodic interchange between green and colorless states. According to Archer (1873), the thecamoeba *Difflugia pyriformis* Perty is infected in spring and summer but not in autumn, and Balbiani (1888) found that *Frontonia leucas* Cl. and L. gradually becomes completely colorless in September. Wesenberg-Lund (1909) and Gelei (1927) also reported periodic appearance of green states. On the other hand, the symbionts not only are regularly imparted to the daughter animals during division but are also usually preserved in the cysts (shelled and naked amoebae, *Actinosphaerium*, ciliates). It is only in *Acanthocystis aculeata* Hertwig and Lesser, according to Entz (1882), that they are relieved of their algae before capsulation.

The adaptations prerequisite to endosymbiosis are clearly revealed in rhizopods and ciliates. Certain species are decidedly incompatible with algae and thus are never infected with them; in other species the algae are present at times and absent at others; and certain species, like many thecamoebae or the colonies of *Ophrydium versatile* Ehrenbg. among the ciliates, never, or extremely rarely, appear in the colorless state.

Of the protozoans which we have just discussed, symbiosis usually involves the *Chlorella* species, and other algae only occasionally. However, many of the rhizopods, flagellates, and ciliates also enter into symbiosis with Cyanophyceae. Verification of tentative opinions of other authors came when Pascher (1929) demonstrated the existence of such extremely interesting "endocyanosis" which leads to a very close relationship between the two partners. In the thecamoeba *Paulinella chromatophora* Lauterborn are always found two sausage-shaped, blue algae, which at first might be taken for chromatophores. During reproduction of the host animal one of these algae glides into the new cell; later a division restores the old number of two in the two daughter animals. In a naked amoeba, on the other hand, Pascher found hundreds of little cyanellae, and in a series of flagellates he found more or less intimately adapted blue algae. For example, the cryptomonad *Cryptella cyanophora* Pascher always has in its center a single, round symbiont which divides simultaneously with the nucleus.

Far more frequent among marine protozoans is symbiosis with zooxanthellae, which usually represent "palmella phases" of non-flagellate cryptomonads. In numerous radiolarians the zooxanthellae are located in the extracapsular soft body; in acanthometrids they are located within the central capsule, where the arrangement and quantity of the symbionts in each case are type-specific and thus indicate extensive regulatory influence on the part of the host organism. When radiolarians multiply, the algae may be taken along by the gametes (Schewiakoff in acanthometrids, 1926) or left behind in the residual bodies which disintegrate after swarm formation as in *Sphaerozoum*. Surrounded by mucus-covered shells, the algae continue to reproduce here in the palmella form or are transformed into flagellate states.

In view of these observations, it is astonishing that Moroff and Stiasny failed to recognize the algal nature of the yellow cells, thus arriving at thoroughly misleading conclusions, though fortunately without influencing other authors. The zooxanthellae of *Acanthometra*, they believed, were trophic nuclei originating from the primary nucleus of the radiolarian (Moroff and Stiasny, 1909, 1910). In *Sphaerozoum*, Stiasny (1910) described the corresponding formations as "schizonts" leaving the mother colony to become new radiolarians. The yellow cells in *Thalassicolla*, he claimed, were early stages of alien radiolarians which had migrated there!

The occurrence of yellow cells in foraminifera is of lesser importance. *Globigerina, Orbitolites, Peneroplis, Polytrema, Trichosphaerium* are among those that function as algal hosts (Schaudinn, 1899; Winter, 1907; and others). More recently W. L. and M. M. Doyle (1940), investigating *Orbitolites*, took into consideration the detailed structure of the algae and studied it under various conditions. In the algae they found oil droplets and calcium oxylate crystals in addition to starch. In marine ciliates and flagellates, yellow cells are found only in isolated instances.

The green coloring of fresh-water sponges has attracted attention from time immemorial. In 1840 Hogg was not greatly surprised to observe that the sponges turned green when exposed to light and remained colorless when unilluminated, for he was still of the opinion that these extremely formless organisms were more closely related to plants than to animals. According to Noll's aphoristic report of 1870, such behavior is logically explained by the algal nature of the green inclusions. Brandt (1881) gave the first documented presentation of this view and successfully established it in the face of considerable opposition (Beyerinck, 1890; Weltner, 1893, 1907; Delage and Hérouard, 1899; Oltmanns, 1904/05; and others). It is hard to understand why according to Minchin's writings (1900) and the *Cambridge Natural History* of 1906 (Sollas) it was still thought that the old concept might possibly be valid.

Van Trigt (1917, 1919, 1920) made a thoroughgoing study of the symbiosis of native spongillids, and there are other reports on exotic relatives to round out our information. On the one hand, the symbiosis is lax, for excess algae are apparently digested constantly and at times may even disappear entirely. On the other hand, even the permanent stage of the gemmules is infected, and the algae, here representing not *Chlorella* but a *Pleurococcus* species, are transferred to the larvae developing in the maternal body (Limberger, 1918), if indeed they have not already been transferred to the egg cells.

In marine sponges we sometimes find zooxanthellae, and sometimes higher algae belonging to the Florideae and the Chlorophyceae.   In the latter case, one can establish all stages from harmful parasitism and harmless commensalism to a genuine symbiotic relationship (M. and A. Weber, 1890; Weber-Van Bosse, 1910; and others).   Sometimes the sponges are so interlinked with the algae that it is difficult to determine which of the two partners actually determines the growth form, a situation recalling lichen symbiosis.

Chlorohydra viridissima Pall., the only fresh-water coelenterate living in close symbiosis with *Chlorella*, represents a classical test subject for the study of algal symbiosis.   As early as the eighteenth century, Rösel von Rosenhof, Schäffer, and Trembley took delight in its lively green.   In 1882 Hamann observed the transfer of algae, limited for the most part to the endoderm, to the growing oocytes, and in 1925 von Haffner studied all aspects of the process in greater detail.   Goetsch, who also contributed to our knowledge of symbiosis in hydras (1924), found that occasionally *Hydra attenuata* Pall. proceeds toward ingestion of algae, in which case, after various disturbances, an equilibrium of the two partners occurs, but it never approaches that of the obligatory symbiosis of *Chlorohydra*.   Goetsch could keep the latter free of algae for months by confining them in calcium-free water kept cold and dark.   In diversified experiments he investigated the return of the green color in de-algaefied *Chlorohydra* and in *Hydra attenuata* by feeding and transplanting and by interchange of the symbionts.

It has been discovered that many marine hydropolyps live symbiotically with yellow cells.   The algae are always found in the endoderm, but the specific areas preferred may vary (*Hydrichthella, Pachycordyle, Aglaophenia, Halecium, Eudendrium, Myrionema*, and others).   The principal studies of these hydropolyps are by Svedelius (1907), Stechow (1909), H. C. Müller (1913, 1914), Light (1913), and Müller-Calé and Krüger (1913).   It was shown that Hydrocorallina like *Millepora* are permeated with zooxanthellae (Moseley, 1881; Hickson, 1900, 1924; Boschma, 1924; and Yonge and Nicholls, 1931).   Among the siphonophores only *Velella* and *Porpita* have been identified as zooxanthellae hosts.   Early observations by Brandt and Geddes were amplified by Kuskop (1920) and Delsman (1923).   Kuskop proved that the algae are usually limited to definite areas of the colony, a circumstance which indicates a state of cooperative living far beyond that of parasitic invasion.   In *Velella*, for instance, the dactylozooids and the central nutritive polyp are avoided entirely, and the so-called hepatic canals of the endodermal canal system are inhabited only in their dorsal section; other areas are closed off in *Porpita*.

During budding, the hydrozoan symbionts are transferred directly to the daughter animals.   When egg cells mature in regressed medusoids and sporosacs which maintain their connection with the colonies, a number of forms are infected almost exactly as in *Chlorohydra*, for example,

Halecium and *Myrionema*.   Among the Hydrocorallina this is also true of
the eggs of the medusoids, which, though radically reduced, are nevertheless in the process of being separated from the stock (Hickson, 1900;
Mangan, 1909).   In medusans which are separating from the *Velella*
and *Porpita* colonies, the radial canals likewise contain zooxanthellae,
but the latter are apparently destroyed before maturation of the egg
cells on the so-called chrysomitras which sink down into greater depths,
inasmuch as the youngest larval stages of *Velella* are at first algae-free
and are dependent upon an infection which never fails to occur during
their rise to the surface.   There is little confirmed information on algae in
hydromedusans, but they are known to occur in *Cunina* (Stschelkanowzew,
1906; compare Buchner, 1930).

Among the scyphozoans it is chiefly the medusans which may act as
hosts to algae.   Whereas among the Semaeostomata zooxanthellae have
been established only in *Linuche* species; here they are situated in the
umbrella joined in little sacs between the gonads (Thiel, 1927).   We have
much information on zooxanthellae among the rhizostomes.   Here Geddes
and Brandt recognized them as such, whereas earlier authors like Haeckel
and Hamann still considered them to be strange gland cells or spermato-
zoan-like formations.   In *Cotylorhiza* the endoderm is colonized, including
the anastomosing canals of the gastrovascular system (Claus, 1884); in
other genera it is usually the mesogloea which the algae prefer and which
is everywhere permeated with them in nests.   Occasionally the mesogloea
remains free and the algae are restricted to accumulations just beneath the
ectoderm, as in *Mastigias* (Uchida, 1926).   In *Cassiopeia* (Bigelow, 1900;
Smith, 1936) they are not only everywhere in the mesogloea and in the
migratory cells, but also, though sparsely, in the endoderm cells and even
in the ectoderm cells.

In the scyphomedusans the egg cells and planulae remain free of algae.
However, some of the hosts contain algae in the polyp phases, and it must
therefore be assumed that they are acquired anew each time with their
food.   Bigelow described in detail how part of the algae are transmitted
during strobilation and how planuloid formations, constricted off from the
scyphistomae, are provided with algae.   Other species have algae-free
polyps and acquire the symbionts only as young medusans (*Crambessa*, von
Lendenfeld, 1888).

Little information is available on the yellow cells of ctenophores.   Chun
(1880) and Moseley (1882) reported them in the meridional vessels and
sexual organs of *Euchlora rubra* Köll.   According to Krumbach, unicellular
algae may be found in the gonads of *Haeckelia*.   Berkeley (1930) described
the red algae in the subepithelial zones of the stomodaeum in *Owenia
abyssicola*, a genus found at greater depths.

Algal symbiosis is extremely frequent among the Anthozoa.   In octocorals algae occur chiefly in many Alcyonaria (*Tubipora*, *Heliopora*, *Xenia*, *Clavularia*), but they have also been reported in several Gorgonacea (*Isis*, *Melitodes*) and in at least one Pennatulacea (Pratt, 1905, 1906; Laackmann, 1908; Hickson, 1916; Yonge and Nicholls, 1931). They are found especially in forms frequenting the shallower water of warm seas, and in such cases the algae are restricted to the endoderm and also are found at large in the lumen of the lacunae lined with endoderm. Forms frequenting the moderate and cold seas tend to be symbiont-free. The strong accumulations of red pigment apparently eliminate reduced illumination as the cause of algal symbiosis. According to Pratt, the rise in the number of algae is accompanied by increasing regression of the mesenteric filaments which furnish the digestive secretions and by discontinued ingestion of formed food.

Numerous algal hosts are known in all orders of hexacorals except Ceriantharia. Infection of the endoderm cells in the mesenteries and the tentacles is more or less general, with more pronounced accumulation of algae in sharply defined zones of the mesenterial filaments, the so-called zooxanthellae strips. Moreover, the algae often appear in the region of the ectoderm, particularly in the Zoantharia (Pax, 1914). Here too the symbionts frequently determine the coloring by almost overwhelming the pigment of the host, but, because of decreasing light with increasing depth, the algae become sparser and sparser and sometimes disappear altogether. Then the animal's own coloring emerges in increasing degree.

Symbiosis with zooxanthellae is also extraordinarily widespread among the Madreporaria. Aside from numerous authors who merely referred to symbiosis incidentally, it was chiefly Boschma and Yonge who dealt seriously with the subject. As director of the expedition to the Great Barrier Reef of Australia (1928–29), Yonge made an unusually thorough investigation, and with his collaborators he contributed valuable data on anatomical and physiological aspects (C. M. Yonge and Nicholls, 1931; C. M. Yonge, J. J. Yonge, and Nicholls, 1932; C. M. Yonge, 1936, 1940, 1944). Boschma (1924, 1925, 1926), investigating thirty-eight species of reef corals of the Indo-Pacific Ocean, invariably found them to contain algae; and Yonge and Nicholls found algae in all species of thirty-five genera. Thus it is certain that all shallow-water corals contain zooxanthellae. That they were lacking only in *Dendrophyllia* merely confirms the rule, since this is a form which originated at greater depths and has only secondarily adapted itself to shallow water. Boschma's finding (1925), that *Astrangia* and *Phyllangia* species sometimes never contain algae and sometimes only occasionally, is not a contradiction, for, although the algae frequent shallow water at times, they nevertheless usually prefer dark to illuminated places.

The algae, surrounded by a rather strong cellulose membrane and filled with oil droplets and starch-like substances, are limited to the endoderm, where they may be found in the interstitial cells as well as in other sites. It is always the locations more exposed to light which are colonized, and, when the corals are weakly illuminated, a lighter coloring, sometimes approaching white, takes the place of the brown coloring caused by the algae. This change resembles the bleaching-out of green spongillids which is caused by a deficiency in the light supply. Attempts to cultivate the algae of reef corals outside their hosts have been unsuccessful.

In all anthozoans, as with fresh-water sponges, the planular larvae appear to be infected before they leave the mother animal or the mother colony. In detailed research on reef corals Marshall (1932) found that the larvae of *Porites*, approximately 0.5–1 mm. in length, contain 1150–7400 algae, and that those of *Poicillopora*, which are substantially larger, contain no less than 25,000. Even the ectoderm may be infected by zooxanthellae, usually, near the mouth, but later the algae disappear from this location and are found only in the endoderm (*Maeandrina*, Wilson, 1888; Boschma, 1929; *Siderastraea*, Duerden, 1904). Some authors believe that even the mature egg cells are infected with algae, but no verified data exist. With unfavorable conditions the transfer fails to take place in aquarium animals, and colorless larvae swarm out instead (Duerden, 1904).

For *Siderastraea radians* Pallas, Duerden presented an excellently illustrated description of the regularity with which the algae are distributed over the septa and tentacles during the development of the first polyps.

The symbiosis of rhabdocoelous and acoelous turbellarians, on some aspects of which we are well informed, is of special interest. On the one hand, there is great variation in the degree of intimacy embodied in the reciprocal relations, but at the same time the maximum intimacy is clearly beyond that of protozoans and coelenterates. We have already mentioned that the green turbellarians were among the test objects used in research on the seemingly mysterious "animal chlorophyll." Schultze (1851) was the first to demonstrate the presence of genuine chlorophyll in *Dalyellia viridis* G. Shaw; Geddes (1879) used *Convoluta* in gas metabolism experiments; and Brandt (1881, 1882) finally recognized the inclusions as algae. Keeble and Gamble (1905, 1907) made the definitive study of the interesting *Convoluta*, and von Haffner (1925) filled in the remaining gaps in our knowledge of rhabdocoele symbiosis.

In rhabdocoele symbiosis it is a case of typical, easily cultivated *Chlorella*, usually located in the interstices of the mesenchyme, thus in an intercellular position, and only rarely in the epithelial cells of the intestines. Strictly obligatory algae hosts, such as *Chlorohydra*, do not occur among fresh-water forms, but, in addition to hosts which are green only occasionally, there are some which are almost always infected, such as the aforementioned

Dalyellia or *Castrada hofmanni* M. Braun.   The egg cells are never infected. After emerging from winter eggs, the colorless females feed on algae covering the breeding container and gradually turn green after several weeks. In this case the algae make their appearance first in the intestinal epithelial cells and later in little embryonic elements located between these cells. It is these elements which then carry the symbionts to the mesenchyme, where later they perish and leave their contents in the gaps between the mesenchyme cells (von Haffner).   This strange process, first described in *Dalyellia viridis*, was verified for *Phaenocora typhlops* in unpublished research carried out in my institute by Gimmler.   When embryos develop within the maternal body from summer eggs, the algae migrate to the pharynx, and from there to the gut and the mesenchyme (Silliman, 1885; Dorner, 1902; Luther, 1904).

The differently constituted symbiosis of acoelous green turbellarians of the sea, *Convoluta roscoffensis* Graff and *Convoluta schultzii* O. Schm., has been studied by these authors: von Graff (1891), Haberlandt (1891, in von Graff), and, in particular, Keeble and Gamble (1905, 1907, 1910). Appearing again in the host in place of *Chlorella* are flagellates which in turn become non-flagellate.   They appear also in the parenchyma and should presumably be classified as chlamydomonads.   This time they are altered so radically in the animal tissues as to become mere irregular formations, without membrane; they cannot be cultivated and no longer come into consideration as starting material for a transfer.   Instead, independent flagellate states, lured apparently by chemotactical stimulants. accumulate and reproduce vigorously on cocoons which contain several eggs.

When the algae, as a result of the new infection, accumulate sufficiently in the body of the host, the host stops its normal feeding entirely and apparently lives exclusively on the food elaborated by its guests.   The animals, having become completely dependent upon their symbionts, take no other food even when infection fails to occur, and thus deteriorate. In such cases they can be kept alive only by prompt artificial infection. But, strange to say, this autotrophic manner of feeding no longer satisfies the aging animals.   To meet the desire for formed protein which is now awakening, they attack their symbionts.   However, their death is merely delayed in this way, not prevented.   After resorption of the symbionts, the animals perish since they are no longer capable of ingesting food independently.

Other marine turbellarians, several acoelous forms and one allocoelous form, live in union with yellow cells.   *Convoluta paradoxa* Örst. was exhaustively investigated by Keeble and Gamble (1908).   In *Convoluta roscoffensis* the larvae slipping out of the cocoons can easily be infected, but with

Keeble and Gamble's species infection is possible only when the larvae are provided with seaweed from the living area of the worms, that is to say, with free-living infection stages. If these stages are lacking—a situation which admittedly might never occur in nature—the colorless larvae ingest food abundantly, for example, diatoms, small crustaceans, and the like, but they do not grow larger; indeed they even become smaller. Nevertheless, as soon as even one of the pale yellow flagellates, to be classified perhaps among the chrysomonads, is taken up into the parenchymatous tissue replacing the gut here, it reproduces vigorously, thus giving rise to a yellow coloring in the host animals. In this case, however, the infection does not cause ingestion of food to be suspended, and, whenever an insufficiency of nutrients is artificially created, part of the algae are digested.

Yellow cells are not lacking in the Echinodera, as we learn from a monograph by Zelinka (1928) on these minute metazoans. Before 1928 they had been overlooked completely. They are found in all subclassifications in the region of the hypodermis. Although they are merely facultative symbionts, they are nevertheless arranged in groups and series in a striking, type-specific manner. They are always absorbed during feeding. Hungry animals attack the algae and then acquire migratory cells laden with their decomposition products.

Many apparent examples of symbiosis in rotatorians must be discarded because they are merely simulated cases resulting from the eating of green algae (cf. Remane, 1929), but some verified symbioses do exist. Obligatory and facultative algae cultivators and pronounced predators can here too be differentiated. Algae which are impervious to digestion proceed from the intestinal epithelium into the body cavity, as is the case with fresh-water turbellarians. Such cases were reported by Harring and Myers (1922, 1924, 1928), Penard (1890), Gelei (1927), and Varga (see Buchner, 1930). Varga found algae also in winter eggs.

The chaetopterids are the only annelids with algae. Brandes (1898) had traced the strange green coloring of the gut in *Chaetopterus* to algae-like inclusions, and Berkeley's research on the same test object (1930) verified the existence of this symbiosis. Not only in *Chaetopterus variopedatus*, but also in *Mesochaetopterus*, *Phyllochaetopterus*, and *Leptochaetopterus*, flagellates are found in the cells of the intestinal epithelium, usually in the palmella phase, without cellulose membrane, chromatophores, or pyrenoids. Some already possess two flagella and are motile when taken fresh from the host animal; in others the flagella make their appearance several hours later.

The algae are especially dense in the outer part of the tall epithelial cells which faces the lumen of the gut; they also settle laterally in a thinner layer, and they avoid the central regions of the cells entirely. They become smaller and less numerous near the anus, and the motile phases accumulate in its direct vicinity, that is, in the younger, regenerative tissue of animals which reproduce asexually through transverse fission. In *Mesochaetopterus* even the eggs are usually covered externally with the algae, and in early larval stages the gut tends to be infected also (Enders, 1909). According to Berkeley, the symbionts should be classified as chrysocapsins.

Among the mollusks two representatives were found living symbiotically with algae. On zooxanthellae in naked marine snails there is much information, some of it uncertain, and some undoubtedly the result of confusion regarding nutritive bodies and inclusions peculiar to the animal. An example of such confusion is Zirpolo's data on algae in the "hepatic caeca" of *Phyllirrhoe* (1923). Neither Fedele (1926) nor I (1930) was able to make verification. Also the old data on green algae in *Elysia viridis* Brandt (1883) cannot be confirmed. In a number of other opistho-branchiates, symbiosis with zooxanthellae is indisputable. This is true also for *Aeolis, Doridoeides, Melibe, Spurilla, Favorinus,* and *Eolidina (Aeolidiella)* species (Hecht, 1895; Eliot and Evans, 1908; Hornell, 1909; Henneguy, 1925; Naville, 1926; Graham, 1938). In all these cases the algae lodge in cells of the intestinal pouches present in the dorsal body appendages, and apparently the algae are always absorbed along with food. Naville and Graham give clear proof for *Eolidina.* Living and feeding on the actinian *Heliactus bellis,* the snail also devours its algae and nematocysts. Reaching the cells of the intestinal caeca intact, the foreign nematocysts are used by the snails along with their own, and the algae reproduce as welcome symbionts. At the same time the algae impart the same coloring to their hosts as to the actinians; and because of this uniform coloration it is barely possible to distinguish between the snails and the actinians.

Apparently the other naked snails mentioned are also carnivorous—in some forms the foreign origin of the nematocysts could be proved—and it is likely therefore that they all acquire their algae in the same way. *Tridacna crispata* Bgh. is an exception. In this case we are dealing with a purely herbivorous form in which the symbiosis differs from that of other opisthobranchiates in other respects also (Yonge and Nicholas, 1940). It possesses no dorsal papillae, is free of nematocysts, and one would look in vain for algae in the richly developed intestinal diverticula. Algae are nevertheless present in an unusually narrow, brownish border in the deeply indented, undulating folds of the leaf-shaped body, and this time they are situated intercellularly in the connective tissue. Apparently the algae are acquired with food, just as with carnivorous snails, but the adaptation is more pronounced, inasmuch as this is a herbivorous form which spurns the adequate diet available and restricts the algae to a narrow, specific area.

At this point we must not fail to mention a type of algal symbiosis the significance of which has been recognized only recently, despite the fact that it is connected with a highly striking animal and achieves a unique degree of complexity. We are speaking of tridacnids, those lamellibranchiates in which excessive growth may at times result in shells one meter in length. Brock (1888) suspected the presence of algae in the cells he observed in their mantle, and Yonge made a detailed study of this exceedingly interesting phenomenon. His report is one of the most valuable in the field of algal symbiosis.

The mantle edge of the tridacnid is thickened and enlarged to such an extent that the shell can no longer be closed entirely and in a position of rest extends over the edge of the shell in folds. To the extent that the mantle is exposed to light, the numerous lacunae of the vascular system contain masses of brown zooxanthellae, each one enclosed within a thereby greatly expanded amoebocyte. Unlike the zooxanthellae in corals, these possess no membrane at all, or a very thin one at best. Oil droplets and much starch are found in their plasma. Excess algae enter phagocytes which accumulate particularly between the greatly reduced intestinal diverticula and elsewhere around the intestinal canal. The algae are found here in all stages of dissolution.

Symbiosis is less highly developed in *Hippopus* than in *Tridacna*. The mantle edge is not so greatly enlarged, yet the shell is opened wider. There are fewer algae and a smaller accumulation of phagocytes in the gut. Above all, *Hippopus* lacks those peculiar organs developed by *Tridacna* in connection with symbiosis. Wherever the mantle of *Hippopus* is spread out toward the light, it has a series of conical protuberances in which one or more nests of transparent cells are embedded. Apparently derived from epithelial cells which sink to the bottom and reproduce there, these cells are surrounded by a covering of pale connective tissue. They are always situated in the densest algae accumulations, and there can be little doubt that their function is to direct the light to deeper areas (Fig. 1). Algal symbiosis causes the formation of light-converging lenses in the animal tissue, just as does symbiosis with light-producing bacteria. This is, indeed, a reaction which is suggested at the very start by the fact that the mantle edge often forms lens-eyes.

We have seen that algal symbiosis achieves a unique degree of complexity in those animals which first made their appearance in the Eocene period. By comparing their organization with that of their progenitors, which were of *Cardium*-like structure, we see also that they have experienced radical modifications through acquiring the algae; and, indeed, that the algae have been the determining factor in every aspect of their unusual form and mode of life.   The unusual extension of the mantle edge over the whole dorsal side, which produced the requisite living space and light supply for the algae, caused a shifting of the umbo toward the front; the siphons likewise shifted frontward from the rear, becoming widely separated from one another during this process and at the same time reduced.

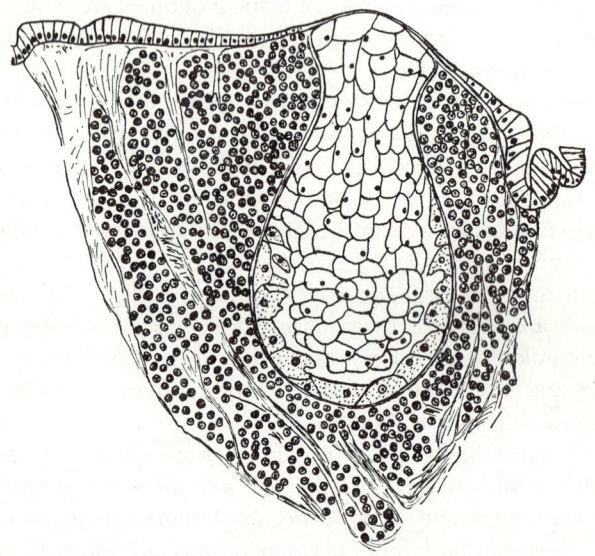

Fig. 1   *Tridacna crocea* Lam.   Lens-like organ of the mantle edge in the midst of numerous zooxanthellae.   180×.   From Yonge.

The rear closure muscle and the rear retractor, though pushed forward less far, were radically enlarged.   Unlike the case in all siphoniferous lamellibranchiates, the anus was shifted as a result of these movements to a position behind the evacuation siphon.   The heart and kidneys were involved in this general change of position, and the mouth and labial palpi were shifted downward.   At the same time the intestinal diverticula, reduced in number and importance, were supplanted by the symbiont-filled phagocytes.   The undigested algae led to considerable enlargement of the kidneys and to extensive accumulations of excreta in them.

It became necessary for the animals to change their mode of life to the degree that they became dependent upon the algae for their food.   They had to remain on the upper surface of coral reefs near the shore where sufficient light was available for their symbionts, and, after burying themselves like *Cardium* in the sand, they learned to assume the position needed for the development of the mantle facing the light. The animals learned this through strong development of the byssus, through changes of the shell form, and through boring into the coral calcium. The question must remain open whether the gigantic growth of the shells is also the result of algal symbiosis. Thiel (1929) thought it possible that the abundance of oxygen, traceable to the algae, might effect an increased precipitation of calcium. On the other hand, Yonge reminds us that algae-free, reef-dwelling forms like *Spondylis* and *Chama* also display marked shell growth; nevertheless he too considers it quite likely that it is the symbiotic relationship with algae which leads to the increased metabolism requisite to such excessive performance.

In *Anodonta cygnea* L. and *Unio pictorum* L. typical *Chlorella* are occasionally found colonizing the tarsus and the posterior sections of the mantle edge and of the gills, in this way coloring them an intensive green, but it would be erroneous to speak of symbiosis here. Nevertheless, this phenomenon, which leads to injury of the shells (studied more minutely by Goetsch and Scheuring, 1926), enables us to understand the tendency of *Chlorella* toward a parasitic mode of life where algal symbiosis takes place harmoniously. The phenomenon also provided Goetsch and Scheuring with a supply of algal material for studying the extent of receptivity on the part of protozoans and hydras.

The only information on chordate animals is based on ascidians. In these animals the greenish blood cells have often been mistaken for algae, especially by older authors. However, it is actually a matter of yellowish or green unicellular algae in the lumen of the cloacal cavity and its branches, as in *Diplosoma virens* Hartmeyer and *Didemnum voelzkowi* Michaelsen, or within the cellulose mantle, as in *Didemnum viride* Herdmann. On the other hand, there is no record of intracellular colonization in true ascidiozooides. Cases like these, applying exclusively to inhabitants of tropical oceans, can scarcely be regarded as examples of true symbiosis (Harant, 1930; Smith, 1935).

The possible advantages of algal endosymbiosis were widely discussed, and numerous morphological and experimental studies were undertaken in clarification. The first was a study by Geddes (1878) on oxygen production in symbiotic algae in *Convoluta*, and reports by others on the same topic soon followed. In 1881 Engelmann demonstrated by his precise bacterial method that even *Chlorohydra* and *Paramaecium bursaria* give off oxygen energetically in the light. Geddes (1882) and Brandt (1883) had corresponding results with the yellow cells of radiolarians and actinians. Geddes was the first to consider this oxygen production as a concomitant symbiotic phenomenon which is useful to the animals. His views were met partially with opposition, partially with support. Brandt had observed that actinians live much longer in the light than in the dark when they are kept with a small quantity of oxygen in a sealed vessel, but he was of the opinion that under natural conditions the animals would not need the oxygen provided by the algae. However, Penard (1902), Gruber (1904), and Doflein (1907) found that green monothalamians and amoebae, as well as *Paramaecium bursaria*, in covered watch-glass cultures were far more resistant than colorless individuals of the same or related forms. Pringsheim, in careful physiological studies on *Paramaecium bursaria* (1925, 1928), kept green and colorless animals illuminated in sealed glass tubes with a limited quantity of oxygen and found that the green animals lived on for months while the colorless ones were soon dead. Algae-containing cultures fared better when illuminated day and night than when they assimilated light only during the day. Hood (1927) found that the oxygen production of algae resulted in improved living conditions for *Frontonia*, a genus living exclusively in stagnant water.

Van Trigt (1919) discovered that new sponges can readily reorganize from green cellular broth of fresh-water sponges in the light but not in the dark. Wilson (1891) stated that *Chlorohydra* can live in stagnant water far longer than species without algae, and Bohn and Drzewina (1928) reported that this genus at first exhibits depression phenomena when kept in oxygen-free water but that these symptoms eventually recede. Trendelenburg (1909), making precise comparisons of the formation and consumption of oxygen in algae-containing actinians in the dark and in the light, concluded that the oxygen production of algae was of great significance to his test objects, since the intensity of their metabolism, as with other lower animals, is dependent upon the quantity of oxygen available. He said that in stagnant or very warm water the sources of oxygen localized in the animals may therefore have great importance.

In recent years a succession of authors have studied the effect of oxygen production by symbiotic algae on host animals. Yonge and his co-workers found that oxygen production exceeded the needs of reef corals which were exposed to daylight for nine hours, whereas they always consumed substantial quantities of oxygen of algal origin in the dark. It is not surprising then to learn, on the one hand, that the intensity of the oxygen production is influenced considerably by the depth at which the algae live and by the variance in light supply during the different seasons, and, on the other hand, that such variations were not observed in the algae-free *Dendrophyllia*.

Marshall (1932) studied oxygen production in the planulae of corals; Verwey (1931) in *Acropora hebes;* Welsh (1936) in the marine turbellarian *Amphiscolops;* Kawaguti (1937) in other madreporarians; and Smith (1939) in *Anemonia sulcata*. Considerable difference of opinion is represented in the views on possible oxygen utilization. Verwey and Smith believe that the oxygen is utilized; Yonge and his collaborators reject this opinion. The latter concede that abundant oxygen-producing plant life might be a vital necessity for the corals, especially where strong currents are lacking, but they see no reason to suppose that the plant life is localized in their interior. In their opinion the animals would utilize the oxygen, if at all, only under certain extraordinary conditions as might arise in small collections of water.

Far more important in judging the advantages of algal symbiosis is the question whether symbiosis may lead to complete cessation of direct nourishment, or at least to a significant decrease. For protozoans numerous data indicate that this is actually the case. Archer (1869–1871, 1873), Nüsslin (1884), and Penard were unable to find ingestion of food in *Amphitrema* species living in obligatory symbiosis; and in other green thecamoebae ingestion is markedly lowered. None of the authors studying the Cyanophyceae-containing *Paulinella* were able to observe an animal mode of nutrition. Lohmann (1908, 1912) described a remarkable marine ciliate, *Mesodinium rubrum* Lohm., which ingests food only in youth. As soon as it is infected with the little red alga, *Erythromonas haltericola*, the cytostome is closed, the mouthpiece upon which it rests degenerates, and all direct ingestion of food ceases. Most of the authors agree that ingestion by Daphnia and other predatory animals plays a much smaller role in *Chlorohydra* than in the colorless Hydrozoa. All the authors mention that no prey is ingested in *Convoluta roscoffensis* after sufficient reproduction of algae. In the highly developed algal symbiosis of tridacnids, Yonge found that the intestine tends to contain only very few exogenous food particles, and that the digestive intestinal diverticula are radically reduced. It is reported that ingestion of food also plays an insignificant role in xeniids (Gohar, 1940) and other Alcyonaceae (Pratt, 1903) and that there is rudimentation of the resorbing mesenterial filaments.

It is dangerous to generalize, however, because other symbiotic protozoans and coelenterates feed constantly just as their algae-free relatives do, but at least it seems clear that curtailed feeding in some cases causes the symbiosis to become obligatory and that in numerous instances the facultative symbiosis must result in varying degrees of improvement in the life situation of the hosts.

Experiments with starving animals confirm this conclusion. Brandt conducted such experiments with actinians (1882). Keeping one specimen of *Anthea cereus* darkened, but in well-aerated water, and another of equal size in the light, he found the first dead after a month, whereas the other was still quite active after six months. It was also shown that darkness as such does not have an injurious effect. The earlier data on the behavior of *Chlorohydra* on withdrawal of food were less decisive. Prings-heim (1915) was able to keep them in nutrient solutions for more than three months, but still the depression phenomenon of knobbed tentacles did eventually appear. A satisfactory experimental procedure was possible only after Goetsch (1924) learned how to make the animals algae-free. When green and white chlorohydras and algae-free *Hydra attenuata* were kept in the same vessel, without food, only the first mentioned were still alive after one and one-half months. Finally they too began to deteriorate. Loss of algae is in itself not injurious, for symbiont-free chlorohydras thrive well with feeding. When algae-containing chlorohydras were allowed to starve, some illuminated and others darkened, the illuminated ones lived almost twice as long (von Haffner). Von Graff had the same experience with *Dalyellia*, and Keeble and Gamble with *Convoluta*.

It was demonstrated again and again that it is impossible for the animals to give up ingestion of food completely and that depression phenomena are inevitable, but it was also demonstrated that algae-containing animals which do not normally undergo complete loss of the algae are able to go without food for a much longer period. Pringsheim (1925, 1928) even claimed that complete withdrawal of food was not injurious to *Paramaecium bursaria*, an animal which ordinarily devours bacteria and yeasts with avidity. Ciliates could be kept indefinitely in a suitable, almost bacteria-free nutrient solution despite the fact that the nutrient vacuoles, the fat droplets, and the excreta known as Schewiakoff crystals disappeared at once. On repeating Pringsheim's experiments, Loefer (1936) was convinced that with complete absence of bacteria this would be true only to a certain extent and that Pringsheim's contrary findings were traceable to the fact that bacteria were still present. Hämmerling reported (1946) that the green *Stentor polymorphus*, though it cannot live indefinitely in the light, nevertheless can go much longer without food than the algae-free variety.

Inasmuch as algal hosts can be nourished completely or partially by their symbionts, the details acquire importance. Are the excess symbionts constantly disintegrated and digested? Do they transmit dissolved metabolites to surrounding tissue? Do both possibilities occur, simultaneously or in succession?

There is much information on dying and resorbing algae, some of which have already been presented. Regarding protozoans we have information on radiolarians (Enriques, 1919; and others), naked and shelled amoebae (Penard, Gruber), and ciliates (*Bursella spumosa* according to Schmidt, 1921; and others). On the other hand, Winter (1907) does not find dying algae in *Peneroplis*, nor does Schaudinn (1899) in *Trichosphaerium*. Van Trigt believes that the algae of the spongillids are constantly being digested;

Beutler (1924) and especially von Haffner (1925) describe the degeneration and resorption of *Chlorella* in hydras. Here most of the algae disintegrate at places of increased metabolism, as in the young buds and in the area of egg formation. Woltereck (1904) actually interprets the yellow cells of chrysomitras as a type of food provisioning. According to von Haffner, the algae of green turbellarians of fresh water are digested, and it has been reported that *Convoluta paradoxa* and *roscoffensis* resorb their symbionts, the first mentioned continually and the second in old age. In *Tridacna* the algae are transported by phagocytes from the mantle to the intestinal tract, where the latter, containing algae in all stages of digestion, are found in especially great numbers in the vicinity of the reduced diverticula. Some of the phagocytes transfer to the intestinal lumen, and here also numerous disintegrating algae are always found.

In such cases it is usually impossible to determine precisely to what extent intact algae are immune to the digestive juices of the host, and whether only those are attacked which are already in the process of degeneration. In their own test objects Pringsheim and von Haffner considered healthy algae to be immune, but actually the question is only of secondary importance in judging the value of algal symbiosis.

The various authors differ on how much importance to attach, in addition to such digestion of whole algae, to the utilization of metabolites passing through the cellulose membrane of healthy algae. Apparently this difference in opinion has arisen because individual algal hosts do not utilize these metabolites in the same way. For example, Winter was firmly convinced that intact algae are responsible for the starch granules found in great quantities in the plasma of *Peneroplis*. W. and M. Doyle were of the opinion that during the division of *Orbitolites* symbionts the formed metabolic products must transfer to the plasma of the host; and Pringsheim, on the basis of his autotrophic *Paramaecium bursaria* cultures, in which resorption of whole algae plays no role, also concluded that the algae continually transfer a large number of their metabolites to the ciliate plasma. This is a viewpoint which is in close agreement with Parker's research (1926) on the same test object. Pütter (1911), who made a thorough study of the metabolism of algae-containing actinians, was actually convinced that the zooxanthellae of these animals can supply the entire requirement of nitrogen, and even the entire requirement of carbon in the presence of light. Keeble assumed that the starch reserves of the symbionts are converted to fat and as such become diffused in the plasma of the hosts. He even went so far as to compare symbiotic flagellates with cells of milk glands. In *Chlorohydra* (von Haffner) and *Spongilla* (van Trigt) it is exclusively through digestion of whole algae that the animal partners acquire their supply of metabolites.

Boschma (1924, 1925, 1926, 1929) thought that the algae of madrepores continually disintegrate in great quantities in the mesenterial filaments, specifically in the area which participates in the digestive process, and that they are thus transferred to the host animals.   Yonge and Nicholls (1931) convincingly demonstrated that these were algae which degenerate under unfavorable conditions, such as hunger, darkness, high temperature, and lack of oxygen, and are then carried by the amoeboid cells to the afore-mentioned area, one which, it is known, also performs excretory functions in madrepores as well as in actinians and scyphozoans (Mouchet, 1930; Yonge, 1931; Smith, 1936).   As dark masses surrounded by mucus, the algae are excreted after their transfer to the intestinal lumen.

Two facts bear out Yonge's conclusion that algal symbiosis is not abso-lutely vital to madrepores and actinarians: (1) in nature and in experi-ments, reef-forming corals can survive in the dark with very few algae; and (2) even *Anemonia sulcata* can be kept indefinitely in the dark if well fed and provided with oxygen (Smith, 1939).   Yonge, who did much meritorious work on the physiology of corals, interprets the benefits of their symbiosis with zooxanthellae quite differently from his predecessors.   When coral are kept illuminated in glass vessels, the carbon dioxide does not accumu-late, since it is used immediately by the algae for photosynthesis.   Con-versely, the carbon dioxide does accumulate in the dark, and also in algae-free *Dendrophyllia*.   Phosphorus determinations of sea water also show complete consumption of phosphorus by the algae and an accumula-tion in *Dendrophyllia*.   When the phosphorus content of water is increased, the algae can use far more phosphorus than is normally at their disposal. Thus everything seems to indicate that the algae act as excretory organs by removing the end products of the host metabolism, that is, carbon dioxide, phosphorus, and nitrogen, or by utilizing them for synthesis. Keeble and Gamble in their time had used the same reasoning to explain the lack of excretory organs in *Convoluta*.

No contradiction is involved when increased development of excretory organs is observed in *Tridacna* in connection with symbiosis, for the symbi-onts in this case are digested in great numbers and consequently there are large quantities to be removed.

Yonge surmises that the metabolism of the host animal is substantially increased through continual removal of end products by the symbionts, and that this in the final analysis explains the unusual intensity of growth of reef corals.

2

Symbiosis with fungi and bacteria

THE PERIOD BEFORE 1910

The existence of algal symbiosis in animals had been clearly recognized by 1880, and the extent of the phenomenon was also beginning to emerge in clear outline. It was to take far longer to establish the existence of bacterial and fungal endosymbiosis, a much larger field involving more complex adaptations.

This is indeed surprising, since here the test objects had been extensively studied and the adaptations were so conspicuous that they had come to the attention of even the older zoologists. However, even where typical bacteria were encountered, authors usually resisted the idea that such bacteria might be harbored in the animal body as regular innocuous or indeed beneficial forms. They spoke only of bacteria-like formations or of "bacteroids." Occasionally one or another spoke out definitely for the bacterial nature of the questionable inclusions, but doubts and false interpretations persisted.

The case of bacterial symbiosis in ants and cockroaches illustrates how difficult it still was to recognize the existence of bacterial symbiosis even long after algal symbiosis had been accepted as fact. When Blochmann (1884) came upon symbionts in *Camponotus* and *Formica fusca* Latr., he first spoke of "plasma rodlets" and of "very conspicuous fibrous differentiation of the egg plasma." In horse ants this plasma is indeed temporarily furrowed in all directions in a most astonishing fashion. In 1888 Blochmann described the same differentiations more precisely, establishing transverse fission and infection of the follicular epithelium, and studying in broad outline their behavior during embryonic development. Although their bacterial nature was becoming more and more obvious, he could not bring himself to take a decisive stand. In a contribution to the *Festschrift des naturforschenden Vereines* at Heidelberg he stated: "At all events, these phenomena deserve our full attention, even though it should finally develop that they are symbiotic bacteria, and rather important general results are in prospect, whatever side the final decision may favor." Adlerz (1890), studying the histology of the gut in *Camponotus*, described the inserted

23

mycetocytes as regenerative elements with secretory cofunctions, and the bacteria thus as a type of ergastoplasm.   As late as 1913 Strindberg also considered the mycetocytes to be "mitosomes."

In 1887 Blochmann gave the first description of those peculiar cells embedded in regular arrangement in the fatty tissue of the blattids.   He was impressed by the great similarity of their inclusions to bacteria, for he found them again outside the cells, on the upper surface of the eggs, discovering that they reproduced there, and he found them during development of the embryo also.   This time he wrote: "In the light of our present knowledge one can scarcely do otherwise than declare these rodlets to be bacteria."   He expressed the conviction that in this case, at least, they did not represent occasional or harmful parasites, but must exercise functions of some type or other and should, therefore, be called symbionts.   Nevertheless, he did not venture to speak with finality, and in 1892 he once more published his observations in *Zentralblatt für Bakteriologie* in the hope that professional bacteriologists would decide the issue.   To Forbes (1892) and Heymons (1895) it also seemed likely that the organisms were bacterial in nature, but Cuénot (1896), Prenant (1904), and Henneguy (1904) interpreted them as metabolic products deposited by the insect, and even such an experienced histologist as K. C. Schneider (1902) was inclined to believe that the inclusions represented some type of "chondria" of still unknown function.   Mercier's detailed study (1906, 1907) dispelled all doubts and verified their bacterial nature.   Meanwhile the designation bacteroid had become so entrenched that its use was unfortunately continued, even by authors concerned with the bacteriological problems of this symbiosis.

It is easy to understand why the symbionts of the lecaniids were also observed very early, for one merely needs to tear one of these scale insects apart to have before one's eyes the considerable formations, frequently with buds, which invade the fatty tissue and the lymph.   It was clear from the very beginning that the formations were alien organisms, but their systematic position remained undecided for a long time.   Leydig (1854), in the first report on the subject, was reminded of the Pseudonavicellae. Later he identified the organism as one alleged by Lebert to be the etiological agent of the silkworm disease.   Other authors, among them Naegeli, interpreted them as unicellular algae; Balbiani (1889) classified them as microsporidians.   Labbé (1899) assigned them to the "animal kingdom," among the sporozoa incerta.   In the meantime Putnam (1880) observed that the mysterious organisms transfer to the egg cells at the upper pole, and Moniez (1887) contributed further details incompatible with their reputed sporozoan nature.   He declared the form living in *Lecanium hesperidum* Burm. to be a fungus and named it *Lecaniascus polymorphus*.

Lindner (1895) felt justified in using the term *Saccharomyces*, and the authors of rapidly amassing reports on symbionts of the lecaniids were also satisfied for a long time with the collective concept of "yeasts." Of course, this conclusion was also unwarranted.

In the case just discussed involving only a small number of the Homoptera, it was beyond doubt from the very beginning that the inclusions were organisms. This was by no means the case with the remaining Homoptera, which today constitute a large part of all symbiont bearers, despite the fact that the sites available to the symbionts are imposingly voluminous and often vividly pigmented. Furthermore, the quantities of symbionts transplanted to the egg for the purpose of transmission to the offspring are often considerable. On the other hand, understanding of the situation was impeded through the circumstance that these organs, customarily called mycetomes today, are not colonized by typical bacteria or by organisms so clearly fungoid as the lecaniid symbionts, at least not in the test objects studied by the older authors. The mycetomes seemed to be filled instead with vitelline spheres and other reserve materials. Thus we can understand Huxley's interpretation in 1858: when he observed the habitat of aphid symbionts for the first time, which had been noted in embryonic stages by Leydig in 1850, he called it a "pseudovitellus," thereby creating a collective term which later was to be applied quite generally to equally mysterious organs in the psyllids, aleurodids, and cicadids.

What a wealth of guesswork this pseudovitellus inspired! The aphids in particular presented themselves as favorable test objects, and the embryos, easy to obtain because they develop in the uterus of the mother, attracted scientists to ontogenetic research. Huxley believed that the contents represented vitelline spheres; Balbiani, in a series of studies devoted to the pseudovitellus of the aphids (1869, 1870, 1872), regarded it as a rudimentary male gonad; Witlaczil (1882), for a time, interpreted it as a kidney replacing the lacking Malpighian tubules. Especially difficult to understand were the devices for transmitting the symbionts from the mother to the young, for in the aphids they are intricately situated and differ in winter and summer eggs. The yolk-rich winter eggs, requiring fecundation, are infected before their development, in the ordinary manner. But, as we are to hear, the transfer of these symbionts to the parthenogenetic generations does not occur until the blastoderm stage or even later. Balbiani mentioned a special petiolate follicle cell at the posterior end of the winter eggs, and emerging from this cell the striking green pigmented infection mass, the "rudimentary testes." Tannreuther (1907) interpreted the symbionts as the nuclei of follicle cells transferring to the egg, then disintegrating and thus forming the spheres of the pseudovitellus.

Processes occurring in the embryos of the summer eggs were also misunderstood. Metschnikoff (1866) described a strange self-amputation of the embryo. Witlaczil (1884) rejected this finding but did not himself arrive at a better understanding, nor did Will (1889), the next author to be concerned with these questions. Henneguy, in *Les Insectes* (1904), had an equally incorrect concept of the first appearance of the pseudovitellus in the embryos; and according to Tannreuther (1907) follicle cells transferred to the egg duct, dividing further here and forming the yolk, which then passed over to the embryos—and this despite the fact that Stevens (1905) in cytological investigations of the aphid egg had already observed the content of neighboring pseudovitellus cells flowing directly into the embryo through an opening in the follicular epithelium. All these authors were at least in agreement that the pseudovitellus did not emerge until the blastoderm stage, but now Hirschler (1912) was to mistake it for the actual yolk, asserting that it was already present in the unfertilized summer egg.

Krassilstschik (1889) had already quite correctly recognized that, in addition to the pseudovitellus, unquestionable rod-shaped bacteria regularly make their appearance in certain aphids as visibly well-adjusted guests, but no attention was given to his work.

As early as 1866 Metschnikoff had made various observations on the pseudovitellus of the psyllids, but of course it was impossible that he should recognize the true nature of those "albuminous bodies," which he observed transferring to the egg from the maternal body, and the course of which he studied in broad outline during their embryonic development. Nor did Witlaczil (1885) advance beyond his predecessor in understanding these unusual phenomena.

The earliest mention of the pseudovitellus in aleurodids is probably by Signoret (1867), who also had observed the orange-colored primary mycetome in the yolk of the embryos. Understandably, the mycetomes of the cicadas were found chiefly in our native spittle cicadas (Porta, 1900). In 1908 they were still called accessory sexual organs by Guilbeau. Heymons (1899), encountering the large botryoidal mycetomes of the singing cicadas in the course of ontogenetic studies, described their genesis in bold outline but was at a loss to explain their function. Regarding the mycetomes of scale insects, Berlese (1893) was alone in discovering the great unpaired organ of *Pseudococcus citri* Risso. He declared it to be a storage organ, one which admittedly differed greatly from the fatty tissue in structure.

Meanwhile symbiosis was seldom observed in other insect families. In connection with the historical account of the discovery of symbiosis in bloodsucking animals, we shall discuss more fully the origin of our information on the mycetome of human lice, which reaches back as far as Hooke (1665) and Swammerdam (1669). It was inevitable that other insect anatomists should also notice the voluminous crypto-sequences of heteropterous bugs colonized by bacteria. We find them mentioned by Treviranus (1809) and by Ramdohr (1811), and a good illustration with the caption "cordons valvuleux" appeared in Dufour's work, "Recherches anatomiques et physiologiques sur les Hémiptères" (1833). Once more it was Leydig who took the lead, establishing vibrio-like organisms in them in 1857; and Forbes later studied them in greater detail. His valuable observations proving regulated cooperative living between many bugs and bacteria did not appear until 1896, and then only in very concentrated form. Forbes had earlier persuaded Burrill (1883) to make a short report on the symbionts of a *Blissus* species.

In 1902 Holmgren had unwittingly described the symbiotic organs of *Apion* and *Dasytes*. In both beetles they are in close relation to the renal organs, being represented in *Apion* by two transformed Malpighian tubules, and in *Dasytes* by masses of mycetocytes which hang between these tubules at the end of the gut. Thus he ascribed excretory functions to them and declared the *Dasytes* symbionts to be excretions arising in the nuclei and transferring to the plasma. The filamentous symbionts of *Apion*, each located in a periblast, he likewise declared to be nuclear derivates which grow out into chromosome-like ribbons. To him, the change in form undergone here by the symbionts represented the individual stages of the maturing excretions.

In another beetle the nature of the discovery was clarified with exceptional rapidity. When Karawaiew (1899) came upon strange proliferation-like ceca in the gut of *Sitodrepa*, he found that their cells were filled with organisms which he first called flagellates. Less than a year later Escherich explained that they were saccharomycetes, asserting that this must be a remarkable case of symbiosis but making no effort to discover the transmission devices.

Another instance of rapid clarification is Petri's research on the olive fly, *Dacus oleae* Gmel. In 1904 the Italian zoologist began a series of reports on the bacterial symbiosis which he had discovered in this economically important pest. In 1909 he summarized his reports in a valuable monograph, discussing anatomical conditions, transmission, and biological problems.

Petri's publication marked the close of an era which represented the tentative beginnings of modern symbiosis research and reached back to the days of Leydig, Huxley, and Metschnikoff. Much had been observed, but it was not yet understood and was subject to strange and misleading interpretations. A few cases had already been recognized as regulated cooperative living with bacteria and fungi, for instance, the symbiosis of blattids, Homoptera, lecaniids, *Sitodrepa*, and the olive fly, but such instances were regarded more or less as curiosities, and indeed as distressing exceptions disturbing the order of things, and were represented in the biological literature in unrelated and seldom read reports.

The same neglect prevailed after the discovery that some animals have excretory organs which are regularly colonized by microorganisms. Lacaze-Duthiers (1874) and Giard (1888) proved the presence of strange fungi in the storage kidneys of *Molgula;* Garnault (1887) was the first to observe the bacteria which are always present in the concrement glands of *Cyclostoma,* and Maziarski (1905) reported on the nephridia of the lumbricoids. But these findings, questioned in some cases, failed to attract the attention of wider circles.

FROM 1910 TO THE PRESENT

The situation changed suddenly when the true nature of the pseudovitellus in the Homoptera was recognized by Umberto Pierantoni in Italy and by Karel Šulc in Moravia simultaneously and completely independently of each other. Now a vast field was opened up to research, rapidly extending beyond the Homoptera, a field in which the few cases that had already been correctly evaluated could be meaningfully placed. Pierantoni and Šulc presented their decisive findings in a series of lectures which began toward the close of 1909 and ended early in the following year. The lectures were published in 1910. Pierantoni spoke before the Società dei Naturalisti in Naples on the symbiotic organs discovered by him in the scale insect *Icerya* (1909), and later on those of the aphids and of *Pseudococcus* (1910). Šulc reported on the symbiosis of the cicada before the Naturwissenschaftliche Gesellschaft of Mährisch-Ostrau in 1909. The following year he presented a report entitled "Pseudovitellus and similar tissues of the Homoptera as sites of symbiotic saccharomycetes" before the Königliche Gesellschaft der Wissenschaften in Prague. In this report he introduced the designation "mycetome" which is now in general use for such closed and symbiotically populated organs.[1]

[1] In view of the importance of these reports to symbiosis research, a few dates showing the independence of the authors may be helpful. On November 5, 1909, Šulc delivered a lecture entitled, "On the biology of yeast fungi and their symbiosis with insects." (A report on this lecture appeared later in 1923 in *Sborník přírodovědecké společnosti v. M. Ostravě,* 2.) He had submitted his paper in October to the Royal Society of Sciences, where it had been discussed in the scientific section, and published in the session reports on March 20. In the same year another publication under the title "Symbiotic saccharomyces of the true cicadas" appeared in the same place. Pieran-

With the publication of these reports it seemed as though a blindfold had been removed from the eyes! Numerous examples of symbiosis in the Homoptera were described in rapid succession, and numerous other important insect families or smaller systematic entities were recognized as symbiont bearers. Soon certain ecologically conditioned areas of distribution were clearly demarcated, indicating the direction in which new instances of symbiosis might be sought. In addition to examples of symbiosis in animals which feed on plant saps came examples of symbiosis established with animals which feed exclusively on mammalian blood. A third, less sharply defined, group comprised insects feeding on wood and other vegetable constituents, especially on cereals. And finally it was discovered that some marine animals are not capable of producing light by themselves, as had been generally assumed until then, but have this ability as the result of symbiosis with luminescent bacteria.

In the following period Šulc contributed only one publication on the scale insect *Margarodes* and one on the fulgorids, but Pierantoni continued his study of this new and entrancing field in collaboration with several of his students (Convenevole, Gambetta, Getzel, Rondelli, Salfi, Tarsia in Curia, and Zirpolo). I too soon fell victim to the spell of this subject, and from 1911 on devoted myself to it to the extent that was possible. I had the assistance of many students, whose collaboration I should like to acknowledge here. Working with me were: Aschner, Breest, Breitsprecher, H. Fraenkel, Glumb, Hecht, Heitz, Herfurth, Jaschke, Kiefer, Klevenhusen, Knop, Koch, Kuskop, Lilienstern, H. J. Müller, Nolte, Pfeiffer, Profft, Rau, Richter, Ries, Rosenkranz, Scheinert, Schneider, Schölzel, Schomann, I. Schwarz, Sell, Stammer, Stier, Tóth, Tremblay, Walczuch, and Zacharias. Some of these students—Aschner, Koch, H. J. Müller, Ries, Stammer, Tóth, and I. Schwarz—continued their study of symbiosis as independent researchers and are now turning over investigations in this field to students of their own. In Ischia I was also able to acquaint Tremblay with the field. We are indebted to him for a number of studies in symbiosis. Other scholars in biology rarely took part in symbiosis research, contributing occasional papers only. Not until recently has there been a consistent growth of interest in our field, and especially in the United States.

toni spoke in Naples on December 19, 1909, on "L'origine di alcuni organi d'*Icerya purchasi* e la simbiosi ereditaria," and on February 6, 1910, on "Origine e struttura del corpo ovale del *Dactylobius citri* e del corpo verde dell' *Aphis brassicae*." The first paper appeared in *Boll. Soc. Naturalisti*, **23** (1909), the volume being published on May 30, 1910. The second paper appeared in the same journal, **24** (1910), being published on May 30, 1911. Pierantoni spoke on *Aphrophora* symbiosis on May 12, 1910. The paper appeared simultaneously with the preceding ones.

It has become almost impossible to survey fully the great diversity of anatomical, histological, and embryological details that resulted from intensive investigation of Homoptera.   As regards plant lice, Pierantoni and Šulc had limited themselves to a few remarks on the true nature of their mycetomes.   In 1912 I described the infection of the winter eggs on the basis of our new knowledge, and Sell in 1919 attempted to clarify the extremely contradictory data of the older authors on the transfer of the symbionts to the summer embryos.   Klevenhusen (1927) augmented our data on individual modes of transfer, which are quite varied in detail, and showed that numerous aphids live with a second symbiont, a phenomenon which Krassilstschik had noted earlier.   Klevenhusen proved that some aphids are even allied with a third symbiont, and that these accessory guests are not nearly so well adapted.   In spite of this work, the process of embryo infection was not yet fully understood, and Tóth (1933) therefore attacked the subject again.   Later (1937) he studied the interesting symbiotic cycle of *Pemphigus*, which gives evidence of amazingly widespread control by the host animal.   Because subfamilies of the adelgines (chermesines) and phylloxerines had been given but scant attention in earlier investigations, I persuaded Profft (1937) to make a study of these groups.

From Pierantoni's institute, even before Klevenhusen's work, Rondelli had published reports (1925, 1928) on several disymbiotic eriosomatines. A series of short papers by Paillot (1929–1933) propounded the question of an additional symbiosis with rod-shaped bacteria, a possibility which was exhaustively presented and then discarded in "L'Infection chez les insectes" (1933).   Symbiosis in the blood louse had been investigated earlier by Peklo (1912, 1916), an author who had also worked with the cultivation of aphid symbionts and was inclined to identify them with azotic bacteria.   Schoel (1934), the first to consider the bacteriological aspect of aphid symbiosis on a broader basis, came to the conclusion that the microorganisms in the form of round utricles in the mycetome could revert to the typical rod shape outside the host.   A publication of Uichanco's (1924) occasioned more confusion than clarity.   Buchner (1958) made the surprising discovery that a number of tropical aphids (hormaphidines) had lost the hereditary symbionts previously found in all other aphids and that these symbionts had been replaced by vigorously budding fungi, very similar to those of the lecaniines, which colonize the lymph and fatty tissues.   This finding was recently confirmed by Kolb (1963) in a study of the entire Thelaxidae family.   Tóth's work on symbiosis in *Pemphigus* was supplemented and elaborated by Lampel (1958, 1960).

The aleurodids, with their unique transferral method, provided a surprise.   Here the symbionts are introduced into the egg together with the cells they inhabit, and these cells, outliving the mother organism, are still preserved for a while in the embryo (Buchner, 1912, 1918).   The shifting of their mycetomes in the course of postembryonic development, which had been reported in 1912, was described in greater detail in Weber's monograph on *Trialeurodes vaporariorum* Westw. (1935).   Recently, valuable supplementary data have been presented by Tremblay (1959).   He was able to show that in *Bemisia* each ovocyte is infected by only a single mycetocyte.

In 1912 I found bipartite mycetomes in psyllids, and Breest demonstrated (1914) that two distinct forms may be differentiated during egg infection.   Profft (1937) more or less completed the collection of data through comparative study of a larger number of species, thereby refuting the opinion held by Salfi (1926) and Tarsia in Curia (1934), who claimed to see in *Troiza alacris* Flor. only a single type of symbiont.   Later Mahdihassan (1947) likewise supported the idea of two types of symbionts, believing that this could be corroborated by cultures.

Showing far greater variety than the aphids, aleurodids, and the psyllids, are the scale insects, which, with their numerous subfamilies, have a more complex organization.   A series of publications, some of them of little importance, were devoted to the lecaniines which were relatively well known even before 1910 (Buchner, 1912, 1921; Breest, 1914; Emeis, 1915; Teodoro, 1916, 1918; Strindberg, 1919; Brues and Glaser, 1921; Brain, 1923; W. Schwartz, 1924, 1932; Granovsky, 1929; Mahdihassan, 1929; Benedek and Specht, 1933; Goux, 1944; Poisson and Pesson, 1937; Tremblay, 1961).   Among these reports, it was chiefly those by Schwartz which represented progress.   For the first time he was seriously devoting himself here to the mycological problems of this symbiosis, and he was now able to show that the microorganisms previously called saccharomycetes, or merely yeasts, were actually the conidia of the ascomycetes, which are so difficult to cultivate.   The data compiled by Benedek and Specht on their alleged symbiont cultures can scarcely be taken seriously in view of the ineffective technique they used.   A considerable advance in our knowledge of symbiosis in the Lecaniinae, which at first glance seems so monotonous, was made by Tremblay (1961), when he established that there are species in which the infected fat cells become polyploid, increase in size accordingly, and are able to form extensive, closed groups.   Steinhaus (1955) presented photomicrographs of several lecaniine symbionts, but he was unsuccessful in his attempts to cultivate them.

Studies of symbiosis in the diaspinines were made by Bennett and Brown (1958) and by Tremblay (1958, 1959, 1960).

Turning to the symbiosis of the pseudococcines, we find that Pierantoni elaborated his early reports on *Pseudococcus citri* Risso. For instance, he added a fuller description of the embryonic development (1913). Shorter contributions by Emeis (1915) and Shinji (1920) followed, and also my own investigations, which included other species and genera, some of which behaved quite exceptionally (Buchner, 1921). Schrader (1923) made the interesting discovery that the nuclei of the *Pseudococcus* mycetomes originate through fusion of polar bodies and segmentation nuclei and thus could not really be compared with the remaining embryonic nuclei. Walczuch rounded out our knowledge of this subfamily of the scale insects from morphological, embryological, and bacteriological points of view in an excellent comparative study showing the diversity of coccid symbiosis (1932). Fink (1952) made successful pure cultures of *Pseudococcus* symbionts and greatly broadened our knowledge of their structure and behavior under a variety of conditions. Steinhaus (1951) presented photomicrographs of several pseudococcine symbionts. Köhler and Schwartz (1961, 1962, 1962) also cultivated successfully the symbionts of *P. citri* and *maritimus* and studied their characteristics.

Studies of *Macrocerococcus* and *Puto* (Buchner, 1955), of several species of *Rastrococcus* (Buchner, 1957), and of the myrmecophilous *Hippeococcus* (1957) demonstrated that symbiosis of pseudococcines is by no means so uniform as it had been assumed to be at first. It was even shown that a symbiosis with yeast-like organisms occurs and that at times the symbiosis disappears completely. Köhler and Schwartz (1961) made experimental studies of symbiosis in *Pseudococcus maritimus*.

The symbiosis of monophlebines was developed chiefly by Walczuch. Pierantoni had followed up his preliminary report on *Icerya purchasi* with a more complete presentation (1912, 1914) in which he gave the first description of the development of specific transmission forms which later were to be observed so frequently. Without knowledge of Pierantoni's reports, Shinji (1920) described egg infection in the same species. Now other varieties of *Icerya* and *Echinocerya* as well as *Monophlebus* and *Monophlebidius* were investigated. It was shown that in this subfamily, in addition to unchecked accessory symbionts, there are also well-adjusted ones, situated in a specific part of the mycetome. Getzel (1936) reported on culture experiments.

Jakubski and his students, Kalicka-Fijalkowska (1928) and Boratynski (1928), investigated the symbiosis of *Margarodes polonicus*. The mycetomes in this species differed markedly from those which Šulc (1923) had discovered in an undetermined species of the same genus. In Šulc's species the mycetomes enter into extremely peculiar relations with the oviduct, and for this reason the infection stages transfer to the oviduct and from there to the egg cells. Mahdihassan (1947) also studied this interesting form.

The unique symbiosis of *Marchalina hellenica* Genn. was discovered by Hovasse (1930), though he did not pursue the subject in all its interesting details. For the first time in the investigation of coccids a colony of intestinal epithelial cells was found, which, closely packed with symbiotic bacteria, assumed tremendous proportions. If infection of the egg cells actually takes place in the face of such conditions, it is surely the only instance among the scale insects and, except for the apionines, among all other symbiont-bearing insects as well.

The ortheziines, so handsomely adorned with wax secretions, I myself investigated only superficially (Buchner, 1921). They were studied very thoroughly by Walczuch (1932), who observed a peculiar egg infection and interesting embryological details. Mahdihassan (1946) claimed to have cultivated *Orthezia* symbionts. However, Köhler and Schwartz (1962) were the first to succeed in this beyond any doubt. They also proved that *Orthezia* can dispense with its symbionts.

Up to this time only Shinji (1920), Richter (1928), and Walczuch (1932) had concerned themselves with the symbiosis of the Asterolecaniae. For the Diaspinae also in the beginning only a few communications were in existence; they were from Šulc (1906), Breest (1914), Buchner (1921), and Richter (1928). However, during the last few years interest in the Diaspinae has awakened, for several reasons. Brown and Bennet (1957), Bennet and Brown (1958), and Tremblay (1958) were able to show that here, too, as in *Pseudococcus*, the mycetocytes arise from the fusion of cleavage nuclei with derivatives of naturation division. Surprisingly it was also established that *Diaspis pentagona* produces two kinds of eggs, white ones from which males hatch, and reddish ones which yield females. Tremblay (1959, 1960, 1961), as well as Brown and De Lotto (1959) and Brown and McKenzie (1962), extended our knowledge of symbiosis and sex determination in the Diaspinae.

The lac scale insects (tachardiines) were investigated for the first time by Mahdihassan (1924, 1928, 1929), one of the studies having been made in collaboration with Sreenivasaya (1929). Walczuch completed their fragmentary data, taking into consideration also the manner of transmission and the embryonic development. Surprisingly, entirely different types of symbiosis were revealed even within this small group. On the one hand, Walczuch observed monosymbiotic forms with free yeasts in the lymph or with isolated mycetocytes situated in the fatty tissue; and, on the other hand, he found di-symbiotic forms in which the two types of organisms were contained in cells of various structure, harmoniously united in little colonies.

For the coccines and chermesines we unfortunately have inadequate data, from Pierantoni (1911), Mahdihassan (1929), and Šulc (1906). The stictococcines, which are limited to tropical Africa and the males of which are born without mouth parts and thus are never able to ingest food of any kind, produce two types of eggs: infected eggs yielding females and symbiont-free eggs yielding males. In one of the two species investigated, yeasts appeared in place of what apparently were the original symbionts. In this connection, the method of transmission and the highly abnormal embryonic development in both forms were directed into new channels (Buchner, 1954, 1955). In a third contribution to stictococcine symbiosis (Buchner, 1963), a species was also described which, in addition to the original symbionts, contains a second bacterium which is transmitted independently of them in both time and location. Another surprising phenomenon appeared during investigation of the apiomorphines which produce galls on *Eucalyptus* species. They develop in a strange manner, but they are without symbionts, apparently because their mode of nutrition differs (Buchner, 1957).

Thus new publications followed in rapid succession, but the picture they presented, though diversified, was far from complete.

Turning our attention to the symbiosis of the cicadas, we find a much smaller number of publications. Nevertheless these studies were more fruitful than those on scale insects. In 1911 and 1912 I was in a position to add to the data on symbiosis in our native spittle insects, and especially in the large singing cicadas, describing in minute detail the transfer of the two kinds of symbionts always present in them. Representing the research of several years, my study on the symbiotic adaptations of the cicadas opened up a veritable fairyland of insect symbiosis, although the thirty-four varieties I listed at the time by no means exhausted all the possibilities (Buchner, 1925). Before this time only a few disymbiotic forms had been discovered. Now a succession of trisymbiotic forms came to light, and it could be shown that all three types are harbored in equally regulated fashion in the egg cells. The cultivation of specific infection forms was seen to be a widespread phenomenon. In one of the most frequent mycetome types, such cultivation occurs in certain segments and is traceable to an unusual method of induction through the larval oviduct, to which I gave the designation "infection mound."

Even with the limited material available at the time, it was apparent that the adaptations of cicadids and fulgorids differ radically from each other. With cicadids the additional symbionts associating with the hereditary form are usually harbored in specific zones of the mycetomes which are already present. The fulgorids prefer to build new, independent living quarters with constant variation in histologic construction. With the acquisition of new material it became possible to study the relations between the symbiosis and the host's system, and therewith the tribal history. Because of greater or lesser dovetailing, the sequence of acquisition in polysymbiotic species could be reconstructed with considerable exactness.

The same year in which I made a preliminary report on my results (*System und Symbiose*, 1924) Šulc published his study of six fulgorids. Our findings were complementary in some respects, inconsistent in others.

A publication by Richter (1928) presented only minor supplementary facts on cicadid symbiosis, based chiefly on material from Formosa. By Rau's monograph on membracids (1943) our knowledge of the subject was markedly extended. For this study I was able to place at his disposal material of ninety species, principally from Brazil. Unexpected variety was revealed, of particular interest because it resulted from a tendency to receive additional symbionts, a tendency apparently fully in effect today. It was noted that these symbionts attach themselves as much as possible to the paired mycetomes already present. Thus we find instances in which two, three, four, five, indeed even six, tenants are more or less harmoniously adapted. Although it is true that some cause serious disturbances, most of them participate in the infection of eggs in rather well-regulated fashion. Rau's work greatly widened our understanding of this gradual step-by-step adaptation and of the underlying principles.

Resühr (1938) contributed an excellent study on the cytology of the delicate, tube-shaped, fluid-filled cercopid symbionts, which apparently represent bacteria in varying degrees of degeneration. Mahdihassan (1939, 1946, 1947), regarding them as remnants of disintegrated host cells, denied their symbiotic nature, but his views in opposition to those of all other authors who have given serious attention to these structures.

H. J. Müller's work (1938, 1940), with more than 300 illustrations, represented a momentous advance in our study of fulgorid symbiosis. His work was based on 186 species, 157 of which were being tested for the first time; 130 of these species, collected in Brazil, could, unfortunately, be classified only to subfamilies even by the specialist. At first glance, one may be confused by the multiplicity of mycetome types and all the possible combinations, for which we refer our readers to the specialized portion of this book. Yet their statistical analysis has demonstrated some surprising parallels, sometimes favoring and sometimes excluding acceptance of certain additional forms. The origin of the transmission devices, the various methods of egg infection, the disintegration of infection masses that usually consist of two or three types, but occasionally of more; their embryonic and post-embryonic development—all were investigated by comparison of numerous forms. Müller's illustrations are excellent. Again and again we are amazed at the regulated, adaptive pattern, reaching down to the last histological and cytological details, details which persist from species to species.

In 1949 H. J. Müller was able to postulate a hypothetical family tree of this complicated cicada symbiosis, using as his source the 369 cicadas already investigated in relation to their symbiosis.   It contained many an uncertainty, but still it gave us a valuable insight into the matter, particularly through the assumption of manifold reciprocal antibiotic effects leading to the elimination of symbionts.

Müller also made an extremely important contribution to the phylogeny of cicadid symbiosis by investigating the symbiosis of a peloridiid, which is doubtless a remnant of the paleolithic era (1951).   Thus he has acquainted us with the earliest cicadid symbiosis, the starting points for all later developments.

Ermisch (1960), a student of Müller, made an important contribution to our knowledge of cicada symbiosis by a renewed, extensive investigation of the fulgoroids with particular consideration of the araeopodids.   Thanks to her work, Müller was prompted to a revision of his first family tree, which led to a considerable simplification and to a good harmonization of morphological criteria of the hosts with the actual symbiont content (1962).

The heteropterous lice, which suck plant juices, cannot possibly display such variety, but today their symbiosis seems far more diverse than indicated by Forbes's investigation.   Glasgow (1914) and Kuskop (1924) acquainted us with a large number of variously shaped intestinal appendages colonized by bacteria; and Schneider (1940) demonstrated that normal mycetomes occur in some genera.   In 1939 Kuskop and Rosenkranz proved that transmission of the bacteria present in the intestinal lumen does not proceed through ovarial infection, as Pierantoni (1932) and Convenevole (1933) had maintained, but circuitously by means of external smearing of eggs during deposit.   In *Coptosoma* Schneider even found a unique transformation of the female intestine into an apparatus for producing bacteria-filled capsules.   These, set between the eggs, were sucked out by the young larvae.   The details of this amazing device were studied in greater detail H. J. Müller (1956), who analyzed the behavior of the animals and the effect of symbiont withdrawal.   In a lygaeid of the genus *Stilbocoris*, Carayon (1963) discovered an interesting preliminary stage of its highly complicated method of transmission.   The larvae are born surrounded only by a delicate cuticle, and an unprotected drop of tenacious gut secretion is placed next to each, which the larvae immediately seek and take up as their first food.   In another instance, complete blockage of the intestinal crypts necessitated formation of an individual smearing organ, as Rosenkranz (1939) had shown for the acanthosomines.   However, wherever true mycetomes are present, the method of ovarial infection is used, as described by Schneider.   Pierantoni reported briefly (1951, preliminary; 1954) on a type occurring in *Tropidothorax* and differing radically in localization and transmission. L. Huber-Schneider, reporting on sym-biosis in *Mesocerus (Syromastes) marginatus,* treated culture and morphology of the bacteria in particular (1957). The same year H. Schorr published a study on *Brachypelta.* In this sand-digging cydnid the newborn larvae remain for days with the mother, sucking up bacteria which emerge from the anus upon tactile stimulation. Steinhaus (1951) and also Steinhaus, Batey, and Boerke (1956) provided valuable information on the mor-phology and culture of the symbionts of several Heteroptera. Bonhag and Wick's statement (1953) that two types of symbionts are present in the male subgenital glands of the lygaeid *Oncopeltus* is probably erroneous.

Since hemipterous symbiosis is apparently linked with the host's special-ized mode of nutrition (corresponding devices were lacking completely in the predatory Heteroptera), it seemed logical to search for symbiosis in other cases of specialized nutrition. This was what first led to my work with lignicolous insects. Resuming my investigation of the anobiids—of which only *Sitodrepa,* clearly a form that has only recently become a stored-products pest, had been recognized as a symbiont bearer—I found that lignicolous forms usually have far more highly developed intestinal pro-trusions filled with yeasts (1921). I also described the behavior of symbi-onts during metamorphosis and showed that these beetles are equipped with special depots on the ovipositor, by means of which the upper surface of the egg is contaminated by the fungi during deposit, so that the emerging larvae are reinfected on devouring a part of the eggshell. Later this type of transmission proved to be widespread.

Breitsprecher, with more material at his disposal, could more fully describe the smearing apparatus; he discovered that it may vary consider-ably from genus to genus (1928). Nolte (1938) showed that the symbiosis of the dorcatomines closely follows that of other anobiids; and Kiefer (1932) studied the relation of colony density to temperature and nourishment in yeasts. Heitz (1927) and W. Müller (1934) applied themselves to culture experiments and mycological problems. Later Pant and Fraenkel (1954) and also Graebner (1954) made successful cultures of anobiid symbionts. Graebner, using the fluorescent microscope, provided additional data on the interrelations of the two partners. Lately, botanists have taken a serious interest in *Sitodrepa* symbiosis (Kühlwein and Jurzitza, 1961; Jurzitza, 1962; Grinbergs, 1961), and we have the zoologist Foeckler (1961) to thank for interesting experiments in replacing the species-specific symbionts with other yeasts.

In 1847 Stein made a comparative study of female sex organs in beetles. On the ovipositors of cerambycids he observed auxiliary glands which were apparently identical with the symbiont depots of the anobiids. Thus it was no surprise to discover similar yeast-containing protrusions and a similar method of transmitting the symbionts.   At my suggestion Heitz (1927) did some preliminary work on cerambycid symbiosis.   I myself then went somewhat more deeply into the subject (1928, 1930), later turning the work over to Schomann (1937) for detailed study.   Examination of 184 varieties showed that relatively few representatives of this family live symbiotically and that the nature of their nutrition is apparently a factor.   The following should be mentioned as authors who experimented in the culture of symbionts of the cerambycids, some of them considering their relation to external conditions: Heitz (1927), Schimitschek (1929), Ekblom (1931, 1932), Kiefer (1932), W. Müller (1934), and Schanderl (1942).   Supplementing our previous knowledge were also Gräbner's studies (1954). The mycological questions were treated in this case also by Jurzitza (1959), Jurzitza, Kühlwein, and Kreger-van Rij (1960), and Jurzitza (1962), and their findings are in hormony with the concept presented by their predecessors, that is, that the yeasts are all species of the genus *Candida*.

In addition to the anobiids and the cerambycids, there is a third insect group that lives largely on wood—the curculionids.   On the basis of considerable material I was able to show that only one type of bacterial symbiosis is involved here and that it is not limited to species living exclusively on wood but includes those living in and on herbaceous plants as well (1927, 1928, 1930, 1933).   In detail the devices are far more varied than those of the anobiids and the cerambycids.   The symbiotic sites appear to be evaginations in the midgut, massive cell quantities lodged on them, cells embedded in the fatty tissue, or Malpighian tubules, which are sometimes greatly modified.   The larval and imaginal sites are often distinct, and transmission is sometimes achieved indirectly either by infection of the nutrient chambers of the ovarioles or by means of special syringe-like organs, which in turn contaminate the eggshell.   Glumb (1933) also investigated the latter transmission method on the basis of new material. Nolte (1937) studied the apionines in particular.   Scheinert, in valuable supplementary work on embryological phenomena (1933), discovered the strange structures which concentrate the bacteria dispersed in the egg yolk; he also observed the guided migrations of the cells to which the bacteria are eventually admitted.   In this way it was possible to trace the infection of the oviducts to an infection of the original sexual cells.

In spite of these contributions great gaps undoubtedly still exist in our knowledge of symbiosis in the diversified curculionid family.   Symbiosis apparently does not occur in all its representatives, for, although it is found in all wood eaters, it is lacking in many herbivorous forms.   *Calandra*, so injurious to grain, farinaceous products, and the like, makes its appearance as a symbiont bearer just as the cereal-loving *Sitodrepa* does among the anobiids.   The unpaired mycetome of *Calandra*, loosely attached to the initial section of the midgut, was still a mystery to Mansour in 1927, but that very same year it was correctly interpreted by Pierantoni and by me. Pierantoni (1927) and Tarsia in Curia (1933) pursued the subject further. Mansour (1930) not only concurred in our interpretation but also provided the interesting information, later to be confirmed by Koch (1939), that the Egyptian variety of *Calandra granaria* L., unlike the European variety and *Calandra oryzae* L., either possesses no symbionts at all or has apparently lost them (1934, 1935). H. Schneider (1954, 1956) resumed his study of the two *Calandra* species including the Egyptian variety, admirably rounding out our information on their symbiosis and its disintegration in the adults. Musgrave and his students have concerned themselves extensively with the symbiosis of species of the genus *Sitophilus* (1956, 1958, 1961, 1962, 1963), and in 1960 Hänsel made a series of observations on South American material which have awakened the desire for more extensive investigations.

The bostrychids are like *Calandra* in that all lignicolous species and *Rhizopertha dominica* F., which develops in grain, clearly possess the mycetomes discovered by Mansour in 1934. In 1954 I made a detailed study of bostrychid symbiosis with *Sinoxylon*, *Scobicia*, and *Apate* as my material. I described the method of transmission, which had gone unrecognized by Mansour, the behavior during embryonic development, and the disintegration of mycetomes in male adults. Huger (1954, 1956) confirmed and augmented my findings with data on the injurious influence of higher temperatures.

Symbionts were found also among the lyctids, the universal wood pests which possess but few species. Gambetta (1927) briefly described mycetomes with two types of organisms; and Koch (1936) gave a complete presentation of their symbiotic cycle, with treatment of the relations of the two forms during egg infection, which takes place in unusual ways, and during embryonic development.

That the ipids also are not exceptions was evident from unpublished observations of Tretzel and from brief remarks of Stammer (1933), who discovered that in several species an infection of nutrient cells effected the transmission. Experiments of my own have shown the wide range and unsuspected diversity of ipid symbiosis. We have a detailed account only for the genus *Coccotrypes*, which houses its symbionts in parts of the Malpighian tubules (Buchner, 1961), but we can say now on the basis of studies embracing a number of forms that, although the mode of transmission is uniform, the localization of the symbionts can be quite varied.

Oryzaephilus (Silvanus) surinamensis L. was the test object for symbiotic studies of the cucujids, that small group of tiny beetles which live beneath the bark of deciduous trees and conifers and which have also adapted themselves to the storage places of men, chiefly tobacco warehouses. It was Pierantoni (1929, 1930) who discovered symbiosis in this species; Koch (1930, 1931, 1936) and Huger (1954, 1956) made an exhaustive study.

In view of such widespread symbiosis among lignivorous insects, it seemed logical to look for it also among the siricids. In the course of the search I observed, to my great astonishment, at the base of the ovipositor those jets, so conspicuous and yet until then unnoticed, which permit the spores of a basidiomycete to pass into the larval tunnel and thence into the wood which surrounds the larvae afterward (Buchner, 1927, 1928). Later the mycelia of the fungus form the chief nourishment of the larvae of wood wasps, but even so this is something quite different from a typical "ambrosial culture." For a time it was not clear how the fungus depots were filled. Parkin (1941, 1942) discovered in some of the larvae strange hypopleural organs which opened outward and were filled with fungi, but it was not until 1957 that Francke-Grosmann could show that they are devices, appearing only in female larvae, which enable fungi to survive in dry wood and at the same time provide the means for infecting the imaginal intersegmental organs.

Because of the importance of siricid larvae to foresters, rather extensive literature devoted particularly to mycological problems has developed: Chrystal, 1928; Cartwright, 1929, 1938; Clark, 1933; W. Müller, 1934; Francke-Grosmann, 1939; Rawlings, 1951; Stillwell, 1960. Among these contributions those of Francke-Grosmann are the most outstanding.

Little was known about the way the hylecoetines, which carry on a typical ambrosial culture, provide for a continuous supply of symbiotic fungus. Therefore I included *Hylecoetus dermestoides* L. in my studies and discovered special devices resembling those of anobiids and cerambycids (1930). Francke-Grosmann made valuable contributions on *Hylecoetus* symbiosis, placing particular emphasis on mycological aspects (1951, 1952, 1961; Batra and Francke-Grosmann, 1961).

At first specific transmission devices appeared to be lacking in the ipids (*Xyleborini* and *Xyloternini*), which cultivate ambrosial fungi on the walls of their passages. In *Anisandrus dispar*, Schneider-Orelli (1911, 1913) reported that fungi in the midgut survive the host's hibernation, but this has recently been proved erroneous. In both groups Francke-Grosmann (1952, 1956, 1963) found glandular pockets and depressions, varying considerably from case to case, by means of which the mother animals infect the walls of the larval galleries they are creating. Males are generally without these devices that evolve in connection with symbiosis, but in certain genera they have taken over this service (Finnegan, 1963; Farris, 1963). Further publications of Francke-Grosmann and Schedl (1960), Fernando (1960), Schedl (1962), and Farris (1963) show the great variety of these adaptations. Francke-Grossmann (1956) was also able to amplify the observations of Strohmeyer (1918) and of Beeson (1917) on the transmission of the fungi cultivated by platypodids.

Even within the curculionids a number of symbiont carriers which feed on fresh plant parts were found. Furthermore, we had Petri's observations on the symbiosis of the olive fly. Thus it was no great surprise that from time to time other instances of this mode of nutrition were found. In examining the remaining trypetids, Stammer (1929) found that *Dacus* is not unique in living symbiotically, for all its relatives feeding on fruits, receptacles, stems, or leaves live in association with bacteria. Thus with regard to localization and transmission, a progression could be set up from extremely primitive states to such perfected ones as *Dacus*. Hellmuth (1956) provided a welcome study on bacteriological and physiological questions of trypetid symbiosis based on thirty-seven species.

Stammer (1929) also discovered bacterial symbiosis in lagriids, which feed on fresh and decomposing leaves; with dry material he was able to point out the auxiliary organs in the ovipositor. Some are extremely simple and others are voluminous and diversely luxurious. Their function is that of encasing the eggs with bacteria-rich mucus at the moment of deposit.

A report which A. Bournier made at the International Congress of Entomology in Vienna in 1960 happily opens up to symbiosis research a new field—the Thysanoptera, which live chiefly on cell sap and fungi, now enter the ranks of symbiont bearers. Only in the species *Candotripes*, which lives on fungi, could the transmission at the posterior pole and the behavior of an unpaired mycetome rudiment in the course of embryonal development be established, while in a few species nourishing themselves on cell sap nothing similar could be observed. It is hoped that the future will bring extensive investigations of this group, despite the difficulties presented by the minute size of these insects.

For our knowledge of bacterial symbiosis in chrysomelids we are also indebted to Stammer (1935, 1936). Symbiosis is by no means universal among them, for it occurs only in the donacids *Bromius* and *Cassida*. Stammer's brief observations on *Trixagus* and the cantharid *Dasytes* (1933) have unfortunately never been completed.

Fermenting tree sap teeming with microorganisms represents an environment of a special kind. An even greater number of symbiont carriers may be involved, to judge from the two forms of endosymbiosis with which we are already acquainted in connection with it. Keilin (1921) encountered bacterial organs in the ceratopogonid *Dasyhelea* in the form of true mycetomes. Several additional facts may be found in Buchner (1930). It is unfortunate that detailed investigation is lacking, for a study of the only diptera known to possess such devices would be of particular interest.   The second group concerns the beetle *Nosodendron fasciculare* Oliv.   The desirability of further research is indicated by what little Stammer (1933) and Öhme (1948) have presented thus far on the mycetomes and the unusual phenomena accompanying the egg infection.

Renewed interest in the long-known symbiosis of ants and blattids developed inevitably with its identification as a link in a longer chain.   I determined the distribution of bacterial symbiosis in the exotic camponotines and described egg infection, here so astoundingly stormy at its start, and the unusual behavior of the symbionts during embryonic development (Buchner, 1918, 1921, 1928).   Hecht (1924) treated the embryonic development in greater detail.   Lilienstern (1932) dealt with symbiosis in *Formica fusca* Latr., which is of special interest because in other species of the genus the symbionts have apparently been lost and now are merely indicated by certain peculiarities in embryonic growth.   Kolb (1957) studied the morphology of *Camponotus* symbionts in fine detail.

Reports on blattid symbiosis are much more numerous.   After reports by Buchner (1912), H. Fraenkel (1921), and Borghese (1946) came those of Gier (1936, 1937) and Koch (1938, 1949), with valuable new details on egg infection and especially on embryonic development.   Little was known about the latter point, and only scanty material was available, from older authors who like Heymons (1895) had been unaware that bacteria were involved.

Again and again attempts have been made to cultivate blattid symbionts.   Mercier (1907) reported allegedly successful cultures, as did Gropengiesser (1925), Glaser (1920, 1930), and Hoover (1945).   However, Javelly (1914), Hertig (1921), Wollman (1926), Hollande and Favre (1931), after unsuccessful cultures of their own, put no credence in the positive claims of their predecessors.   Gubler (1947), in the course of exceptionally thorough research employing a diversity of methods, concluded that no one had yet succeeded in cultivating blattid symbionts.   Yet Keller (1950) believed that he had been successful in cultivating symbionts highly adapted to intracellular life, despite their scarcely altered shape.   De Haller (1955), who had tried in vain to obtain pure cultures, is skeptical about the positive data.

Some authors were concerned with the finer morphological details of blattid symbiosis: Lwoff, 1923; Wolf, 1924; Neukomm, 1927, 1932; Hovasse, 1913; Hollande and Favre, 1931; Tacchini, 1946; de Haller, 1955.   Meyer and Frank (1957) used the electron microscope in experiments.   Brooks and Richards (1955) and Baudisch (1958) reported on the mitosis of the mycetocytes and their polyploidy.   Proof that mitochondria are present in mycetocytes was furnished by Koch (1930), Brooks (1954), and Meyer and Frank (1957). Gresson and Threadgold (1960) and Bush and Chapman (1961) made use of the electron microscope for such studies. The stainability of the blattid symbionts was treated by Fava and Laudani (1960), and Selmair (1962) reported on the lethal changes of the symbionts after the death of the host. In another connection we shall discuss the numerous attempts to sterilize the host animals and to investigate deficiency phenomena.

It was surprising indeed to discover bacterial symbiosis, resembling that of the blattids in minute detail, among the mastotermitids, that is to say, in one of the most primitive and at one time widespread termite families, but represented today by only a single member in Australia. Brief reports of Jucci (1930, 1931) had already suggested the presence of the symbiosis, and Koch (1938) by painstaking comparison clarified it and interpreted its interesting phylogenetic effects.

Whatever is known about the symbiosis of bloodsucking insects has also evolved from sporadic observations and various misunderstandings; however, this information bears a special stamp, for often it was not zoologists who studied the insects, but bacteriologists, hygienists, and physicians interested solely because these insects acted as hosts of pathogenic protozoans and bacteria, or were suspected of being carriers of virus diseases. Hence many occurrences of symbiosis were first discovered by scientists who were not primarily interested in the zoological aspects of the problem and from whom an exhaustive explanation of the findings was not to be expected. They had not the remotest inkling of cooperative living between host animal and microorganisms, and this indeed at a time when the discovery of symbiosis in plant juice-sucking insects should have alerted them to such a possibility. Such unawareness had led to a chain of unusual misinterpretations during the preliminary phase of this aspect of symbiotic theory, a chain reaching down to recent years.

The first to observe the bacteria localized in a special section of the midgut of tsetse flies was no less an authority than Robert Koch. He constantly referred to these bacteria in his conversation as "symbionts," after observing their regular occurrence, and realized that he must ascribe to them a role, as yet unknown, in the bodily functions of the fly. Stuhlmann, the first to make written mention of these organisms in the literature (1907), made reference to such verbal remarks of Robert Koch, but gave only a short description of their occurrence in the adult and remained under the impression that they were of some kind of protozoan.

The symbionts of the hippoboscids, which have so many features in common with the glossines, were discovered in 1917 when Nöller found rickettsiae in *Melophagus ovinus* L. In the following year Sikora also studied these organisms which are of such vital interest to bacteriologists.

In a certain section of the gut containing unusually tall epithelial cells, she found masses of cell inclusions which she considered "similar to the parasites described by Stuhlmann in *Glossina.*" About the same time Jungmann (1918) described the same formations in a report on rickettsiae in the sheep louse, but, in contrast to Sikora, he mistakenly identified them as the exclusively extracellular rickettsiae. Thus these two researchers had been the first to observe the symbionts of one of the hippoboscids, although they were unaware of the full import of their findings.

In 1903 we find the first reference to the filamentous symbionts in the esophageal ceca of a turtle parasite, the hirudinean *Placobdella catenigera* Moquin Tandon, in Siegel's study of the developmental cycle of haemo-gregarines in turtles. Siegel did not recognize the symbionts as such but considered them to be sporozoites of these protozoans. Reichenow (1910) recognized that the structures in question had nothing to do with the haemogregarines, but then made the error of calling them unmistakable cell products.

Always present in the ixodids and argasids are the symbiotic bacteria which had been noted by Robert Koch during his study of *Rhipicephalus*, even before they had been correctly classified, and which have since played an extraordinary role in spirochete literature. The various forms of *Spirochaeta (Borrelia) recurrentis*, which causes relapsing fever, are found in numerous representatives of the two subfamilies of the ixodids; and so they were investigated by a long line of bacteriologists whose primary interests were medical and who were unaware that their test objects sometimes might harbor both spirochetes and similar-appearing symbionts.

Dutton and Todd (1907) were probably the first to describe an argasid symbiosis, though unwittingly. In the cells of the Malpighian tubules of *Ornithodorus moubata* Wheeler, infected with *Spirochaeta duttoni*, they were continually finding small round formations which they unhesitatingly identified as developmental phases of the spirochetes. Leishman (1910) provided much more accurate information on these inclusions in the excretory organs of the same ticks. Finding in the ovaries the same forma-tions, some cocciform and others rod-shaped, he was convinced that they were present in every phase of development. He maintained that even in the embryos and larvae they might be found in the cells from which the excretory organs develop. Yet he too considered them to be develop-mental phases of the spirochetes.

When Balfour (1911) encountered similar formations in a study of *Spirochaeta gallinarum* in *Argas persicus* Oken, he thought he could actually observe all the transitions between them and the spirochetes, and he immediately agreed that they were developmental phases of the spirochetes. Fantham also (1911) confirmed Leishman's observations, and Hindle

(1912) presented an explanation of the developmental cycle of chicken spirochetes that clearly included these symbionts. However, Blanc (1911) had shortly before taken the stand that the doubtful inclusions in the ovaries and Malpighian tubules had no connection with the causative agent of relapsing fever, despite a chance similarity.

The relation of the spirochetes to the ever-present occupants of the renal organs was tested on a broader basis by Marchoux and Couvy (1913). In *Argas persicus* and *vespertilionis*, in a series of amblyommines, and in *Ixodes ricinus* L. they proved conclusively that the Leishman granules have no relationship to the cell inclusions which we now recognize as symbionts. However, they refrained from explaining their essential nature. Instead, they discussed the question whether they were mitochondria or possibly bacteria, without of course having in mind symbionts in our sense of the word and without venturing a decision. Earlier, Nordenskiöld (1908, *Ixodes*) had asserted that they were mitochondria, and Casteel as late as 1916 considered the organisms present in the ova of *Argas* to be such.

Not until publication of my research on *Ixodes hexagonus* (1922, 1926) and of subsequent studies by Cowdry (1925), Mudrow (1932), and Jaschke (1933), which include the argasids, were these bloodsucking animals finally classified as symbiont bearers.

Our information on pediculid mycetomes is much earlier in origin, since, in some of the species at least, we are dealing with a regular organ. Situated on the ventral side of the gut and yellowish in color, it stands out clearly, especially when the louse has just sucked blood, and it is easily seen with a magnifying glass. It was indeed noted by the very first insect anatomists to use the microscope.

The versatile Robert Hooke—physicist, astronomer, and biologist—was the first to mention the mycetome of the human louse in his monograph *Micrographic* (London, 1665). Soon afterwards we find it referred to by Swammerdam in his "Algemeene Verhandeling van bloedloose diertjens" (Utrecht, 1669, also issued as "Historia insectorum generalis" in 1685). Later there is a more detailed presentation of the mycetome in the famous "Bijbel der Nature," published by Boerhave (1737/38) in Dutch and in Latin after the death of the gifted insect anatomist. Since the description is the first and a quite plausible one, a few sentences taken from the German translation in 1792 are given here. We read: "Under the abdomen, somewhat elevated, almost above the middle of the stomach, one sees an organ considered by Hooke to be the liver. I myself should prefer to regard it as a peritoneal gland, if only I had more conclusive evidence. It is actually not white, but verges on lemon yellow. It is not easily detached from the stomach, to which it has grown fast. Under a magnifying glass it is easily seen to consist of many granules, as small, only slightly transparent glands." He describes the tracheae with which they are provided and calls attention to the hardy character of this organ and its varying shapes, which he illustrates in several sketches.

Even Landois (1864), to whom we owe the term "stomach disc" which is still in use today, was unable to discern much more in the organ. Investigating the mycetome of the felt louse, he found it to consist of "cells filled with numerous granules and fat globules which appear to be attached to the interior of the disc." He considered it to be a gland which emits secretions into the stomach during digestion. Graber (1872) believed one might justifiably call the mysterious organ a type of liver, even though it was still impossible at this time to prove a gall-like secretion. Graber also believed that he had observed fat droplets and pigment granules.

J. Müller (1915) interpreted the chambers of the organ as "small fibers, closely packed as a rule." He was the first to describe the "ovarial ampullae," recognized today as filial mycetomes serving to infect the egg cells, although he failed to recognize their full meaning. So did Sikora (1916) one year later in her "Beiträge zur Anatomie, Physiologie, und Biologie der Kleiderlaus [body louse]." In general she saw mycetomes as "irregular nodules and lumps, blisters, and granules"—clearly the abandoned mycetomes of females and old males in which the symbionts finally perish—and only twice did she see them as filaments, often parallel with one another. How close she was to discovering the true situation may be judged from a footnote: the thought that the organisms might be bacteria had of course occurred to her, she said, in view of Blochmann's work on bacteroids in insect tissue; but not for one moment, despite all her pleasure in hypotheses, did she really take the thought seriously. Thus she preferred to interpret symbionts of the swine louse, that is to say, conspicuous parallel tubes surrounding the nucleus, as filaments which originate with the assistance of the nucleus.

It is easy to understand why mycetomes were not observed as early in bedbugs as in lice, for in the former they are inconspicuous and colorless organs in the fatty tissue recognizable only on carefully prepared slides. It is less excusable that they eluded the many bacteriologists in recent years who were studying microorganisms in bedbugs or were engaged in transmission experiments. Attempting to infect bedbugs with the causative agent of Volhynian fever, Arkwright, Atkin, and Bacot in 1921 encountered certain formations in smears of various organs, primarily of the Malpighian tubules, to which they gave the name *Rickettsia lectularia*. For the most part they were not inhabitants of mycetomes at all—the mycetomes themselves had not yet been recognized by the authors—but companion forms which they themselves termed parasites. Cowdry

(1923) also spoke of the rickettsiae of bedbugs without knowing the actual symbiotic sites.

Not until 1919 was the spell broken which all too long had delayed recognition of the true state of affairs in this field. In that year Roubaud published his valuable study on symbiosis in the glossines with analysis of the behavior of their guests, which, to be sure, he still thought were yeast-like organisms in the larvae, pupae, and adults. Hippoboscids, drawn in for comparison but not studied in detail, convinced him that here too it was a matter of true symbiosis.

In the same year the stomach disc in lice was finally assigned its true place among the insect mycetomes. Sikora and I came to this conclusion independently, our reports appearing simultaneously in the same volume of the *Biologisches Zentralblatt*. Sikora treated felt, swine, and rat lice in addition to body and head lice. She was the first to recognize that it is only in young lice that filamentous symbionts occur in the mycetomes, and that irregular clumps replace them after the third molting. The ovarial ampullae, which until then she had considered as a type of phagocytic organ serving to fuse the follicle rests, she now thought might be "ovarial mycetomes," but, failing to recognize their importance in egg infection, she offered no details. My own observations on the detailed structure of the ampullar walls and the transmission of symbionts thence to the oocytes supplemented Sikora's statements. Whereas she considered these ovarial ampullae to be the actual mycetomes and the stomach disc a provisional arrangement, I considered the ampullae to be filial mycetomes with the sole function of transmission, and this interpretation has subsequently been proved for other insects. Similarly, our conceptions of *Haematopinus* symbiosis were not quite the same, for Sikora assumed a disappearance of the mycetocytes, which in this animal are diffusely embedded in the intestinal epithelium, whereas I, along with later investigators, always found them persisting *in situ*.

In 1920 Doncaster and Cannon, studying spermatogenesis in human lice, also chanced upon an accumulation of symbionts at the posterior egg pole. They discussed the possibility that this symbiont mass might have originated from the degenerated nutrient cells. Once again we observe that, to the very present, it has been difficult for some scientists to take full advantage of the contributions of symbiosis research.

A monograph on symbiosis in Anoplura was indeed needed, for Sikora's reports and mine were only fragmentary, although Sikora had added some supplementary facts to hers in 1922. Nor were matters helped much by a report by Florence (1924) on symbiosis in swine lice, for the author made gross errors of various kinds. For instance, she assumed that the eggs are infected through the micropyles and that the symbionts are constantly transmitted to the lumen of the gut.   The first consistent account of this highly interesting Anoplura symbiosis was presented in a publication originating in my own institute, by Ries (1931), who had earlier published a preliminary report (1930) chiefly on the bacteriological aspects.   His study of 1931, based on painstaking investigation of no fewer than eight genera, described the morphology of the sites and the transmission by way of the ovarial ampullae—and these may vary considerably—and explored the behavior of the symbionts during embryonic development in a variety of forms.   Ries gave the first detailed explanation of the complex symbiotic cycle of *Haematopinus* and *Pediculus*, which differs in males and females, and pointed out numerous strange phenomena and astonishing adaptations of the hosts.   A follow-up investigation by Baudisch in 1958 provided important and interesting corrections to Ries's data on ontogenesis in the *Pediculus* mycetome and on details of its structure.   We are indebted to Puchta (1956) for a fuller report on *Pediculus* symbionts and their degeneration in the male sex, but the final word on possibilities for their cultivation, attempted both by Puchta and Kotter (1955), still remains to be said.   Data on the symbionts of *Haematopinus* are found in Bewig and Schwartz (1956).

The Mallophaga, which are similar to sucking lice in several respects, are also closely related to them in their symbiotic devices.   Sikora and I, again independently of each other, discovered the presence of the ovarial ampullae which contain symbionts and effect the transmission; they had been observed long before by Nussbaum (1882), who described the epithelium of the oviduct as unusually thick and as having two nuclei and "a striated protoplasm."   Sikora made her brief report in 1922, and in 1928 I provided the first illustration.   However, we had both overlooked the presence of other nests of mycetocytes distributed in the abdomen. Again it was Ries (1931) who made an exhaustive presentation on the basis of seven genera.   Schoelzel's study (1937) on the embryology of Anoplura and Mallophaga contains supplementary material on the symbiosis.

Soon after the first publications on symbiosis in lice, it was possible for me to report the discovery of paired mycetomes in the abdomen of bedbugs, and to describe the methods of transmission and the ontogenesis of these organs which had hitherto eluded all researchers (Buchner, 1921, 1922, 1923).   Hertig and Wolbach (1924) confirmed my findings in their entirety and greatly strengthened our understanding of the singular diversity of bedbug symbiosis.   We are unable to say this of Kuczinski's studies on the causative agent of Rocky Mountain spotted fever.   Although there is frequent mention of "characteristic rickettsiae" in these studies, the author lacks genuine understanding of symbiosis.   Pfeiffer (1931) in my own institute devoted serious study to bacteriological questions connected with symbiosis in bedbugs. He concluded that a sharp differentiation must be made between actual symbionts and the bacteria occurring also in other organs. A supporting argument for his interpretation was that such bacteria are not found in *Oeciacus*, the swallow bug which was being compared with the bedbug for the first time. Pfeiffer's various attempts to cultivate symbionts were largely unsuccessful. Carayon (1959), on the basis of his studies on the tropical cimicids of Africa and South America, is convinced that the mycetomes, which I had described, have a general distribution. The mycetomes are lacking only in *Primicimex cavernius*, which is beyond a doubt the oldest representative and which lives on bats in caves in Mexico, and the symbionts are found in the cells of the midgut. Unfortunately he has not yet given details about the mode of transmission.

In view of these first proofs of symbiosis in bloodsucking animals, Reichenow forthwith retracted his statements on the nature of the filamentous inclusions in *Placobdella*, and he was soon convinced that these leeches also live symbiotically with bacteria (1922). Somewhat later, my student Jaschke (1933) amplified our knowledge of hirudinean symbiosis and discovered esophageal rugae infested with bacteria in the ichthyobdellids. Reports had been appearing from time to time on symbiotic bacteria, in *Hirudo medicinalis* (Lehmensick, 1941, 1942; Hornbostel, 1941; and others), but for the most part they were inconclusive. In 1953 research by Büsing, Döll, and Freytag provided what is undoubtedly definitive evidence. According to these authors true symbiosis is involved here. In the intestinal lumen they observed independent bacteria which are essential for digestion of the blood and which are transmitted by means of the cocoon fluid. The Corynebacteria, which are present in the urinary bladder, are equally essential.

In research on the haemococcids of lizards, Reichenow found mycetomes in the gamasids sucking them. In 1920 and 1921 he reported briefly on the subject; in 1922, in greater detail. In the same year two brief publications appeared, one by Godoy and Pinto classifying the symbionts of the amblyommines which earlier had often been mistaken for spirochetes, and the other my own report on ixodid symbiosis, the first in the field. Ixodid symbiosis was now treated in other investigations following in quick succession, although gamasid symbiosis, in need of clarification, was studied by only one worker, Piekarski (1935), who contributed little that was new. Cowdry, in a study of rickettsiae in ticks, had also found their symbionts. Though in 1923 he had still considered them to be mitochondria, he described them as symbionts in 1925, on the basis of seventeen species distributed among the ixodids and the argasids. In 1926 I presented in greater detail my own findings on ixodid species, and these were admirably supplemented by Mudrow in 1932.   Investigating the symbionts for the first time during embryonic development, he observed an extremely early infection of the gonad rudiment.   In 1933 Jaschke reported on argasid symbionts.   Thus after many misinterpretations a well-rounded picture of tick symbiosis was at last obtained.   Only Rondelli's data (1925) cannot be reconciled with this picture.   Rondelli, like Roesler (1934), missed the organisms localized in the Malpighian tubules in *Ixodes* and described bacteria from some other location which obviously were unrelated to the symbionts.   In 1958 there appeared in *Acta Tropica* a detailed study by Aeschlimann on the embryonic development of a member of the Argasidae, *Ornithodorus moubata*, the vector of relapsing fever, and shortly thereafter an additional paper by Wagner-Jevseenko on ovogenesis and spermatogenesis of the same arthropod.   Neither author mentions the symbionts, although they were described by Cowdry in *Ornithodorus turicata*, and this author assumed that he had not seen them in *O. moubata* only because of insufficient fixation.   Nevertheless, the careful study of Wagner-Jevseenko represents a welcome addition to our knowledge of tick symbiosis, because those "Feulgen-positive granules," which she interprets as chromatin particles leaving the ovocyte nuclei and whose harmonious transmission she describes and illustrates so well, are without a doubt the symbionts.

Roubaud (1919) had briefly noted that the ever-present intracellular forms in the gut of *Melophagus* were comparable to the symbionts of the glossines.   Arkwright and Bacot (1921) identified them as rickettsiae, as Jungmann had done earlier.   Hertig and Wolbach (1924) adopted the viewpoint of Sikora and Roubaud, whereas Anigstein (1927) again supported the other opposite view.   It was now time for a zoologist to study the complete cycle of hippoboscid symbiosis and reconcile these contradictions.   My student Zacharias devoted himself to this task with conclusive results.   He demonstrated the larval site, the process of metamorphosis, the transmission to the larvae with the secretion of the milk glands.   In *Melophagus* and other louse flies he discovered the disparate mycetomes of *Ornithomyia* (1928).   Aschner's detailed comparative study, based on a variety of forms (1931), fully confirmed the findings of Zacharias and provided various supplementary data, particularly the subtler morphological details of the symbionts and the more or less parasitical companion forms of frequent occurrence.   In addition to the hippoboscids, Aschner included nycteribiids and streblids in his study by way of comparison, determining that the nycteribiids have loosely joined mycetomes but that the only streblids investigated up to that time harbored no unequivocal symbionts. In a later publication on the nycteribiid *Eucampsipoda aegyptica* Mcq. (1946) Aschner acquainted us with another type of symbiosis, one which is extremely peculiar in that it is limited exclusively to females.

Not until now could reduviid symbiosis be linked to like phenomena among other bloodsuckers. Again it was microbiologists and medical scientists who made the discovery through their interest in the predatory tropical parasites which are the exciting cause of Chagas' disease, *Schizotrypanum cruzi*. Apparently Duncan (1926) had already observed these bacteria in the epithelium of the midgut when he found gram-positive bacilli in *Rhodnius*. The actual discovery was made by the Brazilian physician Dias, who was the first to mention symbionts in connection with *Triatoma* (1933, 1934) and who later found corresponding formations in eight other reduviids (1937). In 1935 Erikson recognized that these organisms were actually actinomycetes. More exact information on the site and behavior of the symbionts was given by Wigglesworth (1936). Brecher and Wigglesworth (1944) explained the methods of transmission and undertook experiments of practical significance with animals artificially freed of symbionts. After a while a wave of publications appeared which concerned chiefly studies of the symbionts in light of their capabilities, but which, however, were full of contradictions. From the Swiss Tropical Institute came the work of Geigy, Halff, and Kocher (1953, 1954) and of Halff (1956), who interpreted the easily cultivatable symbionts of *Triatoma infestans* as *Nocardia*, which are similar to those of *Rhodnius*, and who considered it highly probable that folic acid, which is synthesized by the symbionts in exceptional amounts, is indispensable for the normal development of the host. On the contrary, Wigglesworth (1952, 1955) proposed aneurin, Baines (1956) pantothenic acid, pyridoxine, nicotinic acid, aneurin, and another unknown factor. Bewig and Schwartz (1956) considered the problem of the accomplishments of the symbionts as not yet solved, Harington (1960) spoke for folic acid and aneurin, and the last publication of Gumpert and Schwartz (1963) concludes that only pantothenic acid is indispensable for the normal development of the host. In like manner there is disagreement concerning the nature of the symbionts. In 1956 Bewig and Schwartz agreed with Halff that the *Triatoma* symbionts represented a species of *Nocardia*, which is closely related to the one in *Rhodnius*, but they also observed a *Streptococcus* in the gut lumen of their material. As opposed to this, Gumpert and Schwartz (1963) concluded that the Triatominae can house a whole series of diverse microorganisms which all guarantee the development of the hosts. Therefore the symbioses of the Triatominae appear to be poorly developed, a factor also indicated by the primitive method of transmission, which makes possible an infection with additional organisms.

The period just described was a relatively long one, in which again and again it fell to scientists outside our field to observe one or another of the organisms which later came to be recognized as symbionts. Now a second period followed in which, thanks to the efforts of scientists in the field of zoology proper, clarifications and new discoveries followed in rapid succession to form a well-rounded chapter in the new symbiosis research.   Certain parallels soon emerged as guides in the search for new instances of symbiosis in bloodsuckers.   At first glance it may seem surprising to find that such forms as fleas, culicids, and the tabanids are absent from a series which includes lice, bugs, tsetse flies, lice flies, ticks, mites, and leeches, but there is a good reason for this, and no extension of the field can be expected in this direction.   In the search for further occurrences of this kind it was recognized more and more clearly that among the animals which feed on vertebrate blood only those harbor symbionts which throughout their lives never take any other type of food, a circumstance of decisive importance also in understanding the physiological background of such symbioses.

We need hardly mention that it now was impossible to retain the old, frequently repeated data of Schaudinn (1904) on the symbiosis of *Culex pipiens* L.   In the three pouches of the adult esophagus he had quite regularly found yeast-like fungi, and these, he assumed, exerted a hemostatic effect during the act of stinging and thus promoted the ingestion of food. Later authors, primarily Hecht (1928), observed that the fungi were by no means always present and that the sting reactions were not always a specific product of these yeasts or of the gas bubbles produced by them, as Schaudinn had claimed.

Similarly, research on fleas yielded no phenomena comparable to the genuine symbioses in bloodsuckers, even though the authors occasionally spoke of "symbionts" on discovering that bacteria occurred more or less regularly in the intestinal canal (Bacot, 1914; Faasch, 1935).

The possibility of luminescent symbiosis was given serious consideration for the first time in 1914 when Pierantoni reported on *Lampyris noctiluca* L. In the light-producing organs of this species, obviously somewhat similar to many insect mycetomes, he found bacilliform or cocciform inclusions which he considered to be luminescent bacteria.   Evidence for his interpretation was the fact that the ovarial eggs of the lampyrids had long been known to glow in the maternal body, as had been satisfactorily demonstrated by Dubois in *Pyrophorus*.   And, of course, it was logical to recall that most of the symbionts are already present in the oocytes for the purpose of transmission.

As a matter of fact, hypotheses formed by Dubois on the basis of these observations touched closely upon the idea of symbiosis.   In his opinion, the carriers of the luminescence were minute granules which he interpreted as elementary structures of living substance capable of reproduction, and to which he gave the name vacuolids.   They were already present in the egg, he said, and were transmitted to the adults by way of the individual stages of larval development and the pupae, to be returned again to the eggs (1914).

In smears of early developmental phases Pierantoni actually thought he was finding the same bacteria which he observed in the organs. He even reported successful pure cultures of these, but, because they were not luminescent, he later explained that they could play only an indirect role in the process of light production.

These first statements on luminescent symbiosis were generally rejected. Vogel (1922, 1927), who had done much work with the light-producing organs of the lampyrids, did not succeed in detecting bacteria in them, or even in cultivating them from the organs. Vonwiller (1920) declared that the inclusions were mitochondria, and earlier (1914) I too had been unsuccessful in tracing these alleged luminescent symbionts. Negative opinions increased. Dahlgren (1928) disagreed with Pierantoni despite the fact that in *Photinus* he found rod-like inclusions in the light-producing organs of females, and cocciform ones in males. Harvey and Hall (1929) excised the light-producing organs of *Photuris* but found that even without them the adults developed normal luminescent organs. They rightly concluded that this, too, is an argument against luminescent symbiosis, for otherwise one would have to assume the presence of non-luminescent symbiont reserves in other areas of the larval body to take the place of those removed.

Pfeiffer and Koch of my own institute transferred the granules to other insects to see whether these insects would then become luminescent. The experiment proved unsuccessful. Pfeiffer was also unsuccessful in cultivating the granules.

New support for the concept of luminescent symbiosis appeared in Julin's excellent study on the embryonic development of *Pyrosoma giganteum* Les. (1912), and this time the research yielded better results. The extremely simple cell accumulations, in this case representing light-producing organs flanking the ingestion orifice, had caused much difficulty for the older authors, just as with the pseudovitellus of the Homoptera, and had been classified as ovaries, kidneys, or simply "granule accumulations" of unknown function until Panceri (1873) recognized their true purpose. *Pyrosoma giganteum* Les. contained strange plasma inclusions which gave rise to much conjecture. Panceri and Seeliger (1895) spoke of fat droplets; Julin at first considered them to be mitochondria (1909), but later, because of their affinity to nuclear stains, he interpreted them as chromidia originating in the protoplasm through nuclein synthesis.

From Julin's illustrations and his description of the strange embryonic development, anyone familiar with the new knowledge of endosymbiosis in insects would have to conclude that here was an example of photosymbiosis and its curious method of transmission. As a matter of fact, it is only on the basis of endosymbiosis that these embryological processes can be explained at all. When I was attempting to reinterpret Julin's data in 1914 in terms of endosymbiosis, the world situation prevented me from doing research of my own. After the war (1920) I confirmed that Julin's chromidia were unquestionably bacteria. Soon Pierantoni (1921) presented detailed confirmation of this interpretation and offered convincing illustrations of the symbiotic cycle.

Numerous reports on the unusual embryological development of the salpids (Salensky, 1883, 1916, 1917; Heider, 1895; Korotneff, 1896, 1904; Brooks, 1893; Brien, 1928) all noted that the blastomeres contain striking inclusions which follow upon the early cleavage stages. These inclusions were interpreted in most diverse ways, as yolk substances, as remains of devoured follicle cells, as secretions, or even as products of endogeneous budding of the blastomeres. Such guesses call to mind those so frequently made in connection with symbiotic organisms. It had already occurred to Julin in 1912 that the inclusions were similar to those of the light-producing organs in the pyrosomes, later recognized as bacteria. My endeavor (1930) and particularly that of my student Stier (1938), though it did not prove conclusively that they were identical, nevertheless made it seem thoroughly likely and at the same time explained why follicle cells participate extensively in the initial developmental phases of these animals. This phenomenon, found nowhere else in the animal kingdom, had earlier been a complete mystery. Haneda and Tokioka (1954) after experiments with a Japanese salpid (*Pegea*), came to doubt the presence of a luminescent symbiosis.

Several years before definite proof of luminescent symbiosis in the pyrosomes existed, Pierantoni had extended its scope in another direction. He discovered that the accessory nidamental glands of the myopsid cephalopods have nothing in common with the actual nidamental glands, that their apparent secretion consists instead of myriad ever-present bacteria, and that luminescent bacteria are always found in the light-producing organs of the sepiolids. These organs, which are often highly developed and equipped with lenses and reflectors, are obviously onto-genetically connected with the accessory glands (1918, 1934, 1935). The new symbiosis was now treated by a number of other writers. Bacteriologically they were studied chiefly by Zirpolo, Meissner, and Getzel. Zirpolo repeated the successful cultures of Pierantoni and contributed a long series of reports on luminescent bacteria in *Rondeletiola* and *Sepiola* (1918–1938). Meissner (1926) collected additional data on morphology and behavior of light-producing symbionts, comparing these with forms on the surface of the sepias. He also provided serological proof of the specific linkage of the symbionts to their host. Getzel (1934) made a detailed analysis of the inmates of the accessory glands in *Sepia officinalis* L. Pierantoni (1924, 1925) extended his research to *Heteroteuthis dispar* Gray, a species which lives at greater depths and is in many respects disparate; Kishitani (1928, 1932) and Herfurth (1936) compared a number of other forms; and Herfurth also critically analyzed Pierantoni's opinions on the method of transmission. There was some discrepancy, but this could not alter the fact of myopsid symbiosis itself. There is, for instance, the controversy which developed between Mortara (1922, 1924) and Puntoni (1925, 1927) on the one side, and Pierantoni (1923, 1926) and Zirpolo (1924, 1926, 1927) on the other.

Data on luminescent symbiosis in oegopsids were rejected with cause. Pierantoni had supported the concept of symbiosis here, basing his opinion on preparations of *Charybditeuthis* (1919, 1920); and Shima (1926, 1927) had declared the rod-shaped inclusions in the light-producing organs of *Watasenia scintillans* Berry to be bacteria. Yet Mortara (1922) did not succeed in finding bacteria in the light-producing organs of *Abralia verany* Rüppel; and, when Hayashi (1927), Takagi (1933), and Okada, Takagi, and Sugino (1933) tested Shima's experiment, they were obviously correct in concluding that the alleged bacteria were protein crystals.

The same year that Pierantoni's study on luminescent symbiosis in the pyrosomes appeared, a brief report (1921) opening up a third area of clear-cut luminescent symbiosis was published by Harvey, a scientist who has played a prominent role in bioluminescent research. Here two closely related marine fish, *Anomalops* and *Photoblepharon*, were recognized for the first time as hosts of luminescent bacteria. Earlier their strange light-producing organs situated below the eye had indeed been accurately described by Steche (1909) as glandular formations. As has happened so often in the history of symbiosis research, the brief report ushered in a long series of similar discoveries. In 1928 three publications appeared on luminescent symbiosis in Teleostei. Yasaki proved the existence of this symbiosis in *Monocentris japonicus* Houttuyn; Harms in *Equula splendens* Cuv.; and Dahlgren found bacteria in the tentacle organs of *Ceratias*. In 1930 Kishitani reported that the strange light-producing organs of *Physiculus japonicus* Hilgendorf were filled with bacteria. In 1935 Yasaki and Haneda determined that this was also true of a number of macrurans, and in 1936 they studied luminescent symbiosis in *Acropoma japonicum* Günther. Haneda, working alone, pursued the subject further (1950), and he had already worked with *Malococephalus laevis* Lowe in 1938. Even earlier Osorio (1912) had been convinced that bacterial luminescence existed in this macruran, but his observations were too fragmentary to assist in achieving general recognition for what was then a revolutionary idea. Hickling, who had described the light-producing organs of this same fish more fully (1925, 1926) and had used *Coelorhynchus* for purposes of comparison (1931), decisively rejected the new hypothesis despite the fact that Harvey's publication had already appeared in the meantime. Haneda (1940, 1950) was able to show that the discovery of Harms would have to apply to all leiognathids. As a consequence, Haneda and co-workers have dedicated themselves untiringly to the investigation of luminescent animals and plants. This is attested to by the extensive literature references in Haneda's publication, "Luminous organisms of Japan and the Far East" (1955). Our knowledge of the luminescent symbioses of fish is therewith enhanced in many ways. We may now add the Trachichthidae to the already known symbiont-bearing families of the Leiognathidae and Acroponiidae (Kuwabara, 1955; Haneda, 1957). Of general interest is the comparative presentation which Haneda gives of the luminescent organs of five families arising from a common root; they send out in part their own light and in part that of symbionts (1962). Concerning additional contributions the reader is referred to the References (Haneda, 1952, 1956; Haneda and Johnson, 1958, 1962; Haneda, Johnson, and Sie, 1959; Imai, 1942; Abe, 1942, 1951; Kozukua, 1952; Matsubara, 1953).

So much for cases in which the presence of luminescent symbiosis has been proved beyond all doubt. What Molisch had termed a fairy tale to be discarded once and for all had now become an indisputable fact. Still, the application of the principle, when measured against the abundance of luminescent animals, has been rather limited. In the beginning it appeared to some scientists that the new principles might be immediately applicable to all luminescent animals (this opinion may be found, for instance, in Richard Hertwig's textbook on zoology), and many reported cases of luminescent symbiosis, affirmed more or less positively but later discarded for lack of proof, may well have originated from this idea which at first glance is so enticing.

Pierantoni (1924) upheld the hypothesis of symbiotic luminescence in the oligochaete *Microscolex phosphoreus* Ant. Dug., which excretes luminous mucus over its entire body and trails it in its wake. He identified as bacteria certain formations which had been observed earlier in other oligochaetes and which had indeed been recognized as bacteria by some workers. Cerfontaine (1890) was probably the first to point out bacteria-like formations in a diversity of tissues, in the musculature, peritoneum, and lining of the abdominal nerve cord in *Lumbricus*. Cuénot (1898) actually termed them genuine bacteria in reporting on *Eisenia*, but others more or less decisively opposed this interpretation. Willem and Minne (1899) saw the organisms as products of metabolism. K. C. Schneider (1902) studied them exhaustively, described their staining, which was the same as for genuine bacteria, and pointed out the absence of all structure in them; but he was unable to come to the decision that they were microorganisms. Trojan (1919) also firmly opposed the idea that the organisms were bacteria, and he was joined in this stand by Knop, who was again studying these inclusions in my institute (1926).

As a matter of fact, one is strongly reminded of the pseudobacteria in *Watasenia* by their sharp corners, square cross section, lack of structure, and refracting power, and it is clear that with the oligochaetes, too, the formations are not bacteria but crystalloids.

Nevertheless, the possibility existed that it might be the bacteria which cause luminescence at least in *Microscolex*, for in the tissues of this worm Knop had found, in addition to "bacteroids," small nests of genuine bacteria in all areas where, according to Pierantoni, the worm glows. These bacteria are clearly distinct from crystalloids, for they are slender and rod-like, sometimes slightly curved, and stain more vividly at the two rounded ends. Moreover, they do not vary as much as the bacteroids in length and thickness. Although Pierantoni characterized certain formations in the egg cells as luminescent symbionts, it is quite likely that they were the same organisms which we have just mentioned, since the crystalloids are never found in the egg cells of any oligochaete, including this one, and do not make their appearance until in later embryonic stages.

To be sure, it has not yet been possible to prove the presence of either bacteria or bacteroids in luminescent mucus and in the glandular cells producing it. Skowron, in testing Pierantoni's data, rejected his interpretation (1928) and traced the luminescence to small granules in the secretion instead. Moreover, it is hard to understand how the bacteria situated in the mesodermal tissues could transfer to the glands. Obviously, new research is necessary to clarify this case; it will therefore be omitted from a later chapter in this book, which is limited to clear-cut instances of luminescent symbiosis.

The same ambiguity holds true for the ctenophores. Some years ago I found in *Beroe* and *Pleurobrachia*, in the walls of their dorsal vessels where the light is produced, structures that might possibly have been bacteria. Later, when Pierantoni found cells with vacuoles in corresponding locations of *Bolinopsis* (1924), he too considered their elongated or round inclusions to be reminiscent of bacteria. He found the same formations also outside the cells and in the eggs. In *Beroe forskali* he reported similar accumulations in the plasma of the epithelial cells of the dorsal vessels.

At present there is no proof whatever of luminescent symbiosis in ctenophores. One might argue that up to now it has been impossible to prove the presence of luciferin or luciferase and that ctenophores become luminescent in very early stages of their development.   But until new material appears through research, they cannot be classified among the symbiont bearers.

The possibility of symbiosis in the luminescent mycetophilid *Arachno-campa* (*Bolitophila*) *luminosa* Skuse arose when Wheeler and Williams (1915) proved that the site of their light production was the odd, club-shaped thickened ends of the Malpighian tubules.   The situation called to mind the fact that these tubes had often attracted symbiotic bacteria and had been more or less extensively altered for this purpose (ixodids, *Apion*, and other snout beetles, *Donacia*, *Bromius*).   Certain details, however, make such an assumption, shared by Harvey (1940, 1952), seem unfounded. This fly, endemic in New Zealand, lives in mines and caves.   The adults are only slightly luminescent; however, since larvae by the millions give off considerable masses of thread-like secretions, one has the impression of a solid, luminous curtain hanging from the walls of caves otherwise destitute of light.   Indeed, Waitomo Cave, superb photographs of which may be found in Goldschmidt (1948), has for this very reason become a tourist attraction.   Masses of chironomids are caught in the secretion and fall victim to the mycetophilic larvae which by way of exception have adopted predatory habits.   Thus this would seem to be a case of genuine glandular luminescence, similar to the one discovered by Pfeiffer and Stammer in *Ceroplastes sesioides* L. (1930).   In this European mycetophilid the fatty tissue was established as the site of luminescence, yet no bacteria could be observed in it.   Kato (1953) and Haneda (1957) came to the same conclusion regarding Japanese varieties of *Ceroplastes*.

In the beginning the scope of luminescent symbiosis was understandably overestimated.   Today a more cautious attitude has been adopted.   In an overwhelming number of cases it is still a matter of intrinsic animal light, as I emphasized in 1926, and according to all present knowledge it hardly seems likely that our field will be substantially enlarged in this direction.

Research in bacterial and fungal endosymbiosis was restricted to the morphological factors for a longer period than in algal endosymbiosis. This was inevitable in view of its diversity and greater anatomical and developmental complexity.   In addition to the many unsuccessful culture experiments, there were reports of successful ones, some of them already mentioned in this book, but with a few exceptions, like the finding of Glasgow (1914) and Schwartz (1924) and reports pertaining to luminescent symbiosis, they were unconvincing and generally failed to achieve deeper understanding of symbiosis.

Great as the need was for satisfactory cultures of symbionts and analysis of their characteristics, it was recognized that the chief goal of experimental symbiosis research must be to obtain symbiont-free animals and then to study the expected pathological deficiencies. The first work of this kind was Cleveland's classical experiments (1924, 1925), in which termites were freed of flagellates to prove the vital function of intestinal symbionts in the nutritive process. This experiment represented a step in the right direction, but it involved merely a single test animal, one in which the symbiosis could hardly be compared with more complicated types usually linked with intracellular living and with particular habitats.

Aschner and Koch were the first to disrupt symbioses of the latter type and to explain the effects of the loss of symbionts. After surgical excision of the stomach disc in the body louse, Aschner noted pathologic phenomena which suggested a typical case of avitaminosis and soon caused the death of the insect (1932). With Ries (1933) he made careful histologic investigation of symbiont-free insects, and in 1934 he demonstrated that symbionts can also be eliminated by centrifuging the embryos, as the symbionts are destroyed on transmission to other sites. This method was further developed by Puchta (1955) and Baudisch (1958). After Baudisch had determined the exact time of the separate developmental phases and could thereby allow the centrifugal force to act at the decisive moment, he was able to prevent the symbionts from leaving the yolk region, and they had to perish in it.

About the same time Koch reported success in freeing animals of symbionts by a different method (1933). Numerous insects which provide their eggs with symbionts only externally on deposition of the eggs had suggested their elimination through sterile washings (compare Breitsprecher's first attempts in this direction, 1928). Results which Koch achieved by this method in *Sitodrepa panicea* were in closest harmony with those of Aschner. The symbiont-free insects became incapable of further growth, but, just as Aschner was able to prolong the life span of the sterile louse by rectally injected blood containing bacterial filtrates, Koch was able to remove the barrier to growth and to maintain a fairly normal development and sexual maturity in his test objects by introduction of yeast extracts and dry yeast. When Schomann (1937) applied the same method to the larvae of cerambycids, both types of larvae soon died because of greater difficulties in breeding, but the symbiont-containing larvae did nevertheless live longer. Both Aschner and Koch assumed that it must be vitamins of the B-group in particular which the symbionts provided.

Thus long-fostered theories were confirmed. The literature by now listed numerous studies indicating that microorganisms must be included in the diet to ensure the growth of non-symbiotic insects such as *Drosophila*, *Calliphora*, *Stegomyia*, *Aedes*, and that sterile nutrients must consequently be reinforced with yeast or bacterial filtrates, liver extracts, or other vitamin-rich substances. Such experiments had suggested even before

1932 that one of the most important functions of symbiosis is to provide vitamins, and not digestive enzymes as assumed at first (Buchner, 1930; Aschnér, 1931).

This assumption and the results of Aschner and Koch were thoroughly confirmed by experiments of Brecher and Wigglesworth (1944) with the bloodsucking bug *Rhodnius*, and by those of H. J. Müller several years earlier with the plant-sap-sucking *Coptosoma*. Unfortunately Müller's experiments were brought to an untimely end by the war, and detailed presentation of his findings did not appear until 1956. Brecher and Wigglesworth sterilized the eggs of *Rhodnius* by Koch's method. Müller needed only to remove the bacteria-filled capsules attached to the egg deposits before they could be sucked dry by the emerging larvae. In both cases development was greatly retarded, and most of the animals perished before reaching the last molting. The few adults obtained with *Rhodnius* were incapable of reproducing, apparently because of insufficient development of the ovaries. Through later infection with symbiotic bacteria, however, the harm done could be counterbalanced. H. J. Müller was often able to compensate for the deficiencies by allowing the *Coptosoma* larvae to suck the embryos of the host plant, which are richer in growth factors. Similar methods were later used to obtain sterile *Rhodnius* and *Triatoma* (Geigy, Halff, and Kocher, 1953; Halff, 1956; and Bewig and Schwartz, 1956). Halff (1956) and Harington (1959) also sterilized the eggs of *Triatoma* with gentian violet, as Brecher and Wigglesworth had done; Bewig and Schwartz (1956) used 70% ethyl alcohol, and Gumpert (1962) ethanol.

Koch's discovery that in *Oryzaephilus* it is solely the infection stages that are destroyed at a temperature of 36.5°C. (1936; preliminary reports 1931 and 1933) provided another non-surgical method for disrupting even endosymbiosis transmitted by infection of the ovarial eggs. It became possible to obtain symbiont-free animals which would continue to form sterile mycetomes throughout many generations without showing any injury on the standard diet. On the other hand, in symbiont-free cockroaches Glaser observed (1946) that at 39°C. the larval development was retarded and the ovaries disappeared.

This method has since been employed frequently. With blattids, a favorite subject for symbiotic research in late years, it has been used by Noland (reported to the author in correspondence), Brooks and Richards (1955), de Haller (1955), and Selmair (1962). H. Schneider (1954, 1956) was able to obtain symbiont-free *Calandra granaria* by elevating the temperature, and in this way to explain why the variety *africana*, though possessing no symbionts, nevertheless has rudimentary sites. Huger (1954, 1956) used the same method to obtain sterile *Rhizopertha* and

Oryzaephilus. The latter could also be made symbiont-free through the effect of cold. In *Pseudococcus*, Fink (1952) observed extensive disintegration of symbionts under the combined effect of hunger and heat but observed no damage to the mycetomes. Köhler and Schwartz (1962) also successfully used elevated temperatures with *Pseudococcus citri* and *maritimus*, as well as with *Orthezia insignis*. The lethal temperature deviated, with the different test objects used, between 35° and 40°C.

In recent years the astonishing effects of antibiotics and sulfonamides suggested their use in symbiosis research. In 1936 Koch attempted unsuccessfully to inject trypaflavin into the body cavity of the cockroach. Several years later, his pupil Kaudewitz actually caused the disappearance of the mycetocytes in the villi of the midgut by adding ortho-phenylphenol to the diet of the grain weevil *Calandra granaria* (reported in Koch, 1950). Brues and Dunn (1945) now began to inject various sulfonamides into the body cavities of *Blaberus*. They obtained no clear-cut results. Weak doses led to a mere reduction of the bacteria; strong doses destroyed the bacteria completely but also caused the death of the hosts after a few days; the results, however, at least indicated that symbionts are of vital importance to blattids.

Experiments by Glaser, reported only a year later (1946), demonstrated this vital role of blattid symbionts more convincingly. Sodium sulfathiazole was added to the drinking water of adult cockroaches, and the cockroaches themselves were injected with sodium and calcium penicillin. The effect in at least a fair number of experimental insects was complete disappearance of the bacteria, or a reduction in their number, and damage to the ovaries increasing in degree up to complete disintegration. Since then reports have been constantly increasing on the effect of different antibiotics—penicillin, aureomycin, terramycin, chloromycetin, and streptomycin (Musgrave, 1951; Brooks and Richards, 1955; Frank, 1954, 1956; Huger, 1956; Schneider, 1956; Fava and Barbato, 1960; Henry and Block, 1960; Fava and Laudani, 1961; Musgrave, Monro, and Upitis, 1961; Selmair, 1962; Brooks, 1963). The investigators made use of both feeding methods and injection; Köhler and Schwartz (1962) saturated the feed plants with penicillin and in this way obtained sterile scale insects. That Geigy, Halff, and Kocher (1954) had no success with aureomycin and chloromycetin on *Triatoma* is an exception; however, it was to be expected that aerosporin, active only against gram-negative bacteria, would not kill the gram-positive symbionts of *Blatta*.

In a few cases it happened that the elimination was not complete, but that a few symbionts, elusive to observation, remained resistant, prevented the death of the host, and in the course of the following generations again lead to a colonization, even though a sparse one, of the mycetomes (Huger,

1956, with *Calandra*).   Brooks and Richards (1955) and Frank (1956) had the same experience with blattids.   The surprising observation that the progeny of aposymbiotic *Orthezia* could be cultivated for more than two years without the appearance of disability symptoms and without degeneration of the mycetocytes, as is otherwise the case in symbiont loss, cannot be explained at the present time (Köhler and Schwartz, 1962).

It gradually became clear that symbionts provide essential vitamins, and this led to the conjecture that the results obtained might apply also to all symbiont bearers, whether they feed on vertebrate blood, plant juices, wood, or cereals.   The center of interest in symbiosis research shifted accordingly to earlier experimentation on the vitamin requirements of non-symbiotic insects, on the assumption that the substances found in their food would be the same ones which are made available, more or less completely, to the symbiont bearers by their microfauna.

In the beginning there were only a few studies probing more deeply into the vitamin requirement of symbiont-free insects or into the growth factors present in such great numbers in yeast.   Hobson (1933, 1935) studied the vitamins needed in the development of *Lucilia sericata*, and Trager (1935) attempted the sterile cultivation of *Aedes*.   The investigations of van t'Hoog (1935, 1936) provided invaluable information on the vitamin requirement of *Drosophilia*.   Two of Koch's students, Fröbrich and Offhaus (1953), published painstaking studies showing that all factors essential for the growth and development of the non-symbiotic grain pest *Tribolium confusum* are present in yeast.   Since then this easily cultivated beetle has become an indispensable experimental subject in symbiosis research.

Fraenkel and Blewett, utilizing the results of Koch and his pupils, analyzed more fully the vitamin requirement of the symbiont bearers *Sitodrepa, Lasioderma, Oryzaephilus* and compared it with that of *Tribolium* and other symbiont-free insects (1943, 1946; Blewett and Fraenkel, 1944; Fraenkel, 1952).   They made the chief factors of the vitamin-B complex available in pure form to insects which were deprived of their symbionts or were symbiont-free by nature (*Tribolium, Ptinus, Ephestia*) and thus were able to determine whether these factors were indispensable to the insects. In this way Koch's results were confirmed and extended in scope.

All further investigations of vitamin requirements in insects acquired great significance in experimental symbiosis research, even when experimentation involved symbiont-free test objects exclusively, for instance, the work on the nutrition of *Tribolium* by the Basle school (Rosenthal and Reichstein, 1942, 1945; Rosenthal and Grob, 1946; and others).   Using various experimental procedures, some of them new, Koch's school has been investigating the manner in which the symbionts provide their hosts with vitamins. These workers have improved the method used so success-fully by Offhaus, Fröbrich, Rosenthal, and Reichstein for analysis of the vitamins and other growth factors inherent in the yeast; and they have tested the effect of these individual substances upon many thousands of experimental insects (*Tribolium*). A number of doubtful points were clarified and controversies between authors eliminated. It was then chiefly Offhaus and Fröbrich, who, on the basis of investigations in Basel, analyzed in detail the vitamins and other growth factors contained in yeast (Offhaus, 1939, 1952; Fröbrich, 1953, 1954; Fröbrich and Offhaus, 1952, 1953; Koch, Offhaus, Schwarz, and Bandier, 1951). In more recent years Ilse Schwarz has devoted herself with great perseverance to the expansion of our knowledge of vitamin supply by the symbionts, using the *Tribolium* test, which she has greatly improved. In countless series of tests with separate larvae she has established, for example, the necessary quantities of each of the eight B-vitamins which are indispensable for the develop-ment of *Tribolium*, and she has shown that underdosage has as a conse-quence not only a retardation in development, but also, with certain growth factors, the development of characteristic malformations, while overdose-age in no way shortens the developmental time. A large number of her results unfortunately have not yet been published, but some are now in press (Schwarz and Koch, 1962; cf. Koch, 1959).

Using another species of *Tribolium*, Koch's student Naton investigated the interrelationship of incomplete diet and developmental disturbances, and he was able to establish the harmful effect of an underdosage of carnitin (1960, 1961, 1963). Parallel experiments were carried out in America by Fraenkel and co-workers, where the test animal was the well-known *Tenebrio molitor*, likewise an insect not living in symbiosis (Fraenkel et al., 1950; Fraenkel and Chang, 1954; Fraenkel and Leclercq, 1956; and others).

In Koch's institute another new, promising, and relatively simple pro-cedure was adopted to determine the role of the symbionts in providing vitamins. Earlier Graebner had been able to cultivate the symbionts of a cerambycid beetle in such great quantities that a sufficient supply of dried substance could be obtained. Now Reitinger added this substance to a vitamin-free deficiency diet, achieving in *Tribolium* almost the same increase in growth as with the addition of beer yeasts (reported in Koch, 1951). Fink (1952) had similar results in experiments with dried sub-stance of symbionts of *Pseudococcus citri*, symbionts which he himself had cultivated. Similar experiments were made by I. Schwarz with dried substance of *Mesocerus* bacteria, yeasts of *Ernobius* and *Sitodrepa*, and ambrosial fungi of *Hylecoetus* and *Xyleborus* (reported by Koch, 1956).

With this clear demonstration of the causal relationships between the symbiotic devices and the nutrition of the host insects, a new sphere of activity was opened up in symbiosis research. It was necessary now to determine to what extent substances which are essential to the complete development of the insect hosts are lacking or are available only in small amount in the limited diets of symbiont bearers. It was also necessary to prove that these vital substances are present in the nourishment of non-symbiotic animals. Since little such information was present in the available literature, Koch and Schwarz also used the *Tribolium* test in this branch of symbiosis research (Koch and Schwarz, 1952, 1955; Schwarz and Koch, 1954; Koch, 1956). They demonstrated that the necessary growth substances were almost completely absent in various types of wood, plant leaves, grains, and grain derivatives. Analysis of numerous pollen varieties showed that they are extraordinarily rich in essential vitamins, and this explained why no symbionts are found in pollen-collecting insects. Also, *Candida reukauffii*, which lives in nectar, provides the honey bee with pantothenic acid. It was proved that in *Cimex* and *Ornithodorus* the entire riboflavin content of the blood absorbed throughout youth is far less than that present in the mature insect (De Meillon, Thorp, and Hardy, 1947). In spite of such valuable beginnings, much remains to be done in this field. Further research is required on the composition of the sievetube sap, in which the lack of protein is particularly significant, and on the almost complete absence of endosymbiosis in insects developing in galls or feeding on fresh plant parts.

These important investigations are in themselves sufficient to show that in numerous cases the symbionts make the insects independent of vitamin-containing food. But there are many other aspects of endosymbiosis research. Utilization of the end products of animal metabolism by the symbionts was suggested as a possibility by the frequent occurrence of symbionts in the fatty tissue which stores excretions, in the Malpighian tubules of insects and ixodids, and in various types of renal organs such as the storage kidneys of certain snails and the molgulids, or the nephridia of oligochaetes (Buchner, 1912). A series of experiments show that this interpretation was justified. In 1924 Schwartz was the first to demonstrate that lecanid symbionts in pure culture can utilize urea and various purine bodies. Fink's results, in work done in Koch's institute, point in the same direction. Fink succeeded in cultivating the symbionts of *Pseudococcus* by using a "wet-nurse culture medium"; that is, the symbionts were introduced into a culture medium which had acquired growth-producing substances from a streptococcus culture situated below the symbionts and separated from them only by cellophane. The strains thus obtained could be kept alive in a synthetic culture medium containing galactose and urea exclusively, or urea alone in greater quantity.

L. Huber-Schneider (1957) demonstrated a uricase in cultures of the symbiotic bacteria from the heteropterous bug *Mesocerus marginatus;* the organisms were capable of splitting uric acid crystals into NH_3 and CO_2. Frank (1956), as was to be expected, observed an increase in uric acid concretions in the fatty tissue of symbiont-free *Periplaneta.* Thus, even aside from studies in which the purity of the cultures is uncertain—data by Schoel (1934) and Tóth (1951) on the production of uricase and urease in aphid symbionts, and by Keller (1950) on the decomposition of uric acid by symbionts of *Periplaneta*—there are several cases where one can claim that it is the symbionts which decompose the metabolic end products.

Another important point discussed in the new literature is the possibility that symbionts fix atmospheric nitrogen. The question arose fairly early, chiefly from the fact that the symbiotic sites are abundantly provided with tracheae (Buchner, 1921; Cleveland, 1925; Vouk, 1926). Peklo (1912) had lent support to the interpretation by identifying the typical round aphid symbionts as *Azotobacter*. Schanderl (1942) reported that he had actually verified a certain increase in nitrogen in cultures of the yeast-like symbionts in *Rhagium*. Tóth and his co-workers in particular were convinced that such fixing was possible in the symbionts of plant lice, cicadas, and Heteroptera, and they set out to furnish experimental proof (Tóth and Wolsky, 1941; Tóth, Wolsky, and Bátori, 1942; Tóth, Wolsky, and Bátyka, 1944; Tóth, 1940, 1946, 1949, 1950, 1952). Unfortunately an unreliable technique was used. By the micro-Kjeldahl method, a method in itself not infallible, they measured the accumulation of nitrogen chiefly by means of a broth of crushed symbiont-containing organs or even larger body parts mixed with oxalic acid and a nutrient solution. Moreover, as Tóth himself concedes, foreign microorganisms in the cellular pulp often simulate an accumulation of atmospheric nitrogen. And, finally, on the basis of what is known about *Azotobacter*, it must be taken into account that the assimilation of atmospheric nitrogen might possibly be suppressed by the presence of various nitrogen compounds bound in the tissues.

Like Schanderl, Tóth and his co-workers worked with cultures of the symbionts of aphids and spittle insects, claiming the same results as with their former test objects (Tóth, Wolsky, and Bátyka, 1944; Schmeisser, 1944; Tóth, 1946). However, Tóth's latest presentation of the subject (1952) shows that he is no longer convinced that those strains were actually the symbionts.

It is fortunate indeed that Fink of Koch's institute was successful in obtaining good cultures of *Pseudococcus* symbionts and in proving that these symbionts have the capacity to produce amino acids in pure galactose culture media. In harmony with this is the fact that recently Köhler and Schwartz (1962) were able to demonstrate at least a slow growth in cultures of symbionts from *Pseudococcus citri* in liquid media containing no nitrogen compounds.   However, there is little to convince us in Peklo's later pronouncements that *Azotobacter* symbiosis is widespread in insects, and that his cultures of these symbionts demonstrated their ability to assimilate atmospheric nitrogen (Peklo, 1946; Peklo and Satava, 1949, 1950).   Peklo claimed to have found *Torulopsis* and *Azotobacter* species in a great range of insects: in *Tribolium, Drosophila, Limothrips, Ephestia,* the moth, ipids, even in *Lecanium* and *Sitodrepa,* that is to say, among some insects long known to contain foreign symbionts and among others in which symbionts have not yet been detected.   Inasmuch as Peklo's brief publications present no proof to support his statements, they cannot be regarded as serious contributions to symbiosis research.

The direction which research must take in the future is clear.   Evidence exists that symbiotic functions are often complex in character.   For instance, feeding experiments with *Pseudococcus,* already mentioned, showed that the symbionts of this scale insect not only provide nitrogen and the necessary B-vitamins but also utilize the metabolic end products of the host.   Furthermore, when *Tribolium* was fed with the yeasts from the larvae of cerambycid beetles—the same yeasts for which Schanderl made it seem at least plausible that they are able to assimilate N_2—the yeasts almost completely covered the vitamin requirements normally provided by the food.

The problem of determining how the hosts are provided with enzymes for accelerating the disintegration of food received less attention than such problems as provision of vitamins, decomposition of metabolic end products, and assimilation of atmospheric nitrogen.   Nevertheless, earlier observations on the role of bacterial flora in the fermentation chambers of the intestinal tract of insects were now supplemented by important information on their role in mammals and in human beings.   Here research was done less often by zoologists than by hygienists and medical scientists. The latter are indeed becoming increasingly aware of the close interlocking of their field and ours.   We had proof of this when a German medical society invited zoologists as well as medical scientists to contribute to a series of papers devoted to endosymbiotic research (Verhandl. Deutsch. Gesellschaft für innere Medizin, 63er Kongress, München, 1957).

3

Wrong paths in symbiosis research

The preceding historical sketch discloses three errors in particular which appeared now and again, here and there, to retard our progress toward genuine understanding of endosymbiotic living between animals and plant microorganisms. The first was the failure to recognize the symbionts as organisms, and a tendency to interpret them as animal-specific cell inclusions. That this mistake was made in the beginnings of the science is understandable, but, once the nature of the pseudovitellus of the Homoptera and of other phenomena had been clarified, it seems unpardonable that symbiotic bacteria were still mistaken for ergastoplastic structures, mitochondria, or some type of secretion. We shall cite only a few examples. As late as 1916, Sikora was still referring to the symbionts of lice as plasma differentiations; in the same year Casteel considered the symbionts in *Argas* eggs to be mitochondria. In 1919 Trojan still referred to the blattid symbionts as "chondria." In 1924 Wolf was finding it extraordinarily difficult to judge the nature of inclusions in the mycetocytes of *Blatta*. When Mansour presented his first description of *Calandra* mycetomes in 1927, he failed to recognize that they were thickly populated with bacteria. And in 1952, though it seems almost unbelievable, Lanham interpreted the symbionts of aphids as cell-specific structures (Lanham, 1952; Trager and Lanham, 1952). Finally, even as late as 1958, because the symbionts of the argasids give a Feulgen-positive reaction, Wagner-Jevseenko was led to believe that they were chromatin granules.

Occasionally the reverse error was made: in the search for new instances of endosymbiosis, animal-specific structures were mistaken for microorganisms. In such instances it was usually a matter of formations, principally of crystalloid nature, with a close resemblance to bacteria. As discussed in Chapter 2, the sharply contoured protein crystals in the enclosed light-producing organs of cuttlefish were for a time identified as luminescent symbionts (Pierantoni, 1919, 1920; Shima, 1926, 1927). Similar inclusions, rod-shaped or operculiform, which are widely distributed in the tissues of annelids, were mistaken for bacteria, not only by earlier authors like Cerfontaine (1890) and Cuénot (1898), but even by Pierantoni who, searching for luminescent bacteria in *Microscolex*, failed to distinguish between such inclusions and the genuine bacteria present there (see Trojan, 1919; Pierantoni, 1923; Knop, 1926).

Until corrected by Walczuch (1930), I mistook for mycetocytes what were actually cells, distributed throughout the fatty tissue in *Orthezia* and filled with regularly arranged rodlets apparently representing preliminary phases of the wax. The varied inclusions of the oenocytes often assume an especially deceptive form. When Koch was searching for symbionts in *Tribolium confusum*, in which the modus vivendi is similar to that of the symbiont-containing *Oryzaephilus*, *Sitodrepa*, and *Rhizoperta*, he found in the plasma of the oenocytes tube-like structures deceptively like bacteria; it was only the fact that they do not reappear in the egg cells that drew attention to the true situation (Buchner, 1930; Koch, 1940). Nevertheless, Peklo (1946) also fell victim to this superficial resemblance. Similar mistakes were made in interpreting the "staff cells" in the parenchyma of some cercariae. "These cells," Wunder said (1932), "definitely give the impression of being mycetocytes with symbiotic bacteria."

In another case it was apparently the inclusions related to the mitochondria which were mistakenly identified as symbionts. Dendy (1926) thought that bacteria were responsible for the construction of the siliceous spicules in sponges. It was his opinion that these bacteria in the role of "silicoplastids" closely adapt themselves to living in the sponges and, unless they become silicified, play a role in egg infection. The "symbionts" which Peklo (1946, 1947, 1951) still quite recently managed to find in the fatty tissue and ovaries in insects of all kinds—*Ips, Lecanium, Anobium, Limotrips, Tribolium, Drosophila*—are obviously metabolic end products of the fatty compounds and deutoplasmic constituents of the oocytes.

A third error, occurring chiefly in the course of symbiotic studies of bloodsucking animals, was the failure to distinguish sufficiently between the ever-present symbiotic microorganisms and the parasitic organisms which are found only occasionally. For instance, the symbionts of *Placobdella* were regarded as part of the developmental cycle of a haemogregarine in the blood of this leech (Siegel, 1903); and the extracellular rickettsiae in *Melophagus* were repeatedly confused with intracellular symbionts (Jungmann, 1918; Arkwright and Bacot, 1921; Anigstein, 1927). For a long time no distinction was made between the symbionts of ixodids and argasids and the spirochaetes which cause relapsing fever (Dutton and Todd, 1907; Leishman, 1910; Balfour, 1911; Fantham, 1911; Hindle, 1912). Confusion also arose when true symbionts and a *"Rickettsia lectularia"* were both found in bedbugs (Arkwright, Atkin, and Bacot, 1921; Cowdry, 1923).

In another error what was obviously a parasitic invasion was mistaken for symbiosis. This was the case with the "symbionts" of the *Nonagria* grubs burrowing in *Typha*, which are almost always infested to an incredible degree with *Nosema nonagriae*. These microsporidians are also found in the epithelium of the gut, in the musculature, fatty tissue, and even in the nervous system, without in any way influencing the viability of the insects, and they are transmitted by way of the eggs. Portier (1911, 1918) declared such formations to be vital bacteria and made them his point of departure in future hypotheses (Schwarz, 1929). Grandori gave the term "symbiotic sporozoans" to inclusions of the vitilline spheres recurring consistently in eggs and embryos of *Pieris brassicae* and *Bombyx mori*, and he ascribed important effects to these formations (1919, 1924, 1929). It should be emphasized, however, that we have no reason to suppose that sporozoans ever function as true symbionts and that in all probability the supposed protozoan symbionts represented animal-specific structures which occur in connection with resorption of the yolk. Anadón (1944) described a case of intracellular symbiosis in ephippigerines, that is, a group in which no such intracellular symbiosis has ever been found, but once more the symbionts were in all probability being confused with microsporidians. At this point we should mention the strange occurrence of organisms which were described recently by Feldman-Muhsam and Havivi (1962, 1963) as constant guests of species of *Ornithodorus*. These authors considered them to be symbionts and classified them with the Phycomycetes. They live in the accessory glands of the male genital system as round structures, infect the spermatophores and, gradually becoming spindle-shaped, form in them a thick, superficial border. However, they certainly are not symbionts in the true sense of the word.

Symbiosis research was hampered not only by errors involving mere specific cases, but, what is more serious, by theoretical misjudgments which tended to obliterate the limits of what is understood as endosymbiosis today and thus to discredit the results achieved. Again and again there have been authors who insist that endosymbiosis is an elementary principle of all organisms and that structures usually regarded as specific to the plant or animal are in reality symbionts which have attained a high degree of adaptation to the host plasma. Béchamps, in his book *Microzymas* (1875), expounded the theory that the micrococci present in all living organs play an important chemical role in the tissues of which they are a constituent part. These cocci he called microzymes. Readers interested in this understandably much opposed theory are referred to Galippe's publication, *Parasitisme normal et microbiose* (1917), in which Béchamp's theory was resurrected.

Altmann also found such notions convincing. Prior to Béchamps, the Italian zoologist Maggi postulated them in a series of publications beginning in 1868. Altmann actually used the name "elementary organisms" for the granules which he was continually pointing out and which are generally identical with what are termed mitochondria today. By his term "elementary organisms" he meant extremely simple formations, approximately on the structural level of bacteria, which have entered into intimate association with the animal cell (1893). Even so objective an authority on mitochondria as Meves, who had at first rejected Altmann's hypothesis, in the end conceded that mitochondria might quite possibly be well-adapted symbiotic bacteria (1918).

Down to the very present, adherents to the theories of Altmann and Meves have used these views to construct far-fetched hypotheses. Portier, whose misinterpretation of *Nosema nonagriae* we have already mentioned, began the controversy with his book *Les symbiotes* (1918). He claimed that he had cultivated bacteria from organs of all types in a wide variety of vertebrates. He described the characteristics of the individual strains, made attempts to find the organisms again in the cells, and finally classified them as mitochondria. The wild flight of fancy he embarked on knew no bounds. He theorized that the mitochondria which are always present in both animal and plant organisms are destroyed in great numbers in the course of their functions but are constantly being replaced by new ones ingested with food, chiefly with fats. He asserted that plant cells are able to compensate for an insufficient supply of these mitochondria in the seeds by devouring mycorrhizal fungi and that burrowing insects can obtain them from the surrounding plant tissues. When the supply of symbionts is almost depleted in the process of yolk formation, he said, they are transmitted to the egg by the sperms during fecundation and the macronucleus of the ciliates acts as a reserve! The essence of the fecundation consists of the amphimixis of the symbionts; cancer cells were cells that had lost their symbiotic equilibrium; and so on.

Of course, such distorted conceptions have been energetically rejected. The Société de Biologie appointed a commission at the Pasteur Institute to investigate that section of Portier's data which referred to bacteria he had allegedly cultivated from animal tissue. As expected, the commission came to a negative conclusion regarding that portion of the data (Bierry, Marchoux, Martin, and Portier, 1920). Other authoritative voices were raised against identifying mitochondria as bacteria (Regaud, 1919; Guillermond, 1919; Laguesse, 1919; and Levi, 1922). Lumière (1919) published a refutation in *Le Mythe des Symbiotes* in which he disproved Portier's views point by point, and I did this somewhat later also (1921).

Nevertheless, the view that mitochondria and bacteria are identical had not lost its power to fascinate. It was revived by the American anatomist Wallin in a book with the bold title *Symbioticism and the Origin of the Species* (1927) in which he summarized a series of publications beginning in 1922, each one stating his reasons for believing that the two structures are identical. Though his arguments were this time dressed in scientific garb, they were nevertheless unconvincing. It was Wallin's view that it is impossible to differentiate between bacteria and mitochondria by means of staining, and this impossibility he regarded in itself as one proof of their common identity. He was also convinced that he had cultivated mitochondria *in vitro*, and, like Portier, he saw in endosymbiosis a universal principle of far-reaching significance throughout the plant and animal realms. Wallin had two loyal adherents, Hurst and Strong, who claimed (1932) that they too had cultivated mitochondria from the liver of the white mouse, but for the most part Wallin's theories met with dissent, and we need only refer the reader to detailed refutation by Koch (1930).

Faithful followers of Portier and Wallin, however, continue to exist to the present day. Recently their ideas have been invested with new force by late findings in bacterial genetics, by astounding processes occurring with bacterial viruses, and by discovery of the kappa-particle in *Paramecium*. Foremost among their followers is Schanderl, from the field of agricultural botany and bacteriology. Since 1939 he has been waging a desperate battle on behalf of this supposed common identity of mitochondria and bacteria. In a most detailed presentation of his theory, which appeared in a voluminous publication issued in 1947, Schanderl described numerous experiments with many species and various carefully sterilized parts—tubers, roots, seeds, fleshy fruits, and leaves—from which he had repeatedly obtained bacterial cultures, or, as he expressed it "regeneration of the symbiotic bacteroids"; and he cited many predecessors who had obtained similar results with one test object or another. Yet he was, of course, unable to overthrow the accepted dogma of the general sterility of plant tissues. Later (1950) Schanderl included yeast cells and mold fungi in his studies. Just as Enderlein had done before him in his *Cyclogenie*, a book which surely cannot be taken seriously, Schanderl now claimed that he had observed the disappearance of mitochondria and their transformation into bacteria with independent existence. He asserted that he had also observed mitochondria during their disengagement from leuco- and chloroplasts.

Various other authors, who cannot all be cited here, are in agreement with Schanderl. They consider the cells as disintegrating into their original elements, that is, into bacteria or bacteria-like preliminary phases, and they speak of "mitochondria cultures" (Meinecke, 1951; Santo and Rusch, 1951; Socias, Gonzalez, and Ramirez, 1952). Even Lederberg (1951, 1952) is not opposed to such ideas, at least not in principle.

Although Schanderl's "mitochondria cultures" like Wallin and Portier's theories have met with strong opposition, there has also been seeming verification of them in that many plant tissues are frequently, or even regularly, infected with bacteria, some of which do indeed enter by way of the pollen tubes. So reliable a researcher as Miehe (1930) has demonstrated that bacteria can be cultivated from the interior of pumpkin seeds even when their surfaces have been painstakingly sterilized, and similar results have since been obtained with potato tubers, tomatoes, beans, etc. (see Rippel-Baldes, 1952).

However, a finding of this kind does not signify that such germs from plant tissue should be considered mitochondria which have reverted to their original stage of life. In any event, all formations which are described in this book as symbionts are sharply distinct from mitochondria and are coexistent with them in the same cell. Cowdry (1923, 1924; see also Cowdry and Olitzky, 1922) and Duesberg (1923) were the first to prove the coexistence of the two inclusions in tubers of legumes; Milovidov also proved it, again with legume tubers, and presented (1928) excellent illustrations; Meyer (1925) proved coexistence in bacteria-containing cells of the storage organs in *Cyclostoma;* and Koch (1930) proved it in a succession of animal test objects, namely, aphids, *Philaenus, Sitodrepa, Pseudococcus,* and *Blatta.* Sometimes the inclusions are fine granules, as in the mycetocytes of *Pseudococcus;* again, short rodlets, as in *Sitodrepa;* or tiny filaments and nuclei occurring side by side, as in the cockroach. Wherever two types of mycetocytes occur, as in *Aphis rumicis* or *Philaenus leucophthalmus,* the mitochondria are found in both. Brooks (1954) and Meyer and Frank (1957) also demonstrated convincingly the coexistence of bacteria and mitochondria in cockroaches.

With investigation of live material it may not be clear at times whether certain structures are parts of the cell or whether they represent bacteria, but doubtful cases can usually be correctly interpreted after the fixation process, when the bacteria exhibit much greater resistance than do the cell parts. Indeed, most of the symbiotic studies were carried out with fixed material which would have been wholly unsuitable for studying the mitochondria. Of course, Schanderl merely regarded the ambiguity as the result of differences in length of the evolutionary time that the bacteria had been present. To him, all clear-cut cases of symbiosis merely represent much later and therefore more easily decipherable stages of cooperative living.

Such concepts were seductive, attracting not only authors who have played insignificant roles in evolving the science of endosymbiosis, but even a man of stature in the field, Pierantoni, to whom it owes so much. Repeatedly he equates bacteria with mitochondria and other more or less autonomous cell inclusions (1923, 1924, 1925, 1926, 1929, 1942, 1948). He also tends to include under the term symbionts the "vacuolids" of

Dubois, that is to say, minute mitochondria-like structures to which a hodgepodge of cellular functions has been ascribed. In some cases there unquestionably is a connection between symbionts and the formation of pigments; in others, pigment granules appear to originate from mitochondria. Thus Pierantoni is fortified in his idea that, wherever pigment arises, symbionts in the widest sense of the word must be involved. Earlier, indeed, Dubois had traced the purple of *Murex* to his vacuolids, and the pharmacologist Tschirch (1922, 1924), for whom the concept of endosymbiosis also represents a principle with wide application, mistakenly traced the origin of laccaic acid, which colors shellac red, to symbiotic yeasts of the lac scale insects (see Mahdihassan, 1929).

Pierantoni's studies on the vitelline discs of amphibia, birds, and fish all tended in the same direction (1922, 1927, 1928). Golgi (1923), impressed by the capacity of these vitellines for systematic cleavage, concluded that they were structures endowed with all the characteristics of living organisms, and investigations of Nobile (1927) and Migliavacca (1927) served to strengthen the idea that these structures represented not inorganic reserve material but highly organized constituent parts of the egg. Pierantoni even reported successful pure cultures of vitellines with culture media used in bacteriology and was more persuaded than ever of their organic nature (1948, 1949) when Pucher, a pathologist, in 1949, without knowledge of Pierantoni's publications, advanced the theory that bacteria can be cultivated from the yolk of hens' eggs. Pucher regarded the yolk as an accumulation of "micromycetes" which assume amoeboid form in the culture and give rise to gametes. The same organisms, he believed, occur in the cell nuclei, in the plasma of the embryonic cells, and, in great masses, in the gall bladder of the chick, indeed, in animal and plant cells throughout.

Pierantoni was fortified in his conviction of the ubiquity of endosymbiosis by surprising data, still unverified, on bacteria of supposedly regular occurrence in the blood of vertebrates (Ponce de Leon, 1935; Getzel, 1941). Basing his ideas on Béchamps, Dubois, Altmann, Portier, Meves, and Wallin, and in harmony with Schanderl's views, Pierantoni evolved a theory of the symbiotic constitution of the protoplasm, a constitution which enables a large part of the formed constituents of the living substance to have an individual life and which in the last analysis is traceable to originally independent microorganisms. Logical though it may be to assume that organisms, visible or as yet invisible with our present facilities for magnification, can be transformed into symbionts, it is nevertheless a concept which carries the principle of endosymbiosis into infinity and thus is too speculative to use as a scientific tool.

Mitochondria were falsely identified as bacteria, for both have the capacity to divide and have similarity to each other in form, a similarity inevitable in such simple structures. There were also scientists who mistakenly traced the chromatophores and leucoplasts of plant life to symbiotic algae. Schimper (1885) was probably the first to discuss this possibility. To him clear-cut cases of bacterial symbiosis were preliminary phases of mitochondrial symbiosis, and genuine cases of symbiosis with zoochlorellae or zooxanthellae to him manifested the evolution of more intimate associations in which algae were downgraded to the status of organelles of the plant cell. Opinions favoring such concepts are found as early as 1891 in Haberlandt's publication. Mereschkovsky (1905, 1910, 1920) developed this hypothesis further, and Famintzin (1907) elaborated it extravagantly. The desire to subject the ideas of Schimper and his followers to serious review was inspired by the interesting observations of Pascher (1929) on the endocyanoses, discovered by him, in which the division rates of symbiotic *Cyanophyceae* coincided perfectly with the reproduction of the host cell. For example, R. Müller, who had expressed a similar view in 1928, in 1947 unhesitatingly identified the chromatophores with *Cyanophyceae*, which, he said, had become symbionts long ago in the Paleozoic era. The new findings on strains of *Paramaecium aurelia* are also likely to give new life to such views, for in this species the algae are colorless instead of green, as in *Paramaecium bursaria*. Such algae were earlier considered to be the carriers of genes localized in the plasma (Preer, 1946; Altenburg, 1946).

The course of research has been uninfluenced by these bold hypotheses expanding endosymbiosis into a fundamental principle cleaving the uniformity of cells, both plant and animal. Independently of such extravagant concepts, in clear-headed, unassuming work, the science of endosymbiosis has laid stone upon stone to build the structure which is to be described in the following pages and which even in this less grandiose form embodies much that is miraculous. For us who have remained aloof from such speculations, endosymbiosis between animals and plant microorganisms represents a widespread, though always supplementary, device, enhancing the vital possibilities of the host animals in a multiplicity of ways. For us it is but one manifestation of that principle of interdependence prevalent in all animate nature, which is the prime basis of the plant and animal kingdoms and of the existence of humankind.

Specialized section

4

Symbiosis in insects feeding on cellulose, herbaceous plant parts, seeds, and similar substances

FUNGUS-CULTIVATING INSECTS WHICH TEMPORARILY RECEIVE THE SYMBIONTS WITHIN THEIR BODIES

In a work devoted exclusively to endosymbiosis, at first sight it may seem inappropriate to include a discussion of regulated cooperative living between insects and ambrosial fungi, for there the symbionts are cultivated externally. However, in all these cases the continuity of association is guaranteed by a temporary housing of the symbionts within the body of the animals, which produces adaptations very similar to those found in cases of endosymbiosis.

Just before the nuptial flight the Attini queens store viable fungus spores in a pouch on the hypopharynx, in the posterior region of the buccal cavity. As soon as the queens are enclosed in the chamber used in the process of reproduction, they break open the pouch and fertilize the spores in the familiar manner with their excrement. Obviously, these so-called infrabuccal pouches are not new formations created especially for this purpose, for they are found also in other ants and in the Attini workers.

The sexual forms of the termites apparently use the proctodaeum for transporting the conidia of their food fungi during the nuptial flight. With these conidia they infect the excrement, which consists of finely chewed wood particles consumed in suitably abundant quantity before they leave the nest.

Schneider-Orelli (1911, 1913) interpreted the infection of passages newly bored out by females of fungus-cultivating ipids in similar fashion. Finding in the midgut of hibernating females of *Anisandrus dispar* F. pieces of the mycelial mat at a time when the latter is dessicated, he assumed that these mycelial fragments are broken off after hibernation or eliminated with the excrement. Francke-Grosmann (1956) showed that far more intimate transmission devices, in part diversiform, are created in all cases of ambrosial culture among the bark beetles, that is to say, chiefly in the wood-cultivating Xyleborini and Xyloterini. These devices are invariably invaginations of the body surface more or less deeply inserted and closely related to glands. In the various *Trypodendron* species (Xyloterini) prothoracic paired glands empty into a radically sclerotic fissure, usually tightly closed and covered with long hairs. In the oleaginous secretion the oidia of the fungus *Nomilia ferruginea* find a favorable culture medium, according to Mathiesen-Kaärik (1953). An ingenious evacuation mechanism which operates chiefly when the leg muscles are in vigorous movement enables the fungi to emerge from these well-concealed containers which are lacking in the males.

Such prothoracic glands have not been found in the Xyleborini thus far investigated. In *Xylosandrus* and *Anisandrus* the intersegmental membrane between the pro- and mesonotum forms a sac filled with glandular secretion. In females the sac is filled with conidia-like forms of the ambrosial fungus during hibernation but never in summer. In *Xyleborinus* the females have the typical ipid mesonotum. At the base of the elytra Francke-Grosmann found a sharply delineated glandular area covered with long hairs to which the symbionts adhere. In *Xyleborus pfeili* the ambrosial fungi are located under the elytra between glandular hairs of the terminal abdominal segments. Obviously the examples given do not exhaust the diversity of solutions of this vital problem of ambrosia-cultivating ipids. Females of *Xyleborus fornicatus* and *X. mascarensis* carry out with them, on swarming, little clumps of fungal cells in a complicated paired organ, which lies between the labrum and the mandibles and connects with the mouth cavity (Fernando, 1960; Francke-Grosmann and Schedl, 1960). In 1962 there appeared another such study by Schedl on twenty-six species of the family Scolytidae, in which five types of transmission, unknown until then, are described and well illustrated. According to Schedl, eight different possibilities can be differentiated. These are localizea in the head, the mandibular pockets, pharyngeal pockets, pregular pockets; in the region of the thorax are found notal intersegmental pockets, pronotal glands, paired prothoracal tubules, extended coxal grooves of the coxae, and pockets in the thickened base of the elytrae. Almost all these arrangements are found only in the female; at most, only suggestions of such structures are present in the male. The paired tubes in the prothorax are an exception; they are an exclusive arrangement of the males. In view of the extent of this astonishing discovery, it may be supposed that the possibilities are still not all known. The publications of Finnegan and of Farris in 1963 are concerned with structures which come under Schedl's classification.

Among the bark-breeding beetles the cultivation of fungi is never as highly developed as among the wood-breeding beetles, but comparable conditions often develop from various loose associations with fungi. This is true of *Myelophilus minor* and *Ips acuminatus*, in which Francke-Grosmann (1952, 1956) found the fungus depots of hibernating females under the elytra in secretion accumulations of the metathorax, and in the membranous wing under the left elytron. Shifrine and Phaff (1956) have reported on yeasts found in *Ips* and *Dendroctonus*.

An unusual state of affairs exists with some of the platypodids. These beetles, found usually in the tropics and subtropics, all cultivate fungi on the walls of their tunnels. Often extremely odd adaptations are created, obviously to transport symbionts. The epicranium then is so deeply indented that the vertex and eyes appear to be hollowed out. The edges of this indentation are either bare or bordered by rows or groups of bristles curving inward (*Mitosoma, Symmerus*, and others).

The head of *Spathidicerus thomsoni* Chapuis is grotesquely modified. In addition to a cavity on each side of the median epicranial suture, there are mandibles with shovel-like extensions so large as to be useless for gnawing. Tufts of hair emerging dorsally and laterally together with the mandibles encircle a space in front of the epicranium.

Strohmeyer (1918), the principal expert on this group, is undoubtedly right in considering these radical new formations occurring exclusively in females to be devices for transporting fungi. Indeed, he proved the presence of fungal remains in these grooves in dried material. In other cases complicated adaptations are lacking, and the association of the two partners is effected by a rather general contamination of the body surface, which is very rich in glands (Beeson, 1917; Francke-Grosmann, 1956). Doubtless the study of live tropical platypodids would reveal many another specific adaptation. For example, in *Platypus cylindrus* Francke-Grosmann found dense, sticky masses of symbiotic fungi, especially in an indentation in the posterior area of the mesonotum.

Finally, ambrosial fungi are cultivated by the larvae of the lymexylonid *Hylecoetus dermestoides* L., on the walls of tunnels which they bore out chiefly in beech, oak, spruce, fir, and birch (Neger, 1909). This symbiosis is of special interest because it leads to a transmission device vividly reminiscent of adaptations occurring in endosymbiosis. As with platypodids and bark beetles, the tunnels are lined throughout with fungi, appearing to the naked eye like a layer of cream. The septate mycelium passes through elements of the wood singly or branched in bundles, penetrating only superficially, and produces at the free ends quantities of characteristic thick-walled spore-like appendages rich in glycogen. Neger considers them metaspores, but their germination could not be observed

(Fig. 2*a*).   In late autumn the larvae seal the tunnel opening with a stopper of wood borings and then during the winter remain near the blind end.   The mycelial mat regresses during the cold season but, despite this, wall fragments are regularly found in the vicinity of the larvae.   These wall fragments consist of wood particles interspersed with fungal filaments. Schneider-Orelli, who initiated such observations, sees in them an adjust-

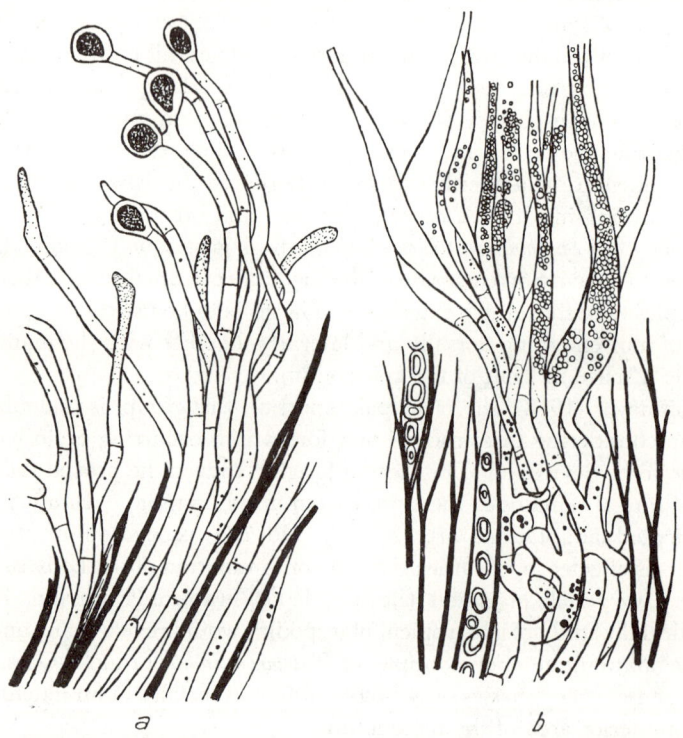

a   b

Fig. 2   *Hylecoetus dermestoides* L.   The fungus, cultivated on the passage wall, (*a*) in its usual growth form, (*b*) in spore formulation.   From Buchner.

ment to hibernation, occasioned by active cultivating on the part of the larvae.   In the spring the mycelial mat again develops luxuriously.

Just how continued association of the two partners is guaranteed at first remained a mystery.   It would be useless of *Hylecoetus* females upon leaving the old tunnels to transport the fungi in the manner of ants, termites, ipids, and platypodids, for they die soon after laying their eggs on the bark and the little larvae swarm apart in all directions after hatching.   An examination of the ovipositor provides the answer to the riddle (Buchner, 1928,

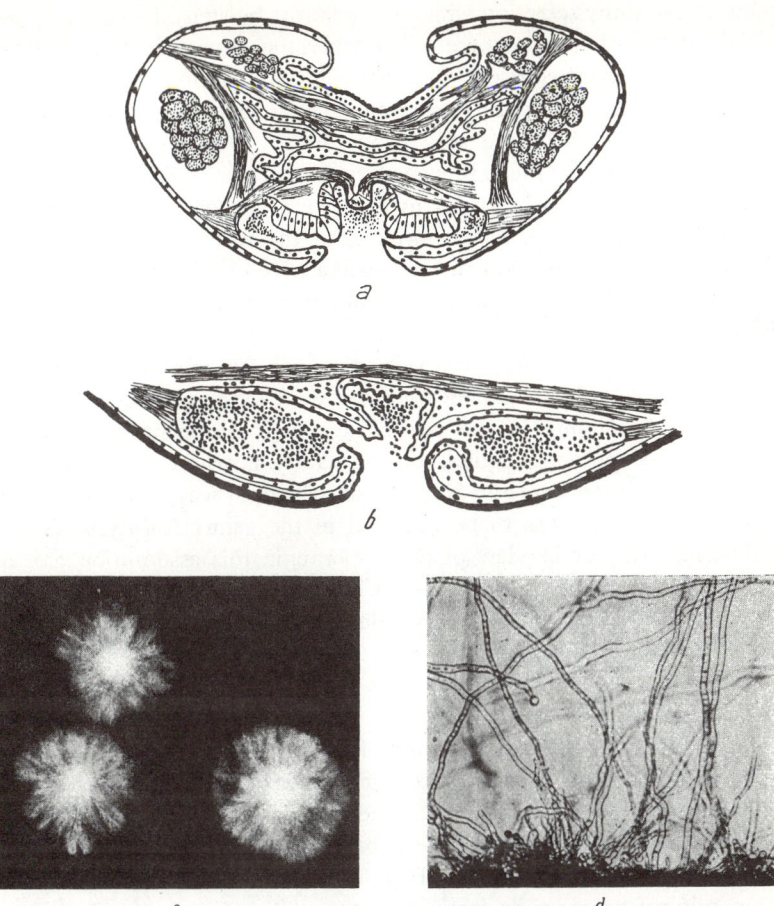

Fig. 3   *Hylecoetus dermestoides* L.   (*a*) Cross section through the egg-laying apparatus; the ventral paired pockets and the median groove filled with spores above the vagina. From Buchner.   (*b*) The pockets more enlarged.   From Buchner.   (*c*) Three eggs in acid malt-gelatin after 7 days.   The ambrosia fungus has developed from the spores. Natural size.   From Francke-Grosmann.   (*d*) The fungus strain developing on the egg shell.   150✕.   From Francke-Grosmann.

1930).   Ventral to the vagina are two spacious pouches opening into a medial groove.   The length of the strands of transverse muscles passing between pouches and vagina can be judged from the total preparation. Groove and pouches are filled with round fungal spores which, particularly because the reservoirs unite with the vaginal mouth, must necessarily reach the upper surface of the eggs during deposit (Fig. 3*a*, *b*).   In fact, they are found in the slimy secretion which abundantly moistens the eggs and frequently causes them to cohere. The emerging larvae are by no means immediately separated from the eggshells but continue to roll about strangely for days, the mass appearing like a mucilaginous ball in which the collapsed eggshells are still recognizable. Obviously this unusual behavior is caused by the existence of the fungal symbiosis and has a purpose, for, when the instinct is finally aroused in the larvae to move on and search for suitable places for boring into the wood, they are so well "lathered" that enough fungal spores adhere to their bodies, particularly in the region of the intersegmental membranes, to guarantee colonizing of the future wall passage.

These spores, which fill the transmission pouches and are carried into the new tunnels, should not be confused with the previously mentioned button-like thickenings of the mycelial ends. They originate in slender bottle-shaped sporangia found predominantly near the outlet and the puparium in the spring (Fig. 2b). Until recently it was assumed that the *Hylecoetus* symbiont was to be classified in the genus *Endomyces* (Neger, Guilliermond), but in view of these sporangia this assumption can no longer be upheld. Similar sporangia have been described for *Dipodascus albidus* Langerh., a genus closely related to *Endomyces* taken from the mucilaginous sap flow of birch.

Francke-Grosmann (1951) was the first to study in more detail the mycological aspects of this symbiosis. She took from material of the tunnels individual cultures originating from the sporangia, the pouches on the ovipositors, from eggs deposited under sterile conditions, or from the trails of the young larvae on agar. Such cultures always yielded the same fast-growing mycelium, which in the beginning smelled of roses and later of over-ripe apples (Fig. 3c, d). That formation of asci never occurred in these cultures provides decisive proof that they are connected with the ambrosial fungi. The terminal swelling of the hyphae occurred only occasionally. The fungus in deciduous or coniferous trees of various types is always the same—only material from Swedish birch differed somewhat in size and form from the German material—yet its growth is of different intensity in the individual wood varieties, and even in the same type of wood the growth is weak or strong in proportion to the depth or shallowness of the tunnels in the bark. In 1961 there appeared still another publication on the subject by Batra and Francke-Grosmann, in which the fungus is classified with the Ascoideaceae and named *Ascoidea hylecoeti*. Here the chlamydospores serving asexual reproduction, the spore-bearing asci, and the hat-shaped spores arising therein are more exactly described.

Unfortunately, these experiments have not yet been applied to other lymexylonids. On examining dried samples of fungi-cultivating *Atracto-*

cerus from Mexico and the Transvaal, I found similar pouches on the ovipositor; and Tanner (1927) provided illustrations of those he found in *Hylecoetus lugubris*. On the other hand, the pouches are lacking in *Lymexylon navale* L., and here also the fungal covering is lacking in the tunnels.

SIRICIDS

Cultivation of fungi in siricids, on which my own reports were the first (Buchner, 1927, 1928), differs in many respects from that of platypodids, ipids, and lymexylonids. With siricids it is not a question of the consumption of a mycelial mat which lines the feed tunnels or of the removal of gnawings from the tunnel system without passage through the intestine. Instead, the mycelia of the symbionts, penetrating more or less deeply, fill the wood around the tunnels and are consumed along with it. Thus the gnawings pass through the intestine and fill the tunnel parts in the wake of the insect. We have seen earlier that ambrosia-cultivators equip the posterior end of their bodies especially for removing gnawed wood particles. In the same way the siricid larvae have a special staff-shaped process for tamping down the excrement behind them. As for devices to guarantee continued symbiosis of the two partners, the siricids are like *Hylecoetus*, but their adaptations are more complicated and the symbionts are found not only in the actual transmission devices of the imagines but also in other organs, the purpose of which had not been understood until just recently. My discovery initiated a whole series of studies (Chrystal, 1928; Cartwright, 1929, 1938; Clark, 1933; W. Müller, 1934; Francke-Grosmann, 1939; Parkin, 1941, 1942). This interest may be attributed in large part to the practical aspect of the relationship between siricids and their fungi.

At the point where the divering arms of the first and second valvulae of the ovipositor curve upwards, two strange organs are present which were overlooked for a long time. In *Sirex*, they represent two piriform evaginations of the membrane connecting the tenth and the eleventh segments. They are supported by a somewhat complicated prolongation of the first valvulae. The side corresponding to the median sternal part of the eleventh sternum is strongly sclerotized basally and is spoon-shaped. A narrow, but clearly defined, prolongation of the median sternal part reaches, on each side, the apex of the sack, where it serves as insertion to a series of thin muscular fibers, which irradiate the whole surface of each evagination (Fig. 4).

The broader section supports a sizable gland, filled with tracheae, which radically constricts the outlet passage.   The gland consists of numerous loosely arranged cells with long evacuation channels ending usually in groups of four and five, especially in the area of the narrower passage, but ending sometimes also in the more extended pouch (Fig. 5).   Except in

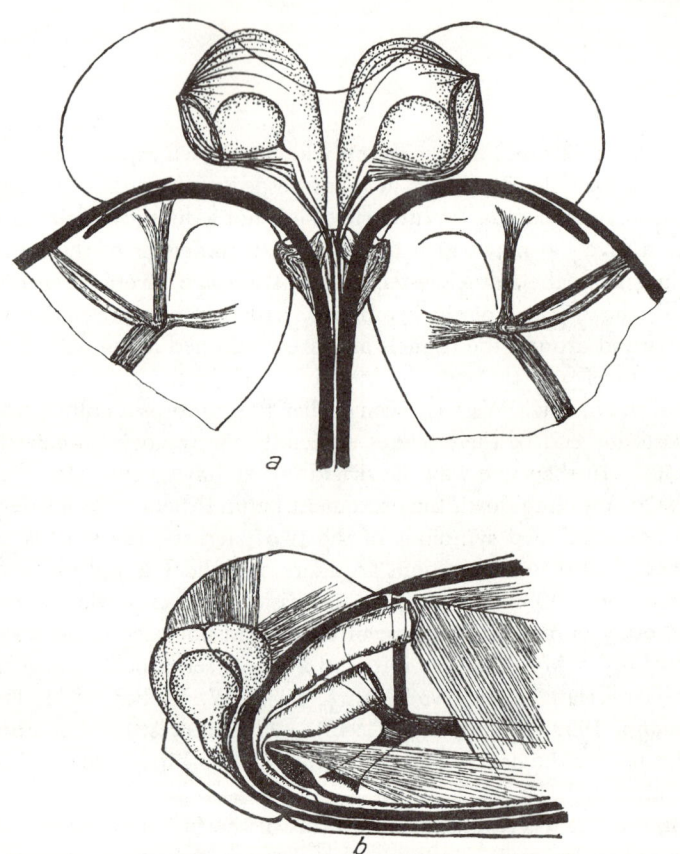

Fig. 4   *Sirex gigas* L.   The pear-shaped intersegmental pockets, filled with fungi, at the base of the egg-laying apparatus, (*a*) seen from above, (*b*) seen from the side. From Buchner.

Xeris spectrum L., which tends to develop in fresh wood, the pouch is filled with numerous one- or two-branched fragments of a mycelium, its numerous "spangles" enabling one to recognize them as the oidia of a basidiomycete (Fig. 6).   Except in the glandular area the chitin forms mono- or multitipped short bristles which are all turned inward, forestalling loss of the fungi.   We may equate these bristles with numerous retention devices found in similar situations in other insects, although some of the devices are more highly perfected than these bristles.

A comparison of the various siricids will reveal differences of many kinds. In *Xeris spectrum* the intersegmental pockets and supporting

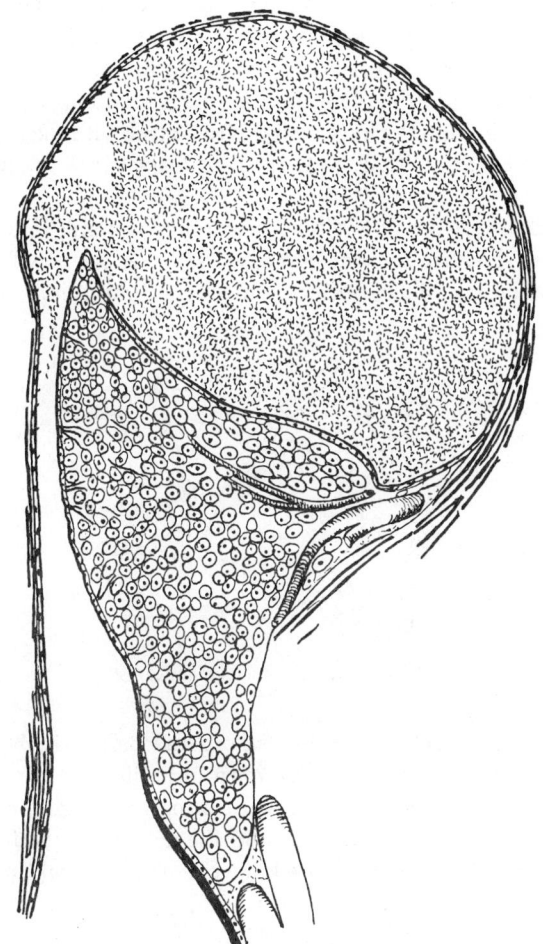

Fig. 5 *Sirex gigas* L.   Intersegmental pocket with oidia and smearing gland, in longitudinal section.   From Buchner.

apparatus, here resembling a simple shovel, are the least developed and, unlike all other species investigated until now, their organs remain free of fungi. *Sirex gigas* L., *S. augur* Kl., and *S. phantoma* F., and also *Paururus iuvencus* L., occupy a middle position in respect to development. In

Tremex magus F. the organs are especially large and extend deep into the body cavity. The chitinous framework of the *Xiphydria* varieties correspondingly shows the greatest deviation from their special systematic position when there is an unpaired shell-shaped part in addition to two lateral leaves, and when these three parts have special sections which apparently serve in the role of the muscle focal points already mentioned. In wood wasps the symbiotic devices are reflected in the chitin, and it is possible to judge their extent with dry material, in which the fungi filling is clearly recognizable at times. We have seen this to be true with the lymexylonids, and similar cases exist. Only the very rare *Konowia* has not been investigated as yet.

It was clear from the outset that the organs opening into the genital orifice on both sides are transmission devices, for we are now familiar with numerous other symbiont bearers which transmit symbionts to their

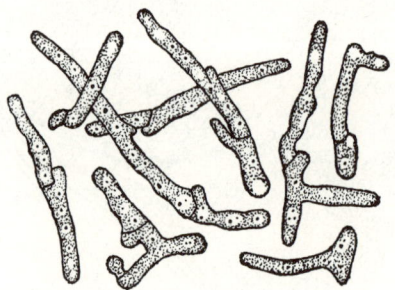

Fig. 6   *Sirex gigas* L.   Oidia from the intersegmental pockets.   From Buchner.

progeny by means of such intersegmental organs localized in the ovipositor. The contraction of their musculature must in itself cause partial elimination of both fungi and secretion and must bring the fungi in contact with the egg which is leaving the vagina; and the activity of the fungal jets is further stimulated by the complicated movements which take place during deposit and often cause the area near the pouches to be squeezed and pressed down. Indeed, when eggs are taken from the wood for examination immediately after deposit, masses of mucus adhere to them, especially at the pole which emerges first during deposit but also at the opposite pole, in which the oidia are embedded. The supply of symbionts must suffice for large quantities of eggs. In *Sirex augur* 1000 and 1100 eggs were found, and in *Paururus noctilio* F. up to 500.

The mucus-filled wall of the tunnel provides favorable growing conditions for the organisms deposited in the wood. These organisms, which then rapidly develop into normal hyphae, do not collect on the surface but penetrate into the surrounding wood at once. Francke-Grosmann showed that, in a moist chamber, luxuriously developed fungal filaments radiate within 24 hours from an egg smeared in this fashion (Fig. 7).

Thus, even before the larvae are hatched, the area of the future tunnels is supplied with fungi-filled wood to nourish them.   It is not surprising, then, to observe the same fungi in the vicinity of older bore tunnels or in

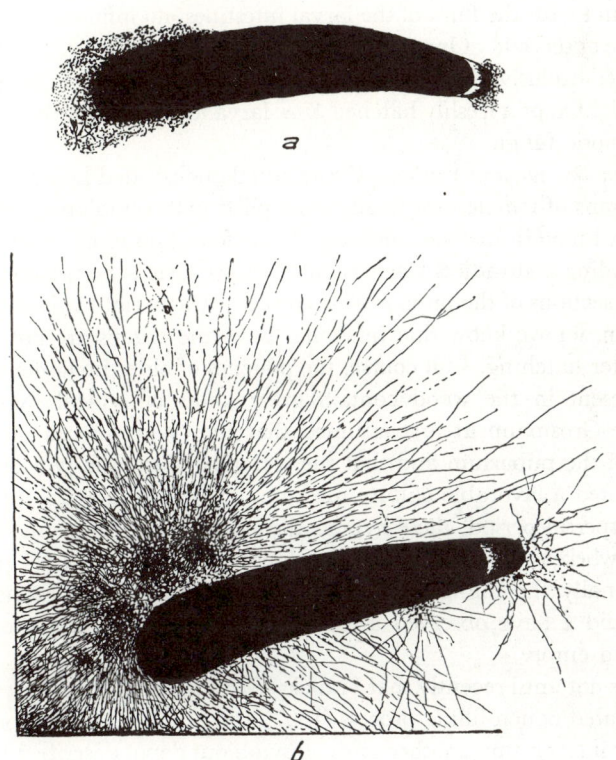

Fig. 7   *Sirex augur* L.   (*a*) Freshly laid egg with oidia-filled mucus on both ends; (*b*) the same egg after 24 hours in the moist chamber, with oidia grown out to mycelia. From photographs by Francke-Grosmann.

the borings which the larvae leave in their wake.   This also explains why "reverse feeding" may frequently be observed in siricid larvae.

Fragments of the mycelia are indeed always dispersed among the wood particles in the intestine.   Their plasmatic content disappears only one hour after consumption, and later even the walls of the hyphae are almost disintegrated.   Although symbiotic fungi break down cellulose and lignin—comparative weighings by Francke-Grosmann of sterile and fungus-filled wood pellets showed a considerable loss of weight in the latter after a rather long period—the intestine of the siricid apparently lacks the capacity for such decomposition. W. Müller believed that it was possible to prove the existence of siricid cellulase by comparative analysis of the borings and the larval excrement, but his experiments are unconvincing in that he failed to take into consideration that even the larval excrement is filled with wood-destroying fungi. In digestive experiments by Francke-Grosmann with the fluid of the larval intestines, no influence on the wood could be observed. One must conclude that in wood wasps the larvae are nourished exclusively by their symbionts and the content of the wood cells. Cartwright kept a freshly hatched *Sirex* larva for three months in a culture of symbiotic fungi.

In *Paururus iuvencus* Francke-Grosmann demonstrated how the transmission organs of females are finally resupplied with symbionts. As long as the pupal membrane was unsloughed, she found no fungi in the pouches. This finding contradicts Cartwright's, who reported the presence of fungi in cross sections of the pupa in *Sirex cyaneus*. Cartwright was undoubtedly mistaken, for we know that in similar cases the organs are not colonized until after hatching. Of course, the pouches of older imagines which are still present in the wood contain great knots of typical hyphae, and Francke-Grosmann at first assumed that they had entered through the lining of the puparium and that the secretion filling the pouches at least represented a favorable culture medium. If the insects are approaching their dispersal period, these hyphae disintegrate into segments of varying length, whereas typical oidia are developed only at the time of egg deposit. Occasionally the filling process is not completed. Both Francke-Grosmann and I have observed a female in which one of the two pouches remained empty.

It was not until recently that Francke-Grosmann clarified the even more complicated pouch-filling process (1957) and at the same time provided a basis for interpreting another unusual symbiont depot described by Parkin (1941, 1942). In studying larvae of various ages in *Sirex gigas* and *S. cyaneus* Parkin had found the symbionts in a location which had escaped all other investigators. In a certain narrowly defined location in the hypopleural folds of the first abdominal segment he found that the chitin forms symmetrically arranged, closely crowded, deep crypts filled with little knots of fungi wound back and forth. The contents are prevented from falling out by sharply tapering thorns distributed over the inner wall and by strange outgrowths shaped like blunt clubs on the outside edges of the indentations. The dimensions of the organ and individual crypts are type-specific. In *Sirex gigas*, for example, the crypt field is about 1 mm. in length and $\frac{1}{3}$ mm. in width. The hypodermis is radically thickened in this location, thus indicating a glandular function (Fig. 8).

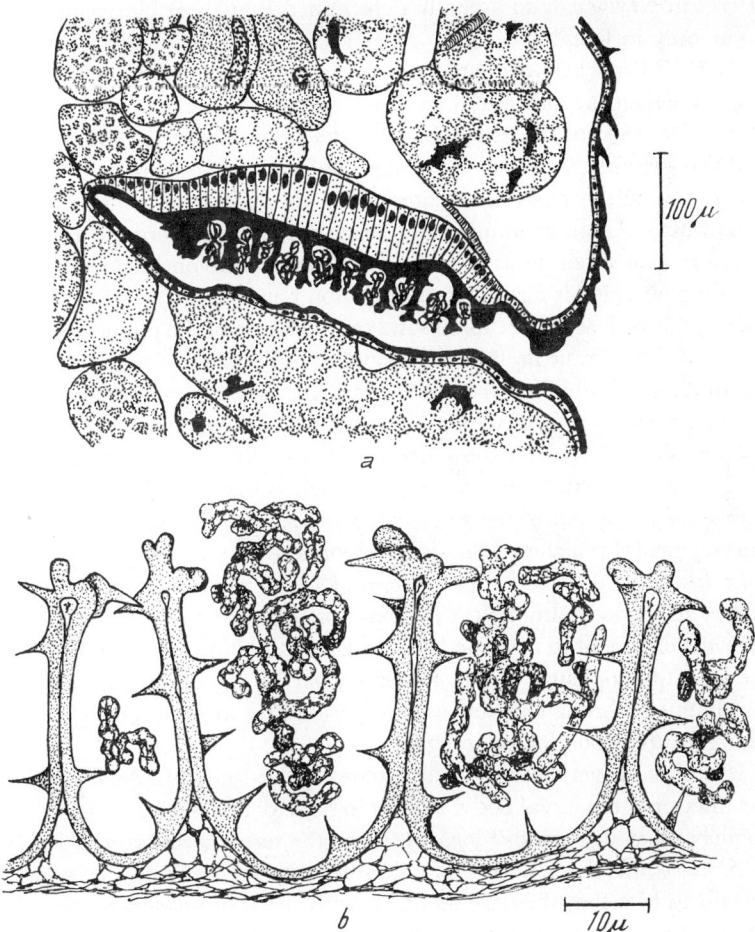

Fig. 8   (*a*) *Sirex cyaneus* F.   Larval symbiont depot in longitudinal section.   From Parkin.   (*b*) *Sirex gigas* L.   Fungus-filled crypts of a 10-mm. larva.   Original, made from a preparation by Parkin.

Parkin searched in vain for these organs in the first larval stage, but he did find them in all the later molting stages. Strikingly enough, they occur in only a fraction of the larvae, and this fraction corresponds quite exactly to the percentage of the female pupae and imagines found in this genus. There can be no doubt that the organ is a specific device of the female larvae, for it is lacking entirely in the pupae and imagines of both sexes. During cultivation of the parasite *Ibalia leucosporoides*, Rawlings

(1951) in New Zealand actually determined that these hypopleural folds occur only in females.

In 1957 Francke-Grosmann was able to interpret the function of these organs, which are not found in other lignivorous insects, and therewith closed the cycle of siricid symbiosis.   Before pupation the folds level off and the two fungal depots then take the form of two wart-like protuberances.   Simultaneously the secretion of the glandular cells reaches its maximum.   Upon examining the larval membranes sloughed off during molting, one often finds masses of delicate symbiont-filled scales, the covering of which is formed from secretion hardened to the consistency of wax.   But such scales are also found on the lining of the puparium and on the pupal membrane itself.   If they are placed intact in a nutrient solution the fungi are unable to germinate, but if they are punctured the spangle-forming mycetes appear within a few hours.   Francke-Grosmann's observations of living insects show decisively that the secretion platelets are broken apart during the movements which young *Sirex* females make in eating their way out of the wood, allowing some of the oidia to reach the intersegmental pouches.   Siricid symbionts stop growing when the wood contains only 20 per cent moisture, and they cannot exist at all in completely dry wood, although it is often possible for the larvae to complete their development in dry wood.   Consequently, it must be the function of the larval hypopleural organs to keep a part of the symbionts capable of development throughout this crucial period and to make possible a reinfection of the wood for the progeny.

It is also not yet clear to what extent regular symbiont depots occur in the interior of the larval body, at least in some of the larvae.   Clark (1933) vaguely refers to "glands" located "near the terminal intestine" of larvae in *Sirex noctilio*, from which he cultivated typical mycelia.   It appeared certain to him that they are identical with the transmission organs of the imago and are carried over through the pupal stage.   Even though such continuity is out of the question, Clark's opinions might have a basis in fact.   Years ago someone sent me cross sections of the larval intestine of an undetermined siricid to which a mass of cells thickly filled with hyphae adhered, but at that time I was not able to pursue the case further.   At any rate, it is not improbable that some siricid or other will one day provide us with a surprise or two in this respect.

The culture of siricid symbionts offers no difficulties and has already been undertaken by several authors, Cartwright (1929) having been the first.   Francke-Grosmann obtained her best results when she used a substrate of malt gelatin supplemented by 0.5 per cent crystallized citric acid to suppress troublesome development of bacteria which always contaminate the contents of the syringe.   W. Müller used both medicated agar and fluid sterilizing agents.   With the medicated agar the fungus of *Sirex gigas* showed great similarity to the symbiotic growth form but did not develop oidia.   With the other agent the submerged hyphae grew out into elongated filaments which similarly formed spangles.

Cultivation is easiest when oidia from transmission organs are chosen as starting materials.   Fungi from the egg surface, the canal, or from the wood bordering on the canals can also be used.   Cultures originating in these places are identical when one and the same population is involved. Representatives of the same species from different locations may yield distinctly different cultures.   Indeed, they may vary considerably with respect to color of the mycelium, its odor, or the character of the air mycelium.   For example, the typical symbiont from *Paururus iuvencus,* which lives in pine trees, smells of fungus and forms a gray mycelium, but all individuals of a population living in fir wood yielded a white mycelium with an odor resembling fennel, regardless of whether transmission organs or wood was used as the starting material.   Similar though slighter differences resulted in symbionts of *Sirex gigas* (Francke-Grosmann).

Cultures are extremely difficult to classify, because in general no sporangia are formed.   Cartwright cultivated a fungus from *Sirex gigas* which he identified with *Stereum sanguinolentum;* from *Sirex cyaneus,* a form closely related yet exhibiting slight constant differences; and from *Xiphydria prolungata,* a fungus similar to *Daldinia concentrica.*   Francke-Grosmann presumes that the symbiont cultivated from the *Sirex* in fir wood is *Trametes odorata.*   The only fungus that form sporangia, and thus is classifiable as *Polyporus imberbis,* also differs from the usual guest.   It was furnished by a *Tremex fuscicornis* population, which by way of exception lives in nut wood.   Usually cultures could not be identified with any of the wood-destroying hymenomycetes identified up to the present time.

Siricids increased in practical importance when it was proved that they live symbiotically with wood-destructive fungi and inject the strains with them during egg deposit.   Since fungal growth is greatly accelerated in moist locations and retarded in dry ones, the wisdom of using wood in mines and cellars is questionable, although it is unlikely that siricid feeding in itself would affect the strength of the wood.   Harmful effects from fungi under dry conditions are negligible (Stillwell, 1960).

Individual siricids are by no means always linked with the same fungus. In spite of the highly perfected mode of transmission, therefore, it is likely that the association between the two partners is a loose one and also that the adaptation is comparatively recent in the history of the breed.   One must keep in mind that originally only one pair of glands was localized in the ovipositor with the function of lubricating the basal parts, whereas another, unpaired gland lubricated the sting to prevent friction with the wall of the tunnel. *Xeris spectrum* still represents this original state, and, although occasionally it does not disdain the fungi on hand, it finds sufficient digestible material in the fresh juicy wood in which it lives. All other representatives of the siricids have learned to live in closer conjunction with wood-destructive fungi, and for this reason the coniferous wood in which the *Sirex* and *Paururus* varieties always live is in general sparsely filled with symbionts. In *Tremex fuscicornis* F. and apparently also in *Tremex magus* F., which like the *Xiphydria* species are limited to deciduous wood, fungal nourishment plays a much larger role. Here the surrounding wood is filled to a far greater extent with rapidly growing fungi and siricid symbiosis most closely approaches the typical ambrosial culture.

PELOMYXA

So striking is the association between bacteria and all *Pelomyxa* species that it was noted even by older investigators like Greeff (1874), Leidy (1879), and Schultze (1875). To some the uncertain inclusions suggested crystals, but to others they indicated bacteria. Penard (1902, 1911) and others verified their bacterial nature. Because the presence of bacteria is so characteristic, Penard even used it as a criterion of the genus, excluding from it all species without bacteria. The bacteria are distributed throughout the endoplasm, but occasionally are found in the ectoplasm. Often they prefer the vicinity of those luminous bodies which are so characteristic of *Pelomyxa* and which, according to Stolc (1900) and Leiner (1924), consist essentially of glycogen. Frequently the bacteria display a partiality for the nuclei which are distributed in great numbers over the plasma of these amoebae, forming a covering all around the nuclei, often in systematic tangential arrangement.

Each *Pelomyxa* species has its specifically shaped bacteria; in *Pelomyxa palustris* Greeff two forms are always present: smaller delicate rodlets 2–3 μ in length, and chains of finely dotted rodlets up to 22 μ in length, their parts held together by a common casing. Penard and Gould (1893) even found much longer chains. Elsewhere only one type of bacteria is found. They are minute in *Pelomyxa prima* Gruber and *P. tertia* Gruber, short and thick in *P. vivipara* Pen., long and thin in *P. caulleryi* Holl. (Hollande, 1945), and in other species they resemble the symbionts of *P. palustris*, which received special study.

The plasma of animals taken in nature tends to be crammed with divers plant parts, sand granules, and other constituent parts of the oxygen-poor moldering detritus in which the animals so often occur in great numbers. Taken in culture, they ingest filter paper or wadding in addition to starch granules and then fill up to an unbelievable degree with the cellulose fibers and, in connection with this, store great quantities of glycogen. They can easily exist on filter paper alone. This unusual ability suggested that they might owe this capacity to a cellulase furnished by the bacteria. Moreover, they live in an environment where cellulose-decomposing microorganisms cannot possibly be absent, and these must necessarily reach the *Pelomyxa* plasma with ingestion of food.

Keller (1949), who entered more deeply into the bacteriological and physiological aspects of this cooperative living, believes that such hypotheses can be confirmed. Like all his predecessors, he sees two types of organisms in the two forms ever present in *Pelomyxa palustris*. He describes their cultivation in agar mixed with sterile *Pelomyxa* broth and reports that he was able to transfer them to ordinary agar after several inoculations. The one organism is described as an elliptical gram-negative coccoid 2 μ in length, with small circular colonies of almost mucilaginous consistency. Filter paper added to a culture in nutrient broth was much loosened in texture after an incubation period of 14 days. The nutrient solution then gave an intensely alkaline reaction. The second organism is described as similar in form but different in its type of growth and its complete inability to produce cellulase. Keller came to the conclusion that the first organism must be classified as one of the myxobacteria which often live in association with genuine bacteria and which favor cellulose-rich substrates. The second organism, he believed, represents an obligate companion form. He named them *Myxococcus pelomyxae* and *Bacterium parapelomyxae*. (Gould, 1905, had declared the stronger form to be a type of Chlamydobacteriaceae.) Keller believed that it was possible to prove through serological testing that the strains he had cultivated and those he was searching for were identical.

Keller's observations seemed to confirm the hypothesis advanced earlier that the unusual capacity of the *Pelomyxae* for utilizing cellulose might be traceable to cellulase production on the part of the microorganisms associated with them.

Research by Leiner, Wohlfeil, and Schmidt (1951, 1953, 1954) cast serious doubt on Keller's findings. It established, first, that even *Pelomyxa* harbors only one bacterium; second, that the broad forms arise when the slender rodlets are folded together in pairs as shown by the longitudinal split. These broad forms then spill out their content in chains of spores. Keller's cultures undoubtedly represent foreign organisms adhering externally to the amoebae, whereas the strains cultivated by Leiner and his co-workers were obtained from thoroughly cleansed animals and have a pronounced resemblance to intracellular states. Although cellulase could not be detected, these authors are convinced that symbiosis is involved, and they conjecture that the bacteria play a role in the protein metabolism of the hosts.

Occasionally it may be observed that under unfavorable conditions amoebae expel their bacteria along with other inclusions, but that in diseased hosts the bacteria gain the upper hand and sometimes continue to reproduce freely even after the death of the host animals. Such observations by no means disprove symbiosis but only emphasize that normally it is the amoebae which control the symbiosis.

Unfortunately it has not yet been possible to study the developmental cycle of the *Pelomyxae* exhaustively, but it may be assumed that, when spores or gametes are formed, a new infection with organisms from the environment must occur each time and that the individual species must make their specific choice.

Finally let us mention that bacteria are always present in the protoplasm of some commensals and independent ciliates. In *Euplotes* it was possible to kill these symbionts with penicillin, but after 10 to 12 days the ciliates also perished (Chatton and Seguela, 1940; Fauré-Frémier, 1952).

FERMENTATION CHAMBERS IN LAMELLICORNS, TIPULIDS, AND TERMITES; SIMILAR PHENOMENA

The guts of insect larvae which feed on diets unusually rich in cellulose often assume strange forms connected with this mode of nutrition. In the phytophagous larvae of melolonthines, the hindgut forms a huge sac-like extension which struck the attention of Swammerdam and was described in a series of forms by the older insect anatomists, such as Ramdohr (1811), de Serres (1813), Dufour (1812, 1824), and De Haan (1835). De Haan also provided excellent illustrations. Later it was chiefly Mingazzini (1889) who studied it anatomically. Even in the living animal this section of the gut, filled as it is with dark material, tends to shine through the tautly stretched dorsal membrane.

Figure 9a shows the intestinal canal of the larva of a rose beetle (*Potosia cuprea* Fbr.) which lives in anthills. Here the intestinal canal is longer than the animal itself by more than one-half. Three rings of glandular ceca encircle the midgut; the hindgut is curved to form an S; a short small intestine first extends toward the posterior, then suddenly joins the huge sac representing the large intestine, and then extends upward toward the head. The narrowed hindgut, which runs caudad on the ventral side, is clasped by two points of the large intestine slightly coalesced below it (Werner, 1926). All related forms investigated, *Oryctes* (Fig. 9b), *Anomala*,

Melolontha, Trichius, etc., vary in development of the midgut evaginations but are largely similar in development of the hindgut.

Lucanid larvae, which feed on dust and moldering wood, are of different construction. In the larvae of *Sinodendron cylindricum* L. a circle of large

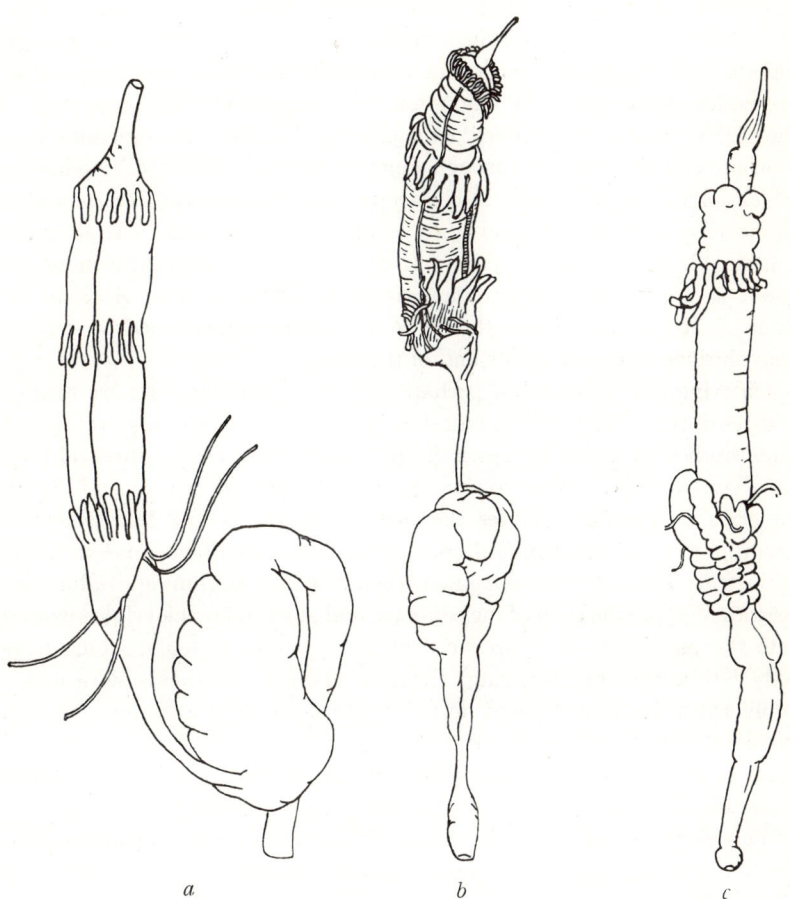

a   *b*   *c*

Fig. 9   (*a*) *Potosia cuprea* Fabr. Intestinal tract with fermentation chamber. From Werner. (*b*) *Oryctes nasicornis* L. The same. From Mingazzini. (*c*) *Sinodendron cylindricum* L. Intestinal tract with corresponding blind sacs. From Buchner.

evaginations, varying in size, is located between the starting points of the Malpighian tubules. No small intestine is formed, and these evaginations are joined directly to a section of the hindgut, laid in folds, which appar-ently takes over the functions of the sac-shaped extension of the melolonthines (Buchner, 1930; Fig. 9*c*). In other species described by de Haan, *Odontolabis alces* F. and a *Eurytrachelus* spec., a sharply defined part hindmost in the midgut is enormously enlarged and has at its beginning a circle of ceca. Doubtless these may be considered to be analogous devices.

In the coprophages, which live for the most part on animal excrement, it is also the midgut which often forms a large, dorsal, hernial sac. Fabre (1891–1897) described this in *Scarabaeus*. As soon as the young larva ingests food, this sac is filled to such an extent that the mid-dorsal region protrudes remarkably. Heymons and von Lengerken (1929), who studied the biology of the "pill rollers" exhaustively, describe it as a gigantic balloon entering the intestine through a narrow section. The resulting hump, which forces the larval body into a deep curve, represents an indispensable supporting organ and an aid in locomotion and ingestion of food in the spherical living and eating space. In the native *Odontophagus* the intestinal canal is similar in structure, and from an illustration of *Aphodius* by de Haan it is apparent that these evaginations are represented by striking convolutions at the posterior end of the midgut.

Differing in character but perhaps performing similar tasks are strange devices in the hindgut of the passalids, described first by Leidy (1853) and later studied in detail by Lewis (1926) and R. and H. Heymons (1934). In these dust-and mold-consuming lamellicorns of North and South America a sphincter divides the hindgut in two. The last section is accompanied by longitudinal rows of sac-like diverticula, and in many species it also has an elongated cecum at its beginning. Chitinous bristles regulate the path of the contents, and at its end special ridges extend into the interior to dam up the contents. The diverticula, which have only a thin chitinous covering, always contain a remarkable flora which proliferates luxuriantly and forms a thick matting on the wall. A phycomycete, *Enterobryus*, is attached to it with suctorial discs, and the end parts of the filaments are detached. Packs of very fine filaments call to mind the *Oscillaria*.

This flora, which varies in detail, is never found in freshly hatched beetles, but it is always present in mature animals. The beetles must therefore ingest spores or permanent states with their food. The larvae are at times weakly colonized and at other times more strongly, a situation which may be possibly connected with the moltings. The larvae also are necessarily infected by organisms from the dust, imaginal in origin. We do not yet know whether the nutrition is influenced by these fungous remains nor how the influence might be brought about, but authors who have studied this remarkable symbiosis are inclined to think that it plays a role in the nutrition and physiology of the larvae. On the other hand, one must not fail to mention that similar fungi called eccrinids are unusually abundant in the intestine of diplopods, isopods, and decapods, as well as of hydrophilids, other lamellicorns, and sundry insects, and yet it would be wrong to assume that symbiosis is involved (unpublished observations of Maessen).

Another comparable case is that of the tipulid larvae, insofar as they feed on cellulose-rich food. Here, too, the hindgut is greatly thickened and equipped with a cecum which is developed in varying degree and, as in all these cases, is filled with food particles. The most extreme development occurs in *Tipula flavolineata* Meig., where the larvae are found in moldering or even quite solid wood. Here the cecum is much longer than the hindgut. In the larvae of *Ctenophora flavicornis* Meig, which live in the same manner, the cecum is less fully developed. In *Tipula maxima* Poda, the larvae of which live among moldering leaves and in the stagnant ooze of rivulets and ditches, we find the first beginnings of this formation. In all species, moreover, the large intestine from which the evaginations spring is unusually widened (Buchner, 1930) (Fig. 10). It is clear that these devices radically alter the larval intestine and often the whole body. Studies on the bacteriological and physiological aspects of the nutrition, based exclusively on melolonthines (Werner, 1926; Wiedemann, 1930) indicate that they are doubtless traceable to an indispensable symbiosis with bacteria which are always present in great numbers between food particles in the areas mentioned.

Werner determined that food takes 3 to 4 days to pass through the larval intestines in *Potosia cuprea*. In the midgut the food is relatively unmixed, but both old food and new food get a thorough mixing in the hindgut, and periodic feeding with filter paper shows that some of the food may remain here as long as 2 months. The bacteria which are necessarily ingested constantly during feeding are sparse in the strongly alkaline midgut but proliferate in the large intestine, which is usually neutral but occasionally slightly alkaline or slightly acid. In this and all related species, the wall of the large intestine contains numerous groups of finely chiseled, pinnate, chitinous bristles which are excellently suited to hold back the food, on the one hand, and to guarantee permanence of the bacterial strain on the other. Such action is important in these species where the food consists of rather firm wood poor in bacteria. Alternating with the bristle fields are areas with a thin chitinous lining and numerous little canals. All authors regard such zones as the seat of resorption, in this instance transferred to the hindgut (Mingazzini; Biedermann, 1911; Wiedemann, 1930).

By Omeliansky's method (1902) Werner proved that cellulose-fermenting forms actually do exist in the intestinal flora of the *Potosia* larvae. The intestinal contents of the larvae were added to an inorganic nutrient solution in which strips of filter paper represented the sole source of carbon.

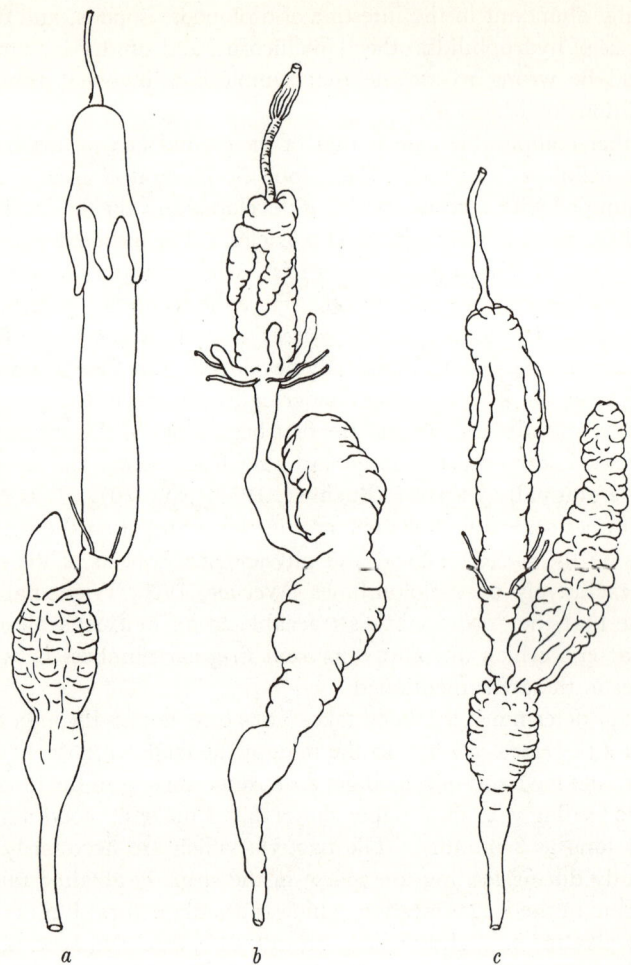

Fig. 10   (*a*) *Tipula maxima* Poda, (*b*) *Ctenophora flavicornis* Meig., (*c*) *Tipula flavolineata*
Meig.   The hindgut bears fermentation chambers differing in degree of development.
From Buchner.

After approximately 50 days only a thin powdery layer of the cellulose was
left on the bottom of the fermentation flask.   Werner was also successful
in isolating from the bacterial mass the only form that attacks cellulose, a
peritrichous flagellate rodlet 1.4 to 4 μ in length.   All other forms isolated
in the usual culture media proved incapable of attacking the cellulose,
being nourished exclusively by decomposition products arising during
cellulose fermentation.

Significantly, temperatures of 33–37°C. proved to be the optimum for this *Bacillus cellulosam fermentans* Werner, that is to say, the same high temperature which prevails in anthills and the one which is also optimum for larval growth. When the temperature falls to 13°C., decomposition of the cellulose is checked and the laboratory larvae stop growing. This corresponds to what occurs in nature, namely, that the animals develop from May to October and then undergo a period of rest during the remaining months, during which the temperature in the anthill falls below 13°C.

Wiedemann's results are in accord with Werner's, and in addition they explain the form in which cellulose decomposition makes its contribution to the insect body. Using larvae of *Oryctes nasicornis* L. and *Osmoderma eremita* Scopoli as well as those of *Cetonia*, Wiedemann found in their intestinal secretion a protease which remains ineffective when its reaction is alkaline, with the result that organisms ingested with the food can pass through the midgut unaltered. As soon as a sufficient amount of acid is produced by bacterial activity to neutralize the intestinal secretion, the protease is activated and the bacteria are digested. The same process occurs in the flagellates which always accompany the bacteria and are nourished by them. Thus it is seen that the host animal can obtain its nitrogen requirement from the bacterial protein, and, since cultures of the bacterial mixture thrive in an Omeliansky solution containing only inorganic nitrogen, one may conclude that the bacterial mixture in the insect's intestine can also utilize the nitrogen which exists in abundance in the ever-present earthy matter.

Inasmuch as these larvae and their symbiotic microorganisms live in an environment differing from that of ant-dwellers, it is not surprising to find that the optimum temperature is also different, specifically much lower. With Werner's ant-dwellers we have seen that 33–37° represented the optimum temperature. With Wiedemann's experimental objects 23° was the temperature sought out by the larvae in containers with temperature gradients; and at 23° the bacterial mixture also flourished best. Thus the greatest increase in larval weight would also be expected to occur at 23°.

The findings of Werner and Wiedemann in connection with the "fermentation chambers" of several melolonthines may undoubtedly be applied to those insect larvae which have the same mode of nutrition and similar extensions of the hindgut, even though they have not yet been studied bacteriologically. The striking differentiations are entirely absent in the imagines, which are indeed adapted to different sources of nourishment, and this in itself clearly indicates that the differentiations are connected with the cellulose-rich, protein-poor diet. In fact, Heymons and von Lengerken (1929) made a statement to this effect with respect to *Scarabaeus*, and Vaternahm (1924), in his studies on the biology of *Geo-*

trupes, had earlier emphasized the nutritional and physiological importance of the bacteria which he always found in the intestinal canal in great numbers. Whether, as Heymons and von Lengerken claimed, the brownish secretion emitted by the larvae of the "pill rollers" actually promotes bacterial activity and thus brings about an extraintestinal predigestion of the dung, must, of course, be given more study.

In this connection it is of further interest that Pochon (1939) was able to establish in the intestinal canal of lignivorous insects (*Tipula oleracea* and *Rhagium sycophanta*) the presence of a bacterium capable of disintegrating cellulose under anaerobic conditions. The form which he called *Plectridium* was said to be very similar to Werner's *Bacillus cellulosam fermentans*.

At any rate, there can be no doubt that all the differentiations which occur consistently in species with a similar mode of nutrition, together with their rich flora, possess nutritional and physiological significance. The fact that Ripper (1930) was unable to find a cellulase in the hindgut of *Dorcus* and *Osmoderma* does not alter matters.

Future research must determine the extent of similar, perhaps less highly developed, adaptations in other insects living on a cellulose-rich diet. For instance, in gryllids the proctodaeum is unusually well developed. The midgut is greatly reduced, but the segment of the hindgut directly following it is greatly enlarged and is provided with strange tufts extending into the interior and forming chitinous bristles and telodendrons in turn. The Malpighian tubules make their entry only at the terminal point of this area, which apparently is also active in resorption, and then a second section follows, in this instance tuftless but likewise wide-lumened. Almost entirely absent in the midgut, the bacteria form a thick matting on the tufts, become sparser again behind the entrance point of the renal vessels, and are present in only very small numbers in the rectum (*Gryllotalpa, Gryllus, Nemobius*, according to Begen, 1948).

Bacterial analysis by Keller (1949) established that a number of organisms, varying from genus to genus, are always present and that other non-permanent organisms are continually making their appearance. Only one strain, *Bacillus subtilis*, was present in all genera. Up to this point it had always been impossible to detect cellulose cleavage in the numerous organisms cultivated, but *Bacillus subtilis* cultures on nitrogen-free media now caused an accumulation, admittedly slight, of nitrogen. At present it seems doubtful that we fully understand the significance of the bacterial flora once again associated with striking morphological adaptations in the grillids.

Comparable in particular to fermentation chambers and similar formations is the enormous extension of the hindgut in termites which feed on more or less decayed wood, or even on fresh wood, or at least on a cellulose-rich diet. Their gut, similar in many respects to that of the blattids, displays a relatively short midgut and an unusually well-developed proctodaeum. Inserted between two narrow sinuous sections of the latter is an enormous sac-like dilation which tends to be largest in the workers, smaller in the sexual forms and soldiers, and absent in the substitute sexual forms which live exclusively on liquid food and in the fungi-cultivating forms.

When the hindgut is opened, a thick milky fluid emerges in which many bacteria and countless numbers of flagellates are found. Indeed, more than a third of the body weight is reported to consist of flagellates. It is particularly the many bizarrely constructed representatives of the polymastigines which interest the protistologists. The plasma body contains numerous wood particles, which, since the cystostome is absent here, must be ingested by means of a groove at the posterior end or by means of pseudopod-like processes. From experiments by Cleveland (1924, 1925), who was first to exclude the protozoan fauna in a variety of ways, and from studies by Montalenti (1927), Pierantoni (1937), Baldacci (1941), Ghidini (1941), Tóth (1945), and Goetsch (1946) we know that the termites have a vital need for these protozoans, which, like *Pelomyxa*, thrive on the pure cellulose diet of the host. The following experiments, alone or in combination, provided proof: temporary increase of temperature to 36°C., hunger, transferral of animals to pure oxygen under a pressure of 1 atmosphere, and feeding of antibiotics. Since diverse flagellates existing side by side in the termite gut are destroyed in part slowly and in part more rapidly, it was possible to determine the various degrees of their dispensability or indispensability in the individual species. Loss of the flagellates always resulted in the speedy death of animals which were incapable of cellulase production, even though other food was offered in abundance. The animals could be kept alive only by reinfecting them in time or giving them humus in which cellulose had been extensively decomposed through fungal and bacterial action. These findings were analogous to those in Montalenti's research on animal-specific enzymes. In the mid- and hindgut he found amylase, invertase, and a proteolytic enzyme but no cellulase (Montalenti, 1932).

All authors agree that in this connection it is primarily the flagellates, and among these in turn the large polymastigines, which are capable of cleaving cellulose and that it is especially the resulting glucose, together with the other water-soluble carbohydrates, which are of benefit to the termites. According to experiments by Trager (1934), other flagellates also come under consideration as useful guests. Trager was able to cultivate *Trichomonas termopsidis* from *Termopsis angusticollis* outside of the host animals, and thus to show that they are incapable of existence without addition of finely pulverized cellulose.   Moreover, it is generally agreed that, as with the bacteria in the melolonthine larvae, a temporary oversupply of flagellates, normally absent in the excrement despite vigorous reproduction, may be digested by the hosts and thus form their chief source of protein.

These close parallels between the symbiosis of termites with flagellates and that of melolonthines with bacteria would in themselves justify discussion of the symbiosis of the termite gut, despite the fact that it involves not a plant but an animal guest.   There is further justification in that other plant guests likewise play an important role in the complicated symbiosis of the termite gut.

For years there had been reports from various sources that bacteria occur regularly in the plasma of the polymastigines (Buscalioni and Comes, 1910, *Trichonympha;* Kirby, 1924, *Dinenympha*, and, 1932, a series of *Trichonympha* species; Powell, 1928, *Pyrsonympha*).   Jírovec (1932) spoke of a symbiosis of unknown benefit when he found always present numerous rodlets and diplococci in *Trichonympha serbica* Georgevitsch from *Reticulitermes*.   Pierantoni (1934, 1935, 1936, 1937, 1940, 1951) expressly directed attention to this strange occurrence.   He theorized that it was not the flagellate plasma but exclusively these bacteria which produce a cellulase and that therefore in the last analysis it is the bacteria that provide the lignivorous termites with living quarters.   Pierantoni studied the *Joenia* and the *Mesojoenia* species in *Calotermes flavicollis* and the *Trichonympha* species in *Reticulitermes lucifugus*.   In *Joenia* the numerous bacteria, usually cocciform but sometimes rodlet-shaped, are crowded together chiefly in the anterior half of the elongated animals behind the nucleus and all around the axis staff, while the space in the posterior half is taken up primarily by the wood particles (Fig. 11).   In *Mesojoenia* the same area is filled with filamentous organisms.   In *Trichonympha minor* numerous minute cocci and coccoids are either distributed uniformly within the whole body or are limited to the denser plasma surrounding the nucleus. In *T. agilis* minute rodlets are located in masses behind the fibrillar basket which closes the anterior nucleus-containing portion of the flagellate body.

From these and numerous similar findings, especially those in the excellent flagellate studies of Kirby (1932, 1938, and other dates) one gains the impression that the microorganisms must in some way be of significance to the flagellates.   They always have a distinct shape and arrangement.   Older authors, insofar as they noticed them, usually mistook them for mitochondria, which indeed are also present.   Especially impressive are cases in which the bacteria are restricted to aggregations so sharply defined, so regularly distributed, and so dense as to give the colonies the appearance of organellae, which are comparable to the mycetomes of multicellular organisms. For example, this applies to the filamentous bacteria surrounding the nucleus of *Trichonympha campanula* as a sharply defined mass; to regular spherical balls of delicate rodlets in many *Caduceia* species; to rodlets which in *Macrotrichomonas pulchra* from a *Calotermes* species are restricted to the so-called capitulum, that is to say, the terminal, swollen end of the axis staff located in front of the nucleus; and to the cocci and diplococci, which form a thin layer under the whole surface in *Pseudodevescovina ramosa*.

Plausible though it may seem that these widespread, highly adapted symbionts in the termite flagellates are beneficent guests, it must not be forgotten that no experimental proof has ever been furnished—Hungate (1938) tried in vain to cultivate them—and that Pierantoni's assumption has by no means met with agreement everywhere (Grassé, 1952).

Nor has any author completely explained the possible symbiotic import of the bacteria occurring independently in great numbers in the termite gut—bacteria which are distinctly differentiated from those living in the flagellates. It was chiefly Grassé and Noirot (1959) who studied their carefully regulated distribution. The foregut, which is considerably enlarged in humus eaters, is bacteria-free. The same is true in general of the midgut, but an exception is found here in the many cases in which a "mixed section" is present anterior to the valvula posterior, whose wall belongs partly to the midgut and partly to the hindgut. In such genera a bacterial mass, practically a pure culture, lies sometimes only between the midgut wall and the food mass surrounded by the peritrophic membrane, and sometimes on the side belonging to the hindgut. Only in certain species in which this section is considerably enlarged does the separation of bacteria from the food appear not so thorough. Where such an unusual

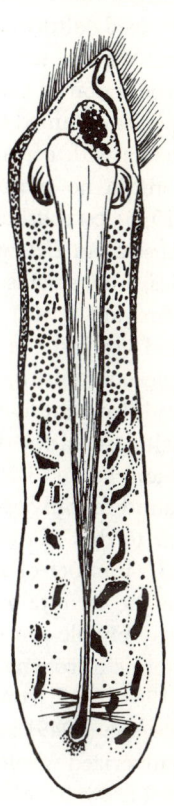

Fig. 11 *Joenia annectens Grassi* from *Calotermes flavicollis.* The anterior half of the axostyle surrounded by symbiotic bacteria. From Pierantoni.

transition zone is lacking, that is, in the majority of termites, the bacteria are limited to the hindgut. There is no such well-defined separation of food from bacteria, but the large mass of bacteria, consisting of various kinds, is collected close behind the valvula and toward the rear ends on the walls of the fermenting chambers. Moreover, we must not forget that, along with these bacteria, cellulose-disintegrating forms are also ingested with the food. Beckwith and Rose (1929) found an analogous situation in a mixed culture of gram-negative rodlets and several micrococci. Baldacci and Verona, sometimes working in collaboration (1939, 1940, 1941), found production of cellulase in pure cultures of a *Cytophaga* and an *Actinomyces* species obtained from the intestine of *Reticulitermes* and *Calotermes*. Experiments with certain termitids, which in general do not live on wood and contain no symbiotic flagellates, have similar implications. Holdaway (1933) established the presence of a rich bacterial flora in the gut of an Australian *Eutermes* species which disintegrates wood in abnormal quantities, and surmised that it played a role in the digestion of the wood. Hungate (1946) kept a related form on a pure wood diet for 18 months and verified the presence in its gut of two anaerobic, cellulose-digesting microorganisms, a *Clostridium* sp. and *Micromonospora propionici*. Misra and Ranganathan (1954) also arrived at the conclusion that in *Odontotermes* cellulase is not derived from the termite but in all probability from the bacteria, and Pochon, de Barjac, and Roche (1959) of the Pasteur Institute found that in *Sphaerotermes sphaerothorax* the bacterial flora is strikingly similar to that of the ruminant rumen.

A number of authors gave different interpretations of the bacterial flora in the termite gut. Cleveland (1925) theorized that they must in some way enable the termites to derive their protein requirement from atmospheric nitrogen; Montalenti and Pierantoni agreed with him. First Tóth (1944/1945; compare also Goetsch, Offhaus, and Tóth, 1944) and then Ergene (1949) attempted to obtain experimental proof. Tóth sometimes pulverized whole *Calotermes* and sometimes only the gut. To the surviving cell broth he added oxalic or succinic acid as a substrate for the amino acid formation and after 24 hours determined the resultant increase in nitrogen by the micro-Kjeldahl method. With isolated hindguts he obtained an accumulation of 0.5% in a solution of 0.5% NaCl + 0.5% glucose which, however, rose to 50% with the addition of small quantities of certain salts. Similar data resulted when whole animals were pulverized (up to 67% with the addition of salts). Surprisingly, a substantial increase in nitrogen (69, 50, or 37%) also resulted with whole animals from which the hindgut, the abdomen, or the thoracic section without gut had been removed. Tóth attempted to explain this unexpected result by saying that other symbionts, capable of binding atmospheric nitrogen and localized possibly in the fat body, had apparently escaped observation. But this must be regarded as an unsatisfactory interpretation and one that is quite likely to cast doubt on the reliability of the method used.

Without knowledge of Tóth's publication, Ergene studied the same problem. He inoculated a nitrogen-free Bortels' solution with an emulsion of the gut of *Calotermes flavicollis*. Despite the fact that atmospheric nitrogen exclusively was available for protein formation, the mixed culture continued to increase in mass even after a series of transfers when possible nitrogen reserves must all have been exhausted. Ergene ascribed nitrogen assimilation to *Bacterium breve aerophilum*, which alone was always present in the mixed culture. That it could not grow in isolation accords with a finding from another source that nitrogen-binding bacteria flourish better on nitrogen-free bases in mixed cultures. A *Coccobacterium calotermitis* Ergene, which invariably appears in the gut whereas *Bacterium breve aerophilum* is said to be present in such small numbers as to escape observation, failed to develop on the nitrogen-free bases, and according to Ergene it might play some other, still obscure role.

We are, however, forced to assume the capability of atmospheric nitrogen assimilation by the symbionts in view of certain experiments showing that termites fed exclusively on nitrogen-free cellulose increased in weight and even reproduced. Lüscher and Clusius (1952), with the aid of radioisotopes, were also able to show that *Calotermes* can synthesize albumin in a bacteriological way through assimilation of atmospheric nitrogen. With normal nutrition, of course, the nitrogen content of wood (about 0.25%) and the fungi which the animals always like to ingest would probably be sufficient to form the bacterial plasma, which together with wood splinters constitutes the diet of the flagellates. For this reason Goetsch also is of the opinion that the questionable bacteria fall back upon atmospheric nitrogen only when other nitrogen sources are insufficient. In spite of all efforts until now, we are still far from comprehending the complicated processes taking place in such a termite fermentation chamber in which the separate regions are characterized not only in regard to the colonization by bacteria but also and chiefly through zonal differentiation in the degree of digestion. Grassé spoke of a cascade of reactions which must take place in this chamber.

Unfortunately it has never been clearly explained how provision is made to safeguard the continuity of this astonishingly complicated symbiosis which unites insects, flagellates, and bacteria living independently in the gut and apparently also symbiotically in the flagellates. On leaving the colony, the sexual forms must carry away in their intestines the various flagellates, which in these forms never reach the stage of cyst formation, just as the queen, in forms establishing fungal gardens, carries away organisms from such gardens. The young animals always ingest the first flagellates with the proctoidal food. Grassé and Noirot (1945) have shown that such food, in contrast to the practically flagellate-free excrement, consists of one drop of intestinal content rich in organisms. Larvae and nymphs must as a rule be reinfected in a comparable manner after molting. Yet cases are also known in which some protozoans survive molting, in the space between the old intima and the new rectal epithelium. The two possibilities discussed in this paragraph apply also to the last molting leading to formation of the imago.

One cannot discuss termite symbiosis without mentioning the interesting results of Cleveland (1934) with *Cryptocercus punctulatus* Scudder. These cockroaches, limited to several areas of North America, are biologically similar to termites. There is clear evidence of the existence of a primitive form of government, and they live in and by wood, often together with termites, and likewise accommodate numerous flagellates in their intestines. Cleveland, in a large monograph devoted to *Cryptocercus* and its intestinal fauna, cites no less than twelve genera with twenty-five species, among them several polymastigines and hypermastigines. They live freely motile in a forward section of the well-developed hindgut, separated from the following section by a type of sphincter. They produce a cellulase and a cellobiase, and after comparable enzymes have again left the host animals they take over the decomposition of the woody food in the manner of termites. Again, it seems to be chiefly the hosts who benefit by the glucose produced in the flagellate plasma, but, as in the case of termites, details are still lacking (Hungate, 1950). As with some termites, the protozoans survive the moltings. Before moltings they are temporarily immobilized in the posterior segment of the hindgut, where they become round and secrete a double membrane. During molting the chitinous lining of this intestinal segment disintegrates, making it impossible for the protozoans to remain. Nevertheless, relatively few leave with the excrement, and these, this time in cyst form, serve to infect the freshly hatched young larvae.

As is well known, birds and mammals feeding on cellulose-rich food press into their service bacteria and protozoans, as do all insects which form voluminous fermentation chambers or in some other way provide space in their intestinal canals for vital microorganisms. Since the discussion of such cases of endosymbiosis would exceed the limits we have set for ourselves, we must be content with brief reference to a surprising analogous phenomenon, namely, the enormous development of the cecum in mammals and even birds, especially woodcock. As with the blind sacs of the *Potosia* larvae and the scarabaei, these are structures in which the food remains for a rather long period.

These devices do not owe their importance solely to an influence which their cellulase-producing bacteria exert on nutrition. It has been demonstrated in considerable research on so-called cecotrophy, that is, the vital necessity of many rodents to consume the fluid content of the cecum (Harder, 1949; Harder, Frank, and Harteler, 1951; Kellner, 1956; and others), that animals soon perish when they are prevented from ingesting the soft excrement. Through polyvitamin concentrates, however, the cecotrophy can be compensated for. Later, in the General Section, we shall show that research on intestinal flora in human beings has been given new impetus through studies of endosymbiosis by zoologists.

The paunch of the ruminants contains a luxurious ciliate fauna as rich in bizarre forms as the flagellate fauna in termites. There is agreement about their widely discussed significance for the host animal, but there can be no doubt that these protozoans also decompose the ingested cellulose fragments in varying degree, reproducing luxuriantly in the paunch but continually being digested upon reaching the rumen (Hungate, 1946, 1950).

THE *HYLEMYIA* TYPE

Like the ambrosia cultivators, the anthomyiids temporarily often accommodate their symbionts within their bodies to ensure continuation of the symbiosis. In all other respects, of course, the situation of the ambrosia cultivators and that of these anthomyiids differ substantially, as shown by Leach's exhaustive research (1926, 1930, 1931, 1933, 1940) on *Hylemyia* (*Phorbia*) *cilicrura* Rond., an important agricultural pest.

The larvae of this fly, which is very similar to the common house fly, live principally in rotting potatoes; but they may be found in all kinds of cultivated plants such as beans, peas, turnips, cabbage, onions, tomatoes, and they always cause decomposition of the surrounding tissue. After hatching from eggs deposited on or in the earth around seed potatoes, the larvae busy themselves for a few moments with the eggshells, which are enveloped by a sticky fluid, before setting out in search of a potato. During the next 24 hours they creep around on the surface of the potatoes, scratching them continually with their two sharp, knife-like mouth hooks until the tissue, which has had no time to form traumatic cork, begins to decay. With rapidly increasing decomposition, they penetrate more and more deeply to feed on the decaying interior. In 2 or 3 weeks the larvae complete their development and undergo pupation in the earth. By then the potato is usually decomposed completely, and the bacteria spread to the stems of the plant, where they cause what is known as blackleg disease. Flies of such origin usually produce a second generation, the eggs being laid on or in these already rotting stems, so that the second generation is of lesser agricultural importance.

The strange behavior of the larvae had already suggested that they infect the plant tissue with putrefactive bacteria during their scratching activity. Through further research this was established as a fact. When an egg is placed on agar after its surface is sterilized with a weak sublimate solution, the hatching larvae cannot grow at all. When bacteria are added after 5 days, the larvae double their size in only 48 hours. Likewise, bacteria-free larvae perish on sterile discs of potato, whereas they develop normally with an addition of bacteria.

Leach investigated the bacterial flora of both larvae and imagines, finding several species quite similar in morphology and physiology. Forms closely related to *Pseudomonas fluorescens* Migula and *Ps. nonliquifaciens* Bergey are most common, whereas *Erwinia carotovora* Winslow is less frequently present. The flora pass unaltered through the intestinal canals of the larvae and imagines and also survive pupation there. During pupation most of them are inevitably eliminated from the body with the chitinous lining of the initial and hindgut, while the others are left behind in the midgut. The eliminated flora represent very diversified forms, whereas those left behind represent but a single form. The latter continue to reproduce in considerable numbers, between the cell fragments of the expelled larval epithelium, until the larvae hatch. Even without specific anatomical devices the chorion, covered with the sticky secretion, must inevitably be contaminated during egg deposit with bacteria from the region of the anus, which is located directly next to the genital opening.

Leach assumes that the bacteria serve solely for the preparation of the food, since the fly larvae thrive equally well when they are given potato tissue which has been decomposed through bacterial activity and then sterilized. When fed on sterile bean germs rich in growth factors, however, they continue to grow, though slowly, and are finally pupated. Thus it is conceivable that in the sterile decomposed potato tissue the active ingredients originating from the bacteria must likewise play a role.

Hylemyia brassicae Bouche causes similar decay in cabbage plants (Johnson, 1930; Bonde, 1930) and *Hylemyia antiqua* Meig. in onions. *Scaptomyza graminum* Fall. and *Elachiptera costata* Leow. lay their eggs on celery leaves, causing rapidly spreading decay and thus destroying the plants. Ogilvie, Mulligan, and Brian (1935) have shown that the decay may be traced to the bacterium (*Erwinia carotovora*) which is inoculated during egg deposit. Their history has not been as thoroughly investigated in these flies as in *Hylemyia*, but it is safe to assume that circumstances are quite similar.

It is known that in other plant diseases the inciting bacteria are transmitted by Coleoptera. The cucumber beetle *Diabrotica vittata* Fabr. causes cucumbers to be infected with *Erwinia tracheiphila* Winslow, which stops up the spiral vessels and leads to complete dessication of the plants.

Leaf lice (*Chaetocnema* sp.) cause a bacterial disease in maize (Lit. in Leach, 1940). Whether the microorganisms live symbiotically with the beetles has not yet been determined, but this possibility must be considered, for in the next chapter we shall learn of bacteria, far more intimately adapted to their trypetid hosts, which cause rotting in the fruits infested by the larvae, just as in *Hylemyia*. Moreover, certain symbionts cause a cancer-like disease in olive trees.

In closing the discussion of *Hylemyia* let us mention another case of bacterial symbiosis, one which has not yet received exhaustive study, that of the tylid *Micropezza*, in which it is not out of the question that the symbionts cause decomposition in the plant tissues serving as food. In this dipteran even the young imagines of both sexes and the mature males contain dense masses of bacteria. In mature females the bacteria reproduce to such an extent that the entire intestinal canal, though containing little formed food, is filled taut with a practically pure culture. Its lining is so delicate and its load so great that it usually tears during dissection and gives the preparation a milky appearance (observations of Stammer, reported by Buchner, 1930).

The ovipositors have no specific lubrication devices, but the great masses of bacteria make them superfluous, as in the case of *Hylemyia*. One may safely assume that during deposit the egg surface is provided with bacteria exclusively from the region of the hindgut.

The development of the fly and thus the habits of the larvae are unknown, but undoubtedly the latter feed on fresh or decomposing plant parts.

TRYPETIDS

The larvae of the trypetids live in various plant parts, some species on fresh, and others on decomposing tissue. Among them are pests of great agricultural importance. The larvae of one group favor the blossoms of composites as a site and devour the immature seeds and receptacles. Occasionally galls are formed as a result, as by *Myopites stylata* Fabr., which is often found on *Inula viscosa*. Other species seek the stems and roots, sometimes causing galls to form in these areas too. In all such occurrences of gall formation, it is chiefly composites which are involved. A third group is comprised of fruit parasites with habitats in olives, cherries, peaches, figs, hips, and many other fruits. A fourth group consists of larval forms which live in the leaves of composites and umbellifers. A fifth group, known only from South America, produces so-called foam galls, which in the dried state have the consistency of pith.

Many years ago Petri made excellent studies of symbiosis in the dreaded olive fly (1904, 1905, 1906, 1907, 1909, 1910). Stammer (1929) later rounded out the description using thirty-seven species from three sub-divisions of the family, the dacines, trypetines, and tephritines, discovering bacterial symbiosis throughout the family. Only two of the ten palae-arctic tribes were omitted. In bacteriological studies Hellmuth (1956) used forty-three species of the subfamilies mentioned.

It was discovered that *Dacus oleae* Gmelin takes first rank in the perfec-tion of the symbiotic devices, and that all other genera can easily be arranged in gradually ascending order. The larvae and imagines of *Dacus* localize their symbionts differently. At the head of the midgut

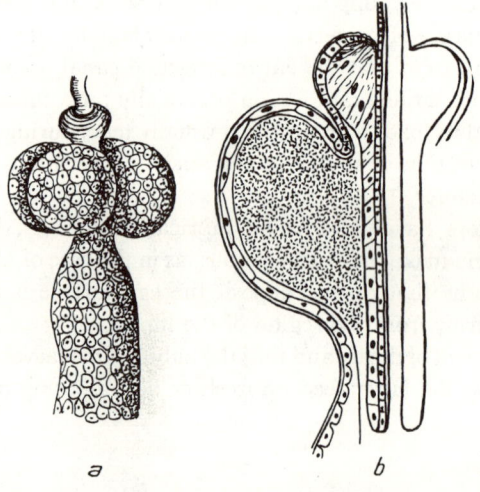

a   *b*

Fig. 12   *Dacus oleae* Gmelin.   The four bacteria-filled blind sacs of the larva, (*a*) total, (*b*) in section.   From Petri.

the larva forms four spacious round ceca, their whiteness shimmering through the membrane. These bacteria-filled ceca are almost closed by intestinal valvules reaching deep into the midgut (Fig. 12). Their always sterile epithelium, much flattened, is differentiated from the remainder of the midgut by the complete absence of structures with secretory function. A weak transverse and longitudinal musculature which passes over the exterior propels some of the contents into the adjoining intestinal section where bacteria are often present in considerable numbers.

Shortly before pupation the bacteria are all suddenly expelled from the ceca. For a while they can still be found in the lumen of other intestinal parts, but directly before pupation most of them are eliminated through the anus. Of the small number left behind, chiefly in the vicinity of the proventricle, some move to the esophagus, survive metamorphosis there, and then toward the end of the metamorphic process settle the so-called cephalic organ of the imago.   This organ is an unpaired, sac-like evagination, at first glandular in function, connected by a short narrow passage with the esophagus (Fig. 13).

The secretion in the evagination apparently offers a favorable culture

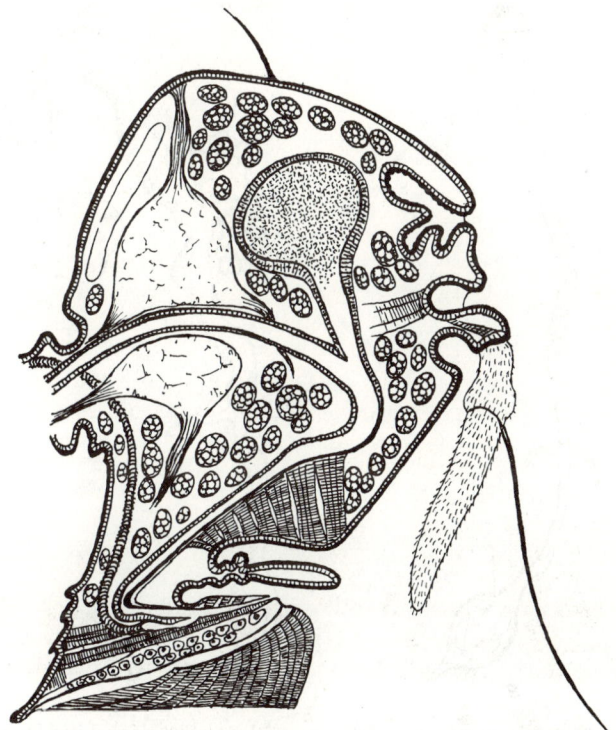

Fig. 13 *Dacus oleae* Gmelin.   Head of an imago in longitudinal section with the unpaired evagination of the esophagus inhabited by bacteria.   From Petri.

medium for the bacteria.   When the imagines hatch, the organ is sparsely settled, yet it is filled with extraordinary masses of symbionts only 30 hours later.   At that point the secretion diminishes, the wall cells become flatter and flatter, and in older animals all glandular function appears to be extinguished.

In the tephritines the symbiotic organs of the larvae greatly resemble those in *Dacus*, except that in older larvae the four evaginations have several subdivisions and are often irregularly constricted (*Tephritis conura*

Loew, *T. heiseri* Frfld., *T. bardanae* Schr.).   The bacteria that were not expelled before pupation now settle the midgut, in the medial section where it is somewhat wider and is covered with short irregular villi (Fig. 14*a*).   In the freshly hatched imago the upper surface of the epithelial cells, which here do not have a brush border, appears irregularly frayed

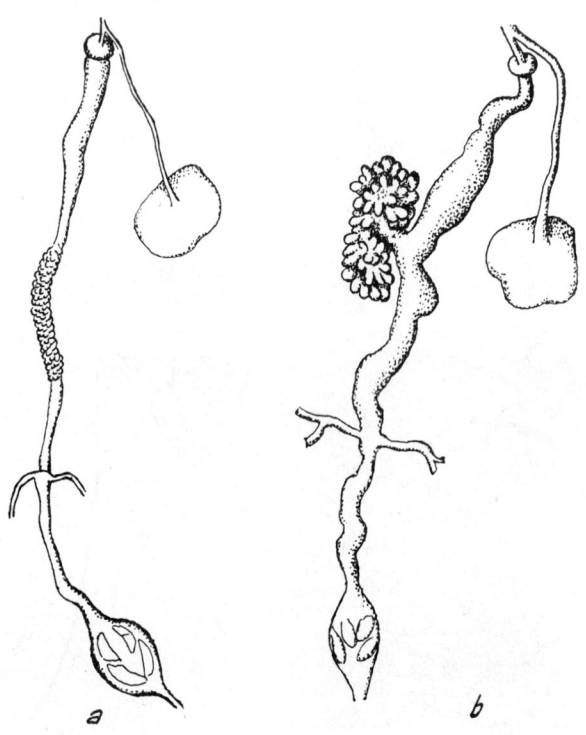

a   *b*

Fig. 14   (*a*) *Tephritis conura* Loew.   Adult gut with crypt zone inhabited by symbionts. (*b*) *Sphenella marginata* Fall.   Adult gut with evagination inhabited by symbionts. From Stammer.

and is covered by a glandular secretion.   In this secretion the first bacteria make their appearance.   As in the cephalic organ of *Dacus*, they multiply so rapidly that eventually they cover the flattened wall cells in an enormously thick layer.   Generally arranged parallel to each other, they change at the free margin into rodlets of varying length and stronger staining (Fig. 15).   In contrast to these masses, the symbionts occurring elsewhere in the intestinal lumen seem to recede into the background. And, just as we found no trace of such an intestinal organ in the imago of

Dacus, so too in the imago of the *Tephritis* species there is no suggestion of a cephalic organ.

In a series of other tephritines—*Trypanea, Paroxyna,* and *Euarestella* species—the same devices exist as in *Tephritis* except that in *Trypanea* imagines the villous zone of the intestine appears to be smooth on its surface. In other species the villi tend to be more sharply set off from the intestine. *Acanthiophilus helianthi* Rossi has a well-separated nest of 20 to 30 broad evaginations along one side of the intestine only. *Sphenella*

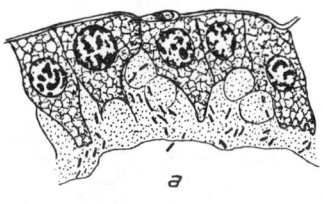

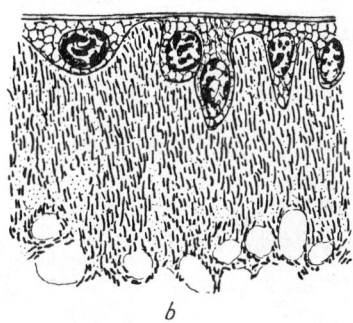

Fig. 15 *Tephritis conura Loew.* Cross section through the region of the midgut occupied by bacteria. (*a*) In a newly hatched animal the first symbionts make their appearance in the secretion; (*b*) the infection in a 9-day-old animal. 950×. From Stammer.

marginata Fall. houses its symbionts in a two-lobed villiferous evagination which is separated from the rest of the intestine only by a narrow passage (Fig. 14*b*).

Of the schistopterines only the larvae of the single palaearctic species living in Egypt could be investigated (*Schistopterum moebusii* Beck). It exhibited subdivided bacteria-containing ceca of the *Tephritis* type. In the remaining tribes of the tephritines (Terellini, Xyphosiini, Ditrichini), as well as in all members of the subfamily of the trypetines, insofar as they have been investigated, far simpler circumstances are revealed than in the dacines and the previously studied tephritines. Stammer found that the four ceca were present at the beginning of the midgut but that they were smaller and free of bacteria.   Hence they are like the rest of the midgut in their histological construction.   Larvae and imagines both lack specific symbiotic organs, and thus their symbionts, all rod-shaped, are found in the intestinal content either in rather large clumps or less densely grouped (*Chaetostomella, Xyphosia, Noeta, Ditrichia, Ceratitis, Rhagoletis,* and others).

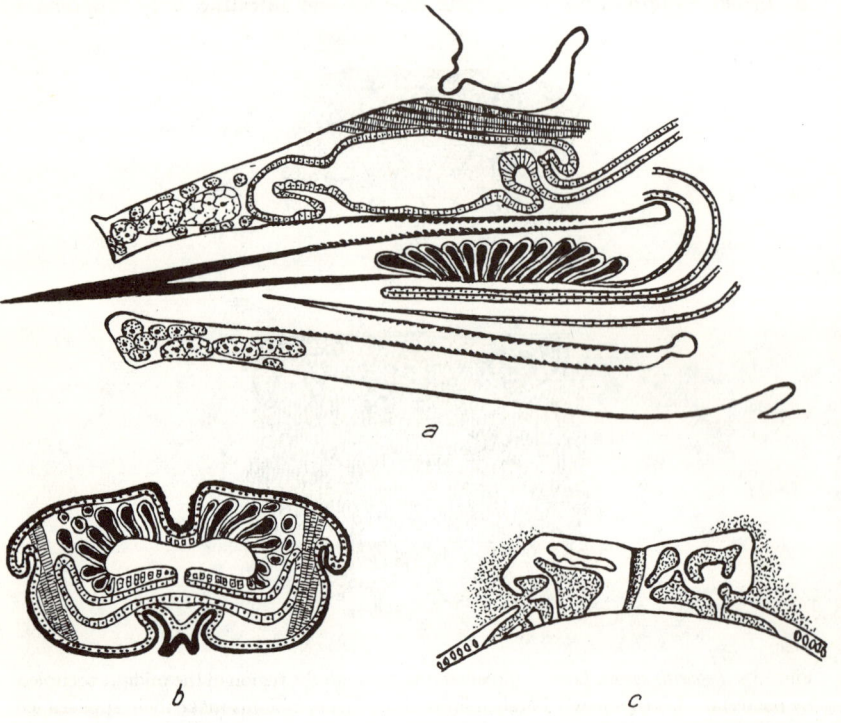

Fig. 16   *Dacus oleae* Gmelin.   (*a*) Longitudinal section through the egg-laying apparatus with the smearing arrangement attached to the hindgut.   (*b*) The cross section shows the connection between vagina and gut.   (*c*) Micropyle infected with bacteria. From Petri.

Stammer had found no cephalic organ in the *Tephritis* species.   Dean (1933, 1935) described one in *Rhagoletis pomonella* Walsh but did not test its content.

In the larvae and imagines of trypetids the transmission devices and other adaptations represent all degrees of perfection, as in their types of symbiont localization.   Again, it is a dacine, *Dacus oleae*, which holds first rank.   As is usual during colonization of the intestinal lumen, the symbionts are given to the egg at the moment of deposit. An ingenious bacterial depot for this purpose is located on the ovipositor. Directly at its mouth, arranged dorsally and on both sides, are approximately twenty finger-like glandular evaginations containing bacteria (Fig. 16a). Inoculation proceeds without difficulty, inasmuch as organisms are constantly gliding past these containers. Their secretion represents a good culture medium, and very soon the bulb-like narrow-necked spaces are filled on all sides. The vagina passes just beneath the hindgut, and in the area of the lubricating apparatus a longish medial slit provides direct communication between the two passages (Fig. 16b). Directly after this the hindgut and the vagina superfluously form a short common outlet. By virtue of such provisions each egg as it passes this area is contaminated with the bacteria. The latter are numerous even outside the crypts of the hindgut, and the area of the micropyle is especially well supplied. This organ has a cap-shaped top filled with numerous irregular hollow spaces, and the mucilaginous secretion for covering the egg on all sides is particularly abundant in this location. The bacteria reproduce luxuriously and fill up the hollow spaces within a short time (Fig. 16c).

For a while egg and embryo remain sterile, but, on the sixth day when foregut and mouth implements are developed, the first symbionts make their appearance in the midgut. Entering through the micropyle, they apparently infect the embryo first, as soon as migration is possible through its oral aperture. By the time the larva hatches, the four ceca are well provided with symbionts.

Aside from a single exception, bacterial reservoirs were found in the hindgut of all tephritines studied by Stammer, though here they are grooves, not sac-like evaginations. Ordinary longitudinal folds of the intestinal walls deepen in the area of the ovipositor to form sharply defined furrows, which widen as they become deeper. In contrast to the crypts of *Dacus*, these transmission grooves develop over the full circumference, though usually more weakly in the ventral part. For example, in *Tephritis conura* Loew. their number ranges between 28 and 32. As in *Dacus*, there is a slit-like connection between hindgut and vagina. Variations in the individual species again represent varying degrees of development: the number of folds may be larger or smaller; toward the hindgut they are more or less well sealed off; deeper in the intestine they may be further subdivided or simply remain rounded; the length of the slit-like communication varies according to the size of body, though it never reaches the length attained in *Dacus*.

A third group, all vestigial forms, are even more elementary. Where specific sites are lacking in the intestine, the special transmission device is also lacking in the otherwise similarly constructed ovipositor, and only shallow folds are formed by the hindgut.   The sole adaptation to symbi-osis is a connection between hindgut and vagina, but this too is consider-ably shorter than in other species.   Nevertheless transmission is ensured, obviously, by a remarkably large supply of bacteria in the hindgut.   In mature females the bacteria may often be found throughout the hindgut up to the anus.

Transmission grooves and slit connection are absent only in *Ensina sonchi* Rob. Desv.   Here Stammer could find no specific sites in the remainder of the intestinal canal, although now and again he found abundant bacterial settlement throughout its extent.

The previously mentioned tendency toward perfection and complication is revealed also in the micropyles.   Among the trypetids it is *Dacus* again that has the richest development of the hollow spaces.   The tephritines have at least a ring of flat hollow spaces in the area of the micropyles, but such spaces are not found elsewhere.

In *Tephritis heiseri* Field. it was possible to study egg deposit in greater detail.   The freshly laid egg is covered with a mucilaginous layer which increases at the inferior pole to form a sizable cap.   In the beginning more bacteria are found there than in the micropyle, but after a few days numer-ous bacteria cover the egg completely, even appearing behind the shell at both poles where the larva does not completely fill out the space.   Thus here too they enter by means of the micropyle.   The ceca are present in the freshly hatched larvae, and as in *Dacus* the colonizing process never fails to set in at once.

Petri, who made a detailed study of the bacteriological aspects of trypetid symbiosis, found that in *Dacus* two forms are distinguishable.   He classifies the actual symbiont as *Bacterium (Pseudomonas) savastonoi* Smith. This rodlet, 2–3.5 μ in length, is rarely linked in chains and is rapidly propelled forward in early liquid as well as solid cultures by means of two or four cilia located at the end.   It is gram-negative and was not encoun-tered in spore formation.   Cultures thrive best on agar with bean broth and were easiest to achieve when the material was taken from the area of the micropyle, from young larvae or from the imaginal midgut.   Hellmuth (1956) cultivated the bacterium from the various phases of the host ani-mals, and, succeeding where Petri had failed, she was the first to obtain clear-cut pure cultures.

Interestingly, it is this bacterium which causes the gall proliferations so frequently found on leaves and branches of olive trees (Petri, 1907, 1909). Indeed, one can produce these galls by inoculating the plant tissue with *Dacus* intestinal evaginations, and, conversely, one can cultivate from the galls an organism identical with *Bacterium savastonoi*.   Where the olive fly does not occur, as in California, this "tuberculosis" of the olive tree is correspondingly less widespread and the infecting agents are borne exclusively by wind and rain from tree to tree (Wilson, 1935). The literature on these bacterial galls, which extends over a long period, is summarized by von Tubeuf (1911).

Allied with this bacterium is another which Petri identified with the widespread *Ascobacterium luteum* Babes. Though lacking in the first stages, it later multiplies considerably, its numbers increasing in the ceca with the age of the larvae. Nevertheless, certain limits are placed on reproduction as long as the insect is healthy, but because of its saprophytic nature it has the capacity to grow luxuriantly after the insect's death or when death is imminent. At times the infection of the larvae may apparently be effected by the lubricating devices, but usually infection occurs later by way of the anus. The almost cocciform capsule-forming rodlet is found also on the galls, on the bark, on the ground in olive groves, and in the decomposition foci of other plants. Hellmuth with good justification believes that the organism belongs to the family of the Rhizobiaceae, specifically to the genus *Agrobacterium*, which occurs pathogenically on roots, and he therefore calls it *Agrobacterium luteum*. Earlier it was usually described as *Phytomonas* or *Pseudomonas*. Hellmuth was also first in cultivating the symbionts of the trypetines and tephritines, always obtaining the same organism, to which she gave the name *Pseudomonas mutabilis* nov. spec. Evidence of its symbiotic nature is, first, that it was cultivated with great regularity even from eggs in a sterile medium; second, that the agglutination test was positive; and third, that the organism could never be obtained from the environs of the larvae.

Free-living symbiotic bacteria also exist among other trypetids and may contribute substantially to larval development. When *Rhagoletis pomonella* Walsh deposits its eggs in the fruit pulp of apples, decomposition sets in at once and then spreads rapidly around the feeding larvae (Allen and Riker, 1932; Allen, Pinckard, and Riker, 1934). The bacteria causing the decay are observable on eggs and ovipositors in imagines of both sexes, and larvae which hatch from eggs with sterilized shells are unable to complete their development in the apple tissue. The female *Rhagoletis* has a pharyngeal bulb like that of *Dacus*, well-defined transmission grooves in the hindgut like those in many other tephritines, a channel between hindgut and vagina, and a richly setaceous micropyle region. (Dean, 1933, 1935.)

When other tephritine species attack fruits such as peaches and figs, as well as when *Dacus* larvae attack the pulp of olives, decomposition quite generally sets in at once. It seemed reasonable then to suppose that the larvae are able to utilize the fruit pulp only after its infection by the symbionts, quite as in the *Hylemyia* type. Accordingly, the reasoning went, trypetid symbiosis must have developed from primitive beginnings, as in *Hylemyia*, and only the more intimate association of the symbionts within the body of the fly had finally made a number of forms independent of such centers of decay, in this manner enabling them to live in the receptacles and other plant parts which remain sterile. Now that Hellmuth has demonstrated that symbionts are lacking in decayed fruit pulp, a more likely explanation is that decomposition is caused by concomitant organisms. Hellmuth was able to cultivate them repeatedly from her test objects.

Stammer directed his attention in trypetids particularly to changes in the form of symbionts, changes which are often rather prominent. Even in *Dacus* the bacteria are much smaller in young larvae than in older ones. The shorter and sometimes somewhat thicker forms assist in transmission. In *Tephritis*, as we have seen earlier, shorter, more deeply staining forms arise at the edge of the imaginal bacterial bolster. The change is most prominent in the symbionts of *Tephritis heiseri*. In the larvae they have the form of plump, short rodlets; in the young imagines the rodlets begin to develop; in somewhat older imagines they assume the form of long filaments; and in mature animals the transmission forms are again short (Plate II,*a–d*). Similar changes in the symbiotic form, more or less pronounced, apparently are found in most trypetids.

ANOBIIDS

Anobiids live chiefly in the wood of coniferous or deciduous trees, some favoring wood far advanced in its decomposition, and others wood which is hard and dry. Some anobiid larvae (*Anobium, Ptilinus, Trypopitys*) attack planks, beams, and old furniture; *Ernobius abietis* Fabr. specializes in the spindles of pine cones; other forms (*Hedobia*) prefer thinner branches to more massive wood. In the forms mentioned it is always dead material that the insects seek out. *Lasioderma* lives in various commercial products, preferably tobacco, as well as in the stems of herbaceous plants. *Sitodrepa panicea* L. departs even further from its original habits by infesting drugs of all types and foods for human consumption, such as farinaceous products, baked goods, rice, and chocolate. The forms mentioned all belong to the same subfamily, the anobiines; the dorcatomines live in part in decaying wood and in part have been adapted in special fashion to tree fungi.

The earliest and best-known instance of symbiosis occurs in the drugstore beetle, *Sitodrepa panicea (Stegobium paniceum)*. Karawaiew (1899) was the first to observe the phenomenon, and Escherich (1900) the first to recognize it as symbiosis. As in all other anobiids in which symbiosis has been examined, its symbionts are housed in the evaginations of the larval midgut. At the point where fore- and midgut meet there are four voluminous ceca, which have numerous subdivisions and are well provided with tracheae (Fig. 17*a*). In cross sections we observe, on the one hand, typical intestinal epithelial cells, relatively few in number, with a brush border and secretory granules, and, on the other hand, numerous extraordinarily enlarged mycetocytes. Sometimes broader than tall, the mycetocytes overshadow the slender pillar-shaped epithelial cells. Their plasma is reduced to a coarsely meshed matrix by the numerous tear-shaped yeast cells which are distributed throughout and are frequently in the process of budding. The centrally located nuclei are correspondingly

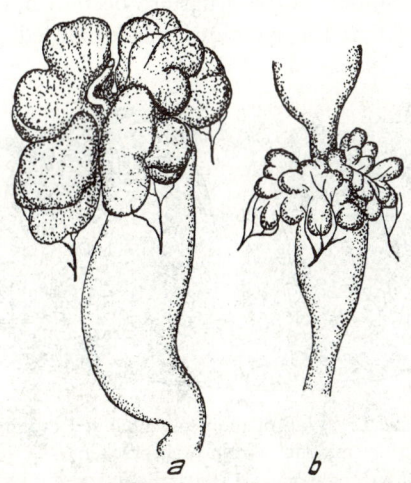

Fig. 17 *Sitodrepa panicea* L. Yeast-inhabited evaginations of the midgut, (*a*) of a larva, (*b*) of an adult. From Koch.

large and often indented; the brush border is absent in the infected cells. At the base of the epithelium, isolated crypt cells are found here and there (Fig. 18). According to Gräbner, some of the mycetocytes always contain dead yeasts (which are continually being expelled into the intestinal lumen) and numerous living symbionts as well. Symbiotic conditions are the same for every mycetocyte, indicating uniformity in the physiological milieu.

Because the symbionts are housed in the midgut epithelium, which disintegrates during metamorphosis, special preventive measures on the part of the host organism become necessary. When the crypt cells increase in old larvae and through fusion form a new intestinal canal, the old epithelium is detached toward the inside and finally disintegrates into tiny fragments which are resorbed during pupal rest.

Disintegration occurs also in the epithelium of the infected ceca. As soon as the larvae are ready for pupation, a mass migration of individual yeasts and whole mycetocytes to the intestinal lumen sets in. When the new imaginal intestine has enveloped the shriveled ceca, individual yeasts are transmitted from the remaining mycetocytes into adjoining young cells and continue to grow vigorously (Buchner, 1921). The imaginal evaginations which originate in this way always remain smaller and more delicate than the larval evaginations, yet they are more complexly subdivided than the grosser larval forms (Fig. 17*b*). The mycetocytes are smaller and thinner in proportion, and again the brush border is lacking.

Karawaiew and Escherich left open the question of the transmission of the ever-present symbionts of the drugstore beetle. Later I was able to offer the first proof that the symbiosis is safeguarded just as effectively

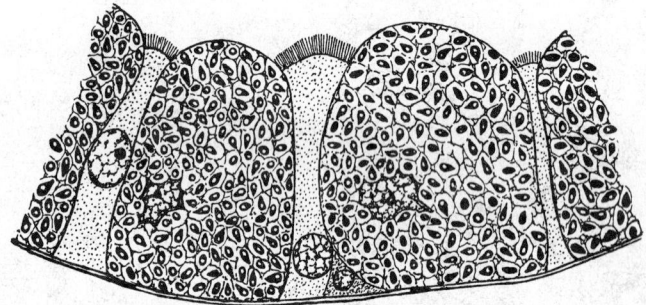

Fig. 18   *Sitodrepa panicea* L.   Wall of the yeast-inhabited evaginations of the larval midgut.   From Breitsprecher.

when the eggs are superficially infected with symbionts and some of the contaminated shell is eaten later, as when ovarial eggs are infected or when the micropyle is penetrated (Buchner, 1920, 1921). Two devices are used to ensure contamination of the egg surface at the moment of deposit. At the point where the retrograde intersegmental membrane and the deeply retracted ovipositor meet in the state of rest, the membrane forms on each side a tubule of considerable size and penetrating deep into the interior of the body, which discharges yeast cells into the sheath of the ovipositor (Fig. 19). Behind the hypodermis of these "intersegmental tubules," lined with a delicate, chitinous lamella, glandular cells cover the entire area except one long, broad strip (Fig. 20*b*), and it is these glandular cells which produce the secretion in which the symbionts are embedded. Fragile longitudinal and transverse muscle strands encircle the tubules, and from the blind end of the latter a strong muscle extends to the spiculum ventrale, the eighth sternite, which has been transformed into a staff shape.

Attached to the sternite are also the muscles which effect withdrawal of the ovipositor.

Two additional yeast-containing pouches located at the end of the ovipositor under the folded vagina are equipped with two imbricate chitinous plates to seal off their valuable content from the outside (Breitsprecher,

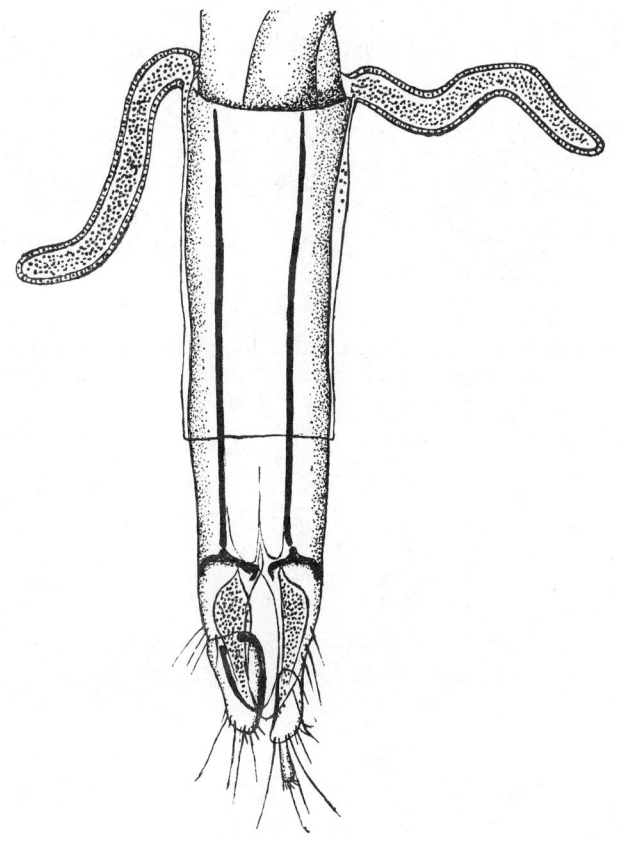

Fig. 19   *Sitodrepa panicea* L.   Egg-laying apparatus with symbiont-filled intersegmental tubules and vaginal pockets.   From Breitsprecher.

1928).   The lumen, which is in direct communication with the vagina, has numerous retention hairs at their juncture to prevent untimely migration of its symbionts into the female genital tract (Fig. 20*a*).

When an egg glides through the much narrower ovipositor, it necessarily presses some of the yeasts down through this apparatus.   In passing from the vagina, the egg dips into a thrombus of yeast cells.   The latter adhere

here and there to the pitted egg surface, either individually dispersed in the secretion or, in some species, in rather large clumps (Fig. 21). It is evidently the function of the intersegmental tubules to act as reservoirs

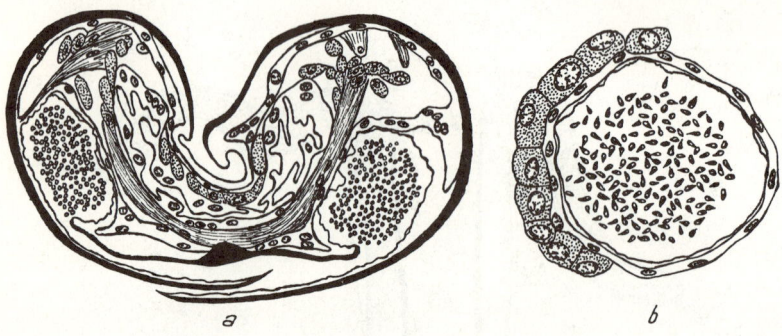

Fig. 20 *Sitodrepa panicea* L. (*a*) Cross section through the egg-laying apparatus with vaginal pockets; (*b*) cross section through one of the two intersegmental tubules with glandular cells and yeasts. From Breitsprecher.

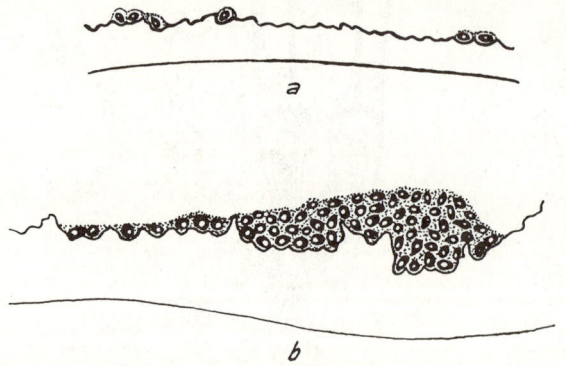

Fig. 21 (*a*) *Sitodrepa panicea* L., (*b*) *Anobium striatum* Oliv. Yeasts on the surface of the egg. From Breitsprecher.

from which the vaginal pouches can be replenished at all times by means of movements executed during deposit.

In the pupal state, pouches and tubules remain empty at first, but preparations for filling them soon take place. When the pupa turns brown, great quantities of yeast and even entire mycetocytes migrate to the midgut to be expelled during the first defecation of the young imago.

Since the hindgut opens into the sheath surrounding the retracted oviposi-tor, the fungi can easily reach the secretion-filled transmission organs. They now reproduce rapidly, and by the time the imago leaves the

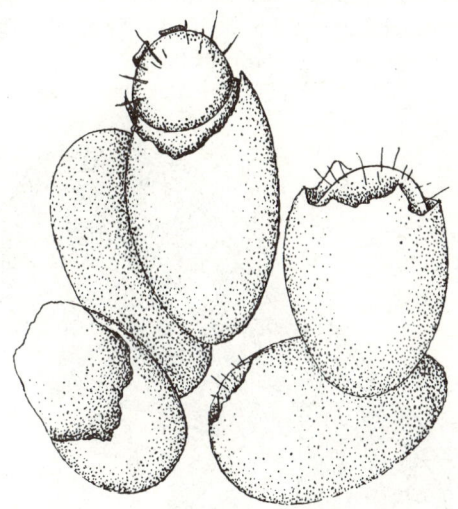

Fig. 22 *Sitodrepa panicea* L. Hatching larvae consume a portion of the shell and thereby infect themselves with the yeasts. From Buchner.

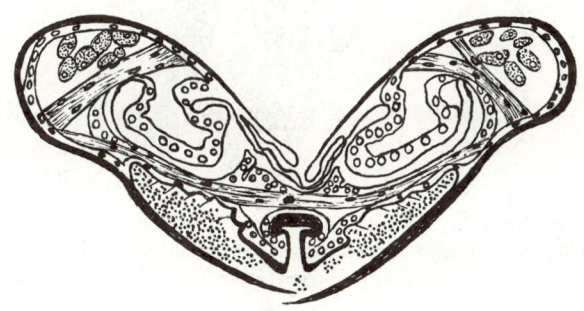

Fig. 23 *Ernobius mollis* L. Cross section through the egg-laying apparatus with yeast-filled vaginal pockets. From Breitsprecher.

puparium the prerequisites for transmitting the symbionts to the progeny are in existence.

While the young larva is hatching, it devours a large part of the egg-shell and thus ingests a number of yeast cells which remain undigested and serve to infect the intestinal epithelium (Fig. 22). Relatively few of the rapidly reproducing symbionts are admitted by the cells existing for this purpose, and these symbionts then in turn begin to infect neighboring epithelial cells. The area of the future ceca is usually contaminated in the first week of larval existence, but at times it is 2 or 3 weeks before an infection of the epithelium can be verified. Interestingly enough, the ceca in such cases have already begun to form.

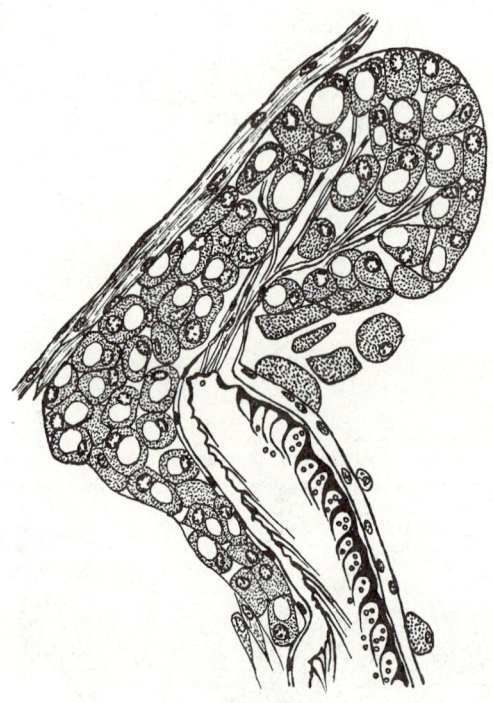

Fig. 24   *Ernobius mollis* L.   Intersegmentally located lubricating glands on the egg-laying apparatus and symbiont-attaching crypt hairs on the outer side of the apparatus. From Breitsprecher.

Research on other anobiids has widened our understanding of their symbiosis. In contrast to *Sitodrepa*, the intestinal evaginations in *Ernobius mollis* L. are much larger in the imago than in the larva, yet the individual mycetocytes are far less developed. The vaginal pockets are essentially like those of *Sitodrepa*, but the strong dorsoventral flattening of the ovipositor demands certain changes in form (Fig. 23). Surprisingly, the intersegmental tubules are absent, another device being used in their stead. The glandular cells form a massive body with proportionately elongated exit channels, necessitating that the yeasts be housed elsewhere. They are now found in zones, adjoining each gland, in which the cuticula of the ovipositor sheath develops strange cryptal hairs.   These are imbricate, scale-like, chitinous formations curving backward, undoubtedly able to hold back a rather large number of symbionts, yet giving the impression of a makeshift device (Fig. 24).   *Ernobius abietis* has similar structures.

Anobiid symbiosis experiences its maximum development in the genera *Anobium* and *Dendrobium*.   In *Anobium striatum* Oliv. the mycetocytes as

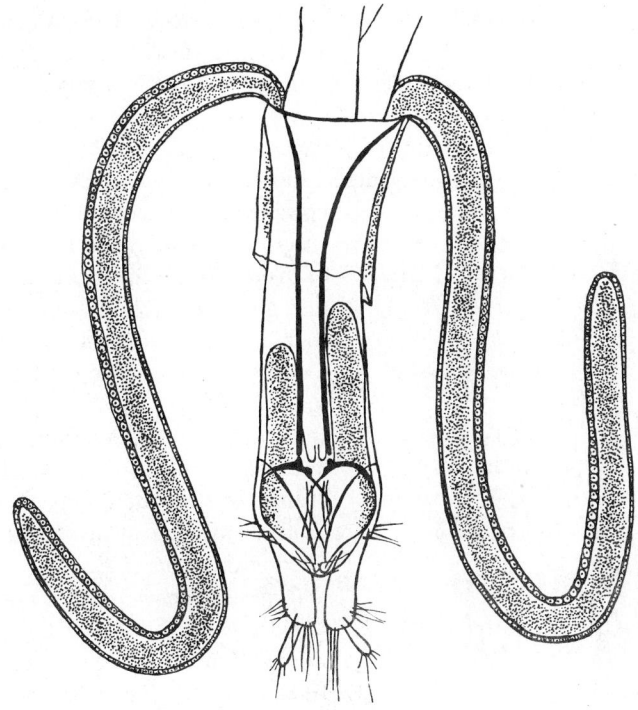

Fig. 25   *Anobium striatum* Oliv.   Egg-laying apparatus with markedly developed inter-segmental tubules and pockets elongated into vaginal tubules.   From Breitsprecher.

well as the intersegmental and vaginal symbiont containers greatly surpass in size and quantity of symbionts those in *Sitodrepa* and *Ernobius*.   The ceca, in this case six, fill out the whole anterior space in the larva and press the fat body backward, but in the imago they are relatively small.   The intersegmental tubules, basically like those in *Sitodrepa*, appear to be vastly elongated, and the vaginal pouches, extending in the direction of the head to form sizable yeast-filled sacs, deserve to be called tubules rather than pouches (Fig. 25).   On the other hand, the longitudinal fissure in the section which is still pouch-shaped is greatly reduced.   The tubule connecting it with the vagina remains, ensuring that masses of symbionts fill the entrance to the vagina.

Other species of *Anobium*, investigated only in chitin preparations, provided no new information, nor did *Anobium rufipes* Fabr., *A. nitidum* Hrbst., *A. fulvicorne* Strm.   It would be desirable to study *Dendrobium* (*Anobium*) *pertinax* L. in fresh material.   As in *Ernobius*, the intersegmental pouches appear to be absent here, but by way of compensation there are other spacious tubules situated in front of the symbiont-filled vaginal pouches within the ovipositor.   The tubules are connected with the space around the ovipositor by slit-shaped openings, and they contain prodigious quantities of fungi (Breitsprecher)!

The intersegmental tubules of *Xestobium rufovillosum* De Geer deviate remarkably from the foregoing descriptions.   The glandular epithelium, completely restricted to one side, continues to extend along the outer side of the ovipositor for a distance.   The non-glandular tubules develop hairs to hold back the symbionts, similar to those in *Ernobius* but much more fully developed.   Bent at right angles, the end parts of the hairs facing the outlet are broken up into a number of little bristles and the spaces arising in this way are always abundantly filled with yeasts (Fig. 26).

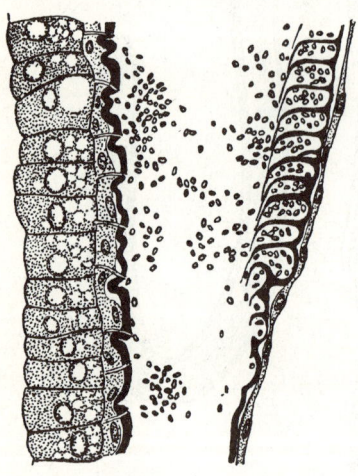

Fig. 26   *Xestobium rufovillosum* De G. Opening of an intersegmental tubule with glandular epithelium on one side and crypt hairs on the other. From Breitsprecher.

Oligomerus brunneus Strm. and *Tryopitys carpini* Hrbst. follow the *Anobium striatum* type; *Lasioderma redtenbacheri* Buch., the type described for *Sitodrepa*. In *Xyletinus ater* Panz., *Ptilinus pectinicornis* L., and *Hedobia imperialis* L., the larvae of which all live in old deciduous wood, Breitsprecher found no transmission devices of any kind with chitin preparations of dry material. But we must not draw the conclusion that these species do not live symbiotically, for at the beginning of the larval midgut in *Ptilinus pectinicornis* we find two yeast-filled evaginations emptying into the gut by narrow passages (E. A. Parkin, 1952; Gräbner, 1954).   Gräbner proved that transmission in this species is also effected by the intersegmental tubules.

Cymorek (1957, 1960) also described the locations and form of the symbionts of some species in the course of taxonomic studies on the genus *Anobium*, and Grinbergs (1962) pictured the gut diverticula and their inhabitants in *Lithraea caustica*.

After having ascertained the existence of similar transmission devices in dorcatomines, I induced Nolte (1938) to study them in detail. Eleven species distributed over five genera were proved to be symbiont carriers, unfortunately determined with dried material almost exclusively. In *Mesocoelopus*, a genus with long intersegmental tubules and vaginal pouches, the ovipositor greatly resembles that of *Anobium striatum*, but in other species the intersegmental tubules are replaced by more compact, indeed even rather small, sac-like invaginations, characteristically braced by a shovel-like elongation, continuing along their wall, of the paired chitinous supports of the ovipositor.

The ovipositors are varied in structure, apparently in connection with the specific reproductive biology in each case. We note all degrees of differentiation, from the extremely long and slender stiletto-like structure in *Dorcatoma dresdensis* Hrbst. to compact short ones (other *Dorcatoma* species, *Caenocara bovistae* Hoffm., *Anitys rubens* Hoffm.). Intersegmental containers of some type are always present, and in most cases also vaginal pouches, although they are occasionally rather inconspicuous. Observations on intestinal organs and the process of metamorphosis have unfortunately not yet been made.

The form of the symbionts may vary considerably from species to species, but it remains constant within any one species. For example, in *Sitodrepa* the symbionts are usually tear-shaped; in *Ernobius mollis*, roundish to oval; in *Anobium striatum*, more elongated with blunted ends; in *Xestobium rufovillosum*, pear- and lemon-shaped; and occasionally tube-like outgrowths occur. A typical cell contains a large vacuole and sharply refracting granules. The buds are arranged terminally or slightly off to one side. The minor differences, which exist between the form of the symbionts in the gut diverticula and in the intersegmental tubules, are illustrated by Jurzitza (1962) with photographs.

Cultural experiments were undertaken by Heitz (1927), W. Müller (1934), Pant and Fraenkel (1950), Gräbner (1954), Kühlwein and Jurzitza (1961), and Foeckler (1961). Heitz and W. Müller cultivated *Ernobius abietis* symbionts with ease on various liquid and solid culture media, but the cultures despite intensive growth could not be induced to form spores. The symbionts of *Sitodrepa panicea* were difficult to cultivate. Escherich claimed earlier that he had had no difficulty in obtaining the growth of such cultures; but, in the experiments under discussion here, their growth on the same media was slight, although unicellular cultures under a glass cover were not difficult to preserve.   After 8 days chain-like groupings occurred.   From the symbionts of other species it proved impossible to obtain cultures at all (*Anobium striatum*, *Xestobium rufovillosum*, *Dendrobium pertinax*).   Pant and Fraenkel obtained cultures of *Lasioderma* and *Sitodrepa* in Hansen's solution, but Schanderl (1942) had no success in cultivating *Ptilinus fuscus* symbionts.

Gräbner had similar experiences with cultures.   In *Sitodrepa* and *Ptilinus* he observed a certain increase of symbionts in the hanging drop, yet single cells never yielded cultures, neither on agar nor in liquid media. The minimum number of cells needed for initiating a colony showed a wide range among the various test objects.   With *Ernobius abietis* approximately 15 cells were sufficient; with *Ptilinus pectinicornis* even an abundance of starting material failed to provide cultures; and with *Sitodrepa* and *Lasioderma*, which occupy a middle position in the variation range, cultures could be obtained only when the ceca were used intact.   Gräbner also found that failures with individual cultures were attributable to the fact that anobiid symbionts require culture media high in growth-factor content.   *Ernobius* symbionts grew rapidly from suspensions only when the agar was first saturated by means of diffusion cultures with growth factors from *Ernobius* yeasts or when it was pretreated with beer yeasts or those of *Sitodrepa*.

Kühlwein and Jurzitza (1961) met with the same difficulties in their attempts to cultivate the *Sitodrepa* symbionts.   Only after they had investigated their amino acid requirements and found that the symbionts need aspartic and glutamic acids were they able to obtain better growth. Both too much and too little amino acid supplement were unfavorable. In regard to vitamin production, they found that the *Sitodrepa* symbionts are heterotrophic for biotin and aneurin.   Foeckler (1961), however, had no difficulty with his cultures on nutrient agar and in nutrient broth with the addition of 2% *Torulopsis*-yeast extract and 2% glucose.

In regard to the taxonomic position of the anobiid symbionts, Müller (1934) declared that the inhabitants isolated by him from *Ernobius mollis* belong to the genus *Torulopsis*, and they are described as such by Lodder and Kreger-van Rij (1952).   Gräbner agreed with this concept and gave the *Sitodrepa* symbionts the name *Torulopsis buchnerii* Gräbner.   According to Kühlwein and Jurzitza, however, these organisms have no relation to *Torulopsis;* but such characteristics as the production of a pigment coloring the cultures red and certain conditions described by Gräbner, which are reminiscent of the germinating conidia of *Taphrina*, speak for membership in the Taphrinaceae.   They therefore renamed the yeast *Symbiotaphrina buchnerii*.   Also, the symbionts of *Lasioderma*, cultivated by Milne (1961), are supposed to be closely related to *Taphrina*.   The facts that the only *Taphrina* now known is the parasite causing leaf deformations and witches'-broom, and that in reinfection experiments of aposymbiotic larvae by Foeckler (1961) only *Torulopsis utilis* was taken up by the sterile mycetocytes and was able to replace the normal symbionts, appear to speak against this new classification.

The reproductive rate of *Sitodrepa* symbionts ranges widely according to temperature and nutritional condition of the hosts. The ceca of hibernating animals are weakly populated, but when taken into higher temperatures they are rapidly filled. Conversely, cooling of summer animals causes a drop in the bacterial population. Heitz had already determined that such lability exists when I persuaded Kiefer (1932) to study it in more detail. In experiments with summer animals of *Sitodrepa* when 70–90% of the cells in the ceca were infected, only 45% of the mycetocytes remained after 41 days at temperatures of 3–5°C. Not only individual yeasts but also entire cells were expelled into the intestinal lumen. At the same time the budding rate sank almost to 0%. In one experiment the rate was 22% when the animals were kept at room temperature, but, when the test animals were subjected to cold for 65 days, the rate was only 3%, and in other cases it was 1%. When the cold animals were returned to warmer temperatures, the budding rate increased after only 4 days to 24%, that is, beyond the normal. When the normal bacterial population was reached, its numbers declined again. With *Ernobius abietis* the findings were much the same. In this species experiments were designed to determine the separate effects of cold and hunger which here are naturally interlinked. It was shown that they are not identical but are summated. With both hunger and cold, reduction in numbers is greater than with cold alone. When hunger and cold influence the host organism unfavorably, incipient mycelial formation occurs in *Ernobius abietis*, just as in pure cultures.

In the General Section we shall discuss the interesting results obtained with *Sitodrepa* species which were rendered symbiont-free (Koch, 1933; Fraenkel and Blewett, 1943; Pant and Fraenkel, 1950; Kühlwein and Jurzitza, 1960; Foeckler, 1961; Jurzitza, 1962).

CERAMBYCIDS

In sites, transmission methods, and nature of the symbionts, the symbiosis of cerambycids greatly resembles that of anobiids. In other respects it is unique: the larval organs disappear during metamorphosis and never return; in the imaginal female remain only reservoirs for the transmission; and in the male the symbionts are totally lacking. Cerambycid symbiosis is also less widespread, being absent in a great number of forms, but symbiotic and non-symbiotic forms are scarcely distinguishable, at least at first glance, because of identical food sources (Heitz, 1927; Buchner, 1928, 1930; Schomann, 1937).

The symbiotic sites of the larvae are all arranged in circles around an enlargement of the initial section of the midgut, but their outward appearance may differ considerably from species to species.   The most impressive formation is the broad girdle of cluster-shaped evaginations in *Oxymirus cursor* L. located slightly beyond the beginning of the midgut.   In *Leptura*

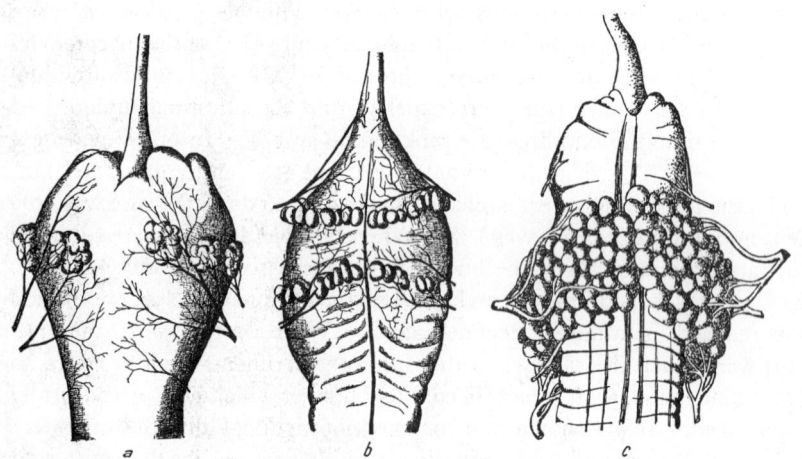

Fig. 27   (a) *Leptura rubra* L.   (b) *Spondylis buprestoides* L., (c) *Oxymirus cursor* L.   Yeast-
inhabited evaginations of the larval midgut.   From Buchner.

rubra L. eight small, non-contiguous rosettes surround the gut, and in *Harpium mordax* Deg. and *H. sycophanta* Schrnk. meager, barely subdivided buds are found.   Other representatives like *Spondylis buprestoides* L. and a *Criocephalus* species have two circles of such bacciform organs, formations abundantly supplied with tracheae and, because of their whiteness, standing out sharply from the gut which is filled with wood particles and therefore brownish in color (Fig. 27).

Schanderl (1949) claimed that in the fatty tissue of all cerambycid larvae he had found paired formations subdivided into chambers and that these were yeast-containing mycetomes.   But the organs he had observed were in reality the young gonads, and whether yeasts actually occur in them remains an open question.

The degree of adaptation achieved between host and symbiont varies in individual cases. The most perfected state of association is embodied in the previously mentioned *Oxymirus cursor*, where gigantic mycetocytes take up most of the space in the highly complex evaginations and slender tear-shaped yeasts appear everywhere in the mycetocyte plasma. As in the anobiines, small groups of slender, unaltered epithelial cells, like pillars, are found here and there, especially at the transit points of the folds. The

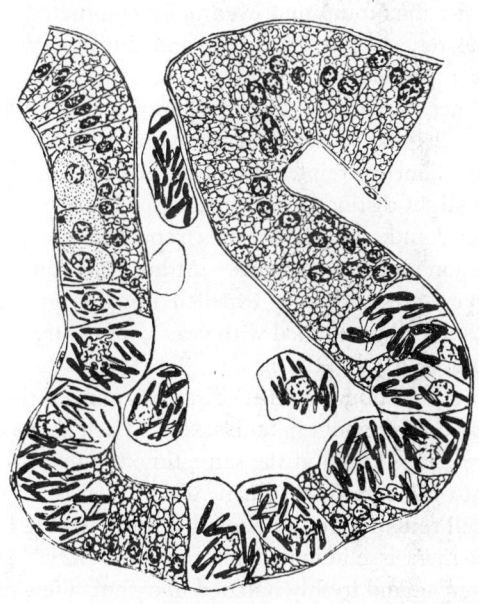

Fig. 28 *Asemum* sp. Section through an infected midgut evagination, new infection of the crypt cells (upper left), expulsion of entire mycetocytes into the gut lumen. From Schomann.

mycetocytes owe their origin to the infection of constantly increasing crypt cells and therefore from the very beginning are without the brush border. The nuclei of the mycetocytes are strongly deve oped in consonance with the enlarged plasma body but retain their roundness and exhibit no degenerative phenomena of any kind. Yeast cells which have been expelled into the lumen are found only in very scattered cases.

The situation is quite different in the *Asemum* species, represented in Fig. 28. Here the mycetocytes do not increase as much in size, the enlarged nuclei assume irregular forms, and a number of expelled mycetocytes, now often in a state of degeneration, may be seen in the lumen. Again it is the crypt cells, usually located in the neck of the as yet sterile invaginations, which replace the degenerating mycetocytes. From adjacent, already populated cells, isolated symbionts are transmitted to their plasma, which up to this point has been compact and deeply staining. These symbionts which produce the infection, and also the yeasts which now multiply abundantly in the newly opened territory, are slender organisms, narrower on one side, which constrict off terminal buds; the cells, filled to the maximum and eventually eliminated, are occupied by much longer forms, shaped like tubes and dumbbells, which have lost their ability to reproduce.

All spondylines, asemines, and saphanines typically undergo such changes in form leading to despecification and elimination and requiring constant replacement by reinfection of crypt cells, although comparison reveals various slight distinctions in the degree of association (Schomann).

On the other hand, not all *Lepturini* represent the perfected stage of mutual adaptation found in *Oxymirus*. Although, so far as is known, complete mycetocytes are no longer expelled into the gut, there is regular abscission of distal cell parts filled with yeasts, for example, in *Harpium* and *Leptura* (Ekblom, 1931, 1932).

Shortly before pupation a gradual disappearance of the larval evaginations can be noted throughout in both sexes. Their sharp contours appear more and more effaced, and at the same time the mycetocytes are almost all expelled into the intestinal lumen, which in this way fills up with a mass of cells, cell remains, and unaltered yeasts. In the last defecation of the larvae the mass is eliminated for the most part. In the intestinal epithelium of pupae and freshly hatched imagines, a few of the yeasts may still be observed in the region of the former evaginations, but after several days they too glide down into the hindgut. In females they reproduce there and fill the transmission chambers of the ovipositor; the males eliminate the yeasts by way of the anus and therewith are rendered symbiont-free.

In cerambycids, just as in anobiids, the intersegmental tubules and vaginal pouches act as transmission chambers. To reach them, the yeasts must first pass from the gut to the sheath surrounding the ovipositor, the transmission being accomplished in the case of a *Harpium inquisitor* larva, for example, 8 days after its metamorphosis. The majority of the symbionts finally leave the body of the host animal from here, but at the same time a relatively small number are transmitted to previously empty spaces and are later used in lubricating the eggs. Reproduction, promoted as usual through secretory action, is so abundant here that the chambers are soon filled completely.

Schomann, author of our most comprehensive study of cerambycid symbiosis, directed his research toward discovery of possible transmission devices. He investigated no less than 184 species, largely with dried material, in which the yeast contents could be recognized each time with great clarity. Schomann discovered that transmission is effected by differentiations which were present before acquisition of symbiosis. He proved that intersegmental tubules are absent entirely in a considerable number of cerambycids and that, where they are represented by saclets,

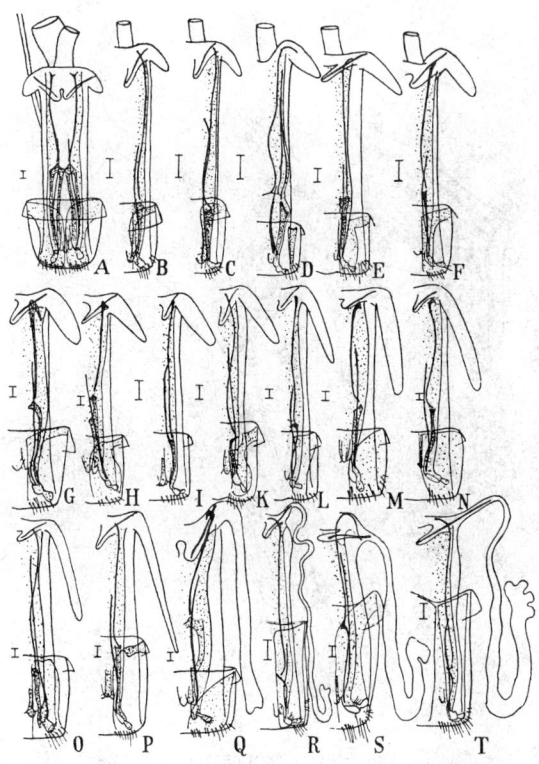

Fig. 29 Egg-laying apparatus of different Lepturini, arranged according to the development of the intersegmental tubules. A to F are symbiont-free, G to T have symbionts. A, *Pidonia lurida* F.; B, *Toxotus vittiger* Rand.; C, *Stenocorus meridianus* L.; D, *Vesperus* sp.; E, *Pachyta quadrifasciata* Blessig; F, *Akimerus Schäfferi* Laich.; G, *Guarotes cyanipennis* Say; H, *Acmaeops tumida* J. Lec.; I, *Leptura virens* Lin.; K, *Harpium inquisitor* Lin.; L, *Cortodera suturale* F.; M, *Nivellia sanguinosa* Gyll.; N, *Grammoptera ruficornis* Fabr.; O, *Alosterna tabucicolor* Deg.; P, *Typocerus attenuatus* Lin.; Q, *Evodinus interrogationis* Lin.; R, *Toxitiades sericeus* Guér.; S, *Rhamnusium bicolor* Schrnk.; T, *Mastododera nodicollis* Klug. From Schomann.

the latter do not necessarily prove the existence of symbiosis. Where symbiosis actually does exist, the saclets become more and more elongated and finally represent true tubules, as shown in Fig. 29. The first six ovipositors in the figure originate from non-symbiotic animals, the

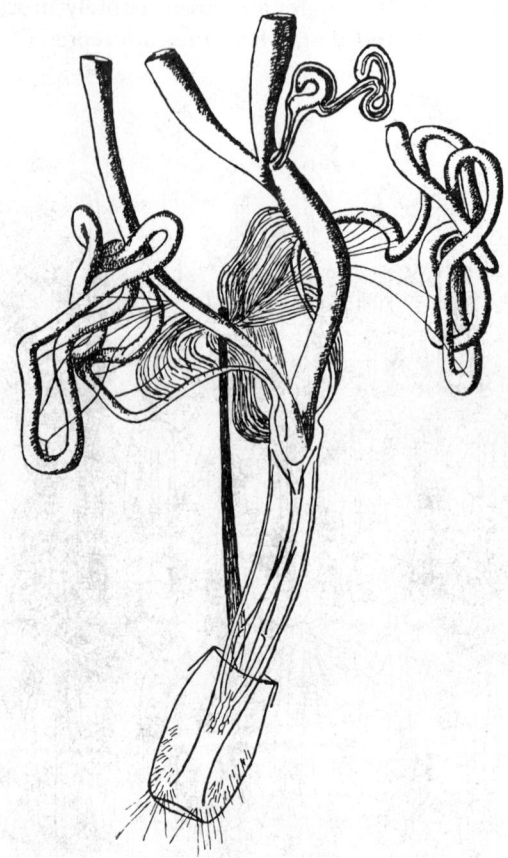

Fig. 30   *Oxymirus cursor* L.   Egg-laying apparatus with maximally developed intersegmental tubules.   From Buchner.

remainder from animals which live symbiotically. Forms like *Oxymirus cursor*, in which the length of the tubules is three times that of the oviposi-tors, and *Necydalis maior* L. show the maximum development (Fig. 30).

At the blind end of each tubule a muscle is inserted, extending to the spiculum ventrale, as in the anobiids, and acting as a support for the various retractors of the ovipositor and other muscles. When the tubules are extraordinarily elongated, as in *Oxymirus*, finer muscle filaments also extend along the whole length. Unless the evaginations are very small, the usual glandular differentiations are present, but they do not by any means involve symbiosis primarily. They are either developed on all sides, and then are unilaminate, as in *Oxymirus*, or on one side only, and consist of several layers. The gland-free parts have chitinous hairs and scales which obviously serve to hold back the yeasts, but they are never so completely developed as in *Xestobium*, and moderately developed ones are also present in species without symbionts.

The tubules, which are far removed from the female genital opening and in addition open freely toward the outside during the act of deposit, as in the anobiids, cannot be relied upon to infect the eggshell with yeasts; hence further symbiont depots in more favorable locations are needed. Again vaginal pouches are used for the purpose. Supported by the so-called gliding plates of the ninth sternite, they open into a ventro-median groove and at the end of the ovipositor are directly connected with the mouth of the vagina.

In addition to the differentiations just mentioned, which are found in all cerambycids with radically elongated ovipositors, there is a third depot in other species, namely, a dorsal pouch on the ovipositor covered by two broad plates and connected with the mouth of the vagina by its lumen (Fig. 31). In the groups which have no symbionts, the pouch is used to protect the more delicate parts of the apparatus. In all Lepturini it fills with yeasts and is then employed as a transmission device. An ample dorsal border of chitinous hairs prevents excessive elimination.

Schomann also discovered the operational procedure of the three differently constituted containers on the ovipositor. Even in mature females the intersegmental tubules are filled with yeasts before egg deposit, but the area around the retracted ovipositor and the vaginal pouches is still free. Unlike anobiid transmission, the vaginal and ovipositor depots are not filled until the female begins the act of deposit with long-drawn-out groping movements of the extended ovipositor. In the process the suitable localities are carefully sought out, the apparatus being repeatedly retracted and projected. The intersegmental tubules, which are folded in a position of rest, are extended when the ovipositor emerges. Muscular contractions cause drops of the contents to escape, and these are pushed forward along the ovipositor by the intersegmental membrane which is in the process of being folded back. Larger quantities of the contents are pressed out principally as the eggs pass the narrow region. The pouches are filled by dehiscence through contractions of the muscles connecting the dorsal and ventral supporting plates near the vaginal mouth, which allow the pouches to spread apart.   Congregated now on all sides around the mouth of the vagina, the symbionts reach the surface of the emerging egg without fail.

As a rule, they are distributed uniformly over the surface, being held fast by the mucilaginous secretion of the intersegmental tubules which function simultaneously as lubricating glands.   Another factor to be considered is that the surface of the egg is often divided into zones by projecting

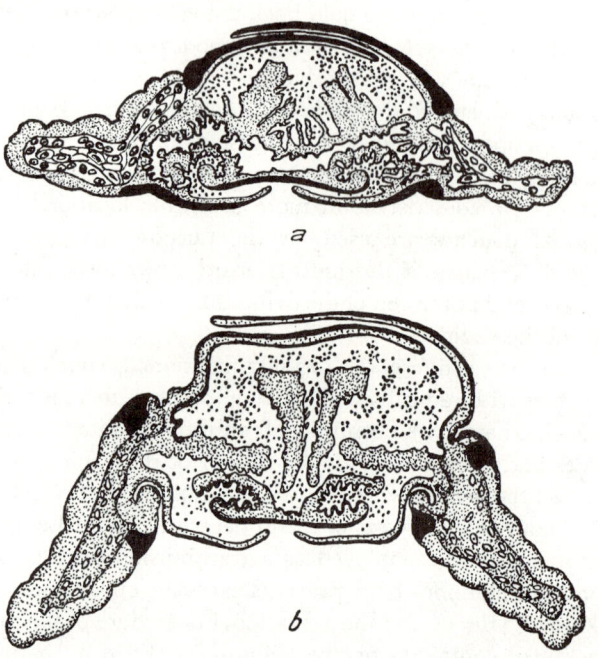

Fig. 31 *Leptura rubra* L.   Cross section (*a*) through the ovipositor with dorsal and ventral vaginal pockets filled with yeasts, (*b*) through the end of the ovipositor, in which both kinds of pockets open into the vagina, which also contains yeasts.   From Schomann.

ridges (*Tetropium, Harpium*, Fig. 32*b*).   It is true that, in *Oxymirus*, long sausage-shaped clumps of yeasts cling to the eggs so firmly as to require force to detach them, but this is a special case resulting from the unusual development of the intersegmental pouches and their glands (Fig. 32*a*). The symbionts housed on the ovipositor must be sufficient for approximately 250 eggs with *Harpium inquisitor*, and for 120 with *Tetropium*.

When the hatching period is at hand, the young larva devours part of the eggshell, which has been softened by a secretion from the inside, and then emerges (Fig. 32).   Free of symbionts up to this point, the gut is now filled with shell fragments to which the yeasts still adhere.   The shells soon disintegrate, and the yeasts rapidly permeate the intestinal lumen.   The zone of the future evaginations can be distinguished now in several respects.   The yeasts lodge in it close against the epithelium, leaving the central intestinal areas free, and now begin to form buds.   The epithelium, in this single instance composed of larger elements with proportionately larger nuclei and without the brush border, sends out plasma processes into the lumen to envelop the yeasts (Fig. 33).   After this admission

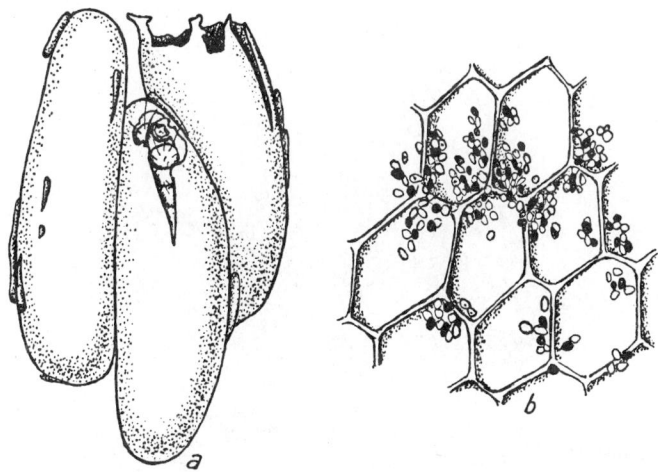

Fig. 32   (*a*) *Oxymirus cursor* L.   The hatching larvae eat a portion of the eggshell, which has been soiled with yeasts.   Combined from Buchner and Schomann.   (*b*) *Harpium mordax* Deg.   Part of the eggshell infected with yeasts.   From Schomann.

period, which begins 13 to 14 hours after hatching, the epithelial cells regain their smooth outlines, and larvae which are only a few days old now possess small and very distinct intestinal evaginations.

The evaginations are modifications necessitated solely by symbiosis, for corresponding ceca are lacking entirely in cerambycids which do not live symbiotically.   But how firmly they are part of the insect's organization, on the other hand, we know from examining sterile larvae obtained through preparation of the prehatching stages or through sterilization of the eggshell.   Even when there are no symbionts, these beetles develop ceca containing typical, greatly distended cells with large nuclei and lacking the brush borders!

We also have reliable information on the mycological aspect of ceramby-cid symbiosis. A comparison based on a large number of species reveals considerable diversity of symbionts, their various forms so typical that they can be utilized as a criterion for classifying the larvae. They are very slender or somewhat broader formations which may take the shape of commas, or of slippers rounded off at both ends, or of ovals, circles, and dumbbells. The budding by which they reproduce takes place usually at the slender end, but occasionally at the blunt end, and even longitudi-nally. Branched yeasts are found rarely in the hosts (Fig. 34*a–c*).

The same shapes occur in the larval organs and in the transmission

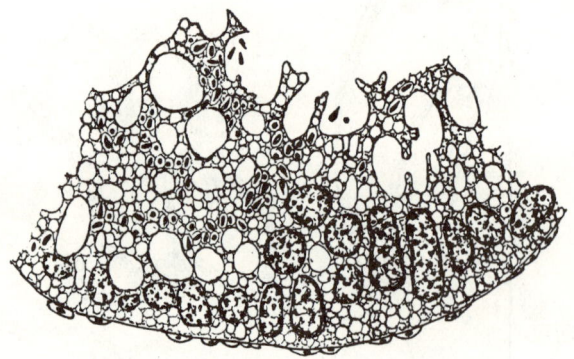

Fig. 33. *Oxymirus cursor* L. The future mycetocytes send processes into the gut lumen of newly hatched larvae in order to take up the yeasts consumed with the shell. Modi-fied from Schomann.

organs of the imagines, but in the latter they tend to be smaller and some-times more or less roundish, apparently as the result of less favorable nutrition. Once again we find an exception in *Oxymirus* with its perfect adaptation of host and symbiont, in that the symboints in the imaginal tubules, with their profusion of secretion, are indistinguishable from those of the larvae.

All this applies especially to the lepturines, to which Schomann paid particular attention in his research. A noteworthy exception is repre-sented by the asemines, spondylines, and saphanines, in which a lesser degree of adaptation is clearly indicated, as mentioned earlier. In their case, with increasing infection of the mycetocytes, the originally slender, tear-shaped symbionts give way more and more to tube-shaped, elongated symbionts blunted at both ends, and only the slender states retain their ability to bud (Fig. 34*d*).

A second indication of the exceptional character of these three groups is even more surprising. In all their representatives, characteristic spores fill the spaces used by the imagines for transmission. These are roundish to pileate formations usually surrounded by a delicate rim and arranged in groups of four (*Tetropium castaneum* L.) or by ones and twos (*Spondylis buprestoides* L.). They develop in the often irregularly shaped yeasts of prepupation larvae even before being transmitted to the intestine, and they are reminiscent of the spores of *Willia* or *Endomyces capsularis* Guill.

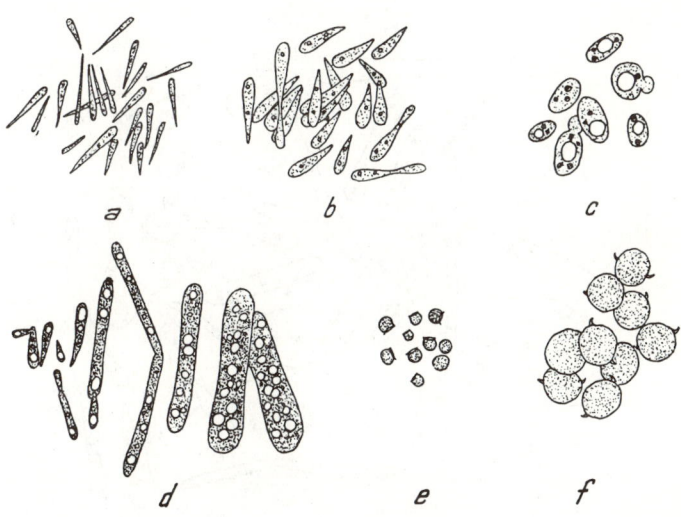

Fig. 34   Different symbionts of cerambycids.   (*a*) From *Rhamnusium bicolor* Schrnk., (*b*) from *Oxymirus cursor* L., (*c*) from *Rhagium bifasciatum* Fbr., (*d*) from *Criocephalus rusticus* L., (*e*) from *Criocephalus ferus* Kr. (spores), (*f*) from *Epipedocera rollei* Pic. (spores). From Schomann.

(Figs. 34*e,f*, 35).   Only in two species of a genus belonging to the tillo-morphines, the *Epipedocera*, could Schomann find roundish, rather large spores with two spurs, in material originating from Formosa (Fig. 34*f*).

The following authors were concerned with culture experiments: Heitz (1927), Schimitscheck (1929), Ekblom (1931, 1932), W. Müller (1934), Schanderl (1942), and Gräbner (1954).   Generally cultures were easily achieved with solid and liquid media (*Harpium* and *Leptura* species), but not in *Oxymirus* (Müller).   Various changes in form occur frequently in the cultures.   *Leptura rubra* symbionts cultivated in medicated media display large branched groupings of 100 to 200 cells, individual parts of which are later detached, but *Harpium* symbionts never showed such groupings (Fig. 36). Often a more roundish to oval form appears in the cultures instead of the slender intracellular form. Cultivated in broth, *Harpium* symbionts develop into long tubes. Gräbner's experiments show that cerambycid yeasts grow unusually rapidly with zylose, that *in vitro* their form tends to be quite different from that in the mycetocytes and transmission organs, and that symbiont forms reappear with pentozylose. When rounded forms, cultivated in medicated media, are injected into the lymph of living caterpillars, they assume their typically slender form again,

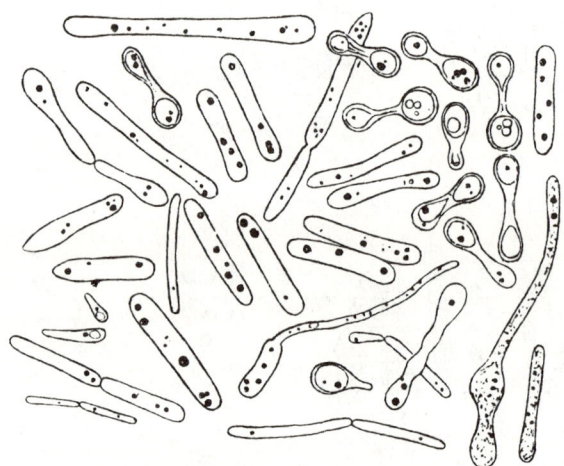

Fig. 35   *Spondylis buprestoides* L.   Symbionts, some of which are in the process of spore formation.   From Buchner.

and after a time tube forms also make their appearance. The yeasts swiftly overwhelm the blood of the animals, and death occurs after 3 days. When the yeasts are returned to ordinary culture media, the round forms appear again (W. Müller). Schimitscheck succeeded in cultivating the drop-shaped symbionts of *Tetropium castaneum* L. from spores, but Gräbner had no success in isolating *Tetropium* symbionts.

Schanderl claimed he had cultivated *Harpium* inquisitor symbionts in bean broth, obtaining a typical mold yeast which produced rapidly growing membranes and numerous asci and spores when transmitted to grape must. He identified his cultures with *Mycoderma bispora* Baltatu and used the term yeast mold to describe the spore-forming symbionts of the spondylines, asemines, and saphanines. Gräbner, on the other hand, classified the pseudomycelium-forming cerambycid symbionts as anascosporogenous yeasts, specifically of the genus *Candida*.

Diddens and Lodder (1942) described the yeast, cultivated earlier by Müller from a species of *Rhagium*, as *Candida tropicalis* var. *rhagii*, and in 1954 Windisch identified Schanderl's *Mycoderma* as *Candida guillermondii*. Independently of these findings, Graebner also saw in the cerambycid symbionts representatives of the anascosporogenous genus *Candida*, a finding well confirmed by the recently begun studies by mycologists on cerambycid symbionts (Jurzitza, 1959; Jurzitza, Kühlwein, and Kreger-van Rij, 1960; Jurzitza, 1962). These authors investigated nine different cerambycids in regard to their symbionts (some of the inhabitants were identical and others were of different species), and they described their characteristics and their chemical capabilities, which showed considerable

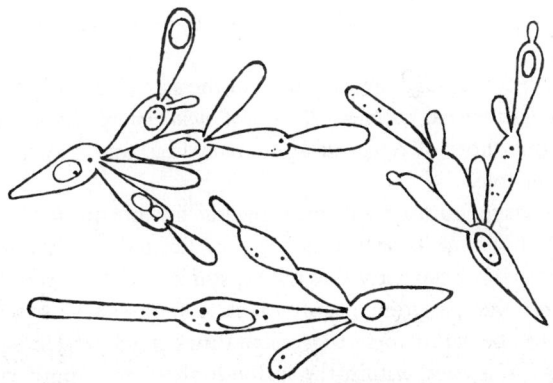

Fig. 36 *Leptura rubra* L. The symbionts form budding groups in wort. From Heitz.

differences also in regard to B-vitamin production. Urea and uric acid appear in general to be utilized, while atmospheric nitrogen cannot be assimilated. In the General Section these results will be considered in detail.

From Schomann's observations on dead larvae of cerambycids it is clear that cerambycid symbionts are rather labile in form, and that such constancy as exists is the result of demands made by the host organism. The symbionts behave therefore just as they do in culture media: they reproduce abundantly in the first days after the death of the host animal: roundish formations take the place of the pointed pear-shaped ones, and branched groupings and tube formations are numerous; and, if the maceration of the host animal continues, the symbionts are also destroyed.

The reproduction rate of cerambycid symbionts is largely dependent upon temperature and state of nutrition, just as among the anobiids. Even in nature a corresponding seasonal change can be determined.

Harpium inquisitor larvae in the winter have ten or more intestinal evaginations widely separated and populated by symbionts; additional humps appear in the spring; and by summer a continuous, broad, aciniform girdle encircles the gut. Kiefer (1932) found a similar change in *Rhagium* and *Leptura* larvae. The number of the budding yeast cells showed a proportionate curve, increasing from 0% in February to 45% in April and then gradually falling to 20% by the end of October.

Where the developmental period of the cerambycids extends over several years, the possibility of proportionate periodicity in the development of the symbiotic habitats must be considered. With other test objects, such as *Harpium mordax* Deg. and *H. sycophanta* Schrnk., which live characteristically in deciduous wood, only an increase in the density of settlement in the individual mycetocytes is noted in summer, but no increase in size of organs, which in these beetles are only meagerly developed.

Hunger and cold retard development of the symbionts in greatly varying degree in the individual species. *Harpium inquisitor* symbionts are strongly influenced, but those of *Rhagium bifasciatum* Fbr. and *Oxymirus cursor* are much less sensitive.

Testing a large number of cerambycids for possibility of symbiosis, Schomann found that its occurrence is very limited in this diverse group. Among test objects from sixty-five tribes, symbiotic devices were found in only the spondylines, asemines, saphanines, a single species of the cerambycines, most of the lepturines, the necydalines, trichomesiines, and tillomorphines. Thus even within the individual tribes no uniformity exists, yet perhaps one day we may find that the gaps in the occurrence of symbiosis among cerambycids, which at first glance seem so surprising, may be explained by differences in feeding. For instance, symbiosis is absent in all cerambycids, the larvae of which live in fresh deciduous wood, either hard or soft, and in herbs. On the other hand, the larvae of all forms in which symbiosis was established—insofar as their biology is known, and this unfortunately is not usually true for exotic forms—are found in living and dead coniferous wood or on dead deciduous wood. Yet this by no means signifies that all species living in that manner have symbionts. For example, all lamiines and prionines are free of symbionts whether they are adapted to coniferous or to fresh or dead deciduous wood.

After such experiences it is not surprising that Grinbergs (1961), who investigated a considerable number of cerambycids from Chile, did not find this typical symbiosis in a single species. It is not clear from his publication whether inhabitants of dead wood were also among his material, which consisted almost exclusively of forms living in deciduous trees. However, he did find in some cerambycid larvae a well-developed gut flora, in which in addition to other yeasts the same *Candida* species, which otherwise live intracellularly, were almost always present. To what extent this may represent a primitive prestage of the more highly developed *Candida* symbioses cannot be said at the present time.

BUPRESTIDS

The paired cecum at the beginning of the midgut in most buprestid larvae and imagines might well be a symbiotic habitat, since it bears an extraordinary resemblance to matching formations in cerambycids, anobiids, and cleonids, all insects with habits quite similar to those of buprestids. These ceca represent smooth, simple, clavate evaginations as in *Trachys minuta* L. or sinuous forms with transverse folds. In *Chalcophora mariana* L. I found ceca studded with numerous little digitate tubules which even cover the actual intestinal tubule at its place of origin, and in the giant larva of the exotic *Megaloxantha bicolor* Fabr. I found proportionately huge moniliform ceca with a ring of coral-like excrescences of the midgut at their root. According to von Gebhardt, who described the ceca in a series of imagines (1929), they may be replaced at times by folds in the initial section of the intestinal tube.

Heitz (1927) reported that he had found bacteria in the epithelium of these appendages in an unclassified larva, and during my own research with live insects I have often thought that his statement must be correct. Unfortunately, we have no thorough study of the matter as yet.

If bacteria actually populate this conspicuous organ of buprestids, their only possible way of transmission would be through contamination of the egg surface, according to our present knowledge. On the ovipositor of a newly hatched *Chalcophora*, thus one with young ovaries, I found minute intersegmental tubules very similar to those in cerambycids. The fact that they were empty does not necessarily exclude the possibility that they play a role in the transmission process, since the filling of corresponding organs in immature females of the cerambycids is gradual.

LAGRIIDS

From Stammer's publication on symbiosis in lagriids (1929), which describes the only research on this subject, we know that their symbiosis differs fundamentally from that of anobiids and cerambycids in localization and nature of the microorganisms, but closely resembles it in type of transmission organs. Its cycle was first investigated in *Lagria hirta* L., which is the only common European species.   Its worldwide distribution and the numerous variants of its egg-lubricating devices were studied with dried material, as in the other two families mentioned.

In *Lagria hirta* and the other European representatives of the genus the imagines feed on shrubs and herbs, the larvae chiefly on the dry and

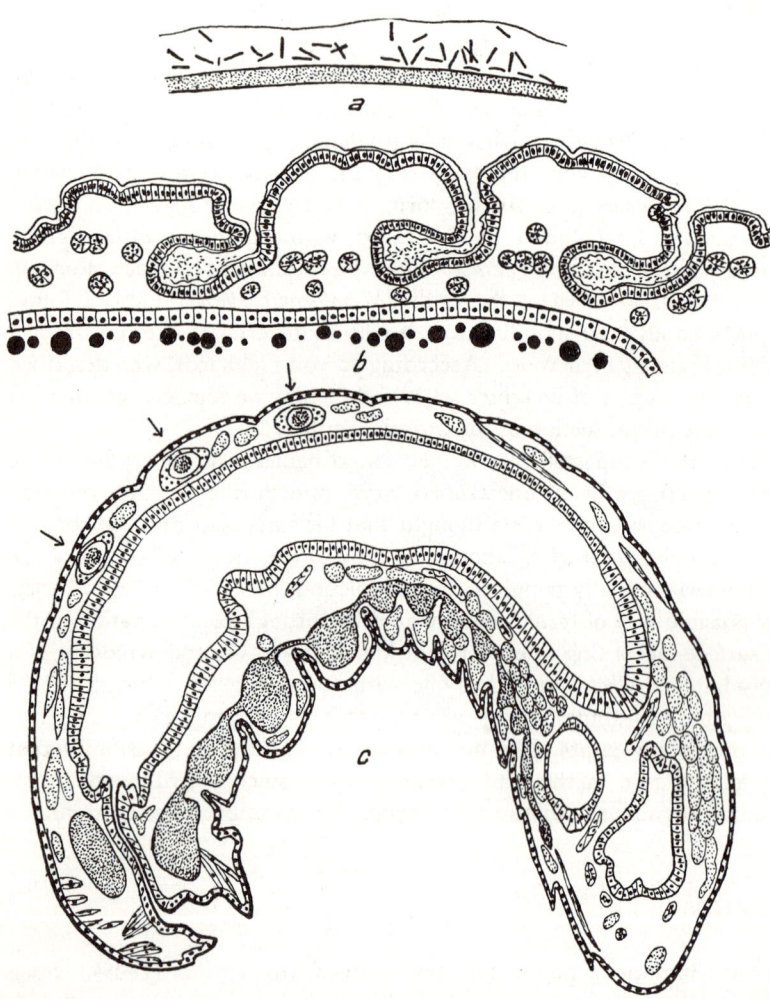

Fig. 37 *Lagria hirta* L. Origin of the larval dorsal organs. (*a*) Eggshell covered superficially with mucus and bacteria. 375×. (*b*) The sacs invaginate and are colonized. 40×. (*c*) Longitudinal section through a larva, after the first molt, with the three completed sacs. From Stammer.

moldering foliage amid which they pupate, and also on fresh leaves. An African species of *Lagria* may defoliate whole rubber trees when it appears en masse and may even attack the green bark. The same habits might possibly be shared by the other symbiotic genera from all parts of Africa, the Far East, and Australia.

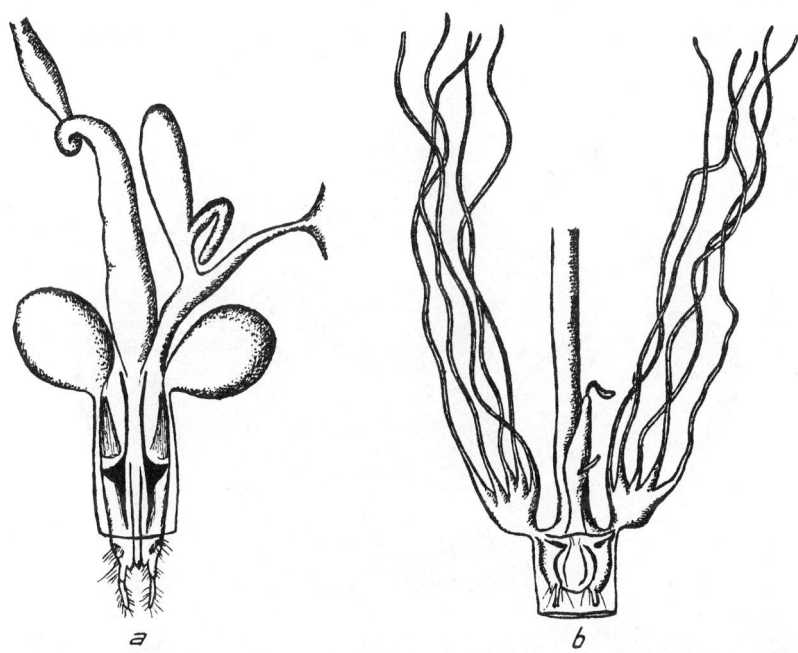

Fig. 38   (*a*) *Lagria hirta* L.   (*b*) *Aulonogria concolor* Blanch.   Egg-laying apparatus with bacteria-filled intersegmental organs, each of which (*b*) is divided into five long tubules. From Stammer.

Lagria hirta larvae have unique symbiotic sites on the dorsal side in the meso- and metathorax and in the first abdominal segment in the sagittal plane. In each of the three locations a dense bacterial population fills a slightly flattened vesicle with a chitinous coating which decreases in strength as it grows older and thus reveals its origin from the dorsal skin! The epithelium of these dorsal organs consists of flat and cubical cells and, at the anterior and posterior ends, an accumulation of vacuolized glandular cells with secretion channels leading into the lumen of the saclets (Fig. 37*c*).

We know that these strange organs are absent in the imagines and that in the imaginal female the symbionts are found exclusively in the transmission organs, but unfortunately there has been no research describing their fate during the course of metamorphosis.   Since symbionts were not found elsewhere in the larval body, one must assume that what happens here is similar to what occurs in *Haematopinus* species of the Anoplura. As we shall see later, the female larvae in this genus develop three dorsal symbiont depots under the hypodermis which are not, however, of hypodermal origin.   During the last molting these depots are transmitted to a

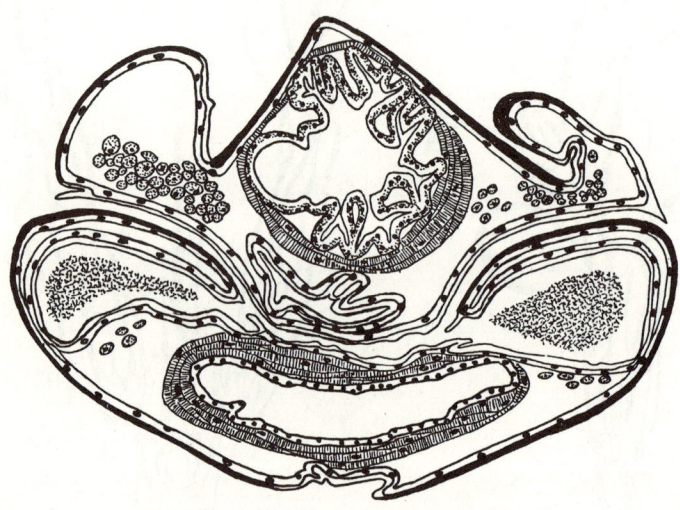

Fig. 39 *Lagria hirta* L. Cross section through the ovipositor; pockets of the egg-laying apparatus and their excretion ducts which open into the vaginal space; above, the gut; below, the vagina. 170×. From Stammer.

liquid-filled space between the larval and the imaginal chitin.   Floating in the liquid, the released symbionts reach this area and pass from it to the wide opening of the sexual rudiment and through the latter to the ovarial ampullae, which later are to be instrumental in infecting the eggs.

It might very well be that during metamorphosis the lagriid symbionts are passed on to the intersegmental tubules and vaginal pockets which are present in their case also.   Stammer investigated on a broad basis the developmental forms which these structures assume from case to case.

At the transition point of the intersegmental membrane and the ovipositor, when the latter is completely retracted in a state of rest, *Lagria hirta* forms, on both sides, a simple, broadly oval sac with a chitinous lining bordered by a unilaminate glandular epithelium (Fig. 38*a*). Strands of muscles connect the sac with the spiculum ventrale and other parts of the ovipositor. The secretion-filled lumen contains a pure culture of very thin rodlets, some of them forming chains with more or less deeply staining sections. Folded in a position of rest, they are extended during egg deposit and allow considerable masses of bacteria to escape into the space around the ovipositor.

In addition to these intersegmental formations there are other symbiotic spaces in the ovipositor proper. Between the dorsal and ventral supporting plates a deep membranous fold sinks on both sides into the interior of the ovipositor to form a bacteria-filled pouch (Fig. 39). In the caudal area the slit-shaped openings in the middle axis communicate with the outside, but toward the head the pouches continue for a distance in the form of folded sacs without direct communication with the outside. By analogy with anobiids and cerambycids it would be logical to interpret the intersegmental sacs chiefly as symbiont reservoirs and to attribute lubrication of the eggs to the pouches of the ovipositor, which are indeed much better suited for this purpose because of their outlet in the adjacent vagina.

As the egg passes through the ovipositor, it squeezes the pouches, releasing the symbiotic bacteria which cling to all parts of the egg in great numbers by means of mucilaginous secretion from the intersegmental sacs. From here they penetrate through the micropyle into the space between egg and chorion, as in *Dacus*.

Only when the young larva is well developed, that is to say, 1 or 2 days before hatching, do the dorsal intersegmental membranes form matching evaginations at the place where later the three mycetomes are found, and quantities of bacteria now migrate to these evaginations (Fig. 37*a,b*). The strange process evidently takes place very quickly, for larvae of the same age show all stages.

Investigating ninety-three other species of the lagriines with dried material, Stammer found that the transmission organs are fundamentally alike in eighty-two of them, but that the lubricating devices show a diversity far beyond that displayed in the anobiids and cerambycids. He differentiates no fewer than twelve types, each stipulated by the variations found in the intersegmental evaginations. In *Lagria hirta* these sacs are still smooth-walled. In other cases they may be moderately lobate, deeply incised, two-pronged, lobate, and folded over, or covered with numerous villi. In still other instances four slender, often very long, tubules are formed, some unbranched, others richly branched, and still others originating from a uniform collecting space (Fig. 38*b*).

Another factor contributing to the diversity is that the elongations of the ovipositor pouches which extend anteriorly, indeed in *Lagria* ending in the head, may be so greatly enlarged as to emerge for some distance, in some species the exposed portion swelling up into club-like sacs. Figure 40 illustrates a case where they are especially large and have at their base additional evaginations which in other test objects may be of considerable

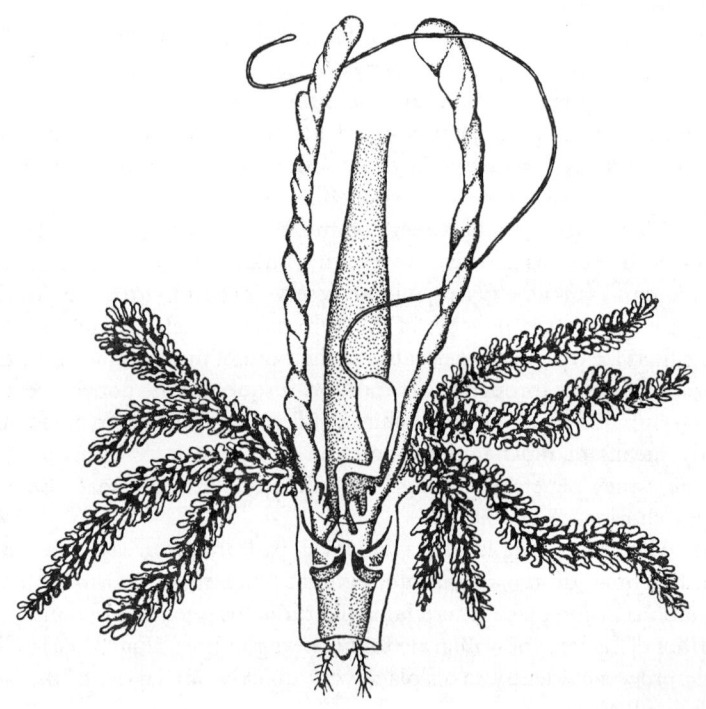

Fig. 40   *Cerogria heros* Fair.   Intersegmental and vaginal organs of transmission in maximum development.   5×.   From Stammer.

size.   These differing formations are combined with those of the intersegmental tubules in varied ways.

Comparatively few lagriines were found to be without such devices and, when viewed in cross section, they also proved to be without the pouches in the ovipositor.   Whether they are symbiotic despite this will, of course, require research using fresh material.   No results were provided by testing of dried material of the remaining lagriid subfamilies, that is, the trachelostines, statirines, chanopterines, and agnathines.

CHRYSOMELIDS

Stammer (1935, 1936) did research on a large number of chrysomelids, testing most of the tribes occurring in Germany for the presence of symbiosis. In only three cases could he establish symbiosis, in *Bromius obscuruᶜ* L., in some *Cassida* species, and in the donaciines.

Fig. 41   *Cassida viridis* L.   Intestinal tract and genital apparatus of a female imago; round dwelling sites for symbionts at the beginning of the midgut, and paired vaginal attachments, divided into three parts, as organs of transmission.   From Stammer.

Cassida symbionts are housed in the same place in larvae and imagines, namely, in evaginations at the beginning of the midgut, as is often the case. In *Cassida viridis* L. there are two pairs of roundish saclets, the partners of which are very close to each other, being separated only by a very thin, though ever-present, membrane. *Cassida hemisphaerica* Hbst. follows this pattern; three other species, *C. rubiginosa* Müll., *C. vibex* L., and *C. nobilis* L., have only two ceca (Fig. 41). A strangely developed, tall epithelium is composed of mycetocytes which sharply separate a large, basal, symbiont-

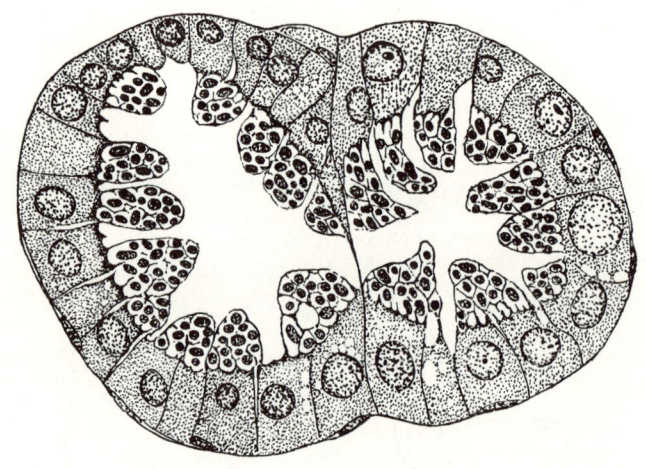

Fig 42. *Cassida viridis* L. Adult gut evagination with symbionts, in section. From Stammer.

free part with dense protoplasm from a distal coarse-meshed part containing roundish and oval microorganisms (Fig. 42).

Stammer did not observe how the organisms reach the ceca of the imago during the metamorphic process which destroys the old ceca, but we may surmise that they are transmitted as in *Bromius*, a genus which appears to have a type of symbiosis closely related in general to that in *Cassida*.

As usual with such localization, the symbionts are given to the egg externally during deposit, but in *Cassida* the host adopts a new device in the form of two glandular symbiont reservoirs opening into the lower third of the short broad vagina. Each reservoir is composed of three curving tubules connected with the vagina by a single delicate outlet. The tubules are thicker in the distal glandular section and are filled tautly with the same microorganisms found in the intestinal appendages (Fig. 41).

The eggs, which are deposited in small packages enveloped in a secretion consisting of numerous lamellae, are firmly attached to the forage plants on which the larvae and imagines feed. Every single egg is topped by a distinct caplet of bacteria at its anterior pole, indicating that the vaginal tubules must enter into well-ordered rhythmic activity during deposit (Fig. 43). On hatching, the larva ingests the entire cap as it eats its way into the open through the eggshell.

The number of tubules varies in the individual species, just as with the intestinal evaginations. Thus two, three, and four tubules may be joined in one outlet.

Strangely enough, Stammer found no symbiotic arrangements of this kind in two species with exactly the same mode of life. In these species, intestinal and vaginal evaginations were accordingly both lacking.

Symbiosis is far more complicated in *Bromius obscurus*. Here the imagines live on leaves of *Epilobium angustifolium*, and the larvae gnaw canaliculate depressions in its roots. The beetle also lives on grapevines, the imagines again feeding on the leaves and young shoots and the larvae on the roots.

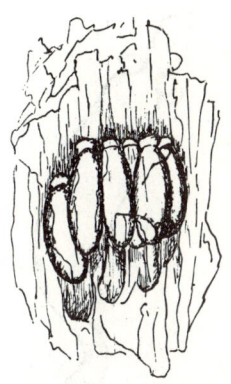

Fig. 43 *Cassida viridis* L. Egg deposit; each egg has a bacterial cap. From Stammer.

At the point of transition of the larval foregut and midgut are two pairs of closely neighboring, roundish sacs which house their symbionts only in cells of the distal part. This time they are more deeply staining organisms, growing gradually and continuously as larvae advance in age and finally developing into large rosettes (Figs. 44*a*, 45*a*). But now still another symbiont variety is found! Numerous rod-shaped bacteria are also present in the lumen of all six Malpighian tubules.

At the same site in the imaginal gut, the four sacs are replaced by a circle of slender villi which are populated by the same symbionts. On the other hand, the rodlets are no longer located in the Malpighian tubules but in an extracellular position, in separate slender ceca covering the hindmost part of the midgut, its remaining portion being covered, as in the larvae, with minute symbiont-free diverticula (Figs. 44*b*, 45*b*).

With *Bromius* it was possible to study more thoroughly the changes in location arising during metamorphosis. Along with the rest of the intestinal epithelium the four ceca are propelled toward the inside by the newly formed imaginal gut. Here the host cells are decomposed, the rosettes disintegrate into their component symbionts, which soon populate the circle of villi which has been forming in the same zone. During disintegration of the symbiotic organs of the larvae, the rodlets reach the intestinal lumen in the same way and from there can easily be transmitted to the evaginations forming in an adjacent location.

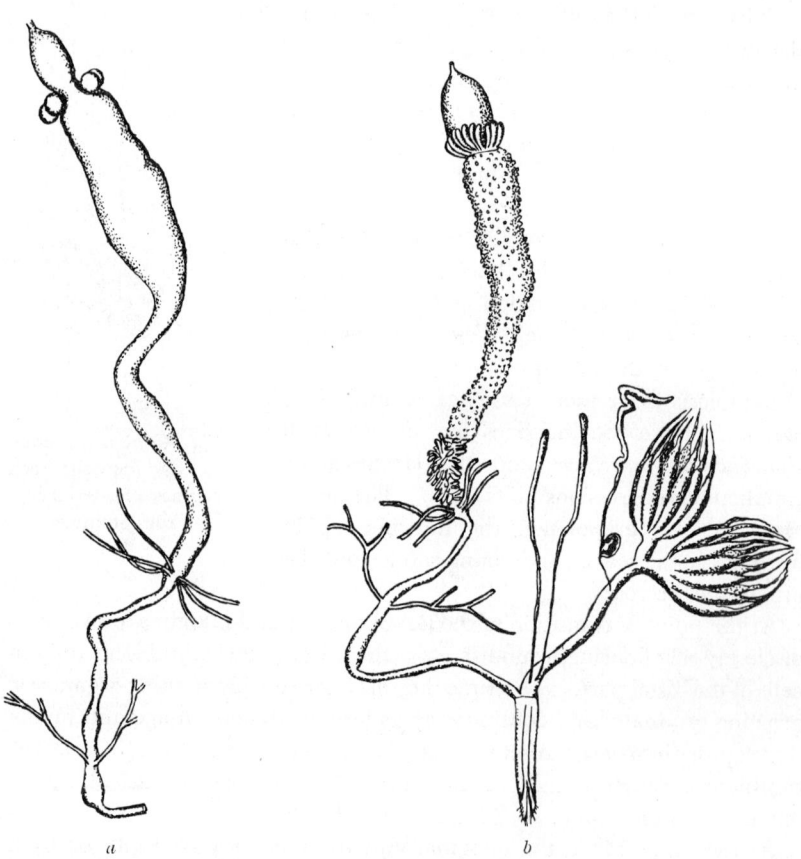

a   b

Fig. 44 *Bromius obscurus* L.   (a) Larval intestinal tract with round bacterial organs, (b) intestinal tract and genital apparatus of a female imago; symbiont sites at the beginning and end of the midgut in the form of finger-like villi; on the vagina two slender, club-shaped organs of transmission.   From Stammer.

The transmission proceeds very much in the manner described for *Cassida*, except that only two gradually swelling vaginal tubules are formed and in their lumen the two symbiotic types are now intermingled (Fig. 44b).   The symbionts are accordingly distributed over the whole egg, individually or in rather large cakes.

Even before hatching, the four midgut ceca are formed. Their cells, all provided with large nuclei, have a strange fibrous structure, oriented toward the lumen and apparently serving to catch the symbionts, which in this case also advance with parts of the eggshell to the larval gut, where they can soon be observed here and there (Fig. 46). In the meantime the Malpighian tubules receive the rodlets which have been untouched in the process.

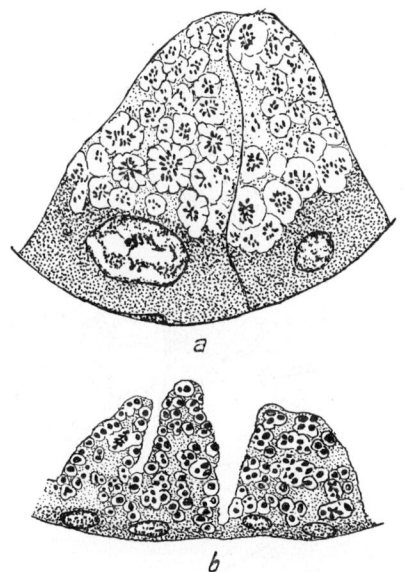

Fig. 45 *Bromius obscurus* L. (*a*) Symbionts in the wall of the larval and (*b*) of the adult gut evagination. From Stammer.

Fig. 46 *Bromius obscurus* L. Rudiment of a not yet infected blind sac shortly before the larva hatches. From Stammer.

The donaciines likewise show many interesting symbiotic traits In Central Europe three genera of this chrysomelid tribe are represented, the genus *Donacia* with its many species, and the genera *Macroplea* and *Plateumaris* with only a few. Stammer (1935), who discovered the symbiosis here, investigated ten *Donacia* species and two of *Plateumaris* and found that they all live symbiotically with bacteria. The imagines of these handsome beetles live on the leaves of swamp and water plants and lay their eggs just below the surface of the water on usually specific forage plants, sometimes in their tissue. The young larvae fall to the bottom of the water and live on sap from roots of forage plants.   At least, the cellular plant parts are never found in their gut.   With stigmata converted into a dagger-like form for this purpose, they bore into the roots to get from the intercellular spaces the air necessary for breathing.   The *Macroplea*

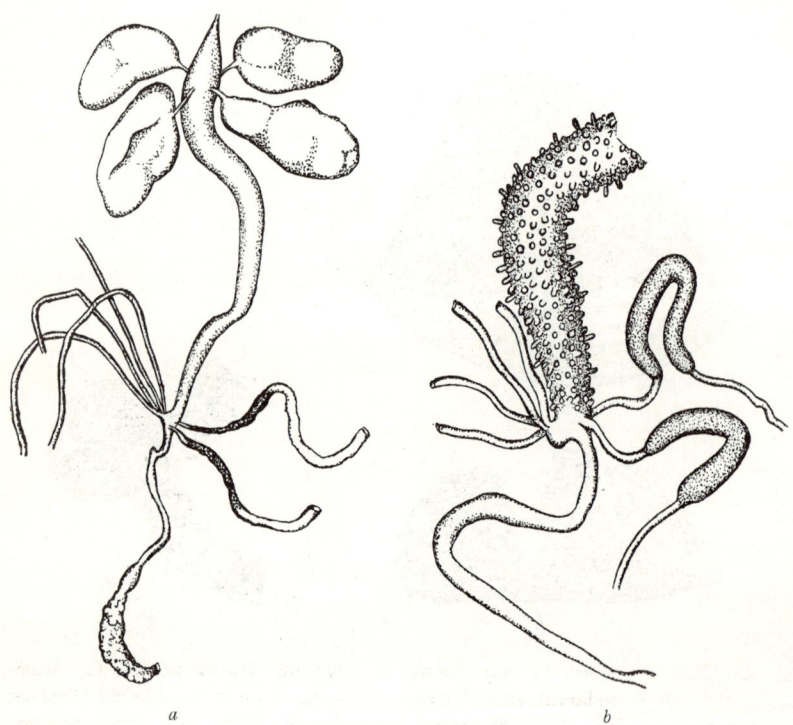

a   *b*

Fig. 47   *Donacia semicuprea* Panz.   (*a*) Intestinal tract of a grown larva with four large symbiont-filled blind sacs; the dark parts of two Malpighian vessels are similarly infected.   (*b*) Intestinal tract of a female imago; a pair of Malpighian vessels house the symbionts in thickened sections.   From Stammer.

species, the symbiosis of which is yet to be investigated, remain under water also in the imaginal form and breathe by intercepting with their strangely altered feelers the oxygen bubbles separating from the plants.

At the beginning of the midgut in adult larvae of *Donacia* are four extraordinarily spacious sacs taking up a large part of the thorax and the first abdominal segments.   They are of varying sizes, very flat, and rather loosely folded.   A thin pipe connects them with the gut (Fig. 47*a*).

Plateumaris sericea L. has similar, but much more protuberant, organs. The epithelium of the sacs is thickly filled with long, filamentous, often parallel bacteria which press the nuclei toward the bottom.

In addition to this basic larval site there is a second symbiotic site in older larvae. When the larvae reach a length of 4–6 mm., two Malpighian tubules are infected by isolated bacteria which migrate from the ceca to the gut. Like all chrysomelids, *Donacia* has six such tubules; four of them joined together open into a small bladder-like swelling, and the remaining two, much smaller, open into it on the opposite side. In the *Donacia* species the shorter tubules have a blind ending; in *Plateumaris* they join the two slings formed by the other four. The bacteria are always housed in the two shorter ones, in a specific section prepared for them. The short initial section is sterile; then follows an infected zone with cells only moderately filled with symbionts, and finally a much longer, again sterile, part. The latter section has greatly enlarged cells with markedly branched nuclei, and it furnishes secretion for the cocoon in which the larva pupates (Fig. 47a). Such secondary colonization of the two Malpighian tubules is found in all *Donacia*, and in both sexes.

After the larva has spun itself into the cocoon, profound changes, which began in the rest period before pupation, bring about the imaginal state of the devices employed in symbiosis. The midgut ceca now experience a strange regression occurring nowhere else in this form. When the larval intestinal cells are expelled and the imaginal intestinal epithelium develops, the ceca do not move to the intestinal lumen but shrink together more and more, and their filamented bacteria undergo a gradual deterioration. Irregularly shaped, deeply staining inclusions appear; gradually they are all expelled into the lumen, which in that way fills up almost entirely. The more weakly staining sections disappear; the more deeply staining ones join in irregular clumps. The remaining pupal bacteria form a more or less homogeneous mass which is no longer connected with the intestine and is surrounded by remains of the lining and its musculature. Finally, these too disintegrate and the organs no longer exist, their removal having been effected exclusively by histolysis without the assistance of phagocytes.

In the meantime important changes in the Malpighian tubules are manifested. In releasing their secretion the two bacteria-filled tubules contract greatly. In female larvae the symbionts now increase extraordinarily, and the cells they populate increase likewise, producing sections light orange-red in color, quite thickened and clearly separated (Fig. 47b).

Two possibilities exist in the case of male larvae. Infected sections may be preserved in rare instances, but they are much smaller and are insignificant, as was the case in three *Donacia* species and in one *Plateumaris* species. In general, the bacteria deteriorate and finally disintegrate, just as in the intestinal ceca, and the resulting bacterial debris is expelled into the lumen of the tube. Comparison shows that infection of male secretory organs is maintained in species in which the infected sections are most strongly developed also in females.

The cells made available to the symbionts in the Malpighian tubules appear enormously enlarged and so crowded with bacteria as to render the protoplasm barely visible (Fig. 48).

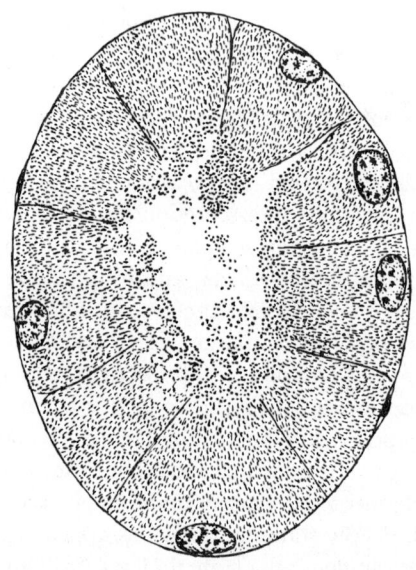

Fig. 48   *Donacia semicuprea* Panz.   Cross section through the symbiont-filled section of a Malpighian vessel from a female imago; development of infective forms in the peripheral regions.   From Stammer.

Apparently no value is placed upon preserving this striking bacterial depot in the males, and this in itself indicates that it is employed in transmission. Indeed, in mature females the bacteria migrate in quantities to the lumen in the two tubules and from there fill the hindgut. Previously they have undergone changes typical of many infection forms, having changed in the peripheral regions of the cells into oval to spherical formations with much deeper staining capacity.

The eggs of *Donacia* are enveloped during deposit by a foamy secretion originating from the vagina and congealing instantly in water. From the hindgut, bacteria in measured quantity are administered to each egg in such a way that they come to be situated where the head of the larva develops, thus where the latter devours the shell to make its way into the open (Fig. 49).

At this time the future symbiont stations have already taken form in the gut and are now available to tenants.   Many years ago Hirschler (1907, 1909) described the development of the larval ceca without realizing their significance.   They originate not simply through a folding of the midgut epithelium, as in anobiids, cerambycids, cleonids, etc., but from an

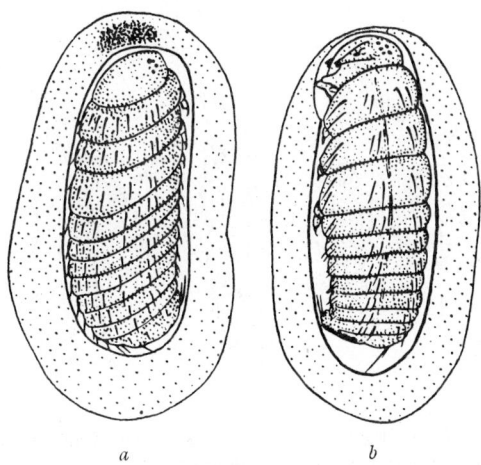

a   b

Fig. 49   *Donacia semicuprea* Panz.   (*a*) Fully developed larva before hatching; mass of bacteria in the jelly of the egg above the head.   (*b*) The hatching larva has consumed the bacteria.   From Stammer.

aggregation of cells associated with a rudiment of the midgut.   This cell mass, which makes an early appearance at the blind end of the stomadaeum, is later cut in two, and then each half again in two, a smaller roundish body and a larger oval one.   In these four groups of cells a hollow space then results which finally breaks through to the midgut.   This produces the reception chambers observable in hatching larvae, in which the sacs still clearly exhibit the variations in size (Fig. 50*a,c*).

Hardly have the symbionts reached the intestine when they become visible in the cells of its lining, and here are rapidly reproduced.   In one case Stammer found that they were amply infected 45 minutes after hatching (Fig. 50*b*).

It appears likely that a causal relationship exists between the unusual origin of these saclets and their tendency to be expelled with the rest of the epithelium when the new gut is forming.

The symbionts undergo remarkable changes in form. In the Malpighian tubules they are broad compact rodlets with rounded ends, 3–4 μ in length and 0.75–1 μ in width, which give rise to the much smaller, short, oval to spherical infection stages. A few hours after transmission to the young larvae, they begin to grow in length, measuring 15 μ or more in older larvae. Staining uniformly at the beginning, they fill up, as the

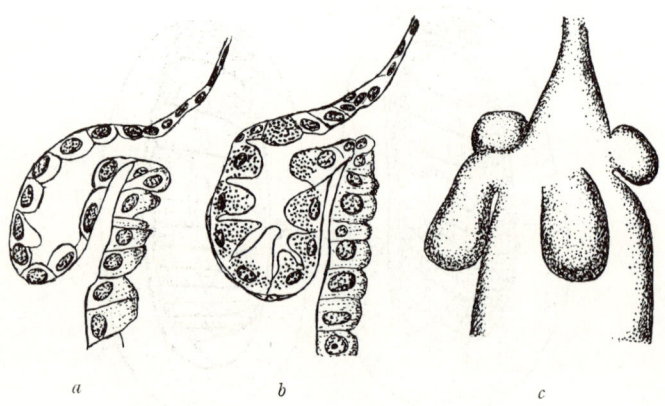

a   *b*   *c*

Fig. 50   *Donacia semicuprea* Panz.   (*a*) Sterile midgut evagination of a larva just before hatching; (*b*) the evagination is infected soon after hatching; (*c*) the unequally large blind sacs in a newly hatched larva. From Stammer.

pupation approaches, with many little granules which finally are fused into larger, deeply staining sections, giving a chain-like appearance to the filaments, which by now have grown even longer. It is these chain-like forms which then disintegrate. Such chain forms are also found in the Malpighian tubules of the larvae. They are rather long at first and resemble those in the gut appendages of younger larvae. In the pupae they disintegrate into the smaller stages with which we began the description of this strictly regulated change in form (Plate I,*a–f*).

CANTHARIDS

Only a single symbiont bearer has been discovered among the cantharids. When Holmgren (1902) described those strange, clavate, deformed

Malpighian tubules of the apionids which were later revealed as habitats of symbiotic bacteria, he simultaneously reported on the equally strange appendicular organs which in *Dasytes niger* L. have their outlets where mid- and hindgut meet.   Once more he had chanced upon symbiotic devices without realizing it.   Much later Stammer investigated the imagines of *Dasytes* (1933) and was able to interpret Holmgren's observation correctly.   Stammer's two reports, presenting no figures, are unfortunately so brief that further research is necessary.   Only a brief description can therefore be given here.

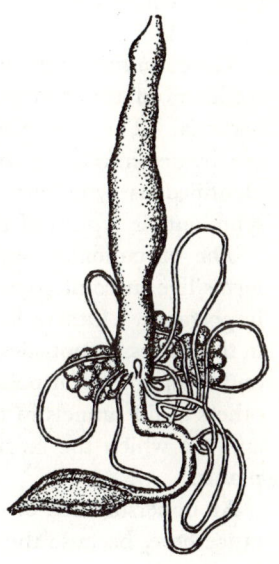

As imagines the *Dasytes* species live on blossoms and feed on their pollen.   The larvae, on the other hand, are said to be predatory and are found inside wood of various types and under bark.

The six Malpighian tubules of *Dasytes niger* are not sealed off but rejoin the hindgut in close succession in two groups of three, and where they first enter the gut they are inserted alternately with three pairs of ducts, which run side by side, and are connected with three morulose cell clumps.   Holmgren interpreted them as aggregations of oeno-cytes.   He had observed them only occasion-ally in connection with the club-shaped

Fig. 51 *Dasytes niger* L. Intestinal tract of an imago with mulberry-shaped my-cetomes opening into it. From Holmgren.

ducts, which had not escaped his notice (three of them are depicted in his illustration) (Fig. 51).

These supposed oenocytes are in reality thickly populated by small oval microorganisms measuring about 5 μ which are transmitted from here to the gut by way of the short ducts.   From another of Holmgren's illustra-tions we conclude that the duct lining is still free of symbionts in the vicinity of the gut, but that its remaining extent, composed of cells rapidly increasing in size, is infected.   Slender mycetocytes apparently belonging to the mulberry-shaped cell clumps extend into the blind end of the passages.   A tunica envelops the strange organs.

Since they are lacking in all males, it is quite probable that they are transmission devices of a special kind which enable the symbionts to travel from the hindgut to the surface of the egg.   Unfortunately it has not yet been possible to study the larvae in which symbionts may be at a different site.[1]

CURCULIONIDS

The curculionids involve a wide field of study, for a great majority of the members of this complex family of Coleoptera live symbiotically with bacteria. The numerous genera—1006 had been discovered by 1871!—are divided into eighty-two tribes, twenty-four of which have already been identified as symbiont bearers. Great variation in localization of the symbionts is typical of the group.

The curculionids feed exclusively on vegetable matter. Numerous forms live in coniferous and deciduous woods and are pests of economic importance. Others live in or on a wide variety of herbaceous plants, in the leaves, receptacles, stems, or roots, eat out the content of seeds, gnaw on bulbs, or live on pollen. Many cause galls or thickening of the stem; others build tunnels in the leaves or ingeniously roll up leaves into packages on which the larvae feed. Usually they specialize in certain food plants.

An understanding of the biology of individual forms is of special importance here, because the symbiotic organs are frequently restricted to the larval states or at least are differently constituted in these than in the imagines.

Nowhere else aside from the cicadas do so many symbiotic sites exist as in this insect family. They have been found in the epithelium of certain evaginations of the midgut, in massive mycetomes encircling its initial section or lodged against it at one side, in cells interspersed throughout its area, in minute syncytia pervading the fatty tissue, in a cell mass similar to the fat bodies and occupying a large part of the abdomen, in cells of the visceral mesoblast, in cells of the Malpighian tubules which are otherwise unaltered, and in strange club-shaped organs which represent transformed Malpighian tubules.

Our information on all these devices rests almost exclusively on my own studies (1927, 1928, 1930, 1933) and on supplementary studies by my students: Scheinert (1933), Glumb (1933), and Nolte (1937). To the symbiosis of *Calandra* these workers also dedicated themselves: Pierantoni

---

[1] In 1952 Stammer withdrew his description of *Dasytes* symbiosis, being then inclined, as Holmgren was once, to interpret the questionable inclusions as secretion granules, but in our opinion this interpretation should not be regarded as conclusive. At any rate, the device has not been found elsewhere and the fact that it is seen only in females is certainly a cogent argument that it is used for transmission of symbionts.

(1927, 1928), Tarsia in Curia (1933), Mansour (1930, 1934, 1935); Mansour and Mansour-Beck (1934, 1939), and Musgrave and his co-workers (Musgrave and Miller, 1951, 1953, 1956, 1958; Musgrave, Monro, and Upitis, 1961; Musgrave and Homan, 1962; Musgrave, Grinyer, and Homan, 1963; Crawford, McDermott, and Musgrave, 1960); while these workers were concerned only with *Calandra* (*Sitophilus*) *granaria*

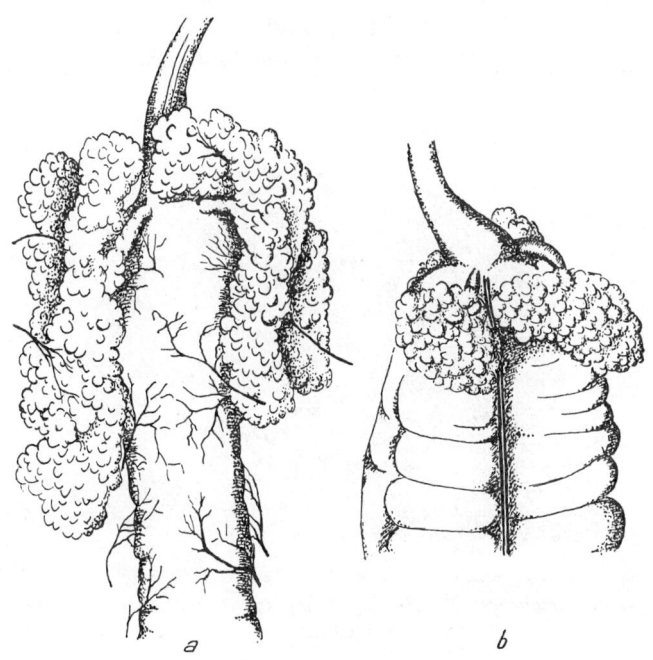

Fig. 52   (*a*) *Lixus paraplecticus* L.   (*b*) *Larinus planus* Fbr.   Larval bacterial organs at the beginning of the midgut.   14×.   From Buchner.

and *oryzae*, Chararas (1956) also investigated the Cossoninae (*Rhyncolus* and *Caulotrupis*) and found in them the same calandrinae-type of symbiosis. In the embryological studies of Murray and Tiegs (1935, 1938) the symbiosis is also considered.   A particular place among all the publications, mostly short ones, dedicated to *Calandra* is taken by the detailed morphological and experimental studies of Schneider (1954, 1956).   Finally, there are two more publications from which it can be seen that many a new symbiosis type is still hidden among the Curculionidae which deserve a new, more penetrating study.   Hänsel (1960) and Grinbergs (1961) have described, unfortunately in too little detail, a type of symbiosis never found elsewhere, in *Empleurodes dentipes* from Chile.

Let us describe first the bacterial symbiosis of a subfamily, the cleonines, which is fundamentally different from that in all other snout beetles, and which suggests that of anobiids, cerambycids, and lagriids in localization and method of transmission.   In the larvae of all representatives which tunnel in the stems of herbaceous growths, be they cleonines or lixines, all have at the initial section of the midgut four variously shaped evaginations which are usually connected with the midgut by only a narrow passage. Such evaginations may be compact protuberant cushions or long acini-

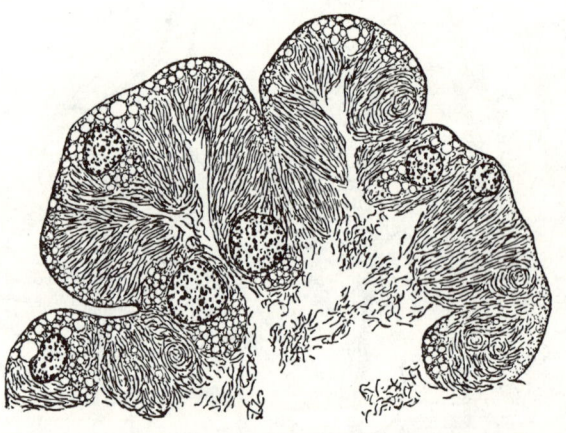

Fig. 53 *Lixus paraplecticus* L. Section from the larval bacterial organ. From Buchner.

form formations accompanying the gut for a distance (Fig. 52).   In their zone the gut epithelium is extensively modified; filamentous bacteria usually arranged perpendicularly to the cell basis fill large cells with greatly enlarged nuclei and also permeate the lumen of these evaginations (Fig. 53).

In view of such a localization of symbionts it is hardly surprising to discover transmission organs which ensure lubrication of the eggshell with bacteria, and yet formerly these organs had always been overlooked in this subfamily.

The extremely compact ovipositor of the cleonids is shielded in the state of rest by the seventh and eighth segments.   In the usual place, that is to say, in the area of the intersegmental membrane, which extends to the ovipositor from the eighth segment with its broad short spiculum ventrale, are situated bacteria-filled, paired formations, the distribution and variants of which were studied in dried material also. In the fifty-seven varieties studied under my supervision, they were always present, and individually they were quite varied in form (unpublished observations of Glumb, 1933; see Buchner, 1933).

Cleonus piger Scop., for example, has two large club-shaped organs attached to the bursa copulatrix by a short bundle of muscles originating at the free end. A well-developed longitudinal musculature extends over the whole surface of each organ, and a sphincter encircles the opening. The epithelium of the sacs is laid in numerous regularly arranged folds dividing the interior into chambers, with brushes to prevent excessive discharge of bacteria. Typical glandular cells, of the kind found in corresponding organs of other Coleoptera, are of course lacking, but the symbionts are nevertheless embedded in a secretion which undoubtedly originates from the plasma-rich epithelial cells (Fig. 54).

Since there are no other symbiont containers housed within the ovipositor and comparable to the vaginal pouches, the symbionts are apparently carried directly to the egg surface by the organs which indeed have been developed in this species to function as regular jets and furthermore are located near the mouth of the vagina. In other cleonines these intersegmental organs may be much longer; among the lixines there are species in which they are identical with those in *Cleonus*, and other species in which they remain quite small and in which the girdling is either only faintly suggested or lacking entirely. *Lixus nitidicollis* Fbr., on the other hand, has very long forked tubes with folded lining, and *Larinus sturnus* Schaller departs from the pattern completely by developing on each side an entire cluster of slender pipelets. Differentiations occur also in the setaceous covering of the invaginated epithelium. Broad-lobed and palmate bristles are occasionally found instead of the single arrangement found in *Cleonis piger*, the palmate bristles often having an imbricate arrangement. In *Larinus sturnus* the bristles fail to take form.

As among the cerambycids, these transmission organs represent the only site in which symbionts are found in the imagines. There has been no research as yet to tell us how the intestinal evaginations regress during metamorphosis, how the male animals become completely symbiont-free whereas the jets are filled in the females, but the processes can easily be reconstructed on the pattern of the cerambycids.

Scheinert actually witnessed symbionts which were embedded in mucilaginous secretion on the egg surface of a *Lixus* species. The four evaginations of the midgut are formed before hatching takes place. The young larva on leaving the eggshell consumes about a third of the same,

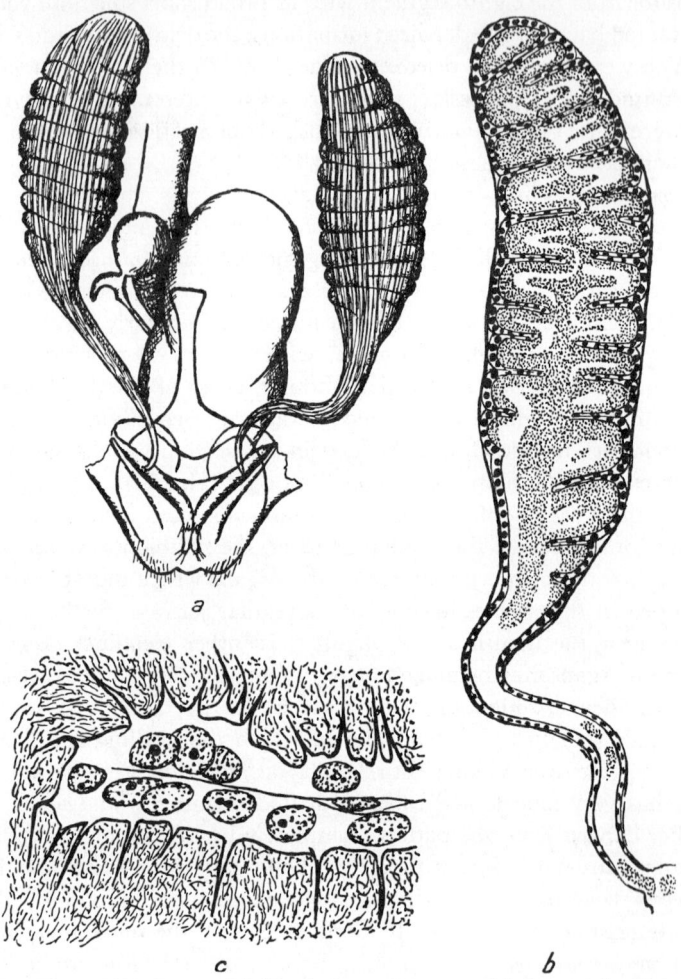

Fig. 54 *Cleonus piger* Scop. (*a*) Egg-laying apparatus with bacterial syringes; (*b*) syringe in longitudinal section; (*c*) wall of a heavily filled syringe with retention hairs. From Buchner.

and soon thereafter bacteria clumps can be found on the wall of the foregut. From there they are easily transmitted to the previously sterile ceca and later infect also their epithelia (Fig. 55).

Another common type of localization is represented by compact organs in the area between fore- and midgut. These organs, which have a superficial similarity to those of the cleonids, have a variety of forms. There are partial mycetomes, completely separate from one another, encircling the gut in fours or eights.   They were found in the larvae of the *Gymnetron* species, which feed on *Linaria* seed capsules or produce fruit galls on *Veronica anagallis;* in *Miarus* species from fruit nodes of *Campanula;* in *Sibinia* species from seed capsules of *Silene;* in *Tychius* species which live in legumes; and in *Bagous* which specializes on *Stratiotes* (Fig. 56*b*).   When the partial mycetomes increase in number, they finally border so closely upon one another that an uninterrupted circle of humps arises.   This is the case

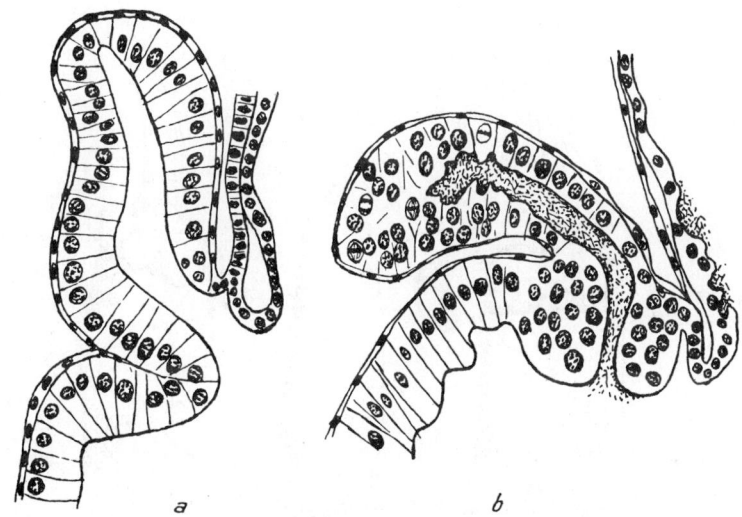

Fig. 55   *Lixus* sp.   (*a*) Gut evagination of a larva before hatching; (*b*) after the shell has been consumed, the bacteria enter it.   200×.   From Scheinert.

with the *Otiorrhynchus* larvae which are all found on the roots of trees and bushes.

The "*Hylobius* type" represents a further development of "*Gymnetron* type" just described.   Externally the mycetomes of the *Hylobius* larvae which tunnel in the splint wood of conifers, and more rarely in deciduous trees, appear to be similar (Fig. 56*a*); yet cross sections show that the individual parts are fused with one another and form a single ring-shaped mycetome.   This applies to the mycetomes of *Molytes germanus* larvae, which live on *Petasites;* to those of *Pissodes* and *Magdalis* species, which are pests of conifers and deciduous trees; and to those of *Cryptorrhynchus lapathi* L., a pest in willows and alders (Fig. 56*c*).

By comparing larvae of different ages it can be shown that even in such states a "*Gymnetron* phase" always precedes and that the partial mycetomes are fused with one another only secondarily.

The method by which these organs are fastened to the gut, based as we shall see on their embryological development, is characteristic in all these cases.   Slightly flattened, the organs rest as though in a socket, on a ring-shaped fold, still part of the ectoderm and therefore lined with chitin, which is formed by the foregut as it joins the midgut (Fig. 57*a*).

Brachycerus represents a third type.   It was a piece of rare good luck that I was able to investigate a larva of this large African stump beetle just as it

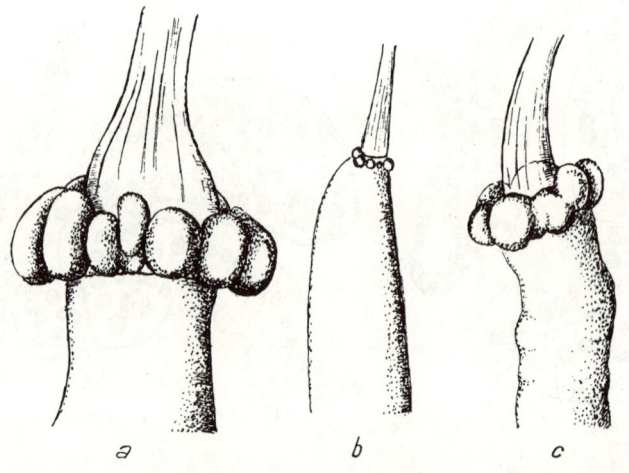

a   b   c

Fig. 56   (*a*) *Hylobius abietis* L.   (*b*) *Sibinia pellucens* Scop.   (*c*) *Cryptorrhynchus lapathi* L.   Larval mycetome between foregut and midgut.   From Buchner.

was about to hatch from the egg.   Already the size of a full-grown *Hylobius* larva, it had a mycetome collar consisting of numerous larger or smaller lobelets situated at the transition of fore- and midgut and only slightly fused with the base.

At this point we must consider *Empleurodes dentripes* Boheman, since the localization of the symbionts follows in part the *Hylobius* type.   Just as in that insect, there lies at the beginning of the midgut a ring of unfused diverticula, whose epithelium is inhabited by long, thin, thread-like bacteria.   Here, too, these mycetocytes, which reach a considerable size, are transported caudad during metamorphosis.   Quite surprisingly, there congregate here on each side, near the opening of the vasa malpighi,

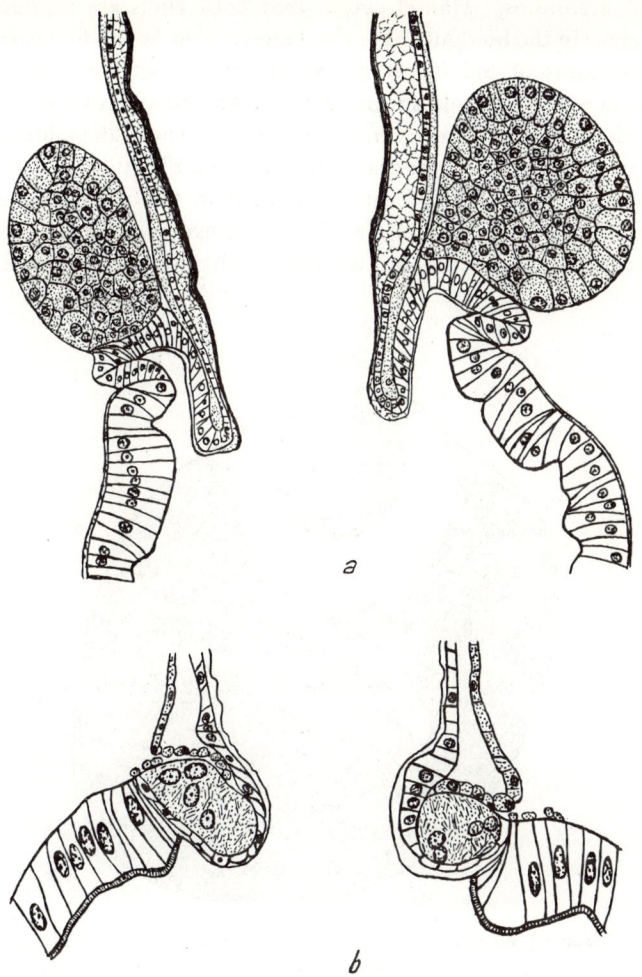

Fig. 57 (*a*) *Cryptorrhynchus lapathi* L. The mycetome rests on a ring fold of the larval stomodaeum. (*b*) *Coryssomerus capucinus* Beck. The syncytial mycetome has sunk in. From Buchner.

three diverticula, which arise close to each other and are filled with organisms of another species, that are much thicker and shorter. Grinbergs presents good photographs of them and the two kinds of quarters housing them. Hänsel has drawn them as rods, which rest upon the epithelial cells and resemble a brush border. According to him the symbionts are very pleomorphic, but both authors are convinced that it is a case of two different organisms.   Hänsel claims that both kinds are represented in these depots in the hindgut.   He also believes that he has found signs that the inhabitants of the larval gut diverticula infect the ovaries.   The diverticula of the hindgut are present in both sexes, but they are said to be considerably less colonized in the male.   In view of these findings, this is apparently the only known case among the snout beetles of a disymbiosis, whereby the guest obtained later is housed in this unusual place.   Its location might suggest at first glance a transmission of the inhabitants by way of smearing of the eggs, but we shall see that also in the apionids there

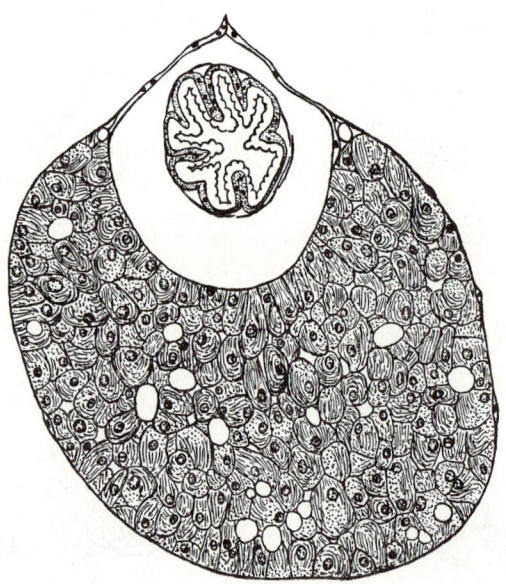

Fig. 58   *Calandra granaria* L.   The larval mycetome is ventral to the foregut.   40×.
From Buchner.

exists a close relationship between the Malpighian tubules and ovarial infection.   When they are treated we shall return again to *Empleurodes*. A more thorough morphological and bacteriological analysis of this object would in any case be very rewarding.

The tropical palm stump beetle *Rhynchophorus ferrugineus* Oliv., one of the calandrines, has a loosely fitting ruff of mycetocytes, differently shaped, it is true, in the usual location.   In all other respects the organ departs from the pattern set by the cases already mentioned.   In the other cases, as will be demonstrated later, the larval organs are completely regressed in the course of metamorphosis, whereas they pass unchanged into the imago in *Rhynchophorus*.

Without exception the mycetomes already reviewed are easily recognizable during preparation of the larvae as spherical formations, rings, or irregular garland-like masses. This is not true of the *Coryssomerus* type, where a closed mycetome ring sinks so completely into the fold formed at one end of the stomadaeum that its existence is revealed only in cross section. With *Coryssomerus capucinus* Beck, the larvae of which gnaw at roots of composites, there is an additional peculiarity in that this ring consists of one or several syncytia, whereas the corresponding organs in other species are built up of uninucleate cells or cells containing at the very most two to three nuclei (Fig. 57*b*). The larvae of the baridines (*Baris* species), which are related to the coryssomerines, have nests consisting of only a few mycetocytes in the same location.

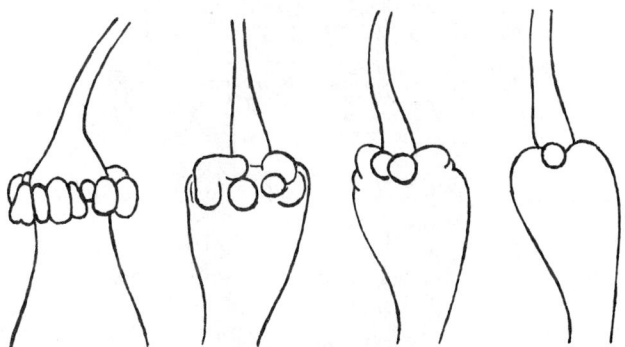

Fig. 59   *Hylobius abietis* L.   Fusion of the mycetome during development of the adult gut.   From Buchner.

The *Calandra* species (*Calandra granaria* L. and *C. oryzae* L.), the larvae and imagines of which infest all types of cereals, as is well known, have been studied by a succession of authors, most recently by Schneider (1956). These species accommodate their symbiotic bacteria in a large unpaired mycetome situated ventrally from the transition point of fore- and midgut. Composed of numerous cells of one nucleus, the organ is spanned by a delicate membrane which runs backward to join a strand which suspends it over the gut (Fig. 58).

Since the mycetomes of all representatives of *Gymnetron*, *Hylobius*, *Brachycerus*, *Coryssomerus*, and *Calandra* types suffer essentially the same fate during metamorphosis, and since nothing similar is encountered among the types still to be described, we shall take up the subject at this point. The mycetomes in all these types disappear at the time of pupation, yet this is not traceable to destruction of the mycetocytes and expulsion of their population, as with the cleonids, but merely to a shifting of the elements composing them.   Even superficial observation of the gut in pupation-ripe *Hylobius* larvae will show that the sharply-defined humps gradually disappear, and that the succeeding section of the gut is clavately thickened instead (Fig. 59).   As may be clearly seen in cross section, the new formation of the gut epithelium has advanced quite far by this time, and the

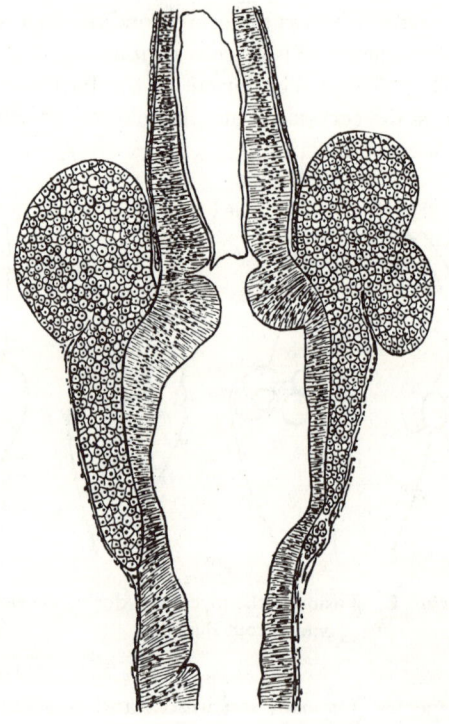

Fig. 60   *Hylobius abietis* L.   The larval mycetome begins to glide posteriorly between the young adult gut epithelium and the muscularis.   From Buchner.

mycetocytes, which are now capable of slipping back and forth against one another, at the same time increasingly glide backward between the imaginal epithelium and its musculature (Fig. 60).   Finally, they invade the entire front, ventricose section of the mid-gut, which is clearly composed of that ectodermal ring-fold of the larval gut, and after metamorphosis are found in small nests enclosing the crypt cells (Fig. 61*a*).

With the other curculionids in this category the general pattern of the process is modified in various ways.   When isolated larval mycetomes are present, they first glide backward as rather dense cell clumps which cause the musculature of the new gut to protrude, until gradually the individual mycetocytes are distributed uniformly over the entire initial section of the midgut.   Sometimes the mycetocytes grow very substantially at this time without corresponding increase in the reproduction of the symbionts, so that we gain a clear picture of what the colonization is like (Fig. 61*b*). The folding of the imaginal intestinal wall may increase until the folds

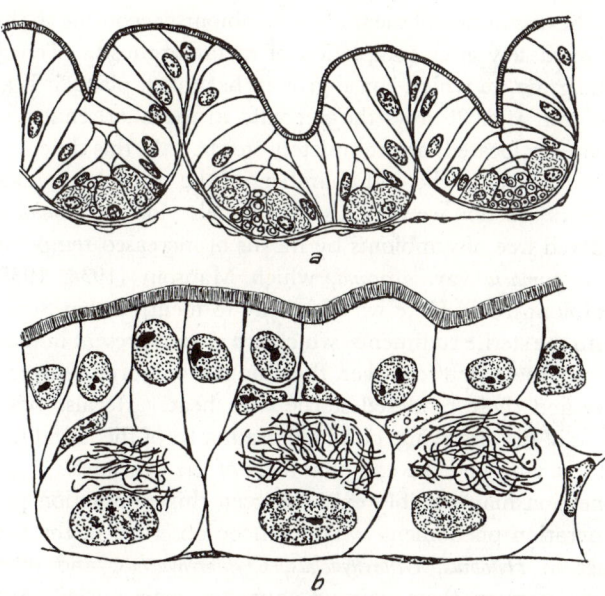

Fig. 61   (*a*) *Hylobius abietis* L.   Midgut epithelium of a young imago with mycetocytes and crypt cells.   200×.   (*b*) *Gymnetron villosulum* Gyll.   Midgut epithelium of a pupa with mycetocytes.   600×.   From Buchner.

have the form of delicate villi.   In such cases the crypt cells and the mycetocytes are always situated at the base.

Calandra develops such villi to an extreme degree.   Here also the large unpaired mycetome slips between the imaginal epithelium and its musculature in connection with the new formation of the intestine, then sends some of its cells into the dorsal region where they finally fill a very large portion of the villi; the latter, about twenty in number, accompany the midgut (Fig. 361).

Rhynchophorus ferrugineus alone has no change of position, as already mentioned.

Oddly, this involved shifting of mycetocytes by no means leads to a state lasting throughout the rest of the imaginal life of the host animals, but merely represents a temporary device whereby the hosts can get rid of most of their superfluous guests.   With older imagines of both sexes one often has to search a long time before finding one or another mycetocyte in the intestinal wall.   This is true of *Hylobius*, *Molytes*, *Miarius*, *Otiorrhynchus*, *Brachycerus*, etc.

The *Calandra* species are no exception in this respect.   Older observations were extended in many ways by Schneider, whose work is well illustrated.   The dissolution of the midgut symbionts, according to this author, sets in immediately after completion of metamorphosis.   Ten days after leaving the grain, in which larval growth has taken place, it is already in full swing, and after 20 days the gut villi, in the blind end of which the mycetocytes are located, are completely sterile.   At this time a very great increase of bacteria in the end chambers of the ovarioles is taking place. In the General Section we shall discuss the behavior of *Calandra* which has been rendered free of symbionts by means of increased temperature, and of *Calandra granaria* var. *africana*, which Mansour (1934, 1935) always found symbiont-free.   Here we wish only to mention that Schneider has drawn both the sterile rudiments, which are always present in the Egyptian variety, and the extension over the midgut of normal mycetocytes of *C. granaria* and those rendered sterile with heat.   He also has excellent illustrations of the emptying, obtained gradually, of the larval mycetomes, the adult gut villi, and the end chambers of the ovarioles.

A connection may possibly exist between this elimination process and the degeneration phenomena which can be observed in the older larval mycetomes in *Hylobius*, *Otiorrhynchus*, *Cryptorrhynchus*, and others.   The mycetocytes vary in their staining capacity; their nuclei appear to be impaired; the filamentous bacteria roll up like snails and finally form little clumps, or by distending at intervals become moniliform, and apparently fall into fragments later.   Especially the cells at the periphery undergo such changes and are usually characterized by increased growth.

The localization of symbionts in mycetomes independent of the gut excludes the possibility of superficial smearing of eggs in this second category of snout beetles.   For the first time we observe a transmission brought about by infection of the oocytes, a method frequently found in other groups.   The special form in which the infection is carried out is rather rare.   At an unusually early time, as we shall see, a basis is laid for the infection of the nutrient cells which later fuse into a syncytial mass at the end of each ovariole.   The bacteria, apparently finding a favorable culture medium here, reproduce rapidly.   With the secretion that passes from the syncytial mass into the growing oocytes, the symbionts also glide into them and are sometimes united there with individual bacteria which penetrate directly into the young oocytes. Mansour gave the first description of this method of transmission in *Calandra*, and later Scheinert and I found it again in a number of other forms. It provides a method by which even pubescent females which have completely disintegrated

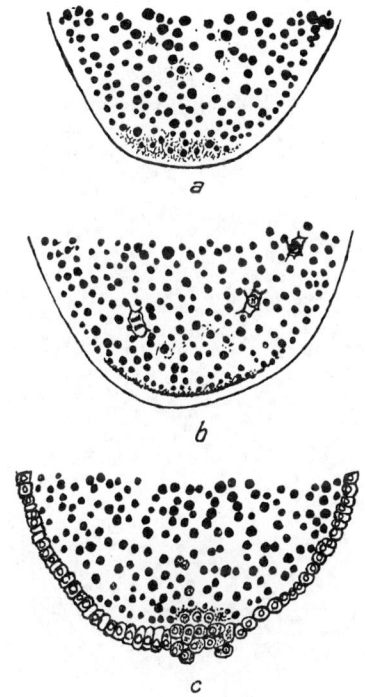

Fig. 62   *Hylobius abietis* L.   (*a*) A newly laid egg with a bacterial mass at the posterior end.   (*b*) Cleavage stage; the bacteria form a thin layer on the germinal membrane of the blastoderm.   (*c*) Development of the primordial germ cells and the uptake of bacteria into them.   67×.   From Scheinert.

symbiotic organs may be tested for presence of symbiosis with smears from their ovariole end chambers.

The study of the embryonic development reveals how the aforementioned early infection of the nutrient cells takes place and how the mycetomes are formed (Scheinert, 1933). Up to the present *Hylobius*, *Liparus*, and *Calandra* have been used as test objects. In the freshly laid eggs the symbionts are distributed over the whole yolk. Whereas the filaments in the mycetomes of *Hylobius* measure up to 30 μ, the short rodlets here measure 1.5–3 μ in length.   Again and again there is a denser accumulation in the vicinity of the anterior pole and a special bacterial zone, shaped like a watchglass, at the posterior pole.   The segmentation nuclei which dip into the accumulation at the posterior pole become the primitive sex cells and without distinction between sexes form a symbiont-containing cell group which stands out from the rest of the blastoderm (Fig. 62).

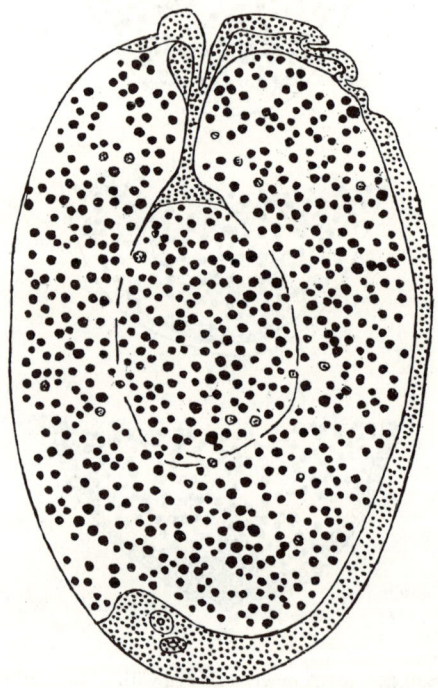

Fig. 63   *Hylobius abietis* L.   Cells of the stomodaeum migrate to the plasma net concentrating the symbionts.   75×.   From Scheinert.

After the germ strip forms and its segmentation begins, a strange process takes place in the egg plasma.   At the free end of the stomodaeum rudiment a plasma radiation occurs; then at some distance from the egg surface a sac-like net suspended from the stomodaeum makes its appearance. The net is connected at all points with the bacteria-containing protoplasm, and now, as it draws together more and more, it concentrates the symbionts in its meshes to such an extent that soon bacteria are no longer detected, either inside or outside the net (Fig. 63).   It is impossible to say to what extent circulation streams or the vicinity of the ever-present yolk nuclei take part in this unusual process.

With increasing constriction of the net, elements of the stomodaeum, thus ectoderm cells, glide toward the net on the connecting plasma rope,

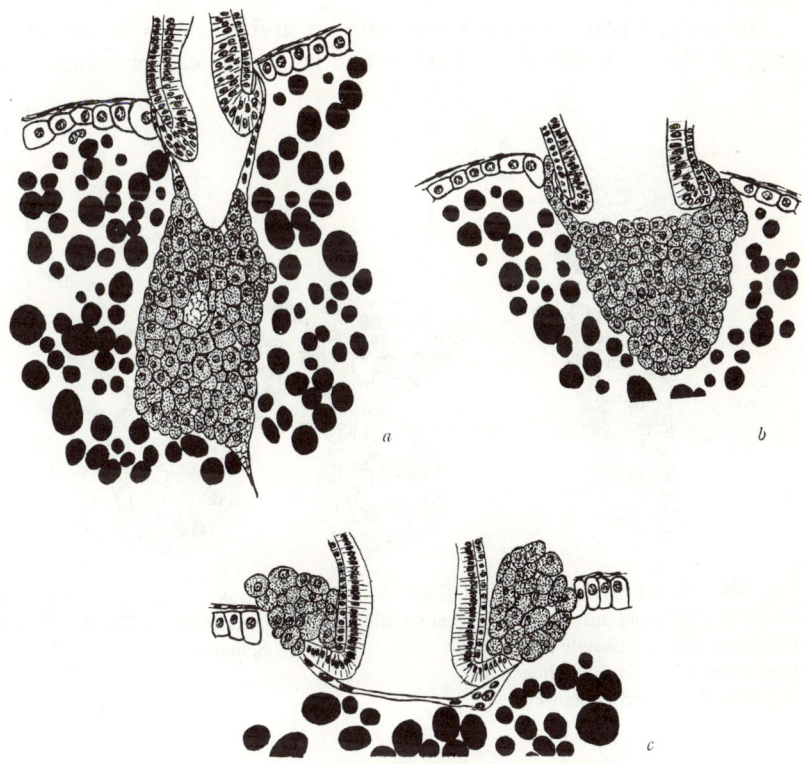

Fig. 64   *Hylobius abietis* L.   (*a*) Mycetome rudiment is attached to the end of the stomodaeum; (*b*) its cells begin to leave the gut lumen; (*c*) the passage through is almost completed.   225×.   From Scheinert.

form a cap upon the latter, then penetrate deeper and deeper into the net which is drawing together more and more, and, now forming clear cell limits, finally assimilate its bacteria.   Finally an embryonic mycetome hangs in the yolk, fastened at the transition of fore- and midgut, and in the next stage, owing to the loose connection of the two epithelia here, it slips between the epithelia into the place assigned to it in the larval organization (Fig. 64).

The mycetomes of the *Hylobius* type are thus ectodermal organs, found nowhere else in this group.   That they are ectodermal is shown not only by their genesis but also by a close topographical relationship to the foregut which is preserved in the larva.   Evidence of a common origin is provided by the reactivation of the cell material in connection with metamorphosis, when it combines intimately with the new (and now ectodermal) midgut.

The male and female genital rudiments are at first infected in the same way, as already mentioned.   Even before both sexes can be recognized

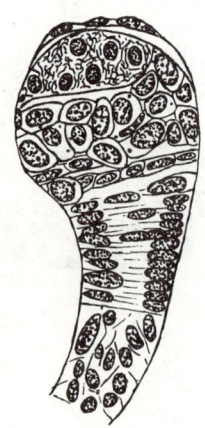 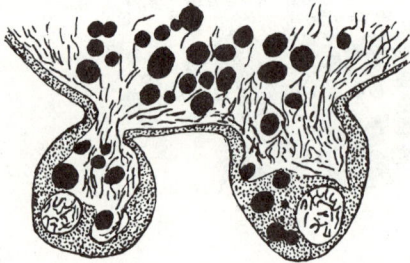

Fig. 65 *Sibinia pellucens* Scop.   Ovariole of a larva ready to pupate; the nurse cells are heavily infected, the ovocytes sparsely infected. 370×.   From Scheinert.

Fig. 66 *Calandra granaria* L.   The bacteria enter the primordial germ cells.   750×.   From Scheinert.

morphologically, it is possible to distinguish between young larvae with clearly infected, primitive sexual cells and those with bacteria in the process of disappearing.   Finally, the bacteria are disintegrated completely in males.   When the quickly growing rudiment of the nutrient cell and the actual genital rudiment, which at first is retarded in its development, are separated in the young female gonads, considerable difference in the behavior of the symbionts can be observed.   In the rudiment of the nutrient cell, rapid multiplication of the symbionts occurs; in the genital rudiment, multiplication ceases (Fig. 65).   This explains why isolated bacteria are already present in the oocytes when the bacteria arrive there from the nutrient cells.

Only slight variations in these embryological and postembryological processes were observed in the various genera available for comparison. In *Calandra* the primitive sex cells together with the bacteria and yolk masses are conspicuously constricted off, and there is no network to concentrate the symbionts (Fig. 66). Instead, the rudiment of the stomodaeum sends out irregular plasma processes between the yolk masses, the nuclei of which swarm out to assimilate the bacteria which were united earlier in the central region. The later clumping and transmission from the midgut rudiment, which here of course occur only ventrally, then take place approximately as in *Hylobius*.

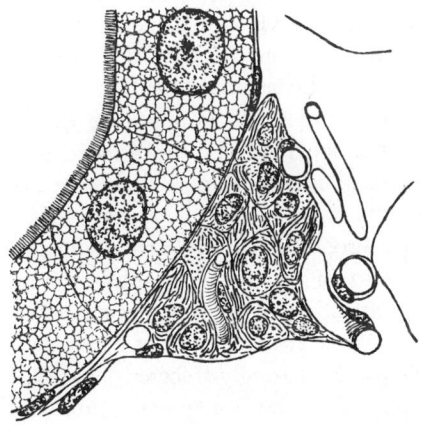

Fig. 67   *Balanius glandium* Mrsk.   Mycetome of a fairly young larva situated between the gut loops.   400✕.   From Buchner.

In some snout beetles extremely minute and therefore easily overlooked infected cell nests were found in the posterior area of the midgut. We shall designate this method of localization as the *Ceutorrhynchus* type. In *Balaninus glandium*, larvae from acorns and, in *Balaninus nucorum* L., larvae from hazel nuts, a tiny nest clearly settled by filamentous bacteria and provided with tracheae is loosely attached to a posterior convolution of the midgut (Fig. 67). With *Smicronyx jungermanniae* Reich., the larvae of which produce galls in *Cuscuta*, I found a little syncytium in the angle of a midgut loop. With *Ceutorrhynchus punctiger* larvae from *Taraxacum* receptacles and with other species of *Ceutorrhynchus* and with *Phytonomus* as well, a rather large collection of small mycetocytes is situated between the convolutions of the midgut in the posterior half of the larval body. Apparently this type of localization, unobtrusive and observable only on a series of slides, is even more widely distributed, and transitions may possibly exist between it and certain cell nests which I found in quantity among the fat bodies of *Dorytomus* and *Elleschus* larvae.

Those primitive-appearing mycetomes are lacking also in the imagines of *Balaninus*, *Smicronyx*, and *Ceutorrhynchus*. Indeed, in these genera they are decomposed rather early. In older larvae mere traces of them are to be found at the very most. The details of this process of dissolution, and the whole symbiosis type in general, are in need of more thorough analysis. With *Smicronyx* the syncytium at first splits up into irregular tatters to which the blood cell nests which assimilate the bacteria are attached. With *Ceutorrhynchus punctiger* Gyll. the structure of the above-mentioned cell accumulation gives way completely, and great numbers of little leucocytes appear between the original mycetocytes, which more and more recede into the background and by pupation time have disappeared entirely. The case is much the same in *Phytonomus*. Again and again, one can observe the close relationship of these mycetomes of the *Ceutorrhynchus* type to the fatty tissue and to blood cells. In *Cionus*, elements of the blood are infected also, although there is no concentration of them and hence no resemblance to an organ. Transmission in all these test objects again takes place by infection of the nutrient chambers. Figure 68, for example, shows an ovariole from a pupa of *Smicronyx*, in which the end filament is followed by the strongly infected few-celled rudiment of the nutrient cells and later by the germ strip and a number of follicles already containing cluster stages.

The profusion of symbiotic devices possible among the curculionids has not been exhausted by our presentation of this *Ceutorrhynchus* type. Other devices, differently constituted and quite surprising, are used by the apionines, those little snout beetles of which more than 1000 species have been described. In research with twenty species I differentiated three different methods of localization, and this number was later increased by Nolte's research based on more extensive material.

Frequently the Malpighian tubules are used in symbiosis. Whereas all other snout beetles have six Malpighian tubules, only four normally developed tubules are found in a series of species of *Oxystoma*, *Erythrapion*, *Protapion*, *Perapion*, etc. In place of the fifth and sixth, the larvae and imagines have two strange club-shaped formations attached by slender pedicles (Fig. 69; see also Fig. 363). Their openings at the transition of mid- and hindgut corresponds precisely to those of the two missing tubules, and in fact these formations represent tubules withdrawn from their original function and radically transformed into symbiont stations. The

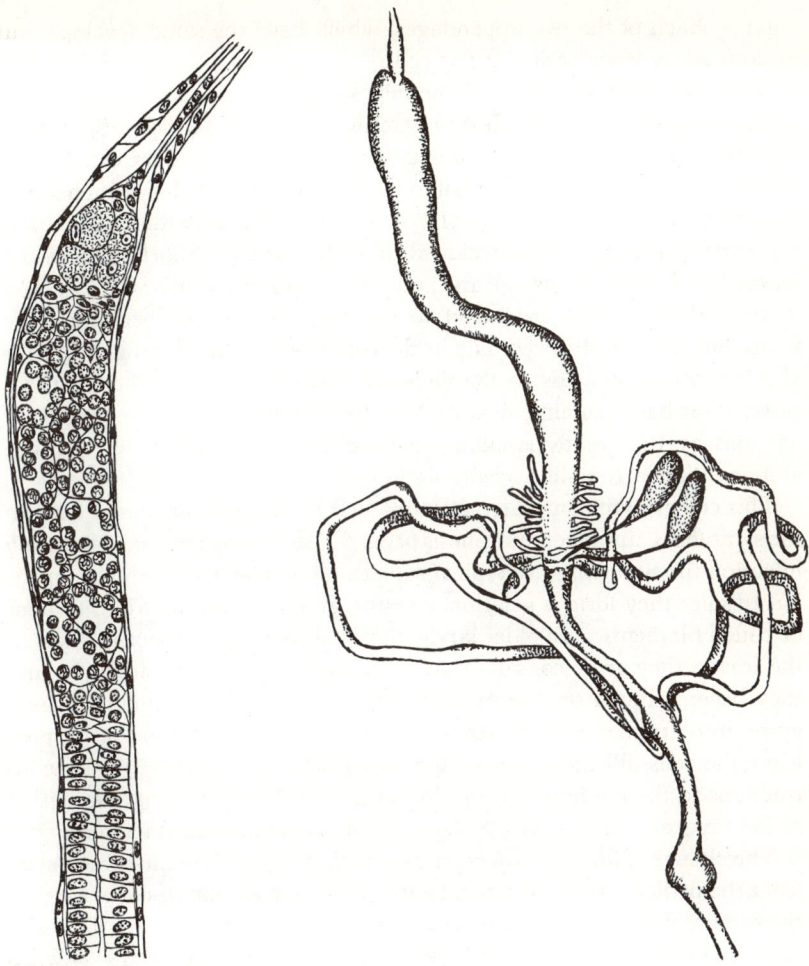

Fig. 68 *Smicronyx jungermanniae* Reich. Ovariole from a pupa, the terminal few celled rudiment of the nurse cells heavily infected. 600×. From Buchner.

Fig. 69 *Erythrapion miniatum* Germ. Adult intestinal tract with two Malpighian vessels transformed into club-shaped sites for symbionts. From Buchner.

pedicle has lost its typical structure and now consists of a few flat cells without a border of rodlets, although the lumen is still preserved, whereas the lumen in the club-like portion has disappeared completely. The pedicle consists essentially of bacteria-filled cells of various sizes, between which little sterile elements are inserted in the longitudinal axis of the organ.   Each of the two appendages, which have the same development in both sexes, has a casing of flat cells.

Nolte and Scheinert studied the genesis of the transformed tubules in *Erythrapion* and *Protapion*.   Both researchers found that the rudiments of six tubules are formed in normal manner as evaginations of the ectodermal proctodaeum, yet two of them fail to develop, remaining short and acquiring a slight clavate thickening.   In the middle of the yolk the symbionts at this time still form a ball enveloped in a fine casing.   Shortly before the larvae hatch, the ball disappears, and the two tubules which were checked in their development are found to be infected.   At the beginning the symbionts may be observed only in the cells close to the opening.   Gradually the colony extends to the elements located deeper, although those situated far back remain empty and are forced aside toward the longitudinal axis by the greatly swelling, infected sister cells and now form the aforementioned rope-like accumulation.

This cell growth continues, accompanied by strange changes of form in the symbionts, until metamorphosis occurs.   In the egg plasma and during infection the symbionts are typical rodlets.   Upon arrival in the Malpighian tubules they form a gelatinous casing and are elongated into sinuous or coiled filaments.   In older larvae the symbionts begin to form knots; in the imago they grow considerably, their casing expanding similarly, and they become much thicker; moniliform states arise which finally disintegrate into fragments together with the casing.   Spherical forms then leave the cells, fill the lumen of the outlet, and are advanced by its contractions to the hindgut.   Hand in hand with this decreasing population of the mycetocytes, a cavity arises anew in the club-shaped section.

The change of form advances parallel with the growth of the mycetocytes from the pedicle toward the blind end and affects all inmates of one cell in the same way.   Consequently, cross sections of the older organs disclose a rather confused situation.   It is puzzling that Mahdihassan (1947) without knowledge of the test object should conclude from Nolte's illustrations that leucocytes, secretion granules, and metabolism products of the symbiotic bacteria were joined here in a hypothetical cycle.

Transmission proceeds in the ordinary manner, although in this case there is nothing to stand in the way of lubrication of the egg surface. Infected nutrient cells send bacteria into the growing oocytes, and long before the Malpighian tubules are colonized the rudiments of the gonads are filled with symbionts.

Such reconverted Malpighian tubules are not found in the other apionines.   With *Aspidapion aeneum* larvae, which tunnel in the stems of mallow, the midgut has moderately enlarged cells thickly populated by bacteria from the visceral mesoblast.   The symbionts are rerouted again in the course of metamorphosis.   The infected cells are detached from the intestine and migrate to the Malpighian tubules, where they reattach themselves. The cells of the lining, being distributed at random over all the tubules, appropriate the bacteria, becoming thickly filled with them on all sides. Some are discharged into the lumen and from here, as one can see in living preparations, are expelled into the gut in balls. There is no striking change of form in the symbionts. Since once more they are situated first in the central regions of the yolk, they must necessarily migrate through the embryonic intestinal epithelium to reach the larval sites. This second type of apionine symbiosis has been found thus far in eight species.

These relationships of the symbionts to the Malpighian vessels, found in some of the apionids, reminds us naturally of the observations made on *Empleurodes*. Since in these apionids, despite the relationships to the hind-gut, no smearing of the eggs takes place, but rather the ovarioles are infected, it would not be wrong to assume such a method of transmission in *Empleurodes*. Even though the behavior of these apionids detracts some-what from the special position of *Empleurodes*, still the symbiosis of the latter remains for the time being unique among the snout beetles.

Omphalapion laevigatum Payk., which may be collected from camomile buds, represents the third type. In the larvae, pupae, and imagines of this species and in one species each of *Taeniapion* and *Diploapion*, a large mycetome-like formation, composed of many small cells thickly populated by bacteria, is situated dorsal and lateral to the gut, enveloping gonads and Malpighian tubules. This time it is clear that the mycetome is a deriva-tive of the fat body, for the convolutions of the latter are mirrored here. Likewise, eosinophilic granules and fat droplets appear simultaneously. Indeed, in older pupae such large fat spheres develop temporarily in the mycetocytes that the bacteria are confined to the plasma ropes that pass in between.

It is, of course, impossible to say whether we have discovered all the methods existing among the apionines for localizing symbionts. At any rate, no relationship exists between symbiosis type and subgenera. The phylogenetic conclusions which may be drawn from these three different methods of localization will be considered in Chapter 15, "Historical Problems." In view of the diversity of curculionids yet to be studied, there can be no doubt that many instances of symbiosis will be unearthed.

IPIDS

Until a few years ago there existed only one statement, brief but sub-stantiated, on the endosymbiosis of bacteria and ipids (Stammer, 1933). In three species each of *Eccoptogaster scolytus* Fabr., *Hylesinus fraxini* Panz., and *Hylastes ater* Payk. Stammer discovered heavy infection of the ovarial end chambers, each time by differently shaped bacteria, and thus, obviously, the same transmission method as in snout beetles.   In *Eccoptogaster* the symbionts were broad rodlets with rounded ends; in *Hylesinus* more delicate, but often slightly swollen, club-shaped filaments; and in *Hylastes* long, often spirally turned filaments.   In other ipids (*Blastophagus, Ips*) Stammer was unable to find the actual symbiotic site in the larval and imaginal bodies, and no infection of the egg tubes could be discovered.

From unpublished research by Tretzel, a student of Stammer, we learn that the ovaries are infected in three *Scolytus* species: (*S. multistriatus* Mrsh., *S. ratzeburgi* Janson, *S. mali* Bechst.), in *Hylurgops palliatus* Gyll., and in another *Hylastes* species (*H. angustatus* Herbst); and that the bacteria are not found in the ovaries of *Pityogenes, Drycoetus*, or in three *Ips* species.

Meanwhile I began to study endosymbiosis in ipids, this area of our field which has been so neglected, and I have reached the conclusion that it is both diverse and widespread.   My observations, of which as yet only those regarding *Coccotrypes* have been published, were made with numerous genera and mostly with living test objects.   Again and again I observed the early ovarial infection noted by Stammer and Tretzel. The localization of the symbionts in the rest of the ipid body may be quite varied.   In some cases scattered mycetocytes occur in limited areas of the fatty tissue, for example with *Ernopocerus fagi* Fabricius, *Scolytus carpini* Ratzeb., *Hylesinus crenatus* Fabricius; in *Crypturgus pusillus* Gyll. the symbionts are found in certain diverticula of the midgut; the *Coccotrypes* species make four of their six Malpighian tubules available; in *Stephanoderes hampei* Ferr., a pest in coffee beans, the rodlets are also located in the Malpighian tubules.   In other cases the symbionts populate cells externally attached to the midgut epithelium, for example, in the *Crypturgus* species, and apparently often populate the intestinal epithelium itself. In many other genera, where I saw symbionts in the ovary, the actual site still needs investigation, for example, in *Leperisinus, Phleotribus, Polygraphus, Hypoborus, Xylocleptes*, and others.

It must still be clarified whether the endosymbionts also occur in the ambrosia-cultivating ipids, in other words, the Xyleborini and the Xyloternini, and in *Myelophilus minor* and *Ips acuminatus* with which Francke-Grosmann (1956) discovered a great diversity of devices for the transportation of fungi.

The form of the symbionts, which in all cases are bacteria, is no less diverse than their localization.   We encounter short rodlets, of variable thickness, delicate, rather long, filamentous forms and clumsier sausage-shaped forms.

Up to the present, *Coccotrypes dactyliperda* F. is the species which I have investigated most thoroughly.   It lives on palm seeds of all kinds and also develops indoors wherever palm seeds are used in the manufacture of buttons.   In 1961 the first publication of the planned ipid series presented an analysis of its striking symbiosis.   In the advanced embryos the symbionts are found between the yolk spheres in dense nests of varying size; in the prolarvae they concentrate in the posterior region of the anterior ventricose midgut; in the first larval stage they are broken up into dense groups and glide in the form of delicate stafflets into the adjoining, narrower section of the gut; next, some transfer to four of the six Malpighian tubules, while others glide past their opening, collect in front of the anus, and permeate the convoluted excrement globules.   They are admitted into the cells during the course of this first larval stage.   As the larva develops, a vast reproduction of the symbionts and a corresponding swelling of the host cells take place, and the latter lose their brush border after infection (Fig. 70).

On comparing the symbiotic population in individual sections of a Malpighian tubule, a number of constant regional differences may be established.   Adjacent to the place of origin, at the meeting point of mid- and hindgut, a short section is symbiont-free.   In the next zone the bacteria are clumsier in form and are situated extracellularly, at first sparsely and then in great numbers.   Now comes an abrupt transition to the thick and by far the longest section with innumerable bacteria in the form of more slender rodlets.   Shortly before the tubule reaches the hindgut, it again becomes more slender and suddenly is free of symbionts again. Finally, becoming even narrower, it forms convolutions and covers the rear section of the hindgut in partnership with the five other tubules.   In connection with pupation the symbionts are temporarily reduced in numbers.

Still awaiting clarification is the transmission of some symbionts to the gonad rudiment during early embryonic development.

The great similarity between ipid and curculionid symbiosis already emerges clearly to confirm from an unexpected direction the close relationship which has always been assumed to exist between the two families. The typical transmission method in most snout beetles is the same as in the bark beetles, and the same diversity of localization prevails.   In the curculionids too there is the same availability of certain renal vessels and an infection of the fatty tissue, midgut appendages and of the cells which are attached to the intestinal epithelium; and the parallels will undoubtedly emerge even more clearly on further investigation.   In both cases one has the impression that the symbiosis is markedly polyphyletic in origin.

Peklo's data on ipid symbionts, which are discordant with my own observations, in all probability rest on mistaken identity of diverse

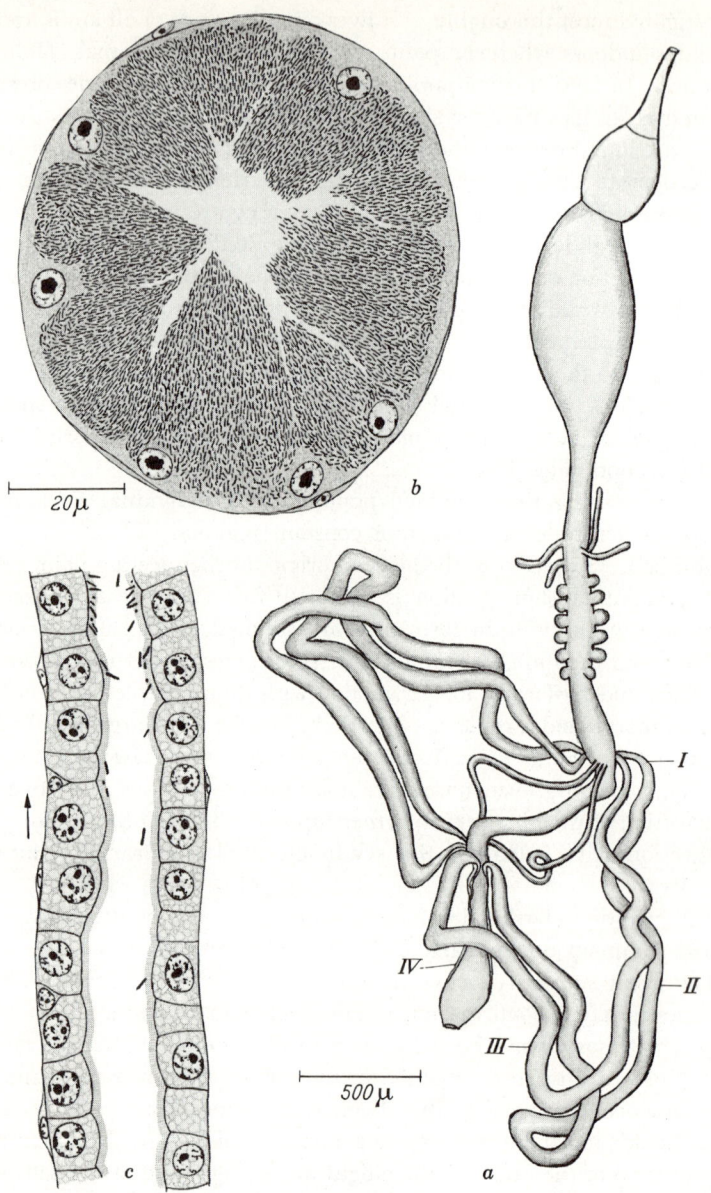

Fig. 70 *Coccotypes dactyliperda* F. (*a*) Intestinal tract with modified Malpighian tubules; (*b*) cross section through a maximally infected tubule; (*c*) first migration of symbionts into one of the predestined tubules. From Buchner.

metabolic products, which are so often described as "symbiosis" by this author. With various *Ips* species (*I. typographus* L., *I. chalcographus* L., *I. amitinus* L.), as well as with *Eccoptogaster rugulosum* Ratz., he claimed that the symbionts are present in great numbers in the peripheral areas of the fatty tissue and then are transmitted to the gut where they are digested and in that manner decrease in the course of development to the point of being present only in the egg cells. In 1946 he called them *Azotobacter;* in 1949 he reported that they were chiefly a *Torulopsis* species, although he had also isolated numerous colonies of *Candida* and bacteria of all sorts, among them types similar to *Azotobacter.*

SILVANIDS

The inconspicuous representatives of the silvanid family live generally on plant products; some live under the bark of deciduous and coniferous wood, and others transported from tropical lands often develop into troublesome pests in warehouses throughout the world. *Oryzaephilus* (*Silvanus*) *surinamensis* L., recognized as a symbiont bearer by Pierantoni (1929, 1930) and later studied exhaustively by Koch (1930, 1931, 1936), feeds mostly on cereals and various products manufactured from them, but also on seeds, flower bulbs, nuts, and dried fruits. In fact, these little beetles are not at all finical and do not even disdain yeast, tobacco, or sugar, in this respect reminding one very much of *Sitodrepa panicea.*

Larvae and imagines have four well-separated round to oval mycetomes. In the former, two dorsal adjacent mycetomes are situated above the gut in the fourth and fifth segments and two ventral mycetomes on each side of the abdominal nerve cord in the third and fourth segments. During pupation the dorsal ones are shifted to the second abdominal segment, the ventral ones to the first. In the female imagines the dorsal organs are even situated in the metathorax and in the first abdominal segment diagonally behind one another between the ovarioles; and in male imagines between the testes. At the same time there are close topographical relations with the Malpighian tubules in that both organs are enveloped by branches of the same tracheae.

The remarkable histological structure of these four organs is unique among the numerous mycetome types found thus far. In the larval states each mycetome is divided into 12 to 15 chambers, each chamber representing a little syncytium with one large, angular, central nucleus and a number of much smaller rounded nuclei. Crowded together in their plasma are curved, tubular symbionts 15–30 μ in length, clearly honeycomb in their structure; one symbiont is situated in each mesh of the greatly reduced host plasma.   A well-developed envelope tissue surrounds the syncytia and pushes forward between them into the center where an especially large nucleus is embedded in a rather large plasma collection, whereas smaller nuclei permeate the outer lining, being found in pairs wherever one of the radial septa branches off (Fig. 71).

In the pupae these complex organs begin to change.   The casing temporarily becomes much larger, the central nucleus grows considerably,

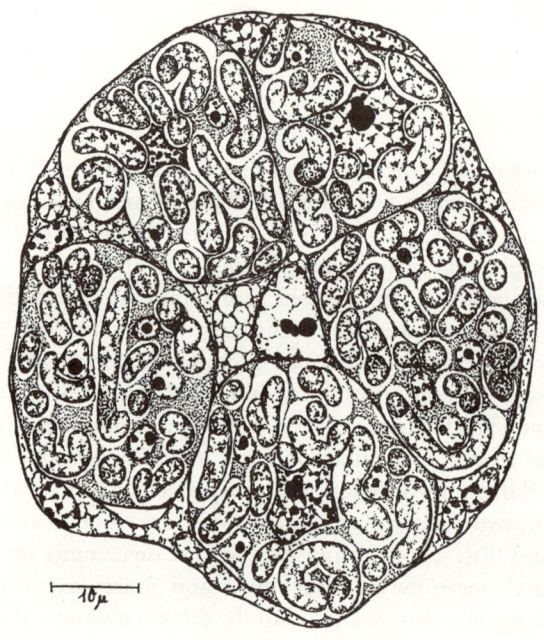

Fig.   71   *Oryzaephilus surinamensis* L.   Larval mycetome.   From Koch.

the little nuclei of the syncytia are everywhere divided mitotically, and the symbionts achieve their maximum length of 60–70 μ (Plate VI$g,h$). Here and there between the normal states, highly branched formations with coarsely honeycomb protoplasm lead to more deeply staining clumped states which are clearly degenerative (Figs. 72, 75d).

These changes introduce the enormous growth which now begins in the imago and often proceeds to the extent that a single partial syncytium swells to the size of the entire larval or pupal mycetome.   The casing of connective tissue is thus expanded to maximal extent; the gigantic nuclei of the syncytia, deeply indented on all sides, appear almost homogeneous and filled with great clods of nuclear substance, thus increasing the contrast between them and the much smaller nuclei interspersed between the symbionts.   The symbionts themselves have by this time divided into countless formations now only 3–6 μ in length (Fig. 72*a*).   With this increase of the symbionts there is considerable development of the female mycetomes which now must furnish sufficient material to infect numerous eggs.   In one female Koch counted 285 eggs!

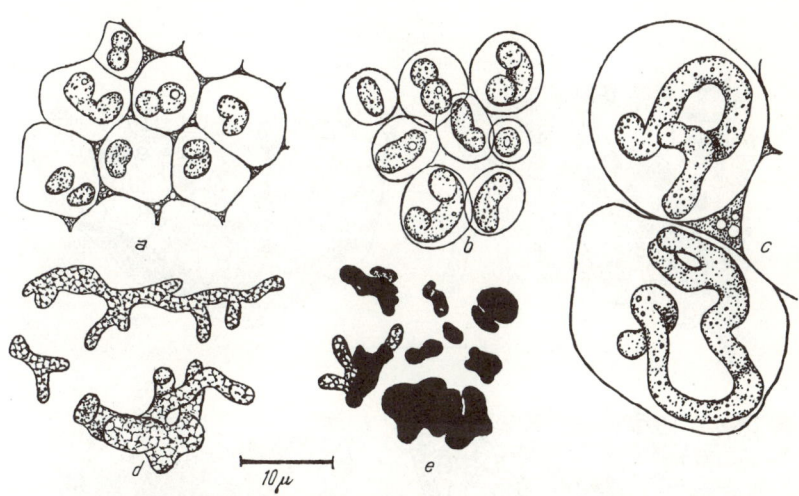

Fig. 72 *Oryzaephilus surinamensis* L.   Changes in form of the symbionts: symbionts (*a*) from the mycetome of a mature female, (*b*) from the polar infection mass of the egg, (*c*) from the mycetome of a larva, (*d*, *e*) dissolution forms from adult mycetomes.   From Koch.

Huger has shown (1956) that the mycetomes of male imagines undergo far-reaching degenerative processes soon after hatching.   The symbionts do not increase as in the females.   First assuming round or pleomorphic shapes, they are sometimes fused and often disintegrated.   In this case the nuclei migrate to the margin of the mycetome chambers and coalesce into a compact group.   Finally, the borders of the syncytia are obscured and only a few encased remains are left.

Appearing in the syncytia when the time of ovarial infection approaches are nests of symbionts differentiated from the others only through stronger staining and representing transition forms.   When the casing bursts, they reach the body cavity, where they may be observed here and there in little groups.   In adult eggs the place of infection is the posterior pole, a location of easy transmission to the interior.   Crowding between the follicles and the pedicle of the ovariole, they find a number of entry points where the follicle cells diverge sufficiently (Fig. 73*a*), at a strictly defined spot with the sylvanids.   In the case of *Lyctus*, as we shall see later, this process involves the entire egg surface.   The infection stages first extend into the

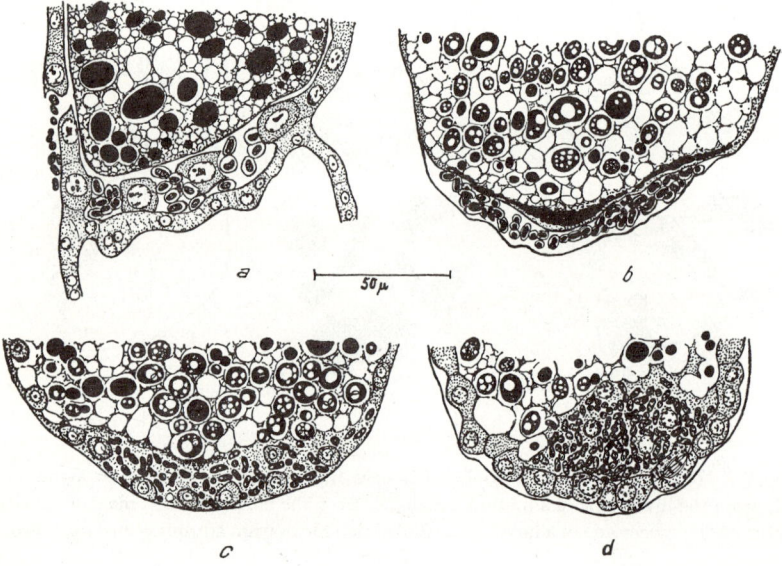

Fig. 73   *Oryzaephilus surinamensis* L.   (*a*) The symbionts force their way, at the posterior pole, between the follicle cells into the egg; (*b*) symbionts at the posterior end of a newly laid egg; (*c*) early blastoderm stage; (*d*) conclusion of blastoderm formation, development of a transitory mycetome.   From Koch.

yolk for a short distance, then glide back again and are united in a cap-shaped mass between the blastoderm and the vitelline membrane which has formed meanwhile (Fig. 73*b*).   Not until now is the yolk-free plasma zone of the egg surface impregnated to a proportionate extent with strongly staining substances which, like similar structures also occurring in this position in other cases, later accompany the germ tract.

After the end of egg deposition only symbionts in the process of degeneration are left in the mycetome.

An understanding of the complicated histological structure of the mycetomes was not possible until Koch made his exemplary presentation of the embryological processes. When the blastomeres dip into this rearmost area of the egg, some fill up with the chromophilous inclusions,

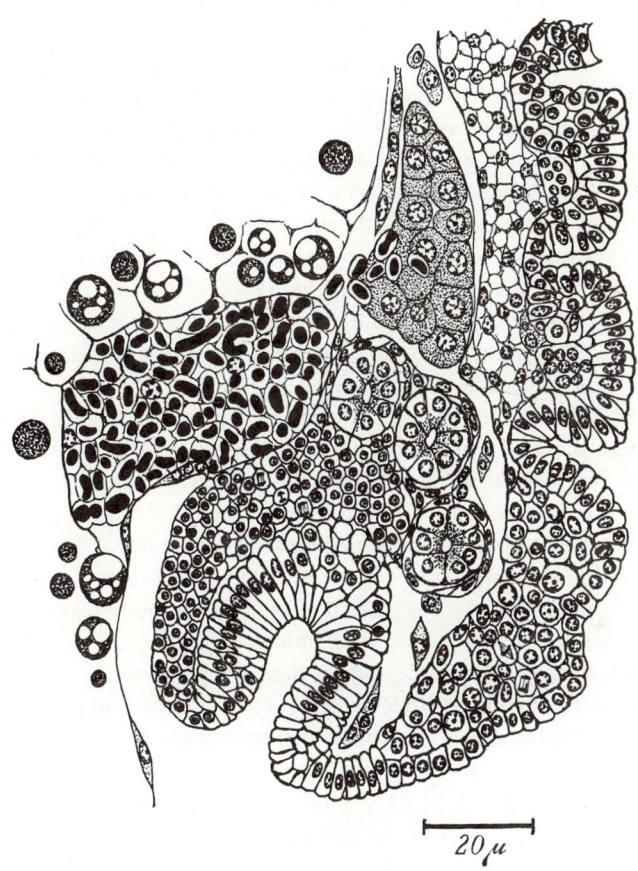

Fig. 74 *Oryzaephilus surinamensis* L. Sagittal section through the posterior end of an embryo; the symbionts are transmitted to the rudiment of the definitive mycetome. From Koch.

therewith becoming the primitive sexual cells; others enter and traverse the zone of the symbionts, arranging themselves in front of these and, forming the blastoderm, include them in the interior of the embryo. In the next stage the symbionts form a group and constantly acquire additional nuclei, thus giving rise to a provisory mycetome. This organ, which continues to receive nuclei in substantial numbers, can be easily recognized in the rearmost region of the blastoderm stage as a conspicuous sharply defined spherical body (Fig. 73*c,d*).

When the germ band develops on the ventral side of the yolk, the ball of symbionts acquires intimate relationships with the posterior amnion fold. The latter is invaginated just where the ball borders on the blastoderm and thus pushes it along. Since the two are slightly coalesced, the symbiont ball is forced to undergo the same shifts in position. Thus the ball is pushed into the interior of the egg by the developing posterior end of the germ band and then terminates its relationships with the amnion fold, sinking back into the posterior regions of the yolk as a closed cell group or in a long procession. At this time the nuclei which supply the symbiont ball visibly succumb to degeneration phenomena and decrease markedly in numbers. When the organization of the embryo makes substantial advances and the rudimentary stomodaeum and proctodaeum are deeply invaginated, the symbionts are once more situated close against the germ band dorsal from the hindgut.

Now an interesting process takes place. Near the hindgut and the place where the rudimentary Malpighian tubules have taken form by this time, the host organism has meanwhile created the beginnings of four spherical mycetomes, and these are now taken over by the symbionts which leave the provisory, now degenerate, mycetome. Some traverse the thin epithelium demarcating the embryonic rudiment and the yolk, and others make use of gaps therein (Fig. 74). Finally each cell of the four rudiments is thickly filled with microorganisms, and the definitive mycetomes occupy their final position described above.

Soon after the hatching of the larvae further histological differentiation begins. Disappearance of cell limits and development of the casing layer had already occurred in the young mycetomes. Now the nuclei of the casing grow considerably and reproduce mitotically. In the process one particular cell pushes in between the symbionts, its nucleus swelling considerably, and is constricted off. This cell then becomes the characteristic center of the organ. At the same time the syncytial symbiont mass begins to divide into an increasing number of smaller syncytia. The final state is achieved at the close of the process when one nucleus in each syncytium has undergone special development. The symbionts have sacrificed the strong staining capacity typical of the infection stages after population of the definitive mycetomes, and now they begin to grow out into the longer sausage-shaped formations which prepare the way for vigorous reproduction in the future (Fig. 75).

Baudisch (1958) recently investigated the number of chromosomes of these nuclei, which differ so greatly in size, in the mycetomes of *Oryzae-philus* and found that the central nucleus is 128-ploid in the last larval state, in the pupa, and in the imago, whereas the nuclei of the mycetome chambers are only 64-ploid. The smallest nuclei, other than these, are

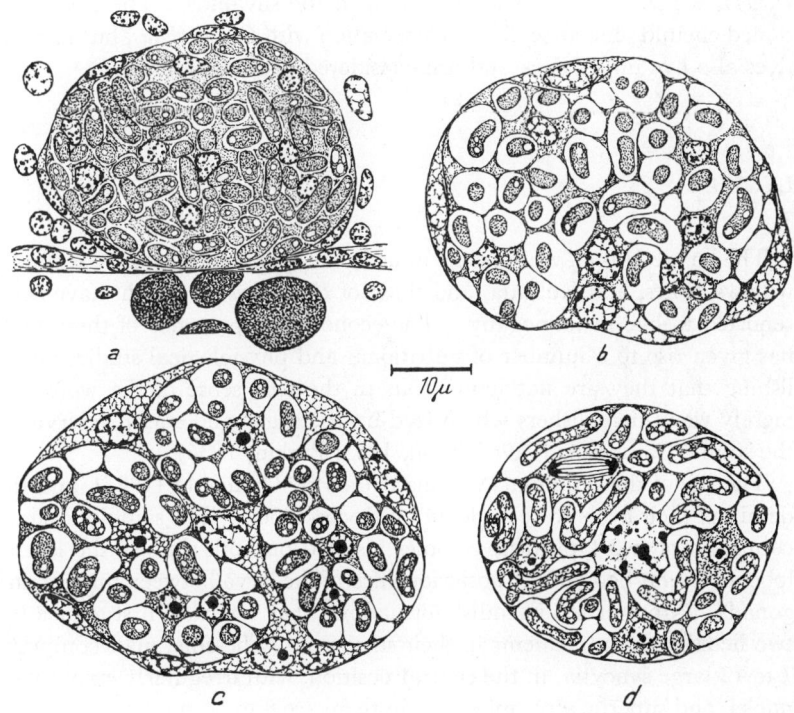

Fig. 75   *Oryzaephilus surinamensis* L.   Differentiation of the mycetomes: (*a*) syncytium with envelope cells; (*b*) immigration of the central nucleus; (*c*) septa formation and division into several syncytia; (*d*) a syncytium from the mycetome of a pupa with mitoses of the small nuclei.   From Koch.

diploid; that is, they contain 14 chromosomes in the female, and 13 in the male.

These mycetomes are obviously new acquisitions of the host animals and yet they have entrenched themselves firmly in the organizational plan of the insect, for, when egg infection is prevented by experimental means, thus producing sterile animals, the mycetomes are formed nevertheless not only in the embryo but throughout numerous generations.   Koch's findings on this subject (1936) will be summarized later in the book.   His material was later augmented by Huger in his experiments on reactions to heat, cold, and antibiotics (1956).

Silvanus unidentatus Fabr., the nearest relative of *Oryzaephilus*, which lives under bark and feeds on rotting substances there, has no symbiotic organs of any description, nor do the cucujids *Laemophloeus ferrugineus* Steph. and *Uleiota*, which were formerly united with the silvanids.   The first-mentioned cucujid, of course, lives in association with *Oryzaephilus*, but its relatives also live under bark and are considered predatory forms.

LYCTIDS

The larvae of the *Lyctus* species mine in the splint wood of oaks, walnuts, willows, ashes, and Robinia, and those of *Lyctus pruneus* Steph. have been reported to live in mahogany.   The economic importance of these pests has given rise to a number of nutritional and physiological studies establishing that they are not lignivorous in the true sense of the word but merely wood demolishers which feed on the sugar and starch reserves of the splint (Campbell, 1929; Wilson, 1933; Parkin, 1936).

The site of the symbionts was discovered by Gambetta (1927), and the details of their cycle were set forth by Koch (1936).   The symbiotic site is composed of two typical mycetomes inserted on both sides of the lateral lobes of the fat body in the posterior third of the larva between midgut and gonads.   More or less roundish formations, they reveal at first glance the two heterogeneous elements in their structure.   The chief mass comprises 7 to 12 large syncytia, in the central position, with irregularly emarginate nuclei, and superficially embedded in them are 8 to 14 smaller ones, often protruding like tumors, with roundish nuclei.   The mycetomes are well provided with tracheae and are surrounded by very flat epithelial cells pushing in between the syncytia (Fig. 76).

Corresponding with these two zones in the symbiotic site are two types of symbionts: those in the medullary layer are weakly staining, roundish or sausage-shaped, slightly curved organisms, and those in the cortical layer are strikingly shaped rosettes with stronger staining capacity.   These will be described later in more detail.

The mycetomes survive metamorphosis without change, or at most take on a rather lobular appearance and increase somewhat in size.   At the same time they are shifted from the third, fourth, and fifth abdominal segment to the metathorax and the first abdominal segment.   In the mature female the mycetomes, now loosening somewhat in texture, are deformed by being pushed against the hypodermis by the enlarging ovaries.

Such intimate connections with the oviducts facilitate transmission, which proceeds through infection of the oocytes. When the latter have

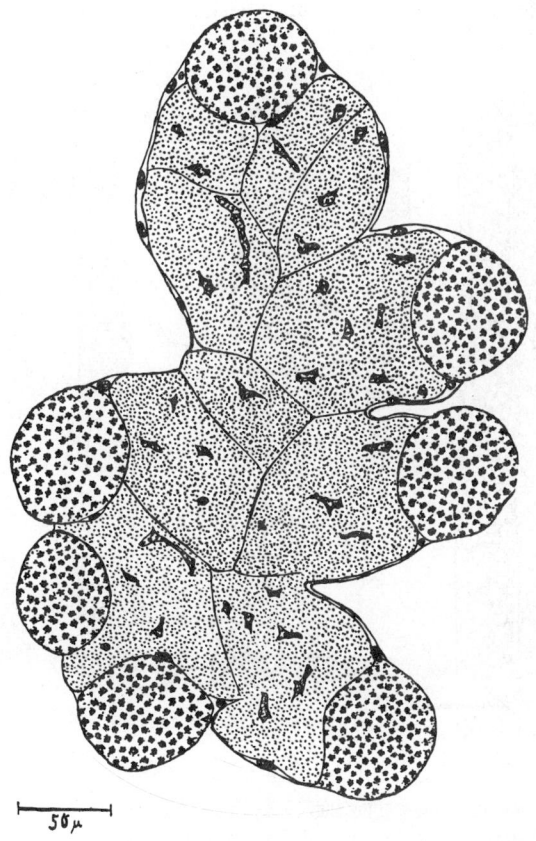

Fig. 76 *Lyctus linearis* Goeze. Mycetome of a pupa with two kinds of syncytia corresponding to the two types of symbionts. From Koch.

attained maximal size, the follicle cells, which up to this point have been joined together closely, now diverge over the whole surface of the egg to form invasion points for the multitudes of symbionts of both types which are now leaving the syncytia bursting apart here and there. Since 14 egg cells are always at the same stage of readiness, considerable numbers of infection forms must be discharged periodically. Hardly have the intermingled representatives of cortical and medullary layers sunk into the uppermost layers of the egg plasma when the gaps close in the follicle cells, which at this time are still connected with one another by delicate plasma bridges, and the formation of the chorion begins (Fig. 77).   The myce-tomes of old females appear to be markedly exhausted, and some are com-pletely detached from the syncytia.

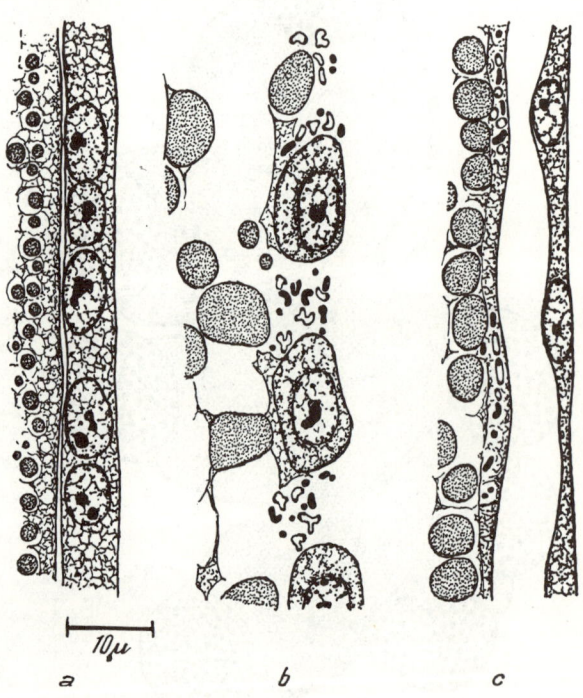

Fig. 77   *Lyctus linearis* Goeze. The two types of symbionts infect the egg cell: follicle (*a*) before, (*b*) during, and (*c*) after the infection.   From Koch.

Koch painstakingly investigated the complicated changes in form under-gone by *Lyctus* symbionts in the individual phases of the life of their hosts. The rosettes, greatly resembling those in the similarly constructed *Bromius* symbionts, are built up of 5 to 15 spherical or tear-shaped elements held together by short plasma pedicles about 1 μ in diameter and surrounded by a hyaline envelope which is difficult to recognize.   Occasionally they fall into individual components, though they are still held together by the envelope.   Before entering pubescence they assume a completely different shape. Y-shaped forked tubes grow lateral processes of several types and turn into knotted, densely packed formations with much more deeply staining free ends. Apparently each of these still encased states, which now take part in the egg infection, is derived from a component of the rosettes. During embryonic development the old rosette shape is again assumed (Fig. 78). The males do not develop specific infection stages, but in addition to the rosettes they have involution forms which apparently disintegrate later. They are deep-staining, notched voluminous masses which result through fusion of the rosette parts.

The large, round or cylindrical, slightly curved symbionts of the medullary area, measuring 0.5–1.5 μ, likewise secrete a gelatinous envelope, and

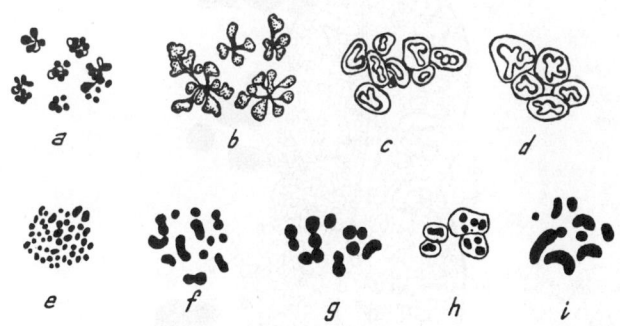

Fig. 78 *Lyctus linearis* Goeze. Morphology of the symbionts. (*a–d*) The symbionts of the cortical layer: (*a*) from a larva, (*b*) from a pupa, (*c*) from a sexually mature imago, (*d*) from an egg. (*e–i*) The symbionts of the medullary zone: (*e*) from a larva, (*f*) from a young female imago, (*g*) from a sexually mature female imago, (*h*) from the egg, (*i*) from a germ-band stage. From Koch.

in the female they assume a different shape at the beginning of pubescence, being transformed into larger, much deeper-staining, oval or sausage-shaped infection stages which are easily distinguished from those originating from the cortical zone. In both sexes there is no lack of states which have been interpreted as involution forms, and especially in the males there are some which decidedly resemble the degeneration phenomena found in the male mycetome of *Pediculus*.

We are also well informed on the course of embryonic development. During formation of the blastoderm the symbionts are still situated in the superficial plasma zone which passes over into the blastoderm, even appearing here and there in the free state under the chorion, but at the end of blastoderm formation they are located in the border area of the yolk.   From here they push into the interior of the egg in swarms, remaining there in little groups between the yolk masses until the germinal band is formed.   Then they sink toward the lower third of the egg and are stationed mainly at the caudal end of the embryo by the time mid- and foregut develop.   Most of the symbionts then collect dorsal from the

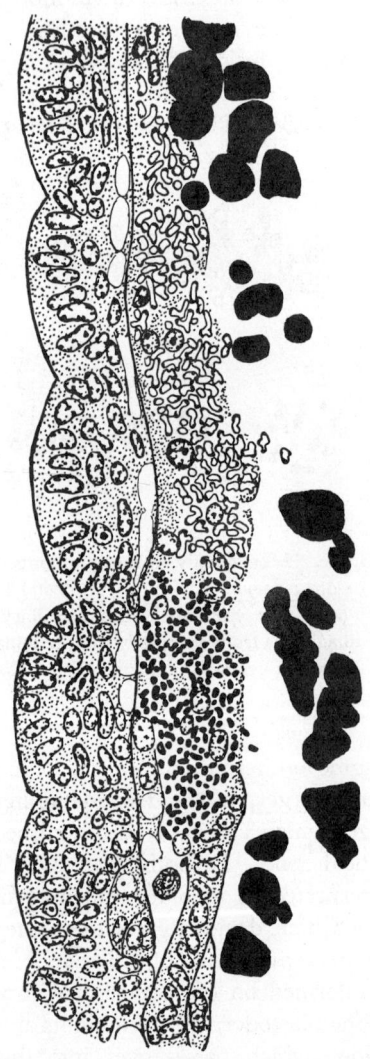

Fig. 79 *Lyctus linearis* Goeze.   Sagittal section through the posterior end of the germ band; the two types of symbionts have separated.   From Koch.

rudiment of the proctodaeum and push into the body cavity of the embryo. In the process they use chiefly the open connection which still exists at this time between the cavity and the yolk in the median plane of the germ band.

In the lateral regions, where the coelom is already closed off against the yolk side, a transverse fissure in front of the proctodaeum provides an outlet. Migrating cells float in the body cavity and between the symbionts which have not yet migrated there. These cells, some of which are laden with symbionts of both types, apparently assist in the transmission process. When all the symbionts have left the yolk, the two types begin to separate (Fig. 79), the front group receiving exclusively symbionts of the future cortical area; a two-fold group following directly behind, exclusively those of the future medullary area. Both groups now have strikingly large nuclei which Koch believes may be yolk nuclei. The anterior group is temporarily disintegrated into uninucleate mycetocytes while paired, roundish mycetomes are forming behind them to accommodate the medullary symbionts in syncytia containing 4 to 5 nuclei.

With dorsal closure of the embryo, the anterior mycetome rudiment is also torn into a right and a left half, and finally its individual elements are inserted into the surface of the two remaining mycetomes, which in the meantime have thrown cell limits around their nuclei. The definitive syncytia of the two zones are formed later by amitosis. Vigorous multiplication of the symbionts and strong growth of the mycetomes finally bring about the state of maximum development that will be encountered in the adult larvae.

BOSTRYCHIDS

From a report by Mansour (1934), unfortunately not very thoroughgoing, we know that the bostrychids (apatids) are among the beetles which live endosymbiotically. This family, of which about 50 genera with 350 species have been classified thus far, has developed particularly in the tropics·and subtropics, where forms measuring several centimeters occur. The larvae and imagines are lignivorous and are found in diseased and dead trunks, branches, and rootstocks of hard, deciduous trees. Some genera have adapted to dry grapevines, to blades of rice plants, bamboo, or grain; and still others, like *Sitodrepa* among the anobiids or *Oryzaephilus* among the silvanids, have given up their original habits to become storage pests in drugs, rice, and other cereals.

Mansour studied two lignivorous species, *Sinoxylon ceratoniae* L. and *Bostrychoplites zickeli* Mars., and the pernicious *Rhizopertha dominica* F., known the world over as a stored-products pest. His explanation of the symbiotic devices of these Coleoptera, which are pests of economic importance, lacked discernment and contained statements which at the very outset sounded unconvincing. For example, he asserted that the symbionts are lacking in ovarial eggs, but infect the testes in the form of long-

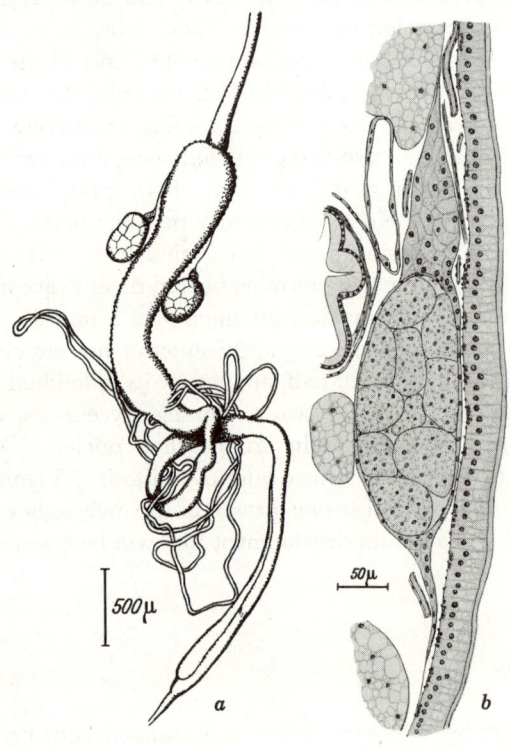

Fig. 80 *Scobicia chevrieri* Villa. (*a*) Intestinal tract of a male imago with paired mycetome and suspensories; (*b*) mycetome of a female larva with bilateral suspensory.
From Buchner.

growing filaments, cause abnormalities here in sperm formation, are transmitted during copulation to the bursa copulatrix, and finally reach the micropyles of the eggs with the sperm. A complete description of bostrychid symbiosis, which is extremely odd in some of its aspects, was unavailable before my own detailed study in 1954. My subjects were the *Rhizopertha* species already used by Mansour, and in addition *Synoxylon sexdentatus* Oliv. and *Scobicia Chevrieri* Villa from dead branches of *Ficus carica* and *Ceratonia siliqua*, *Apate degener* Murray from coffee plantations in Uganda, and *Apate monachus* L. from coffee plantations in French Guinea.

All bostrychids investigated thus far have paired mycetomes adjacent to the midgut, differing from case to case in form and histology. In *Rhizopertha* they are round or oval. A thin epithelium surrounds large multinucleate syncytia, and situated close to the latter is a cell group which at first seemed very mysterious (Fig. 80). In all other relatives the mycetomes are provided with strange hanging devices set together from a club-shaped, sterile cell mass and a long thread joined to the midgut. Such anchorage, occurring either at the anterior pole or at both poles, is more or

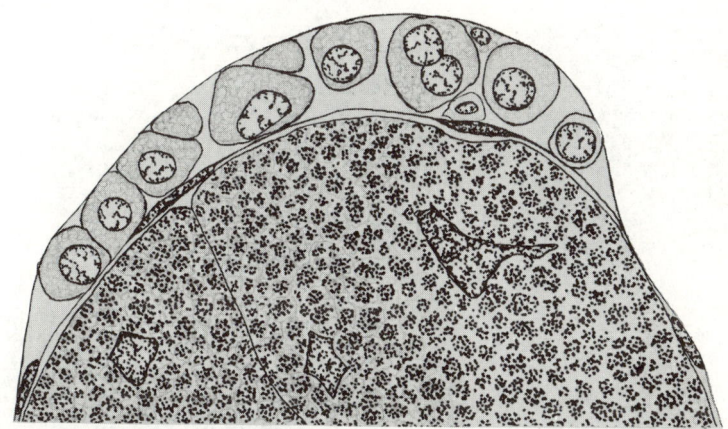

Fig. 81   *Rhizopertha dominica* F.   Section of mycetome of female larva with the accessory cells.   From Buchner.

less well developed, and in connection with metamorphosis it is usually much reduced (Fig. 80b). In *Synoxylon* the imaginal mycetome is a very elongated and irregularly convoluted organ with uninucleate mycetocytes; in *Scobicia* oval mycetomes are composed of multinucleate syncytia, and in the two *Apate* species sausage-shaped organs divide into a number of irregular fragments in the course of development but usually still hang together apparently by means of long thin strands of the epithelia.

The symbionts are extraordinarily diverse in shape. In general, they form loose or very dense rosette-shaped groups with small coccoid components held together by delicate threads (Figs. 81 and 82a), but occasionally very drawn-out branched groupings occur instead (Fig. 85b). *Apate*, in which the rosette formation may be replaced by sausage-shaped or filamentous formations, represents an exception, according to present knowledge. Under various cultural conditions (Huger, 1956) and in connection with the degeneration of the mycetomes in the male pupa and imago, which will be referred to again later, this polymorphism continues to increase substantially and with respect to diversity shows great similarity to bacterial cultures *in vitro* (Braun, Berg, Kessler, and Mavroidi, 1954).

Transmission is achieved by infection of the ovarial eggs, as in all other insects with symbionts and mycetomes. Special transmission forms arise

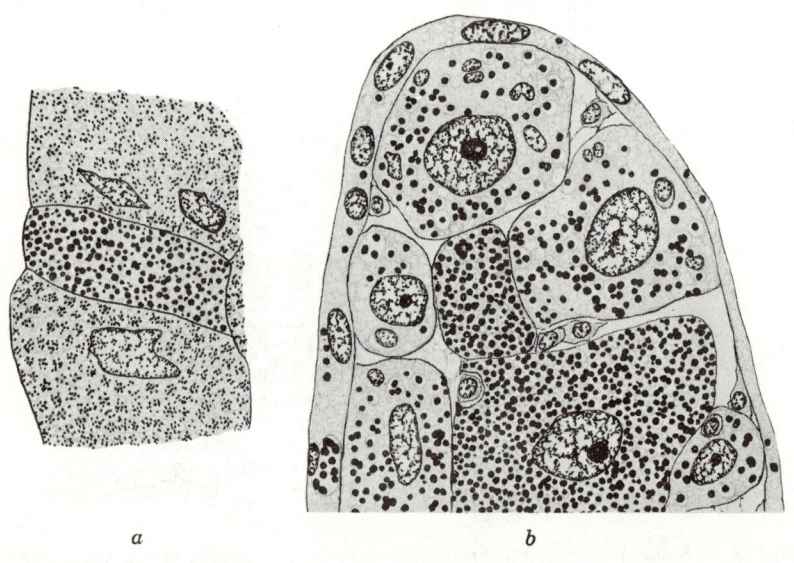

a   b

Fig. 82   (*a*) *Rhizopertha dominica* F.   From the mycetome of a female imago; in one-cell transmission forms.   (*b*) *Sinoxylon sexdentatus* Oliv.   Formation of transmission forms; small sterile cells have migrated into the mycetocytes.   From Buchner.

in the female imagines, a process at first limited to a few cells or syncytia but gradually transforming the whole mycetome. In place of rosettes there are now denser, heavily staining, and more loosely arranged formations which often show that they have several components. They traverse the mycetome epithelium in this state and transfer to the body cavity (Fig. 83). In the follicle of the oocytes, now in a state of readiness, changes occur such as we have already seen in *Lyctus*. Its cells, closely joined until now, diverge on the whole surface to the extent that passages are created for the symbionts, and when the latter reach the surface of the egg, the gaps in the follicle close again. After formation of the chorion the symbionts are located between the yolk membrane and the denser periplasma in a light, finely constructed, marginal layer (Fig. 83c). *Scobicia* and *Synoxylon* are like *Rhizopertha* in this respect.

In *Rhizopertha* I was able to observe what happens to these organisms in the course of embryonic development. During formation of the blastoderm they come to lie under the base of the cells (Fig. 83d). At first almost uniformly distributed, they are later grouped in a large mass under a rudiment of the germ band. The rudiment, consisting of taller cells, at first occupies the entire length of the embryo and later is drawn along during its invagination. When germ band and amnion reach their full

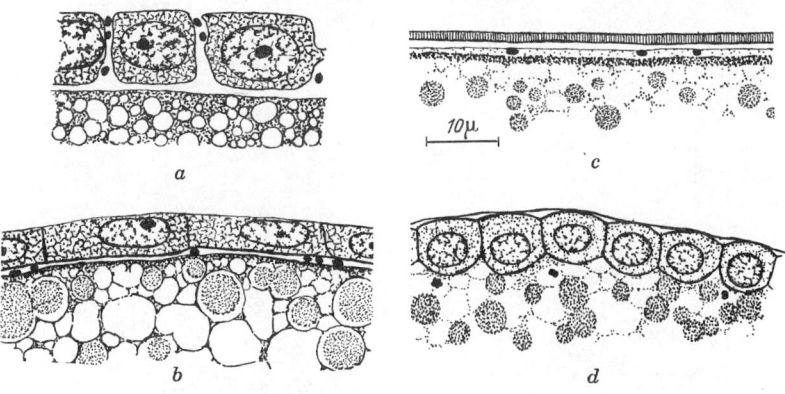

Fig. 83 *Rhizopertha dominica* F. Transmision of symbionts to the egg cells: the symbionts (*a*) in gaps of the follicle, (*b*) between follicle and egg, (*c*) on the outer surface of a mature egg, (*d*) behind the blastoderm. From Buchner.

development, the symbionts are lying dispersed between the yolk nuclei over the entire space occupied by the germ band, but in the stage when the mouth parts and extremities begin to form and the future midgut epithelium gradually surrounds the central yolk, they have already departed through a gap still remaining in the epithelium to an area situated behind it and are progressively received here by conspicuous, already paired elements with large nuclei. Some of these cells which branch off from the lower lamella have no part in building the actual mycetomes but remain sterile or take up the few remaining bacteria which failed to reach the mycetomes. The mysterious "accessory cells" of *Rhizopertha*, illustrated in Fig. 81, and the strange suspensoria of the other genera are derived from these cells. Any symbionts taken up by them naturally lose all ability to reproduce yet retain unaltered the shape typical of transmission

forms.   In view of the close relationships existing between the bostrychids
and the lyctids, the latter with two types of symbionts, it seems reasonable
to interpret these accessory cells as a vestige of a second symbiont type

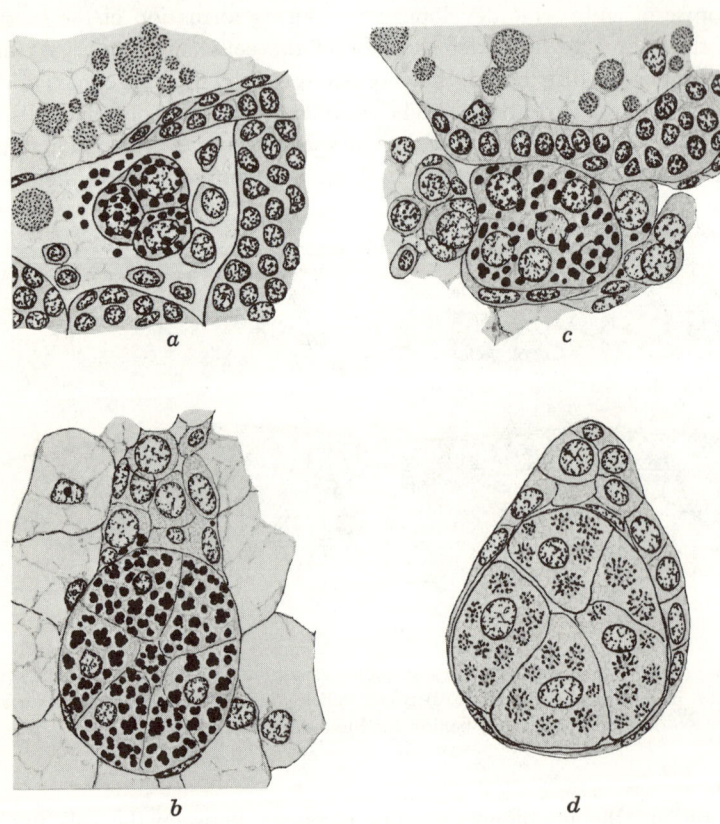

Fig. 84   *Rhizopertha dominica* F.   Formation of the mycetomes: (*a*) the symbionts are
taken into the cells after they have left the embryonic midgut; (*b*) an embryonic mycetome with partially infected accessory cells; (*c*) mycetome of a prolarva, loosening up
of the symbionts; (*d*) mycetome of a young larva, rosette formation.   From Buchner.

which has been lost.   Where these cells are used in the formation of suspensoria, a change of function will occur, whereas *Rhizopertha* will show a
behavior more nearly like the original.   Figure 84 illustrates several stages
of the early development of the *Rhizopertha* mycetomes: (*a*) and (*b*) refer to
embryos and illustrate the entry of the symbionts into one of the two cell
groups, and (*c*) and (*d*) each represents a mycetome from a young larva.

At this time the separation of the cells which remain sterile comes to a decisive end, the fact that the transmission forms are composed of several parts emerges very clearly, and soon the typical rosettes begin to form.

Another strange quality of the bostrychid symbiosis, as yet unexplained, concerns the small sterile elements situated between the mycetocytes in both sexes in *Synoxylon, Scobicia,* and *Apate,* though not in *Rhizopertha.* Some of these interstitial cells are transmitted to the mycetocytes and degenerate there between the symbionts. In *Synoxylon* and *Scobicia,* when the transmission forms are in the midst of their development and some of the mycetocytes have already been extensively emptied, there occurs a second invasion of small sterile cells which at first still keep to their limits but later take up a position in the alien plasma as naked nuclei (Fig. 85*b,c*). In all this there is nothing to suggest phagocytic functions in these migrating cells.

The multiplicity of symbiosis types among the bostrychids is particularly increased in that the male mycetomes in *Rhizopertha* and *Synoxylon* are completely or almost completely disintegrated, whereas in *Apate* only partial degeneration takes place and in *Scobicia* males the mycetocytes remain unchanged. In *Rhizopertha* symbiotic degeneration begins in the pupal state, although at times the first symptoms may be observed in the old larvae. Deterioration may attack a syncytium in isolated patches, and in occasional cases a syncytium here and there may be attacked on all sides. Zones still untouched may exist side by side with greatly reduced syncytia in which the nuclei become involved likewise. Some bacteria disintegrate and others mass together in balls of varying size until only an insignificant remnant of the former symbiont site is left. The accessory cells, suffering no damage, even undergo considerable growth and, in addition to nuclei which have grown proportionately and have often become lobular, still contain some of the transmission states which they acquired during formation of the mycetome (Fig. 85*a*). Figure 85*b* represents the mycetome remains of a *Synoxylon* male in which only one mycetocyte has survived, and the remnants of all the remaining ones are situated as undefined masses at the two poles, where the suspensoria attach. In *Apate monachus* the male mycetome is permeated by cells in which all the bacteria have gradually disappeared and only a few plasma remnants surround the shriveled polymorphic nucleus together with several nuclei stemming from migrating interstitial cells (Fig. 85*c*).

In 1956 Huger published the results of a study, originating in Koch's laboratory, concerned primarily with the influence of high and low temperatures and the effect of aureomycin and terramycin on *Rhizopertha* symbionts. His study not only confirmed my own observations but also amplified them, particularly with respect to the diversity of symbiotic

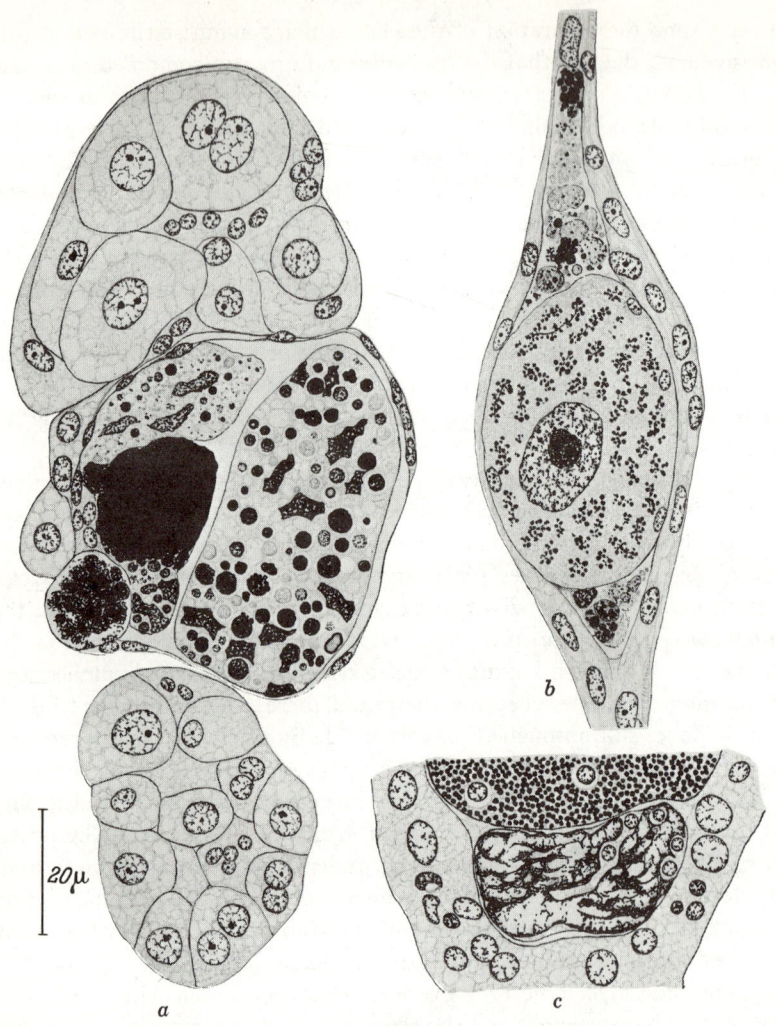

Fig. 85 Degeneration of the mycetomes of male bostrychids.   (*a*) *Rhizopertha dominica* F. Final degenerated remainder with two groups of accessory cells.   (*b*) *Sinoxylon sexdentatus* Oliv. Fragment of a degenerated mycetome, a mycetocyte with conspicuous budding groups.   (*c*) *Apate monachus* L. One mycetocyte contains only its degenerated nucleus and small nuclei which have migrated into the cell.   From Buchner.

degeneration phenomena. As will be explained later in more detail, Huger succeeded in rendering the mycetomes almost or completely free of symbionts. Yet even in the latter case, it seems that some heat-resistant symbionts survive outside the mycetomes, the progeny of animals with sterile mycetomes having slightly populated mycetomes in the first, second, or third generation. It seems fairly certain that this unusual

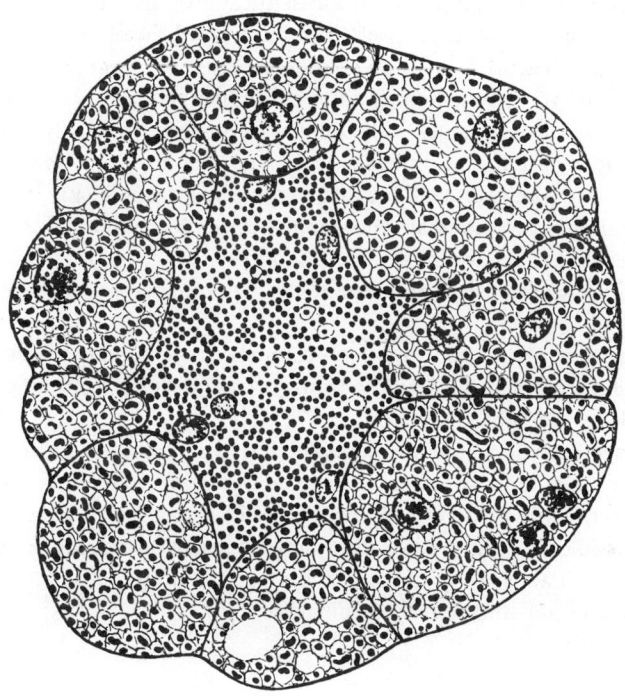

Fig. 86 *Throscus dermestoides* L. Spherical mycetome of a female imago with two types of zones. From an unpublished drawing by Stammer.

resistance, particularly of the transmission forms, to elevated temperatures is connected with the conditions of the tropic and subtropic homeland of the host animals. Interestingly enough, the symbionts which occasionally are situated in the accessory cells often also prove to be more heat-resistant than those in the mycetomes. The antibiotics likewise led to far-reaching degeneration of the mycetomes, but here, too, a remnant of resistant symbionts survived to preserve the quality of resistance in the following generations.

THROSCIDS

Stammer (1933) proved that *Throscus (Trixagus) dermestoides* L. lives symbiotically, but unfortunately lack of material made it impossible to pursue the subject in all its details. This is particularly regrettable, because apparently three types of symbionts are housed in this beetle. The latter lives on fruit trees, in forest meadows, and on the fringes of forests; and the larvae in all probability are wood dwellers.

A paired, spherical to oval mycetome consists of a large, uniform, central syncytium surrounded on all sides by smaller syncytia. Round nuclei are distributed throughout. The central syncytium is thickly filled with small round symbionts, and the cortical layer with numerous longish, oval, often curved organisms, each located in a plasma vacuole (Fig. 86). The corresponding mycetome pair of the males is smaller and of different structure. Because there are far fewer marginal syncytia, the inner zone, subdivided into several syncytia, frequently approaches the surface of the organ.

Furthermore, two broad oval mycetomes are embedded in the fatty tissue on both sides of the sexual organs. They develop from fat cells and contain long filamentous bacteria.

Stammer considers it quite possible that the symbiotic forms in both outer and inner zones represent two states of the same symbiont, and thus that the symbiosis involves only two types of organisms. However, it is just as possible that three types of organisms are involved, all three being, at any rate, bacteria. The stages that came under observation were unfortunately not sufficient to confirm this.

5

Symbiosis in animals which live in tree sap

CERATOPOGONIDS

In 1921 Keilin made a brief report on the interesting symbiosis in *Dasyhelea obscura* Winn., one of the chironomids (a subfamily of the Ceratopogoninae). Using preparations made by Stammer, I later supplemented his material by a few observations made with two other species (Buchner, 1930). The larval forms of the three species investigated live in tree sap; those of other species live in fresh and salt water in a variety of places: in algae fields, decaying seaweed, spray water pools on the seacoast, water collecting in the leaf axils of *Dipsacus*, and in rotting roots.

Mycetomes thickly filled with bacteria are present in larvae, pupae, and imagines. *Dasyhelea versicolor* Winn. and *obscura* have two pairs of spherical organs, each composed of a single syncytium with few nuclei, filling a space in the thorax between the salivary glands and the hypodermis. Another unclassified species of the same genus has a syncytial mass in the last thoracic segment and the two adjoining abdominal segments. Larger at the back and formed possibly through fusion of three mycetomes, one behind the other, it borders on the gut laterally and ventrally (Fig. 87). *Dasyhelea longipalpis* Kieffer, on the other hand, appears to be without organs of this kind.

Unfortunately we have no detailed information on transmission. Keilin merely states that it takes place by ovarial infection and that hatching larvae are already provided with mycetomes. Since other diptera with mycetomes and ovarial infection are not known to exist, we may be quite sure that new devices are present here and that further research in *Dasyhelea* symbiosis would be rewarding.

NOSODENDRIDS

The symbiotic organs of *Nosodendron fasciculare* Oliv. were discovered by Stammer (1933) and somewhat more fully investigated by Öhme (1948). Larvae and imagines of this beetle live in the fermenting sap flow of elms, or sometimes of chestnut and other trees, and lay their eggs under scales of the bark in the vicinity of the wounds.

On each side in the imagines of both sexes are yellow-tinged, dumbbell-shaped mycetomes in the vicinity of the gonads. The mycetocytes are enveloped in a delicate casing with flattened nuclei the mycetocytes at the

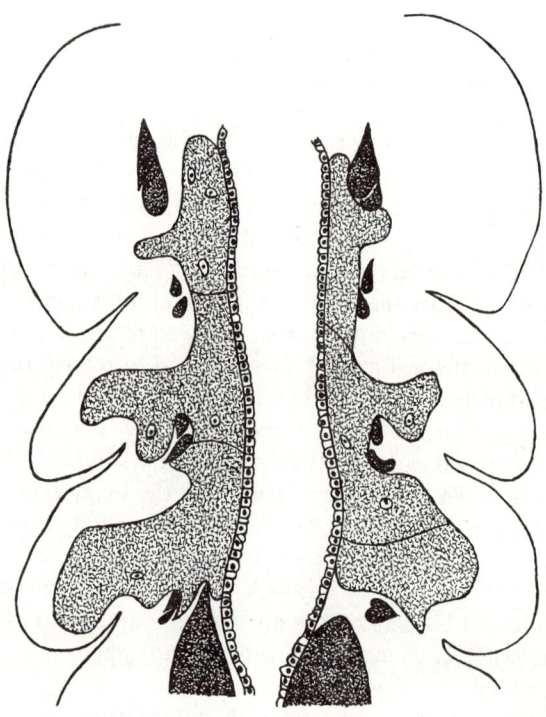

Fig. 87 *Dasyhelea* sp. Frontal section through the last thoracic segment and the adjoining abdominal segments of an imago; a syncytially constructed mycetome surrounds the gut. From Buchner.

periphery being of moderate size while those in the central area develop into giant cells with proportionately larger nuclei. The symbionts are roundish or somewhat polygonal formations so thickly crowded that they are recognizable only on smears. In females the mycetomes are abundantly supplied with tracheae and are connected by means of the latter with the ovarioles. They are also shaped like dumbbells, but with the approach of pubescence they are cut into two completely separate parts, one of which fuses with the germ layer of the ovarioles. It is in these parts exclusively that transmission devices are formed, chiefly in the marginal areas, when much smaller, denser, and therefore more deeply staining particles are constricted off at the periphery of the irregularly shaped organisms.

Unfortunately it was not possible to provide a detailed explanation of the infection, the transmission forms being so hard to recognize in cross-section preparations. They are found in the free state in great quantities in the body cavity and apparently penetrate even the youngest oocytes and adjoining sections of the ovarioles. In the deposited egg they are larger and therefore more easily recognized. In the male imagines they are in a state of advanced decay. The mycetomes retain their old form but take on a glassy appearance and contain exclusively slender spindle-shaped formations which Öhme interprets as degeneration products of the symbionts.

We also have very little information about embryonic development. At first several symbiont groups are distributed at random in the yolk; later they form two aggregations provided with nuclei. The larval mycetomes are then located in the area of the mesothorax and the metathorax in a position ventrolateral to the gut; in the younger stages the mycetomes are differentiated from those of the older larvae and imagines in that the smaller marginal mycetocytes are still lacking, the whole organ consisting of giant cells with one large central nucleus and a number of smaller peripheral nuclei. Apparently, a part of the peripheral protoplasm around the smaller nuclei is later constricted off to form the smaller elements of the cortical zone which are typical of the imaginal mycetomes.

Very probably these *Nosodendron* symbionts are modified bacteria, but nevertheless it would be desirable to study more thoroughly their change in form as well as the unusual development of several transmission states on the surface of each symbiont. It would also be profitable to study the details of the strange conversion of one particular organ part into a transmission device and the behavior of the symbionts during egg infection and embryonic development.

6

Symbiosis in animals which suck plant juices

HETEROPTERA

From observations of older insect anatomists (Treviranus, 1809; Dufour, 1833; Leydig, 1857) as well as from newer research (Forbes, 1892; Glasgow, 1914; Kuskop, 1924; Rosenkranz, 1939; Steinhaus, 1951; Steinhaus, Batey, and Boerke, 1956; Huber-Schneider, 1957; and Schorr, 1957), there was reason to believe that in all heteropterous bugs living primarily on plant food, the symbiotic bacteria are localized in a special section of the midgut, but it is a comparatively recent discovery that plant-sucking Heteroptera may be supplied with regular mycetomes just as bloodsucking members of the order are (Schneider, 1940). In keeping with its more primitive character, let us review first the type of symbiosis linked with the intestinal canal.

The insect species involved, in contrast to those without symbionts, all have a longer midgut by virtue of an additional section of considerable size. First, there is a thin-walled, greatly widened ventricose part followed successively by a slender tube, a smaller widening either round or oval, and a sharply defined end section in which symbiotic habitats are developed. In forms which do not have symbionts, this final section is lacking or at the very most only faintly suggested, and in pronounced predators the first three zones are often not so sharply separated.

The evaginations in the fourth section are extraordinarily diverse in form. In the blissines and the related aphanines they are long slender tubes developing near the end of the midgut and often bearing a certain resemblance to the neighboring Malpighian tubules. For example, *Gastrodes abietis* L. has one or two of these sparsely ramified, filamentous appendages on each side (Fig. 88a); in *Tropistethus holoscenicus* Schltz. they are more ramified, and in other cases there are four to six somewhat flat ceca (*Drymus silvaticus* F.).

As the number of ceca increases, flat leaf-shaped formations arise through lateral concrescence. They are more or less deeply cleft and consist of two or three palmate lobes or of ten to fifteen tubes, at times even more. In some forms the places of derivation may be distributed over a rather large section of the gut and the two sides may develop quite differently, so that even within the same species some degree of diversity may be found, for example, in *Myodocha serripes* Oliv. and the *Aphanus* species (Fig. 88*b*); and the berytids are quite similar in this respect.

More frequently and in far greater number, we find very short and uniformly developed crypts extending a great distance over the fourth section of the midgut, and in all cases they extend almost to the outlet of the

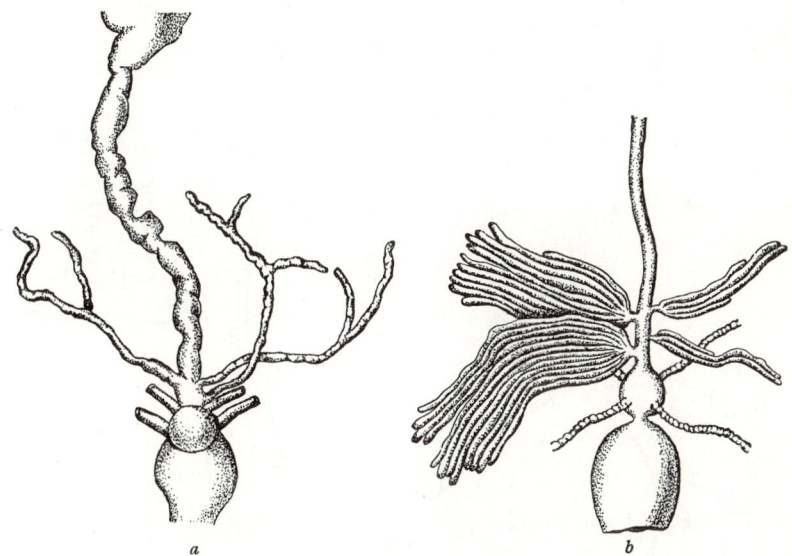

a   *b*

Fig. 88   (*a*) *Gastrodes abietis* L.   From Kuskop.   (*b*) *Myodocha serripes* Oliv.   From Glasgow.   Tubular outgrowths of the midgut as sites for symbiotic bacteria.

Malpighian tubules.   They are arranged in two rows in the cydnids, lygaeids, acanthosomines, plataspids, and many coreids (Fig. 89), and in four rows in the pentatomines and scutellerines (Fig. 90).

There are also many secondary variations.   The number of evaginations may be low, as in the lygaeid *Oedancala dorsalis* Fabr. in which there are approximately thirty-five clearly defined, pyriform evaginations on each side.   Another example of such primitive states is *Thyrecoris unicolor* Pal. in which the sacs are still well separated but already quite numerous. The maximum development is found in the pentatomids, in which many hundreds of closely crowded crypts form four long garlands (Fig. 90).   In

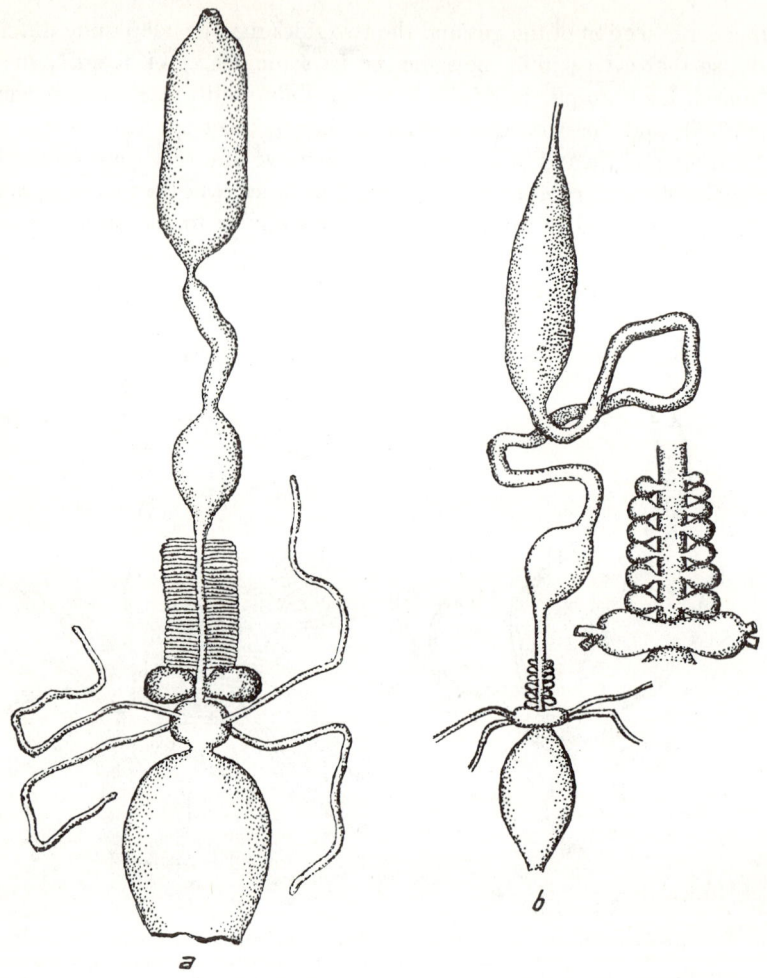

Fig. 89   (*a*) *Peliopelta abbreviata* Uhl.   (*b*) *Dysdercus suturellus* Scha.   The female midgut has blind sacs arranged in two rows; some of them are enlarged for purposes of transmission.   From Glasgow.

one of the pentatomids their number was estimated at 1400! Since the evaginations toward the blind end tend to become broader, the rows—whether two or four—are forced into diverse convolutions when evagination development is especially luxuriant. The musculature passes smoothly over the crypts but sends out longitudinal and transverse strands, which penetrate between them and entwine each single one. Tracheae also develop in profusion. The second- or third-last stigma pair sends out strong branches to the crypts and provides them abundantly with capillaries.

The epithelium of the crypts frequently consists of binucleate cells which are sometimes thick and rich in plasma, sometimes quite flattened, and in many cases, chiefly in pentatomids, contain a pigment, not to be confused with the yellow pigment which often fills the matrix cells of the tracheae in species of the Homoptera. Its color is specific for every genus: red in *Pentatoma*, orange-red in *Eurydema*, and pink in *Carpocoris*. The intensity is subject to strong variations and is greatly dependent upon temperature. In winter animals the pigment may be absent entirely, as in *Carpocoris pudicus* Pd. *Pentatoma rufipes* L. is only faintly colored at 15°C., but when placed directly in the sunlight the crypts change to an intense red by the next day. *Dolycoris baccarum* L. generally has colorless crypts; but after particularly hot days a red dyestuff appears. The reactions determined by Rosenkranz permit the assumption that the coloring agents are prestates of melanins, very similar to the material which so intensively dyes the testes of the Heteroptera.

The bacteria are restricted to the lumen of these variously shaped evaginations except in the aphanines, in which the symbionts occur extracellularly or within the lining cells, and in the lygaeid *Tropidothorax leucopterus* investigated by Pierantoni (1951), in which the symbiosis is also exceptional. According to Pierantoni, the cells and the ceca of the much folded and spherically distended prestomach are filled with bacteria, as is the much thinner epithelium of the whole remaining midgut.

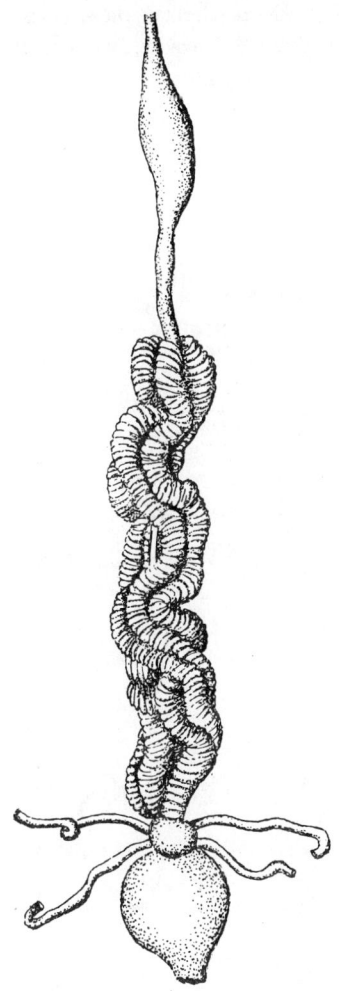

Fig. 90 *Carpocoris fuscispinus* Boh. Adult intestinal tract with outpocketings arranged in four rows and inhabited by bacteria. From Kuskop.

Even where typical crypts are developed there is always a connection, however narrow, between their cavities and the gut, and therefore symbionts may always be found in the intestinal canal in varying numbers.   The acanthosomines are the sole exception.   In their case Rosenkranz determined complete obstruction of passages originally connected with the gut.

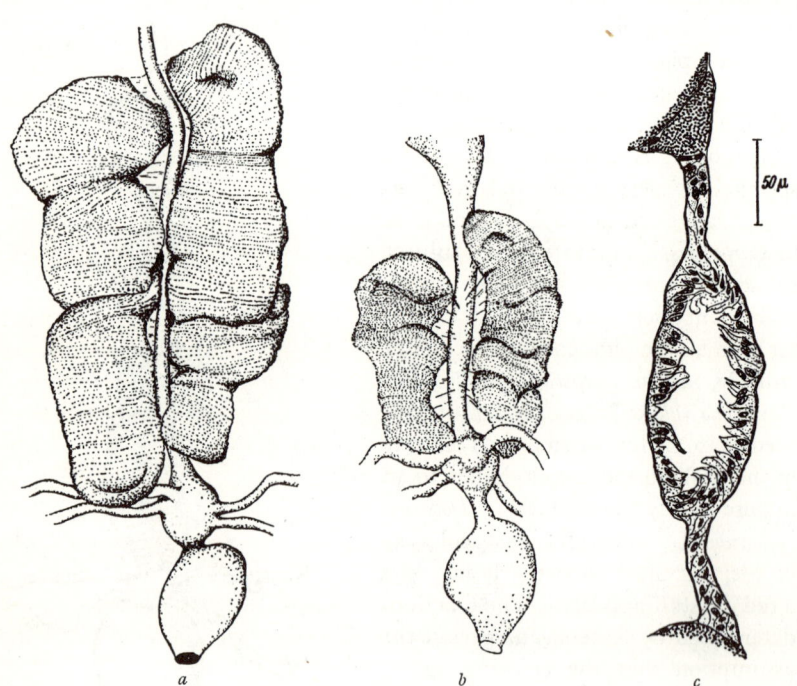

Fig. 91   *Acanthosoma haemorrhoidale* L.   Cryptic gut (*a*) of a female imago, (*b*) of a male imago; (*c*) cross section through the gut; the openings of the crypts into the gut are sealed up.   From Rosenkranz.

Figure 91 illustrates *Acanthosoma haemorrhoidale* L. in which these formations, fastened only by a delicate membrane, approach the character of regular mycetomes.   Here the symbionts are completely sealed off, and it would be lost labor to look for them in the intestinal tract.   In general, the crypts are fundamentally the same for both sexes, but in males they tend to be smaller.   The crypts of *Acanthosoma* are also less developed.

The pyrrhocorids require special discussion.   In this family, in which animal diet obviously still plays a substantial role, there are no conspicuous symbiotic organs. It is true that in females of *Pyrrhocoris apterus* there are several small evaginations varying greatly in size and number, of a type found nowhere else among Heteroptera which lack symbionts. However, they contain no bacteria. On the other hand, the lumen of the first ventricose midgut section is always populated by great numbers of bacteria, always in the same form, which are crowded together in great knots or permeate the food pulp on all sides. Often they are anchored by the free end to the intestinal epithelium, as is true so frequently of these bacteria in the ceca. In another pyrrhocorid, *Dysdercus suturellus* Scha., it was only in females that Glasgow discovered, at the same place on both sides of the gut, bacteria-populated ceca which were essentially more developed but grew smaller and smaller toward the front (Fig. 89*b*). An investigation of the pyrrhocorids based on more extensive material may provide information on the phylogeny of Heteroptera symbiosis.

The method of transmitting these symbionts seemed obscure at the beginning. Glasgow thought that the eggs themselves were infected, for he had discovered that infection of the larval intestine of pentatomids and coreids occurs only 24 to 48 hours after hatching, at a place where the crypts ordinarily develop, and he had repeatedly cultivated the symbionts from deposited eggs. At the same time, he attributed a certain role to supplementary infection in various later developmental stages, for he had often observed in these later stages the habit of sucking up all drops of fluid, even symbiont-containing excrement of fellow members in the species. Examining primarily *Graphosoma italicum* Müll. in which superficial smearing of the eggs during deposit seems plausible, Kuskop found that the crypts closest to the hindgut are substantially enlarged in the females and accordingly contain many more bacteria than those farther away. The epithelium is then quite flattened, the bacteria are much smaller, thus forming a specific transmission form, and are embedded in a richer secretion. Moreover, a particular physiological state of the end crypts is indicated by a diffuse yellowish coloring of the crypt content; this must not be confused with the general granular pigment of the epithelial cells, which appears even in forms where pigment is otherwise absent in the crypt gut. All such differentiations are lacking in males.

In connection with the differentiations, numerous symbionts occur in the hindgut of pubescent females. However, in *Graphosoma italicum* Müll. and in *Palonema* and *Carpocoris* species, Kuskop observed the bacteria even before the hatching of the larvae in the anterior regions of the midgut— or thought she saw them? From here, she said, they glided into the area of the future fourth section in the course of the next few days. Thus she was convinced that they are transmitted to the egg yolk through the micropyle after depositing the eggs, as actually is the case with *Dacus oleae*.

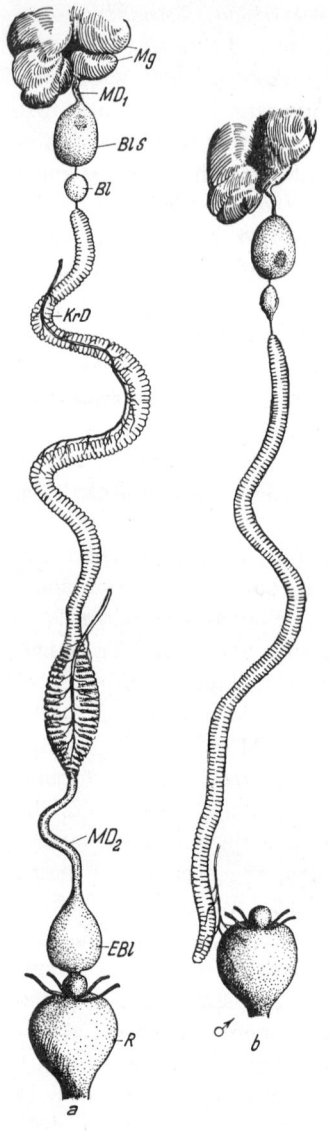

Fig. 92 *Coptosoma scutellatum* Geoffr. Intestinal tract (*a*) of the female, (*b*) of the male. *Mg*, stomach; *Md*₁, short midgut section; *BlS*, blind sac; *Bl*, small blister-shaped section; *KrD*, cryptic gut; *MD*₂, tubular section; *EBl*, terminal bladder; *R*, hindgut with Malpighian vessels. From Schneider.

Pierantoni (1932) and his student, Convenevole (1933), upheld the theory of ovarial infection, but this method seems improbable in a symbiosis localized in the intestinal lumen.

Rosenkranz (1939) and Schneider (1940) made observations which clearly indicate a smearing of the eggs at the moment of deposit. Rosenkranz was able to confirm Kuskop's information on the differentiation of special end crypts in the female and even found their number to be type-constant. *Palonema prasina* L. and *Pentatoma rufipes* L., the test objects he used most often, have ten and eight differentiations, respectively; *Peribalus vernalis* Wlff. and *Graphosoma* have only five. Moreover, Rosenkranz actually found irregular clumps of tightly crowded bacteria in the secretion which surrounds the eggs after deposit, but he did not find them in embryos and larvae which had been fixed at the moment of hatching.

In fact, the behavior of the young larvae, now long familiar to experts on Heteroptera but formerly not understood, seemed to indicate that the symbionts are acquired after leaving the eggshell. The solution to the mystery involves the so-called rest period, during which the young larvae grope around on the empty eggshells with their snouts, resisting all efforts to detach them forcibly. By the time they leave of their own accord, symbionts in great number can be clearly seen in the midgut.

Instead of several widened end crypts, there may also be a single pair of much larger saclets, differing to an even greater degree from the typical crypts. Such symbiont reserves, apparently quite rare and lacking in males, undoubtedly assist in the smearing of the eggs. Glasgow describes them in *Peliopelta abbreviata* Uhl. (Fig. 89), and Carayon in *Phlegyas*.

An astonishing variation of this transmission method was found in *Coptosoma scutellatum* Geoffr., the only representative of the plataspids found in Central and Southern Europe, though other species occur throughout the Eastern Hemisphere, from Russia and Turkey to Japan and Australia. It sucks principally at *Coronilla varia* but also at other Papilionaceae, and because of its unique transmission device, discovered by Schneider (1940), it is excellently suited for use as a test object in experimental symbiosis studies. The intestinal canal of this bug deviates from the previously described type in a number of aspects (Fig. 92). After the first ventricose segment of the midgut there follows a short narrow tubule enlarging quickly and directly into an almost spherical cecum which contains a homogeneous liquid and always a dark firm mass. The digestive intestine ends with this section, for only a delicate filament formed from the tunica propria still leads to the voluminous adjoining part of the midgut, which now is used exclusively in the symbiotic process and thus recalls *Tropidothorax*. Such an interruption of the digestive tract had previously been found among the Heteroptera only in *Ischnodemus sabuleti* Fall., and in *Brachypelta aterrima* Först., to be described later, and in the coccids (*Lepidosaphes*) and phylloxerids. In *Coptosoma* the posterior section begins with a new, small, colorless swelling, already filled with bacteria, and thereafter it has the form of a long slender tube bordered on each side by a series of crypts. We do not find the usual narrowing of the section connecting them with the central intestinal lumen. Indeed, it is now superfluous, for here the crypts open into it along their whole breadth, being formed from a fold in the intestinal wall rather than by evagination of saclets. The epithelium is extraordinarily flattened, but the musculature and tracheae are of the usual type.

In the males the crypt gut ends blindly and is attached to the rectum only by tracheae. In the female another important transmission device follows upon the section just described, and therefore the connection with the hindgut is accordingly maintained (Fig. 92a). In the following zone, lacking in males, the crypts, filled with small transmission states, are much larger, and the tracheae are regularly distributed throughout (Fig. 93a). Beyond the crypts, the tube becomes narrower and the epithelium forms only shallow folds. An eosinophilic substance, secreted in abundance, fills the lumen and envelops the numerous infection states. Following this section is an end vesicle more or less pear-shaped. The epithelium again is laid in deep folds and produces fibrous ropes of secretion which unite toward the middle to form a resistant covering around the porridge-like bacterial mass (Fig. 93*b*).   A ring musculature clasping the outlet of the end vesicle provides for rhythmic dissolution of the previously coherent mass, so that regularly formed, bacteria-filled capsules are sent first to the hindgut and then expelled from the body.

When the female sets out to deposit her eggs, this complicated apparatus

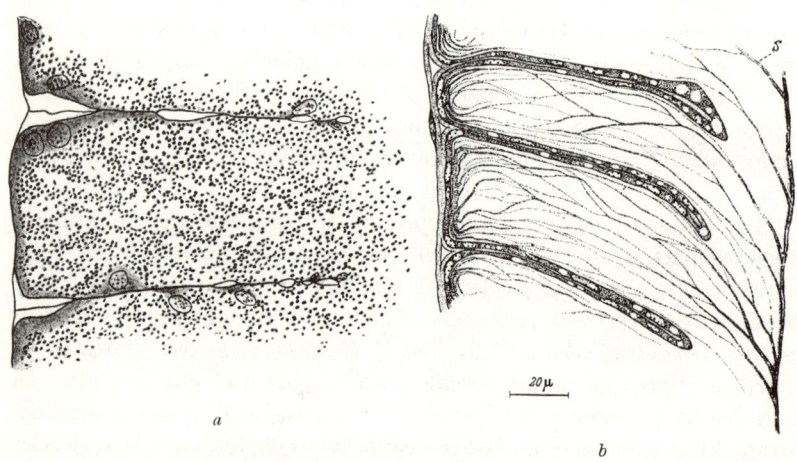

a

b

Fig. 93   *Coptosoma scutellatum* Geoffr.   (*a*) Terminal crypts with numerous infection forms; (*b*) section through the terminal bladder, in which the capsule wall is secreted and the cocoons are formed.   *S*, secretion.   From Schneider.

begins to function: The eggs are laid in two lines, diagonal to the longitudinal axis of the animal, hence arranged like spikes.   Through constant groping movements of the abdomen, the animal keeps oriented with regard to the deposition.   When the first two eggs are laid, it places close to the eggs in the angle formed by them, in the longitudinal axis of the spike, a small oval formation with a brownish content, the first of the bacteria-filled capsules.   After a short pause the third and fourth eggs appear and another capsule is placed.   This activity continues until ten or twelve eggs have been deposited.   Occasionally one or another cocoon may be missing (Fig. 94*a*,*b*).

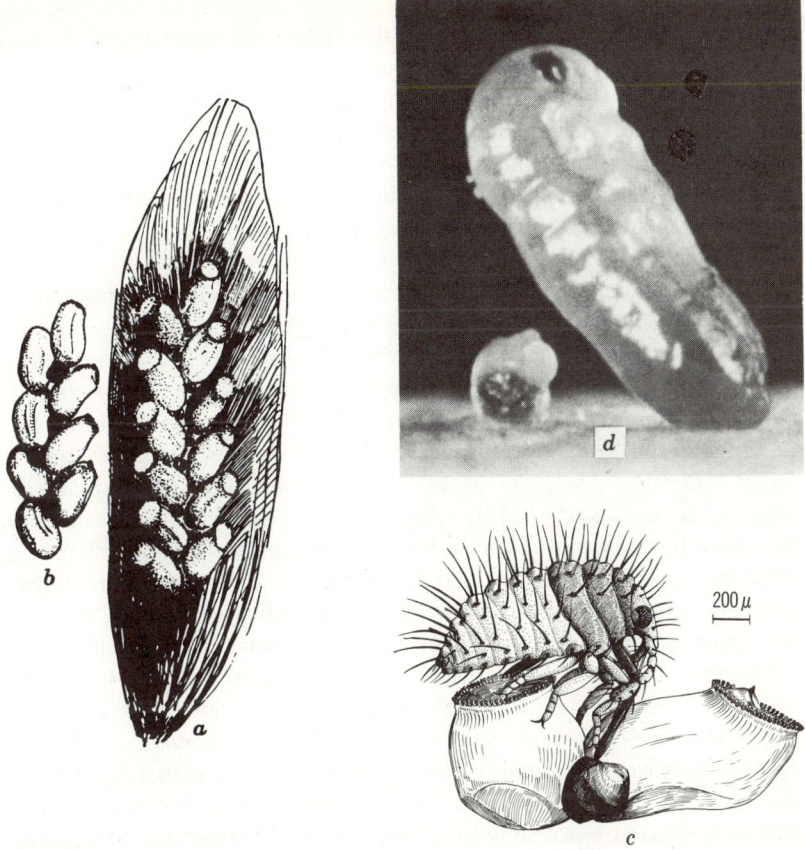

Fig. 94 *Coptosoma scutellatum* Geoffr. (*a*) Eggs on a vetch leaf; the bacteria-filled cocoons are between the eggs, (*b*) eggs from below, (*c*) a newly hatched larva sinks its proboscis into one of the symbiont-containing capsules, from H. J. Müller. (*d*) *Stilbocorus natalensis*. The mother has deposited a portion of symbiont-containing content of the gut beside the non-infected larva, from Carayon.

When the larvae finally break the cover of the eggshell with their egg teeth, they undergo the so-called rest period, which in actuality is spent in sucking out the capsules. H. J. Müller (1956) described the behavior of the newborn larvae in detail. After resting for 10 or 15 minutes, they move about on the deposit in greater and greater agitation, palpating the eggs with their rostra, and making deep punctures between them. Surprisingly quickly they come upon one of the hidden packages of symbionts.   After 30 to 105 minutes, during which they suck unceasingly, they leave the deposit to recuperate from their labors on the leaves of the host plant (Fig. 94c).   The capsules are empty, the punctured places can be recognized, and the symbionts are now found in the larval gut.   If the capsules are removed, the larvae unceasingly poke around on the eggs for hours until they are exhausted.   We have here a case in which new anatomic creations and highly specific instincts are uniquely combined to safeguard the symbiosis of hosts and guests.

I was able to confirm the existence of similar capsules also in several unclassified Javanese forms closely related to *Coptosoma*, but surprisingly the tropical plataspids living on tree-shaped Papilionaceae in the Cameroons revealed a very different picture (Carayon, 1949).   Here also a moderately developed bacteria-settled crypt zone is followed by a strongly developed glandular section of the midgut, but this time the latter furnishes material for a complicated ootheca measuring up to 40 mm. in length (*Plataspis flavosparsa* Mont., *Niamia bantu* Schout.; compare the illustration in Poisson, 1951).   Approximately 60 eggs are arranged in two rows, but it is not yet known how they acquire the symbiotic bacteria. *Coptosoma* evidently represents a type in which the formation of the ootheca has been abandoned and the glands which originally secreted them are now engaged exclusively in the service of symbiont transmission.

Recently Carayon (1963) discovered a no less original and interesting method of transmission, although this solution is much simpler and perhaps less efficient.   He established in *Stilbocoris natalensis* Distant, a lygaeid of Africa belonging to the Rhyparochrominae, that the females are viviparous and bring forth developed larvae, which are surrounded not by a chorion but simply by a thin embryonal covering.   The larvae are born separately, and the mother deposits next to each one a drop of symbionts from her highly developed crypt gut (Fig. 94d).   Shortly thereafter the larva sheds its covering, feels about with its proboscis, and remains for a long time with the newly found and vital material.   Only when it has found its regular food, the fruits of figs, does its proboscis begin again to function.

From studies by Rosenkranz we learn of another clear-cut case of externally transmitted symbionts which illustrates again that organisms are able to react according to the new requirements imposed upon them by the symbiotic process.   As already mentioned, this author found no communication whatsoever between crypts and intestinal tubes in the acanthosomines.   Of course, absence of such a connection would preclude smearing of the eggs by the anus, as in all other pentatomids.   Indeed, there are no enlarged end crypts, and the host organism is compelled to find a new method.   It does not merely take the solution closest at hand, namely, an infection of the oocytes, but establishes an entirely new symbiont depot in the ovipositor, similar to vaginal pockets and other devices in anobiids and cerambycids. It had previously escaped notice that *Acanthosoma haemorrhoidale* L. and its numerous relatives have a paired transmission organ on the ventral side of the posterior end of the body.

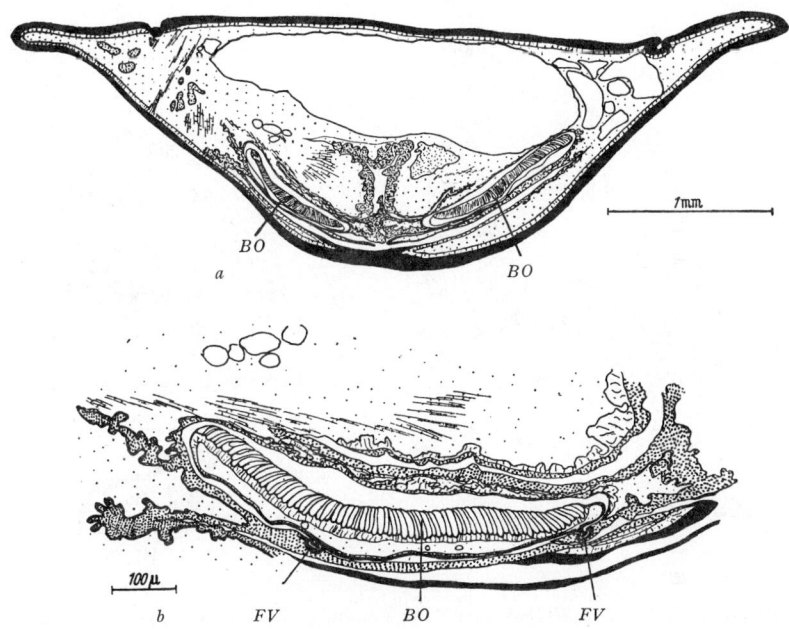

Fig. 95 *Acanthosoma haemorrhoidale* L. (*a*) Transverse section through the posterior end of a female imago with both lubricating organs; (*b*) one lubricating organ greatly enlarged, with chitinous tubules and pleated toothing. *BO*, lubricating organ; *FV*, pleated toothing in transverse section. From Rosenkranz.

This organ, pyriform and vividly colored, has a chitinous lining with numerous bacteria-filled tubelets (Fig. 95).

Well hidden by a complicated pouch formation, this organ is arranged in such a way that eggs gliding down must necessarily squeeze out a portion of its contents into the vagina, which communicates with the lubricating organ. A pair of dovetailing chitinous ridges on each side prevents gaping of the fold covered with symbiont containers when the genital segments are tautly stretched during egg deposit. The closely crowded tubelets have only a small opening facing the vagina and must be thought of as having originated from bundles of hair, of the type we have already observed in the devices of anobiids (Fig. 96).

All other acanthosomines investigated have fundamentally the same organs. Since they are clearly defined in chitin preparations, it was possible to use two species from Chile and seven palaearctic species.

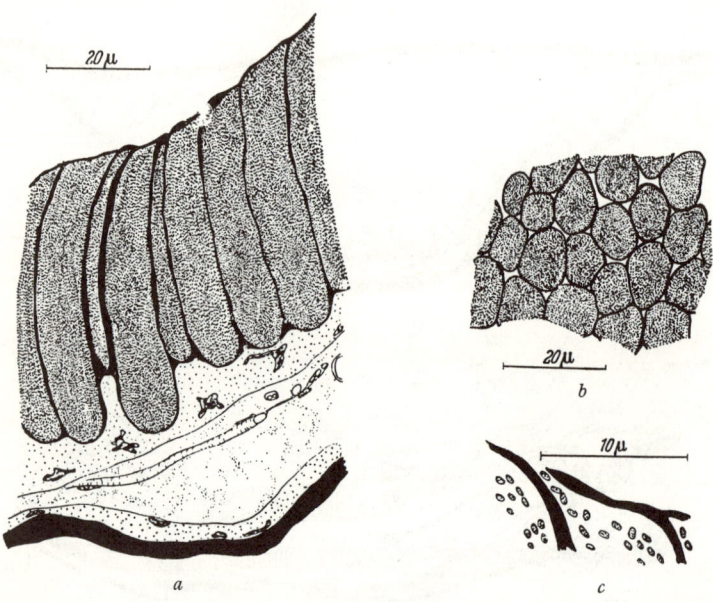

Fig. 96   *Acanthosoma haemorrhoidale* L.   The chitinous tubules of the lubricating organ are filled with symbionts: (*a*) in longitudinal section, (*b*) in transverse section, (*c*) their opening.   From Rosenkranz.

As would be expected, the larvae remain for a while on the eggshell and are infected with the bacteria which cling to its surface.

Unfortunately, lack of material made it impossible to observe how these organs are formed and then colonized. In the second larval state the crypts are still connected with the intestine. Thus their obstruction must take place during the third to fifth larval state, and some of the symbionts must previously have infected the ovipositor spaces, which have already begun to form, by way of the hindgut and the anus.

From observations made by Pierantoni (1951, 1954) with *Tropidothorax*, at first only with fixed animals, these lygaeids also appear to have adopted exceptional methods of transmission. Here the four Malpighian tubules are converted into long formations with sudden constrictions and frequent convolutions, and their epithelia and lumina are likewise thickly populated with bacteria. In this case, which is unique among the Heteroptera, in all probability the renal organs are used to smear the eggs, as in *Donacia*, but this unusual object is still in need of more thorough investigation.

According to the statements made on pentatomids and acanthosomines there can be no doubt that the Heteroptera symbionts in an overwhelming majority of the cases are transmitted externally at the moment of egg deposit. That other paths may also be chosen is seen from the splendid observations of Schorr (1957) with *Brachypelta aterrima* Först., a member of the cydnid family, thus the only Heteropteron burrowing deeply in loose sand. The imago alone moves also on the surface. Roots and leaves of the *Euphorbia* species are their favorite food. No external smearing of eggs with symbionts is observable when the eggs are deposited, 30 to 40 at one time. After hatching, the young larvae remain close together for 6 to 8 hours until instinct impels them to cling stubbornly to all parts of the maternal body within reach. For 8 to 9 days the mother lives with her young in this manner, and during this period the larvae have as their sole food the watery drops, filled with dense balls of symbionts, which emerge from the maternal anus. Larvae isolated within the first 2 days are unable to survive, and normal development is possible only after an association of 60 hours.

Very interesting is Schorr's comparison of the intestinal development in the two sexes. The formation of crypts begins shortly before the first molting; after they are completely formed, symbionts densely populate the evaginations, tightly crowded and arranged in two rows, and the intestinal lumen which is now very narrow. The nuclei of the epithelial cells have been pushed to the wall by the symbionts. In both sexes, exactly as in *Coptosoma*, the gastric part is constricted off from the bacteria-containing part, which is now used exclusively in the symbiotic process. In male imagines this section is reduced to an extent observable in no other heteropterous bug, until finally it is a mere filamentous formation practically devoid of content. In female imagines this unusual transmission device is impressively developed. In an animal containing ripe eggs the posterior crypts are already quite large, although they have almost completely disappeared from the anterior section. During brood care the anterior crypts are also enormously enlarged, and by virtue of prolific bacterial reproduction all crypts are once again abundantly populated.

Not only such complicated devices but also quite primitive transmission methods may be found, as shown by Huber-Schneider (1957) with *Mesocerus (Syromastes) marginatus* L. In this test object the eggs were not conspicuously smeared before deposit.   The larvae leave before they are colored, the typical palpating of the eggshells appears not to occur, and the mortality of isolated young larvae is very high.   In these animals living in the open in family groups it is apparently the habit of sucking up the bacteria-containing excrement of their fellows that guarantees the symbiosis which has become indispensable to them.   Brecher and Wigglesworth (1944) arrived at a similar concept regarding *Rhodnius* and *Triatoma*, two bloodsucking reduviids which will be discussed later.

Malouf (1933) was of the opinion that the symbionts of the pentatomid *Nezara viridula* L. are localized not only in the usual place but also in the appendages of the male sexual apparatus and thus infiltrate the eggs with the sperm, but what he saw was in actuality the secretions occurring so frequently among the Heteroptera and often bearing a resemblance to bacteria.   Woodward (1949) was misled by the same similarity into speaking of bacterial symbiosis even in the case of nabids, predatory forms which never possess symbionts (Carayon, 1951).   Nor can we take seriously the opinion of Bonhag and Wick (1953) who mistakenly thought they had observed symbionts of two types in the epithelium of the "subgenital glands" in *Oncopeltus fasciatus* (Dallas), a species in which Steinhaus, Bathey, and Boerke had made a fruitless search for bacteria-containing intestinal evaginations.

Rosenkranz, like Schorr, provided accurate data on the formation of crypts during larval development.   In *Palonema prasina* L. they begin to develop 2 or 3 days after hatching.   On all sides of the corresponding intestinal section, already filled with bacteria, there first appear small well-separated evaginations.   During the second larval state and its concomitant growth, they are arranged in two rows and border closely on one another.   In the third state they are more sharply set off from the gut by a longitudinal constriction of the latter.   The fourth state marks the appearance of a second longitudinal fold which brings forth the four transverse crypt rows.   The final condition is achieved in the fifth larval state.   The typical pigmentation of the crypt-gut sets in only after the symbionts have colonized it.   In females the enlarged end crypts appear for the first time, not in the pubescent animals but in the larvae.

With *Coptosoma* the symbionts quickly reach the posterior section of the midgut which remains free of yolk and is filled with secretion.   Those which remain in the yolk-filled section disintegrate, and, when the yolk is resorbed, intestinal abscission sets in at once.   The enlarged end crypts of the female are already present in the third larval state, whereas the two end sections reach their full development only in the imaginal form.

Kuskop, Pierantoni, and Convenevole established that a certain polymorphism of the symbionts prevails within any one species, and later Rosenkranz and Schneider provided a detailed study of the basic cyclic changes. In *Palonema prasina* the pigmented end crypts used in transmission are filled with short rodlets only 1 μ in length, and the first larval state accordingly has very small symbionts. When the crypts increase in size in the second state, the symbionts lengthen substantially, so that filaments up to 16 μ in length occur. In the following stages the symbionts continue to reproduce but again become shorter. In the fifth larval state they usually measure only 4 μ. When this state is reached in autumn, the two sexes take different paths. In the male one sees renewed growth up to an average length of 10 μ; in the female further disintegration leads to forms of approximately 2 μ. Not until the following summer at the time of egg deposit does a new wave of division occur in the enlarged end crypts, and the resultant forms now measure only 1 μ.

The situation is the same for other pentatomines and scutellerines. The lubricating organs of the acanthosomines are also filled with infection forms of small size, and the crypts with much longer rodlets and filaments. Heteroptera symbionts often grow into formations which are far removed from the typical bacterial form. For instance, Glasgow found gently turned tubes up to 50 μ in length in *Peribalus limbolarius* Stål, and completely irregular, corkscrew-like involuted forms, usually measuring 10–30 μ, in *Murgantia histrionica* Hahn, although there were among them some giants 100 μ or longer (Fig. 97). Steinhaus (1951) gave more exact information on the variable morphology of gram-negative *Murgantia* symbionts and furnished photomicrographs for the very divergent symbiont forms specific for the individual Heteroptera (*Chlorochroa, Chelinidea*). Steinhaus, Bathey, and Boerke (1956) again studied the morphology of several Heteroptera symbionts, describing especially those of *Chelinidea vittiger* Uhler and *Euryophthalmus cinctus californicus* (van Duzee) as new flagellated species of *Pseudomonas*.

The *Coptosoma* symbionts also show clear differences in form in the two sexes. The infection stages are coccoid; in the young larvae they are only moderately developed, and some become sausage-shaped, arched, and rather bloated. In the females they retain their size, with 3.5 μ the maximum; in the males they grow into long filaments and filamentous chains with links measuring up to 17 μ. *Brachypelta* symbionts differ radically from the other Heteroptera symbionts. According to Schorr, they are roundish oval formations resembling four beans and are surrounded by a closely fitting membrane; thus they resemble the symbionts of some cicadas or of *Pseudococcus*. In the various stages, now one and now the other condition predominates. The degeneration of symbionts in the male has been described in exact detail. A *Cydnus* species was included in the study; in spring it had filaments 10–70 μ in length, which in the females disintegrated into rodlets 2–4 μ long by June (Plate VI*d–f*).

A more or less adapted companion form accompanies the actual symbionts rarely. For example, among the deep-staining short rodlets of *Stegonomus pusillus* H. S., an additional form was encountered in the shape of long, very delicate, and weak-staining filaments. In cultural experiments by Huber-Schneider an organism distinct from the symbionts and

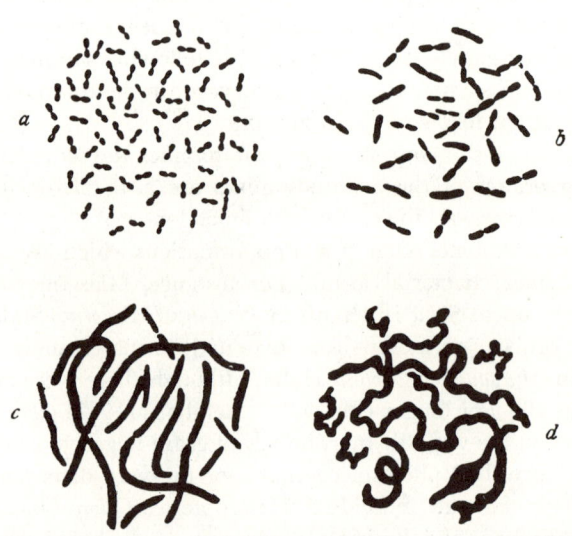

Fig, 97   (*a*) *Anasa tristis* De G., (*b*) *Euchistus servus* Say., (*c*) *Peribalus limbolarius* Stål, (*d*) *Murgantia histrionica* Hahn.   Symbiotic bacteria from the crypts of the midgut. From Glasgow.

of possible importance to the host animal was obtained with great regularity from the third midgut section.

As early as 1914 Glasgow had experimented with cultures. With some species, in which the symbionts were obviously highly adapted and modified, as in *Murgantia* and others, his efforts were in vain. With other species, such as *Anasa*, *Alydes*, and *Metapodius*, cultures were successful when a decoction of pumpkin stems and leaves was added to the bouillon. With such an addition the *Anasa* symbionts, which were always immotile when taken fresh from the organ, showed great motility. Glasgow classifies them in the large group of fluorescent bacteria. He could prove serologically that it was actually the symbionts that were cultivated. An immunization serum obtained by intraperitoneal injection of dead symbionts in guinea pigs yielded instantaneous coagulation of the cultivated organisms, even with radical dilution. Steinhaus (1951), though unsuccessful in cultivating *Murgantia* symbionts, had success with two species of *Chelinidea*, although he used ordinary media such as nutrient agar with and without addition of glucose. The minute gleaming white colonies grew very slowly at first but developed more rapidly after subculture. In 1956, in collaboration with his co-workers, he reported successful cultures of several Heteroptera symbionts, and one year later Huber-Schneider described the various forms which *Mesocerus* symbionts assume with the media used (Plate VI*a–c*). The gram-negative double rodlets, which tend to form chains in a liquid medium, yielded clear nuclear structures with Robinow staining.

In 1923 I discovered paired mycetomes situated in the area of the fat body completely independent of the intestinal canal, thus confirming the existence of symbiosis in the bedbug. This finding, disturbing the unity of Heteroptera symbiosis, was surprising at the time, but today we know that many phytophagous bugs have typical mycetomes and send their symbionts into the oocytes. With the blissines, a subfamily of the lygaeids, the organs in question are situated in the abdomen on both sides beneath the hypodermis. There are two species among them which occur only in Central Europe, namely, *Ischnodemus sabuleti* Fall., which sucks at all types of water plants, and *Dimorphopterus spinolae* Sign., which lives on *Calamagrostis* in sand dunes.

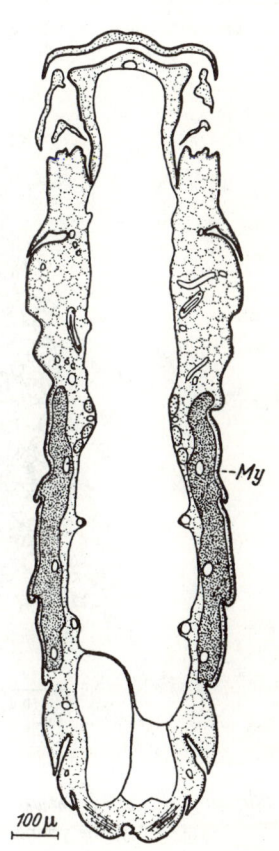

Fig. 98   *Ischnodemus sabuleti* Fall. Frontal section through a female imago with paired mycetome accompanying the gut. *My*, mycetome. From G. Schneider.

Only the first could be investigated. In three places dorsoventral bundles of muscle pass through the transparent organs embedded in the fatty tissue (Fig. 98). Large syncytia, some with bizarrely shaped nuclei, are thickly

populated with bacteria and are surrounded by a flat epithelium.   The
end branches of numerous tracheae which supply the organ penetrate
the syncytia to some extent.   The inmates are multiform bacteria which
appear originally as slender rodlets and filaments but may later assume
the distended form of vesicles and tubes (Schneider, 1940).

Transmission is achieved by early infection of the gonads.   In the

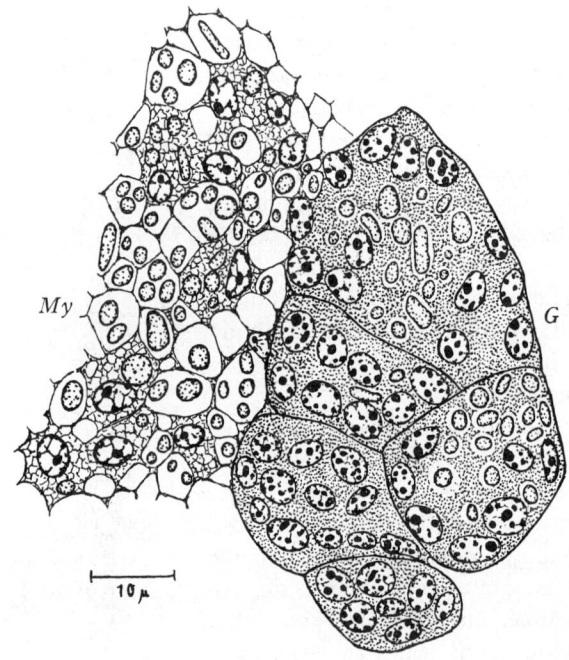

Fig. 99   *Ischnodemus sabuleti* Fall.   Early infection of the female embryonic gonad.
My, the young mycetome; *G*, gonad.   From G. Schneider.

female embryo the distended vesicles and sausage-shaped forms can be
plainly seen as they are transmitted from the young mycetomes to directly
adjacent locations in the ovaries, evenly permeating the cell material,
in which at this time it is not yet possible to distinguish between nutrient
cells and oocytes (Fig. 99).   Later the symbionts invade particularly the
nutrient cells at the end, collecting throughout in nests.   In the plasma of
very young oocytes which do not as yet have contact with the nutrient cord,
one occasionally finds isolated symbionts with great variations in form and
staining.

With the flow of secretion from the nutrient cells to the oocytes, the symbionts finally glide in great numbers into the egg plasma, and yet one cannot speak of specific infection stages. Later the plasma growth is fostered at the expense of symbiont reproduction to the extent that only a cap-shaped aggregation of rodlets is left at the posterior end.

Accordingly, the blastoderm cells occupying that area are the first to be infected when the embryo begins to develop. Originally arranged in a single layer, these young mycetocytes, beginning to increase even before invagination sets in, finally form a cellular mound which closes the amnion hollow and then is pushed along almost to the opposite pole by the germ

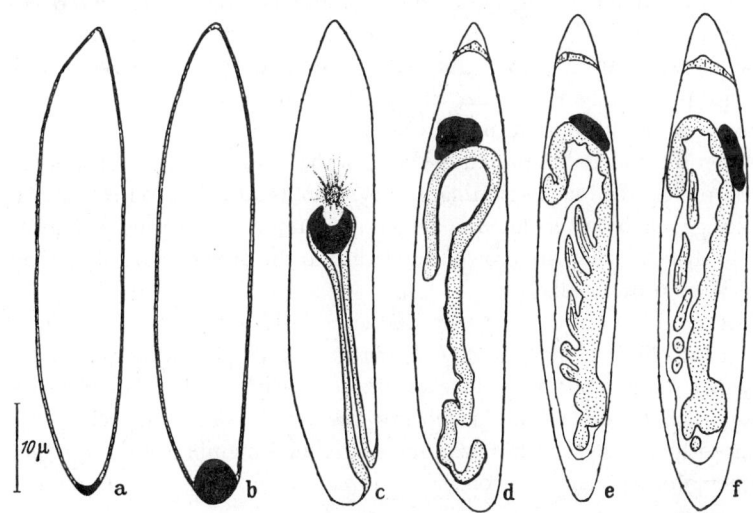

Fig. 100   *Ischnodemus sabuleti* Fall. Embryonic development. Behavior of the mycetome rudiment.   From G. Schneider.

band invaginating in this location. In the process a strange plasma radiation forms on top of the invagination, giving the impression that it is about to drag the latter behind it. Something similar is found occasionally in certain scale insects, cicadas, Anoplura, and mallophages, yet we can draw no definite conclusions about the significance of this striking structure, as it also occurs without reference to symbiosis, although far more rarely. When the posterior end of the germ band doubles back to form an *s*, the still unpaired mycetome rudiment is detached and, now flattened off, borders on the abdominal region of the germ band. The final state is then brought about through abscission into a right and a left half (Fig. 100). The *Ischnodemus* mycetomes thus develop very similarly to those of *Cimex lectularius* and numerous Homoptera.

Surprisingly, not all members of the subfamily of blissines have mycetomes.   Carayon reports in correspondence that *Blissus* and *Dimorphopterus* house their bacteria in crypts of the midgut.

Paired mycetomes were found in several species of *Nysius*, a genus of the lygaeines.   The oblong or oval organs, situated close to the gonads and composed of syncytia or of uninucleate mycetocytes, are colored an intensive red by a pigment occurring between the thickly lodged bacteria. Again the symbionts are transmitted to the young oocytes, but in all other details *Nysius* differs from *Ischnodemus*.   Following close upon the sterile nutrient chamber where the oocytes begin to form one behind the other, the ovarioles have a sharply delimited zone which has the same red color of the mycetome (Fig. 101*a*).   Examination of cross sections shows that these ovarioles are a type of filial mycetome and that particularly at the anterior pole the symbionts are transmitted directly to the oocytes without recourse to the fibrous strands (Fig. 101*b*).

When the oocyte has passed the zone of the filial mycetome, its infection is complete.   In the beginning the symbionts are distributed over the whole egg, each one enclosed in a vacuole, but with increasing yolk formation in this case they are crowded together in the area of the anterior pole in a dense round ball.

The symbionts are rodlets or filaments 3–11 μ in length, frequently linked into little chains, as in *Ischnodemus*.   A satellite form is found, in varying numbers but nevertheless quite regularly.   It is a smaller, weak-staining bacterium which penetrates the ovaries and egg cells and is apparently on the road to becoming a qualified member of the symbiotic partnership.   A similar coccoid companion form is encountered in *Ischnodemus* but not with such regularity.

Schneider also found mycetomes in representatives of a third subfamily of the lygaeids, the cymines.   *Cymus claviculus* Fall. and two *Ischnorrhynclus* species were investigated.   In the former no symbiotic devices could be discovered; the two *Ischnorrhynchus* species, *I. resedae* Pz. and *I. ericae* Horv., have an unpaired red racemose organ, which extends along the ventricose part of the intestine in the medial position or slightly to one side.   Long, thickly matted, filamentous chains, some of them branched, fill dinucleate mycetocytes surrounded by a flat epithelium and are often accompanied by a less adapted short rodlet.   As in *Nysus*, a girdle-shaped infected zone in the ovarioles is the instrument of symbiont transmission.   Although it is not red like the mycetomes, but orange, the symbionts finally form a ball at the posterior egg pole and may be recognized immediately by their pigmentation.

Finally, we must mention that Schneider chanced upon a widespread bacterial invasion of the abdominal fatty tissue in a representative of a fourth subfamily of the lygaeids, the geocorine *Geocoris grylloides* L., but was unable to decide whether this was a true case of symbiosis. Other species of the same genus showed nothing comparable, but this in itself would not exclude its regulated occurrence here, for in many other subfamilies of the

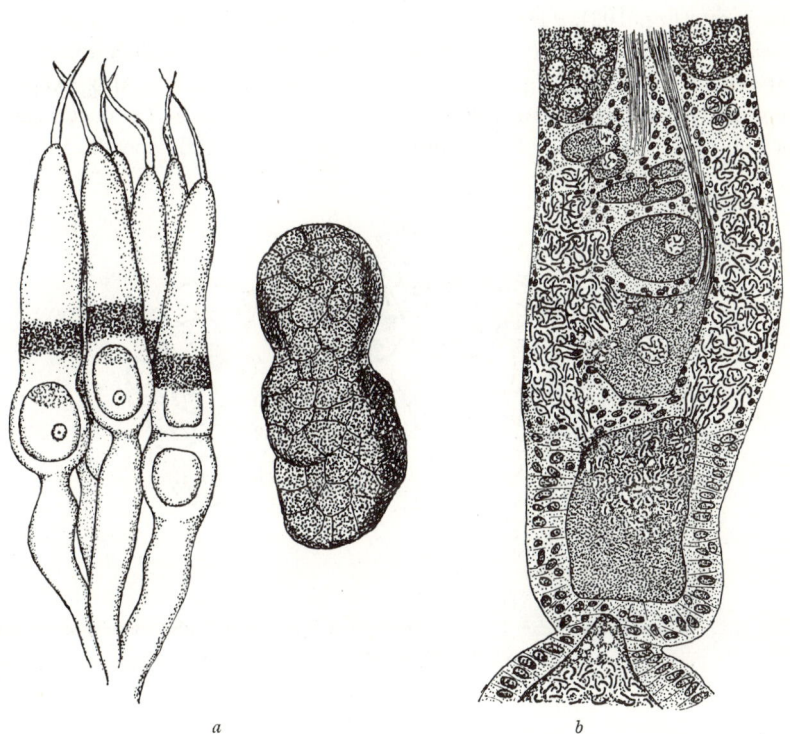

a   *b*

Fig. 101   *Nysius senecionis* Schill.   (*a*) Ovarioles with pigmented infection zone and the neighboring, similarly pigmented mycetome; (*b*) the pigmented zone of an ovariole in section. The ovocytes are infected by way of the symbiont depot. From G. Schneider.

Heteroptera the symbiont bearers exist side by side with non-symbiotic species.

    Later we shall analyze the relations between system and symbiosis, but here we shall only say that they are quite obviously based on the same ecological, that is to say, nutrio-physiological factors which are decisive for the presence or absence of symbiosis.

COCCIDS

LECANIINES.   The frequently investigated symbionts of the lecaniines have been familiar to us for almost 100 years.   In all the many species studied they float independently in the lymph or live intracellularly in the elements of the fatty tissue.   The fat cells undergo slight changes at the most (Fig. 102).   However, Tremblay (1961) showed in *Sphaerolecanium prunastri* Fons. and *Eulecanium coryli* L. that through the infection of the fat cells a considerable growth of the same can take place, also, accompanied by polyploidy resulting in giant nuclei and often by the occurrence of two nuclei per cell.   The symbiont form is completely specific for the

Fig. 102.   *Lecanium hesperidum* L.   Symbionts free in the lymph and in cells of the fat body (fresh, teased preparation).   Original.

individual species of the host animals and shows considerable diversity. The forms may be oval, clavate, guttate, or lemon-shaped, or they may be slender and cigar-shaped with abruptly tapering ends.   Sometimes they seem rather knotted, and occasionally they grow out to form longer tubes. Sharply refracting inclusions are almost always present, sometimes, in small numbers; at other times they form dense groups, preferably around the two poles; and at still other times they permeate almost the whole plasma in great numbers.   Occasionally fat droplets and vacuoles filled with cell sap are found in the plasma (see Buchner, 1921, plates).

All previous information referred exclusively to the tribe of the Lecaniini. But two other tribes, the Aclerdini and the Micrococcini, behave in fundamentally the same way, according to observations with *Aclerda berlesei Buffa,* in which the formation of giant cells with polymorphic nuclei takes place quite often, and a species of *Micrococcus.* Whether finer differences exist must be determined by future research.   At any rate, the facts found in connection with the symbiotic devices are sufficient to confirm the classification of *Micrococcus* among the lecaniines (Balachowsky, 1948), although Leonardi (1920) still classified it among the pseudococcines.

Reproduction occurs exclusively through budding.   It may occur apically or laterally, at one or both poles.   When reproduction is intensified, other buds may form behind the bud which is not yet detached and

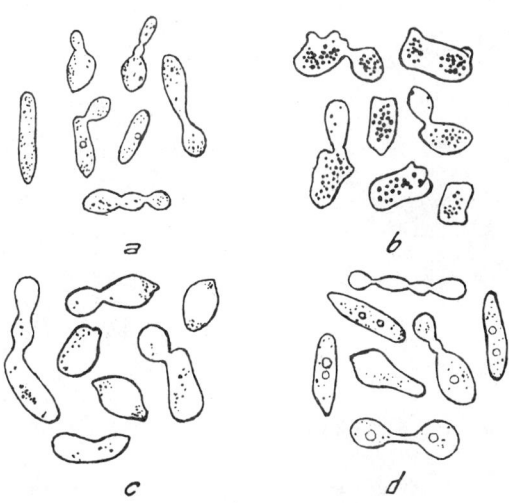

Fig. 103   (a) *Lecanium tessellatum* Sign., (b) *L. longulum,* (c) *L. hemisphaericum* Targ., (d) *L. hesperidum* L.   Symbionts inhabiting the lymph and fat cells.   From Buchner.

in this way give rise to little chains of three or four links, from which occasional tube forms, most of which still have numerous constrictions, may easily be derived (Fig. 103).

Transmission takes place by ovarial infection, as with all Homoptera. Where the few large nutrient cells of the still young oocytes are attached, relatively few symbionts (approximately 10 to 15) appear in the follicle cells connecting nutrient cells and oocytes.   Before long they are advanced to the space around the nutritive cord, where they wait until the nutrient cells are mere degenerating rudiments still affixed to the now yolk-rich full-grown egg; then they glide toward the egg along the cord which is also regressing and in that way come to lie between the follicle and the egg surface. The egg, forming a little scrobe at this place, retreats, and, when the symbionts sink into this furrow, the plasma closes over it again (Fig. 104). The egg at this time is in the stage of the first maturation division.

We have very little information about the behavior of the symbionts

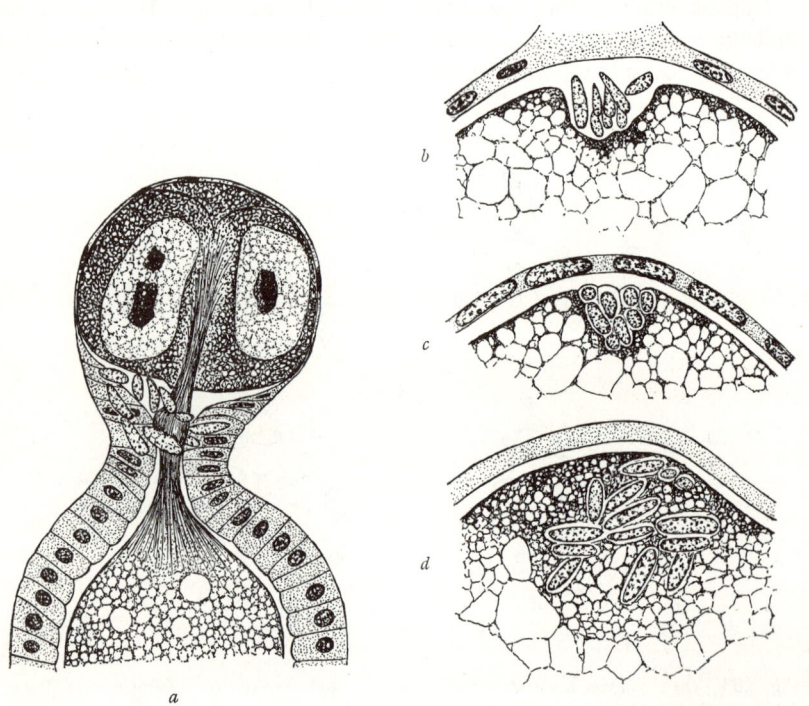

Fig. 104   *Lecanium corni* Bouché.   (*a*) The symbionts penetrate the follicle; (*b*) they sink into the receiving groove of the egg; (*c, d*) the latter closes behind them.   From Buchner.

during embryonic development.   Breest (1914) reported that the symbionts of a *Lecanium* species do not disperse at all but temporarily build a nest under the blastoderm, which is then penetrated by cells apparently representing blastomeres left behind in the yolk.   A well-defined cellular construction arises that survives the involution of the embryo but then disintegrates and makes space for a profuse settlement of lymph and fatty tissue. We shall use the term "transitory mycetome" for these temporary mycetome-like structures which are found among other coccids and lead, as in this case, to rapid degeneration of the mycetocytes or to their dispersal. According to Tremblay (1961), who also describes the infection in the form already discussed, the symbionts of *Sphaerolecanium prunastri* always remain at the border of the upper pole without coming in contact with the cleavage cells during the development of the germ band, which is pushing the primoidial sex cells before it.

As long as the young lecaniines are still motile, the number of symbionts remains rather limited, but, when the period of increased growth of the insects begins some time after they have settled down, the symbionts reproduce rapidly and sometimes invade the host organism to a terrifying degree. Later the intensity of the sprouting appears to decrease, and clear indications of degeneration may occur. Particularly during and after egg deposit it may be observed that some of the symbionts elongate into the already mentioned tube forms, which are also typical of moribund and dead animals (Schwartz, 1932).

According to observations by Poisson and Pesson (1937) with *Pulvinaria mesembryanthemi* Vallot, the symbionts in this test object are often housed in juvenile wax cells which originate from leucocytes and still maintain their phagocytic qualities for a while. Such cells cannot develop into mature wax cells and are finally destroyed, but the yeasts in them increase and finally are returned to the blood, the authors state. Since nothing similar has been reported as yet, a confirmation is desirable.

A number of authors report successful cultures of the lecaniine symbionts. Berlese's statements (1905) that the symbionts of *Ceroplastes rusci* L. can be cultivated easily, and those of Conte and Faucheron as well (1907) are probably of historical interest only. Later Brues and Glaser (1921) reported success with *Pulvinaria innumerabilis* Rath. in approximately 50 per cent of the inoculations, potato agar with and without addition of maple sugar proving to be especially suitable. However, the question of cultivability was not thoroughly tested until Schwartz's experiments in 1924. He had far greater difficulties than his predecessors, but this was undoubtedly due in large measure to his more fastidious methods. Peptone-cane sugar solutions were particularly favorable culture media for the symbionts of several *Lecanium* species. Cultures also grew well in caterpillar lymph, but the addition of extracts of the host plant had an unfavorable effect.

Generally speaking, extensive assimilation of the organisms more or less different within the host animals took place in all these cultures. Just as the milieu of the host influences the morphology of the symbionts in nature, so does the type of culture medium in these experiments. On malt agar, yellowish gleaming discs of mucilaginous consistency, in which the yeasts sprouted almost entirely as individual cells, appeared around the place of inoculation; on agar with sodium hippuric salt (0.1%) and cane sugar (6%), the central disc remained small and from it grew a richly branched mycelium with clumps of individual cells at the tips.

In Schwartz's experiments the *Lecanium* symbionts were retarded in their growth and had to be precultivated by the drop method, but Benedek and Specht (1933) reported far hardier cultures of the *Lecanium corni* symbionts which could be continued indefinitely in subcultures. They claimed that they obtained vigorous reproduction without difficulty in a 5% glucose bouillon and that from such cultures they could cultivate others on various solid media. In approximately half of the cultures another bacterium appeared, a sporogenous one closely related to *Bacterium megatherium*, which they found again in the scale insects and termed an auxiliary symbiont. However, their interpretation of these organisms as a *Torula* species scarcely deserves serious consideration, for they merely rinsed the animals in distilled water, cut them in two, and then inoculated them, so that it is very probable indeed that they were merely cultivating organisms adhering externally to the animals.

Lindner's theory (1895) concerning the systematic position of the lecaniine symbionts, namely, that they are true saccharomycetes, can no longer be upheld. Schwartz comes to the conclusion that they are ascomycetes resembling *Dematium pullulans*, apparently pyrenomycetes, which usually are reproduced in the host animal only in the form of conidia. He tested his cultures with respect to fermentation physiology and established the presence of amylase, saccharase, emulsin, trypsin, lecithinase, lipase, and urease, as well as the capacity for splitting hippuric acid.

KERMESINES (HEMICOCCINES). The kermesines (hemicoccines), which formerly had a certain economic importance as dyestuffs, comprise only the one genus *Kermes*, of which approximately 40 species, occurring exclusively on oak trees, have been classified. Šulc's data on the symbionts in *Kermes quercus* L. (1906) were very brief. He found the body cavity invaded by small saccharomycetes which were elongated to form little staffs tending to be blunt at one end and pointed at the other. Vejdovsky, to whom Šulc's material was available, found them in quantity chiefly in old females, and he theorized that they had the task of destroying the tissue of females and thus of creating the necessary space for the embryos (!). I investigated *Kermes quercus* and *K. vermilio* Blanch. In the latter I saw no trace of such organisms; in the former I found enormous masses of yeast-like fungi finally developing into a luxurious mycelium in the brood space. Here they penetrate the embryos and may even overwhelm them. They are lacking in the eggs and normal embryos. Thus they are evidently saprophytes of frequent occurrence. I have never been able to detect clear-cut symbionts in either of the two test objects (Buchner, 1955).

ASTEROLECANIINES.   The symbiosis of the asterolecaniines was investigated by Shinji (1919), Richter (1928), Walczuch (1932), and Mahdihassan (1933, 1951).   Their symbiotic devices have certain similarities to those of the lecaniines but nevertheless are distinct from them.

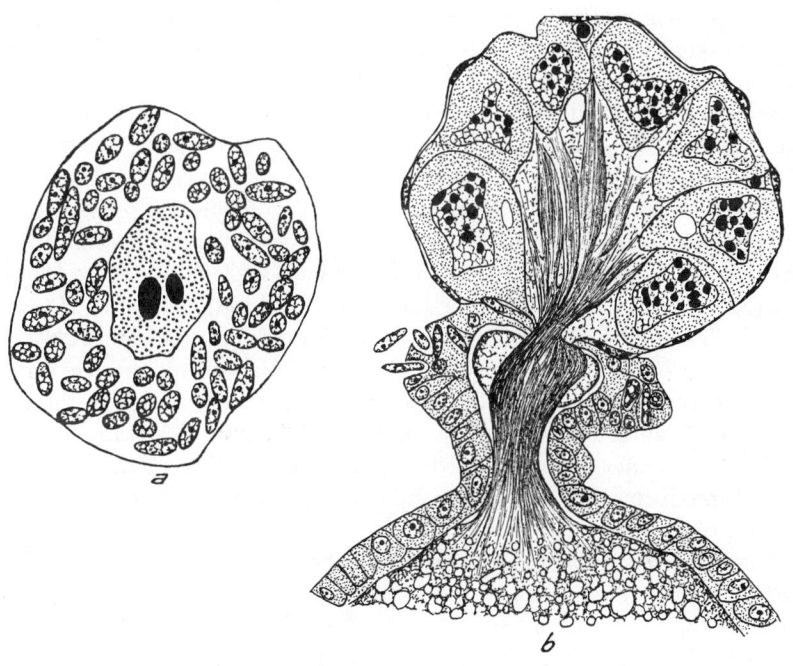

Fig. 105   *Lecaniodiaspis africana* Newst.   (*a*) Mycetocyte from the fat tissue, 750×; (*b*) the symbionts infect at the anterior pole of the egg, 450×.   From Walczuch.

The habitats are specific cells, evenly distributed in the fatty tissue, with dense protoplasm and with no inclusions of fat substances.   With *Lecaniodiaspis africana* Newst., Walczuch's test object, the cells of one or two nuclei vary considerably in size and the symbiotic population of their plasma varies even more.   Some cells are filled to bursting, others contain very few symbionts, and still others have none at all.   In the last case they are easily differentiated from the fat cells by their plasma constitution (Fig. 105*a*).   Individual mycetocytes display clear signs of degeneration: their plasma is harder to stain; the surface appears ragged; they drop their contents, so that very frequently one sees free symbionts in the lymph which bring about an infection of mycetocytes still sterile.

In form they are similar throughout to the yeast-like symbionts so typical of the lecaniines: they are spindle-shaped; they have a frothy protoplasm; a strongly chromatic chondrium is always present; and they reproduce through budding.   Thus it is safe to interpret them as the conidia of an ascomycete.

The manner of transmission is also the same as in the lecaniines.   The fungi appear in the follicle between the nutrient cells and the oocytes, then pass through the follicle into the space around the fibrous nutrient cord, which here too is surrounded by egg plasma (Fig. 105*b*).   In another species not yet classified, I observed yeast in a free state, particularly in the body cavity, as it was transmitted to the egg at the same place as in *L. africana*.

On the other hand, Shinji reports on the basis of defective preparations that in *Lecaniodiaspis pruinosa* Hunter the corresponding specific cells of the fat body are populated by rod-shaped symbionts and that these are transmitted to the ovarioles at the same place at which the yeasts of *L. africana* are.   This finding is in consonance with Richter's statement that *Asterolecanium variolosum* Ratz., which lives on oak trees, has little rod-like symbionts which populate cells, in this case of one or more nuclei, so thickly that their form can be recognized only with difficulty.   The same is true of *Cerococcus ornatus* Green, in which Mahdihassan found rod-shaped formations reproducing through transverse division and forming forks or little chains, obviously bacteria.

Asterolecanium aureum Boisd., a form brought to our hothouses from India, again occupies a special position with regard to symbionts.   In young animals the mycetocytes are so crowded that they give the impression of being two lobate organs, yet later they are dispersed in the fatty tissue in the manner typical of the subfamily, so that the term "transitory mycetome" is applicable again.   In the beginning the organisms which thickly fill the cells are small and usually dumbbell-shaped; in the older animals they are replaced by oval and round forms more loosely arranged, for the most part, and considerably distended.   Appearing again at the upper pole, they are transmitted to the egg not at the time of the maturation divisions but in a subsequent quadrinucleate segmentation stage.

The multiplicity of the phenomena and the complete lack of information on embryonic development led me to make a rather large collection of asterolecaniines in Java, although the arrangement of the material is still in its early stages.   With *Cerococcus, Pollinia*, and a number of *Asterolecanium* species, I always encountered round or oval symbionts in the diffusely distributed mycetocytes, but never rodlets. The infection always took place at the upper pole, where later a group of mycetocytes provided with yolk nuclei awaits the germ band which is growing forward.

Only the previously mentioned *Lecaniodiaspis* species seems to behave at all differently. During segmentation a marginal proliferation occurs which, as in *Pseudococcus*, is traceable to the nuclei of the polar globules, and in all probability their derivates later join the symbionts by a strangely circuitous route.

A sporadic appearance of yeast-like symbionts as in *Lecaniodiaspis*, has also recently been observed with other coccids (*Rastrococcus*, Buchner, 1957; *Stictococcus*, Buchner, 1955), and even with aphids (Buchner, 1958) and no doubt is caused by a secondary displacement of older bacterial symbionts (see pages 269, 282).

DIASPIDINES. The diaspidines, which are subdivided into six tribes and represent the subfamily of the coccids which is richest in species, appear quite monotonous with respect to their symbiosis. Small uninucleate mycetocytes are distributed at random throughout the fatty tissue in all representatives of the Parlatoriini, Lepidosaphini, and Aspidiotini investigated thus far (*Parlatorea, Lepidosaphes, Chrysomphalus, Chionaspis, Aspidiotus, Pseudoparlatorea*).

In 1906 Šulc made the first symbiosis study in this group with a species of *Lepidosaphes;* I studied *Chrysomphalus, Pseudoparlatorea,* and *Aspidiotus* species in the live state (Buchner, 1921); and Breest (1914) reported on embryonic development in *Aspidiotus hederae* Sign. Later Richter (1928) specialized in this group and broadened the base of study by including additional species.

The mycetocytes are usually uninucleate and occasionally binucleate; they frequently divide mitotically; and the organisms which they contain are roundish, oval, or elongated, but never very numerous. Strongly refracting grains and droplets which are usually situated in the plasma of the mycetocytes are colorless, yellow-green, or orange, and tint the cells correspondingly. In *Chionaspis salicis* L., Richter found that the symbionts of youthful animals have the same shape in both sexes but that pubescent females contain more numerous mycetocytes and larger symbionts than do adult males (Fig. 106).

Egg infection follows the pattern of the lecaniines and asterolecaniines, subfamilies to which the diaspidines are clearly related in other respects. Individual symbionts which now are more slender and thus resemble the lecaniine yeasts, pass through the follicle into the space between the latter and the nutrient cell cord until they cause considerable protrusion there and finally glide toward the egg and sink into it during the first maturation

division.   *Aspidiotus piri* Lcht. is an exception in that it collects numbers of
motile mycetocytes in a ruff-like aggregation around the follicle connecting
the nutrient cells and the egg, and only then does it discharge the symbionts.

Breest and Richter provided the first data on embryonic development.
According to these authors the symbionts are originally grouped together
at the superior egg pole and during formation of the blastoderm are shifted
to a position close behind the latter, where residual blastomeres join them

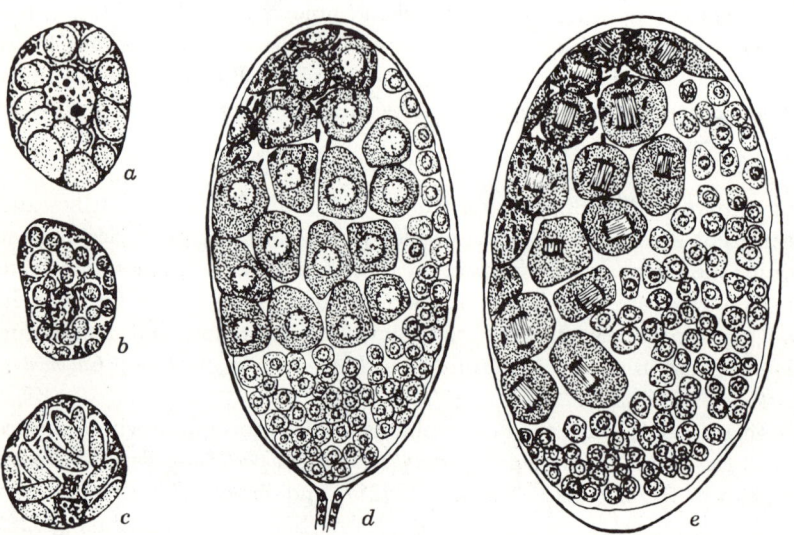

Fig. 106   *Chionaspis salicis* L.: mycetocyte (*a*) from a female, (*b*) from a male, (*c*) with
the more slender infection forms; from Richter. *Diaspis pentagona* Jarg.: (*d*) the nearly
developed polyploid, in which the transmission forms of the symbionts are beginning
to sink, (*e*) progressive penetration of the future mycetocytes which are now undergoing
synchronous division; from Tremblay.

immediately (*Chionaspis*) or after dispersal of symbionts in the yolk
(*Aspidiotus*).   For some years now interest in the Diaspidines has greatly
increased for many reasons; in America, Brown and Bennett (1957),
Bennett and Brown (1958), Brown and De Lotto (1959) and in Italy,
Tremblay (1958, 1958, 1959, 1960, 1960, 1960) have concerned themselves
with *Diaspis* (*Pseudaulacapis*) *pentagona* Targ.   It is of particular significance
for our field that these works have shown that the mycetocyte nuclei of
this object, as well as of *Pseudococcus citri*, as we shall learn later, arise from
a fusion of the nuclei of the polar bodies, that is, from elements which really
do not belong to the embryo, with a cleavage nucleus.   Of the two polar-

body nuclei, one has 8 chromosomes and the other 16. The cleavage nucleus, fusing with these two, contains 16 chromosomes. Thus a nucleus with 40 chromosomes arises from which, this chromosome number being maintained, a considerable region of correspondingly larger cells is formed; and this region is sharply separated from the small cleavage cells and blastoderm cells. Bennett and Brown (1958) assumed, on the basis of analogy with a similar occurrence long known in *Pseudococcus*, that here too it was a case of prospective mycetocytes, and Tremblay (1959) not only confirmed this assumption but also demonstrated the mycetocytes in detailed and impressive pictures (Fig. 106*d,e*). At the same time he investigated the early development of eggs yielding males, which, as in all diaspidines, are differentiated from those producing females in that a set of 16 chromosomes is eliminated during cleavage, and he found that the mycetocyte nuclei arise exactly as in the eggs bearing females and each contains 40 chromosomes. He also investigated the transmission of the spindle-shaped yeasts and the shifting and later loosening of the young mycetocytes without finding any significant deviations. The male larvae nourish themselves normally in the first stage and are well provided with symbionts. Later, however, an increasing degeneration of the mycetocytes sets in and leads finally to complete disappearance, a process which also takes place in other coccid males.

Tremblay calls attention to the close topographical relationships which also exist in the female between the loosely distributed mycetocytes and the oenocytes, and which are so intimate that they allow deep indentations to arise in the latter. He believes that he has occasionally seen a complete transmission of intact mycetocytes into the oenocytes, but I consider this an error caused by sectioning.

Another discovery made by the authors who worked with *Diaspis* does not concern symbiosis directly, but rather sex determination, which, however, in certain objects, as we shall learn later, can be related to symbiosis. Two kinds of eggs are formed by *Diaspis*, white eggs, which produce males, and reddish ones yielding females; consequently, essential differences in the trophical arrangements of the ovocytes correspond to this difference in eggs.

TACHARDIINES. The lac-producing tachardiines, limited to the tropics and subtropics, represent a subfamily of the scale insects of more than thirty known species. In contrast to the subfamilies already described, the tachardiines are quite heterogeneous with respect to symbiotic devices. Three of the four known genera, namely *Tachardina*, *Tachardiella*, and *Lakshadia* (*Tachardia*), were investigated first by Mahdihassan (1924, 1928, 1929) and later more extensively by Walczuch (1932).

All species of the genus *Lakshadia*, a name introduced by Mahdihassan to designate the only shellac producers which are of technical importance, live symbiotically with yeasts in the manner of lecaniines.   The yeasts are housed in the fat cells or float independently in the lymph.   Each host species has specifically fashioned symbionts; usually they are large spindle-shaped formations constricting off terminal buds which frequently proceed to chain formation (Fig. 107*a*).   Vacuoles and deep-staining granules permeate their plasma.

Symbionts of the *Lakshadia* species do not choose the method of the lecaniine symbionts for infecting the egg cells, however.   In young females

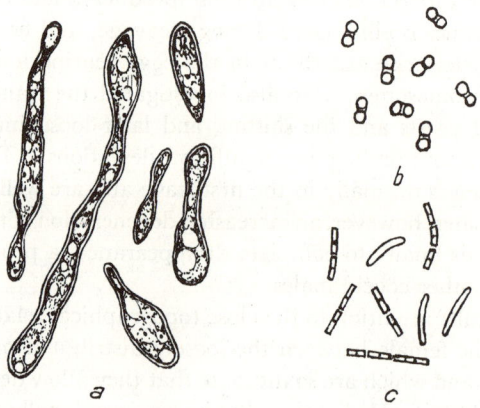

Fig. 107   (*a*) *Lakshadia communis* Mahd., (*b*) *Tachardina lobata* Chamb., (*c*) *T. silvestrii* Mahd.   The symbionts.   800×.   From Walczuch.

rather large symbiotic masses may be observed developing near the ovaries. When the oocytes are mature, the symbionts in these species are transmitted at the posterior egg pole through the follicle cells into a space arising between the follicle and the egg (Fig. 108).   Remaining there until embryonic development begins and the blastomeres rise to the periphery, they then push between the blastoderm cells into the interior of the young embryo, where they are temporarily closed off in a dense group by the cleavage nuclei but, with progressive invagination of the germ band, are dispersed over the whole yolk.   They reproduce freely and frequently grow out into chains, penetrating the fat cells only when the embryos are ready to hatch.

The genus *Tachardiella*, investigated by Walczuch on the basis of *Tachardiella cornuta* Cockerell, embodies a completely different type. Here the symbionts are typical bacteria in the form of minute delicate

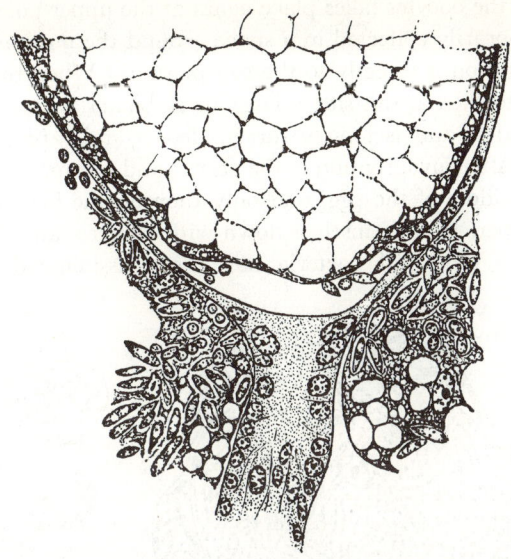

Fig. 108   *Lakshadia communis* Mahd.   Infection of the ovocyte at the posterior pole. 450×.   From Walczuch.

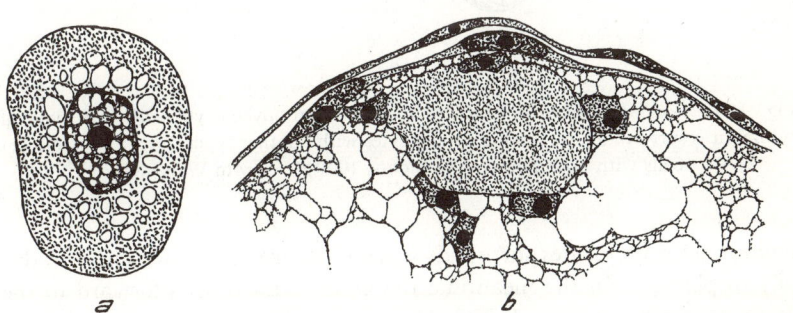

Fig. 109   *Tachardiella cornuta* Cockll.   (*a*) A mycetocyte with small rod-shaped bacteria, 800×; (*b*) blastoderm stage; the cleavage nuclei migrate toward the symbionts, 666×.   From Walczuch.

rodlets which are accommodated exclusively in genuine fat-free mycetocytes. The marginal parts of the mycetocytes are so thickly populated that the plasma structure is not recognizable. A vacuolized space around the symbiont-free central nucleus is usually left untouched (Fig. 109*a*). The mycetocytes are distributed throughout the fatty tissue, individually or joined in little nests.

Infection of the oocytes takes place again at the upper pole. The bacteria are temporarily collected in a space around the nutrient cord, then they glide along on it directly to the egg plasma. When the very small blastoderm is formed at this site, a large thick bacterial clump is situated directly behind it and is loosely surrounded by migrating blastomeres (Fig. 109*b*). After invagination of the germ band sets in, the clump sinks down to the middle of the egg, and only then do the blastomeres, reinforced in number, load themselves down with bacteria and the clear cell limits appear. When all bacteria have been assimilated, a circle of

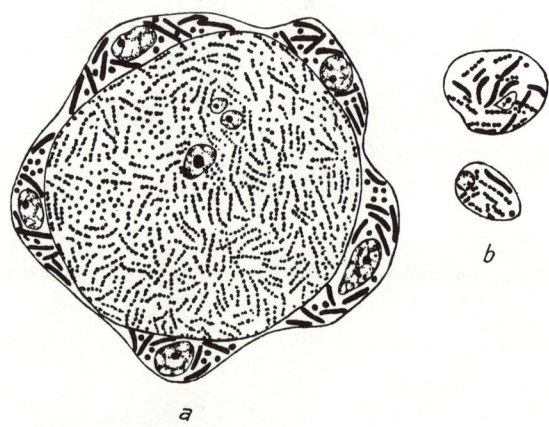

Fig. 110 *Tachardina silvestrii* Mahd. (*a*) A central mycetocyte with one type of symbiont is surrounded by smaller mycetocytes with another type, 750×; (*b*) transport cells with both types of symbionts, 1000×. From Walczuch.

embryonic mycetocytes surrounds a symbiont-free space which is gradually displaced. In the meantime the germ band presses forward to the embryonic mycetome and, intimately fused with it, transports it toward the upper pole. When the germ band curves into an *s*-shape, it grows out beyond the still adjacent mycetocyte group. Shortly before revolution the mycetocyte group is divided into two transitory mycetomes situated on both sides of the thoracic region. It has not yet been possible to observe how these mycetomes disintegrate into scattered and multiplying cells. Thus we see that the same tendency toward embryonic mycetome formation which we observed earlier among the lecaniines and in *Lakshadia* is in this case especially pronounced.

Even within its genus, *Tachardina* displays two symbiosis types which are very different from each other or from *Tachardiella* and *Lakshadia*. *Tach-*

ardina silvestrii Mahd., outwardly very little different from T. *lobata* Chamberlin but revealed as a separate species on the basis of its divergent symbiosis, is classified among the rare scale insects which live in association with two types of symbionts (Walczuch). The mycetocytes are not inserted individually in the fatty tissue but in strange groupings consisting of a large central cell and a number of smaller cells at the periphery (Fig. 110*a*). The nucleus of the central cell is so deeply lobate that the latter appears multinucleate in cross section. The plasma is densely packed with chain-forming symbionts which Mahdihassan interpreted as actinomycetes similar to *Nocardia*. Working exclusively with smears, he failed to observe the smooth, tubular, and slightly curved organisms which live in the smaller, similarly uninucleate elements.

Transmission is effected here both by symbionts floating free in the lymph and by combinations of the two types in the small transport cells which may be observed frequently near infection-ripe oocytes (Fig. 110*b*). Once again, infection takes place at the upper pole. Both types of symbionts penetrate between the follicle cells at the usual site and are collected in a furrow which forms all around the nutrient cord as the egg plasma retreats. When the nutrient cells and cord degenerate, the symbionts fill the space between egg and follicle and gradually spread in a thin layer almost to the equator of the egg.

In the meantime embryonic development begins without actual infection, the latter process being deferred until after formation of the blastoderm. Then the two symbiont types enter the embryo between the blastoderm cells, which here are loosened in texture, first forming a thin layer behind them (Fig. 111*a*). The germ band has been pushed close to the grouped symbionts by the time the latter are released from the blastoderm and are filled with yolk nuclei. Separation of the two symbiont varieties now begins, and during its course the homogeneous rodlets gradually are grouped around one part of the nuclei and the chains around the other. In Fig. 111*b* the mysterious process has already taken place.

At the time the extremities begin to form, the unpaired transitory mycetome which arises in this way is shifted to the area of the third thoracic segment, where it is temporarily located at the left side of the embryo. About this time the fastening of the one type of mycetocyte to the periphery of the other begins. After revolution the mycetome is paired. The final loosening of its structure occurs during the larval period, enabling the cell groups to penetrate the fat body.

Tachardina lobata Chamb. possesses only one symbiont type. The symbionts fill uninucleate cells which are distributed in the fatty tissue, as in *T. silvestrii*, so thickly that the host plasma is barely visible. Their shape is roundish; when isolated, they frequently appear to be joined in biscuit-like pairs (Fig. 107*b*). Within any one mycetocyte they are approximately 'the same in size, but there are definite differences from cell to cell.

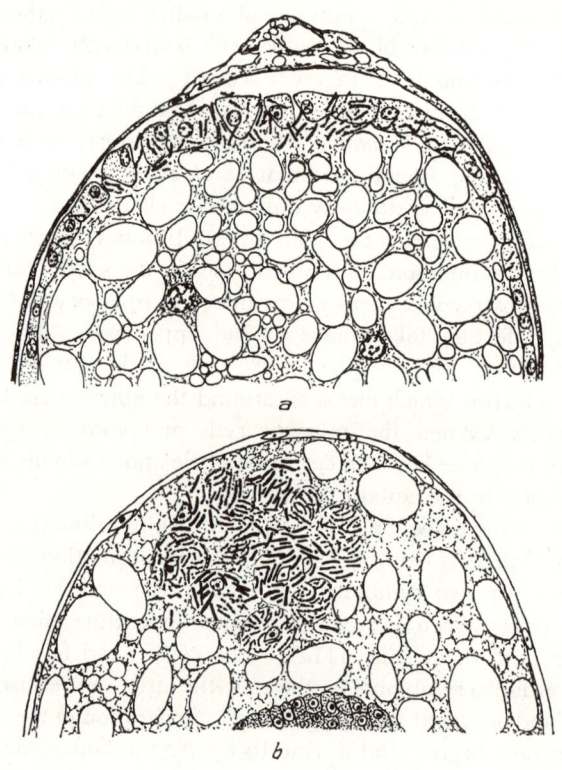

Fig. 111   *Tachardina silvestrii* Mahd.   (*a*) Both types of symbionts are transmitted between the blastoderm cells into the yolk; (*b*) they are distributed to different cells. 555×.   From Walczuch.

Infection of ovarioles and late transmission to the blastoderm stages conform to the pattern described for *T. silvestrii*.   The unpaired transitory mycetome is replaced by a paired organ, which in older embryos again becomes unpaired and is then situated, surprisingly close to the outer surface, on the ventral side between hypodermis and ventral nerve cord. The cleavage and dispersal occur after hatching, during somewhat older stages than with *T. silvestrii* (Walczuch).

ORTHEZIINES. The subfamily of the ortheziines, distinguished by rich and handsome wax secretions, has approximately sixty species. Its symbionts are typical bacteria contained in cells dispersed in the fatty tissue, sometimes supplemented by a general infection of the lymph. Šulc had made brief reference to the existence of bacteria-like organisms in 1910. My own report on ortheziine symbiosis was incomplete and in part even erroneous (Buchner, 1921). In 1932 the subject was exhaustively studied by Walczuch with *Orthezia insignis* Dougl. from Africa and India and in Germany with *O. urticae* L. commonly found on stinging nettles.

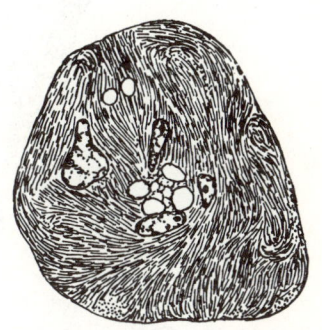

Fig. 112   *Orthezia insignis Dougl.*
Multinucleated mycetocyte.
750×.   From Walczuch.

In *Orthezia insignis* the long, thin, and extraordinarily crowded rodlets are found in cells, with usually one and sometimes several nuclei. The central area is occupied in most instances by symbiont-free vacuoles. In this species the lymph contains no bacteria except those which are migrating to the ovarioles (Fig. 112).

Egg infection, an exceedingly strange process here, takes place at the posterior pole where the bacteria often collect in dense groups. In a limited ring-shaped zone the follicle cells near the egg secrete small caps which quickly grow out into long tubules (Fig. 113). The symbionts enter this secretion, which is so constituted that it reduces their staining extraordinarily. They still stand out fairly well outside and in the plasma of the follicle cells, but, since the tubules of *Orthezia insignis* appear fully homogeneous, in contrast to those of *O. urticae*, it cannot be said with certainty whether the secretion formation begins, as is quite plausible, before the onset of infection. The oocytes retreat a distance from these large formations but through lack of space are often forced into curved shapes. When the maximum number of tubules is achieved, up to forty having been counted, these large parts are detached from the follicle cells, the chorion presses in between and, closing like an iris diaphragm, dislodges the stately infection mass from the follicle.

Now placentiform, the infection mass is situated in a hollow of the egg plasma to await embryonic development. During the formation of the blastoderm some nuclei with denser plasma attach themselves at the surface of the symbiont mass (Fig. 114). Through progressive formation of the blastoderm the infection mass is dislodged but slightly at first

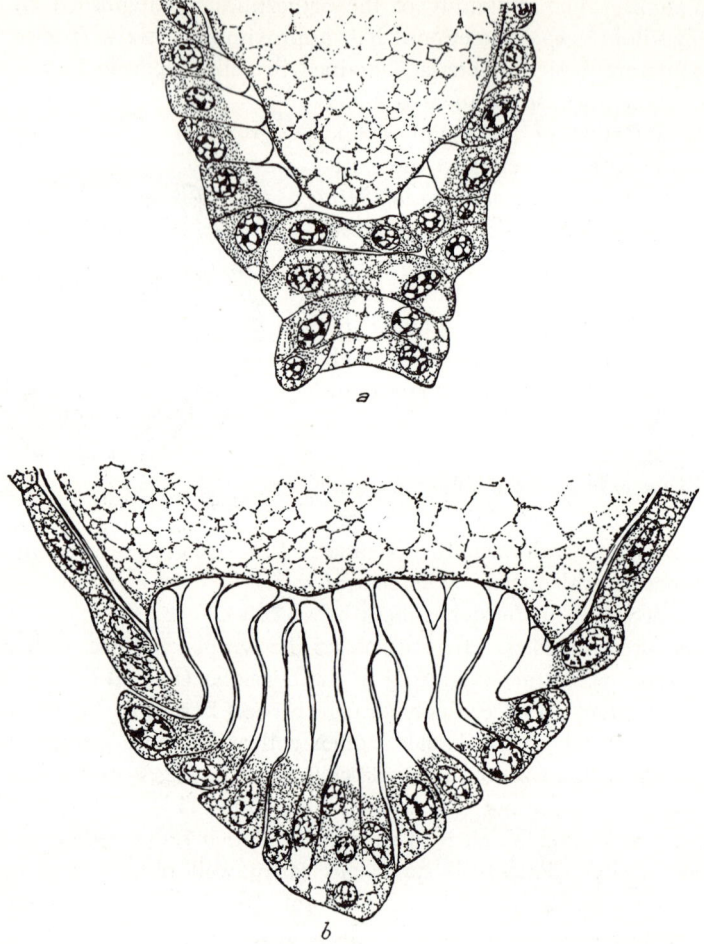

Fig. 113. *Orthezia insignis* Dougl.   Infection of the follicle cells at the posterior pole of the egg.   (*a*) Beginning of the formation of secretion caps, (*b*) the secretory tubules, filling up with bacteria, at the height of development.   555×.   From Walczuch.

toward the inside, together with the primordial germ cells that are separating nearby, but, when the germ band develops, its free end pushes symbionts and gonad rudiment into the opposite area of the embryo, just as so often occurs in similar cases.   During this process a strange differentiation becomes noticeable in the egg plasma.   Cords appear which place themselves close against the symbiont package, which is now in an enclosed casing, and partially envelop it.   Later the cords, dispersed in

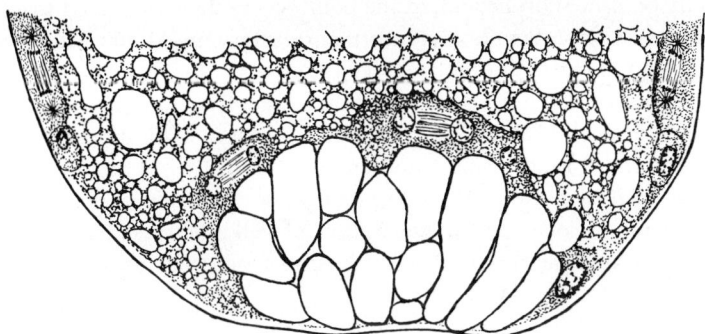

Fig. 114. *Orthezia insignis* Dougl. Blastoderm cells surround the bacteria-filled tubules. 555×. From Walczuch.

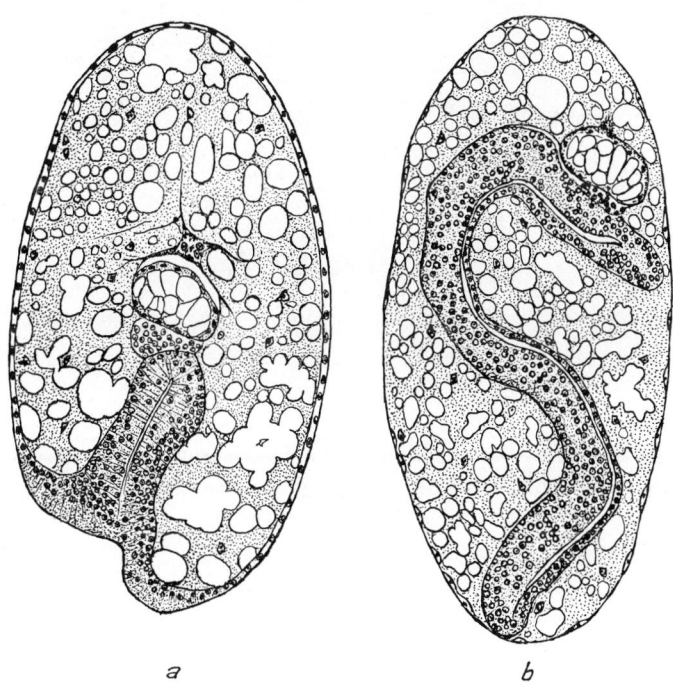

a   *b*

Fig. 115 *Orthezia insignis* Dougl.   (*a*) Invagination of the germ band, which is pushing the genital rudiment and the symbionts forward; a plasma radiation is present in front of the moving elements; (*b*) the transitory mycetome has become separated from the peak of the germ band.   From Walczuch.

the yolk, become shorter, and, at the point where they are pulled together, a small radiation settles close against the transitory mycetome but degenerates before the extremities form (Fig. 115*a*).

When the germ band assumes an *s*-shape, the symbionts are situated off to one side (Fig. 115*b*). The outlines of the tubes disappear, and the bacteria form a uniform mass and again stain easily. The nuclei of the envelope migrate between them but soon show clear signs of degeneration. At the same time a second generation of cells, of mesodermal origin, migrates toward the organ, encases it once more, and partly penetrates

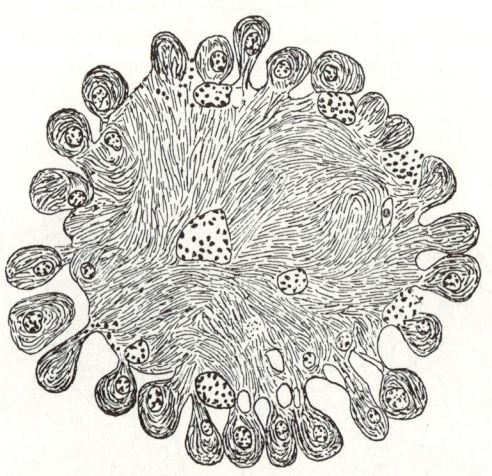

Fig. 116   *Orthezia insignis* Dougl.   The symbionts of the transitory mycetome are taken over by the definitive mycetocytes.   666×.   From Walczuch.

into it. As soon as the symbionts come in contact with these cells, they enter the plasma bodies of the latter. The division of the bacterial supply, still permeated with the old degenerating nuclei, begins at the surface and progresses to the definitive mycetocytes. The sufficiently infected cells tend to rise toward the periphery, temporarily connected with the central remainder only by thin cords (Fig. 116). In the meantime revolution of the germ band proceeds and, before hatching, a loose collection of infected cells, situated ventral from the abdominal nervous cord, replaces the transitory mycetome. Later a separation into two aggregations takes place, the mycetocytes not being dispersed in the fatty tissue until the larval stage.

Remarkably, Walczuch quite regularly found great quantities of minute, rickettsia-like rodlets also in the intestines and Malpighian tubules of the insects from India and Africa. In the hindgut, their chief site, they form a regular thrombus, and in the midgut they are also situated intracellularly. They are not transmitted to the egg cells. In larvae which have not yet fed, the intestines are symbiont-free, but the organisms soon appear in the esophagus and later are no longer observed there. Apparently there must be some type of regulated transmission after all, perhaps by the excrement of the mother, and it may be possible to interpret these organisms as additional symbionts.

Orthezia urticae differs from *O. insignis* in that the minute symbiotic rodlets populate not only certain uninucleate cells embedded in the fatty tissue but also occur free, in varying numbers, between the fat cells and in the lymph. The peripheral zone of the fatty tissue consists of strange cells, which at first glance give the impression that they are populated by bacteria, as, indeed, I once considered them to be. Their rodlets, which are arranged in impressive, usually regular rows, are actually crystals of some type, probably preliminary stages of the wax secreted in such great quantity. An analogous, specially constituted marginal zone of the fat body is also found in *O. insignis*.

The type of egg infection and the circumstances during embryonic development correspond in principle to what has been reported for *Orthezia insignis*. Secretion tubes, but far fewer and smaller than those of *O. insignis*, are formed by certain follicle cells at the posterior pole. The tubes disintegrate much more quickly once they are taken over into the egg, and the accumulation of little rodlets is surrounded during blastoderm formation by a denser, sharply defined, yolk-free plasma zone. When the primitive sex cells and this symbiotic area, provided with yolk nuclei, are pushed into the interior of the egg by the germ band, a strange reaction of the central plasma takes place. A net-like plasma sac fused to the front of the embryonic mycetome becomes more and more concentrated as it approaches the symbionts, and is drawn together (Fig. 117). As the yolk masses fall between its meshes, the surface of a small completely yolk-free area forms and sends out a strong radiation, which resembles a similar one in *Orthezia insignis*, and the narrowing plasma sac resembles reactions found in the embryonic development of snout beetles. Both may safely be interpreted as an expression of the spatial sealing-off of the symbionts. Before and shortly after revolution the radiation disappears. As in *O. insignis*, the symbionts are absorbed by the definitive cell material, and the accompanying nuclei degenerate, but, in contrast to *O. insignis*, some of the symbionts remain free and are dispersed between the fat cells.

Mahdihassan (1946) reported symbiotic cultures of *O. urticae* and *O. insignis* in which the colonies assumed the same greenish color as the symbionts.   Quite recently Köhler and Schwartz (1962, 1962) studied more thoroughly the symbionts of *Orthezia insignis*, which they described as gram-negative rods, 7–8 μ long and 0.8 μ in diameter.   Only in older animals do longer forms sometimes appear.   They too experienced no

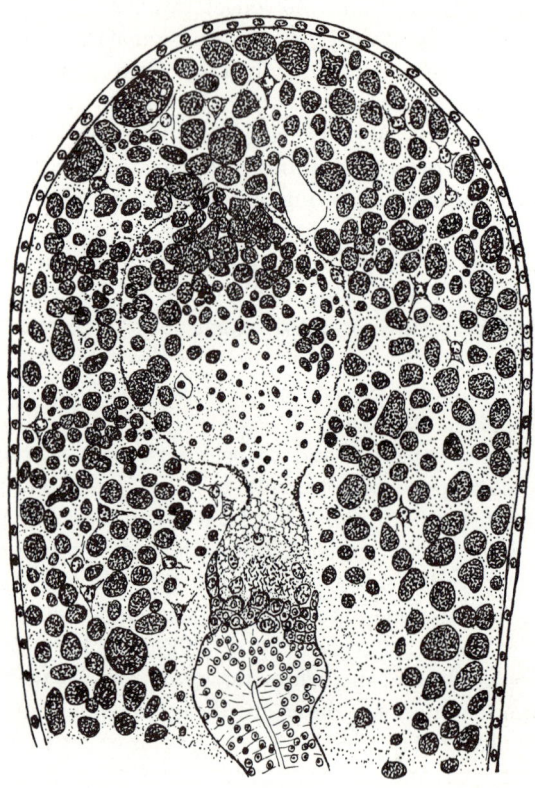

Fig. 117   *Orthezia urticae* L.   At the end of the germ band a plasma sac arises in front of the symbionts.   From Walczuch.

difficulties in cultivating the organisms.   The rods in pure culture had a length of only 1.5 μ, and the greenish pigmentation which had been observed by Mahdihassan appeared.   Both authors would classify the organism in the genus *Achromobacter*.   In their second publication they report on the possibility of killing the symbionts with progressive, inter-mittent heating.   The mycetocytes remain as large empty regions.   The only harmful effect that could be recognized after cultivation of the aposymbiotic animals for more than 2 years was a small delay in development which occurred after a few generations.

ERIOCOCCINES. Of the eriococcines only *Cryptococcus* and *Eriococcus* have been investigated, by Walczuch, to whom we are indebted for so much of our knowledge of symbiosis in scale insects. In *Cryptococcus fagi* Dougl. the symbionts are small rod-like bacteria which populate cells greatly resembling the surrounding fat cells in size and profusion of vacuoles, but nevertheless of different origin (Fig. 118). The isolated symbionts, which are occasionally found extracellularly, may possibly be those which are on route to the ovarioles. A specific entry site for the symbionts is afforded

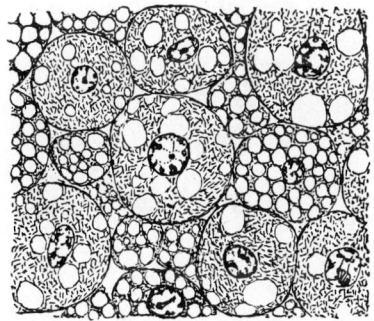

Fig. 118 *Cryptococcus fagi* Dougl. Mycetocytes in the fat tissue. 750×. From Walczuch.

by the unusually tall follicle cells between the nutrient cells and the oocyte. Around the nutrient cord behind them, a thick bacterial mass gathers and later spreads in the shape of a cap over the anterior egg pole. In connection with blastoderm formation the symbionts later are transmitted to the young embryo, where they sink down in thick cords to the middle area of the yolk. Here, meanwhile, a strange, completely yolk-free plasma area forms and sends out cords like suspended ribbons, toward all sides to serve as access roads for the yolk nuclei. The entire bacterial force now sinks into this plasma (Fig. 119); a thick bacterial mass forms around each yolk nucleus; cell limits appear; and a looser collection of mycetocytes arises, to be distributed later throughout the fatty tissue.

Eriococcus spurius Lindinger, with thicker rodlets as symbionts, is very similar to *Cryptococcus*, but the central plasma area of the embryo is absent and the assimilation of the symbionts by yolk cells takes place later, although here also it leads to the formation of elements distributed in the yolk.

PSEUDOCOCCINES. In Balachowsky's new system (1948) three tribes—the pseudococcines, the eriococcines, and the chermesines—are joined in a

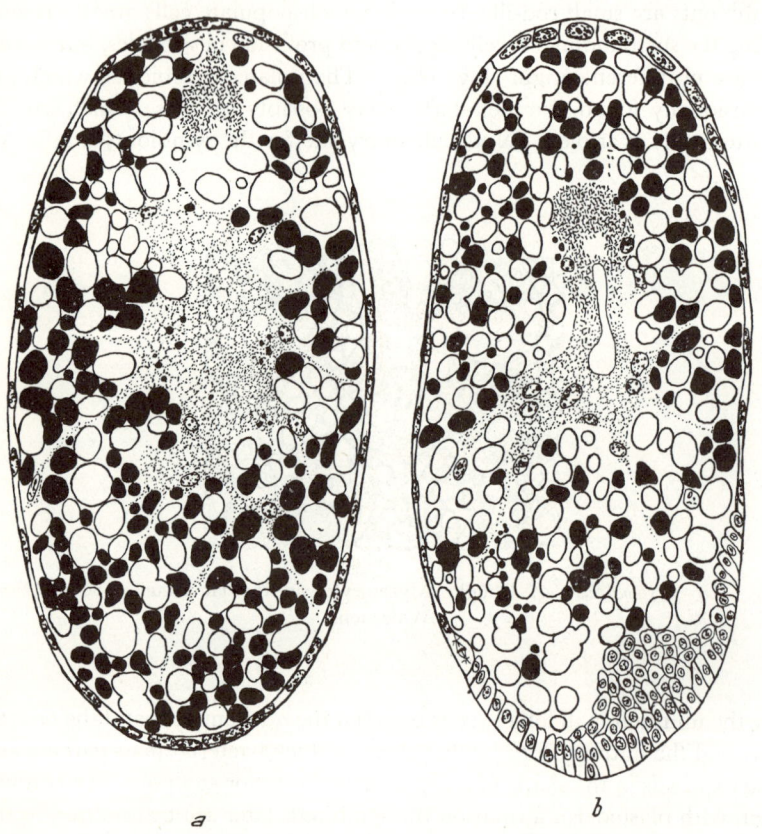

Fig. 119 *Cryptococcus fagi* Dougl. (*a*) In the center of the yolk a plasma region is developed; (*b*) the symbionts sink into it. About 400×. From Walczuch.

subfamily, the Cherminae, although the three groups are completely different in their symbiosis. In *Chermes*, as already mentioned, the yeasts appear to be localized in the fatty tissue and in the body cavity, the bacteria of eriococcines in true mycetocytes, and those of pseudococcines in a highly characteristic unpaired mycetome. The symbiosis was studied mainly in *Pseudococcus (Dactylobius) citri* Risso and *adonidum* L., but I was able to prove to my own satisfaction that many other pseudococcines, for example, *Ferrisia*, *Trionymus*, *Rhodania*, *Antonina*, and *Antoniella*, have the same symbiosis type. The mycetome, situated ventrally from the intestine and a vivid yellow in color, is so striking that it caught the attention even of early authors. In *Pseudococcus citri* it measures approximately one-third of the total body length in females; in males it is much smaller. Intensive in coloring, it shimmers through the body wall of the female, especially

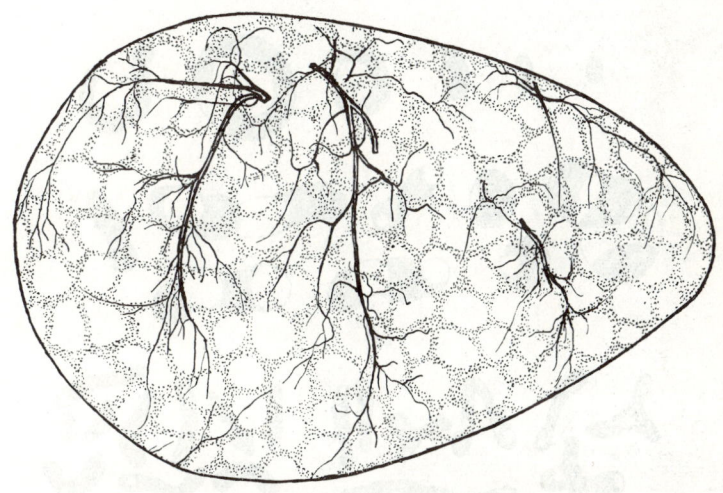

Fig. 120 *Pseudococcus citri* Risso. The unpaired mycetome, side view (fresh preparation). From Buchner.

before the ovaries are developed. The organ is abundantly supplied with air from branches of the last pair of tracheae, their end tubules penetrating it deeply (Fig. 120). A flat epithelium containing yellowish pigment granules surrounds the mycetome and, accompanied by the tracheae, sinks into the interior among the mycetocytes. The latter, unusually large, have a large central nucleus. In the plasma, around the centrally located, sizable nucleus of the mycetocytes, are roundish or longish mucilaginous globules in which the symbionts are thickly embedded (Pierantoni, 1910, 1913; Buchner, 1921; Walczuch, 1932; Fink, 1952).

Such organs are found in the very same manner in *Pseudococcus adonidum*, *P. diminutus* Leonardi, and *P. nipae* Maskell, but the morphology of the symbionts differs considerably. Older information on *Pseudococcus citri* has been corrected in many respects by more thoroughgoing studies by Fink (1952). Fink points out particularly that with normal development of the wall each of the sausage-shaped symbionts closely adjoins a vacuole, being thus forced into a curve, and that the vacuole, absent only in the smallest symbiont forms, is developed from a liquid secreted unilaterally in the symbionts in gradually increasing quantity (Fig. 121). Such states can then constrict off small, again vacuole-free symbionts. That these casings actually represent hypertrophic vacuoles characteristic of an

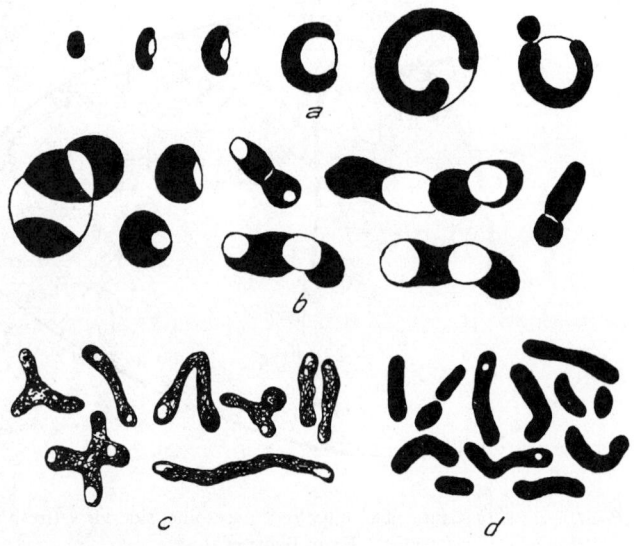

Fig. 121   *Pseudococcus citri* Risso.   (*a*) Normal symbionts, (*b*) cold-hunger forms (both fresh preparations), (*c*) smear from a primary culture, 1900×; (*d*) cultured symbionts (fresh preparation), 1600×.   From Fink.

important form of cicadid symbionts, the so-called *a*-symbionts, is shown with great clarity in states with abnormal vacuole formation such as occur under the combined influence of hunger and cold (Fig. 121*b*). Sometimes degenerate forms occur in which the protoplast swells, becoming lighter in the process, and gradually displaces the vacuole completely (Fig. 122*c*).

Each of the three or four globules in a mycetocyte is similarly surrounded by an invisible membrane which causes the globule to burst under pressure, and the symbionts in each tend to be in the same state of development (Fig. 122). According to Fink, the little oval or roundish states with homogeneous plasma, frequently in the process of dividing, represent the reproductive phase, and the long, variously expanded, granulated forms represent the degenerative phase. The reproductive phase achieves its maximum from egg infection to hatching during the first larval state and during the ovarial development; the degenerative phase is typical for the remaining time and increases suddenly when ovulation terminates.

A final instance of the varied morphology may be seen in pubescent imagines which form specific stages to accomplish the transmission. Such stages, bent into a *u*-shape, are shorter and thicker; they consist of dense protoplasm and are enveloped in a covering of their own. These casings,

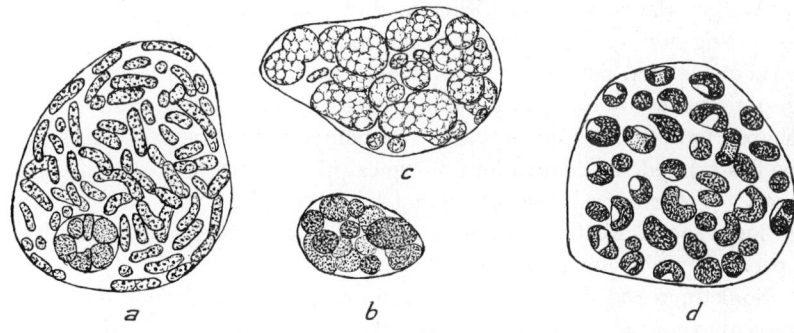

Fig. 122 *Pseudococcus citri* Risso. (*a*) Mucus packet with typical inhabitants; (*b*) forms with dense plasma; (*c*) degenerating symbionts; (*d*) mucus packet with infection forms. From Buchner.

which I had the opportunity to observe earlier, are most clearly visible between the two ends. The tendency toward transformation is shared by each symbiont in the bundle, and only in rare cases can one be found which has escaped the process (Fig. 122*d*).

The symbionts of *Pseudococcus nipae* assume various forms. They may be thick rodlets with blunt ends and constricted like biscuits or rodlets cut off like dumbbells and then frequently reversed with both halves situated close together, often tapering at both ends or even becoming lemon-shaped. In *P. nipae* all these states are found side by side in one and the same gelatinous ball without special stages to effect transmission. The symbionts of *P. diminutus* have a thick, moderately curved, often transverse-dividing sausage shape with completely homogeneous protoplasm, which in this form can also effect transmission (Plate III*b*).

Differing from both *P. nipae* and *P. diminutus* is another species, unfortunately not identified, in which delicately filamentous symbionts in mucilaginous bundles secrete special casings into which they are fitted in curved form or rolled up spirally, depending upon their own length (Plate IIIa). Smaller states composed in the same way serve to infect the oocytes. The *P. adonidum* symbionts, also radically divergent, are for the most part straight slender, filamentous forms which often form chains of 2 to 4 links through frequent transverse division. For infection they use mucilaginous balls of symbionts of the same type, but these disintegrate into shorter rodlets in the young larvae and grow out again only in the older larvae (Plate IIe,f). The symbionts of *P. citri* and the very divergent ones of *P. maritimus* are illustrated in Plate V d,e.

At first glance it is surprising to find such heterogeneous symbionts occurring side by side in mycetomes which have the same rigidly specific structure and which occur in representatives of the same genus, but the situation is easily clarified if we interpret all these sausage- and bladder-shaped formations, and we shall be encountering them very often, as more or less expanded modifications of typical filamentous or rod-shaped bacteria. Transitions between original bacterial forms and extreme involutional forms are being found continually with increasing research on species in such related groups.

Sometimes rod-shaped or cocciform organisms, on the one hand, and typical expanded symbionts, on the other, will occur in one and the same animal, in the mycetome as well as in the eggs and embryos. During research on the relationship of *Pseudococcus brevipes* Ckl. to a disease of pineapple plants characterized by green spotting of the leaves, Carter found that the disease makes its appearance only when animals which contain both these forms suck at the plants, whereas those without the rodlets proved to be harmless. He was convinced that it is the symbionts in rodlet form which produce the poisonous substances and that the latter reach the host plant by way of the salivary glands (Carter, 1933, 1935, 1936, 1937, 1939, 1942).

Carter later gave up this hypothesis on finding in Africa and Singapore *Pseudococcus brevipes* which produced the same disease despite the fact that their mycetomes did not contain these additional bacteria (see also Leach, 1940; and Zimmerman, 1948).

In the peripheral zones of the mycetome in mature females the individual bundles with infection stages disappear, and occasionally whole mycetocytes are expelled, which then disintegrate and transmit their content to the blood stream.

In all *Pseudococcus* species, the egg cells are again infected at the superior pole. The mucilaginous globules appear at the favored place between nutrient cells and oocytes and sink into a groove in the egg which closes behind them. By this time the symbionts have begun to increase, and some of them leave the bundles to become starting points for new aggregations.

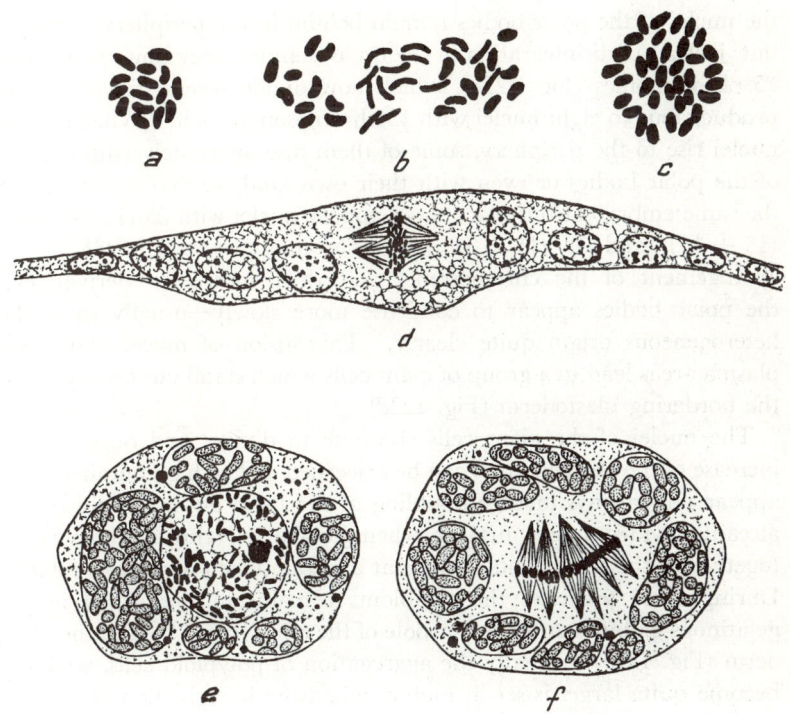

Fig. 123 *Pseudococcus citri* Risso. (*a*) Equational plate of the united polar bodies, (*b*) fusion of the chromosomes of the united polar bodies with two cleavage nuclei (15 + 10 + 10), (*c*) equatorial plate of a giant cell (35), (*d*) the giant cells in association with the blastoderm; from Walczuch; (*e, f*) mycetocytes from a larval mycetome, prophase and mitosis with more than 200 chromosomes, from Buchner.

It was no small surprise when Schrader (1921, 1923) established that in *Pseudococcus citri* and *P. maritimus* the mycetocyte nuclei arise from a fusion of the nuclei of the polar bodies with the cleavage nuclei. Pierantoni considered the derivatives of maturation division to be yolk nuclei. In the meantime Walczuch confirmed Schrader's presentation on the same object and the study of the Diaspinae uncovered a parallel case, as we have already learned, which, however, shows no such extensive variability and constantly increasing polyploidy as in *Pseudococcus*.

The normal number of chromosomes is 10. At the end of the second maturation division, which is the actual reduction division, the pronucleus, needing fertilization, therefore receives 5 chromosomes; the first polar body receives 10; the second, 5. As so often occurs among the insects, the nuclei of the polar bodies remain behind in the periphery of the egg, but, instead of disintegrating in the usual manner, they unite in groups of 15 chromosomes (Fig. 123*a*), which now divide several times and thus produce four to eight nuclei with 15 chromosomes each. When cleavage nuclei rise to the periphery, some of them fuse alternately with derivates of the polar bodies or even with their own kind, so that side by side in the same embryo at this place there appear nuclei with 25 (15 + 10), 30 (15 + 15), and 35 (15 + 10 + 10) chromosomes (Fig. 123*b,c*). The arrangement of the chromosomes and their form—the derivatives of the polar bodies appear to condense more slowly—usually show their heterogeneous origin quite clearly. This fusion of nuclei and related plasma areas lead to a group of giant cells which stand out clearly against the bordering blastoderm (Fig. 123*d*).

The nuclei of the giant cells continue to divide, and once more an increase of the chromosomes can be discerned. The high numbers which appear now always involve doubling of the basic numbers arising when already divided nuclei fuse or when just-halved chromosomes remain together in the mother nucleus, or at any rate when division is retarded. During these processes the symbionts remain untransformed in their gelatinous globules at the upper pole of the embryo just behind the blastoderm (Fig. 124*a*). Finally the aggregation of polyploid cells, which has become quite large, is set in motion when the invagination of the germ band begins, and it migrates in open formation toward the symbionts, which glide toward the cells a short distance. Once the cells have come in contact with the symbionts, they gradually flow around the individual globules and hence become the definitive mycetocytes in which the symbionts continue to increase (Fig. 124*b*).

Even before Schrader, I found that the mycetocytes in the larvae and imagines contain unusual numbers of chromosomes. In accordance with the postembryonic growth of the mycetomes, there is no lack now of cases where mitosis proceeds normally, thus increasing the number of cells. In other cases mitosis terminates with renewed nuclear joining, further increasing the number of chromosomes. Thus I was able to find nuclei with more than 200 chromosomes (Fig. 123*e,f*; for the stage of Fig. 123*e* there are several other sections with additional chromosomes). In such cases multipolar or otherwise disturbed spindles often occur, and occasionally two daughter cells lie so close to each other that it may be assumed that they are fusing.   The nuclear growth thus produced results

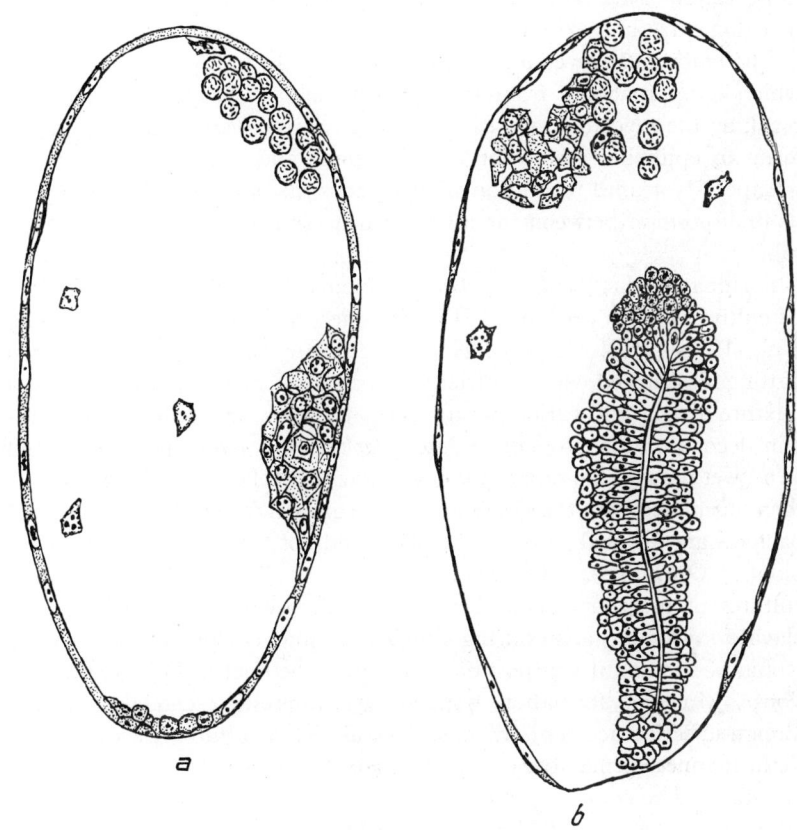

Fig. 124   *Pseudococcus citri* Risso.   (*a*) Embryo with giant cells stored on the side; (*b*) the giant cells have migrated to the symbiont packets lying at the anterior pole and are beginning to take them up.   From Schrader.

in a corresponding increase in the protoplasm and consequently serves to enlarge the mycetome.

Until just recently fusion of this kind had never been found elsewhere in symbiosis.   It resembles a similar fusion of the polar bodies in the honey bee, where, however, the derivatives are destroyed immediately.   Actually, it has greater similarity to the remarkable behavior of the polar bodies discovered in chalcicids by Silvestri.   Here a great part of the egg plasma is given over to the nuclei, which then increase extraordinarily by amitosis.   In the form of a trophamnion the nuclei and their plasma encase the many embryos arising through abscission of the cleavage stages. Observations even remotely resembling this genesis of the pseudococcine mycetome are non-existent.

The embryonal mycetome with its nuclei sinks toward the germ band which is approaching from the opposite pole.   The germ band, after reaching the mycetome with its vertex, gives its mesodermal elements to form an epithelial casing for the mycetome and doubles back into an *s*-shape.   Not until revolution takes place is the organ finally brought to its final position between the abdominal nervous cord and the intestinal tube.

Fortunately the *Pseudococcus citri* symbionts are among those which can be cultivated with certainty.   The first successful cultures were made by Fink (1951, 1952).   He used Kanz's celloplate method in which cotton batting is saturated with a nutrient solution of Merk-Standard I, that is, a mixture of agar and various amino acids, peptides, and other soluble protein decomposition products.   After *Staphylococcus aureus* has been added as a "wet nurse," the mixture is covered with a cellophane sheet which is then inoculated with the symbionts.   The growth factors produced by the bacteria are diffused through this sheet and act beneficially on the symbionts.   Only under such favorable conditions did vividly pink round cultures of more or less arched tubes with occasional branching and clavate thickening arise during Fink's experiments (Fig. 121*c,d*).   With Robinow staining it was possible to discover the nuclear bodies and divisions.   Mucilage formation stopped, yet, interesting enough, when a foreign infection accidentally took place in such a culture, it was resumed in the manner normal to the mycetome and it produced the usual symbiont globules.   The color is connected with lipoid inclusions usually accumulating at the ends of the tubes.   The organism is strictly aerobic and decidedly gram-positive after a series of transmissions.   Its appearance changes according to cultural conditions.   With intensive growth it develops shorter and thicker tubes; under unfavorable conditions extremely long forms appear in increasing degree (up to 35 μ).

In his cultures Fink also tested the characteristics of the symbionts.   He investigated the utilization of separate carbohydrates and found that they produce a urease and can synthesize amino acids from both atmospheric nitrogen and pyruvic acid.   On the basis of their cultural, pigment-producing, and fermentation characteristics, Fink classified the organisms under the Corynebacteriaceae [*Corynebacterium dactylopii* (Buchner)].   Later Köhler and Schwartz (1962, 1962) also succeeded in cultivating the symbionts of *Pseudococcus*, but only after they had worked with nurse cultures and had enriched the primary medium with various nutrients. Their results harmonized in general with Fink's; however, they found that their strains became motile after long cultivation, a capability which could not be realized in intracellular life but nevertheless was not lost. The finding of this characteristic, in addition to others, strengthened Köhler and Schwartz in the belief that the organism is a *Corynebacterium*. They were also the first to culture the gram-negative symbionts of *Pseudococcus maritimus*. These organisms were non-motile in cultures and under certain conditions grew out into long threads which reached a length of 80 μ. Their morphological and physiological characteristics speak for membership in the genus *Flavobacterium* (Plate V*d,e*). We shall report in Part 3 on the behavior of different species of *Pseudococcus* which were rendered symbiont-free with both antibiotics and increased temperature.

The genus *Phenacoccus* investigated by me and by Walczuch represents a type of symbiosis clearly different from that of *Pseudococcus* and yet related to it. Here the symbiotic device is an unpaired mycetome similar to the *Pseudococcus* type with respect to shape and histology. It is milk white, or at the very most weakly pigmented. In *Ph. piceae* it retains its oval shape throughout its existence; in *Ph. aceris* it becomes much more voluminous in the course of embryonic development, and, where the segmental dorsoventral muscles pass, it has deep indentations which give it the outlines of an oak leaf on frontal sections.

The nuclei of the mycetocytes are again very abundant in chromosomes, but the symbionts, no longer situated in separate bundles, permeate the plasma uniformly. They are extremely expanded formations, round or oval, and sometimes notched. The symbionts used in transmission are often noticeably larger than those remaining in the mycetome. With the host cell, they leave the organ and are then released (Fig. 125). In other respects they are like *Pseudococcus* in the manner of their transmission to the egg, their acquisition of nuclei, and their behavior during embryonic development. I have also found this type of symbiosis in *Heliococcus, Centrococcus, Ripersia, Eumyrmecoccus*, and in *Ceroputo* Šulc (that is, the true *Ceroputo*, to which so many non-related forms were later assigned).

Until recently it was still assumed that with these two types the subject of pseudococcine symbiosis had been exhaustively presented, but my own investigations based on *Macrocerococcus* and *Puto* (1955), the genus *Rastrococcus* (1957) newly set up by Ferris (1954), and the myrmecophilous *Hippeococcus* (1957) showed that symbiotic devices in this subfamily are far more diverse than they had seemed to be.

Macrocerococcus and *Puto*, it is true, at first have an unpaired, ventrally situated mycetome which is sharply defined. However, the mycetocytes, which were still homogeneous and filled with filamentous symbionts in the female first larval stage separate into two distinct types with increasing ovarial development (Fig. 126*a*).   Those of one type maintain their ability to divide and furnish numerous cells which crowd in between the rudiments of the ovarioles and are thereby reduced (Fig. 126*b*); those of the other type no longer reproduce, although they continue to grow and acquire bizarrely shaped nuclei.   The symbionts themselves differ even more radically in their behavior.   Their original rodlet form is preserved

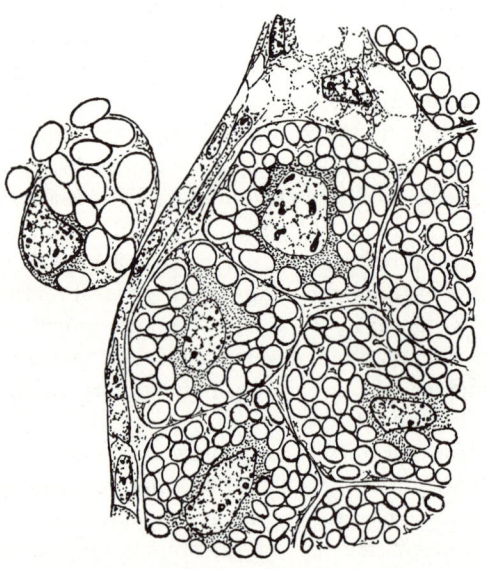

Fig. 125   *Phenacoccus* sp.   An intact mycetocyte has left the mycetome and now releases the symbionts serving egg infection.   750×.   From Walczuch.

only in the mycetocytes which retain their capacity to divide.   In the others, as we shall see later, the bacteria very soon begin to degenerate into small roundish formations.   One would infer then that it is the smaller mycetocytes alone which assist in transmission.   These exceedingly interesting test objects, however, also have a very unusual transmission method. When the ovarioles are sufficiently developed, the mycetocytes at first collect around the cervix connecting the egg and the nutrient cells, then sink into deep niches forming in the follicle, and from there are transmitted without alteration of any kind to the oocytes (Fig. 127).   These mycetocytes are highly viable cells, for here and there one happens upon mycetocytes which continue to divide mitotically during and shortly after the sinking process (Fig. 127*c*).

Such transmission through intact mycetocytes had been observed previously only in the aleurodids (Buchner 1912, 1918), where the migrating cells are destroyed in the course of embryonic development and new cell material appropriates the symbionts.   It is different with *Macrocerococcus*

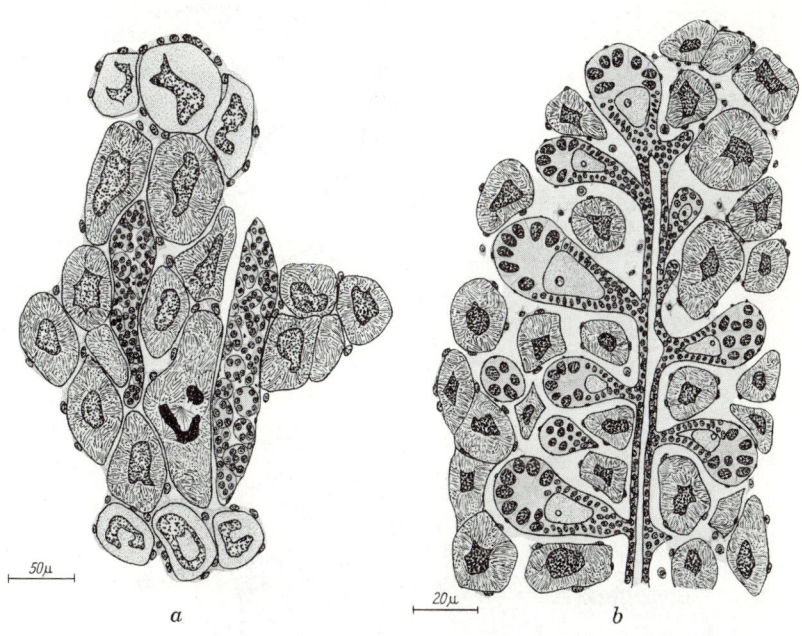

a   *b*

Fig. 126   *Puto antennatus* Sign.   (*a*) Mycetome and gonads of a female larval stage; primary and secondary mycetocytes; (*b*) cranial portion of a young ovary; the secondary mycetocytes penetrate between the ovarioles.   From Buchner.

and *Puto*, in which the symbiosis differs only in unimportant details. Nowhere in their case is there anything to indicate that the implantation of mother cells into the daughter animals is excluded.   In a compact mass of only a few cells, they move toward the germ band, which is growing in their direction, and glide toward the dorsal side of the embryo when the extremities sprout.   During this shift two types of nuclei appear: small ones, originating from the lower layer, which are to supply the epithelial casing of the mycetome, and larger yolk nuclei, each at first merely clinging to the mycetocytes but later actually passing over into them and

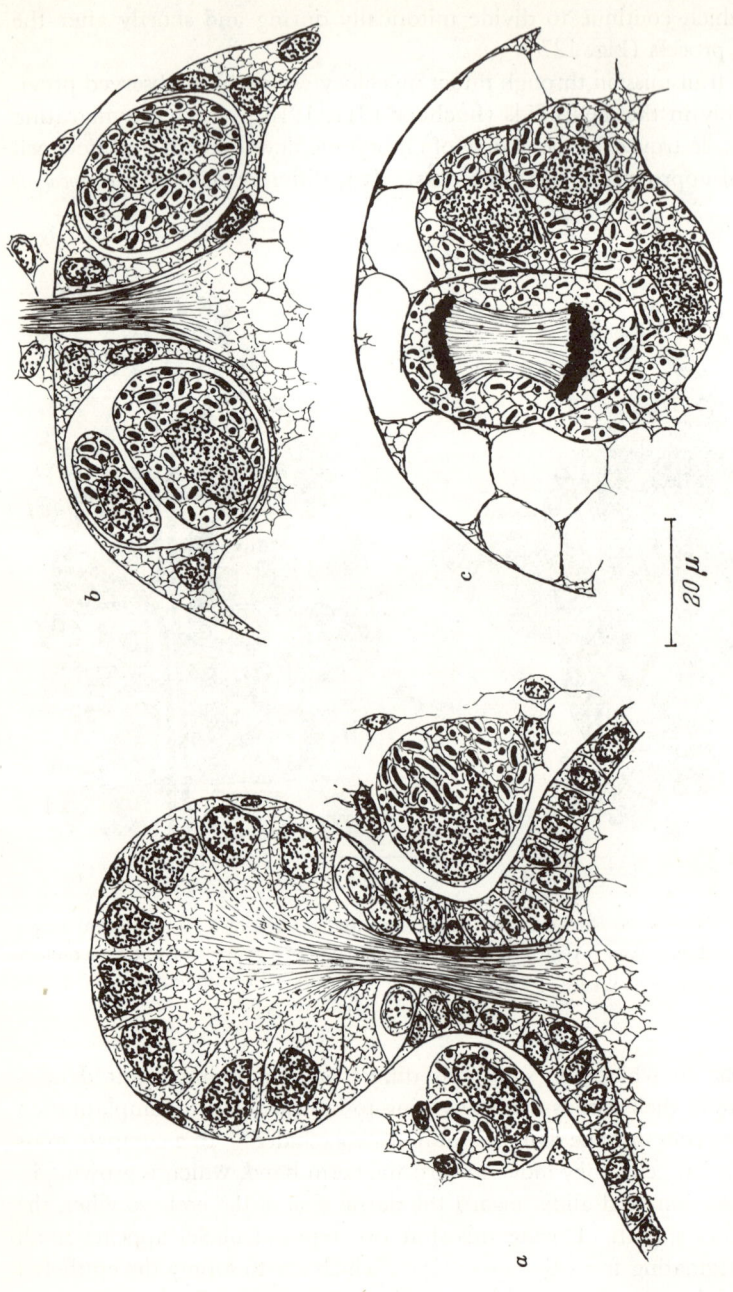

Fig. 127 *Macrocerocus superbus* Leon.   Three stages of the transfer of intact mycetocytes to the egg.   From Buchner.

20 μ

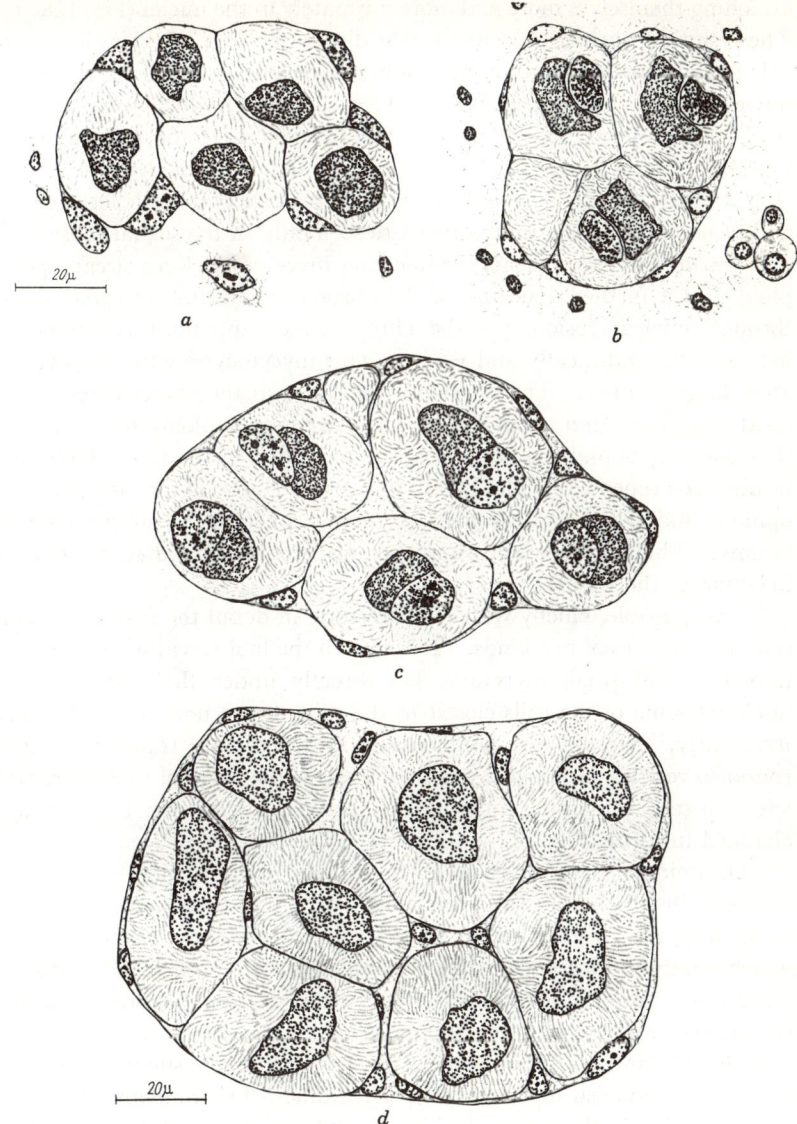

Fig. 128 *Macrocerococcus superbus* Leon.   Fusion of each of the maternal mycetocytes
with a separate yolk nucleus.   (*a*)   The yolk nuclei are in contact externally; (*b*) the
yolk nuclei have penetrated; formation of the mesodermal membrane; (*c*) both kinds
of nuclei before fusion; (*d*) "somatic fertilization" is completed.   From Buchner.

attaching themselves more and more intimately to the nuclei (Fig. 128*a,b*). The resultant nuclear pairs are at first distinct because of their characteristic nuclear structure but fuse into a homogeneous, proportionately larger nucleus when the larva is born! The symbionts, which were shorter, thicker, and fewer in number in the infected mycetocytes, increased markedly in the meantime and now take the form of delicate rodlets (Fig. 128*b,c*). Here one is observing an actual case of somatic fertilization and of potential immortality otherwise seen only in the sexual cells.

It is true that the nuclei of the infecting mycetocytes were already polyploid and a further doubling of the chromosome numbers came about through nuclear fusion, yet the chromosomes apparently continue to increase endomitotically and lead to giant mycetocytes with proportionately large mitoses. The divisions giving rise to the mycetocytes which assist in transmission again reduce this extreme polyploidy somewhat, so that one may actually speak of somatic reduction of the type which may be observed repeatedly in body cells during the course of the insect's development and which is analogous to certain phenomena in the field of botany. Finally the whole supply of mycetocytes is exhausted during infection of the numerous eggs.

It was possible, chiefly with *Puto*, to study in detail the first separation into mycetocytes of two kinds. It begins in the first larval state when the unpaired, still small mycetome lies directly under the midgut. The nuclei of some of the cells closest to the midgut are now split extremely irregularly. Most of the symbionts in these cells are transformed into roundish vesicles, although a few small nests of the original rodlets are still left. In the remaining mycetocytes neither nuclei nor bacteria are changed in form (Fig. 129).

This unique sharp division of the viable and degenerate symbiont material, in which apparently the neighboring intestinal epithelium plays an inciting role, vividly recalls the so-called *x*-symbionts of the fulgorids which continue to populate the mycetome, transformed in high degree, whereas the non-degenerate specific states used in transmission were transmitted very early to the "rectal organs."

Although in males of both genera there is a symbiotic site in the form of an unpaired longish mycetome, the mycetome is in an advanced state of decomposition in the imagines, the only state investigated, and in the end apparently disintegrates completely. If they have not wholly disappeared, the symbionts are spherical and rarely are remnants of the original tube forms encountered. It may be assumed that in earlier stages the mycetocytes do not split into two types, a process which has significance indeed only for the female sex.

Although the symbiosis of these two genera differs from that of *Pseudo-*

coccus and *Phenacoccus*, certain relationships nevertheless exist through the unpaired ventral mycetome and the nature of its symbionts. However, investigation of several pseudococcines, described earlier as *Ceroputo* or *Phenacoccus* but recently assigned to the new genus *Rastrococcus* by Ferris (1954), has demonstrated that the symbionts in this subfamily are extremely heterogeneous (Buchner, 1957). Indeed, Ferris himself had assumed that this, his own genus, would need further subdivision once a larger number of species had been identified, and the validity of his assumption is affirmed by the numerous provisional names I had to invent during research with five species from Java.

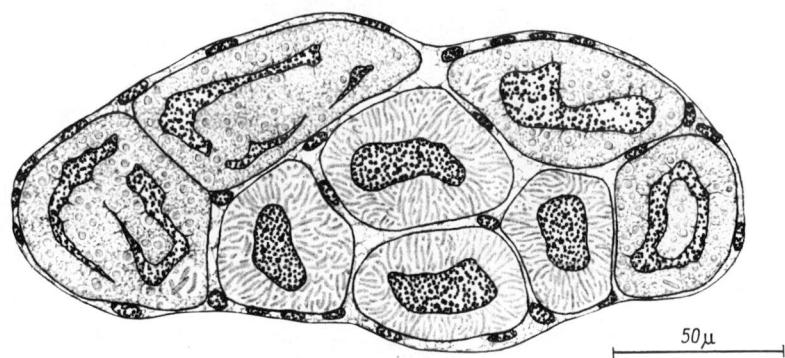

Fig. 129 *Puto antennatus* Sign. Mycetome of a first-stage larva; differentiation into primary and secondary mycetocytes. From Buchner.

In contrast to all previously investigated pseudococcines, *Rastrococcus* species do not have an unpaired mycetome. *Rastrococcus spinosus* (Rob.) houses its rod-shaped bacteria in loosely distributed mycetocytes, great numbers of which permeate the ventral abdominal region particularly while others extend as far as the cephalic area. As in all species of the genus, the symbionts enter the oocyte at the site typical for the entire subfamily, namely, between the follicle and the nutrient plasma cord. Here they collect in great masses and form on the exterior of the egg a cap-shaped aggregation before sinking into the interior as a group. When the cleavage nuclei then move in between the symbionts, a mycetome-like structure is formed (Fig. 130), but of course the components are dispersed in the prolarva.

Rastrococcus "franssenii" is surprising because of the presence of symbiotic yeasts! They live in the multilobular fatty tissue, which is built up

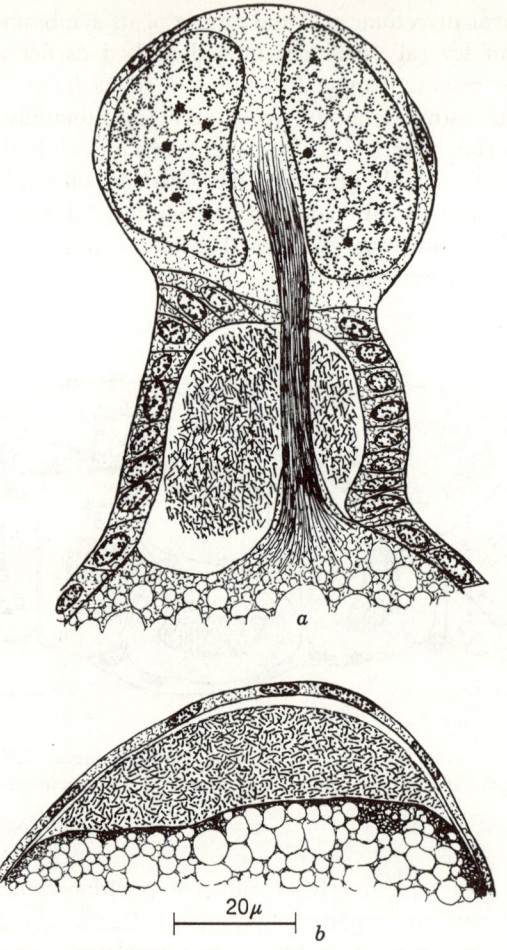

Fig. 130   *Rastrococcus spinosus* Rob.   Two stages of transmission of symbiotic bacteria. From Buchner.

syncytially and is therefore unlike that of pseudococcines studied earlier. Little uninucleate fat cells surround huge fat-containing terrirories in which two types of nuclei can be differentiated: larger, irregularly shaped ones which always attract numerous yeasts, and smaller ones dispersed throughout the symbiont-free areas (Fig. 131).   During egg infection the yeasts follow the course described for the bacteria of *Rastrococcus spinosus*. There has been no research as yet on embryonic development.

Rastrococcus iceryoides Green has a similarly composed fatty tissue with uninucleate small fat cells forming an almost continuous circle around great areas which are without cell limits, but these areas are without nuclei throughout and they contain in addition to sparse fat droplets numerous staff-shaped bacteria, in the manner of a pure culture.   Yeasts are always present, though in much smaller numbers than with "franssenii," but they are loosely dispersed in uninucleate fat cells, where they bring about

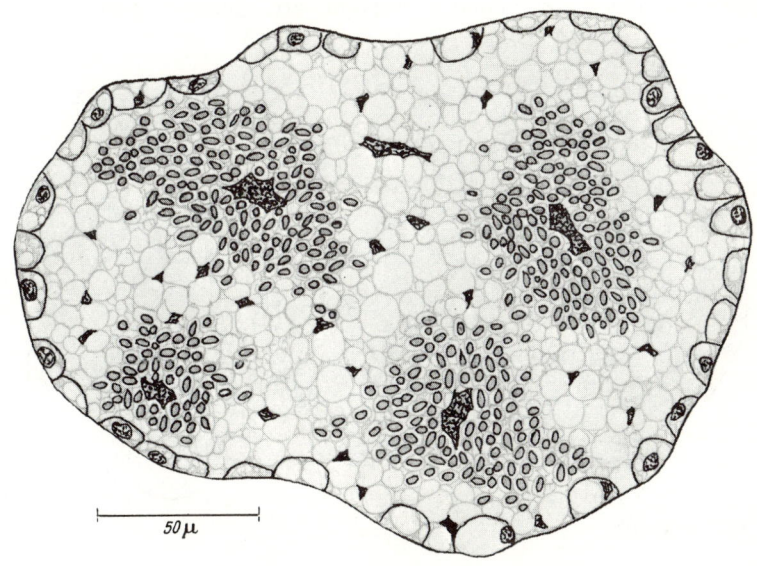

50 μ

Fig. 131   *Rastrococcus* "franssenii."   Fat tissue occupied by symbiotic yeasts.   From Buchner.

moderate growth and formation of polymorphic nuclei (Fig. 132).   Both types of symbionts infect at the same place, thus forming in the mature egg a roundish ball at the upper pole (Fig. 133*a*).   When the germ band approaches them, the two partners, which were united until now, separated and it is interesting to see how differently the host treats each of them.   First it shows interest only in the yeasts.   While the yolk nuclei migrate toward them and soon penetrate them on all sides, the bacteria, which in the meantime have markedly increased, glide toward the periphery of the embryo in irregular groups, yet without allying themselves with nuclei.   Yeasts and bacteria are soon clearly separated, as seen in Fig. 133*b*, which illustrates a late stage of the separation process.   By the time the extremities sprout, cell limits have been formed around the yeasts, whereas the bacteria groups, unchanged, are situated here and there in the plasma.

With respect to further development we are forced to hypothesize.   It will have to be assumed that here and there the bacteria get into the young fatty tissue and infect its cells—infected fat cells are seen occasionally even in older animals—causing them to disintegrate and thus giving rise to the great nuclei-free areas.   The providing of the yeasts with nuclei, on the other hand, may represent only a temporary arrangement; apparently they

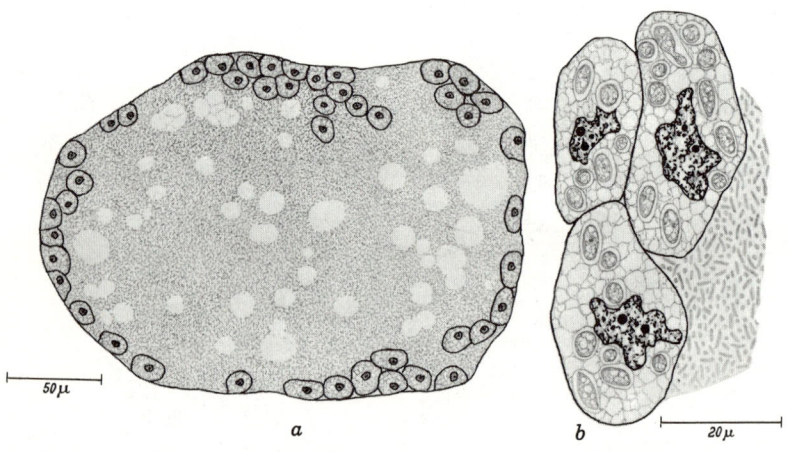

Fig. 132   *Rastrococcus iceryoides* Green.   (*a*) Section of fat tissue flooded with bacteria; (*b*) fat cells infected with yeasts.   From Buchner.

are released again later, drift around in the lymph, and are sporadically transmitted to the fat cells.   Occasionally such free-lancing yeasts are still found in mature animals, and in very rare cases they are found in the midst of the bacterial masses.   Also among the lecaniines there are cases where nuclei are temporarily given to yeasts which later populate lymph and fat cells.

Another variant occurs in *Rastrococcus* "pseudospinosus."   Again the fatty tissue is built in the typical way, that is, it consists of well-limited cells of one nucleus or, more rarely, two nuclei, but there are occasional fat cells, some of them very large, which contain relatively few yeasts.   Bacteria drift about in the lymph in inconspicuous little groups but nevertheless are transmitted to the egg simultaneously with the yeasts.

It would be interesting to study the symbiosis of other tropical species of this genus, since it is clear from the foregoing observations, even though they are based on scanty material, that loss of symbionts and new acquisitions of them are no rarity here. We have seen that typical pseudococcine

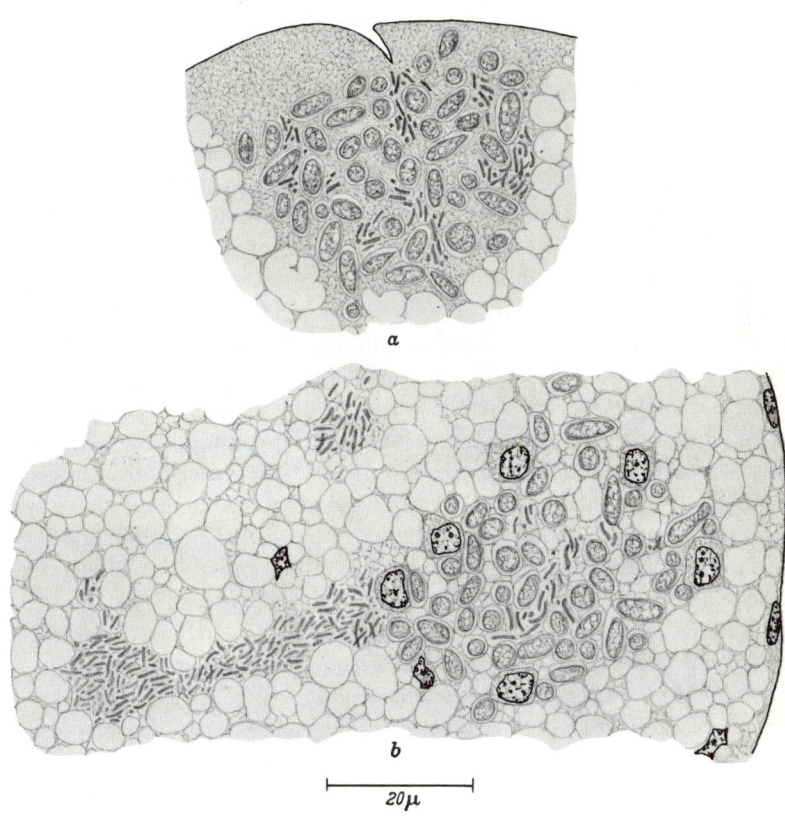

Fig. 133 *Rastrococcus iceryoides* Green. (*a*) Bacteria and yeasts after egg infection; (*b*) separation of bacteria and yeasts at the beginning of blastoderm formation; the yeasts are provided with nuclei. From Buchner.

symbionts are frequently replaced by yeasts and that these yeasts are often allied with more or less well-adapted bacteria; yet for the present it must remain undecided whether *Rastrococcus spinosus* represents an original state or whether any conclusions on symbiont exchange can be drawn from a symbiosis which is admittedly unusual even for pseudococcines.

Even old well-organized cases of symbiosis are not irrevocable.   Let us take as an example the interesting *Hippeococcus wegneri* Reyne.   This genus which is known at present in three species from Java, and at first glance does not suggest a pseudococcine at all, lives on various host plants in close relationship with *Dolichoderus* species.   Numbering in the hundreds, they are guarded by the ants and when endangered are brought to safety after climbing several at a time on the backs of the ants.   When order has once more been restored, they are carried back to the feeding place.   When necessary, the foster children are even transported to new feeding plants (Reyne, 1954).

On studying *Hippeococcus* in its native habitat, I found to my great surprise that it has no symbionts and yet develops a voluminous mycetome rudiment, which remains sterile, in a manner greatly resembling *Pseudococcus!*   In both genera a lateral proliferation of large polyploidal cells, which later are detached from the lining, takes place when the blastoderm forms.   Though far greater in *Hippeococcus*, the increase in both genera is brought about in all probability through a fusion of the nuclei of the polar bodies and those of the cleavage nuclei.   In *Pseudococcus* the prospective mycetocytes travel to the anterior egg pole and assimilate the symbionts there; in *Hippeococcus* a larger body, rounding itself off, glides backward and, remaining sterile, is pushed to the center of the egg by the tip of the germ band which is not invaginating.   Remaining there while the curving germ band develops further, it broadens out on the dorsal side to the area of the head when the extremities bud, in the manner characteristic of numerous insects which are infected at the posterior pole (Fig. 134). The sterile mycetocytes mingle with the fat cells, which are separating at this time, and, showing greater and greater resemblance to fat cells, are now apparently transformed into fat cells themselves.   Only a small number of mycetocytes which have already degenerated are phagocytized by the fat cells when revolution sets in.

On examining imagines taken from the host plants I discovered that their ovaries are always very retarded in development and that embryos are completely absent.   It is in this stage that the ants carry the females to their nests, and it is only here that the females are found in great numbers with maturing ovaries and an abundance of embryos.   It is almost certain that the foster parents feed the animals with a food sap containing the growth factors which are necessary for sexual maturation and which are lacking in the plant food.   We have as yet no corresponding observations on synthetic nests, but, on the one hand, the scale insects obviously do not suck at roots in the ant nests; and furthermore an aphid which always lives in association with ants, *Paracletus cimiciformus* v. Heyd., actually takes up liquids from the mouth of the ants with its snout, the tip of which is

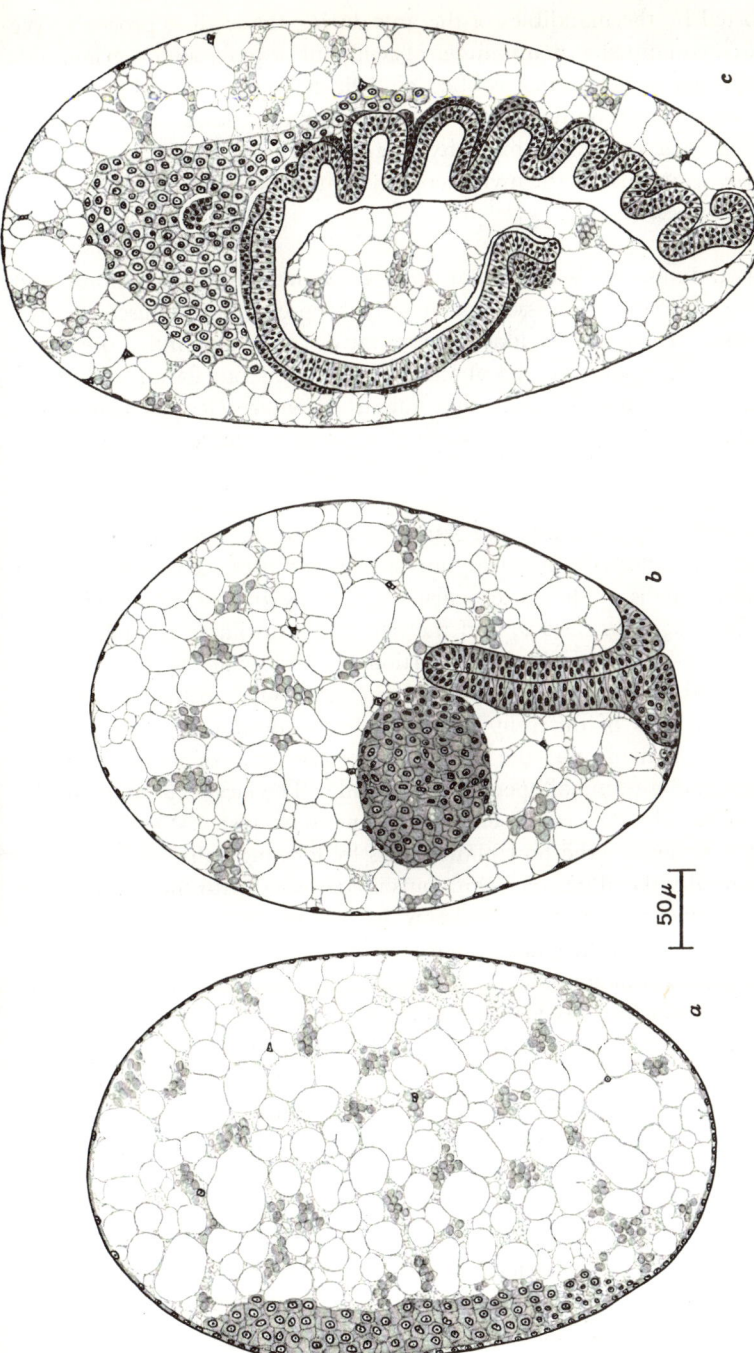

Fig. 134 *Hippeococcus wegneri* Reyne. During egg development a mycetome rudiment which remains sterile is separated off. From Buchner.

supported by the mandibles of the ants during the feeding process.   We are also acquainted with a number of symbiont bearers, such as triatomids and blattids, which, when robbed of their symbionts, are able to complete their cycle of development, though more slowly, and without bringing their ovaries to maturity (see the General Section).   Thus in *Hippeococcus* it must be the myrmecophily that causes the breakdown of the endosymbiosis.

APIOMORPHINES.   It is clear that *Hippeococcus* has lost its symbionts and that the embryonic' processes just described represent a reminiscence phenomenon.   However, in four representatives of a subfamily, the apiomorphines, I established the absence of symbionts but found nothing in the embryonic development that might be interpreted as primordial phenomena.   The question whether this group is symbiont-free from the beginning, or whether we are dealing with a secondary state, must therefore remain open.

The apiomorphines, a group restricted to Australia and New Zealand, feed almost exclusively on eucalyptus species and produce on them highly diversiform galls.   The latter vary enormously in size, some of them achieving huge dimensions, and the size of the turbinate compact occupants ("peg-top coccids") is proportionate.   Antennae and extremities are reduced.   The males, which are much smaller than the females and develop in galls of different formation, also outnumber them by far. Copulation is made possible when the posterior end of mature females emerges from a terminal opening in the gall.   The larvae are born in the gall and leave by the same opening after a period of rest.   When they suck at shoots or leaves, the host plant reacts by forming galls.

Apparently the absence of symbionts is connected with the fact that the apiomorphines are the only coccids which live in galls, and it is certainly no accident that in the gall-forming Hymenoptera and Diptera or in most of the insects which mine in leaves and stems, symbionts are always lacking.   Unfortunately we have as yet no research on the special way in which the apiomorphines absorb their food or on the growth-factor content of galls.

Interestingly, the special position of these scale insects is emphasized also by their sometimes very exceptional early embryonic development.   The small yolk-poor eggs of *Apiomorpha urnalis* Tepper exhibit a development determinate to an extent encountered nowhere else among the insects. The 16-cell stage consists of 12-yolk nuclei and 4 blastoderm cells.   Three of the latter lose their capacity for dividing, grow enormously, and constitute the serosa, which always retains this 3-cell structure; the fourth blastoderm cell forms the rudiment of the germ band and amnion.   In the other three species I investigated, the early development follows more or less the usual pattern. Serosa and follicle epithelium take over the metabolic transport to the yolk-poor embryos.

STICTOCOCCINES.   The stictococcines, representing a subfamily with but a single genus, *Stictococcus*, has many interesting aspects.   All known species

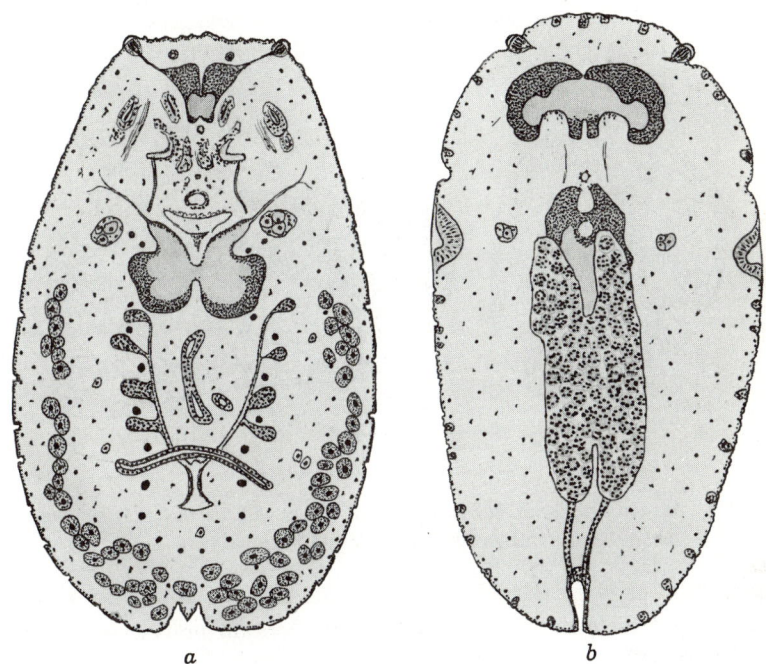

a   *b*

Fig. 135   *Stictococcus sjoestedti* Cock.   Frontal section (*a*) through a first-stage female larva with peripherally situated mycetocytes, (*b*) through a first-stage male larva without mycetocytes.   From Buchner.

—in 1936 the list included only ten—live in tropical Africa in a region in which temperature and moisture tally exactly with what is suitable for the planting of *Theobroma cacao*.   Not less than eight species have been found on *Theobroma*, but the coffee bush and many other plants also serve as food plants.

Two factors have been of special interest to systematists: first, the sexual dimorphism, in contrast to that of all other coccids, becomes manifest in varying degree in the first instar; second, the nymphal states of males ingest no food from the moment of birth onward, for their mouth parts are mere functionless rudiments at the most or have regressed completely.

Investigation of the symbiotic devices (Buchner, 1954, 1955, 1963) revealed additional pecularities.   In the six species available to me, only the females possessed symbionts!   Figure 135 illustrates in cross section the prolarvae of both sexes in *Stictococcus sjoestedti* Cock.   In the female it can be seen that the mycetocytes are rather scanty in this early stage and appear individually or in groups chiefly in the marginal regions of the fatty tissue.   The oviduct and rudiments of the ovarioles are formed like antlers.   In the male of the same age the mycetocytes are lacking com-

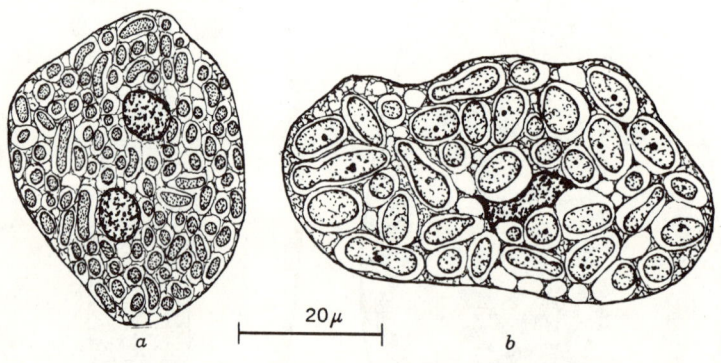

Fig. 136   Mycetocyte (*a*) of *Stictococcus sjoestedti* Cock. with bacteria, (*b*) of *S. diversiseta* Silv. with yeasts.   From Buchner.

pletely, but the voluminous testes already contain numerous rosettes of spermatogonia.

In regard to egg formation it was found that oocytes which produce males are symbiont-free from the beginning.   The mycetocytes of *Stictococcus sjoestedti*, with one or two nuclei, have a dense population of rather thick sausage-shaped microorganisms (Fig. 136*a*).   When the ovarioles develop further, they frequently reach the zones permeated by the mycetocytes and are infected from there outward.   In the process a number of cells collect around the posterior area of the oocytes, and only in these oocytes does a thoroughgoing transformation of the bacteria into large spherical infection stages result, a transformation patently caused by contact with this narrowly defined zone of the follicle.   In the mycetocytes which adjoin the follicle at other points the symbionts retain their original form (Fig. 137*a*).   The oocytes remain very small and completely free of yolk material.   When transformation into infection states is completed, the symbionts are transmitted to the egg through gaps arising in the posterior ring-shaped zone in the follicular epithelium. Figure 137*a* illustrates the beginning of infection; in Fig. 137*b* most of the symbionts simulating vitelline spheres have already transferred. They are now

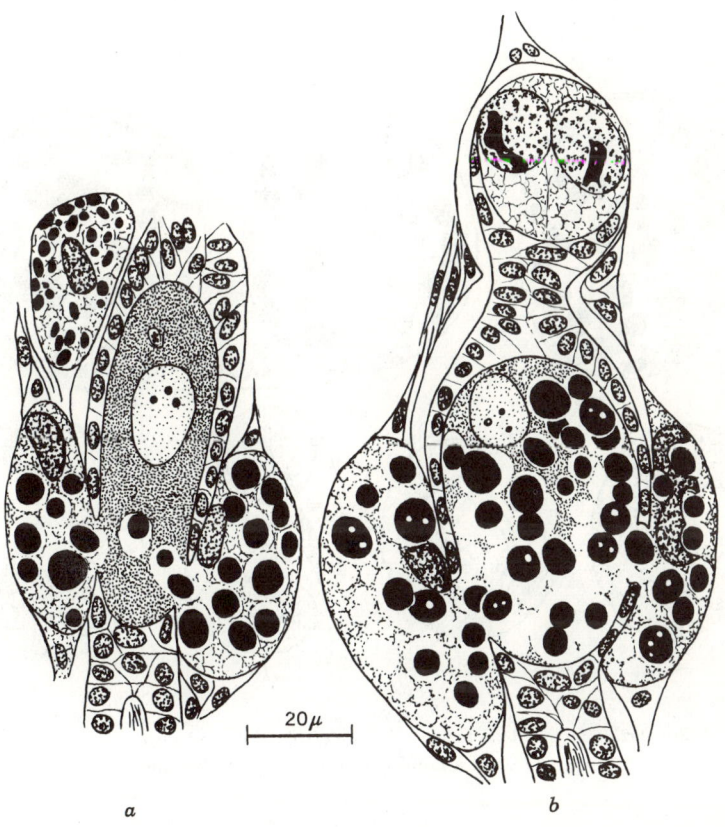

a   b

Fig. 137   *Stictococcus sjoestedti* Cock.   Two stages of egg infection.   From Buchner.

situated in meshes of the egg plasma and displace all the latter except a small part around the nucleus, which is now approaching its maturation division. The egg cell is inseminated by the time infection begins; the chromosomes of the spermatozoans may be recognized in both illustrations in the vicinity of the egg nuclei. When oocytes attain the necessary maturity in regions in which no mycetocytes exist, infection fails to occur and the egg plasma is not vacuolized.

These two egg types, very different yet both inseminated, are illustrated in Fig. 138, which also illustrates several following stages of fertilization and cleavage with and without symbionts.   In both types the egg plasma is concentrated before and during the maturation divisions to form a bridge which connects the two poles and in the middle of which the union of the pronuclei takes place.   The first of the two polar globules floating about in the liquid-rich area forms a spindle which rarely leads to division.

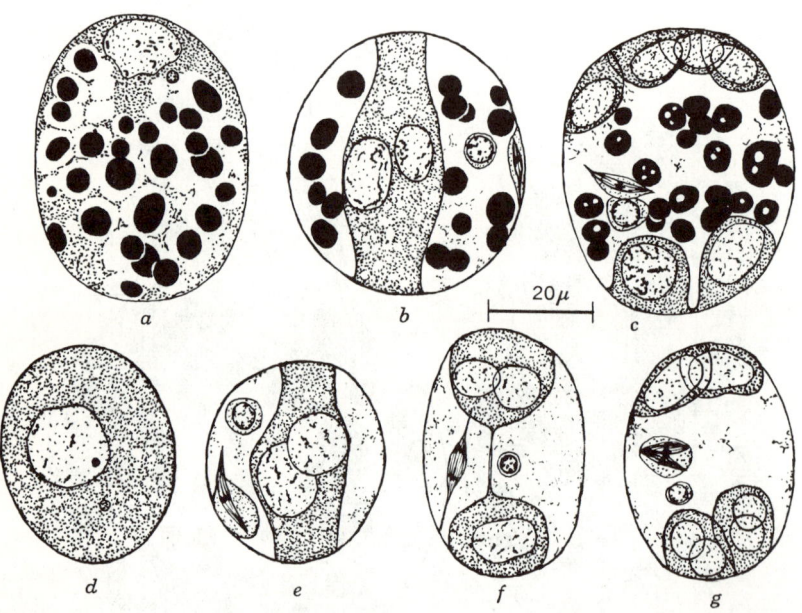

Fig. 138   *Stictococcus sjoestedti* Cock.   (*a–c*) Fertilization and cleavage of an infected egg; (*d–g*) of a non-infected egg.   From Buchner.

Distribution of the plasma to the two poles occurs in connection with the first cleavage.   The blastomeres are still located superficially during later cleavages, but formation of a connected blastoderm does not take place. The cleavages are not synchronized, and they reveal a varying gonomery.

Complete lack of reserve substances and the fact that development begins exceptionally early suggested from the very beginning an alimentary device of the type found elsewhere in diverse forms under similar conditions (compare Buchner, 1957).   The device used by *Stictococcus sjoestedti* is again unique!   In a few-celled stage, several cells of the rudimentary blastoderm leave the area of the former egg at the anterior pole and arrive at a type of placenta which the follicle forms between the egg and the degenerating nutrient cells. In this way a basin-shaped formation arises which represents the rudiment of the germ band. In its interior the follicle nuclei and the remains of the nutrient cells finally disintegrate (Fig. 139). At this time one of the blastomeres is detached from the lateral

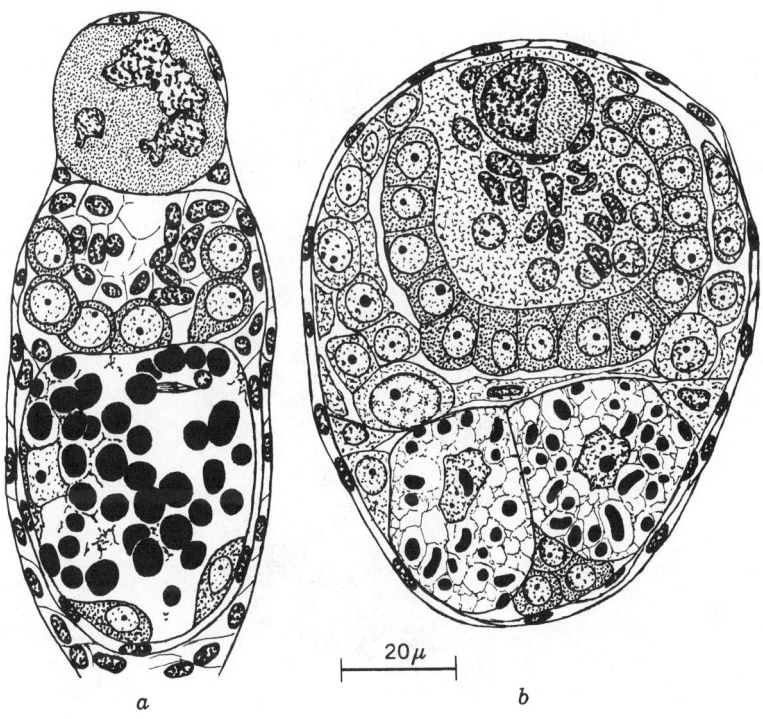

Fig. 139 *Stictococcus sjoestedti* Cock.   (*a*) Extraembryonic rudiment of the germ band in an infected female embryo; (*b*) advanced extra-embryonic formation of the germ band; in the remainder of the embryo, the two first mycetocytes.   From Buchner.

wall, becomes amoeboid, and enlarges its nucleus. It now ingests all the symbionts, forms a cell limit around the latter, and finally is divided in two (Fig. 139). After this division the symbionts, spherical up to this point, again assume tubular shape, and at the same time the embryo and the germ band, situated so unnaturally outside, are again united into a harmonious whole.

Comparison of male embryos shows that they develop exactly as the females do, except that dimensional differences heighten the contrast between the migrated rudimentary germ band and the small, now symbiont-free, egg derivate situated behind it (Fig. 140).　For all intermediate stages of these processes which sound so improbable, the reader is referred to the wealth of illustrative material in the original work.

When I had the opportunity to study specimens of *Stictococcus diversiseta* Silv. for purposes of comparison, I found to my great surprise that,

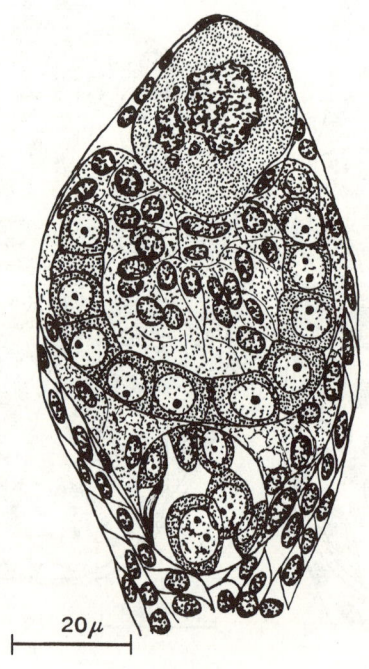

20μ

Fig. 140　*Stictococcus sjoestedti* Cock.　Extraembryonic formation of the germ band in a non-infected male embryo.　From Buchner.

although males remain symbiont-free in this species too, typical "yeasts" again, instead of bacteria, populate cells dispersed in the body (Fig. 136*b*); that embryonic development and alimentary devices are highly exceptional; and that female embryos are infected only at an advanced stage of germ-band formation (Buchner, 1955).　Here also the egg cells remain very small and yolk-free.　The yeasts are transmitted individually to a cellular casing, at first very thin, and from there to the area of the nutrient cells, and occasionally they even settle among the blastomeres of the egg, which is now in the state of total cleavage (Fig. 141*a*).　However, the

incipient symbiont transmission regresses completely.   The morula now
taking form is always free of symbionts.   No terminally situated placenta
of follicular origin exists, and the embryo is enveloped instead by a very
large extrafollicular casing in which giant cells appear in a layer.   These
giant cells now receive the yeasts and advance them to the posterior area

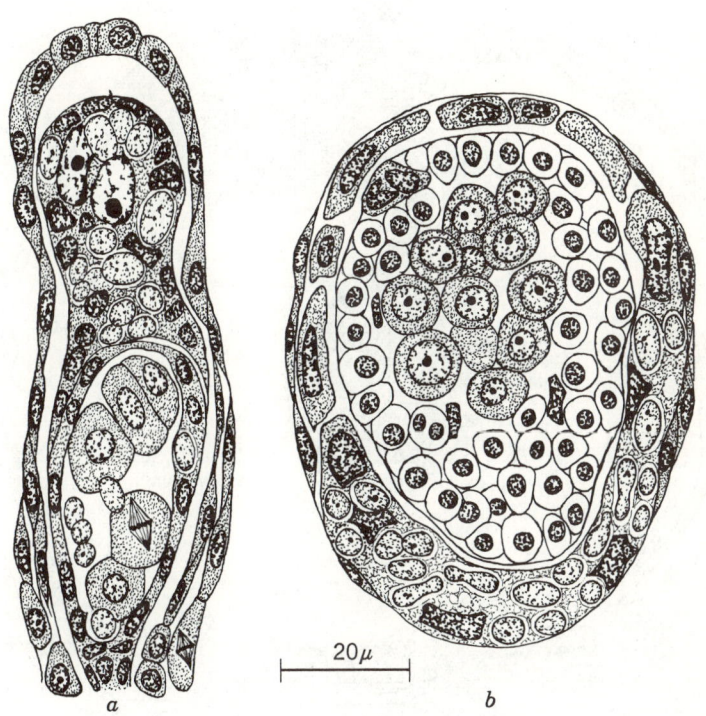

Fig. 141   *Stictococcus sjoestedti* Cock.   (*a*) Cleavage stage with temporary infection
of the nutrient cells and embryo with symbiotic yeasts; (*b*) older cleavage stage with
rounded follicular cells and with extraembryonic giant cells infected with yeasts.
From Buchner.

of the embryo.   In the meantime the follicle cells loosen and round off and
are then situated around the morula which as yet consists of only a few
cells (Fig. 141*b*).

   Through a type of delamination, a cell layer which represents the rudimentary serosa is now delimited around the morula; absorption of considerable quantities of liquid causes it to stand out from the other cells
which are arranged in the form of a rosette and from which the germ band

and the amnion later develop.   At the same time several vitelline spheres appear in the space between the two rudiments.   When signs of separation into germ band and amnion appear in the rosettes curving sinuously and suspended at the upper pole of the serosa, the first yeasts at the opposite

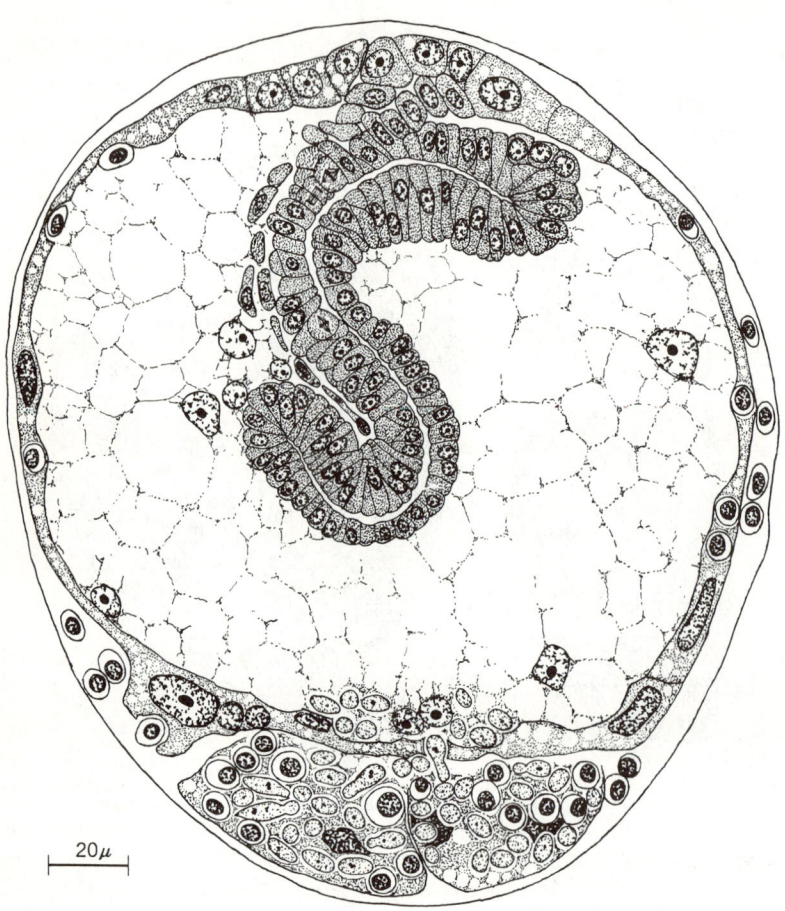

Fig. 142 *Stictococcus diversiseta* Silv.   Female embryo with advanced rudiment of the germ band; the yeasts begin to pass over into the embryo at the posterior pole.   From Buchner.

pole are transmitted through gaps in the serosa to the body cavity of the embryo, which in the meantime has grown much larger (Fig. 142). Development of eggs which produce males takes the same unusual course except, of course, that the symbionts are lacking in all stages.

Everything points in the direction that *Stictococcus sjoestedti* represents the original state of symbiosis in this subfamily and that in *S. diversiseta*, as in the genus *Rastrococcus* among the pseudococcines, the bacteria are displaced by yeasts. Examples of this type have indeed accumulated to such an extent recently that a different interpretation is scarcely permissible. This applies particularly to the symbiosis of several aphids with yeasts which we now describe. That endosymbiosis may very well appear and disappear is demonstrated in a particularly drastic way precisely by the stictococcids, for there can be no doubt that the lack of symbionts in the males represents a secondary state which is caused by loss of their capacity to absorb food. Several reminiscence phenomena in males prove that they also lived symbiotically at one time. The origin of the free space at the time of fertilization and cleavage in *S. sjoestedti* must be considered a reminiscence phenomenon which, indeed, was caused originally by the presence of symbionts. With *S. diversiseta*, cells serving the transport of yeast are differentiated during development of males as well as in the female embryo and finally are destroyed without function. And the fact that the male embryos of *S. sjoestedti* form the germ band outside the embryo also, might thus be explained, for one may assume that the strong loading of the little egg with the foreign organisms supplies the reason for this device which is found nowhere else. Opposed to the assumption that the small size of the egg alone makes it necessary is the fact that the symbiont-free *Platygaster* egg is approximately only 16 μ in length and 7–8 μ in width and yet is able to develop without recourse to such radical devices, as well as the fact that the egg of *S. diversiseta* which at the beginning develops without symbionts does not form the germ band outside the egg area.

The fact that two species should behave so differently understandably awakened the desire for an extension of this knowledge. In 1963 I was able to report on four additional species of *Stictococcus*. The most surprising finding was with *Stictococcus acaziae* De Lotto. In this case there is, in addition to the original symbionts, another bacterium. Unlike the former, this bacterium is not taken up into the ovocytes but is transmitted only in an advanced embryonic stage, like the yeasts in *S. diversiseta*, after this later acquisition has multiplied considerably in elements of the peritoneal covering which become giant cells. Although the acquisition of a yeast caused the loss of the original symbionts, there is no such antagonism between the two kinds of bacteria. Other species follow essentially the two known symbiosis types. An interesting variation is offered by *S. formicarius* Newstead; here merely one nucleus of the two-cell stage leaves the germ plasm and becomes the extraembryonic rudiment of the germ band. In *S. acaziae*, which was infected with Hymenoptera larvae, different kinds of disturbances caused thereby, and also influencing the transmission of the symbionts, were discovered.

The stictococcid symbiosis has also thrown a certain light upon the problem of sex determination which is singular in the coccids. Research by Hughes-Schrader and Schrader (see summary by Hughes-Schrader, 1948) has shown that in a series of primitive scale insects the original heterogamous state of the male sex still prevails but that the *x*-chromosomes have been lost and that sex determination was consequently transferred to the females. What constitutes the deciding factors after this change, however, is still unclear. Experiments on the numerical proportion of the sexes, which has been proved to vary widely, and its dependence on factors of the milieu suggest that a sex-modifying function is involved. For *Stictococcus* this seems to be proved in a completely unexpected form. One may evidently assume that in this case the very different content in growth factors in the eggs with and without symbionts is the deciding factor. The removal of the heterogamous state of the male sex, in view of the short-livedness of the so very feeble males, appears to be an ecological necessity.

MONOPHLEBINES. An as yet unpublished investigation by me, which encompasses a wealth of material, concerns itself with the fifteen tribus of the family Margarodinae, including the primitive steingeliines and xylococcines, which live without symbionts. This study will enhance our knowledge of the coccid symbioses in many ways; however, the results obtained thus far unfortunately cannot be evaluated in this text. The data on monophlebines published to date treat only a few representatives of the genera *Icerya*, *Echinocerya*, *Monophlebus*, and *Monophlebidius*.

Icerya purchasi Mask., a common pest on various citrus plants, plays a role in the history of symbiosis research because it was one of the test objects in the first experiments to explain the true nature of the homopterous mycetomes (Pierantoni, 1910, 1912, 1914). Its paired mycetomes, which occupy substantial portions of the abdomen, are deeply cleft formations, often subdivided into small mycetome parts and held together by tracheae. Large mycetocytes with irregularly shaped nuclei are covered by a well-developed epithelium which also pushes itself between them. Small sterile elements in the vicinity are constantly being infected, usually without developing to the size of the mycetocytes, and apparently they assist in replenishing the symbiotic material consumed during egg infection. Oenocytes with thick plasma and large nuclei are often inserted between the mycetocytes.

The oval to bean-shaped symbionts which fill all parts of the host cells contain glycogen and metachromatic granules in their liquid-rich protoplasm. In view of the many transitional states in *Pseudococcus* between typical bacteria and similarly shaped involutional forms, it is only natural that one is reminded of highly despecified bacteria. Of course, this concept would not harmonize with Getzel's interpretations of *Icerya* symbionts (1936). He reported that he had made successful cultures on glucose agar in Pierantoni's laboratory and that he had observed in them an organism of yeast-like appearance with oval, elliptically elongated, or irregular cells. On solid media the organism continued its yeast-like growth but in potato water was transformed into a pseudomycelium with only a few branched arms which were thick-walled and flattened against one another. Once cultivated, it could easily be recultivated on almost

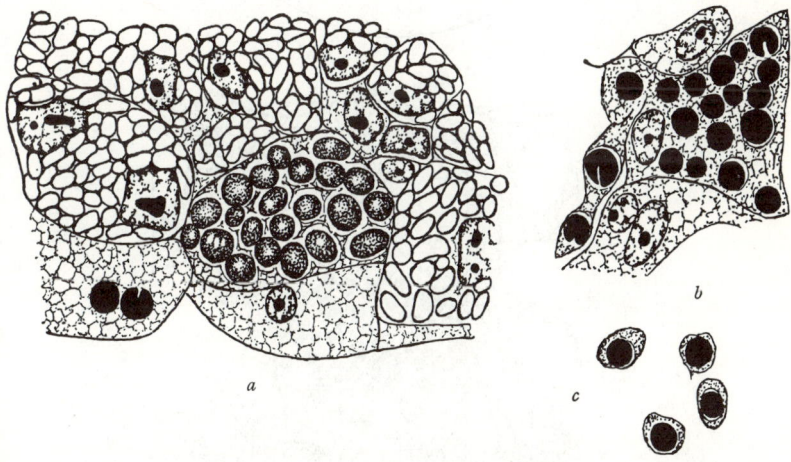

Fig. 143 *Icerya aegyptica* Dougl. (*a*) Formation of transmission forms, 750×; (*b*) their exit from the mycetome, (*c*) transmission forms with plasma remnants in the lymph; 666×. From Walczuch.

all nutrient media. Getzel classifies it as one of the Mycotorulae in the family of the Torolopsidiaceae, assigning to it the name *Geotrichoides pierantonii*. Retesting of his findings would be highly desirable.

In observing *Icerya purchasi*, Pierantoni was the first to discover the formation of specific states for transmission to the progeny. Within the mycetocytes, symbionts distinguished by increasing affinity for nuclear dyes appear individually or in small groups. In *I. purchasi* they usually retain the original form; *I. aegyptica* Douglas, investigated by Walczuch, they become spheres and increase in volume. A fissure reminds one of the more elongated shape of the original. As these infection states leave the mycetome, they usually carry away a part of the host plasma, which later disappears (Fig. 143).

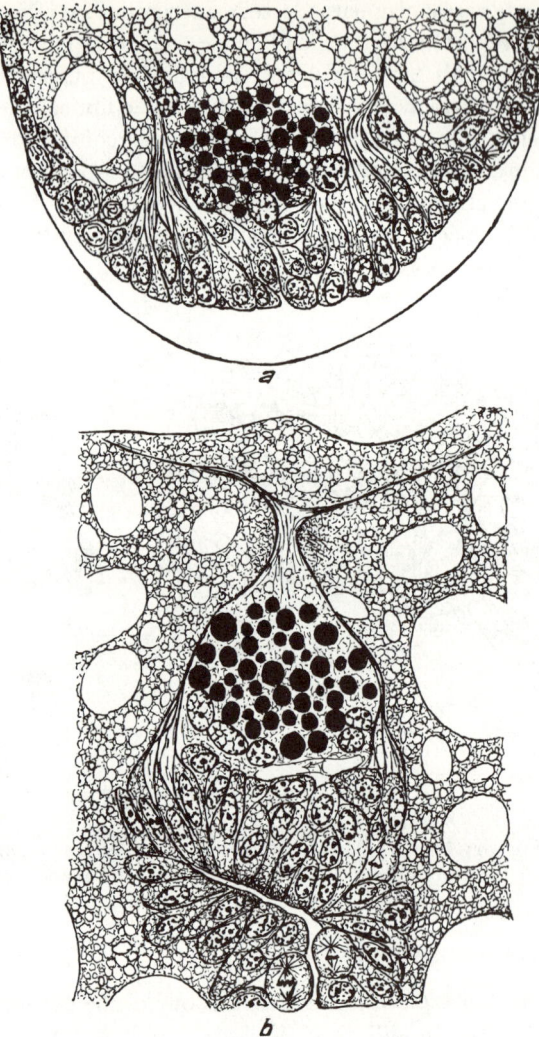

Fig. 144 *Icerya aegyptica* Dougl. (*a*) Blastoderm stage; the symbionts grasped by fibrous processes of the blastoderm cells, 333×; (*b*) the symbionts at the end of the germ band, which has penetrated to the middle of the egg, enveloped in a fibrous sac, 400×. From Walczuch.

As with *Lakshadia*, egg infection takes place at the posterior pole. Gliding between the follicle cells, the symbionts are transmitted to a space formed through retreat of the egg plasma. The finely honeycombed coagulation in which they are embedded here originates from follicle cells.

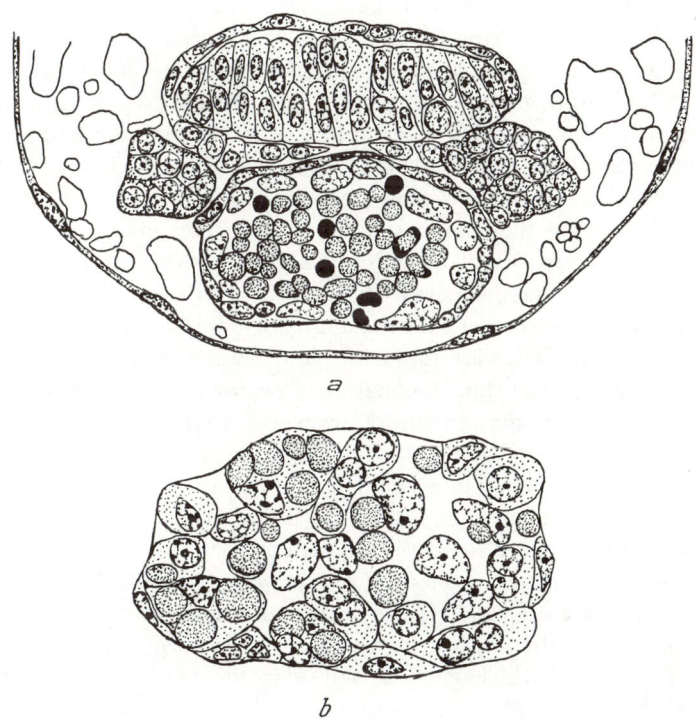

Fig. 145 *Icerya aegyptica* Dougl. (*a*) Frontal section through an embryo before revolution; above, the germ band; below, the mycetome surrounded by mesodermal cells; on both sides, the rudiment of the gonads; 450×; (*b*) formation of the mycetocytes; 666×. From Walczuch.

The symbionts are ingested by deglutition after the maturation divisions cease.

The fate of the symbionts during embryonic development is described by Pierantoni and Walczuch. Ascending cleavage nuclei sink between them, increasing slowly at first without forming cell boundaries. The germ band, which is invaginating at the same place, pushes along, in the usual manner, the embryonic mycetome flanked by primordial germ cells.

In *I. aegyptica* a strange formation envelops the symbionts, somewhat as in *Orthezia*. In the yolk of the blastoderm stage the neighboring cells grow radiating fibrillous processes which are soon united to form a regular casing over the symbionts (Fig. 144).

When the symbiotic group has almost reached the opposite pole, it detaches itself from the germ band. Pierantoni had established that mesodermal elements of the lower lamella which are added at this time furnish material for the epithelial casing, but not until much later did Walczuch recognize that the cleavage nuclei are also replaced by such elements. First, a layer of mesodermal cells rich in nuclei develops around the syncytium to take the place of the disappearing fibrous casing. The larger nuclei of the syncytium are superficially located, and its symbionts gradually lose their strong staining capacity and again assume their typical form. Then the yolk nuclei are collected in the central part of the primary mycetome and the mesodermal nuclei are separated, some remaining on the surface as a casing but the majority penetrating to the syncytium. Here plasma boundaries are formed around the symbionts each time as they are captured, and thus the final mycetocytes arise. During involution one can still see the remains of degenerated cleavage nuclei within the young mycetome (Fig. 145).

After the mycetome becomes paired, large numbers of oenocytes migrate to the mycetome and become located on the surface. Further cleavage of the organs takes place shortly before hatching.

Walczuch investigated species of *Icerya aegyptica* from Africa and India. In the mycetomes of all individuals of African origin she found a second symbiont type, in the form of filamentous bacteria which are obviously acquired later and which are less well adapted (Fig. 146*b*). In large masses they push the mycetocytes apart and in certain places stretch the epithelial casing so taut that the aggregations permeated by small nuclei border on the surface of the organ. They are also found, rolled into spirals, in the egg intermingled with typical symbionts.

Examination by a taxonomist proved that the form in question is not merely another species of *Icerya*, and certainly one may conclude that a special type of symbiont exists, for Walczuch's findings with other species of *Icerya* and of *Echinocerya*, *Monophlebus*, and *Monophlebidius* show us that the occurrence is not a mere parasitic invasion lacking significance. In all the species mentioned Walczuch found a more or less extensively adapted second symbiont of the kind described; comparison of their forms affords an insight into the gradually progressing perfection of the symbiosis, just as in many cicadas which are still to be described.

Icerya littoralis Cockerell has the most primitive type. Here the second symbionts, usually in the form of thick knotted tubes, are located outside the mycetomes, though always nearby, around and in small-celled elements of the blood.   In *I. monteserratensis* Ril. How. and *Echinocerya anomala* Morr. the additional symbiont populates the mycetomes, which are quite lobate but not constricted off; and with respect to the mycetome it follows the pattern of behavior described for *I. aegyptica*.   In *I. monteserratensis* these symbionts are longish, uniformly shaped formations; in *Echinocerya anomala* they are organisms with strongly vacuolized protoplasm which assume clavate or knotted forms and put forth lateral processes (Fig. 146).

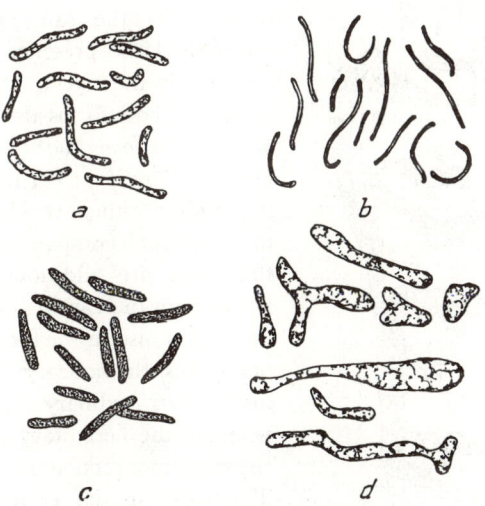

Fig. 146   (*a*) *Icerya littoralis* Cockll., (*b*) *I. aegyptica* Dougl., (*c*) *I. monteserratensis* Ril. How., (*d*) *Echinocerya anomala* Morr.   Additional symbiont forms.   800×.   From Walczuch.

With *Monophlebus dalbergiae* Green and *M. tamarindus* Green and with *Monophlebidius indicus* Green, the monophlebine symbiosis reaches its perfection.   Enormously developed mycetomes extend through seven segments and are constricted correspondingly but do not form partial mycetomes.   Tracheae envelop them closely and even permeate the interior; but ventrally and toward the intestine their borders remain smooth.   The hereditary symbionts live in giant mycetocytes which have polymorphous, often paired nuclei and which are arranged like an epithelium and are permeated occasionally by oenocytes.   The accessory symbionts live in a large syncytium which has small polygonal nuclei, is penetrated by many tracheae, and is surrounded by the giant mycetocytes.

They are coccoid forms or short rodlets, thick or very slender, and are type-specific (Fig. 147). Unfortunately Walczuch's material had no infection-ripe ovaries, but it is clear nevertheless that the two forms are transmitted to the egg cells in the regular manner.

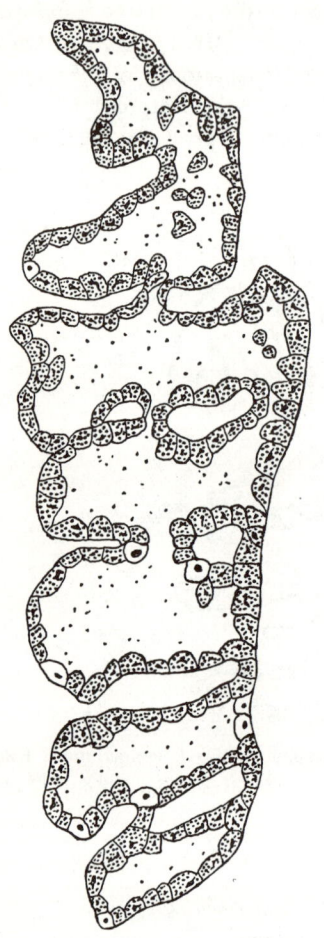

From data by Kitao (1928), who, to be sure, failed to evaluate the situation correctly, one would have to conclude that the genus *Warajicoccus* (*Drosicha*) behaves in the same way. Later I established the presence of companion bacteria in the mycetome and the transmission of both forms also in *Aspidoproctus* and *Walkeriana* and other monophlebines. Thus it is clear that this subfamily, unlike other coccids, is inclined to accept such companion forms. Proof that they are additional guests exists not only in the varying degrees of adaptation but also in the form of the microorganisms, the accessory symbionts having an extraordinary number of forms, whereas the hereditary symbionts have the same monotonous shape throughout. The close spatial relationships of the oenocytes and the mycetomes are typical of all monophlebines. For details of my observations the reader is referred to a report to be published shortly.

Fig. 147 *Monophlebus dalbergiae* Gr. Frontal section through the left mycetome in which two kinds of symbionts are living. 50×. From Walczuch.

MARGARODINES. Today the margarodines are separated from the monophlebines and are classified as an independent subfamily. This is entirely justifiable from the standpoint of symbiosis, for what little we know of them harmonizes little or not at all with our knowledge of the monophlebines. Šulc (1923) was the first to investigate an unclassified species of *Margarodes*, not described in detail, which sucks at the roots of *Festuca ovina*. Its paired mycetomes, in an extremely odd manner seen nowhere else, are so intimately bound up with the oviduct that they form its boundaries for a certain distance. Muscles of the oviduct radiate into the mycetome at the place of attachment and serve to anchor it. The mycetocytes, arranged in parallel rows oriented toward the lumen of the oviduct, begin distally with small cells which become larger and larger as they approach the

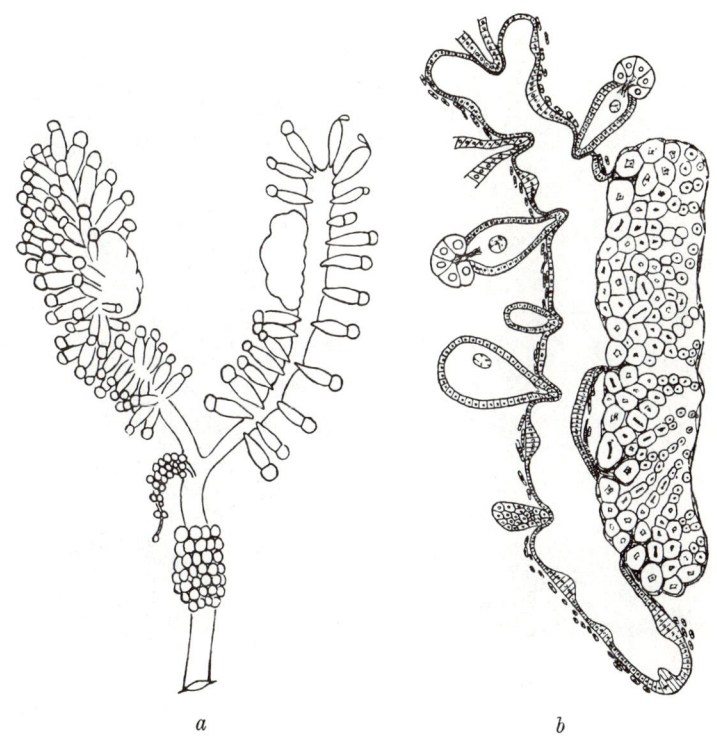

a   b

Fig. 148   *Margarodes* sp.   (*a*) Female sexual apparatus with the fused mycetomes, (*b*) section through both.   From Šulc.

oviduct.   A flat pigment-free epithelium encases the free sides of the mycetomes which, according to Mahdihassan (1947), develop from originally unpaired organs connecting the two ovaries (Fig. 148).   The structure and site of the male mycetomes are unknown.

The manner of transmission is as strange as the symbiotic site itself. As the symbionts approach the oviduct, they become increasingly smaller and each develops a central granule.   In this state they are transmitted to the lumen of the oviduct.   The follicle of the egg cells develops an extremity, covered by the epithelium of the oviduct and projecting into the latter, which must be interpreted as a specific adaptation to the unusual type of transmission.   The symbionts, their central granules having disintegrated into many particles, pass through the epithelium of the oviduct into the extremity which is provided with a transmission canal.   Through this

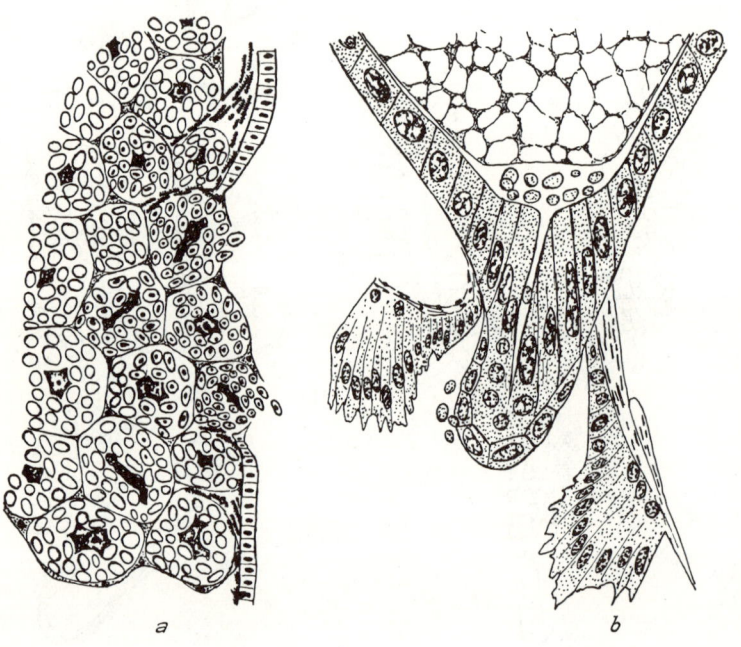

Fig. 149   *Margarodes* sp.   (*a*) Development of infection forms and their transmission to the oviduct, (*b*) infection of the egg through the terminal follicle peg.   From Šulc.

passage they reach the space arising in the usual fashion at the posterior end of the oocytes (Figs. 148*b* and 149).

Also available are data of Jakubski and his students on *Margarodes polonicus* Ckll., a species which is of cultural and historical interest because it produces a red dye formerly in wide use.   The mycetomes differ greatly in location and structure from those of the species available to Šulc.   The paired organs, elongated bodies built up of large, polygonal, uninucleate mycetocytes, accompany the sharply reduced intestine through five or more segments of the abdomen.   In females ready to deposit eggs the organs loosen, and groups of mycetocytes are freed and scattered between fat cells and ovarioles. According to an illustration furnished me by Jakubski, infection of the egg cells takes place at the posterior pole, somewhat as in *Icerya*. But his student, Boratynski (1928), who describes the mycetomes as two irregular masses composed of fat cells, states that infection takes place by way of the oviduct, that is, as in Šulc's test object; but Boratynski was not able to establish how the symbionts reached there.

Kalicka-Fijalkowska (1928) provides data on the symbionts in *Margarodes polonicus* during embryonic development. As usual, the germ band pushes the first mycetome rudiment toward the front together with the primordial germ cells. The rudiment then slides to the rear, stopping when it reaches the vicinity of the fourth or fifth abdominal segment. Later it is shifted toward the front and, now bipartite, attaches itself to both sides of the stomodaeum.

The above-mentioned investigations, not yet published, have shown that there exists a much greater multiplicity in margarodin symbiosis, in which the acquisition of additional symbionts also plays a large role.

COELOSTOMIDIINES. The only coelostomidiine studied symbiotically is *Marchalina hellenica* Genn., a scale insect common on the Anatolian seacoast and in the Dardanelles. It lives chiefly on the Aleppo pine, hidden under the bark but revealing its presence through protrusion of its luxuriant wax secretions. Hovasse (1930), the only author to investigate the species, came upon devices which are unique among the diverse symbiosis types of scale insects, but she failed to describe the cycle in its totality.

The symbiotic site is the intestinal epithelium, a localization found in no other scale insect. The midgut, following upon a long, slender esophagus with an oval bulb at its end, has initial and end parts of the normal narrow construction and an extremely uncommon middle section curved into a *u*-shape. The cells of this section, which are the only infected ones, form convexities which extend into the body cavity and are so large, with diameters of 0.5 mm., that they are recognizable with the unaided eye. Their nuclei, proportionately large and observable with an ordinary magnifying glass, are often cleft and distorted excessively. Interrupted here and there by short sections of normal, low epithelial cells, they arch into the intestinal lumen causing various kinds of buccae which in turn may extend deeply into them. The tracheae are not only attached externally but also penetrate to the interior of these giant cells. The remainder of the midgut section in the area where it is joined to the rectum admits the separate outlets of the three Malpighian tubules.

Hovasse is apparently in error when she states that slight aggregations of smaller rod-shaped microorganisms are present even in the distal sections of the epithelial cells, which remain flat in the initial and final parts of the midgut. The symbionts which fill the enormously grown cells of the middle part on all sides are markedly elongated filaments generally about 20 μ but somtimes as long as 50 μ. The filaments are gram-negative non-motile forms which may be straight, waved, or closely spiral, and they are frequently split up into rodlet chains.

It is known that transmission takes place indirectly by way of the nutrient cells, but the details are not yet clear. Hovasse was of the opinion that symbionts present in the young gonad rudiment originated from neighboring midgut cells. Opposed to this interpretation is the fact that delicate infection stages are involved. According to unpublished observations, it is even before separation of the ovarioles, let alone the differentiation of egg cells and nutrient cells, that an aggregation of infected somatic cells, surrounded by the epithelium of the future oviduct, is found behind the oogonia. When the ovarioles take form they press between the oogonia, and when the latter are separated into egg cells and nutrient cells, they are situated in quantity beneath the follicle wedged between the nutrient cells. Whether the symbionts then glide with the secretions on the nutrient cord to the oocytes or arrive there later in connection with the resorption of the degenerating nutrient cells remains to be investigated.

In the blastoderm stage the symbionts are found in larger and smaller aggregations scattered throughout the yolk. They are accompanied by several yolk nuclei, and there are more in the posterior area of the embryo at the start of invagination. With advanced invagination, denser symbiont nests are found in the direct vicinity of the midgut rudiment and also in that of the primitive sexual cells. Despite lack of information concerning details, it is practically certain that it is from these nests, that is, from the outside, that the intestinal epithelium is infected and that the somatic elements of the gonad rudiment are provided with symbionts.

COCCINES. This subfamily, with its single genus *Coccus* and a small number of species occurring in Mexico, India, and North Africa, is still awaiting research. Pierantoni stated in 1910 that in the familiar cochineal, *Coccus cacti* L., the body cavity of young females in particular, and of adult females to a lesser extent, is filled with a large-celled specific tissue containing inclusions which might possibly be symbionts, yet he ventured no final judgment. With respect to *Coccus indicus* Green we have, on the one hand, Mahdihassan's statement (1929) that it contains roundish colorless symbionts and, on the other hand, the finding of Walczuch, who is known as a painstaking researcher, that she had hunted in vain for symbionts in this test object.

APHIDS

As stated in the historical section, the older zoologists studied the pseudovitellus of aphids anatomically and especially embryologically, but their attempts to understand the unusual phenomena linked with this organ were fruitless (Huxley, 1858; Leydig, 1850; Metschnikoff, 1866; Balbiani, 1869, 1870, 1872; Witlaczil, 1884; Will, 1889; Henneguy, 1904; Flögel, 1905; Tannreuther, 1907). When Šulc (1910) and Pierantoni (1910) ranked the pseudovitellus among the mycetomes, it became necessary to study the field again in all details; hence with some of my students I set myself to the task (Buchner, 1912, 1921, 1958; Sell, 1919; Klevenhusen, 1927; Tóth, 1933, 1937; Profft, 1937). Others have helped, more or less successfully, to develop our knowledge of the complicated symbiosis of aphids (Peklo, 1912, 1916; Webster and Phillips, 1912; Uichanco, 1924; Rondelli, 1925, 1928; Paillot, 1929, 1930, 1931, 1932, 1933; Paspaleff, 1929; Schoel, 1934; Mahdihassan, 1947; Schanderl, Lauff, and Becker, 1949; Lanham, 1952; Trager and Lanham, 1952; Maillet, 1957; Lampel, 1958, 1960; Kolb, 1963).

With the exception of certain hormaphidines, in which the original symbionts were replaced with yeasts living free in the lymph (Buchner, 1958; Kolb, 1963), the numerous aphidines, pemphigines, and adelgines (chermesines) investigated all have mycetomes which are well provided with tracheae and usually consist of two longish strands passing through a number of abdominal segments. Some have broader lobes connected by a transverse bridge situated in the posterior section of the organ, approximately in the sixth abdominal segment, or in exceptional cases on the border between the first and second segments. Toward the rear the mycetomes tend to terminate in an unpaired point (Fig. 150), although in older females the strong development of the ovaries and of the embryos developing in the maternal body in summer generations generally leads to extensive cleavage and distortion of organs which in younger animals were formed normally, as in the psyllids and aleurodids.

Sparse, very flattened cells develop a thin casing around the mycetomes and their partial forms. The individual mycetocytes are large, polygonal, or rounded cells with plasma permeated evenly by numerous roundish microorganisms. Strongly refracting granulae and glycogen droplets as well as mitochondria (Koch, 1930) may be observed between them. In addition, there is a diffused specific coloring of the mycetocytes: in *Callipterus* species, pale green; in *Aphis saliceti* Kalt., yellowish; in *Aphis sambuci* L., dark green; in *Pterochlorus roboris* L., brownish green; in *Pemphigus spirothecae* Pass., rather grayish. The coloring is said to vary

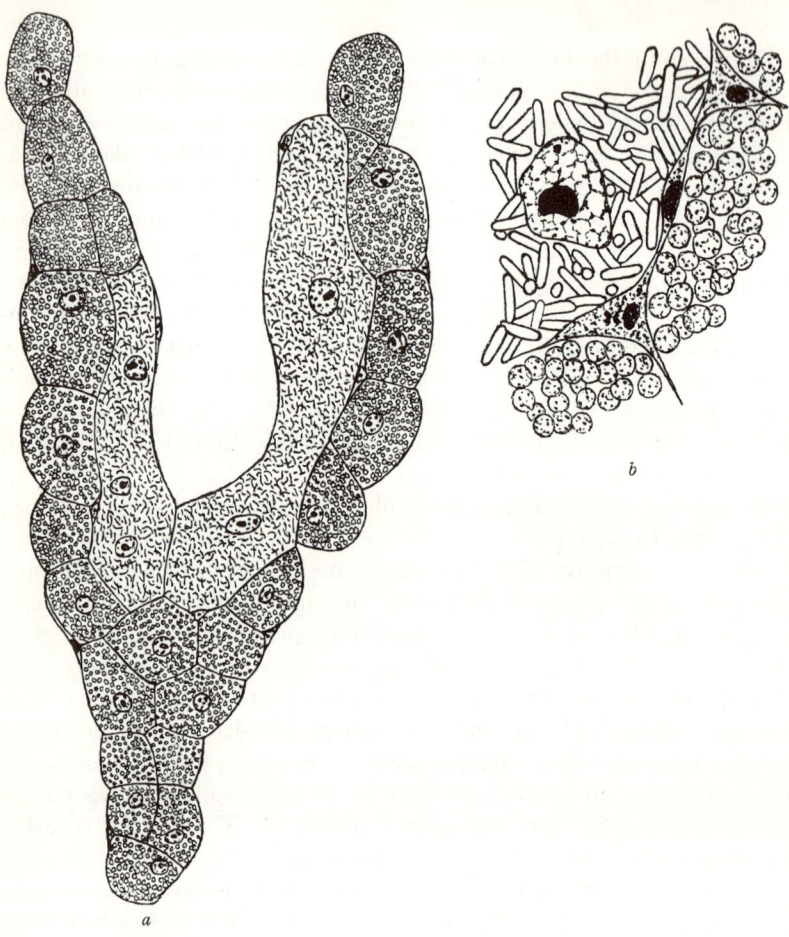

a

b

Fig. 150   *Macrosiphum jaceae* L.   (*a*) Frontal section through the mycetome housing two kinds of symbionts; (*b*) the two kinds, separated by envelope cells, at greater magnification.   From Klevenhusen.

from colony to colony within species, depending on the various host plants.

The organisms populating these uninucleate mycetocytes have an extraordinary resemblance to one another, but there are a few exceptions among the adelgines which are to be described later.   These organisms are gram-negative roundish formations 2–4 μ in diameter, and they resemble extremely small nuclear vesicles. As soon as they were recognized as independent organisms, they were studied by a number of authors, some of whom made cultivation experiments. The concepts formed during these studies differed radically. Within the host cell their multiplication undoubtedly takes place by dumbbell-shaped abscission, but the daughter cells are not always detached immediately and may be seen in groups of three to six or more individuals. In the center of these vesicles one can usually detect a strongly staining granule enclosed by a vacuole which is given to both daughter individuals in the cell division, a circumstance suggesting that it might be a nucleus.

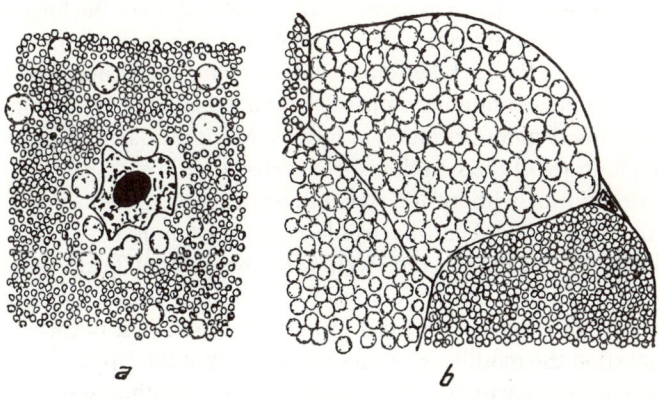

Fig. 151 *Aphis* sp. (*hederae* Kalt.?). Locally limited growth in size of the symbionts (*a*) occurring sporadically, (*b*) involving entire mycetocytes. 750×. From Klevenhusen.

In general, the symbionts in all cells are of equal size, although quite frequently they attain considerable dimensions, having diameters ten times larger than usual. Such giant symbionts may be scattered among the normal symbionts or may occupy mycetocytes of their own. I called attention to this phenomenon years ago, and Klevenhusen discovered it too in his extensive research with aphid symbiosis, especially with *Macrosiphum tenaceti* L. and a species of *Aphis*. In the latter the neighboring cells may have three different sizes (Fig. 151). Some may be degenerative; in others the mycetocytes appear to be quite sound; and, as we shall see later, only the small original states are used during transmission.

These organisms, which occur in aphids in untold numbers throughout the world, were called saccharomycetes by Pierantoni, and schizosaccharomycetes by Šulc, but today the majority of authors agree that they are bacteria. Peklo declared with great emphasis that they were a type type of *Azotobacter* and reported luxurious cultures of the symbionts from *Pterocallis tiliae* L., *Aphis papaveris* Fabr., *Pemphigus* and *Schizoneura lanigera* Hausm. He asserted that the vesicle-shaped symbionts of the *Schizoneura* species are gradually transformed into small rodlets and may experience gigantic growth in nitrogen-poor nutrient liquids. According to Paillot, this would not be a type of *Azotobacter* at all, but an originally coccoid or rod-shaped bacterium modified under the influence of the host organism, for in teased-out preparations of numerous forms he had observed states which he considered to be transitions between the round and rodlet forms of the organisms. Klevenhusen, Tóth, Rondelli, and Buchner did not observe such transitional forms, and today there is scarcely a doubt that the species on which Paillot based his theories belong to a group, of which we shall hear later, containing both the round hereditary symbiont and an additional guest more primitive in form. Mahdihassan (1947), who also reported cultures of the symbionts, concurred that in such cases there were two distinct forms without intermediate transitional states.

On the other hand, Schoel's detailed reports on aphid symbiosis, if confirmed, would harmonize very well with the data of Peklo and Paillot. According to Schoel, the symbionts of *Cavariella aegopodi* Scop. have a low degree of motility when first freed from the mycetocytes, but with increasing separation the motility increases to the point of the circular movements of the still united partners. Separated from one another, some of the now oval daughter cells become spherical and fall into their former state of almost complete immotility, but much more frequently they elongate more and more during division, so that after 3 hours their length is 3 μ and their thickness only 0.8 μ. In their final state they are normal rodlets, rounded at the end and measuring up to 3.5 μ in length and 0.4 μ in width. Accompanying this change in form taking place under the cover glass is a change of locomotion, the rotary movement being replaced by a rectilinear one.

Schoel also reports successful cultures on bouillon-agar plates and in nutrient solutions. Again, vesicle-shaped states retreat in favor of actively motile homogeneous rodlets which are flagellate on all sides, but the pleomorphism, Schoel claims, goes even further. There are chains with 8 to 10 links, measuring up to 16 μ in length. In older cultures the long rodlets are increasingly replaced by short rodlets, and the latter in turn are replaced more and more by oval cocci with active rolling movement. The cocci may again grow out into short rodlets or change into vesicle-shaped states identical in appearance with the inmates of the mycetocytes. On comparing *Megocerca viciae* Buckt. and *Doralis saliceti* L., Schoel found their symbionts so similar even with serological testing that he proposed the same name, *Bacterium aphidinum*, for all three species.

Tóth (1933) arrived at an entirely different concept. In the center of the symbionts on his preparations he repeatedly found quite a large body evenly blackened with iron hematoxylin and interpreted it as the actual bacterial body, declaring the remainder to be an envelope formation taken along at the time of cleavage. When cleavage takes place rapidly, the development of the envelope cannot keep pace, Tóth said, and in rare cases, as with *Macrosiphum carnosum*, the envelopes are lost entirely and the oval coccoid bodies are left naked. The sole cause of these occasional enlarged symbionts, he argues, is a swelling of the envelope. Tóth's observations are perhaps traceable to a fixing which caused the central plasma to ball up or to an impregnation of the vacuoles with the dyestuff.

Whatever the reason, a new and thorough study of aphid symbionts is desirable.[1] Even Schoel's data need verification in view of the surprising ease with which he obtained his cultures. Uichanko is quite skeptical about the results of his own culture experiments and those of his predecessors. Schwartz (1924) and others had tried in vain to cultivate these organisms.

The symbionts of the adelgids occupy a special position. Two genera, *Adelges* (*Cnaphalodes*) and *Sacchiphantes* (*Chermes*), have mycetomes which are fundamentally alike and both also have uninucleate mycetocytes, but the latter contain not round vesicle-shaped symbionts but inmates of a different shape. In the *Adelges* species the organisms with their tapering ends resemble caraway seeds, and during transverse cleavage they fold together in the middle and thus come to lie in bundles parallel to one another. In *Sacchiphantes* the organisms are blunt-ended tubes which behave in a similar way during division (Buchner, 1921; Profft; Fig. 152). Also different, though to a lesser degree, are the oval or round organisms of the genera *Pineus* and *Dreyfusia*. That some symbionts of the adelgids are so different from the typical guests of aphids is to be explained by the fact that they are a very old side branch of the aphid family tree.

The symbiotic devices of the aphids already described typify the original form of general occurrence. In a large number of species the symbiosis is even more complicated by the presence of additional symbionts displaying

---

[1] *Note added in proof:* While this volume was in press, an important electron microscope study on aphid symbionts by Vago and Laporte (1965) appeared. They found that these microorganisms consist of much nuclein but very little cytoplasma and evidently represent modified bacteria which are very close to the *L*-forms and especially to the *A*-type. Their gram-negative character is probably a consequence of the lack of a wall, and their scanty cytoplasma may be the reason why it is impossible to cultivate them.

varying degrees of adaptation.   Usually one additional guest is adapted in addition to the hereditary form, but occasionally two extra forms may be taken in.

The first observations on this aspect of symbiosis are those of Krassilst-schik (1889) at a time when the nature of the pseudovitellus was still mis-understood.   He repeatedly found colonies of bacteria in a number of species within a narrowly defined area adjacent to the pseudovitellus and called the organisms biophytes to indicate their apparent usefulness to the host animal.   Further information was supplied by Peklo with respect to *Schizoneura lanigera*, in which he regularly found a cultivatable rod-shaped bacterium which in his judgment was unrelated to the round symbionts.   He gave no information about the symbiotic site.   Paillot also observed typical bacteria in many species but decisively rejected the interpretation that they represent a second type of symbiont, and he declared all statements to this effect to be erroneous.   In his opinion the aphid symbionts which are shaped like the original ones, as already mentioned, represent the actual hereditary type from which the round vesicle-shaped states are continually replenished.   His concept was based on his observation of the spontaneous appearance of parasitic bacteria after introduction into the body cavity of butterfly larvae (see Chapter 14).   However, what he again and again interpreted as transitional forms in a great variety of aphids are possibly accidental shapes arising through smearing, or they may be degeneration forms of the original symbionts.   Nevertheless, in our own opinion and in that of others as well, he made valuable contributions to the knowledge of accessory aphid symbionts through his extensive research, despite his basic misconception.

Fig. 152   Symbionts of (*a*) *Adelges strobilobius* Kalt., (*b*) *Sacchiphantes abietis* L.   From Buchner.

The di- and trisymbiotic aphids remained unclarified until the appearance of thorough studies of a series of aphids by Rondelli, Klevenhusen, Tóth, and Profft which were not based merely on teased-off preparations as most of Paillot's were.   Rondelli was the first to find two symbionts among the eriosomatids; Klevenhusen, independently, discovered them in the same tribe and also in *Macrosiphum jaceae* L., *M. tanacetum* L., and *M. tanaceticulum* Kalt.; Thornton encountered the disymbionts in a species of *Aphis* (*hederae?*), in *Pterocallis juglandis* Frisch and *Chaitophorinella testudinata* Thornton, and in working with *Pterochlorus roboris* L. and *Stomaphis quercus* L. he was the first to confirm the existence of the trisymbionts. Tóth supplemented the list of disymbiotic forms substantially by adding species from *Chromaphis, Mycocallis, Doralis, Hydaphis, Colopha*, and *Byrsocrypta*, for the most part without furnishing details, and by verifying trisymbionts in three *Stomaphis* species. From Paillot's data it may be seen that *Eriosoma, Tetraneura, Drepanosiphum, Siphonophora*, and *Chaetophora* also possess two types of symbionts. Among the adelgines Profft established two genera with disymbionts, *Dreyfusia* and *Pineus*. We see then that the tendency of aphids to have additional symbionts is characteristic, and that it is observable in roughly half the species studied thus far and is manifested in the same manner in a series of tribes among the aphidines, pemphigines, and adelgines without indication as yet of any systematic connection.

These additional guests and the nature of their sites differ more or less from case to case, but in almost all these genera it is clear that they are of later adoption. In general, there is a tendency to house them within the framework of the mycetome already on hand. In *Macrosiphum jaceae* the homogeneous, slightly curved tubes, approximately 8 μ in length, increase through transverse fission and live in few-celled syncytia inserted unpaired in the angle of the two shanks of the mycetome and thus surrounded laterally, dorsally, and posteriorly by the uninucleate mycetocytes (Fig. 150). The envelope cells in part pass over the syncytia, in part between the two zones, thus providing a firm connection.

The disymbiotic adelgine genera, *Pineus* and *Dreyfusia*, differ in some respects from other aphids. With the exception of *Pineus pineoides* Cholodk., to be considered later in view of its special position, in which the mycetomes consist of two large wings with a broad connecting bridge, the mycetomes are built up half of syncytia and half of uninucleate mycetocytes. In both sites the symbionts are round to oval microorganisms, but they are easily distinguished, those of the syncytia being somewhat larger and basophilic, those of the uninucleate mycetocytes smaller and eosinophilic. Because of the histological character of the syncytia, Profft considered the forms there to be the accessory symbionts, but for a number of reasons, and principally because of circumstances in *Pineus pineoides* still to be described, it seems far more probable that this subfamily, which is exceptional in other respects, is exceptional also in not adhering to the tendency of the aphids to accommodate the additional symbionts in syncytia, and that in this case the occupants of the uninucleate mycetocytes should be interpreted as the accessory forms. In Klevenhusen's unclassified *Aphis* species there are the usual round symbionts and delicate slender tubes accommodated in a special unpaired syncytium, just as in *Macrosiphum jaceae*. In both species the syncytium has the same location, but in *Aphis* species it is without an epithelial covering and in general lacks close relationship with the original mycetome. In older animals the syncytium may be situated on either side of the intestine when the myce- tome is deformed.

The accessory symbionts of *Pterocallis juglandis*, filamentous and approxi- mately 15 μ in length, are definitely part of the mycetome system. Like the original symbionts, they colonize uninucleate cells which tend to be superficially located but are often inserted deep between other mycetocytes (Fig. 153). Very frequently the long filaments arrange themselves parallel to one another and thus force the nuclei into a corresponding elongation. In older animals the mycetome is constricted off into fragments which contain both types of cells.

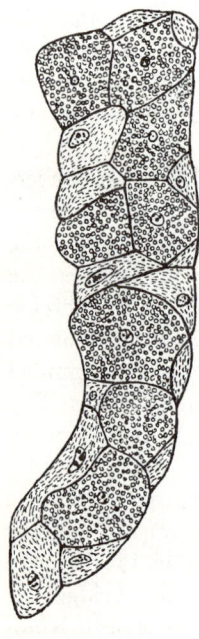

Fig. 153 *Pterocallis juglandis* Frisch. Mycetome with two kinds of symbionts. 400 ×. From Klevenhusen.

The accessory symbionts of *Chaitophorinella testudinata* are cocci 1.5–2 μ in diameter situated in two syncytia, one to the left and one to the right. The organ complex is broken down very early into fragments in which, nevertheless, the old space relationships are preserved (Fig. 154).

In all the cases discussed the symbiosis is quite firmly established, but there are other cases in which its development is more or less incomplete. Not all *Macrosiphum* species, for example, stand on the high level of *M. jaceae*. In *Macrosiphum tanacetum* the filaments, usually 20 μ long but sometimes as long as 30 μ, are housed in two syn- cytia which may be situated anywhere in the chief organ and, permeated as they are by large vacuoles, are unevenly settled. *Macrosiphum tanaceticolum* has only one syncytium which maintains the relationship with the mycetome only in occasional cases. When fissuring sets in, some of the syn- cytial fragments are without nuclei and their symbionts are united loosely, but even the fragments containing nuclei are often surrounded by an area of isolated rodlets. In oviparous females the body cavity contains a large dense mass of bacteria. In *Macrosiphum tanacetum* uninucleate parts are split off from the syncytium which then come to lie between the mycetocytes or free in the body cavity. Frequently some of the rodlets are taken into the blood cells and apparently are digested there. However, these are all irregularities which are not observed in the forms investigated so far.

Tetraneura ulmi De Geer houses a delicate filamentous guest in an unpaired syncytium which is always located in the mycetome above the gut in young animals. Later it is distorted, and the symbionts are transmitted in great numbers to the body cavity, with the result that, at most, little remains of the original syncytium in the old animals. According to Rondelli, in *Tetraneura rubra* Licht. the reduction of the syncytium proceeds without such loss.

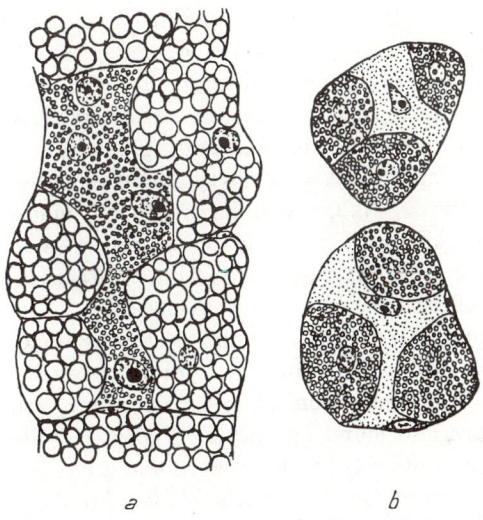

a   b

Fig. 154 *Chaitophorinella testudinata* Thornton. (*a*) Part of the embryonic mycetome with two kinds of symbionts, 750×, (*b*) fragments of the mycetome from older insects, 250×. From Klevenhusen.

States of this type lead to an extensive invasion of the body, as in *Schizoneura lanigera*. In the younger animals both slender rodlets and filaments up to 25 μ in length are found in a syncytium, according to independent, concurring descriptions by Rondelli and Klevenhusen; in older animals little of the syncytium remains. Its inmates have been released, and they permeate the body cavity in the form of rodlets now only 6 μ in length, forming small aggregations between the fat cells throughout the body but particularly large ones in the cephalic region, including the proboscis. Moreover, elements of the blood which are occupied more or less densely by the rod-shaped bacteria are found everywhere.

Pineus pineoides shows comparable arrangements. The mycetome consists solely of syncytia with round and oval symbionts, but many fat cells contain sharply defined nests of rodlet-shaped bacteria. Again, the fat cells are found in great numbers in the cephalic region and are present also in the proboscis. According to Profft's theory which we rejected above, the inmates of the syncytia would be the accessory symbionts, but then the original symbionts, always found everywhere else, would be lacking. It would probably be wiser, therefore, to interpret the inhabitants of the syncytia in the *Pineus* and *Dreyfusia* species as the original symbionts, which, it should not be forgotten, also differ from those of the other aphids in the case of the monosymbiotic adelgines.

Unique thus far is Paillot's observation with *Rhopalosiphum vitis* L. In addition to the original ever-present states he found a second type, usually large coccobacilli but often, depending on the origin of the material, curved organisms with tapered ends or very small cocco-bacilli; he found the small ones only in isolated cases but in far greater numbers.

Among the numerous aphids with two symbiont types there are several that also have a third. As far as is known today, only lachnines are involved. The aphids investigated for the presence of a third symbiont are *Pterochlorus roboris* L., *Stomaphis quercus* L., *St. longirostris* Fabr., and two other species of the same genus as yet unclassified. Klevenhusen and Tóth, each studying different species, found the third symbiont in all cases. The genus *Lachnus* has unfortunately not yet been tested. With *Pterochlorus roboris* all three forms are united in the mycetome, although the two accessory forms are also found outside it. Figure 155 illustrates a site where smaller cells with cocci-like formations and a larger one with delicate rodlets are wedged between mycetocytes with the typical round guests. The cells with the cocci either press between the primary mycetocytes or are situated on the exterior of the organ; location and size, as well as processes of embryonic development yet to be described, indicate that they are infected casing cells. Cocci and rodlets are also found in cells which tend to collect on the tracheae, or in cells which are situated in the fatty tissue or drift about in the body cavity. Blood cells apparently come in contact with them there and may perhaps resorb some of them. Unlike the cocci, of which the chief mass remains in the mycetome, often only fractions of the rodlets are left behind while the overwhelming majority form larger or smaller aggregations in various other places in the body, chiefly on the intestine. This "disorder" is increased when two types occur in the same cell. In such cases, moreover, they usually occupy different areas; or one of the two forms may be represented in scarcely perceptible quantity or, in a minority of cases, may be enclosed in special vacuoles.

In the case of *Stomaphis quercus* L., which sinks its unusually long proboscis into the trunk of oaks at the foot of the tree under moss, the older embryos readily reveal that they are colonized by three types of symbionts. Between the mycetocytes with vesicle-shaped symbionts are numerous smaller, uninucleate elements thickly filled with tube-shaped formations 6–8 μ in length (Fig. 156a). In several large syncytia which are enclosed by the mycetome and intestine the inmates are variable in form, occurring as straight rodlets approximately 3 μ in length or as forms bent into hooks or acute angles or turned like corkscrews. The same organisms are also observed in several small cells in the middle of the mycetomes as well as

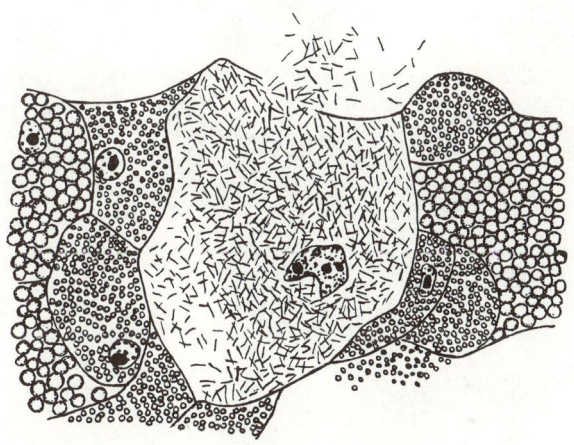

Fig. 155   *Pterochlorus roboris.*   Section from an old mycetome with three kinds of symbionts.   750×.   From Klevenhusen.

free in the body cavity and in spaces between the fat cells, where they may form rather large aggregations. In older animals in which the syncytia have completely disappeared, this variety of symbiont can be discovered only after some effort.

Later phases of embryonic development show that an originally uniform syncytium has been fragmented. Symbiont-containing plasma is cut off around its nuclei, as has been noted also in *Pterochlorus*, so that independent little cells arise. Free tube-forms in the vicinity join the remains without nuclei and, thus intermingled, both are taken up by blood cells which build nests between the fat cells. Moreover, rodlets may infect mycetome cells which contain only tubes at first and, conversely, tubes may infect cells which contain rodlets (Fig. 156b,c). Finally, Klevenhusen reports that the rodlets are also ingested by large yellow cells, apparently representing oenocytes, which lie scattered under the hypodermis.

In view of such chaotic conditions it may seem too venturesome to apply the term symbiosis here, but we are justified by observations of the ordered manner of transmission, still to be described, which prove that the di- and trisymbiotic forms among the aphids are recently acquired guests which are still in need of adaptation.

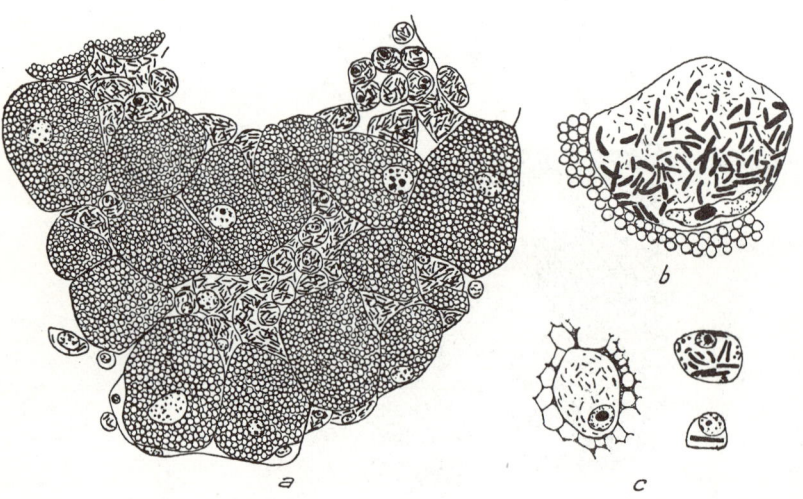

Fig. 156 *Stomaphis quercus.* (*a*) Section from the mycetome of a younger insect, 385 ✕; (*b*) mycetocyte of an older insect with mixed infection, 720 ✕; (*c*) infected blood cells, 720 ✕. From Klevenhusen.

Schanderl, Lauff, and Becker (1949) report that they find cultivatable *Mycoderma* yeasts in the intestinal epithelium of *Mycodes, Doralis,* and *Aphis* species and that the same organisms, in this case uncultivatable, are found in *Phylloxera vastatrix* L., which, as will be described later, has no pseudo-vitellus and apparently no other symbionts. However, the description may be based on a mistake, as undoubtedly also are the "symbionts" which Haracsi (1937) found in the salivary glands of the aphid *Prociphilus.*

The series of adapted forms already described, never observed so impressively elsewhere except among the membracids, lent special interest to aphid symbiosis, but the interest was still increased by the unusual events occurring during transmission. This is a process in which symbionts, and at times di- and trisymbionts, must be furnished to two generations growing in very different fashion—the progeny of the oviparous sexual forms arising from large yolk-rich eggs (winter eggs), and the numerous summer generations which develop parthenogenetically in the maternal body from much smaller, almost yolk-free eggs.

Infection of winter eggs proceeds in the manner thoroughly familiar to us today. Let us use as an example *Drepanosiphum platanoides* Schrk., the test object I used in giving the first description of this infection in 1912. Long before yolk formation has reached its peak, the oocyte becomes detached from the nutrient cord and protrudes somewhat at the posterior

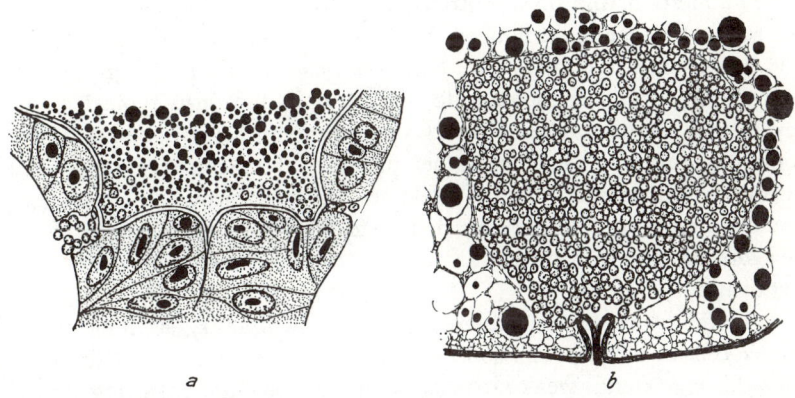

a   b

Fig. 157   *Drepanosiphum* sp.   Infection of the winter egg: (*a*) the symbionts are transmitted in a ring-shaped zone to the ovocyte; (*b*) the infection is ended.   From Buchner.

end in a sharply defined ring-shaped zone. Here the bordering follicle cells form gaps permitting the roundish symbionts, which have here and there left the mycetocytes and are drifting about in the body cavity, to be transmitted unaltered to the egg plasma without being hindered by the rather well-developed chorion (Fig. 157*a*). Symbionts are occasionally observed in the plasma of neighboring follicle cells, but they are mere stragglers which do not gain admittance and are destroyed there. Along with changes in the egg form during the growth process a narrowing of the entry passage takes place by a progressive concentration of the follicle cells in the direction of the posterior pole.

Through continual replenishment and probably also in part through reproduction now beginning in the egg, there arises an aggregation which is so typical of the winter eggs of aphids and is so large that it did not even escape the attention of older authors (Fig. 157*b*). Lenticular in form and attached to the chorion, or rather roundish and enclosed on all sides by yolk and egg plasma, it may reach astonishing proportions—in the eggs of *Pterochlorus roboris* it has a diameter of 0.38 mm.! These symbiont balls of the winter eggs are dark green, light green, or yellowish, depending on the coloring of the mycetome; they are rarely colorless, as in *Chaetophorus populi* L.

The behavior of the other aphids during winter-egg infection is essentially like that of *Drepanosiphum*, but a more detailed comparative study would probably unearth some variations. Where two types of symbionts are present, both appear at approximately the same time at the follicle zone in question, and they are transmitted together to the oocyte through the gaps. Klevenhusen noticed the double infection in a series of forms, and he was able to establish in *Macrosiphum tanaceti* and *Pterocallis juglandis* that the additional symbionts do not infect in the form typical of the mycetomes but are broken down through transverse cleavage into shorter rodlets for this purpose. Profft then furnished the pictures of such mixed infections still needed. In the adelgines, his test objects, all generations are oviparous, and his observations refer therefore to the eggs which develop parthenogenetically as well as to those which develop after fertilization.

As with the remaining aphids, there is a temporary swelling of the follicle cells in the infection zone. Their texture loosens, and at the same time the egg extends weak protrusions into the resultant gaps (Fig. 158*a*). *Pineus pini* has a combination of smaller eosinophilic and somewhat larger basophilic round and oval symbionts which finally form a lens-shaped body at the posterior end; *P. pineoides* has a combination of roundish and rodlet-shaped symbionts (Fig. 158*b*).

An example of infection of winter eggs by three types of symbionts is unfortunately not available.

We have been well informed about the shifting of the symbionts during development of winter eggs, although most of the authors were not aware that they were describing the genesis of a mycetome. Balbiani (1888), Flögel (1905), and Tannreuther (1907) made studies, and Webster and Phillips, in a study of the biology of *Toxoptera graminum* Rond., furnished good microphotographs and a brief description. Klevenhusen and Paillot made observations on the subject, and it was Profft especially who furnished the data on egg development in disymbiotic forms.

The large foreign body of the symbionts at the posterior egg pole at first prevents blastoderm formation in this region, with the result that it is finished everywhere else before the first cleavage nuclei sink into it (Fig.

159*a*). When the germ band invaginates, the blastoderm around the
symbiont ball grows inward, thus cutting the latter away from its environment and shoving it along ahead to act as a closure, for in the beginning
the invagination is still open at the front (Figs. 159*b* and 160). With

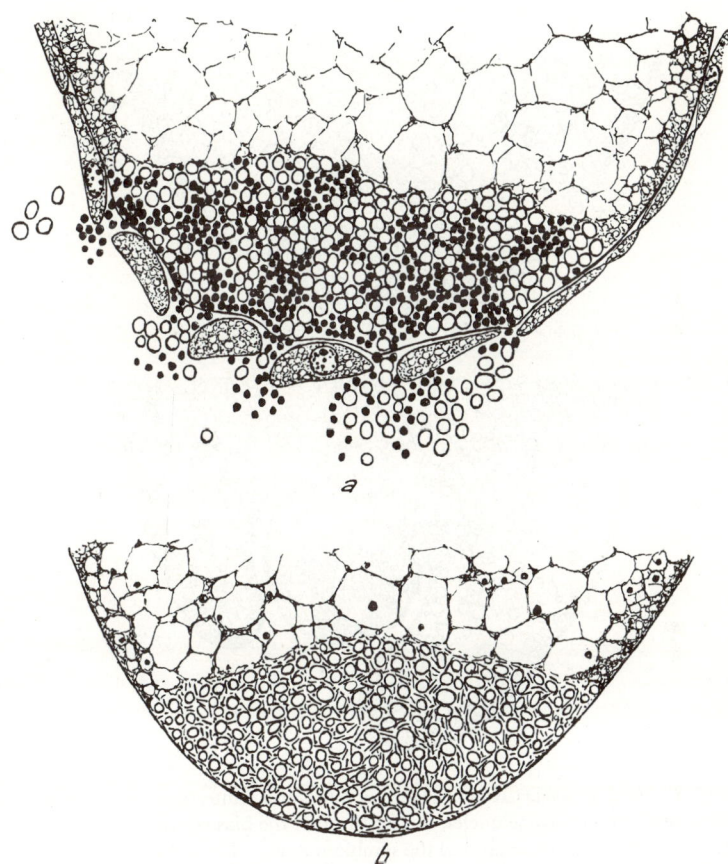

Fig. 158   *Pineus pini* Macqu.   (*a*) Two kinds of symbionts transfer between the follicle
cells into the egg.   (*b*) *Pineus pineoides* Cholodk.   The mass composed of two kinds of
symbionts after completion of infection at the posterior pole.   From Profft.

further growth the ball is displaced by the amnion hollow, as is usual in
such cases.   At this time amnion and germ band are clearly differentiated,
and the rudiment of the gonads makes its appearance beside the mycetome
rudiment.   Simultaneously cell boundaries are formed around the indi-

vidual nuclei of the embryonic mycetome, and the extremities develop on the sinuously curved germ band.   By the time stomodaeum and procto-daeum take form, the mycetome is already situated in the abdominal region as an unpaired oval body.

It is in this stage that winter eggs customarily await the beginning of

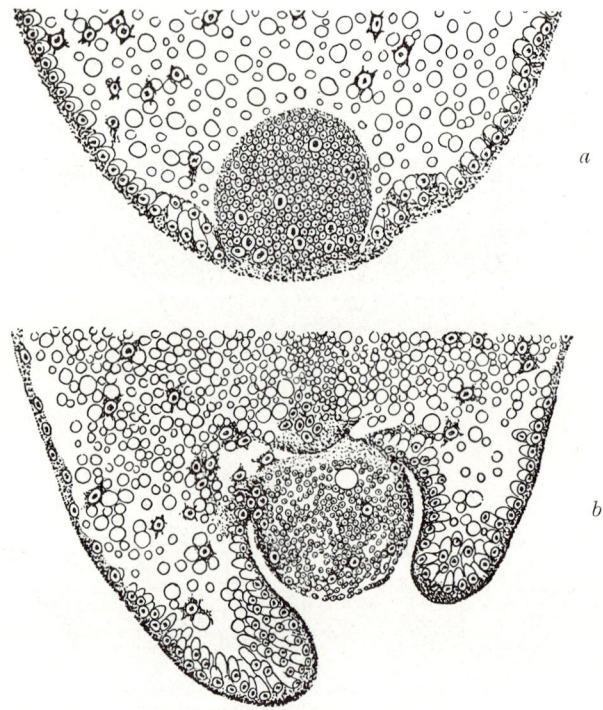

a

b

Fig. 159   *Toxoptera graminum* Rond.   Development of a winter egg: (*a*) the symbiont ball, permeated with cleavage nuclei, breaks through the blastoderm; (*b*) beginning of invagination of the germ band around the symbiont mass.   From Webster and Phillips.

the warmer season.   With the development of the midgut the mycetome is constricted off to form two partial ones, to be reunited later by a bridge laid across the intestine.   Involution of the embryo is without significant effect on the mycetome, which is now waiting in its place.

Profft studied egg development in the adelgines.   The monosymbiotic forms behaved in the manner just described for *Toxoptera*.   A symbiont ball provided early with cleavage nuclei, on the surface of which a number of flattened envelope cells adhere, is shifted into the interior of the embryo by the germ band. Division into uninucleate cells with simultaneous migration of envelope cells does not take place until after revolution. With disymbiotic forms, separation of the two types begins during invagination. In *Pineus* the basophilic symbionts, which at this time are growing

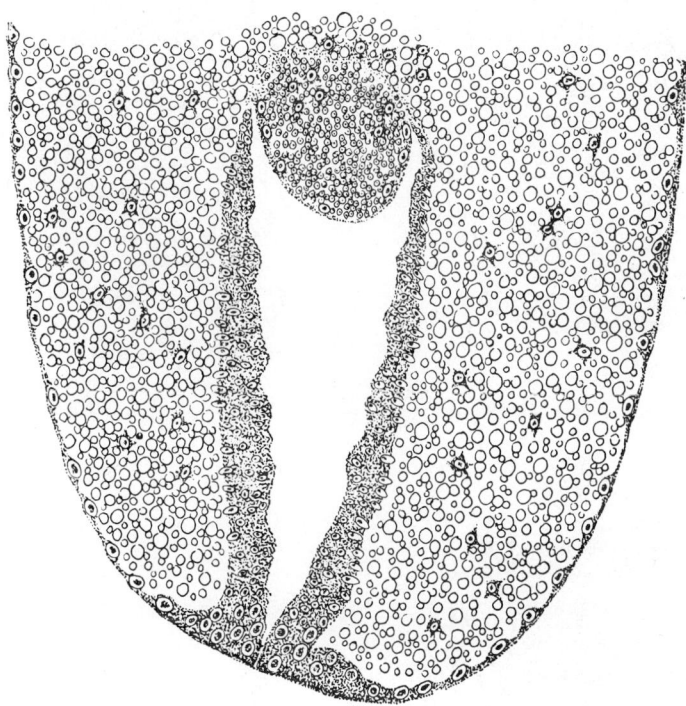

Fig. 160 *Toxoptera graminum* Rond. Development of a winter egg. Progressive invagination of the germ band. From Webster and Phillips.

considerably, tend to collect in the peripheral regions of the provisory syncytium, but no dissociation of any depth occurs. Even before the close of invagination some of the envelope nuclei are transmitted to the syncytium to permeate its peripheral regions (Fig. 161a).

In a succeeding stage all basophilic, hence phylogenetically younger, symbionts suddenly congregate around the envelope cells and are caught in the still small uninucleate mycetocytes by cell limits which now arise. Thus, in monosymbiotic species, cells that remain sterile are assigned to the latecomers (Fig. 161*b*)! In the same manner cell limits are next thrown around the central nuclei which enclose the eosinophilic symbionts (Fig. 161*c*).   Later both types of mycetocytes are intermingled, and from

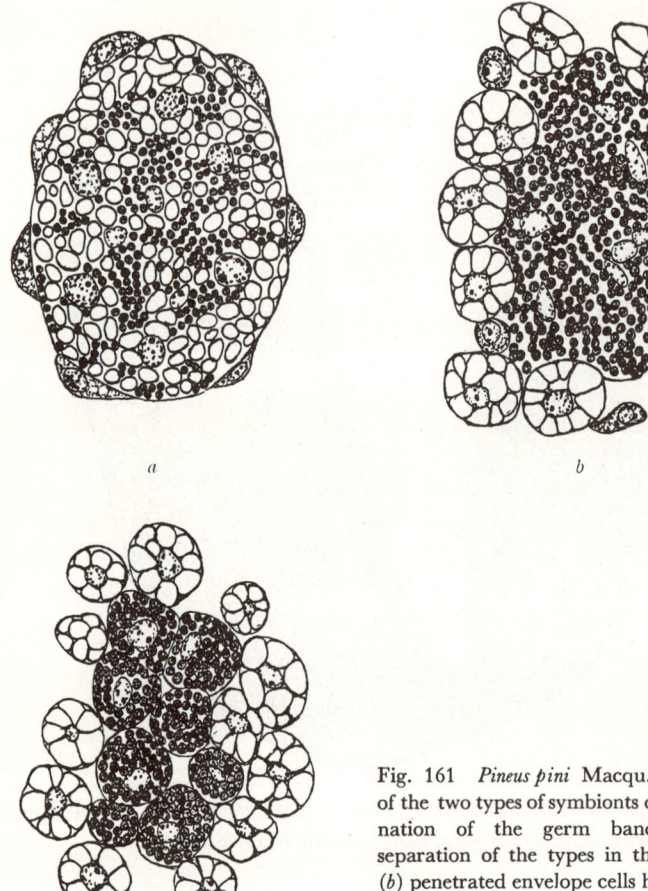

a

b

c

Fig. 161 *Pineus pini* Macqu. Separation of the two types of symbionts during invagination of the germ band. (*a*) Early separation of the types in the syncytium; (*b*) penetrated envelope cells have taken up one type; (*c*) the second type has also been taken up into cells.   From Profft.

those of earlier origin the little syncytia are formed by amitotic nuclear abscission.

Paillot reviews the development of winter eggs in a disymbiotic form also, but his interpretations are obviously based too rigidly on his conception that the two symbiont types are identical.   According to him, the rodlets, that is, the accessory symbionts, would be transformed in great numbers into the vesicle forms in the embryonic mycetome.

It was far more difficult to explain the transmission of symbionts to summer generations developing in the maternal body. Because of investigations by Sell, Rondelli, Klevenhusen, Paillot, and Tóth, we understand the subject much better in general, but some difference of opinion still exists with regard to minor details. The difference in behavior between the two generations is based in the last analysis on a conflict stemming from a device created early in phylogenesis: eggs of virginipara begin to develop at a very early stage of growth when they are almost yolk-free; but the adaptations which regulate infection of the completely developed yolk-rich eggs experience no corresponding change in the parthenogenetically reproducing generations. In remarkably rigid fashion the follicle temporally and spatially holds fast to traditional methods, and thus, when it opens, the symbionts do not come upon an infection-ripe egg but upon an embryo. The many contradictory interpretations given to these repeatedly investigated processes, especially in the older literature, may be explained for the most part by variations which make it possible for symbionts to come upon a blastoderm stage or a younger or older form of the germ band, depending on the stage of developmental progress of the embryo and the duration of symbiont influx in the individual species.

According to a description of *Aphis sambuci* L. by Sell, the blastoderm of the young blastula in this species is only weakly developed, and the vacuolized central plasma, which is poor in yolk and still contains a few remaining cleavage nuclei, borders partially upon the surface. Several strikingly large nuclei which occur laterally in the posterior quarter, according to Sell, are primitive sex cells. In a succeeding stage the part of the blastula situated behind these nuclei—Metschnikoff's "cylindrical organ," which later authors had declined to accept—is separated in strange fashion from the rest of the embryo by a groove. Its cell limits disappear, and from it a syncytium develops with several large nuclei which are conspicuous because of their different structure. This large body closes the gap in the blastoderm which develops in the process and represents a mycetome rudiment opportunely made available (Fig. 162*a,b*). Only now do certain changes connected with infection begin in the follicle. As already described, in winter eggs the symbionts are transmitted through a gap in the follicle cells. In summer eggs the gap develops only on one side. It is surrounded, like a fertilization cone, by especially well-developed cells, and through it the mycetome rudiment sends out a conical process to the body cavity (Fig. 162*c*), that is, in the same way that the plasma of winter eggs protrudes into the follicle gaps in the winter eggs of *Pineus*.

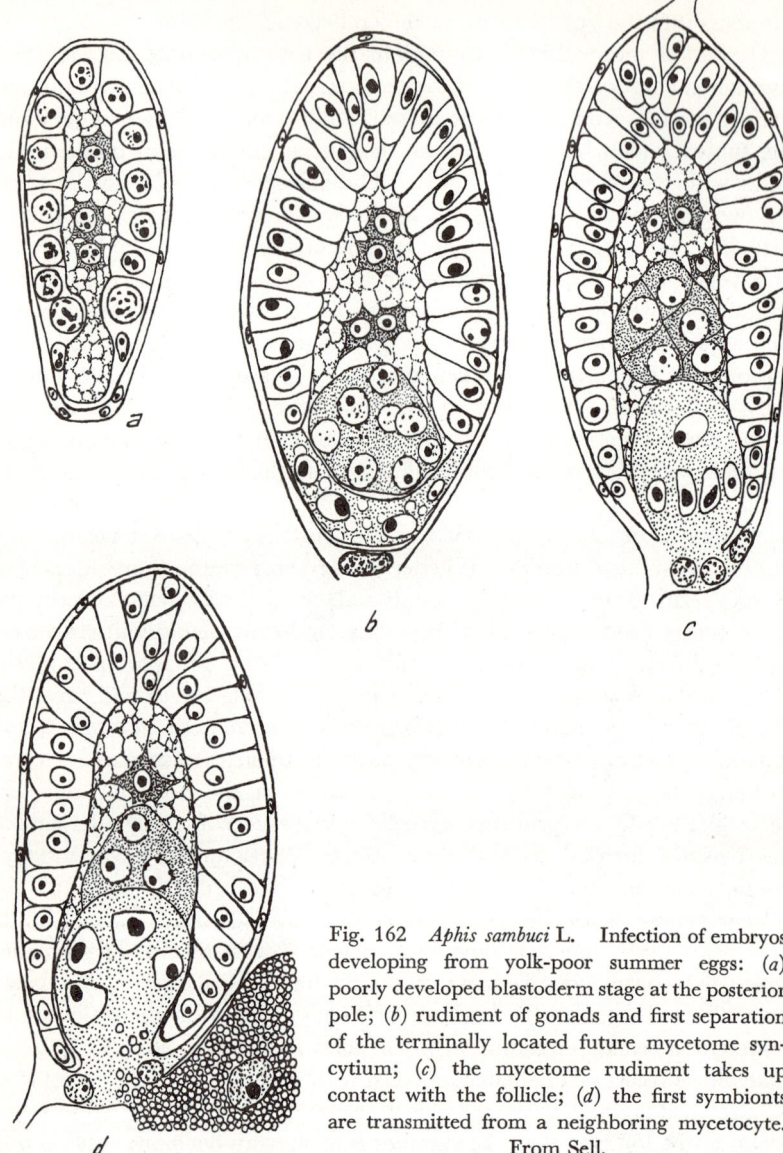

Fig. 162 *Aphis sambuci* L.   Infection of embryos
developing from yolk-poor summer eggs: (*a*)
poorly developed blastoderm stage at the posterior
pole; (*b*) rudiment of gonads and first separation
of the terminally located future mycetome syncytium; (*c*) the mycetome rudiment takes up
contact with the follicle; (*d*) the first symbionts
are transmitted from a neighboring mycetocyte.
From Sell.

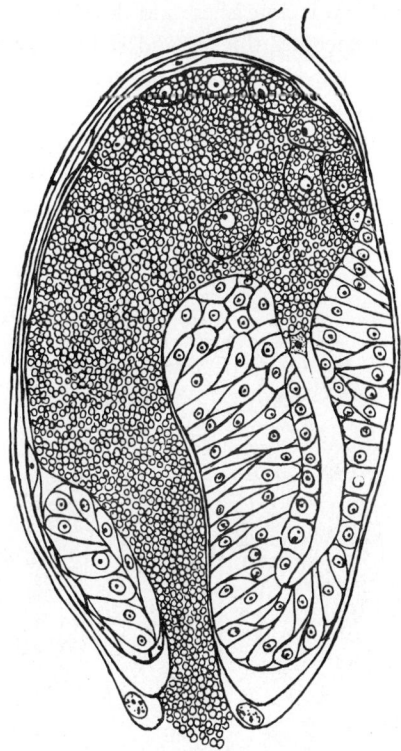

Fig. 163 *Aphis sambuci* L. Infection of the summer embryos. The symbionts stream through an opening of the amnionic cavity into the embryo. From Sell.

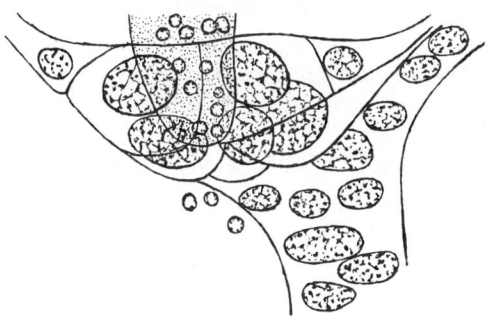

Fig. 164 Aphid of *Cineraria*. Infection of summer embryos. Follicle swelling and infection of the peg of mycetome rudiment penetrating them (fresh preparation). From Buchner.

Meanwhile the embryo is nourished and developed by maternal secretions; the blastoderm becomes thicker and richer in cells; and the rudiment of the gonads forms a sizable body between the mycetome rudiment and the few existing yolk cells.   At this stage infection sets in.   Usually it is symbionts circulating free in the blood which are transmitted to the follicle peg, but occasionally some are transmitted directly from a

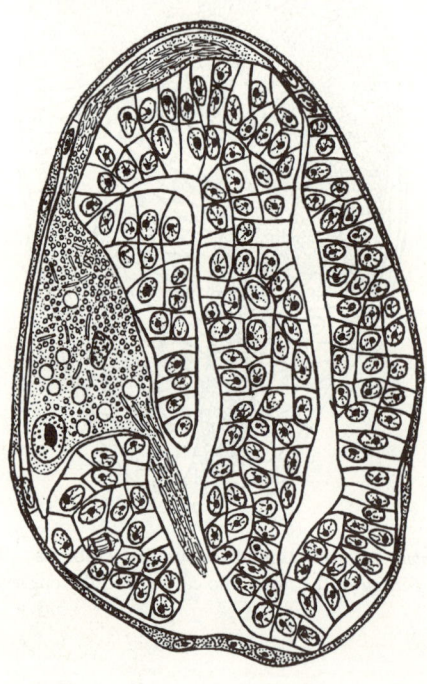

Fig. 165   *Chromaphis juglandis.*   Infection of the embryo with both types of symbionts has ended; displacement of the portal of entry.   From Tóth.

neighboring mycetocyte (Figs. 162*d* and 164).   Very gradually the symbionts take possession of the place made available and thus force the nuclei toward the upper regions.   In this way a considerable mass of foreign organisms eventually fills a large part of the embryo and necessarily influences the course of further development.

During transmission of symbionts to the embryo the germ band begins to invaginate near the invasion point and, inasmuch as the influx has not yet ended, a wide opening is necessarily left toward the inside (Fig. 163). When the infection is prolonged, as in this case, the germ band develops unequally and the opening is shifted toward the side which has been retarded in its development. There is only a slight indication of this in our test object, but it may sometimes lead to a considerable displacement, so that the opening eventually is found not at the top of the invagination but close to its place of origin (Fig. 165). When the infection ends earlier,

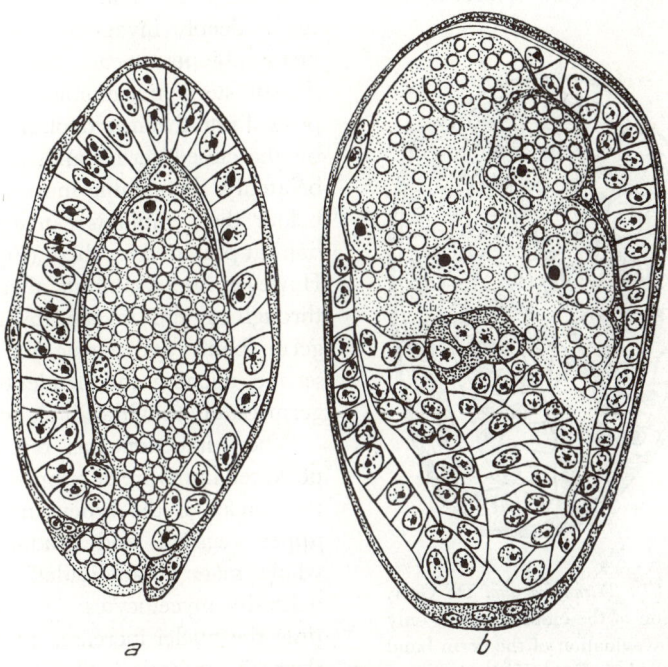

Fig. 166 *Hydaphis* sp. Infection of embryos before invagination of the germ band: (*a*) the round symbionts pour in; (*b*) infection by both types ended, appearance of cell boundaries in the syncytium, formation of the envelope, tetranucleated rudiments of the gonads. From Tóth.

the symbiont mass is closed off without occurrence of such asymmetry. For example, in a *Hydaphis* species the available space in the blastula is entirely filled with symbionts even before invagination begins (Fig. 166).

Such behavior, which may evidently be regarded as original, readily discloses that winter and summer eggs develop in basically the same way. Even in winter eggs invagination takes place near the embryonic symbiont-containing syncytium, which temporarily represents closure of the amnion hollow (compare Fig. 160). When the symbiont mass has reached the desired size, in both kinds of eggs it is separated from the germ band and the latter is finally closed.

All variations have a temporal basis: the infection periods in individual species may vary in duration, or the time at which they begin may vary. Early assertions of Hirschler (1912) that infection sometimes takes place in the unfurrowed egg are probably erroneous. At any rate, transmission begins as a rule before invagination, although in some test objects the egg is deeply invaginated when the first symbionts appear. In such cases the still sterile mycetome rudiment is pushed into the interior of the embryo by the closed germ band, and now, belatedly, an opening must be made before the symbionts can enter the habitat prepared for them (Fig. 167). However, the symbionts always pass through a gap in the invaginated germ band and never through a second opening independent of the germ band, as Witlaczil claimed.

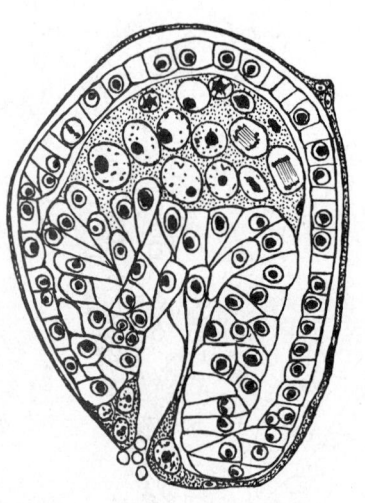

Fig. 167   *Tetraneura ulmi* De Geer. Infection of the embryo begins cnly after invagination of the germ band has set in.   From Klevenhusen.

When infection has reached its peak, cell limits are developed around the nuclei of the syncytium in the upper regions, and gradually the whole mass is subdivided into the definitive mycetocytes. At the same time the nuclei increase, apparently through amitosis. The yolk cells, flattened radically at this time and nestling close against the surface of the mycetome, insert delicate processes between the new mycetocytes to provide the indispensable envelope (Figs. 166b and 172b).

It is not yet entirely clear when the primitive sex cells first appear. It is generally thought that they are not separated from the germ band until after invagination begins. Figure 166b, for example, shows a quadrinucleated gonad rudiment. On the other hand, as mentioned before, Sell described a much earlier separation and later Paspaleff (1929) supported the idea that separation occurs during the blastoderm stage. Apparently there are temporal variations at this stage also, and Tóth is wrong in his opinion that the gonad rudiment in Sell's description represents yolk cells.

The follicle opening used by the symbionts during transmission is closed when infection ends, and the surrounding cells resume their former appearance.

Paillot, comparing infection of embryos in young females and aging animals, made the interesting discovery that in the older animals its course is abnormal in the terminal regions of the ovarioles. The development of the embryos is checked; blastoderm stages are formed, and their interior is filled by round symbionts. This observation made with *Schizoneura lanigera* and other species should be supplemented, as it promises interesting glimpses into the mechanism of embryo infection, which in this case again approaches that of the older ovarial infection.

In forms with two or three symbiont types, infection of embryos is basically the same. The accessory guests, more or less intermixed or temporally separated, are transmitted from the point of entry to the mycetome rudiment, which is at first homogeneous. Curtailment of the filamentous or tube-shaped symbionts, such as is occasionally observed during infection of winter eggs, does not take place with summer embryos. Klevenhusen and Rondelli described the process in a number of forms, and Tóth added further details. Paillot also refers frequently to double infection of embryos, but we shall not enter into these here since his statements are based on what we believe to be a mistaken conception. For us, the intermixing of the forms is proof that they always exist side by side; for him it is proof of their identity.

At first *Macrosiphum jaceae* has only primary symbionts. The rod-shaped ones which follow are soon collected in the center of the syncytium, which is permeated throughout by nuclei but lacks cell limits at the start. Eventually the middle area contains the secondary symbionts exclusively, and then cell limits are thrown around the peripheral nuclei to form the mycetocytes, the definitive habitats of the round symbionts. At the same time migrating envelope cells invest the mycetocytes and the central syncytium (Fig. 168). It is impossible to determine morphologically the forces involved during symbiont separation, just as it was impossible to determine them from the morphology of the strange phenomenon which occurred during the development of winter eggs. Subsequently enormous growth of mycetocytes and syncytium sets in, but mitoses or incipience of the process is never observed. Nevertheless, one is inclined to accept the interpretation that a polyploidy accompanies this enlargement of the nucleus. *Macrosiphum tanacetum* behaves approximately the same as *M. jaceae*. In *Chaitophorinella* the coccoid secondary guests are mixed with the primary ones when they penetrate and are soon collected again approximately in the central area of the mycetome rudiment.

In *Aphis* sp. (*hederae?*) the delicate tubes infect at the last moment and are pushed into a narrow area at the margin by the far larger mass of round symbionts. The syncytium in which they are located is at no time involved during formation of the envelope. In *Pterocallis juglandis* also there is almost no intermixing of the two types. Again the tubes penetrate first and are then displaced by the oncoming mass of primary symbionts in the area of the blastoderm. Only a few stragglers arrive intermingled with the later arrivals. With *Hydaphis* species the rodlets appear later and form the central aggregation (Fig. 166).

When two accessory symbionts join the primary form, all three proceed

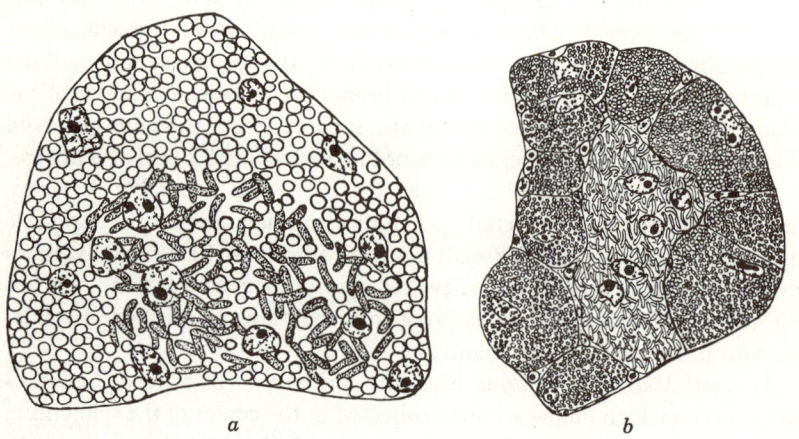

a   *b*

Fig. 168   *Macrosiphum jaceae* L.   Separation of both kinds of symbionts in the embryo developing parthenogenetically: (*a*) separation beginning, 750×; (*b*) peripheral mycetocytes with round symbionts, and the central synctium with rod-shaped ones; lower magnification.   From Klevenhusen.

to infect the embryos. In *Pterochlorus* the primary symbionts take possession of the syncytia and then are pushed to all sides by a flux of intermingled cocci and rodlets which finally accumulate in front of the blastoderm (Fig. 169*a*). Straying round symbionts which enter the stream belatedly may succeed in rejoining the others of their type, or they are expelled from the mass and degenerate. The central vein of the symbiont mixture finally passes into the polar aggregation.

Segregation of the two symbiont types in the polar aggregation extends throughout embryonic development, and different methods are employed (Fig. 169*b*). Some of the cocci are transmitted to neighboring envelope cells and thus maintain their association with the mycetome. Others reappear in the proximate fat cells, and still others form small rodlet-free aggregations around one nucleus on the periphery and withdraw from the mycetome as little independent mycetocytes. The rodlets are transmitted individually to the body cavity or are taken up by little elements in it. Occasionally mixed infections are involved in such cases. In this way the rodlets may disappear completely from the mycetome, or a small remnant, now unmixed, may be left behind.

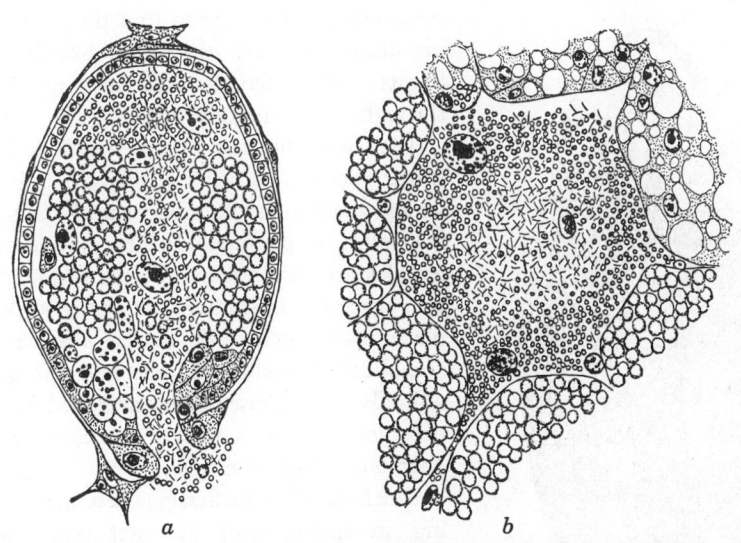

a   *b*

Fig. 169   *Pterochlorus roboris.*   Infection of embryos with three kinds of symbionts: (*a*) a mass of cocci and bacilli compresses the primary symbionts, already transmitted over, to the side; (*b*) separation of the cocci and bacilli begins.   750×.   From Klevenhusen.

From such extensive lability we see clearly the still inferior adaptation of the two symbionts which are acquired later.

Stomaphis quercus behaves in a similar fashion. As already mentioned, in this species tubes and rodlets join the round symbionts, and again they arrive intermingled after most of the other symbionts have been incorporated. However, segregation occurs early with detachment of the tube-shaped forms near the edges of the central infection strand, and they are ingested by the original envelope cells. The rodlets settle down almost entirely in a larger syncytium, which is fissured later, and infect the small number of elements lingering behind in the mycetome (Fig. 170). It occasionally happens that round symbionts get into the rodlet syncytium by accident and degenerate.

That aphid symbiosis is basically well controlled is impressively shown in Tóth's description of the interesting regulation phenomena which occur during rudimentation of the sexual forms, sometimes affecting only the males and sometimes both sexes. In *Stomaphis* species the females are normal. The males are very small and short-lived, because they molt only once after birth, they ingest no food and their mouth parts are accordingly reduced. When males are tested for symbiosis, it is discovered that they have no mycetomes and are completely free of symbionts! The females behave normally throughout, as already stated.

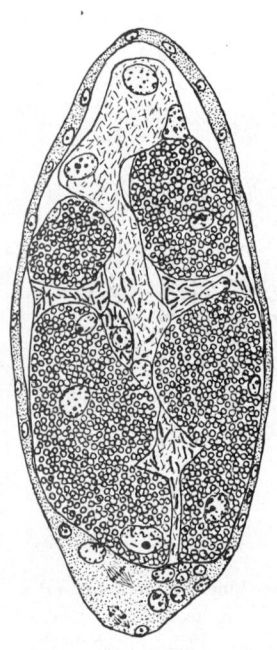

Tóth was the first to investigate a *Pemphigus* species, namely, *P. spirothecae* Pass., which causes the familiar spiral galls on leafstalks of poplars. In this species the females are also affected by the rudimentation process. In studying the changes taking place in the symbiotic devices during the complicated cyclic reproduction of these animals, each phase of which has a differing need for symbionts, one is surprised by the harmony which prevails.

The females of the first generation, appearing in spring with the first buds, suck themselves fast to young leafstalks after 2 or 3 weeks and deposit their progeny in the galls after four-fold molting. Each fundatrix produces approximately 75 living young parthenogenetically. Since all are infected as embryos in the manner described, the mycetomes of this first generation are completely normal. As usual among aphids, the future third generation takes form here in the unborn embryos, yet before the birth of the second generation it does not proceed beyond the still sterile blastoderm stage (Fig. 171*a*).

Fig. 170 *Stomaphis quercus.* The young embryo is infected by three kinds of symbionts. From Klevenhusen.

The second generation remains in the gall, depositing in it a substantially smaller portion of offspring. Each of these virgins bears only approximately thirty young furnished with symbionts from the supply of the maternal mycetome. These are the so-called sexuparae, of particular interest as producers of the rudimentary sex forms (Fig. 171*b*).

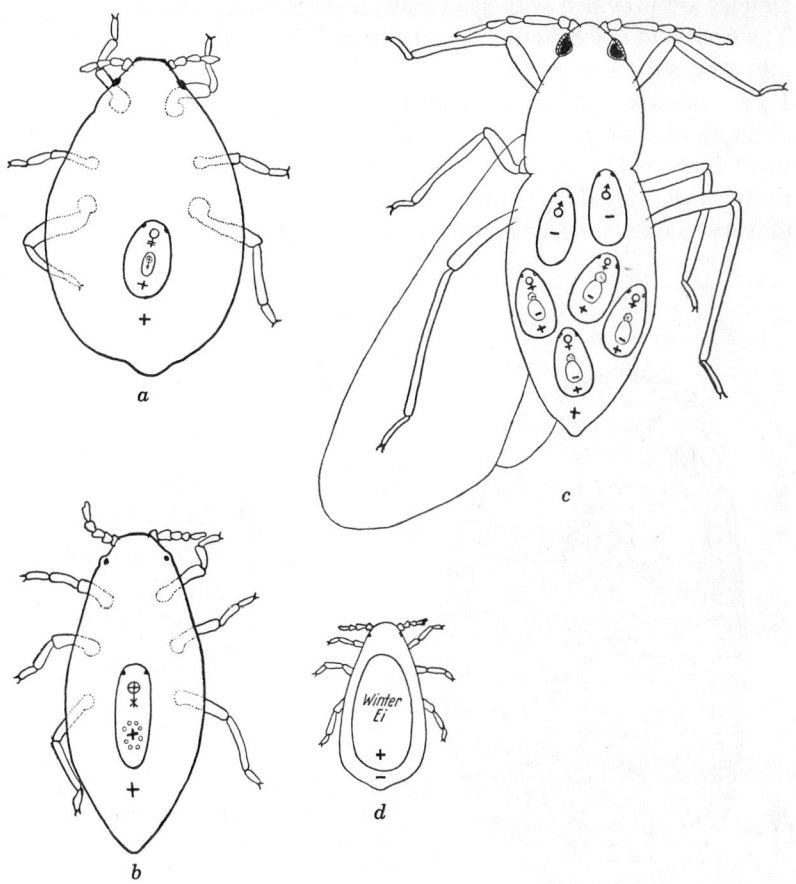

Fig. 171 *Pemphigus spirothecae* Pass. Schematic presentation of the yearly cycle. (*a*) First generation (fundatrix) with an embryo of the second generation, in which an embryo of the third generation is already present; (*b*) second generation (virgo) with the embryo of the third generation, which develops eight oviducts; (*c*) third generation (sexupara) with two sterile male embryos; (*d*) fourth generation; sexualis-female with the single infected winter egg. Crosses indicate the presence of symbionts; dashes, the absence of symbionts. About 20×. From Tóth.

The alate third generation leaves the gall and produces only six offspring. Originally there were eight ovarioles, but two habitually degenerate and the others bring forth only one embryo each. The two that are closest to the head become males, and the other four become females. In the two males the usual infection fails to occur in the embryonic stage; the four females are provided with symbionts in the normal manner (Fig. 171*c*). The frontmost embryos thus lack the usual syncytium which later develops into a mycetome and which is colonized normally in the other four (Fig. 172).   In view of the small number of symbionts required, the mycetome of the third alate generation is much less developed than that of the first two.   The sexual forms which arise in this manner are short-lived, relatively small animals ingesting no nourishment whatever.   Proboscis and salivary glands are lacking; antennae and eyes are rudimentary.   The

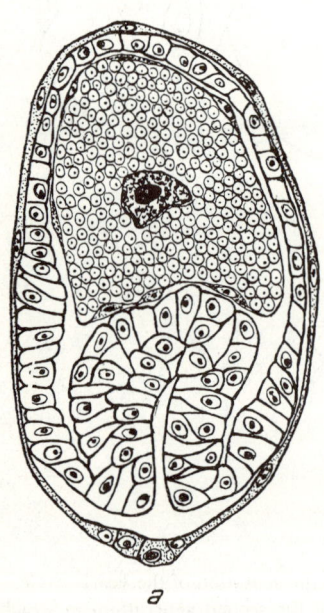

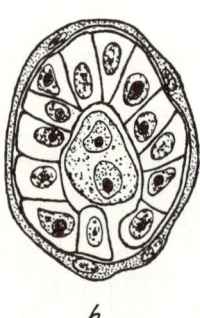

b

Fig. 172 *Pemphigus spirothecae* Pass. Embryo (*a*) of a female sexual insect with mycetome rudiment, (*b*) of a male sexual insect without mycetome rudiment. From Tóth.

a

females deposit on unopened buds a single relatively large egg from which the aforementioned fundatrix develops (Fig. 171*d*).

The rudimentary female has a mycetome, but it is extraordinarily reduced and just barely suffices to provide for the single egg.   Consisting of only a few mycetocytes, it lodges close against the egg at its posterior end and encompasses the oviduct.   When infection begins, gaps arise between the follicle cells situated there, allowing the symbionts in the mycetocytes to transfer by several routes which converge behind the follicle.   The entire supply of symbionts is exhausted, although sometimes a small remnant is left behind to degenerate.   The reproductive intensity of this modest quantity of symbionts imparted to the winter egg is enormous, as may be judged from the fact that the fundatrix developing from it has approximately 9000 infected progeny in one year.

Tóth gives only a general description of these interesting conditions, and he does not touch upon the chromosome cycle, a companion cycle to those of the generations and symbionts.   It is therefore all the more satisfying that Lampel (1958, 1959) renewed studies on three species of *Pemphigus*, not only with the monoecious *P. spirothecae*, but also with the two heteroecious species (*P. bursarius* L. and *P. populi-nigrae* Schrk.), and he filled many gaps in our knowledge.   Among other things, he followed more exactly the further development of the mycetome of the generations arising parthogenetically from the yolk-poor ovocytes and showed that, from the one or two nucleated mycetome rudiment, the separate small nucleated mycetocytes arise through amitotic constricting off of nuclear buds, which migrate to the periphery and there cause a type of cleavage of the plasma. In such a way twenty mycetocytes, on the average, are finally formed, which do not further increase.   Infection of the winter animal was described more exactly.   In contrast to other aphids, there is no formation of plasma buds which break through the follicle and form the paths for the invasion.   At the most, there is only a small remainder of symbionts in the mother's body.   Strangely enough, a degeneration of the centrally located symbionts and nuclei takes place in the rudiment of the mycetome.

Of particular interest, moreover, is the detailed comparison of the male embryos, which develop without symbionts.   In the same way as for female embryos, which contain symbionts, nuclei are prepared for a future mycetome.   However, they soon degenerate and the yolk nuclei, which must form the mycetome envelope, are likewise given up into the blastocoele and suffer the same fate.   Also, the interruption of the blastoderm at the posterior pole, which serves the transmission of the symbionts to the female embryos, is formed unnecessarily.

After the discovery of these striking relationships between the absence of egg infection and the formation of male larvae in the stictococcines, it might seem appropriate to relate closely to this the suppression of the infection in eggs producing male aphids.   However, the similarity of these cases apparently lies only in the principle of thriftiness, which appears so often in endosymbioses.   In addition, the situation is different in each case, since in aphids only females are produced from the non-fertilized eggs as a consequence of the loss of spermatozoa, which are necessary to produce males, as was confirmed by Lampel also for *Pemphigus*, whereas in *Stictococcus* all eggs are fertilized and both sexes are produced.

The impressive equilibrium between the symbionts required and the quantity on hand is probably brought about primarily through promotion or stoppage of their reproductive intensity.   Another possibility mentioned by Tóth is disintegration of symbionts by phagocytosis, which at times may be quite considerable. He assigns the responsibility for this to the oenocytes, which in general have the tendency in aphids to attach themselves externally to the mycetocytes. Here it might be a question of purely topographical relationships lacking significance. Yet Klevenhusen observed in *Stomaphis quercus* that symbionts may be transmitted to oenocytes, but he was not able to supply details of their later history. Profft never observed symbionts in *Pineus*, where the oenocytes are adjacent to the mycetome and are unusually regular in their arrangement. Tóth believes in general that his preparations show a resorption of the symbionts in the oenocytes.   Oenocytes with clear protoplasm but slightly tinged with yellow, he says, detach the divisional cell membranes in the locality, and the round symbionts then are transmitted in great numbers to the oenocytes, where they often turn a uniform dark brown (*Stomaphis*). They do not actually disintegrate and are observed in this condition even in older animals.

Tóth's description of the decimation of symbionts in *Pemphigus* differs somewhat.   Particularly in the sexuparae, normal oenocytes attach themselves to the mycetocytes in the process of degeneration. Their plasma fills with secretion and drops of the latter are transmitted to the mycetocytes.   The nucleus thereupon degenerates, the symbionts are loosened somewhat, and they stain more intensely.   The oenocyte sinks more deeply into the cell, its nucleus grows, secretory activity continues to increase, and the symbionts finally disintegrate completely.

In view of certain cases in which the entire symbiont supply or a part of it is destroyed at certain times through lytic processes, we must keep in mind from the very beginning that such decimation may possibly exist among the aphids, but, inasmuch as the occurrence is rather isolated and details are lacking, a verification of Tóth's observations by someone else would be desirable.   Lampel is among those who discount any relationship between aphid symbiosis and oenocytes.

Lastly, we have to report on a form of aphid symbiosis which falls completely out of the framework of those symbiotic arrangements known up till now in leaf lice.   I came upon it while I was investigating in Java a series of Hormaphidinae belonging to a subfamily of the Thelaxinae.   I found that in some of the twenty-three species examined, which belong to fourteen genera, the usual mycetomes were lacking and, differing from species to species like budding yeasts, inhabit the lymph and infect fat cells without causing increased growth of the latter.   This surprising occurrence, however, was limited to some genera of the Oregmini and was lacking completely among the Hormaphidini (Figs. 173 and 174). In *Cerataphis freycinetiae*, slightly bent, often long delicate bacteria were present as companion symbionts between the yeasts.   That this is a case of displacement of the older symbiont form by yeasts, which is observed time and again, is quite evident and is confirmed beyond question by the method of transmission to the embryos.   The yeasts use the ever-present port of entry described by me in 1921, a fact which has not been duly

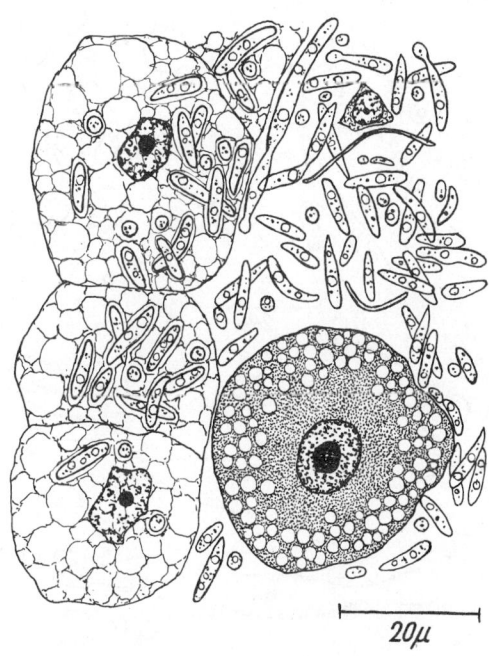

Fig. 173   *Cerataphis freycinetiae* v.d. Goot.   Symbiotic yeasts in lymph and fat cells; in the former, also thread-shaped companion symbionts; one oenocyte symbiont-free. From Buchner.

recognized by many later authors (Fig. 175).   Another proof of the loss of the old device is the fact that the characteristic large nuclei destined for the mycetome are given up into the inside by the blastoderm, and there they disintegrate without being put to use (Fig. 175).   Finally, it is of general interest that in two genera, *Glyphinaphis* possessing yeasts and *Hamamelistes* containing its original symbionts, the embryos developing into males remain symbiont-free in the sexupara generation, which is widely suppressed in the tropics.   This is a phenomenon which was known formerly only in the Pemphiginae.

Kolb (1963), using chiefly my material, investigated thoroughly the yeast-bearing genera, taking into consideration the embryonic development, and she also included in her studies representatives of the Anoecinae and Thelaxinae. She came upon another deviating form of symbiosis. *Hamamelistes* and other genera possess symbionts which are differentiated

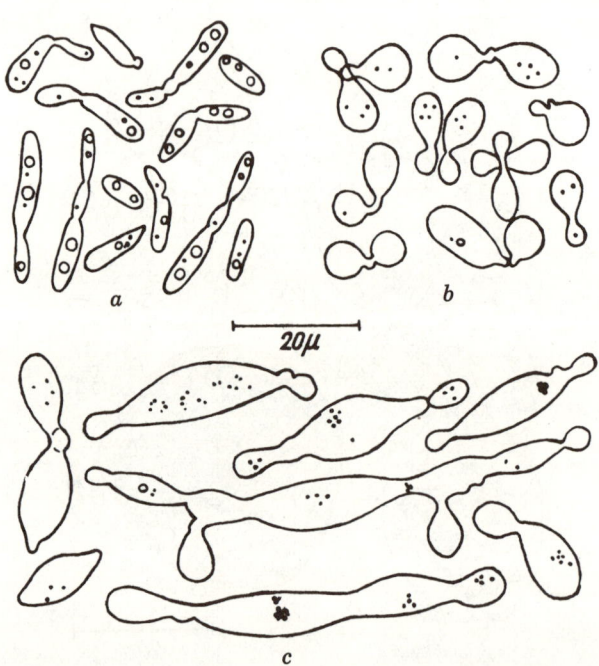

Fig. 174   Symbiotic yeasts from three representatives of the Oregmini; fresh mounts. (*a*) From *Cerataphis variabilis* R.H.L.; (*b*) from an undescribed genus; (*c*) from *Glyphinaphis freycinetiae* v.d. Goot.   From Buchner.

from the tropical ones in that in cross section they sometimes appear round, and other times as irregular clods, forming rows of chains.   Infection of the embryos takes place much earlier, and there are other differences. Unfortunately no investigation has been made on living forms of these unusual symbionts.

In our description of aphid symbiosis we have had no occasion to mention the subfamily of the phylloxerines, for it is the only aphid without a homologue of the pseudovitellus.   Some authors, of course, have considered it possible to verify an analogous organ.   Grassi in his long

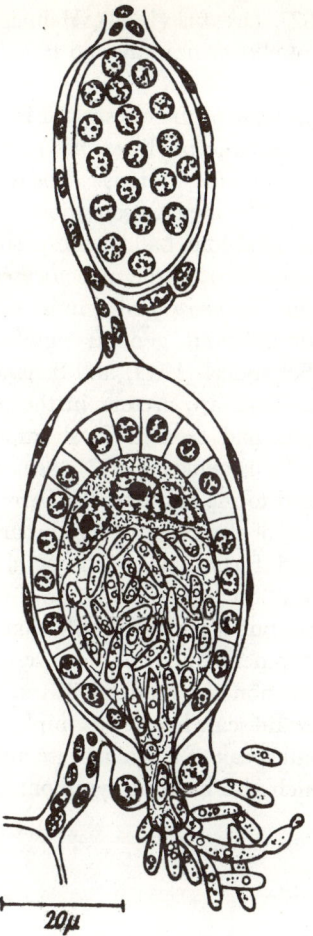

Fig. 175   *Cerataphis freycinetiae* v.d. Goot.   Portion of an ovariole (fresh mount); above, the protruding rudiment of the infection mound; below, transfer of the yeasts into the approaching plasmatic processes.   From Buchner.

Phylloxera monograph (1912) frequently pictures the "pseudovitellus" of the vine louse in the form of cell groups, loosely arranged and partially connected, and is of the opinion that they might possibly represent habitats of symbiotic organisms.   Peklo (1916) even asserted that he had made successful pure cultures of *Phylloxera vastatrix* Pl. symbionts and I (1930) described curved rod-shaped organisms which I had found scattered diffusely in *Phylloxera quercus* Boyer.   On the other hand, several authors, such as Witlaczil (1882), Dreyfus (1894), Henneguy (1904), and Portier (1918), asserted that phylloxerines have no pseudovitellus.   Profft came to the same conclusion.

After a painstaking review of the question, Maillet (1957) comes to the apparently final result that the phylloxerines have no pseudovitellus and evidently no other form of endosymbiosis.   For the history of the problem we refer the reader to his detailed presentation.   It is beyond doubt that the easily recognizable symbiont ball at the posterior end of the eggs is lacking as well as an equivalent of the mycetocytes.   Thus the organisms I observed years ago in *Phylloxera quercus* must represent an occasionally occurring phenomenon without general significance.   Maillet rightly rejects statements by Schanderl (1949) and Schanderl, Lauff, and Becker (1949) that symbiotic yeasts are present in the midgut epithelium, and indeed these statements had seemed most improbable from the very beginning.   Maillet, like Fraenkel (1952), assumes that the absence of symbionts may be traced to the fact that *Phylloxera* is not fed by sieve-tube sap but by the contents of the cells, which apparently contain materials lacking in the sap and furnished by the symbionts in other aphids. Whether it is a question of a primary state or of loss of symbionts cannot be decided at this time, but it is conceivable that careful investigation of early embryonic development might disclose vestiges of a symbiosis once present.   In this connection the reader will recall the no less surprising lack of symbionts in certain scale lice, for example, the apiomorphines; and in the following we shall speak of a similar case among the cicadas, that of the typhlocybines, which also have no symbionts.

ALEURODIDS

Our knowledge of aleurodid symbiosis is based almost exclusively on my own investigations in 1912 and 1918.   Despite the many symbiosis types which have come to our attention since then, its method of transmission was until a short time ago without parallel.   Weber (1935, 1935) contributed several supplementary details in his publication on *Trialeurodes vaporariorum* Westw.

All aleurodids have relatively small, paired, roundish, or oval mycetomes which stand out even with low magnification by virtue of their vivid orange coloring.   At first situated cephalad in front of the gonads, they glide backward in females of the fourth larval stage and sink between ovarioles arranged in clusters.   At the same time the mycetocyte unit is loosened and is distributed over the pedicles in such a way that finally only the end parts of the ovarioles protrude (Figs. 176 and 177).   A study of the trans-

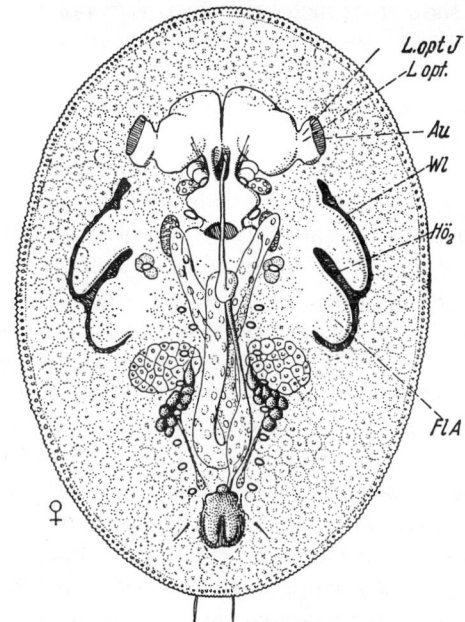

Fig. 176   *Trialeurodes vaporariorum* Westw.   Female larva of the fourth stage.   FlA,
wing rudiment; Wl, epithelial wall; Au, eye; L. opt., lobus opticus; Hö₂, middle
developmental cavity.   From Weber.

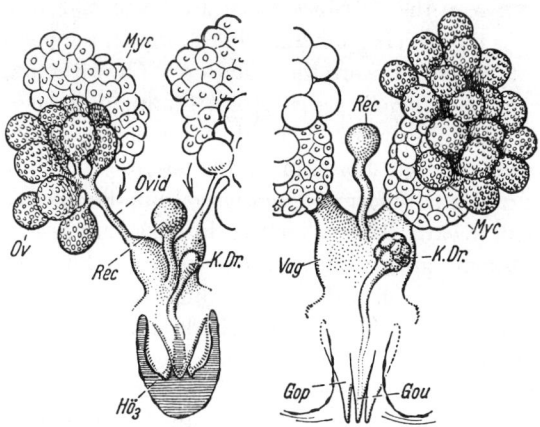

Fig. 177   *Trialeurodes vaporariorum* Westw.   Shifting of the mycetome to between the
ovarioles in the course of the fourth larval stage.   Ov, ovarium; Myc, mycetome; Ovid,
oviduct; Rec, receptaculum seminis; Vag, vagina; Gop, Gou, paired and unpaired
gonapophyses; K.Dr., cement gland, Hö₃, posterior developmental cavity.   From
Weber.

mission method shows that such behavior is very practical, yet it is hard to understand why the mycetomes shift backward at the same time in males and thus come in closer contact with the testes, which indeed are sometimes surrounded by some of their mycetocytes in the manner of an epithelium (Buchner, Weber).

The individual mycetocytes are always uninucleate and relatively small.   Their inmates are roundish or oval organisms often located in a single layer around the central nucleus.   Occasionally abscission of buds can be observed in them.   Steinhaus (1951) found long rod-shaped forms in *Tetraleurodes stanfordi* (Bemis), sometimes in chain formation and sometimes growing out in long filaments 25–30 μ in length.   The yellow pigment granules are inserted in the meager host plasma, collecting particularly around the nucleus and in the spandrels between the symbionts.

Often the mycetocytes are observed in mitotic division whereby the pigment is chiefly concentrated at the two poles of the spindles.   Although the mitosis rarely exhibits disturbances of any kind, here, as in *Pseudococcus* and probably in some other cases, a later fusing of the daughter cells must come about from time to time, for Schrader's study of sex determination in *Trialeurodes* (1920) showed that there were approximately twice as many chromosomes in the mycetome nuclei as in other body cells.

In the aleurodids not only symbionts migrate from mycetome to egg cells for transmission to the progeny, but also completely intact mycetocytes, as was found recently in some coccids (*Puto, Macrocerococcus*) (Buchner, 1955).   Since the latter may be found in the mature female dispersed everywhere between the ovarioles, there are no special difficulties.   The place of entry is the follicle epithelium bordering on the young oocyte and narrowing progressively toward the posterior.   The mycetocytes, numbering nine or ten in *Aleurodes aceris* Geoff., gradually push between cells, which are very flat there, and, when they arrive at a space behind them which has arisen through shortening of the oocytes, they spread them still farther apart and finally form a longish oval bundle there.   The bundle is now pushed toward the egg, where it sinks into a furrow partially enclasped by the margins.   The mycetocytes group themselves in a rather spherical formation (Fig. 178).   According to studies of the genera *Bemesia* and *Aleurolobus* by Tremblay, a single strongly pigmented mycetocyte infects the egg.

The follicle cells which gaped to provide an entrance passage meanwhile move close together again to restore the thick epithelium.   The canal which became free when the mycetocytes shifted forward is narrowed; the cells bordering it secrete a strong, chitinous wall around it and form a thin-walled vesicle at its end, giving rise in this manner to a pedicle, temporarily squeezed into an *s*-shape, by which the egg is later attached to

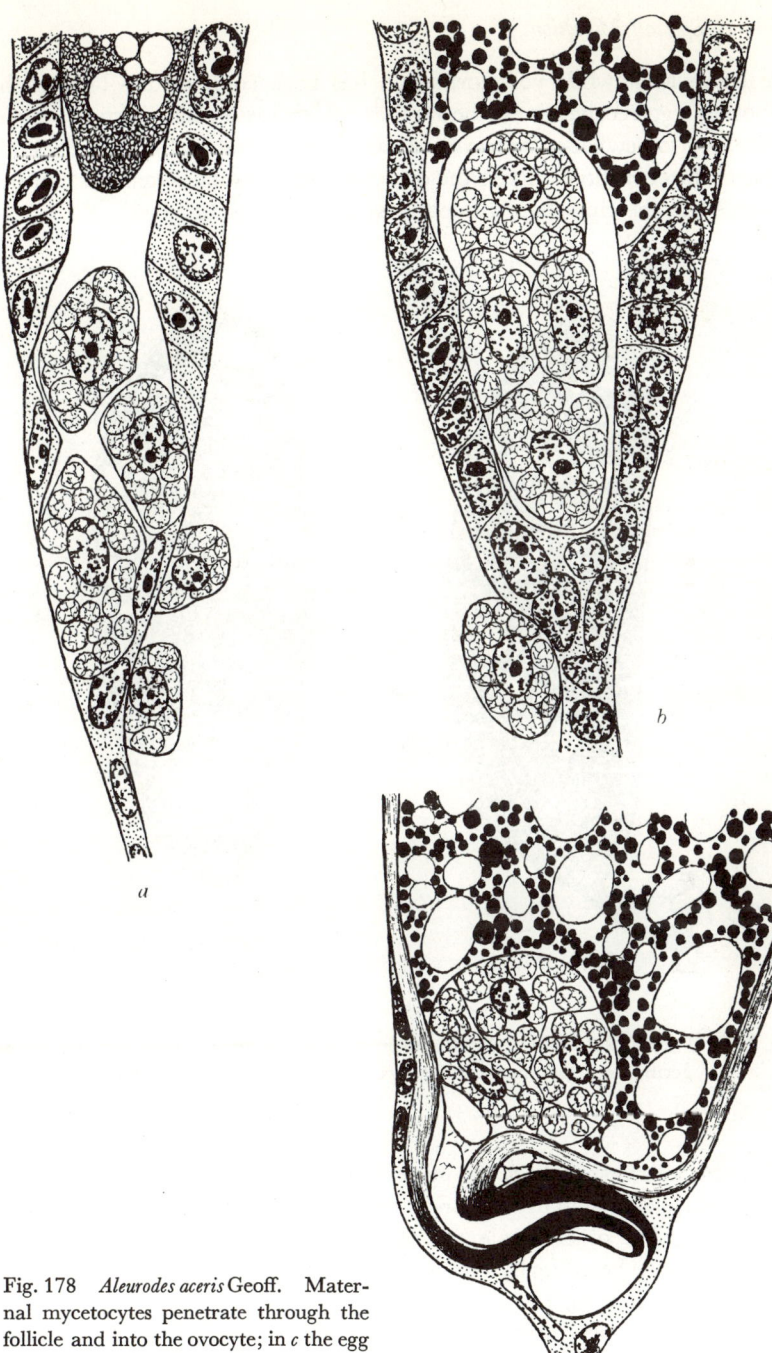

Fig. 178 *Aleurodes aceris* Geoff. Maternal mycetocytes penetrate through the follicle and into the ovocyte; in *c* the egg stalk, still pressed together, develops. From Buchner.

the leaf. Meanwhile yolk formation has been completed. Thus each deposited egg receives from the mother mycetome a small portion which is situated somewhat eccentrically at its posterior end and which bears no signs of degeneration. Even the original pigmentation is kept, as can be observed in the ring-shaped deposit with a magnifying glass.

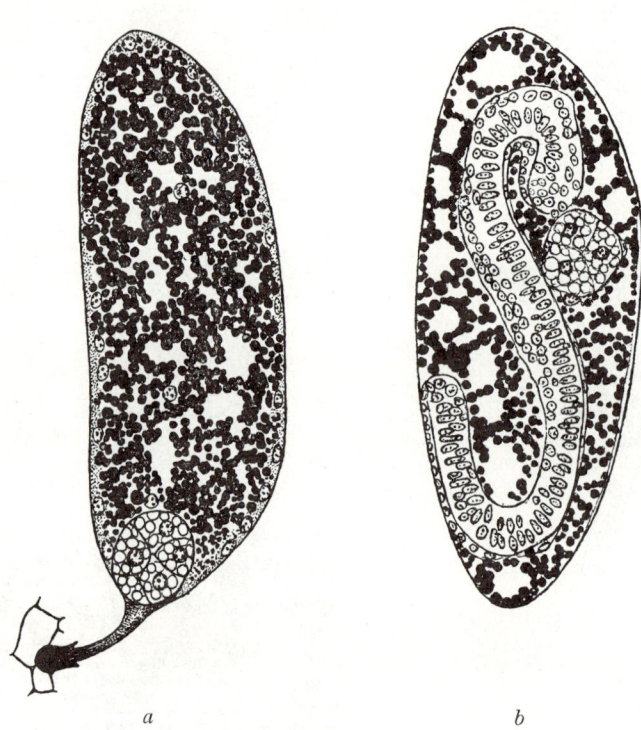

a   *b*

Fig. 179   *Aleurodes proletella* L.   (*a*) Blastoderm stage; (*b*) embryo with s-shaped, bent germ band; the maternal mycetocytes persist.   From Buchner.

The backward shifting of the mycetomes and their cleavage in the area of the ovarioles is actually the first step toward their infection. The corresponding shift of the male mycetomes must perhaps be regarded as aimless, purely hereditarily conditioned repetition of behavior which in the females is purposeful.

How will the new organism deal with this implantation of maternal body cells during embryonic development? Will the strangers meet with early destruction or even build up the new mycetome? It would appear so at first, quite as if these cells were preserved by transplantation from the destruction befalling the mother animal itself very soon after deposition of the eggs. During cleavage the cells remain unchanged; the place they occupy is left free by the ascending nuclei. When the blastoderm is formed, it surrounds the spherical cell group, approximately at the equator, and the group is moved forward by the inverting germ band, as so often in such cases, without entering into closer relationship with it. When the germ band forms the *s*-shape, the ball glides back somewhat, stopping when it reaches the ventral side and settling close to the amnion. During revolution it keeps its position and thus reaches the dorsal region of the abdomen (Fig. 179).

Not until revolution, sometimes indeed just before hatching, do embryonic elements act as a replacement for the maternal cells. Earlier, several large nuclei attached themselves to the mycetocytes of the cleavage nuclei or the lower lamella; now the old cell limits are obliterated, and the mother nuclei migrate to the symbiont-free plasma area arising in the interior. This is the signal for the peripheral nuclei to penetrate with their meager protoplasm between the deserted symbionts (Fig. 180*a*). At this time the young organ is already broadening out and displays a medial furrow caused by the placement of the intestine. The new nuclei divide several times, and the symbionts, which are reproducing freely at this time, thus gradually form cell limits, the interior ones last. The old nuclei, still located in the central plasma, discard the angular outlines acquired through pressure of the symbionts and resume their round form. Apparently they even continue to be constricted off amitotically (Fig. 180*b*). But various signs of degeneration appear at last, and the result is that the ring of new mycetocytes on the dorsal side opens temporarily to discharge plasma and nuclei of the mother animal into the vicinity, where they disintegrate.

Occasionally the implanted elements show enough resistance to prevent the substitute cells from penetrating and thus necessitate formation of the multinucleate epithelium. A closer study of these variants and the behavior of the symbionts during transmission would be of interest.

After the larvae hatch, the mycetome is completely divided into paired sections. Vigorous mitotic cell divisions lead to an enlargement of the organs, causing the latter to envelop some of the dorsoventral muscles.

Particularly in *Aleurodes proletella* L. there are individuals which give evidence of a deep-seated disturbance in the symbiotic equilibrium. The mycetomes appear more voluminous in such cases and often split off into unequal fragments which are passed on in the body to penetrate as far as the thoracic region at times. In connection with this dispersal there is a substantial increase in pigment, making the sick animals easily recogniz-

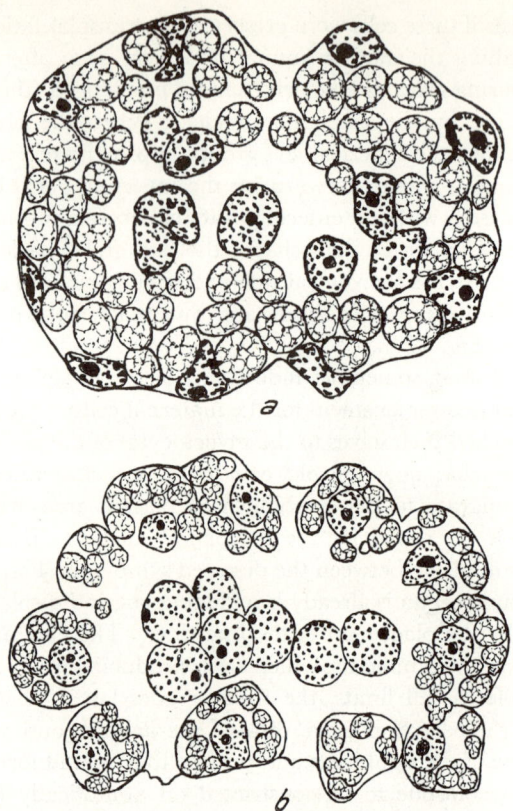

Fig. 180 *Aleurodes proletella* L. (*a*) The maternal mycetocyte nuclei concentrate in the center of the provisional mycetome, the definitive nuclei crowd toward the periphery; (*b*) the definitive mycetocytes are bounded off; maternal plasma and nuclei still persist in the center. From Buchner.

able. The nervous system, the matrix cells of the tracheae, and the body epithelium in particular are filled with it. The obvious connection between the unusual behavior of the mycetomes and this increased formation of pigment merits a thorough study.

PSYLLIDS

Just as with the pseudovitellus in aphids, the vividly colored mycetomes characteristic of all psyllids were observed by early insect anatomists but were not recognized as symbiont bearers until the time of Pierantoni and Šulc. The psyllid mycetomes are basically unpaired, bilaterally symmetrical organs which undergo radical deformation and cleavage when the gonads develop in the pubescent animal. There may be a disproportionately large center with relatively small lobes and humps on both sides, as in *Psylla buxi* L. (Fig. 181), or two separated wings may become so large that the center is like a thin connecting bridge, as in *Psylla*

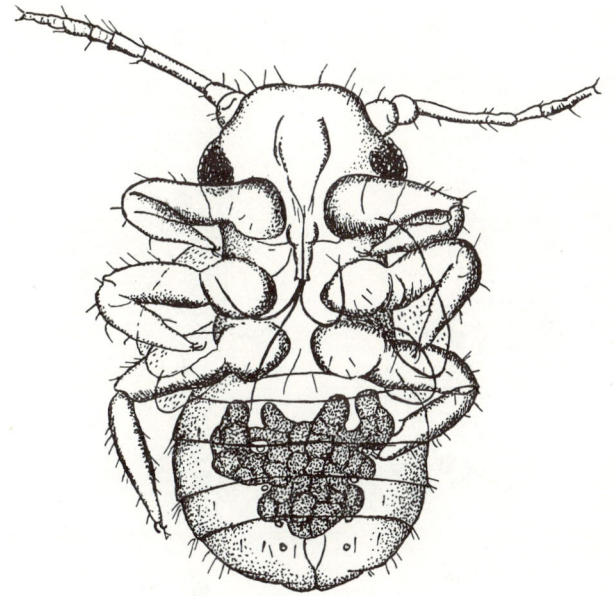

Fig. 181   *Psylla buxi* L.   Nymph with mycetomes.   From Buchner.

pirisuga Frst. (Fig. 182).   Segmental muscles then cause corresponding furrows in the organ.   In the imagines the gonads tend to force the mycetomes forward and divide them in half.   In females the cleavage is so great during egg infection that fragments of the mycetome are found between the ovarioles.

Three components can always be differentiated in these organs: uninucleate cells which are round or are flattened at the edges in the manner of an epithelium; a syncytium not always the same in size; and an envelope composed of much flattened cells encasing the whole.   The distribution of cells and syncytium is varied.   The uninucleate cells may form a more or less connected marginal layer, relinquishing the center to the syncytium, or they may penetrate to the syncytium and collect here and there in nests. Finally, there are species in which the whole syncytium, uniformly penetrated by the cells, borders the top surface in all places (*Psylla alni* L.). The yellow pigment is found in the plasma of both zones but is particularly abundant in the syncytial zone, which is richer in plasma.   Profft's reaction tests (1937) show that it is not a typical lipochrome and yet is not readily classifiable as a melanin either.

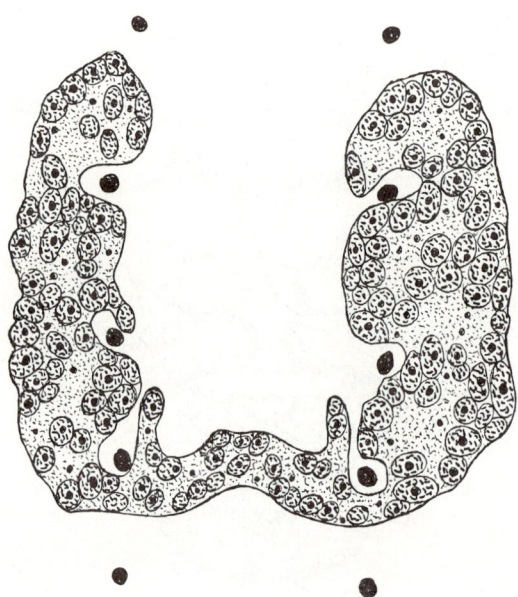

Fig. 182   *Psylla pirisuga* Frst.   Mycetome composed of mononucleated cells and syncytia.   96× From Profft.

In accordance with the two histologically different zones, there are two types of symbionts.   Throughout the uninucleate zone there are very similar elongated, delicate tubes with irregular, thickened places and swellings, which are connected by thread-like sections.   They are so densely packed that their true form can be revealed only by teased-out preparations.   The symbionts of the syncytia are much more varied. Sometimes they differ so much from those of the uninucleate cells that the difference is indisputable.   Profft discovered a number of such cases during his comparative study of eighteen psyllids.   For example, in *Trioza salicivora* Reut. there are loosely arranged slender rodlets with intensely staining granules at both ends (Fig. 183*a*); in a Brazilian *Psylla* species, filamentous bacteria; in *Psylla alni* L. and *Aphalara nebulosa* Zett., thin tubes (Fig. 183*b*). Because in other species the differences are slighter, it is possible that they sometimes may go unrecognized. Šulc asserted that there were two types of symbionts for *Aphalara calthae* L., a view upheld by me (1912) and by Breest (1914). Salfi (1926) and Tarsia in Curia (1934) opposed the concept that psyllids are disymbiotic. However, even where the two symbiont types are very similar in shape, they may be differentiated by structure and staining. For instance, the inhabitants of the mycetocytes assume chiefly the plasma colors, whereas those

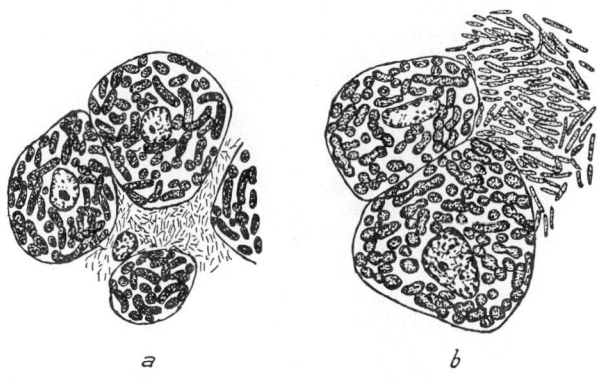

a   *b*

Fig. 183   (*a*) *Trioza salicivora* Reut.   (*b*) *Aphalara nebulosa* Zett.   Different kinds of symbionts live in the mycetocytes and in the syncytium.   855×.   From Profft.

of the syncytia show a pronounced affinity to basic colors, and this differentiation is observable even with low magnification.

Profft's material contained two species in which the symbiosis type diverges radically from that of other psyllids. In *Strophingia ericae* Curt. and a *Trioza* species from Brazil, the uninucleate mycetocytes are colonized in the usual way, but the syncytium is limited to small areas and is free of symbionts. In the *Trioza* species the additional organism, in the form of longish ovals or tubes located in cells of the fatty tissue next to the mycetome, is revealed as a true symbiont by its regular transmission to the egg cells. *Strophingia* never has more than the one ever-present symbiont of the usual shape, but there is every indication that a second symbiont once existed and for some reason was eliminated. For additional instances of such loss in other species the reader is referred to Chapter 15. The symbionts in the fatty tissue have the same color as those in the syncytium of the other psyllids.

At any rate, comparison of the symbiont forms in the two zones shows that those of the syncytia, which often retain the original bacterial form, are the late-comers and that only some of them assume those bizarre growth forms as a result of long exposure to intracellular living.

We find only one reference in the literature to successful culture of psyllid symbionts, a statement by Mahdihassan (1947) that he cultivated

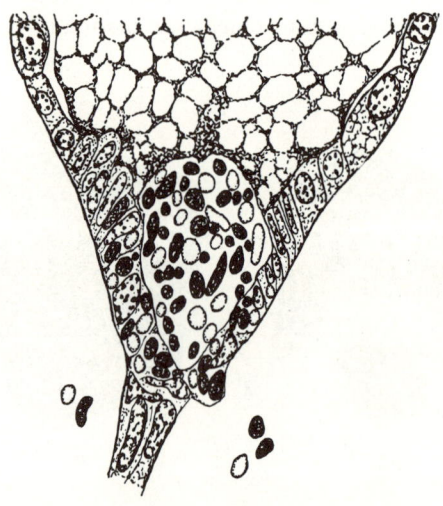

Fig. 184   *Trioza* sp.   The two types of symbionts infecting the egg.   From Profft.

both symbiont forms from *Psylla mali* on apple juice and obtained a relatively thick comma-shaped organism, which produced a light-green pigment, and a very short delicate rodlet.

That two distinct forms are actually present is also confirmed by a study of the transmission method, if indeed more proof were needed. Irregularly shaped tubelets in the mature females become roundish or oval formations through transverse cleavage.   Where the inmates of the syncytium still possess the original bacterial form, they proceed in this form until the infection of the egg cells.   Transformation and emigration take place irregularly here and there in the mycetome.   Transmission to the egg cells takes place at the posterior pole, where the psyllid eggs form a peg-shaped constriction surrounded by unusually distended follicle cells. Both types meet on the exterior but can be distinguished from each other in spite of great similarity by differences in form, color, and structure. Little processes, which can be observed occasionally on the follicle cells, receive the symbionts at random in the manner of amoebae, so that arrivals from both areas may be found in the same cell.   Meanwhile, the process of the egg is withdrawn, and the symbionts are transmitted to the

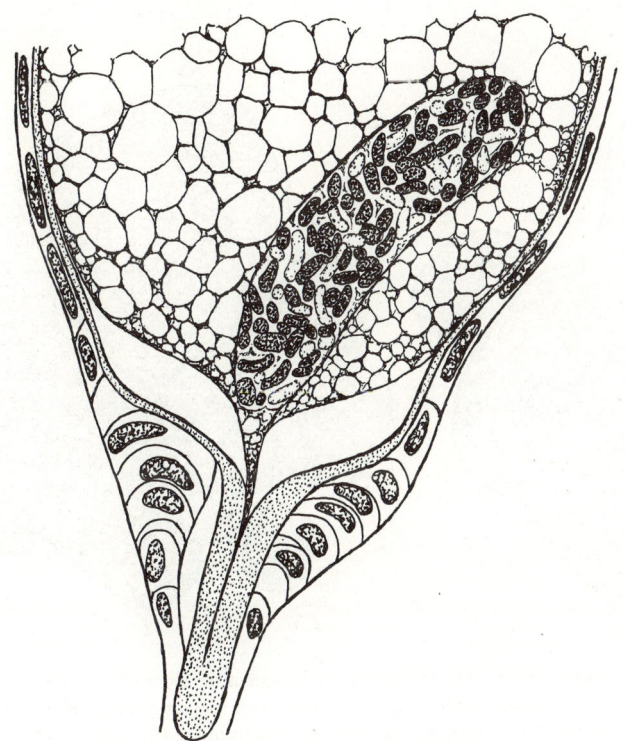

Fig. 185   *Psylla alni* L.   Infection of the egg with the two types of symbionts is ended. From Buchner.

resulting hollow (Fig. 184).   Finally the egg cell enclasps the symbiont ball formed in this manner and draws it in (Fig. 185).   In the *Trioza* species from Brazil not only occupants of the mycetome are involved in the infection, but also, as already mentioned, those of the fatty tissue.

The same yellow pigment that colors the mycetome and sometimes disappears there in pubescent females is also found in the ovary.   Diffusely distributed at the beginning, it is later collected in the extension at the posterior egg pole and after infection is completely concentrated around the roundish or longish symbiont ball. In place of the former extension, there is a pedicle by which the eggs are fastened to the base, as in the aleurodids.

When Profft studied egg development in *Psylla alni*, he found a certain lack of singleness of purpose in the genesis of the mycetomes which at first seems surprising. With termination of blastoderm formation a number of

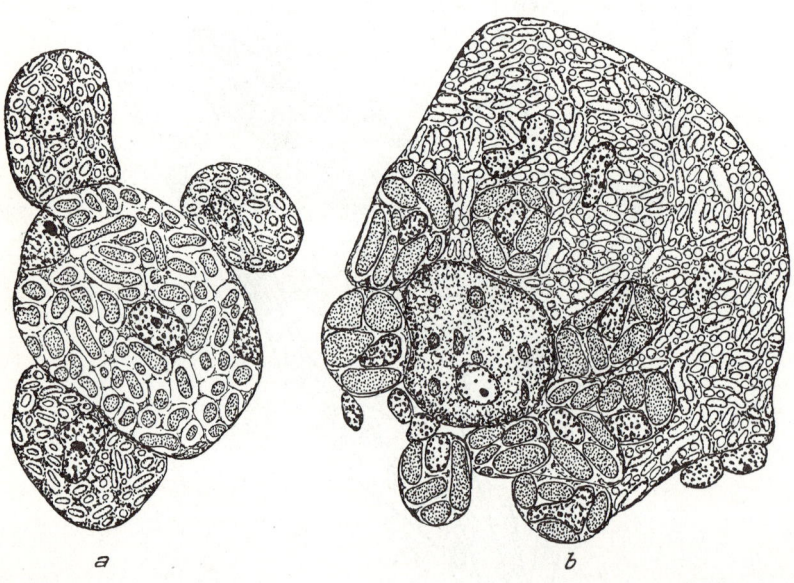

a   *b*

Fig. 186 *Psylla alni* L. (*a*) Mycetome rudiment after separation of the two types of symbionts; (*b*) dividing of the syncytium into mycetocytes, joining of the provisional mycetocytes to the syncytium, and rearrangement of the two zones. From Profft.

cleavage nuclei are transmitted to the symbiont ball; others are attached externally as envelope cells. Gradually the future occupants of the syncytium are collected, not as one would expect in the center of the ball but at its periphery, and the future occupants of the cortex now occupy the center. In a succeeding stage the symbionts which are later located in the interior are collected without exception around the envelope cells, and the latter now become uninucleate mycetocytes attached like a bud to a round syncytium with the cortical symbionts! The yellow pigment is then abundant in the peripheral cells and sparse in the syncytium.

During the following phase, which gives a rather chaotic impression, the nuclei of the provisory syncytium multiply greatly. Several large symbionts collect around them, and little uninucleate mycetocytes originate through free cell formation. At the same time the cell limits of the peripheral elements are lifted, and the latter then merge to form the definitive syncytium. The definitive mycetocytes pressing more and more toward the surface finally surround the entire syncytium (Fig. 186). We have no data on the origin of the envelope cells which form again during this strange process and this time persist. Only one explanation appears possible for the apparent lack of singleness of purpose the first time: the otherwise sterile envelope cells were used for the new cell material necessitated by the acquisition of the second symbiont, just as with the disymbiotic adelgines; but, since this measure contradicted the necessity expressed again and again for shifting accessory guests to the interior of the mycetome, secondary shiftings of the newly infected cell material occurs in both cases, to an extent which varies from species to species.

Salfi and Tarsia in Curia also studied the embryonic development of their test objects. According to the latter, only the symbionts of the cortex are transmitted to the eggs, and after involution some of the inmates of the newly formed mycetocytes are transmitted to the area arising in the center through fusion. According to Salfi, the converse is true: the mycetocytes are split off from the primary syncytium and are provided with symbionts. These interpretations, based on the hypothesis that only one symbiont type exists, are both untenable.

The species investigated thus far are members of the subfamilies of the aphalarines, psyllines, and triozines, yet their symbiotic devices show no relationship to the system.

CICADIDS

In the cicadids a thoroughly confusing diversity of symbiotic devices goes hand in hand with unparalleled wealth in forms and colors. When Šulc and Pierantoni made their first reports on the mycetomes of *Aphrophora* and *Philaenus* (1910) and therewith correctly interpreted what had already been observed in 1899 by Heymons in *Cicada septendecim* L. and by Porta (1900) and Guilbeau (1908) in foam cicadas, it was not yet realized that these were the first insights into an area of unprecedented complication. Later Šulc (1924) made more detailed statements on fulgorid symbiosis, and Resühr (1938) studied the morphology of certain aphrophorine symbionts. Since then the most important studies have been my own (Buchner, 1912, 1923, 1924, 1925), and those of my students

Richter (1928), H. J. Müller (1940, 1949, 1951, 1962), and Rau (1943). Two students of Müller also did their part in extending our knowledge of this subject. Strübing (1956) in her studies on the oviduct and the laying of eggs in the Delphacidae gave information on the symbionts, and Ermisch (1960) investigated thoroughly a large number of additional fulgoroids. At this point we should not forget the findings of Sander (1955) on the fate of the symbionts of *Pyrilla* during embryonic development. The observations by Mahdihassan (1939, 1941, 1946, 1947) and Naidu (1945) cannot be considered serious contributions to the subject.

My comparative study of 1925 was based on more than 100 species; thanks to my successors this number has grown to 405, of which 188 species belong to the Cicadoidea and 217 to the Fulgoroidea. Ermisch studied 19 additional new species. However, it must be remembered that, according to Melcalf's estimate of about seven years ago, 30,000 species of cicadas, which are distributed in 3500 genera and 45 families have been described. Only 52 of the 74 subfamilies differentiated by Haupt (1929) have been tested for symbionts. As the latter are not generally uniform in constitution, there is much virgin territory awaiting exploration. The Brazilian material I was able to obtain for the research done by Müller and Rau was extensive, but the need for further work particularly on exotic fauna still exists.

The great symbiotic diversity of these animals is the result of a marked tendency to take up additional symbionts. This hunger for symbionts is so great that monosymbiotic forms are a rarity. Aside from the typhlocybines, which have no symbionts, only 22 of 405 species investigated are monosymbiotic; 224 (55%) are disymbiotic, 120 (30.5%) trisymbiotic, and 17 (4.2%) tetrasymbiotic, 6 (1.5%) quintasymbiotic, and 2 sexasymbiotic. The colorful character is further heightened by certain sympathies and antipathies which bring about various combinations of the symbiont types. In order to present them and to understand them statistically, it is necessary to use letters to designate the forms and organs that proved to be different. Thus one speaks of *a*-, *b*-, *c*-, *d*-, *e*-symbionts and of *a*-organs, *b*-organs, and so forth. Sometimes there are so many that it is necessary to use Greek letters.

Comparison reveals that age and degree of adaptation of multisymbiotic forms can be most diverse at times just as in other Homoptera with several symbionts and that four categories are easily distinguishable. One is inclined to follow H. J. Müller in designating as chief symbionts of cicadas those which, unlike most of the other guests, may occur alone and are not dependent upon the companionship of other types. Until recently only "yeasts" were considered possible as chief symbionts among both *Cicadoidea* and *Fulgoroidea*, and the so-called *x*-symbionts among the latter. However, when H. J. Müller (1951) had the opportunity of investigating among the peloridiid *Hemiodoecus*, a representative of the unique Coleorrhynchae, which are customarily considered a special division as opposed to that of the Auchenorrhynchae (= Cicadidae) and the Stenorrhynchae (= Phyllidae, Aleurodidae, Aphidae, and Coccidae), he found that it resembled the cicadids in its symbiosis, but that it possesses only *a*-symbionts, that is, a form which had never before been encountered alone in the cicadas. In treating the peloridiids as cicadoids or fulgoroids rather than as representatives of a third division of the Homoptera, we are following H. J. Müller, who does not consider it absolutely necessary to create a special division for them and prefers to classify them on the basis of Auchenorrhynchae development.

The organisms which he calls auxiliary symbionts are of widespread occurrence and are also highly adapted to symbiotic living but are found only in connection with the chief symbiont or in companionship with other auxiliary symbionts, for example, the *f*- and *t*-symbionts. Companion symbionts are found only in association with chief and auxiliary symbionts; although they are well integrated, they have only a slight distribution. They are clearly distinct from the fourth category, the so-called accessory symbionts, which are only imperfectly assimilated and not completely constant in their appearance. Their transmission to the egg cells may display irregularities, and they usually cause disturbances in the histological structure of the mycetomes, the better-adapted symbionts frequently being severely damaged. Thus it is clear that they are the symbionts which were acquired last.

PELORIDIIDS. The rare peloridiids, found in the humid forests of South Australia, New Zealand, and Patagonia, are primitive Hemiptera which were classified with certain predatory shore Hemiptera (sand bugs) until their homopterous nature was recognized. An unusually primitive appearance is lent by peculiarities in head structure and especially by paranotal prothoracic pouches such as occur in no other living insect and are otherwise known only from fossil forms of the Paleozoic era. Especially in the nymphal states, these pouches are similar to the pterothecae of the meso- and metathorax, and the pronota are provided with a tracheal system very similar to that of normal wings (Evans, 1941).

Indeed they have been referred to as living fossils and thus have been equated with *Peripatus*, *Anaspides*, and similar forms. They are closely related to the Palaeorrhynchae, known from the Australian Permian period, which also have paranota. In all probability they represent relics of insect fauna which once populated the Southern Continent, the widely dispersed remains of which are found today in South America, Australia,

South Africa, and Antarctica.   The peloridiids have been preserved down to the present, because they are the only living Homoptera which suck exclusively on mosses, in other words, plants which have survived millions of years, whereas most Palaeozoic Homoptera were condemned to extinction through disappearance of their host plants.

It is obvious that investigation of these animals was of very great interest in the phylogeny of the homopterous symbionts and that of their hosts. Even though specimens of *Hemiodoecus fidelis* Evans available to H. J. Müller (1951) were insufficient for studying minute details, they nevertheless afforded sufficient insight into the symbiosis type, for it might well be exactly the same in other peloridiids, although, of course, they too should be investigated.

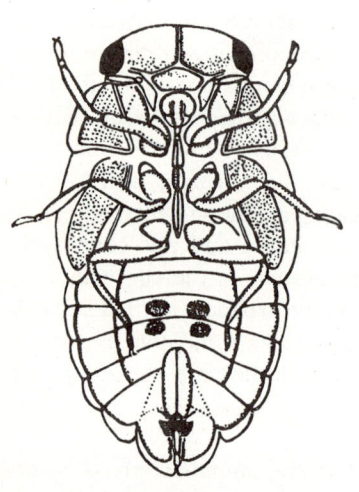

Following closely upon each other in the abdomen of males and females are two orange-colored, more or less spherical, mycetomes on both sides of the median line (Fig. 187).   In larvae they are situated in the fifth and sixth abdominal segments; in pubescent animals they are pushed more to the outside and are deformed by the development of ovaries, or of testes and the appendicular glands in males.   Each partial mycetome has a very flattened epithelium with very few nuclei and a single central syncytial mass originating presumably through fusion of smaller complexes. The compact tube-shaped symbionts are located in the interstices.   Here and there irregularly shaped nuclei are found in the islands of the host plasma, which is permeated by pigment granules.   Numerous fine tracheae and tracheoles, originating in each case from a strong branch, traverse the epithelium and penetrate to the marginal parts of the syncytium.

Fig. 187 Peloridiid. Older male larva with four *a*-mycetomes as the only symbiotic arrangement. From J. W. Evans.

Transmission takes place at the posterior egg pole.   Yet only states were found in which transmission through the follicle had already taken place and in which a flat cap of more deeply staining symbionts permeated by granules lay between yolk and chorion.

Even though typical envelope membranes could not be recognized in the fixed material, they were without doubt symbionts of the *a*-type, which are so widespread among cicadas and which are to receive further discussion later. Additional corroboration is afforded by the fact that in *Hemiodoecus* the specific transmission forms originate in the manner typical of these symbionts. Proliferations of epithelial nuclei characteristic of the infection cones of *a*-symbionts were found in certain zones of the mycetomes, where the tubes are transformed into more compact structures with denser plasma, which then leave the organ in that form and infect follicle

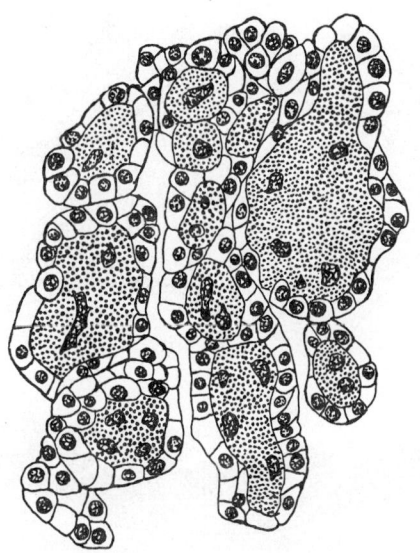

Fig. 188   *Ledra aurita* L.   Fat tissue with giant cells colonized by yeasts and syncytia. From Buchner.

and oocytes in an adjacent area. As always in such cases, the tubes in the more centrally located regions are transmitted to these uninucleate elements, where the transformation begins and progresses gradually toward the surface. Later we shall describe the origin of these epithelial proliferations in other test objects through contact of larval mycetomes with young oviducts. Indeed, the finding of such intimate contact in *Hemiodoecus*, justifies the conclusion that it was the cause of the proliferations.

A closer relationship to the fulgoroids is indicated by the structure of the mycetomes and particularly by the situation that these transformation places are not protrusions as in the cicadoids, but sunken areas as in the

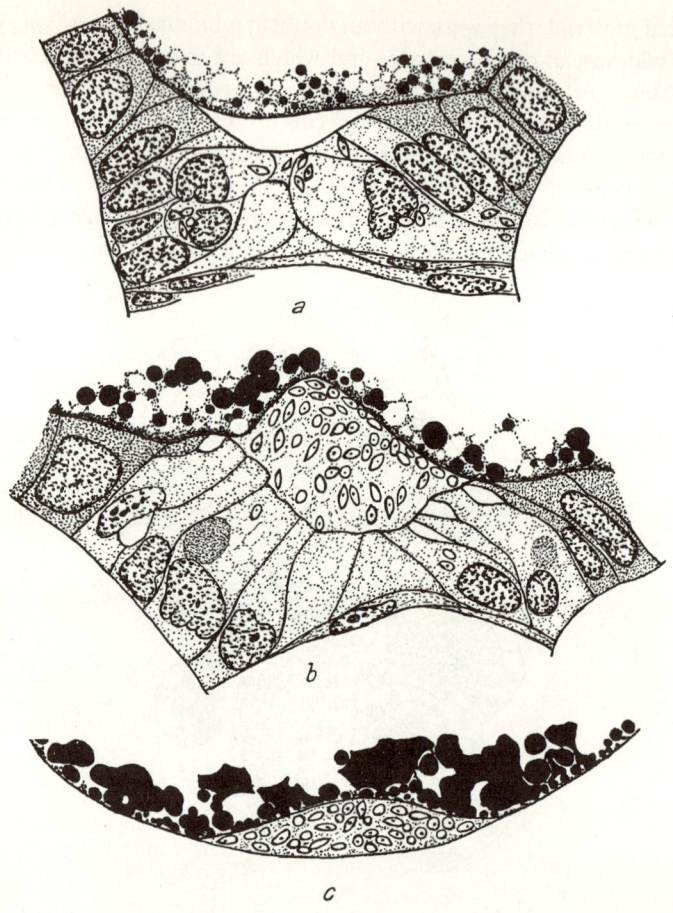

Fig. 189   *Ledra aurita* L.   (*a*) The first yeasts appear in the enlarged cuneate cells; (*b*) the yeasts leave the cuneate cells; (*c*) conclusion of the infection process.   From Buchner.

fulgoroids.   Only the very weak development of the epithelium, which appears rarely in this mycetome type, is exceptional.   Primarily, however, this type of cicada symbiosis, though thoroughly typical in all other details, is an exception unique among the cicadoids and fulgoroids in that only the *a*-symbionts occur here.   For discussion of the significance of this fact in the phylogeny of cicada symbiosis, the reader is referred to Chapter 15.

CICADOIDS.   The division of the cicadas into the two subfamilies of the cicadoids and fulgoroids, carried out first by Kirkaldy (1906), is thoroughly justified from the point of view of symbiosis. We shall discuss the somewhat simpler cicadoids first.

Of the cicadoids tested, only two species possess a single symbiont of the type characterized as chief symbiont. In both cases they are those yeast-like organisms which among the cicadas too should probably be interpreted throughout as conidia of ascomycetes. With *Ledra aurita* L., the only representative of the ledrids investigated, these yeasts are lemon-shaped or more elongated structures in uni- to multinucleate cells or syncytia scattered in the fatty tissue and surrounded on all sides by its cells as though by a unilaminate epithelium (Fig. 188). The infection of oocytes proceeds, as almost always with the cicadas, at the posterior pole, where, as with the psyllids, a ring of follicle cells, approximately 3 cells thick, for which Šulc introduced the term "cuneate cells," swells up to a greater extent even before the appearance of the first fungi. Unchanged in form, the yeasts are transmitted to it alone and later reach the space originating behind it through retreat of the oocytes; in the mature egg they are situated beneath and close to the surface in the form of a small lenticular aggregation (Fig. 189).

The second case concerns *Pyrgauchenia breddini* Schmidt, the only representative of the terentiines in the membracid family that it was possible to study. Here the yeasts sites are paired lobed organs without epithelial envelope, which fill out the largest part of the abdomen and are constructed from numerous large syncytia (Fig. 190). The fatty tissue, to which in this case there are no

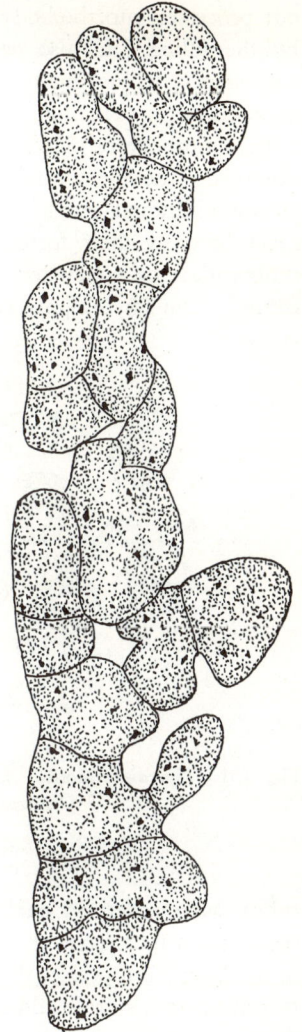

Fig. 190 *Pyrgauchenia breddini* Schmidt. One of the two mycetomes colonized by yeasts. From Buchner.

closer relationships, is limited to sparse cords. The approximately cigar-shaped yeasts are again taken up in two rows of cuneate cells which send out processes into the body cavity. These are probably not only caused by the filling, but also perhaps serve to capture the symbionts. The aggregation which finally is again found at the posterior end of the egg is much larger this time.

In a number of cicadoids the aforementioned *a*-symbiont, which we observed as the sole symbiont in the peloridiid *Hemiodoecus,* joins these yeasts as an additional guest. These microorganisms can always be clearly recognized by their form, the structure of their sites, and the strange differentiations connected with the production of specific transmission forms. The *a*-symbionts have the form of oval, sausage, or elongated

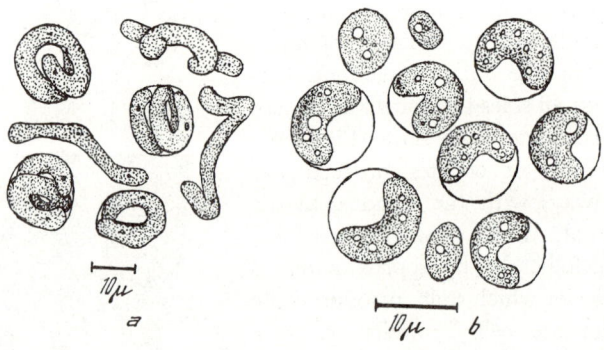

Fig. 191   The *a*-symbionts.   (*a*) *Philaenus spumarius* L.   From Resühr.   (*b*) *Cixius nervosus* L.   From H. J. Müller.

tubes, surrounded in each case by a membrane, not always easy to recognize, which forces the sausage-shaped states into *u*-shaped curves and in so doing often presses the ends close together and thus causes the tubes to roll up into spirals and loops of many types. Within this membrane there is a division into two that can be followed also in live species and a cleavage into a larger number of partial pieces (Fig. 191). Rough handling of the mycetomes and use of non-isotonic media often caused deformations of various kinds in these symbionts and once misled Šulc into an incorrect description of their shape. Resühr and H. J. Müller in particular studied the physical chemistry of the very sensitive protoplasm of these *a*-symbionts and the finer details of their morphology and reproduction; and they corrected earlier errors. Šulc had interpreted them as saccharomycetes, but Resühr and Müller agree that they are markedly degenerate bacteria, rich in fluid, as are many similar cicada symbionts.

In a number of reports Mahdihassan expressed the opinion that they were not microorganisms at all but remains of disintegrating host cells. However, it is hardly necessary to enter into a serious dispute in view of the overwhelming factual evidence against such an opinion.

a-Symbionts occur as the sole companions of yeasts in a number of jassids, membracids, euscelids, and cicadids as well as in the only aethalionid tested. They are always housed in well-defined paired mycetomes, provided with tracheae, which generally consist of numerous large syncytia but often of uninucleate giant cells and are surrounded by a well-developed epithelium which sends out narrower processes among the syncytia. The form of these organs varies a great deal in detail. Often they are radically elongated and are more or less deeply constricted chiefly by the dorsoventral muscular fascicles and stronger tracheal trunks; for example, in *Selenocephalus griseus* F. (euscelids), with *Oncopsis* (*Bythoscopus*) *scutellaris* Fieb. and *O. lanio* L. (jassids); and with some membracids, for example, with several *Heteronotus* species (Fig. 192*a*). Sometimes the sectional parts are held together merely by a narrow bridge, as in a number of *Hoplophora* species (membracids), where a chain of five segmentally arranged partial mycetomes are embedded bilaterally in the fatty tissue. Occasionally the sections are held together only by the common epithelium. In other species the constrictions are slight and unilateral, as in the smaller organs consisting of uninucleate mycetocytes in *Grypotes puncticollis* H. S. and *Opsius heydeni* Kbm. (euscelids) (Fig. 192*b,c*), or the similarly built *a*-organs of *Ulopa reticulata* F. (aethalionids).

The *a*-organs of the cicadids, which are found in conjunction with yeasts, are very different in their structure. Like all other *a*-organs, they are traceable to originally homogeneous paired organs, yet in this case the cleavage beginning during postembryonic development does not lead to segments placed one behind the other but to racemose arrangements and finally to a group of completely independent spherical partial mycetomes. With some of the gaeanines (*Mogannia*, *Huechys*, and *Scieroptera* species) the constructions are so extreme that the central connection can be recognized only by a detailed study of a series of sections; in others— *Tettigia orni* L., *Tibicen haematodes* Scop., *Cicadetta montana* Scop.—complete cleavage has started, and only a strong tracheal branch holds the parts together like a cluster of grapes.

Although all the gaeanines tested had *a*-organs and yeasts, only some of the other large singing cicadas have them. Among the cicadines the combination was found only with *Dundubia mannifera* L. and *Rihana ochrea* Walk. Histologically these conspicuous organs are also somewhat exceptional. The interior of the spherical forms is occupied by relatively few giant syncytia with numerous roundish nuclei which apparently, as with

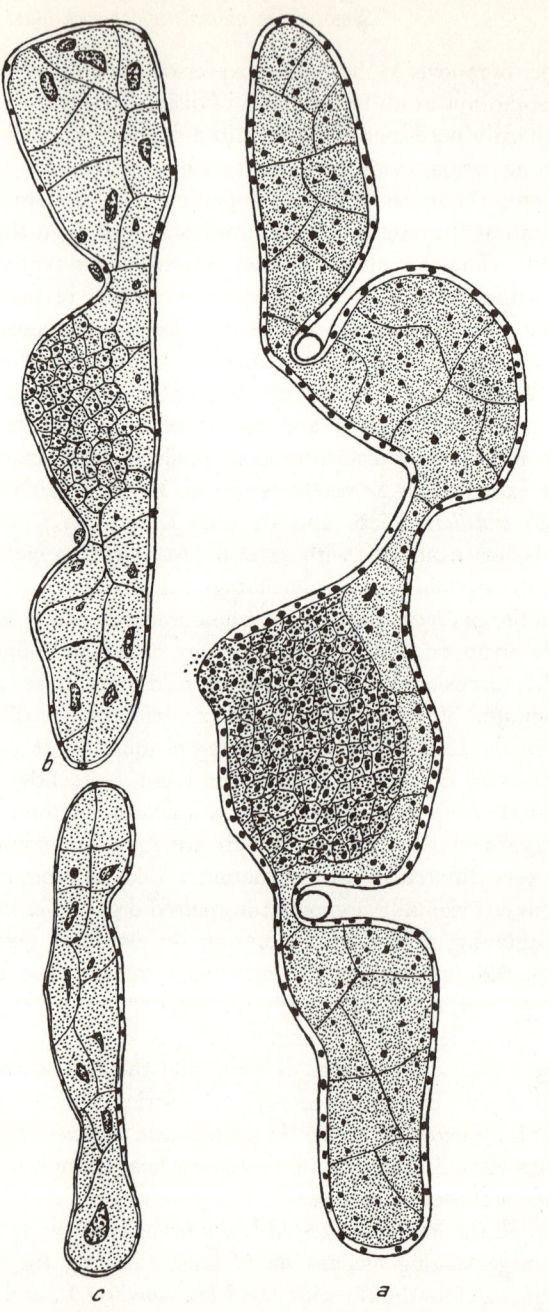

Fig. 192 (a) *Selenocephalus griseus* F., (b) *Opsius heydeni* Kbm. Female mycetomes, occupied by a-symbionts, with infection mounds. (c) *Opsius heydeni* Kbm. Male mycetome without infection mounds. From Buchner.

many other cicadas, originate through secondary fusion of several smaller syncytia and therefore more strictly represent "synsyncytia." Their plasma is penetrated throughout by tubes which are sometimes very long and frequently interlaced. The exterior of the thick mycetome epithelium has a remarkable border of rodlets not observed elsewhere, and numerous light-refracting inclusions in the form of ringlets, rodlets, and threadlets, such as are occasionally seen in the epithelium of other cicadoid mycetomes. Large tracheal branches do not merely penetrate to the epithelium but, in the manner usual with such *a*-organs, send finer branches deep into the interior of the organs.

Localization of the yeasts in these $H + a$ combinations may be quite varied, ranging from diffuse inundation of fat cells and lymph to well-defined organ-like sites. The first is the case with *Penthimia nigra* Goeze (jassids) and *Notocera brachycera* Irm. (membracids); also with *Selenocephalus griseus* the multinucleate cells contain at the same time fat in which the yeasts occur. *Oncopsis scutellaris* and *lanio* (jassids) as well as *Micrutalis ephippium* Burm. (membracids) behave like *Ledra* in that completely fat-free polynuclear cells are inserted here and there in the fatty tissue. With *Macropsis scutellata* Fieb. the syncytia are already developed into extensive complexes no longer encased by fat cells.

With a species of *Aconophora* (membracids) a very thin organ-like layer of yeast-populated syncytia surrounded by an epithelium of fat cells is situated between the *a*-organs and the hypodermis, the *a*-organs being accompanied throughout their expanse and clasped at the margins by this layer; in *Ulopa reticulata* a very large yeast organ takes up almost the whole space between the two *a*-organs. It is constructed of giant polygonal cells which usually have several small nuclear fragments but sometimes only one large bizarre nucleus. In this case relationships to the fat bodies still exist in spite of such extensive independence, for typical fat cells furnish an epithelium-like casing, although one that is frequently interrupted. An isolated yeast cell may sometimes be found in these fat cells, but never outside the organ.

A number of other membracids have evolved a completely different and interesting method. With *Lycoderes, Omolon,* and *Sphongophorus* species the yeasts are admitted into the epithelial envelope of the *a*-organs, which in this case are broken down into four partial mycetomes. At the same time the limits of the epithelial cells disappear, and a voluminous multinucleate zone arises which, like the bridges extending between the syncytia in typical *a*-organs, continues into the interior and pushes aside the syncytia or deforms them into irregular stellate complexes (Fig. 193). These "epithelial organs," which we shall see again in other combinations, actually go back to the original unilaminate epithelium, for they are enclosed solely by a delicate membrane without nuclei.   The yeasts are ordinarily confined to the epithelial organ, but with *Omolon laporti* Grm. they occur occasionally also in the fat cells and are especially numerous in the body cavity; in such cases they frequently grow out in long tubes.

Finally we should mention a unique combination of *a*-organs and yeasts which I observed in *Cicadula* (*Thamnotettix*) *4-notata* F. (euscelines) and Rau in a species of *Cymbomorpha* (membracids).   Both forms have the typical *a*-organs, but the yeasts are found in the epithelium of the midgut.

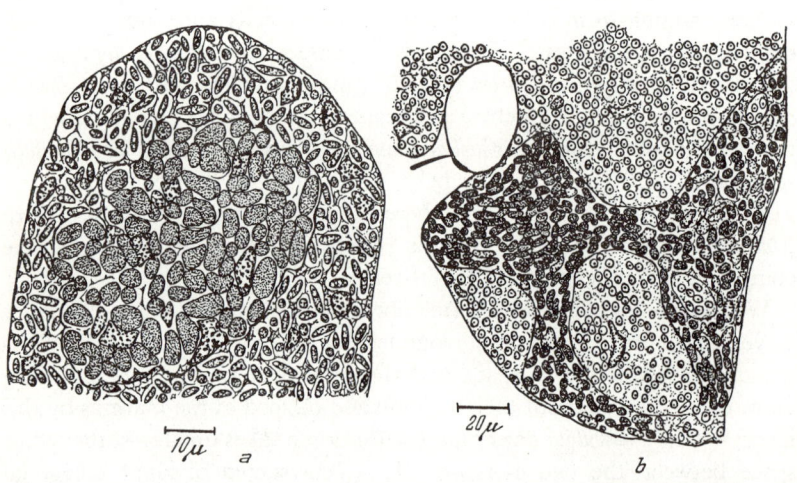

Fig. 193   (a) *Lycoderes galeritus* Less.   (b) *Omolon laporti* Grm.   The epithelium of the *a*-mycetomes is taken over by yeasts; in (b) the syncytia of the *a*-symbionts is reduced to irregular complexes; nevertheless, the infection mound reaches the surface.   From Rau.

With *Cymbomorpha* the latter is filled tautly in its whole expanse and is distended and swollen to an extreme.   The cell limits have disappeared; the nuclei are amitotically increased and become angular; and the completely altered intestine takes up a substantial part of the abdomen. In addition, isolated yeasts are found in the fat cells, in the body cavity, and occasionally even in the cells of the Malpighian tubules.   In the females, yeasts rolled up into clumps are found free in the hindgut.   In all five animals investigated the infection was the same (Fig. 194*b*).   With *Cicadula* the settlement of the intestinal epithelium is much less dense, and consequently such syncytial deformation does not occur (Fig. 194*a*). No yeasts were discovered outside the epithelium, and, although intestinal infection was found in all thirteen females tested, it was not observed in the majority of the males.

With respect to transmission, the yeasts occurring in connection with *a*-organs behave like those that appear without them, except that naturally they are found on and in the cuneate cells with the *a*-symbionts and are

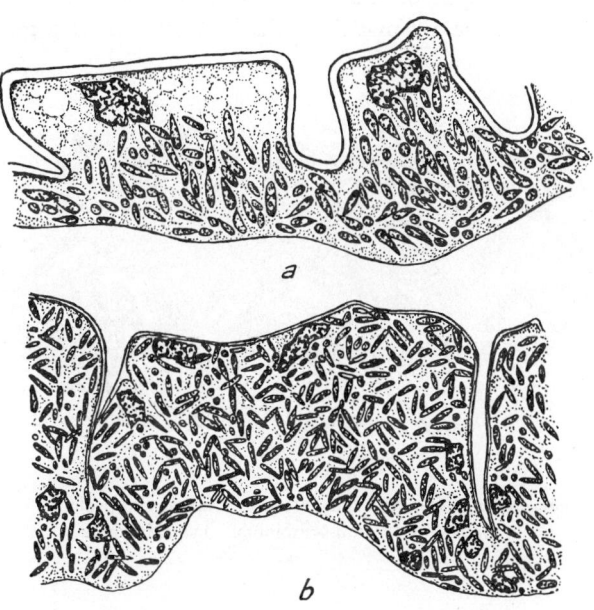

Fig. 194   (*a*) *Cicadula 4-notata* F.   From Buchner.   (*b*) *Cymbomorpha* sp.   From Rau. The midgut epithelium is occupied by yeasts and thereby, in *b*, is transformed into a syncytium.

intermingled with the latter in the egg (Fig. 195). They never form specific states, and, where they are bound in an organ-like fashion, they apparently can burst out to infect the egg from any point. The yeasts settling the intestinal epithelium in *Cicadula* and *Cymbomorpha* are lacking in the egg, which contains only *a*-symbionts, but in view of their regular occurrence, especially in *Cymbomorpha*, some regulated type of transmission seems probable. Perhaps the yeasts travel from the intestinal lumen to the surface of the egg and, like the yeasts of cerambycids and anobiids or the bacteria of the Heteroptera, are taken in by the larvae after hatching.

The situation is far more complicated with the *a*-organs.   Here it is thoroughly characteristic that special infection states, cultivated in zones which are of a special histological structure, are used exclusively for transmission to the progeny.   These are the "infection mounds," which I was the first to describe in a series of forms and which have already been alluded to in our description of the peloridiid *Hemiodoecus*.

On comparison of male and female *a*-organs we note not only that the male organs tend in general to be smaller but also that at one or several fixed places the syncytia or giant cells are replaced in females by smaller cell material of one or two nuclei (cf. Fig. 192*b,c*).   Among the cicadoids

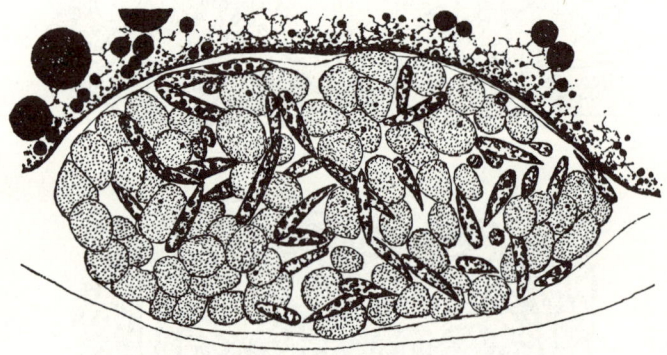

Fig. 195   *Selenocephalus griseus* F.   The infection mass at the posterior end of an egg with yeasts and *a*-symbionts.   From Buchner.

these altered areas first form more or less pronounced protrusions and sometimes very conspicuous, sharply set off appendages; among the fulgoroids, as we shall show later, they are usually more submerged and cause no change in the exterior form of the mycetomes.   Always, however, this particular zone reaches the surface of the organs and here at the peak of development pushes apart for a distance the flattening epithelial cells which remain sterile (Figs. 192*a,b*; 193*b*; 197; 199*a*; 201).

By studying the genesis of these mounds, which present a contrast to the usually vividly pigmented organs because their own coloring is rather weak and fades more and more toward the top, we observe that they are traceable to a nest of sterile epithelial cells, which is formed in the larval mycetome by active cell division at the point where the mycetomes are temporarily slightly fused with the oviduct.   When such contact exists in more than one place, as it frequently does with *a*-organs in fulgoroids, several of these cell nests are induced. However, vacuoles tend to emerge from the base to the surface and to all appearances contain an active ingredient which causes changes in the symbionts whereby they become stages capable of infection. At first only a few of these at the basal zone are transmitted to the specialized epithelial cells, increasing there, becoming more and more compact, and containing a denser and therefore more deeply staining protoplasm. Gradually the whole mound fills from the base upward with stages capable of infecting. Increase of the symbionts leads to formation of cyst-like nests in which the cells are distended more and more and their nuclei are pressed against the wall. Usually they are then fused into a syncytium in the apical area, and the infection stages are transmitted to the body cavity at the top when the ovaries attain the necessary maturity (Fig. 196).

Obviously the active ingredients arising here are diffused into the neighboring areas of the mycetomes, for at the base the specific cell nest is not sharply set off against the surroundings but shows that an influence, lessening with increasing distance, is exerted on the environment. When organs with uninucleate mycetocytes are involved, the vicinity of the cell nest has a growth-retarding effect, as is illustrated so well by the mycetome in *Penthimia* (Fig. 197). When syncytia are involved, they remain smaller in the area of the infection mounds because here the usual fusions are apparently thwarted. Such actions operating at a distance are manifested in changes which sometimes take place in the adjacent symbionts even before transmission to the epithelial cells. For instance, we have a description of an *Aconophora* (membracid) telling how the *a*-tubes become shorter, more compact, and more deeply staining in adjacent syncytia and how, the nearer they are situated to the cells to which they finally are transmitted, the more strongly they are affected by this transformation.

Where the usually sterile epithelium of the *a*-organs is taken over by yeasts and the *a*-symbionts are consequently forced away from the surface, as in *Lycoderes* and *Sphongophorus*, these *a*-symbionts must find a way toward the outside under difficult circumstances. The areas corresponding to infection mounds push through the cortical areas occupied by the yeasts, and thus at least the tips reach the surface of the organ. In *Omolon*, which has a sharply cleft *a*-organ, the cell nest producing the infection states borders the top surface along a broad front but is still connected with the other areas populated by the *a*-symbionts by means of a pedicle-like process (Fig. 193*b*).

The spherical partial mycetomes of the cicadids form infection mounds of their own, yet they are all situated in such a way that they apparently are traceable to an originally homogeneous cell nest not subdivided until cleavage takes place.

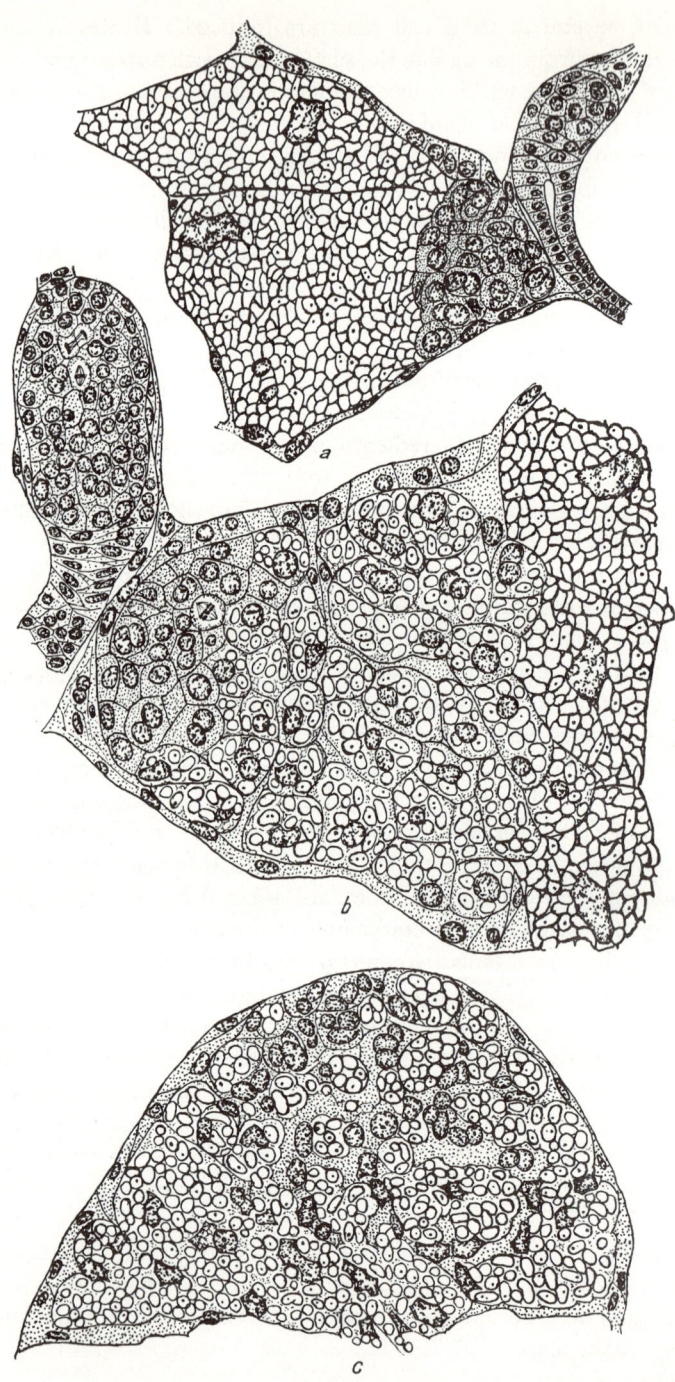

a

b

c

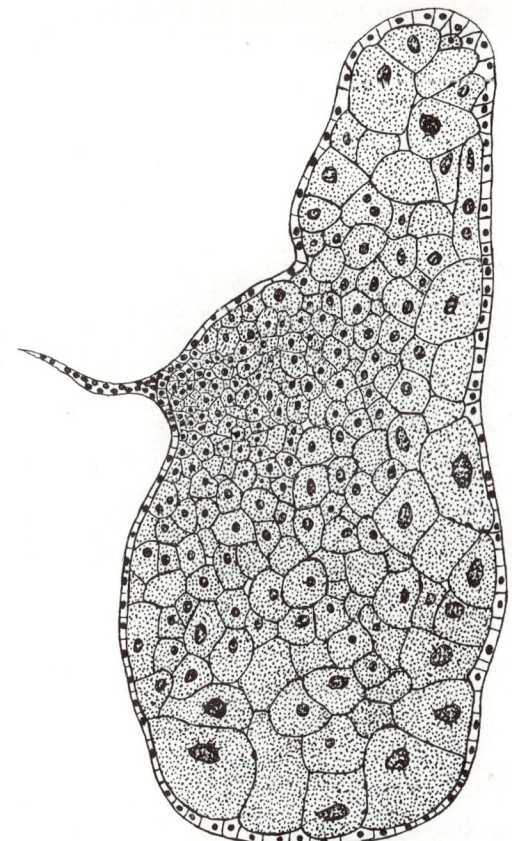

Fig. 197   *Penthimia nigra* Goeze.   Mycetome of a female with infection mound; the growth-inhibiting effect dies off with increasing distance.   From Buchner.

The *t*-symbionts constitute another well-characterized form of auxiliary symbionts widespread among the cicadoids but absent completely in the fulgoroids.   In the disymbiotic combination $a + t$, which is the only one that we shall treat here, they occur much less frequently than in the more complicated tri- and polysymbiotic combinations.   Jassids, euscelids, and membracids furnish examples.

Fig. 196   *Aphrophora salicis* DeG.   Development of the infection mound.   (*a*) Sterile nest of cells at the point of fusion of the mycetome with the oviduct; (*b*) progressive infection of the mound; (*c*) peak of the infection with mature infection forms.   From Buchner.

The organisms designated *t*-symbionts are relatively large formations, which are housed solely in uninucleate cells and are so crowded in these cells that it is often difficult to recognize their shape. Some are roundish to oval, others are lobed. Sometimes in place of roundish lobes there are more slender processes extending toward all sides and resulting in the

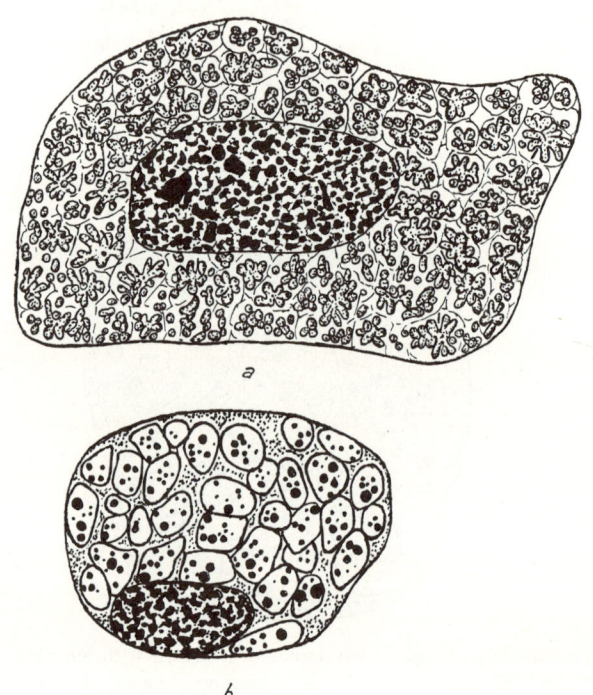

Fig. 198   *Euacanthus interruptus* L.   (*a*) Mycetocyte with *t*-symbionts; (*b*) the same as a migratory cell with infection forms.   From Buchner.

formation of delicate rosettes (Fig. 198*a*). As with the *a*-symbionts, a membrane surrounding each individual organism can be observed in such forms. The plasma of these *t*-symbionts is usually very rich in eosinophilic granulations. In spite of the overloading, the mycetocytes populated by them quite generally retain their capacity for mitotic division. The equatorial plates appearing in this connection often display unusually high chromosome numbers and thus remind one of the polyploidy found with *Pseudococcus* and *Aleurodes*.

This symbiont type never develops mycetomes for itself alone, but always in more or less intimate association with *a*-symbionts. Three progressive stages of amalgamation can be differentiated. *Euacanthus interruptus* L. (jassids) represents the most primitive. On both sides of an elongated *a*-organ a somewhat shorter, but equally slender, *t*-organ nestles toward the outside and is enclasped by it at the margins as by a hollow. The *a*-organ is surrounded on all sides by the customary well-developed epithelium; the *t*-organ has only a very delicate casing membrane with very few minute nuclei. The fact that the dorsoventral muscles pass between the two sections shows how loose their alliance is (Fig. 199*a*).

The connection of the two parts is much more intimate in a large number of other jassids and in euscelids. The *t*-section is firmly inserted into the *a*-section spanning it on three sides in all species investigated thus far of *Euscelis* (*Athysanus*), *Deltocephalus*, *Macrosteles*, *Paramesus*, *Eupelix*, *Strongylocephalus*, and *Thamnotettix*. To the extent that the two sections are in contact, the thick epithelium originally surrounding the *a*-organ on all sides is lacking and, where the *t*-symbionts reach the surface, the epithelium passes over into its delicate boundary (Fig. 199*b*).

Such incomplete spanning of the *t*-symbionts is also occasionally found in membracids, for example, *Cyphonia trifida* I. Far more often we find the most perfect stage of the double mycetome in which the *a*-zone encases the *t*-zone on all sides. *Gargara*, *Cyphonia*, *Ceresa* are among the genera which furnish examples of this (Fig. 206*a*).

We have already seen that the formation of transmission states in special epithelial nests of their own was typical for the *a*-symbionts. Now we see that the *t*-symbionts also form states serving the egg infection in a very special, though different, manner. In the central areas of the female mycetomes, and only there, when the egg cells mature it may be observed that an occasional mycetocyte, polygonal up until that point, is rounded off and that its entire supply of symbionts experiences certain changes. These symbionts then are more clearly set off from one another and, if they were lobed, they now gradually assume a roundish shape. The eosinophilic granulations at the same time occasionally flow together to form larger drops (*Euacanthus*). In this state the usually elongating mycetocytes migrate between the unchanged mycetocytes toward the surface of the mycetome and finally leave it (Fig. 198*b*). Places preferred for exit are the sections not surrounded by the *a*-cortex and, where the cortex goes all the way around, the constricted area on the inner side of the organs. Gaps arising through migration are rapidly closed, and cell divisions provide the necessary replacement. Outside the mycetome the mycetocyte disintegrates and discharges into the bloodstream the infection-ripe states which are frequently deformed during migration and only now

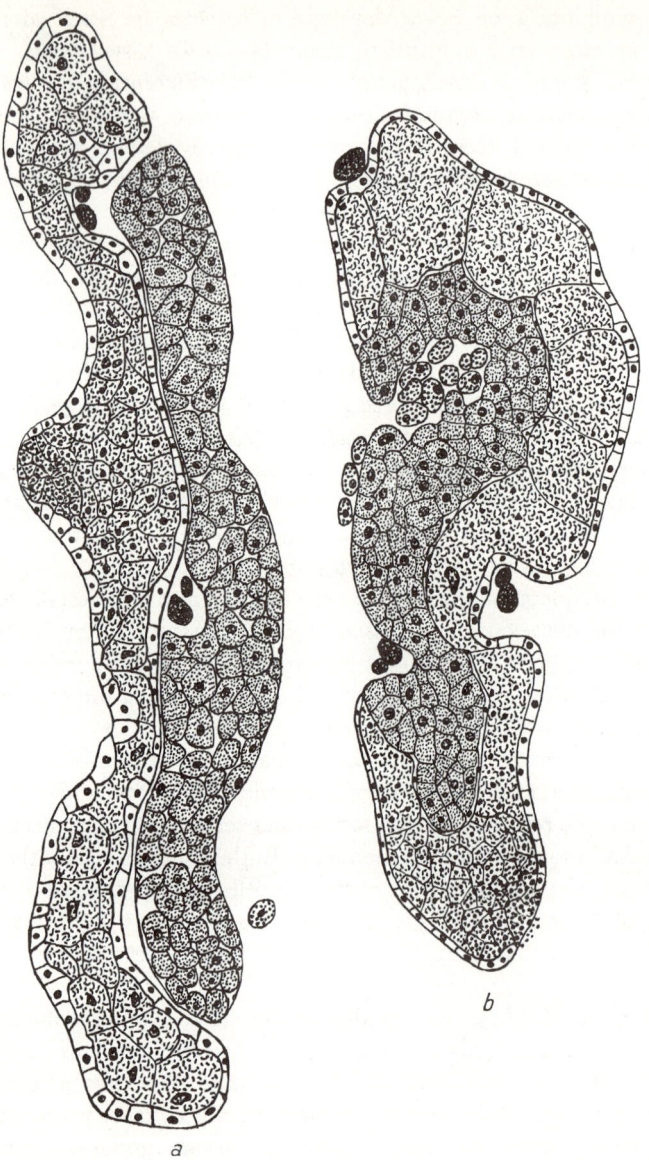

a

b

Fig. 199   (*a*) *Euacanthus interruptus* L.; mycetome of a female with loosely joined *a*- and *t*-sections, and the infection mound of the former.   (*b*) *Paramesus nervosus* Fall.; *a*- and *t*-sections are intimately united, the infection mound is located terminally.   From Buchner.

can be rounded off again (Fig. 205*b*). The two symbiont types are united in the usual manner in cuneate cells, which sometimes form large proliferations, and are finally transmitted to the egg (Fig. 200).

In addition to *a*-, *H*-, and *t*-symbionts there are several types of companion symbionts and numerous accessory symbionts to complicate the picture among the cicadoids. Very characteristic of the cercopids are the *b*-symbionts, which are found in no other cicadoids and among the fulgoroids are limited to some of the cixiines. They are related to the *a*-symbionts with which they are always allied, even among the fulgoroids, but do not form such long tubes. Instead, their shape is that of roundish,

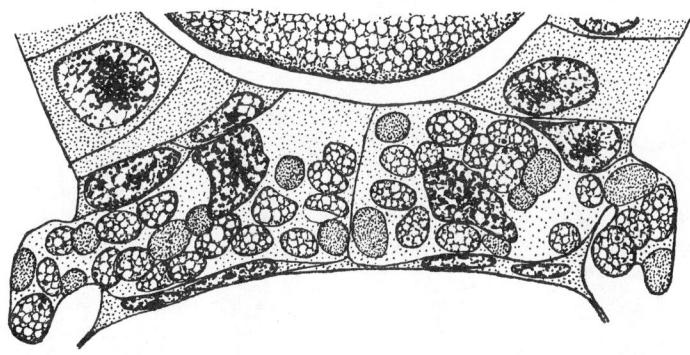

Fig. 200   *Paramesus nervosus* Fall.   Infection of the cuneate cells by *a*- and *t*-symbionts. From Buchner.

oval, or shorter, uniformly thick, *u*-shaped sausages, surrounded by a membrane and cut off behind it into two or more fragments. Like the *a*-symbionts, they are found in uninucleate giant cells or in syncytia. The well-defined mycetomes in which they live, in contrast to the *a*-organs, are always surrounded only by a flat cellular envelope and are always situated in the vicinity of the *a*-organs. Sometimes one or two roundish or oval organs loosely attached to the exterior of the *a*-mycetomes or inserted in niches (many *Neophilaenus* species and other species); at other times a regular double mycetome, resembling that of *Euacanthus*, is created; then the *a*-section clasps the *b*-section closely but leaves its exterior side free and takes no part in the epithelial enclosure (*Cercopis sanguinolenta* L., *Lepyronia coleopterata* L., Fig. 201*a,b,e*).

Infection mounds are never formed for transmission purposes in the *b*-organs. Instead, isolated symbionts, slightly changed in shape and

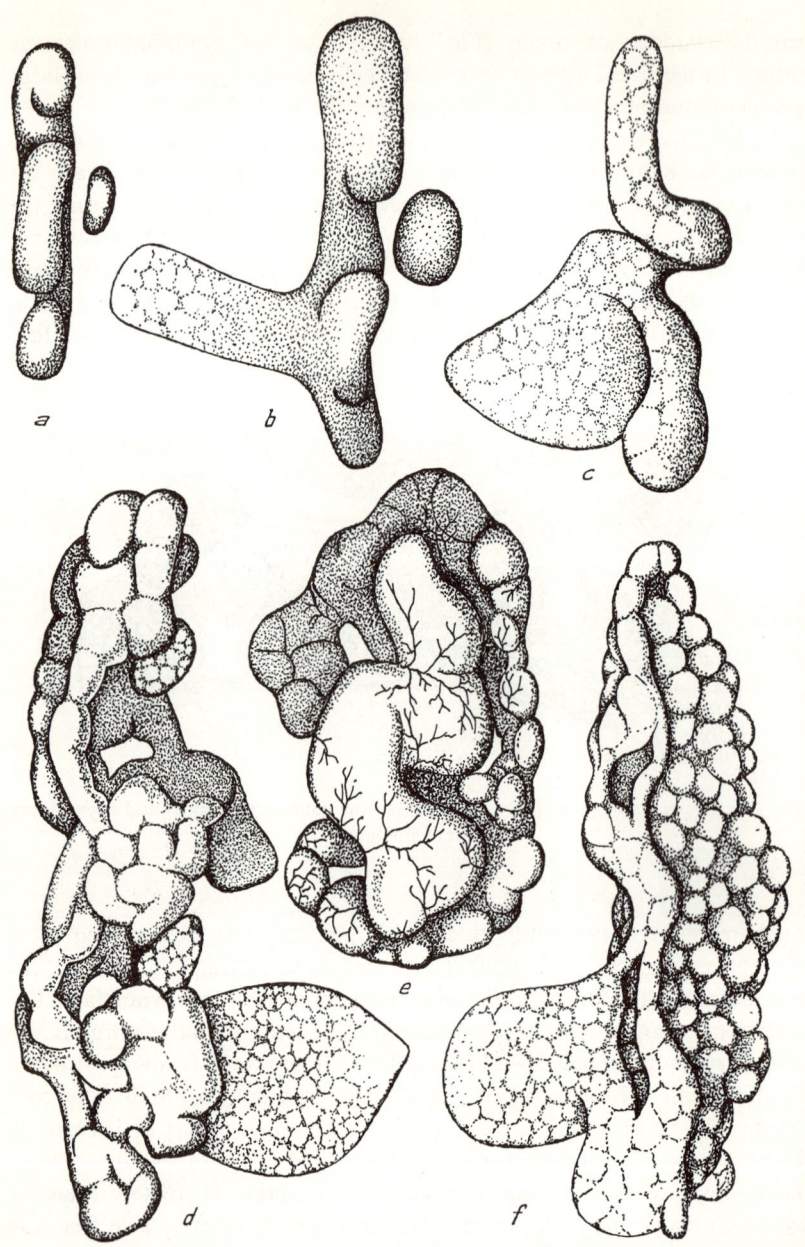

Fig. 201 Various Cercopidae mycetomes (fresh preparations). (a) *Neophilaenus lineatus* L. ♂; (b) *N. lineatus* L. ♀; (c) *Philaenus spumarius* L. ♀; (d) *Aphrophora salicis* DeG. ♀; (e) *Cercopsis sanguinolenta* L.; (f) *Aphrophora alni* L. ♀. (a to d, f) Side view; (e) surface view; (b–d, f) with infection mound. From Buchner.

often distinguishable from transmission forms of the *a*-symbionts only on very close examination, migrate from the mycetome to the body cavity and reach the oocytes in association with the *a*-symbionts.

Another isolated occurrence of a companion symbiont allied with *a*-symbionts is found in some of the cicadines. The combination of a yeast with *a*-symbionts typical of the gaeanines is apparently relatively rare with them, whereas the combination of $a + w$-symbionts is found repeatedly (*Cicada plebeja* Scop., *C. septendecim* L., *Rihana* sp., *Poicilopsaltria octopunctata* Fab.). In such cases too there are racemose combinations of spherical partial mycetomes, but these now consist of a single giant central syncytium in which numerous little nuclei are usually limited to radial plasma routes and in which the smaller *a*-tubes are housed, and a marginal layer also built up syncytially and containing the larger *w*-tubes which are richer in inclusions. These strange double mycetomes penetrated by strong tracheal branches are now closed off toward the outside by a well-developed epithelium.

In this case too specific transmission forms are provided in a manner so characteristic of the type of symbiosis. In the medullary zone, nests of roundish stages arise in one section of the partial organ corresponding to the infection mounds but not protruding toward the outside; the same process takes place in the bordering area of the cortical layer, and, since at the same time the separating membrane is dissolved, both types of transmission forms are intermingled and are transmitted together to the body cavity (illustrations in Buchner, 1912).

Before turning to the tri- and polysymbiotic cicadoids, we must mention several cercopids and a jassid in which, in addition to the ever-present *a*-symbionts, filamentous, rod-shaped, or spherical companion symbionts take the place of the *b*- or *w*-symbionts. *Aphrophora salicis* DeG. has very conspicuous mycetomes colored an intense red by the pigment filling the epithelium. These mycetomes are irregular, lobed formations with fenestrate interruptions through which the dorsoventral muscles pass. In addition there are two smaller mycetomes, diffusely tinged light yellow, inserted on the inner side of the organs in depressions (Fig. 201*d*). The chief mycetome containing the *a*-symbionts consists of giant cells which are usually binucleate; the smaller mycetomes are constructed from uninucleate cells smaller in size and give a less balanced impression. Their slender straight rodlets fill the individual cells in varying degrees of density. Smaller mycetocytes clasp other rounded mycetocytes or, quite flattened, close off the mycetome toward the outside. In addition this companionate symbiont type, which is not yet well adapted, occurs here and there in the fat cells or, in elongated form, both in the fat cells and in blood cells and oenocytes and occasionally even in the epithelium of the *a*-organs. Nevertheless, these rodlets take part in egg infection in a thoroughly regulated manner.

Among the cercopids, *Tomaspis tristis* F. is another exceptional case. Whereas *T. rubra* L. has *a* + *b*-mycetomes, *T. tristis* F. has very conspicuous *a*-organs forming the large infection mounds, but in place of the *b*-organs there is a loose irregular encasement by polynuclear cell groups, similar to fat cells, with symbionts that once more have the typical bacteria shape. *Philaenus spumarius* L. (*leucophthalmus* L.) has its own special behavior. In addition to the usual *a*-organ it has small, roundish, spherical organisms which are often encountered in the process of division and resemble aphid symbionts. They are housed in polynuclear giant cells surrounded by fat cells; avoiding the more distant regions, they group themselves around the mycetome (Fig. 201*c*).

In addition to *a*-symbionts the jassid *Cicadella* (*Tettigoniella*) *viridis* L. also has filamentous and rod-shaped bacteria, housing them, as does *Aphrophora salicis*, in two smaller organs loosely attached to the exterior of the chief mycetome. Here the two mycetome types are tinged light yellow. The large polygonal uni- or binucleate cells of the additional mycetomes, which do not have the irregularities of *Aphrophora salicis*, are thickly filled down to the last corner by long slender filaments braided like hair. The behavior of the filaments with respect to the infection mound of the *a*-mycetome is almost unique, for the mound is also infected by them. In a middle zone of the well-developed mound, which even in the living specimen has more intensive pigmentation, a number of large round cells, uninucleate to trinucleate, can be observed without difficulty. In these are the companion symbionts, now of course usually in the form of much shorter rodlets between which one of the larger filaments and several scattered *a*-symbionts are situated occasionally (Fig. 202*a*). Moreover, they also occur in the form of individual short rodlets in the areas bordering on the infection mound. Toward its tip they form little bundles, probably held together by a gelatinous substance, such as are never encountered in the actual mycetomes, and it is only these which leave the place of their formation together with the transmission forms of the *a*-symbionts and take part in egg infection. In this case therefore the formative stimuli localized in that cell nest influence the two types of symbionts at the same time, each in a different manner.[1] In the simple circle of cuneate cells which swell and are vacuolized even before the symbionts arrive, several

[1] Mahdihassan (1947) evolved completely different concepts of symbiosis in *Cicadella viridis*. He considers the *a*-symbionts to be remnants of disintegrated host cells and finds in the accessory mycetomes a larger and a smaller symbiont variety; and he declares that the filaments appearing punctiform in optical transverse section in my illustrations are the variety acquired later!

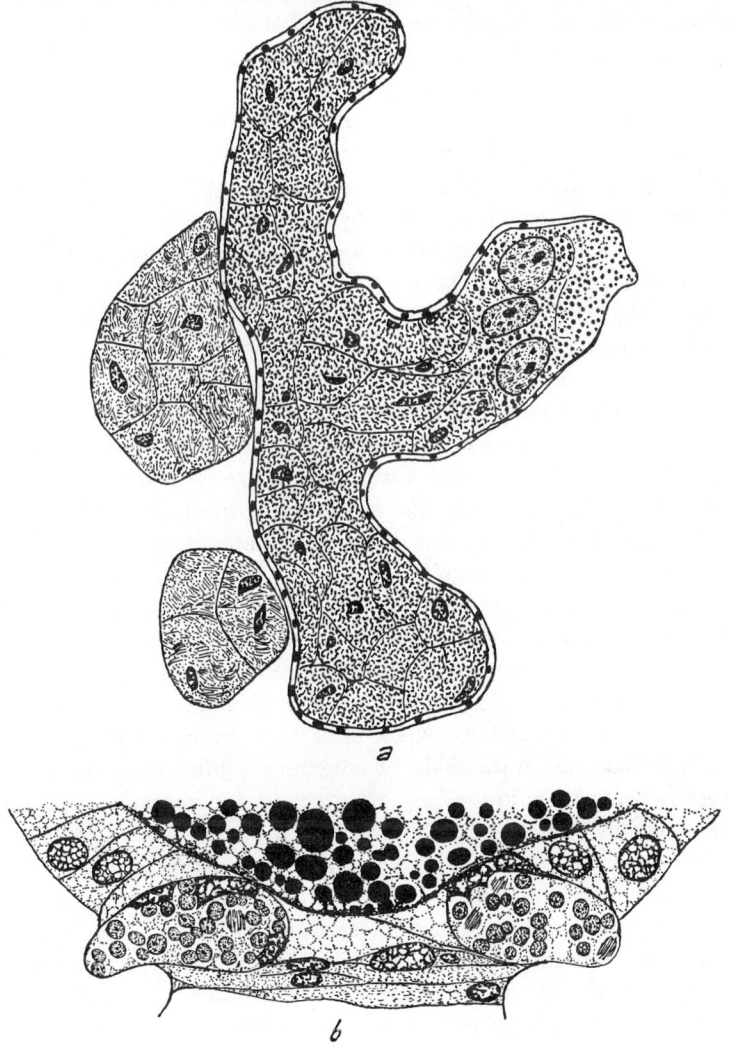

Fig. 202 *Cicadella viridis* L.   (*a*) The *a*-mycetome with two additional, bordering mycetomes in which live thread-like symbionts which have infected three large cells of the infection mound and have formed bundles here (partly schematic); (*b*) a wreath of cuneate cells was infected with both types of symbionts.   From Buchner.

a-forms appear first followed by the little bundles which were transmitted in smaller numbers. The nuclei, round at the beginning, take various shapes in connection with the increasing filling but resume their original character after their discharge (Fig. 209*b*).

By acquiring much less well-adapted companion symbionts, *a* + *b*-, *a* + *t*-, and *a* + *w*-combinations can become trisymbiotic. The first possibility is embodied in *Aphrophora alni* Fall. (*spumaria* L.). A lobed red *a*-mycetome here seizes incomplete hold of a large quantity of roundish rose-colored *b*-syncytia along three sides (Fig. 201*f*); moreover, several inconspicuous nests of uninucleate cells are situated between these two zones with plasma thickly filled by strong, slightly curved filaments. In the symbiont ball, gigantic in this case and occupying the whole posterior region of the egg, the three types are therefore intermingled. They occur in quantities proportionate to the size of their sites, and the *b*-symbionts are the most numerous and the filamentous form is the least numerous.

With the cicadine *Platypleura kämpferi* E. there is in addition to the *a* + *w*-symbionts a third guest, also a bacterium, this time in cells which, situated in the fatty tissue, join into a complex closely fitted to the racemose double mycetome. The combination *a* + *t* + filamentous or rodlet-shaped companions occurs often with jassids and euscelids; among the jassids it even attains a high degree of adaptation in *Idiocerus* and *Agallia*, whereas other cases give the impression of later, less well-regulated adaptation. With *Aphrodes* (*Acocephalus*) *bicinctus* Curt. and *A. trifasciatus* DeG. there are even rodlets which with their transmission forms live primarily as parasites in the midst of numerous *a*-tubes, occasionally also scattered between them in the form of little nests, but never in other zones of the mycetomes or in other tissues. Between the rosette-containing *t*-cells in *Paropia* (*Megophthalmus*) *scanica* Fall. are other random uninucleate elements of changing dimensions with thick masses of rod-shaped bacteria. In *Idiocerus stigmaticalis* Lewis and *I. cognatus* Fieb. and in *Agallia venosa* Fall. this third cell and symbiont type are inserted as another connected mycetome part between the cortical zone of the *a*-symbionts and the closed mass of *t*-cells (Fig. 203). All three parts are uniformly enclosed by an epithelium, but the latter as usual remains thin on the outer side which lacks a cortical zone. Occasionally a group of these smaller bacteria-containing cells may be observed wedged between the *t*-cells, or a nest of their occupants may be found in the neighboring mycetome epithelium, reminding us that adaptation is not yet completely stabilized.

In the founding of the previously described three-fold symbiosis an isolated step was taken here and there in the cicadoid system, but with the membracids, thoroughly tested by Rau with material representing ninety species and nine of the twelve subfamilies, we find a much more widespread inclination toward acquiring companion symbionts and additional, less well-adapted, accessory guests. Thus only in this family do types of symbiosis with more than three distinct forms occur among the cicadoids.

Once more it is the never-failing *a* + *t*-organs which represent the starting point. We shall consider first the instances where only a third symbiont is housed in more or less well-regulated form. In this category are the only membracids which are native to Germany, and consequently they were tested years ago by me; they are *Centrotus cornutus* L. and *Gargara genistae* F. In the first species uninucleate cells which contain numerous, deeply staining small tubes appear between the cortical and medullary zones and within the medullary zone. Here and there they are wedged between the *a*-syncytia at points where the tracheae enter and where later the cells laden with *t*-symbionts take their departure (Fig. 204). Although they were regularly found in material of heterogeneous origin, they cause clear-cut disturbances. The mycetocytes lose their sharp delimitation wherever they come in contact with the organisms which must be classified as accessory symbionts, and occasionally they show clear signs of degeneration. When these addition forms appear now and then in the interior of *t*-cells, they are presumably those states which are used in egg infection, as the migrating cells which are leaving always contain them. At first strongly staining, they very quickly lose this property and as pale formations can be detected only with great difficulty, but, since

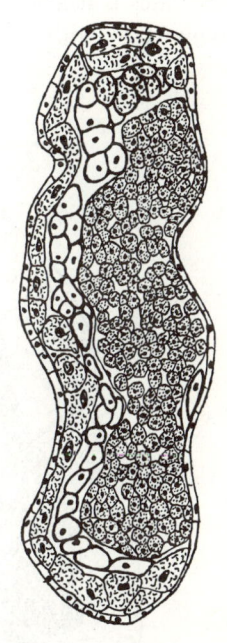

Fig. 203 *Idiocerus stigmaticalis* Lewis. Mycetome with three zones, each of which is inhabited by one type of symbiont. From Buchner.

they occur again in all clarity in the course of embryonic development, there can be no doubt about their regulated transmission.

With *Gargara genistae*, in place of the readily stainable small tubes, there are roundish accessory symbionts, sometimes in racemose formations. They enclose only a few nuclei, which are in the process of disintegration and therefore disappear entirely over rather long areas, and again cause severe disturbances in the neighboring mycetocytes. Many of these are disintegrated or are infected by the still ill-adapted guests, but here too the latter are taken along by the migratory cells and thus the hereditary union of the three mycetome inmates is protected.

We shall cite only a few of the many additional examples given by Rau to prove the presence of a third symbiont in membracids. Only in very rare cases is it a yeast cell, and indeed yeasts do not generally occur in conjunction with *a*- and *t*-symbionts. In a smiliine they were found

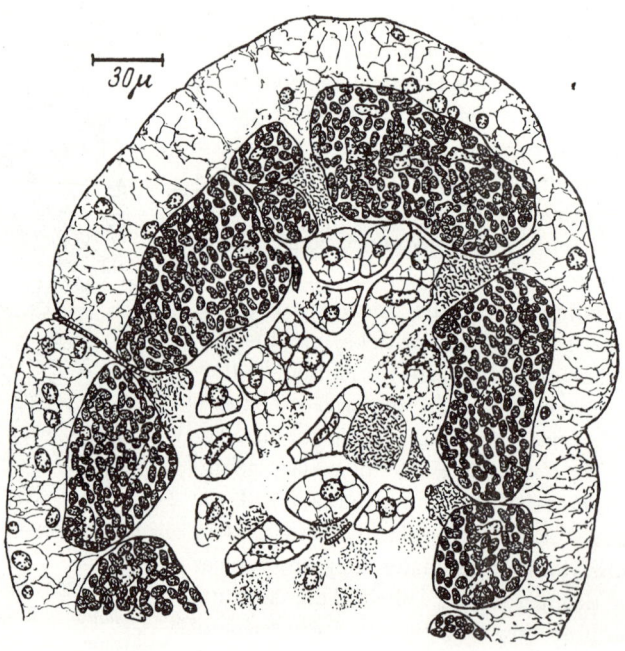

Fig. 204   *Centrotus cornutus* L.   Section from the mycetome with three kinds of symbionts: *a*-symbionts (tubular), *t*-symbionts, and accessory bacteria.   From Rau.

in the fatty tissue, free in the abdomen, and in the testes; in two polyglyptines they were taken up into the mycetomes and, at least in the females, formed more or less closed nuclei-containing complexes in the area of the *t*-cells or between them and the *a*-syncytia. Numerous hollow spaces and fissures in the *t*-organ indicated deep disturbances. In part, the migratory cells remain free of them and in such cases contain sharply described and healthy infection states; or they may acquire numerous yeasts, in which case the *t*-symbionts can no longer be clearly recognized; that is, they are apparently severely damaged. In the symbiont ball of the egg all three forms are again represented.

The third symbiont type in the *a*-organ of *Hypsoprora coronata* J. takes the form of slender filaments. Where these are not very numerous the older inmates suffer very little, but wherever they appear in great numbers the older inmates are indeed injured by their presence. The *a*-tubes then grow out in large round or irregular formations which soon disintegrate into several parts. Their behavior in relation to infection mounds in itself shows extensive adaptation. The threadlets distributed individually between the *a*-tubes until that time, now gather together more and more in small bundles—like those we saw in still more regular form in *Cicadella viridis*—the nearer they lie to the infection mound. As such, they are transmitted with the *a*-symbionts to the still sterile cells of the young infection mound and later again leave with them. In the cuneate cells they meet the other two types again, and reunited the three are situated in the symbiont ball of the egg. With more extensive material it was possible to determine that only the intensity of the population varies from animal to animal, chiefly in the males; in all other respects the behavior of the third guest is always the same.

A stricter localization is found in certain very long, slender, and pale tubelets, found by Rau in *Sundarion, Eualthe,* and *Hemikyphta* species and designated *v*-symbionts.[1] They avoid the cortex of the *a*-symbionts completely and settle primarily in the area between it and the mycetocytes of the *t*-symbionts, and to a lesser degree in the midst of the latter in uninucleate cells. They are found nowhere else in the cicada body. Of course, a certain lack of balance is revealed in the formation of large empty spaces in the *t*-organ and the sharp reduction of its cells, which are not otherwise injured. The infection mound is infiltrated by these additional guests; they may now be designated companion symbionts, and in all probability, together with the two primary forms, they infect the eggs in the form of small states not accessible to observation.

Also the *p*-symbionts of *Adippe alliacea* Grm. appear to be well adapted. This time very small but distinct cocciform organisms occupy exclusively nuclei-permeated spaces between the *t*-cells, in which the symbionts are scarcely altered even when several accessory symbionts occasionally push in between them (Fig. 205*a*). Possibly we have again a case of future migratory cells, for these now serve quite regularly for the transportation of the *p*-symbionts to the egg. After leaving the mycetome, the latter are first transmitted to a certain place in the body cavity and later the same thing happens with the *t*-symbionts, which are rounded off again in the process (Fig. 205*b*). During transmission to the cuneate cells and to the egg these third symbionts lag behind and therefore occupy the posterior cortical zones of the symbiont ball (Fig. 205*c*).

[1] Rau used Greek letters to characterize the accessory symbionts.

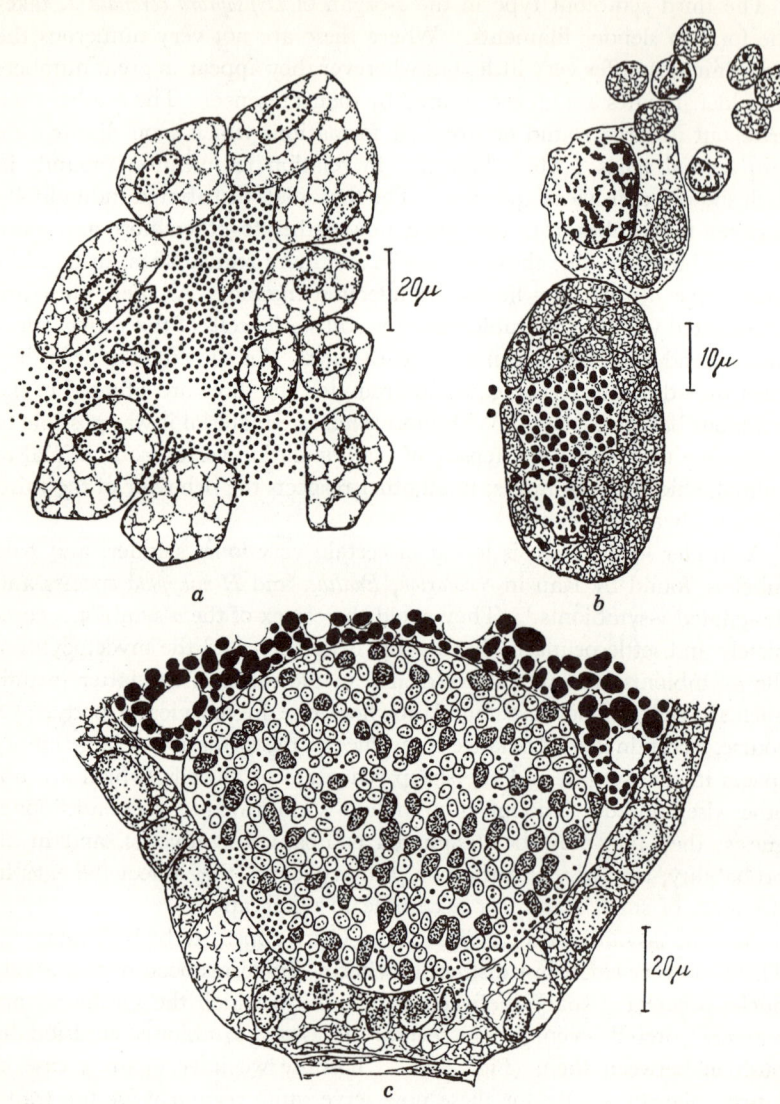

Fig. 205   *Adippe alliacea* Grm.   (*a*) Section from the *t*-organ with ρ-symbionts; (*b*) disintegrating migratory cells releasing the *t*- and ρ-symbionts; (*c*) infection of the egg with *a*-, *t*-, and ρ-symbionts.   From Rau.

The σ-symbionts of *Bolbonota* species are also limited to the *t*-section of the mycetomes, being transmitted to the migratory cells in great numbers and infecting the egg cells along with the remaining symbionts.

A tolaniine, which could not be further classified, separates its auxiliary symbionts, in the form of pale compact tubes, almost completely from the area assigned to the *t*-symbionts and houses them around the latter in an intermediate zone scantily provided with nuclei.   Only here and there is this order interrupted by a small nest inserted among the *t*-mycetocytes (Fig. 206*b*).   The infection mound is reserved for the *a*-symbionts, and the migratory cells again carry the third guest along to the ovary.

The membracids with four different symbionts house two additional symbiont types in the typical *a* + *t*-mycetome.   In each instance one can clearly determine, by the often unbridled behavior, which of the two is the symbiont acquired last.   Whereas the third symbionts may be the most varied kind, the fourth are almost always the so-called η-symbionts, that is, small, and extremely small, rod-shaped bacteria.   Only in a single polyglyptine are there large strong tubes forming a narrow border between the *a*- and the *t*-organs.   In general, this form represents an exception among membracids with four symbionts, since it is a yeast that appears in the mycetome as third guest, of the type that has already been observed with a number of polyglyptines (Figs. 206*a,e*; 207).

In *Bolbonota corrugata* Jrm. such yeasts are found free in the body cavity, more rarely in the fatty tissue, and only in isolated instances in the mycetome, where the fourth accessory symbiont is a minute η-form.   Enclosed in vacuoles of the *a*-syncytia, such symbionts cause the latter to become hypertrophied and in the area of the *t*-symbionts damage these symbionts and cells severely.   Moreover, they permeate the infection mound and from there apparently seek the ovarioles.

The mycetomes of a species of *Campylenchia* also contained *a*-, *t*-, and η-symbionts, the last being found in all the mycetome parts.   Indeed, the η-symbionts not only occur between the *a*- and the *t*-symbionts but also fill the otherwise sterile epithelial cells of the organs in thick masses and penetrate the migratory cells in quantity.   They are present also in wide areas of the fatty tissue and in the oinocytes (Fig. 208*a*).   Furthermore, the parasitic filamentous λ-symbionts, which are also encountered elsewhere, infect some of the *a*-symbionts earlier and at times bring about their disintegration.

All four forms find their way to the egg cells, the *t*-symbionts in the ordinary manner by means of the migratory cells; the *a*-symbionts, as always, indirectly by way of the infection mounds.   At the same time the λ-symbionts are passively brought along by some of the *a*-symbionts.   In the form of little bundles, that is, in a specific infection form, the λ-symbionts

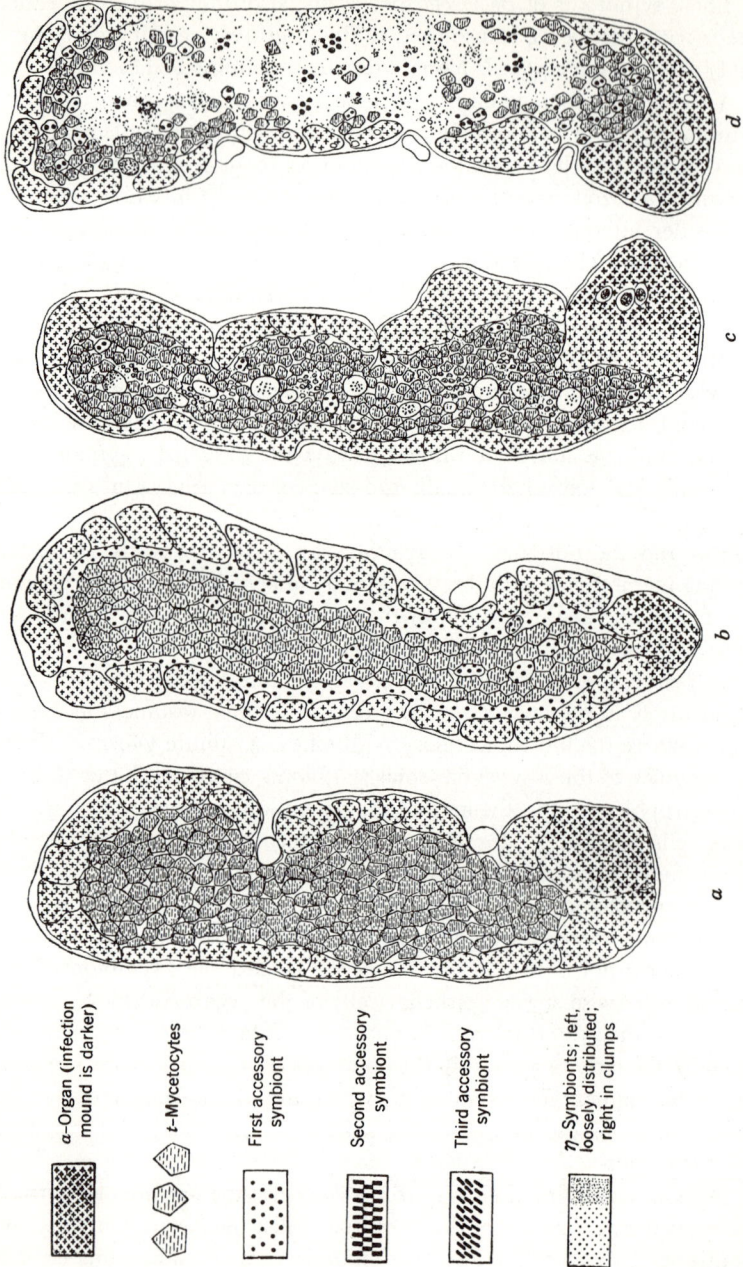

a–Organ (infection mound is darker)

t–Mycetocytes

First accessory symbiont

Second accessory symbiont

Third accessory symbiont

η–Symbionts; left, loosely distributed; right in clumps

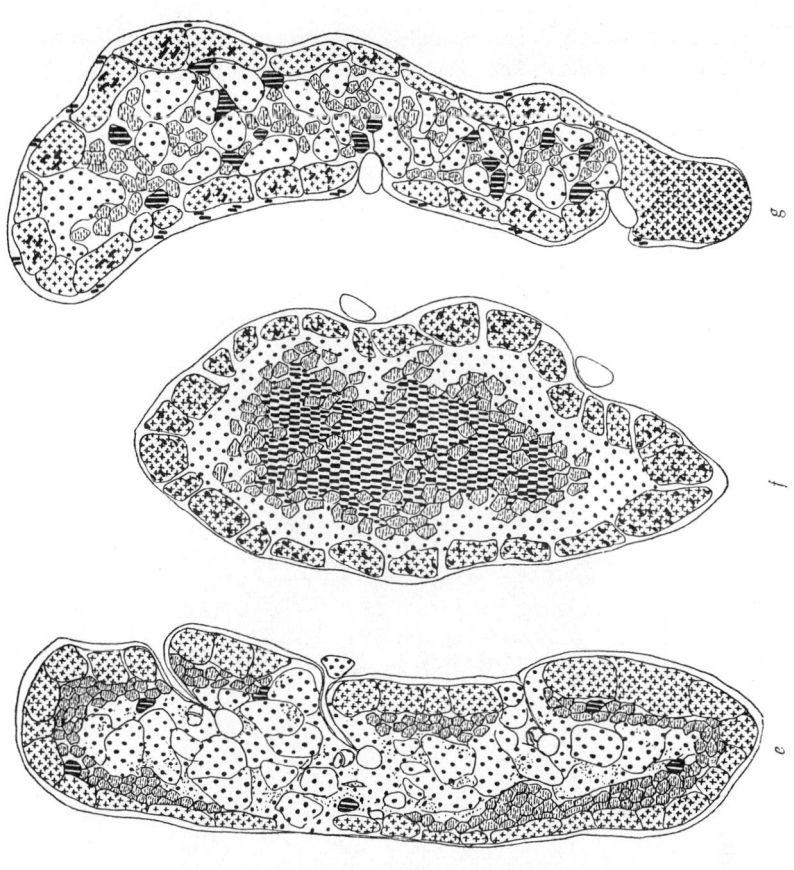

Fig. 206 Frontal sections through membracid-mycetomes with (*a–d*) two to four different symbionts and (*e–g*) five to six different symbionts. Only the designations for the symbionts are drawn schematically. (*a*) *Ceresa* sp., (*b*) Tolaniine, (*c, d, e*) Polyglyptines, (*f*) Tragopine, (*g*) *Enchophyllum 5-maculatum* Jrm., 150 to 300×. From Rau.

377

fill out almost entirely.   Intermingled, the three types are acquired by the
ring of cuneate cells.   The η-symbionts take a different route.   Even
before the cuneate cells discharge their contents into the oocytes, they form
a cap-shaped accumulation at the posterior end of the latter (Fig. 208*b*).
Apparently they even infect the youngest oocytes which are still in the
end chamber; this could be clearly observed in another case in which there
were very young symbionts also.

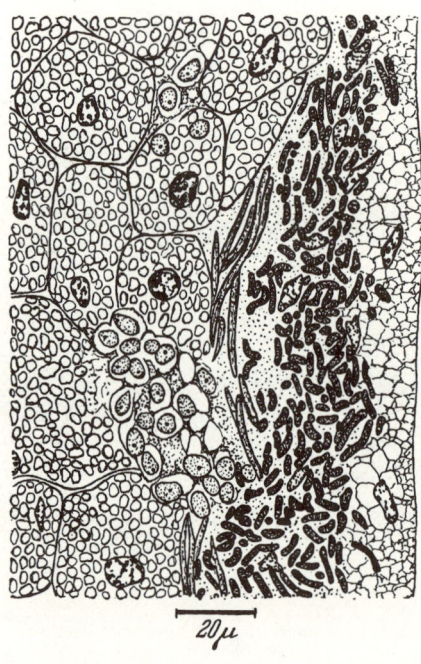

Fig. 207   Polyglyptine.   Section from the mycetome with *a*- and *t*-symbionts and two
additional guests.   From Rau.

Figure 206*c* shows the localization of four symbionts in the mycetome of
an undetermined polyglyptine.   This time the η-symbionts are already
limited exclusively to the *t*-area of the mycetomes, are situated in large
vacuoles, and cause damage to the neighboring mycetocytes, bringing
about their disintegration and even penetrating in quantity the interior of
the *t*-symbionts, which are then in the process of dissolution.   In the
mycetome there is an additional type of symbiont in very small quantity
between the *t*-mycetocytes.   This is a tube form similar to the one occurring in *Centrotus* as a companion symbiont.   Some of the η-symbionts are

carried into the egg as intracellular parasites of the *t*-symbionts, and some apparently are transmitted to the young oocytes, for they are also found at the posterior end of the egg between yolk masses, as in *Campylenchia*.

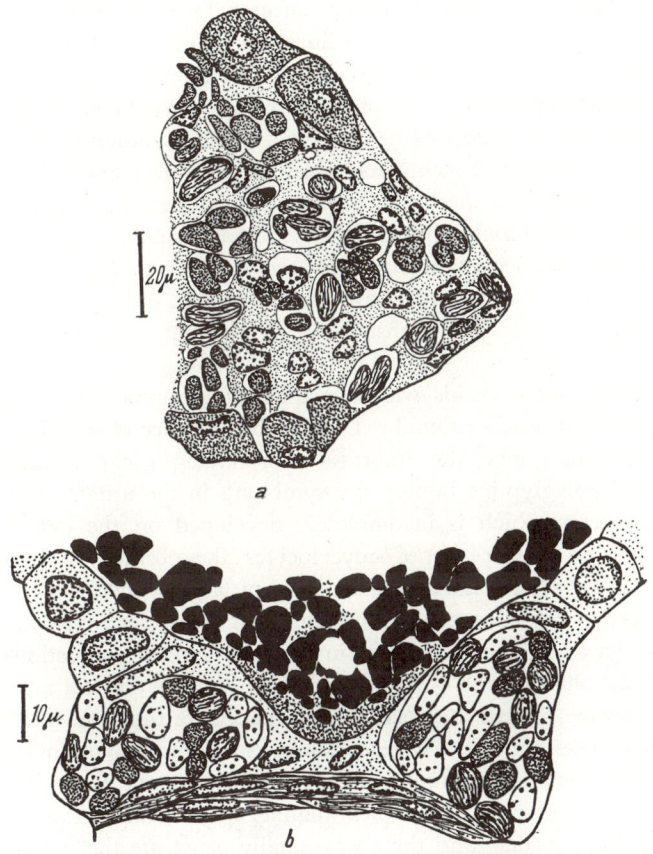

Fig. 208   *Campylenchia* sp.   (*a*) Infection mound; the *a*-symbionts are partly burdened with the λ-symbionts, the epithelial cells are heavily infected with η symbionts.   (*b*) The *a*-, *t*-, and λ-symbionts are joined in the cuneate cells, the λ-symbionts are already collecting at the posterior pole of the egg.   From Rau.

In the highly disrupted mycetome of another polyglyptine, the *t*-cells and with them the third guest in the form of large tubes are driven back; its remaining part is filled with an irregular coagulation of plasma, which here and there houses colonies of the η-symbionts and thus appears thoroughly devastated (Fig. 206*d*).

With a third tetrasymbiotic polyglyptine, Rau found a clear difference in the regulating capacity of the two sexes. The mycetomes of males appeared exceptionally disorganized; those of females seemed much more balanced in spite of damage of various kinds.

As the last membracid-mycetome infected by four symbionts, let us describe that of *Hypheus erythropterus* Burm., which is also characterized by extensive displacement of the *t*-mycetocytes yet gives the impression that a useful new arrangement of symbiont distribution has taken hold beyond the pioneer state represented in Fig. 206*d*. The *t*-symbionts form a connected layer behind the *a*-syncytia, which are only developed unilaterally, and the remainder of the organ is essentially taken up by syncytia with tube-shaped *v*-symbionts. Only the *η*-symbionts, which are still unbridled and not very numerous, disturb the order by occurring between the *a*-syncytia, in the *t*-mycetocytes and the *v*-syncytia, even penetrating the *t*- and *v*-symbionts. The transmission of the four types could not be studied in this species.

Of the five membracids with five different symbiont types we shall describe three in greater detail. Despite the presence of so many guests, their mycetomes give the impression of having great balance. An unclassified polyglyptine houses its *a*-symbionts in the usual manner in a syncytial cortex which is incompletely developed on the exterior and behind which a meager layer of *t*-mycetocytes takes position. The interior is taken up by a medullary zone built up of large syncytia filled with a tube-shaped, well-adapted, companion symbiont type (*ψ*-symbiont). Long slender *ι*-symbionts are found in smaller complexes, sometimes without nucleus, between the *a*- and *t*-sections and in the cortical area of the medullary zone, where they occasionally penetrate the syncytia. Obviously these accessory guests were acquired as the fourth symbiont, for the small *η*-symbionts are undoubtedly the ones that came last. They are distributed irregularly in the typical manner in more or less large groups in the medullary zone, and they occasionally penetrate the syncytia and their inmates, but nevertheless avoid other areas of the mycetome (Fig. 206*e*). The ovaries were too young for the study of egg infection, but interestingly enough it could be determined at least that the *ψ*-symbionts, well inserted as they are in uninucleate cells, leave the mycetome at the preferred places and behave exactly like the *t*-symbionts (Fig. 206*e* in the middle to the right).

A membracid and a tragopine, both unfortunately unclassified, acquaint us with a somewhat modified pattern and furnish supplementary information on egg infection. Here a mycetome composed of four zones appears to be in formation (Fig. 206*f*). In both animals small, roundish, pale formations are found in another narrow area, totally without nuclei, which forms on all sides between the *a*- and *t*-zones. The center of the organ is occupied by slender filaments with few nuclei. The youngest symbiont type is represented by similarly shaped bacteria which behave in a corresponding way. They are found primarily in the syncytia of the *a*-symbionts, usually penetrating the interior, but are found here and there occasionally in the otherwise sterile epithelium. In the membracine the bacteria also invade the fatty tissue, but in the tragopine, where regional separation in the mycetome is more advanced, they are found only occasionally.

All five varieties together infect the egg cells: some of the *a*-symbionts leaving the infection mound are infected by the bundles of the fifth symbiont; the two other additional guests make use of the migratory cells of

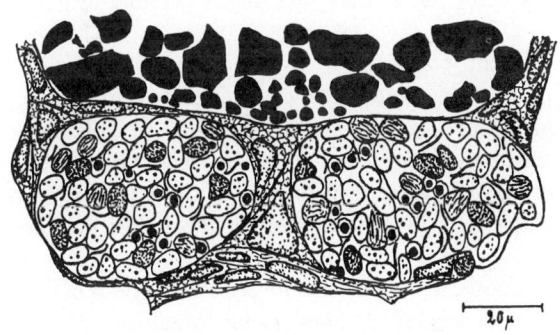

Fig. 209   Membracine.   The cuneate cells contain five types of symbionts.   From Rau.

the *t*-symbionts, which may vary greatly in appearance, depending on the number of "blind passengers" of the one or the other type which have been transmitted to them according to their position (Fig. 209).

Membracids with six different symbionts represent the maximum membracid "hunger for symbionts," a hunger bordering on the fabulous. We have two examples of this. Like the aforementioned species, *Enchophyllum quinquemaculatum* Jrm. houses five types in the mycetome and moreover contains a yeast in the fatty tissue; and an unclassified membracine finds place for all six guests in its mycetome. In *Enchophyllum* the solidarity of the *t*-mycetocytes is surrendered and a mixed zone is formed which, surrounded by a layer of *a*-syncytia, houses three forms in lively confusion. Between the *t*-cells, long, very matted tubes (ω-symbionts) take up much space chiefly in uninucleate or polynucleate associations, whereas small, uninucleate cells, settled by rodlets (λ_1-symbionts), which are also inserted here and there, are fewer in number but nevertheless fill the epithelial cells in uniform density.   A fifth form, composed of some-what coarser and thicker rodlets (λ_2-symbionts), was found also within the mycetome loosely distributed in the *a*-organ, but it occurred chiefly in the fatty tissue, in the oenocytes, and in the Malpighian tubules. Avoiding the fatty tissue, the sixth symbiont, a slender yeast, is found exclusively in the body cavity and only in very small numbers (Fig. 206*g*).

a-, *t*-, and *ω*-symbionts and the yeasts meet in the cuneate cells.   The first two originate, as always, from the infection mound; the *ω*-symbionts are taken along by the migratory cells of the *t*-symbionts; the yeasts find their way alone.   The two small rodlet types are transmitted again to the nutrient chamber and even in early stages are situated there in the plasma of the glandular cells, in the fiber cord, and in the youngest oocytes (Fig. 210*a*).   The more delicate form leaves the mycetome also with the help of the migratory cells; the coarser one might possibly originate from the fatty tissue.

The two different points of invasion result in a separation, never observed before, of the symbionts which are finally united at the posterior pole.   The four types originating from the cuneate cells form an apical cushion, whereas the youngest symbionts, carried to the posterior by cur-rents of the egg plasma before transmission of the others, form a spherical aggregation which rests in a depression of the larger symbiont cushion (Fig. 210*b*).

As mentioned above, a membracine unites all six guests in its mycetomes. An extremely well-developed filamentous organism drives apart the greatly diminishing *t*-mycetocytes and occupies a large central area. Intermingled with the filaments is a fourth, roundish, homogeneous sym-biont type in sparse, small aggregations.   The fifth type is represented by small pale tubes found here and there in the *a*-organs and frequently in the midst of the *a*-symbionts.   The ordinary, less well-adapted *η*-form is found loosely distributed in the epithelium of the mycetome, but more densely again in the fatty tissue, in the Malpighian tubules, and in the oenocytes. Five forms were found in the ovary, only the insignificant *η*-symbionts not being accessible to observation.   The migratory cells must serve as usual to transport the two other inmates of the inner zone; the fifth form is trans-ported in the interior of the *a*-symbionts.

With the aid of our native German membracids it was also possible to study the embryonic and larval development of the mycetomes.   When several cleavage or blastoderm nuclei penetrate the symbiont ball, the *a*-symbionts are the first to gather around them and are soon enclosed by the developing cell membranes.   Whereas flat nuclei which remain at the surface form an envelope around the developing mycetome, the *t*-symbi-onts and the additional, less well-adapted guests still represent a homogene-

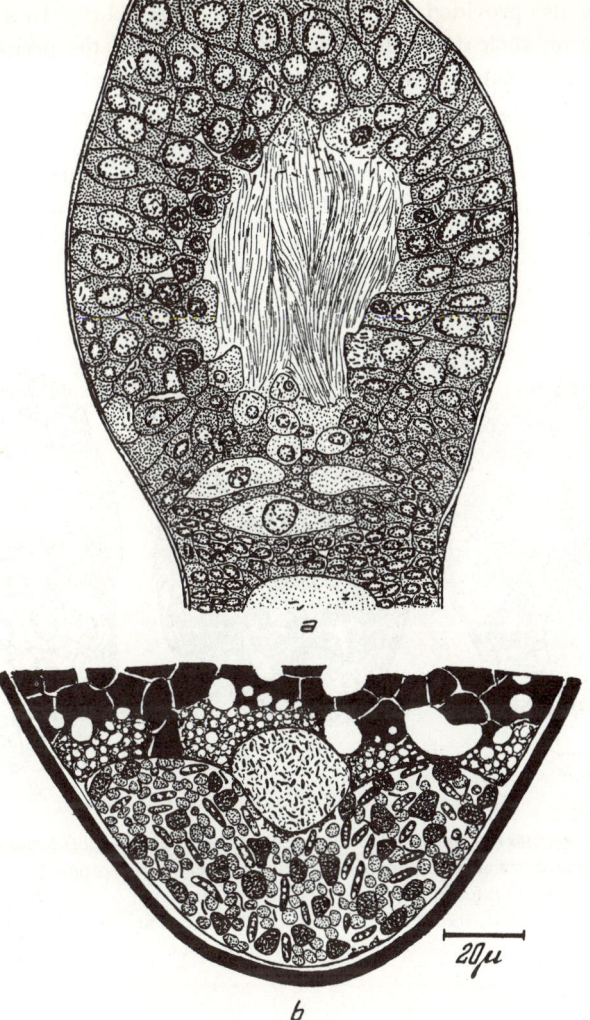

Fig. 210 *Enchophyllum 5-maculatum* Jrm. (*a*) End chamber of an ovariole with λ_1- and λ_2-symbionts, which entered the young ovocytes with the secretion. (*b*) Posterior end of an egg after conclusion of infection; the λ-symbionts rest on a cushion of *a*-, *t*-, *ω*-, and yeast-symbionts which penetrated over the cuneate cells. From Rau.

ous mass also provided with several nuclei (Fig. 211*a*).   In a succeeding
stage the uninucleate *a*-cells begin to travel toward the periphery of the
embryonic mycetome where meanwhile a circle of plasma-rich cells
apparently originating from the previously adjoining blastoderm takes
form (Fig. 211*b*).   Since the *a*-cells become more and more numerous in
connection with a strong increase in their inmates, without occurrence of
mitosis, and since these sterile cells at the same time make way for a sparsely

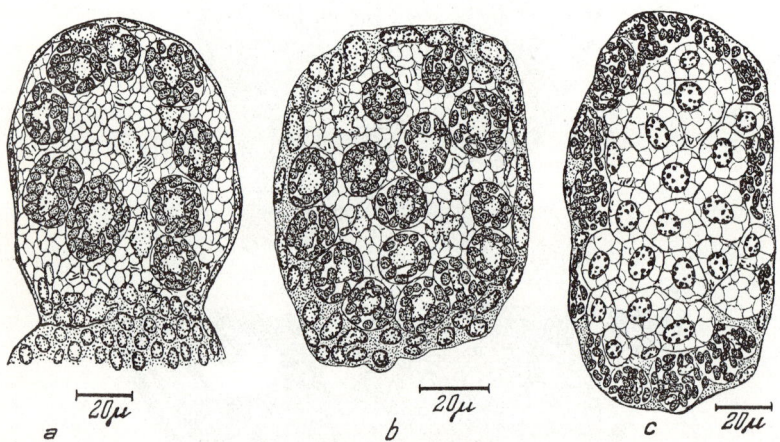

Fig. 211   *Centrotus cornutus* L.   Development of the mycetome: (*a*) formation of a first
set of *a*-mycetocytes around penetrated cleavage nuclei; (*b*) reproduction of the *a*-sym-
bionts and formation for a secondary mesodermal envelope; (*c*) the latter also infected,
both zones separated, the *t*-cells bounded off, the small accessory bacilli between the
t-symbionts.   From Rau.

populated envelope, it may be assumed that they are progressively infected
and become *a*-cells.   When a continuous cortex is formed by the latter,
cell limits grow up around the other free and also increasing nuclei, giving
rise in this way to *t*-cells.   Symbionts of the third type, still few in number,
are enclosed along with them (Fig. 211*c*), but in this early stage the *t*-cells
are still able to defend themselves by forcing the plasma buds filled by the
unwelcome guests into gaps in their own ranks.   Not until much later are
they reinfected and harmed in the manner already described.   Thus
the temporal sequence of the acquisition of the three symbiont types is
clearly reflected in their progressive elimination.

During these events the young mycetome is carried forward in the manner usual with Homoptera by the germ band developing at the posterior end of the egg and is divided in two before hatching. The nuclei of the *a*-mycetocytes cannot divide mitotically during further growth of the organs; amitosis unaccompanied by cell division leads instead to formation of typical syncytia. The *t*-cells present a different picture. In spite of the strong loading of their plasma, they continually maintain the capacity for normal mitotic division, and thus even mature animals are able to compensate for losses resulting from withdrawal of migratory cells.

Finally we must mention the still unclarified symbiotic devices of the typhlocybines. At this time it is certain only that there are no mycetomes and no immediately conspicuous symbiotic settlement of the fat or of the fluid of the body cavity. Accordingly, the posterior end of the egg does not have the accumulation of symbionts which ordinarily is easy to establish. On investigating a series of species some years ago, I frequently came upon enormous masses of small rod-like bacteria, principally in a ventricose section of the midgut, which were not only free in the lumen but also formed a thick border along the epithelial cells. It is of course rather doubtful that they may be interpreted as symbionts of the typhlocybines transmitted superficially during egg deposit. Observations by Stüben (reported by Buchner, 1948), according to which one would have to take a settling of the Malpighian tubules into account, remain unconfirmed. It seems more than likely that the typhlocybines, which are also exceptional in other respects, are the only cicadas that have no symbionts. In all probability, as Müller (1949) believed, their disparate behavior is explained by the fact that they do not live by sieve-tube sap but suck out cells, and thus absorb nourishment of higher quality.

The only statements on culture of cicadoid symbionts are those by Mahdihassan (1939, 1947) and Tóth (1946, 1951). Without giving details of technique and culture media, Mahdihassan reports that he succeeded in developing two different kinds of bacteria from *Cicadella viridis*, a red form apparently producing carotene and establishing colonies, and a form tinged yellowish green. In research on assimilation of atmospheric nitrogen, Tóth used cultures of organisms which he had cultured from *Aphrophora salicis*, *Philaenus spumarius*, and others, but he did not furnish proof of their identity. Resühr tried in vain to cultivate the symbionts of *Philaenus* and *Cicadella*.

FULGOROIDS. The symbiotic devices of the fulgoroids are as diverse as those of the cicadoids but totally dissimilar. Organ types lacking in the cicadoids are characteristic of the fulgoroids, and their entirely different character is the result of their tendency to house the individual symbionts in separate paired and unpaired organs, instead of uniting them in one mycetome pair as in cicadoids. Moreover, the symbiont types acquired later are inserted in the body of the animal much more harmoniously for the most part than is usually the case with cicadoids, and an accumulation of half-parasitic, accessory symbionts of the kind observed in the membracids is never found.

Greatly complicating the symbiosis in fulgoroids is the addition of a third, very odd, chief form, usually designated by the letter x, to the a- and H-symbionts. Although the latter occur frequently in association with auxiliary and companion symbionts, they occur rarely alone. They also manifest a clear, mutual disinclination toward one another, so that, with a single negligible exception, they never make their appearance at the same time. As far as we know today, this disinclination has been overcome in the genus *Issus* and, according to the new findings of Ermisch, also in *Jassidaeus*.

Yeasts have been found as sole symbionts only in *Nisia atrovenosa* Leth. (meenoplines) and an unclassified derbine. In both cases they are embedded on both sides of the fatty tissue in irregularly defined areas in which syncytia take the place of uninucleate fat cells. The yeasts have the typical shape. With both species lack of material made it impossible to observe the infection of the egg cells.

Exclusive occurrence of the x-symbionts was ascertained in a number of other derbines, only one of which could be determined to be a *Mysidia* species, and in the meenopline *Paranisia*. These cases are especially important because they hold up to view a remarkable and at first incomprehensible phenomenon in the simplest form conceivable, uncomplicated by cooperative living with other types. In both sexes of the derbines in question insignificant paired mycetomes, elongated or bean-shaped, are shifted in various ways by the developing sexual glands. Behind an envelope of greatly distended cells are syncytia in which more or less distorted nuclei are found primarily in larger, central, plasma islands. The symbionts are strikingly large, polygonal, or round formations, few in number and distinguished by various dark lumps, or sometimes by intensively eosinophilic granules. Today it can no longer be doubted that these and the bizarrely shaped structures, often even larger, which are found in still larger corresponding mycetomes of other species, are extremely degenerate bacteria. The mycetomes are much smaller in males than in females (Fig. 212a).

In addition to these x-organs only the females have an odd unpaired mycetome, a so-called rectal organ, in this case inserted behind the valvula rectalis (pylorica) between the intestinal epithelium and the muscularis and protruding like a hernial sac into the lumen of the hindgut (Fig.

212*b*).   It consists of a few large mycetocytes, each of which has two irregularly shaped nuclei and numerous small compact tubes.   At the posterior end of the mature egg cells, there is a symbiont group, and, despite the fact that a great many different symbionts are found in the two

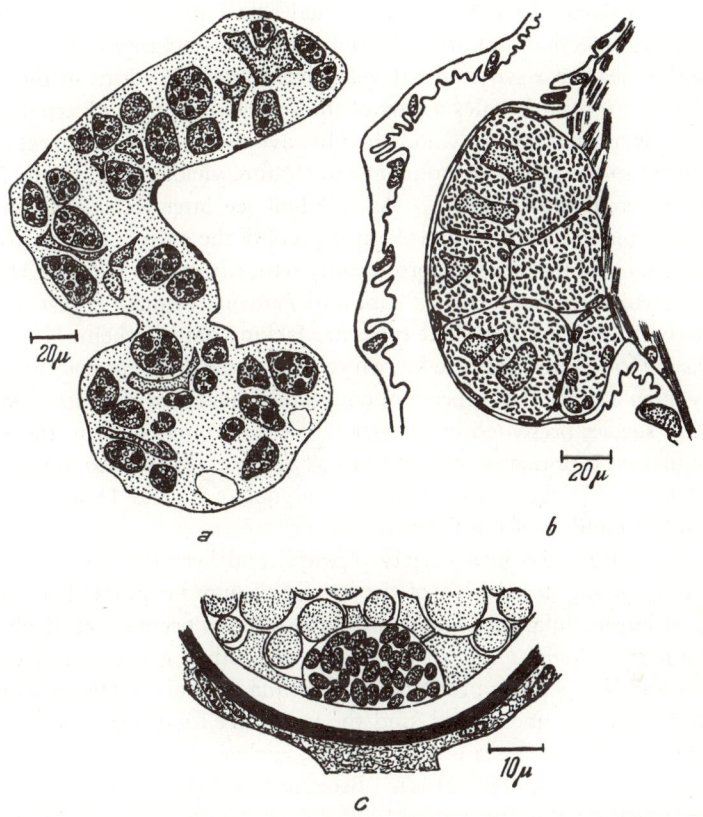

Fig. 212   Three different derbines.   Monosymbiotic forms with *x*- and rectal organs. (*a*) Part of the *x*-organ; (*b*) rectal organ; (*c*) posterior end of the egg with a single symbiont type stemming from the rectal organ.   From H. J. Müller.

organs, this group doubtless contains only a single type, namely, small oval deeply-staining formations such as can be differentiated here and there in the rectal organ as specific infection states.   Clearly larger than the normal symbionts, they are situated between the mycetocytes (Fig. 212*c*).   Therefore, although no representatives from the *x*-organ take part in egg infection, these organs and their symbionts always appear in exactly the same manner in the progeny of both sexes. When Šulc (1924) came upon x-organs and rectal organs in conjunction with other symbionts for the first time in *Cixius* and *Fulgora*, he overlooked the fact that the one organ type is always present in the female only, and he also believed that he had found the x-symbionts again in the egg. In the same test objects I proved the lack of rectal organs in males and the lack of x-symbionts during egg infection and drew the only possible conclusions, namely, that in both sexes the x-symbionts develop from infection forms of the rectal organs and that in females a part of the symbiont store is diverted before this hypertrophy sets in and, thereby avoiding such degeneration, is advanced to the filial mycetome at the rectum, yielding exclusively infectious states (Buchner, 1925). As we shall see later, H. J. Müller was actually able to furnish embryological proof of the correctness of this idea in the course of his basic fulgoroid study with *Cixius* and *Fulgora; Mysidia*, other derbines behaving like them, and *Paranisia* confirm it just as decisively because other symbiont types are lacking only in them.

Disymbiotic fulgoroids are known to us in three different combinations. Very often a yeast (*H*) appears in conjunction with *f*-symbionts. Among the 217 species presented by Müller (1962) in tabulated form, the *H* + *f* combination is present in no less than 88 species; 54 cases are found in the delphacids, 9 in the issids, 14 in the dictyopharids, and 11 in the flatids and other families of the flatines.

Yeasts are housed in a variety of ways, and here too this points to a relatively young acquisition of them. They may be situated intercellularly or intracellularly. The first type of position occurs exclusively with fulgorines (*Fulgora, Pteroplegma, Nersia,* and other genera) and certain issines and is especially pronounced in young animals. The uninucleate fat cells then are uncrowded, and the yeasts are frequently enclasped by pseudopod-like processes of the fat cells, exclusively in the fissure spaces between these cells (Fig. 213*a*). Because later the fat cells are usually closer together, they no longer give the symbionts the same latitude as originally.

Phalaenomorphids and flatids handle their yeast symbionts in exactly the same way. The uninucleate fat cells are infected over wide sections, and, then dissolving their cell limits, they flow together to form syncytia. Density of settling and sharpness of separation, as compared with non-infected zones, are subject to very strong variations, indicating a still relatively loose relationship between host tissue and symbionts. Only in a species of *Ormenis* was there clear-cut separation of infected, but still irregularly distributed, fatty tissue lobes that appeared to mark the beginning of an organ-like separation.

The issines belonging here behave in general like the flatids, but forms

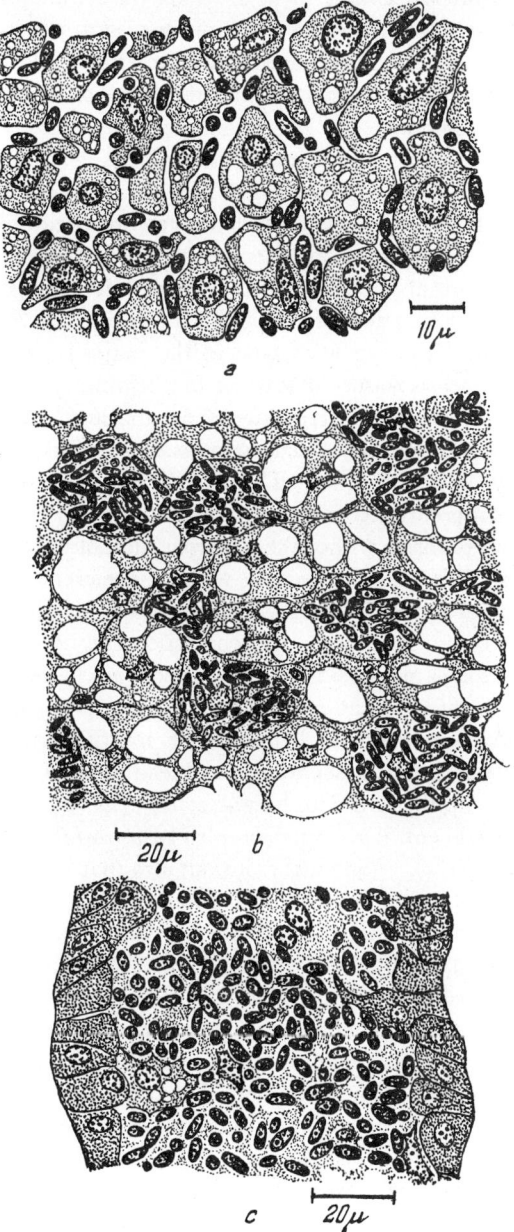

Fig. 213 (a) *Fulgora nodivena* Walk. (b) Megameline. (c) *Liburnia aubei* Perr. Different forms of colonization of the fat tissue by yeasts in the presence of *f*-organs. From H. J. Müller.

are also encountered among them in which the symbionts live intercellularly. With stronger infection they appear also to penetrate the fat cells and cause them to undergo syncytial amalgamation. With some megamelines and delphacines we find another new type of localization of yeasts, namely, in individual cells which remain uninucleate and usually are distributed at random over the fatty tissue lobes (Fig. 213*b*), but even in these subfamilies a flowing together of such cells into syncytial associations usually occurs. In this case a tendency toward organ-like, indeed even mycetome-like, concentration of the yeast cells can be distinguished when a rather large number of species are compared. Ermisch, who studied the kinds of colonization of the yeasts in numerous *H-* and *f-*bearing araeopids, found the syncytial type in the majority; only in *Megamelus notula* are the yeasts found at least at the beginning in uninucleated fat cells, which are sharply differentiated from the sterile fat tissue. Chiefly in females, dissolution of the cell boundaries often takes place in the end and extensive syncytia arise. The two types of symbiont housing can scarcely be differentiated in these animals.

In addition to forms without sharp separation of infected syncytial and sterile zones of the fatty tissue and without preference for specific body regions, there are other forms in which the yeasts avoid the peripheral cell layers of the fatty tissue lobes. A progression can be set up among them leading from forms in which the cortical zones vary in their spatial development, and the infected and non-infected zones are imperfectly separated; through forms in which unilaminate or multilaminate envelopes are usually sterile and sharply separated; and leading finally to states in which a unilaminate epithelium with fungi-free fat cells regularly surrounds the densely infected central syncytia completely, as in *Liburnia, Conomelus,* and *Chloriona* (Fig. 213*c*). We must point out also that in the last instance the elements of the epithelium can be differentiated from the other non-infected fat cells by their abundance of dense plasma, their roundish nuclei, and often also by their size, and, by the fact that branched processes extend out between the syncytia (as they so frequently do in mycetomes).

Because the *f-*organs allied with these yeasts are rather insignificant mycetomes, they were sometimes overlooked in the past. At times these organs are unpaired, long, extremely thin tubes which wind transversely through the posterior part of the abdomen; they are then often entwined and form balls, especially at the ends, as in flatids and phalaenomorphids and in some fulgorines. At times the organs are paired and greatly elongated, or short and compact, as in some fulgorines (compare Fig. 229*a*). They are paired or unpaired in the issines; in the delphacids they are very small and insignificant tubes, always paired, which tend to be situated in the vicinity of the gonads. An epithelium, often somewhat flattened, surrounds mycetocytes which, depending on their size, sometimes have only sufficient space in the narrow tube to be situated one behind the other and at other times have enough space for several to be situated side by side (Fig. 214). Rarely are their inmates so clearly recognizable as spherical formations as in *Liburnia;* usually they are much smaller and so delicate that in the fixed test object they stick together in pale granular coagulations; but apparently they are always spheres or short filaments.

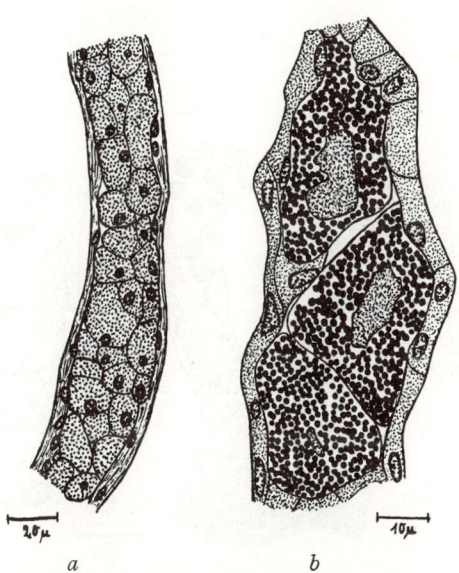

a   *b*

Fig. 214   (*a*) Phalaenomorphid.   (*b*) *Liburnia fairmairei* Perr.   The *f*-organs in the presence of yeasts in the fat tissue.   From H. J. Müller.

The tracheae are very abundant, and sometimes the branches penetrating the interior are so strong that the mycetomes are deformed.

During egg infection it is always easy to study the appearance of the yeasts in the cuneate cells and their transmission to the egg.   On the other hand, the *f*-symbionts cannot usually be recognized with the desired clarity, and they appear as in the mycetomes as a fine coagulation which often makes up a substantial part of the symbiont ball after termination of infection and encloses relatively few yeasts.   The states accompanying transmission are of special interest not only because the cuneate cells are differentiated before the arrival of the symbionts, as happens frequently in other cases, but also because the egg cells in both megamelines and delphacines make elaborate preparations. Appearing long before the onset of yolk formation in the posterior region of the still very young oocytes is a slender, oval, hollow space resembling an empty symbiont ball. As growth proceeds, it breaks through behind and is joined with the hollow enclosed by the cuneate cells which have been differentiating in the meantime. A ring of such cuneate cells approximately four cells in thickness admits the yeasts and *f*-symbionts, and transmits them without delay to the interior, relatively few symbionts ever being found in the

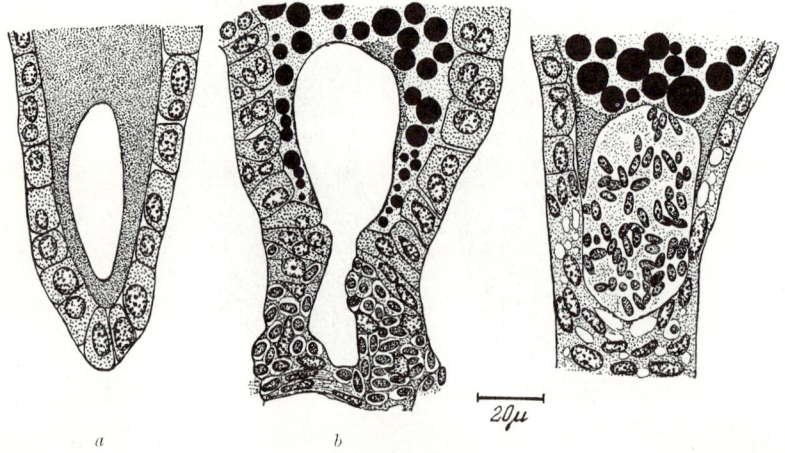

Fig. 215 Megameline. Infection of the egg with yeasts and *f*-symbionts. (*a*) Forward cavity in the ovocyte; (*b*) infection of the wedge cells; (*c*) crossing over of the symbionts into the egg. The *f*-symbionts appear as a fine coagulum. From H. J Müller.

follicle. Whereas the egg cell usually tends to swallow the entire mass of symbionts all at once, here the corresponding space fills gradually. In Müller's words, it is a "gliding infection." Finally the egg cell is completely sealed off in the usual manner by a drawing together of the free edges in the manner of an iris diaphragm (Fig. 215).

It was possible also to study the transmission of the flatid symbionts in detail. A hollow, open at the posterior from the very beginning, appears in this case also here in the oocytes; the cuneate cells are extremely elongated and, joined in a syncytium, finally hold back a surprisingly large number of yeasts (flatid type, Müller).

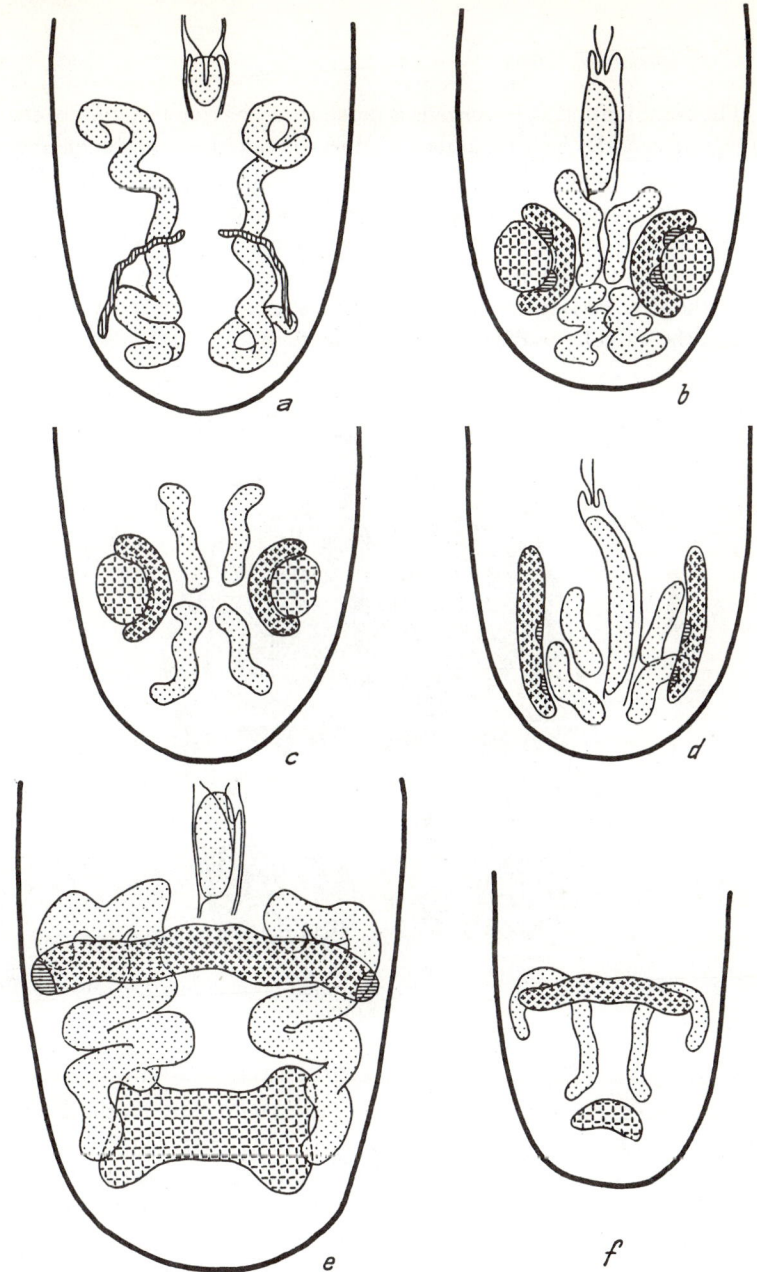

Fig. 216 Symbiotic arrangements of different fulgoroids I. (*a*) Tropiduchine ♀.
(*b*) *Cixius nervosus* L. ♀ (*c*) *C. nervosus* L. ♂. (*d*) *Myndus musivus* Germ. ♀. (*e*)
Caliscelis bonellii Latr. ♀. (*f*) *C. bonnellii* Latr. ♂. Schematic, rectal, and *x*-organs,
dots; *a*-organs, crosses; *f*-organs, vertical lines; organs with companion symbionts,
cross-hatched; infection mound, horizontal lines; filial mycetomes, black. From
H. J. Müller.

393

The combination $x + f$-organs is much rarer, having been discovered in only thirteen species distributed among the achilines, derbines, meeno-plines, and tropiduchines. The x-organs of the achilines are paired as always and also so split up on both sides that two partial mycetomes, some-times differing in shape according to sex, come to lie one behind the other. The associated rectal organ is situated not behind the valvula intestinalis, as in the derbines described above, but, as usual in the remaining cases, within the valvula, with six large binucleate mycetocytes in strict radially

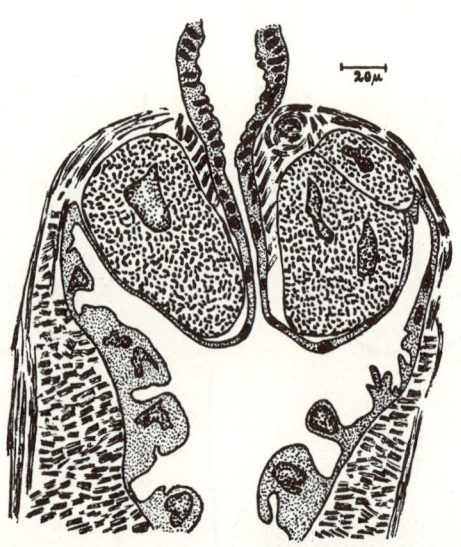

Fig. 217   Meenopline.   Ring-shaped rectal organ opening into the valvula.   From
H. J. Müller.

symmetrical arrangement forming a regular circle between the two walls of the intestinal fold (Figs. 216a and 217). The f-organs in these achilines are also paired and small, or unpaired and elongated, and their inmates are not set off clearly. The remaining subfamilies follow the same pattern except for variations of a secondary nature. Egg infection among the tropiduchines deserves special study because it embodies a simplified and primitive-appearing type (tropiduchine type, Müller). The derivates of the rectal organ are first congested in great numbers in front of the cuneate cell follicle, and then are quickly transmitted with the f-symbionts at the posterior end of the egg to the small hollow space which is of such recent origin that the cuneate cells are scarcely differentiated as such.

The third disymbiotic combination possible among the fulgoroids is represented by *x*-organs with *a*-organs, in other words, a mycetome type which also plays a great role among the cicadoids. It has been discovered in fifteen species in six subfamilies, that is, about as frequently as the *x* + *f*-combination. With *Myndus musivus* Germ. (cixiines) the paired *a*-organs

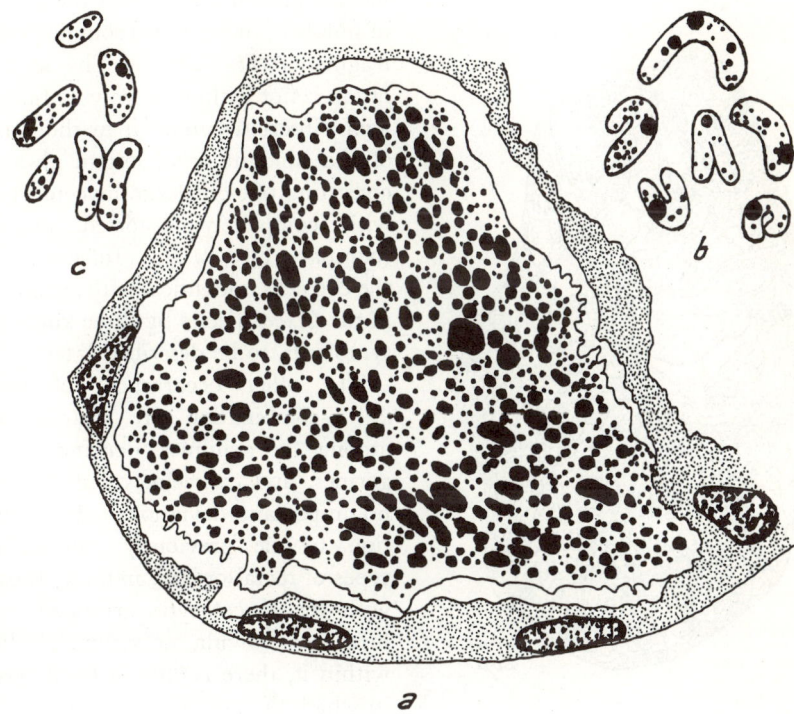

Fig. 218 *Myndus musivus* Germ. (*a*) A giant symbiont from the *x*-organ; (*b*) infection forms of rectal symbionts; (*c*) symbionts from the ovocyte (at the same magnification). From Buchner.

extend over the exterior of the *x*-organs in the form of long tubes well provided with tracheae. Yellow-brown to orange pigment in the embryonic cells of the tracheae gives them a corresponding coloring. A cubical epithelium encloses only a few large syncytia, in which almost all the nuclei nestle flat against the wall and only here and there occur in little plasma islands in the center. They originate through gradual amalgamation of primary syncytia. The symbionts are short compact tubes characteristic of the type and only in certain places are transformed into smaller infectious states; however, the mycetome sections do not protrude like cones but are inserted cell nests with the usual origin and construction. Two are present in each organ, since the temporary adhesion to the ovary occurs in two places in the cixiides (Fig. 216*d*).

The *x*-organs appear on each side as subdivided, strong, yellowish tubes consisting as usual of syncytia with parietal nuclei; the giant symbionts located in their meshes have the shape of unlobed, more or less rounded off, fragments (Fig. 218*a*). The rectal organ is not within the valvula but is again inserted behind it in the wall of the hindgut as an organ always consisting of eight large mycetocytes with completely distorted nuclei (compare Fig. 216*d*). Infection of the egg cells is identical with infection of the *Cixius* species by three kinds of symbionts, to be described later.

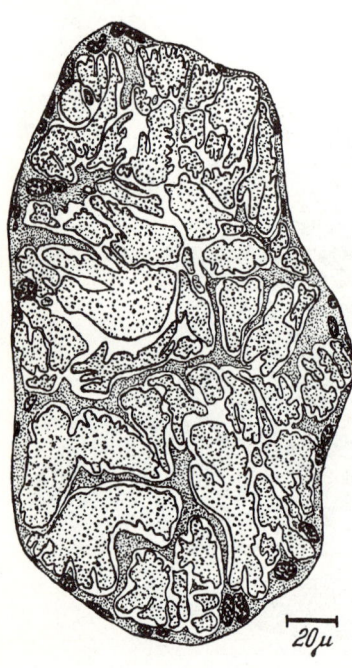

Fig. 219   Achiline. Syncytial *x*-organ with nuclei lying on the wall, and greatly tattered symbionts. From H. J. Müller.

The other occurrences of this *x* + *a*-combination are distributed among the achilines, derbines, fulgorines, and caliscelines and display no important deviations. In addition to the previously mentioned types of rectal organs, namely, those with the mycetocytes arranged behind the valvula or symmetrically within it, there is often a third type in which the symbionts are arranged asymmetrically within it, so that the intestinal lumen passes obliquely through the organ (compare Fig. 216*b*). The *x*-symbionts frequently have the deeply split form observed in *Cixius*, *Fulgora*, and many others (Fig. 219). The calisceline *Bruchomorpha*, an exception in that the *a*-organ passes transversely through the abdomen, has an unpaired huge tube with an infection mound at each end (compare Fig. 216*e*).

Quite often three different kinds of symbionts are found in the fulgoroids. In an overwhelming number of cases the two chief symbionts *x* and *a* were united with a companion or accessory symbiont. Rarely is an additional symbiont combined with the *x* + *f*-pair (one meenopline and seven delphacines). The combination $H + f$, though not rare in itself, adds a companion symbiont in only one fulgorine species. Also very rare is the combination of a third form with two chief symbionts x and H, which indeed, as already emphasized, manifest a pronounced mutual antipathy. This possibility exists for only two *Issus* species, which thus acquire an exceptional position within their group.

On the other hand, in a large number of species, bacteria recognizable as specific forms in section join the $a + x$-complex. In the very uniform cixiines it is a case of b-organs such as occur in the cercopids. Figures 216b and 216c show the location of eight or nine mycetomes side by side in the abdomen of the *Cixius* species. On both sides the x-organs are divided into two tubes placed one behind the other; the paired horseshoe a-organs hold the b-mycetomes which are also paired, spherical, and orange-colored. The few large syncytia of the latter, which show the frequent tendency to fuse into syncytia, are surrounded by a very flat epithelium with but a few nuclei. Tracheal branches penetrate between the syncytia, and tracheoles pass over into the syncytia. The symbionts are spherical to oval, like those encountered in the cercopids, and each is surrounded by a closely fitting envelope. After being transformed here and there in the cortical areas of the organ into slightly changed infection stages, they are transmitted as such to the body cavity. The a- and x-organs have the construction described for *Myndus*. The rectal organ is not situated within the valvula, but behind it, as in *Myndus* (Fig. 220).

The three symbiont types, which can be easily distinguished according to their place of origin, meet during egg infection. As occurs so often, the slender cuneate cells are clearly distinct from the remaining cubical follicle cells even before the symbionts appear. In the broad cell cushion which they form after infection it is easy to distinguish between the slightly curved and strongly staining short tubes of the rectal organ, the lighter, small, round or oval infection stages of the a-organs, and the very pale, finely punctate representatives of the b-organs; however, one would look in vain for a representative of the x-organs. In this cushion of cuneate cells a lively increase of symbionts now begins. The vacuoles originally enclosing only one symbiont now flow together to leave large hollow spaces; the cell limits disappear; and a syncytium arises in which the nuclei are deformed through the increased filling and are pressed toward the periphery. Not until then does a hollow space form between the posterior end of the egg and the syncytium, which fills up to the extent of its size with the emerging symbiont group. After being admitted to the egg, the remains of the cuneate cells shrink together and are destroyed along with symbionts which are occasionally left behind in them.

Because the representatives of the various mycetomes do not appear

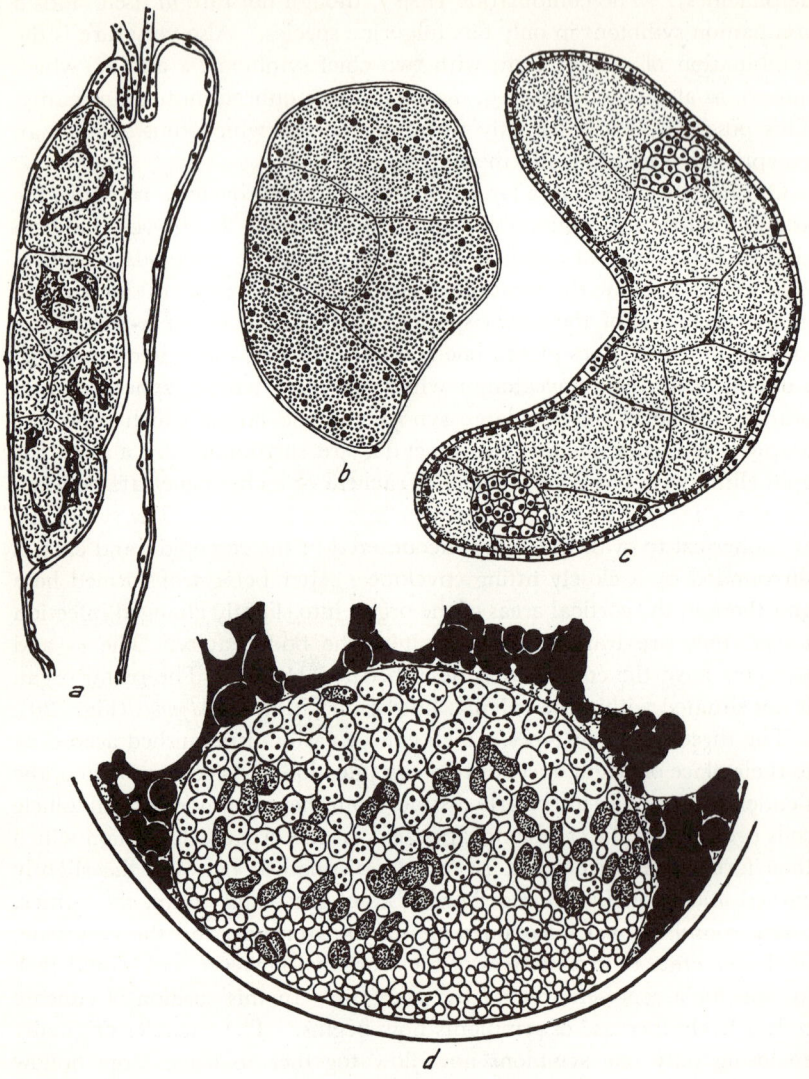

Fig. 220   *Cixius nervosus* L.   (*a*) Rectal organ; (*b*) *b*-organ; (*c*) *a*-organ with the rudiment of two infection mounds; (*d*) three markedly different types of symbionts in the egg.   From Buchner.

in the cuneate cells all at the same time, approximate separation of the types occurs and is adhered to during exit from the cuneate cells and transmission to the egg (Fig. 220*d*).

Establishment of an unpaired *k*-organ, radically varying in shape, is typical of the poiocerines, a subfamily of the laternariids. Usually it is a flattened and laminar form which may have a considerable circumference and lobes of different types. Sometimes it has the form of a broad flattened transverse band; it is rarely small and spherical. A flat epithelium surrounds numerous large syncytia and synsyncytia. The symbionts, very delicate short tubes, are thickly crowded and, since they have a tendency to swell, resemble a honeycomb.

The infection stages are developed in the *k*-organs in a peculiar way. Inserted in the mycetome under the epithelium in isolation or in groups, little syncytia are spread out one after another over certain sections of the whole surface like a continuous integumental zone. Their few nuclei are pressed toward the wall and contain somewhat larger and lighter symbionts. In all probability, sterile epithelial cells are disengaged in an earlier stage from their association here and there, are infected like cell nests of young infection mounds, and are transformed into syncytia in connection with amitotic nuclear divisions.

Among the poiocerines the *x*-organs are filled with extremely lobate giant symbionts. The *a*-organs, populated by elongated organisms and forming two, four, or six infection mounds according to the species, are extremely long and slender tubes, with many entwined windings, in a voluminous mass on both sides of the posterior area of the abdomen (compare Fig. 225*e* which illustrates a poiocerine with four types of symbionts and a small oval *k*-organ).

Though very little is known about the larger pyropsines, and most laternariids, it was possible to determine that three *Pyrops* species, in addition to an *x* + *a*-complex, also have an unpaired mycetome, the *i*-organ, in appearance like a loaf of bread, which is situated transversely behind the *a*-organ and consists of uninucleate cells or cells with only a few nuclei with slender taut filaments.

Symbiotically the fulgorines fall into two groups: one, as already mentioned, has yeast in the fatty tissue and *f*-organs; the other lives trisymbiotically and contains a so-called *m*-organ besides its *a*-organ and *x*- or rectal organ. Figure 221 depicts the symbiotic apparatus of *Fulgora europaea* L., the largest and most striking of the Central European fulgorids, and illustrates, in addition to the tube-shaped primary organs, the strange, unpaired, bowl-shaped formation present in both sexes which enclasps the convolution of the midgut loop from below. It is loosely constructed of little syncytia, and it lacks a cellular envelope. The rodlets which it con-

tains are small and slightly arched.   In some of the other fulgorines of
this type, the paired *m*-organ is situated in a different position and is in
part infected by longer delicate filaments.   In *Fulgora confusa* Stål it
contains small roundish symbionts.   The *x*-organs are filled with extremely
large and deeply lobate giant forms, in the living state with thin fluid

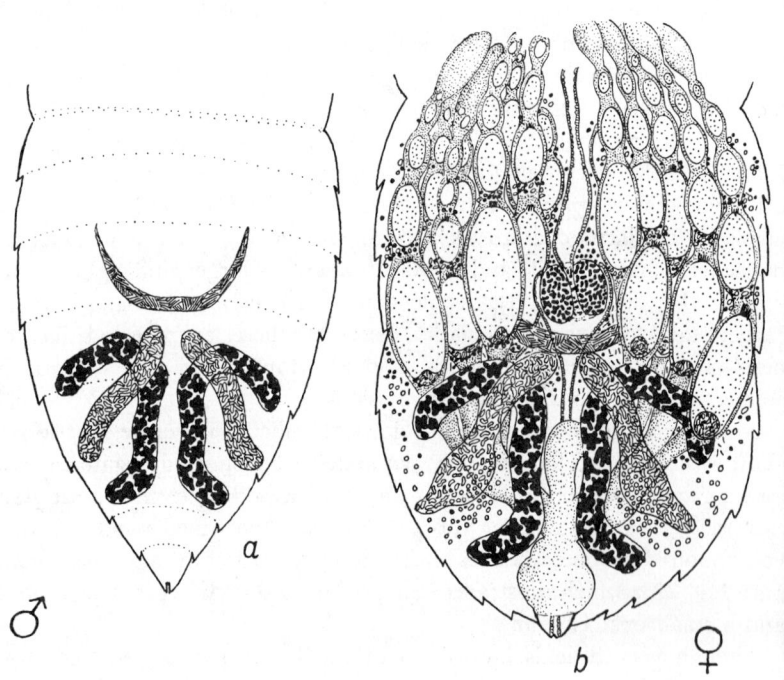

Fig. 221   *Fulgora europaea* L.   (*a*) Male with paired *a*- and *x*-organs and the unpaired
m-organ; (*b*) female which possesses in addition the rectal organ and has three kinds of
symbionts in the body cavity and in egg infection.   From H. J. Müller.

plasma densely filled with markedly light-refracting eosinophilic inclusions of the most varied size and shape (Fig. 222).

The transmission method of the trisymbiotic fulgorines is especially
interesting.   With *Fulgora confusa* and another unclassified species the
inmates of the *m*-organ do not enter the egg by way of the cuneate cells,
but indirectly by constructing a filial mycetome in the upper section of each
of the twelve ovarioles.   Inserted between the apical nutrient chambers
and the young oocytes, an almost spherical organ tends to cause a slight

swelling of the ovarioles.   An epithelium flattened like a membrane with very sparse, correspondingly shaped nuclei encloses mycetocytes which are at first uni- to dinucleate but gradually merge to form a syncytium with nuclei escaping toward the margin where the original division into cells is recognizable for the longest period of time (Fig. 223*a*).   The symbionts correspond to those of the *m*-organs.   This peculiar organ is

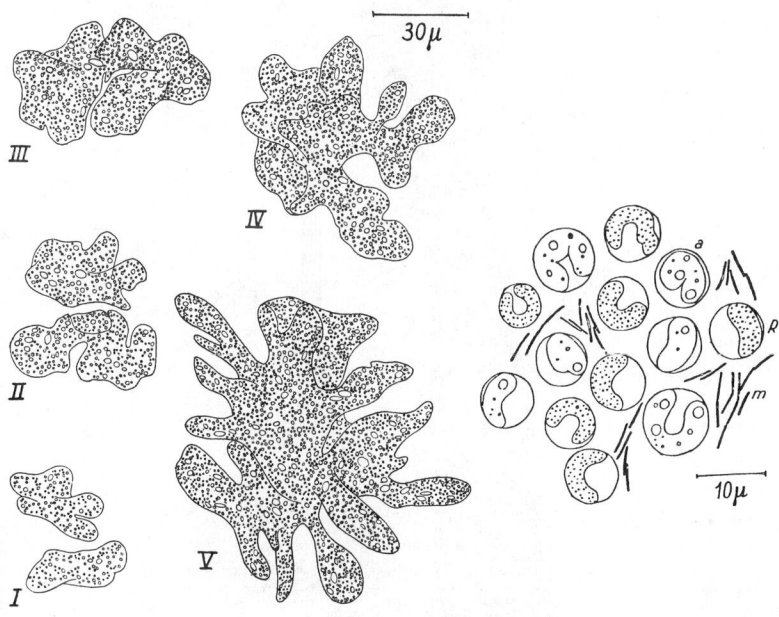

Fig. 222   *Fulgora europea* L.   (*Left*) Giant symbionts of the *x*-organs during larval development (I–V larval stages).   (*Right*) The three types of symbionts from the infection clumps of the egg.   *a*, *a*-symbionts; *R*, rectal symbionts; *m*, *m*-symbionts.   All are fresh preparations.   From H. J. Müller.

penetrated by fiber cords which conduct the secretions of the nutrient cells to the oocytes, which are now used as a route by the symbionts.   It can be observed that they are transmitted to the cords and, enclosed by vacuoles, glide with the secretion stream to the egg cells (Fig. 223*b*). Having arrived there, they first remain in the upper areas.   Unfortunately, because of lack of material, it was impossible to trace their later course.   The route used here brings to mind the fact that ill-adapted accessory symbionts among the membracids also avoid the cuneate cells and infect the zone of the nutrient cells and the youngest oocytes without,

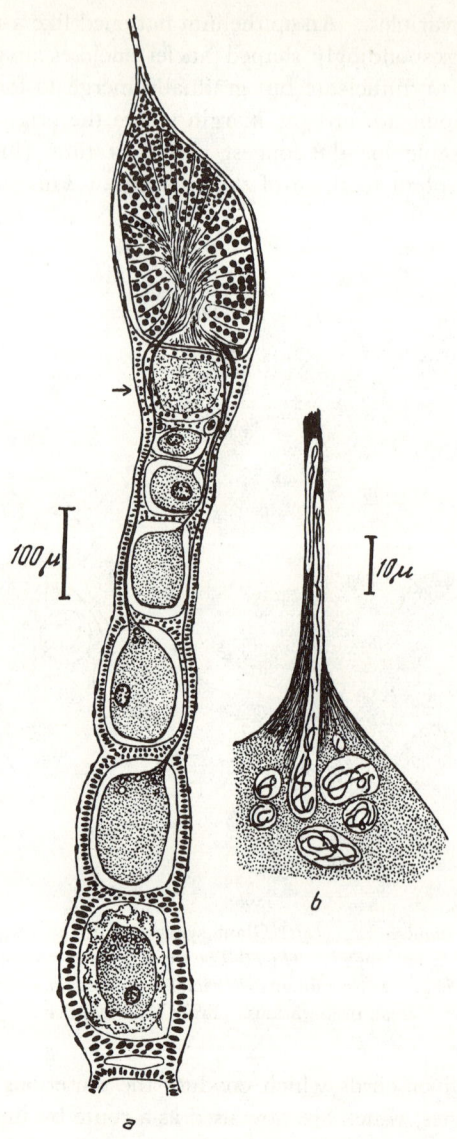

Fig. 223　Fulgorine.　(*a*) Ovariole with a filial mycetome serving transmission of the *m*-symbionts; (*b*) the thread-like bacteria glide in the nutritive cord to the ovocytes. From H. J. Müller.

of course, establishing regular filial mycetomes in the manner of these fulgorines or *Bladina fraterna* Stål, which is to be described later.

The two phylogenetically older *a*- and *x*-symbionts of these animals naturally use the route over the cuneate cells.

Accordingly, one would expect the same behavior of *Fulgora europaea*, which has *m*-symbionts that give the impression of lesser adaptation, but

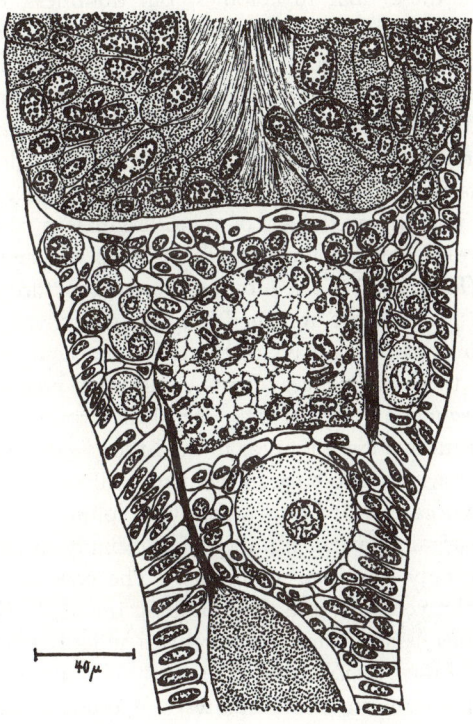

Fig. 224   *Fulgora europaea* L.   Rudimentary filial mycetome, which remains sterile, behind the nutritive chamber.   From H. J. Müller.

surprisingly all three symbionts are found here in the cuneate cells and in the symbiont ball (Fig. 222).   On studying the ovarioles, however, one discovers at the place where *Fulgora confusa* erects the filial mycetomes a very similar sharply defined but symbiont-free formation which clearly betrays, through marked vacuolization of the plasma, through nuclei which are in part pyknotic and disintegrating, and through incompletely executed syncytial union of the cells, that it is in the process of degenerating (Fig. 224).   The only possible explanation for this strange occurrence is that in *Fulgora europaea* the filial mycetomes were originally erected for the *m*-symbionts and that the host organisms still reproduce them, although today the youngest guest also reaches the egg by the simpler route over the cuneate cells.   Study of the postembryonic development showed that this always sterile site is formed in the third nymphal state, although no signs of degeneration could be determined at first.

The fulgorine material permitted no conclusions with respect to genesis and infection of these filial mycetomes, but doubtless, on leaving the mycetome, the symbionts seek the organ rudiment and infect it individually.   In the ricaniid *Bladina fraterna* Stål and in another unclassified nogodinine, the corresponding organs are at first glance similar; however, they are formed in a completely different manner.   In *Bladina* one finds voluminous paired *x*-tubes and an *a*-organ, which is developed unusually in that it consists of a transverse unpaired section and one extending backward.   Behind the unpaired part of the *a*-organ is a roundish *n*-organ which has no epithelial envelope and is built up of syncytia in various stages of fusion (Fig. 225*a*).   The inmates present a picture never observed before in cicadas.   They are arranged in little groups, varying in number and size of organisms but always of the same character within any one syncytium.   By these differences also one can often recognize the subsequent fusing of syncytia.   Apparently each little group may be traced to the increase of a mother symbiont, as in the little rosettes which are found in *Euacanthus* and which also vary in size from cell to cell.

Whereas the *a*-organ forms a total of four very voluminous, inserted infection mounds, *n*-organs have a number of externally attached cells with larger, more strongly staining, symbionts.   The cells are obviously not part of the actual mycetome, and the symbionts definitely give the impression of being transmission forms.   That this is actually the case is proved by examination of the ovarioles, for once again filial mycetomes are found there with the very same roundish inmates.   Without remotely approaching the high degree of organization of those established in the fulgorines, they are situated, in the form of an indefinitely limited cushion of binucleate, loosely organized mycetocytes, behind the nutrient chamber and are penetrated by the fiber cords of the chamber.   Transmission of symbionts to this chamber apparently takes place indirectly by way of an infection of the anucleate nutrient plasma where they are also enclosed in vacuoles (Fig. 226).   These vacuoles are affected by a current passing through the fiber bundles and must be elongated accordingly.   Having arrived at the oocyte, they first sojourn in the area of the entry point, but later are uniformly distributed throughout the whole egg.   It was not possible to study their further history.

Acquisition of the *a*- and *x*-symbionts by the egg takes place amid

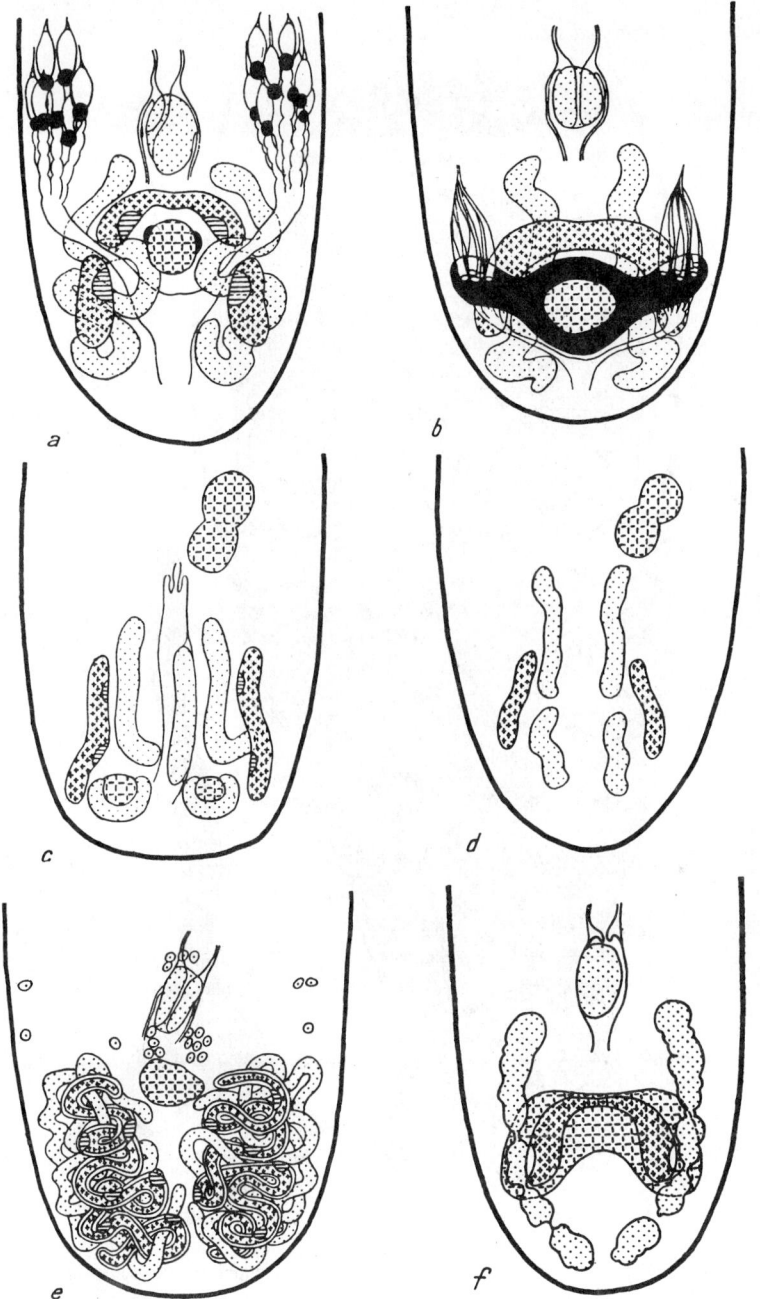

Fig. 225 Symbiotic arrangements of different fulgoroids II. (a) *Bladina fraterna*
Stål. (b) Nogodinine ♀. (c) *Oliarius villosus* F. ♀. (d) *O. villosus* F. ♂. (e) *Crepusia*
nuptialis Gerst. with a companion symbiont in the epithelium of the *a*-organ. (f)
Tettigometra atra Hagenb. For explanation see Fig. 216. From H. J. Müller.

405

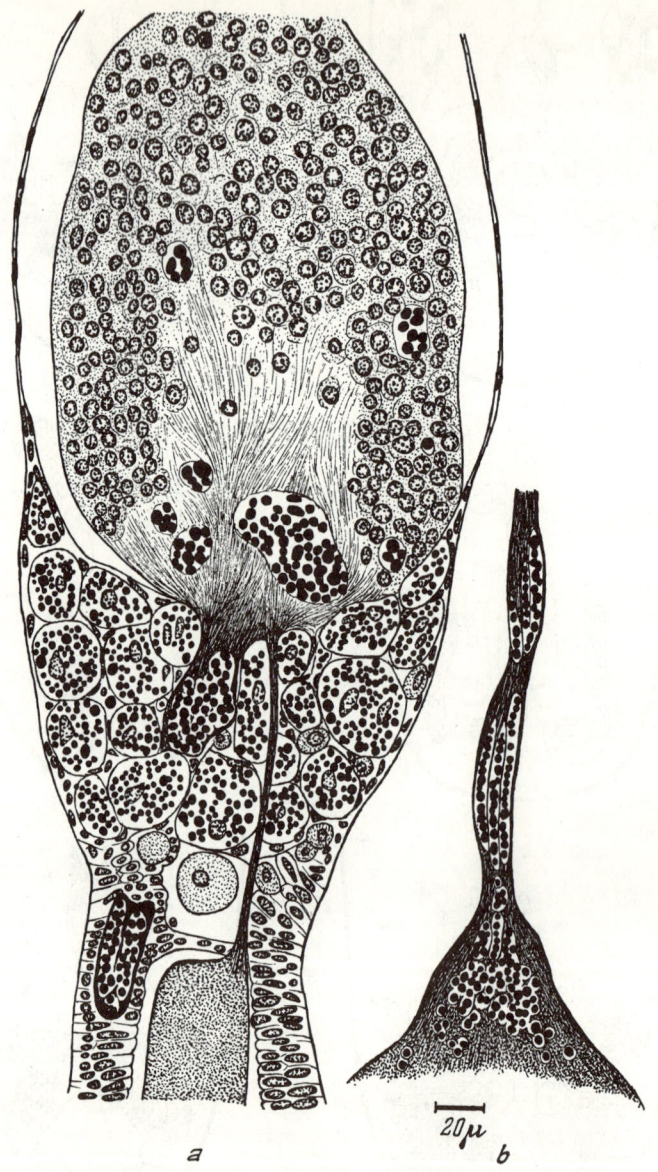

Fig. 226   *Bladina fraterna* Stål.   (*a*) Upper end of an ovariole; the symbionts in the plasma of the nutritive chamber and in a filial mycetome; (*b*) the symbionts in the strands leading to the young ovocytes.   From H. J. Müller.

phenomena found in no other cicadas. Whereas only cuneate cells are usually differentiated, here there is a separation of another type of follicle cells, which are elongated into slender clubs and toward the egg constrict the hollow space enclosed by the cuneate cells like a bottleneck. When the symbionts leave the cuneate cells, they first remain in the space enclosed by these "roof cells," which are undergoing regulated changes of position, apparently because the oocyte is as yet incapable of receiving them, and not until later do they pass into the oocyte through the mouth enclosed by them (ricaniid type). In the group the darker symbionts of the rectal organs are present in exceptionally small numbers.

A younger unclassified nogodinine larva fortunately reveals some details regarding the formation of these filial mycetomes. Its *n*-organ is located at the same place as in *Bladina* and has the same structure. In addition, it is surrounded by a thick layer of large dinucleate mycetocytes, which continues on each side into a huge lobe extending to the ovaries and, somewhat loosened there, surrounds the still very young ovarioles (Fig. 225*b*). The mycetocytes of these processes thus finally are situated at exactly the same spot on which the fusion takes place between the ovarioles and the still compact oviducts which are growing out toward them; in other words, the laterally developed and unusually voluminous infection mounds are penetrated by the oviducts and the ovarioles which are trying to reach each other (Fig. 227). Later in all probability whole mycetocytes are transmitted to the zones in question, for the components of the filial mycetomes have the same structure and, even before their transmission, changes take place in the young ovarioles which apparently are caused by the adjacent mycetocytes and presumably have the goal of acquiring the requisite space for the latter. The nuclei clump together and disintegrate in this area, the plasma is vacuolized, and the cells are finally disintegrated. With the further growth of the ovarioles, the connection with the mother mycetome tears apart after transplantation has been completed, and finally only a few remains of the infection mounds are attached to it, as in *Bladina*.

In addition to a group characterized by yeasts plus *f*-organs, the issids have another with three symbionts. In this group an *o*- or an *l*-organ is added to *a* and *x*, as in *Hysteropterum* and *Acrisius*, respectively. The bean-shaped, syncytially built *o*-organs form a special sharply set off zone, which is comparable to the infection mounds but is traceable not to an originally sterile nest of epithelial cells but to mycetome-derived sections with symbionts developing in a different direction. The *l*-organs, on the other hand, are very large, broad, elliptical formations with a confused network of thin filaments in a single giant syncytium.

Occupied by spherical symbionts, the unpaired *p*-organ of *Asiraca clavicornis* F. (delphacids) is significant because of its extreme sexual dimorphism. In the fifth nymphal stage it is of approximately equal size in males and females. Later it continues to grow substantially in females but decreases more and more in males. Concomitantly the symbionts, which appear first in the form of pale indefinite spots, are completely decomposed and the nuclei clump together and disintegrate. In May when the female organs are at the peak of their development, only a few

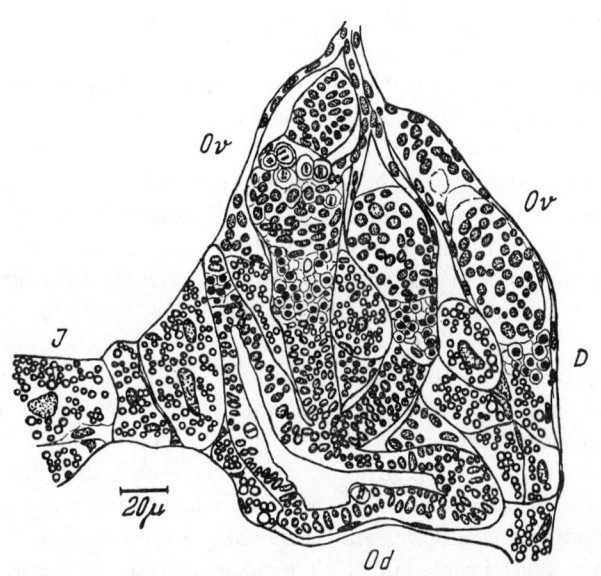

Fig. 227 Nogodinine. The cells of the infection mound of the unpaired *n*-organ penetrate between the rudiment of the ovarioles and the oviduct and thus establish the filial mycetomes. From H. J. Müller.

final remains of "symbiont debris" can still be discerned, and with difficulty, in the male mycetomes!

Also the *q*-organ of *Stenocranus* and *Kelisia* (delphacids, megamelines), like the *p*-organ of the related *Asiraca*, is decomposed except for mere traces in males.

As the last form that contains *x* + *a*-organs joined with a third symbiont type, let us consider the genus *Oliarius* (cixiines). We place it at the end of the progression for, owing to another complication entering here, it occupies an exceptional position in comparison with other fulgoroids housing a third guest with the *a*- and *x*-symbionts. A female of *Oliarius villosus* F. contains no less than ten mycetomes.   Thus Šulc, the first to investigate a species of *Oliarius* (*cuspidatus* Fieb.), concluded that five different symbiont types were present, all taking part in egg infection.   Soon thereafter the giant forms of the *x*-symbionts were found to be identical with the inmates of the rectal organ, and the number was reduced to four. Apparently, in spite of the large number and the diversity of the organs, the symbiosis here involves only three forms, for Müller succeeded in showing that the stock of mycetomes is fewer by far in males, which had not been studied until then, and that in males not only the rectal organ is lacking but also another organ which thus is equivalent in value to a filial mycetome used in transmission.   Figures 225*c* and 225*d* show the mycetomes of the two sexes.   The paired *x*-organs, populated by deeply lobed giant forms, are divided on both sides into two partial mycetomes, the posterior one in females holding the spherical *d*-organs as though in a dish. The rectal organ is situated behind the valvula and is composed of twelve to fifteen large mycetocytes, each with two highly bizarre, deeply cleft nuclei.   The tube-shaped *a*-organ extending along the exterior on each side consists of a single large syncytium and forms two infection mounds on each side.   The already mentioned *d*-organs are constructed from many uninucleate mycetocytes enveloped only by a thin anucleate membrane and are occupied by confused balls of long thin filaments.   Like the rectal organ, the *d*-organ cannot be found in male animals, but the fifth organ type, the *c*-organ, is still represented in both sexes.[1]   This is a broad unpaired formation situated far forward in the abdomen, usually asymmetrically to the rectum.   In its structure it shows great similarity to the *x*-organs; the nucleus-containing host plasma is reduced to a marginal border and to numerous septa which lead into the interior; a non-cellular membrane surrounds the syncytium.   The form of the obviously very delicate and feeble symbionts cannot be determined with certainty. Either they are confused balls of fine filaments or large states similar to the giant symbionts of the *x*-organs, but much more delicate and extremely tattered, with finely granulated protoplasm.

   Neither Müller nor Ermisch, who also investigated two species of *Oliarius*, was able to establish four kinds of symbionts in egg infection.   In all probability only representatives of the *a*-organs and rectal organs as well as thin filaments of the *d*-organ make their appearance.   The *c*-organ is actually a depot of degenerate *d*-symbionts, for it is lacking even in younger males and is histologically similar to the *x*-organ.   Furthermore, like the *x*- and rectal organs, *c*- and *d*-organs are never found alone.   The

[1] Unacquainted with Müller's work, Mahdihassan (1947) declared the *c*-organ of *Oliarius* to be testes (for this see Buchner, 1947) although it occurs in both sexes, possesses no exit of any kind, and differs completely in structure.

final answers to the question can only come from a study of the symbionts during embryonic development.

We have seen that cases in which the combination $x + a$ is extended by addition of a third form are extremely numerous. Now we shall see that cases in which a third symbiont is added to the $x + f$-organs are as rare as the former are numerous. There is just one instance, that of a

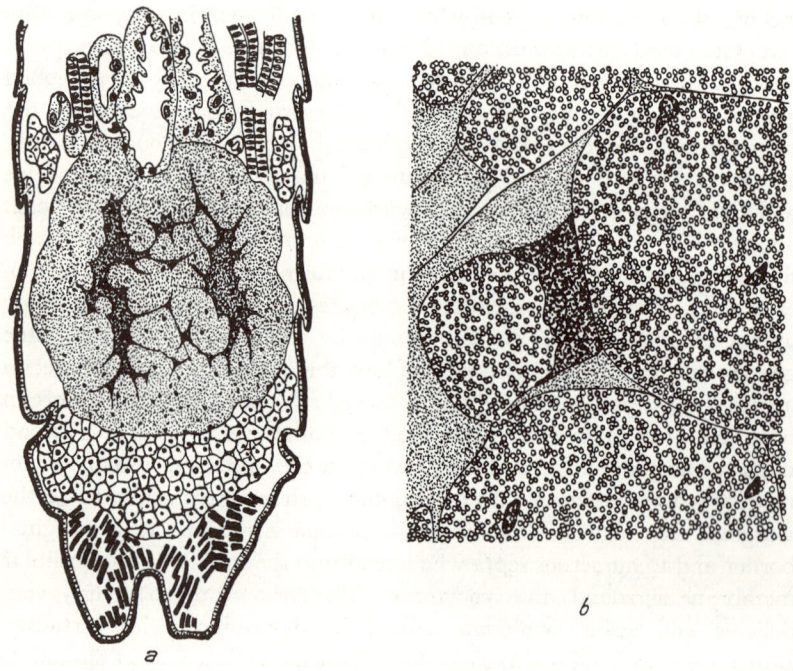

Fig. 228   Megameline.   (*a*) Abdomen with giant *r*-organ; (*b*) a section therefrom at higher magnification.   From H. J. Müller.

meenopline, in which besides the $x + f$-organs there are paired, but not sharply set off, areas of the fatty tissue which are occupied by pale tube-shaped symbionts and which have acquired a syncytial construction. Somewhat more common, on the other hand, is an enrichment by *r*-symbionts, as found in some megamelines and delphacines. These *r*-symbionts live in mycetomes which acquire a very unusual size and which, in proportion to the size of the host animals, are possibly the most voluminous of all cicada mycetomes. Figure 228 shows a megameline in which the *x*-organs are subdivided on both sides and the paired *f*-organs are very small. The *r*-organ, on the other hand, extends in laminae in all directions, sends out lobed processes, and stretches out toward the dorsal side with lateral bulges. Its construction is no less unusual. An anucleate membrane sets the organ off against the fatty tissue which is usually adjacent, and in general the organ gives the impression of being an infected, giant fat-cell complex which has been elevated to the status of a mycetome. As is usual with such a heavy load of protoplasm, the original syncytia are extensively fused and the protoplasm withdraws essentially into two symmetrical central zones from which processes, anastomosing in many ways, pass toward the periphery, where little plasma islands have been left behind here and there. The behavior of the nuclei is as remarkable as the construction of these unique organs as a whole. In the large plasma accumulations there are only a few giant lobed nuclei, but between the symbionts there are numerous uniformly distributed, very small nuclei. During separation of plasma and symbionts the nuclear development apparently is continued, but it differs this time in that some nuclei develop with suppression of division, whereas others continue to divide amitotically and become smaller. In the form of tiny spheres the symbionts fill up all the plasma-free spaces in enormous quantity and form no specific infection states. Tracheae and tracheoles span the organ and penetrate its peripheral parts.

The combination of yeast symbionts in the fatty tissue plus *f*-organs, encountered in some fulgorines, is extended only in a single species, belonging to the dictyopharids, through the acquisition of an *m*-organ, which replaces the usual *x*- and *a*-symbionts of this subfamily.

In a very few issines a three-fold symbiosis takes place which is distinguished in very unusual fashion by the joint appearance of the two chief symbionts, *x* and *H*, which usually occur vicariously.

In formation of the *x*-organ and the rectal organ, *Issus coleoptratus* Geoffr. is like those issines which unite *x*-organs with *a*- and *l*- or *a*- and *o*-mycetomes except that there is also a horseshoe-shaped slender *f*-organ with the usual indistinct filling of bacteria and an infection of the fatty tissue by yeasts. The latter ordinarily avoid the marginal zones of the fat lobelets, although there the separation is less sharp than with the delphacids, especially since the symbiont-containing areas are still penetrated by fat globules. In addition, there are groups of giant cells harboring yeasts exclusively which perhaps might be interpreted as incipient yeast mycetomes (Fig. 229).

On the other hand, *Issus dilatatus* Oliv. also contains primarily *x*-organs and yeasts but houses the latter in a well-defined unpaired mycetome, which, though not enveloped cellularly, represents a single giant syncytium. The tendency toward organ-like separation of infected fatty-tissue areas, observed again and again, in this case leads to a peak performance not attained elsewhere. As a third symbiont, in place of the *f*-organ which ordinarily tends to occur with yeasts, a paired *a*-organ is found,

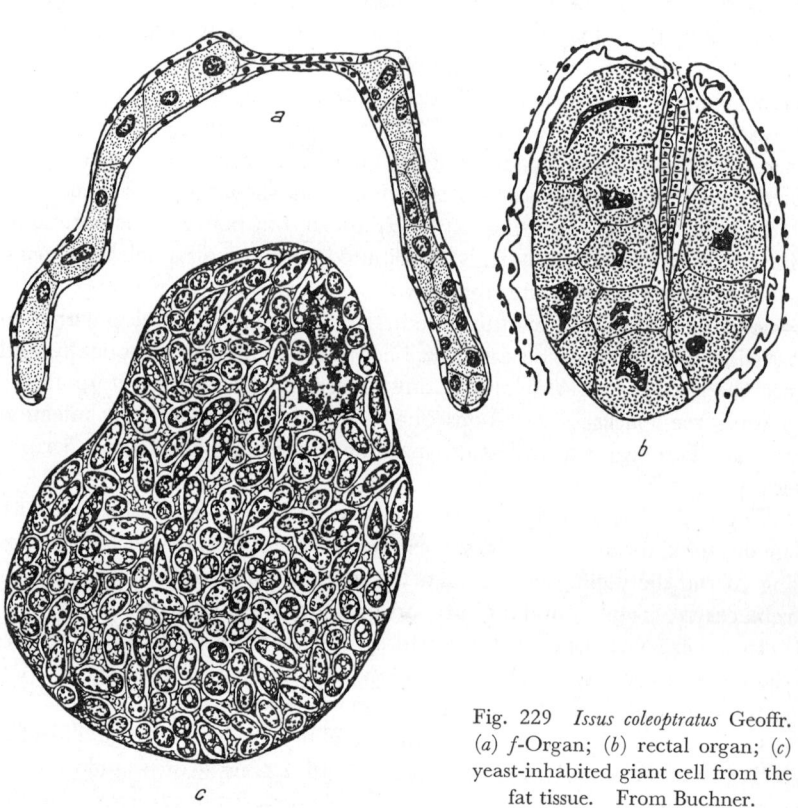

Fig. 229 *Issus coleoptratus* Geoffr.
(*a*) *f*-Organ; (*b*) rectal organ; (*c*) yeast-inhabited giant cell from the fat tissue. From Buchner.

that is, an organ type that ordinarily excludes the simultaneous presence of yeasts and is never compatible with *f*-organs. In a later discussion of preferred or always avoided combinations of symbiont types we shall return to these two interesting *Issus* species.

Only three species of fulgoroids have four symbionts. A form which is assigned to the genus *Oliarius* or at least belongs somewhere near it taxonomically—only larvae were available for observation—has the complete symbiotic apparatus described for *Oliarius* and, in addition, six or seven mycetocytes in the fatty tissue in one large loose group or in two or three groups. These are giant cells with highly bizarre nuclei and plasma thickly filled with minute cocci (Fig. 230).

In similar fashion two *Cixius* relatives have acquired a paired *e*-organ in addition to the mycetomes typical of this genus. On each side of the exterior of the *b*-organs a large syncytium, in the form of a broad oval sac, originates through joining of originally dinucleate cells and is loosely filled

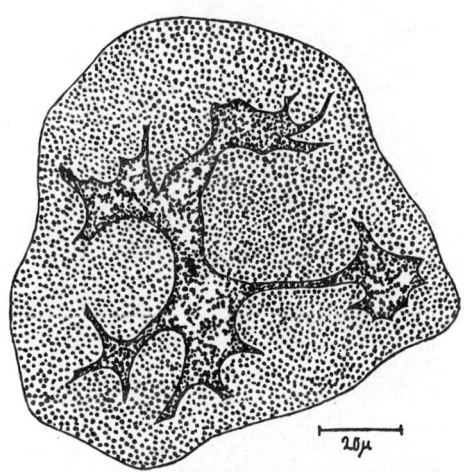

Fig. 230 Form belonging to the *Oliarius* group. One of the giant mycetocytes, irregularly embedded in the fat tissue, with coccus-shaped symbionts. From H. J. Müller.

with slightly curved rodlets. In this case it could be proved that all four types participate in the egg infection, which follows the pattern of *Cixius* exactly.

An unusual solution of the housing situation is afforded finally by some of the poiocerines (laternariids) which have associated an additional fourth symbiont with the already described *x*-, *a*-, and *k*-organs (Fig. 225*e*). Among the membracids we have already seen an unregulated infection of the mycetome epithelium and a more regulated one which affects the cells uniformly and extensively. Now *Crepusia nuptialis* Gerst., a species of *Poiocera*, and others furnish examples of completely developed "epithelial organs" of this kind.

The epithelial cells which surround the syncytia with their long thin a-tubes are thickly filled on all sides by delicate, long, filamentous bacteria (*Crepusia*) or by compact short tubes (*Poiocera*); their nuclei increase amitotically, and the little syncytia originating that way tend to fuse together into larger entities (Fig. 231). In this manner a thick mantle, enclosing the a-symbionts on all sides, originates which the development

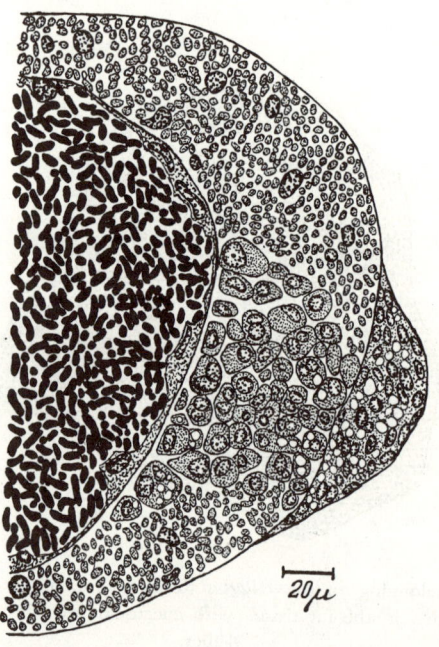

Fig. 231 Poiocerine. Section from the tubular a-organ with an additional symbiont inhabiting the mycetome epithelium. A separate infection mound is established for each of the two forms. From H. J. Müller.

of the infection mounds formed chiefly by the a-organs must take into account. And indeed the symbiont-containing margin is interrupted by a nest of sterile cells, and when these have been infected and the customary transformation to infection stages has taken place, there is nothing to prevent them from leaving the mycetome in the usual fashion. Moreover, infection stages are also formed in the case of the symbionts of the marginal zone, which are clearly acquired much later. Above the place where the rudiment of the infection mound of the a-symbionts sinks in, there extends a broad mound of additional, more or less syncytially fused and markedly vacuolized, cells. Later this mound stretches out in the form of a shield over a larger section of the epithelial mycetome and receives the symbionts of the margin, thus transforming the filaments into shorter, thicker, and more deeply staining structures.

In view of such a well-developed colonization of the *a*-mycetome epithelium by a younger acquisition, a special meaning is taken on by the observations which Ermisch made on *Trypetimorpha fenestrata* Costa, a tropiduchid. In addition to *a*- and *x*-symbionts, it possesses still another accessory inhabitant of the *a*-mycetome which disturbs its construction and inhabitants to a considerable degree. The typical epithelial covering of the mycetome disintegrates markedly, so that only a few of its nuclei lie between masses of delicate, only faintly staining bacteria. The latter not only form in this way a covering, interrupted only by the infection mound of the *a*-symbionts, around the original guests of the mycetome, but they also penetrate the syncytia inhabited by them. The result is pyknosis of the nuclei and, according to the degree of invasion, at first hypertrophy and finally disolution of the symbionts. In the infection mounds between the transmission forms of the *a*-symbionts are also found similar forms of the invader. Even though the latter could not be demonstrated with certainty in the thick symbiont balls of the egg cells, their transmission cannot be doubted, because they occurred in all insects examined. It should be noted that this new acquisition is strictly limited to the *a*-mycetome. It apparently represents a prestage to the harmonious incorporation of the epithelial organs of the Poiocerinae.

Ermisch was also able to demonstrate among the Araeopidae a form with four symbionts, in the interesting combination $x + f + r + H$ (*Issidaeus lugubris* Signoret). The slender *x*-organs, paired at first, became constricted off into four parts through the action of the *r*-mycetome, which contains numerous slightly spiral rods and swells up considerably. Each of the few paired, very small *f*-organs consists of about ten mycetocytes with minute spherical organisms. The yeasts occupy the fat tissue diffusely; although only sparsely present at first, later they are located in the female within almost all fat cells of the abdomen and free between them. The dissolution of the symbiosis in the male, which has been so often established, is in this case quite pronounced. When the yeasts in adult males become so few that they are easily overlooked, the nuclei of the *f*-organs of old males show pyknosis and few symbionts are left in the mycetocytes. The *r*-organ ceases its growth and attains about one-fourth the volume of the corresponding female organ.

The systematic position of the only fulgoroid known to contain five different symbionts is not quite certain. Although it has been classified among the derbines, Müller is more inclined to classify it as a cixiine. The symbiotic formula of the animal in question is: $x + a + g + h + \beta$. In other words, to the well-known combination of an x- and a-organ are added two previously unknown, unpaired mycetome types, and a fifth, also new, form which is housed in the fatty tissue.

The g-mycetome consists of three concentric zones. A marginal layer of large, more or less rectangular syncytia follows upon a flat epithelium; the interior consists of a single large synsyncytium with nuclei of various sizes. Brighter, symbiont-poorer lines in it seem to suggest former delimintations. The marginal layer is interrupted by the rudiment of a large infection mound. The symbionts are apparently the same in both zones, but they are delicate densely packed forms which suffer on fixation; hence that nothing definite can be said about them. Apparently they have the shape of spherelets or short tubes. Because of certain observations concerning the development of the x-organs, to be discussed later, Müller surmises that the unusual stratified construction may perhaps come about through gradual degeneration of the older central syncytia and appropriation of their content by a second peripheral cell generation.

Close behind the g-organ lies the h-organ, which also does not recur in the long series of cicada mycetomes. It is a large syncytium, surrounded by a very flattened cellular envelope, in which the flattened nuclei are lodged in the inner surface of a sharply set off marginal plasma and which is penetrated by long filamentous bacteria arranged in whorls and arches and filled with eosinophilic granules.

The fifth symbiont of this interesting form, a spherico-polygonal organism with dense protoplasm, populates smaller and larger areas of the fatty tissue, in which the cells merge into syncytia as a result of infection. Usually the syncytia are situated then in the vicinity of the h-organ in greatly variable numbers. This organism, which has not yet established itself firmly, is apparently the symbiont acquired last.

Our enumeration of the various combinations of fulgoroid symbionts known at present, which of course does not even begin to exhaust the actual diversity, is now drawing to a close. Though at first glance perhaps boring, it reveals an astonishing unique phenomenon, posing a series of general questions. In ever-changing forms, new types of independent sites are created for additional symbionts; the tendency to reserve special areas for them in the mycetomes already present plays no role at all, aside from the exceptional case of the epithelial organs of the *Crepusia* type. The more or less parasitical-appearing accessory forms in the interior of the mycetomes, observed often among the cicadoids and extremely often among the membracids, almost never occur among the fulgoroids. Only the tettigometrids constitute an exception if we disregard the Trypetimorpha just described, and for this reason they are being considered last, a decision justified by the completely isolated position of this family in the fulgoroid system.   Its very uniform symbiotic arrangements are quite in keeping with its exceptional position.

All tettigometrids studied thus far contain: an *x*-organ more or less clearly divided into partial mycetomes, together with the associated rectal organ; paired or unpaired *a*-organs; and a third large unpaired *s*-organ, oined sometimes by an additional fourth, half-parasitic guest which appears in the *s*-organ (Fig. 225*f*).   In the rectal organs the unusual number of the mycetocytes united here, up to forty, is striking.   The inmates of the *x*-organs are differentiated in a very remarkable manner from all others known at present.   Originally roundish polygonal fragments, they undergo the usual widespread serration and pinnate cleavage, but after that there is more or less complete division into little tube-shaped or compact, almost roundish symbiont fragments (*Tettigometra atra* Hgb., *obliqua* Panz., and others).   That it is not a regular cleavage process is evident from certain irregularities which are always observable.   The phenomenon manifests itself in degrees of intensity varying not only from species to species but also even from individual to individual, and it does not always affect the symbionts of one mycetome in the same way. Individual giant forms may finally be surrounded by numerous smaller structures which greatly resemble the *a*-symbionts.   Thus it is justifiable to interpret this behavior, which is important in evaluating the unique giant symbionts, as another proof of the extreme degree of degeneration that changes them so much from their original bacterial nature.

The *s*-organ, surrounded by an anucleate membrane, consists of sparsely nucleated mycetocytes which show slight inclination toward fusion, and sometimes also of cells which remain uninucleate or at most become binucleate, with spherical or oval symbionts.   In three species there were intruders, limited to this voluminous organ, which may evidently be interpreted as accessory but still half-parasitic symbionts.   As heavier compact rodlets or very small rodlets, they fill the spaces between the loosely set mycetocytes or, where fusion occurs, crowd among the symbionts in huge aggregations or narrow tracks corresponding to former limits.   Some *s*-symbionts show clear marks of degeneration, and the mycetome as a whole also gives the impression of being disturbed.

The embryonic and larval development of the symbiotic devices of the fulgoroids, investigated until now almost exclusively by Müller on the basis of *Cixius nervosus* and *Fulgora europaea*, affords interesting insights, first, into the formation of the various mycetomes and their population by symbionts which are intermingled in the egg and, second, into the unusual origin of the rectal organs.   My own data, acquired earlier on the basis of several

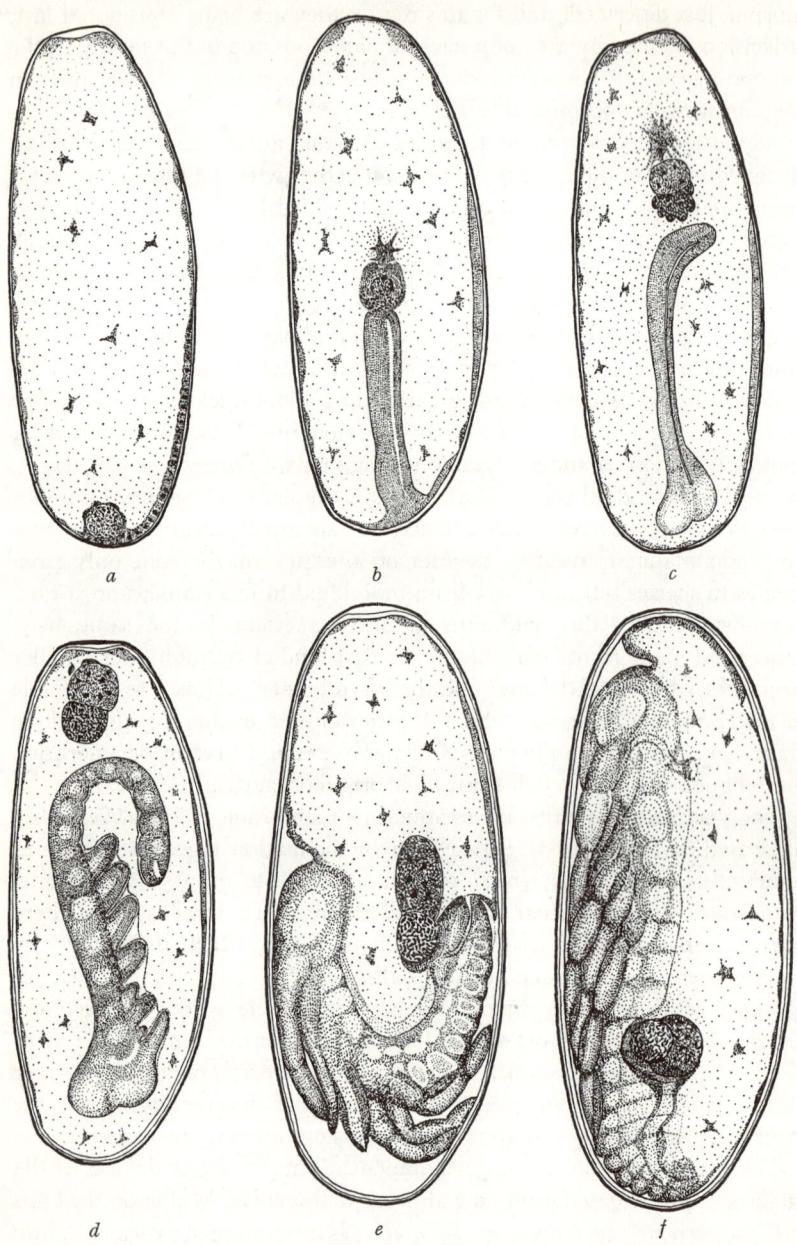

a   b   c

d   e   f

embryos of *Eurybrachys*, represents only a very modest supplement to Müller's work and is of value only because it involves an entirely different symbiosis type. Only Sander and Ermisch have made recent contributions to this phase.

During cleavage of *Fulgora europaea*, blastoderm cells from below and vitellophages from above approach the symbiont balls and increase on its surface to form a connected envelope. Then the embryonic blastoderm in the posterior part of the dorsal region thickens, thus being sharply set off against the extra-embryonic blastoderm, and invagination of the germ band begins, by which, as usual with the Homoptera, the symbiont ball is shifted to the front (Fig. 232*a,b*). During grouping of the blastoderm cells and especially during the first phase of invagination, the envelope cells on the symbiont ball are increased substantially, the nuclei becoming larger and the plasma body swelling. At the same time their two-fold origin again becomes apparent, for the vitellophages now loosen in structure and send out amoeboid processes to take up yolk masses while the less vacuolized elements situated farther back retain their character as blastoderm cells. With progressive invagination this polarity is considerably increased. A club-shaped accumulation of yolk reduced to small pieces and held together by the vitellophages arises anteriorly; in front of this is a plasma radiation, such as is found in Heteroptera, Anoplura, and other test objects; and posteriorly the envelope cells begin to fill with the dark emigrants from the *a*-organ. As in similar cases, the forces at play in such dissociation cannot be judged from the histology. While the *a*-symbionts approach the posterior envelope cells, the smaller rectal symbionts reach the front and enter elements of vitellophagic origin which sometimes even now also process the yolk. Unaltered, the small *m*-bacteria retain their middle position. Thus, in a period from the ninth to the eleventh day after egg deposit, a primary collecting mycetome arises; this mycetome consists of a roundish syncytium with rectal symbionts and, attached behind, a grape-like cluster of mycetocytes with *a*-symbionts, and its interior contains the still anulceate bacteria of the future *m*-organ (Fig. 233).

In this state it is detached from the germ band now forming an *s*-shaped curve (Fig. 232*c*). Observation of these states in life shows that *a*- and rectal symbionts are both in the process of lively increase at this time. Also, the number of *a*-mycetocytes still increases visibly until, by the end of

Fig. 232 *Fulgora europaea* L. Embryonic development: (*a*) development of the germ band; (*b*) the invaginating germ band shifts the primary mycetome forward; (*c*) after separation of the *a*- and rectal symbionts the primary mycetome frees itself from the germ band; (*d, e, f*) the primary mycetome before, during, and after revolution; in (*e*) and (*f*) secondary covering with mesodermal cells. Partly schematic. From H. J. Müller.

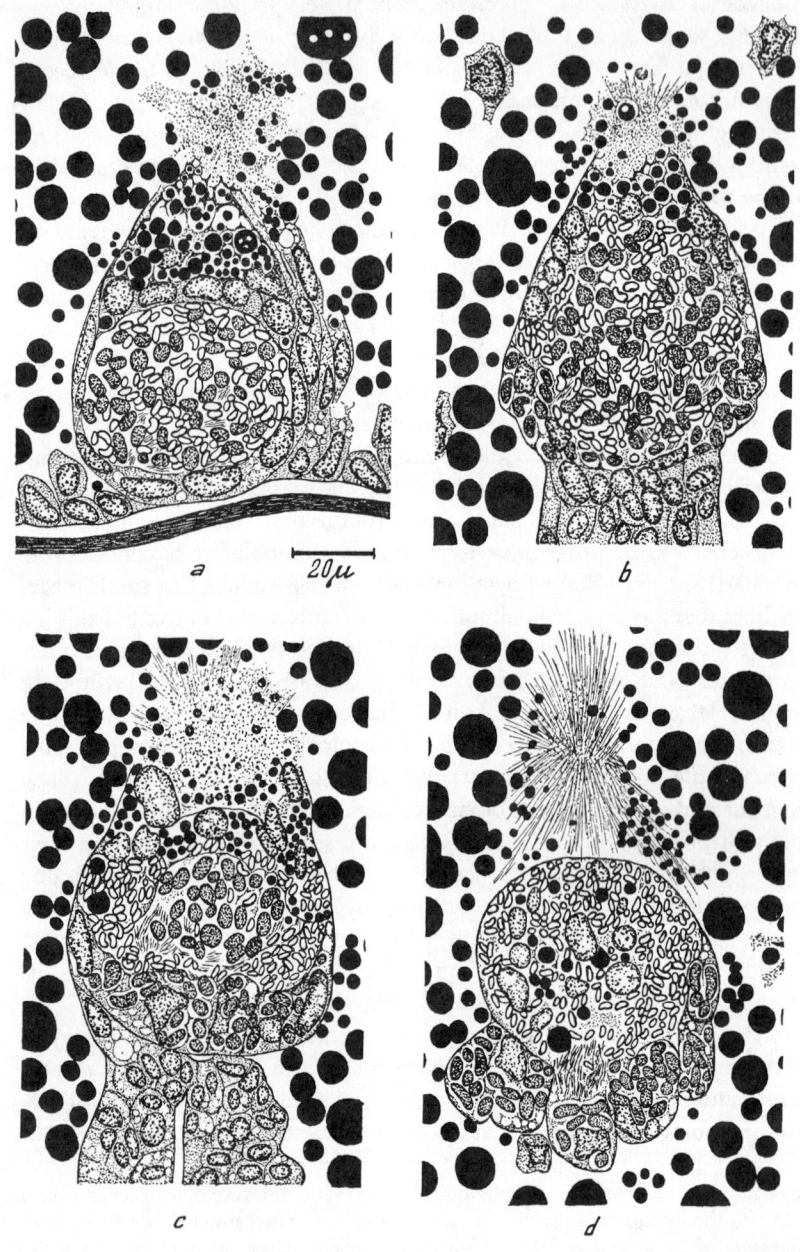

20μ

the invagination period, on the fifteenth to the twentieth day, the two symbiont types are increased about four-fold and populate a formation shaped
like an opened bun which is now located close to the upper pole. The
smaller bacteria, by this time having been forced toward the outside, are
loosely attached to the rectal syncytium in small anucleate heaps held
together by a gelatinous envelope. The plasma radiation, which had
hovered like a star at the tip of the germ band, is also preserved for a while
after detachment of the collecting mycetome. Indeed, it shows its
strongest development during the process, and Müller considers it quite
possible that it participates in the development and then draws the symbionts even farther to the top (Fig. 232*d*).

Sander (1956), studying the behavior of the mycetome rudiment during
embryo formation in *Pyrilla,* found strange relationships to progressive
yolk cleavage which remind one of observations made by Baudisch (1958)
in *Pediculus.* An ever-decreasing yolk part not yet affected by cleavage
("core" of yolk) here enters into close relationship to the radiation, on
which finally a large cluster of yolk-free cells is attached.

Ermisch studied the behavior of the symbionts of *Conomelus* and *Stenocranus* during embryonic development. In *Conomelus* he demonstrated a
great similarity to the behavior which Müller found typical of *Fulgora* and
Lixius, but *Stenocranus* differed in many aspects. Here the blastoderm
cells are located some distance from the symbiont clump and therefore do
not become a part of its covering. Vitellophages, however, surround it
and resorb clumps of yolk with the branches proceeding outward, while
those moving inward form a net which encompasses the symbionts. The
nuclei, which later penetrate between the yeasts, are also derivations of
the vitellophages. The *f*-symbionts could not be recognized with certainty. Like Sander, Ermisch is convinced that the radiation, which is
more poorly developed in *Conomelus* than in *Stenocranus* and *Fulgora,* serves
the purpose of releasing the embryonic collecting mycetome from the germ
band.

It has already been mentioned that in *Cixius nervosus* the symbionts are
extensively sorted out even before the end of egg infection (Fig. 220*d*); correspondingly, the final separation takes place in a much less impressive
manner. The hindmost blastoderm envelope cells open toward the side

---

Fig. 233   *Fulgora europaea* L.   Development of the primary mycetome and separation
of the symbiont types: (*a*) the symbiont mass surrounded by epithelium shortly before
invagination of the germ band; (*b*) beginning of invagination, first uptake of the
a-symbionts into cells; (*c*) development of the syncytium of the rectal symbionts; (*d*)
a- and rectal symbionts separated, *m*-symbionts not yet in cells.   From H. J. Müller.

turned to the symbionts, gripping the mass of small *a*-symbionts already concentrated there with branched and raveled processes, and overgrowing them. Simultaneously the rectal symbionts populate the other cells, again of vitellophagic origin, which do not participate in the disintegration of the yolk to the same extent as in *Fulgora*. With plasma processes they fish, as it were, the last rectal symbionts from the remaining *b*-symbionts and later merge to form a syncytium.

As with the little bacteria in *Fulgora*, the *b*-symbionts are not provided with nuclei at first, yet, unlike the bacteria, they make up a substantial

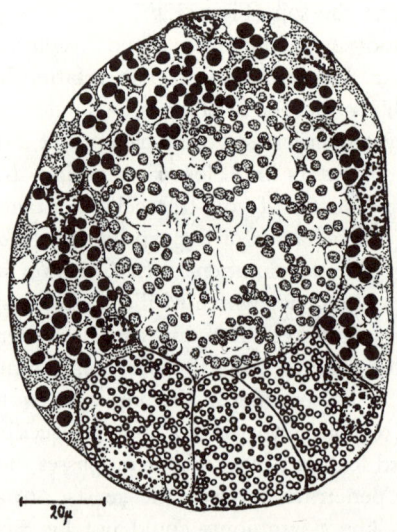

Fig. 234   *Cixius nervosus* L.   Primary mycetome after separation of the three kinds of symbionts.   From H. J. Müller.

portion of the collecting mycetome, which in *Cixius* is extremely imposing and remains round (Fig. 234). There is only an intimation of the development of the plasma radiation in *Cixius*, in which the germ band shoves the symbiont ball to the anterior end, where it is more or less mechanically held back.

Even before involution begins in the *Fulgora* embryo, the collecting mycetome is reattached to the embryo near the posterior end of the germ band. Thus cells which apparently belong to the lower lamella or at any rate are of mesodermal origin are added, and their arrangement resembles a shallow bowl. During revolution the mycetome maintains its position, thanks to this fusion, and passively reaches the area of the fourth segment from the last (Fig. 232*e,f*). During and after displacement, the mesodermal cells still increase substantially and finally not only do they form an epithelial covering over the whole mycetome but also they are inserted among its symbionts and thus initiate their separation.

At the same time the two symbiont types begin a period of marked growth and structural change. The *a*-symbionts grow in length and are forced into curves and spiral coils by a spherical envelope surrounding them. Their vacuoles are increased to such an extent that the long tubes are swollen in places and temporarily appear to degenerate. The rectal symbionts, which until now have been small slender tubes, also develop considerably, and the texture of the formerly dense and strongly staining plasma loosens to the extent that it stains very lightly. At the same time the plasma is enriched with strongly eosinophilic granules, and here and there irregularly lobed and emarginated forms are observed which almost appear to be miniatures of the giant symbionts of the future *x*-organs.

The collecting mycetome, which naturally becomes much larger with the growth of its inmates, is decomposed into its components soon after revolution. In the process the *a*-organ remains unpaired, whereas the organ populated by the forms from the rectal organ divides into a right and a left half, thus revealing even more clearly that it is the rudiment of the *x*-organs, which are indeed also paired. During these processes the *m*-symbionts, clearly documenting their much poorer adaptation, are all dispersed in the body cavity, usually joined in small clumps. Occasionally they appear to have been taken in already by mesodermal cells. When they prefer the region of the midgut slings, this is the first indication of their future position.

Much more significant, of course, is the history of the *b*-symbionts of *Cixius*. After revolution when the rectal syncytium and the cells with the *a*-symbionts begin to withdraw from the collecting mycetome, the *b*-symbionts, as yet not provided with nuclei, slip forth through a gap which makes its appearance. They are immediately taken up by mesodermal cells, and the young mycetocytes, remaining uninucleate, are arranged into a paired organ which at first has no cellular envelope, since the *b*-symbionts were still located in the interior at the time when young mesodermal elements formed an epithelial covering around the collecting mycetome.

The further history of these envelope formations occurring in both test objects is as strange as it is significant. It can be studied most clearly in the *x*-organs. Here the cells are almost cubical in the beginning but, after apparently losing their mitotic capacity for division, are markedly elongated, owing to growth of the organ, and flattened off. Then they develop a strange polarity, just as in the case of the blastoderm cells surrounding the symbiont ball, in that they remain smooth toward the outside but lose their limits toward the inside and reach into the peripheral area of the mycetome with branched protoplasmic processes to entwine the nearest symbionts.   At the same time the old primary mycetome nuclei and the plasma areas surrounding them are destroyed.

The original syncytium is thus replaced by a second garniture of cells which also assume a syncytial character.   Comparable transmission of symbionts from primary to secondary mycetocytes occurs in other test objects, but no other symbiont bearer manifests this "stealthy" narrowing of the syncytium with simultaneous preservation of outer contours.   In the *a*-symbionts, on the other hand, secondary envelope cells are conversely employed to enlarge the mycetome.   Here the central mycetocytes do not degenerate, but the intensity of increase in symbionts appears to weaken temporarily.   At the same time individuals among them transfer at the periphery to the envelope cells and increase there so quickly that the cells are soon densely filled.   Beginning in the center, disintegration of cell limits, and hence formation of syncytia, take place toward the end of embryonic development.   According to Müller, the final sparse envelopment of the *x*- and the *a*-organs may be traced to the peritoneal epithelium.

Thus *x*-, *a*-, *b*-, and *m*-organs develop from the collecting mycetome. The forms from the maternal rectal organs clearly become *x*-symbionts, but the origin of the rectal organ itself, which is limited to females, still remains an enigma.   Only after the midgut begins to develop in females does it enter into remarkable relationships with the young *x*-organ which lead very circuitously to formation of the rectal organs.   In *Fulgora* the rudiment of the midgut, at first in the form of a solid cell cord, grows obliquely from the blind end of the proctodaeum to the anterior, then quickly curves back and approaches a point where the two halves of the *x*-organ are connected merely by a narrow symbiont bridge which is penetrated by a little canal.   The anterior end of the midgut cord, which at first consists of only a few loosely connected cells, now attaches itself, with simultaneous manipulation, more and more closely and firmly to this bridge of the *x*-organ (Figs. 235*a*; 237*a*,*b*).

Through this contact exceedingly strange alterations occur in some of the organisms predestined to become giant symbionts.   The secondary mesodermal envelope cells, at this time about to receive them, send processes out between them, as we stated above.   Where contact is made with the midgut rudiment, these protoplasmic collecting arms extend very deep into the syncytium, thus surrounding the symbionts there more intimately than elsewhere.   Thus on both sides, at the point where the connecting tube is inserted, well-limited mycetocytes are formed; here the considerably grown *x*-symbionts quickly fall into smaller fragments; they are also conspicuous because of their deeper staining, which takes place immediately (Fig. 237*b*).

Meanwhile the tip of the midgut rudiment advances toward the front and passes into the yolk. In many places it begins to form a lumen, but in the contact zone it remains massive at first. Now the moment arrives when the mycetocytes, laden with the smaller symbionts, become migratory cells. They force their way from right to left through the narrow bridge canal which opens straight into the intestinal rudiment, travel to

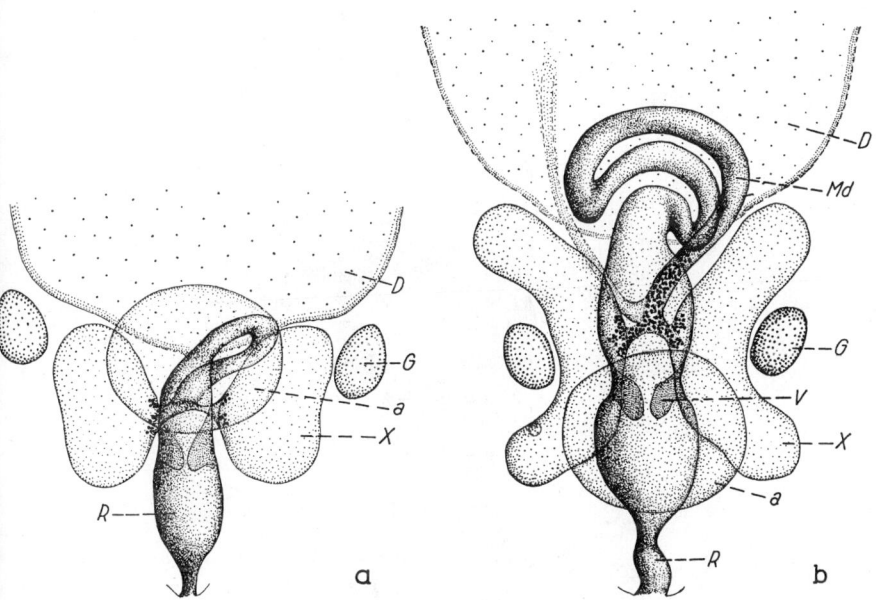

Fig. 235 *Fulgora europaea* L. Development of the rectal organ I: (*a*) the midgut rudiment locates itself with its tip on the bridge of the *x*-organ, wherein migratory forms of the inhabitants are developed; (*b*) development of the midgut has progressed, the migratory forms develop in it a provisional gut organ. *a*, *a*-organ; *G*, gonad; *D*, yolk sac; *R*, rectum; *X*, *x*-organ; *Md*, midgut, *V*, valvula pylorica. Schematic. From H. J. Müller.

the rudiment, and, gliding forward, are arranged in an irregularly delimited cell mass which can be called a transitory intestinal organ (Fig. 235*b*). Shortly before hatching, its cells begin to disintegrate and its nuclei to degenerate. The symbionts are freed, and within a few hours, shortly before the freshly hatched larvae ingest their first food, they are propelled by peristaltic movements, now beginning, through the complicated convolutions of the midgut in the direction of the rectum. Without adhering to the intestinal wall, they approach the valvula rectalis which looks like a large swelling even before the midgut rudiment is full-grown (Fig. 236).

Meanwhile remarkable preparations are made during the decay of the intestinal organ. The intestinal cells located just above the valvula

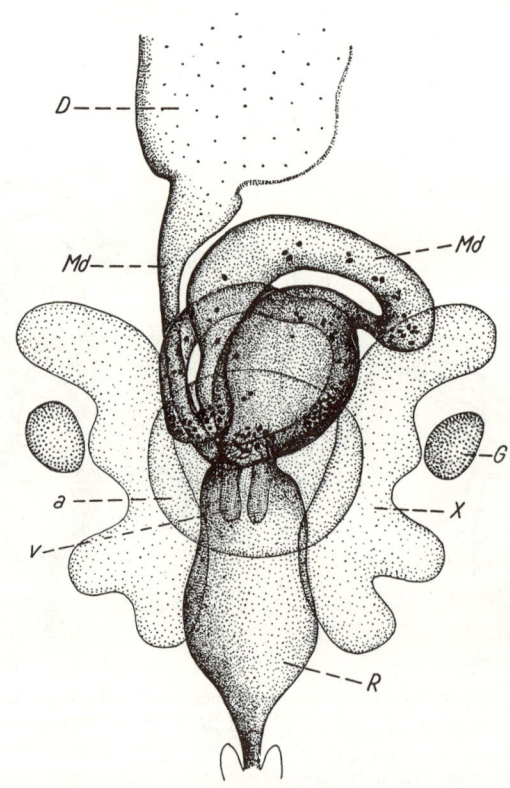

Fig. 236   *Fulgora europaea* L.   Development of the rectal organ II.   The gut organ is degenerating, the symbionts, moving posteriorly, are caught up on the valvula by a grid.   Schematic.   From H. J. Müller.

rectalis sprout long, protoplasmic, branched processes resembling weir-baskets and send them into the greatly extended lumen.   Arranged in two or three layers, they form an apparently sticky network to which all the symbionts without exception adhere as they float up.   At the base of the "weir-basket" cells a ring-shaped slit in the intestinal epithelium is opened and leads into the ring fold, which appears at all clearly only now, and into

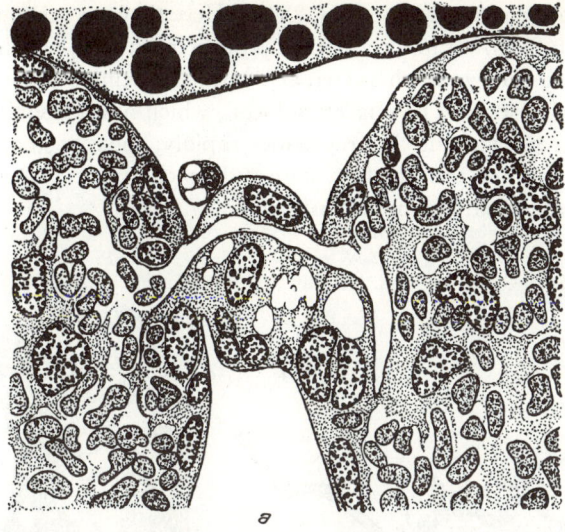

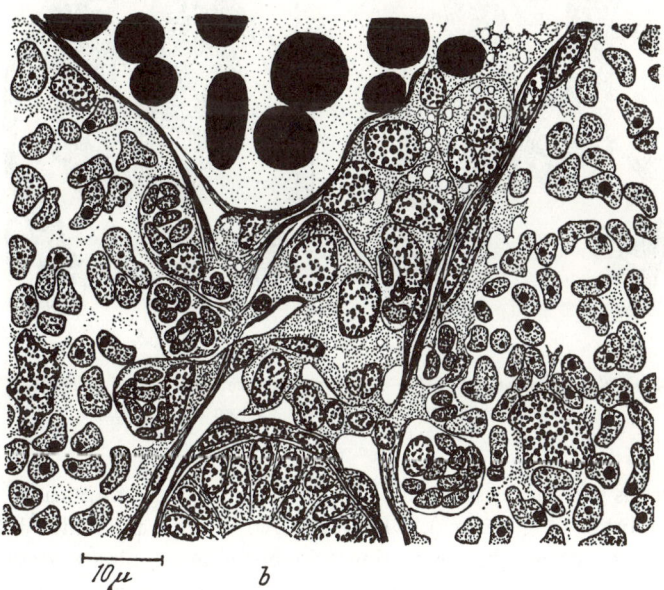

Fig. 237   *Fulgora europaea* L.   (a) Cells of the midgut rudiment begin to grow around the bridge between both halves of the x-organ; (b) growth has progressed, the young giant symbionts have disintegrated into smaller, darker migratory forms in specific epithelial cells.   From H. J. Müller.

which all the symbionts glide (Fig. 238).   Here they are taken up by cells belonging to the lower and retrograde portion of the fold, which still bears a pronounced embryonic character, and at the same time lose their capacity for mitotic division.   The migrating forms, which now cease their travels, lose their strong staining and reproduce rapidly; the mycetocyte nucleus is severed once, and thus the final state of the rectal organ with its numerically fixed binucleate giant cells is started.

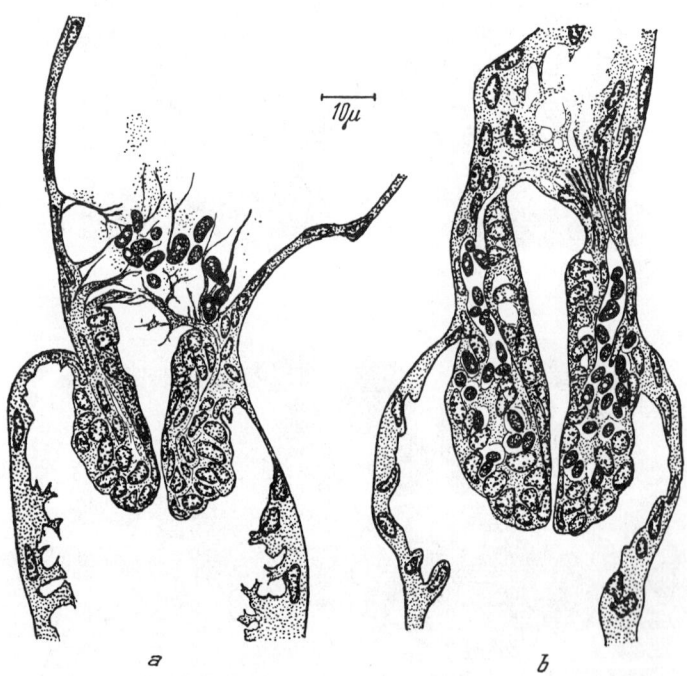

a   b

Fig. 238   *Fulgora europaea* L.   (a) The symbiont on the valvula rectalis; (b) the symbionts pass into the valvula.   From H. J. Müller.

There is not the least trace of all these astounding adaptations in the male embryos (Fig. 239a,c).   The unpaired x-organ is divided into two mycetomes without the connecting bridge; the developing midgut grows past them but there is no fusion of any kind; the secondary envelope cells take over the symbionts, but they do not become mycetocytes; and conversion of intestinal epithelial cells into "weir-baskets" does not occur at all.

Cixius differs in several respects from *Fulgora*.   A broad connection remains between the two halves of the x-organ, and the developing midgut attaches its end to the unpaired section, as though with a sucker.   The

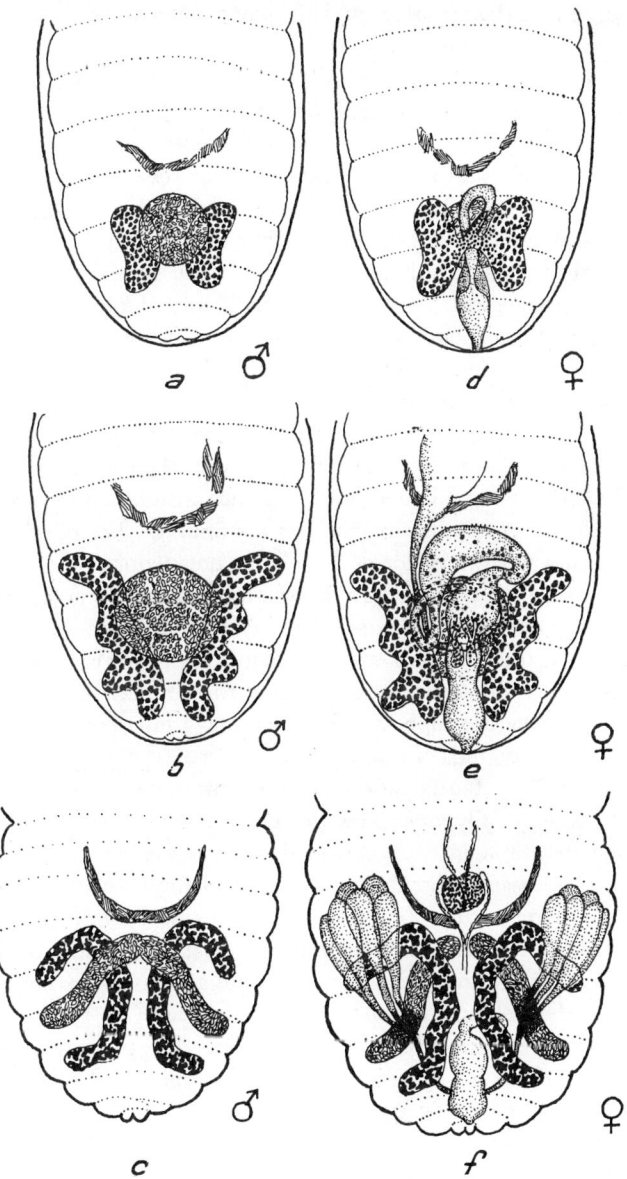

Fig. 239   *Fulgora europaea* L.   Postembryonic development of the symbiotic organs:
(*a*, *b*, *c*) in the male; (*d*, *e*, *f*) in the female; in (*e*) development of the rectal organ; in
(*f*) induction of the infection mound by contact of the *a*-organs with the oviduct.
Schematic.   From H. J. Müller.

envelope cells differentiating at this place form a very characteristic accumulation extending deep into the mycetome, and the migratory cells are taken into the intestine in closed formation.   The provisory intestinal organ is much more sharply defined, and it distends the midgut correspondingly.   The weir-basket, probably because the intestinal section in question is much narrower here, consists of less conspicuous, almost antler-like processes.   The symbionts fill the valvula only unilaterally at first and later glide backward still farther in the young larvae, quite in accord with the final position of the rectal organ behind the valvula.

Thus embryological research has brilliantly confirmed my earlier hypothesis of the genetic connection of *x*- and rectal symbionts.   The extensive degeneration of the *x*-symbionts, although it would be wrong simply to assume at the start that they are "unwanted," makes them unsuited at any rate to act as starting material for egg infection, and in females it forces the host to transplant some of the symbionts to a place where the environmental conditions stop such a situation in time, that is, before the giant growth has reached its maximum.   For this purpose the host cultivates migratory forms which do not acquire until later their typical shape in the rectal organ, becoming small, oval, or short tubes in *Fulgora* and longer tubes in *Cixius*.   Finally, they undergo a second change of form which, though small, renders them suitable for egg infection.   In this sense the rectal organ is comparable to the filial mycetomes of the Anoplura and the Mallophaga which serve egg infection, or comparable to the infection mounds of the *a*-organs.

It would take us too far afield to enter into detailed discussion of the changes, already fully described by Müller, which occur during postembryonic development.   As expected, all events concerned with growth, potential subdivision, and shift in position of the mycetomes take place strictly according to rule (Fig. 239).   In each case a very definite increase in size of the individual symbiotic organisms is in proportion to the growth of the host animals; in other words, the rate of increase in the symbionts during larval development is also regulated precisely by the host animals.   As proved statistically, only in the *x*-symbionts does progressive growth of symbionts leads to the enlargement of their organs.   The splitting into partial mycetomes and in general the regulated shifts in position of the symbiotic organs apparently always represent passive processes caused by increased volume and shifting of the various organs in the host animal.

Of greater general interest among postembryonic phenomena is the genesis of the infection mounds of the *a*-organ, a process which could be followed more accurately in *Fulgora* and which proceeds entirely in accordance with observations made with *Aphrophora*.   In the very earliest nymphal states the arched paired oviducts attach themselves closely to the *a*-organs, although this close contact, which in some cases amounts almost to fusion, has no immediate visible effect (Fig. 239*f*). It is not until the beginning of the fifth nymphal stage that a slight thickening, apparently traceable to cell divisions, appears in the area of the accumbent oviduct in the epithelium of the *a*-organ. Because of animated mitotic divisions, this place later swells rapidly to a mound which is flat at the beginning but soon forms a hemisphere and finally is inserted into the mycetome. This cell nest always stands out in sharp relief from the section of the mycetome populated by the symbionts, but only in the imaginal state is it more clearly differentiated from the bordering epithelium, from which it doubtless takes its origin. The oviducts usually give up their immediate contact with the mycetome toward the end of the fifth nymphal stage; the cells of the infection mound which become dinucleate through amitosis begin to be populated in the young imago. In *Cixius* it could be shown that, correlative with the paired infection mounds on each side, double contact with the oviduct actually does occur temporarily.

Fulgora and *Cixius* provided valuable insights into the development of the symbiotic devices, but in view of the enormous diversity of fulgoroid symbiosis we must not forget that they afford only a sectional view of an area which abounds in interesting details. For instance, one would like to have actual knowledge of the embryonic development in *Oliarius* instead of being forced to the mere assumption that another mycetome pair is connected with the *x*-rectal organ association! It would also be interesting to study at greater depth the behavior of the yeast symbionts in fulgeroids during embryonic development, now that several observations with *Eurybrachys* embryos show that in this case also the nucleated symbiont ball is first thrust forward by the germ band and is advanced during involution, adhering to the embryo, just as in *Fulgora* and *Cixius*, to the posterior pole, where it gradually seems to disintegrate. At any rate, yeasts infect extensive areas of the fatty tissue of embryos about to hatch. One is reminded of similar tendencies in coccids toward temporary organ-like binding of yeasts which are otherwise scattered diffusely in the fatty tissue.

THYSANOPTERA

At the International Congress in Vienna in 1961 Bournier reported for the first time a symbiosis in Thysanoptera. The insect concerned was *Caudotrips buffai* Karny, one of the largest European forms which sucks out fungi. Although the entire cycle is not yet known, this short publication is sufficient to give an approximate picture of the type of symbiosis, which shows a striking similarity to that of species of *Pediculus*. From the beginning of yolk formation on, a spherical clump, about 20 μ in diameter and consisting of very delicate thread-like bacteria, is located in the posterior region of the ovocytes. Only the oldest ovum of the ovariole is infected. Even though the transmission of symbionts could not be observed, certain facts lead us to the assumption that egg infection takes place in a fashion similar to that in *Pediculus*. Each oviduct ends in an ovarial ampule, which in the living state is bright red and which is reminiscent of that in *Pediculus*. Also, the little information we have concerning the later fate of the symbionts fits in well with *Pediculus* symbiosis. *Caudotrips* belongs to the few ovoviviparous forms. In the beginning only two to four nuclei are found in the symbiont clump. During invagination of the germ band the symbionts are pushed to the upper pole. In the course of further development the mycetome slides again to the rear, as is usual in such cases, until it has reached approximately the middle of the embryo. Often it appears then to be divided into two or three fragments of unequal size. During further development it comes to lie on the ventral side and loses its spherical form. At the same time its nuclei become constantly smaller. At this stage the egg is laid. Nothing has as yet been published concerning the further fate of the mycetome in larvae and adults.

The extent of symbiosis in the Thysanoptera remains to be investigated. This may be no easy task, considering the minuteness of the insects. Bournier has already studied a number of additional genera, without coming upon a similar arrangement; however, it has been entirely a case of forms which pierce the cells of higher plants. Therefore Bournier assumes that such species represent a counterpart to the Thyphlocibinae. Since there are also species which form galls, which suck nectar, and which live as predators on small insects and mites, there might possibly be interesting relationships between nutrition and symbiosis. It would be no less interesting if a more penetrating investigation of *Caudotrips* should still more definitely demonstrate relationships to the *Pediculus* symbiosis, since Anoplura and Thysanoptera are closely related phylogenetically.

7

Symbiosis in animals sucking vertebrate
blood and feeding on corneous substances

HIRUDINES

RHYNCHOBDELLIDS. It was recognized by Reichenow (1921, 1922) that the "esophagus gland" of *Placobdella catenigera* Moquin-Tandon, which feeds on the blood of turtles, is erroneously named and actually is always infected by symbiotic microorganisms. The gland consists of two evaginations of the intestinal canal, swelling like clubs toward their blind ends and originating between proboscis and stomach (Fig. 240a). Two types of elements take part in the building of their walls: ever sterile, slender, pillar-like cells and large, similarly uninucleate cells with plasma densely filled with filamentous bacteria intertwined with one another in all directions (Fig. 240b). The same organisms, which in life are coiled and appear partly rolled in a spiral, are found also in the lumen of the ceca.

Shortly after a leech sucks, the organisms are found also in great quantity in the initial part of the gut. When digestion is more advanced, they are situated even in the posterior regions, some individually and some in balls, and the red blood corpuscles around them appear to be dissolved into a pulp, according to Reichenow, whereas in other places they are still well preserved.

Even young animals still living on the yolk of the egg possess the evaginations in question, but at first the epithelium still resembles that of the esophagus. However, since the symbionts appear before the first ingestion of food, there can be no doubt that they are transmitted systematically by the mother animal to the daughter animals. It has not yet been possible to determine how this transmission takes place, but observations made with other leeches indicate that, as with lumbricids, they are transmitted through an infection of the cocoon fluid surrounding the developing animal and that the bacteria are transmitted with the fluid to the intestine after the mouth has been pierced. Indeed, when the cocoon is formed, the anterior end of the grub is drawn through its rudiment, which is at first ring-shaped, and thus the organisms here can easily be pressed from the esophagus and added to the content, which represents a favorable culture medium.

It is not known whether other glossosiphonids have created similar devices. "Glandular appendages" in the foregut have been specified for *Protoclepsis tesselata* O. F. Müller and *Hemiclepsis marginata* O. F. Müller, but this designation is erroneous. They are actually the four frontmost

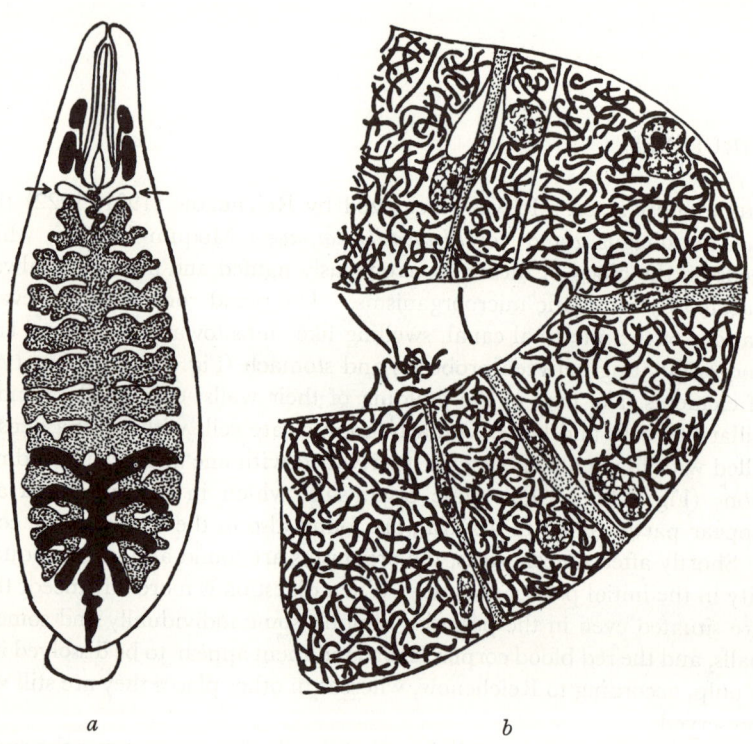

a   *b*

Fig. 240 *Placobdella catenigera* Moquin-Tandon. (*a*) Digestive apparatus with evaginations of the esophagus occupied by symbionts; (*b*) section through a portion of the esophageal evaginations.   From Reichenow.

evaginations, histologically exactly like the adjoining cecal sacs and different only in their more delicate shape, and no symbionts are found in them (Jaschke, 1933).   On the other hand, Withman (1891) specified cecal sacs for a form described by him as *Clepsine plana* and which, according to Kovalevsky (1900), is closely related to *Placobdella*.   These ceca resemble those of the latter so greatly that here, too, colonization by symbionts must be taken into consideration.

Strangely it went unobserved for a long time that similar formations occur in ichthyobdellids. It was not until 1933 that Jaschke found that *Piscicola geometra* L., our common fish leech, has two diverticula of a particular type on the foregut, in which the lumen is always filled in the same way with a homogeneous bacterial culture. They are separated from the adjoining intestinal ceca, which serve solely for the storage of blood, by the ducts of the sex glands, the spermatophore pouch, and the associated glandular bundles and are easily distinguished from them because they never contain blood. Formed like clumsy, pear-shaped sacs, they are connected with the intestinal canal merely by very narrow, short canals (Fig. 241). The walls are very similar to those of the following diverticula; they have no glandular cells and are somewhat more flattened than those of the diverticula.

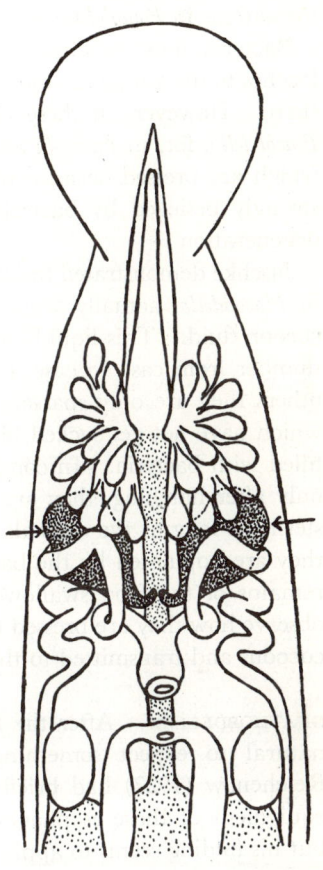

In contrast to the symbionts of *Placobdella*, the bacteria are situated only in the extracellular position. With animals from Silesia and Bavaria, the same rodlets, approximately 3 μ in length and 0.7 μ in width, always appeared.

Jaschke found the same devices in the North American *Piscicola punctata* and in *Cystibranchus respirans* Troschel living on our barbels. *Cystibranchus* has curved rodlets and somewhat different positional relationships between the ceca containing them and the loops of the vasa deferentia. In *Branchellion torpedinis* Sav., a marine leech and a parasite of the electric ray, the evaginations,

Fig. 241 *Piscicola geometra* L. Anterior section of the animal with esophageal diverticula occupied by bacteria. From Jaschke.

which do not contain blood, were noted by Sukatschow (1913). They are folded and divided into several spaces by dorsoventral muscles, but they are not set off from the gut so extremely and in general have a greater resemblance to the other evaginations. Moreover, the bacteria, which again are somewhat different in shape, also occur in the first pair of diverticula situated behind the genital openings. Apparently it is not a question of new formations but of ceca which originally also stored blood and which were specialized subsequently, to an extent which varies among the individual species. Jaschke did not find corresponding formations in *Pontobdella muricata* L., a large parasite of rays.

Bacteria must inevitably transfer through the contractions of these leeches to the gut again and again, but as yet they have not been detected there. However, in these cases they also work hemolytically, as with *Placobdella*, for, in *Piscicola* and particularly in *Branchellion*, red corpuscles which are pressed occasionally into the ceca containing symbionts were strongly besieged by bacteria and were found to be in the process of degeneration.

Jaschke demonstrated finally that transmission, as we already suspected in *Placobdella*, actually takes place circuitously through infection of the cocoon fluid. This liquid contains many microorganisms which vary in number from case to case, for at times it is teeming with them and at others they are only sparsely represented. However, in young animals which have not yet sucked blood, the diverticula in question are already filled with bacteria. In contrast to the remaining ceca, which originate only after hatching, they are formed in the cocoons but usually remain sterile as long as the animals are still enclosed in them. After hatching they are populated by the bacteria which are abundantly present in the remains of the food swallowed in the cocoon. It was not possible to observe how they are pressed from the diverticula during formation of the cocoons and transmitted to the secretion filling them.

GNATHOBDELLIDS. After the findings made with *Placobdella*, it was only natural to expect something similar in the medicinal blood leech. Reichenow (1922) had briefly reported that he actually had found an equivalent of those esophageal ceca, although a radically different one, but his finding seems to apply to the filamentous bacteria living regularly in the ampullae of the nephridia, to be reported on later, for this occurrence apparently corresponds to similar colonization of excretory organs in other nonbloodsucking annelids (lumbricids, glossoscolecids). Zirpolo failed to furnish a more detailed presentation of his brief statement (1922) regarding organs which he claimed were directly connected with the gut, and it may be concluded that he later recognized his error.

The reports of Lehmensick (1941, 1942) and his student, Hornbostel (1941), must be taken more seriously. In the gut of the blood leech kept for medicinal purposes and in that of a strain occurring endemically in Germany, they constantly found a short, gram-negative rodlet 1.5–2 μ in length, which they assigned in the broader sense to the *coli* group, although, of course, mistakenly.   Because of polar ciliation it is actively motile and can be cultivated without formation of spores in bouillon with dextrose or on agar mixed with a suspension of blood in dextrose solution.   After 24 hours the agar is filled with the colonies to its full depth.   Their hemolytic effect appears clearly on the blood plates.

Büsing (1951) and Büsing, Döll, and Freytag (1953) examined the bacterium more fully and proved that it is a *Pseudomonas* species (*P. hirudinis*) which grows luxuriantly on all the usual culture media.   No companionate forms were found.   Since it had been impossible up to this time to prove that an animal-specific proteolytic enzyme is present in the blood leech, the idea occurred that it is only this bacterium which makes blood digestion at all possible in *Hirudo*.   In fact, Lehmensick and Hornbostel themselves were convinced of this by their observations, but it was the data of the three former authors that actually established proof of this symbiotic achievement.   They also determined that the corpuscular components of the blood are attacked *in vitro* and that one can also make the *Hirudo* intestine completely germ-free with chloromycetin and thus stop the blood digestion.   The establishment of antibiotic properties in the symbiotic bacterium further clarified the absence of other organisms.

Transmission to the progeny takes place as in *Piscicola* by infection of the cocoon fluid.   When the cocoon surrounding the region of the clitellum in the shape of a ring is stripped off over the anterior end of the grub, its fluid is infected with the microorganisms on the surface, and with *Pseudomonas hirudinis* in the area of the esophagus.   The symbionts are then transmitted through the esophagus to the young animals.

This form, which is harmless to man, should not be confused with a bacterium which Schweizer (1936) described as living in the slime of the surface of *Hirudo* and without justification called a symbiont.   The testing of Schweizer's data by Büsing, Döll, and Freytag proved instead that it is completely unstable.

GAMASIDS

In *Liponyssus saurarum* Oudemans and *Ophionyssus natricis* Berlese which suck on reptiles, and in *Ceratonyssus musculi* C. L. Koch living on the blood of mice, mycetomes are found that reveal intimate relationships with the gut.   As we shall show later with *Ornithomyia* and *Pedicinus*, they are situated between the intestinal epithelium and the muscle layer surrounding it. With *Liponyssus* and *Ophionyssus* it is a mycetome pair associated with the left and right ceca of the intestine, and a third unpaired organ, larger than the first two, on the ventral side of the middle cecum (Reichenow, 1922;

Piekarski, 1935; Fig. 242*a*).   The great consumption of epithelial cells of the gut during digestion causes the mycetomes to border on the intestinal lumen, sometimes even before pubescence but at any rate in older animals.

The structure of the mycetomes is very simple for they are merely accumulations of a few uninucleate mycetocytes growing considerably in the course of development, with plasma densely filled with microorganisms (Fig. 242*b*).   Of *Ophionyssus* it is stated that the paired organs consist of

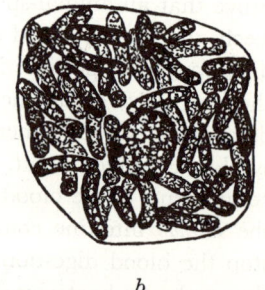

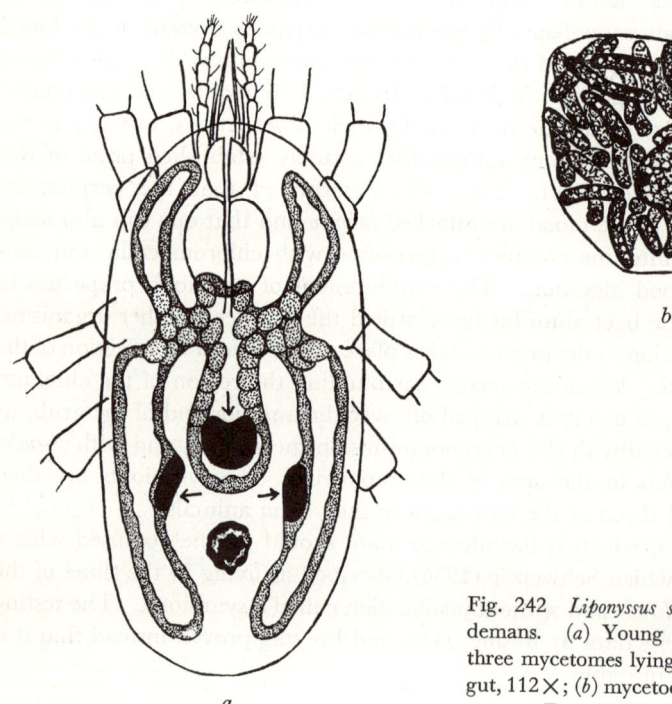

Fig. 242   *Liponyssus saurarum* Oudemans.   (*a*) Young female with three mycetomes lying close to the gut, 112×; (*b*) mycetocyte, 1440×. From Reichenow.

a

b

only two to four cells, whereas ten to twelve may be found in the unpaired organs.   *Ceratonyssus* is an exception in that it contains a single, correspondingly larger mycetome situated closer to the head and dorsal to the gut but in all other respects similar.

Reichenow made the astonishing discovery in *Liponyssus saurarum* that not one type of symbiont may be present but no less than six different types, usually only one at a time in the same host.   Occasionally two types are found together, but both forms are always distributed over different cells.   The specimens collected in Spain contain one bacterium most frequently, a clumsy slightly curved rodlet with rounded ends resembling *Periplaneta* and *Blatta* symbionts.   With some mites there is an additional slender rodlet, more rarely a considerably longer form, and sometimes an additional extremely long thin filament.   Animals from Rovigno had symbionts unlike those of the Spanish mites.   Instead they had bacteria, some of which were shaped like spindles, others like filaments and thereby somewhat reminiscent of the Spanish forms.   Piekarski also encountered two different symbiont types in *Ophionyssus*, rodlets 5–10 μ in length, and slender spindles, but the two forms never occurred side by side in the same host.

Such variability, found in no other instance of endosymbiosis, demands research on a broader basis.   Since *Liponyssus* occurs on diverse lizard species and even in snakes, one might be tempted to think that some mutual relationship exists between the blood-giving host and the symbiont types of its parasites; on the other hand, the *Ophionyssus* specimens of Piekarski all originated from the sand viper (*Viper ammodytes* L.).

Here it is clearly a case of differentiated organisms and not of various states of the same form, as indicated by their transmission to the progeny. Sometimes the ovaries also have close spatial relationships to the gut and, like the mycetomes, may even border on its lumen as the result of extensive consumption of intestinal epithelial cells, as in *Liponyssus;* or they may avoid such connections and instead pass through the body cavity in the form of paired cords, as in *Ophionyssus*.   In both cases the eggs finally transfer individually to the body cavity, where, situated directly beneath the central mycetome and between the lateral ones, they continue to develop.   Thus the symbionts can easily migrate from their sites to the egg, which is still without an envelope.   When such an egg moves to the uterus, the symbionts have already penetrated deeply into the yolk and are then situated within the yolk spheres, a peculiarity occurring in no other case of symbiosis (Fig. 243).   This is also true of the embryos in which symbionts frequently continue to increase in the vitelline spheres.   The various symbiont types present in the mycetomes can be recognized easily.   Only the long slender filaments seem to infect in somewhat briefer form, and then grow out again.

During embryonic development, which takes place partly in the maternal body, changes in the symbionts are rarely observed.   Indeed, the six-legged mite larva has no gut on leaving the egg and the symbionts are always embedded in a mere unorganized yolk mass.   Only in the eight-legged nymph, which arises after 1 or 2 days, is the intestinal epithelium constructed and the remains of the yolk resorbed.   This is also the period when the mycetomes are formed, a process which, according to Reichenow, apparently uses material equivalent to the intestinal cells, although he was

unable to observe the stage, apparently of short duration, when the symbionts transfer from the gut to the cells which are becoming mycetocytes.

There is not yet sufficient material on the extent of such symbiosis in leech mites. We are only certain that it does not occur in all.
Piekarski was unable to trace it in a chicken mite *Dermanyssus gallinae*

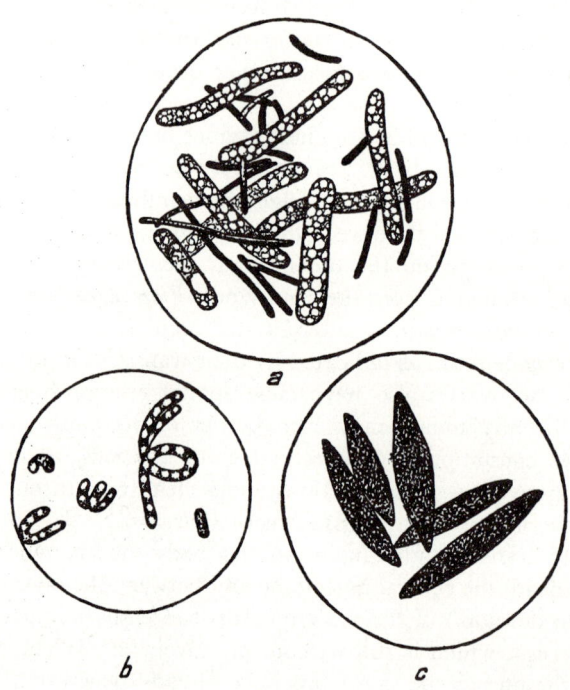

Fig. 243   *Liponyssus saurarum* Oudemans.   Yolk clods with symbionts: (*a*) with two
different forms, (*b*) with a third type, from Spain, (*c*) with a fourth, from Istria.   250×.
From Reichenow.

De Geer, and Reichenow found no mycetomes on sections of a non-classi-
fied species of the same genus living on doves.   On the other hand,
Marchoux and Couvy (1913) report that on smears of the gamasid *Laelaps
echidninus* they found rod-like organisms which may possibly originate
from mycetomes.

At present it is impossible to decide whether the rickettsia which have
been repeatedly reported in bird mites (Nöller, 1920; Reichenow, 1922;
Sikora, 1924) may be interpreted as a replacement for the symbionts.   At
any rate, further investigation of gamasid symbiosis would be desirable.

IXODIDS AND ARGASIDS

The multiform ixodids are distributed over two tribes differentiated through the position of the anal furrow: the poststriates or amblyommines and the prostriates or ixodines.   Among the first mentioned are numerous genera, such as *Amblyomma, Hyalomma, Boophilus, Rhipicephalus, Dermacentor, Margaropus,* and *Haemaphysalis,* whereas the prostriates comprise only the one genus *Ixodes.*

All these genera live symbiotically with microorganisms which are always localized in the Malpighian tubules and transmitted to the progeny by way of the egg cells (Buchner, 1922, 1926; Cowdry, 1925; Mudrow, 1932; Jaschke, 1933).   The extent of the cells infected in the excretory organs varies greatly.   In *Ixodes* they are appropriated almost throughout their length.   Only the last section extending from the posterior margin of the animal to the rectal bladder remains free, whereas all or at least most of the remaining cells of the tubes extending in complicated windings into the head area are infected.   The colonization itself may vary greatly (Buchner, 1922, 1926).   In *Ixodes hexagonus* Leach it is frequently a case of slender, often much elongated filaments which on sufficient differentiation usually reveal clearly that they are composed of granules set one behind the other.   These filaments are more or less perpendicular to the cell basis, a situation of general occurrence which is determined by the structure and the direction of the operation of the host cell.   A narrow basal zone and a broader distal plasma zone are customarily free of symbionts.   Frequently there is a tendency to form bundles.   The filaments joined in this way may be wound about each other in spirals, and at certain places such pairs pass over into one another at one end.   During cross-cleavage the two halves apparently double over, and the repetition of this process leads to bundles resembling tufts, which in some cells are much intertwined (Fig. 244).

There are also cells filled with thick masses of numerous and much smaller bundlets.   They take up a large zone of distal coarsely vacuolized plasma.   Besides these distinct types of colonization are various transitional types.   Occasionally individual cells or whole sections remain completely free or are only sparsely populated, thus giving rise to a great variation in the host cells, which moreover are extremely variable in form.   The differences in behavior may apparently be explained by differences in origin of the material used.   In Mudrow's specimens of *Ixodes hexagonus* she occasionally found small symbiont bundles only in the anterior blind end of the vessels, and in the remaining sections chiefly great quantities of tiny rodlets. In *Ixodes ricinus* the symbiont filaments are somewhat shorter, but they are also interbraided in diverse ways (Buchner, 1926).

Differences of many kinds are seen on comparing the infection in the amblyommines. Where the relationships were investigated in greater detail, as by Mudrow in *Rhipicephalus sanguineus* Latr., *Dermacentor reticulatus*

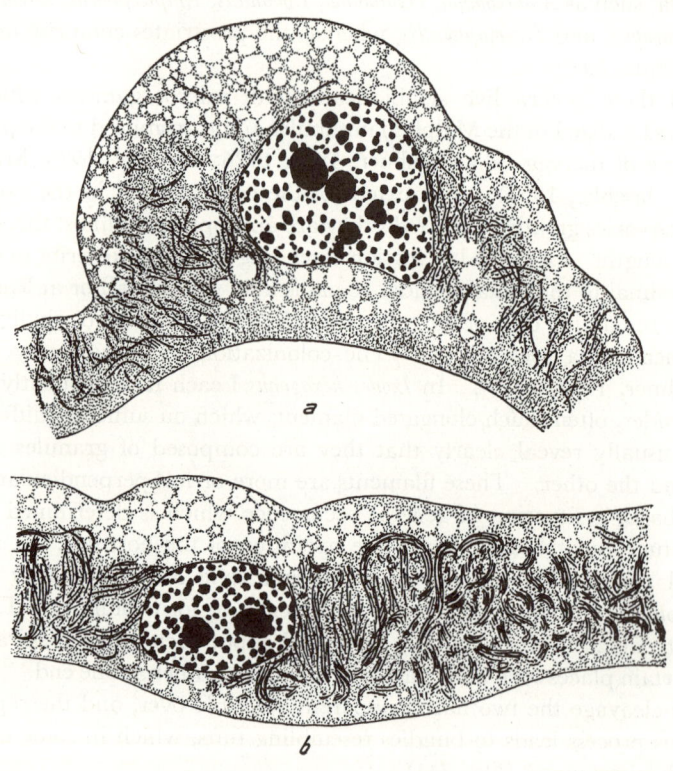

Fig. 244   *Ixodes hexagonus* Leach.   Two cells from the Malpighian vessels with symbionts.   From Buchner.

Fabr., and *Boophilus annulatus* Say., it was found that the symbionts were usually limited to the zone of the blind ends in the vessels. Figure 245*a* illustrates a *Rhipicephalus* female in which at the very most one-fourth of each of the two vessels contains symbionts; the same is true of *Dermacentor;* and in *Boophilus* the colonized section is even much slighter and sometimes sharply set off from the sterile part. The infected cells are markedly enlarged and form a terminal button which is no longer so distinct in females which have sucked their fill. Here the unpopulated cells are still in the process of considerable growth, and the vessels swell as they fill with excretions (Fig. 245*b*).

The symbionts of the amblyommines, which usually permeate the infected cells at random in all directions, have various shapes but are also strictly type-specific. This was already emphasized by Cowdry (1925). After a study of fourteen species he was convinced that they could be definitely classified on the basis of their guests. The origin of the host animals has no influence on the morphology of the symbionts, for he

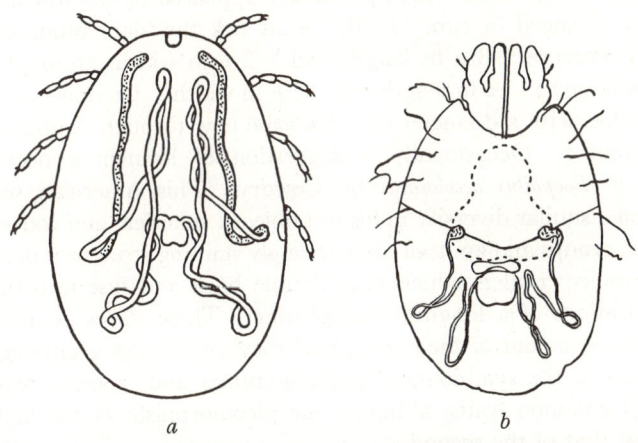

a *b*

Fig. 245 (*a*) *Rhipicephalus sanguineus* Latr. A long terminal section of the Malpighian vessels houses the symbionts. (*b*) *Boophilus annulatus* Say. A small swelling at the end of the Mapighian vessels is colonized. From Mudrow.

found it to be the same in material originating both from Hawaii and from South Africa at an altitude of 4000 feet.

Mudrow describes the symbionts of *Rhipicephalus sanguineus* as straight or slightly arched, short rodlets measuring 0.5 μ in width and composed of tiny granules in varying numbers. Occasionally they occur in little nests, each apparently representing the progeny of one symbiont. There are also bundles and intertwined longer filaments consisting of longer rows of rodlets and granules. Formation of groups and bundles occurs to an increased extent when the cells of the Malpighian tubules develop markedly and divide more animatedly in connection with the ingestion of food. The symbionts of *Dermacentor reticulatus* are more compact and lie side by side in larger and smaller roundish groups, which reveal that they are composed of countless individual little granules only when the host animals are plucked apart.   Most protean in form are the symbionts of *Boophilus annulatus;* their changes in form parallel the development of the host animal.   In addition to "symbiont granules" which are like those of *Dermacentor*, Mudrow finds in the larvae thicker filamentous states which far exceed the symbionts of all other amblyommines in length and thickness.   They too are composed of individual granules, but it is easy to recognize the more weakly staining, basic embedding substance, which probably has the mucilaginous consistency of other gelatinous envelopes of bacteria.   In the nymphs these symbiont colonies continue their growth, and the granules which previously appeared in one row are now frequently arranged in two.   In the adult tick the filamentous inmates ordinarily measure 5–8 μ in length and 1–2 μ in width, although some may be as large as 15 μ in length and 2.5 μ in width.   Compact and disc-like structures with a diameter of 3–5 μ arise now through doubling back and shortening.   Occasionally a suggestion of branching occurs, as observed in *Boophilus decoloratus* by Cowdry.   This American scientist came upon a similar diversity in his test objects: spherical and rod-shaped organisms; groups and knots of more strongly staining structures in a paler basic substance; rodlets which may double back and fuse into rings or show indications of a longitudinal splitting.   These states do not have individual movement of their own, and they are always gram-negative. Comparison of the symbionts of amblyommines and ixodines reveals a number of common traits, although the pleomorphism of the first goes far beyond that of the second.

Transmission of the tick symbionts is accomplished by egg infection, as is already clear from the findings of authors who saw them without knowing what was involved, and whose work was discussed more fully in Chapter 2.   Later Godoy and Pinto (1922) and Cowdry (1925) found the organisms now recognized as regular guests in the ovaries and in the eggs, but these authors were not concerned with the interesting details of behavior.   With ixodines (Buchner) and amblyommines (Mudrow, Jaschke) such details are essentially the same.   The moniliform ovary of *Ixodes* is quite like that of the spiders in structure, for the germ epithelium is a pipe on which the oocytes protrude exteriorly like berries with increased growth and at the same time push along before them a cell-free membrane.   In the process a broad, socket-formed pedicle is developed upon which the egg rests and which also develops from the germ epithelium.   The undifferentiated elements of the germ layer in themselves are more or less strongly infected over wide sections, and the same is true of the cells which are revealed as young oocytes by their increased growth (Fig. 246).   In such cases the symbionts are characteristically grouped at the two poles of the nearly oval cells.   The elements of the egg pedicle also contain symbionts, though in small numbers, but these do not influence the animated mitotic divisions, although they are themselves visibly prevented from increasing at this point.   The oocyte attached to the pedicle gradually assumes a bean shape, and at this time the symbionts are more and more forced into the zone which is turned away from the pedicle until they form a dense ball with approximately the same circumference as the nucleus.   Vaguely delimited at the start and still surrounded by various strays, it now is set off sharply against the plasma, which in the meantime has been enriched with vitelline spheres (Figs. 247a,b; 248).

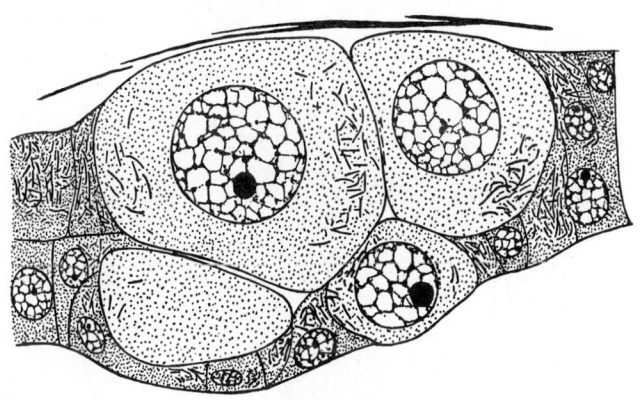

Fig. 246   *Ixodes hexagonus* Leach.   Germinal epithelium and young ovocytes with symbionts.   From Buchner.

This state of maximal balling represents merely a transition state.   The oocyte has not reached its final length, and, as it continues its growth, the strange sphere disperses again, so that its components are scattered very inconspicuously here and there in the now sparse plasma.

The phenomenon is accompanied by form changes and lively increase of symbionts.   The filaments found in the germ layer are much shorter than the forms found in the Malpighian tubules; they do not form bundles or even simple twists; that is, they barely increase.   On the other hand, they clearly reveal that they are constructed of granules placed one behind the other.   This arrangement remains unchanged even in the young oocytes.   Only when the balling process begins does an especially increased reproduction set in and give rise to numerous bundlets.   When this concentration has reached its maximum, this period of increased

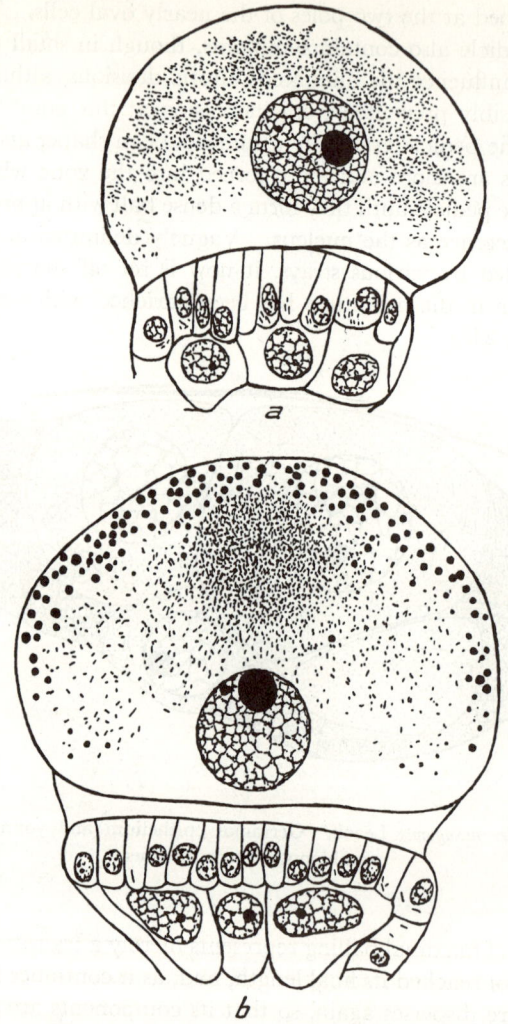

Fig. 247 *Ixodes hexagonus* Leach.   The symbionts in growing ovocytes: (*a*) beginning
of the shifting; (*b*) advanced stage of agglomeration.   From Buchner.

reproduction ceases and the symbionts are very small, usually weakly
curved rodlets, a form which is retained later.   Shortly before egg
deposit a second process of concentration occurs, this time in the narrow
yolk-free marginal zone of the mature egg close to the posterior pole and
in sparse plasma trains protruding toward the inside.   The behavior in
Ixodes ricinus is similar in every regard to that in *I. hexagonus*.

Mudrow, the first to investigate the corresponding processes in the amblyommines, obtained results with *Rhipicephalus*, *Dermacentor*, and *Boophilus* which were similar to those described for *Ixodes*.   In the species mentioned there are also four stages: bipolar grouping, central balling, dispersal, and renewed grouping at the margin of the egg.   The primary merit of Mudrow's work is that she studied the embryonic development of her test objects and found that infection of the ovaries takes place at the

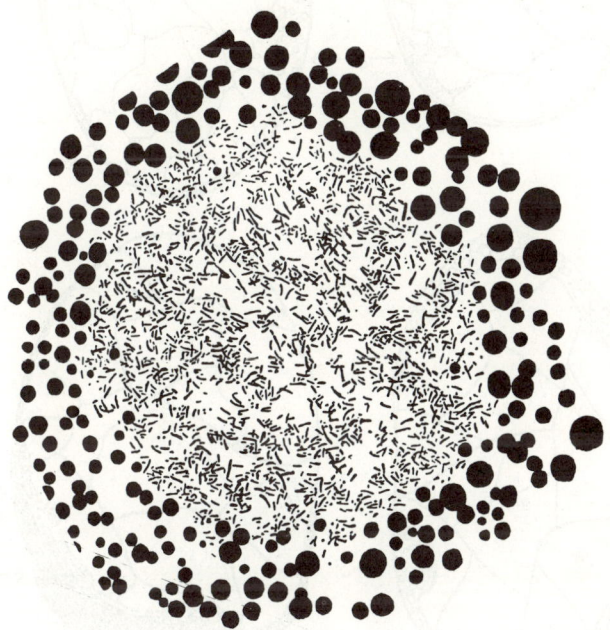

Fig. 248   *Ixodes hexagonus* Leach.   The symbiont clump in an older, yolk-rich egg.
From Buchner.

earliest imaginable stage.   When the blastoderm nuclei appear in the sparse plasma border on the surface of the egg, they also serve the symbionts congregated there.   At this same place a proliferation representing the rudiment of the endoderm originates and pushes them more to the inside (Fig. 249*a,b*).   The germ band forms and gradually leaves only a part of the dorsal area free.   The symbionts are waiting at the same place and are now situated in the so-called caudal lobe (Fig. 249*c*).   Here in a following stage an inner infected cell mass is separated, which furnishes the Malpighian tubules, and also a peripheral layer of endodermal cells, which

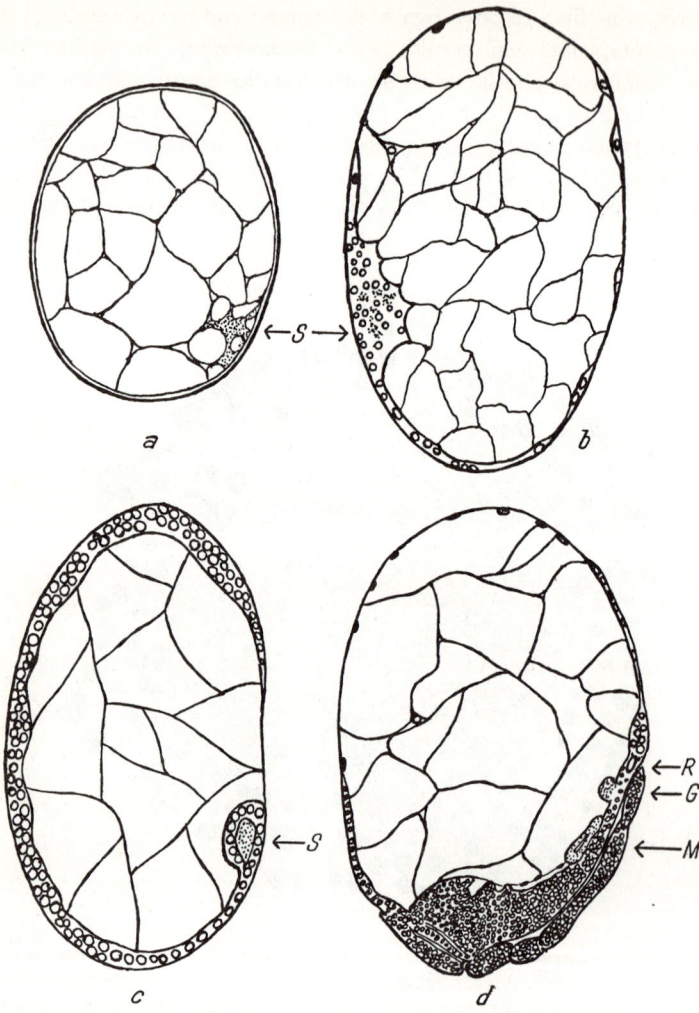

Fig. 249 *Rhipicephalus sanguineus* Latr. Development of the egg and early infection of the Malpighian vessels and gonads. (*a*) Egg ready to be laid, 265×. (*b*) Beginning of ingrowth of the endoderm; (*c*) rudiment of the infected tail flap; (*d*) rudiment of the Malpighian vessels and sex glands; 125×. *S*, symbionts; *R*, rudiment of the rectum, *G*, of the gonads, *M*, of the vasa malpighi. From Mudrow.

in all probability give rise to the sexual cells that at first are situated close to the rudiment of the excretory organs. When the sexual cells can be identified with certainty, they appear to be provided with symbionts. Even before the larvae hatch, it is possible to recognize the sex of the young gonads and thus establish that only the female sexual glands are

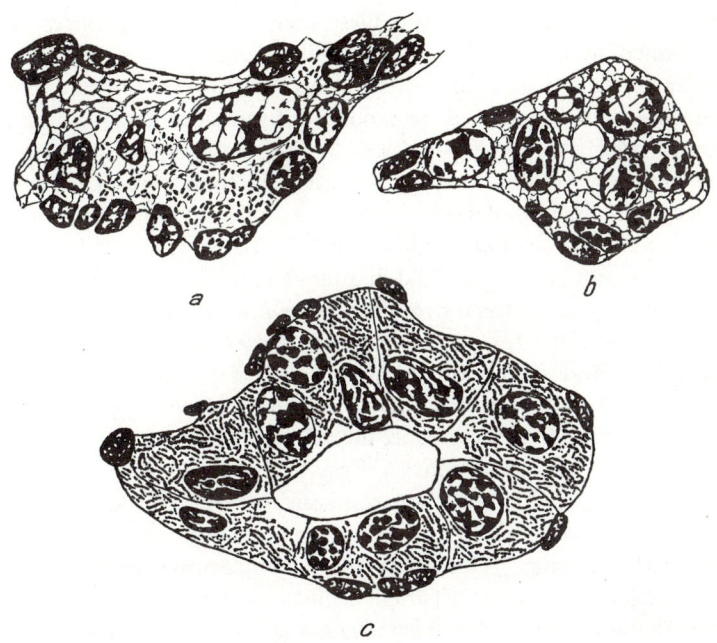

Fig. 250 *Rhipicephalus sanguineus* Latr.: (*a*) female embryonic gonad with symbionts; (*b*) male embryonic gonad without symbionts. (*c*) *Boophilus annulatus* Say.: young female gonad heavily infected. 1350×. From Mudrow.

infected (Fig. 249*d*). It is impossible to tell whether this situation exists because the symbionts are destroyed in the males immediately after the rudimentary testis is infected or whether such an infection fails to occur at all. The latter seems much more probable. Figure 250*a,b*, showing sections through the still undeveloped female and male sexual glands of a *Rhipicephalus* larva, graphically illustrates that only the female glands are infected. In Fig. 250*c* showing gonads of the same age in the larvae of the much more rapidly developing *Boophilus*, the germ epithelium of the more fully developed female gland is so thickly filled with symbionts on all sides that the host plasma is just barely visible.

Unfortunately *Ixodes* has not yet been compared with the amblyommines in this respect, but, in view of analogous behavior of the symbionts in the mature ovaries in both genera, there is little doubt that an extremely early safeguarding of symbiotic living takes place in *Ixodes* also.

Limitation of the symbionts to more or less large distal sections during development of the Malpighian tubules might be explained by the fact that the cell divisions take place chiefly in the section near the rectal bladder, so that the cells, infected and now no longer capable of division, are pushed more and more toward the front. In young Malpighian tubules and in the germ epithelium, cases of mitosis are not rare, however.

Since so many genera of the ixodids are symbiont bearers, it is all the more surprising that Mudrow failed to find symbionts on smears and sections of the ovaries and Malpighian tubules in *Ixodes ricinus* L. from Mecklenburg, and that Roesler also (1934) was unable to find them in the same species, although he had material from Germany, Italy, and Greece; yet symbionts were present as usual in animals investigated by Marchoux and Couvy and by me. In this connection we must refer to a publication by Rondelli (1925). In an unclassified *Ixodes* species from Sardinian cattle she found no symbionts in the cells of the Malpighian tubules, and yet she found numerous short rodlets in the lumen of the esophagus and initial section of the gut, in the salivary glands, and in parts of the lumen of the excretory organs. Eggs and pedicles were also infected. On the other hand, the "ergastoplasm" which Nordenskiöld (1908) found in the Malpighian tubules of *Ixodes reduvius* might be identical with our symbionts. However, only a comparative study based on a large amount of material of varied origin could indicate the significance of their occasional absence in the genus *Ixodes*.

At any rate the bacteria described by Rondelli may be merely occasional guests, if indeed they are not to be identified with developmental stages of piroplasmas as Mudrow suspects. Such organisms actually do appear, for in the Malpighian tubules of a *Dermatocentor* female Mudrow found numerous, extremely delicate, intracellular rodlets in sections avoided by the symbionts.

Among the argasids also, the symbionts are found only in the excretory organs and in the ovaries. In *Argas persicus* Oken the former are infected almost to their full extent, a few cells near the rectal bladder remaining sterile. In *Ornithodorus moubata* the blind end remains sterile, and the infected zone is limited to an adjoining section which comprises approximately one-fifth of the total length of each tube; even during preparation it is conspicuous as a whitish thickening.

As described in Chapter 2, the symbionts were observed by a number of authors before being recognized as such by Cowdry (1923, 1925). When they were studied later by Mudrow (1932) and Jaschke (1933), they proved to be far less polymorphic than those of the ixodids. Longer filaments and chains are completely lacking; instead there are minute gram-negative cocci and short rodlets approximately 0.4 μ in diameter and 1 μ in length at the maximum. Embedded in a stroma evidently self-produced, they form spherical grouplets in which as many as forty components may be counted. Such nests of organisms seem to correspond to the rows and groups of granules so common among the ixodids. In digesting and starving animals of both sexes they fill the cells of the Malpighian tubules in such quantity that scarcely more than a basal border and the immediate vicinity of the nucleus are left free. *Argas, Ornithodorus,* and *Otobius* are practically alike in this respect. Steinhaus (1946) tried in vain to cultivate *Argas* symbionts in the embryonic tissue of chickens. Roshdy (1961) studied the symbionts of *Argas persicus* with the electron microscope.

The method of transmission is understandably similar to that of the ixodids. The still undifferentiated germ epithelium is already infected, although sparsely, and there are no later reinforcements from the symbiotic sites; the symbionts are completely lacking in the egg pedicles; in the growing oocytes an animated periodic increase takes place, but not the regular shiftings and aggregations which are characteristic of the ixodids. In a later stage, when the egg cell has already reached three-fourths of its final size and is filled on all sides by clumps of yolk, the bacteria, forming characteristic little groups as always and up to that point diffusely distributed, suddenly concentrate in a yolk-free hollow, forming such a thick layer on the latter that only a small symbiont-free area remains; in it the plasma forms a weak radiation. Other symbiont nests are then found only between the bordering vitelline spheres. This concentration, caused apparently by a centric current, undoubtedly corresponds to the balling in the older *Ixodes* egg.

However, in *Ixodes* the aggregation is disintegrated again during further egg growth; in *Argas* it is preserved until fertilization and deposit, indeed, even during early embryonic development. When the blastoderm forms and the yolk cells are differentiated, the symbionts, guided by these yolk cells, press forward between the vitelline spheres to the area of the future caudal lobe, where they come upon the rudiment of the endoderm, which is in the process of separation. Some of them gradually are drawn in by it, infecting the paired rudiments of the Malpighian tubules and at the same time temporarily giving up the formation of little groups. Jaschke, who compiled these data, believes that the remaining symbionts are destroyed, but we must also consider the possibility that they serve to infect the gonad rudiments, which appear much later here than with the amblyommines and were found to be symbiont-free at first by Mudrow.

She assumes that the symbionts in this case transfer from the neighboring Malpighian tubules.

Recent publications by Geigy and Wagner (1957), Wagner-Jevseenko (1958), and Aeschlimann (1958) furnished a welcome addition to our knowledge of symbiont transmission in *Ornithodorus moubata*, although of course the authors did not realize they were making a contribution. Misled by the circumstance that in this case, as with some other symbionts, the presence of desoxyribonucleic acid could be detected in the structures in question by the Feulgen method and that the nucleus was not found in the mature egg, the authors declared the symbionts in the oocytes to be chromosome substance which had migrated from the nucleus. However, their size and shape as well as their regulated increase and displacement during egg growth which lead to a dense peripheral aggregation in the mature egg correspond to what is known at present about symbionts in argasids and ixodids. In *Ornithodorus* as in *Ixodes* one finds in the older oöcyte the socket bearing it, the nucleus, and the aggregation of the symbionts in an axis. This finding and the good pictures furnished by Wagner-Jevseenko of the various stages of egg development fill a gap in our knowledge of argasid symbiosis.

GLOSSINES

The midgut of the imagines of all glossines is divided into three separate parts. The anterior section has a low weakly staining epithelium composed of pillar-like cells; the center section has tall, club-shaped, deeply staining cells; and the end section is composed of cubical cells which lack such indications of increased secretory activity. Approximately in the middle of the first section one's attention is drawn to a short, sharply set off, and quite thickened area which corresponds to the symbiotic site. The site was described for the first time by Stuhlmann (1907) and in more detail by Roubaud (1919) (Fig. 251).

The infected cells, arranged in two longitudinal bands, are enormous in comparison with the neighboring symbiont-free cells. In hungering animals they protrude like cushions into the intestinal lumen, and, when the latter is filled with blood, appear, on the other hand, to be extremely flattened. The nuclei are situated in the distal region; the plasma is thickly filled with longitudinally oriented rodlets 3–10 μ in length, in which sections differing in degree of staining can be differentiated (Fig. 252*b*,*d*). The organisms divide transversely or bud-like end sections are constricted off, a situation that caused Roubaud to consider them to be yeast-like organisms, whereas Wigglesworth (1929), doubtless with good reason, declared them to be gram-negative bacteria (Fig. 252*e*).   The infected section is well provided with tracheae, with branches penetrating deeply into the infected cells.

According to Roubaud, the symbionts, especially after ingestion of food, often transfer to the intestinal lumen in such quantities that they give the impression of being a thriving pure culture, and the disintegration of red corpuscles then begins in the area of such aggregations.   On the other hand, Wigglesworth observed only one clear case of transmission of symbionts from the cells through the peritrophic membrane, and he asserts that the blood digestion always takes place several millimeters behind the symbiont zone in the second midgut section; only here were large quantities of a tryptase secreted.

The transmission of symbionts to progeny is closely connected with the strange reproductive method of the glossines.   As is well known, their larvae develop to the point of pupation in the maternal body and are fed exclusively on the secretion of the so-called milk glands.   At first Roubaud could merely prove that the egg cells are still symbiont-free and that the larvae already contain symbionts, and he therefore supposed that the guests reach the young larvae with the chyme.   Once Wigglesworth did find the cells and lumen of the milk glands abundantly infected.   Moreover, the assumption is doubtless correct, for this method of transmission has been satisfactorily demonstrated in the Pupipara which feed their larvae in similar fashion and create symbiotic devices which in general resemble those of the glossines.

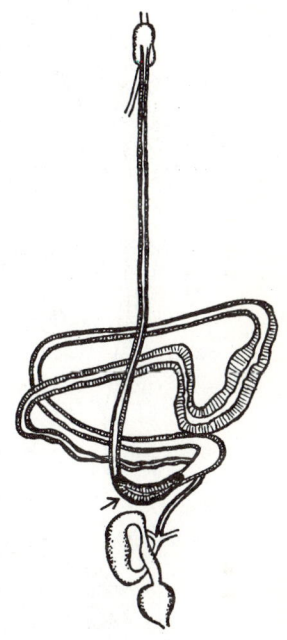

Fig. 251   *Glossina* sp. Adult intestinal tract with the thickened section containing the symbionts. From Wigglesworth.

The remarkable shape of the intestinal canal in glossine larvae is the result of their unusual method of nourishment.   A proventricle with a deeply inserted valvula follows a swallowing apparatus and a narrow stomodaeum; the proventricle passes over almost directly into another shapeless sac filled with chyme which ends blindly and takes up most of the larval body (Fig. 253*a*).   The bacteria, ingested with the "milk," colonize the frontmost midgut cells of the proventricle, again in a sharply set-off area.   In this ring-shaped zone the epithelial cells are somewhat larger than in the bordering section and are always free of the fat droplets which are stored abundantly in the remainder of the midgut epithelium (Fig. 252*a*).   According to Roubaud, free symbionts are always present

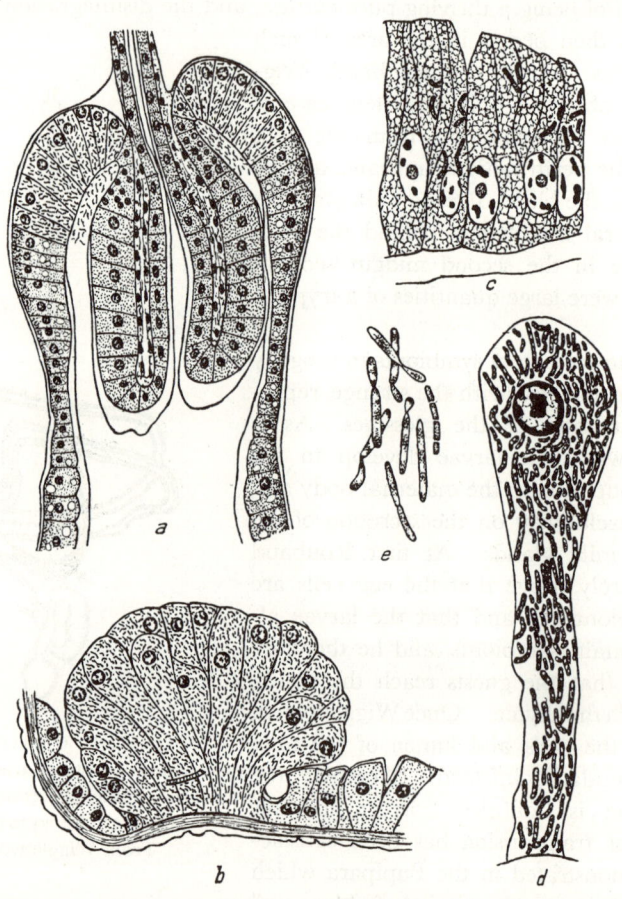

Fig. 252   *Glossina palpalis* Rob.-Desv.   (*a*) Proventriculus of the larva with symbionts, 340×; (*b*) infected giant cells of the adult midgut, 350×; (*c*) the rudiment of the adult site is colonized by the first symbionts, 1350×; (*d*) mycetocyte of the adult, 1150×; (*e*) symbionts, 1700×.   From Roubaud.

in the intestinal lumen, and he assumes that they are transferred to the following intestinal part where it is impossible to observe them.

During pupation the infected zone of the epithelium is cast off into the intestinal lumen.   Little spherical cell masses, laden with symbionts, are

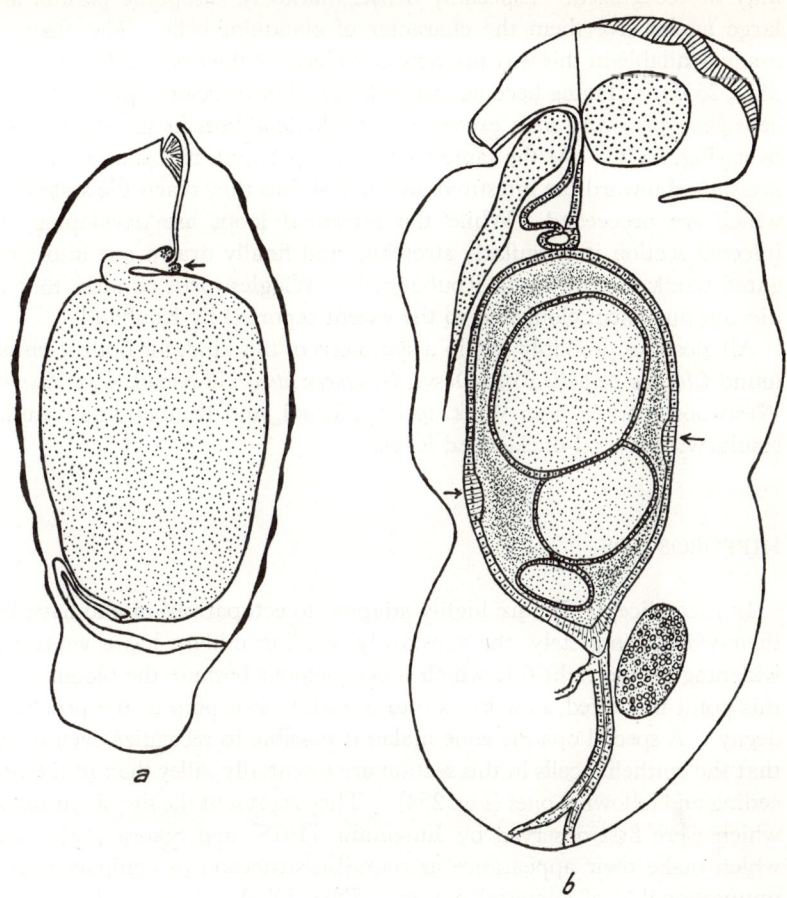

Fig. 253   *Glossina palpalis* Rob.-Desv.   Sagittal section (*a*) through a larva; (*b*) through a 4-day-old pupa; the symbionts are housed in the proventriculus. In the middle of the adult gut lies the larval gut; (*a*) shows the protruding sites of the symbionts, which at this time are located between the epithelial linings of the two guts.   30×.   From Roubaud.

pinched off from it and begin to disintegrate in the lumen.   On the fourth day of pupation, which takes place outside the maternal body, the new midgut is already completely formed, but in its interior is still the large remnant of the larval gut filled with a dense protein-rich mass, and between the two epithelial the now freed symbionts can be detected.   At this place in the middle region of the sac-like intestine a sharply delineated group of much larger cells, which represent the future sites of the bacteria, may be recognized.   Especially dense, markedly basophilic plasma and large nuclei give them the character of glandular cells.   The elements made available in this way are now colonized by the first symbionts (Figs. 252c, 253b).   Having become intracellular, they increase rapidly; the host cells keep pace and soon exceed the sterile neighbors in size many times over (Fig. 252b).   At the same time their nuclei, at first situated basally, are shifted toward the intestinal lumen and thus they reach the state from which we proceeded.   While the intestinal loops are developing the infected section is irregularly stretched and finally drawn out into elongated bands, according to Roubaud, but Wigglesworth remarks that he did not find them elongated to the extent reported by Roubaud.

All glossines probably live in a symbiosis of this kind, for both scientists found *Glossina palpalis* Rob.-Desv., *G. submorsitans* Newstead, *G. tachinoides* Westwood, and *G. fusca* Walk. to be infected, and Roubaud had similar results with several unclassified forms.

HIPPOBOSCIDS

In sheep lice, which are highly adapted to ectoparasitism and have lost their wings completely, the constantly winding midgut has a ventricose widening on the right side which is conspicuous because the blood, up to this point unaltered, now turns into a dark brown pulp in the process of decay.   A special opaque zone makes it possible to recognize even *in vivo* that the epithelial cells in this section are essentially taller than in the preceding and following ones (Fig. 254).   They represent the site of symbionts which were first observed by Jungmann (1918) and Sikora (1918) and which make their appearance as roundish structures or compact rodlets unquestionably of bacterial nature.   They fill the slender cells densely, although they often more or less avoid their distal zone.   The nuclei are barely distinguishable from those of the flatter, non-infected cells, but their brush border is much less developed and may even disappear completely in some places (Fig. 255).

Zacharias (1928) explained the transmission of these ever present organisms, which, Mahdihassan (1946) reports, were cultivated on plum agar with the addition of liver extract and traces of iron chloride.   As Roubaud had already suspected in the case of glossines, the "milk glands" serve both as transmission medium and as food for the larvae, which develop singly in the maternal body until pupation.   Figure 256 illustrates an early stage of an embryo located in a very widened uterus.   The right ovary which produced it is in the process of regression; in the left ovary, which is to produce the next larva, the young oocytes are visible.   The

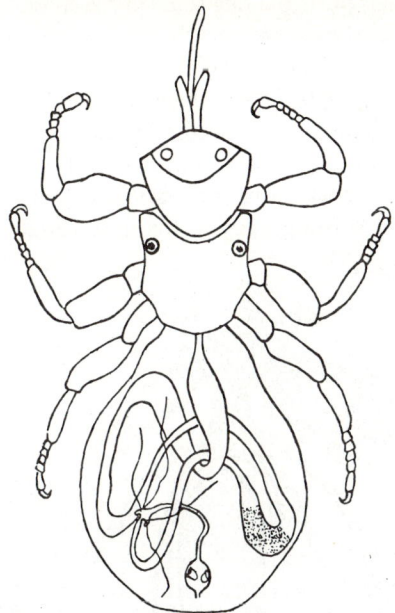

Fig. 254   *Melophagus ovinus* L.   Schematic presentation of the intestinal tract; the symbiont-containing zone is dotted.   From Zacharias.

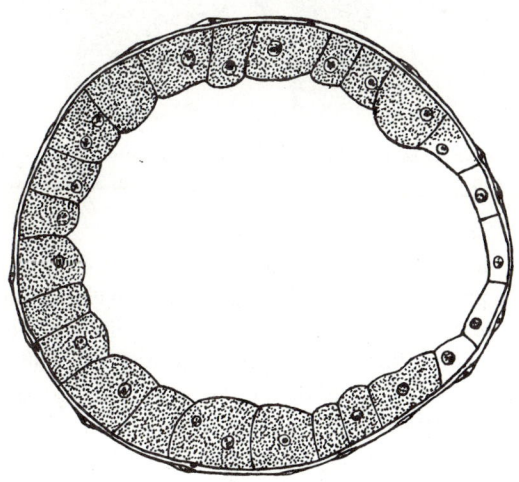

Fig. 255   *Melophagus ovinus* L.   Cross section through the adult gut in the region of the transition to the infected section.   From Zacharias.

down-gliding egg has pushed along some of the sperm originating from the seminal receptacle.   Opening into the uterus are two small glands, the function of which remains obscure and which are always free of bacteria, and a pair of milk glands, which develop enormously during gestation. The symbionts, detectable in the secretion-filled lumen, are transmitted to the larvae as they absorb this nutrient material.   The symbionts appear to vary in number according to the gestation state of the mother animal and are not distributed uniformly in the little canals but are localized in

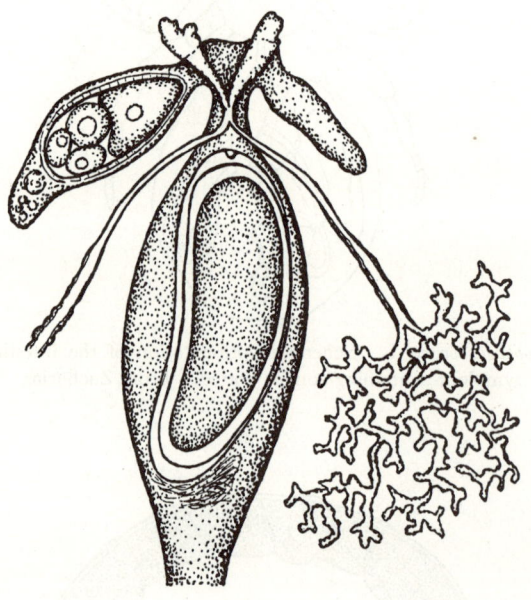

Fig. 256   *Melophagus ovinus* L.   Female sexual apparatus with milk glands and embryo in the uterus (fresh preparation).   From Zacharias.

groups.   Division stages are not rare.   The ducts of the milk glands open into the uterus precisely where the mouth of the larva will later be situated. The mouth develops a special muscular "sucking lip" and regular swallowing movements, approximately 45 per minute (Pratt, 1893, 1899).

Arriving in the larval midgut, the symbionts take possession not of the section which corresponds to that occupied in the imago, but of one which follows directly on a short esophagus and which has a narrow lumen and cells that have been enlarged long before colonization (Fig. 257*a,b*).

When the imaginal discs of the larval gut build up a new intestinal epithelium in connection with pupation, the old one is still completely

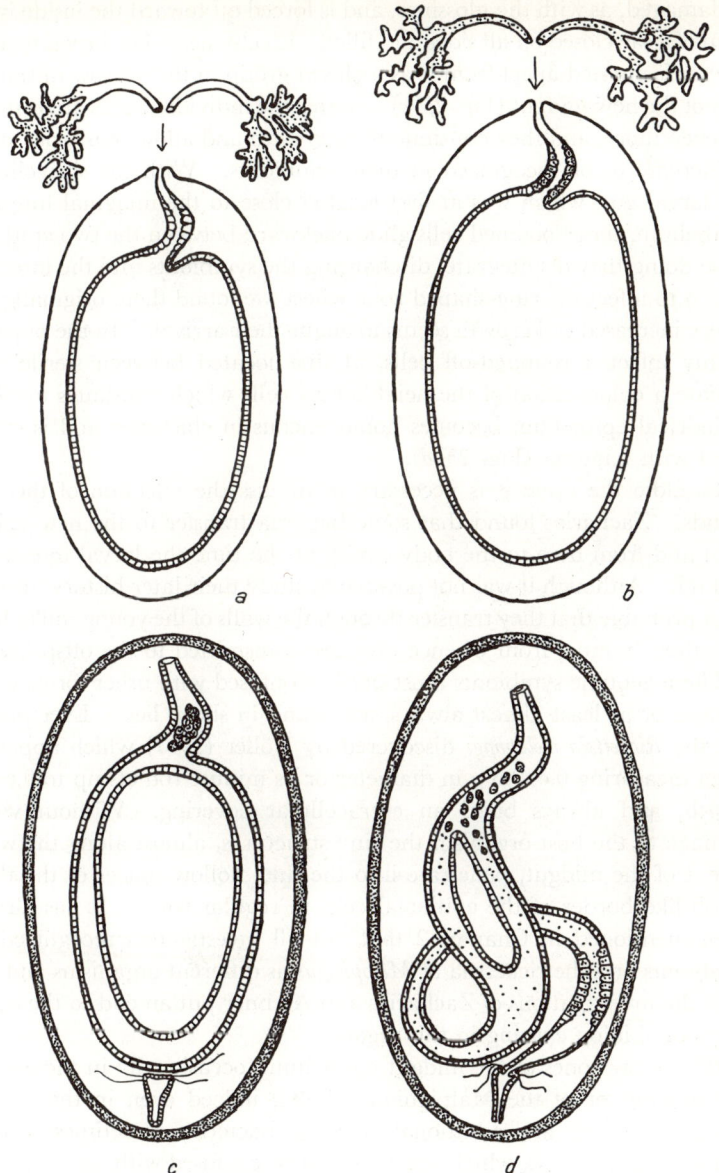

Fig. 257 *Melophagus ovinus* L. Infection of the larva and relocation of the symbionts during pupation. (*a*) Larval site still sterile, milk glands infected; (*b*) larval site infected; (*c*) larval mycetocytes cast off into the gut lumen during pupation; (*d*) infection of the adult site in the pupa. Schematic. From Zacharias.

undamaged, as with the glossines, and is forced off toward the inside in the form of a sac closed on all sides and filled with chyme. The larval myceto-cytes are situated apart from it as a closed group in the narrow initial section of the new midgut (Fig. 257c). Later the individual cells are rounded off even more, and they continue their growth and allow numerous stages of increase to be recognized in their symbionts. With the shriveling of the larval gut, which was at first located close to the imaginal intestinal epithelium, these loosened cells glide backward between the two epithelia. In so doing they disintegrate, discharging the symbionts into the intestinal lumen to infect the ring-shaped zone where we found them originally and where increased cell growth seems to augur their arrival. In the pupa the freshly infected rounded-off cells, at first located between sterile cells, initiate a colonization of the neighboring cells which continues until the cylindrical epithelium becomes homogeneous in character and is evenly filled with bacteria (Fig. 257d).

To close the cycle it is necessary to discuss the infection of the milk glands. Zacharias found that some bacteria transfer to the new epithelium and from here to the body cavity at the time the larval intestine is cast off. Although it was not possible to study their later history, it seems most probable that they transfer through the walls of the young milk glands into their lumina, from whence they are transmitted to the offspring.

These genuine symbionts must not be confused with other forms which always, or at least almost always, are found in sheep lice. Ever present are the *Rickettsia melophagi* discovered by Nöller (1917) which appear as cocci measuring 0.4–0.6 μ in diameter or as minute rodlets up to 0.6 μ in length, and always build an extracellular covering. Without visible damage to the host organism the tiny structures, almost along the whole extent of the midgut, penetrate into the fine, hollow spaces of the thick, brush-like border of the epithelial cells in regular rows. It has already been mentioned in Chapter 2 that not all investigators recognized the symbionts and the rickettsia of *Melophagus* as different organisms and that only the investigations of Zacharias and Aschner put an end to the differences of opinion existing in this regard.

In the low zones of the midgut epithelium, occasionally in the walls of the hindgut or of the Malpighian tubules, indeed even in the seminal receptacle there are additional delicate bacteria, sometimes forming rather long filaments, which also must not be confused with the symbionts. Whereas the rickettsia are apparently transmitted by ovarial infection, these companionate forms, like the symbionts, reach the larvae with the secretion of the milk glands and can be detected in the free state among the fat droplets of the larval gut. In the course of pupation they transfer through the dying intestinal epithelium to the imaginal intestinal lumen where they are still found in the young hatched animals.   Only later are they admitted to the epithelial cells.   Thus these occasional guests interestingly show a series of traits recalling the history of the symbionts.

In the remaining hippoboscids tested for symbiosis—*Lipoptena cervi* L. and *L. caprina* Austen, *Lynchia maura* Bigot, *Hippobosca equina* L., *H. capensis* Olfers, and *H. camelina* Leach, as well as *Ornithomyia avicularis* L.—the devices, though fundamentally like those in *Melophagus*, display rather extensive variations of the basic form.   In the imagines the symbionts are almost always housed similarly in the interior of the intestinal epithelial

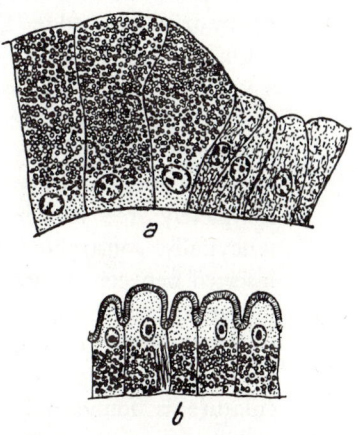

Fig. 258   (*a*) *Lynchia maura* Bigot: sterile and symbiont-containing gut epithelial cells of the adult.   (*b*) *Hippobosca camelina* Leach: symbiont-containing adult gut epithelial cells, one of which also with thread-shaped companion bacteria.   From photomicrographs by Aschner.

cells, but the extent of the infected section and the manner in which the individual cells are infected may vary widely.   In *Lynchia maura* the bacteria-containing zone is very short, comprising only one-twentieth of the total length of the midgut, but the cells far exceed those of all other hippoboscids in size; hence the symbiont-containing section is conspicuously set off, especially when the intestine is filled with blood.   The symbionts fill the greatly expanded cells very densely up to the apical pole, avoiding only the narrow, basal part in which the nuclei are embedded (Aschner, 1931; Fig. 258*a*).

In the *Lipotena* species the symbionts, as in *Melophagus*, occupy approximately one-eighth to one-tenth of the intestinal length; they fill the cells uniformly, and the nuclei are in a central position.   In the *Hippobosca* species the infected area is much larger, taking up six-tenths to seven-tenths of the length in *H. camelina* and five-sixths to six-sevenths in *H. equina*

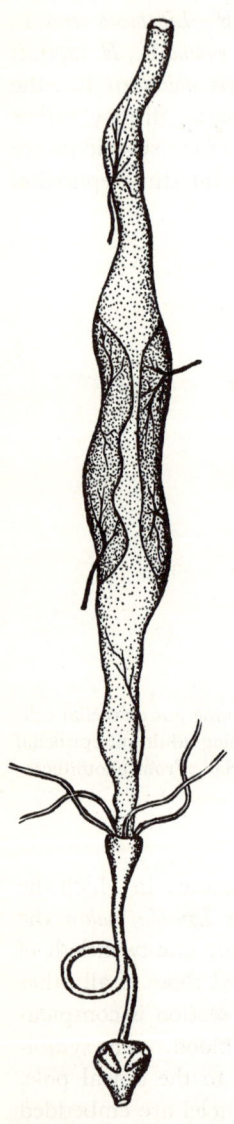

Fig. 259 *Ornithomyia avicularis* L. Gut of an adult just emerged from the pupa with paired mycetomes. From Zacharias.

and *H. capensis!* In *H. equina* the symbionts are located only at the cell basis, and the free part containing the nucleus extends like a tuft into the intestinal lumen. The same behavior is seen in *H. camelina* and *H. capensis* where the symbiont-containing part of the cell plasma is so sharply set off from the symbiont-free part that Aschner, who is responsible for all these data, compares the infected section with an evenly trimmed hedge (Fig. 258*b*).

On the other hand, *Ornithomyia avicularis*, which sucks on swallows, behaves in a completely different way, according to Zacharias. On both sides of the ventricose, thickened intestinal section, in which the cells remain completely free of symbionts, an elongated, syncytially constructed mycetome is firmly inserted between the intestinal epithelium and the musculature and is surrounded and penetrated throughout by a net of muscle fibers. Such rich development of the intestinal musculature is found only in the area of the mycetomes and elsewhere is restricted to the usual sparse fiber tracts. Abundant tracheae, some even penetrating into the interior of the mycetomes, further complicate the histological picture of this device, one which is thus far unique within the Pupipara, but with respect to position and the behavior of the intestinal musculature suggests the mycetomes described by Reichenow for *Liponyssus* (Figs. 259, 260).

Aschner points out that a certain proportion between the population strength of the individual epithelial cells and the extent of the infected sections exists in the cases cited. The more densely the symbionts fill the cells and the larger the latter become, the more limited is the populated zone of the intestine. *Lynchia maura* represents one extreme, and the *Hippobosca* species the other. In the latter the cells are infected only at the base, but almost the whole midgut is infected.   It is undoubtedly a sign of improved organization when larger and larger sections are gradually given back their real functions and when that high degree of concentration is achieved which causes one short section to appear like an organ.

It was also Aschner who provided the detailed morphology of the hippoboscid symbionts.   The diversity is striking and the variations are such that the individual hosts can be recognized easily with the aid of

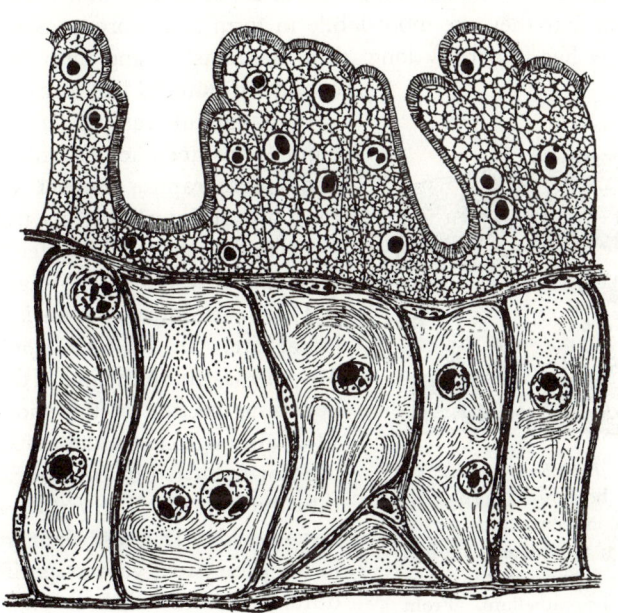

Fig. 260   *Ornithomyia avicularis* L.   Part of the mycetome embedded in the muscularis of the gut epithelium.   From Zacharias.

smears (Plate III, *c–l*).   As we have already seen, the symbionts of *Melophagus* are spheres or compact rods; those of *Lipoptena caprina* are curved, short, sausage-shaped rodlets; and *L. cervi* occupies a middle position in that it has both spherical and sausage-shaped forms.   The symbionts are short straight rodlets in *Hippobosca equina;* disc-shaped or bladder-shaped in *H. camelina;* and very polymorphous in *H. capensis*, which has both filamentous and spherical structures.   The filaments, usually irregularly curved and sinuous, have various lengths.   In addition to those which resemble the symbionts of *Melophagus* and *Lipoptena*, there are some, often as long as 25 μ, which display irregular constrictions and sinuations or are separated into fragments held together by thin connections; there are also round, oval, pear- and disc-shaped structures some of which clearly originate through balling of filaments. What causes the formation of such filament balls is a mystery at present. They are not necessitated by intracellular living, as they are also found in the fluid secretion of the milk glands. At any rate, this great diversity, which we shall encounter again in the symbionts of the bedbugs, undoubtedly arises from various phases of one and the same organism.

Like *Hippobosca camelina*, *Lynchia maura* contains disc- and vesicle-shaped symbionts, but they are more labile in form and more weakly staining. *Ornithomyia* has uniformly long, thin filaments passing through the syncytium in bundles like locks of hair.

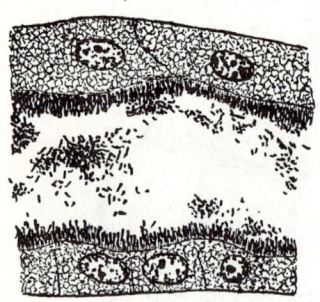

Fig. 261 *Hippobosca equina* L. Unfolded beginning portion of the symbiont-containing larval gut section; the bacteria lie free in the lumen and vertical to the surface of the epithelium. From a photomicrograph by Aschner.

In a careful analysis of these symbiotic forms documented with numerous photomicrographs, Aschner states that they are all interlinked by transitional forms and in spite of all differences may very probably be closely related.

There are interesting variants of the larval site. In *Lipoptena caprina*, as in *Melophagus*, the epithelium of the narrow beginning section of the midgut is the site, although here the infected cells are not cubical but protrude like tufts into the lumen. The *Hippobosca equina* larvae evidently house their symbionts in the same area, yet the specific details differ. Behind the esophagus, the midgut forms first a bellied and then a narrow section before passing over into the sac-shaped chyle gut. The wall of the widened part protrudes toward the inside in folds; the cells themselves have cone-like protrusions, and the symbionts, now extracellular, form a thick layer on the diversely organized surface. The short rodlets are all placed perpendicular to it, so that the protrusions and incisions are evenly covered by a matting which ends abruptly shortly before transition to the sac-like section of the gut (Fig. 261).

Hippobosca capensis behaves in a similar way except that the intestinal section in question has no folds, and protrusions of the individual cells serve to enlarge the upper surface and to retain the symbionts more efficiently. The rod-shaped elements tend to be anchored at the bottom of the fossae in dense tufts perpendicular to the surface, and the roundish ones form a thick layer of caps on the protrusions.   Occasionally clumps of them are found in the free state in the lumen.   We may safely assume that in both test objects a smaller number of the symbionts are always carried to the chyle gut with the secretion and in that position are withdrawn from observation or soon disintegrate.

Surprisingly *Hippobosca camelina* does not behave like the other two *Hippobosca* species.   Symbionts can be detected neither extracellularly nor intracellularly in the beginning section of the midgut, which has no anatomical adaptation to their presence.   Since they are present in the milk gland secretion they must also reach the chyle gut.   In fact, certain racemose associations in that location appear to be identical with them, yet the shape of the symbionts is too similar to that of the numerous yolk and protein spheres to permit positive statements to be made.

The situation is similar for the larvae of *Lynchia maura*, where symbionts are also lacking in the initial section of the midgut and in all probability are found only in the chyle in the free state until they take possession of their imaginal sites.

Zacharias was unable to investigate larval material of *Ornithomyia avicularis*.   However, the symbionts are undoubtedly housed first in the gut, free or intracellularly, for in the pupal stage balls of the filamentous structures are observed passing through the young intestinal epithelium and sinking into an adjacent zone where the musculature has been proportionately thickened and in which the syncytium which they are to populate is prepared.

In all these test objects transmission takes place when the symbionts are fed to the larvae with the secretion of the milk glands.   Aschner always found correspondingly shaped organisms, sometimes in sizable aggregations, in the lumen of the glands.   He was less concerned with the processes taking place during metamorphosis, but what he observed substantiated the hypothesis of Zacharias.

As in *Melophagus*, additional extremely diverse microorganisms occur in the other hippoboscids.   Some apparently are parasites, but others display traits which suggest the beginnings of a symbiotic relationship. Aschner, who made exhaustive studies, differentiates extra- and intracellular rickettsiae in addition to the rod-shaped bacteria which Zacharias found in *Melophagus*.   In *Lynchia maura* Aschner almost always found bacteria which are similar to those in *Melophagus*.   The population density varies widely as do the tissues in which the bacteria are found.   Bacteria often appear free in the chyme as well as between the intestinal epithelium and the peritrophic membrane surrounding the chyme.   Sometimes the musculature passing around the intestine is infected to such an extent that the muscle cells are completely interpenetrated and surrounded by the filaments, thus producing a state somewhat resembling the strange mycetomes of *Ornithomyia*. The cells already taken over by the symbionts may be infected by additional forms. Like the symbionts, they are also found in the lumen but never in the cells of the milk glands and transfer with the symbionts to the chyle gut of the larvae.

Hippobosca camelina houses a fellow traveler which is much better adapted and acts almost like a true symbiont. It is rarely absent, shows a constant infection strength, and is limited strictly to the intestinal epithelial cells and the lumen of the milk glands. Joined in chains, the delicate thin rodlets, 1.5–2 μ in length, are found mainly in the area made available to the symbionts; joined in thick bundles and placed perpendicular to the base, they, like the symbionts, populate only the basal zone of the epithelial cells (Fig. 258*b*). Although interwoven among the symbionts, they do no visible damage to the symbionts or to the cells themselves. The symbiont-free intestinal cells are attacked more rarely. In such cases the whole cell body is occupied by them, as in *Lynchia maura*.

Extracellular rickettsiae were found in *Lipoptena caprina*, *Hippobosca capensis*, and *H. equina*. In *L. caprina* they greatly resemble *Rickettsia melophagi;* they form a thick covering on the exterior of the intestinal epithelial cells and, although they cannot be detected in the milk glands because of their small size, they are nevertheless transmitted with their secretion, according to Aschner. Pupae and freshly·hatched imagines, at any rate, are already infected, as in *Melophagus*. In *Hippobosca capensis* a similar organism appears in far fewer numbers on the surface of the intestinal epithelial cells, and in *H. equina* the organisms appear so infrequently that they could be detected only at times after being allowed to accumulate in intestines which had been removed under sterile conditions and placed in suitable culture media.

In *Lynchia maura* only an intracellular *Rickettsia* could be detected. The tiny organisms never appear in great quantity and are difficult to stain. They are found chiefly here and there in the intestinal epithelial cells and at times in the cells of the fat body or of the milk glands. Transmission takes place not by feeding but by egg infection. The organisms permeate the entire extent of the follicular epithelium of young oocytes rather densely, transferring from there to the egg cells, a particularly abundant accumulation being found at the posterior pole, in other words, where the genuine symbionts are found in so many cases. Thus in this hippoboscid three types of microorganisms are found together: the symbionts; the rod-shaped companion bacteria which are transmitted by the milk glands; and the intracellular rickettsiae which are transmitted by the egg cells.

NYCTERIBIIDS

The bat-sucking nycteribiids differ widely from the hippoboscids in their symbiotic devices. As far as is known from Aschner's studies, special zones of the larval or imaginal gut are never used as symbiotic sites.

Nycteribia biarticulata Herm. and *N. blasii* Kol. house their symbionts in mycetomes. In females of *N. biarticulata* a racemose aggregation of mycetocytes flanks each side of the sizable hindgut in the most posterior

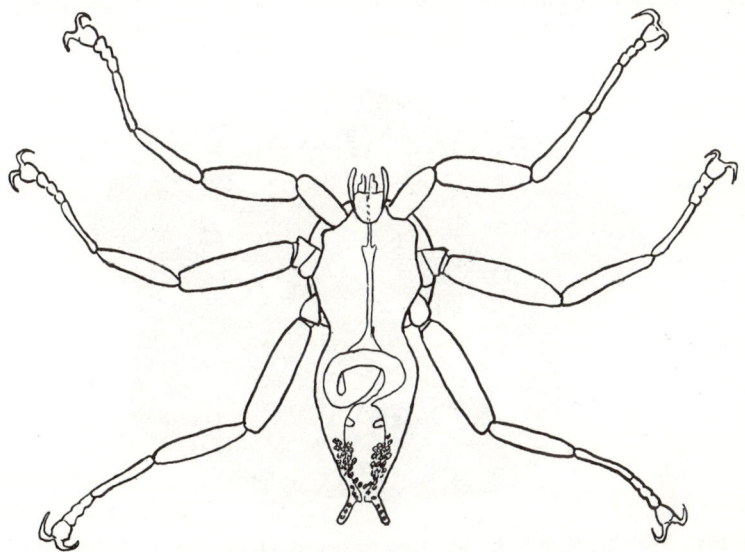

Fig. 262   *Nycteribia biarticulata* Herm.   Two clustered mycetomes in the posterior region of the abdomen.   From Aschner.

section of the abdomen, and each of the two accumulations, connected above the anus by a loose bridge of mycetocytes, sends out a smaller cell cord into the styloid processes which form at the posterior end of these strange wingless animals (Fig. 262). The arrangement is less regular in the males. The mycetocytes are not concentrated in organ-like formations but are scattered individually or in little groups in the fatty tissue and are found occasionally in abdominal areas more toward the head. *Nycteribia blasii* females also lack connected mycetomes, having merely three groups of mycetocytes arranged on each side in cords. In both species the symbiont-containing cells are not held together by a special envelope but are closely entwined by the branches of the milk glands. The symbionts settle the cells thickly and are pasted together, so that it is difficult to recognize their shape even on smears.   In *N. biarticulata* they are polymorphous structures which make their appearance as minute cocci, slender rodlets, and in every conceivable transitional form in between.   The symbionts of *N. blasii* are somewhat larger and less variable, usually roundish or oval, and tend to force the cell nucleus to the periphery (Fig. 263).

Transmission is again undertaken by the milk glands.   In *N. biarticulata* the lumen is filled by a granulated mass which apparently consists essentially of the forms found also in the mycetocytes.   On the other hand, the

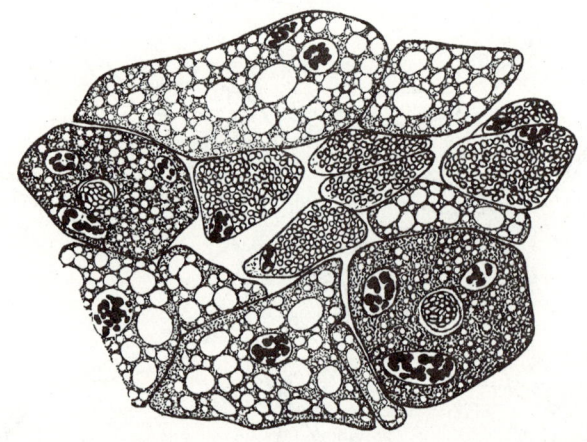

Fig. 263   *Nycteribia blasii* Kol.   Mycetocytes between fat cells and tubules of the milk glands; in the lumen of the latter are symbionts.   From Aschner.

secreting portion is weakly developed compared with that in hippoboscids, and there are no signs of increased secretory activity in its cells.   Aschner believes therefore that the symbionts serve not only in transmission but also as larval food.   Figure 263 illustrates that the lumen of the milk glands in *N. blasii* also has a dense population of symbionts.   Owing to lack of material, it was only possible to determine that typical mycetocytes and extremely polymorphic symbionts are present in *N. blainvillei* Leach.

For the same reason information is lacking on the development of the mycetomes and the localization of the symbionts in the larvae.   At any rate, these symbionts must traverse the intestinal epithelium soon after ingestion of food begins or during metamorphosis, just as in *Ornithomyia*.

In *Eucampsipoda aegyptica* Mcq., Aschner (1946) established another interesting type of nycteribiid symbiosis. There are no mycetomes or intestinal sites, the symbionts being found in the milk glands exclusively. The latter have been adapted in remarkable fashion. The lumen of the ducts is, in general, very narrow and is lined with a strong wavy intima. In the terminal areas of the individual tubules it widens substantially, and the lining becomes delicate. These numerous, slender, club-shaped spaces are thickly filled with symbionts, which in this case are long thin filaments (Fig. 264).

The same organisms transfer through the esophagus to the larval chyle gut, where they are found abundantly at the mouth. Aschner is convinced that here also they constitute part of the larval food. In the

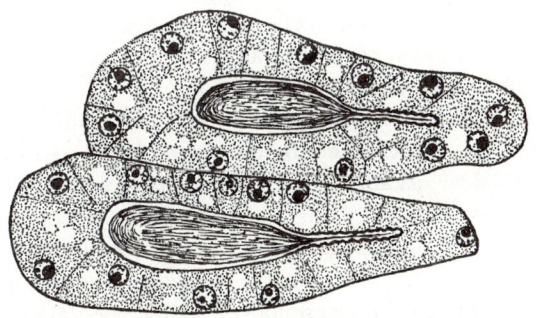

Fig. 264   *Eucampsipoda aegyptica* Mcq.   The ends of two tubules of the milk glands with broadened, symbiont-containing ducts.   From Aschner.

female larvae or pupae the milk glands must also be infected by them in some way or other; in the larval gut of males, which naturally are without milk glands, all symbionts are apparently disintegrated, for the imagines of this sex prove to be symbiont-free, a situation indicating that the Pupipara symbionts may not be necessary for blood digestion.

In numerous specimens of *N. blasii*, Aschner found a fellow traveler, a very small punctiform organism which he had already established in *Lynchia maura* and characterized as a rickettsial form. In *N. blasii* too, it tends to populate the intestinal epithelium and cells of the Malpighian tubules but may be found also in fat cells and in cells of the milk glands. As in *Lynchia*, it appears also in the follicle cells of young oocytes and from here penetrates into the oocytes themselves, forming a particularly strong aggregation at the posterior pole.

STREBLIDS

Aschner was able to investigate only one of the streblids, *Nycteribosca kollari* Frfld. He established that, in relation to the symbiosis, this genus differs substantially from the hippoboscids and the nycteribiids. Intestinal epithelium and milk glands are free of the typical Pupipara symbionts, and mycetome-like structures are lacking, but it is possible that the rickettsiae, which are mere occasional fellow travelers in *Nycteribia blasii* and *Lynchia maura*, here carry on the work of symbionts. They are exclusively intracellular, appearing most frequently here and there in the intestinal epithelial cells where they are arranged in their characteristic manner in loose cords or little morular groups; moreover, as in the other two forms mentioned, they appear also in the Malpighian tubules, the milk glands, and the fat body. Often the cells accompanying the heart tube are infected, but transmission occurs through infection of the follicle and the oocytes, as in the other cases.

Doubtless such an isolated occurrence would not justify the assumption that a symbiosis is present. Yet all other forms designated as "Pupipara" create symbiotic devices ranging from the very simple to the complicated, and it is very possible that in this case Aschner is correct in interpreting the presence of these guests of *Nycteribosca* as a still primitive stage of "symbiont replacement." Among the trypetids, Aschner points out, a similar situation obtains, for occasionally a simple population of the lumen occurs in addition to highly developed forms of symbiosis. As an isolated occurrence, he adds, such population cannot be regarded as symbiosis without further evidence.

TRIATOMIDS

The bloodsucking triatomids (Wigglesworth, 1936, 1943, 1952, and co-worker Brecher, 1944; Geigy, Halff, and Kocher, 1953, 1954; Goodchild, 1955; Halff, 1956; Bewig and Schwartz, 1956; Gumpert, 1962; Gumpert and Schwartz, 1961, 1962, 1963) have many symbiotic traits in common with Heteroptera which suck plant juices. Like the majority of plant-sucking Heteroptera, these predatory bugs, important in tropical South America as the carrier of Chagas' disease, house their symbionts in the area of the midgut; however, they do not use the hindmost section but the endodermal section of the proventricle adjacent to the valvula intestinalis, as the glossine larvae do (Fig. 265). Authors are in striking disagreement on the exact localization. Wigglesworth at first stated that the symbionts in *Rhodnius* are intracellular and transfer to the intestinal lumen by means of distally constricted off cell parts. Later (1952) he withdrew this explanation, asserting that they are found only in the lumen of the gut or between the cells of the proventricle. Halff, at the Tropical Institute of Basel, similarly found the symbionts of *Triatoma* solely in

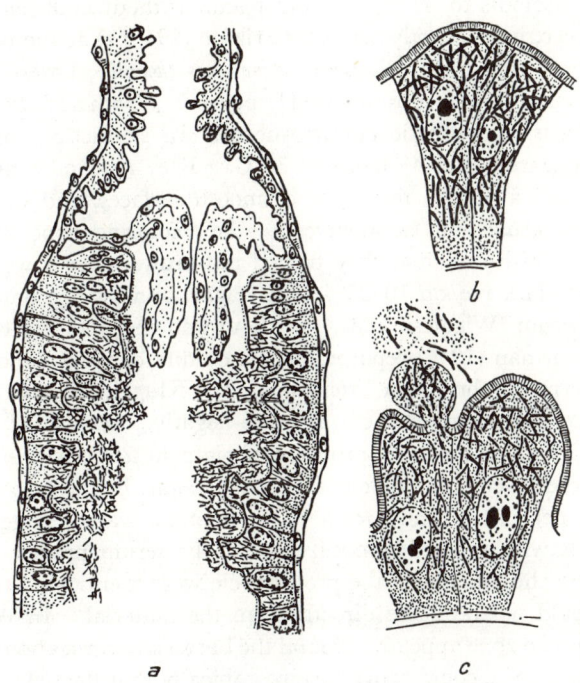

Fig. 265 *Rhodnius prolixus* Stål. (*a*) Beginning of the midgut of a second nymphal stage, 3 weeks after feeding; the symbionts are intra- and extracellular; (*b*) gut epithelial cells of a first nymphal stage; (*c*) same stage, 7 days after feeding; the symbionts move into the gut lumen. From Wigglesworth.

the lumen, but in animals from California she also found them in vacuoles of the epithelial cells (1956). Bewig and Schwartz reported that they occur in both sites in *Rhodnius* and *Triatoma* and provided a photomicrograph of *Triatoma infestans* which coincides completely with Wigglesworth's original pictures of *Rhodnius*.

According to Wigglesworth's first and evidently still valid presentation, the intestinal lumen of the young larvae is bacteria-free before the first feeding, and from the infected epithelial cells, which function at the same time as glandular cells, distal symbiont-containing plasma buds are constricted off. Released, the originally rod-shaped formations grow out into richly branched filaments in the still undigested blood within 6 to 10 days according to temperature. Later they are decomposed again into rodlets and cocci.

Wigglesworth (1936) had at first declared the gram-positive rod-shaped *Rhodnius* symbionts to be diphtheroid bacilli without knowing that they had been recognized shortly before by Erikson (1935) as actinomycetes and had received from him the name *Actinomyces rhodnii*. Later (1943) she accepted Erikson's well-substantiated hypothesis. Both authors cultivated the symbionts in synthetic culture media. In synthetic agar the first colonies appeared after 48 hours at 30°C. They reached a diameter of 1–2 mm., had a smooth margin and smooth surfaces, and were slightly arched. As usual with actinomycetes, their coloring depended on the medium. Whitish at first, they turned red-brown after a week on 5% blood agar, dark red on 10–25% blood agar, and orange on agar with Loeffler's serum (Wigglesworth). Without knowledge of her two predecessors, Weurman (1946) reported on the culture of the symbionts which had transferred to the gut in *Triatoma infestans* Klug. Classifying them as *Pseudomonas*, she describes them as gram-negative, very small, flagellate rodlets which show the same property of pigment formation in that they produced cream-colored colonies in peptone agar, but were colored rose in certain peptone-gelatin media or in peptone water. Agglutination experiments with rabbits immunized with the serum proved that these cultures and the inmates of the proventricle were identical, although the cultures could not be made directly from the material. In Weurman's cultures a green zone appeared around the bacterial aggregations in blood-glucose agar. She could prove spectroscopically that the oxyhemoglobin content at first lessened and then disappeared completely. Thus she theorized that the green zone represented verdo-hemochromogen, a compound transitional between hemoglobin and biliverdin. Wigglesworth (1943, 1951) rejects the possibility that *Rhodnius* symbionts participate in the dissolution of the blood, and Bewig and Schwartz come to the same conclusion. Weurman's classification of *Triatoma* symbionts is rejected by other authors, who generally use the term *Actinomyces* (*Nocardia*) in speaking of *Rhodnius* and *Triatoma*. Only Goodchild (1955) considers the symbionts of the two genera to be essentially different and is inclined to place the diphtheroid symbiont of the triatomes in the vicinity of *Corynebacterium hofmannii*. Bewig and Schwartz describe a companion coccus in addition to the *Actinomyces rhodnii*, but, since the coccus was found by no other author, Halff believes that a mixed infection may have occurred as a contamination from contiguous cultures. At any rate, it would be desirable to amplify our knowledge of the triatomid symbionts on the basis of a larger number of species, using experiments on interchangeability wherever possible.

Regarding symbiotic transmission to the progeny, Wigglesworth at first was of the opinion that the oocytes are already infected and that in freshly hatched animals the symbionts are in the intestinal epithelium, but she revised her hypothesis (1944) on the basis of studies by Rosenkranz (1939), who observed superficial smearing of heteropterous eggs at the moment of deposit and subsequent admission of symbionts through the mouth of newly hatched larvae. Wigglesworth now obtained experimental proof that the egg plasma was sterile in her test objects also and that symbionts were present in the chorion of the deposited eggs, which, unsterilized, had yielded the typical colonies in bouillon with 0.2% glucose. Correspondingly, the larvae were always symbiont-free when they hatched from sterilized eggshells and were kept sterile. However, the symbionts on the eggshell were so few in number that they could be detected only through culture; special devices to increase smearing were absent; and it was impossible to observe the newborn instinctively palpating the eggshells with the mandibles. Thus Wigglesworth concluded that not only the eggshells but also the symbiont-containing excreta of fellow animals represents an important source of infection. The animals are indeed inclined to suck the fluid feces of one another, and sterile larvae were easily infected when provided with such feces or when simply brought together with normal larvae. In this connection it should be recalled that Huber-Schneider (1957) arrived at similar ideas on transmission in the plant-sucking bug *Mesocerus*. Actually, all authors agree with Wigglesworth that the acquisition of symbionts is guaranteed under normal conditions by means of feces consumption.

In harmony with such a primitive arrangement are additional findings of Gumpert and Schwartz on the symbionts of *Rhodnius prolixus* and *Triatoma infestans*. From both insects they cultivated not only a *Nocardia* (*Actinomyces*) species, which was not identical with *Nocardia rhodnii*, but also a *Corynebacterium*, a *Pseudomonas*, and a *Mycobacterium*, which appeared in two forms. When they infected animals, easily rendered symbiont-free by preventing their normal infection, with these strains, it was found that each one had the value of a genuine symbiont and guaranteed normal development. Reinfection with certain companion forms, which also were regularly isolated, proved to be without effect. This surprising plasticity of *Triatoma* symbionts could explain the divergence of findings of different authors.

In the General Section we shall discuss the differing concepts of the problem of the usefulness of symbionts.

In addition to *Rhodnius prolixus* Stol, six *Triatoma* species and *Eutriatoma flavida* Neiva and *Psammolestes coreodes* Bergroth were investigated; all were found to be living symbiotically with an *Actinomyces* (Dias, 1937; Liem Soei Diong, 1938; Weurman, 1946).

In the final chapter we shall describe the interesting experiments of Wigglesworth and successors with sterile animals.

CIMICIDS

In both sexes of *Cimex lectularius* L., the glassy, whitish mycetomes may be overlooked at first glance because of their resemblance to fat lobes (Buchner, 1921, 1922, 1923). The paired mycetomes are situated near the third abdominal segment in females; strange to say, they are slightly fused with the vasa deferentia of the testes in males. The testis of *Cimex lectularius* consists of seven oval chambers arranged in an arch, and the approximately oval mycetome, resembling a strayed eighth chamber, is suspended on the concave side of the testis (Fig. 266). At first the mycetome is more voluminous than the testis rudiment, but, although it grows steadily and reaches a length of approximately $\frac{1}{2}$ mm. in mature males, it is overtaken in size by the testis rudiment during postembryonic development.

Histologically these organs are easily distinguished from the fatty tissue. They consist of giant cells with three to five chromatin-rich nuclei, four to five cells being placed crosswise and eight lengthwise. A delicate cellular envelope with sparse flattened nuclei encases the organs and, accompanied by tracheoles, pushes in between the mycetocytes, in which the plasma is so thickly filled with microorganisms that it appears finely granulated on sections.

The morphology of the inmates is revealed only through study of the crushed organs and of smears of them. Side by side in the imaginal mycetomes are slender straight rodlets with sinuous individual motility, convoluted filaments, cocci, and strange disc-like states. Despite their varied appearance, these forms evidently represent one and the same organism.

The occurrence of such mycetomes in a heteropterous bug came as a surprise, but today it is no longer considered exceptional, for Schneider (1940) later found similar well-defined organs in several bugs which suck plant juice. The same thing may be said about the once unique finding that the symbionts are transmitted indirectly by infection of the nutrient cells. Even in the early larval states it is possible to observe in the plasma of the binucleate nutrient cells not only dense clumps of smaller, strongly staining states but also rodlets which stain even more strongly.   Smears yield the same shapes found in the mycetome.   They are carried by secretory flow in the fiber bundles to the young oocytes.   Moreover, very early stages are found, even "bouquet" stages, with plasma containing easily stainable rodlets (Fig. 267).   At first permeating the egg plasma

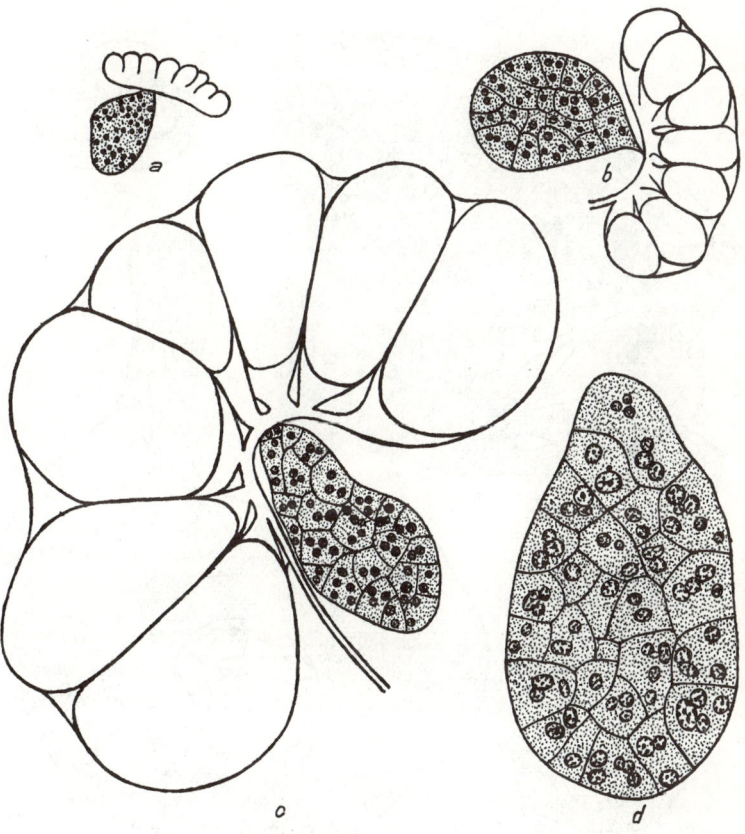

Fig. 266   *Cimex lectularius* L.   Mycetome and testes during postembryonic development.   From Buchner.

on all sides, the symbionts are finally collected, as in so many test objects, at the posterior egg pole in a thin inconspicuous layer.

In the ovoviviparous bedbug, embryonic development is still associated with the ovarioles until the germ band assumes an *s*-shape.   Cleavage nuclei rising to the periphery appear in the symbiont zone at the posterior end of the egg.   Later the blastoderm cells infected in that way increase and form a multilayered cushion protruding into the yolk. Since invagination of the germ band sets in round about this cushion, the latter is again shoved inward and for a while seals off the amnion hollow like a cork. Only after the germ band curves into an *s*-shape is the now spherical

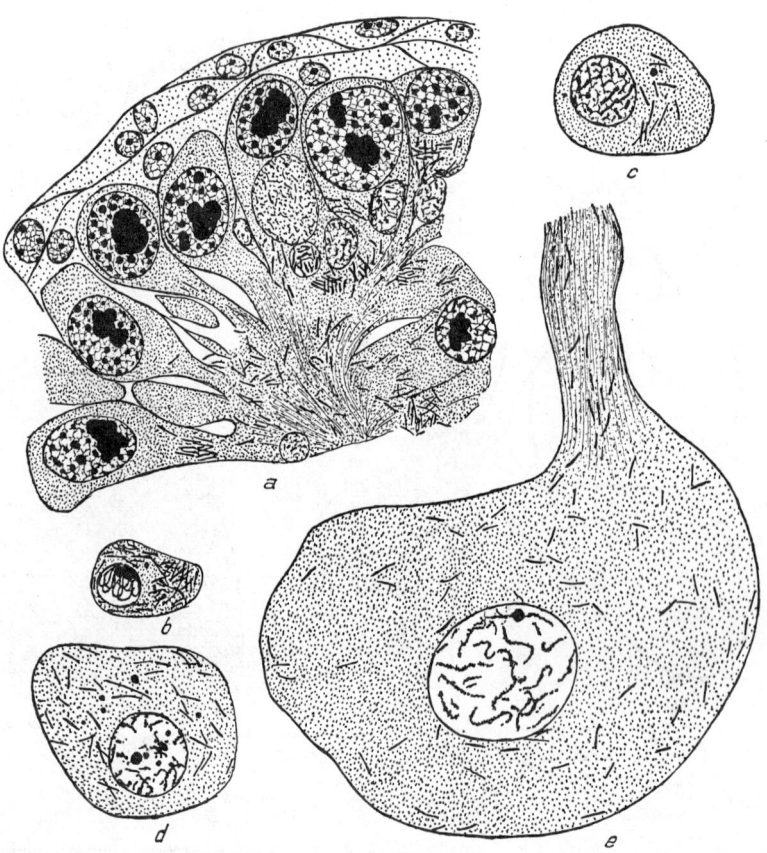

Fig. 267   *Cimex lectularius* L.   Transmission of the symbionts.   (*a*) Infection of the nutrient cells; (*b, c, d*) young infected ovocytes; (*e*) the symbionts pass through the nutritive cord into the egg.   From Buchner.

infected cell mass finally separated. While the germ band is still in growth, the mass remains in its position; when the extremities sprout, it flattens and again settles closely against the germ band which thus transports the young mycetome during revolution. After revolution it is con-

stricted off in dumbbell shape (Fig. 268). Finally the two young mycetomes come to rest between the intestinal epithelium and the hypodermis
on one side and two dorsoventral muscle bundles on the other. Whereas
the settlement of the embryonic mycetocytes is very sparse at the beginning
and some are perhaps entirely symbiont-free, a lively increase of symbionts
begins before revolution and leads to a filling so dense that the form of the
organisms is barely distinguishable. This whole process of embryonic

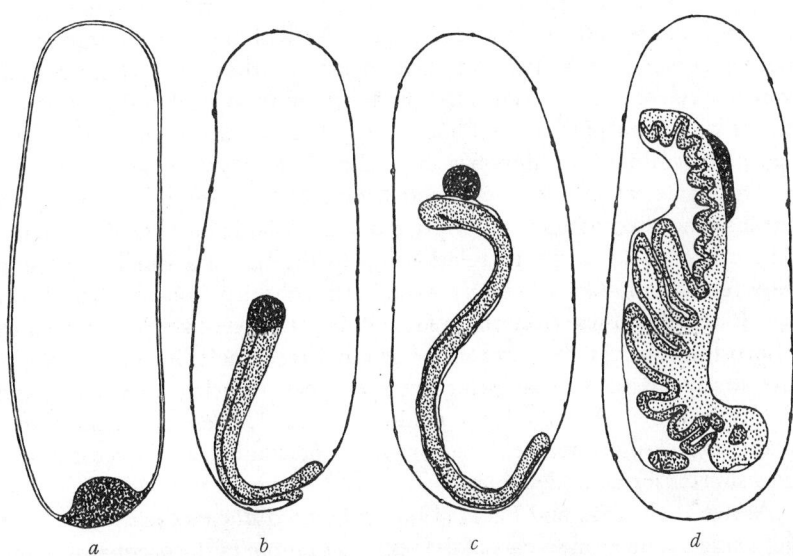

Fig. 268 *Cimex lectularius* L. The development of the mycetomes. In (*d*) the mycetome is considerably constricted, and the section is therefore placed more laterally.
From Buchner.

development is extraordinarily similar to what Schneider found in the
heteropterous bug *Ischnodemus* (Fig. 100). Until the flattening process
takes place, the embryonic mycetocytes divide mitotically; then mitosis
is arrested and replaced by amitosis, with resultant polynuclearity and
gradual gigantic growth (Fig. 266).

Whereas I had used only sections in most of my research, Pfeiffer also
used smears in studying (1931) the behavior of symbionts during embryonic and postembryonic development. Thus he determined certain
cyclic form changes of the symbionts which in part had already struck the
attention of Hertig and Wolbach (1924), who had been the first to confirm

my observations. In the early larval states Pfeiffer found convoluted rodlets of different sizes which were constricted and often were held together only by thin bridges, in that way giving rise to roundish states. Even this early the strange "filament balls" may occur that were also observed by Hertig and Wolbach; according to them, these "balls" are caused by an invisible envelope which prevents elongation. Later there are disc-shaped formations, which had also attracted the attention of these American authors. Round or angular in their boundaries, they are also present in the imagines and are formed perhaps through merging of the filamentous balls. Hertig and Wolbach also mention ring-shaped or *c*-shaped stages and round and oval granules with a denser margin and a lighter center. It is doubtless safe to assume that these states are all genetically connected. Thus the organism under consideration is exceptionally pleomorphic, resembling in many respects the similarly polymorphic symbionts of hippoboscids and some amblyommines (*Boophilus*).

The bacterial problems connected with bedbug symbiosis are by no means exhausted with the description of this variable mycetome occupant. There is need to clarify its relationship to the bacteria which are found very regularly in *Cimex lectularius* even outside the mycetomes. Some time ago I identified these bacteria in fat cells in leucocyte-like elements which resorb the superfluous sperm in the seminal receptacle, in similar cells of the strange Ribaga-organ which receives sperm during copulation, and finally in the epithelium of the oviduct. Usually these additional occupants are easily stainable, longer rodlets and filaments and occasionally also shorter forms in dense nests.

Arkwright, Atkin, and Bacot (1921), who used smears exclusively, were only able to catch, in smears of the testes, a glimpse of the occupants of the mycetomes, the latter still being unknown to them. They also found that isolated and sometimes considerably swollen cells of the Malpighian tubules are infected regularly with organisms ranging from minutest cocci to filamentous-shaped forms, the latter disintegrating sometimes into tiny structures. They proved that the intestinal epithelial cells themselves are often not spared. The designation "*Rickettsia lectularia*" which they introduced into the literature refers solely to this finding.

I had only tentatively put forward the idea that the typical occupants within the mycetome and the organisms outside of it were identical, but Hertig and Wolbach joined them in one cycle of forms without hesitation. Pfeiffer decisively upheld the hypothesis that they are strictly separate organisms and that the term *Rickettsia lectularia* should be applied only to those ill-adapted companion forms which are subject to variations in their occurrence. He also found them within the mycetome side by side with the symbionts and distinguishable from them, and he proved that the companion forms also infect the egg cells, using the same route as the symbionts. He is of the opinion that the easily stainable rodlet, which is found even in the youngest oocytes, should also be considered a companion form and not a symbiont.

Pfeiffer's conception gains in probability when it is recalled that such non-pathogenic fellow travelers may occur very regularly among the hipposcids and some other symbiont bearers. Another argument in its support is that the swallow bug *Oeciacus hirundinis* Jen., which has mycetomes of exactly the same type, does not have such companion forms but does have the coarser rodlets, filament bundles, and disc shapes which are characteristic of bedbug mycetomes. In the mycetome smears neither fine filamentous forms nor minutest cocci are found, an indication that these, when they are present in the bedbug mycetome, should be classified among the additional guests which do not stop before the site of the legitimate symbionts.

When one considers that these cimicid symbionts assume shapes not observed before in clear-cut rickettsia, such as balled-up filaments, discs, and the like, and that true rickettsia have never been found elsewhere as symbionts of mycetomes, one will do well not to apply the term *Rickettsia lectularia* to the regular inmates of cimicid mycetomes.

Further investigations are needed. Such research would have the goal of separating the two forms as sharply as possible and of focusing attention on their presence side by side during embryonic development. Years ago it struck me that not all rodlets and filaments in the posterior area of the egg are drawn into the embryonic rudiment of the mycetome but that some, together with several yolk nuclei, still adhere to the exterior during formation of the embryo. It would be necessary to determine whether these forms on the exterior are companion forms and thus serve to infect the remaining tissue, or whether they might possibly be the symbionts which later infect the female gonads where settlement apparently also begins early, as described in *Ischnodemus* by Schneider (Fig. 99).

Attempts to cultivate these microorganisms have been made repeatedly. Arkwright, Atkin, and Bacot (1921) and Kuczynski (1927) were unable to report any degree of success, although they used many different culture media. Pfeiffer's diversified experiments, which included inoculation into tissue cultures, were also unsuccessful. Occasionally he found a limited possibility of development in both symbionts and rickettsiae, especially with an addition of embryonic extracts, but the forms had no stability when transferred to the same culture media. Steinhaus (1946) reported the appearance of a diphtheroid organism in his culture experiments, but its identity with the bedbug forms being described here must remain an open question.

That the latter are not pathogenic has been proved by the investigations of Arkwright, Atkin, and Bacot and by experiments of my own. The first three authors mentioned inoculated two experimental persons with pulverized bug intestine without reactions of any kind. Using an artificial puncture canal, I inoculated myself several times with mycetomes and achieved merely the type of swelling occasioned by a bedbug bite or by introduction of other protein bodies originating from sterile insects. Implantation of pulverized bugs in mice, guinea pigs, and rabbits also produced no reaction.

So far our presentation has referred almost exclusively to phenomena occurring in *Cimex lectularius*, which has understandably been the chief test object. It has already been mentioned that the swallow bug *Oeciacus* possesses very similar symbiotic devices. It is also possible to assert today that all the other bloodsucking cimicids very probably have their symbionts too. An illustration of the male sexual apparatus of the tropical and subtropical *Cimex rotundatus* Signoret in Patton and Cragg's *Textbook of Medical Entomology* (1913) shows a relatively smaller, roundish formation at the origin of the vas deferens which the authors call an "accessory lobe," but which is doubtless identical with the mycetome of our bedbug (Fig. 269). Furthermore Carayon reports by letter that he found corresponding mycetomes in other tropical relatives. In males of *Ornithocoris uritui* Lent. and Abalos and *O. toledoi* Pinto, two South American forms, he also found a round mycetome attached to the seven-chambered testis at the point where the vas deferens has its origin. The same applies to *Leptocimex boueti* Brumpt, a bug of tropical Africa, which is as easy to cultivate on the human being as are our bedbugs. By courtesy of Carayon, I know the situation here in part through personal observation. The very insignificant mycetome of females, situated on each side independently of the gonads close beneath the skin, is composed of small uninucleate cells, its plasma being filled with long, interbraided, filamentous symbionts. Thus in shape and clarity they are clearly distinguishable from the *Cimex lectularius* symbionts.

The manner of transmission resembles that of *Cimex lectularius* in that very young oocytes are infected but differs considerably in that they are overwhelmed by the microorganisms to a degree observed elsewhere only in *Camponotus* species and *Formica fusca*. In such cases their plasma is limited to a sparse network, at least in *Leptocimex*, where the symbionts now appear to be more compact and more deeply staining, and only after the oocyte is connected with a powerful nutrient cord does its growth gradually gain the ascendancy. Finally the symbionts are concentrated at the posterior end of the older oocytes in a flat layer. As with *Cimex lectularius*, the blastoderm then forms a terminal, symbiont-containing, thickened zone which is shifted forward during invagination.

Of particular interest is Carayon's finding in his study of *Primicimex cavernius* Barber. This Mexican bug which lives on bats is a representative of the subfamily of the Primicimicinae, which in turn represent the most primitive cimicids in existence. Carayon's test object is revealed as such

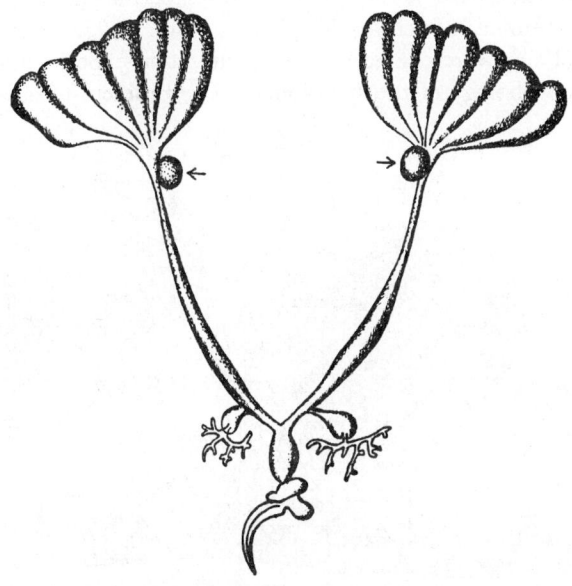

Fig. 269   *Cimex rotundatus* Signoret.   Male sexual apparatus with both mycetomes. From Patton and Cragg.

not only in its morphology and insemination devices but also in its localization of symbionts! For here, in contrast to all other forms investigated, the symbionts live not in mycetomes but in the epithelium of the midgut (cf. Carayon, 1959, p. 96, footnote).

In all species of *Leptocimex*, Carayon found in the fat body, and in females also around the ovarioles, companionate bacteria which resemble those of the bedbug but are lacking in the two American species, as was the case also with *Oeciacus*. It is to be hoped that in the not too distant future Carayon will publish a more detailed comparative study of cimicid symbionts which will contribute essentially to our understanding of a subject which is still unclear in several aspects.

ANOPLURA

When Sikora and I, independently of each other, discovered Anoplura symbiosis in 1919, each of us submitted a brief report.   In 1931 the subject was exhaustively studied by Ries (1931) and became one of the most interesting chapters in symbiotic theory.   Valuable supplementation was provided by Baudisch (1958) in his studies of the embryology of the *Pediculus* mycetome.   Puchta (1954, 1955), Kotter (1955), and Bewig and Schwartz (1956) were concerned with cultivation experiments and with the finer morphology of the symbionts.   The problem of usefulness was

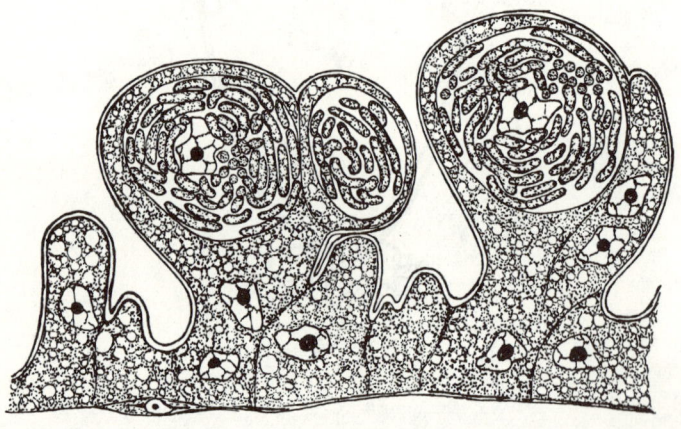

Fig. 270   *Haematopinus eurysternus* N.   Gut epithelium with mycetocytes.   1200×.
From Ries.

investigated successfully by Puchta (1955) on animals made sterile by centrifugation.   His results are found in the General Section of this book.

The symbionts may be localized in very different ways.   The simplest conditions are found within the genus *Haematopinus*.   In *H. eurysternus* Nitzsch of cattle, *H. suis* L. of swine, and *H. macrocephalus* Burm. of horses, numerous mycetocytes are scattered over the epithelium of the midgut, except for the proventricle which is peculiar to this genus.   They are so intimately inserted in the epithelium that it often appears as if they were lying within the cells, but the embryological history of the gut reveals that they are merely wedged between the cells and surrounded by them (Figs. 270, 272*b*).   The mycetocytes are roundish; the tube-shaped symbionts, which are clearly honeycombed in structure, are doubtless bacteria;

arranged parallel to each other and sometimes forming chains, they surround the central nucleus; the plasma of the host cells is barely visible. The symbionts are never transferred to the intestinal lumen, contrary to the statements of Florence (1924).

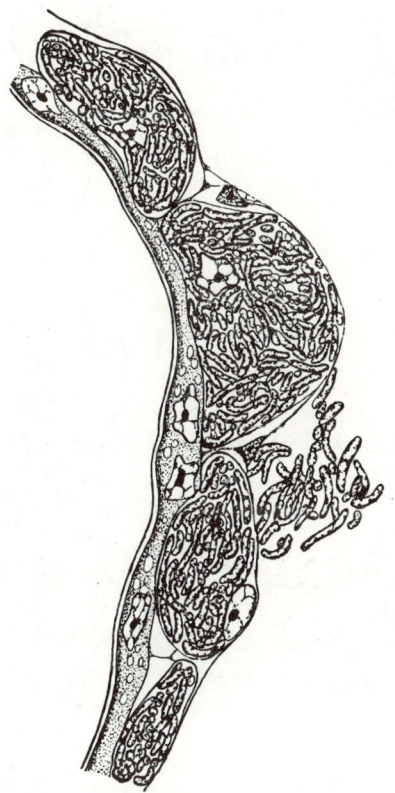

Fig. 271   *Pedicinus rhesi* Fahrh.   Section from the zone of the midgut occupied by mycetocytes.   1800 ×.   From Ries.

With the genus *Pedicinus* living on catarrhine apes, Ries found a new type of localization.   Between the intestinal epithelium and the basal membrane with its circular and longitudinal musculatures, a closed ring of mycetocytes sharply constricts the stomach gut approximately at its middle (Fig. 272c).   The membrane, which is provided sparsely with flattened nuclei, is forced away from the epithelium and now is connected with it only by fine fibers passing between the mycetocytes.   The nuclei are

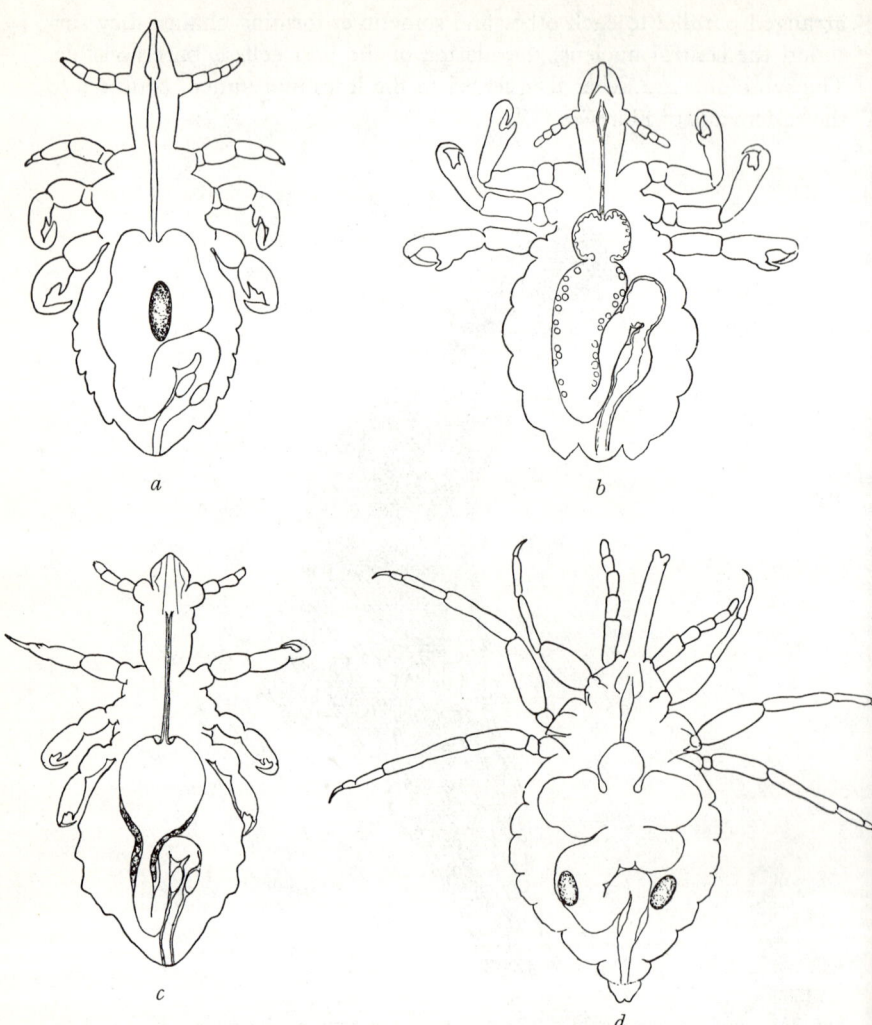

Fig. 272   (a) *Linognathus tenuirostris* Burm.   Grown male with unpaired mycetome.
About 22×.   (b) *Haematopinus suis* L.   Male, mycetocytes in midgut epithelium.
About 18×.   (c) *Pedicinus rhesi* Fahrh.   Grown male, mycetocytes between gut epi-
thelium and basal membrane.   About 22×.   (d) *Haematomyzus elephantis* Piaget.
Grown male with paired mycetomes.   About 22×.   From Ries.

lobate and are usually placed eccentrically in the mycetocytes; the sur-rounding plasma is filled with short granulated tubes passing criss-cross. The intestinal epithelium shows no peculiarities of any kind in the zone under discussion (Fig. 271).

Most frequently, however, the symbionts of lice are housed in well-defined mycetomes which may vary in several respects in structure and location. Usually they are unpaired, oval, or roundish organs situated

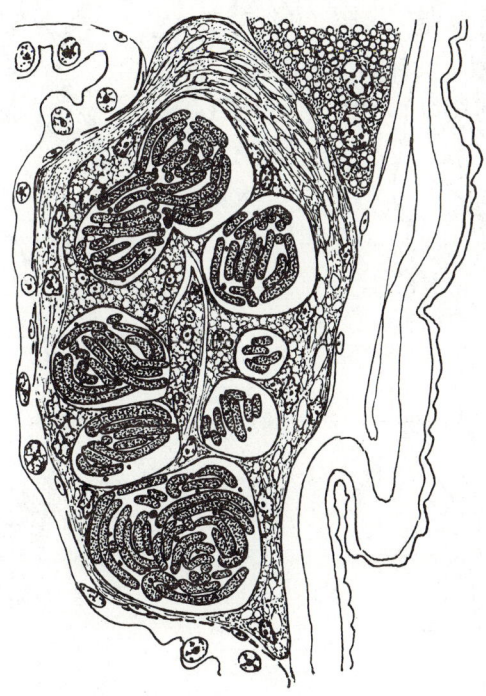

Fig. 273   *Pediculus capitis* DeG.   Sagittal section through the stomach disc of a male nymph before the third moult; left, the gut epithelium; right, the hypodermis.   770 ✕.
From Ries.

ventral to the stomach and ordinarily are without close relationships to the latter (*Linognathus*, *Polyplax*, Fig. 272a), although at times they may be connected with it somewhat more intimately (*Pediculus*, Fig. 274h).

In the simplest case these organs consist merely of a sharply defined aggregation of mycetocytes flattened off toward one another without envelope formation, as exemplified in *Linognathus tenuirostris* Burm. and *Polyplax spinulosus* Burm.   The cells of the first are filled with typical fila-

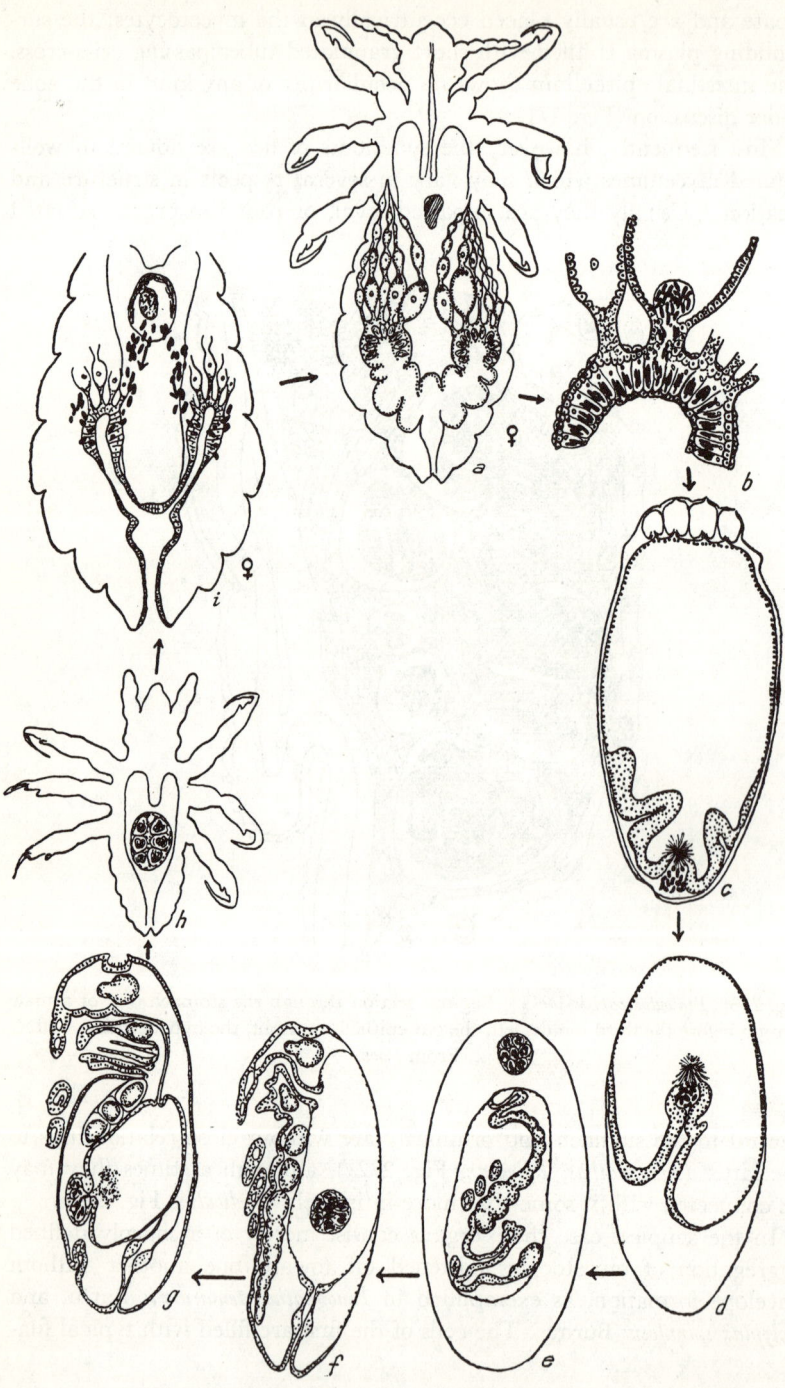

ments, and the extremely simple *Polyplax* mycetome with small, irregular, granulated vesicles. More complex in structure are the "stomach discs" of the *Pediculus* and *Phthirus* species. Unlike the colorless mycetomes of *Linognathus* and *Polyplax*, they are whitish yellow formations which stand out clearly against the blood-filled, dark red gut. In *Pediculus capitis* De Geer, a syncytium surrounds 10–16 radially arranged chambers containing sausage-shaped symbionts which are fully homogeneous *in vivo*. When the mycetome reaches its maximum development, as in the larvae before the third molting, three zones can be distinguished: a central, weakly vacuolized cell mass; a second similarly constructed mass in which the chambers are situated; and a third, more fibrous, coarse envelope layer surrounding the organ. Irregularly shaped, frequently lobate nuclei are distributed in these syncytia. The chambers taken up by the symbionts never contain nuclei and thus are not cells but merely well-defined and regularly ordered vacuoles of the syncytium (Fig. 273).

The opaque appearance of these stomach discs is produced by numerous granulations inserted in the exterior envelope. These minute, oval, strongly light-refracting granules which appear in early embryonic stages are dissolved in diluted mineral acids and thus are rarely visible in the microscopic preparation. Between the symbionts in the vacuoles are strongly staining, yellow granules. These are metabolism products, which Kotter (1955) studied under the fluorescent microscope and classified as pterines because of their light blue fluorescence. At the place of contact these organs cause the intestinal epithelium, with which they are slightly fused, to protrude inward like a cushion, although there is no connection with the intestinal lumen.

Anoplura symbionts are always transmitted to the progeny indirectly by infection of special organs between the ovaries and the oviducts, the so-called ovarial ampullae, but their histological structure, and especially the methods by which they are filled with symbionts, may differ widely from species to species. In *Pediculus* certain changes begin in the mycetome with the approach of the third molting. The symbionts increase more markedly, the chambers fuse, giving rise to an isolated central cell group which looks as though it had been sintered, and to a peripheral, strongly vacuolized envelope. The symbionts swarm in thick clumps

---

Fig. 274 *Pediculus capitis* DeG.   Symbiotic cycle.   (*a*) Female imago with evacuated mycetome and infected ovarial ampullae.   (*b*) Infection of the ovocytes by way of the ampullae.   (*c*, *d*) Invagination of the germ band; the symbionts are carried into the interior of the embryo.   (*e*) Primary mycetome in the yolk.   (*f*) The transitory mycetome has arrived in the embryonal gut.   (*g*) Abscission of the stomach disc.   (*h*) Youngest nymphal stage with mycetome.   (*i*) The symbionts transfer from the mycetome into the ovarial ampullae.   Partly schematic.   From Ries.

through gaps in the outer connective tissue and glide backward on the ventral side of the stomach (Figs. 274*i*, 275*a*).   It is not clear just how this movement takes place, for motility has never been observed in the symbionts.   At any rate, the latter are not scattered throughout the body cavity but advance toward the ampullar rudiment only.   At this time the

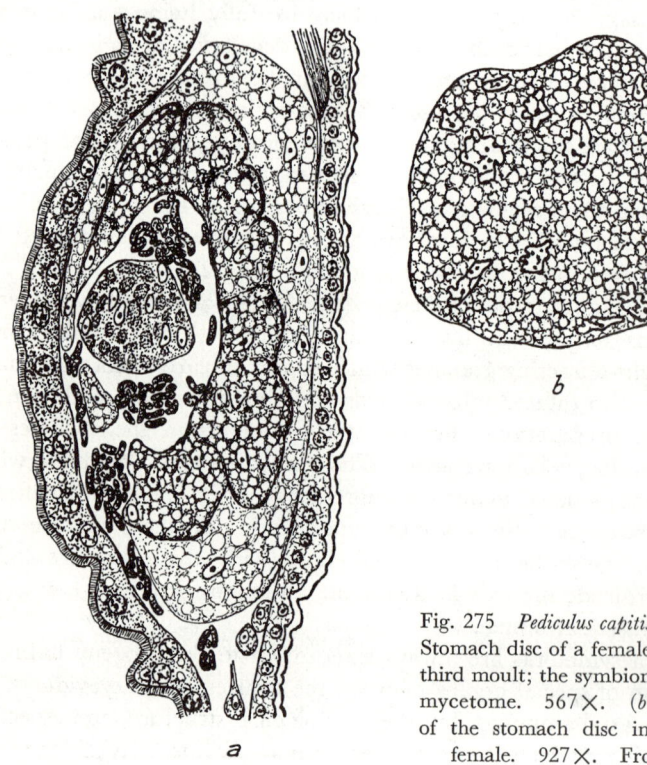

Fig. 275   *Pediculus capitis* DeG.   (*a*) Stomach disc of a female during the third moult; the symbionts leave the mycetome.   567×.   (*b*) Remnant of the stomach disc in a mature female.   927×.   From Ries.

rudiment is merely a moderately thickened section of the oviduct epithelium adjoining the young ovarioles.   At this place exclusively the symbionts transfer to the oviduct epithelium, thereby temporarily causing the otherwise homogeneous tunica to be roughened irregularly, and then are usually situated individually in vacuoles of the cell plasma.   Gradually a tetra-laminate infected section develops which is characteristic of most ovarial ampullae.   Behind the tunica, which has regained its smoothness, cells pushing backward from the egg follicles force the infected epithelium inward.   On the other hand, cells pushing forward from the non-infected part of the oviduct furnish the inner lining of the ampulla (Fig. 276).

Finally, mesodermal elements attach themselves to the exterior of the tunica and thus bring about the final state of the ampullar wall (Fig. 277).

Through this migration the stomach disc becomes completely symbiont-free; in the grown female it represents only an irregular cell group, often ramified and still strongly light-refracting, with shriveled nuclei (Figs. 274*a*, 275*b*). Swammerdam's illustrations, mentioned in our historical introduction, refer to such devastated and distorted organs.

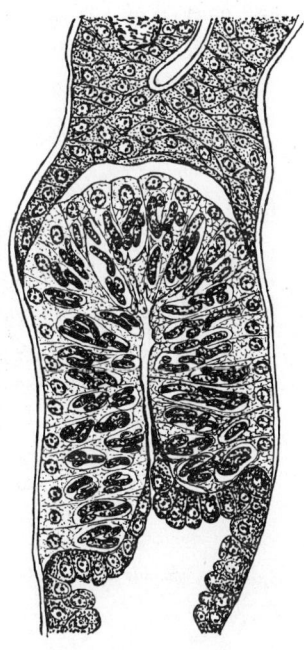

When the ovarioles attain a certain age and when the oocyte situated behind each reaches the infectious stage, some of the symbionts, joined in vacuoles, transfer from the ampullar epithelium to the egg pedicle. From here the bacteria glide into a depression in the oocytes and, when the latter are sufficiently enriched, they close behind the symbionts. A seam leading to the symbiont bundle recalls even later the erstwhile invasion place (Fig. 274*b*).

Strange to say, at the time when the migration of symbionts is prepared for in females, degenerative changes also take place in the male mycetomes. The well-ordered division into chambers is blurred; the symbionts which are still tube-shaped begin to be colored in a highly irregular fashion. All sorts of deviating shapes, roundish and elliptical forms, appear. In the grown male the chamber walls at length disappear almost completely, and a thin fibrous envelope surrounds a symbiont clump, the components of which are no longer clearly delineated.

Fig. 276   *Pediculus capitis* DeG. Development of the four-layered structure of the ovarial ampulla. 510✕.   From Ries.

Ries therefore uses the term "symbiont debris." Yet in some pubescent males the mycetome still has the appearance of the larval stomach disc.

This description based on *Pediculus capitis* applies also to *P. vestimenti* Leach, although it is by no means certain that the latter represents an individual species; and the situation in *Phthirus pubis* Leach is very similar.

In *Linognathus tenuirostris* Burm., the ovarial ampullae are settled in much simpler fashion. Shortly before the third molting, the less complex mycetome is taken over in its entirety into the ectodermal genital rudiment, which is not yet joined to the ovarioles.   At the point where the lateral sections of the oviduct arise, a bundle of strange cells of fibrous structure is differentiated, breaks through the wall of the embryonic oviduct, and advances toward the tube-shaped mycetome.   Like drawn fibers, the cells attach themselves closely to the mycetocytes, in which the cell limits are beginning to blur, and shortly thereafter all bacteria are situated within the genital rudiment although the preparations did not reveal whether a pulling action or merely a path breaking is involved (Fig. 278).   At any rate, the task of the fiber cells is concluded; they disappear while the oviduct joins the ovarioles on both sides.   The nuclei of

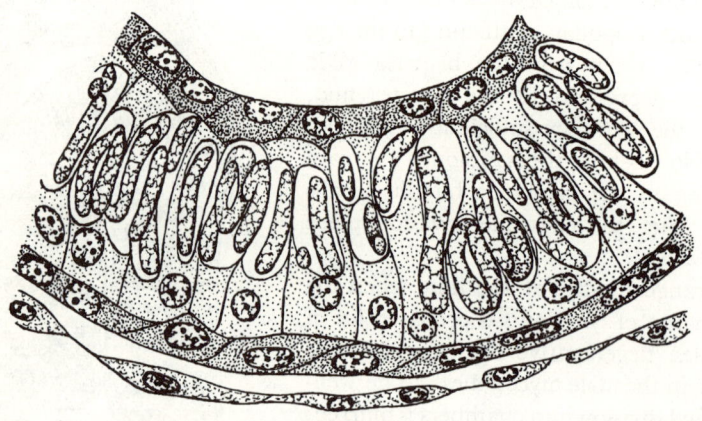

Fig. 277   *Pediculus capitis* DeG.   Histology of the completed ovarial ampulla.   From Buchner.

the old mycetocytes degenerate, the abundantly increasing symbionts diverge toward both sides to advance toward the ovarioles.   On arriving at the blind end of each oviduct they penetrate its wall, which still represents a syncytium.   Some of the nuclei of this syncytium are used to form a sterile basal cell layer; the remainder are used for the mycetocytes which are now being constructed again.   A cell layer terminating within is furnished as in *Pediculus* by the non-infected epithelium of the oviduct, the latter functioning in its later course as a cement gland.   In this way a similar stratification of the ampullar wall arises, but no follicle cells are used.   Unlike the male mycetomes in *Pediculus*, those of *Linognathus* exhibit no degeneration phenomena.

In *Polyplax spinulosus* Burm., the rat louse, the whole mycetome is also taken over into the rudiment of the oviduct, but special fiber cells are not developed.   Apparently the epithelium of the ectodermal invaginations is tensely stretched where the mycetome obstructs development, tearing finally and closing behind the transferred organ.   The old nuclei of the mycetocytes degenerate, and the symbionts are distributed over the two wings of the genital rudiment.   A new feature is that elements grow out

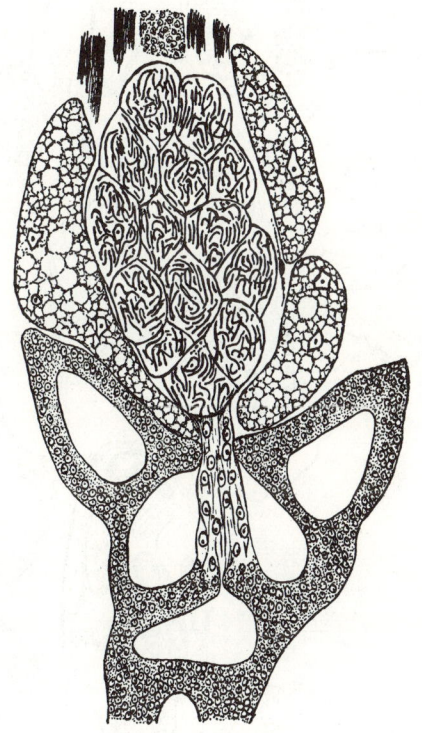

Fig. 278   *Linognathus tenuirostris* Burm.   Frontal section through the rudiment of the female genital apparatus; the mycetome is united with the oviduct by fibrous cells. 360×.   From Ries.

from the closed cell formation of the egg pedicles and advance toward the newcomers (compare Fig. 282).   They receive the symbionts, which have the same roundish shape that they had in the mycetomes, and thus give origin to the syncytium which constitutes a substantial part of the ampulla, the latter otherwise being constructed in the typical manner.

After the end of egg infection in this test object, the symbionts are not situated in roundish bundles in the yolk but nestle in a densely packed

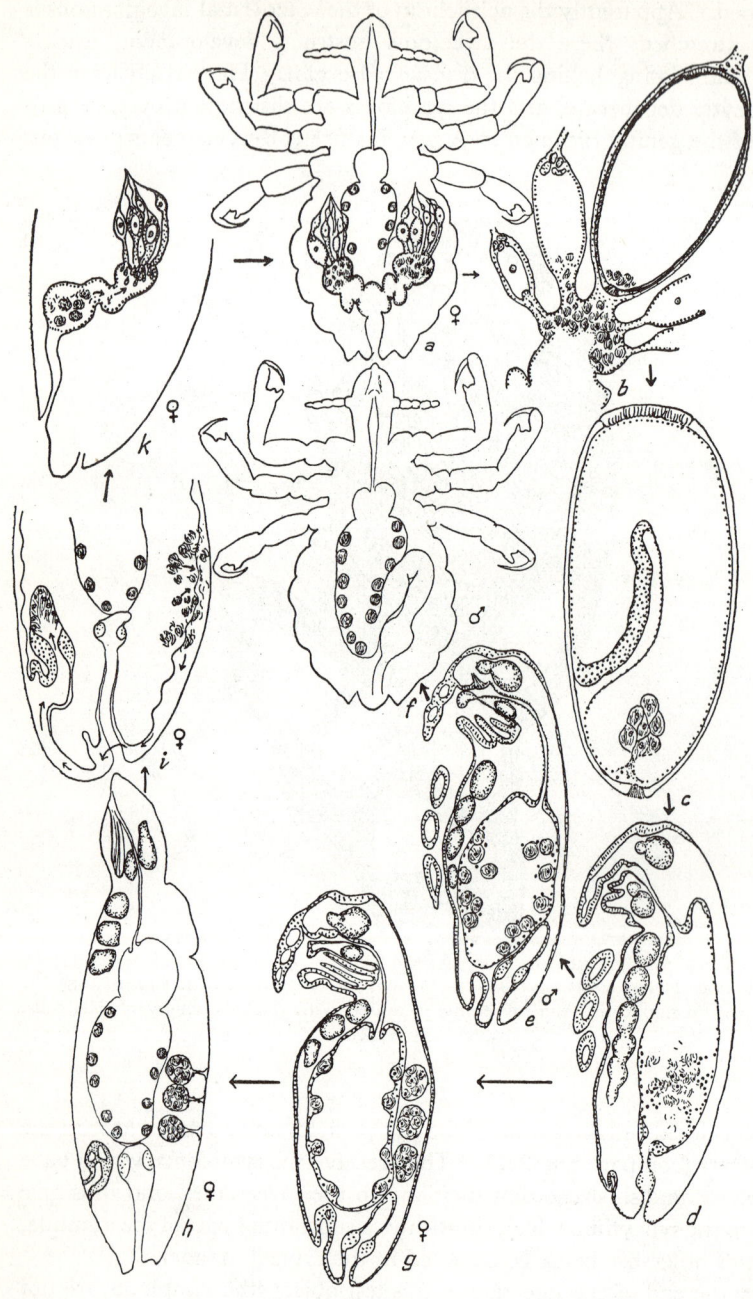

flat mass behind the vitelline membrane close to the blastoderm and thus, strictly speaking, remain outside the egg at first.

In *Pedicinus rhesi* Faluh., only in males do the mycetocytes retain their position between gut and musculature; in females the symbionts all continue to migrate and, when they reach the place where the rudiment of the ampullae divides, they throng into the ampullae, as in *Pediculus*.   Thus a refreighting also takes place here, and the appearance of the multi-laminate structure finally is approximately that found in *Pediculus* and *Linognathus*.

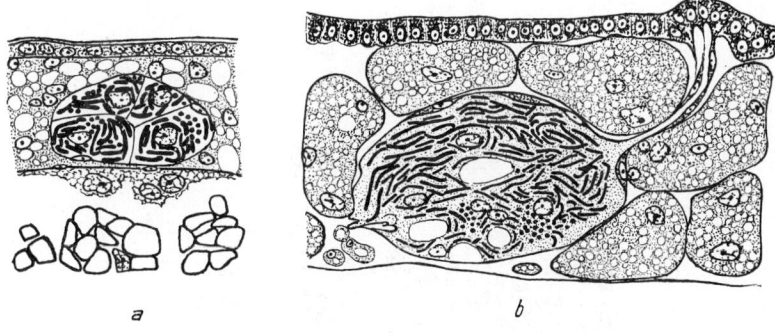

a   *b*

Fig. 280   *Haematopinus eurysternus* N.   (*a*) One of the three depot mycetomes between gut and wall of the back; (*b*) depot mycetome shortly before decomposition.   824×.
From Ries.

A completely different situation exists in *Haematopinus*, a genus in which the complications involved in filling the ovarial ampullae reach the maximum.   Preparations for their filling and thus for egg infection begin to take place even during embryonic development.   As will be described later, in the female embryos only some of the secondary mycetocytes which form in the interior of the intestinal rudiment and are derived from it are installed again in the epithelium of the midgut, the others passing through it in its dorsal area.   In this way three unpaired aggregations of myceto-

Fig. 279   *Haematopinus suis* L.   Symbiotic cycle.   (*a*) Female imago, mycetocytes in the gut epithelium, ovarial ampullae infected.   (*b*) Infection of the ovocytes by way of the ovarial ampullae.   (*c*) Invagination of the germ band, development of a transitory mycetome.   (*d*) Degeneration of the same.   (*e*) Uptake of symbionts in midgut cells.   (*f*) Male with infected midgut cells.   (*g*) Three depot mycetomes are formed in the female embryo.   (*h*) Female with infected gut epithelial cells and the three completed depot mycetomes.   (*i*) The symbionts leave the latter and enter the exuvial fluid.   (*k*) Infection of the ovarial ampulla.   Partly schematic.   From Ries.

cytes are formed between gut and dorsal hypodermis in the still diffuse fatty tissue. They might be called depot mycetomes, for their task is to harbor the symbionts which later fill the ovarial ampullae. Mesodermal cells form a delicate envelope around them, and in later stages they even hang like bell clappers in the body cavity from the back by means of a hollow cellular cord on the hypodermis (Figs. 279, 280).

Far-reaching changes in the depots set in shortly before the third molting. The localization of the mycetocytes is given up, fluid vacuoles appear, the envelope begins to become unclear and in some places disappears entirely, and the suspension cord is dissolved. When the last

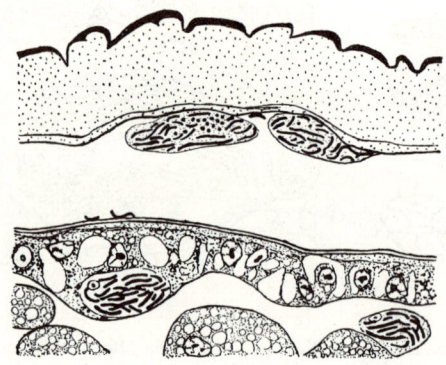

Fig. 281   *Haematopinus eurysternus* N.   Passage of the symbionts through the hypodermis.
824×.   From Ries.

larval skin is sloughed off and the imaginal skin begins to have the appearance of a delicate border, great numbers of symbionts, usually crowded in clumps around a shriveled nucleus and originating from the depot mycetomes, pass through the hypodermis into the space filled with molting fluid between the old and the new cuticula (Figs. 279*i*, 281).

Now the only invasion place in this stage of development takes form in a wide opening of the genital rudiment. Strange, fibrous, conducting cells sprouting at the distal end of the rudiment are set off sharply from their more deeply staining surroundings; the symbionts glide with the molting fluid to that opening, apparently very rapidly, pass into it, rise to the fiber cells, and are soon appropriated by them (Figs. 279*i*, 282*a,b*). It is impossible to decide whether pumping movements of the host animal or active movements of the symbionts are involved and whether in this case a chemotactic telekinesis operates.

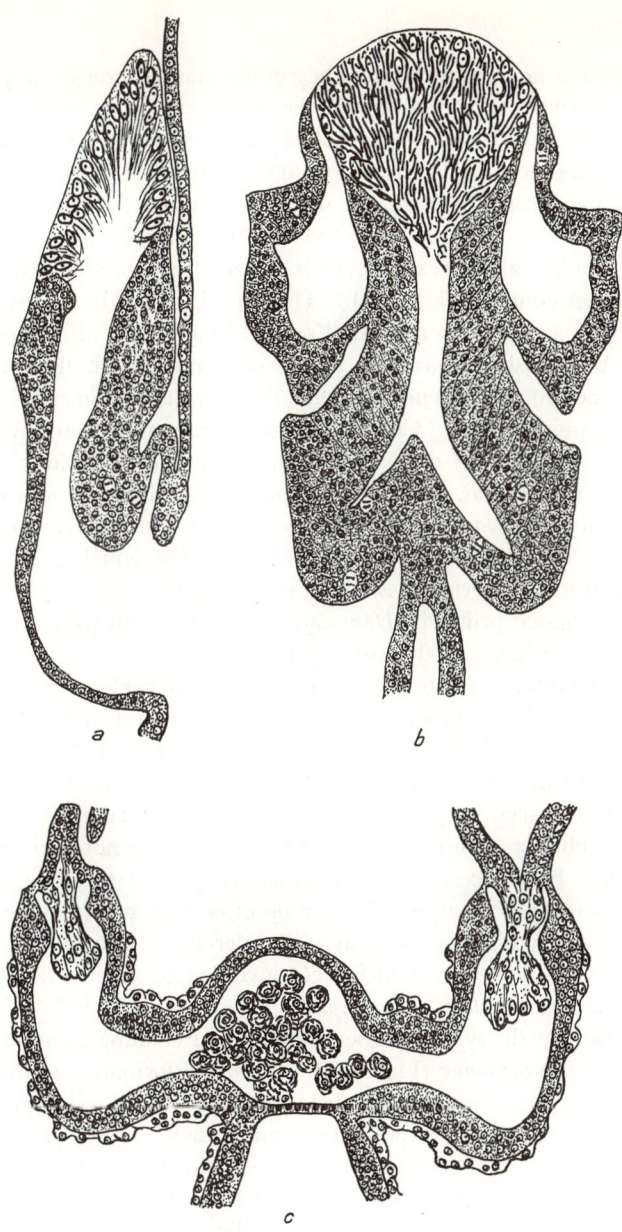

Fig. 282   *Haematopinus eurysternus* N.   (*a*) Sagittal section through the rudiment of the sexual passages before colonization by the symbionts; fibrous transport cells await the symbionts.   400×.   (*b*) Frontal section through the sexual rudiment after colonization of the transport cells.   400×.   (*c*) Frontal section through a later stage of the sexual passages; the primary transport cells are rounded off, secondary transport cells wait at the site of the future ampullae.   210×.   From Ries.

Now the two lateral sections of the rudimentary excretory apparatus protrude toward the top and fuse with the pedicles of the ovarioles advancing toward them to form the ampullar rudiments. The infected conducting cells are meanwhile released from the place of their formation and migrate in two groups toward the ovarioles without forming clear cell limits (Fig. 279*k*). As in *Polyplax*, each of the ovarioles differentiates a second group of paler prehensile cells which are extended like hands toward the newcomers (Fig. 282*c*). The nuclei, which have just joined the symbionts, are not deserted until several hours later, after they have completed their task and as soon as the symbionts reach the ampulla. The abandoned nuclei degenerate, while the new prehensile cells become the definitive mycetocytes.   Cell limits appear in a mass which is syncytial at first, and this infected cushion is pushed between ovarioles and oviducts.   As with *Pediculus*, the ampullar wall finally consists of four layers: a mesodermal muscular tunica; a basal cell layer which grows out from the ovarioles; the infected zone; and an epithelium which is delimited toward the inside and originates from the oviduct.

Ries investigated primarily *Haematopinus eurysternus* but convinced himself that *H. suis* behaves in the same manner.

To close our description of the different symbiotic cycles, we must consider the embryonic phenomena.   These were also studied by Ries, and supplementary material was contributed by Schölzel (1937).

The different methods used by the Anoplura in housing the symbionts or settling the ovarial ampullae clearly demonstrate that they represent solutions which are found from case to case, and are not derived from one another.   It will come as no surprise, then, that the study of the embryonic development affords further proof of such polyphyletic mode of origin, for the symbionts are dealt with differently during egg development in each of the genera which were investigated in detail: *Haematopinus*, *Linognathus*, *Pediculus*, and *Polyplax*.

In *Haematopinus*, the symbionts between egg and vitelline membrane are unaffected by the cleavage (Fig. 283*a*), but during formation of the germ band, which invaginates at some distance from the posterior pole, the symbionts are sucked in by the embryo and provided with yolk nuclei. When the ball sinks deeper into the yolk—by this time the embryo has formed mouth parts and extremities—a cellular pedicle develops between it and the serosa.   The pedicle then sends fibrous processes between the bacteria to form individual mycetocytes around each yolk nucleus (Figs. 279*c*, 283*b*).   During revolution of the embryo the transitory mycetome, which up to this point has been inactive at the end of the egg (Schölzel gives excellent pictures of this), reaches the central regions of the yolk, where it displays clear signs of degeneration when the future midgut epithelium grows out from the fore- and hindgut.   The mycetocytes are like swollen bladders; their limits begin to disappear.

At this time cell cords migrate toward the disintegrating mycetome from the embryonic intestinal epithelium; the symbionts, now completely freed, gather around some of the cords; new cell limits are formed; the symbiont-free sister cells transport processions of the new mycetocyte generation back to the intestinal epithelium again, where they are once more inserted

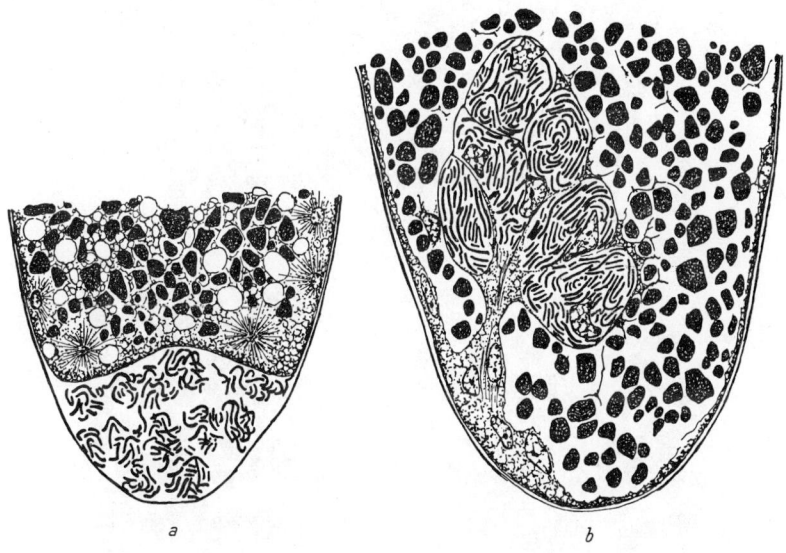

Fig. 283   *Haematopinus eurysternus* N.   (*a*) Early blastoderm stage; the symbionts still lie outside of the embryo.   (*b*) The transitory mycetome is detached from the serosa. 660×.   From Ries.

here and there.   The yolk nuclei freed in this way and placed temporarily in the service of symbiosis apparently do not degenerate immediately but at the usual time along with the other yolk nuclei (Fig. 284).

It is clear that the mycetocytes inserted so intimately in the stomach gut in the *Haematopinus* species are equivalent in value to actual intestinal epithelial cells.   This fact cannot be obscured by the considerable difference, presently manifesting itself, between the nuclei of the two cell types, a disparity resulting from their different functions.

As already mentioned elsewhere, in the females, in connection with the return migration, a substantial number of the secondary mycetocytes pass through the intestinal epithelium into the body cavity and there build up three depot mycetomes.

In *Linognathus* the symbionts are already in the interior of the egg when cleavage begins, but the yolk cells are not provided until the germ band forms.   As in *Haematopinus,* the seam which appears during the swallowing act gives origin to a fibrous pedicle, the branches of which play a role in

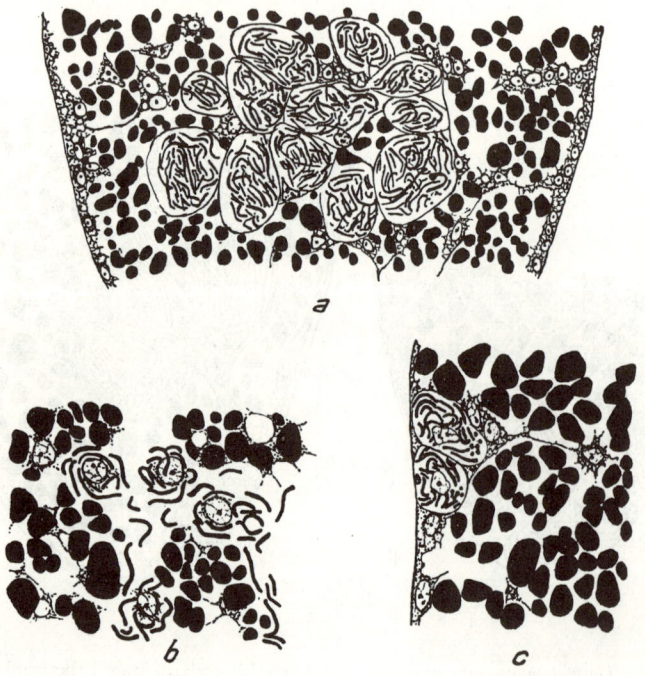

Fig. 284   *Haematopinus eurysternus* N.   (*a*) Section from the rudiment of the midgut; gut cells migrate to the degenerating provisional mycetocytes.   500×.   (*b*) The symbionts, having been freed, are reloaded.   824×.   (*c*) The definitive mycetocytes, migrating back, become embedded in the embryonic gut epithelium.   824×.   From Ries.

formation of cell limits.   The young mycetome reaches the central yolk during revolution, but this time without refreighting it glides out from the yolk and away to its definitive position before formation of the intestinal epithelium.   Thus the genus *Linognathus,* a member of the haematopinids, has some traits in common with *Haematopinus,* but its behavior is much simpler.

The *Pediculus* and *Phthirus* species as repr esentatives of the pediculid

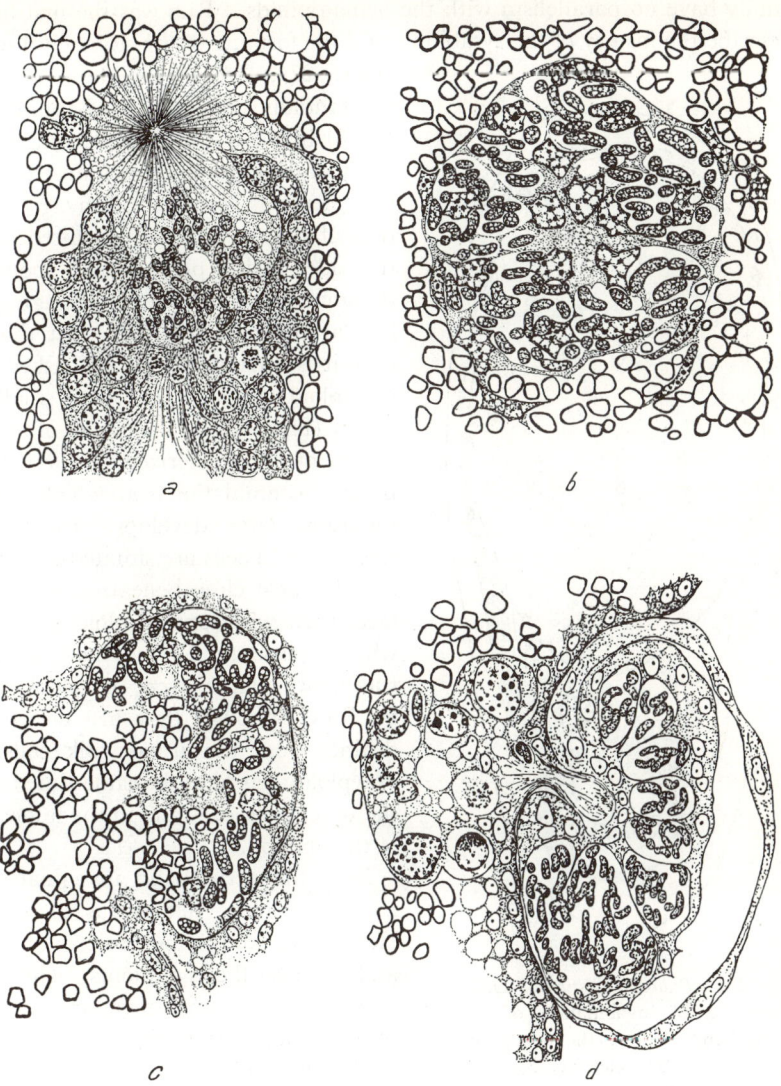

Fig. 285   *Pediculus capitis* DeG.   Development of the stomach disc.   (*a*) Upper end
of the germ band with symbionts and plasma radiation.   700×.   (*b*) The transitory
mycetome shortly before the inversion of the germ band.   770×.   (*c*) The transitory
mycetome pushes the gut epithelium outward.   770×.   (*d*) Development of the
definitive mycetome and degeneration of the provisional mycetome.   770×.   From
Ries.

family have no parallelism with the hematopinids. Ries was the first to describe the history of the symbionts and the forming of the stomach disc (Figs. 274, 285). His data were substantially corrected by Baudisch (1958) in a retesting which included study of the chromosome numbers of the elements involved. Invagination of the germ band is shifted to the side, taking place round about the small group of symbionts hanging free at the posterior pole. At this time it is surrounded by a "basket" of diploid cleavage cells with small nuclei (Fig. 286). The place of attachment is characterized by a denser plasma accumulation from which the radiation later develops. In this stage 12 to 14 cells are situated at the anterior pole close beneath the surface, each with four 16-ploid nuclei which represent the actual mycetome rudiment! They were overlooked by Ries, although they are clearly distinct from the usual yolk cells. Polyploidy sets in at the same time in the other yolk cells and in the nuclei of the amnion.

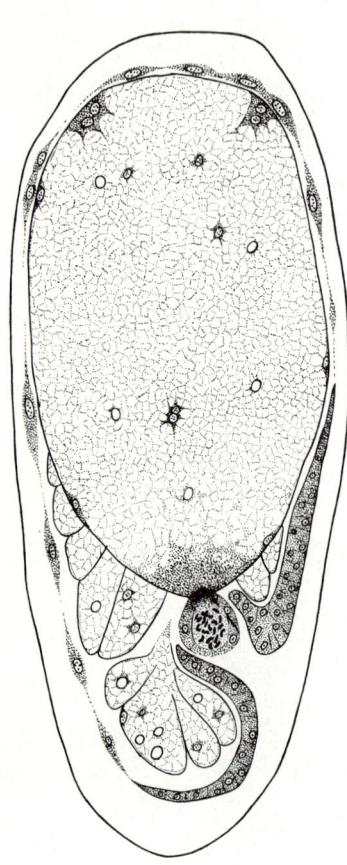

Fig. 286   *Pediculus vestimenti* DeG. Beginning of invagination of the germ band; below, the symbionts surrounded by cleavage nuclei; above, two prospective mycetocytes, each of which has four nuclei. From Baudisch.

Now the germ band invaginates more and more deeply, as in the Homoptera and some Heteroptera, and pushes the symbionts, which prevent its closure, together with their envelope proportionately to the front. As in the two fulgoroids, *Ischnodemus* and *Orthezia*, a clear radiation appears now in front of the symbionts. Ries and Sander (1955) interpreted the radiation as a locomotor center (Figs. 274c,d; 285a; 287a,b). According to Baudisch, it originates in connection with the progressive yolk cleavage and may be traced back to a yolk membrane (which encloses an unfurrowed space in the process of growing smaller and smaller) and its derivates.

Through its contraction the symbiont-containing basket is shifted toward the anterior pole, where it finally comes upon those yolk free, tetranucleated cells which attach themselves to it like little sacs (Fig. 287a). Concentrically arranged, they are now opened toward a

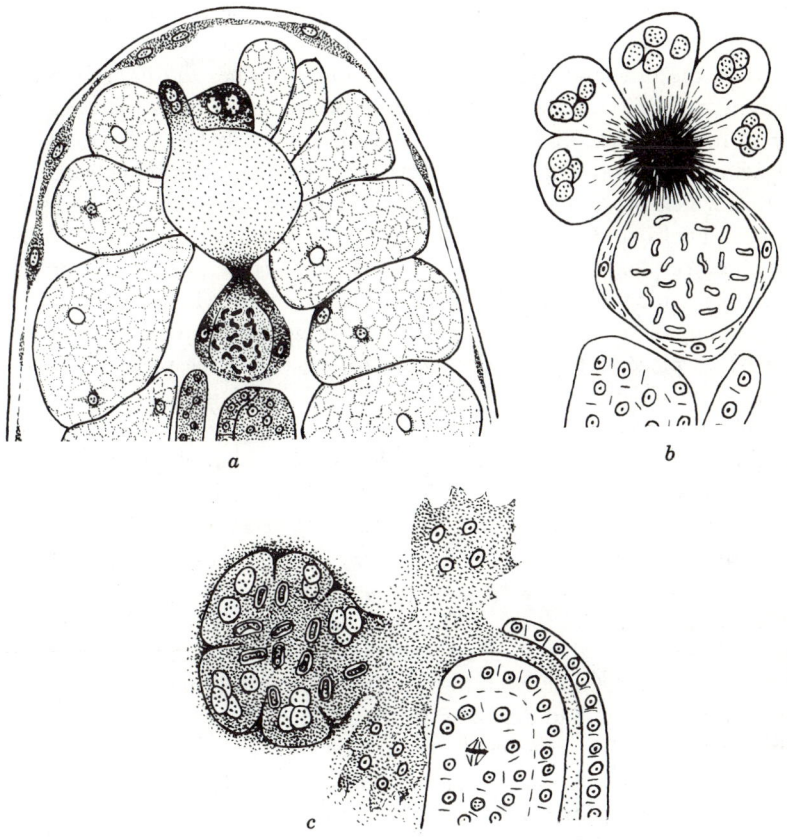

Fig. 287 *Pediculus vestimenti* de Geer. (a) The symbionts have arrived at the upper pole; yolk cleavage completed, the prospective mycetocytes lie next to the remainder of the yolk membrane. (b) The prospective mycetocytes and the symbionts are united by a large monaster. (c) Formation of the provisory mycetome, degeneration of the envelope cells. From Baudisch.

central plasma with a powerful radiation and gradually receive the transferring symbionts (Fig. 287b). After transfer, the radiation disappears, the little basket opens completely, and its former provisory envelope is now a mere socket rapidly disintegrating between the germ band and the embryonic mycetome (Fig. 287*c*). Ries supposes that the symbionts are received by a syncytium derived from the germ band, which until then lay inactive above the symbionts, and the latter are then united in vacuoles without formation of cell limits (Fig. 285*a*,*b*). What is actually taking place, however, is much more complicated and harmonizes better with the behavior of the other Anoplura, in which the symbionts are also always housed in elements derived from the yolk nuclei.

As in the other lice, this young mycetome reaches the interior of the rudimentary midgut and must therefore be advanced to its final position by a more or less violent act. Beneath the germ band far from the mycetome a mesodermal plate has originated earlier from binucleate cells within which two types of elements are differentiated: one type retains the diploid nuclei and consists of dense, darkly staining plasma, the other becomes 16-ploid, nuclei and plasma swelling correspondingly and the latter becoming coarsely foamy and rich in fluid (Fig. 287*b*). After revolution, these cells representing the rudimentary mycetome envelope glide to a place in the intestinal wall toward which the provisory mycetome is meanwhile approaching. In its cells, still in rosette form, the nuclei now migrate to the center with much protoplasm whereas the symbionts keep their distal position (Fig. 288*a*). The cell plate up to this point has lain flat on the intestine, but now it becomes bowl-shaped and the intestinal wall with the embryonic mycetome sinks into this depression. Next the intestinal evagination is narrowed like an iris diaphragm whereby the empty primary mycetocytes are pushed into the interior of the intestine (Figs. 285*c*,*d*; 288*b*). The diploid cells are installed in the wall of the larval mycetome; the 16-ploid cells keep their position between the mycetome and the body wall to form that fibrous cell mass with coarsely lobed nuclei which had appeared in the pictures provided by Ries. The mycetome is divided into anucleate chambers by septa which extend from the intestinal wall and are preserved during the abscission.

Using living specimens, Baudisch determined precisely the duration in each case of this complicated and unique chain of events and acquired additional insights by centrifuging developmental stages. He discovered that the mycetome envelope sometimes reveals a certain degree of autonomy. If it arrives too late or at the wrong place, a mycetome-like formation arises from its rudiment without participation of the intestinal wall. Further investigation is necessary to determine whether there is an additional connection between the conspicuous unique formation of a radiation accompanying invagination and the behavior of the yolk membrane and of yolk cleavage. Sander's observations (1955) with *Pyrilla* and those of Ermisch with *Conomelus* and *Stenocranus* seem to indicate that this may be the case.

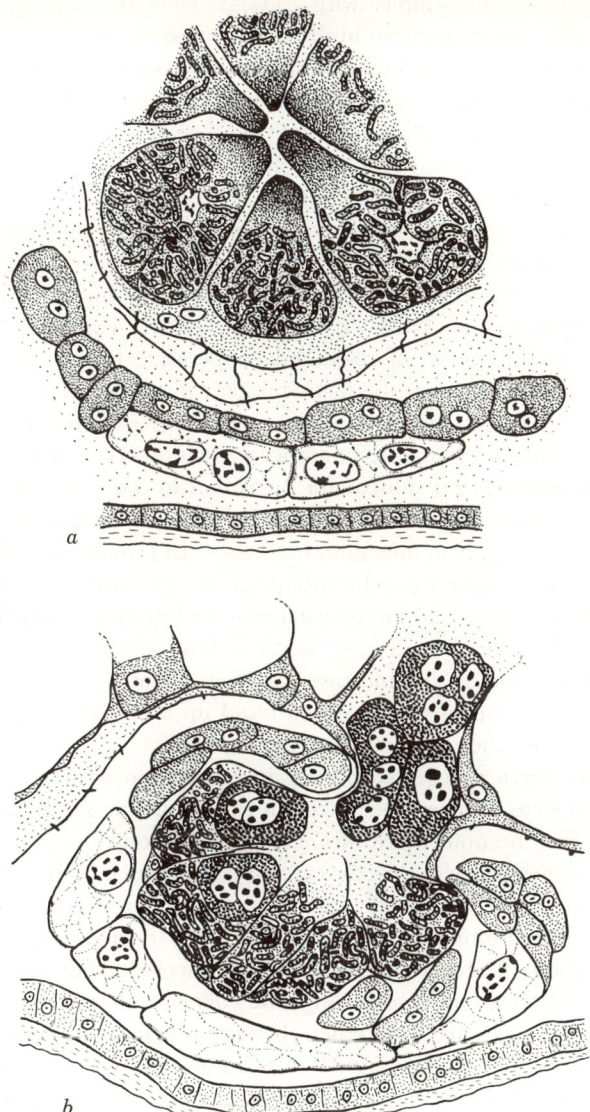

Fig. 288   *Pediculus vestimenti* de Geer.   Formation of the definitive mycetome.   (*a*) The rudiment of its enveloping membrane, consisting of two types of cells, lies between the gut and the integument.   (*b*) Evacuating the provisory mycetocytes, the symbionts leave the gut and are engulfed by the envelope cells.   From Baudisch.

The situation is far simpler with *Polyplax*. Here the symbionts, which are not received in the embryo until after the germ band has invaginated, are provided with yolk nuclei; the embryonic mycetome emerges from the intestinal rudiment in its simple form and reaches its definitive place apparently without reloading.

Unfortunately it has not been possible as yet to study embryonic development in *Pedicinus*.

It is implicit in what has been said here and there in our discussion that these various symbiotic cycles are accompanied by well-regulated phases of increase and by changes in the form of the bacteria, but there is need for detailed presentation. The increase phases clearly correspond in each case to the requirements of regulated cooperative living. In *Pediculus* three waves of increase can be distinguished: one begins after abscission of the definitive mycetome from the gut and furnishes symbionts in quantities proportionate to its final size; another wave is observed at the approach of the third molting and subsequent migration to the ampullae; a third, after the ampullae are occupied, and the large new area must be provided with a sufficient supply of infectious bacteria. In *Haematopinus* the symbionts also begin to increase substantially after they are received by the secondary mycetocytes and are inserted into the intestinal epithelium; and an additional wave of transverse fission follows after the ampullae are settled. In *Linognathus* the increase also runs parallel with reception of the symbionts by the yolk cells, which build up the definitive mycetome, and sets in anew when the ampullae are infected.

Form changes are observed chiefly in connection with egg infection. When an oocyte reaches the infection-ripe stage in *Haematopinus*, the symbionts first become more strongly stainable only in the bordering region; in connection with this, the cells containing such preparatory states disintegrate in order to allow the symbionts to transfer in groups to the egg pedicle. It is interesting that increased affinity to the dyes reappears every time the symbionts change over again to extracellular life. This occurs when the inmates are discharged into the yolk cells by the disintegrating primary mycetocytes and later when they are transported by the molting fluid to the female genital opening. In *Haematopinus suis* the symbionts in the ovary are only half as thick as those in the mycetocytes, according to Bewig and Schwartz (1956).

Frequently the stages used to infect the egg are distinguished by special chromatic deposits. In *Linognathus* the filaceous symbionts of the ampullae possess a central, strongly staining granule which disappears after they transfer to the egg again (Fig. 289*a,b*). The symbionts of *Pediculus*, investigated by Ries with sections only, were more fully studied later by Puchta (1955) and Kotter (1955) with living specimens and smears.

Particularly from the illustrations of the first mentioned author, we see the highly diversiform character of these microorganisms which are not identifiable with any other known bacterial type.   In the mycetome of a young larva just hatched they have the form of tubes 25 and 30 μ in length and 2 μ in width and increase through terminal abscission of smaller fragments, now only spherical in shape.   In the second larval state the symbionts continue to increase in the same way, becoming substantially longer in both sexes (ca. 50 μ).   The occupants of the ovarial ampullae in imagines of different ages and the organisms found in freshly laid eggs are similar to the mycetome occupants of the third larval stage, although a

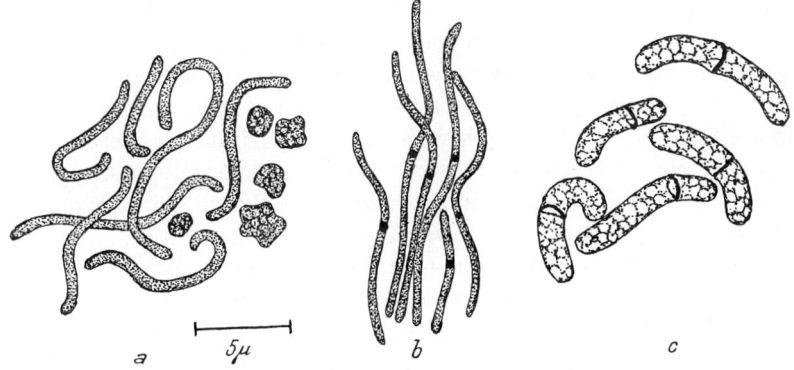

Fig. 289   (*a*) *Linognathus tenuirostris* Burm.   Symbionts from an embryo, in part involution forms.   (*b*) *L. tenuirostris* Burm.   Forms serving egg infection.   (*c*) *Haematomyzus elephantis* Piaget.   Symbionts serving egg infection.   2800×.   From Ries.

tendency immediately manifests itself in the latter to form more compact shapes, spheres, and dumbbells.   According to Puchta, only round symbionts are found on the second day, but on the third and fourth days the tendency to grow out into tubes breaks through in increasing degree.   In male animals the deviation is revealed in the third larval stage in which the growth in length considerably exceeds that of the female symbionts. Filaments 100 μ or longer arise; they are still homogeneous at the beginning, but in the imagines they gradually form chains of darker thickenings and knots and finally are constricted off here and there, although the fragments may still hang together by thin threads.   The older the males, the more irregular become the symbiont fragments which vary in size and are rich in dense inclusions and which Ries earlier called "symbiont debris."   Kotter used the most diverse methods in studying the histology of the mycetomes and the structure of the symbionts. Nevertheless he was unable to prove the presence of chromatin in the symbiont plasma with any of the usual methods.

In the freshly laid egg in *Linognathus*, odd, club-shaped, and lobed states appear in addition to tubes (Fig. 289); in *Polyplax*, the usual filaments and tubes are replaced throughout by small irregular vesicles. All this points to the unusual plasticity of the Anoplura symbionts, a quality which enables them to adapt harmoniously to the complicated symbiotic cycle.

Cultivation of Anoplura symbionts was attempted by Puchta (1956) and Kotter (1956) with *Pediculus* and by Bewig and Schwartz (1956) with *Haematopinus*. The latter report that they were unsuccessful; Puchta obtained in not quite 4% of the cases a pleomorphic organism of diphtheroid appearance which he interprets as the symbiont; Kotter, who cultivated an identical organism only five times in 453 cases, cannot accept Puchta's interpretation. Since organisms highly adapted to intracellular living are involved, the prospect of cultivating them *in vitro* is very slight, and one would be inclined to share Kotter's skepticism. In the General Section we shall describe the methods used to obtain symbiont-free lice, the pathological deficiencies of the latter, and the elimination of the deficiencies by a diet rich in growth factors.

Our comparative analysis did not permit us to describe the interconnections of the individual cycles, but we shall at least briefly survey the matter as presented schematically for *Pediculus* and *Haematopinus* in Figs. 274 and 279. In *Pediculus* the symbionts undergo a three-fold change of cellular sites: reception by a transitory mycetome, transferral to a new syncytial organ, and complete resettlement in the ovarial ampullae. In *Haematopinus* the cycle is still more complicated. Here the symbionts are received by a transitory mycetome and are transferred not only to the intestinal epithelial cells but also to the already mentioned strange depot mycetomes, and by way of the molting fluid to the genital rudiment. On arriving here they are received by another cell type before they populate the definitive sites and then disintegrate. Thus there are five successive sites of cellular construction, and on two occasions the symbionts are housed extracellularly: in the exuvial fluid, and later in the space between egg and chorion. Furthermore, there are differences conditioned by sex aside from the presence or lack of the ovarial ampullae. In *Pediculus* the female mycetome is emptied completely, whereas the symbionts disintegrate in the male stomach disc. In *Haematopinus* the symbionts are preserved in both sexes, but the aforementioned depot mycetomes are formed only in females.

Well informed as we are today on Anoplura symbiosis, no study has been made as yet of any representative of the family of the echinophthiriids or of the subfamily of the euhaematopinines, belonging to the haematopinids. Undoubtedly new devices would be found in these groups. We must also consider the possibility that some forms may be without symbionts. For instance, Ries was unable to discover symbiotic organs in two representatives of the linognathines, *Haemodipsus ventricosus* Denny of the rabbit and *Hoplopleura acanthopus* Burm. of the field mouse. In both of them the swellings characteristic of the oviducts were also lacking.

MALLOPHAGES

The symbiotic devices of the mallophages, of which Sikora (1922) and I (1928) observed only the ovarial ampullae, were thoroughly studied by Ries (1931) in connection with his Anoplura studies. He found that these devices, although far less complex than those of lice symbiosis, nevertheless appear to be closely related to them. Thus they are being included here even though the hosts feed primarily on corneous substances—feathers, quills, scales of the skin—and only occasionally use blood as a supplementary item of diet.[1]

The sites made available to the bacterial symbionts of the Mallophaga are the simplest conceivable. Closed organs are never formed. The symbionts live exclusively in individual mycetocytes. In the very slender, almost rod-shaped *Columbicola columbae* L., these mycetocytes form loose nests on both sides of the abdomen close beneath the hypodermis (Fig. 290*f*); in representatives, which are much broader in size, such as *Sturnidoecus sturni* Schrk., they are found deeper in gaps between the fat lobelets; in *Turdinirmus merulensis* Den., they are forced chiefly into the folds of the deeply notched segments. Sometimes these mycetocytes are inserted

---

[1] In the text we have used the nomenclature employed today, but for the reader's information we are listing below the equivalent terms appearing in Ries.

Ries:	Modern:
Lipeurus baculus Ntz.	*Columbicola columbae* L.
L. lacteus Gbl.	*Anaticola tadornae* Den.
L. frater Gbl.	*Falcolipeurus frater* Gbl.
Docophorus leontondon Ntz.	*Sturnidoecus sturni* Schrk.
Nirmus merulensis Den.	*Turdinirmus merulensis* Den.
N. fuscus Ntz.	*Kélerinirmus fuscus* Ntz.
N. subtilis Gbl.	*Brüelia subtilis* Gbl.
Goniocotes compar Ntz.	*Campanulotes compar* Ntz.
Goniodes damicornis Ntz.	*Coloceras damicornis* Ntz.
G. dissimilis Ntz.	*Oulocrepis dissimilis* Ntz.
G. falcicornis Ntz.	*G. pavonis* L.
Menopon biseriatum Piaget	*Eomenacanthus stramineus*

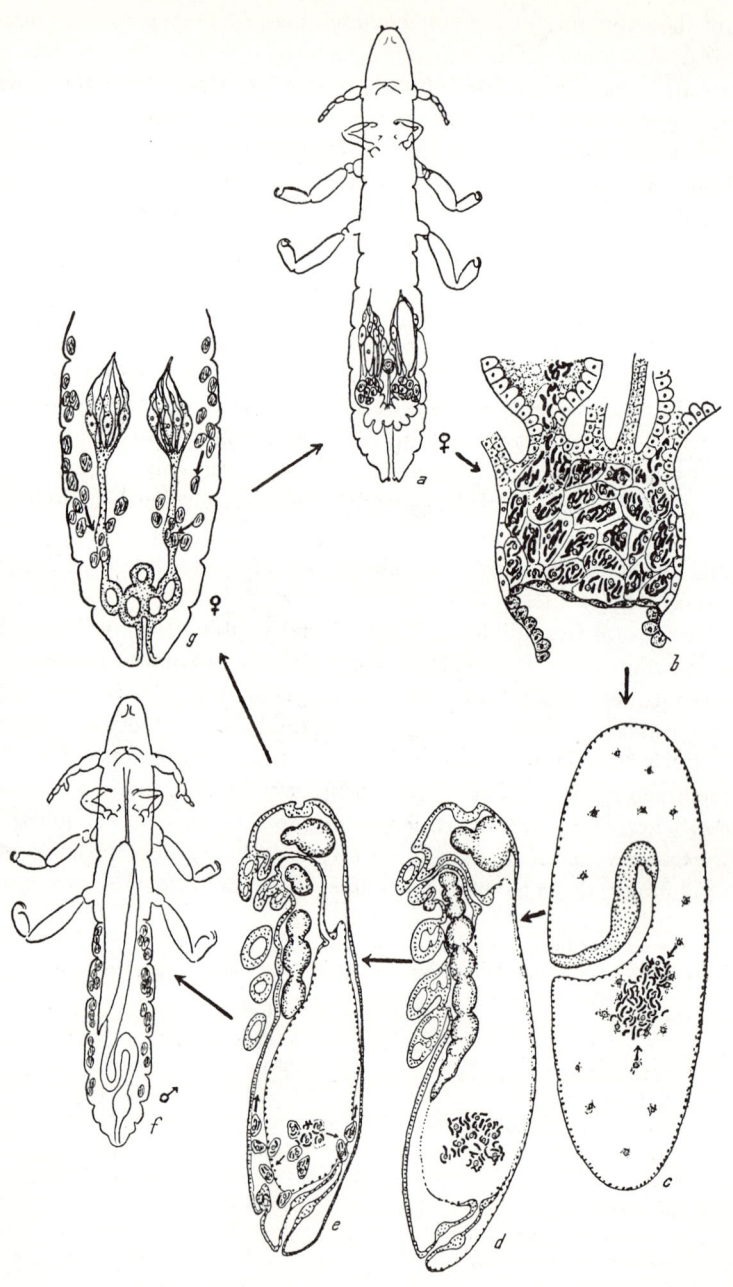

deeper in the fatty tissue and thus, except for their much less regular arrangement, resemble the mycetocytes of blattids. The genera *Goniocotes* and *Campanulotes* initiate this form of housing. It is best developed in *Coloceras damicornis* Ntz. of the dove (Fig. 291), whereas a "*Goniodes* sp." of the peregrine falcon houses the mycetocytes merely in the gaps of the fat body.

The size of the mycetocytes and the manner in which they are filled may vary widely. In *Columbicola columbae*, the most thoroughly investigated of the bird lice which live on the dove, the mycetocytes are small

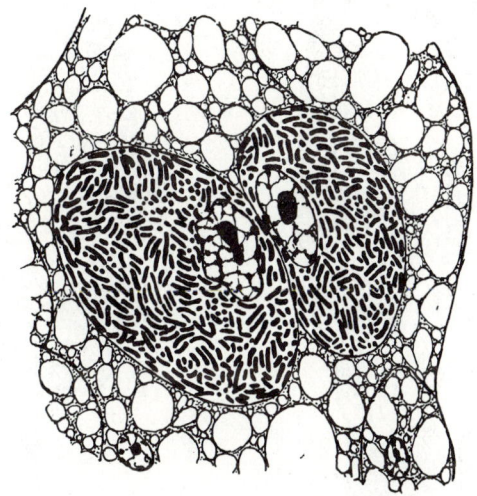

Fig. 291   *Coloceras damicornis* Ntz.   Mycetocytes in the fat tissue.   1600×.   From Ries.

and contain only a few sausage-shaped symbionts which stain so weakly that they eluded the predecessors of Ries (Fig. 292); on the other hand, the cells occupied by symbionts in *Anaticola tadornae* Den. are larger than those found in the Anoplura and the other Mallophaga. Their inmates are more oval to roundish. The symbionts of *Sturnidoecus sturni* Schrk. of

Fig. 290   *Columbicola columbae* L.   Symbiotic cycle.   (*a*) Mature female, the symbionts in the ovarial ampullae only.   (*b*) The symbionts infect the young ovocyte by way of the ovarial ampulla.   (*c*) The symbionts are located in the yolk and are provided with nuclei during invagination of the germ band.   (*d*) The symbionts in the lumen of the embryonic gut.   (*e*) The mycetocytes transfer to the body cavity.   (*f*) Males with mycetocytes beneath the hypodermis.   (*g*) Female; the mycetocytes transfer to the ovarial ampullae.   Half schematic.   From Ries.

the starling, which populate very small elements and are difficult to recognize, are extremely fine, thin filaments apparently bacterial in nature. The same situation prevails in *Turdinirmus, Kélerinirmus,* and *Brüelia,* but in *Goniocotes* and *Campanulotes* the mycetocytes are large with thick, completely unstructured, short tubes. Those of *Coloceras damicornis* are even larger, and their plasma is crowded with numerous short rodlets. The other species, designated "*Goniodes* sp.," is more like *Columbicola* (Fig.

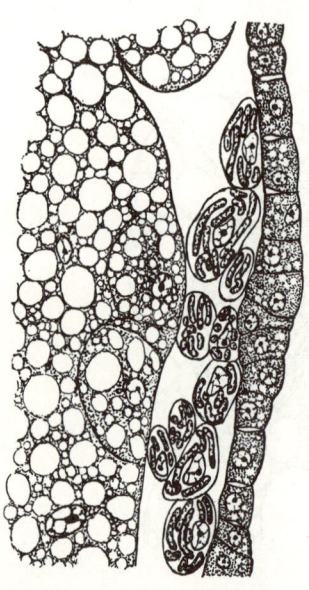 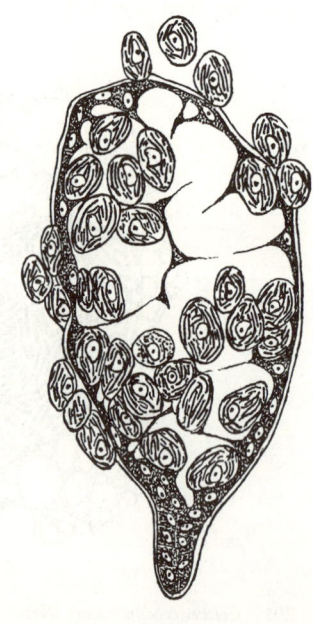

Fig. 292 *Columbicola columbae* L. Male larva; mycetocytes between fat tissue and hypodermis. 927×. From Ries.

Fig. 293 *Sturnidoecus sturni* Schrk. N. The mycetocytes penetrate the developing ovarian ampulla. 927×. From Ries.

295). Here the mycetocytes have but a single nucleus, which may be round or lobed but is always type-constant in form and structure.

The symbionts are almost always transmitted indirectly by an infection of ovarial ampullae, as in the Anoplura. In the majority of cases the mycetocytes are merely transmitted unaltered to the ampullar sites. In females they begin to migrate for this purpose before the third and last molting and collect in the area of the oviduct (Fig. 290g). Here they attach themselves to the cell cords which, especially elongated in the

mallophages, connect the ovarioles with the oviduct and gradually transfer to its vacuolated interior (Fig. 293). After its filling, the cords are
markedly shortened; the young epithelium of the oviduct forms a thin
limiting layer against the infected section and is set off clearly against it
by stronger staining. The ampullar epithelium, on the other hand, is
supplied by the sterile follicle cells and is accompanied in the direction of

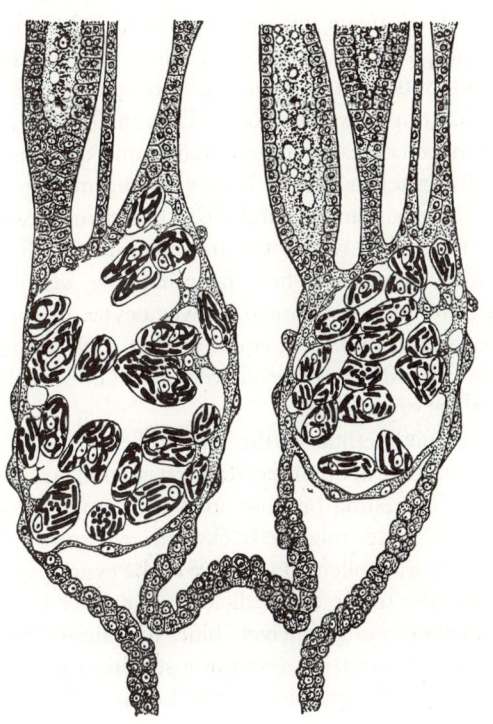

Fig. 294   *Columbicola columbae* L.   Sexual apparatus of a female immediately after the
third moult with filled ampullae.   567×.   From Ries.

the body cavity by mesodermal elements which gradually attach themselves externally to form a tunica externa which is lacking in the ovarioles
and the oviduct.   The end result is one which is thoroughly typical of the
lice.   Figure 294 represents the two ampullae of *Columbicola*, just filled, at
a time when the mycetocytes, which later interlock more densely and form
a multilaminate cushion, are still situated more loosely and thus disclose
a peculiarity found only in this species.   The previously uninucleate
mycetocytes become binucleate in the ampullae, apparently through
fusion of pairs.

In females of *Columbicola*, all mycetocytes transfer to the ampullae (Fig. 290*a*); in *Anaticola*, nests of free mycetocytes still remain between fatty tissue and hypodermis and the mature females consequently look more like mature males.

In *Turdinirmus merulensis* Den. of the blackbird, the mycetocytes also transfer to the area adjoining the ovarioles without undergoing further changes. In other cases the difference between the mycetocytes in the body cavity or fatty tissue, on the one hand, and the infected ampullar cells on the other, is so clear-cut that it can be explained only by a reloading of the symbionts such as customarily occurs among the Anoplura in connection with ampullar infection. Thus in *Goniocotes* and *Campanulotes* the ampullar mycetocytes are much smaller and only barely defined, with a small roundish nucleus, which is poor in chromatin, instead of a large lobed one. The abundant granulations found in between indicate that the reloading took place within the ampulla and that it is not a matter of previously freed symbionts first infecting the genital rudiment. In *Coloceras* the difference between the mycetocytes in the fatty tissue and those in the ovarial ampullae is equally great. Here the large symbiont-rich cells are replaced by numerous very small ones with proportionately smaller quantities of symbionts.

The transmission method of the unclassified *"Goniodes"* species differs substantially from that of all other Mallophaga and Anoplura investigated. The situation is interesting because no ovarial ampullae are formed. To be sure, the mycetocytes migrate backward, even during preparations for the third molting, and collect at the base of the ovarioles. From here they transfer individually to the egg pedicle and advance in it toward the egg. In the process their nuclei shrivel, blur, and finally disappear entirely. The egg, as so often happens, develops a secretion-filled reception groove which closes behind the symbionts after being filled with the same (Fig. 295).

This more primitive type of transmission, which to a certain extent resembles the transmission of intact mycetocytes to the egg in the aleurodids, explains how the ampullae typical for both groups came to be developed and at the same time warns against drawing from their absence premature conclusions that symbiosis does not exist.

As in sucking lice, one sometimes observes in biting-lice symbionts, which infect the ampullae and egg, certain changes suggesting the specific transmission states so common in other cases. In *Columbicola* the bacteria of the ampullae contain a fine but clear granule which disappears again in the egg and is always lacking in the mycetocytes. The same thing applies for the symbionts of *Turdinirmus*, which moreover undergo an animated period of increase before migrating to the ampullae. The ordinarily pale symbionts of *Columbicola* become intensively colored after they leave the ampullar cells, but they lose this stronger staining again as soon as they transfer to the egg cells.

The symbionts were studied during embryonic development only in *Columbicola*. Here, too, analogies with the Anoplura were found. Again the symbionts remain inactive in the plasma until the germ band invaginates, and then they are drawn somewhat deeper into the yolk. Yolk nuclei throng between the symbionts without forming cell limits at first.

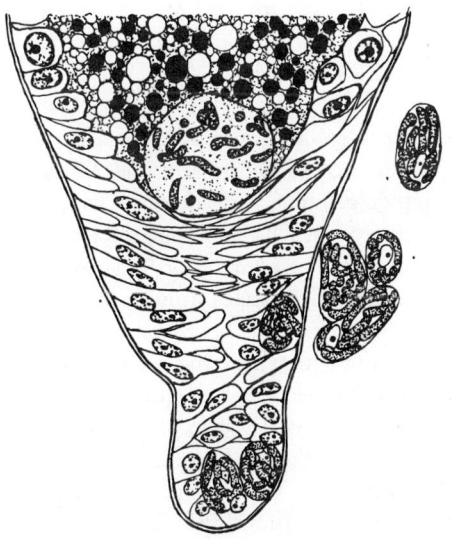

Fig. 295 *Goniodes* sp. Mycetocytes carry the symbionts into the egg stalk and to the egg without ampulla formation taking place. 824×. From Ries.

As with the sucking lice, it is only after the symbiont group reaches the central yolk with revolution that the cell limits take form, beginning at the periphery and gradually progressing toward the interior. When the subdivision is finished, the newly formed mycetocytes swarm apart and penetrate the fatty tissue individually through the still poorly nucleated intestinal epithelium and gradually advance toward the places being made available to them (Fig. 290*c,d,e*).

We have seen that in several species of sucking lice it was impossible to determine symbiosis, and this is true in increased degree for the mallophages. Apparently all trichodectids, of which Ries investigated four species, lack symbionts, and the same applies to the family of the liotheids, where eight genera were tested, some to be sure with deficiently fixed material, without yielding evidence of symbiosis. Also in *Oulocrepis dissimilis* Ntz. no indication of symbiotic devices could be discovered.

With *Eomenacanthus stramineus*, Ries found that the chitinous lining of the crop was accompanied in all the cases studied by an uninterrupted border of minute rodlets, which he calls rickettsiae although they might very well be a type of replacement for the symbionts. He is inclined to interpret the strange folding of the crop, in the area where it passes into the esophagus, which holds particularly large quantities of these organisms, as an adaptation useful for such replacement (1931).

RHYNCHOPHTHIRINES

The symbiosis of the elephant louse, *Haematomyzus elephantis* Piaget, with traits suggesting both the Anoplura and the Mallophaga, does not fit harmoniously in either group. Indeed, quite aside from the findings of Ries, who was the first to study its symbiotic devices, this insect has always been considered exceptional by virtue of its strange, snout-like anterior end, and has been classified at times among the Anoplura and at other times among the Mallophaga. Its behavior with regard to its symbionts, which is an effective gauge in evaluating related aspects, harmonizes best with the conception of Weber (1939), who creates a special suborder of the rhynchophthirines for *Haematomyzus*, placing it with the Anoplura and Mallophaga in the order of the Phthiraptera.

Unlike other Anoplura and Mallophaga, the elephant louse has paired mycetomes which have no relationship to the gut (Fig. 272d). In males they retain their original position between the testes and the ventral hypodermis among the lobes of the fatty tissue. Their histological structure also differs from that of other Anoplura mycetomes in that they consist of a well-developed groundwork of connective tissue which forms a plasma-rich envelope and more delicate septa between the individual large mycetocytes densely filled with short tubes.

As with the Anoplura and the Mallophaga, egg infection in *Haematomyzus* emanates from ovarial ampullae erected especially for this purpose, but the manner in which the ampullae are filled constitutes another exception. The mycetomes, which in this case are paired from the very beginning, are so situated that they inevitably come in close contact with the developing genital rudiments, and yet they are all taken over almost unchanged into the sections of the oviducts, which are indeed always preformed for this purpose. Unfortunately Ries was unable to observe the process in detail, but even in mature females the typical arrangement of the mycetocytes and the characteristic syncytial envelope are retained in the new location, while on the other hand the usual tetralamination is lacking.   However, since some of the mycetocytes seem to have become larger and multinucleate, it must be assumed that there is some fusion of the primary mycetocytes, as was the case with several mallophages (*Columbicola*).

In *Linognathus*, it will be recalled, the unpaired mycetome is taken over into the genital rudiment in its entirety, and later its cells degenerate and the symbionts are resettled.   In a number of mallophages the mycetocytes remain intact during translocation but are isolated in the process.   Here, then, the elephant louse has traits of both suborders.   In the *Haemato-myzus* symbionts, as with some lice in which specific changes are found in the infecting stages, a strongly staining, ring-shaped growth develops temporarily around the middle of each bacterium after transmission to the ampullae (Fig. 289c).

It has been impossible as yet to study the embryonic development of this interesting form.

8

Symbiosis in insects living on mixed diets

BLATTIDS

All blattids live in symbiosis with bacteria. Since Blochmann's initial discovery of bacterial symbiosis in *Blattella (Phyllodromia) germanica* L. and *Blatta (Periplaneta) orientalis* L. (1887, 1892), an unusual number of authors have been engaged with this phenomenon, but relatively few have studied the symbiotic cycle (Mercier, 1907; Buchner, 1912; Fränkel, 1921; Gier, 1936; Hoover, 1945; Borghese, 1946; Koch, 1949). The majority of the authors were concerned chiefly with bacteriological problems, the cultivation of symbionts in particular, and their numerous reports, some of them quite brief, did not always contribute appreciably to our information. In recent years the emphasis has been placed increasingly on problems of metabolism and physiology and on the use of antibiotics to obtain symbiont-free animals. Through this shift in emphasis the cockroach has become the most widely used test object in symbiosis research (Frank, 1954, 1956; de Haller, 1955; Brooks and Richards, 1955, 1956; Brooks, 1957, 1960, 1962, 1963; Henry and Block, 1960; Fava and Barbato, 1960; Fava and Laudani, 1960, 1961; Selmair, 1962). The significant results of this research will be evaluated in the General Section.

Up to the present not less than twenty-five classified and some unclassified species from sixteen genera have been tested for the presence of symbiosis: *Blatta, Periplaneta, Blattella, Parcoblatta, Blaberus, Eurycotis, Cryptocercus, Pycnoscelis, Ectobia, Heterogomia, Epilampra, Nauphoeta, Homalo, Derocalymma, Platyzosteria,* and *Nyctobora.* Remarkably similar results were obtained in all instances. Specific cells lodged within the fatty tissue are always used as the bacterial sites, although more detailed observations frequently reveal slight type-constant differences in number and arrangement of the mycetocytes which are always located in the interior of the lobelets. In *Blatta orientalis* and a number of other species these mycetocytes are always arranged in a single row (Fig. 296), in *Nauphoeta* they are arranged in two, and in *Blattella* in three to four. In *Blatta aethiopica* Sauss. and *Pycnoscelis surinamensis* the infected cells are irregularly scattered, and in *Homalo demascruralis* the lobelets are almost completely

516

taken up by enormous closely crowded accumulations.   In *Ectobia lap-ponica* L., only a narrow border is reserved for the sterile fat cells.   *Crypto-cercus punctulatus* Scudd. reveals dispersed mycetocytes exclusively, each surrounded by fat cells.

The symbionts do not permeate all areas of the fatty tissue, for they are found only in the visceral sections and never in the thorax or the parietal lobes of the abdomen.   According to Koch, they are also limited to the seven anterior segments in *Blattella germanica*.   Furthermore, mycetocytes are always found in the connective tissue enveloping the gonads in both sexes.

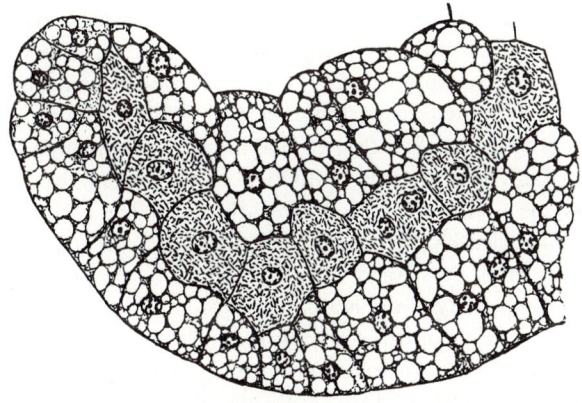

Fig. 296   *Blatta orientalis* L.   Fat lobe with mycetocytes.   280×.   Free from Blochmann.

Among the blattids the mycetocytes retain their capacity for mitotic division throughout the development of the host animals.   It had been assumed at first that only amitotic divisions occur, but Brooks and Richards (1955) determined that toward the close of the individual developmental stages numerous divisions by mitosis take place throughout the area of the fatty tissue.   According to Baudisch (1958), the nuclei finally become octoploid during larval development through fusion of daughter nuclei of mitotic origin (*Blatta orientalis*).   The presence of mitochondria in the mycetocytes, pointed out by Koch (1930), was confirmed by Brooks and Richards and by Meyer and Frank (1957).   The latter authors used the electron microscope to study the fatty tissue and the symbionts of *Blatta orientalis* and discovered rod-shaped inclusions which apparently represent reserve material.   In the fat cells they are found in masses, and in the mycetocytes in smaller numbers. Gresson and Threadgold (1960) and Bush and Chapman (1961) also used the electron microscope to study blattid symbionts.

The non-motile symbionts are gram-positive but not acid-fast (Plate V,*a–c*). Sometimes they are relatively compact forms and at other times more slender filaments and rodlets. They tend to be slightly curved, and the ends are usually rounded. There is considerable variation in length. For *Blatta orientalis*, lengths of 2.5–5.3 μ are cited; for *Cryptocercus punctulatus*, 2.5–8.1 μ; and, for *Heterogomia*, 5.3–9 μ. The symbionts of *Blatta aethiopica* measure only 1.6 μ, and those of *Blattella germanica* 3 μ. These various forms are all rigidly type-specific. Nor is the microstructure always the same: some are structureless; others reveal a vacuolated inner structure. Sparser zones may alternate with denser cross bands. Granules which are apparently metachromatic in nature are sometimes found at the periphery (Neukomm, 1927). In *Blatta orientalis* a central granule, probably differentiated from those of the periphery and usually surrounded by a light area, is interpreted as a nucleus by Hollande and Favre (1931). In fact, this granule is divided in two when the rodlets begin their transverse division. In so doing, they develop a transverse membrane, for which Lwoff (1923) and Neukomm have already provided more detailed data, and thus form sometimes a very clear ligament. Also, Frank (1956) in a study of the symbionts of *Blatta orientalis* found inclusions which might possibly consist of desoxyribonucleic acid, to judge from their staining. He describes them as formations which occur in pairs in the rodlets and are distributed to each daughter cell during transverse cleavage. Through cumulative transverse divisions, little chains of three or four links are sometimes formed, a tendency especially marked in *Parcoblatta* but also reported for *Blattella germanica* (Lwoff). Formation of spores has never been observed in living animals, but Lwoff reports that strongly staining inclusions which might conceivably be regarded as spores appear in killed animals after 48 hours (Plate V,*a–c*).

Although the bacterial nature of these organisms is beyond all doubt, Neukomm (1932) needlessly proved serologically that they are not animal-specific differentiations. Guinea pig sera obtained by injecting flotations of bacteria or bacteria-free fat cells yielded a complementary formation very specific for bacteria and fat. Neukomm also proved that blattid symbionts react much like bacteria to ultraviolet rays. When Koch (1930) demonstrated that mitochondria as well as symbionts are present in the mycetocytes of roaches, the concept of K. C. Schneider (1902), who thought that the inclusions in question might possibly represent chondriosomes, was further undermined. It seems inexplicable that

Wolf should still be uncertain about the bacterial nature of the blattid symbionts as late as 1924.

Although the symbionts vary but slightly from species to species, they are nevertheless always rigidly type-constant. Individuals originating in the most varied localities and subjected to the most varied conditions always yielded the same forms. Unbalanced or insufficient nourishment, complete withdrawal of food, extreme temperatures, injection of yeasts or bacteria into the body cavity, treatment with x-rays or ultraviolet rays—none of these factors, alone or in combination, had the slightest effect on them (Gier). In starving *Cryptocercus*, Hoover found an increase of the mycetocytes in the individual fat lobes, but this may have been an illusion due to simultaneous reduction of the fat cells.

According to Mercier and Gropengiesser (1925), yeast-like organisms would sometimes replace the typical bacteria in the mycetocytes. It is reported that the organisms displace the regular guests to a varying extent. Gropengiesser even reports that his material was infected exclusively by oval forms resembling *Torula*, but this phenomenon, never encountered by other investigators, was doubtless merely simulated through insufficient fixation and faulty staining. In hypotonic solutions the blattid symbionts actually swell up into yeast-like forms, and in such cases the unequal transverse partitions, which are also frequent under normal conditions, resemble buddings (Gier).

The symbionts are transmitted in all blattids principally in the same way, with only minor deviations distinguishing the individual species. As already mentioned, more or less numerous mycetocytes are inserted without fat cells in the encasing connective tissue of the ovarioles and thus make possible a direct transfer of the symbionts to the same. Without temporary infection of the follicle cells, such as Mercier and other older authors still assumed, the bacteria glide one by one among the follicle cells and thus come to be situated between the oocytes. At the same time the youngest areas of the germ layer always remain free of symbionts; but otherwise the time of transfer varies. In *Periplaneta americana* it occurs at a time when several oocytes are still situated side by side; in *Blattella germanica*, *Blatta orientalis*, and others, it is deferred until these oocytes are stationed one behind the other (Figs. 297, 298).

Unlike the other authors, Gier believes it to be very probable that even the rudimentary sex glands in the embryos are infected by isolated bacteria, and that these bacteria may constitute the source of all the symbionts which he found appearing for the first time approximately in the third month and then in greatly increasing numbers in the ovarioles. For mechanical considerations he rejects the possibility that these immotile organisms transfer from the surrounding mycetocytes to the ovarioles through the tunica, and yet such transmission is often observed in other symbionts and is doubtless to be explained on the basis of their lytic capacities.

Without penetrating the egg cells themselves, the symbionts later increase abundantly in the space between these cells and the follicle. In this way an unbroken layer of bacteria originates on the surface of the oocytes. In the beginning, the bacteria tend to be situated tangentially, yet as they increase in number they sometimes form palisade-like and even

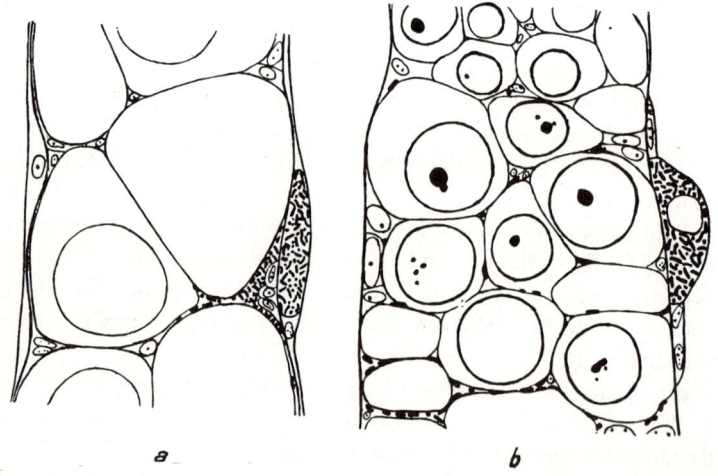

a   *b*

Fig. 297   *Periplaneta americana* L.   (*a*) The bacteria leave one of the mycetocytes attached to the ovariole and penetrate the latter.   1000×.   (*b*) Bacteria everywhere between the young ovocytes.   1100×.   From Gier.

multilaminate casings, as in *Periplaneta americana* and other species. In such instances there is a pronounced increase at the two poles accompanied by a varying degree of local folding on the surface of the egg. In *Periplaneta americana* it begins when the egg has attained a length of approximately 0.1 mm. and continues to increase until it measures 1 mm. The bacteria first accompany the folds in uniform thickness, filling out every recess of the follicle cells until they protrude like tufts, and finally penetrate deep into the yolk-free marginal area of the eggs in the form of sharply set-off tubes (Fig. 299*a*,*b*).

In *Heterogomia aegyptica* the deep folding of the egg surface is even more extreme (Fränkel). Also, in *Blatta orientalis* the increase is so exceptionally strong that half-spherical bacterial masses protrude into the egg plasma

(Gier). In *Periplaneta australasiae* L., the symbionts again form a ring-shaped growth at the two poles of the egg (Koch; Fig. 300).

Other species, as for example *Cryptocercus punctulatus*, show only very weak polar aggregations. With increased egg growth, the bacteria occupying the lateral areas can no longer keep pace. The initially solid lining undergoes interruptions here and there, and finally there are only scattered nests of symbionts.

Reception by the egg itself usually takes place at the last possible moment. When the yolk membrane forms shortly before its deposit, the bacteria sink into the surface protoplasm and, giving up the earlier arrangement in furrows, form a disc-like, yolk-free aggregation at the anterior and posterior ends (Fig. 299c). In *Periplaneta americana* it is about 15 μ thick at the center and 150 μ in diameter, whereas in *Blatta orientalis* it is not very much wider but may acquire a thickness of 40 μ.

We have varied information on the activity of symbionts during embryonic development from research by Wheeler (1889) and Cholodkowsky (1891), and especially from research by Heymons (1895) which had other goals. Gier and Koch provided important supplementary information. From the reports of the authors mentioned above, it is clear that the exceptional uniformity of blattid symbiosis is similarly preserved during this period, but that interesting differences may occur, especially with regard to formation of transitory mycetomes.

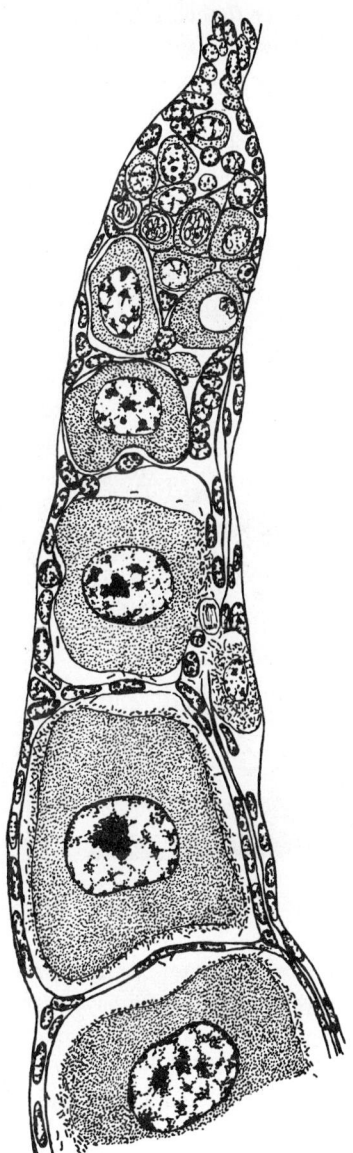

Fig. 298 *Blatta orientalis* L. Upper end of an ovariole at the time of infection; symbionts on the surface of the ovocyte and in an externally attached mycetocyte. From Koch.

Gier describes the complicated processes in *Periplaneta americana* in the following way: while the cleavage nuclei are in the process of rising to the surface of the egg, some of the same are sinking, especially at the two poles, into the marginal areas which house the symbionts, thus creating a break in the blastoderm which otherwise forms on all sides. At first 5 to 10 nuclei unite with the symbionts at each pole; in a 4-day-old embryo the number has risen to 40 nuclei each, but only some of them retain their original size and become nuclei of the primary mycetocytes. The rest, apparently under the influence of the surrounding bacteria, swell enormously and, during the shifting of the mycetocytes now taking place, are

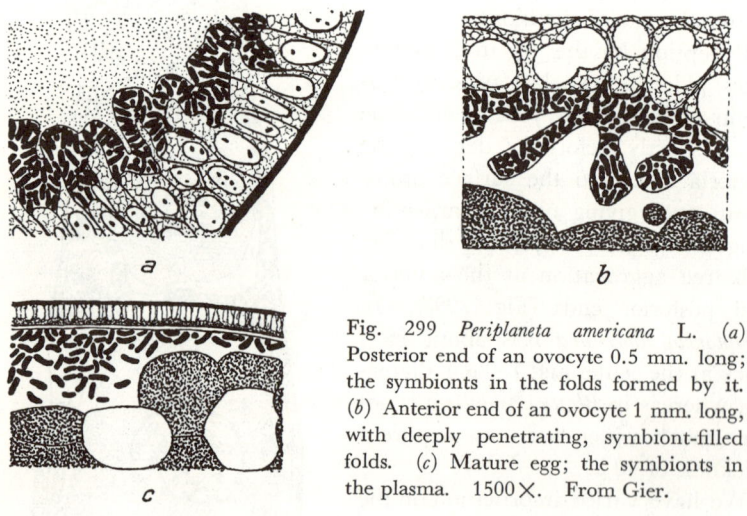

a

b

c

Fig. 299 *Periplaneta americana* L. (*a*) Posterior end of an ovocyte 0.5 mm. long; the symbionts in the folds formed by it. (*b*) Anterior end of an ovocyte 1 mm. long, with deeply penetrating, symbiont-filled folds. (*c*) Mature egg; the symbionts in the plasma. 1500×. From Gier.

dissociated from them; their further history could not be followed. About this time the mycetocytes at the posterior pole form an almost spherical aggregation, and now the latter takes up a position at the end of the germ band which has been developing meanwhile. In an embryo 3 days old it begins to glide forward and ventrally between germ band and yolk, either as a compact mass or divided into individual parts. Arriving approximately in the middle of the embryo, it turns aside toward the egg center and is joined here at the close of the eighth day with the parts which are simultaneously moving forward and with the negligible fractions which until now have been situated laterally. Thus a well-described transitory mycetome is formed; it consists of a rosette of wedge-shaped, uni- or binucleate cells thickly filled with bacteria (Fig. 301). Around this regular organ-like formation is a circle of vitellophages to which the liquefaction of the surrounding yolk must be ascribed.

The symbionts remain in this provisory site for about 3 days before abandoning it. The rosette is elongated, it disintegrates, and the bacteria reach a position among the yolk pieces again. Some transfer then to the body cavity where the intestinal epithelium is absent or still incomplete; others collect at places where the cells are more crowded, increase there for a while, and then are apparently rapidly resorbed from the eighteenth

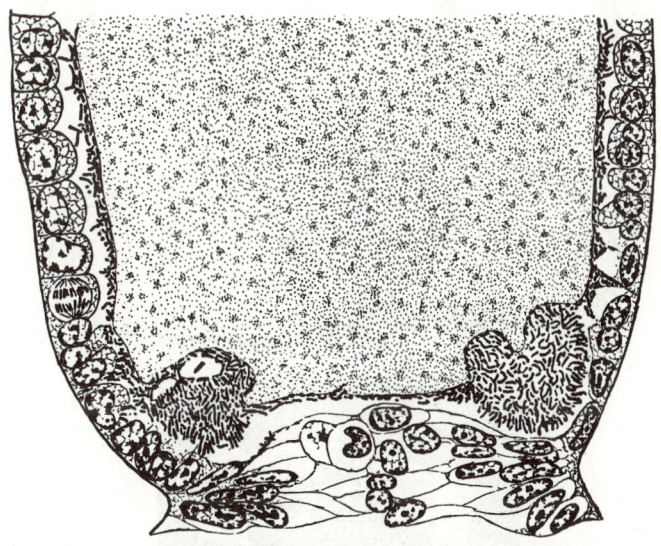

Fig. 300 *Periplaneta australasiae* L. Posterior end of an egg with ring-shaped bacterial pad. From Koch.

day on. The nuclei of the disintegrated mycetome can no longer be distinguished from the remaining yolk nuclei, whose fate they share. The bacteria floating in the body cavity are finally received by peripheral cells of the lateral fat lobes.

Blattella germanica apparently develops somewhat differently, according to Koch. Instead of collecting, as usual, at the two poles, the bacterial symbionts are distributed uniformly and loosely over the entire egg surface, at least at the time of the maturation divisions. Only after separation of the egg casings do they tend to shift more and more toward the future ventral side. During cleavage and blastoderm formation they group around the individual yolk nuclei and for the most part are carried by these to a position directly behind the rudiment of the germ band

(Fig. 302*a*).   Those which are situated more to the side and dorsally keep their position at first but soon show a tendency throughout to concentrate in the yolk-rich middle of the embryo, where masses of bacteria provided with yolk nuclei now collect in an irregular aggregation (Figs. 302*b*,*c*).

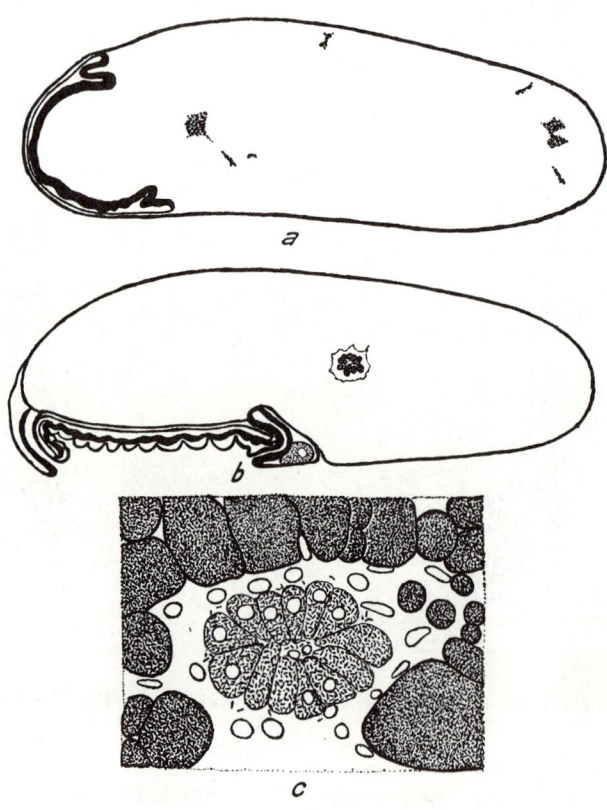

Fig. 301   *Periplaneta americana* L.   (*a*) Sagittal section through a 6-day-old embryo with bacterial nests scattered irregularly in the yolk, 55×.   (*b*) 11-day-old embryo with transitory mycetome in the yolk, 55×.   (*c*) Transitory mycetome of a 10-day-old embryo, 300×.   From Gier.

A regularly built, organ-like transitory mycetome is thus not formed in this case, yet after a while the symbionts here too leave the nuclei with which they are associated and press ventrolaterally through the embryonic midgut epithelium to the body cavity.   Koch observed no bacteria being destroyed in the gut during the process, yet according to him it is possible

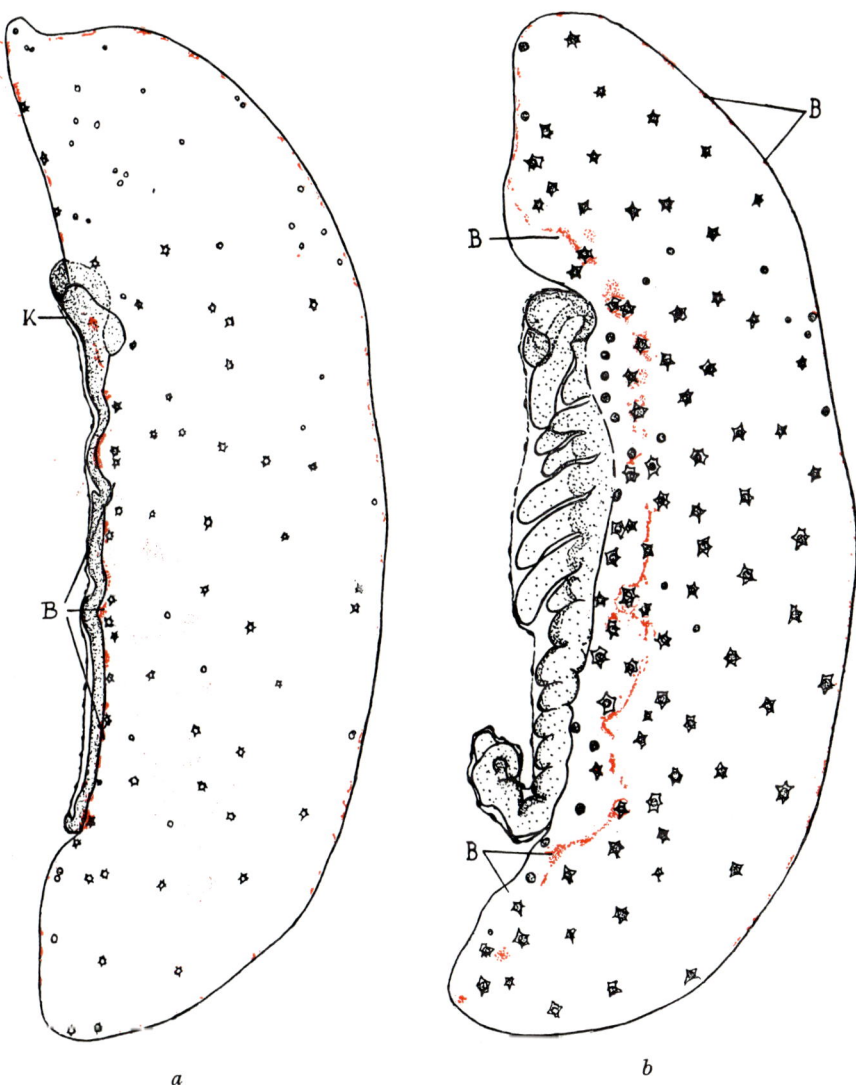

Fig. 302 *Blatella germanica* L.   Embryonic development, I.   (*a*) Early stage of germ-
band formation; the bacteria (B) still lie at the periphery of the egg; K, head fold.   (*b*)
Somewhat older embryo; the bacteria sink into the interior.   From Koch.

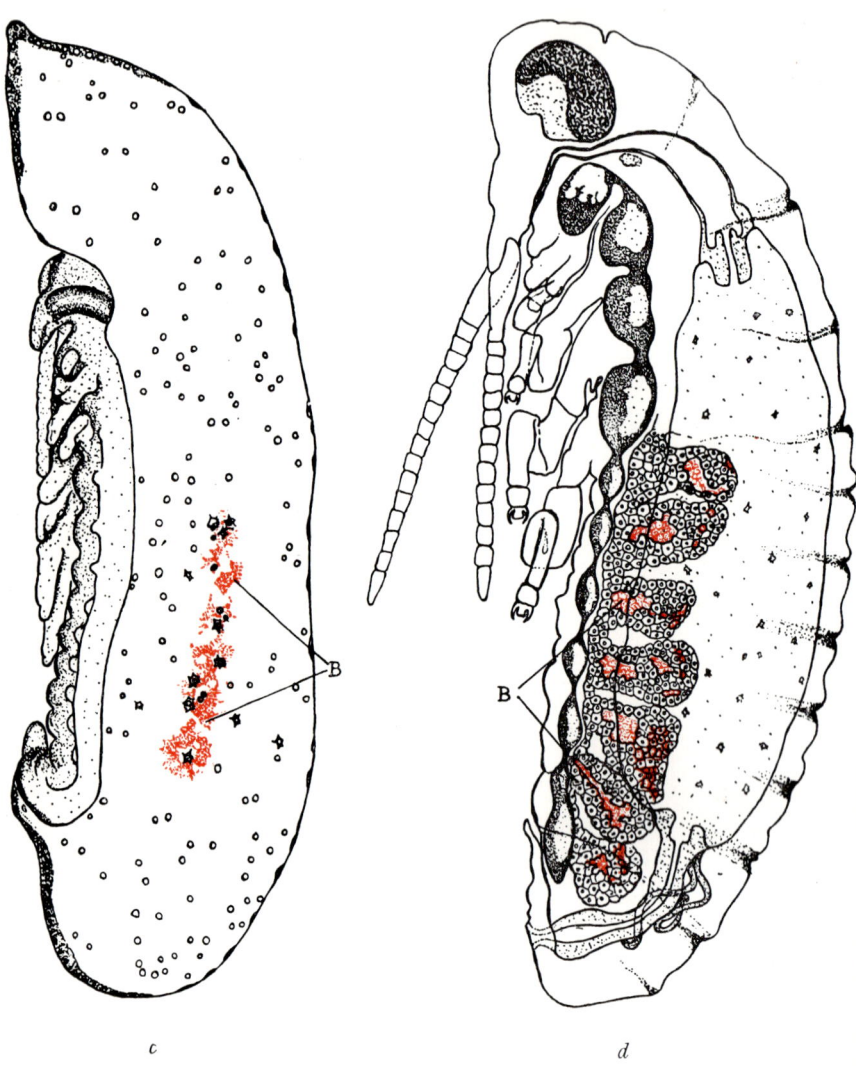

c   *d*

Fig. 302 *Blatella germanica* L.   Embryonic development, II.   (*c*) The bacteria have
assembled in the center in union with yolk cells.   (*d*) Embryo shortly before hatching;
the symbiont-filled cells are in the interior of the fat lobes.   From Koch.

even before infection to distinguish the future mycetocytes from the fat cells in *Blatella germanica* by the striking size of the nuclei and the completely homogeneous plasma (Fig. 303). Only after their settlement do the nuclei sink deeper into the fat lobes, and it is here that strong increase of bacteria and parallel amitosis of the host cells bring about the end result (Fig. 302*d*).

To judge from Heymons' report, *Blatta orientalis* occupies a midposition with respect to behavior of symbionts in the midgut rudiment. The

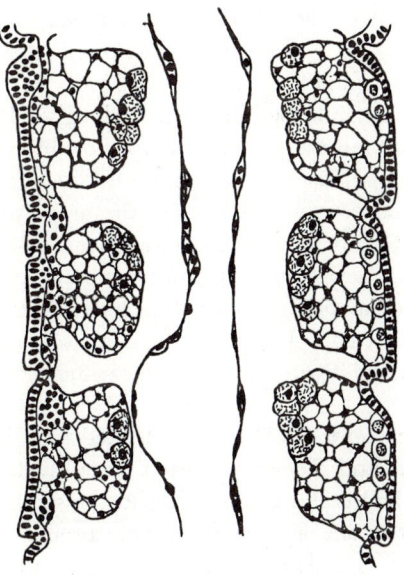

Fig. 303   *Blatella germanica* L.   Three embryonic segments of the abdomen with the just occupied mycetocytes, which have not yet sunk into the fat lobes.   From Koch.

bacteria increase strongly, and the numerous yolk nuclei which permeate them on all sides also undergo animated amitotic divisions and thus give rise to a large, sharply defined syncytium. When its nuclei begin to flow together to form large, irregular chromatin groups, multitudes of bacteria glide toward the intestinal epithelium, which is already developed on all sides, and through it reach the body cavity. *Ectobia livida* also forms a voluminous, sharply set off, transitory mycetome (Heymons).

In no other case of insect symbiosis has so much effort been expended in the cultivation of symbionts as with blattids. Reports of successful cultures and unsuccessful cultures followed each other in turn, and yet it remains impossible even now to be certain that blattid symbionts have been successfully cultivated.

It is hardly surprising that the older authors who were setting out to make cultures, for example, Blochmann (1887), Krassilstschik (1889), and Forbes (1892), had no success. Mercier followed with detailed data on positive results, but the motile, completely ciliated, spore-building organism which he cultivated and named *Bacillus cuenoti*, despite a certain similarity to the symbionts of *Blatta orientalis*, was doubtless a regularly recurring contamination such as was inevitable with his starting material, namely, the content of the cocoon, which is hard to sterilize. Javelly (1914) and Hertig (1921), who in part followed Mercier's method, also obtained mere contaminations of various kinds, but never strains with which the symbionts could be identified. Hovasse (1913) and Wollman (1926) had to admit failure in their attempts to cultivate them. In 1920, Glaser obtained a *Spirillum* which, though not identical with the *Bacillus cuenoti* of Mercier, was said to represent the organism sought, but soon thereafter he was forced to retract his statement. Before he could come forward again with new claims of positive results, Gropengiesser (1925) published a study on an organism which he had cultivated from cocoons on ordinary nutrient media. It was very similar to Mercier's, and without hesitation he identified it with the symbionts without presenting serological proof.

In his new experiments (1930), Glaser started with the fat body. From *Periplaneta americana* and *Blattella germanica* he cultivated three strains of a bacterium which had certain similarities to the symbionts. They could indeed be delimited one against the other through agglutination with specific rabbit sera, but their identity with the symbionts could not be proved by a complement fixation, since he was unable to produce sufficiently dense and clean suspension of the inmates of the fat body. In the opinion of his followers, his *Corynebacterium periplanetae*, a diphtheroid rodlet, was an atmospheric contaminant. On the basis of their own failures, Hollande and Favre (1931) rejected the results of Mercier, Gropengiesser, and Glaser. Bode (1936) was unsuccessful in obtaining drop cultures from the usual blattids kept in the laboratory, although with starting material of large quantities of fat bodies in meat bouillon he obtained pure cultures of a bacterium which to him seemed identical with that of Mercier and Gropengiesser. He was uncertain whether the bacterium represented the symbiont.

In 1937, Gier also reported failure in cultivating the symbionts of *Periplaneta americana* and various other blattids. He could merely keep them alive for a few days in an explant of the fat body, but he did not observe an increase. Transferral to crickets and chicken embryos also led to no positive result. Steinhaus (1945) repeated Glaser's experiments without success. When Hoover, in the same year, was making smears of the fat body of *Cryptocercus* on blood-agar plates, she obtained cultures after 12 to 14 days in which, like Glaser, she could differentiate three types and which she too identified with the symbionts.

In 1947, Gubler published a careful study of his diversified attempts to cultivate the symbionts of *Blatta orientalis* and *Periplaneta americana*. Forms which bore a certain similarity to the symbionts did occasionally grow on the various culture media or with transferral of the bacteria to the yolk sac of chicken embryos, but serological proof showed that they were neither identical with one another nor did they correspond to the symbionts. Apparently they were slow-growing Corynebacteria originating from the air, which were indeed often barely distinguishable from the symbionts on the basis of morphology and staining.

Keller (1950) reported successful cultures. He was the first to provide a culture medium equivalent to the milieu of the symbionts by mixing agar with the fat body of the host pulverized in Ringer's solution. Such plates were unfavorable to atmospheric organisms and other contaminations but after approximately 7 days yielded, in the places inoculated with the symbiont-containing fatty tissue, very delicate, transparent colonies which could be further inoculated on fat-body agar though not on ordinary bouillon agar. Because the blattid fat body is also a storage organ for a number of metabolic end products, among which uric acid plays the most essential role, it occurred to Keller to transfer the strains obtained in that way to a culture medium which consisted of 0.2 gm. of uric acid, 0.3 gm. of NaCl, 100 g. of water, and 2% agar and therefore practically eliminated the danger of a foreign infection. The bacteria grew in the form of exceedingly thin, smooth-edged colonies, although more slowly than on the fat-body agar. With increasing transinoculations their shape was transformed in proportion; the rodlets grew shorter, approaching more and more the form of coccobacilli and cocci. According to Keller, they are related to the symbionts of the legumes, to which there are also physiological parallels, and he therefore calls them *Rhizobium uricophilum*. Keller's assumption that he had actually cultivated the symbionts has been strengthened by experiments using the extremely sensitive method of Cannon and Marshall with the serum of immunized rabbits, but until these data are confirmed by some other source it is better to reserve judgment.

It is beyond understanding that several authors interpret the blattid symbionts as rickettsiae or organisms related to them. Glaser actually identifies them as rickettsiae; Gier believes that they are related organisms at the very least; and Hoover, that they are in the process of becoming rickettsiae. Hertig and Wolbach emphasize the intrinsic difference between the disproportionately large, gram-positive symbionts and the far smaller, gram-negative rickettsiae; and in a description of true rickettsiae from the chicken louse *Eomenacanthus*, Ries rightly warns against such an extension of the term rickettsia.

MASTOTERMITIDS

Jucci (1930, 1932) discovered that the only termite with bacteria-laden cells in its fatty tissue, in the manner of blattids, is *Mastotermes darwiniensis*

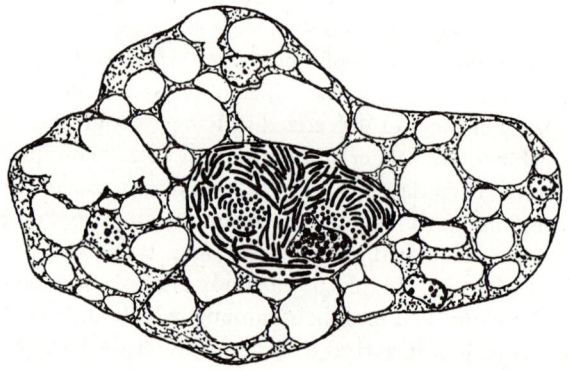

Fig. 304   *Mastotermes darwiniensis* Froggatt.   Mycetocyte, surrounded by fat cells, from a worker.   From Koch.

Froggatt, sole surviving member of the primitive and once widespread family of the Australian mastotermitids, and this finding acquired increased significance when Koch demonstrated (1938) that the similarity extends down to the last details of transmission.   For this reason *Mastotermes darwiniensis* will be discussed here in spite of the fact that it feeds not on a mixed diet but on living and dead wood.

As with blattids, the mycetocytes are limited to the visceral parts of the abdominal fat body.   These parts extend to the dorsal surface and in addition send out several tips into the metathorax from the first dorsal segment; here, too, the mycetocytes are always completely surrounded by fat cells; the structureless gram-negative bacteria, 3–6 μ in length, fill all parts of the uninucleate cell body but are not scattered at random, being usually arranged in parallel bundlets (Fig. 304).

The histological details of the fat body of the various states differ with respect to relative frequency of the mycetocytes and the microstructure of the fat cells. In adult workers, Koch found only one to three mycetocytes, completely separated by fat cells, in the individual lobes of a weakly developed fat body. In nymphs and the winged young imagines of both sexes a voluminous fat body pushes the settled cells so far apart that they may easily escape observation. In pubescent imagines the distribution of mycetocytes remains the same; but the surrounding fat cells are now filled with giant urate concretions, whereas fat vacuoles and protein granules are decidedly reduced or are completely absent. Conversely, the fatty tissue of soldiers is, of course, urate-free but is nevertheless laden with pigment granules and so shriveled that there is often room for 15 to 20 mycetocytes in one lobelet. Apparently these easily observable differences are due chiefly to differences in age. Whether there are differences caused by caste would require comparative study of the fat body in all developmental stages.

The details of egg infection are so similar to those observed in blattids that the pictures illustrating them might apply to any cockroach. As the egg begins its growth, only isolated rodlets appear at first in the space between the young oocytes and the follicle; thereafter follow states with surfaces completely covered by tangentially arranged bacteria (Fig. 305). Later they increase to such an extent that they almost completely fill the space behind the follicle several layers deep. With growth of the egg, there are progressive devastation of the lateral areas and an increase in the symbionts collected at the two poles.

Since the shift is so rapid, Koch believes that it cannot be explained merely on the mechanical basis of failure of the symbionts to keep pace with the egg in growth and that movements of the bacteria must also be involved.

Finally, such enormous bacteria caps form at the poles that they are clearly recognizable at the lowest magnification. They are drawn into the egg plasma earlier than with the blattids. When the eggs, which may attain lengths of 1.25 mm., reach

Fig. 305 *Mastotermes darwiniensis* Froggatt. Ovariole from the ovary of a young queen. From Koch.

approximately one-third of their ultimate size, thick masses fill an area 50–60 μ in width and indistinctly defined against the bacteria-free plasma. Older stages were not available to Koch.

In the ovaries of the young winged queen, penetration of the ovarioles by symbionts has already ended. A thick matting of tracheae envelops the individual ovarioles and surrounds the whole organ like a cloak. Mycetocytes and fat cells are not represented in the peritoneal envelope. The ovary of a female nymph, on the other hand, presents a completely different picture. Here the now maximally developed fat body is attached close to the tufts of the ovarioles, which are not as yet so abundantly supplied with tracheae and between which the mycetocytes are scattered everywhere. The latter are usually flattened off and nestle closely against the young ovarioles. The symbionts, discharged one by one, now pass through the wall of the still very flat follicle into the oocytes which are still in the "bouquet" stage (Fig. 306). Such mycetocytes are also attached in the same way to the younger areas of the germ layer which contain only oogonia, yet these areas always remain bacteria-free. On the other hand, older oocytes which are already infected occasionally receive reinforcements from adjacent mycetocytes. Since such mycetocytes are lacking in the ovaries of mature queens, one must conclude that the symbionts are all discharged at the end of metamorphosis and that the abandoned mycetocytes degenerate.

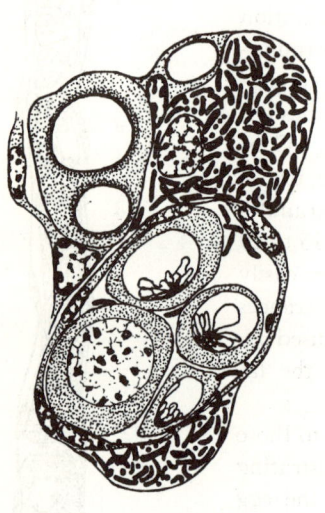

Fig. 306 *Mastotermes darwiniensis* Froggatt. Cross section through the ovariole of a nymph; the symbionts transfer from two attached mycetocytes to the ovocytes in the bouquet stage; tracheae are visible in three locations. From Koch.

As among the ants, the ovaries of the termite workers show more or less vigorous beginnings of regular infection. The ovaries develop in differing degree, as with the ants. Apparently a considerable period of growth tends to occur before degeneration sets in. In other cases the gonads are checked so early in their development that it is impossible to recognize the sex. *Mastotermes* occupies a middle position in that typical ovarioles are formed which do not contain synaptic stages but merely oogonia. Nevertheless the first steps toward infection are also taken here, for mycetocytes appear everywhere among the young ovarioles. They send no bacteria into the adjacent germ layer, but this should be no surprise, for analogous areas in the ovaries of queens also remain sterile.

Unfortunately it was not possible to study the embryonic development of *Mastotermes darwiniensis*, but there is little doubt that the behavior of the symbionts during this process follows the pattern of the blattids, for in the preceding description we find parallel after parallel in the two families: the physical appearance of the mycetocytes, their localization in the fatty tissue, their appearance in the immediate vicinity of the ovarioles, their role during infection, the time of transmission to the ovarioles, and the immunity of the germ layer revealed thereby, the way the symbionts traverse the follicle without sojourning in the plasma of the follicle cells, distribution of the symbionts over the egg surface, secondary concentration at the two poles, and late transmission to the egg plasma—all these are peculiarities which recur in both families.

The significance of this surprising observation for the historical problems of symbiosis in general and for the phylogeny of blattid symbiosis in particular will be explained elsewhere.

FORMICIDS

As far as is known today, only a negligible fraction of the ants live endosymbiotically with bacteria, namely, the *Camponotus* species and *Formica fusca* Latr. The other *Formica* species have no symbionts, although the embryonic development of some of them reveals that symbiosis was once present and then abolished (Blochmann, 1884, 1886, 1887; Buchner, 1918, 1921, 1928; Hecht, 1924; Lilienstern, 1932; Kolb, 1957, 1959).

The sites differ in the two cases. In *Camponotus*, mycetocytes resembling interstitial cells are inserted over the entire midgut between the secreting and resorbing epithelial cells. Some of these mycetocytes are club-shaped and of very considerable size, and others are of more modest dimensions. They all rest on the basal membrane. Their plasma is filled thickly and evenly with slender filaments and rodlets (Fig. 307). Blochmann observed these cells only in *Camponotus ligniperda* Latr.; however, I found the histology to be the same, aside from variations in size, in *C. senex* Smith from Mexico, *C. rectangularis* Em., and *C. maculatus* F. in three subspecies from various areas of Africa. Recently Kolb made a thorough study of the finer morphology of the *Camponotus* symbionts and the different ways in which they develop according to the developmental state of the hosts. The same organism may appear as little rodlets, thicker sausages, or tennis rackets. They may grow out later into branched or unusually long filaments. Also, amorphic spherical states occur. The presence of desoxyribonucleinic acid can be proved by the usual method. The arrangement of the acid is also subject to a certain change, and it disappears in connection with other degeneration phenomena of the symbionts occurring in the old host animals (Plate IV,*a–f*).

In *Formica fusca* the somewhat shorter, more or less curved, symbionts in the imagines are situated in cubical cells arranged behind the midgut in a loosely-drawn-out, usually unicellular layer. In the larvae they are smaller and have a dense even filling of bacteria; in the pupae they

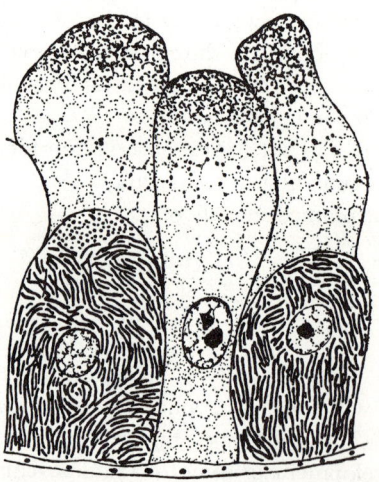

Fig. 307 *Camponotus ligniperda* Latr. Gut epithelium of an imago with mycetocytes. From Lilienstern.

increase greatly in size, are permeated by vacuoles, and then temporarily appear much less densely settled (Lilienstern; Fig. 308).

Larvae, pupae, and imagines of queens, males, and female workers behave in the same way in both *Camponotus* and *Formica*.

Transmission is effected by an extremely early infection of the ovarioles. As soon as the young oocytes are arranged one behind the other and the nutrient cell nests are added to one of them in each case, the first symbionts appear in the ovarioles of *Camponotus ligniperda* and quickly inundate the plasma of the follicle cells. Strictly avoiding the nutrient cells, they soon transfer to the plasma of the still very young oocytes, where the first little "accessory nuclei" now develop around the true nucleus, which is still in the synaptic stage.

The bacteria, sparse in the beginning, are later augmented through reinforcements, multiplication, and through their growing out into long filaments in the new milieu, which apparently constitutes a favorable medium.   The oocyte is thus permeated throughout and now appears like a single ball of threads; its plasma is limited to sparse walls separating the symbionts (Fig. 309).   The follicle cells meanwhile increase in number

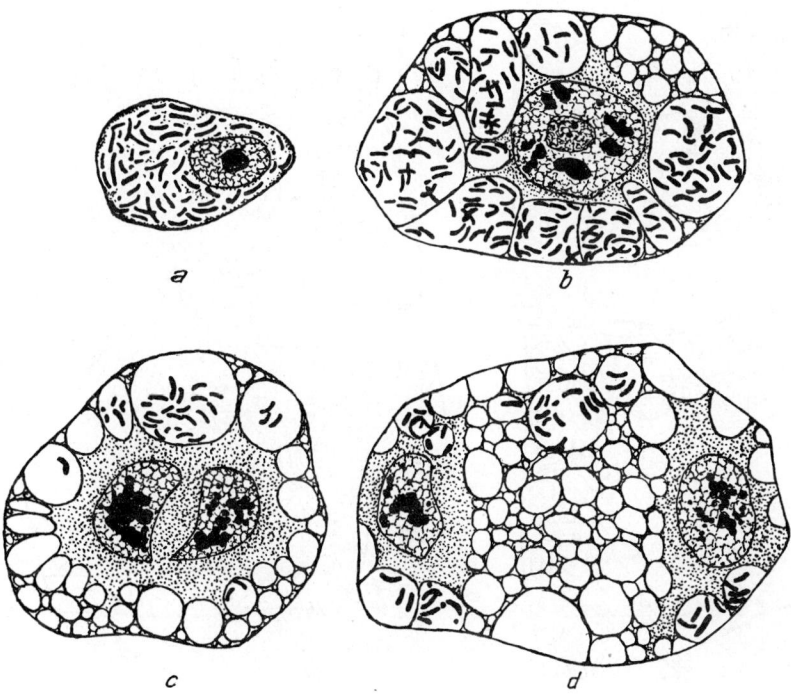

Fig. 308   *Formica fusca* Latr.   Mycetocyte (*a*) from a larva, (*b*) from an older pupa, (*c*), (*d*) in amitotic division.   From Lilienstern.

in proportion to the growth of the egg.   Their symbionts do not keep pace with the increase; only here and there does one still come upon an infected cell, and even its inmates are finally transferred to the egg.   Gradually the equilibrium which had clearly been temporarily lost in the egg is restored. Symbiont-free protoplasm, in the building up of which the nutrient cells take part, wins more and more ground; the bacteria, which stop multiplying when the yolk begins to form, are forced back increasingly.   In the egg about to be laid they are found only at the posterior pole, where they form a cap between the blastema of the germ membrane and the plasma carrying the yolk.

In *Formica fusca* the eggs are infected in much the same way.   As Fig. 310 effectively illustrates, shorter rodlets first permeate the follicle cells in huge masses yet never transfer to the nutrient cells, although the latter are

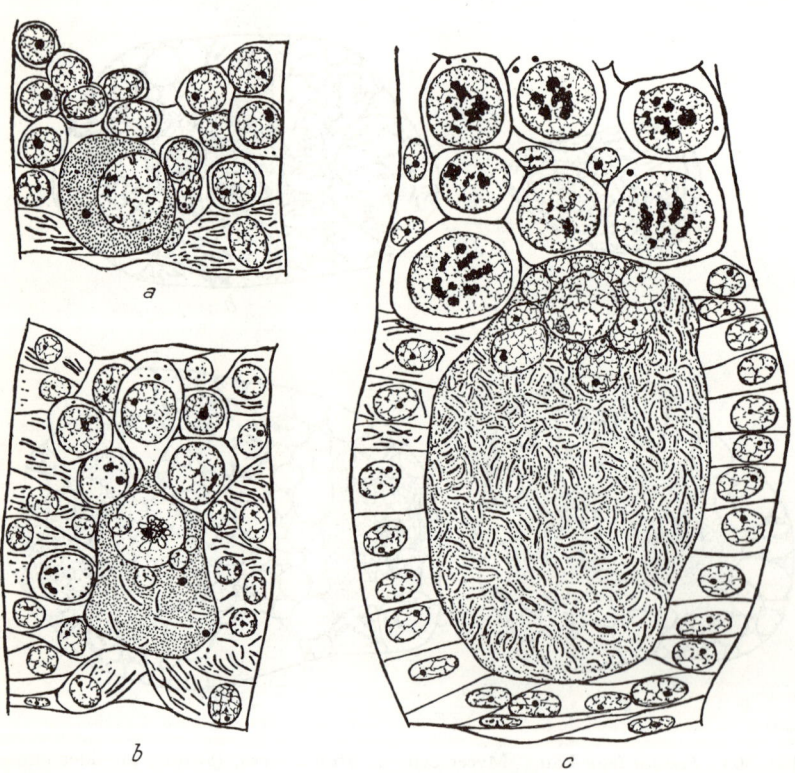

Fig. 309   *Camponotus ligniperda* Latr.   Three stages of infection of the follicle and the young ovocytes.   From Buchner.

sister cells of the oocytes.   This time, universal permeation is followed by a phase in which a peripheral symbiont-free zone occurs, but soon the entire protoplasm is again as extensively inundated as in *Camponotus*.

The ovaries of *Camponotus* workers, normally remaining rudimentary and forming only young oocytes which soon disintegrate again, are infected exactly in the same way as in the queens (Buchner, 1928).   In *Formica fusca*, transfer to the ovaries is additionally facilitated through the fact that here and there, as in blattids and *Mastotermes*, mycetocytes are directly adjacent; but in general only an infection of the follicle cells apparently takes place, and oocytes which would already be maximally inundated in the case of queens are symbiont-free.   However, even in this test object the eggs must be infected to some degree, for some of the larvae which develop occasionally from the egg of workers when no queen is

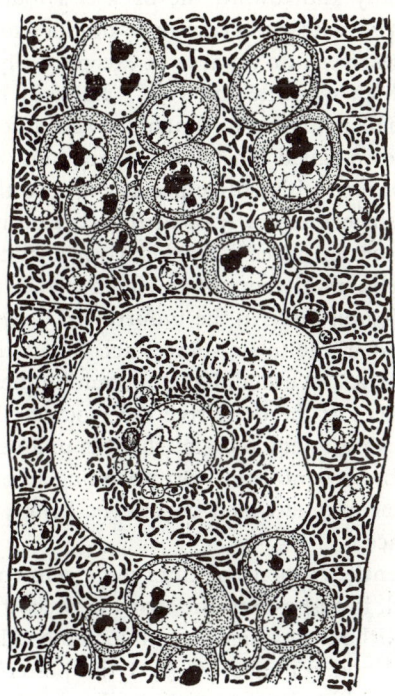

Fig. 310   *Formica fusca* Latr.   Infection of the follicle and of a young ovocyte.   From Lilienstern.

present contain mycetocytes which are at least in part sparsely infected (Lilienstern).

The history of the symbionts during embryonic development of the *Camponotus* egg is very complicated (Strindberg, 1913; Buchner; Hecht). When cleavage nuclei enter the symbiont-containing marginal areas, they cause the delimitation of tall cylindrical cells in which the bacteria are restricted essentially to the distal sections richer in plasma.   The nuclei of these primary mycetocytes now form a number of accessory nuclei, just as those of the oocytes did previously. Aside from this symbiont-covered section, a number of other zones which were differentiated unusually early can be observed in the completed blastoderm stage. At the anterior pole, vacuolated, yolk-rich cells, later rounded off, are dissociated from the embryo complex and now possess only trophic functions. Joining them ventrally and laterally is a section with extremely regular plasma-rich and yolk-poor cells which represent essentially the actual embryonic rudiment. Ventrally and toward the back is a fourth blastoderm area which again comprises extraembryonic cells destined to decay, and without distinct limits passes over into an area of exceptionally large, cubical and polygonal cells which engirdle the yolk. Bacteria have been received here too, but in smaller numbers. Later the cells fuse to form only a few giant elements, the so-called blastoderm syncytium. At first the dorsal side is deficiently covered with cells, especially toward the front (Fig. 311*a*). In the blastoderm stage a narrowly defined nest of minute, deeply staining cells has already been separated. Earlier I was inclined to call them the primordial germ cells, an interpretation which seemed all the more plausible since at an analogous place in the unfurrowed *Camponotus* egg a body is found which is comparable to substances accompanying the germ course in other Hymenoptera. We shall have to mention these cells again, since Hecht interprets them as the definitive mycetocytes already prepared.

Interestingly, this rudiment mosaic which is unique in insect development can already be recognized in corresponding separations in the blastema of the germ membrane which begin even before the cleavage nuclei reach the egg surface.

As the above-mentioned plate of embryonic cells continues to develop into the germ band, it pushes into the space between the embryo and the yolk membrane, without forming an amnion, the cells which are being anteriorly attached and rounded off. At the posterior pole where the germ band joins the blastoderm syncytium, a second group is dissociated from the embryo complex. Further development is now accompanied by progressive shifts of position. Gliding toward the dorsal side, the cells of the blastoderm syncytium, sometimes fused into a single, gigantic, multinucleate whole, push the symbiont-laden cells along. These cells of the blastoderm syncytium are also dissociated from the epithelial complex, are rounded off, and extend from the dorsal area to which they were thus transported into the sides and the ventral areas of the yolk, the latter already having been enveloped by a poorly developed midgut epithelium (Fig. 311*b*).

After the mycetocytes, easily recognizable by their accessory nuclei, have been brought in this way to the vicinity of the gut, the symbionts transfer to the many smaller, angular, bacteria-free elements which appear

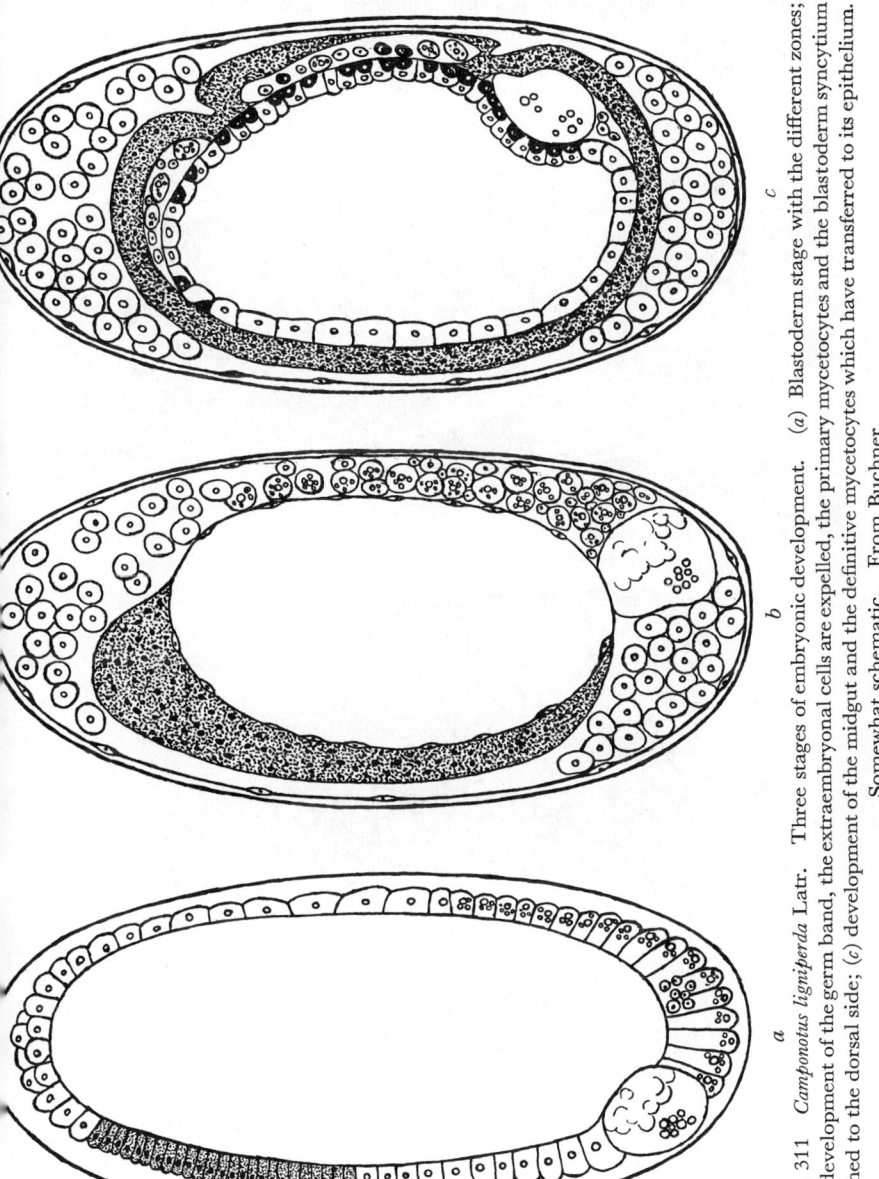

Fig. 311 *Camponotus ligniperda* Latr. Three stages of embryonic development. (*a*) Blastoderm stage with the different zones; (*b*) development of the germ band, the extraembryonal cells are expelled, the primary mycetocytes and the blastoderm syncytium pushed to the dorsal side; (*c*) development of the midgut and the definitive mycetocytes which have transferred to its epithelium. Somewhat schematic. From Buchner.

everywhere between the mycetocytes.   Figure 312*a* shows the three distinct cell varieties which, now intermingled, surround the central yolk: the flattened, embryonic intestinal epithelial cells which contain the vitelline spheres; the huge transitory mycetocytes; and one of the future

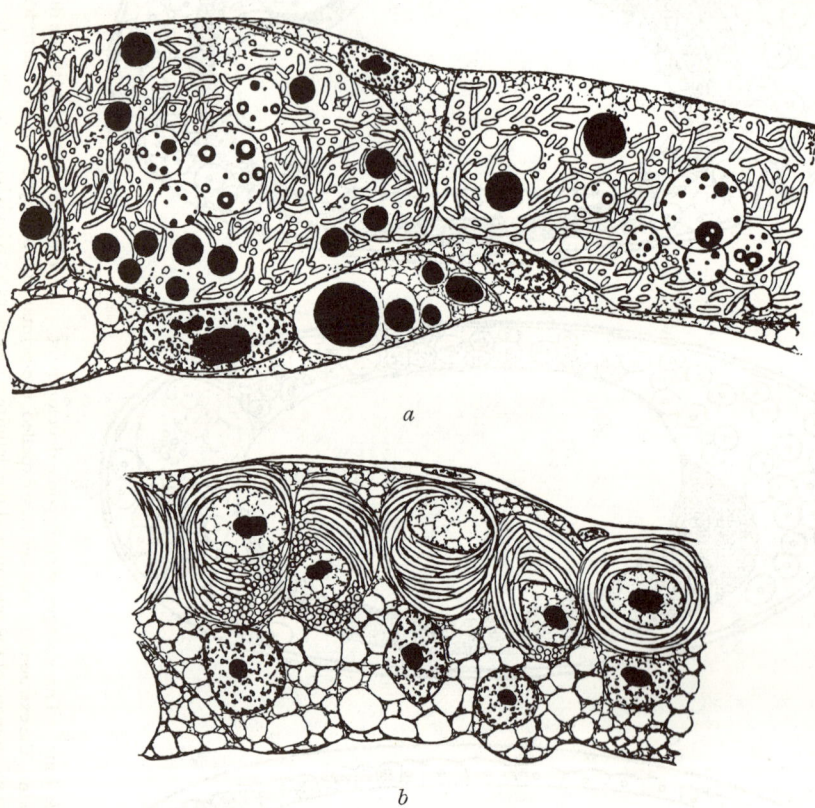

a

b

Fig. 312   *Camponotus ligniperda* Latr.   (*a*) Epithelium of the embryonic midgut with transitory mycetocytes and interstitial cells lying posteriorly; (*b*) midgut epithelium of a larva ready to hatch, with inserted mycetocytes.   From Buchner.

mycetocytes wedged between them.   Figure 313*a* represents the two latter types as seen from the surface.   The moment of refreighting was not observed by me or by Hecht, but there can be no doubt of the fact itself, for soon thereafter the uninucleate interstitial cells have grown considerably and are thickly filled with bacteria, whereas the original mycetocytes still contain symbionts only here and there and exhibit clear-cut signs of incipient disintegration.   Their nuclei become hyperchromatic, and the nuclear membranes disappear with the result that the nucleoles are now situated in the plasma; the latter becomes more and more homogeneous and shrivels into irregular masses which are finally destroyed in the yolk. The young mycetocytes push between the opening intestinal epithelial cells and therewith reach their ultimate position (Figs. 311*c*, 312*b*).

The question of the origin of the definitive mycetocytes has not yet been completely answered.   Strindberg's assumption that some of the original mycetocytes assume the character of the definitive ones is automatically

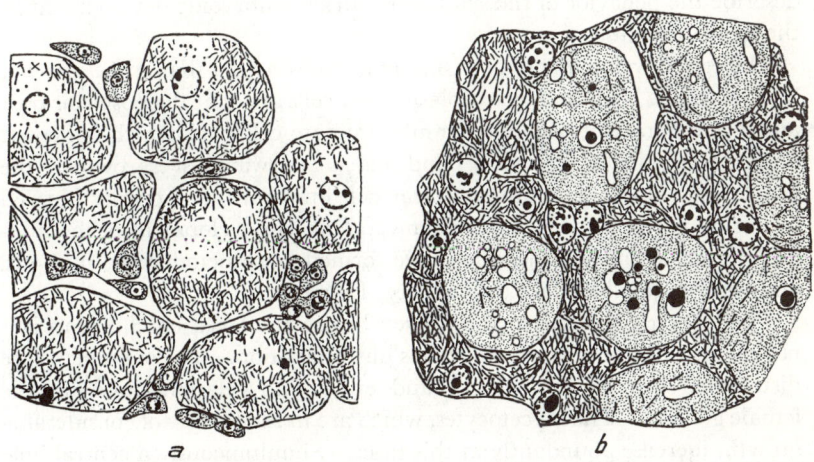

Fig. 313.   *Camponotus ligniperda* Latr.   Transitory mycetocytes and sterile interstitial cells (*a*) before relocation, (*b*) after relocation; the now symbiont-free mycetocytes degenerate.   From Hecht.

eliminated in view of the basically different structure of the two cell types· Objections can also be raised to Hecht's theory that they are derived from the cell group which appeared early in the blastoderm stage.   Quite aside from the fact that this group greatly resembles the primordial germ cells, an insertion of foreign elements into the intestinal epithelium has never been observed elsewhere, and in *Haematopinus* the mycetocytes which are inserted there in very similar fashion are clearly derived from the embryonic intestinal cells which also take over the symbionts from transitory mycetocytes.   For these reasons we must consider again the possibility that those once sterile interstitial cells similarly represent material from the midgut rudiment.

The history of the blastoderm syncytium is very strange. Here, also, the symbionts increase to sizable masses and, along with this increase, numerous accessory nuclei are formed, sometimes in great swarms. With progressive dorsal closure, the syncytium, like the other mycetocytes, is overgrown by the germ band and reaches the space between the midgut and the posterior end of the embryo (Fig. 311c). Like the primary mycetocytes, which are analogous in many points, the syncytium is destroyed after it divides into two masses situated on both sides of the hindgut and loses its symbionts, which apparently are also taken over by the definitive mycetocytes and transported to the intestinal epithelium. We shall discuss the possible significance of this strange cell mass after we describe the behavior of the symbionts during embryonic development in the genus *Formica*.

The eggs develop much more simply in *Formica fusca* than in *Camponotus*. Again a single layer of infected blastoderm cells arises at the posterior pole and is later transformed into a multilaminate cell mass which is dented during formation of the germ band and is afterwards overgrown by the latter (Fig. 314). The initially clear cell limits are temporarily blurred in this unpaired embryonic mycetome. During development of the mid-gut the mycetome reaches the angle formed by proctodaeum, mid-gut, and ventral ganglia chain (Strindberg, Lilienstern).

The mycetome is flattened off, even before the larva hatches, but it is not divided into two loose cell groups until pupation. Its components are distributed around the midgut and enter into relationships with the female gonads. The mycetocytes, which are in the process of considerable growth, increase abundantly at this time. Simultaneously a central hole appears in the nucleus, breaking through first on one side and then on the other, and the two sickle-shaped halves diverge without severing the dumb-bell-shaped nuclear constrictions which are otherwise characteristic of amitosis (Fig. 308c,d). In older pupae a renewed increase of the symbionts regains the lost territory.

Lilienstern also studied the development of unfertilized eggs of female workers, which, as is well known, yield only males, and found that such eggs, whether infected sparingly or not at all, always develop an unpaired mycetome, just as do the eggs of queens which are provided normally with symbionts, but the mycetome remains completely sterile or at most poorly settled.

This interesting situation throws light on a fact which Strindberg had discovered in the course of his research but naturally was unable to explain. He discovered that, early in the development of *Formica rufa* L. and *F. sanguinea* Latr., a cell mass which bears an extraordinary resemblance to the mycetome rudiment in *F. fusca* is delimited. Lilienstern retested

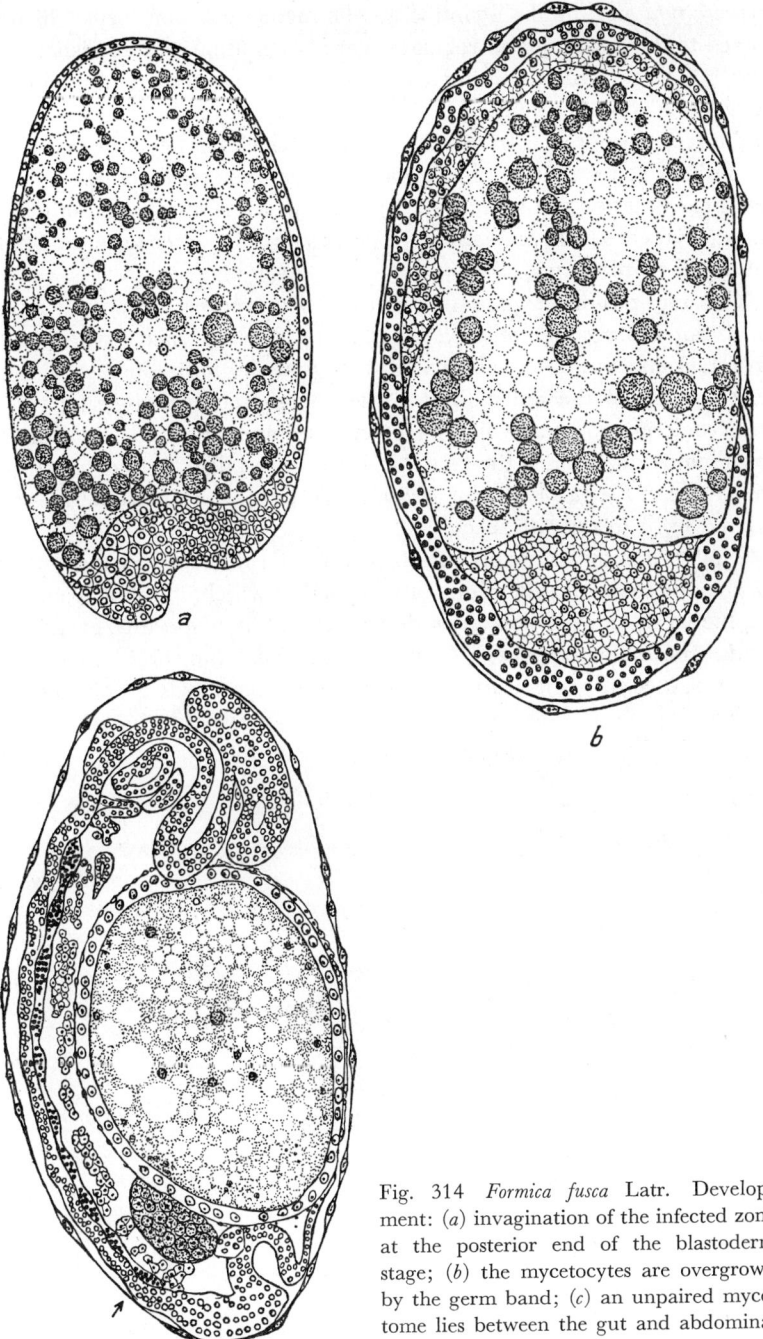

Fig. 314 *Formica fusca* Latr. Development: (*a*) invagination of the infected zone at the posterior end of the blastoderm stage; (*b*) the mycetocytes are overgrown by the germ band; (*c*) an unpaired mycetome lies between the gut and abdominal mesoderm. From Lilienstern.

Formica rufa and, unlike Strindberg, who merely mentions cells which are forced to the inside, found regular polar invagination and abscission leading to a large formation which resembles the mycetome rudiment in *F. fusca* to the point of confusion in situation and structure but lacks symbionts and is later regressed.   In *Formica sanguinea*, which has never been retested, Strindberg states that the invagination consists merely of a few cells and leads to formation of a small roundish mass which is situated exactly as the mycetocytes are in *F. fusca* and disappears some time after formation of the intestinal epithelium.

The only possible explanation of these occurrences, which are apparently not observed in other ants, is that symbiosis once existed in the ancestors and was later eliminated, and that the hosts, just as with the development of the unfertilized *F. fusca* eggs, retain vestiges of this in embryonic phenomena which have since become meaningless.   Tending in the same direction is the fact that Lilienstern failed to find bacteria in the larvae of *Formica fusca* var. *glebaria* Nyl. Em., which is sometimes considered to be a separate species.

Finally, it is by no means out of the question that the strange and perplexing blastoderm syncytium of *Camponotus*, which, though it contains symbionts, is nevertheless later destroyed, ought to be interpreted as a reminiscence phenomenon.   One might also take into consideration that, here too, the symbionts were housed in paired mycetocyte groups until the present-day site was evolved and that this cell mass with similar behavior should also be interpreted as a reminiscence of this.

9

Symbiosis in luminous animals

CEPHALOPODS

Symbiosis with bacteria has been demonstrated satisfactorily thus far only in the myopsids, both loliginids and sepioids, almost all of which live in shallow water. The decisive discoveries were made by Pierantoni (1917, 1918, 1924, 1925, 1934, 1935); Zirpolo (1918–1938), Mortara (1924), Meissner (1926), and Getzel (1934) worked on bacteriological problems; Kishitani (1928, 1932) and Herfurth (1936) made anatomical and bacteriological contributions. In spite of these numerous studies, the field lacks the high degree of clarity which in general distinguishes the research on luminous symbiosis of pyrosomes and fishes.

Surprisingly, the so-called accessory nidamental glands, devices which had long been known to exist in all myopsids, acquired new significance as bacterial organs. The actual nidamental glands are limited to females. As paired sacs accompanying the ink bag, they are conspicuous because of the laminated structure which has developed from profuse infolding of the glandular epithelium. The substances which they produce, together with those secreted in the oviduct glands, leave the pallial sinus to form envelopes around the eggs, after which the latter are pasted to a base of some kind. The accessory nidamental glands, usually located close in front, may appear alone or in connection with the luminous organs but at any rate do not in general themselves emit light perceptible to the eye. As the name indicates, it was formerly assumed that they assisted in the construction of the eggshell. This premise was seemingly buttressed by their location, their outlet (to be described later), and their apparently glandular structure, but now that it has been proved that they do not perform this function, we shall merely speak of them as "accessory glands," fully aware that this name also is not ideal.

The diversity of the symbiotic devices of the myopsids demands organization of the material into four groups. Let us consider first the numerous cases in which the accessory glands are limited to females. Classic examples are furnished by the sepiids, of which chiefly *Sepia elegans* D'Orbigny and *S. officinalis* L. were studied in detail. Here the accessory glands, situated directly in front of the nidamental glands and provided with medial branches of the visceral nerves, are stately structures, originally paired, which collide along the midline in full breadth and then are fused outwardly and inwardly into an unpaired organ (Fig. 31*a,e*).   In

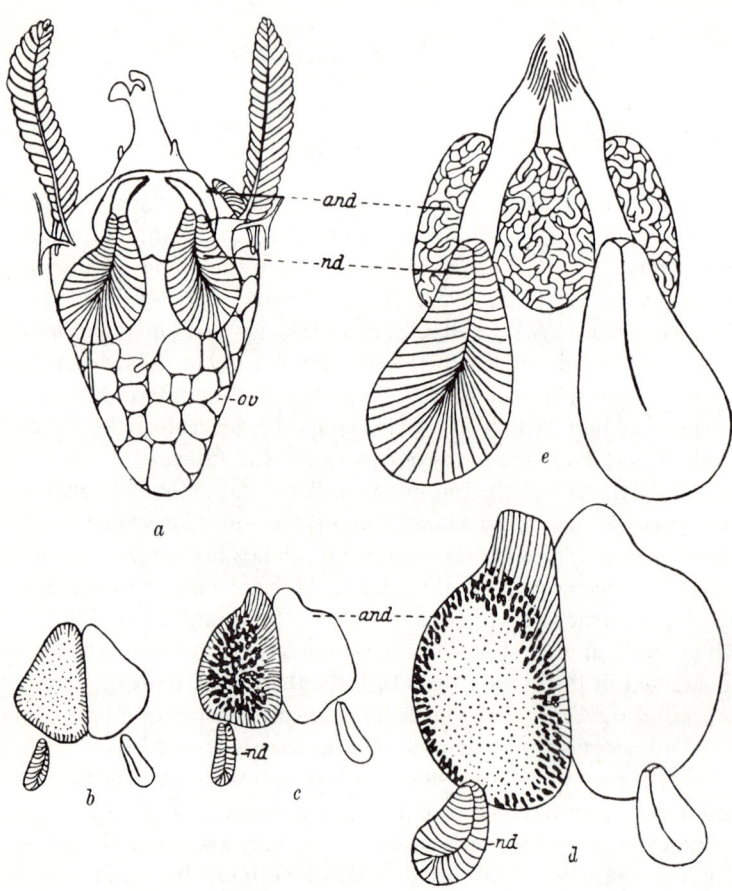

Fig. 315   *Sepia elegans* D'Orb.   (*a*) The abdominal organs; (*b* to *e*) development of the nidamental glands and the at first essentially larger "accessory nidamental glands." *ov*, ovary; *nd*, nidamental glands; *and*, accessory nidamental glands.   From Döring.

sections they give the impression of being tubular glands.   A cushion of connective tissue is filled with numerous, greatly branched, and closely interbraided tubes which open into the pallial sinus through narrower ducts.   Their openings are collected on two longish, clearly distinct fields, the posterior ends of which are directly opposite the opening of the nidamental glands, a position which makes an intermixture of the contents inevitable. The entire field occupied by the invaginations is ciliated, and the ciliate cover continues into the interior of the ducts. *Sepia officinalis* differs from *S. elegans* only in minor points.

Strange to say, these tubules are now colored differently. White, yellow, and orange-red tubes can be distinguished, and these three types are intermingled throughout the organ (Fig. 316). Their interior is not

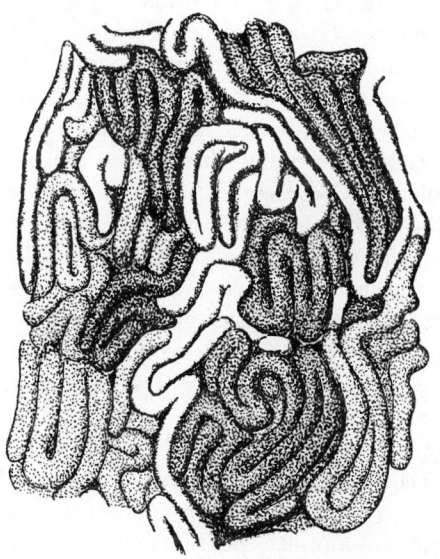

Fig. 316 *Sepia officinalis* L.   Section from the accessory glands with white, yellow, and orange-colored tubules colonized by different bacteria (fresh preparation). From Pierantoni.

filled with a secretion, as the earlier authors assumed, but is always populated by a large number of bacteria. In the histological preparation it is possible to recognize even with low magnification the differences in the development of the tubular epithelium and also the variations in the structure of their contents which are analogous to the various symbiont types represented in them. In *Sepia officinalis*, Pierantoni found slender rodlets in the white tubes, coccus bacilli in the yellow, and minutest cocci in the orange-colored, and he determined further that each host species has slightly different symbionts which should not be confused with one another.

Their culture in sepia bouillon caused no difficulty whatsoever. The colonies cultivated from red and white tubes gave no evidence of light capacity; those cultivated from the yellow were said to be always luminescent. The symbionts of the luminous organs of other sepioids were soon to be subjected to thorough bacteriological testing in connection with the new viewpoints, but Getzel (1934) of Pierantoni's institute first studied the luminous organs of the accessory glands of *Sepia officinalis*. As culture media he used chiefly agar with a bouillon of *Sepia* musculature, a mixture with which Zirpolo meanwhile had had good success with symbionts of other cuttlefish; for studying pigment formation he used a mixture, recommended by Pergola, of sea-water, glycerin, and egg yolk. In these media he found no less than five forms, each accompanied by specific pigmentation and each detectable in smears of the organs. They were gram-positive throughout, possessed no cilia, and formed no spores. *Micrococcus nidamentalis albus* is an exceedingly small round coccus 0.5 μ in diameter, yielding white colonies; *Coccobacterium nidamentale rubrum*, a scarcely larger, elliptical coccus with colonies which appear a vivid brick red; *Diplobacterium nidamentale pallidum*, a delicate, short double rodlet 0.9 μ in length, is characterized by yellowish ocher pigment. *Staphylococcus nidamentalis croceus*, the largest of the symbionts, is spherical; its diameter is 1.2 μ and its cultures are ocher-colored. The fifth symbiont, *Staphylococcus nidamentalis malus*, is smaller, measuring only 0.8 μ in diameter, and produces a greenish yellow pigment.

None of these five organisms had the capacity of bioluminescence. Thus in this respect also it was necessary to correct Pierantoni's data. The organisms are obviously specific forms adapted to this living area, for, as shown by their non-motility, positive behavior with respect to gram-staining, odd pigment production, lack of capacity for luminescence, and other characteristics, they are sharply distinct from all the forms obtained from the surface of the sepias and studied in culture by Zirpolo (1917) and Meissner (1926).

The same type is often represented in the family of the sepiolids. Using *Sepietta obscura* Naef for purposes of comparison, Pierantoni found that, unlike other sepiolines, it behaves like the sepias except that the orange-red tubes, which predominate in numbers, fill out the entire central area of the likewise secondarily unpaired organ, and the two other types cause white and yellow peripheral spots. Again there is a separate symbiont type to correspond to each of the three types of tubes. Herfurth also determined in *Sepietta oweniana* Naef the same arrangement of the varicolored tubes. In this genus, the only sepioline which follows the *Sepia* type throughout, the orificial conditions guarantee in high degree an intermingling of the symbionts and the secretion of the nidamental glands, for the flat groovelets into which the symbiont-populated tubes empty are not only situated close to the openings of the nidamental glands but are also connected with them by special membranous folds (Fig. 317).

Among the sepiolids, almost all rossiines behave like *Sepietta*. In *Rossia macrosoma* Owen, the species given the most detailed study, the tubules of the accessory glands, which remain paired, are much more

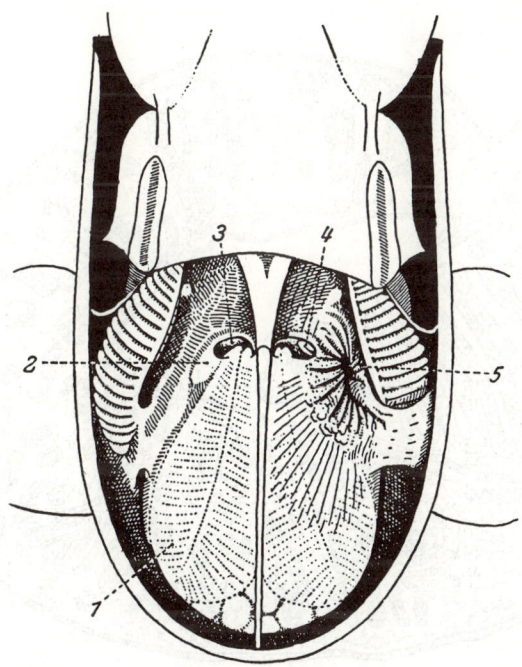

Fig. 317   *Sepietta oweniana* Naef.   After removal of the ventral portion of the mantle. 1, Nidamental gland; 2, accessory gland; 3, common opening of both; 4, opening of the oviduct; 5, bursa copulatrix.   From Naef.

weakly developed than usual and do not always permeate the whole area of the relatively much more developed connective tissue.   Döring (1908) actually speaks of connective tissue degeneration in the organs, and Naef (1923) also concludes from their behavior that they have a tendency to retrogress.   Herfurth investigated smears and this time determined only two types of symbionts.   The openings of the tubes are again in especially close relationship to the opening of the nidamental glands with which the accessory glands are fused laterally.   Between the two organs a well-defined space takes form, and to this both secretions and symbionts transfer (Fig. 318).

Here may also be added *Spirula*, the one genus of the family of the spirulids, which, as is well known, is a form living at greater depths.   In *Spirula spirula* Hoyle, the accessory glands, limited to the female, are well developed and easily visible.   They are situated in an arc around the openings of the nidamental glands, and each of three different bacteria

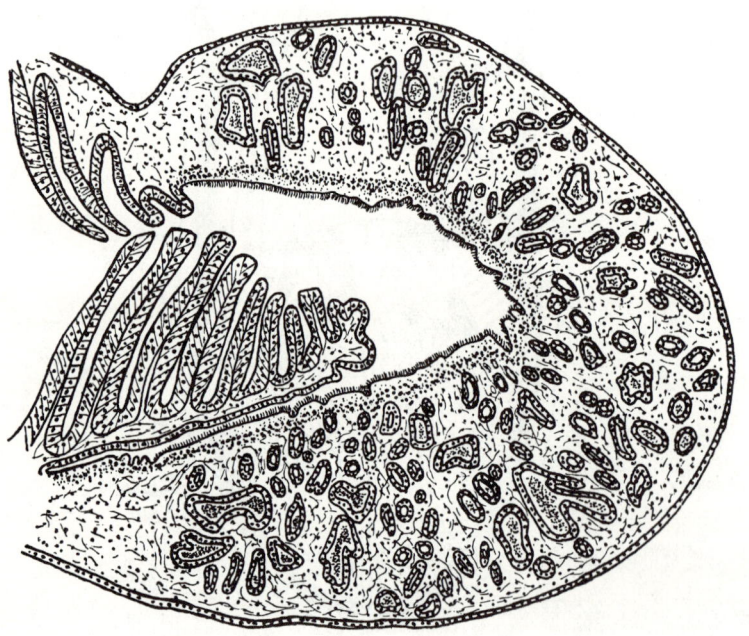

Fig. 318   *Rossia macrosoma* Owen.   Accessory gland and nidamental gland open into a
common cavity.   From Herfurth.

types corresponds to one of the three colors of the tubes (Herfurth).   The luminous organ at the rostral end in both sexes, however, is a formation which has no counterpart among the other myopsids.   Herfurth could detect no bacteria here.

In the family of loliginids the accessory glands, limited to females, also represent in general the only symbiotic device.   They are paired cushions which are covered to a great extent by the nidamental glands and give the impression of being smaller.   The tubules are shorter and less branched than with *Sepia*, the ducts short and narrow.   Their openings extend over the whole surface (*Loligo vulgaris* Lam., *marmorae* Verany). These organs reach maximum development in the young female and then no longer keep pace with further growth of the animals. Herfurth also did research on *Alloteuthis media* Naef, where the accessory glands are also more deeply set, but this time did not find their openings in closer topographical relationships with those of the nidamental glands. Here, too, this shift in position, the marked reduction of the tubes, and the strong development of the connective tissue constitute evidence that the organs are in the process of retrogression (Naef). In the tubes, three types of inmates are found: larger and smaller rodlets and small cocci. More exact data on the bacteria are available only for *Loligo forbesi* Steenstrup, but this is an exceptional case necessitating the establishment of a second category.

As determined by Wülker (1913) and later by Naef (1923), *forbesi* Steenstrup is the only *Loligo* species in which the males have a rudimentary

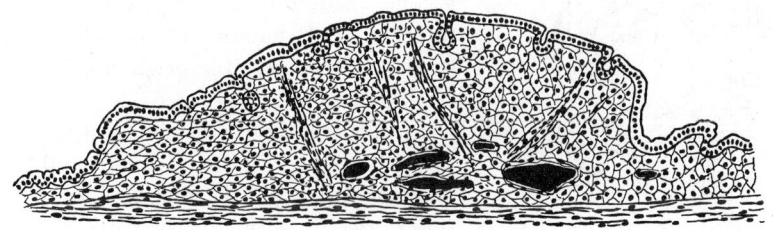

Fig. 319   *Loligo forbesi* Steenstrup.   Rudimentary accessory gland of a male animal.
From Wülker.

form of the accessory glands that are ordinarily restricted to females. The glands are situated in front of the kidney openings, not behind them as in sepiids and sepiolids. The resemblance between male and female glands is closest in half-mature animals. Their structure is fundamentally the same as in the *Sepia*, but the epithelial invaginations remain much shorter, less branched, and less interbraided, and as a result the fibrous part appears much more richly developed. The accessory glands are quite poorly developed even in females, and in males they not only are checked prematurely in their development but even retrogress. The tubes disappear again in part; the connective tissue condenses and is repeatedly furrowed by muscle fascicles (Fig. 319).

Wülker had still interpreted the content of the tubes as a secretion, but Pierantoni was able to prove that here too it consists of rodlet-shaped bacteria. Generally the accessory glands of the *Loligo* species are also an orange-red; for *Loligo edulis* Hoyle, it is expressly stated that three types of color tones occur irregularly intermingled, and thus one may conclude that several symbiont types are present here also.

The third category embraces those myopsids in which only females have accessory glands and both sexes have genuine bacteria-populated luminous organs. This type, more or less perfectly developed, is particularly widespread among the sepiolines. The luminous organs of *Rondeletiola minor* Naef represent a more primitive stage. Directly in front of the nidamental glands in the female is a bialate accessory gland, larger on the right side, where the roundish, unpaired, luminous organ is inserted. This organ not only sends a greenish white light shimmering through the pallium, but can also at the same time emit its contents into the surrounding sea water and thus fill this with a fiery mist. Paired outlet fields and a fine membranous fold, which passes through the middle of the luminous organ, still suggest the original two-sided rudiment.

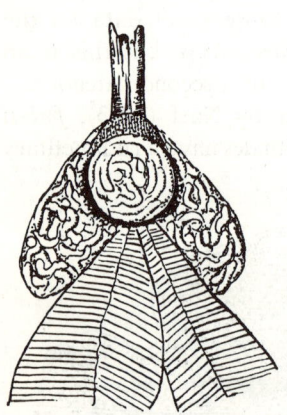

Fig. 320 *Rondeletiola minor* Naef. The luminous organ of the female with the flanking accessory glands and the posteriorly located nidamental glands. 10 ✕. From Pierantoni.

Even in the living specimen it may be seen that the luminous organ and the accessory glands are composed of spiral tubes, the area of the first being light yellow, the narrow, ring-shaped, adjacent area a vivid orange, and the rest of the accessory glands whitish (Fig. 320). Cross sections of the organ complex reveal that the luminous organ is sharply delimited against its surroundings. The cuplike ink sac directly below receives the invaginations, which are few in number but widen out like bottles toward the bottom and act as a pigment shade. The musculature passing on its inner side has a stronger capacity for light refraction than elsewhere and functions as a primitive reflector. On the other hand, the connective tissue, which is provided with numerous, fine, vascular, branches, has an extremely compact construction at the place where it is penetrated by the more slender ducts and is exceedingly light-refractive, actually having the value of a lens, although one that is still imperfect. In general, the tubes of the accessory glands are not swollen like clubs and drain into the peripheral areas (Fig. 321).

The luminous organ of the males is much larger than that of the females, but its structure is the same except that the accompanying tubes of the accessory gland are naturally lacking.

Both *Sepiola,* which was studied in a long series of species, and *Euprymna* are differentiated from *Rondeletiola* in that the luminous organs, now located far from the accessory glands, are laterally superimposed upon and inserted into the ink sac as paired, roughly auriform structures (Fig. 322*a*). From the standpoint of bacterial symbiosis, Pierantoni studied primarily *Sepiola intermedia* Naef; Kishitani (1932), *S. birostrata* Sasaki; Herfurth, *S. robusta* Naef, *S. atlantica* D'Orbigny, *S. ligulata* Naef,

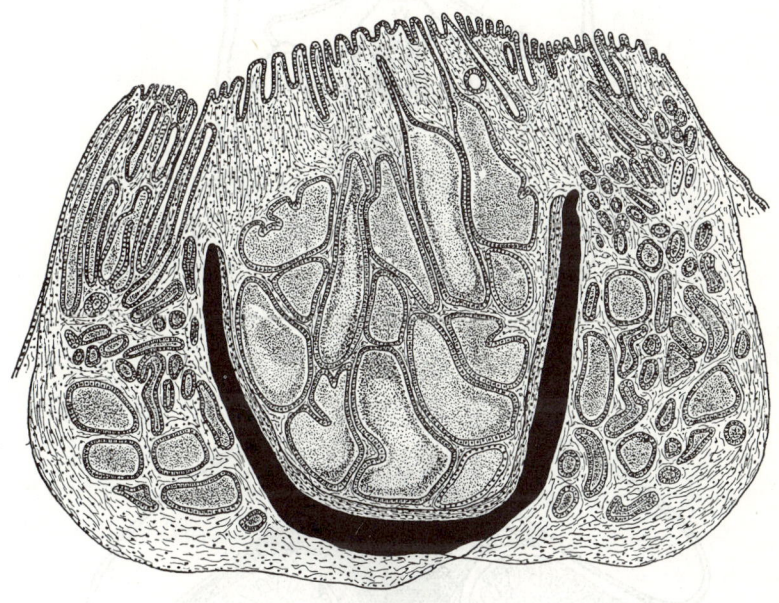

Fig. 321   *Rondeletiola minor* Naef.   Section through the luminous organ and the accessory glands of a female.   Partially schematic.   Combined from Pierantoni and Herfurth.

S. rondeleti Steenstrup, and *S. affinis* Naef.   Haneda (1955, 1956) provided brief data on *Sepiolina nipponensis* Berry, which, if correct, would disturb the order usual among the sepiolines.   Indeed, Haneda interpreted the content of the luminous organs as a secretion and made vain attempts to obtain bacterial cultures.   Kishitani used *Euprymna morsei* Verrill for purposes of comparison, but except for slight differences all these species have luminous organs and accessory glands of the same structure.

The exterior of the first is always covered by an elliptical, sharply elongated, and highly iridescent lens.   It doubtless represents a consider-

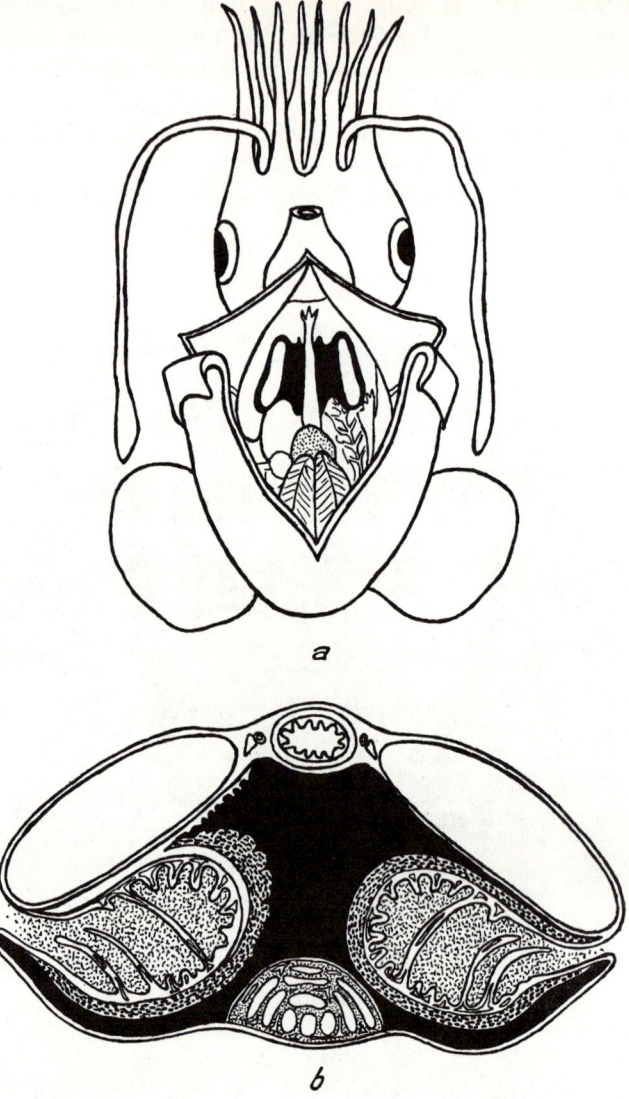

Fig. 322 *Euprymna morsei* Verrill. (*a*) Male animal with opened mantle, seen from the ventral side; the ear-shaped luminous organs are embedded in the ink sac; the unpaired accessory gland is anterior to the leaf-like nidamental glands; (*b*) schematic cross section through the paired luminous organs, and their lenses, ink sac, ink gland, and rectum. From Kishitani.

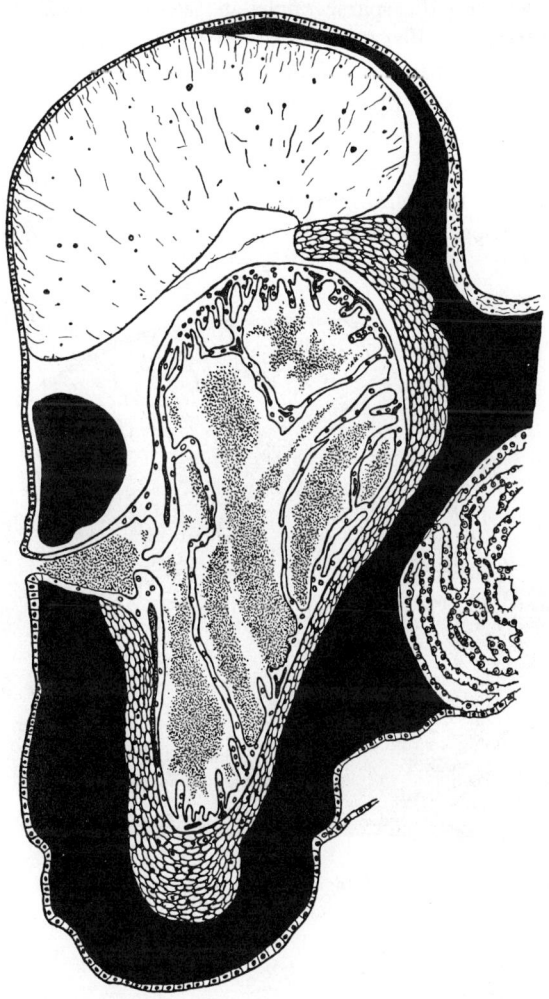

Fig. 323   *Euprymna morsei* Verrill.   Longitudinal section through the luminous organ
with lens, reflector, ink sac, and ink gland.   From Kishitani.

able and constructive advance when the ducts of the wide-lumened
epithelial sacs situated behind the lens no longer penetrate it individually,
but advance together toward a single outlet situated laterally to it on a
little mound and, shortly before reaching it, drain into a common space
(Fig. 323).   In life the lens is glassy in character and consists of connective
tissue which is rich in fibers and vacuoles and is penetrated by many
capillaries with a few small nuclei.   The reflector is now substantially

perfected.   Between the sparse nuclei in *Sepiola intermedia* are masses of strange oval leaflets.   Pierantoni is doubtless correct in considering them to be derived from the muscle fibers, which in *Rondeletiola* are still found in the same layer in slightly changed form.   In other cases they are large, spindle-shaped cells situated parallel to one another.   The ink sac, which in *Rondeletiola* has the form of a cup shielding the light laterally and toward the inside, in *Sepiola intermedia* functions simultaneously as a dimming device.   Especially with fixed specimens, it often reaches far around

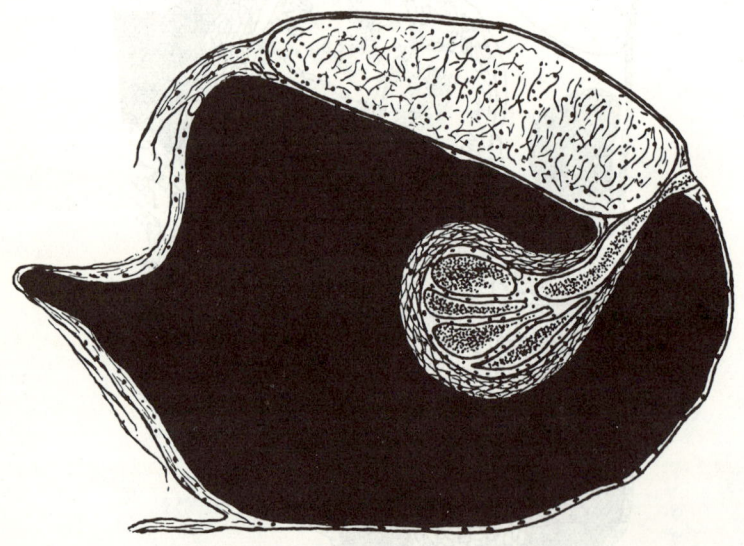

Fig. 324   *Sepiola ligulata* Naef.   Longidutinal section through the luminous organ with lens, reflector, and ink sac.   From Herfurth.

the light source and even pushes between it and the lens (Figs. 322*b*, 324). Everything indicates that the living animal, on which at times no luminescence is perceptible (Kishitani), can make use of this device at will. On the other hand, the bacteria can be also expelled into the sea water (Pierantoni, Skowron, 1926).

In consonance with the high development of these luminous organs, they are excellently provided with capillaries.   On each side a branch of the aorta aboralis, accompanied by a vein, penetrates the capsule from the rear, branches off in the connective tissue, penetrates the septal walls, and supplies the lens with the previously mentioned capillaries.   In the same way a branch of the nervus visceralis serves each of the two organs, which are bound together by a transverse cord of connective tissue. An illustration by Kishitani gives a graphic representation of the intimate relationship of these organs, which represent the apex of cephalopod symbiosis, with the body of the host animal.

The pigment distribution follows that in *Rondeletiola*. The yellow pigment is lacking throughout in the accessory glands where they separate completely from the luminous organs; a regular border is white; and a large central area is colored a vivid orange. In addition to Pierantoni, the following have studied the contents of the tubes in detail: Zirpolo, Mortara, Meissner, Herfurth, and Kishitani. The luminous organs have organisms which are very similar to each other: *Rondeletiola* contains a *Coccobacillus pierantonii* Zirpolo, the dimensions of which vary between 1 × 1 μ and 1 × 2 μ and which is in part immotile and in part moves by means of one or several apical flagella; *Sepiola intermedia* contains a rather variable organism called *Vibrio pierantonii* Zirpolo, which appears sometimes as a coccus and sometimes as a rodlet 2–4 μ in length and which is also equipped only in part with one to three flagella situated at one pole. From the luminous organ of *Sepiola birostrata*, Kishitani describes a *Micrococcus sepiolae*, and, from that of *Euprymna morsei*, *Pseudomonas euprymna*, the rodlets of which have unipolar flagella and are 1.5–2 μ in length. Among the *Sepiola* species Herfurth repeatedly found organisms which are very similar to the luminescent symbionts of *Sepiola intermedia*.

The symbiotic population of the accessory glands provides a far more colorful sight. *Rondeletiola* and *Sepiola intermedia* have white tubes with short rodlets and orange-colored tubes with coccobacilli. In other species, minute or coarser cocci and coccobacilli appear in association with longer rodlets or spindle-shaped formations with variously arranged, strongly staining inclusions. Figure 325 illustrates the diversity of these forms, a diversity which is always constant within the species. The accessory glands may have two or three inmate types or even four, as in *Sepiola rondeleti* and the luminescent symbiont represents an additional guest.

Pierantoni, Zirpolo, Meissner, Mortara, and Kishitani constantly cultivated the luminescent symbionts without difficulty. Meissner used as a solid medium a bouillon of *Sepia* or flounder musculature and sea water with addition of 1% Witte peptone and 2% agar or 15% gelatin. By adding an egg and 40 cc. of glycerin to a liter of this mixture, Zirpolo grew cultures which were by far the most intense in light production. Morphologically the various strains obtained from *Rondeletiola* or *Sepiola* were comparable throughout, yet when grown on agar and gelatin they exhibited greater or lesser differences in sugar fermentation and luminescent capacity. Comparison with all other luminescent bacteria previously cultivated showed that in both cases new forms are involved. In particular, they

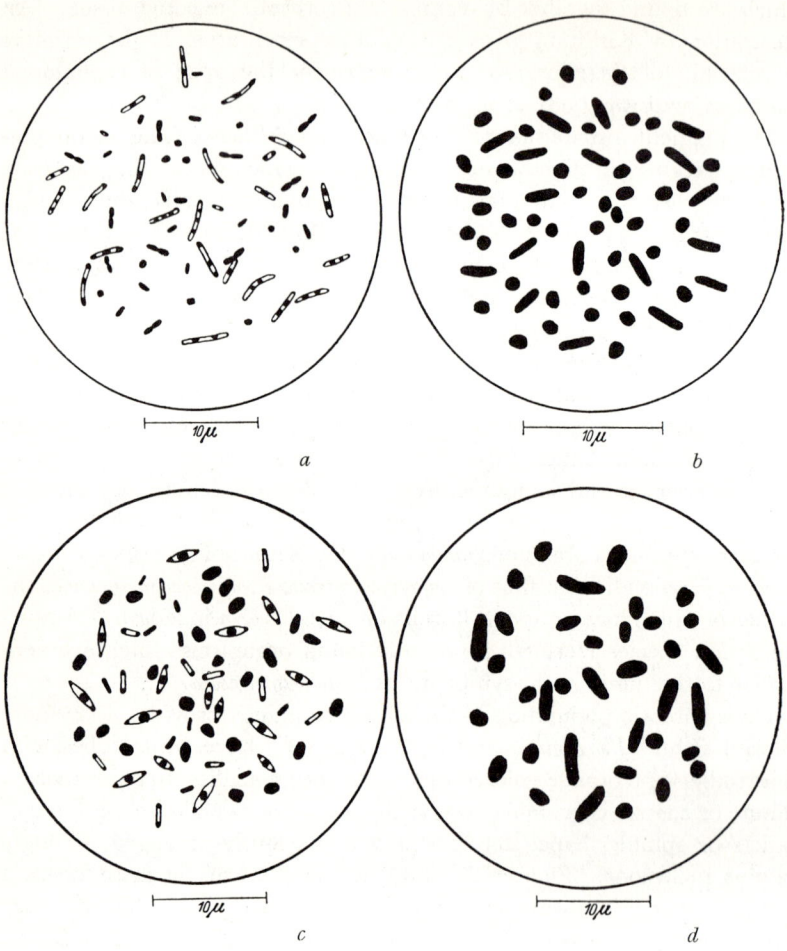

Fig. 325 *Sepiola ligulata* Naef.   (*a*) Three kinds of inhabitants of the accessory glands; (*b*) the coccobacillus from the luminous organ.   *Sepiola rondeleti* Steenstrup.   (*c*) Four kinds of inhabitants of the accessory glands; (*d*) the coccobacillus of the luminous organ. From Herfurth.

are clearly distinct from all cultures obtained by Meissner and Zirpolo from the skin and musculature of dead cuttlefish (*Sepia officinalis*).   The latter, bacilli and vibriones, did not show the racial specificity of the symbionts but were revealed in each case as an identical or at least serologically uniform species.   Through this carefully executed research, corrections were made in Mortara's data, according to which the occupants of the luminous organs would be identical with the luminescent bacteria clinging outwardly to *Sepiola* and inside various body cavities, including the anterior eye chamber. The reader is referred to controversies on the subject between Mortara and Puntoni and between Pierantoni and Zirpolo.

It is not surprising that the luminous organs of *Loligo edulis*, described by Okada (1927) and Kishitani (1928), constitute exceptions, as the

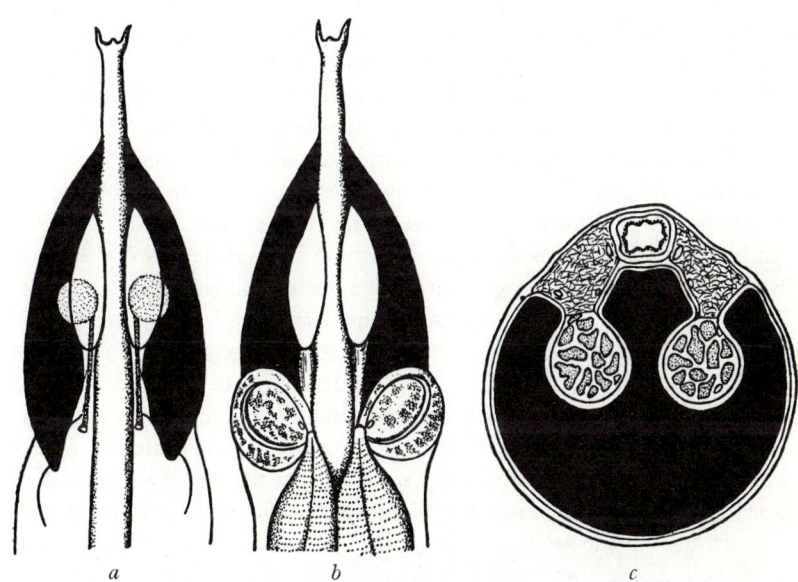

a   *b*   *c*

Fig. 326 *Loligo edulis* Hoyle. (*a*) Male animal with spindle-shaped lenses, round luminous organs, and excretory ducts; (*b*) female animal with the same kind of lenses, accessory glands, and nidamental glands; (*c*) cross section through the trunk with the luminous organs embedded in the ink sac and the lenses which flank the rectum and are penetrated with fibers.   From Kishitani.

organization of the loliginids is also exceptional in many other respects. Flanking the hindgut on each side is a slender, spindle-shaped lens approximately 30 mm. in length and 10 mm. in width penetrated in all directions by coarse fibers.   Below it is a smaller, spherical complex of bacteria-filled invaginations of the epithelium, which is parallel, by and large, with the longitudinal axis of the lens and opens into a duct arising at the posterior end.   This passage penetrates the lens and leads, in females, to the accessory glands and, in males, opens at the analogous place into the pallial sinus (Fig. 326).   The wall of the passage is lined with cilia and contains the same gram-negative, motile, short rodlets which are found in the luminous organ and which are so easy to cultivate.

Because of more ventrally situated openings into which the duct of the luminous organs also drains, the accessory glands are connected with the pallial sinus. Again three types of inmates can be differentiated in the tubes: a *Vibrio*, a *Coccus*, and a long rodlet. From *Euprymna morsei* and *Sepiola birostrata*, Kishitani had always cultivated only the non-luminescent occupants of the accessory glands; but, when he used the content of the accessory glands as starting material, he also obtained numerous luminescent forms which were identical with the occupants of the luminous organs and doubtless originated from the common outlet area.

In this third category we must also discuss the heteroteuthines, another subfamily of the sepiolids in which the luminous organs are highly developed in both sexes. Here it is a question of forms which have adapted to greater depths and thus differ in several respects from the sepiolines, from which they apparently originated. The only species tested as yet for symbiosis is *Heteroteuthis dispar* Gray, which lives at depths of 1200–1500 meters. The luminous organs are large formations measuring 0.5–1 cm. situated between the gills. Resembling large beads, they are enclosed by the cup-like ink sac in its inserted part and are arranged in a slight arc on the free side. Internally they are divided into a right and a left membranous sac with folded ciliated lining. Each of the two spaces filled with luminescent inclusions has an independent outlet on a raised wart. Inserted between them and a reflector at the rear is a tremendously developed reticulated area the significance of which is not yet clear. Toward the outside are masses of connective tissue. These are penetrated by muscle fibers, they resemble the reflector in structure, and evidently they serve only secondarily as converging lenses (Fig. 327).

On stimulation, *Heteroteuthis* also expels its luminescent content (Harvey, 1940; and others). Leaving the organs in the process are diversiform but chiefly ribbon-like masses which consist of roundish, oval, or longish structures, usually 3 μ in diameter and 6–7 μ or more in length, embedded in a basic substance. These inclusions, which correspond to the symbionts of the luminous organs in other myopsids and which always have the typical bacterial shape, are increased through separation into two equal or unequal parts, through multiple division and especially through a type of budding. They contain a strongly staining formation which is either roundish or rod-shaped, depending on their own shape, which Pierantoni, the first to study these inclusions in detail (1924, 1925), was inclined to interpret as a nucleus but which might just as easily represent the bacterial body surrounded by a gelatinous envelope. According to all indications, at any rate, symbiotic microorganisms are involved and also, as in so many other cases, a degeneration of the original bacterial shape which is admittedly unusual for luminescent bacteria. Unlike Mortara (1922), Harvey (1940) does not seem disinclined toward such an interpretation. Skowron (1926) leaves the question open but believes that self-luminescence is more probable. In the luminous organs of *Heteroteuthis* as well as of other sepiolids, neither luciferin nor luciferase could be detected. This fact, and above all the systematic position of the heteroteuthines, is consistent with Pierantoni's hypothesis (Fig. 362).

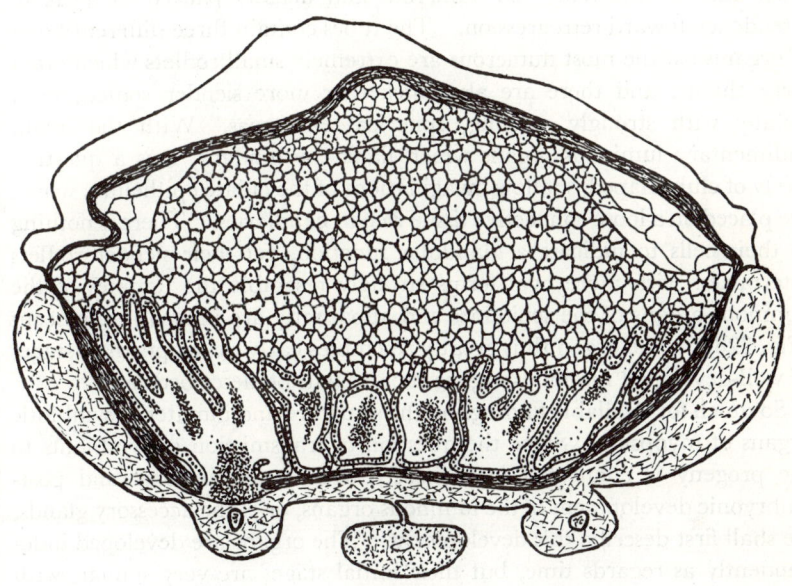

Fig. 327 *Heteroteuthis dispar* Gray. Cross section through the luminous organ. From Herfurth.

It must also be considered that research on the accessory glands, present here too in females, provided results in complete agreement with those for the other myopsids. In the heteroteuthines they are intimately fused with the nidamental glands and are in part overlapped by their opening. In some respects their structure differs from the norm; for example, a gelatinous pallium found only here calls to mind that the tissues of the deep-living oegopsids often assume a similar character. Two spatially separated types can be differentiated by the tubes: one thick-walled with numerous large and slender rodlets, and the other thin-walled with numerous small cocci (Herfurth).

The last category comprises animals with well-developed accessory glands in females, rudimentary ones in males, and rudimentary luminous organs in both sexes.   According to observations of Chun (1910), the only representative of the category is *Rossia mastigophora* Chun, which is thus distinct from the other *Rossia*, which admittedly are in need of further research.   In testing females for their bacterial content, Herfurth found partial fusing of accessory and nidamental glands, with the openings of the latter again enclasping the former in an arc.   The part containing connective tissue is by no means as conspicuous as in *Rossia macrosoma*, and the tubes are too numerous and densely packed to speak of a tendency toward retrogression.   The tubes contain three different types of organisms: the most numerous are extremely small rodlets which often form chains; and there are also thicker or more slender rodlets, both usually with strongly staining, spherical end-areas.   With the small, rudimentary luminous organs which flank the hindgut, it is a question solely of club-shaped, shallow invaginations of the ciliate epithelium which are placed on a loose connective tissue rich in capillaries.   There is nothing in their cells to indicate a glandular function, yet their slender rodlets intermixed with more oval forms strongly resemble the inmates of the luminous organs of the sepiolines and apparently are indeed luminescent (Fig. 328).   The organ thus greatly resembles early developmental stages of well-developed luminous organs such as will be described later.

So much for available findings on the structure and inmates of symbiotic organs of myopsids.   Since the question of transmission of symbionts to the progeny is connected most closely with the embryonic and post-embryonic development of the luminous organs, or of the accessory glands, we shall first describe this development.   The organs are developed independently as regards time, but their initial stages are very similar, with which, in the case of the luminous organs, the additional differentiation of the various auxiliary devices is also associated.

We are indebted chiefly to Döring (1908) for our information on the development of accessory glands.   He found their first rudiment in *Sepia elegans* in embryos which were 9 mm. in length.   They thus appear later than the nidamental glands, but greatly outdistance them at first in their development and later are outstripped themselves.   On both sides of the ink sac is a roundish, epithelial thickening, characterized by cubical, large-nucleated cells, which later increases on all sides, and in post-embryonic states, 11 mm. in length, begins to develop an extremely characteristic ridge system radiating from a central point.   These ridges, in the formation of which the connective tissue situated below does not take part, are low at first and gradually become higher.   When all the rudiments have grown to the point that they touch each other medially, the margins of the grooves grow together in such a way that they are replaced by rows of little groovelets. This process begins from the center and advances toward the periphery, and therefore the radial striation is still preserved only at the periphery (Figs. 315*b,c,d;* 329*b,c,d*). Gradually, then, the connective tissue, which has meanwhile been permeated by numerous capillaries, pushes between the invaginations which are already growing out into short tubes in the midfield and splitting down below. Only now do the invaginations develop at their blind ends the actual "glandular tubules," first in the form of spherical swellings (Fig. 329*d*).

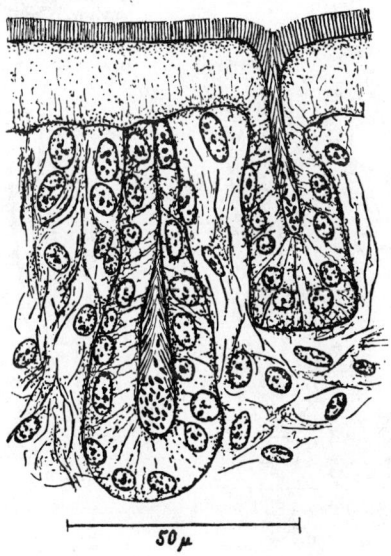

Fig. 328   *Rossia mastigophora* Chun.   Section through the rudimentary luminous organ of a female with bacteria in the poorly developed tubules.   From Herfurth.

At this time the surface of the rudiment gradually swelling into a flat cushion is covered with a ciliate covering which also lines the narrower primary and secondary invaginations, but not the tubes, which are sharply set off from them and later filled with symbionts. This process of tube formation encroaches more and more on the peripheral areas in which the organs are still developing and therefore still contain radial grooves. Finally, by concentration of the ducts, two outlet fields converging front-ward are formed (Fig. 329*e*).

According to Wülker, the tubes begin to form exceedingly early in *Loligo forbesi.* In an embryo with a pallium 1.5 mm. in length one can

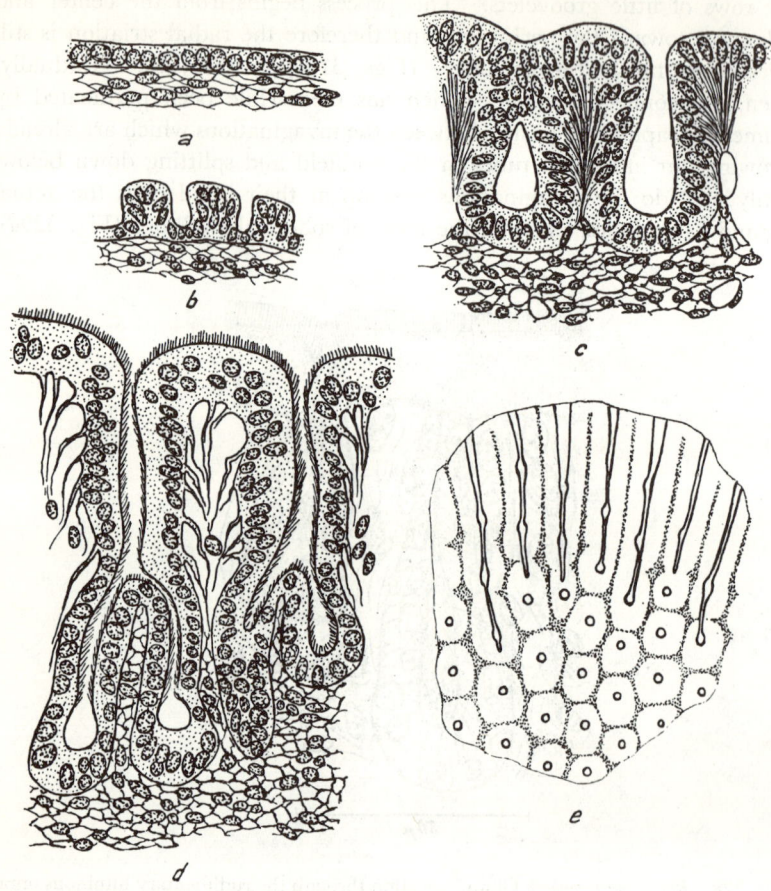

Fig. 329 *Sepia elegans* D'Orb. Development of the accessory glands. (*a*) First thickening of the mantle epithelium; (*b*) beginning of cleat formation; (*c*) progressive cleat formation, partial fusion of the groove walls; (*d*) later developmental stage, formation of the definitive tubules and appearance of cilia; (*e*) formation of pits from the original grooves, seen from the surface. From Döring.

recognize in the area of the future organs a locally defined, slight thickening of the still unciliated epithelium, which also forms widely opened, radially arranged groovelets from which the tubes develop. These grow downward to a small extent and are ciliated in their initial portion. When young stages of other species were used in research, it was found that the accessory glands were formed in much the same way.

When both luminous organs and accessory glands are present, the former develop far more rapidly (Fig. 330). Pierantoni found their rudiment in *Sepiola intermedia* in an embryo 15 days old, thus 12 to 14 days before hatching. In stages which have just left the egg capsule they are prepared in all essential parts, whereas the accessory tubes in the same test object begin to develop only in animals which have already attained a length of 1 cm. Figure 331 shows a cross section of an embryo which is almost ready to hatch. The future luminous organ is already visible as a paired proliferation of the connective tissue, also covered with cilia, in which the first folds of the epithelium are invaginated; the cushion of the future lens begins to differentiate, and the musculature resting on the ink sac already reveals the laminate structure of the future reflector. In *Rondeletiola*, where the luminous organ in females is represented only by a section of the accessory gland, the rudiment of the symbiotic organs of both sexes is developed in the same way in the form of radial furrows and only subsequently is there a separation into luminous organ and accessory gland in females (Naef).

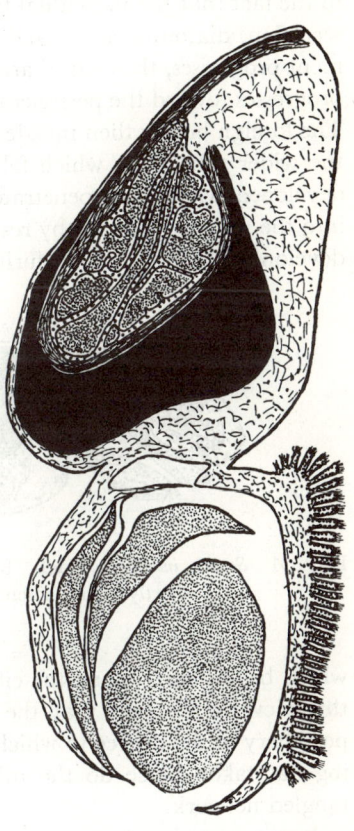

Fig. 330 *Sepiola intermedia* Naef. The luminous organ is well developed in the young animal, whereas the rudiment of the accessory glands is in an early stage of development. 20×. From Pierantoni.

Turning to the question of the route by which the new settlement of the symbiotic organs is achieved, we enter a field which is not fully clarified. First it is necessary to decide when the always type-specific bacteria populate the rudiments of the accessory glands and of the luminous organs. These rudiments open sometimes before hatching into the perivitelline substances, and sometimes at a later period into the free sea water. In this connection we must speak of a very remarkable process, one which has been described especially accurately by Döring for *Sepia* but is quite similar in other genera.   When the pallium of a young animal achieves a length of 16 mm., the rudiments of the accessory glands turn brownish to blackish except in the marginal area, a change which may be traced to the fact that the individual tubes are filled taut with sea sand, pigment secretion, diatomin shells, and other detritus (Fig. 315*c,d*).   As development progresses, the central areas lose their color again, and the pollution progresses toward the peripheral areas which at first were still fairly clean. The foreign bodies then invade the ducts which are splitting up but avoid the wider tube parts which follow.   Because the ciliate lining is not yet developed when they penetrate, Döring surmises that they are pressed into the narrow passages by respiratory movements of the pallium.   The development comes about during the filling and provides for a cleansing

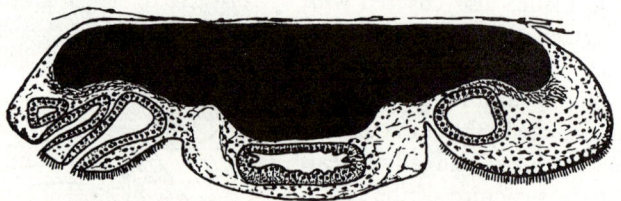

Fig. 331   *Sepiola intermedia* Naef.   Rudiment of the luminous organs in an embryo shortly before hatching.   160×.   From Pierantoni.

which begins early when the cilia are turned outward.   Also supporting this view is the fact that the cleansing process advances toward the periphery to the degree to which cilia are formed.   Only after the cleansing has taken place do the tubes develop intensely and give rise to a tangled network.

As numerous organisms are doubtless introduced with the "garbage," it would be plausible to assume that this strange occurrence serves to infect the tubes, and indeed quantities of symbionts do begin to fill the young tubules in connection with the cleansing.

We are strengthened in this view when we hear that Pierantoni actually observed such an infectious process in *Loligo forbesi*, where Wülker had already described corresponding filling with foreign bodies (Fig. 332). In young animals Pierantoni found that the rod-shaped bacteria representing the final symbionts are still mixed with much detritus, but that the latter gradually disappears and gives way to an even settlement.   Apparently the bacteria in the tubes find a favorable culture medium, originating perhaps from the surrounding connective tissue.   Using sections treated with bacterial dyes, Herfurth investigated the tubes in *Sepia officinalis,* where the temporary pollution is equally great, and determined that infection by rodlets and cocci first occurs in animals with pallia exceeding 25 mm. in length. In animals where the pallia are 40–45 mm. long the tubes are still polluted only in part, some having a uniform content and others containing a mixture of cocci, rodlets, and foreign bodies. Such gradual development of a uniform population is consistent with the difficulty encountered in obtaining pure cultures of symbionts from young animals (Pierantoni, 1934).

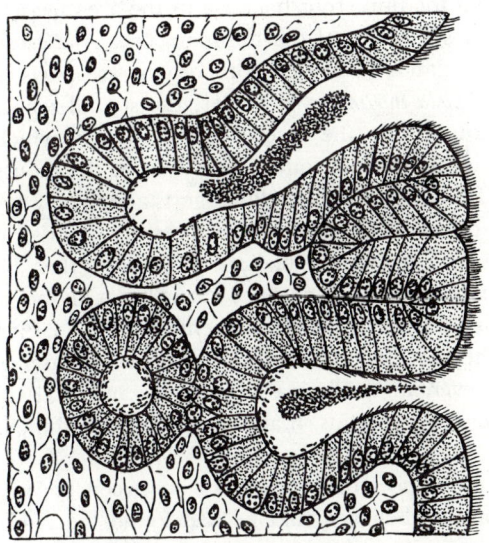

Fig. 332   *Loligo forbesi* Steenstrup.   Infection of the accessory glands from the outside. From Pierantoni.

All these facts at first glance suggest that rudiments, originally empty, are infected for the first time by sea water. No corresponding observations on temporary pollution are available for luminous organs, but the situation here is so similar, for instance in the matter of temporary ciliation, that one could hardly expect two basically different methods of infection. On the other hand, the accessory glands, as they are limited to the female sex and open more or less in common with the nidamental glands, give the impression of being transmission organs! The substances surrounding the egg after deposit originate from the glands of the oviduct insofar as they are located near them; and adjoining these oviduct glands are the nidamental gelatins furnished by the nidamental glands and the outer envelopes which are also derived from them and are blackened in part by the secretion of the color sac.

In *Sepia*, Pierantoni found the inmates of the accessory glands again chiefly in the inner gelatinous parts of the capsule wall, and he states, moreover, that they are present in great numbers in the direct vicinity of the egg. In the light orange peripheral areas of the egg capsule in *Sepiola intermedia*, which was being used for purposes of comparison, he came upon the same forms which produce an identical pigment in the accessory gland. He also found that the inmates of the luminous organs were predominant among the numerous bacteria in the area near the egg.

Obviously the symbionts of the luminous organs are united with those of the accessory glands wherever both organs are parts of a connected complex. In *Sepiola intermedia*, however, the spatial separation is rather extensive. Here we cite the case of *Loligo edulis* in which the distantly located luminous organs send out a long duct to the mouth of the accessory glands. This behavior strongly suggests a tendency to unite the symbionts of these organs with those of the accessory glands.

Herfurth, testing egg capsules of *Sepia officinalis* for their bacteria content with dab preparations, could convince himself that all five of the symbiont types differentiated by Getzel are represented. Unlike Pierantoni, he discovered that they are strictly limited to the nidamental gelatins, which they permeate even in the bursa copulatrix, and that they are always lacking in the oviduct gelatins, that is to say, in the immediate vicinity of the eggs.

Far greater differences than these exist between the views of the two authors. Pierantoni believed that he had found the symbionts again in the young embryos, if not in the egg, of *Sepia officinalis*, and he describes a diffuse infection of the connective tissue and epithelium in the young tubes of the accessory glands. According to him, these constitute the source of infection for colonization of their lumen. Thus he makes external contamination responsible for infection of the *Loligo forbesi* organs but ascribes no significance to it in other cuttlefish, although he is unable to deny that other wild forms penetrate the young organs of the *Sepia* from the outside.

Even with specific methods, Herfurth, however, was unable to detect bacteria in the tissue of the embryos up to the point of hatching. With good reason he considers the symbionts nests found by Pierantoni in the epithelium of the tubules to be nuclei in the prophase stage, and for this reason he rejects the possibility of infection at the moment of hatching. Owing to a protease secreted by the Hoyle organ, the egg membrane is obstructed locally by mucus and the remaining body is then pulled along.

Thus only external contamination would be possible at best, but, according to Herfurth, even young animals hatched in distilled water are still free of symbionts; and Pierantoni himself confesses that in young *Sepiola intermedia* which had already formed accessory glands and luminous organs he never could be sure that he had actually found bacteria.

Less important objections to Pierantoni's concept had been raised earlier. Kishitani, Skowron, and Mortara also believe that infection after hatching is far more plausible. Kishitani points out that the easy cultivation of all myopsid symbionts would suggest a new infection in each instance. Skowron frequently found cases of *Sepiola intermedia* where no light-producing capacity was observable and where the luminous organs had few bacteria. He found one specimen completely free of bacteria 3 days after its capture, a period which would allow for replenishment of the symbionts in cases where a temporary oversupply had been expelled. He believes that such findings indicate that filling is effected from the outside but occasionally fails to take place. Mortara supports the idea of such filling because she, certainly without justification, admits no difference between specifically adapted symbionts and wild bacterial flora occurring on the exterior of cuttlefish.

Finally it should not be forgotten that transmission by eggs has not been observed as yet in organisms which house their symbionts extracellularly, be it in the intestinal lumen or in any of its appendages, in integumentary organs, as in luminous fish and lagriids, or in excretory organs, as in oligochaetes. Instead, the symbionts in such cases are taken in through the mouth, through excretory pores opening into the cocoon fluid, or through glandular apertures opening into the sea water.

It must not be concealed that the hypothesis of a new infection by sea water is not always easy to apply to cephalopods. At first glance it is hard to picture how the various forms which are so specific for each species find their way into their hosts, but we must consider that the symbionts are continually being overproduced in these symbiotic organs and are continually being expelled into the environment. Some must also be freed each time they transfer from the accessory glands to the pallium and the bursa copulatrix. The organisms necessary for new infection are therefore never lacking in the environment of the individual species. Although undesired forms must also reach the organ rudiments at the time they are contaminated, such forms would undoubtedly soon retreat in favor of those which find the living conditions adequate and, through selection, the type-specific settlement would gradually be achieved. Special conditions in the various histologically distinct tube types apparently favor the development of this or that form without necessarily having the fixed objective in each case of a pure culture in the bacteriological sense.   Colonization in this form would best explain the diversity of the types in the accessory glands, in that an initially limited symbiont supply might gradually be augmented by co-optation and the organs in this way might become a refuge for all possible organisms.   Finally, what explanation could exist for this strange, indeed unique, occurrence of regulated pollution and recleansing of delicate organ rudiments which is so strictly synchronized with the appearance of the symbionts, if not that of an intended infection?

When a symbiont stock is renewed from generation to generation, it is not simply commonplace saprophytic bacteria which are taken in, as Mortara claimed.   This is not only clear from comparative study of the morphological characteristics of the symbionts, but it has also actually been proved by serological research, chiefly by Meissner and later successfully repeated by Kishitani.   In numerous agglutination experiments with *Sepiola, Rondeletiola, Euprymna,* and *Loligo,* both authors found that immune sera, obtained by injecting rabbits intravenously with living or dead symbionts, invariably yielded fundamental differences between symbiotic and non-symbiotic forms.   Moreover, the former revealed such high racial specificity that only the homologous injection strain was highly agglutinized, whereas strains cultivated from other individuals of the same species were agglutinized slightly if at all.   In Pfeiffer's experiment, specific bacteriolysins for the symbiotic *Vibrio* or *Coccobacillus* were determined in guinea pigs and sharks.   In this case also the bactericidal effect of an immune serum specific for the symbionts was directed solely toward the injection strain, whereas other strains from the same host species, but originating from other individuals, were uninfluenced (Meissner).   Such fine racial differences would also indicate discontinuity of symbiosis rather than hereditary linkage.

Whichever viewpoint one accepts, questions still remain to be answered. If, as it seems, the accessory glands do not function as transmission organs, a viewpoint supported by the fact that their inmates are again limited to the organ specific to the female, they must play some other role in the life of cuttlefish.   Although cultures of their inmates have not as yet been luminescent, we shall have to consider the possibility if luminous organs of a special kind are involved.   It is well known to fishery zoologists that intense light is emitted from the ventral side of *Sepia* females during the estrual period, and Pierantoni observed their luminescence in the aquarium during copulation.   He traces the luminescence to bacteria in the pallial epithelium, but it is just as likely that it originates from the accessory glands.   Naef, for instance, assumes that the latter are occasionally luminous in *Loligo forbesi.*   We must also consider the possibility that they radiate a light which is imperceptible to the eye, for Pierantoni

(1934) found that cultures of their inmates evoke clear-cut changes on highly sensitive photographic plates.

Obviously, the close spatial relationships of accessory glands and nidamental glands bring about a mingling of their contents, and it must therefore be significant that the egg membranes are provided with symbionts. What is the significance, however, if, as the majority of authors contend, it does not involve the infection of eggs and embryos? No satisfactory answer exists, for one will scarcely be satisfied with the bold hypothesis of Pierantoni (1935) that the symbionts send out rays which have a favorable influence on the development. Perhaps the clarity lacking at present will be provided by new and admittedly difficult research on the possible infiltration of tissues of embryos and young animals by symbionts, on the history of the bacteria in the egg membranes, and on the first colonization of the symbiotic organs.

Enigmas of a different kinds are posed by myopsid symbiosis. The apparently random occurrence of the variously constituted devices, the tendencies manifested here and there toward retrogression, and indications that symbiosis is eventually eliminated force us to phylogenetic considerations. Naef had this in mind in his long monograph on cephalopods, and such speculations acquire increased interest through the new insights into bacterial symbiosis which can no longer be denied but which Naef surprisingly ignores. We shall not enter into details until we present the phylogeny of symbiosis. Here we shall merely discuss a question connected with the significance of the accessory glands. Pierantoni and Naef agree that phylogenetic relationships exist between these glands and the highly developed luminous organs. Pierantoni proceeds from the striking tri-division of the accessory glands expressed in the variegated pigmentation, and assumes, though he later retracted the assumption, that the contents of the yellow tubes are luminescent, that this section gradually separates from the white and orange-colored areas and is raised to the status of a luminous organ, an organ which in the case of *Rondeletiola* is still imperfect and is embedded in its old environment but which is finally equipped with all devices and is more sharply separated spatially. Without considering the differentiation of the tubes, Naef also assumes a hypothetical stage in which the accessory glands constrict off a luminescent bud which is then shifted farther toward the front; but, in the behavior of *Rondeletiola*, for reasons which we cannot go into here, he sees an atavistic retrogression to older states. At any rate, such deliberations also assign to the accessory glands the status of old organs with a specific function of their own beyond their utility as simple transmission organs.

There is little positive to report on symbiosis in oegopsids, which develop an abundance of closed luminous organs.   Pierantoni was of the opinion (1919, 1920) that the rodlet-shaped and granular inclusions in the luminous organs of *Charybditeuthis maculata* Viv. represent bacteria highly adapted to intracellular life, and Shima (1926, 1927) held the same view with respect to the sharply defined, rodlet-shaped formations which appear as extremely conspicuous inclusions in the luminous center of the brachial, eye, and skin organs of *Watasenia scintillans* Berry.   However, these data have not been confirmed; indeed, they have been decisively rejected.   In a retesting of *Watasenia*, Hayashi (1927) demonstrated decisively that these homogeneous, sharply limited structures with four edges in cross section are cristalloids; Okada, Takagi, and Sugino (1933), who made exact microchemical analyses, were also convinced that they are protein crystalloids which dissolve within several months with animals fixed in formalin; Takagi (1933), who described the mitochondria located between the crystalloids, and Kishitani (1928) reject the assumption of bacterial symbiosis, and on the basis of these original Japanese preparations I can only agree.   The cultures which Shima claims to have obtained from the luminous organs in *Watasenia* were doubtless merely cultures of organisms adhering to the surface.   Mortara (1922) studied *Abralia veranyi* Rüppel without finding evidence of luminous symbiosis.

Thus we shall have to accept, first, that in oegopsids animal-specific light is produced in all these wonderful, diversiform, luminous organs located in such a variety of body parts and, second, that a contrast exists between the forms of shallow waters and those of deep waters, just as in the case of luminous fish.

In view of the apparent connection existing between the two arrangements, it is perhaps no accident that the oegopsids have neither luminescent symbionts nor organs corresponding to the accessory glands.   Yet here we must speak of a remarkable exception.   As far as we know, it is simply and solely in *Ctenopteryx siculus* Pfeffer that females possess an accessory gland!   Naef provides an illustration of a young female in which the gland, in the form of a large unpaired organ behind the ink sac and in front of the renal orifices, is far ahead of the nidamental glands in development, just as in the myopsids.   Interestingly, the furrows are also present here; however, they do not radiate as usual from two formation centers, but are arranged in the manner of a fan, with the result that fusion goes one step beyond the ordinary.   Research is lacking, but obviously a symbiotic site is involved here.   The degree to which the organ is developed in the growing female is unfortunately not yet known, but Naef assumes that its development is still more or less incomplete. It is possible that weakly developed rudiments of the organ were overlooked in the related bathyteuthids.   The large, flat, luminous organ on the eye bulb of *Ctenopteryx* (Chun, 1910) is a closed one, as in the other oegopsids. When we treat the phylogeny of cephalopod symbiosis, we shall discuss the significance of this isolated occurrence more fully.

TUNICATES

PYROSOMIDS. All pyrosomes investigated have exceedingly simple luminous organs which are almost uniform in structure. Each individual of a colony, which may be composed of thousands of individual animals, displays on both sides of the ingestion opening a mesodermal cell group which is fastened loosely to the exterior wall of the peripharyngeal blood space and radiates an unusually intense light despite complete lack of auxiliary devices (Fig. 333). Each of the plates, usually unilaminate but occasionally dilaminate and without special envelope formation, is composed of twenty to several hundred oval cells. The plates have no innervation at all. Only in *Pyrosoma agassizi* Ritter, is there occasionally a race in which several interconnected strings of luminous cells surround the oral aperture instead of the two roundish cell plates. *P. spinosum* Herdman and *P. agassizi* are also exceptional in possessing not only the generally distributed cell groups but also two additional ones constructed in the same way and located to the right and left of the cloacal opening (Farran, 1909; Neumann, 1913). This deviation is in harmony with the fact that these two forms, for various reasons, are set apart from other pyrosomes, under the designation Pyrosomata fixata.

The nucleus of the luminous cells tends to be attached to the cell wall; their plasma is taken up by numerous sausage-shaped, more or less convoluted structures which were so perplexing to earlier researchers and today are known to be bacteria (Buchner, 1914, 1919; Pierantoni, 1921, 1922, 1923). Approximately 10–30 μ long and 2–3 μ wide, they can easily be recognized even in life; their plasma has a definite stromatic structure; the more strongly staining granulations lodged in it prefer the nodal points and the marginal area particularly (Fig. 334a).

Some of the luminous cells contain additional bacteria which concentrate the chromophilous substances at one or two places, especially at the ends while the remainder gradually lose their staining and are finally destroyed. Thus arise oval, intensively staining spores which transfer to the peripharyngeal blood sinus (Fig. 334b,c).

This occurrence of spore formation, exceedingly rare among the symbionts, is connected with the transmission of the luminous symbionts to the progeny. Carried by the bloodstream, the spores reach the genital sinus, where the only egg produced by each of the hermaphrodite indi-

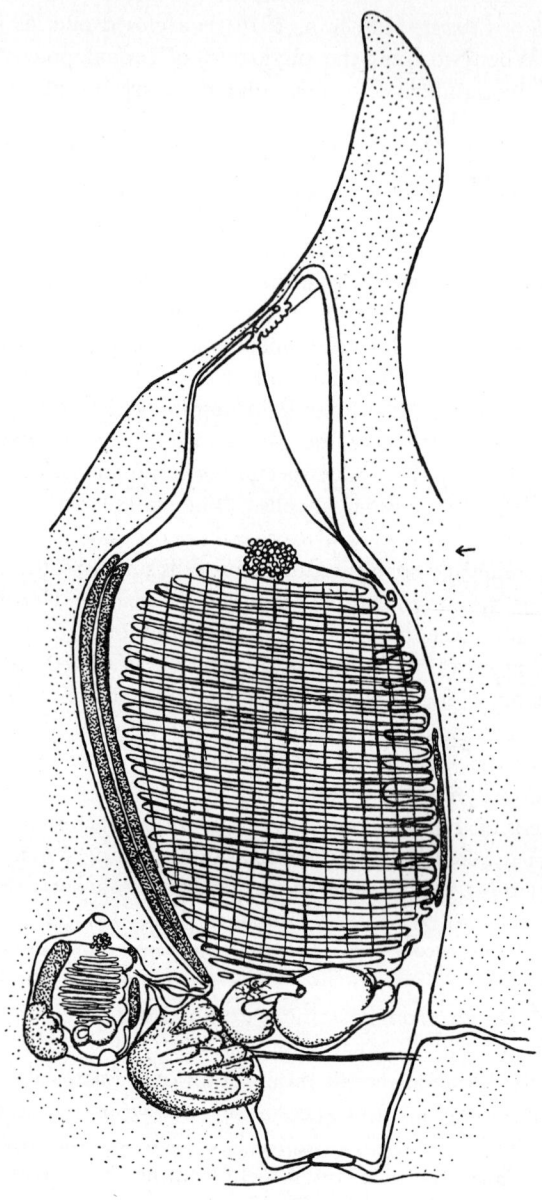

Fig. 333   *Pyrosoma* sp.   An individual of the colony in side view with the luminous organ situated above the gill-intestine; the young animal, developed on the stolon, also possesses a luminous organ.   From Seeliger.

viduals of the colony is developing.   Since Pierantoni often found them
intracellularly, it is quite likely that they are also able to penetrate cells
during this migration.   The complicated processes which now take
place in the egg and in the embryo, for the clarification of which we are
indebted chiefly to Julin (1909, 1912) and Pierantoni, reveal a surprisingly

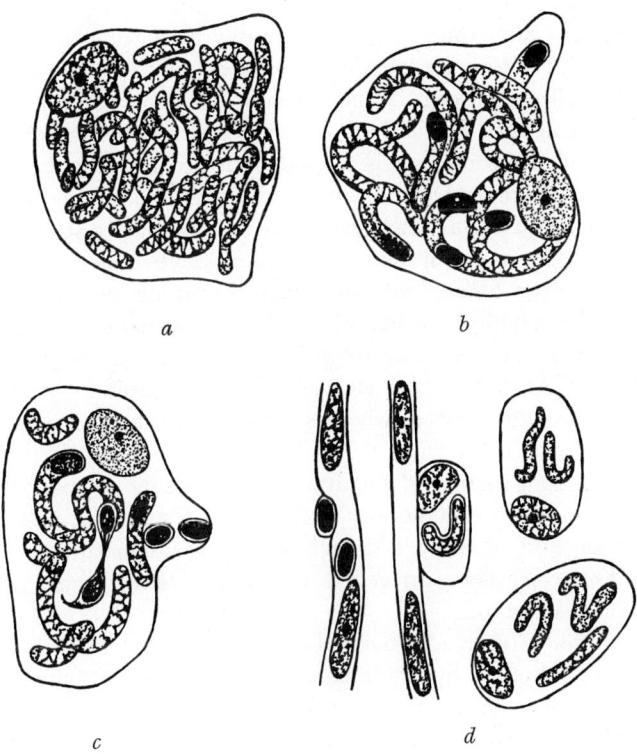

a   *b*

c   *d*

Fig. 334   *Pyrosoma giganteum* Les.   (*a*) A mycetocyte of the luminous organ; (*b*, *c*)
mycetocyte with spore-forming symbionts; (*d*) infection of the follicle by spores and
separation of the testa cells containing the reproducing symbionts.   From Pierantoni.

extensive adaptation of the luminous symbiosis to the phenomena, here so
highly specialized, of embryonic as well as asexual development.

As with other tunicates, the originally unilaminate follicle of the pyrosome egg is divided into two layers, the secondary follicle, and the testa
cells migrating eggward; but the latter lose their trophic functions and,
used only for transportation, enter into close relationships with the
luminous symbiosis.   Toward the end of egg growth, the spores appear

in the primary follicle (Fig. 334*d*). Each follicle cell provided with such a spore divides by amitosis into a secondary follicle cell and a testa cell which withdraws from the association. Assigned to the testa cell, the spore begins immediately to assume its original form. Later this one bacterium is increased again through cross-division and gradually imparts to the testa cell the character of those of the luminous organ.

Infection of the follicle and thus the origin of the testa cells, which are at first not very numerous, take place mainly at the upper pole of the telolecithal eggs. With the ascidia, these soon sink into the superficial areas of the still uncleft egg; in the pyrosomes it is only in the course of development and very gradually that they transfer to the embryo. The cleavage of the pyrosome egg is discoidal, in accord with the unequal distribution of its yolk. During the first divisions the infected testa cells glide into the constrictions which take place as the plasma splits, and are imprisoned there, so to speak. With progressive cleavage they sink deeper and deeper among the numerous dwindling blastomeres until blocked by the adjacent uncleft yolk. At first forming regular chains between the cleavage cells, they are gradually distributed more and more among them. At the end of cleavage few testa cells may be found between follicle and embryo, whereas it is possible to find as many as 40 to 50 inside the latter (Fig. 335). The older authors who inevitably observed this participation of somatic elements in the formation of the germ layer assumed active motility of the testa cells. Julin, who studied their behavior most carefully, although he did not fathom the true nature of their plasma inclusions, arrived at the opinion that they are passively drawn into the embryo.

With the end of cleavage, the follicle infection and transfer of testa cells by no means cease, but the area of their formation is shifted. Even the testa cells left in the embryo at end of cleavage do not all arise from the original polar field of formation. Some take rise from a new ring-shaped area analogous to the circumference of the germ disc and therefore take up a more peripheral position without traversing the disc. During gastrulation, the germ disc enlarges, flattens, and epibolically enclasps the yolk syncytium which corresponds to the endoderm, thereby pushing the testa cells, which are still interspersed in the germ disc, toward the periphery and strengthening the ring zone which up to this point was less clearly defined.

Now the nervous system, pericardial organ, and peribranchial spaces take form, the cloaca is invaginated and gives rise to the short-lived, incomplete oozooid, the first and only sexually produced individual of the future colony. Yet there is no essential change in the distribution of the testa cells. The oozooid still lacks the well-defined luminous organs, but already it contains a substantial part of the cell material destined to build up the eight luminescent plates of the four primary ascidiozoids which bud asexually on the oozooid.   Their origin is introduced by the sprouting of a stolon which is immediately joined by some of the organ rudiments of the oozooid (Fig. 336).   Evaginating on its ventral side, it leads to an interruption of the testa cell ring.   The follicle ceases to give off further testa cells at this place, but by way of compensation forms them laterally

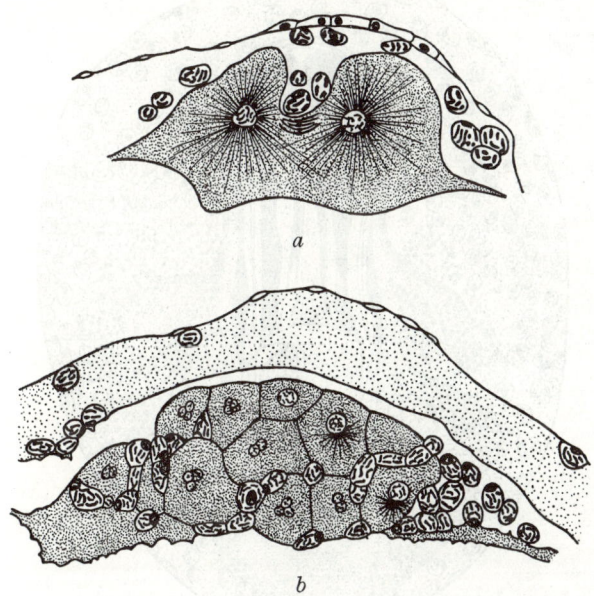

Fig. 335   *Pyrosoma giganteum* Les.   (*a*) First cleavage division with infected test cells; (*b*) advanced cleavage stage; the infected testa cells between the blastomeres and in the border zone around the germ disc.   From Julin.

to an increased extent.   Indeed, it is only now that the provisioning of the germ layer with migrating luminescent cells reaches its maximum, the number having risen to approximately 400.

Julin and Pierantoni agree that no division of testa cells ever takes place during this period and that they all owe their origin to infection of a follicle cell.

Originally straight, the stolon places itself diagonally around the oozooid and is constricted off into the rudiments of the first ascidiozoids. When these acquire six to eight pairs of transverse gill slits, the testa cells begin to circulate and build up their luminous organs, which are identical with those described at first, except that their cell material is not specific to the animal possessing them but represents an implant of grandmother elements!

With construction of these eight luminous organs, the cells which have survived so long are finally destroyed.   A new stolon forms now at the pos-

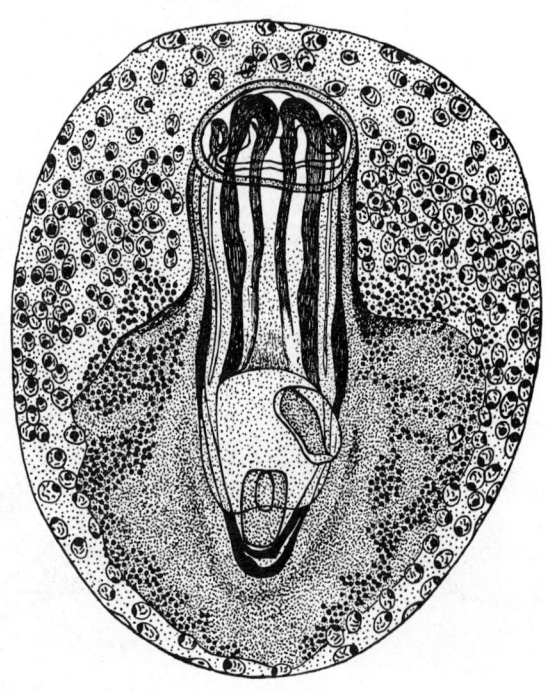

Fig. 336   *Pyrosoma giganteum* Les.   Formation of the stolon on the oozoid.   The infected testa cells in the latter are arranged in horseshoe shape.   From Julin.

terior end of the pharynx in each individual and constricts off the ascidiozoids which determine the further growth of the colony and gradually push in between the older individuals; according to Pierantoni (1923), their luminescent cells are furnished by the so-called dorsal organs of the mother animal.   Individual elements of this mesenchymatous cell group situated in the dorsal sinus are detached from it, take up bacteria which are migrating from the luminescent cells, and transfer to the stolon. Pharynx, pericardium, nervous system, sexual rudiment, and peribranchial pouches in the form of excised cords are extended directly into the stolon. Such migrating elements of the dorsal organ then unite to form luminescent plates which at the time of pubescence will again send spores to the eggs for transmission to new colonies.

The inclusions of the migratory testa cells are identical with those of the luminous organs, for observation of living developmental stages in the dark shows that the points which become luminescent in the embryos on addition of ammonia always correspond exactly to their arrangement.

In general, the pyrosomes illuminate only on stimulation and then remain dark for a while. Under natural conditions it is mechanical stimuli which are most prominent. Pyrosomes set into motion by storms or the ship's propeller are intensely luminescent, but even the gentlest contact with the tip of a colony is sufficient to evoke the phenomenon. In such cases the light usually travels from one end of the colony to the other in a few seconds, although certain sections may be bypassed occasionally or the stimulus may be conducted very slowly. An English zoologist reports that he could write his whole name in fiery letters on a large colony! The luminescence can also be induced chemically by alcohol, ether, ammonia, fresh water, etc., and even by induction currents (Polimanti, 1911; and others).

Contrary to expectation, the organs also respond to light stimuli. Burghause (1914) showed that one pyrosome colony can be ignited by the light of another in a neighboring glass jar and that the same effect can be obtained through other luminous animals or by the light of a match. Presumably the light stimulus is received by the primitive ocellus next to the ganglion and then is conducted to the luminous organs. It is not clear precisely how this comes about. As mentioned, neither a colonial nervous system nor an innervation of the simply constructed cell group has as yet been demonstrated. On the other hand, the cloacal muscles are connected with one another by strands of fibers which appear well adapted for transmitting a contraction stimulus. Such stimulus will necessarily result in more rapid circulation of the respiratory water and in a greater supply of oxygen to the light organs which produce the luminescence in the symbionts.

Many impressive experiments show the amount of oxygen required by luminescent bacteria and indicate that minimal quantities suffice to cause temporary luminescence. We shall merely cite that of Beijerinck (1889). He proved that the light furnished by striking a match is enough to produce luminescence in a culture to which chlorophyll-containing extract is added after the available oxygen has been consumed. In this manner a highly sensitive indicator is made available for detection of minute quantities of oxygen produced by algae.

Surprisingly, a sustained luminescence is often produced by dying animals already in the process of decay and therefore certainly poor in oxygen. One is inclined to conclude from such cases that in healthy pyrosomes it is not only the lack of oxygen but also other unusual conditions which prevent even weak, continuous glimmering unless the organs are stimulated by more animated respiratory movements.

Pierantoni, in cooperation with Zirpolo, reported on cultures of *Pyrosoma* symbionts on agar mixed with *Sepia* broth. Admitting that his test object is one which is difficult to sterilize, he is nevertheless convinced that the organisms cultivated were identical with the symbionts. They had the shape of coccobacilli and thus were more similar to the strongly staining spores.

SALPIDS. As is well known, the pyrosomes are not the only luminous tunicates, for the salpids, Appendicularia, and doliolids are also luminescent. The latter light up on the whole surface, according to what little information exists, and the light emitted by the Appendicularia appears to be animal-specific. At any rate, there is every probability that luminous symbiosis is present among the salpids.

Well-defined luminous organs are found exclusively in the genus *Cyclosalpa*, which are distinguished from other salpids through the wreath-shaped arrangement of the sexual animals. It is a matter here of elongated, sharply described sections of the blood vessels lodged on both sides of the body wall, either singly or in a chain divided into five cords. At times the solitary salpids and the chain salpids of the same species are different in this respect. In *Cyclosalpa pinnata* Forskal, in which Chamisso discovered metagenesis and which has received detailed study only with regard to possible luminous symbiosis, the solitary form has two light organs, each with five divisions, and the chain form has two short, undivided light organs. They are penetrated by columns of connective tissue and at the same time represent centers of blood formation.

In the remaining salpids the light is emitted mainly from the so-called nucleus, that is, the *u*-shaped or convoluted section of the gut situated behind the pharynx and presumably is also connected with the larger accumulations of blood cells forming there.

In the salpids, therefore, the luminous and hematopoetic organs are not so sharply divided as in the pyrosomes, where blood-forming centers are created in the lateral organs.

Histologically, the luminous organs of *Cyclosalpa* present a heterogeneous picture. Julin (1912) had noted that among the variously structured elements of the blood, there are some in which the inclusions have an extraordinary resemblance to the testa cells which infect the embryos in pyrosomes. Their numerous inclusions, which may be roundish or longish but always small, push the cell nucleus to the wall, exactly as in the pyrosomes (Fig. 337).

When it was discovered that the structures found in the luminous organs and the infecting testa cells of the pyrosomes were bacteria, it was natural to use this knowledge to trace the symbiotic cycle in salpids, for, as with the pyrosomes, unexpected findings in embryonic development were available here which might conceivably be explained as phenomena resulting from symbiosis. Unfortunately, I must say in advance that, in spite of my own efforts (1930) and those of my pupil Stier (1938), the subject of luminescence in salpids, though its existence can no longer be questioned, has by no means been eluci-dated in all its aspects.

All the numerous authors who dealt with the development of salpids had observed the strange inclusions occurring in all species in the blastomeres of early cleavage stages. Such inclusions were interpreted as plasma fragments fore-shadowing degeneration, as daughter blastomeres arising through endogenous budding (Salensky), as remains of con-sumed follicle cells (Heider, Brooks), as vitelline spheres (Korotneff), and as secretions condensed into plasma vacuoles as a result of intensive nourishment of the embryo. Such bewilderment brings to mind the various interpretations given

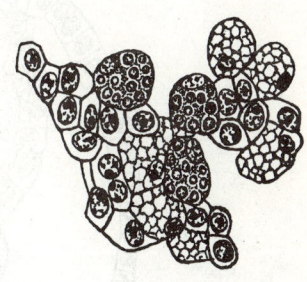

Fig. 337 *Cyclosalpa pinnata* Forsk. Blood cells and three cells from the lateral organs filled with the supposed lumi-nous symbionts. From Stier.

to the contents of the luminescent cells of the pyrosomes, the pseudovitellus of the aphids, and to many other structures which we have finally come to understand as symbionts or sites of symbionts.

Like the pyrosomes, the salpids are viviparous. The first cleavage stages of the egg, which is generally formed singly by each individual of the chain, are surrounded on all sides by a well-developed follicle and are anchored only at one specific place with a cushion consisting of especially tall cells of the unilaminate epithelium. The cleavage of the yolk-poor eggs is one which is bilaterally symmetrical throughout and strictly deter-mined with respect to size and position of the blastomeres and to their divi-sion rhythm. A six-cell stage is followed by a twelve-cell stage, and only then do the bodies in question emerge for the first time. Eight specific cells always contain the bodies, whereas the other four remain free. In my opinion there is every indication that it is around this time that they penetrate the blastomeres from the blood sinus surrounding the follicle.

Apparently this process takes place very rapidly, for it has been impossible to trace it in the preparations. The earliest stage discoverable showed several of these inclusions in only four blastomeres close to their surface (Fig. 338). For the sake of brevity, these inclusions will be designated as symbionts from now on. At any rate, they do not appear spontaneously in all cells in the ultimate quantity, and this is indeed exactly what one would expect in the event of an infection. Roundish or oval in shape,

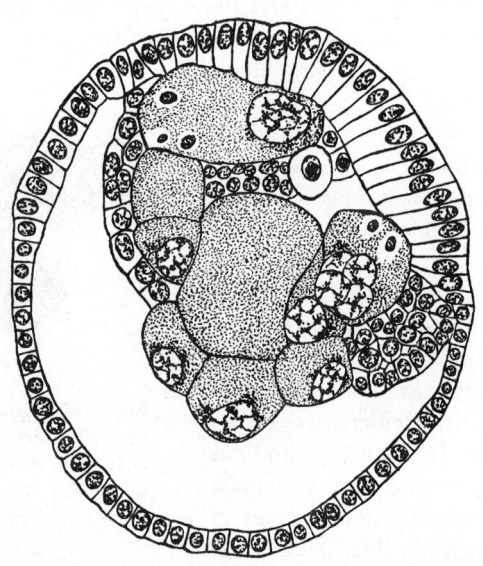

Fig. 338 *Cyclosalpa pinnata* Forsk. Sagittal section through a 12-cell stage. The first symbionts appear in the blastomeres; the follicle cells begin to grow around the blastomeres and to penetrate between them; upper right, the three polar bodies lying against the follicle cushion. From Stier.

permeated by more strongly staining granules, and only slightly defined at first, they are always situated in a lighter plasma area and bear a strong resemblance to the spores of the pyrosomes which swell in connection with infection of the testa cells.

By the time all eight of the blastomeres are infected, the symbionts have become somewhat elongated and now, usually arranged like a fan, occupy a substantial part of the chiefly pear-shaped cleavage cells in which the nucleus is situated at the plasma-rich, tapered end (Fig. 339).

During the extremely slow progress of cleavage the follicle cells increase rapidly and, as so-called calymmocytes, make their way between the blastomeres into the interior of the morula, forcing it apart and surrounding it in the form of easily distinguishable, small elements. A continuous covering of follicle cells is formed also on the surface. Thus the originally spacious follicle hollow, in which insemination took place, is more and more reduced and eventually disappears completely. In the form of a "uterus sac," its epithelium then is attached to the embryo, which is penetrated and covered by somatic elements (Fig. 340).

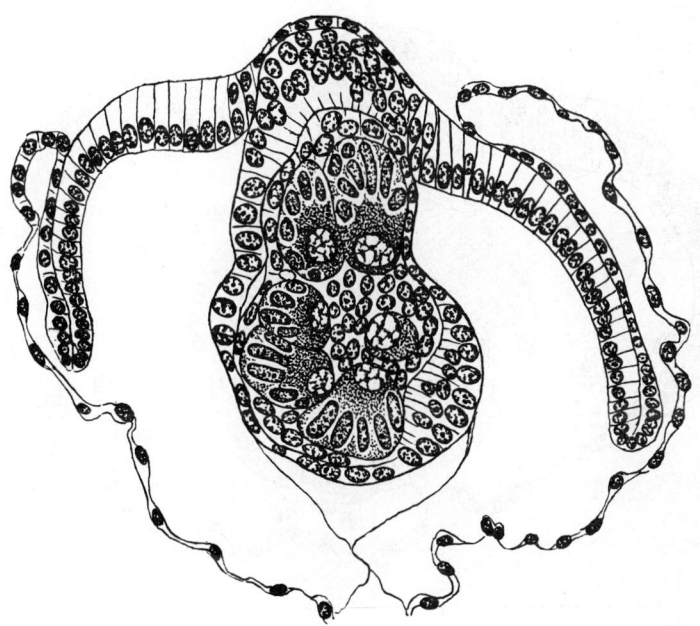

Fig. 339 *Cyclosalpa pinnata* Forsk. Sixteen-cell stage. Numerous follicle cells between the infected blastomeres; folding of the respiratory-cavity epithelium. From Stier.

In addition to these changes there are simultaneous changes in the cloacal epithelium bordering the surrounding blood sinus. A section already distinguished by taller cells, which protrudes into the cloaca as a so-called epithelial mound, is more sharply set off by a ring fold and, like a bell and its clapper, surrounds the germ layer, called the embryonic massif (Brien), which at this time is still at the sixteen-cell stage (Fig. 340 before formation of the ring fold; Fig. 339, early stage of its formation). In this way a placenta begins to develop, which we shall discuss later.

There are various opinions about the further development of the "embryonic massif" synthesized from heterogeneous materials. Salensky

(1882, 1883) was convinced that the largest part of the salpid body is constructed from maternal calymmocytes and that the blastomeres are destroyed in the course of development and serve the embryo as food ("follicular budding"). Heider (1895) conversely interpreted the calymmocytes as being resorbed by the blastomeres, their inclusions, as already mentioned, as remains of their nuclei, and elements which generally are assigned to the follicle as micromeres originating from the egg. Brooks

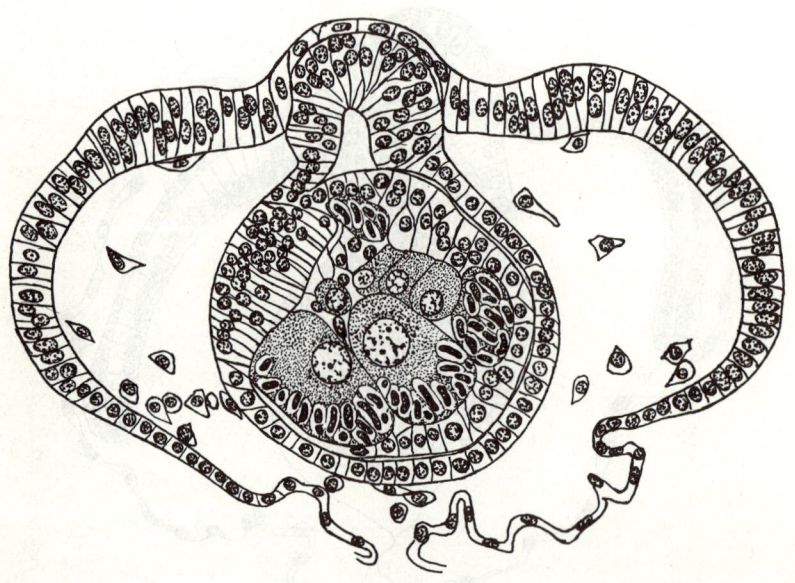

Fig. 340   *Cyclosalpa pinnata* Forsk.   Sixteen-cell stage with advanced infection.   The follicle cells surround the blastomeres on all sides; the follicular cavity has disappeared; the respiratory cavity epithelium lies over the follicle.   From Buchner.

(1893) also arrived at the conclusion that the embryo is synthesized in all its essential parts from the follicle cells. According to his view, the increase of the blastomeres is an unusually long process during this period, but they too are eventually situated in small numbers at the sites of future organs, where they increase rapidly and build up the final tissue. Meanwhile the follicle cells, which, as it were, had made a quick sketch of the future organ, are replaced cell by cell and finally resorbed. Brooks also interprets the mysterious inclusions of the cleavage cells as remains of consumed follicle cells.

Salensky (1916, 1917, 1921) also adopted this concept but rejected the assumption that the blastomeres have normal capacity for division. According to him, the daughter blastomeres are hampered by the surrounding calymmocytes and originate through an unusual endogenous budding. The plasma, he said, is cut to pieces in this way and the fragments, our bacteria, are provided with nuclear parts. In this way a few large blastomeres come to possess numerous small ones which later become free and displace the previously dominant somatic elements. Davidoff (1928) accepted this hypothesis on the basis of Salensky's preparations.

Korotneff (1896), who also believed that the new organism originates solely from blastomeres, interprets the framework of the follicle cells not as actual forerunners of the final organs, but as a framework which merely simulates these organs, and Brien (1928) speaks of a kind of "follicle gall" (blastophore), in and from which the derivates of the egg form the embryo. He correctly rejects a phagocytosis of the follicle cells by early cleavage cells as well as an endogenous budding on their part, but he makes the error of interpreting the mysterious inclusions as vitelline spheres. It seems from the first most unlikely that such vitelline spheres should suddenly appear in a certain group of blastomeres during cleavage.

When I persuaded Stier to reinvestigate the history of the blastomeres, I considered it possible that at least a part of the embryo would be supplied by the follicle cells. After finding that the cleavage of the salpids is highly determined in character and thus has only a limited regulation capacity at the very most, it seemed plausible that somatic cells would compensate for the deficiency which results when the symbionts appropriate a substantial part of the blastomeres. That something of this nature is theoretically possible and can be induced by the existence of symbiosis is shown by the pyrosomes, in which the eight luminous organs of the first four ascidiozoids are unquestionably constructed from the follicle cells.

Stier's observations failed to confirm my hypothesis. In fact, she too came to think that a provisory somatic organ structure is replaced finally by derivatives of the egg. On the other hand, she studied the individual cleavage steps of the blastomeres and the history of their inclusions much more exactly than her predecessors did and thus contributed substantially to our understanding of the symbiotic cycle.

As already mentioned, the blastomeres divide very slowly during proliferation of the follicle cells. The twelve-cell stage with four non-infected cells is followed by a fourteen- and a sixteen-cell stage in which ten or twelve cells are infected and four cells are always free of inclusions. The latter are also distinct from other blastomeres because of their nuclear structure and in all probability represent sexual cells. In the twenty-four-cell stage, twelve cells are infected and twelve are symbiont-free. The first are composed of two large "intestinal blastomeres" which separate early and remain undivided for a long time. Such blastomeres represent the rudiment of the epithelium of the respiratory cavity, of the gut, of the gills, of four cells equivalent to the pericardial rudiment, and of six cells the derivates of which participate in building the nervous system. Eight of the elements of the neural rudiment remain sterile in addition to the four which are probably the original sexual cells. This relationship of infected and non-infected blastomeres is also preserved later. In a forty-two-cell stage, for example, twenty-two cells have inclusions and twenty do not. As always, the four germ cells stand out particularly.

There is nothing to indicate that additional supplies of symbionts are received from without, the enrichment of the infected blastomeres taking place by division of those which are already filled with symbionts. The non-infected elements are increased not only by division of sterile cells but also through bud-like separation of sterile daughter cells from infected blastomeres, a process which is facilitated by the eccentric position of the nuclei in the bacteria-free plasma area (Fig. 341a).

While the blastomeres are slowly increasing, other complicated changes take place in the environment. They will be described here only in brief; for details the reader is referred to Ihle (1935), who considered the existence of a light symbiosis entirely possible. On the ventral side of the germ layer, a space, which is separated by the plafond of the placenta from the placental cavity beneath it, arises through renewed gaping of the erstwhile follicle cavity, which has been reduced to a virtual slit. At the same time the calymmocytes are arranged much like an epithelium; cavities appear in the mass of the germ layer which was solid up to this point; and, deeply indented, folds open into the newly arisen follicle cavity (supra-placental cavity). Blastomeres in the form of nests are now distributed over the framework of somatic cells which forms in this way. The intestinal blastomeres are added to the cavity which is arising within the calymmocytes and is analogous to the future intestinal lumen; other cell groups correspond to the rudiment of the pericardium and of the nervous system, etc. (Fig. 342a). Since the germ layer is still connected with the former embryo sac only in its dorsal region, horizontally placed sections provide the picture so typical of salpid development, namely a cross hovering free in the secondary follicle cavity (Fig. 342b).

This is not the place for a detailed account of the further development of the placenta that takes place at the same time. A ring-shaped proliferation of the lateral wall of the secondary follicle cavity, thus of calymmocytes, usually gives rise to the so-called placental plafond. Growing into a voluminous formation, it greatly contracts the follicle cavity which is separated into two spaces. As a result of close relationships with the maternal circulation, the greatly enlarged placenta, which is still attached to the already well-developed embryo as a voluminous body, finally becomes an important nutrient organ replacing the gradually decaying calymmocytes.

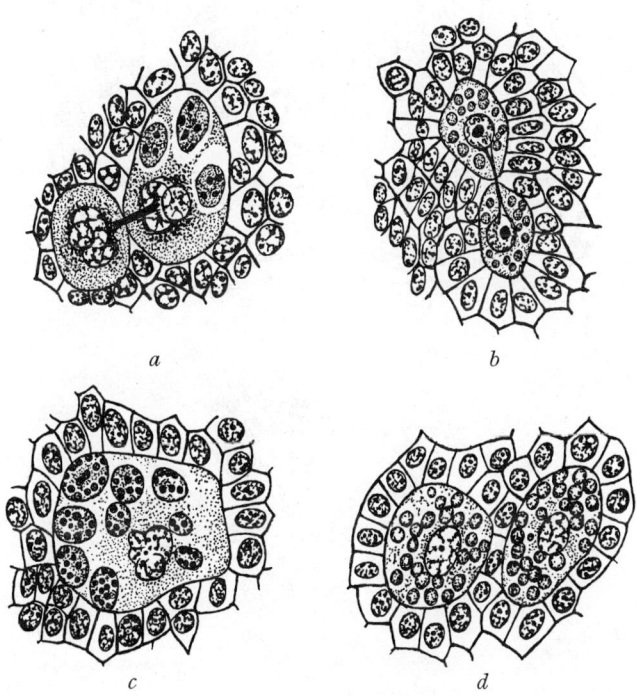

Fig. 341 *Cyclosalpa pinnata* Forsk. (*a*) One sterile and one symbiont-containing blastomere arise through division; (*b*) an infection blastomere distributes the symbionts, broken up into small vesicles, to the two daughter cells; (*c*) the symbionts of a blastomere guide the decomposition products into small vesicles; (*d*) blastomeres containing vesicular-shaped symbionts. From Stier.

Since the calymmocytes are markedly eosinophilic and the cells derived from the blastomeres are basophilic, the history of the two cell types can be easily traced. At the surface of the embryo the provisory ectoderm cells furnished by the follicle give way in increasing degree to the definitive cells. The task of lining the cavity which appears in the massive cloacal rudiment is also gradually taken over by embryonic elements. The remaining embryonic rudiments similarly develop at the cost of the calymmocytes and more and more form a connected whole (Fig. 343). Hand

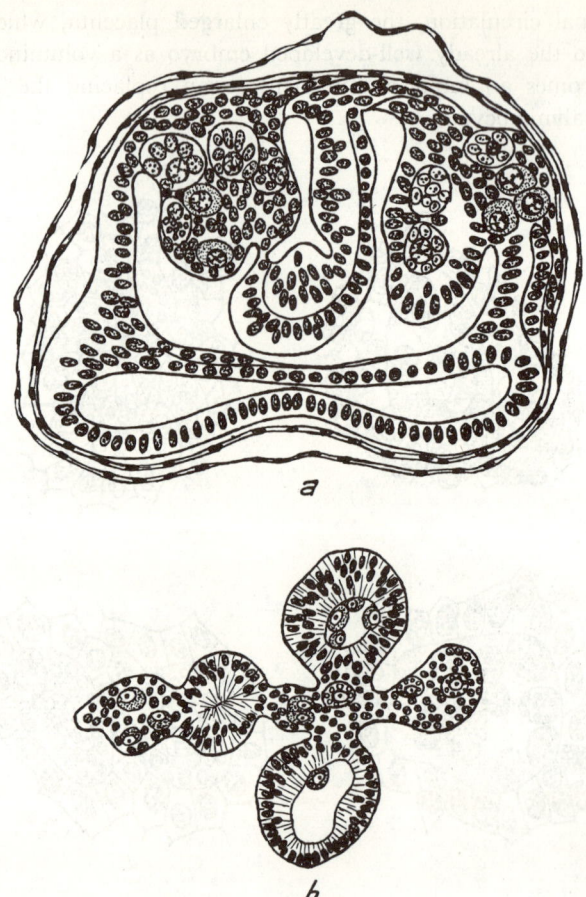

Fig. 342   *Cyclosalpa pinnata* Forsk.   (*a*) Longitudinal section; the secondary follicular cavity is developed, the blastomeres appear distributed over the germinal framework which forms the folds and cavities and which consists of follicle cells; (*b*) the same stage; a horizontal section shows the cross formation; the blastomeres represent (from right to left) the rudiments of the pericardium, the gut, and the nervous system.   From Stier.

in hand with this, the structure of the calymmocytes loosens, cavities arise in them, and at the periphery they are gradually detached while forming amoeboid processes, and drift around individually in the follicle cavity.

From our point of view, the behavior of the conjectural symbionts is of special interest.   Approximately at the thirty-cell stage a substantial change takes place.   Previously of great size and small in number, they are transformed by a type of endogenous increase into many small, roundish

vesicles surrounding the nucleus on all sides (Fig. 341*c*,*d*).   They give
the impression of being thoroughly sound and do not resemble products of
decay at all.   Apparently a second, more intensive wave of infection sets
in later, for in a slightly more advanced stage all the previously sterile cells,
with the exception of the sexual cells, contain such vesicle-shaped formations.   They are stored up gradually, and cells in which only a single or

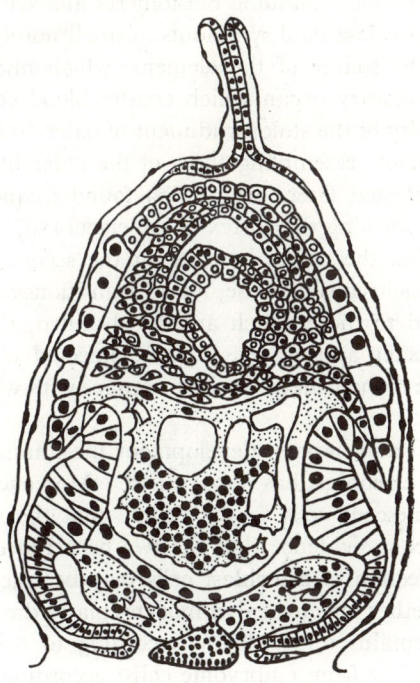

Fig. 343   *Salpa fusiformis* Cuvier.   Cross section of an embryo, in which the replacement of follicle cells (with black nuclei) by embryonic cells (with white nuclei) is taking
place; ventral to the embryo are the greatly swollen and vacuolated placental cover,
the supraplacental cavity, the placental "plafond" and the placenta.   From Heider.

very few inclusions are present strongly suggest the first infection stages
of the cleavage cells.   Stier, whom we are following in all these details,
was unable to decide whether additions from the maternal blood are
involved or whether the inmates of the blastomeres themselves represent
the source.   At any rate, it will be recalled that the luminescent bacteria
of the pyrosomes also transfer to the embryo over a protracted period.

   This temporary state of a general infection is followed by a phase in which the embryonic cell material is completely freed of symbionts. According to Stier this is achieved in various ways. Some infected cells transmit their contents to neighboring calymmocytes, while others divide into an infected and a non-infected daughter cell, the former falling sacrifice to degeneration. Also, anucleate symbiont-containing plasma fragments are often found among the follicle cells.

Among the calymmocytes floating free in the follicle cavity, one finds not only the cells which consume blastomeres and symbiont remains but also some which enclose vital symbionts in small numbers. These are in all probability the source of the elements which finally appear in the elaeoblast, a temporary organ which creates blood cells and now takes form in the vicinity of the stolon rudiment of older embryos. Cells, with inclusions completely resembling those of the older blastomeres and the calymmocytes infected later, are at first found frequently between the usual blood cells forming the cortex of the elaeoblast, predominantly and significantly where the organ borders on the secondary follicle cavity. Apparently the inclusions increase, for all transitions are found from cells sparsely provided to those which are thickly filled, the latter therewith acquiring the exact appearance of those infected elements which are found among the blood cells of luminous organs with which our discussion began.

In both sexual and asexual development the luminous organs do not appear until the elaeoblast has retrogressed. Moreover, they fill up with blood and luminescent cells to the degree that the elaeoblast shrivels. Thus there is strong evidence that the two occurrences are interrelated. Whether the infection of elaeoblast cells may be traced to free floating symbionts or whether infected calymmocytes play a direct part in building up the organ remains undecided. According to Brien, the elaeoblast originates exclusively from embryonic cells; according to Salensky, from embryonic and follicle cells jointly. Salensky's hypothesis encompasses the possibility that at least the luminescent cells of the solitary salpids, like those of the first four individuals of a pyrosome colony, may be derived from maternal body cells.

Painstaking complicated research is needed to corroborate all aspects of the symbiotic cycle as presented here. This is especially true for phases in which the inclusions first appear in the blastomeres and those in which they are later eliminated, as well as for the apparent role of the elaeoblast as an organ which supplies symbionts to the luminous organs, or, where the latter are lacking, to the blood. The course of the symbionts in the stolonial budding should also be investigated. Although it is apparent from figures available in the literature that the same inclusions are found in the blastomeres of all salpids which lack an actual luminous organ, they should nevertheless be compared once more from the stand-point of symbiosis. Certain differences would at least come to light. For example, for *Salpa zonaria*, Korotneff specifies rod-shaped inclusions, in other words, forms similar to typical bacteria.

Now that it has been proved that in the last analysis the new organism is constructed from derivatives of the egg, we are more than ever convinced that the unusual quality of the phenomenon has its basis in a symbiosis with luminescent bacteria. We know of no other case in which the blasto-meres of an egg in total cleavage are appropriated by the symbionts, and it is obvious that in a test object with determined cleavage and extensive infection this appropriation of the blastomeres must cause severe disturb-ances in the course of development. The cleavage cells do indeed appear paralyzed; as so often occurs when embryonic cells are laden with symbi-onts, the tempo of their division is maximally reduced; somatic cells of the maternal body, which generally possess extensive reproductive capacities in tunicates, step into the breach and build a provisory substrate of the future organism. The substrate is not replaced until the embryonic cells are in a position to transmit their inmates to the elements of the blood which constitute their definitive site and thus make it possible for them to assume their actual organ-building tasks.

TELEOSTEI

ANOMALOPIDS. The two genera *Anomalops* and *Photoblepharon* represent the first Teleostei in which luminous symbiosis was determined. The structure of their strange organs was described in detail by Steche in 1909, but the bacterial nature of their luminescent content was not recog-nized until much later (1921, 1922) by Harvey. Abe (1942, 1951) made a study of *Anomalops*. *Anomalops katoptron* Bleeker and *Photoblepharon palpebratus* Boddaert, the only known species of the two genera, have the typical dark-brown coloring and the appearance of deep sea fish but actually live in shallow atolls of the Banda Islands where the organs, which continue to be luminescent for a while after death, are used by the natives as fish bait. *Anomalops* is also found in other areas of the South Seas: in Celebes, the Fiji Islands, the New Hebrides, etc.

In both cases there are astonishing, highly conspicuous organs, ap-proximately bean-shaped, which are situated close to the lower margin of the orbit and take up one-eighth to one-tenth of the total length of the little fish (Fig. 344*a*). Tinged light yellow, they are sharply set off from the darkly pigmented surroundings. The continuous light emitted by them can be arbitrarily dimmed. *Anomalops* continually rotates the

organs around a cartilaginous pedicle toward the bottom of the eye socket
to the extent that they are luminous for 10 seconds and dark for 5; *Photoblepharon* merely draws an underlid fold over the organ occasionally.

The actual luminous body consists of numerous glandular tubes, perpendicular to the surface and separated by a small amount of connective

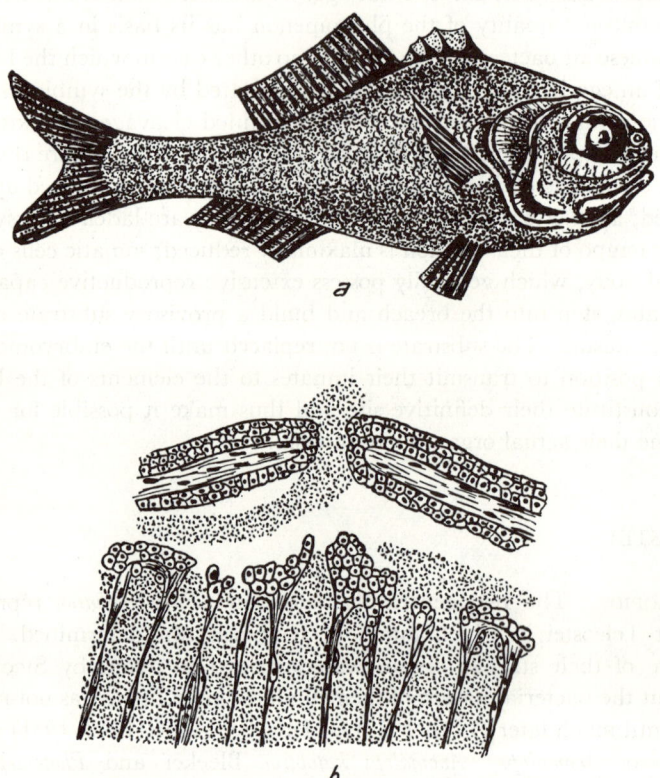

Fig. 344   *Anomalops katoptron* Bleeker.   (a) Total view with luminous organ under the
eye; (b) section from the organ colonized by bacteria.   From Steche.

tissue filled with capillaries.   Subdermal, flat, collecting spaces receive
a number of such tubes, and each is connected with the outside by a pore
(Fig. 344b).   The otherwise unilaminate epithelium which lines the tubes
passes into the multilaminate epithelium of the skin.   Posteriorly a bowl-
shaped reflector composed of connective tissue and guanin-containing
cells encloses the entire light body.   The organs are well provided with
blood.   At regular intervals a major artery extends branches of first

and second degree. Also a rather prominent nerve from the trigeminus-facialis complex approaches it.

Harvey had the perspicacity to recognize that the luminous filling of the glandular tubes consists of numerous rod-shaped, motile bacteria, some of them linked into spore-forming chains. He achieved good growth on peptone agar but, as occasionally happens with other luminescent bacteria also, the light production ceased. When fresh-water or other reagents with cytolytic effect are added to flotations of the symbiotic bacteria, they are dimmed immediately. As usual with luminescent bacteria, their requirement for oxygen is very great. When remoistened after drying out, the organs light up weakly. Luciferin and luciferase could not be detected.

Migration of bacteria from the apertures of the luminous organs was observed neither by Steche nor by Harvey, but it may very possibly take place under certain circumstances and compensate for an overproduction of the symbiotic microorganisms. Haneda (1943, 1953, 1955), at first finding no ducts, doubted the existence of bacterial symbiosis; finding them later, he acknowledged its existence. His reports on the subject are accompanied by photomicrographs.

Unfortunately we know no more about the development of the luminous organs than we do about their infection by the symbionts. Harvey merely established that eggs taken from mature females are not luminescent and cannot be made so by any of the usual stimulants.

MONOCENTRIDS. The *Monocentris japonicus* Houttuyn, a knight fish common in Japanese coastal waters, is about 12 cm. in length and strongly compressed laterally. At the anterior end of the lower jaw, two oval luminous organs, touching in the median line, give the underlip a swollen appearance (Fig. 345*a*). They are transversely placed, cushion-shaped protrusions approximately 4 mm. in length. These organs have usually escaped the attention of ichthyologists because their surface is densely set with dark-brown, pigmented papillae and is therefore only slightly set off from the surroundings. The construction has been described in detail by Yoshizawa (1916) and preeminently by Okada (1926). As with the anomalopids, it is a case of two complexes of numerous, slender evaginations opening toward the outside (Fig. 345*b*). The tubes, opening into special collecting spaces and approximately nine in number, are separated by septa which in turn are penetrated by trabeculae. Each has an individual narrow passage to the outside. A transversely placed slit contains the individual pores with which these canals end and from which the luminous material is pressed out (Fig. 346).

A tall plasma-rich epithelium forms the glandular distal sections which

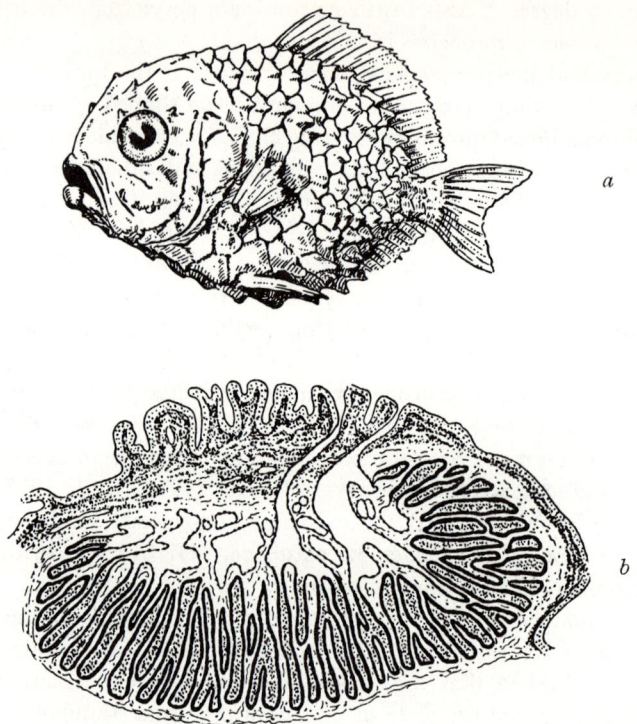

a

b

Fig. 345   *Monocentris japonicus* Houttuyn.   (*a*) Total view with luminous organ at the anterior end of the lower jaw; (*b*) section through one of the two organs.   From Buchner.

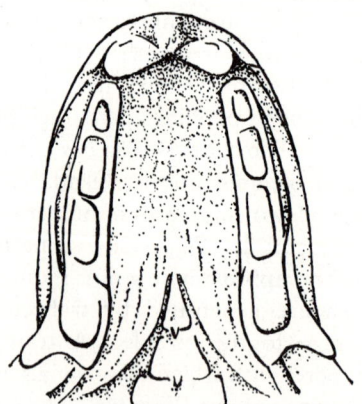

Fig. 346   *Monocentris japonicus* Houttuyn.   Interior view of the lower jaw with both luminous organs and their slit-shaped openings.   From Okada.

furnish the secretion for embedding the bacteria; the collecting spaces, as with the anomalopids, are furnished by the multilaminate skin. Here too a blood capillary penetrates the connective tissue separating the tubes and a reflector provides for intensification of the luminous effect. The connective tissue, centrally located and extending around the margin for a distance, is always permeated by numerous crystals which appear opaque in incidental light and are so arranged that the glow of the light is directed toward the mouth. Yasaki's observations (1928) confirm this. As with *Anomalops*, the light is continuous but occasionally is stopped for a few minutes. If in such cases one opens the mouth, it can be seen that the light radiates through the mucous membrane into the buccal cavity. The dimming mechanism has not been discovered.

Spontaneous migration of luminous matter through the narrow pores has not been observed and yet, as mentioned, the fiery rays can be made to emerge through pressure.

The symbionts limited to the glandular lumen are slightly curved, gram-negative rodlets 1.5–3 μ in length which are rendered extremely motile by one to three unipolar flagella. Spore formation could not be observed, but round, spindle-shaped, or filamentous involution forms occur. When the glandular content is removed by a bacterial filter, only the remainder glows, affording additional evidence that it is actually the bacteria which are luminescent and not the animal-specific secretion in which they live. Luminous cultures were easily obtained from all 79 animals investigated in a culture medium which was composed of 500 g. of meat, 30 g. of NaCl, 10 g. of peptone, 20 g. of agar, and 1000 cc. of distilled water and which showed a pH of 7.4. Transferred to various warmblooded and coldblooded animals, the *Monocentris* symbionts exhibited neither luminescent phenomena nor pathological effects.

The method of transmission again is unknown. Another *Monocentris* species, *M. gloria-maris*, also has luminous organs next to the mouth, and presumably they are filled with bacteria.

LEIOGNATHIDS. In the same year that Yasaki found luminous symbiosis in *Monocentris*, Harms (1928) was successful in furnishing analogous proof for the genus *Leiognathus* (*Equula*), of the family of the leiognathids, several species of which are caught on the Javanese coast and dried in quantity to be served as a garnish for rice. Harms investigated *Leiognathus splendens* Cuv. The site of the luminous organ is a very odd one. In the form of a thick ring, the organ surrounds the esophagus at the point where it passes into the stomach. On the dorsal side it extends into the air bladder. The greenish blue light, which is present day and night and at times is intermittent, shines through the abdominal wall and also makes its appearance at the posterior margin of the gill openings as though projected like a searchlight. With fish 10–12 cm. in length, the ring measures about 4–5 mm. in diameter and 1.5–2 mm. in width.

The actual light-body consists of a mass of finger-shaped, glandular tubes usually parallel to one another and radial to the longitudinal axis of the esophagus. Delicate connective tissue filled with capillaries is inserted between them. Several such tubes open into a duct of the first order and then pass into a collecting reservoir from which, in turn, ducts of a second order open into the esophagus between the mucous villi (Fig. 347a).

The lumen of the ceca is lined by tall cuboidal cells which grow flatter and flatter toward the outlet. At the blind end are groups of cells each of which has a strong light-refracting inclusion with a possible lenticular effect (Fig. 347b). Although the tubes contain a fine granular secretion, their prime content consists of thickly crowded, motile, staff-shaped bacteria. In general, these bacteria avoid the ducts of the first order, but those which do pass over into them are converted to longer filaments tending toward sprout formation.

A reflector with especially strong ventral development surrounds the esophagus and is necessarily penetrated by the ducts. Toward the outside a pigment mantle which extends around the strange organ is interrupted in the vicinity of the air bladder, so that light must strike into it. Acting in its turn as a reflector, the bladder then throws light into two tips elongating toward the front to produce two lateral luminescent cones at the gill slits. The light which illuminates the whole abdomen weakly originates from the ventral part of the ring which does not extend into the air bladder. Again it is the peritoneum, also densely permeated with guanin, which acts as a reflector. Harms also describes odd lentiform structures which are said to throw light into the abdominal cavity and into the air bladder. Figure 347a shows their position along the wall of the latter. Breaking the pigment envelope here and there they have in the center a pigment collection which expands to cut off the lens action.

Luciferin and luciferase are not detectable here. Dried organs when remoistened will illuminate dimly; cytolytic media extinguish the light.

Unfortunately we know nothing in this case about the devices which safeguard continuity of cooperative living. Harms could merely determine that a very young specimen barely 1 cm. in length already had infected esophageal glands which were still paired.

Without knowledge of Harms's discovery, Haneda (1940, 1950) also came upon the strange luminous organs of the leiognathids and recognized their symbiotic nature. In addition to the species available to Harms, he investigated ten other *Leiognathus* species, *Gazza minuta* Bloch, and two *Secutor* species and always found the same devices in this material origi-

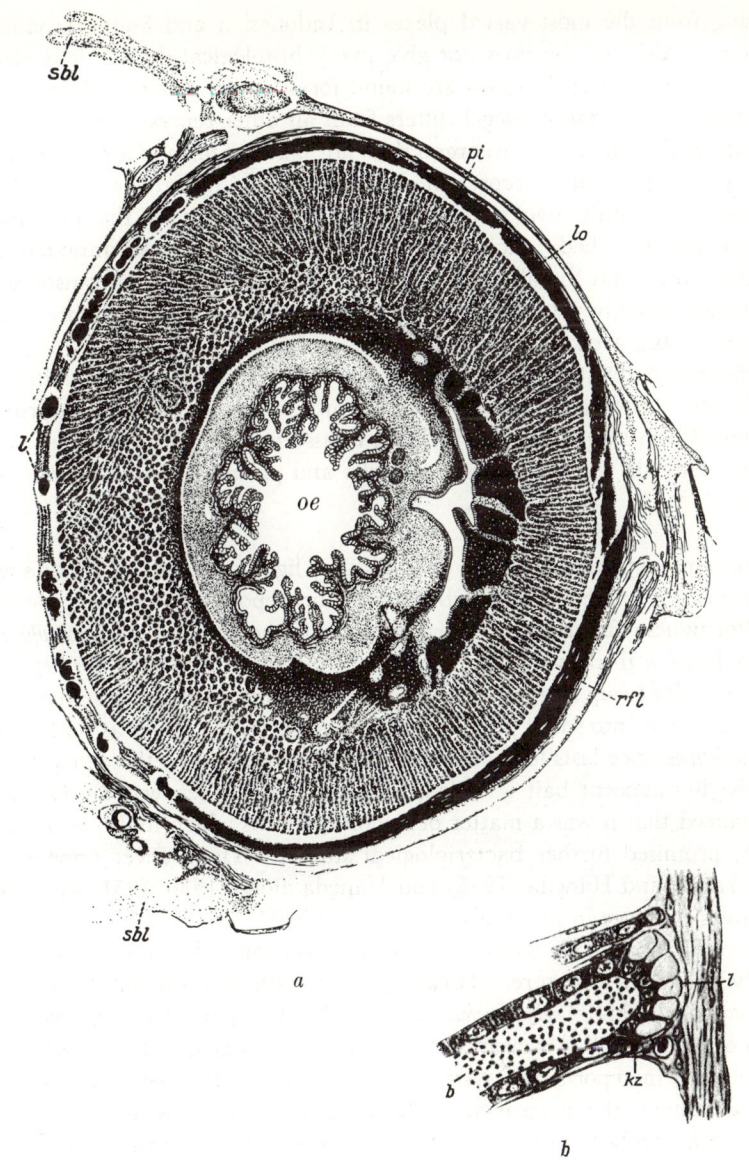

Fig. 347 *Leiognathus splendens* Cuvier. (*a*) Transverse section through the esophagus and the luminous organ surrounding it; (*b*) end of one of many glandular tubules. *oe*, esophagus; *lo*, luminous organ; *rfl*, reflector; *l*, lenses; *pi*, pigment; *sbl*, swimming bladder; *b*, bacteria. From Harms.

nating from the most varied places in Indonesian and South Japanese waters. Although he does not give many histological details, it is clear that the same related organs are found for all three genera. *Leiognathus rivulatus* Temm. and Schlegel differs from all other representatives of the family in that the luminous organ does not take the form of a ring around the esophagus but is represented merely by a dorsal swelling. For *Gazza minuta* only two symmetrically situated ducts of the luminous gland are described. According to Haneda, the ventral musculature of the thorax and the abdominal area is milk-white and semitransparent and thus actually acts as a lens. The blue-white light is therefore most intense in this area.

Haneda cultivated the symbiotic bacteria in various media without difficulty. They are non-motile or slightly motile coccobacilli approximately 3 μ in length and are inclined to assume filamentous form in culture. He calls them *Coccobacillus equulae* and promises further reports on them.

MACRURIDS AND GADIDS. The possibility of light symbiosis was first suggested in 1912 when Osorio published a short report on a strange practice of Portuguese fishermen. They press from the anus of *Malacocephalus laevis* Lowé a thick, yellow, luminescent fluid and rub it into the flesh of *Scyllium, Pristurus,* and other fish. Made luminescent in this manner, the fish is cut into small pieces and fastened to fishing hooks as bait. Its phosphorescence lasts for hours and is renewed each time these "candil," as the luminescent bait is called, are cast into the sea water. Osorio, convinced that it was a matter of microorganisms which multiply on the flesh, promised further bacteriological studies. These never appeared, but Yasaki and Haneda (1935) and Haneda alone (1938, 1951) supplied detailed evidence in proof of Osorio's hypothesis.

Investigating a long series of forms, they compared the luminous organs in their detailed structure. Hickling had conducted a similar study with *Malacocephalus* and *Coelorhynchus* (1925, 1926, 1931) but had rejected the idea of symbiosis. It is clear from the findings of Yasaki and Haneda and from those in reports of a systematic nature that such organs are widespread among the macrurids. The Japanese authors found them not only in a number of other species of the genera *Malacocephalus* and *Coelorhynchus* but also in *Abyssicola, Hymenocephalus, Nezumia,* and *Coryphenoides,* and they were convinced that all contained bacteria. Parr (1946) described their outward appearance in *Ventrifossa, Grenurus,* and *Trachonurus.* Vinciguerra (1932) studied the luminous organs of *Hymenocephalus italicus* Gigl.

When *Malacocephalus laevis* is viewed from the inferior side, two spots are observable: a small roundish one in front of the anal opening between the ventral fins, and a larger one, transversely placed and slightly luminescent, in front of the first (Fig. 348*a*). Both spots are black but have no scales; the connective tissue behind them is of a glassy consistency. These spots constitute lenses, and their melanophores, according to Haneda, contract and expand as a result of various stimuli to allow the light to shine or grow dim. A reflector rich in guanin cells is situated above the anterior larger lens, and a muscle extends between both lenses. Behind the anterior lens and above the posterior one is the strange luminous body, a sac-shaped organ consisting again of a number of glandular tubes with a duct extending backward. The mouth of the latter curves around the anal opening and a number of ceca are formed round about it (Fig. 348*b*).

A double fibrous pouch with many chromatophores, which is permeated by a network of nerve cells and blood vessels, covers the gland and at the same time separates the individual tubes. The luminescent material is extruded by contraction of the smooth muscle cells which extend over the organ.

According to the presentation of the Japanese authors, the bacteria, limited to the lumen of the tubes, are enclosed in roundish sacs (gelatinous globules?) which hang like grape clusters from the glandular epithelium (Fig. 349). If the envelope tears, the symbionts drift free in the lumen of the gland. In my opinion Hickling errs in his belief that a granulated luminescent secretion arises in the epithelial cells between nucleus and surface and is freed only through degeneration of the cells and even then is still united in clumps.

From Hickling's description it appears that the generally similar luminous organ of *Coelorhynchus coelorhynchus* Risso differs from that of *Malacocephalus* in several respects. Here the organ is also an anal gland, its location marked by a fleck of pigment which may possibly constitute a shallow lens. However, a second spot free of scales is lacking. In this case the gland itself consists of far more numerous, perpendicularly placed, narrow tubelets which open into a system of collecting spaces, and these in turn give rise to a duct which extends to the anus and has blind ceca in the vicinity of the latter (Fig. 348*c*). The fibrous envelope, reflector, and blood supply correspond to the situation in *Malacocephalus*, but the musculature which effects the expulsion of the luminescent substance is lacking. Instead, the fibrous envelope has more intimate relationships with two special median pieces of cartilage in the pelvic arch which are connected by a ligament and so constituted that a bending of the two shanks expels the "gland content." Hickling compares the effect of the cartilage pieces with that of a nutcracker.

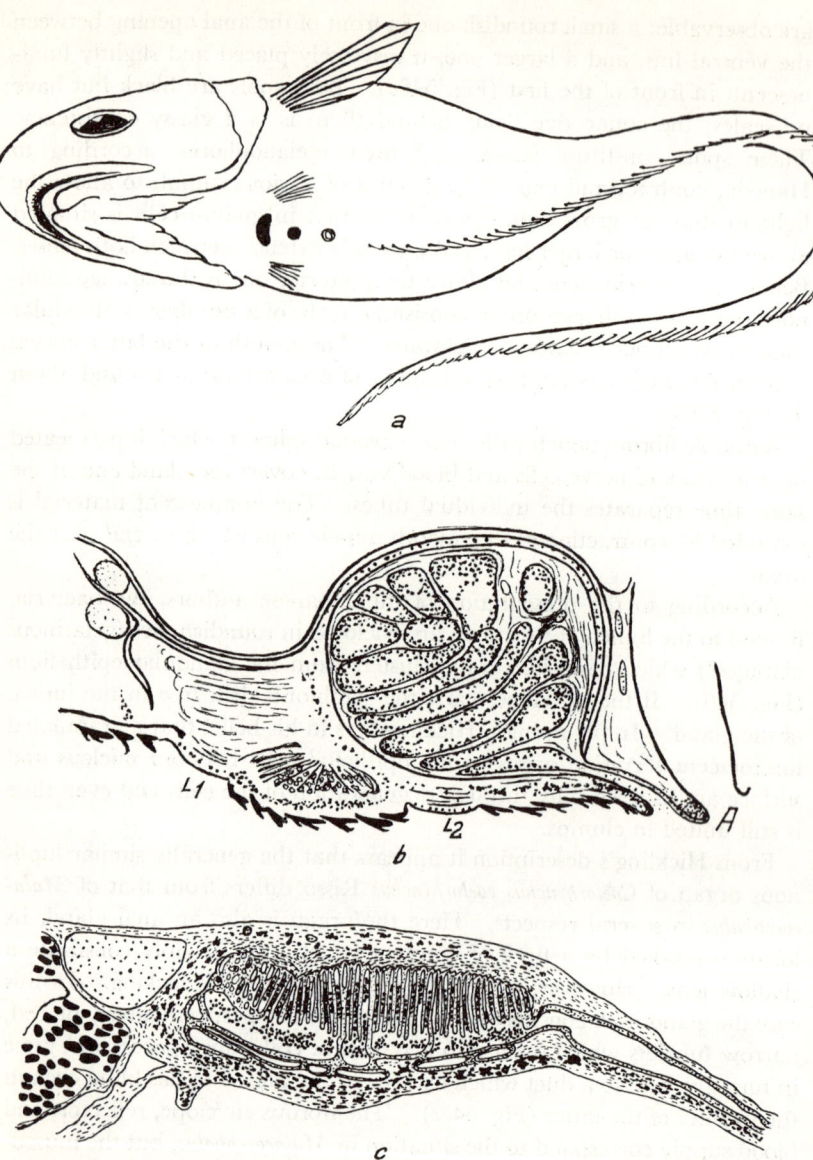

Fig. 348   *Malacocephalus laevis* Lowé.   (*a*) Ventral view.   Between the abdominal fins in the region of the luminous organs are two scale-free, darkly pigmented lenses, and behind these the anal papilla.   (*b*) Sagittal section through the luminous organ; L_1, L_2, the two lenses; *A*, anus.   From Haneda.   (*c*) *Coelorhynchus coelorhynchus* Risso. Sagittal section through the luminous organ.   From Hickling.

The most striking difference is that the luminous organ of *Coelorhynchus coelorhynchus* retrogresses as the animal grows. It is best developed in young animals; in older specimens the light-producing capacity decreases to the extent that it is difficult to extract even a thin, barely luminescent emulsion. Finally, animals occur in which the glandular tubelets cannot even be detected on sections.

Further variants are discovered on comparing the remaining forms. In *Coelorhynchus anatirostris* Jordan and Gilbert, the organ is very small, is situated directly in front of the anus, has neither lens nor reflector, is externally just barely visible, but nevertheless contains luminescent

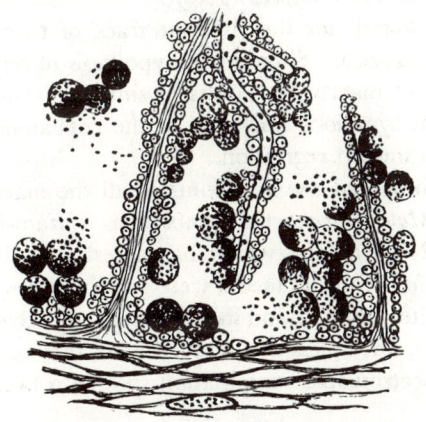

Fig. 349 *Malacocephalus laevis* Lowé. Section through the "luminous gland." From Haneda.

bacteria. In *Coryphenoides* species the lens and the scale-free spot are at least indicated, and the duct is very short. These species represent transitions to the well-developed state found in *Abyssicola*, *Nezumia*, and two Japanese *Coelorhynchus* species in which rather large luminous glands, reflectors, and lenses are located behind a scaleless spot with a thin transparent membrane. The canal extending to the anus also varies in length, and the lenses and scale-free places may now occur in pairs, as with *Malacocephalus laevis*. There are cases in which the bacteria-filled canal is very long and is equipped, along its full extent, with a reflector layer and a lens, admittedly a weak one, so that in young fish a luminescent line between the pectoral fins extends as far as the anus. In older animals the passage is covered by melanophores which screen the light from its area. In *Coelorhynchus hubbsi* Matsubara, Haneda found an extremely long, thin duct in which the two ends are transformed into luminous organs with all appurtenances. In *Hymenocephalus* species the duct is also greatly elongated, but only the blind end contains a luminous organ. The latter, however, has a double lens (Vinciguerra, Haneda).

In view of such diversity, the question is raised whether the tendency is toward continual improvement of the devices or gradual reduction. The organs of young animals are often more fully developed than those of adults, and indeed, in adults of *Coelorhynchus coelorhynchus*, Hickling was unable to find them even in sections. These two facts would point toward regression and toward the possibility that there are species which have lost the organs completely. In fact, they are not mentioned in the description of the young forms of *Trachyrhynchus trachyrhynchus* Risso living in the deep sea, and Hickling found not the slightest trace of them when he was studying adults in sections. Should the hypothesis of regression be confirmed in the case of macrurids, points of similarity might be revealed between their light symbiosis and that of the cephalopods, which also manifest tendencies toward regression.

Haneda easily cultivated the symbionts of all the macrurids included in his study. In *Malacocephalus* the symbiont is a gram-negative coccus or short rodlet 2.2–3.6 μ long which has a terminal flagellum and is therefore very motile. Culture media were gelatin plates, agar, bouillon, and cooked egg white. Suspension in sea water shows that the symbionts need a lot of oxygen, for, unless they are agitated, they soon light up only at the surface. Spectroscopic examination showed a broad band in the area of 638 to 430.

The hypothesis of the two Japanese authors was verified shortly after Haneda's report was published when Kishitani (1930) demonstrated that similar luminous organs populated by bacteria occur in the gadids, which, with the macrurids, constitute the Anacanthini or Gadiformes, but unlike the macrurids live in shallow water and can therefore be studied in aquariums. Kishitani studied *Physiculus japonicus* Hilgendorf, a fish living near the coast, in which Franz (1910) had found opening into the anus a ventral gland the function of which he was unable to explain.

Outwardly the luminous organ of Kishitani's test object looks like a small, round disc, black and free of scales, as in *Malacocephalus* and its relatives. In a specimen measuring 37 cm. it has a diameter of 4 mm. The somewhat larger, heart-shaped gland beneath it is built similarly, being divided into tubes which become shallower toward a duct permeating the abdominal musculature and opening into the rectum in the direct vicinity of the anus. The covering of the gland again consists of two layers of fibrillous connective tissue permeated by chromatophores. Here, too, the connective tissue between the luminous organ and the peritoneum is particularly transparent and thus resembles the primitive lenses of *Malacocephalus*.

Because of the strong pigmentation in this area Kishitani was undecided about interpreting it as a lens formation, especially because he never observed light shimmering through the peritoneum. Here the observations of Haneda (1938, 1951) provide supplementary material. Haneda saw the animals light up in the aquarium and found that the light is withheld by expansion of the melanophores. Furthermore, when Japanese fishermen observe this fish, which they call "Dongo," light up in the depths, they are accustomed to say "Dongo is yawning."

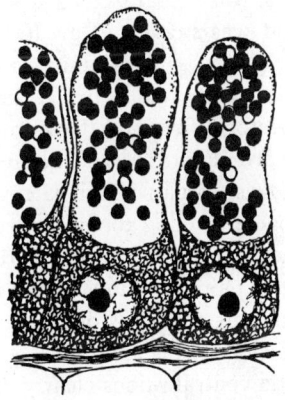

Fig. 350   *Physiculus japonicus* Hilgendorf.   Cells of the wall of folds in the luminous organ with luminescent bacteria.   From Kishitani.

Approximately as in *Malacocephalus*, the cocci measuring 1–1.5 μ are enclosed in "plasma sacs" which by no means are part of the cubodial epithelial cells on which they rest but represent an excretory product of the symbionts and secondarily enter into this positional relationship (Fig. 350).

These cocci light up under the microscope but possess no spontaneous movement. Observations on their oxygen requirement correspond to those made with other fish. On Hattori's culture medium for luminescent bacteria or on meat broth agar and gelatin with 3% cooking salt, thirty specimens in all sixty experiments yielded intensely green luminescent pure cultures in which the shorter rodlets gradually took the place of the cocci. Like Kishitani, Haneda (1951) had no difficulty in cultivating the inmates of the same bacteria-containing organs in *Lotella phycis* Temm. and Schl.

Micrococcus physiculus Kish. does not liquefy gelatin. Dextrose, mannose, galactose, maltose, and levulose are fermented in all strains; lactose and saccharose are not.

Kishitani tested also the contents of the intestinal tube and found that there is always a small number of luminescent bacteria in the rectum near the opening, but that only water vibrios clearly distinct in shape and behavior grew in inoculations of the other sections. Agglutination experiments with rabbit serum demonstrated a pronounced specificity of the symbiotic strains and their differentiation from the saprophytic vibrios.

ACROPOMATIDS AND APOGONIDS. *Acropoma japonicus* Günther, a small fish living in Japanese waters, was recognized as a symbiont bearer by Yasaki and Haneda (1936) and was studied in greater detail later by Haneda (1950) and Matsubara (1953). The luminous organ, shaped approximately like a tuning fork, is situated between the ventral fins, which are shifted in front of the pectoral fins, and flanks the anal opening with the two ends turning backward (Fig. 351*a*,*b*). Numerous tubules are formed by the wall of the yellowish organ which is enveloped by a capsule of connective tissue; the bacteria cling to the surface of the epithelial cells or make their appearance joined in clumps. A duct which sends numerous end branches into the paired sections of the luminous organ extends between these to the hindgut and ends opposite the opening of the urogenital apparatus in the cloaca (Fig. 351*c*).

The musculature of the ventral side is clearly distinct from the remaining musculature, not only in the direct vicinity of the luminous organ but also as far away as the head and tail areas. Of milky transparency, it acts as a lens, as with the leiognathids, and radiates like a frosted electric light bulb in which the light source is not visible. Behind the modified musculature is whitish, opalescent tissue which divides it from the upper lateral and dorsal musculature and thus plays the role of a reflector. The light is continuous, but special melanophores of the corresponding regions of the abdominal membrane appear to be able to vary its intensity.

When Haneda later investigated a greater amount of material of *Acropoma*, he discovered that what was previously called *A. japonicum* apparently represents two different species. In addition to the one on which the preceding description is based, there exists another, previously unknown, which differs chiefly with regard to the luminous organ, but also in coloring, scales, location of anus, and so forth. Its luminous glands, consisting of two much longer tubes occupying almost the whole ventral side of the fish, are not connected cephalad, but are instead reversed at the posterior end and extend again for a distance toward the front.

The symbionts are cocci or short rodlets measuring 0.8–2 μ which are rendered very motile by a flagellum. In all experiments they form colonies radiating a bluish green light on the culture media usual for luminescent bacteria. Two days after the death of the fish the organ is still luminescent. Oxygen requirement and other behavior of the luminescent substance are the same as in other luminescent bacteria. Aggluti-

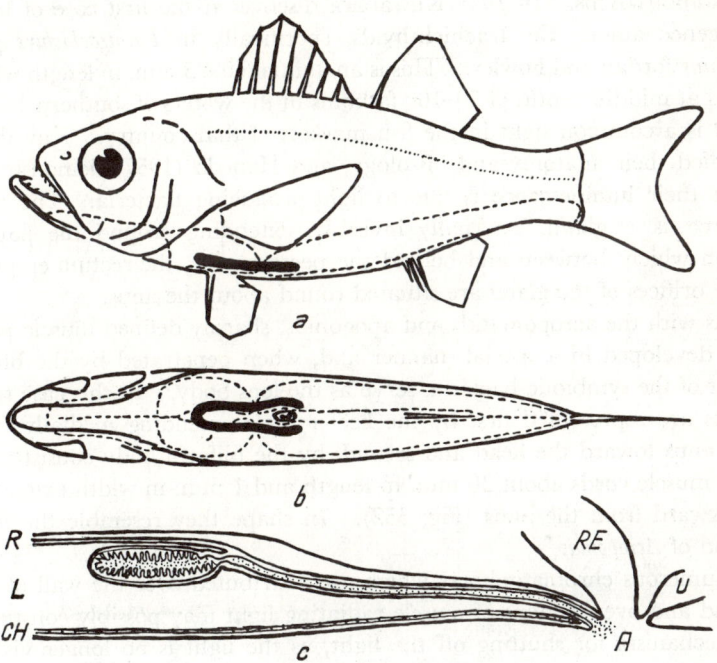

Fig. 351 *Acropoma japonicum* Günther. (*a*) Side view. (*b*) Ventral view. (*c*) Medium longitudinal section through the unpaired part of the luminous organ and its excretory duct; *R*, reflector; *L*, lens; *CH*, chromatophores; *RE*, rectum; *A*, anus; *U*, opening of the urogenital duct. From Yasaki and Haneda.

nation experiments showed that each of the two species has specific non-banal forms, on which Haneda is planning to make further reports.

According to a Japanese report by Kato (1947), *Apogon marginatus* Döderlein, a representative of the apogonids, has luminous organs similar to those of the acropomatids. However, according to Haneda and Johnson (1962) this is not symbiosis. Both authors present a family tree of the teleosts in which the trachichthyids, leiognathids, and acropomids possess luminescent symbionts, while the luminous organs of the pempherids and apogonids are bacteria-free. On the other hand, the great similarity of both organs is so evident from the comparison of types with and without symbionts that one could assume that the symbiosis arose through the subsequent infection of sterile luminous organs which were already present. In this connection it is of interest to note that Kozuka (1952) cultured luminescent bacteria from the gut of luminous fish.

TRACHICHTHYIDS. In 1955 Kuwabara discovered the first case of luminescence among the trachichthyids, specifically in *Paratrachichtys prosthemius* Jordan and Fowler. This is an animal 60–75 mm. in length which lives at middle depths of 50–100 fathoms in the waters of southern Japan, and is a common sight in the fish markets of that country. Kuwabara studied their anatomy and histology, and Haneda (1957) demonstrated that their luminescence is due to light-producing bacteria. The light source is a gland, externally invisible, extending around the papilla upon which, between and behind the pectoral fins, the rectum empties. The orifices of the gland are situated round about the anus.

As with the acropomatids and apogonids, sharply defined muscle parts are developed in a special manner and, when penetrated by the bluish light of the symbiotic bacteria, serve as the lens body. In this case these parts are represented first by the keel muscle extending medially from the anus toward the head and second, by the filiform body consisting of two muscle cords about 30 mm. in length and 1 mm. in width extending backward from the anus (Fig. 352). In shape they resemble the light gland of *Acropoma*.

Numerous chromatophores which are distributed over the wall of the gland and over sections of muscle radiating light may possibly constitute a mechanism for shutting off the light, as the light is no longer visible when the chromatophores are flattened out after the death of the animal. The light gland is filled with numerous gram-negative cocci which can be cultivated with ease and are identical in all specimens.

PEDICULATES. According to a report, unfortunately very brief, by Dahlgren (1928), the pediculates are among the fish which owe their light-producing capacity to symbiosis. Dahlgren states that in *Ceratias* he found bacteria populating the spherical organ located in the terminal swelling of the long tentacle which extends beyond the mouth like a fishhook. Dahlgren's discovery is not surprising in view of the fact that the organ involved is an exposed gland-like formation. Again, septa of connective tissue protrude from the periphery into the interior, and a reflector and a pigment mantle surround the organ. It is plentifully supplied with blood vessels.

Further testing for evidence of symbiosis is desirable not only for ceratiids but also for the other pediculates, that is, gigantactids, antennariids, and malthids, since in all these families the luminous organs have an external opening and are similar in anatomy and histology. There are many variations, of course, in the number and site of the organs. In addition to the bizarre tentacle organs at the end of isolated motile

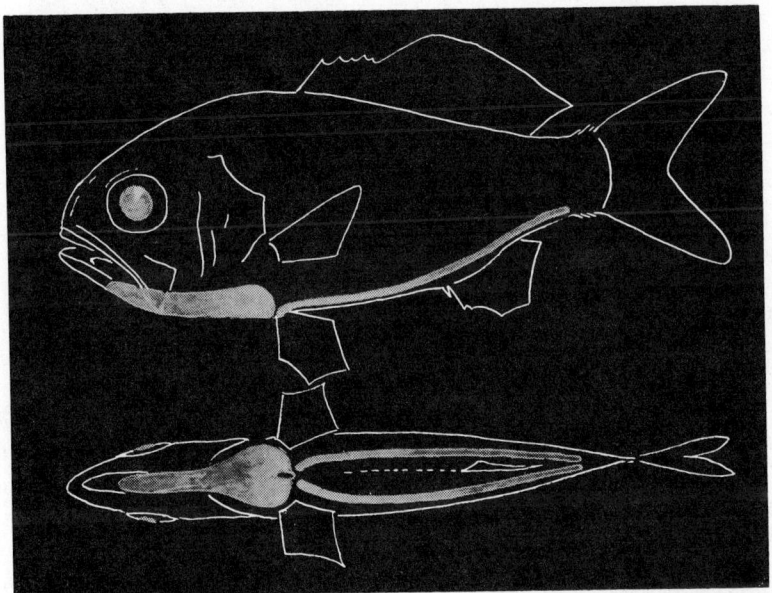

Fig. 352   *Paratrachichthys prosthenius* Jordan and Fowler.   Lateral and ventral views with muscle parts beaming the bacterial light.   From Haneda.

rays of the dorsal fin which at times may be shifted to the snout, some ceratiids have at the beginning of the dorsal fin three additional short, club-shaped formations, the so-called caruncles, which, according to Brauer (1906, 1908), are quite like the tentacle organs in their structure. In *Chaunax* and the malthids the tentacle is very short and lies free on the forehead, but the organ in it is again an open gland; in other cases the tentacle is situated in a frontal sinus.   Apparently all these organs are sites of luminescent bacteria.

10

Cases of symbiosis localized in excretory organs

ANNELIDS

LUMBRICIDS. The complex segmental excretory organs of the lumbricids are regularly populated by bacteria in a strictly defined area. Maziarski correctly discerned their nature in 1905, but his explanation was more or less questioned by the older authors. It was not until 1926 that Knop eliminated all doubt by his presentation of the complete symbiotic cycle. Pandazis (1931) reported on the physiology of the organs.

Following upon the nephrostome of a segmental organ of *Lumbricus terrestris* L. are the initial canal and the trilooped, narrow nephridial canal. The latter continues into the ciliate canal, the lumen of which is lined with cilia. At the end of the second loop the ciliate canal passes over into the ampulla, which in turn opens into the section called the gland canal. After complicated convolutions the latter ends in the urinary bladder, which empties outward. The bacteria populate only the ampulla, the unciliated wall of which is strongly vacuolized and is free of the granules found in the gland canal (Fig. 353).

In the form of rodlets 3–5 μ in length and more strongly staining at both ends, the bacteria are arranged in a broad border on the cell surface, usually perpendicular to it. Where they are less dense, it is clearly recognizable that they are embedded in a sticky secretion which prevents them from being washed in greater quantities into the adjoining parts of the nephridium, and indeed they are never found there except during periods of reproduction (Fig. 354).

It seemed obvious from the completely regular occurrence of the bacteria and their strict spatial limitation that well-functioning transmission devices might be present, and such devices actually were discovered by Knop. The egg cells remain symbiont-free, a fact which coincides with the general experience that other methods of transmission are used when the symbionts are localized in the body cavities. If one investigates the ampullae of worms during the reproductive period, April through August

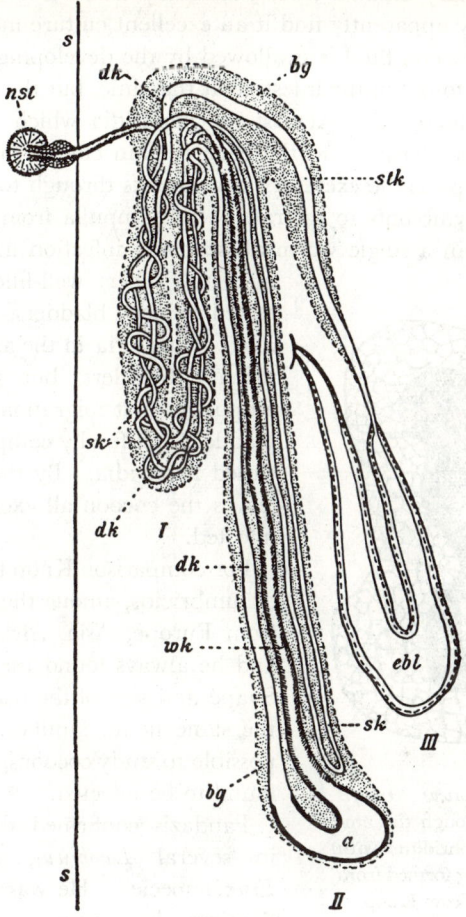

Fig. 353 *Lumbricus terrestris* L. Schematic presentation of a nephridium. I–III, the three chief loops; *a*, ampulla; *dk*, glandular canal; *ebl*, terminal bladder; *sk*, loop canal; *wk*, ciliated canal; *nst*, nephrostome. From Maziarski.

in the case of *Lumbricus terrestris*, one frequently comes upon some which show more swelling than others. In such ampullae the fluid is permeated with symbionts as soon as the cohesive substance imprisoning them is dissolved. When this is the case, the symbionts transfer to the urinary bladder in great numbers. Clearly, this change is preliminary to infection of the cocoon fluid, for here also the bacteria are sparse at first but soon are observed in great numbers evenly distributed in the fluid. When the worm later withdraws its body from the secretion ring, organisms from the urinary bladder must necessarily transfer to the future cocoon fluid and they apparently find it an excellent culture medium.

Since the cocoon fluid is swallowed by the developing worms, the bacteria are also found in the intestine at this time, but the ampullae are not infected by this route. At first the nephridia which are forming progressively from the front to the back remain closed, but when they are perfect in all parts the excretory pore breaks through to the outside, permitting the symbionts to migrate to the ampulla from the surrounding fluid. Thus in a single worm all stages of infection may be found one after the other: well-filled ampullae, infected urinary bladders, and first appearance of bacteria in the ampullae; infected urinary bladders, but still bacteria-free ampullae; first migration into the urinary bladder; and finally completely sterile, still closed nephridia. By the time the worm leaves the cocoon all excretory organs are infected.

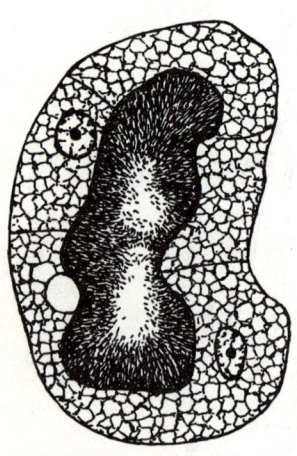

Fig. 354 *Lumbricus terrestris* L. Section through the ampulla of the nephridium with the wall covering formed from symbionts. From Knop.

For comparison Knop used thirty species of lumbricids, among them representatives from Europe, Asia, Africa, and America, and he always found the same conditions. Shape and size of the bacteria were about the same in all families; whenever it was possible to study cocoons, they were always found to be infected.

Pandazis confirmed the same bacteria in several *Lumbricus, Lumbriculus,* and *Eisenia* species. He was successful in cultivating the bacteria. As the culture medium he used 2% agar with ascitic fluid or cattle serum, and he found that a weakly acid reaction was advantageous. Colonies of gram-negative rodlets 1.5–12 μ in length with bipolar staining and without individual movement appeared after 20 hours. They often formed smaller or larger chains but never spores. In older cultures numerous involution forms appeared.

In a study of the physiological properties, it was shown that they split albumin into albumases and peptones and change higher fats into fatty acids. They attack no type of sugar, are not acid-forming, and reduce nitrates. Injected into the body cavity of the worms, they cause a strong increase of the phagocytes, which also fill immediately with the bacteria. Groups of weakly staining bacteria appear later in the wall cells and in the lumen of the nephridia. With an increase in the number of organisms introduced, disturbances become noticeable in the worms; large doses bring about their death. The bacteria in the ampullae always remain unchanged under these conditions.

GLOSSOSCOLECINES. Knop pursued the question of the extent to which similar regulated cooperative living with bacteria occurs in other families of the oligochaetes. With material originating from many countries he proved that it does not occur among the megascolecids, which also live terrestrially, or among the criodrilines, hormogastrines, and micro-chaetines of the family of the glossoscolecids which live in part on land and in part in fresh water. He found regular colonization of the excretory organs only in the subfamily of the glossoscolecines, which are representatives of the non-indigenous lumbricids in tropical and subtropical South America. Furthermore, the colonization differs in the details of its development and in some other respects as well.

At first, forms are found which still resemble the lumbricids insofar as the bacteria are restricted to the ampulla, but in contrast to the lumbricids there are scattered bacterial aggregations in the plasma of the cells forming the ampullar wall. *Enandriodrilus* and *Diachaeta* are examples of this. *Andiorrhinus* and *Andiodrilus* are somewhat different, as their ampullae develop markedly, at the expense of the ciliate canal, and they also have cilia; the wall consists of large, secretion-laden, glandular cells. Symbionts fill these cells in great quantities, being particularly abundant in the secretion globules (Fig. 355*b*). In *Andiodrilus* the usually slender rodlets are replaced by thicker and shorter forms. *Thamnodrilus* has an even larger ampulla but otherwise is the same. Conversely, the ampulla is smaller in *Pontoscolex* than in the lumbricids, and the wall and lumen are weakly infected, whereas the epithelium of the urinary bladder is thickly settled (Fig. 355*a*). In *Glossoscolex* the ampulla is lacking entirely, but in its place not merely the area corresponding to it but also all the other parts of the nephridium, looped canal, ciliate canal, and urinary bladder are populated by numerous, almost filamentous, elongated micro-organisms. Only the genus *Aptodrilus* appears to have no symbionts.

The symbiosis of glossoscolecines differs therefore from that of the lumbricids through a general tendency toward intracellular settlement and through the inclination to take possession not only of the ampulla but also of the remaining sections of the nephridia. Although only fixed material was studied, it is safe to assume that the cocoon fluid and the worms developing in it are also infected.

HIRUDINES. Bacteria which invite comparison with those of the lum-bricids and glossoscolecines are also found in the urinary bladder of the nephridia of *Hirudo medicinalis*. As with the lumbricids, there was no agreement in the interpretations after attention had first been drawn to them, in this case by Leuckart. Thickly crowded and attached to the surface of the epithelium by one end, the bacteria form a matting which, especially since they execute weak, sinuous movements, simulates a ciliate epithelium.

Jaschke (1933) became convinced of the regularity of the phenomenon during the course of research. Moreover, he came upon bacteria, joined in nests, in the epithelium of the urinary bladder. These are clumsy,

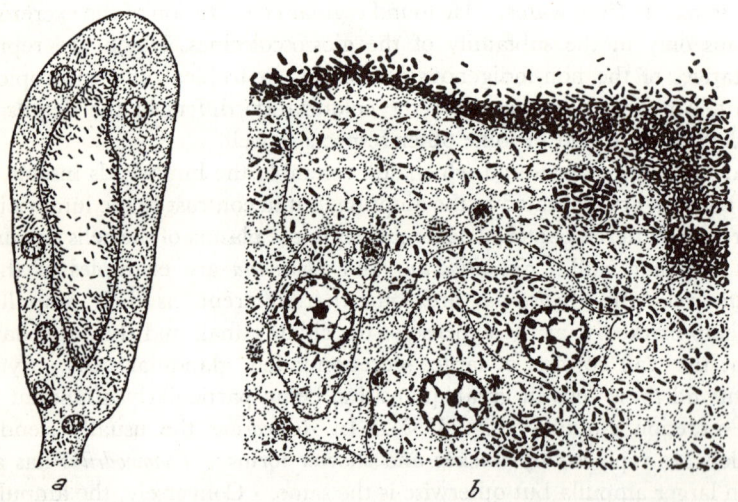

Fig. 355   (*a*) *Pontoscolex corethrurus* Fr. Müll.   Section through the extra- and intra-cellularly colonized ampulla.   (*b*) *Andiodrilus affinis* Mich.   Section through an extra- and intracellularly infected nephridium.   From Knop.

slightly arched coccobacilli, apparently always present (Fig. 356).   In the horse leech, *Haemopis sanguisuga* L., Jaschke always found bacteria not only in the urinary bladder but deep in the urethra as well.

In their productive research on *Hirudo*, Büsing, Döll, and Freytag (1953) reported both on symbionts of the intestine and on the bacteria of the urinary bladder.   The authors were successful in proving the concept, not generally accepted, of the bacterial nature of the much longer filaments attached to the epithelium.   They were able to cultivate these organisms, which are completely non-motile according to their observations, and to analyze their biochemical properties.   On the basis of their biochemistry and their morphology, these organisms were classified among the Corynebacteriaceae under the name *Corynebacterium vesiculare*. In addition to this form in the urinary bladder of the blood leech, the authors established another regular guest appearing free in the lumen. To this much shorter, clumsy rodlet, which is easily cultivated, they gave the name *Corynebacterium nephridii*. The authors do not discuss the possibility that this organism is identical with that found intracellularly by Jaschke in the bladder epithelium. Indeed, they do not even mention Jaschke's organism.

The authors' demonstration that both types of bacteria can break down low-molecular protein fragments explains a long-known fact,

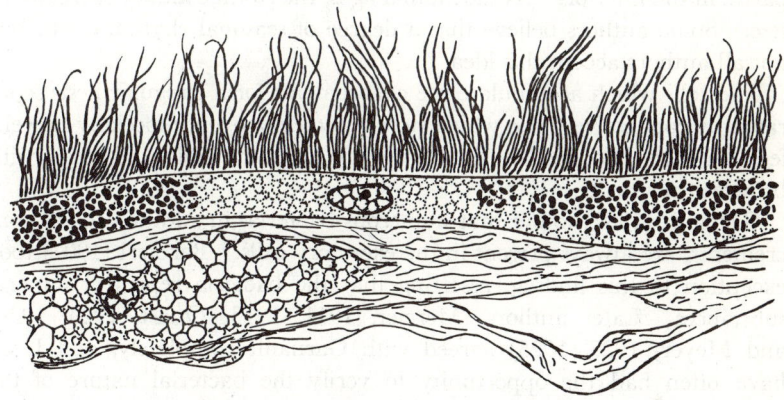

Fig. 356 *Hirudo medicinalis* L. Wall of the urinary bladder of a nephridium with extra- and intracellular bacteria. From Jaschke.

namely, that the urinary bladder has an unusually high ammoniac content. A considerable part of this content, at least, is expelled, not by action of the nephridia but by action of the bacteria, as proved by ammonia determinations of the animal in which flora of the urinary bladder were destroyed by injecting a mixture of penicillin, streptomycin, and cattle serum into the intestine. The authors consider it quite likely that even the occupants of the urinary bladder are symbionts, but proof is still lacking.

In all probability these two types of bacteria also infect the cocoon fluid, thereby guaranteeing the symbiosis, for it is not merely in the reproductive period that the blood leech on stronger stimulation discharges cream-like droplets, representing bacterial masses, after the watery content of its bladder has run out.

CYCLOSTOMATIDS AND ANNULARIIDS

It has long been known that bacteria regularly populate the strange ductless storage kidney of the cyclostomatids, a family of the Proso-branchiae. The first research was done with *Cyclostoma elegans* Drap., a moisture-loving snail found under leaves and underbrush, often in huge masses, in the Mediterranean countries and warmer areas north of the Alps. In this animal, the storage kidney or "concrement gland," the meandering windings of which shimmer more or less clearly through the shell, is usually pure white and is close to the gut, partly on the surface and partly in the interior. As the animal ages, the storage kidney increases in size. Some authors believe that a degree of seasonal rhythm exists, but not all authors accept this idea.

The cells which accumulate the concrements form irregular nests separated by connective tissue which stores much glycogen, especially in well-fed animals. Lodged in the tissue are calcium cells and cells with brownish pigment.

Garnault (1887) found that numerous bacteria, in addition to the concrements, are always present in the excretory cells. Barbieri (1907) took exception to this finding, asserting that the bacteria represent mineral substances. Later authors (Mercier, 1911, 1913; Quast, 1923, 1924; and Meyer, 1923, 1925) agreed with Garnault completely, and I too have often had the opportunity to verify the bacterial nature of the forms.

According to Mercier, these kidney cells are at first completely free of special inclusions. Later, small globules fluctuating in number and consisting of concentric layers are found in a large vacuole arising in the protoplasm. With increasing production of the excretion they all transfer to a single body which is lamellate in construction. Around this time, isolated symbionts make their appearance in the previously bacteria-free plasma. Multiplying rapidly, they soon fill all parts of it. Further growth of the stratified concrement, abundant proliferation of the bacteria, and swelling of the cells to considerable dimensions follow (Fig. 357). The protoplasm is limited in places to a thin border, and the nuclei assume irregular forms. Then a phase of decrease sets in. The numerous amoeboid elements between the kidney cells play the role of phagocytes and resorb the cells, together with the concretions and bacteria. Vacuoles emerge with bacteria which are joined in clumps and gradually digested. The excretory bodies are also liquefied and resorbed. Since the concrement cells are often enveloped by entire groups of phagocytes, epithelioid associations of amoebocytes originate after dissolution of the cells. Meyer agrees essentially with Mercier, whereas Quast, apparently without justification, believes that the concrements are situated extracellularly.

Quast and Meyer find that the concrements are composed of 36–50% uric acid. Since they also contain xanthine and other purine bases (hypoxanthine, adenine), Meyer proposes therefore that they be called purinocytes.

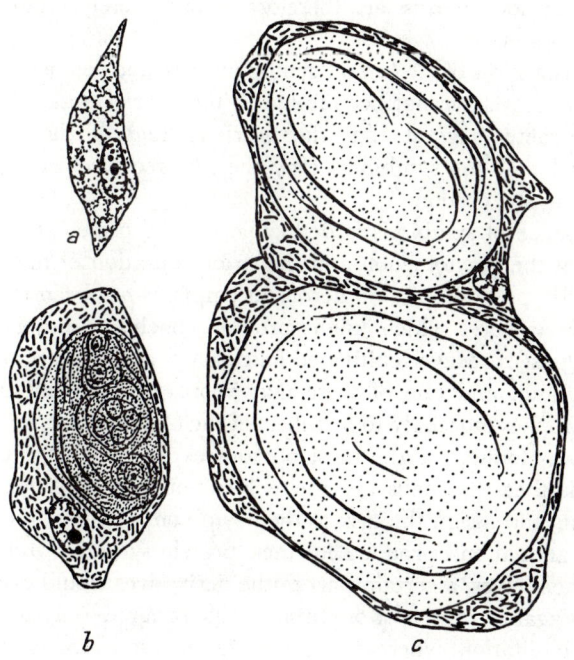

Fig. 357 *Cyclostoma elegans* Drap. (*a*) Purinocyte before storage of the excretion and infection by symbionts; (*b*) beginning of concrement formation; the plasma is infected; (*c*) advanced concrement formation. From Mercier.

The bacteria are non-motile, gram-negative, slender rodlets, 2–5 μ in length, which are usually curved and occasionally grow out into longer filaments.

Meyer compared other cyclostomatids and found essentially the same conditions in *Cyclostoma lutetianum* Bourg. from France, *Cyclostoma mauretanicum* Plry. from Algeria, and *Leonia mamillare* Link. from North Africa. He found a bacteria-free individual only once, in *Cyclostoma sulcatum* Drap.

Meyer also included a number of the closely related annulariids in his research. In *Tudora putre* Pfeiffer, from Cuba, he found concretions and bacteria; in *Chondropoma subreticulatum* Maltz, from Haiti, he found large cocci resembling *Torula* instead of the typical bacteria; in other species no microorganisms could be observed.   In a *Glossostyla* species from Manila the storage kidney is lacking completely, the purines being secreted through the nephridium; *Chondropoma dentatum* Say., although it has extracellular concrements, appears to have no bacteria.

From the point of view of symbiosis, the annulariids, formerly classified among the cyclostomatids, are therefore in part closely related to them, and in part different.

The transmission of *Cyclostoma* symbionts has not yet been explained. At any rate, they have not been found in the ovarial eggs.   Experiences with oligochaetes suggest that the nutrient fluid of the egg may be infected and that the symbionts may then be received by the embryos developing in the cocoon.

The presence of symbiosis is deducible for the present only from the regularity of the phenomenon and the strict limitation of the bacteria to a single cell type.   Despite numerous attempts, it has been impossible to obtain pure cultures which might provide conclusive proof (Garnault, Mercier, Quast).   Meyer cultivated four groups of gram-negative rodlets from the gland, and one of these may represent the symbionts.   This statement applies especially to members of the *C. fluorescens* group in which a uricase was found, for the most obvious explanation of cyclostomatid symbiosis is that the snails are themselves not able to break down the concrements and assign the task to their symbionts.   Since the symbionts utilize the animal end products for their protein synthesis and are finally resorbed themselves, a certain part of the derivatives would eventually be transferred again to the host organism.   Quast agrees only in general on a useful cohabitation, whereas Meyer supposes also a practical benefit in this direction.

MOLGULIDS

The molgulids are the only tunicates with ductless storage kidneys. The organ, a bean-shaped sac filled with varying amounts of urates, is situated unilaterally in an intestinal sling in the direct vicinity of the heart.   The lumen is the site of concrement formation, and it always contains fungi of a size which would forbid intracellular living.   The occurrence is so striking that it did not escape the notice of early authors like Lacaze-Duthiers (1874) and Giard (1888).   Giard describes the fungi in *Molgula socialis* Alder, under the name *Nephromyces*, as a fine, filamentous mycelium spanning the concrements and occasionally interspersed with thicker and irregular states. From time to time, sporangia with round, long flagellate spores are formed and also zygospores surrounded by a finely granulated envelope. *Anurella roscovitana* L-D. harbors another very similar fungus. In a *Listonephrya* species the kidney

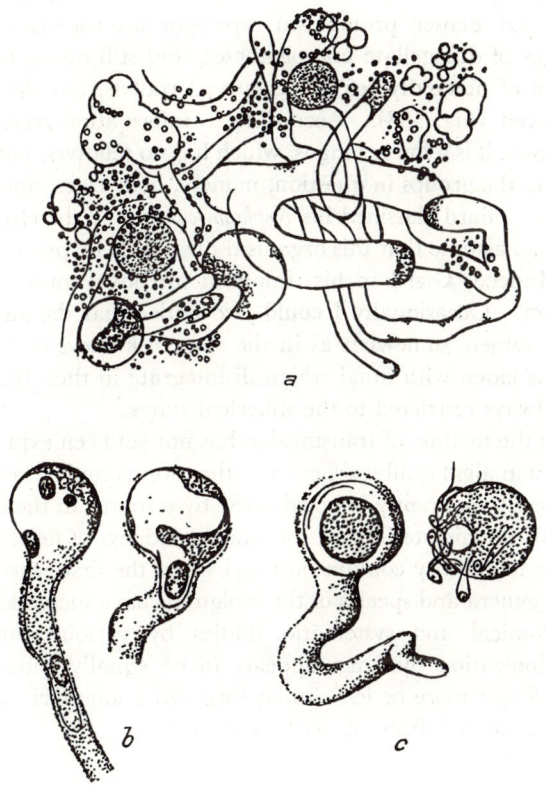

Fig. 358 *Molgula impura* Heller. (*a*) The fungi in two forms penetrate the excretion of the storage kidney; (*b*) tubules rounding themselves off; (*c*) macrogamete and copulation. In (*b*) and (*c*) the excretion covering, from which the fungi have withdrawn, is partly visible. From Buchner.

is taken up almost entirely by a single concretion so large that little space is left for the fungi, which in this species are characterized by pyriform spores.

I published a detailed account of the diversified states of the kidney symbionts in *Molgula impura* Heller (Buchner, 1930). The hyphae of various lengths are here slightly branched, and occasionally they constrict off conidia-like ends. Septa are lacking entirely. Among the hyphae are unusually thick and fluid-rich states with thicker plasma only at the ends. Many hyphae are incrusted with light-refracting granules which originate from excreta accumulating in the kidney and at times seem to cause their death. Some hyphae are transformed into spherical stages, which may again assume a tube shape. Some spherical states with envelopes and denser protoplasm represent macrogametes, others the initial stages of diflagellate microgametes, and still others become spores in the form of markedly refracting curved rodlets, on which no flagella were observed (Fig. 358). According to information received by letter from Claussen, it is a lower fungus, which has no relatives, not even distant ones, among the groups in question, namely, oömycetes, ancylistines, and chytridines. Giard classified his *Nephromyces* among the chytridines.

It is safe to assume that this organism lives at the expense of the ascidia excreta. Indeed, Giard in his time had already expressed his opinion to this effect. Occasionally I could also notice that the amoebocytes in the kidney lumen, somewhat as in the storage kidneys of *Cyclostoma*, are more or less laden with fungi which disintegrate in their interior. Such cases are always restricted to the spherical states.

Here too the method of transmission has not yet been explained. As so often occurs in algal symbiosis, even in the most complex cases, the young animals are all apparently infected anew by a source in the sea water, for the egg cells remain sterile and the storage kidneys of firmly fixed larvae born in the laboratory contain no fungi when the first excreta appear.

That all genera and species of the molgulids have such a kidney is clear from anatomical and systematic studies by various authors. Their strange colonization by fungi appears to be equally widespread, for I invariably found more or less similar forms in a long series of molgulids, preserved in alcohol, from all parts of the earth.

General
Section

11

Localization of the symbionts

The interior of the gut is unquestionably the foremost site in the housing of plant guests in multicellular animals, for in the overwhelming majority of cases the plant guests must inevitably have reached it on some occasion with food. In many instances they are acquired anew through the mouth from generation to generation, either because free-living organisms are widespread and ever present in the milieu of the hosts or because special devices exist to furnish the larvae with the desired organisms as their first food. Accordingly, other invasion points recede in importance. Nevertheless, the light-producing organs of many cephalopods and teleosts, which attach themselves as membranous invaginations of glandular character and later carry bacteria, constitute a strong argument that the symbionts are acquired without intestinal participation, and in the nephridia of the bristle worms the bacteria also enter anew through excretory pores. Just as the luminous anal glands of macrurids and gadids point to the anus as the place of entrance, perhaps the sweet excrement of the Homoptera may also have led occasionally to anal acquisition of symbionts, especially since it is known that the rickettsia use this route at times. Unique among the insects are the dorsal organs of the lagriid larvae. Formed as glandular membranous sacs and therefore settled from outside, these organs presumably reflect the manner in which the symbionts were first acquired. Since Fekl (1954) showed that the tracheae of insects often contain bacteria generally coinciding with those of the milieu, and that the hemolymph may also contain the organisms without injury to the animals, it is possible that symbionts were also acquired in this manner, although no experimental proof exists. No relationships appear to exist between the intestinal flora and the forms found in tracheae and lymph.

The simplest form of intestinal settlement is naturally intermixture with the chyme, more or less evenly diffused or restricted to special areas, with no resultant anatomical or histological changes of any kind. Because it tends to elude observation, a symbiosis of this type requires special circumstances to bring it to attention and to make it possible to distinguish between symbionts and other intestinal inhabitants. For example, in the case of trypetids, which as larvae and imagines contain numerous bacteria diffused in the midgut, we justifiably used the term symbiont, for the nearest relatives provide the bacteria with specific appendicular organs of the gut as sites and certain new formations for smearing the eggs exist even with this low stage of adaptation. Similarly, comparison with other Heteroptera which suck plant juices indicates that the bacterial masses in the midgut of *Pyrrhocoris* are to be equated with those housed by the former at great expense, and one can say with certainty that the bacteria diffused in the milk-gland secretion in the larval chyle gut of *Hippobosca camelina* or *Lynchia maura* correspond to more strictly localized symbionts in the other hippoboscids, especially when we consider that they are taken up in special cells of the intestinal epithelium during metamorphosis. In the case of bacteria intermingled with the chyme in the *Hylemyia* gut, it could be shown by sterilizing the egg surface that the bacteria are vital to life. Another possibility for obtaining information about the utility or non-utility of such localization is provided by cultivating the microorganisms outside the host and by proving that they have capacities which must be welcome to the host. However, aside from the intestinal flora of the vertebrates, very few data exist with respect to this matter.[1]

Where the import of cooperative living exists as a direct influence of food from the symbiont enzyme, such diffuse settlement must guarantee the best utilization of it. Obviously, on the other hand, it may also cause certain incompatibilities, thus necessitating improved fixation of symbionts, or preparation of more or less sealed-off sections of the gut in cases where a longer time is required to decompose the food. Such tendencies will naturally be more pronounced wherever the symbiosis lies in another direction and wherever housing of the guests in or on the gut, which indeed in the case of insects frequently undergoes complete renewal during metamorphosis, will involve unnecessary complications and dangers for its continuation.

The chitinous retention devices, observed very frequently on the wall of spaces which are derived from membranous folds, do not occur in the area of the midgut. A similar effect may be achieved here through

---

[1] *Tenebrio molitor* illustrates that symbioses which are limited to the intestinal lumen may develop from preliminary stages for which the use of the term symbiosis is not yet justified. Indeed, Busnel and Drilhorn (1948) and later Kaudewitz (1953) proved that intestinal bacteria of this tenebrionid synthesize substances which make a certain degree of growth possible even in the presence of deficient diet. When the intestine is sterilized by administering antibiotics (sulfaguanidin, prontosil), growth ceases with deficient diet.

formation of tufts, and indeed in *Hippobosca equina* numerous folds are formed on the wall of the convex widening of the larval gut, which is the sole symbiotic site, and every cell has a plug-shaped process, so that a complexly organized relief of sinuses and protrusions, to which a thick border of bacteria is anchored, arises.   With *Hippobosca capensis* the folds are lacking, but plugs are formed in the individual cells and the symbionts are deposited in great quantity in the resultant niches.

Occasionally it is noted that symbiotic bacteria quite regularly cling to the cell surfaces in palisade form.   Thus in the urinary bladder of *Hirudo* they were long considered to be animal-specific cilia.   In addition, with the extracellular rodlets in the Malpighian tubules of *Coccotrypes,* such polar fixation is also found on the cell surface.

Where symbionts live in sections of chitin-lined insect gut, as in the hindgut of tipulids, crickets, and lamellicorn beetles, variously ramified bristle formations serve admirably to fix the bacteria.

Much more frequently we find that symbionts are limited to intestinal evaginations which are more or less closed off and thus represent, as it were, quieter spaces.   At the same time we must distinguish between formations which were present before symbiosis took place and others which owe their existence solely to symbiosis.   Among the first are the four roundish ceca at the beginning of the midgut in trypetid larvae, which are found as empty evaginations also in forms with diffuse settlement of the intestinal lumen.   Their appropriation by the symbionts then causes considerable growth, sometimes also subdivision, and simultaneously quiets the secretory activity of the wall, an activity which is recognizable in sterile saclets by vacuolization of the plasma.   Another site independently established before symbiosis consists of the little tufts accompanying the posterior area of the midgut in the imagines of *Bromius.* During metamorphosis the bacteria housed until then in the Malpighian tubules transfer to the lumen of some of them and thus cause an essential lengthening of the appropriated tufts.

Otherwise, as far as we know, new formations caused by symbiosis are always involved.   The so-called fermentation chambers exemplify the increasing degree to which the organism satisfies such requirements. In the hindgut of crickets there is no corresponding separation of the bacteria-populated section at all; with tipulid larvae all transitions are found up to the formation of sharply set off, voluminous ceca; in lucanid larvae a circle of little chambers is formed; in those of the melolonthines a section of the hindgut is inflated to a huge balloon; and in *Scarabaeus* a narrow cervix is the only connection between a large cecum and the hindgut.

The trypetid imagines offer to their symbionts in part completely new evaginations in the midgut which represent various degrees of separation. An effort to keep the symbionts from the intestinal content is clearly revealed by a zone of shallower folds enclasping the gut on all sides; by deeper evaginations more sharply set off, unilaterally placed, emptying in a broad front into the gut; and finally by an appendage connected to it by a narrow passage. The peculiar cephalic organ of the *Dacus* imago, which has not been observed in other trypetids, must also be regarded as a new formation.

New formations of a special size, through which extracellular intestinal symbionts can also be removed extensively from the lumen, are found especially among the Heteroptera which suck plant juices. Here may be found a few slender, sometimes branched tubes or a greater number of rather long and lightly fused tubes, but chiefly hundreds of little crypts arranged in two or four rows. In acanthosomids the narrow passages by which they are connected with the intestinal lumen have grown together, and thus there is no connection at all between the two types of spaces.

In the next significant step revealing the same tendency toward more intimate union and simultaneous withdrawal from the intestinal lumen is the admitting of symbionts to the interior of the intestinal epithelial cells. That such admission is not difficult, is proved by numerous cases where the transfer is repeated again and again before our eyes, sometimes in limited areas and sometimes throughout the entodermal intestinal system, as often occurs in algal symbiosis. This step is not surprising when one considers that intracellular digestion is widespread both in protozoa and in lower, multicellular animals such as coelenterates and worms. Wherever bacteria and fungi are incorporated like food fragments in intestinal epithelial cells which usually receive only dissolved substances, it may be said that the insects are reaching back to capacities rooted in their phylogeny.

Sometimes the transfer to intracellular life does not represent complete surrender of free living, and sometimes the transfer occurs desultorily with individual representatives of a larger group holding fast to settlement of the intestinal lumen. In heteropterous bugs, for example, only among the aphanines is there extra- and intracellular settlement of intestinal appendages, in this case elongated and filamentous. Of course, blood-sucking triatomids in part house their bacteria in slightly enlarged cells which form a zone adjacent to the valvula cardiaca, yet they also transfer continuously to the intestinal lumen and are found there in large accumulations mixed with blood in the process of dissolution. With the glossines there are differences between larvae and imagines caused by holometabolism. In the larvae the symbionts occupy a very short section of the intestinal epithelium adjacent to the valvula cardiaca but cause no enlargement of the cells worthy of mention; in the imagines they occupy two sharply defined zones, which are situated somewhat farther back and protrude markedly because cell growth has become considerable. In both states, bacteria transfer still to the intestinal lumen, though to be sure in smaller numbers.

In all curculionids except the cleonids the symbionts occur in the most varied places but always intracellularly and never in intestinal epithelial cells. In the cleonids the symbionts occur both intra- and extracellularly in variously formed evaginations, which are always situated at the beginning of the midgut and which also must be considered to be new formations. Both methods of living are encountered in similar form among the hippoboscids. In some species both larvae and imagines house the symbionts intracellularly, even though at different places in the intestinal epithelium. Other species do not open their cells until metamorphosis; and as larvae they house the bacteria in the lumen of the midgut, throughout its whole extent or in an enlarged anterior part. However, the infected cells, which usually react with increased growth and development of a weaker brush border or none at all, are always sharply set off against the immune cells and display a species-specific increase in dimension.

Several corresponding stages of adaptation are found among the hirudines. *Hirudo medicinalis* appears to diffuse the symbionts in the intestinal lumen; *Piscicola* concentrates them in the hollow of special esophageal evaginations; and *Placobdella* copiously infects the epithelium of corresponding organs, without giving up its extracellular mode of living.

An unusual situation prevails among the cleonids, where symbionts are formed both inside and outside of cells in the intestinal evaginations which were doubtless formed expressly for symbiosis, for, with all other new formations now known, their settlement precludes simultaneous occurrence in the lumen, for instance in the anobiids and cerambycids as well as in *Cassida, Bromius,* and *Donacia.* Throughout there are evaginations at the beginning of the midgut. This is a location which seems predestined for the purpose, and therefore such evaginations are sometimes formed even when no symbionts are present. We can prove that they are actually new formations by comparing them with symbiont-free relatives without such differentiations. The forms may vary: roundish and tube-shaped saclets may occur side by side, or smooth saclets along with buckled or tuft-covered ones. Sometimes there are only two or four evaginations, and sometimes a whole circle of them. With the cerambycids the circle is set somewhat farther back and in several species is joined by a second. Larvae and imagines may both have the same organs at different stages of development, as in anobiids, buprestids (?), and *Cassida;* or their organs may differ considerably one from the other as in *Bromius;* or the organs may be lacking in the imagines (then they are not replaced by other sites), as in long-horned beetles and, within certain limits, in *Donacia.* Besides bacteria there are also yeast-like inhabitants (cerambycids, anobiids), that is to say, organisms which up to the present have never been found as regular settlers of the intestinal lumen.

The same transition from extracellular to intracellular settlement may take place in the Malpighian tubules, which are used in a number of cases as symbiotic sites and in the process are more or less transformed. An extracellular confiscation taking place uniformly over all vessels has been observed thus far only in *Bromius* larvae, the symbiotic bacteria of which are transplanted during metamorphosis to the midgut tufts mentioned above. By a transformation emanating from the body cavity certain apionines, *Aspidapion* and others, infect a portion of the cells which make up the vessels, without limitation to certain areas.

The same applies to *Ixodes,* although with other ixodids only certain zones, recognizable from the outside as thickened sections, of the Malpighian tubules are settled. In *Rhipicephalus* and *Dermacentor* only the fourth section of the two vessels might at most contain symbionts, and in *Boophilus* radically enlarged cells are set off from the sterile areas in the form of small, terminal buttons. The situation is similar with the argasids, where *Argas persicus* opens the vessels throughout; *Ornithodorus,* on the other hand, places only a limited thickened section at the disposal of the symbionts.

With *Donacia* the narrowing goes one step farther, for in older larvae of both sexes only two of the six Malpighian tubules are occupied for a distance. Yet, strictly speaking, this is merely an early settling of future transmission organs, which in males leads to reduction of the symbionts or even to their complete disintegration. A unique occurrence is the hypertrophy undergone by four of the six Malpighian tubules in the ipid *Coccotrypes,* for they yield almost throughout their extent to symbiotic bacteria and are in this way withdrawn from their original functions. The most radical reshaping of the Malpighian tubules is observed in some of the apionines. The two tubules, here likewise placed exclusively in the service of symbiosis, are transformed into little clubs in which the lumen is now limited to the short pedicle and which otherwise consist essentially of infected giant cells.

We were able to follow clearly the transfer of symbionts from intestinal lumen to intestinal epithelium in a whole series of test objects. However, transfer of the symbionts from the gut to the space between the intestine and hypodermis is less easily demonstrated because we cannot observe the process so clearly, at least with bacteria and fungi, for a site of this kind always leads to transmission by ovarial infection, thus precluding a possible repetition of the passage from the intestine into these regions.

However, when algae are found in the mesenchyme they also have reached it indirectly by infection of intestinal cells. A small number of the green Turbellaria still harbor the Chlorella in the intestinal epithelium exclusively, but in a larger number, although they are also received there at first, are later all transferred to tissue behind the epithelium. The algae of Anthozoa, which live in the lumen and epithelium of the endodermal cavities and in gaps of the mesogloea, probably also transfer circuitously by way of the intestinal epithelium.

Certain Pupipara represent the only forms allied with bacteria in which a corresponding migration is repeated again in each generation. The swallow louse *Ornithomyia* differs from other hippoboscids in that the imagines do not house their symbionts in the intestinal epithelium but in a mycetome-like, sharply defined syncytium inserted behind the latter in the musculature of the intestine. However, the larvae carry their symbionts as usual in the intestinal part adjoining the deglutition apparatus, and during pupation these glide backward. Then they force themselves, in the form of balls without nuclei, through the intestinal epithelium into the mycetome rudiment, which has been prepared in advance and apparently is derived from the mesoderm.

With the nycteribiids, where symbiosis developed much as it did among the hippoboscids, the symbionts are also localized in the area of the mesoderm. Indeed, Aschner had discovered regular, imaginal mycetomes or mycetocyte groups, in this case situated far from the intestine. In spite of this, transmission is executed by the milk glands, linking the larval symbiosis unfailingly with the gut and thus causing these exceptional states. Unfortunately, neither larvae nor pupae of these animals have as yet been made the subject of an investigation which would actually prove that the migration of the symbionts takes place through the intestinal epithelium.

In this connection we are, of course, reminded of the two cicadas, *Cicadula 4-notata* and *Cymbomorpha* sp., which carry yeasts in the intestinal epithelium and partly also in the area of the body cavity without taking them up into the egg cells.

Even in a group generally housing its symbionts in the space between intestine and hypodermis, occasional cases are found where well-adjusted symbionts are localized in the intestinal epithelium or in derivations of it, without infection of the egg cells, however. Whether it is a case here of an original condition or of a stage of gradual penetration cannot be decided.

In two other cases, in which well-adapted symbionts infect definite sections of the gut epithelium, it is a question of objects of considerable phylogenetic significance. *Marchalina hellenica,* representative of a special tube of the margarodids, is the only known scale insect which localizes its guests in a well-circumscribed section of the midgut, but nevertheless its symbionts, like so many other, less well-adapted coccid symbionts, infect still young ovocytes by way of the nurse cells. A similar condition is found in *Primicimex.* Also in this most ancient representative of the cimicids are the symbionts localized in the epithelium of the midgut. Even though in this case too nothing is known of the method of transmission, one may still be certain that it takes place through the egg cells. The fact that in almost all other Heteroptera the symbionts are united with the gut is in this relation of special significance. At this point we must also call to mind the Anoplura, for with them too reminiscences of an older localization are not lacking. Indeed, the mycetomes of *Pediculus* and *Phthirus,* which are located beneath the gut, can be traced back to an abscission of the same, and this is true for the mycetocytes of *Pedicinus,* which are inserted between the gut epithelium and the local membrane.

It is not true that settlement of the intestinal epithelium might perhaps occur frequently where the adaptation of additional, still more or less undisciplined, bacteria is visibly in full swing behind the intestinal epithelium. Thorough research on the membracids has not yielded one single example of this; and for the fulgoroids there has been only one observation, although a very striking one, and that was in a cixiine where only in the area of the rectal organ was the intestinal epithelium swollen into a thick layer and heavily filled with filamentous bacteria, instead of being flat and spread out. We must also recall that in *Orthezia insignis,* both in European and in African material, masses of small bacteria which should possibly be regarded as incipient symbionts were found in the intestinal lumen, in the intestinal epithelium, and in the Malpighian tubules.

Thus everything indicates that this transferral of the symbionts from the intestinal lumen to the spaces behind it usually took place very rapidly and often probably without lasting settlement of the epithelial cells. More or less shadowy reminiscences of such a route during symbiont acquisition may have been preserved in embryonic development, for in its course the symbionts often come to be situated in the midgut rudiment still filled with yolk, until a kind of flight from it sets in. An especially primitive type of behavior may be involved among the bostrychids when the transferral from the embryonic midgut to the body cavity takes place without previous addition of yolk nuclei.

The manner in which symbionts are housed in the space between hypodermis and intestinal tract is extraordinarily varied, the individual devices being almost too numerous to be surveyed. We shall discuss first the yeast-like organisms which appear in this area exclusively with Homoptera, and even among these Homoptera only within the scale insects, cicadas, and aphids. In part they behave in a very primitive manner and usually cause organ-forming reactions at a low level, if at all. They are found only extracellularly in some of the fulgorines, where they are situated in gaps between fat cells, in such regularity that a honeycomb arrangment results. Later the affinity to the fatty tissue usually becomes more pronounced. Its next step is illustrated among the scale insects by the lecaniines, the genus *Lakshadia,* and certain *Rastrococcus* species, and among the cicadas by the cicadines and certain membracids, such as *Notocera.* In these cases numerous free yeasts populate the lymph; other yeasts sometimes press in masses into the interior of the fat cells, apparently without influencing their functions. Recently it was discovered that several hormaphidines carry yeasts in the lymph and fat cells in place of the hereditary symbionts.

This condition leads us to far more numerous cases where the yeasts, aside from a few exceptions and from symbionts on the way to egg infection, are situated in the interior of the fat cells thereby withdrawing the cells from their original function completely or at least to a very high degree. Heavy encumbrance of the protoplasm leads to production of giant cells with polymorphic nuclei that often constrict off buds and, especially, to lifting of the cell limits and thus to formation of syncytia. At the same time the origin of the fat cells is usually clearly recognizable, for transitions with weaker infection may be found and occasionally one or another fat cell may still be inserted in the syncytia.

Generally, sterile fat cells still surround these yeast-occupied zones, which may be distributed at random or restricted to certain areas of the tissue. In issids, phalaenomorphids, and flatids, as well as a number of cicadoids, such as the ledrids, jassids, and euscelids, this fat-cell envelope is still multilaminate and of varying strength; in delphacids the organ-building tendencies are more pronounced, the fat cells having a marked epithelial character in the infected areas which at the same time are often compressed. In addition to species with envelopes which are already unilaminate but still supplied by typical fat cells, there are species in which the syncytia, enlarged by fusion, are surrounded by a regular plasma-rich epithelium, in which no fat droplets are observable. Transitions of the most varied kind document the path which leads from diffuse settlement of the fatty tissue to this final state. The fatty tissue in *Rastrococcus* "franssenii" displays great similarity to that organ-like development which in cixiids is caused by localization of yeasts in the fatty tissue. A fatty syncytium contains numerous small nuclei interspersed with large nuclei around which masses of yeast cells are grouped, and the individual lobelets are surrounded by a single layer of fat cells.

There is also a tendency toward organ-like organization of another kind, culminating in a much more rigid separation of the yeasts in sharply defined sites, paired or unpaired. In *Ulopa reticulata,* the only aethalionid investigated, are found two large, irregularly shaped formations, situated toward the inside against the paired *a*-organs and still surrounded by a simple layer of fat cells that sometimes is incomplete. In addition to its other symbiotic organs, *Issus dilatatus* develops a large, unpaired yeast organ consisting of a single syncytium without an envelope. Among the membracids, which house their yeast guests in such varied ways, we also find such organs, in addition to all possible earlier devices. Indeed, we find not only diffuse permeation of the lymph and fat cells and spatially defined synctia formation with very sharp separation from the fatty tissue, but also paired mycetome-like sites, which, like those of *Ulopa,* have a unilaminate covering of fat cells (microtalines, *Aconophora*) or, as with *Pyrgauchenia,* take the form of two elongated lobed syncytial complexes without a cellular envelope of any kind. Ultimately they also localize the yeasts in their *a*-organs; hence they will be discussed again later.

Similar sites established for bacteria at the expense of the fatty tissue are rarer and only occasionally do they achieve a formation suggesting mycetomes. The fatty tissue of the beetle *Trixagus* contains paired bacteria-populated cell groups which are apparently derived from it and indeed contain fat cells here and there in the interior. Whether the large cell mass, consisting of uninucleate mycetocytes and taking up a good part of the abdomen of *Omphalapion,* also represents modified fatty tissue must remain uncertain until embryological data are obtained.

Again it is primarily the Homoptera in which phenomena of this sort occur. A *Trioza* species represents a special case among the psyllids. Here a symbiont type, which has replaced one of the two forms which elsewhere are housed together in a mycetome, prefers to live in these neighboring fat cells. In the cicadine *Platypleura kaempferi* E. a third accessory symbiont is housed in the same way near the mycetome. The *a*-organ of *Tomaspis tristis,* consisting of giant uninucleate mycetocytes, is surrounded by numerous, small, bacteria-populated syncytia, the fat-cell nature of which is still clearly recognizable. The multinucleate giant cells of *Philaenus spumarius,* which contain little globules and accompany the *a*-organs, do not seem to have the nature of fat cells but in spite of this one will have to take into account that they may have been derived from them.

Among fulgoroids there is no lack of similar phenomena. For instance, a meenopline houses coarse tubes in a right and left zone of the fat body, and an asiracine houses long, tube-shaped organisms in somewhat laxer fasion. In both instances such housing results in growth of the cells and their irregularly developing nuclei, as well as in fusions. In a derbine the primitive histological character of certain bacteria-containing cells is still clearly recognizable, and in the laternariid *Lycorma* the fatty tissue is permeated on many sides by bacteria-containing large syncytia. In *Rastrococcus* "franssenii" we have seen that the yeasts greatly modify the fatty tissue. There is also a high degree of modification when *Rastrococcus iceryoides* houses bacteria. In this species the spaces which are somewhat less regularly bordered by fat cells contain no nuclei but only enormous masses of bacteria enclosing numerous fat droplets.

Not infrequently it happens that ill-adapted symbionts localized in mycetomes may also appear in the fatty tissue. *Aphrophora salicis,* in addition to the *a*-mycetomes, has organs in which filamentous or rod-shaped bacteria are housed; yet these bacteria are still quite undisciplined and may also be found in great numbers in and among the fat cells, which are thereby stimulated to multinuclearity, and in the oenocytes and blood cells as well. The membracids also afford examples of such appearance in the fatty tissue, and here too the ill-adapted accessory symbionts may cause considerable growth of the elements attacked (*Darnis, Enchophyllum,* and others).

Something similar is true of the supplementary symbiotic bacteria in many aphids. Especially in older animals the symbionts reveal their deficient adaptation when they occur not only in the mycetomes but also in lymph and fatty tissue and even in blood cells and other elements. *Schizoneura lanigera, Pterochlorus roboris,* and *Stomaphis quercus* are examples. The latter two are trisymbiotic species in which the two accessory symbionts—cocci and rodlets—transfer to lymph and fat cells.

The methods of populating the fatty tissue thus far observed, whether strictly regulated or more or less wild, are alike in that the symbionts populate cell material which is already unilaterally determined and therefore for the most part differentiated. Thus these sites differ widely from the great mass of the sites to be treated at this point, for the latter, be they isolated cells or cell nests inserted in the fatty tissue or well-defined organs, are always derived from cleavage cells or at least from undifferentiated elements of the later mesodermal rudiment.

It thus seems appropriate to designate organ-like formations found in *Pyrgauchenia, Ulopa,* and *Aconophora* as pseudomycetomes, and to extend this concept to all the other mycetomes which also consist of cells but at first had a different function; examples are the stomach disc of *Pediculus* which is derived from an endodermal intestinal saclet, the ectodermal organs at the beginning of the midgut in the snout beetle, the three dorsal membranous formations in the lagriids, and the luminous organs composed of blood cells in the pyrosomes.

At first glance, mycetomes embedded in fatty tissue suggest genetic connection with the fatty tissue, but embryological research usually shows that they have nothing to do with it but represent instead genuine mycetomes which have, so to speak, been dissipated. Examples include the little uninucleate cells filled with minute bacteria in *Tachardiella;* the similarly shaped elements permeating the fatty tissue of *Cryptococcus* and *Eriococcus* in great quantity and invariably traceable to cleavage cells or yolk cells; and the sometimes larger and multinucleate mesodermal mycetocytes in *Orthezia insignis,* limited to certain zones and thickly filled with fine filaments. *Orthezia urticae* is different in that there are also quite large, extracellular symbiont masses. In *Tachardina lobata* the symbionts are paired biscuit-like organisms; in the diaspidines, they are usually round forms, but during transmission they are spindle-shaped, and presumably are closely related to the symbionts of the Lecaniinae; the cells populated by them are also situated without transitions between the fat cells and are derived directly from cleavage cells. The same is true of the yeast-populated mycetocytes of asterolecaniines and of the diffusely distributed bacteriocytes of *Rastrococcus spinosus* and the *Stictococcus* species, in which it is at times a matter of bacterial symbiosis and at others of yeast symbiosis.

Among the blattids too, one never finds transitions between the fat cells and the uninucleate mycetocytes inserted in type-specific manner among them. This is in consonance with Koch's finding that, when the bacteria leave the yolk of the embryonic intestine, they are admitted first to special plasma-rich elements, different totally in nature from the embryonic fat cells, and only afterward are submerged in the individual fat lobelets. In a number of mallophages the minute cells populated by filamentous or swollen bacteria, which may be found between the hypodermis and fatty tissue or between gaps in the latter, are derived more or less clearly from cleavage cells, and even if they are shifted in the manner of blattids to the interior of the fat lobelets, as in the mallophage *Coloceras,* these foreign elements are definitely involved.

Again, groups of mycetocytes with low organization and populated by bacteria, and occasionally groups of syncytia too, scarcely deserving the term "mycetome," are found in a number of snout beetles, usually in the vicinity of the midgut, but here a good many things point to partial relationships to blood cells and to fatty tissues, so that some of these devices, still in need of more thorough research, are presumably pseudo-mycetomes (*Balaninus, Cionus, Ceutorrhynchus, Smicronyx,* and others).

Occasionally there are closer topographical relationships to the intestine. We have already discussed the strange imaginal organs situated in the area of the musculature in Ornithomyia. In the larvae of *Aspidapion* small bacteria-filled cells adhere to the midgut, and the symbionts of *Formica fusca* live in cells forming a simple closed layer directly behind the intestinal epithelium. The *Camponotus* species, the only ants among which a bacterial symbiosis has been discovered, go one step farther by arranging the corresponding cells within the epithelium of the midgut, even though it is unlikely that they have a genetic connection with it.

In general, the symbionts do not maintain relationships with the intestinal epithelium after they have passed its boundary. This is true for all the variants of genuine mycetomes which we are about to discuss. Let us consider first only mycetomes populated by a single symbiont type. Very seldom is it a matter of simple accumulations of uninucleate mycetocytes, which are held together by an anucleate membrane such as the unpaired formation resting on the midgut in *Polyplax,* the rat louse, and in *Linognathus,* the dog louse, are loosely inserted or sometimes dissipated in nycteribiids. Similarly uncomplicated are the paired oval organs of the aleurodids, whose uninucleate mycetocytes are shifted in female animals to a place between the ovarioles during postlarval development. The two mycetomes on each side of the *a*-organs in *Cicadella viridis* also consist of a few large, uninucleate cells, thickly filled with bacteria and held together by a very delicate membrane.

Much more frequently they are the mycetomes covered with an epithelium, which is sometimes well supplied with plasma and nuclei, and at other times remains low and is provided with markedly flattened nuclei. Chiefly the cicadas and scale insects furnish here countless examples. Frequently cells of this envelope, both in greater and in lesser numbers, glide into the interior of the mycetomes and form an interstitial network between the mycetocytes. This is true both of the monosymbiotic mycetomes, e.g., the unpaired mycetome of *Pseudococcus,* and also chiefly of the cases in which the central region of the mycetomes makes room for secondarily acquired guests.

Mycetomes of this type are not lacking among the cicadas. Where they occur among the cicadoids they are also always surrounded by a well-developed epithelium abundantly supplied with plasma and nuclei. The *a*-organs of the cercopids, jassids, and euscelids provide numerous examples. Seldom are they so compact as in *Penthimia;* as a rule, they are elongated and are situated on both sides just below the hypodermis and, like the mycetomes of the majority of the Margarodidae, the segmental muscles undergo constrictions which are greater or lesser in depth and which increase in number with their length (*Opsius, Grypotes, Bythoscopus, Philaenus,* and others). With the fulgoroids, mycetome types consisting of uninucleate or, at most, dinucleate mycetocytes are very rare. Here we must mention first the insignificant, but widespread, usually paired *f*-organs, slender, sausage-shaped organs surrounded by a well-developed epithelium the mycetocytes of which, often arranged behind one another in a single row, are the site of minute cocci. The unpaired *q*-organ of *Stenocranus* and others has no epithelium and consists of a few giant mycetocytes with a single, strangely shaped nucleus in the center, around which the characteristic symbionts are grouped.

Usually syncytia occur instead of the uninucleate mycetocytes. Amitotic divisions may lead to formation of a moderate number of nulcei and cause commensurate growth in cells, for example, in the simply constructed paired mycetomes of the bedbug, where polygonal syncytia are spanned by an envelope consisting of only a few nuclei, or in the mycetomes of the related *Ischnodemus* which sucks at plants. Of similar uncomplicated construction are the mycetomes of the bostrychids and those of the *Dasyhelea* species. The cicadas especially have a great diversity of such syncytial mycetomes. In the *a*-organs of the cicadoids, syncytia very rarely take the place of mycetocytes and generally reach no unusual dimensions (*Selenocephalus, Cicadula*). The only exceptions are the gaeanines, where the otherwise uniform *a*-organs are cut into a whole group of spherical partial mycetomes, each consisting of a giant syncytium with numerous nuclei. As may be clearly recognized again and again in the fulgoroids, such large nuclei-rich masses are not derived from uninucleate mycetocytes but represent so-called synsyncytia. That is, they originate through later fusion of several syncytia and thus represent a type of plasmodium.

In these fulgoroids the *a*-organs bear an entirely different stamp and should apparently be considered synsyncytia. Sometimes the tendency to fuse goes so far that there is but a single gigantic territory, lacking all subdivision, which tends to be surrounded by a well-developed sterile epithelium of cubical cells. The plasma of the syncytium then is limited chiefly to a narrow marginal area and contains only the correspondingly flattened nuclei. As far as shape is concerned, they may be unpaired roundish or broadly oval organs, sometimes lengthened into transverse tubes; they may be bent into horseshoes; they may retain only a narrow bridge; or they may be developed in pairs, in which case they exhibit all transitions from spherical formations to coiled tubes of extreme length.

Another type of syncytially constructed mycetome, of special significance for fulgoroids and showing in its structure a certain similarity to that of their *a*-organs, is represented by the *x*-organs, those paired, tube-shaped mycetomes which are often divided into two or more partial pieces each and serve as a site for the giant symbionts in their process of growing to such an unusual size. Here too the protoplasm is usually restricted to the marginal zone containing the nuclei. In this case only septa-like processes project from the marginal protoplasm to the inside to form the niches in which the symbionts are situated. In the interior a cavity results from decomposition of the primary mycetocytes in which the still somewhat normally formed bacteria had first been housed. In addition to this type which is widespread among the fulgorines, cixiids, and others, there is a second type in which, apparently because such degeneration phenomena do not occur, the plasma fills the whole organ in uniform meshes, although the nuclei avoid the periphery and are situated everywhere in the plasma wedges instead (derbines, delphacines, meenoplines, *Myndus,* and others).

In addition to these most widespread syncytial fulgoroid mycetomes, there are a multitude of other mycetomes, established each time for different symbionts and more or less histologically distinct. Consisting of syncytia without genuine epithelium are the *c*-organs of some cixiines, the *o*-organs of issines and *Caliscelis,* the *s*-organ of the tettigometrids, the *m*-organs of *Fulgora europaea* and other fulgorids, and the *n*-organs of the nogodinines. Nuclei-containing epithelia span the *k*-organs of the poiocerines, the *b*-organs of the cixiids, and the *h*-organ in one derbine. The latter builds for its symbionts a single large syncytium the plasma of which is limited almost exclusively to the margin. The flat nuclei are situated just inside this border and in part are drawn along into the interior with the whirls of bacteria arranged in parallel bundles.

Another strange organ is the unpaired *r*-mycetome which is spread out flat and occupies a large part of the abdomen in some megamelines and delphacines. It lacks an epithelial covering, and it consists of giant syncytia, their plasma being united in the center and sending out bizarre, nuclei-containing processes on all sides. An organ type found in a single derbide suggests at first glance that two types of symbionts are housed here, for behind a strong epithelial envelope there is a circle of middle-sized syncytia and within this a large synsyncytium. In actuality this arrangement is explained by the fact that the secondary fusing affects only the center. The *c*-organs of *Oliarius*, unpaired tubes with anucleate covering, like the *x*-organs, represent a single syncytium from the plasmatic marginal border of which radial septa are projected between the fine filamentous symbionts.

This may suffice to illustrate the diversity of the cicada mycetomes insofar as they are populated by a single symbiont type.

The syncytially built mycetomes of *Oryzaephilus* originate in a very unusual fashion. After the cell limits are dissolved in embryonic mycetomes, a number of multinucleate chambers are separated in which a central nucleus grows to a special size each time, whereas the others around it remain small and retain their capacity for mitotic division. Also, the envelope-cell nuclei are arranged and differentiated in remarkably regular fashion. Two each are situated where the dividing walls protrude toward the inside, and a particularly large nucleus is situated in the center of the whole organ, which in no other instance is so perfectly constructed. While this nucleus was found to be 128-ploid in the imago, the central nuclei of the symbiont-containing syncytium are 64-ploid and the small nuclei lying between the symbionts are diploid.

Another device without counterpart is found in the species of *Margarodes* investigated by Šulc. This animal builds its paired mycetomes, surrounded by thin cellular envelopes, into a fenestra of the oviduct, so that the section comprising the transmission forms borders directly on its lumen and anchors them there with the help of a musculature surrounding the oviduct. Also, the *Nosodendron* mycetomes appear to represent an anomolous type, but more thorough embryological investigation is needed to clarify their structure.

The diversity of devices increases considerably when additional forms are admitted into animals which originally established an organ for a single symbiont type. Mycetomes with several symbionts are found with the lyctids, trixagids, coccids, aphids, psyllids, cicadines, and gamasids. With *Lyctus* and *Trixagus* both zones of the paired organ are built syncytially; in *Lyctus* a number of smaller syncytia are situated surperficially on several larger central syncytia, or they may be more or less submerged; in *Trixagus* a single central syncytium is surrounded on all sides by smaller syncytia. In coccids the tendency to incorporate additional symbionts is diversely developed. According to my results, as yet unpublished, it is found that chiefly in the margarodid family, in which, when three still completely symbiont-free tribes are not considered, only the Callipappini, Marchalini, and Coelostomidiini are throughout monosymbiotic, but the remaining tribes, either for the most part or completely, contain very differently shaped bacteria in addition to the always similarly formed hereditary symbionts. The former always live in the central regions of the mycetocytes, sometimes in the mononucleate cells and other times in syncytia, and they relinquish only the border zones to the hereditary symbionts, which are often forced drastically into the background. According to all reports presently available, only in the Iceryini is such an intimate union lacking, circumstances permitting. As far as the other coccid families are concerned, everywhere in the fatty tissue of *Tachardina silvestrii* are numerous little associations composed of a central cell with a strongly cleft nucleus for the one type, and isolated, superficially situated uninucleate cells for the other type. The genus *Rastrococcus,* as we have already reported, comprises species which contain only bacteria or only yeasts as well as species which have both types of symbionts, yet housing in mycetomes is never achieved; instead, fat cells and lymph alternate in this function.

Among aphids, disymbiotic species are not rare and occasionally there are even trisymbiotic species. There is a great variety of methods for incorporating supplementary areas into mycetomes, which are invariably present and are occupied by typical small, round symbionts. An unpaired syncytium may be inserted between the shanks of the mycetome as with *Macrosiphum jaceae,* or a paired syncytium may be firmly built into the two halves as with *Chaetophorinella testudinata.* With *Pineus* the additional guests populate superficially situated uninucleate cells and the hereditary symbionts are housed instead in syncytia; and with *Pterocallis juglandis* both types are probably units in mononucleate cells.

Where two types of additional symbionts are present, they probably both find a habitat in the mycetome. *Stomaphis quercus,* for example, localizes a tube form in mononucleate cells and a rod-shaped symbiont in several larger syncytia, inserting both between the cells of the original mycetome. The additional symbionts may also occur in the area of the fatty tissue, in the lymph, and in the blood cells; thus they reveal their incomplete adaptation. When, in addition to the yeasts present in lymph and fat cells, supplementary bacteria occur in hormaphidines, these bacteria are only found drifting free in the lymph.

In this respect the disymbiotic psyllids stand on an essentially higher niveau. On their mycetomes, originally unpaired but very distorted and cleft with increasing growth, two sections can always be differentiated aside from the well-developed epithelium: a syncytially built section with the one symbiont type, and a section consisting of mycetocytes with the other type; and the mycetocytes either form a well-separated, connected margin around the syncytia or penetrate the same, distributed evenly or collected in nests.

Cicadas vary remarkably in housing additional symbionts. When it is not a matter of yeasts, the cicadoids in an overwhelming number of cases, as with aphids, psyllids, margarodines, make spaces available in their heriditary mycetomes for the symbionts or at least loosely install them there, whereas the fulgoroids are almost always disinclined toward such concentrations and prefer to erect specific housing for each kind. For the diverse manner in which these various organ types can be combined with the di-, tri-, and polysymbiotic species of the fulgoroids, the reader is referred to the detailed presentation in the corresponding chapter.

When several forms are united in one organ in cicadoids, it is always the *a*-organs which reveal such readiness to receive. In this case it is chiefly the *b*- and *t*-symbionts which are so obstinately mutually exclusive, yet each is handled in very different fashion. The *b*-symbionts are frequently found in completely independent mycetomes, and, even when they enter into close topographical relationships with the *a*-organs, they still have considerable independence insofar as the *b*-symbionts are grasped by the *a*-organs patelliformly; but toward the hypodermis they always remain naked and are never closely connected with the *a*-organs by a common envelope formation. Their mycetomes, as long as they remain independent, tend to be composed of large, mononucleate mycetocytes (many *Philaenus* species), but in company with *b*-organs they are syncytial in structure (*Cercopis sanguinolenta, Lepyronia coleoptrata,* and others). *Philaenus alni* possesses such mycetomes too, but in addition to the *a*- and *b*-symbionts a third tube-shaped guest populates smaller nests which also come to be situated loosely in gaps between the two mycetome parts.

The situation is different when it is a question of housing *t*-symbionts, which are more widespread among the membracids, jassids, and euscelids and which live in mononucleate mycetocytes exclusively. Here a pronounced disinclination to erect independent organs comes to light, and the various cases show a clear sequence of ever firmer rooting in the *a*-organs. *Euacanthus interruptus* represents a very rare primitive state, for the complex of *t*-cells is still inserted only loosely in the *a*-organ in the manner of the cercopids, the two epithelial envelopes remaining completely separated, and even the dorso-ventral muscles pass between the two organ parts. Otherwise the cellular envelope covers the two territories and unites them more intimately even when the *a*-section remains unilateral (*Eupelix, Strongylocephalus, Aphrodes, Paramesus, Deltocephalus,* and others). The next significant step takes place when the *a*-organ finally encloses the *t*-cells on all sides and therewith furnishes the entire epithelial envelope. Such concentrically built double myce-tomes are thoroughly characteristic of the membracids.

These mycetomes composed of *a*- and *t*-symbionts are extraordinarily inclined to attract other microorganisms, usually bacteria and, more rarely, yeasts; and, especially in the membracids, they display all con-ceivable stages of an effort toward harmonious adaptation limited to certain areas. For details the reader is again referred to the Specialized Section. A tendency is manifested here again and again to reserve a third zone between the areas of the *a*- and *t*-symbionts for another symbiont type; in other cases a special mycetocyte type is intermingled among the *t*-cells; or both tendencies appear side by side, since frequently not only a third symbiont is housed in sites of this type, but even a fourth and at times a fifth. Although additional receptions occur far less frequently in jassids and euscelids, yet even here they occur occasionally, sometimes in a very balanced manner, as with *Idiocerus* and *Agallia*, where sometimes a cleanly divided zone of mycetocytes is inserted between the areas taken up by the two other symbionts.

A third type of double mycetome is found in a long series of cicadines, such as *Cicada plebeja* and *septemdecim;* each individual, round, partial mycetome consists of one central syncytium and of one forming a spheri-cal disc, and one of these is given over to the *a*-symbionts and another to tube-shaped symbionts which are not identical with *b*-symbionts.

Observations of many kinds show that the epithelial envelopes of the cicada mycetomes are used sometimes to house symbionts. In a membracine wih six different symbionts the still extremely undisciplined *η*-symbionts appear here and there in the epithelium of the *a*- and *t*-organs; in *Enchophyllum 5-maculatum* another accessory bacterium is found which also lives in the *t*-section, this time thickly and uniformly. The invasion of the epithelial cells by *n*-symbionts seems to be more regular with *Campylenchia,* where they already have the total character of uniformly densely occupied mycetocytes, although here the same guest takes possession of the *t*-organ also, indeed to a very great extent.

An apparently long-standing occupation of the mycetome epithelium, leading to an extensive transformation of the organ type, is found also in *Lycoderes, Omolon,* and *Sphongophorus.* Here yeasts localized in the epithelium of the *a*-organs change the character completely. The epithelium swells into a giant syncytial mass which cleaves the original mycetome in various ways. In place of the epithelium disposed of in this manner, a mere anucleate membrane now envelopes the organ.

The mycetome epithelium is used also in several fulgoroids (laternariids). Again it is the *a*-organ which makes its epithelium available to delicate filaments in *Crepusia nuptialis* and several of its relatives. Once again the epithelium reacts with syncytial formation, being transformed into a thick marginal zone which extends all around and is covered by an anucleate membrane. With another poiocerine, a *Lystra* species, it is the epithelium of the *k*-organ which is colonized in similar fashion by filamentous symbionts and is transformed into a uniform syncytium. Just how, in general, an unregulated infection may have preceded such a harmonious housing of additional symbionts is shown by the cicada *Trypetimorpha,* which today still remains at this stage.

All these mycetomes, the epithelia of which were sterile at the start but were raised to the status of symbiotic sites, have unfortunately been grouped under the term epithelial organs, but one must not forget that they were actually extremely diverse formations occurring here and there spontaneously.

Over and against these formations involving true mycetomes, the pseudomycetomes are of lesser importance. In consonance with their heterogenous origin, they afford no points of comparison, merely representing solutions found here and there on occasion within a narrower systematic association. We have already spoken of the beginnings of mycetome-like formations for the purpose of enclosing the yeast symbionts. Much more clearly delimited are the compact pseudomycetomes, individually so different in shape, which numerous snout beetles form at the border of the foregut and midgut from presumptive ectodermal cells and which clearly reveal their exceptional nature during metamorphosis by dissolving into their individual components and being distributed as such over the midgut epithelium. Another unique site is found in the lagriid larvae. In the dorsal region three glandular membranous saclets, one behind the other, are constricted off after the symbiotic bacteria transfer to them and are submerged below the hypodermis. The structure and manner of origin of the stomach disc in *Pediculus* and *Phthirus* species are equally unique. As we have learned, mesodermal cells in two forms participate in constricting off the midgut epithelium, although the symbionts remain extracellular and are united in chambers.

Among the symbiotic luminous organs, only those of the pyrosomes represent mycetome-like formations, but these accumulations of mycetocytes, without envelopes and simply constructed, can be traced back in the first four individuals to testa cells and later to blood cells and therefore must also be classified among the pseudomycetomes. However, the infected blood cells in the lateral organs of the salpa appear loosely intermingled among the non-infected constituents of the blood, so that it is decidedly inappropriate to speak of organ-like sites. In all other cases of light symbiosis there is an extracellular colonization of invaginations of the integument or evaginations of the intestinal canal which have no points of contact with the sites observed previously but nevertheless arouse special interest because they sometimes cause the host organism to form auxiliary devices of many types, such as lenses, reflectors, and dimming devices, and occasionally influence development of nerves, muscles, and blood vessels as well as pigmentation and scale formation.

Convergence phenomena are involved when the Teleostei establish symbiotic luminous organs resembling membranous glands in the orbital region, in the lower jaw, or in specially differentiated fin rays, and when other forms convert glandular structures on the esophagus or hindgut to act as luminous organs. On the other hand, the various bacteria-containing luminous organs of the myopsids are derived from a common root, as will be thoroughly discussed in the chapter dedicated to the historical problems. In this case they are always membranous invaginations which, like the luminous organs of fish, are colonized from outside, some retaining the primitive structure of the accessory glands and others being separated from them and developing into highly complex organs with all conceivable auxiliary devices.

We have already learned that the Malphigian tubules of insects and ticks harboring symbionts undergo either no change or at most only small changes, such as a thickening or shortening. The bacteria-infected purinocytes of the cyclostomatids and annulariids are not influenced in any way thereby, and in molgulids the symbiotic fungi drift in the lumen of the fluid-filled storage kidney; at most, a special adaptation to the bacteria may be observed among the oligochaetes, namely, in the ampullae of the nephridia, which are sharply set off and highly distinct in their histology.

12

Methods of transmission

The study of transmission devices is doubtless one of the most stimulating fields of symbiosis research. It is interesting enough to discover the never-flagging inventive skill revealed in creating the sites, but it is even more fascinating to discover the astonishing variety and precision in the devices used to preserve the symbiosis.

Surprisingly, the research of the last decades has demonstrated that even ambrosia-cultivating insects create distinctive devices in the structure of their bodies for transplanting the symbiotic fungi to passages newly laid by the mother animal, or for supplying them to the young larvae. Interestingly, the adaptations vividly resemble those caused by endo-symbiosis. In *Hylecoetus*, spaces which open at the vaginal mouth are filled with fungi spores; in siricids, there are paired intersegmental pouches and the females have also complex devices to prevent desiccation of organisms destined for transmission. The fungi-cultivating ipids have taken other paths, and we know today of eight different solutions to the problem of transmitting the vital fungus strains to the offspring. Some of the platypodids, which unfortunately have not been so thoroughly investigated behave in a very similar manner.

Nevertheless, even in cases of endosymbiosis the acquisition of new guests is sometimes left to chance. This is true in numerous cases among protozoa, coelenterates, and worms, where algal symbiosis frequently is not essential to life and where its continuation is not at stake even when symbiosis is more or less obligatory symbiosis, because the algae are clearly present in quantity in the milieu of the host animals and can be easily encompassed by the pseudopodia or taken up through the mouth opening. In similar fashion the useful bacterial flora of many insect larvae living on a cellulose-rich diet, as found for example in the so-called fermentation chambers, and the bacterial flora of higher vertebrates are usually taken in anew with food by the young animals.

Our present knowledge of Teleostei indicates that colonization of luminous organs, which indeed always open directly toward the outside or into the intestinal canal, is also traceable to the constant availability of abundant bacteria, in spite of the astonishing degree of organization achieved sometimes. When the organs open into the esophagus, as with *Equula*, the symbionts migrate through the mouth; if it is a question of the appendicular organs of the hindgut, the symbionts may possibly reach them each time anew through the anus, for it has been shown that the luminescent symbionts do not thrive in other intestinal sections. With the symbiotic luminous organs of cuttlefish, which are likewise connected with the outside, there has been no confirmation of the assumption, at first very striking, that the accessory glands are transmission organs which infect the cocoon fluid, and everything seems to indicate that even these luminous organs are always entered from the outside as with the fishes. The same mode of infection was clearly demonstrated for the accessory glands of sepiolids. As with luminescent fishes, this presupposes the presence of more complicated adaptations, for the organism must of course possess the capacity for stopping the development of undesirable intruders. With the accessory glands of the cuttlefish, a ciliation of the invaginations serves to remove the particles of dirt, sand, and the like that were admitted simultaneously; and with cephalopods and Teleostei, glandular separations of the organs in question promote settlement and rapid increase of benevolent guests. The never-failing inmates of the storage kidneys of the molgulids, which are not yet present in eggs and larvae, must similarly be taken up from the environment through the ingestion opening. A certain safeguarding of the new infection exists in *Convoluta roscoffensis* by the circumstance that the freely motile flagellate stages are lured on by the cocoon gelatins and find favorable conditions for increase, so that they are available immediately to the hatching young animals in sufficient quantity.

In other cases the infection of the cocoon fluid occurs within the mother animal. This method has been proved for bacteria in the ampullae of the nephridia in the oligochaetes and for intestinal symbionts of the hirudines, and it is presumably the same for symbionts of cyclostomatids and annulariids.

Such arrangements lead to numerous cases in which the mother animal gives a little store of symbionts to each egg during deposit, arranging the supply in such a way that it is the first food ingested. For this purpose there are special smearing apparatuses, jet-like organs, and additional symbiont depots on the ovipositor. Sometimes the egg surface can be sufficiently provided with symbionts by abundant removal of symbionts with the excrement, and in such cases bacteria are always involved.

Primitive conditions such as exist in *Hylemyia* are apparently not at all rare, yet they may easily escape our notice. Here abundant but loosely adapted bacteria flora fills the gut and the chorion is contaminated by it during egg deposit without the necessity for special devices. The fact that a number of trypetids quite regularly provide the egg surface with symbiotic bacteria originating from the hindgut without special devices being recorded in the anatomy might easily have been overlooked if it were not for the fact that other representatives of the same family possess special devices, some of them highly complex.

Buried in the hindgut of a number of relatives are longitudinal folds, always sealed rather well toward the lumen, but sometimes still subdivided at depth, which now serve as special bacterial reservoirs. The most primitive forms took the first step toward adaptation by devising a connection, between intestine and vagina. In later forms this connection grew more complex. The end of this developmental series is seen in *Dacus* with flask-like invaginations arranged around the hindgut and with a particularly long passage connecting the latter with the vagina. In addition to these improvements on the ovipositor, the micropyle region of these animals is equipped with cavities for reception of symbionts in a manner not observed elsewhere. This arrangement is of special significance here, because in contrast to all other bacteria given to the eggs in this manner, those of the trypetids, like spermatozoa, penetrate directly through the micropyles into the embryos and, after the formation of the intestinal tract, are admitted through the mouth.

With the Heteroptera there are four distinct devices by which bacteria localized in the intestinal lumen are transmitted in more and more highly perfected manner to the progeny. *Rhodnius,* and evidently also other bloodsucking triatomids, lack such devices which might guarantee an abundant supply of the bacteria although the seat of the symbionts is situated far from the anus, at the beginning of the midgut; in fact, the smearing of the eggs here is scanty but counterbalanced by the habit of the larvae of sucking at the excrement of their fellows. As a rule Heteroptera, which suck plant juice, make better provisions. Only in *Mesocerus (Syromastes) marginatus* does the smearing of the egg shell seem to be very slight and does the eager palpation of it by the young larvae seem not to take place, but as with *Rhodnius* the sucking up of the excrement appears to be significant. One is justified in speaking of primitive transmission organs in reference to some forms in the female sex where a certain number of the crypts situated directly next to the hindgut are radically enlarged, are thus filled more abundantly, secrete fluid in increased degree, and finally at the time of egg deposit send their content in masses to the lumen of the hindgut. Symbionts reaching the egg shell in this way are then sucked up by the young animals during the so-called rest period.

Where the crypts are sealed off completely from the intestinal lumen, more radical measures must be taken and in fact the representatives of all acanthosomines have established on the ovipositor a unique complex organ which is so constructed that the eggs gliding down press out part of its content. Characteristic of this organ and sometimes seen in similar devices of other insects are the following: special bacteria-filled chitinous tubelets on the wall of the depot; a connection of this depot with the vagina; hair formations which prevent undesired transmission to the vagina; and a clever chitinous serration closing the space toward the outside. An equally certain way is taken by the lygaeid *Stilbocoris,* which lays larvae ready to hatch and deposits next to each of them a clump of its bacteria-containing feces, which becomes the first meal after rupture of the chorion. *Coptosoma* shows us how this solution of the problem can be developed still further. In these plataspids the posterior section of the female gut is separated completely from the rest of the midgut and now serves only to manufacture those special symbiont-filled capsules set between the eggs in a regular arrangement. Joined here in perfected manner to ensure continued cooperative living are morphological differentiations and newly acquired instincts of the mother animal and of the hatching larvae which suck the capsules.

A method which is completely different but no less original is taken by *Brachypelta.* For days the freshly hatched larvae suck up drops of bacteria emerging from the maternal anus, and during this time of brood care the crypts of the intestine swell enormously. If Pierantoni's hypothesis, namely, that the eggs are smeared in *Tropidothorax* by end sections of the Malpighian tubules which are settled by symbionts and thereby modified, be confirmed, there will be another transmission method of Heteroptera symbionts localized in the gut.

All cleonids have an entirely different method for providing their progeny with the bacteria, which as larvae they house in evaginations of the frontmost midgut. At the boundary between the seventh and the eight abdominal segments, in other words, where the ovipositor passes into the sheath surrounding it in the position of rest, the cleonids have created on each side an invagination, usually club-shaped, with a greatly folded, bristle-covered epithelium and a well-developed, longitudinal musculature. Functioning like a regular jet, it allows some of the content to reach the eggshell each time during the moment of egg deposit and thus resembles similar formations in the siricids, which drive the oidia of their symbiotic basidiomycetes into the puncture canal. The same service is rendered in other species by more slender tubules, sometimes even clusters of them. As with the two following insect families allied with yeasts, the hatching larva is infected by consuming a part of the eggshell.

Anobiids and cerambycids have symbiont-filled containers in the same location, but they apparently act as reservoirs for filling additional spaces which have closer connection with the vagina. The anobiids have slender intersegmental tubules in place of the jets in the cleonids. Such tubules have a weaker transverse and longitudinal musculature and moreover are well provided, in contrast to the organs of the cleonids, with glandular cells which furnish the secretion surrounding the yeasts. Joining them at the end of the ovipositor are two pockets which are sealed toward the outside by overlapping chitinous plates but are connected with the vagina. From the pockets the symbionts reach the vagina and the egg-shell in quantity (*Sitodrepa*). In some species the intersegmental tubules are enormously lengthened (*Anobium, Dendrobium*), and the vaginal pockets moreover increase their capacity through sac-shaped extensions. When much more voluminous sites in the intestine correspond to these more efficient transmission devices, this parallelism confirms the rule, also confirmable elsewhere, that the circumference of the sites and the quantity of the symbionts placed at the disposal of transmission are balanced.

The cerambycids, although they are otherwise not closely related, be-have in a similar fashion, showing that certain possibilities are, as it were, dormant in the constitution of the hosts and that, when the need appears, these possibilities are realized. Again and in the same place, one finds glandular, intersegmental tubules and vaginal pouches, the latter not being filled through the former until after metamorphosis. The tubules are excessively elongated with many species; the pouches are stiffened and protected by plates. Some species ensure prolific egg smearing by an additional yeast-containing space dorsal to and connected with the vagina. A mucilaginous secretion covers the egg surface after deposit to keep the symbionts fixed to the surface, and sometimes an additional system of ridges on the chorion protects their fixation. The intestines of the newly hatched larvae contain mixed fragments of shell and symbionts.

Although the lagriids have formations similar in principle to those of the anobiid and cerambycids, the former reveal an inclination, not seen to the same degree in any other group, toward modifying those in the most varied manner and frequently even surpassing them. The native *Lagria hirta* has merely simple, oval, intersegmental sacs and vaginal pouches sufficient to guarantee complete coating of the eggs with bacteria-containing mucilage, but its many exotic relatives show varied serration and subdivision of the intersegmental organs, in part enormously enlarged, and possess large, sac-shaped, appendages of the vaginal pouches, of which only modest beginnings are present in *Lagria hirta*.

The method by which the symbionts are then taken up into the animal body is completely unique. One to two days before hatching, the intersegmental membranes of the dorsal region form three glandular invaginations, situated one behind another, to which the symbionts transfer and which are then closed behing them. In view of the bacterial reservoir on the ovipositor, sometimes so enormously developed, one may conclude that these odd organs, limited to the larvae, are sometimes much larger than with *Lagria hirta*.

Smearing of eggs by way of the hindgut, a method most highly developed in Heteroptera and trypetids, or by means of chitinous, membranous invaginations on the ovipositor, as developed by anobiids, cerambycids, cleonids, lagriids, and perhaps even the buprestids, does not exhaust the possibilities of such transmission methods, as the chrysomelids demonstrate. With *Bromius* and *Cassida* the host organism develops on the vagina slender glandular bacteria containers, lacking in all the symbiont-free relatives, from which originates the bacterial caplet which is placed on the head of every single egg precisely where it cannot fail to drop into the mouth of the hatching larva. The *Donacia* use an organ already present, developing it expressly for transmission purposes. Two of their six Malpighian tubules are withdrawn from their original function, their lower section being converted into considerably swollen containers for bacteria and their distal section into a gland furnishing the secretion of the egg cocoons. Again, they embed a well-defined clump of bacteria in the foamy secretion which surrounds the eggs and after deposit rigidifies at the place where the larva later gnaws its way through the shell.

Almost all insects which transmit symbionts to their eggs in some way are monosymbiotic insects. As far as we know today, only *Bromius* has two types of symbionts, each housed in places spatially separated but meeting in the vaginal tubes and together finding their way to the eggs.

Finally there is one other completely different method by which symbionts are "fed" for transmission purposes to the larvae. This method is used by the Pupipara and the glossines which feed the larvae developing individually in the maternal body with the secretion of their milk glands. Their symbionts are located in these glands and are sent with the secretion to the larvae in the act of swallowing and are then settled in specific places. The only exception is the nycteribiid *Eucampsipoda aegyptica* which does not link the symbionts to the intestine but houses them solely in the milk glands and in their secretion, in the way that the imagines of the cerambycids restrict the symbionts to transmission organs of females. By way of compensation the blind ends of the glandular lumina, especially widened for this purpose, are transformed into reservoirs richly filled with the delicately filamentous bacteria.

All these transmission devices involving a circuitous path by way of the outside world, or at least intra-uterine feeding, have one factor in common, namely, the localization of the symbionts in the lumen or in the wall of the gut. With such settlement the symbionts, even when housed intracellularly, are expelled through the anus, continually or at least in connection with metamorphosis, and thus it is easily understandable that such habits have led from accidental smearing of the egg surface to more regulated devices, the lubrication originating from the anus only or from spaces supplied through the anus which bring the symbionts into a closer relationship to the vagina. In lagriids it might seem at first glance that an infection of the egg cells by way of the dorsal organs is an extremely simple device of this sort; yet a complicated detour is preferred by renewed expulsion of symbionts during pupation, presumably also by migration in the exuvial fluid to the ovipositor and thus by inoculation of the intersegmental organs and vaginal pouches located there. All this is merely a clear indication that the host animals strive to keep the egg cells immune when their symbionts are not housed in extraembryonic or mesodermal cells.

Only a few cases are known in which this rule of insect symbiosis is violated. The ovaries remain sterile both in *Ornithomyia,* where symbionts populate a mycetome embedded in the musculature, and in the nycteribiids, where mycetome-like sites are formed. This is easily explained, however, through the close relationships of the Pupipara symbiosis to the milk glands. Conversely, it may seem more striking at first glance that egg infection is permitted in a number of Anoplura (*Pediculus, Phthirus, Pedicinus, Haematopinus*), although in their case the symbionts populate endodermal cell materials. But it is already clear from the preceding chapter that, where intestinal epithelial cells are used, the symbionts in these cases in all probability have first been housed in extraembryonic cell material, and thus a type of behavior which at first seems unusual is seen to be phylogenetically the older type. Of course, such an explanation might not be applicable when ipids and apionines allow their symbionts, which are housed in the intestinal epithelium or in the Malpighian tubules, to reach the egg cells, or when those snout beetles which erect pseudomycetomes for their symbionts analogous to the stomodaeum choose the same path, indeed the one which is usual for curculionids. Possibly in all these cases the unusual behavior comes about because it is not the occytes which are infected here, as they usually are, but the primordial germ cells. Finally, it is easy to understand how the rectal symbionts reach the egg cells, although they are localized in derivations of the intestinal epithelium, for doubtless they were originally true mycetome dwellers which merely moved to this unusual location later.

Many algal hosts behave differently and, in spite of localizing symbionts in the intestinal epithelium, use egg cells for transmission, but this is not surprising in view of their completely different organization.   Indeed, *Chlorohydra* is the genus for which infection of oocytes was first observed. A local severance of the supporting lamella and of the egg surface here ensures transmission of the egg cells for a longer stream of Chlorella without appearance of further complications.   Where migrations of the oocytes occur in the medusoids or sporosacs, they lose their immunity only when they reach the place of their maturity (*Aglaophenia Halecium, Millepora*).

Concerning egg infection, so widespread among insects, there are available numerous cases which may be organized under the rubric apolar, unipolar, or bipolar according to the place where the symbionts transfer to the egg cells.   With bipolar infection the symbionts are received at both the anterior and the posterior ends of the egg cells; the unipolar infection, the type most frequent by far, may take place at the anterior or posterior pole.

Ants and lyctids use apolar transmission.   With the first (*Camponotus, Formica*) there is a very early, extraordinarily abundant infection of the follicle cells and immediately thereafter a universal transfer to the still very young oocytes surrounded by them.   A stormy increase sets in at once, causing an extraordinary inundation of the protoplasm. The apolar infection in *Lyctus* is very different, for it does not take place until the eggs reach their ultimate size.   The follicle cells which at first were set closely together now diverge on the whole surface, as though by military command, to form a regular network with meshes through which the symbionts migrate.   As soon as these symbionts submerge in the marginal zone of the egg plasma, the follicle cells again push close together.   In very similar fashion the bostrychid symbionts in later stages transfer on all sides to the egg cell through gaps in the follicle cells.

Bipolar infection by one and the same symbiont type is known only among blattids, where it is generally widespread, and we may also expect it with certainty in *Mastotermes*, where the symbiosis is indeed similar in all other respects to that of the blattids.   Numerous bacteria-laden cells interspersed in the fibrous envelope of the ovarioles prepare for transmission into the ovarioles themselves.   Very early, individual bacteria approach the young oocytes, forcing their way through the follicle cells, and multiply on their surface until the simultaneously developing egg cell is surrounded on all sides by an extracellular covering of bacteria. When the increase of the symbionts can no longer keep pace with the enlargement of the egg, they are gradually restricted to the two poles. The symbionts are admitted to strange fold formations, varying from species to species, and when these formations are constricted off in the fully developed egg, the symbionts are eventually admitted to the latter.

Infection limited to the posterior pole is by far the most common method, and it is constantly being varied. The symbionts may at times reach the egg by passing through the follicle cells or by colonizing them temporarily, and at other times they may require the establishment of special ovarial mycetomes from which to feed the oocytes. Interfollicular infection is found with aphids, with several coccids, and in a different form with a *Goniodes* species and the aleurodids. With the aphidines it is only the winter eggs, and with the adelgines (chermesines) also the eggs that develop parthenogenetically, where in a narrowly defined follicle zone bordering the posterior end of the eggs, the cells swell at a certain time and draw apart so that direct access to the egg is made possible for the symbionts. Although the space between follicle and egg is not immediately filled, as it usually is with infection at the posterior pole, the symbionts transfer directly to the egg plasma which reaches the surface between the gaping follicle cells, and indeed in the adelgines even lightly protrudes. The migration lasts until the standard symbiont quantity required has transferred and the gaps in the follicle have closed again.

As already presented in detail in the chapter on specialized forms, the remarkable events, modified in many ways among the individual species and occurring during transmission of symbionts to parthogenetically developing summer embryos, are easy to explain in light of a once-present infection of yolk-rich eggs. In place of numerous gaps in the follicle, a single gap is formed. Through it a plasma process, formed early and surrounded by enlarged follicle cells, extends into the body cavity to receive the symbionts which transfer in a continuous stream to the space to be infected. The sole difference is that the symbionts come not upon unfurrowed egg plasma but upon embryos at various stages of development, for embryonic development sets in at a very early stage of egg growth but the follicle nevertheless maintains the old time schedule with regard to receptiveness. In isolated hormaphidines where yeasts appear in place of the original aphid symbionts, these much larger organisms nevertheless use the same place of entry.

With many scale insects (Margarodidae, *Lakshadia*) the symbionts also approach the egg at the posterior pole between the follicle cells, yet their transmission method is more like the one which is especially characteristic of cicadas, for they gather first in a space formed between follicle and egg and then are received all at one time in a groove formed at the posterior end of the egg. Also the orthezines are numbered among the scale insects which use the posterior pole, but they behave thereby with marked variation. A limited number of follicle cells, about forty, form a receptive girdle in the corresponding area. Animated formation of secretion begins, giving rise in each cell to a large thrombus turned toward the egg, to which the bacteria collecting outside in considerable masses transfer. Finally these thrombi are all constricted off and are taken into the egg.

In *Oryzaephilus*, too, gaps finally arise between the follicle cells that are developing particularly at this place, and through these gaps the symbionts next reach the space between the egg and the yolk membrane.

A *Goniodes* species (Mallophaga) is very strange in its behavior. Whole mycetocytes, such as were situated previously as normal sites between the fat cells, transfer to the ovarioles after collecting at the time of the third molting between the pedicles of the latter and migrate toward the egg between the cells. In so doing they degenerate, the nucleus disappears, and the freed symbionts gather in a groove at the posterior egg pole. The infection method which is universal among aleurodids, and until recently believed to be unique, is only one step removed from such use of whole mycetocytes, a subject to which we shall have to return in the discussion of the ovarial ampullae. Again intact mycetocytes of the dissipated organs appear at the posterior end of the young ovarioles and force themselves between the follicle cells. This time, in part through retreat of a narrowed process of the oocyte and in part through gaping of the follicle cells, a space originates which receives the intruders which unite to form a type of mycetome, the intruders being finally pushed forward and the free symbionts, just as always, being received by the egg as unaltered cells.

In two pseudococcines, *Macrocerococcus* and *Puto,* a counterpart of this method of infection was recently discovered. Here too, although at the anterior pole, several mycetocytes, which occasionally are in the act of dividing, transfer to the egg, but the fate of the implant of mother cells in the embryo differs occasionally. In the aleurodids they are replaced later by embryonic cells, and with these two scale lice a highly remarkable "somatic fertilization" of the mother cells take place through embryonic yolk nuclei.

The psyllids behave as very many cicadas do, in that the egg first develops a terminal process and the follicle cells surrounding it are separated as cuneate cells with the sole function of reception. The symbionts transfer to these cells and soon are transferred to the space which forms behind through reduction of the process. When the transfer comes to an end after a while, a groove of the egg receives them and closes after them.

With the cicadas this type of infection is modified with the help of cuneate cells in various ways. However, among the fulgoroids, on which more research is available than on the cicadoids, this manner of infection permits differentiation of a series of types which generally are connected closely with the individual systematic details. It is therefore quite possible to classify an unknown form with some certainty in the correct subfamily according to characteristic traits of egg infection. The zone of the cuneate cells can be of various heights; in the most extreme and admittedly rare cases it is merely a ring consisting of a single cell layer as with *Cicadella viridis;* usually there are two, three, or more rows of cells which for the most part undergo certain changes before the symbionts arrive. They usually develop to some degree, vacuoles occur in their plasma, and very often they protrude above their neighbors. All this points to changes which, in contrast to the neighboring cells, make them capable of reception. The transferred symbionts are enclosed in vacuoles and then, as with psyllids, take their departure on the other side of the wedge cells, sometimes without sojourning for any length of time. The result then is approximately a steady moderate filling of cells which serve solely as a passageway, so that it is appropriate to use the term "gliding manner of infection." This method is characteristic of the delphacids but is also found in tropiduchines, *Ledra aurita, Pyrgauchenia,* and others.

With this gliding type, the cuneate cells understandably undergo very few radical changes, but this is not true when the symbionts first collect in them and multiply. In such cases the microorganisms are no longer situated individually in the vacuoles but are united in nests; the plasma is soon permeated by them; the nuclei are deformed, and the cells are swollen, now more, now less. Moreover, there may be passive protrusions which go far beyond those which appear in connection with preparatory changes and which definitely have nothing to do with the reception of symbionts. As a rule, strong loading of the cells causes their limits to be dissolved and a bolster-like syncytium to be formed, which after maximal filling quickly pours the whole supply of symbionts into the space forming between egg and follicle or directly into the reception groove of the first. The empty cuneate cells then collapse and begin to degenerate together with any remaining symbionts. Representing this type are cicadids, cercopids, membracids, jassids, euscelids, and, not less, many fulgoroids, among them the cixiines.

Investigations with fulgoroids have shown impressively the great number of variations found in this method of circuitous transmission by infection of the cuneate cells. The number of wedge cells, the quantity of symbionts, the place of their multiplication, the tempo of transmission, the formation of the cavity arising behind the cuneate cells, the manner in which the egg cell approaches the reception—all this is modified from group to group in a manner dependent on the specific constitution of the subfamilies. In addition, there are specific form changes of the cuneate cells, displacements they undergo during the infection process, recourse to particular "roof cells," and many other peculiarities. The reader is referred to the Specialized Section for the constitution of these various types, such as flatid, fulgorine, tropiduchine, derbide, nogodinine, issine, and delphacid types.

Among the forms which send their symbionts to the egg cells at the posterior pole are the Anoplura, the rhynchophthirines and mallophages, but the methods they use are different from all others. For each of the two ovaries they erect a filial mycetome between the ovarioles and the oviduct, a so-called ovarial ampulla, from which the egg infection proceeds. The filling method, which varies greatly with the individual forms, is highly peculiar. The *Goniodes* species, already discussed, opens the way for erection of such ampullae. There the mycetocytes which transfer to the oviduct disintegrate immediately; in other mallophages they are preserved and as such fill the ampullae. With *Haematomyzus,* the elephant louse, paired mycetomes are transplanted in toto to the places in question; with *Polyplax* the ectodermal sex rudiment also swallows the intact, here unpaired, mycetome, but then its cells disintegrate, and the free symbionts migrate from the inside to the rudiment of the two ampullae. In *Pediculus* the symbionts leave the mycetomes earlier one by one and penetrate from the body cavity into the rudiment. *Pedicinus* behaves in the same way. *Haematopinus* species, however, behave in the most unusual way. In addition to their infected intestinal epithelial cells, they establish in females three other symbiont depots from which they supply the ampullae circuitously by the exuvial fluid and by migration through the sex opening. Although with these modes of origin the source of the cell materials may differ greatly in detail, the wall of the ampullae has the same more or less complicated tetralaminate construction.

In the case just discussed, the total supply of symbionts is finally displaced in this or that way to safeguard its transmission to the sexual apparatus. Similarly, a *Margarodes* species, contrary to all its relatives, incorporates its paired mycetomes into the oviducts in such a way that they border on the lumen, thus enabling the symbionts to transfer to it directly. This time the individual eggs extend into the oviduct by means of special processes of the follicle equipped with canals through which they receive their symbionts.

Thus far we have discussed methods of transmission now known which take place at the posterior pole. A third category exists in which the point of transmission is at the anterior pole. It is primarily the pathway which many coccids prefer and likewise all curculionids with exception of the cleonids, which carry their symbionts in the intestine, the ipids, which have not yet been sufficiently investigated, as well as mycetome-containing heteropterous bugs and several polysymbiotic cicadas and coccids; among the latter some symbionts use the anterior, and others the posterior pole.

The scale lice, insofar as they choose the anterior pole, use very simple methods. The bacteria or yeasts penetrate chiefly between the follicle cells, which combine nutrient and egg cells, and after a short intracellular life transfer to the space between follicle and nutrient cord, where they accumulate to a greater or lesser degree. Much rarer are the cases in which the symbionts prefer the path between the follicle cells. Occasionally the follicle cells swell in a very striking way (*Lecaniodiaspis*) at this place, as in most cases of infection at the posterior pole, and the symbiont-filled space around the nutrient cord may swell considerably. However, the symbionts invariably glide toward the egg without penetrating the cord, where, just as in cases occurring at the opposite end, they are admitted by a groove or remain behind the yolk membrane until gaps occur between the blastoderm cells through which they can transfer (Lecaniines, asterolecaniines, diaspidines tachardiines, eriococcines, pseudococcines).

In the case just described, the nutrient cells, which are made up of only a few large cells, were untouched by the symbionts. When, on the contrary, a much larger nuclei-rich syncytium is developed, from which nutrient cords passing to several oocytes situated one behind the other arise, an infection of the syncytia often takes place and the nutrient cords are used for transportation (Heteroptera, curculionids, ipids). With the Heteroptera and the snout beetles, the infection of the gonads is put into effect extraordinarily early. In the snout beetles the primordial germ cells, which are already separating at the blastoderm stage, are infected without regard to future sex from the supply imparted to the egg. The symbionts which reach the male gonad rudiments are then forthwith destroyed; in the females they are increased, and they fill the cells of the future nutrient chambers even in the youngest ovarioles. With the Heteroptera the symbionts still transfer during embryonic development, but at a much later time, from the young mycetomes exclusively to the female gonad rudiments which have not as yet displayed a separation into egg and nutrient cells. As a result of this, even after their separation several individual bacteria are always found in the young oocytes before the greater number transfer from the nutrient cells (*Ischnodemus*). This first infection of the gonads has not been described for *Cimex lectularius* and the ipids, but presumably it also takes place early, especially since even in very young animals of the first mentioned species a weak infection of the oocytes is observed before they enter into connection with the nutrient cells. Invariably the bacteria appear to find favorable conditions in the glandular nutrient cell plasma and increase abundantly in it before gliding with the secretion through the fiber cords into the egg cells. With the tropical bedbug *Leptocimex*, the infection of the nutrient cells seems, on the other hand, to play only a small role and the inundation of the youngest oocytes appears to set in independently.

In the cases cited previously, all symbionts transfer to the egg together in the same way wherever several live in the same host. Among the forms which infect at the upper pole there has been but a single disymbiotic species, *Tachardina silvestrii*. Here the two types arrive at the follicle individually or united in migratory cells and reach the egg together through the follicle. The same has been shown recently for the *Rastrococcus* species, about which we have already learned that some live with yeasts and bacteria. Also among the disymbiotic lyctids, a mixture of the two symbionts passes everywhere through the meshes of the follicle epithelium. However, in numerous cases two symbionts transfer simultaneously at the posterior egg pole, and in many cases three or even four and five. Examples are provided by all psyllids, the disymbiotic aphids and probably also the trisymbiotic one, the majority of the disymbiotic margarodids among the scale insects, and especially by a multitude of cicadas. It makes no difference whether the symbionts in question are housed in various divisions of the same mycetome or in separated organs; at the appointed time they all meet in front of the zones of the follicle and pass over to the egg together, sometimes using gaps in the follicle, but usually by temporarily infecting the wedge cells. Only occasionally, with certain cicadas (*Cixus, Stenocranus*), may slight temporal differences be found which lead to an approximate sorting of the various symbiont types in the egg ball.

It is all the more striking that among the cicadas and coccids several forms have been found in which, as mentioned, this rule is broken, transmission of the symbionts taking place at both the posterior and the anterior poles. The former are members of the cicadoids or ful-goroids. In all the membracids the *a*- and *t*-symbionts are admitted at the posterior pole, and additional guests, which occur in this group in quantity, usually follow the same route. But some symbionts char-acteristically prefer other paths and in so doing reveal that they are later acquisitions. *Enchophyllum* 5-*maculatum* harbors, in addition to its four symbiont types, two small rodlet types which live undisciplined inside and outside of the mycetomes. Both types infect the end chambers of the ovarioles, including the central fiber masses and the very young oocytes situated in the vicinity. After lively multiplication in the grow-ing eggs, they are all carried finally to the posterior pole, appearing there as a narrow border just beneath the surface, at a time when the other symbionts are still in the follicle cells. When the other four types transfer, they form a well-separated, spherical accumulation above the border. The η-symbionts of a *Campylenchia* species behave in a similar way, and with a polyglyptine the same symbiont variety leaves the mycetome in clumps and likewise reaches the egg cells by detour over the nutrient chambers.

From such random behavior, devices have developed at three places which represent a substantially higher stage of organization. The bugs *Nysius* and *Ischnorrhynchus* create for themselves, in a ring-shaped zone behind the layer of the youngest oocytes, a filial mycetome of sorts which strikes the attention readily because of the same vivid orange coloring which is peculiar to the mycetomes. When the oocytes glide in sequence through this infected cell girdle, the bacteria transfer to them without reverting to transportation by the nutrient cord and, after having been first diffusely distributed in the egg, are finally united in a ball, situated, by way of exception, at the anterior end.

Only the follicle cells were infected here, but *Fulgora confusa* takes an additional step. This cicada installs in the ovarioles behind the nutrient chamber a regular little mycetome consisting of a secondary syncytium in the center surrounded by a layer of flat envelope cells. It is a device which invites comparison with the ampullae erected at the opposite end of the ovarioles in Anoplura and Mallophaga. It is filled by the filamentous *m*-symbionts, which otherwise live in a primitive cell group on the intestine, and some of these get into the cords which pass through the organ from the nutrient chamber and thus are carried into the young egg cells. Of course, it has not been possible to observe how this filial mycetome is colonized by the *m*-symbionts, but presumably it is by penetration of individual bacteria which float free through the body cavity.

Finally, a similar formation is found with *Bladina fraterna*, a ricaniid, but here it could be shown that its origin is completely different and that, in the final analysis, a fragmented infection cone is involved. For this reason we shall return to this formation during the discussion of these devices which serve to produce specific infection stages.

Finally, my investigations of the margarodid symbioses, which are not yet completed, have uncovered here too within the iceryines and llaveiines a series of cases in which additional, less intimately adapted symbionts are transmitted over the nurse cells, as opposed to the hereditary symbionts. The fact that the symbionts of the *Marchalina* with their very primitive localization also choose this path is thereby understandable.

It is not surprising that symbiont transmission in bloodsucking arachnoids should display traits not found among the insects. Among the gamasids the eggs which transfer to the body cavity come in direct contact with the mycetomes, and the symbionts then migrate from mycetomes to eggs. Here, and this is found nowhere else, they are admitted to yolk spheres, in which they may still be detected in the embryos. In ixodids the first rudiment of the gonads has already been infected simultaneously with that of the Malpighian tubules; the symbionts are found subsequently in all cells of the germ layer and need only to be supplemented in the growing oocytes. The argasids behave similarly; with them, too, there is no further addition of symbionts from the Malpighian tubules at all, and even the undifferentiated germ layer is infected, though sparingly at first.

Understandably, the tunicates also use entirely different methods in transmitting the symbiotic, luminescent bacteria to the progeny. From the mycetocytes of the pyrosomes, spores emerge which infect the follicle cells of the egg, just beginning its furrowing, and are later increased there. The cells, converted thus into transportation media, are dissolved from their epithelial association and as testa cells force their way between the blastomeres. During subsequent development the embryo is penetrated by constantly migrating and regularly arranged cells of this sort until they have grown to approximately 400. Then, without degenerating, they form the eight luminous organs of the first individuals which bud immediately on the oozoid. This unique behavior resembles the migration of intact mycetocytes to the eggs in *Aleurodes* and bears an even stronger resemblance to *Puto* and *Macrocerococcus*, for in these two genera the maternal mycetocytes serving transmission are not destroyed as in *Aleurodes*. Thus, once again, it is seen in particularly impressive fashion that adaptation of foreign organisms in the animal body can lead to extremely odd consequences.

In the development of the salpa, which falls completely outside the expected, a transitory framework of organ rudiments composed of maternal cells is replaced step by step by embryonic material, and this development is traceable ultimately to the circumstance that here a good number of the blastomeres of a strictly determinate system are occupied by symbionts for purposes of transmission and thus are withdrawn at least temporarily from their function. At any rate, no other animal has been found in which eggs are totally furrowed and at the same time serve to transfer the symbionts.

The manifold form changes accompanying transmission to the egg cells and the specific propagation periods of symbionts will be discussed in a later chapter, which deals with the reciprocal relationships of both partners; however, at this point we cannot avoid raising the question of the nature of the forces in operation as the symbionts are transported from their sites to the ovarioles and transferred to follicle cells and egg cells.

Sometimes the path which the free organisms must traverse is shortened by cleavage and displacement of the mycetomes, which bring the symbionts to the direct vicinity of the ovarioles (aleurodids, aphids, lyctids). In certain cases mycetocytes specialized for this purpose build a kind of envelope around the young ovarioles (blattids, *Mastotermes*). With *Aspidiotus piri* a number of mycetocytes, usually permeating the fatty tissue, gather around the follicle zone which later allows the symbionts alone to pass, and with the disymbiotic *Tachardina* regular migratory cells carry a mixture of organism types to the same place.

The task is even simpler for the symbionts of a *Goniodes* species where intact mycetocytes penetrate the follicle and degenerate in it, and for symbionts of the Mallophaga and Anoplura, which transfer mycetocytes or even mycetomes to the ovarial ampullae, and thus to the immediate vicinity of the oocytes which are to be infected. However, the culmination of this tendency is found in an unsurpassed shortening of the free path of the symbionts among the aleurodids and the few pseudococcines in which we saw the maternal mycetocytes even transferring to the egg and the embryo.

Yet all these are special cases, either the symbionts are in general given over to the lymph and isolated or, at most, united in bundlets or in mucilaginous packages, and they must find their own way among the organs. An active motility and therefore a somewhat controlled migration do not come into question here. At most, one or another accessory bacterium, still primitive in shape, may perhaps retain the capacity for independent locomotion, but this is completely lacking in all the more adapted forms and naturally in all the yeast-like organisms. In fact, the transmission forms are often found in various crannies of the body cavity, and it is scarcely conceivable that they would appear again between the follicle cells with a uninterrupted regularity, strictly controlled temporally and spatially. At times it is the follicle of still very young ovarioles which is infected over wide areas, as with the ants; and at other times it is a very limited number of cells at the posterior end of more or less mature eggs, the area between egg and nutrient cells, or only a ring-shaped zone of the follicle situated in the posterior third, as with *Stictococcus sjoestedti*. Sometimes the whole follicle of older eggs is capable of reception. In other cases, transmission is guaranteed by an infection of very young gonad rudiments. Apparently the resistance prevailing everywhere else is cancelled occasionally at these locations, which vary from case to case, thus providing the necessary invading points for symbionts carried to the vicinity by the blood stream.

Two explanations are possible for this constantly observable transfer to the follicle cells. Either the follicle cells are active participants and incorporate the approaching organisms with cell processes in the manner of phagocytes; or the symbionts as former parasites still have the ability to liquefy the cell walls with the help of lysines, and thus without a home and, as though knocking everywhere but finding only one door here, they sink to the cell interior. In the case of wedge cells, the cell growth preceding the infection and the occurrence of fluid vacuoles indeed may indicate a special preparation that might very well be related to the cessation of resistance; but in other cases the infection of the follicle proceeds just as certainly within strict local limits without appearance of morphological changes. Regular grasping processes of the kind that are ordinarily found during embryonic development and, when symbionts are occasionally transferred, during postembryonic development, are scarcely ever seen. One might apply this interpretation to a slight protrusion sometimes observable with some cicadas in still empty or almost empty wedge cells.

Although a certain activity must inevitably be ascribed to the symbionts during egg infection, it is obviously the host organism that always directs the flow and controls the amount by limiting the function of resistance temporally and spatially, by closing routes created by divergence, or by forcing the receptive wedge cells from the body cavity, after they have performed their service, through complicated shifts of position.

Just as the follicle becomes receptive over wide areas or at narrowly localized places, so, too, does the egg cell itself. With the ants, with heteropterous bugs, and others, very young oocytes permit transfer of the bacteria even before the yolk is formed. With *Lyctus* the two symbiont types which are infecting cooperatively sink over the whole surface into the yolk-filled egg; with aphids and psyllids the symbionts are accepted only in the narrow zones bordering on the gaps in the follicle; with most of the other Homoptera and many other animals the posterior egg pole admits the symbionts; with still others, the anterior pole; and with blattids we see the bacteria infecting both poles simultaneously. Sometimes the egg cell is closed entirely to the symbionts, entrance not being given to them until furrowing or possibly even during the blastoderm stage, or indeed after invagination of the germ band. At this point we must remind the reader of the striking infection method of the aphid embryos, which extend a receptive plasma process into the body cavity through a preformed opening of the follicle, and of the two types of eggs, side by side in aphids and stictococcids, one of which is infected and the other of which remains sterile in significant fashion in connection with rudimentation of males. Everywhere the same picture is presented of a high degree of temporal and spatial limitation.

Wherever symbionts transfer to the oocyte without any change of the latter in shape, as with ants, aphids, lyctids, etc., it is undoubtedly appropriate to think again of a local dissolution of surface limits by symbionts. In numerous other cases, however, there will doubtless be regular deglutition by the egg, suggestion phagocytosis. Shallower or deeply incised fossae soon appear at the anterior or posterior ends and are closed up behind the symbionts. The blattids represent an isolated variation, for they draw in the bacteria, not through a single groove but through a varying number of folds.

Whether such interplay of local dissolution of the resistance of the host cells, on the one hand, and a general effort of the symbionts for intracellular life, on the other, is sufficient to allow the free-drifting symbionts to reach the desired places each time cannot be answered here. One often has the impression that increased numbers collect in the direct vicinity of the invasion points, and that then case factors still unknown but nevertheless acting as a directive force come into play. In *Orthezia,* for example, the freed symbionts are occasionally found in great quantities in the direct vicinity of the follicle cells which are to be infected, and in a tropiduchine species there is not only a clear grouping at an equivalent point but also a multiplication of the symbionts; and similar things are observed elsewhere too. When *Pediculus* symbionts leave the stomach disc to infect the ovarial ampulae, the situation is basically similar to the infection of the wedge cells, and in this case too it is reported explicitly that the bacteria proceed in well-regulated fashion from one organ to the other. Indeed, when only half of the mycetome is available to the symbionts as a result of artificially induced partial elimination, they also infect only the corresponding side of the ovarial ampullae.

13

Embryonic and postembryonic phenomena

To compare phenomena of embryonic and postembryonic development we must consider insects exclusively, since other known symbiont bearers have a different organization and thus offer practically no points of comparison. Wherever symbionts are present before egg development, or at least transfer during early stages, it can be determined in most cases that the host organism very soon attempts to provide them with nuclei and to admit them to cells or syncytia. However, the phenomena involved do not always lead resolutely to the development of final sites; for in many cases the housing is merely provisory, and it is clearly recognizable that the host animals are primarily concerned with sealing off their guests and making them maneuverable. Pointing in this direction are also the membrane-like plasma compactions which are placed around sharply defined symbiont balls which form in the oocytes, for they may attain considerable firmness and, when they are isolated in cicadas, a good deal of pressure is necessary to make them burst under the coverglass.

In a number of cases certainly the symbionts are not first brought into intimate contact with nuclei. The symbionts of *Rhizopertha* are forced behind the blastoderm during its formation and still permeate the entire yolk while the germ band is forming, and only after they leave the yolk at the posterior end and transfer to the body cavity are they taken up by cells which join to form mycetomes. Similarly, the symbionts of *Coccotrypes,* in dense accumulations, are found in the area of the yolk and do not loosen until the first larval stage; they then glide into the yolk-free part of the midgut and into the Malpighian tubules without previously entering into relationship with yolk nuclei. The snout beetles also provide their symbionts very late with nuclei, and a smaller number reach in this case the primitive sexual cells and are thus reserved early for future transmission. A greater number, however, are taken up by cells which belong to the rudiment of the stomodaeum. In the most unusual case of *Stictococcus* the tendency to delimit symbionts in cells does not awaken until after the strange extra-embryonic rudiment of the germ band is separated.

Retarded provision of nuclei may be caused also by deficient adaptation of additional symbionts, as shown by *Rastrococcus iceryoides* where the yolk nuclei first collect only around the yeast symbionts and the symbiotic bacteria are ignored for a rather long time. *Lakshadia* temporarily and superficially envelops the yeast symbionts with yolk nuclei during cleavage, but during invagination these symbionts are disseminated in the area of the yolk and do not infect the fat cells until the larvae are about to hatch.

Elsewhere one always finds more intimate and longer-lasting provision with nuclei. It makes no difference whether the symbionts are present in the egg at the beginning of cleavage, or only enter during development: with negligible exceptions, it is always segmentation nuclei on their way to the egg surface, those which have already become blastoderm nuclei, or those which remained in the yolk—yolk nuclei or vitellophages—which are released from their original functions for this purpose. The nuclear materials are basically identical and are differentiated only by the number of the mitoses undergone; this explains why many types of transition exist here and why sometimes even blastoderm nuclei and segmentation nuclei participate jointly in provisioning the symbionts, as in the fulgoroids, or why even, as appears to be the case in the winter egg development of the aphidines, segmentation nuclei and blastoderm nuclei penetrate the symbiont ball and yolk nuclei finally form the envelope of the mycetome. Adelgines, psyllids, membracids, and silvanids use segmentation nuclei exclusively to build their symbiotic organs. This applies also to *Orthezia,* but with all other scale insects investigated for embryonic development—*Tachardiella, Tachardina, Icerya, Eriococcus, Lecanium, Rastrococcus, Stictococcus,* etc.—it is always yolk nuclei which are added to the symbionts. The same plan is followed by aleurodids, whereas blastoderm nuclei are made available by blattids, by formicids, and, among the Heteroptera, by *Ischnodemus* and *Cimex.*

There are many variants. When, as so often happens, the symbionts form a sphere or lenticular accumulations situated directly at the posterior end of the egg, they represent an obstruction for the nuclei migrating to the egg surface, and this often leads to temporary interruption of the blastoderm. Examples are: the winter eggs of aphids; the aleurodids, with which this place is taken up by the maternal mycetocytes; cicadas (*Cixius*); and *Pediculus.* Sometimes the blastoderm cells are heaped around the obstruction into a swelling (*Cixius, Gargara*) and, closing the gap, gradually force the ball off to the egg interior. Frequently the germ band, when it invaginates at this place and thus pushes the mass of the symbionts ahead, still remains open for a long time at the anterior end or is interrupted by the embryonic mycetome (aphids, *Cimex, Ischnodemus,* and, for a shorter time, also *Oryzaephilus*). Quite exceptional, according to new investigations by Baudisch, is the genesis of the unpaired *Pediculus* mycetome. After a cellular envelope has been formed around the symbionts by cleavage nuclei, the germ band pushes the provisory mycetome forward, bringing it into close relationship with twelve to fourteen cells, prepared long before, with four 16-ploid nuclei each. The symbionts transfer to these, the envelope cells soon being destroyed. Also, the definitive envelopment of the mycetome is derived from a mesodermal cell group formed far from the symbionts.

Where the symbionts form a loose superficial layer, the cleavage nuclei tend to submerge easily, and, if cell limits are formed, they are transformed into nuclei of the mycetocytes (*Formica, Camponotus,* Heteroptera, blattids). With *Oryzaephilus* the cleavage nuclei first traverse the infected area, form behind it a sterile blastoderm, and push the symbionts toward the inside where the cleavage nuclei arriving later take possession of them; yet here this somewhat unusual behavior is apparently caused by the fact that a substance accompanying the germ path is situated, by way of exception, at this place close beneath the egg surface and must necessarily reach sterile elements. On the other hand, where it is the part of the symbiotic plan that primitive sexual cells separating at the posterior pole are infected, as with some snout beetles, the nuclei destined for this purpose transfer easily to the symbiont-containing marginal area. With *Cimex, Ischnodemus,* and *Formica* the infection of the blastoderm cells causes a striking increase there even before the germ band invaginates, so that a large cell cushion penetrates inward.

When yolk or cleavage nuclei provide the symbiont aggregations, these nuclei usually all force their way in at once and transform the aggregations into a syncytium; but sometimes they linger first at the surface, as with *Lakshadia,* only deciding after a while to penetrate more deeply. This is true to an extreme degree in *Orthezia,* evidently because the symbiotic bacteria at this time are still united in large balls held together by secretion. However, as soon as these balls break up, the nuclei take possession of the bacteria. Sometimes a retardation occurs, obviously because the arriving vitellophages must first serve in the transfer of the symbiont mass (*Tachardiella*). Occasionally some of the nuclei sojourn at the surface, either forming the envelope or, with disymbiotic forms, serving to house an additional symbiont type, as with adelgines and psyllids.

A late penetration of symbionts into already developing eggs, such as is observed chiefly with coccids and Anoplura, by no means excludes the use of yolk nuclei. Use of the latter is not at all surprising when the infection takes place earlier, during blastoderm formation, as with *Tachardina, Cryptococcus, Phenacoccus,* and others, or even in very early cleavage stages, as with *Asterolecanium.* Yet even where a blastoderm stage is infected at the upper pole across a broad front, reception never occurs through the cells that are set together like an epithelium and border the surface. Instead it is always yolk nuclei which first approach the symbionts after they reach the space behind the blastoderm. In *Haematopinus* and *Polyplax,* however, the embryo forms the germ band and the extremities earlier, at the time of the infection, but even this has no influence on providing the symbionts with nuclei. Since the invagination in these animals begins laterally and the bacteria are nevertheless admitted at the posterior end, they come upon an area which is still constituted exactly as during blastoderm formation. With viviparous aphids, on the other hand, in which the symbionts are also given only to embryos whose invagination has progressed to greater or lesser degree, the same cell materials are made available for admission of symbionts that had served the same purpose during the development of the already infected winter eggs. Although in many stictococcines the original symbionts are replaced by yeasts, the egg cells remain closed to them and the new guests first enter the embryo in an advanced stage, where they are likewise taken up by the yolk cells. The same is true for the disymbiotic *Stictococcus acaziae,* in which the bacteria, which formerly were the only ones present, still infect the egg cells, while the symbionts added later, which are also bacteria, behave like the yeasts.

Only a few exceptions violate the rule that the first provisioning with symbionts devolves upon cleavage, yolk, and blastoderm nuclei. We have already learned that with *Pseudococcus* the polyploid cells which move toward the symbionts entering at the upper pole originate through fusion of cleavage nuclei with derivations of the polar bodies, although the reason for this unusual measure is unknown. A very similar process has been observed recently in *Diaspis,* where polyploid, prospective mycetocyte nuclei arise from the fusion of a cleavage nucleus with the nuclei of the two polar bodies. The reader will recall that in *Pediculus* the symbionts are finally received by very specific yolk cells in the process of separation far from the place of infection. With lyctids and bostrychids the nuclei are provided much later, the reason doubtless being that the symbionts are still distributed in the fat over the whole surface when development begins and must first be concentrated during its course. It could not be decided definitely whether the cells used are still yolk cells or were derived from the rudiment of the fat body, as seems more likely to an expert on lyctid symbiosis. If the latter should be the case, such unique behavior might be explained by the no less unique late housing of the symbionts.

When one attempts to arrange the numerous observations available today on the origin of the various sites, one sees that the great diversity derives in large part from a tendency to keep the symbionts united in mycetome-like, closed associations during embryonic development even when the final housing is of a different kind. We have contrasted such transitorily formed mycetomes, which are dissolved earlier or later, with mycetomes of the persisting kind. Cases where the embryo forgoes all organ-like concentration are very rare and are probably found only within the scale insects. With *Lakshadia,* as already mentioned, the symbionts are not provided with nuclei at all, but the fact that they are surrounded by a cellular envelope after collecting suggests that we have here a transitory mycetome. It is different with *Eriococcus,* where the symbionts in the form of short rodlets are distributed in the egg yolk and are admitted here and there by yolk cells only when the extremities are being formed and consequently are housed from the very beginning in mycetocytes which are diffusely distributed. Several diaspines (*Aspidiotus, Chionaspis*) behave in the same way. Yeasts admitted at the upper pole are gradually propagated and dispersed until they are seized by yolk cells.

For the rest, it is always transitory mycetomes which are formed in scale insects which distribute their mycetocytes in the fatty tissue; in this case, to be sure, there is wide variation in the sharpness of the delimitation and in the organ-like behavior. *Cryptococcus* forms only a loose conglomeration of mycetocytes which, though separating early into groups, does not disintegrate into completely individual cells until the larval state. Some of the diaspines (*Lepidosaphes*) group their mycetocytes first in paired accumulations; *Tachardiella* creates a transitory mycetome from densely packed, wedge-shaped cells, and it is later paired and then finally dissolved anyway. Other lac insects go one step farther by temporarily concentrating their symbionts. *Tachardina silvestrii* unites two symmetrically arranged cell groups into a very sharply defined and unilaterally arranged embryonic organ which is loosened only in the larva. Curious to relate, *T. lobata* locates its voluminous transitory mycetome, also originating from two cell groups and afterward falling apart, in the space between the abdominal nervous cord and the hypodermis. With mallophages a well-defined cell assemblage enters the yolk which is surrounded by the midgut; later its components swarm apart, traverse the embryonic intestinal epithelium individually, and are then distributed in gaps between the organs. *Formica fusca* first groups the embryonic mycetocytes in an unpaired aggregation, then in a paired, well-contoured one where there is nothing to indicate that its cells will finally be spread in a single layer behind the midgut.

There are provisory formations even with the lecaniines, where symbionts populate lymph and fat cells in a very undisciplined manner. Unfortunately, very little research has been done with their embryonic development. However, with one *Lecanium* species it was discovered that yeasts situated at the upper pole and provided with yolk nuclei are transferred to the abdomen during involution as a sharply defined formation, just as are the rudiments of a definitive mycetome, and that not until then does the inundation of the body cavity and the fatty tissue set in. It may be assumed that this is also true for at least some of their relatives. Accordingly, degeneration of the primary mycetocytes must occur here and likewise a reloading of the symbionts, recalling many similar phenomena occurring with replacement of primary mycetomes by secondary ones.

That equivalent events take place in numerous cases in which yeast-like symbionts of cicadas live diffusely in the fatty tissue or in pseudomycetomes may be concluded from individual facts already known about the fulgoroid *Eurybrachys*. Yolk nuclei penetrate the symbiont balls which are first located at the posterior end of the egg and then are pushed forward by the germ band; during revolution the ball is again drawn away from the caudal zone and is then loosened up, thereby making free the path to the body cavity for the yeasts. Once more the nuclei of the transitory mycetome will simultaneously degenerate.

Previously, either the mycetocytes of the transitory mycetomes were preserved as such, or the syncytia initially housing the symbionts degenerated, and in their place lymph and fatty tissue were populated. In other cases transitory mycetomes are evidently decomposed again, but the symbionts are also assigned to cells which are lodged in the fatty tissue but which originate elsewhere. *Orthezia* first provides its symbiotic bacteria with cleavage nuclei and then handles the unpaired syncytium arising in that way exactly as if it were to become a definitive mycetome. Nevertheless, when the extremities are formed, cells of the lower lamella, that is to say, elements derived from the mesoderm, migrate to the syncytium, are distributed over its surface, and are gradually filled with the bacteria until finally only the degenerating remains of the former cleavage cells are left. The loose cell mass which arises is first divided into a right and a left one, and not until later do its components spread over the entire area of the fatty tissue.

The blattids are comparable, displaying the tendency to form transitory mycetomes to a greater degree. *Blattella germanica* has only a loosely elongated aggregation of bacteria, grouped around former blastoderm nuclei, inside the yolk at the time the extremities are formed; *Blatta orientalis* has a somewhat more sharply defined complex of very numerous nuclei and bacteria without cell limits; and *Periplaneta americana* and *Parcoblatta pennsylvanica* have a genuine little mycetome composed of wedge-shaped cells. However, with all blattids these nuclei or cells degenerate, and the bacteria, freed anew, swarm through the embryonic intestinal wall to the body cavity where numerous specific mesodermal cells, which later sink into the fatty tissue, are held in readiness for them.

Four possibilities may be distinguished when permanent mycetomes are created: the primary mycetomes may remain unchanged; they may be replaced by secondary mycetomes; mycetomes may be built from maternal mycetocytes which are replaced relatively late by embryonic cells; or the maternal cells used in transmission are not replaced by such cells but enjoy the same potential immortality that the sexual cells do.

The primary mycetomes are by far the ones most frequently kept. Anoplura, Heteroptera, coccids, aphids, psyllids, cicadas, lyctids, bostrichids, and curculionids afford numerous examples. With some Anoplura (*Linognathus, Polyplax*) these mycetomes are small cell groups, without envelopes and of the simplest construction conceivable, which are preserved in the form in which they arose unpaired from the symbiont ball through migration of the yolk nuclei. Usually, however, the organ-like character is substantially increased by development either of delicate envelopes provided with a few flat nuclei, or of epithelia rich both in plasma and in nuclei which may be extra-embryonic or mesodermal in nature, and finally by abundant provision with tracheae. To clarify the extent and significance of what are undoubtedly the most primitive mycetomes, it will suffice merely to refer to the wealth of genuine cicada mycetomes, which when considered individually are so very different and yet, except for the *x*-organs, preserve their primary state, according to present knowledge.

It is much more rare that primary mycetomes are replaced by secondary mycetomes. In such cases the definitive cell material may be mesodermal or endodermal. It is mesodermal in *Icerya*, and thus undoubtedly also in other relatives, as well as in *Oryzaephilus*; and it is endodermal in certain Anoplura. In detail, each case bears its own stamp. In *Icerya*, mesodermal cells loosening from the germ band form a continuous envelope, even before revolution, around a primary mycetome permeated with yolk nuclei; then the nuclei penetrate the mycetome and take over the symbionts, beginning in the marginal areas, as always with this type of transmission, whereas the old supply of nuclei degenerates. Several cells which are left behind flatten off and furnish the envelope for the future paired mycetomes. *Oryzaephilus* presents a unique case insofar as its primary mycetome, which until that point had adhered to the germ band, is dissolved at the time the extremities are formed, and symbionts and nuclei are transferred to the posterior pole. The nuclei gradually degenerate, the symbionts come upon four rudiments of the definitive mycetomes, which are already surrounded by a delicate envelope, and then immediately take possession of the mycetomes.

The shifts found in some Anoplura are entirely different in nature. As with the blattids, many curculionids, the mallophages, and other Anoplura, the symbionts enter the yolk surrounded by the midgut rudiment, but, whereas they usually make their way to the definitive site in the area of the mesoderm (blattids, *Polyplax*, *Linognathus*) or are seized by prospective stomodaeum cells (curculionids), this time cells which were actually destined to participate in building up the midgut epithelium become mycetocytes. With *Haematopinus* the cells of a loosely joined mycetocyte group degenerate, and cells of the embryonic intestinal epithelium migrate toward them through the yolk and then, traversing the same path with symbionts, are again inserted in the intestinal epithelium. With female animals, closed groups of such infected endoderm cells also migrate through the intestinal wall, giving rise to the three depot-mycetomes which later, when the last larval membrane is separated, discharge their inmates into the exuvial fluid and thence to the ovarial ampullae. With *Pedicinus*, the ape louse, the endodermal secondary mycetocytes are not localized in the intestinal epithelium but directly behind it, and thus all must traverse it together. With *Pediculus* a different method is used. The nuclei of its primary mycetome, which is more highly organized and is surrounded by envelope cells, are also destroyed, but this time a sharply defined out-pocketing of the intestinal epithelium is constricted off and reformed into a syncytium in which the symbionts are later located. Mesodermal elements form at last an evelope round about this secondary mycetome, which is so unusual in its manner of formation.

As a third possibility for replacing primary mycetomes by secondary ones we mentioned cases where the primary ones are supplied by maternal cells and do not give way to animal-specific organs until later. As we know, this unique method has been used only by the aleurodids. The embryos have already completed revolution when the replacement of the old cells takes place. Elements, which either have the value of vitellophages or originate from the lower lamella, encircle the primary mycetome and take the symbionts from the margin, as with *Icerya* and

Orthezia. Several years ago it was found that also in certain pseudo-coccines (*Puto, Macrocerococcus*) intact maternal mycetocytes carry symbionts to the egg cells, and it was surprising indeed to find that they are not destroyed in the course of embryonic development, as with aleurodids. In unique fashion a yolk cell unites with every mycetocyte, and the two nuclei fuse with one another, as in fertilization, so that the term somatic fertilization can actually be used. At the same time the closed complex of the maternal mycetocytes is maintained; mitoses lead solely to further growth, and smaller yolk nuclei furnish an epithelial envelope.

Finally in this connection we must mention another remarkable phenomenon taking place in the *a*- and *x*-organs of the fulgoroids and leading to a very irregular type of transferring symbionts in the *x*-organs. With *Fulgora* and *Cixius* the embryonic *a*- and *x*-organs are covered during and after revolution by a plasma-rich, mesodermal epithelium. In the case of *a*-organs built up from individual mycetocytes, the symbionts, which remain small, also transfer immediately to these epithelial cells, and without degeneration of the primary mycetocytes a second additional layer of them arises, so that uniform organs result which house their symbionts partly in former blastoderm cells and partly in mesodermal elements. In all probability this unique type of enlarging of mycetomes occurs in all *a*-organs, which indeed are extraordinarily widespread with cicadoids too; for, among the membracids also, the only cicadoids studied as yet for mycetome embryology, a layer of mesodermal cells originating from the end of the germ band covers the still unpaired young mycetome after separation of *a*- and *t*-symbionts, and these cells are apparently also infected by the *a*-symbionts adjacent to them except for a few which are transformed into fat envelope cells.

On the surface of the syncytially built *x*-organs, additional epithelial cells become active at the same time. They break away toward the inside, extend processes between the symbionts, which are moderately augmented up to that point, and envelope them. This time, however, it is the old, centrally located nuclei which are destroyed, and thus, since cell limits disappear in the epithelium also, a new syncytium unobtrusively replaces the provisory one without essential change in form or structure.

Comparison of the origin of various cell materials supplying symbionts in the course of egg development, as we have presented it up to now, will afford insight into the multiplicity of available methods and into the immutable regularity of their execution, although it will by no means exhaust the cellular and histophysiological phenomena reflecting even better the intimate reciprocal actions between the growing organism and its guests.

Frequently it is possible to observe correlations between the manner of egg infection and behavior of symbionts in the growing oocyte on the one hand and later embryological phenomena on the other. Even before development begins, a tendency is repeatedly manifested to concentrate the symbionts in a sharply defined zone, which thereby paves the way for the formation of transitory or final mycetomes. Considered from this viewpoint, the limited receptivity of the follicle or of the egg surface seems extremely purposeful and the spherical shape of the apical symbiont ball seems to be the ideal device. Frequently therefore, even where the oocyte is first infected more or less diffusely, a tendency may be observed to bring about this concentration later, and it is certainly not by chance that even an infection of the egg cells at the anterior pole frequently ends with the symbionts collected at the posterior pole, exactly as in cases where the eggs were penetrated only at this pole.

For example, the *Ischnodemus* bacteria enter the egg through the nutrient cord, permeate it at first in all directions, and nevertheless finally form a cap at the posterior end. The symbionts of *Camponotus* and *Formica* are united at the posterior pole after first inundating the oocytes. As the reader has seen, accessory symbionts of membracids frequently infect the youngest oocytes or transfer through the nutrient cells and yet are strictly concentrated in the mature egg, in the direct vicinity of other symbionts which had entered at the posterior pole from the very start! With *Nysius* and *Ischnorrhynchus,* the embryonic development of which is unfortunately unknown, the symbionts are crowded together at the anterior pole after having first taken possession of the whole young oocyte, but we do not know whether they do not after all glide backward before the beginning of cleavage.

That these symbiont movements are not accidental but are executed with a view to embryonic development is proved with great clarity by the behavior of the snout beetles which form ectodermal mycetomes on the stomodaeum and at the same time infect the primordial germ cells when their development begins. These beetles keep the majority of the bacteria in the anterior area in which they are needed primarily and send a smaller delegation to the posterior end of the egg where the primordial germ cells are separated!

No less surprising are the complicated shifts undergone by tick symbionts during egg growth. Diffusely distributed at first, they are next collected into two groups and then neatly separated from the yolk-rich egg plasma at a certain place in a single round ball. This time, to be sure, it happens not with a view to the future development but probably more from a need for isolation, inasmuch as the symbionts are distributed again throughout the space toward the end of egg growth. However, before the egg is deposited, the symbionts are concentrated again in a narrowly defined place in the periplasm and in the bordering yolk, a location corresponding exactly to that where the development sets in and where the Malpighian tubules and the primordial germ cells which are to be infected take form!

With all these movements in space, active symbiont migrations probably do not play an essential role any more than they did during egg infection. On the contrary, everything seems to indicate that it is chiefly currents of the egg plasma that are involved in such shifts. The results of many observations show that they prevail in the developing insect egg in the required direction. An especially impressive parallel exists in the movements of the yellow pigment in psyllid eggs. Distributed over the whole surface at first, it is all collected before infection in the cone-like process of the egg forming at the posterior end and, after the transfer of the symbionts, it is even collected all over the surface of the ball formed by them. In very similar fashion, substances accompanying the germ path, mitochondria, and other specific inclusions gather at the posterior pole of insect eggs, and repeatedly one has the impression that they are blocked here on their journey from the front (cf. Buchner, 1918). Like the symbionts, parasites sometimes are finally concentrated after diffuse settlement at the posterior pole of the eggs, as may be observed in the rickettsiae occurring occasionally with hippoboscids and *Nycteribia*. This is also proof that such transport possibilities exist quite generally in these egg cells and therefore may be utilized correspondingly when symbiosis is established. In the last analysis, therefore, the symbionts are handled in the growing egg exactly like "organ-forming substances" of the mosaic eggs of ascidians, molluscs, etc., which also owe their final arrangement primarily to currents prevailing in the plasma of oocytes or cleavage cells.

A more penetrating cytological study of all factors of egg growth would doubtless deepen our knowledge of these conspicuous symbiont transfers and concentrations, which obviously may be influenced not only by currents but also by plasma growth traceable to activity of the follicle and nutrient cells.

Significant shifts of symbionts not yet provided with nuclei also take place after development begins. As to be expected, they are especially conspicuous with *Lyctus* where, by way of exception, there is no concentration of the bacteria before development. From all directions of the egg surface, they drift, united in swarms, toward the posterior area of the germ band and there transfer from the yolk to the body cavity.

When the same method of transmission and arrangement of the symbionts was found later in bostrychids, it could be shown that here also the symbionts, first widely scattered, are gradually collected at a narrowly defined place from which they reach the body cavity. In *Cryptococcus* multitudes of symbiotic bacteria sink at the blastoderm stage from the anterior egg pole toward the middle of the egg where the nuclei intended for them are in readiness. Similarly, the symbiont balls of *Pseudococcus* advance toward the future mycetocytes which are moving in their direction. The symbionts of those apionids which are housed in two Malpighian tubules which early develop in a specific manner first reach the center of the yolk-filled midgut but nevertheless find their way to the opening of these two vessels. In a quite similar manner the symbiotic bacteria of the bark beetle *Cryptococcus* arrive from the yolk-filled stomach at the four Malpighian tubules reserved for them there. In neither of these two cases do they ever stray into one of the other vessels! With *Oryzaephilus* it was confirmed that there is a well-directed movement of the symbionts, freed again after degeneration of the primary mycetome, in the direction of the secondary mycetomes awaiting settlement; and something similar occurs among the blattids, where the bacteria must find their way through yolk and midgut epithelium to the body cavity after the transitory mycetomes are dissolved.

In addition to such symbiont shifts in the developing germ there are other highly interesting reactions of the embryonic cells and of the egg plasma, which serve to seal and concentrate symbionts which had been more or less scattered. Both functions can naturally go hand in hand, yet the sealing off may predominate; thus, when *Icerya aegyptica* concentrates its symbionts from the beginning at the posterior pole of the egg, the blastoderm cells arising there form fibrous processes which never occur otherwise, and these are united above the symbionts and the cleavage nuclei which have been added to them. In this way they are closed off on all sides during invagination of the germ band and at the same time are anchored with long fibrillar processes in the egg yolk. Only when this closing-off process becomes superfluous through formation of an epithelium by migrating mesodermal cells do the fibers regress. When the germ band invaginates in *Orthezia urticae,* a plasma sac arising at its free end encloses the symbionts and their yolk nuclei. This sac is voluminous in the beginning and includes numerous yolk spheres, but later it pulls together more and more toward the symbionts, thus allowing the yolk pieces to pass, and finally encloses a narrow, completely yolk-free plasma area containing only the rudiment of the transitory mycetome. About this time a considerable radiation goes out from it which is not reduced until the mesodermal cells get ready to take over the bacteria conclusively. This, too, obviously involves a reaction of the egg plasma which is released by the presence of foreign organisms and does not involve concentration but rather is the expression of a need for sealing off. With *Orthezia insignis* no sac is formed; instead, plasma fibers develop which radiate into the yolk on the one hand and enclose the symbionts on the other, and later, after shortening, make way for a weaker radiation.

Comparable in snout beetles are the strange plasmatic structures initiating formation of mycetomes situated on the stomodaeum. This time a radiation appears at the end of the embryonic initial section of the gut and from it the formation of a plasma net proceeds. When maximally developed, this net encloses the whole central yolk permeated by symbiotic bacteria. Again, it is progressively narrowed, allowing the yolk spheres to glide through its meshes but at the same time drawing together to collect the bacteria in the manner of a fishnet and uniting them finally in a saclet suspended on the stomodaeum.

In *Cryptococcus* the origin of the large, yolk-free plasma congregation must be caused by centripetal currents which arise in the blastoderm stage and obviously exert an attractive force on the still nuclei-free bacteria, as on the yolk nuclei.

At the time the germ band invaginates in *Ischnodemus, Pediculus, Fulgora,* and *Cixius* there appear plasma radiations which rest upon the nuclei-provided symbionts and thus precede invagination but which cannot have the function of gathering or isolating symbionts even though they may seem to resemble radiations appearing with *Orthezia* and snout beetles. Their significance has been interpreted in different ways. Some have seen in them a type of locomotor center responsible for the movement of the germ band (Ries, 1931; and others), and H. J. Müller (1940) considers them a device to facilitate detachment of the embryonic mycetome. With good justification, apparently, the phenomenon has recently been interpreted as being closely related to yolk cleavage, that is to say, to division of the yolk into individual nuclei-containing areas, proceeding from the yolk membrane and progressing forward from the posterior part. The observations and considerations of Sander (1956) relate to the fulgoroid *Pyrilla*, those of Baudisch (1958) to *Pediculus.* Sander concludes that the radiation in his test object represents the product, on the one hand, of centripetal contractions of the net-shaped plasma causing the yolk cleavage and, on the other hand, of longitudinal contractions of the plasma structures which cause and control the invagination and which have close relationships to the yolk membrane. Baudisch, who was unacquainted with Sander's publication, writes in the same vein in stating that the long rays directed to the anterior egg pole doubtless represent sections through the yolk membrane and their derivations. Like Sander, he asserts that the yolk cleavage increases more and more, from the posterior forward, that the part spanned by the yolk membrane thus becomes smaller and smaller, and that ultimately the relationships to the former egg surface are lost entirely. During this time the symbiont-filled "little basket" invariably hangs at the posterior pole of the yolk membrane and accordingly is drawn forward more and more during the progressive narrowing process. In fact, the end stages of this process have a surprising similarity in two insects which are otherwise quite different. At any rate, in considering these structures in the future, more attention will have to be given to their relationships to yolk cleavage. Supporting the interpretation that they are not caused by presence of symbionts but merely enter into secondary relationships with them is the circumstance that a radiation occurs at the top of the invagination in the mallophage *Trichodectes scalaris,* which plainly has no symbionts, and in *Haematopinus eurysternus,* where the symbionts are located in another area (Schoelzel, 1937).

In connection with spatial shifts undergone by free microorganisms during embryonic development, one must think primarily in terms of passive transportation, but with respect to certain directed movements of sterile and even infected elements one actually has the impression of active migration. Especially convincing in this respect are, first, the embryonic midgut cells of *Haematopinus,* which at a given moment move toward and receive the symbionts being freed in the yolk and then retreat in processions accompanied by cells that remain sterile; and, second, the cells of the stomodaeum rudiment which in *Hylobius* and its relatives glide along the suspension fibers to the bacteria, infect themselves, and then force their way through at the boundary between the ectodermal and endodermal intestinal epithelium. The polyploid prospective mycetocytes of *Pseudococcus* also reveal by their outlines that, like amoebae, they strive toward the symbionts, soon enclosing them with their pseudopodia. The symbionts of *Diaspis pentagona* behave just the opposite. First they are concentrated at the anterior pole, then move posteriorly and thereby gradually permeate the future mycetocytes.

More complicated migrations must be undergone especially by the bacteria of the blattids, which, accompanied by yolk nuclei, come from various regions of the embryo, chiefly from the anterior and posterior zones, but also from zones situated laterally and dorsally, and, gliding along paths strictly adhered to, are finally united in the middle of the yolk to form the transitory mycetome. Heymons thought in terms of active symbiont motility, but his followers are right when they consider the escorting vitellophages to be the actual transporters. According to Gier, the migrating bacterial mass in *Blatta americana* is often shaped very irregularly at its anterior end, giving the impression of a body which is being forced against the yolk spheres. In other cases the yolk spheres appear liquefied on the route traversed by the bacteria. Even with the smaller groups coming from the sides, the accompanying yolk nuclei are never lacking. Gier, on the other hand, is inclined to consider the current of liquefied yolk passing to the germ band responsible for the return migration of the bacteria from the yolk to the body cavity. With mallophages, after loosening of the transitory mycetome erected in the yolk, all the mycetocytes creep one by one to the body cavity through the intestinal wall which is taking form; in *Camponotus* the definitive mycetocytes transfer to the midgut epithelium along a broad front.

Many more cases of this kind could easily be cited, but the decision as to how far active motility or passive transportation is involved cannot be made with certainty purely on the basis of morphological findings. On the other hand, where shifts of well-described transitory or definitive mycetomes are concerned, it will always be necessary to consider the second possibility, that of passive transportation. Above all, we must recall the widespread significance of the invaginating germ band as a means of transportation, by which not only symbionts but also often primordial germ cells are similarly moved from posterior to anterior poles (aphids, psyllids, aleurodids, cicadas, many coccids and Heteroptera, lycitids, *Pediculus*). This close relationship between symbionts and invagination process is doubless of deeper significance and should not be regarded as a chance occurrence resulting from infection at the posterior pole. A clear proof is the circumstance, to which little attention has been paid, that with rare exceptions (orthezines) the relationship always goes hand in hand with formation of mycetomes, whereas the position at the upper pole, which excludes such relationships to the germ band, with the exception of *Pseudococcus, Puto, Macrocerococcus,* always result in settlement of lymph and fat cells or in the origin of diffusely distributed mycetocytes. The significance of the contact of the symbionts with the developing germ band can be clearly seen from those cases where the germ band is formed at a distance from the symbionts. This is the case with the formicids, where insertion of the embryonic rudiment along a broad front guides the development of the mycetocytes to completely different paths. Also in blattids the germ band is completely lacking in relationships to the symbionts, the latter being located in a great variety of places and, after extremely involved migrations, finally coming to rest in cells embedded diffusely in the fatty tissue.

Other untoward consequences are observed with Anoplura and Mallophaga where a well-described symbiont aggregation is formed at the posterior end of the eggs but the invagination sets in laterally. As with blattids, the symbionts reach the embryonic midgut without coming in contact with the invagination and, in their later behavior, which is unusual and differs from case to case, apparently again reflect the effort of the host animals to compensate for this "bad situation." Polyphyletic devices represented here are: a refreighting to the midgut epithelium not observable elsewhere, migration of isolated mycetocytes derived from yolk cells to the body cavity and later dispersal in it, or migration of intact, very primitive mycetomes to it. It is only in *Pediculus* that there occurs a meeting of the symbionts with the end of the germ band and therewith their transportation forward, but these symbionts, as in the relatives, nevertheless get to the midgut. The complicated phenomena recently discovered by Baudisch suggest all too clearly that these are emergency measures.

It is true that lyctids and bostrychids represent an exception and do finally form mycetomes, although no relationships to the germ band exist, yet this merely confirms our hypothesis that mycetome formation is fostered in high degree by such relationships, for the path taken in these two forms, especially in the lyctids, is unusually protracted and intricate.

The transmission of embryonic mycetomes through the germ band is of decisive significance for their ultimate localization in the host body. First they are removed from the future cephalic area, in which they would only create disturbance, and are transported to the caudal region of the embryo. Before revolution, however, they tend to enter into more intimate contact with the dorsal region of the embryonic rudiment. With cicadas and elsewhere, mesodermal cells provide for a connection; with aphids originating from winter eggs, the posterior end of the germ band actually encircles the mycetocytes; and often the mycetomes extend above the embryonic rudiment like a cake and thus achieve better support. When revolution takes place, they follow the movements of the germ to reach the place assigned to them in the organization of the animal.

From such considerations there results the surprising and significant relationship between the most frequently used infection method and embryonic development, and all the special precautions which are connected in some way with infection at the posterior pole or with a secondary concentration there of symbionts which transfer at other places appear to have been made with a view to future housing in well-individualized mycetomes.

Finally, we must deal with another interesting phenomenon, one which presents great obstacles to understanding the process: separation of the various symbiont varieties infecting jointly during embryonic development. The processes in question have been investigated with di- and trisymbiotic insects (aphids, psyllids, lyctids, *Tachardina, Rastrococcus,* and aphids, fulgoroids, and membracids). For all these cases, we have at hand a more or less uninterrupted series of the individual stages of the process demonstrating that the symbionts, at first mingled in confusion, are suddenly cleanly separated. Nevertheless, nothing certain can be deduced about the factors in operation here. At times one has the impression that the sorting comes about because the individual cells each time exert a certain attractive force on a certain variety, and at other times it looks as though they had simply fished out one or the other from a mixture.

The first is true, for example, of *Tachardina silvestrii,* a lac insect. Here, after the blastoderms have finished forming, a loose symbiont mixture is permeated by yolk nuclei of the very same constitution, and now staff-shaped symbionts increase in the plasma of the one and chain-shaped symbionts in that of the other, without formation of clear cell limits at first. Subsequently the settlement of the originally uniform cell material has various effects: the cells infected by rodlets remain small and retain their mitotic capacity to increase, whereas those infected by chains grow out into giant cells with a single strongly lobed nucleus.

The situation is very similar with *Lyctus*. Again no differences in mesodermal nuclear material are detectable; nevertheless it seems that the symbionts, this time not sought out by the nuclei but drifting in turn toward them from the yolk, feel attracted by the one or the other nuclear group and each time, as in the other case, they cause a specific reaction when the one sort is immediately enclosed in uninucleate mycetocytes and the other in syncytia.

The separation occurs in a somewhat different manner where the symbiont mixture forms a thick ball as with aphids, psyllids, and cicadas. In this case the various types are assigned to nuclei already clearly differentiated. With aphids and psyllids the one is provided with cleavage nuclei, the other is taken up by yolk cells. With *Fulgora* and *Cixius* the *a*-symbionts arrive at former blastoderm cells, and the rectal symbionts at former yolk cells; with the membracids, nuclei which stem from the blastoderm join the *a*-symbionts, and those joining the *t*-symbionts are undoubtedly derived from yolk nuclei. For the third accessory symbiont, the nuclear material is of a different kind.

There are some other differences. With adelgines and psyllids, only cleavage nuclei enter the symbiont mixture at first and envelope cells are attached outwardly to this mixture as sterile formations. In *Pineus pini* a certain degree of separation occurs even at this stage, for the smaller and more strongly staining symbionts in uninucleate cells of the mycetome begin to group more closely around the nuclei of the syncytium; then the envelope cells penetrate into it on all sides and suddenly form cell limits around larger, paler organisms which later populate syncytia. Not until now do the central nuclei follow their example. The final stage is brought about later when the mycetocytes which were the first to arise become multinucleate and in part sink between cells of the middle, which remain uninucleate. With *Dreyfusia* the envelope cells which are being infected sink to a greater depth from the beginning. The only aphidine studied behaves quite like *Dreyfusia* in this respect. The same thing is true of psyllids. Here also the type which are later found in more or less central syncytia are first assigned to the envelope cells, thus bringing about a shifting of the two infected territories to varying degrees.

With *Fulgora* the two cell varieties surround the symbiont ball in a sterile state at first. Derivations of the blastoderm form the posterior part of the envelope, and yolk cells form the anterior part. As soon as the germ band begins to form, before their infection, a difference becomes noticeable in the cells. First the increasing cells generally acquire more plasma; vacuoles appear next in the lower ones; and the upper ones, in accordance with their vitellophage nature, loosen toward the middle of the egg to receive and reduce the yolk spheres. Differences in the nuclear structure accompany this differentiation. The striking bipolarity of this stage is increased by the plasma radiation which appears around this time, even before the symbiont ball does. Then, as though at a secret command, the large, dark *a*-symbionts move downward and enter the former blastoderm cells, whereas the smaller rectal symbionts glide forward and fall victims to the vitellophages which are still receiving yolk spheres also. The third type, the little *m*-bacteria, which are still not very well adapted, keep their place in the middle.

Cixius follows *Fulgora* in principle, but an approximate pre-embryonic separation of the symbionts prepares the way for an embryonic sorting. In place of the *m*-bacteria, after formation of the two polarly situated mycetocyte varieties, the *b*-symbionts are left over again as a third kind in the middle, in great quantity and without nuclei. In *Conomelus* with its $H + f$ symbionts, the infection clump lies some distance from the polar blastoderm and only vitellophages surround it on all sides. At the beginning of invagination a lively increase in yeasts takes place, the vitellophages penetrate and distribute themselves homogeneously among them; and thereby the small, scarcely recognizable f-symbionts are separated from them. *Stenocranus,* on the other hand, behaves like *Fulgora* and *Cixius.* First the a-symbionts separate off and fill several large mycetocytes in the posterior region. In the meantime the q-symbionts make their way to the opposite pole and are provided with nuclei, as are also the descendants of the rectal organ, which lie in the middle, so that four days after the beginning of invagination there is present a collecting mycetome in which not a single symbiont is in the wrong place.

With the membracids, no envelope is formed to prepare for separate reception. Blastoderm nuclei immediately enter the polar group of symbionts, are scattered there, and forthwith are surrounded by the large dark a-symbionts. Free cell formation leads rapidly to uninucleate mycetocytes which are now set like little islands in the mass of the remaining t-symbionts. Although the mass is already permeated by nuclei which are different in structure and, at least in *Gargara,* also much smaller, and which correspond apparently to yolk nuclei, there is no separation into the t-mycetocytes until much later. Meanwhile the a-mycetocytes, rising to the surface, reach their definitive location, and, as we have already noted, they increase substantially through subsequent infection of mesodermal envelope cells. The accessory symbionts, which here represent the third variety, at first share the fate of the t-symbionts and are enclosed with them in cells.

In the aphidines the situation is reversed by belated infection of the parthenogenetically forming embryos. It is not the nuclei which approach the symbionts but the symbionts which stream toward a syncytium made ready for them. When di- or trisymbiotic types are involved, a sorting usually occurs which is more or less already prepared for in that the symbionts make their appearance at different times, a differentiation corresponding to the pre-embryonic sorting in *Cixius.* With disymbiotic kinds only the round primary symbionts usually stream in at first to take exclusive possession of the more peripheral areas and their nuclei; than a mixture follows in the sections left over for it in the middle of the syncytium. Thus once again the same mysterious process takes place by which all primary symbionts are eventually situated peripherally, and all accessory symbionts centrally. The first are enclosed in uninucleate cells, the latter sojourn in syncytia (*Macrosiphum jaceae* and many other species). With an *Aphis* species the accessory symbionts transfer only at the very last, shortly before the possibility of infection is cancelled, and are provided not with nuclei of the syncytium but with others situated in the vicinity. Moreover, these ill-adapted and tube-shaped organisms, separated in this manner at the start, are not at any later time ever firmly installed in the mycetome. *Pterocallis juglandis* represents the third possibility. Here the accessory tubes appear first, forming a peripheral layer supplied with nuclei of the syncytium, and are joined later by a few which arrive belatedly with the primary symbionts.

With the trisymbiotic aphids the situation is more complicated. *Pterochloris roboris* first sends the primary symbionts to the syncytium; then a mixture of cocci and rodlets appears which breaks through the mass and is blocked at the anterior end of the embryo. Primary symbionts arriving belatedly at this time are still able to join those of their kind, but the cocci infect envelope cells of the mycetome and several other tissues, and the rodlets, if they do not also transfer to the body cavity, are housed in cells which are not a part of the original syncytium. The two accessory symbionts of *Stomaphis quercus* also infect later than the primary symbionts; the tube form in part seems to collect around nuclei of the syncytium which are actually destined for the primary symbionts, and in part it again infects envelope cells; the other form, a smaller rodlet, it also housed partly in uninucleate cells and partly in a syncytium; in addition, these two varieties, which are still quite undisciplined, transfer to the body cavity and to other tissues.

By surveying the mass of facts available at present, one gains interesting insights into the relationships between the sorting process and the age of the symbiosis. The better the companionate symbionts and other additional guests are adapted to the host organism, the more well-traveled are the paths of their sorting. Especially well-adapted additional forms have the same cell material made available to them as those originally present. Examples are not only *Tachardina* and *Lyctus* but also *Fulgora* and *Cixius,* for when the latter two distribute both blastoderm nuclei and yolk nuclei to the *a-* and rectal symbionts, the material is fundamentally equivalent to that of monosymbiotic species. *Centrotus* and *Gargara* do not quite reach the same standard as the fulgoroids, preferring clearly the older *a*-symbionts, a preference which is expressed particularly in the much later division of the *t*-syncytium into cells. The behavior of the aphids and psyllids makes an even less perfected impression in assigning envelope cells, which first had a different task, to the symbionts acquired secondarily. In this respect it is informative to compare a monosymbiotic adelgine, like *Sacchiphanthes viridis,* and a disymbiotic one, like *Pineus pini.* The providing of the symbiont ball with nuclei is entirely the same in both, but in one case the envelope cells remain envelope cells and in the other they take possession of the additional symbionts. An even deeper understanding of the working method of the symbiotic hosts will be gained by including in the comparison the embryo infection of those aphids which have already adapted the second symbiont to a high degree (*Macrosiphum jaceae* type). In this case the additional symbionts now streaming to the syncytium come in immediate contact with its nuclei and are thus provided with them in the parthenogenetic generations and are not received by envelope cells as during winter-egg development! We must also call the reader's attention to *Rastrococcus iceryoides,* where the bacteria being received in addition to yeasts are at first neglected and are not provided with nuclei.

Inadequacies are seen wherever a third guest must be provisioned. The *b*-symbionts of *Cixius* remain longest without nuclei, and only during involution are they provided with embryonic cells about the origin of which there is no certainty; the much less well adapted *m*-bacteria of *Fulgora* are likewise not included when the *a*- and rectal symbionts are provided with nuclei and are either attached externally to the double mycetome in little bundles or are enclosed by it to be admitted by loose, insignificant host cells only after revolution takes place; the accessory symbionts in the two membracids are first caught along in the *t*-mycetocytes and emigrate from them much later. Even greater confusion is seen in the behavior of a number of other accessory symbionts during embryo infection of the aphids. Here it is also envelope cells which often bridge the gap and sometimes even mixed infections, observed nowhere else, which end with the expulsion or dissolution of one of the two types.

During the discussion of the impressive persistence with which symbionts transfer to egg cells, we expressed doubt that it might be explained solely by local functioning of a generalized resistance and the universal tendency of free symbionts to pass over, if possible, to intracellular life, and these doubts increase when we consider the directed migrations of free symbionts during embryonic development and the symbiont movements preceding the sorting-out process.

It is true that, wherever egg cells or early developmental stages are infected, the various sites attain their essential form and are settled by the end of embryonic development. However, they frequently undergo further change during postembryonic development. Many modifications may still be made in the organs in location, form, and histological structure. Unpaired mycetomes may be cut into pairs and sometimes may undergo further subdivision, as explained particularly by H. J. Müller in many cases among the fulgoroids. Much more rarely, mycetomes which are separated at first may be fused later, as with symbiosis of the *Hylobius* type among snout beetles. Particularly in females, in later stages a more or less extensive cleavage is observed—and, indeed, complete disintegration of compact mycetomes which are paired or unpaired up to that point and which are usually brought into closer topographical relationship to the ovaries in this manner. Other shifts in position are more complicated but are based, all things considered, on development and shifts of the other organs in the host animals.

Histological changes are brought about chiefly by frequent fusing of the syncytia, development of mycetocytes into giant cells, and similar phenomena, but may also affect the envelopes. The growth of the mycetomes sometimes causes a flattening of the epithelia as with *Oryzaephilus* and some cicada organs; on the other hand, in the *a*-organs of the latter the reverse situation occurs when the envelope, at first very insignificant and composed of flattened cells, is gradually transformed into a strong epithelium composed of almost cubical cells and well provided with tracheae (*Fulgora*).

In another connection we have already mentioned the origin of the ovarial ampullae, which of course also falls in the postembryonic period, and in the next chapter we shall speak of the development of the infection mounds, for which the same thing is true. For other transformations of lesser importance which fall in the same period, we must refer the reader to details given in the specialized section of our book.

Completely different, of course, are the relationships of the postembryonic development to symbionts when the latter are not admitted until hatching of the larvae or even later. When the symbionts live in the lumen of the whole gut or in evaginations independent of it, they simply populate the corresponding spaces and no special measures are necessary on the part of the host animals (some of the tephritines, all trypetines, *Dacus,* some hippoboscids, the one of the two *Bromius* symbionts which is housed in the lumen of the Malpighian tubules). The same thing applies to the intracellular population of certain sections of the intestinal epithelium which are not essentially changed thereby (triatomids, glossines, *Melophagus,* and other hippoboscids).

Where new formations entailed by adoption of symbionts are made available, they originate partly during embryonic development and partly during larval development. The symbionts of the chrysomelids and cleonids come upon the receptive rudiments of the corresponding intestinal evaginations. In the larvae of *Lagria* the bacteria given to the egg during deposit transfer to the three dorsal membranous sacs formed several days before hatching. The four Malpighian tubules infected later in *Coccotrypes* are already differentiated in the prolarva from the symbiont-free ones by their width. In the same way the two club-shaped Malpighian tubules, which are used symbiotically in some of the *Apion* species, take this variant form even in the sterile state, and the fermentation chambers of the lamellicorns, coprophages, and tipulids are evidently also formed before hatching. In the cerambycids a specific histological character is observed, at least in the hatching animal in those places in the midgut where the ceca are differentiated after the yeasts are taken over, and, even though it is not yet discernible at this time in the anobiids, the fact that the evaginations of the gut develop even when infection is stopped artificially demonstrates that cell groups already determined during embryonic development are involved. The crypt formation on the Heteroptera intestine also begins one or two days after arrival of the symbionts, extending through the major part of the larval development, but in this case we do not know whether this process is also fixed by heredity to such an extent that it takes place when the bacteria are not present.

Such housing of the symbionts in the lumen or in the cells of intestinal evaginations offers no difficulties of any kind in the hemimetabolic insects. In fact, the crypts of the larval gut of the Heteroptera or the infected intestinal epithelial cells of *Rhodnius* are simply taken unchanged into the imaginal organization. Nevertheless, difficulties demanding special measures for safeguarding the stability of the symbiosis are required where holometabolism is present and where the larval intestinal canal, more or less radically fused during pupation, is replaced by a newly built canal composed of embryonic cell reserves.

The methods used may be divided into three groups. A radical solution occurs when the host organism uses the opportunity of getting rid of his guests, which meanwhile have become superfluous, except for a small part serving transmission. This is done by cerambycids, cleonids, and *Donacia*. When the imaginal midgut forms behind the larval gut in cerambycids, the mycetocytes are also expelled into the intestinal lumen, fall victim here to general disintegration, and with other cell fragments are removed from the body during the last defecation of the larva. A remnant of yeasts surviving metamorphosis then also reaches the out-side with male imagines of cerambyrids, whereas in the females it serves to inoculate the intersegmental tubules and vaginal pouches. The dis-appearance of bacteria-containing intestinal organs in cleonids may take place in a similar way. In their case, too, it is only the transmission organs of females that still contain symbionts. The ceca of the *Donacia* larvae, which are connected with the rest of the midgut only by an extremely thin passage, can apparently for this reason no longer reach its interior and therefore, in connection with the renewal of the gut, are destroyed on the spot by lytic processes.

In all other known cases the host organism keeps its symbionts, even though the larval sites are fused, and either erects at the same place new imaginal organs which more or less resemble the larval organs, or metamorphosis compels the host to transplant the symbionts to other places and sometimes to create completely different organs.

The anobiids use the first method. In their case also, the majority of the mycetocytes are expelled to the intestinal lumen and pass to the outside, but some of the yeasts, at a time when larval and imaginal epithelia are still in contact, transfer to the latter, where only somewhat smaller ceca are again formed for them. The *Cassida* species probably act in the same way. Larvae and imagines have similarly formed evaginations in the corresponding place, although it has not yet been possible to observe the transfer of the symbionts to the organs of the sexual forms. The *Bromius* larvae also expel the two roundish ceca at the beginning of the midgut during pupation; in the process the rosettes formed by the symbionts fall into their component parts, and these now infect a circle of slender tufts which is located at the same place but differs greatly in form. The second symbiont type is forced by metamorphosis into a more radical change of residence, for the staff-shaped bacteria, which also populate the Malpighian tubules, reach the interior of the gut during fusion, transferring from there to certain tufts which take form toward the end of the midgut.

The fate of these bacteria leads us to the third group of phenomena which all have in common the fact that the holometabolism causes the hosts to house their symbionts in completely different places in the larval and imaginal states. In general, the imaginal sites are also situated on the intestine but occasionally it happens, as we have seen, that spaces behind the gut are populated during pupation. The imagines of *Dacus oleae* transplant the symbionts to the unpaired evagination of the esophagus; other trypetids transfer them from the same ceca situated at the beginning of the larval midgut to evaginations of the midgut established specially for this purpose, which are more or less highly differentiated and are situated more to the posterior. In both cases the symbionts are limited to the lumen. Indeed, during metamorphosis Diptera tend in general to fix the symbionts, which in the larval state had been housed in the area of the proventricle, in areas of the midgut situated more toward the posterior. In this case they may be intracellular in the larvae and the imagines, or they may be extracellular in the first and proceed to intracellular life in connection with the transformation. The first possibility occurs with glossines and some of the hippoboscids (*Melophagus, Lipoptena caprina*), whereas others furnish examples of the second possibility (*Hippobosca equina* and *capensis*). *Hippobosca camelina* and *Lynchia maura*, as larvae, do not limit their guests to more narrowly defined sites but carry them free in the chyme of the whole intestinal lumen and, during pupation, decide after all to admit them to midgut cells. *Ornithomyia* is forced to take much more decisive measures through fusion of the larval gut and gives up housing in the intestinal epithelium in the imaginal state by building a mycetome-like formation outside the gut.

At first glance, the reverse seems to be true among those curculionids which as larvae have compact mycetomes at the juncture of the fore- and midgut but as imagines carry the symbionts in mycetocytes which are distributed like crypt cells over the whole midgut epithelium. Yet we have seen that such mycetomes are built at the expense of the rudiment of the foregut and that during metamorphosis, which leads to a renewal of the midgut by ectodermal-cell reserves, recall their origin, so to speak, loosen up, and take part in building up the imaginal gut. Thus there is no transfer in this case to other cells but only a complicated shift of larval mycetocytes (*Hylobius, Otiorrhynchus, Calandra,* and others).

We are unable to tell whether the mycetomes of *Rhynchophorus,* which are situated in the same place and by way of exception survive metamorphosis unchanged, are of different origin and therefore behave differently.

From a consideration of the various methods used by host animals to rectify an awkward situation, we once more build up a portrait of a superior organizer who understands how to direct the symbionts to other parts of the body, how to transplant them from one cell type to another, how to build new and different types of sites, and thus how to safeguard the endangered stability of the symbiosis.

14

Correlations between host

organism and symbionts

Every page of the section on specialized forms—for instance, the comparative chapter on sites, methods of transmission, and embryonic phenomena—indicates with commanding clarity that with genuine endosymbioses the host animals are in all respects master of the situation. Thus, in general, discussion of correlations between the two partners must necessarily revolve about this important part of our investigation. The symbiotic host regulates in the same fashion the degree and special form of propagating for his guests; it may give them a certain shape, regulate their size, and it even has the capacity for getting rid of them on occasion.

Even in algal symbiosis an astonishing control is found in the rate of increase. Thus the number of yellow cells in the radiolaria is often greatly limited. Among the Sphaerozoa there are species in which almost 100 algae are grouped around each individual, others in which 1 or 2 or at the very most 3 or 4 symbionts lie near the central capsule, and still others which occupy a midposition with respect to numbers. With the acanthometrids the number of algae also varies from species to species and according to developmental stages. Only a few contain 6 to 10, or approximately 20 to 40 zooxanthellae; usually there are 60 to 100, and with the dorotaspids and hexalaspids they are in groups of 8 to 16 around every skeletal ray, so that in all they amount to 300. The greatest number is found among the stauracanthids and the dictyacanthids, in which large associations surrounded by special envelopes fill almost the whole central capsule.

Even more surprising is the high degree of dependence of propagation in some symbiotic Cyanophyceae on adaptation among the rhizopods and flagellates. Indeed, *Paulinella chromatophora* always contains only 2 blue algae, of which one glides into the daughter animal during division of the host organism, whereupon both divide again. In other cases the number is restricted to 1 to 6 cyanellae, whereas certain amoebae have more than 100.

The remaining symbionts also propagate within certain limits, for again and again it may be observed that the individual sites always display a specific size and density of population in the various stages. Apparently this constraint goes so far in some cases that mycotomes are built up of a constant number of cells. To be sure, we do not have detailed research, which indeed would be laborious in the case of larger organs, and we must be content with an approximate estimate. In one case, however, such precise measuring may easily be determined in very impressive fashion. Among the fulgoroids, the few-celled rectal organ consists frequently of a specific number of cells in individual species. *Cixius nervosus* and *Oliarius horisanus* always build it up from 9 cells; other *Cixius* species, from 4, 6, 7, or 11 cells; *Kelisia praecox* Hpt., from 7; *Kelisia vittipennis* I. Shlb., from 8; and a meenopline, from only 3. Among the issids there are rectal organs with 11 and with 20 mycetocytes; in other cases one does of course come upon slight variations, yet the numbers always vary within the same limits. Only the tettigometrids have rectal organs with higher cell numbers, for example, *Tettigometra fusca* Fieb. with 25, *T. obliqua* Panz. with 30, and *T. atra* Hgb., with 35 to 40 cells. Occasionally there are cell divisions not provided in the general plan of the organ, but in such cases the surplus elements are forced out from it and are resituated in similar form in fissures of the abdomen, close by or distant, as the case may be. In *Crepusia nuptialis* Gerst., where rectal organs normally consist of 25 cells, 21 such expelled cells were found by H. J. Müller, to whom we owe all these observations, so that one may conclude that all embryonic mycetocytes have inadvertently divided another time in this case.

Detailed examination soon shows that the rate of increase of the symbionts varies at different times, and that the host organism, according to momentary requirements as it were, tightens or loosens the reins from time to time. In the course of the symbiotic cycles three types of increased reproduction can be differentiated: temporally limited periods of increase during embryonic development; gliding, longer-lasting mass increase during postembryonic development; and spurts of increase serving special purposes and usually limited to special zones.

Increases during embryonic development, with sharp time limits, are etched very clearly in *Fulgora* and *Cixius* embryos, which have been investigated so accurately. When the germ band begins to invaginate, the *a*- and *x*-symbionts awaken their rest after division, which usually appears after the end of egg infection, and increase approximately four-fold until its end, that is, up to the fifteenth to the twentieth day of development. The the period of division is replaced by one of growth in the two types. There is statistical proof that later the *x*-symbionts do not increase at all, but a second wave of increase occurs for the *a*-symbionts after revolution. The third guest of these two fulgoroids, the *b*-symbionts of *Cixius* and the *m*-symbionts of *Fulgora,* increase far less than their two companions during this whole time.

Another striking increase in the embryonic mycetomes, populated meagerly at first, sets in before revolution, for example, in *Cimex lectularius.* An animated increase of the two types is also confirmed in *Lyctus* in direct connection with the occupation of the sites, hence again before revolution. The same thing is found in aphid development, in the early embryonic enlargement of the area of the *a*-symbionts by infection of the envelope cells in membracids, and in many other cases. Understandably, the intensity of the embryonic symbiont increase is by no means the same everywhere but is adjusted to each particular situation. For example, the increase of the symbionts during embryonic development is only slight in *Pediculus.* Ries counted 150 to 250 symbionts in the deposited egg and 290 to 330 in the stomach disc before hatching. One may conclude that for the most part not every symbiont divides once in this case.

Quite different in nature is an increase in the quantity of symbionts which may be confirmed step by step in the course of postembryonic development. It usually runs parallel to the growth of the individual developmental stages ensuring that the proportion fixed by heredity is maintained between the size of the host animals and their mycetomes. Müller demonstrated very graphically for *Cixius* that the relation between the body size of the individual nymphal stages and the total volume of the various mycetomes always remains about the same. For the most part, no penetrating studies of this kind are available for late stages of development, but we can judge from statements made in the literature that many animals behave in a similar way. Examples are well-defined mycetomes like those of *Pseudococcus citri, Cimex lectularius,* or of the aphids, and settlement of scattered mycetocytes. Also, where the symbionts drift free in the lymph, as in the lecaniines, it was possible to observe a gradual increase in population strength during larval development. Where symbionts populate the intestinal epithelium their increase tends to be seen less in the increase of the individual cell than in the extent of the populated zone; and *Marchalina* also represents a unique case in that the infected intestinal-epithelium cells swell extraordinarily in consonance with excessive increase of symbionts during postembryonic growth. Symbionts living in the intestinal lumen are also subject to a proportional increase controlled by the growing host animal, as proved by the trypetid larvae, by the Heteroptera, where the crypt gut becomes larger from molting to molting, and by other cases.

The degree of symbiont increase and consequently of mycetome development is proportionately regulated in cases within the same species where generations follow upon one another and the increase in symbionts differs according to the different body sizes and intensity of reproduction. This is shown especially clearly in *Pemphigus* where the fundatrix produces about 75 progeny, the virgo 30, the sexupara 6, and the sexual form has only a single egg to provide with symbionts (Tóth), but even with the adelgines there is a similar gradation corresponding to the individual generations (Profft).

Occasionally there is no proportional growth of the host body and of the symbiotic sites, but all such deviations evidently have their special reasons. For example, in *Pediculus*, the embryonic mycetome grows only about two-fold during postembryonic development and is relatively largest in the young stages, but this may be connected with the later transfer of the symbionts to the ovarial ampullae.

In comparing the sites of male and female animals, considerable differences may be observed in the relation of their size to that of the host body. For example, in the *Acanthosoma* female, the crypt gut amounts to about one-half the body length, and in the male to one-fifth of it. No less impressive is the comparison of the crypt gut of a mature male with that of a mature female in *Brachypelta*. In view of the considerable differences in size between the sexes in the coccids, the size difference of mycetomes increases here in proportion. The unpaired mycetome of *Pseudococcus*, which in the female constitutes about one-third of the body size, is so insignificant in the male that it is easily lost sight of during dissection. Also, the male mycetomes of *Puto* and *Macrocerococcus* or of *Margarodes polonicus* are much smaller than those of the females, and it has been stated about *Chionaspis* that the diffuse mycetocytes in the female sex are relatively more numerous than in the male. With *Oryzaephilus* large increase of the symbionts in the mature host is restricted essentially to the female. The study of the cicada *Asiraca* reveals drastic differences in the behavior of the symbionts in the two sexes. In the male the reproduction of the inmates of the unpaired *p*-organ disappears completely in the autumn with the attainment of pubescence and there is only a slight enlargement of the symbionts, but in females animated reproduction of symbionts and decrease in their size sets in during the same period. The same is true for the *a*-mycetomes of this cicada. Their volume in females exceeds that in males ten times, and this is also true for some of the other cicadas. The so-called yeasts are also subject to retardation linked to the male sex, as shown in the delphacids, where they usually are loosely distributed and, in case of syncytial formation, take over·a smaller number of fat cells. The bacteria which in *Donacia* transfer in the larval state to two of the Malpighian tubules, and take possession of a limited section with a view to future provisioning of eggs, later multiply considerably in the female pupa. Not only does such multiplication fail to occur in males, but disintegration of the symbionts may even set in.

These conspicuous sexually caused differences suggest a link to increased need for symbionts on the part of females. However, with the previously mentioned spurts in increase, the relationships to transmission requirements are usually clear beyond doubt. With some cicadas, for example, the wedge cells of the follicle are infected by a relatively small number of symbionts, and these then undergo a short, sharply limited period of multiplication which ends at the moment when the cell cushion arising in this way is filled to the maximum. With other cicadas the multiplication does not occur in the wedge cells but begins instead when the symbionts transfer from these to the space forming behind them; and in still others the symbionts are not increased until admitted to the oocytes. This is seen very clearly in *Eurybrachys* where only a very few yeasts are situated at first in the spherical symbiont balls between the accumulation of the fine accompanying bacteria, but these yeasts become much more numerous even before development begins. Also in *Oryzaephilus* a clear division period is observed in the symbionts when they are still situated between egg and yolk membrane; and in *Ixodes* there are especially conspicuous waves of increase even during egg growth. It is obvious that the production of transmission forms undergoes a perceptible relief in this way.

In general, a conspicuous spontaneous increase of the symbionts frequently goes hand in hand with egg infection. With ants we observed that the follicle is greatly inundated in the area of the young oocytes and that subsequently an actually stormy increase occurs there. In blattids and *Mastotermes* a very animated reproduction of the bacteria on the egg surface led to a thick covering over it.

In all these cases the quantity of symbionts needed for transmission is covered by increase of those which have already arrived at the ovarioles. In other cases the same provision is already made in the sites proper and may extend to the whole mycetome or be limited to narrow zones. In *Oryzaephilus*, after the first increase period fills the maturing organs in the usual way, a universal enormous increase of symbionts suddenly begins in the female imagines, during which the previously tube-shaped inmates are transformed into the small transmission forms. There is a similar situation in *Lyctus* in the rosettes living in the marginal zone of the disymbiotic organs. In females they fall into fragments with sexual maturity, first sporadically and finally everywhere, and these fragments are transformed later into very characteristic infection stages. In the stomach disc of *Pediculus* the symbionts clearly undergo an increase before they all leave the organ to transfer to the ovarial ampullae; and among the viviparous aphids, in which infection of the numerous embryos causes a particularly strong demand for symbionts, the needs are met by epidemic waves of division (Tóth).

When Fink investigated the symbiotic cycle of *Pseudococcus citri* in greater detail, he also came upon a rhythm visibly corresponding to varying needs. Until hatching, the predominant reproductive phase of the symbionts is one characterized by oval and roundish shapes and frequent division stages. As long as the larval stages are motile, regulated ingestion of food does not take place, reproduction lags, and the degenerative states accumulate. The symbionts become long and more or less swollen, or they may be extremely large, roundish forms which stain badly. From the moment of establishment until egg infection begins, a new reproductive activity of the symbionts predominates; during egg deposit it sinks radically and the stages of degeneration increase again, until finally, after egg deposit is completed, the mycetomes of the gradually dying animals are filled with extremely degenerate symbionts with greatly enlarged vacuoles. Unfortunately, no comparison of the male animals has been made, but such a comparison would doubtless document even more clearly the close relationship of the symbiotic cycle to the demands of the host animals and their physiological state.

The special depots which are built for transmission also indicate purposeful and intensified multiplication among the symbionts. This applies equally to the cavities used in egg smearing, such as intersegmental tubules and vaginal pouches in cerambycids, anobiines, etc.; it also applies to the equivalent rectal formations in the trypetids, and to filial mycetomes with symbionts located intracellularly. The ovarial ampullae of *Pediculus*, according to Ries, contain 3000 to 6000 symbionts. Approximately five eggs each with 150 to 250 symbionts are fed daily by them without visible decrease in density of settlement. This means that in the same period of time as many as 1000 symbionts must be supplied subsequently through divisions. We must also recall to the reader's mind the division wave, sharply limited in time, in the Malpighian tubules of the female *Donacia*. Where eggs are not provided with symbionts and, by way of exception, the nymphs must first take them for days from the maternal anus, as in the bug *Brachypelta*, an enormous increase of the bacteria and enlargement of the crypts housing them begins only in the "nursing mothers."

Locally limited spurts in increase connected with the mycetomes are primarily caused by the infection mounds which are always formed by *a*-organs of cicadas. These mounds are supplementary cell nests, sterile at first, which originate in certain parts of the mycetomes, sometimes resting on them in the form of more or less sizable elevations, as with the cicadoids, or inserted in them and thus not changing the shape of the organs, as is usual in the fulgoroids. The cicadoids always form only one such mound, which makes its appearance laterally or apically; in the fulgoroids, especially when the organs are elongated, several of the corresponding cell nests are often present at the same time. With the issines or with *Fulgora europaea,* for example, there are two each; with the poiocerines the number changes according to the species, increasing sometimes to six or more with tube-shaped mycetomes. These cases should not be confused with the rare instances in which a subsequent increase of infection mounds is brought about through growth or cleavage of the mycetomes in question. The first is the case with *Asiraca* is which the *a*-organs are drawn out to an extreme length; the latter takes place during division of cicadine mycetomes into racemose associations.

The symbionts multiply animatedly in the small uninucleate or dinucleate elements—which may be traced back to locally active increase of the epithelial cells in the mycetomes and which are always contrasted sharply with the symbiont-containing zones—as soon as the elements are infected from these zones at a fixed moment. At first embedded individually in vacuoles, they soon fill cyst-like spaces and progressively take possession of the whole area toward the top until finally the infection stages which are enriched in that way are given to the lymph stream at the highest part of the mound.

Corresponding devices are rarely observed in other mycetome types. In the unpaired *g*-organ of a derbine a corresponding cell nest behind the epithelium is pushed between the syncytia, but an early non-infected stage here not has been observed as yet. An exceptional case of this kind is represented in the so-called epithelial organs of certain poiocerines, in which the *a*-organs, which form as such an infection mound, make their epithelium available to an additional guest. A second, differently constructed, sharply separated cell nest is also made ready for this guest; and with the cicadines which form a marginal layer with *w*-symbionts around the *a*-mycetomes, without recourse this time to the mycetome epithelium, the two types, each in a cell nest, undergo increase and transformation for the purpose of transmission.

Finally, an excessive development of a formation, which is evidently comparable to the infection mounds and is linked again to a considerable stimulus to divide, is found in the ricaniid *Bladina* and another nogodinine which has not been classified. The unpaired *n*-organ is enclosed on all sides by a thick layer of large dinucleate mycetocytes. Processes resembling the often markedly developed infection mounds of the aphrophorines extend from the layer to the two ovaries and here enter into surprising relations with the young ovarioles. Cell degenerations in them make way for the implantation in each of a number of mycetocytes which develop into filial mycetomes, and from the latter first the nutrient cell plasma and then the egg cells are later infected. With the transplanting into the first there is once more lively multiplication of the symbionts, which then glide on the fiber cords to the eggs. Thus with transplantation into new areas—sterile cells of the infection mounds or nutrient cells—two well-separated increase waves furnish the symbiont quantities necessary for continuation of cooperative living.

Quite differently constituted is the division impulse, which is limited to a few cells, and which precedes formation of rectal organs in the fulgoroids. A small group of nuclei of the *x*-syncytium, again derived from original envelope cells of the mycetome, provides in this case the impetus for cell delimitation of several symbionts already quite well developed. In them, growth ceases forthwith and its place is taken by division and increase in the small migratory forms which fill the provisory mycetome now in the process of separating. As long as the latter remains incorporated in the intestinal lumen, renewed division-rest prevails but, after the final site in the valvula pylorica is settled, it is replaced by a new period of animated multiplication which leads to the filling of the new mycetocyte generation which at first had been only meagerly provided for.

From all this we see clearly that the multiplication of the symbionts under normal conditions is controlled chiefly by the host organism itself. Only where several symbiont types are cultivated side by side is it necessary to take into consideration the possibility that stimulating reciprocal relationships exist between them. Supporting this idea are, first, several observations with various mixed cultures of non-symbiotic microorganisms in nature and in the laboratory (satellism) and, second, the frequently proved production of growth-promoting vitamins by symbionts, a subject we shall discuss in detail in the closing chapter. On the other hand, the antagonism asserting itself in certain combinations, which will play a considerable role in our discussion of historical questions, makes it appear possible that in plurisymbiosis one symbiont may sometimes check the increase of another.

However, before explaining the means available to the host for regulating the quantity of symbionts, it seems opportune to furnish corresponding proof that the host controls also their shape, for both types of influences are frequently found side by side and sometimes may be traced to the same or similar causes.

Symbiosis research today has available considerable material on the problem of pleomorphism in plant microorganisms, especially in bacteria. Although the great plasticity shown by the symbionts was a fact forced at every turn upon the attention of zoologists working in this field, bacteriologists ignored their discoveries, treated them with skepticism, or expressed doubt that the symbionts were living organisms at all. The bacteriologists themselves were all too prone to doubt the purity of their cultures when they found surprising pleomorphism therein; and it was completely overlooked how favorable for such investigations those cases of endosymbiosis are in which the homogeneity of the individual states, thanks to localization in specific cells and organs of the egg, can be studied continuously to the death of the individual without necessity of analyzing the cultures. Thus animals were encountered in which the bacteria, originally simple in construction, were modified to an extreme extent, but, in view of the conservative bias of the professional bacteriologists, a certain courage was needed to interpret the shapes in question as bacteria. My school of thought, however, supported this viewpoint from the very start.

All those roundish, sausage-shaped, tube-shaped, rosette-shaped or gnarled formations were explained by us to be modified bacteria. Indeed, the comparison of symbiont shapes in closely related host animals would of necessity lead one to this concept. For example, among the Heteroptera symbionts which settle the ceca in the various hosts, one finds typical bacterial shapes, larger and thicker forms, and, in *Murgantia,* gigantic irregularly convoluted and locally swelling tubes are observed. The various *Pseudococcus* species usually have sausage-shaped, thickened organisms, but *Pseudococcus adonidum* harbors in the very same organs the rod-shaped and filamentous bacteria joined in gelatinous balls typical of the genus. The bacterial nature of these is no longer in doubt. Also, with the mallophages one comes upon all transitions between short rodlets, slender filaments, and roundish forms as well as oval and very swollen ones; and, with Anoplura, in addition to swollen and elongated, sometimes completely homogeneous shapes, there are granulated vesicles, as with *Polyplax,* and only slightly modified formations. Even the gigantic x-symbionts of the fulgoroids, which may measure 200 μ and more, emanate, as we have seen, from states which were still very close to typical bacteria and must be regarded as grotesquely deformed bacteria.

The microbiology of the last decades has meanwhile completely justified this concept. At first it was chiefly the technical mycologists who frequently observed considerable plasticity of the bacteria and were thus in a position to support the symbiosis expert in his concept. Today, however, the material on pleomorphism in bacteria is so extensive that we briefly remind the reader how significant these findings are for the understanding of our own results. Again and again, with the various environmental changes brought about experimentally, modifications resulted in the bacterial shape, in the cytological microstructure, and in the manner of multiplication. For example, the influence of lithium or cesium salts radically changes the form of the bacteria, and an extensive literature exists today on this subject. Little rodlets, for instance those of *Paratyphus-b* or *Escherichia coli,* become huge, structureless protoplasmic clumps which may attain a diameter of 20 μ without losing their capacity for multiplication (Maassen, 1904; Eisler, 1909; Sander, 1938; Klieneberger-Nobel, 1949, 1951; L. Dienes and H. T. Weinberger, 1951; J. Piekarski, 1952; L. Carrère and T. Roux, 1952; H. Braun, St. Berg, S. Kessler, and P. Mavroidi, 1954; H. Bender, 1957; and others). A counterpart to research in this direction is that on the effect of organic substances, principally the various antibiotics, on the shape and multiplication of the bacteria (*Proteus vulgaris, Escherichia coli, Paratyphus-b, Staphyloccus, Streptococcus,* and others). In their case extreme longitudinal growth, giant forms, club shapes, branchings, and mycelia-like networks were obtained (A. D. Gardner, 1940; A. Fleming, A. Voureka, J. R. H. Kramer, and W. H. Hughes, 1950; R. S. V. Pulvertaft, 1952; F. Grieder, L. Neipp, and R. Meier, 1954; and others).

Of special interest to the zoologist, who must be concerned chiefly with shapes that bacteria assume in the insect body, are the somewhat earlier findings of Paillot (1922, 1933) on transferral of parasitic bacteria to the body cavity of other insects. *Bacterium pieris liquefaciens alpha,* obtained from the blood of *Pieris brassicae,* where it has the form of a coccobacillus, remains almost unaltered in the caterpillar of *Vanessa urticae;* in the caterpillar of *Vanessa polychloros* it becomes somewhat longer; in *Lymantria dispar* it grows out into more or less convoluted filaments 40 to 50 μ in length. *Bacterium melolonthae liquefaciens gamma* produces a short rodlet in the May beetle; in *Agrotis segetum* big filaments occur; and in *Lymantria dispar* they become much thicker and their ends are conspicuously pointed.

In other cases the growth forms are fantastic. *Bacillus liparis* is normally a gently curved rodlet; when cultivated in certain sugars, it grows out and swells like a club at one or both ends; when transferred to the blood of *Pieris brassicae,* it grows gnarled branches resembling the "rosettes" appearing in *Centrotus* and *Euacanthus. Bacillus melolon-*

thae liquefaciens gamma rapidly swells into a tube shape in *Lymantria dispar;* in advanced stages of infection there are median or polar swellings to which gradually the whole bacterial substance transfers. The strongly staining constituent parts collect into balls in the center or permeate the whole organism which is simultaneously becoming immotile, so that a certain similarity with infection stages of cicadas results. After the death of the grub, these forms disappear again and only the typical coccobacilli float around in it. When *Bacterium lymantricola adiposum* is isolated for a rather long time and is then transferred back to *Lymantria,* all sorts of proliferations and thickenings are formed, resulting in mycelial-like stages such as are found in so many symbionts.

In general, it can be seen that radical changes of form in endosymbionts occur, with negligible exceptions, only in forms living intracellularly between intestine and hypodermis. Conversely, the more primitive shapes are displayed chiefly by inhabitants of the body cavities emptying to the outside, or by inhabitants housed in the epithelium of the gut or of the Malpighian tubules. This is true for symbionts of trypetids, glossines, triatomids, and most Pupipara, for bacteria in luminous organs of fish and cephalopods and in accessory glands of the latter, as well as for the inmates of the nephridia of certain oligochaetes. Throughout there are rod-shaped or filamentous formations, and it is no accident that it is precisely these that can be cultivated outside the host body without difficulty. Among the few exceptions are the unusually shaped symbionts of *Murgantia* among the Heteroptera, the inmates of the Malpighian tubules in *Apion* species which are reshaped into clubs, the rosettes in the intestinal organs of *Bromius,* and the inclusions in the luminous organs of *Heteroteuthis.*

We know, moreover, that all the more or less undisciplined accessory symbionts keep their original forms, although they usually live intracellularly and at times even in one and the same cell with deformed older comrades, nevertheless in addition they often drift free in the lymph. Here the degree of swelling and deformation actually becomes one of the criteria for determining the relative age of the adaptation in polysymbiotic forms. At any rate, the deformation is generally not caused spontaneously through transfer into the lymph of the host body, as in experiments of Paillot cited above, but is revealed, but by no means invariably, as a gradually appearing companion phenomenon of intracellular life. That it may fail to occur in spite of the old age of the symbiosis is proved by the symbionts of blattids.

Much less plastic are the symbionts designated yeasts even when housed in quite mycetome-like formations. Even though a good many of the morphological peculiarities in each case are apparently traceable to environmental influences which differ from species to species, nevertheless these peculiarities fluctuate within modest limits.   On the other hand, there is almost complete stoppage in formation of long tubes, tuft-like associations, or mycelium-like states, in other words, the growth form peculiar to them by heritage and making its appearance in cultures. Although tube-like outgrowth is scarcely ever found in the pseudomycetomes and the mycetocytes of intestinal evaginations, it still occurs occasionally in healthy animals when the yeasts live in lymph and fatty tissue.

An influence quite generally found with bacteria and yeasts concerns general stoppage of the capacity for building spores.   At first glance this is surprising, as one would think that they in particular would furnish suitable states for symbiont transmission to progeny.   In fact, spores do serve this purpose in the two exceptions known today, namely, the spores of luminescent bacteria infecting the egg follicle of the pyrosomes and the spores which fill the lubricating organs on the ovipositor of a number of the cerambycids.   In the latter case the basis for the exception might be that the still conspicuously ill-adapted yeasts in the groups in question soon develop enormously after each infection of new intestinal cells and at the same time lose their capacity for budding.

In view of such plasticity in shape of symbionts, it is not surprising that arriving and departing impulses toward their multiplication accompany the individual cycle of the host animals, in the course of which the symbionts are also often transferred to other places in the body. Moreover, the impulses also cause cyclical form changes, and these changes frequently act as sensitive indicators of the environmental changes taking place in the course of the individual developmental stages. In our section on specialized forms such changes were mentioned at every turn.   For example, they were particularly pronounced in the *Donacia*.   In their Malpighian tubules, settled for the purpose of egg smearing, are broad stout rodlets; from them arise much smaller infection stages, which are short oval to spherical forms that may still be found in the newly hatched larvae.   But after 12 hours they begin to elongate again in the now colonized intestinal appendages, and in the course of larval development they grow out into long filaments.   Evenly colored at the beginning, they gradually fill with more deeply staining granules which finally flow together to form larger sections and give the still elongating filaments the appearance of chains.   Whereas the majority of the symbionts are disintegrated during metamorphosis in the intestinal organs, some infect two of the Malpighian tubules and grow shorter there.

More or less pronounced cyclic form changes are found also with the trypetids, where they reach their climax in *Tephritis heiseri*.   The symbionts of the larvae have the form of short plump rodlets; in young imagines they tend to elongate; in older ones very long filaments are found; in mature animals short transmission forms are found again; but even with *Dacus* the bacteria are much smaller in the young larvae than in the older. The apionine symbionts in the Malpighian tubules pass through a series of such varied states that Holmgren, who still considered them to be cell-specific inclusions, interpreted them as individual stages in maturity of secretion droplets. In reality, very small rodlets grow out into sinuous filaments, produce a gelatinous envelope, and finally fall into fragments together with the envelope. Of particular interest in this respect are observations on *Coccotrypes* symbionts. In the egg cells they are found in the form of very small rodlets; in the first larval state they are somewhat larger and transfer to the gut and four of the six Malpighian tubules; and here, in the individual sections, which are also differentiated histologically, they assume extremely varied shapes. Moreover, these states are subject to other transformations in the larval, pupal, and imaginal states. The most penetrating bacteriological analysis of such cyclical changes is one of *Camponotus* symbionts. Using smears and sections, Kolb (1957, 1959) carefully described the changes undergone by the essentially rod-shaped and filamentous organisms living in the intestinal epithelium in the individual stages of post-embryonic development. She took into consideration the various division tempos each time, directing her attention especially to the behavior of the chromatic substances which are subjected to considerable change. Previously, the only positive statements regarding proof of desoxyribo-nucleinic acid in symbiotic bacteria were those of Rizki (1954) and of Frank (1956) with *Blattella germanica* and *Periplaneta orientalis*, respectively. Kolb now proved their presence in her test object and studied the sizeable changes which these nucleoids undergo during the individual phases of the cycle.[1]

The *x*-symbionts of the fulgoroids undoubtedly undergo the most extreme change in form as yet encountered by the symbiosis investigator.

---

[1] Kolb's observations are in consonance with those made in studies of modifications caused by antibiotics or lithium chloride. In connection with her investigations of *Camponotus* she also tested several other symbionts and found desoxyribonucleinic acid in those of *Oryzaephilus* and *Ceratitis*, whereas she found no nucleoids in those of *Pediculus* and in the chief symbiont of *Aphrophora salicis*. Her findings correspond to the negative results of Resühr (1938) with *Philaenus*, of Lanham (1952) with aphids, of Bewig and Schwartz (1956) with *Haematopinus*, of Kotter (1955) with *Pediculus*, and of Nicol (1958) with *Pyrosoma*. Therefore it does not seem justifiable to doubt for this reason the bacterial nature of these symbionts. Even in the egg nuclei themselves, Feulgen-positive reactions can sometimes not be demonstrated.

During embryonic development they still are very normal short tubes and increase through cross-division, but, shortly before the embryo hatches, they suddenly lose their capacity for division and grow out into structures which exceed all previously known symbionts in size by many times. They may retain their polyhedral shape and a more or less smooth surface, as is the case with the derbine type where growth is kept within certain limits, or they may assume extremely strange forms and be deeply slit, as with the fulgorines, poiocerines, *Bladina*, and others. Appearing simultaneously in the very thin protoplasm are countless, highly refractive eosinophilic granules which always undergo changes in the same way during larval and imaginal life; they develop into larger, elongated, flowing structures, and sometimes they swell into irregularly shaped clumps in older animals.

Sometimes the behavior of the symbionts reveals that various conditions prevail at one and the same time in the individual cells or syncytia of mycetomes without any morphological signs whatever. The symbionts show then a different character from cell to cell, but all are in the same stage of transformation. This is often true of aphids, for between mycetocytes containing the typical vesicle-shaped primary symbionts there are other mycetocytes containing symbionts which are all of a different order of magnitude. In the t-mycetocytes of *Euacanthus* also the rosettes are very different in character, and an analogous situation occurs among the same symbionts in membracids, although the phenomenon is not connected with cultivation of transmission forms. Fulgoroid symbiosis also provides us with examples of such behavior. In the individual syncytia of a-organs in poicerines there are organisms quite different in size and shape: in one species, for example, many smaller and more slender tubes on the one hand, and relatively few, larger, almost egg-shaped structures on the other. If, then, such syncytia flow together, as so often happens, the former areas can still be recognized by the regionally different sizes. In *Bladina*, multiplication grouplets of these symbionts occur in the individual syncytia, with components different in size and number. In apionines, the club-shaped transformed Malpighian tubules have a checkerboard appearance, conspicuous even with low magnification, which may be traced to the fact that each cell is filled by a different developmental stage of the symbionts; and striking differences in size are also found from cell to cell in *Tachardina*.

Just as considerable differences in intensity of multiplication were found in comparing the two sexes, so too do the symbionts react sometimes in formational respect to different environments. For example, in the fulgoroids there is often dimorphism of the x-symbionts. With *Caliscelis* they are larger and deeply incised in the broad tube-shaped organs of females; in the more slender mycetomes of males they are polygonal in shape and smoothly delimited; in *Asiraca* they are complexly pinnate in females and coarser and less deeply lobed in males. Conversely, the *a*-symbionts as a rule develop more markedly in males, in which, as we have already seen, in contrast to female symbionts, they lose their capacity for dividing and become two to three times as large as in females. It is the same with *Issus* and some derbines. Concerning the diaspine symbionts, it was noted that the symbionts of females are larger than those of males in certain species. In a smiline Rau found certain yeasts which gave the impression of being parasites and had the usual slender tear-shaped form in the lymph and fat cells of females, whereas in the males they had the same form in the seminal vesicles but otherwise appeared as tubes which often were ten times longer.

To what extent these changes in form are of significance for the host and to what extent they are only automatic reactions to changes in milieu cannot always be determined with our present knowledge. It is, of course, quite possible that increases in achievement of some kind go hand in hand with the often considerable increase of the symbionts. H. J. Müller interprets the giants of the *x*-symbionts as organisms over which the host animals have lost their power, so to speak, and which, since they have lost their capacity for division, increase in this uncontrolled manner. But one might argue that is precisely the genesis of the rectal organs which shows with great clarity that the host has means at its disposal to cancel growth and cause division. What one is inclined to consider deformed states may very well be connected with certain achievements. For instance, *Pseudococcus* symbionts in N-free nutrient solutions almost always change into the Y- and cross-shaped states typical of the bacteroids of the legumes, and at the same time also here an assimilation of atmospheric nitrogen sets in.

Significant correlations of such general form changes with the needs of the moment are clearly revealed when the symbionts as a whole are converted in this way into a state which prepares the way for origin of forms suitable for transmission. This is the case when the rosettes in *Bromius* and *Lyctus* suddenly fall into their component parts; when in *Oryzaephilus,* this time in both sexes, the tubes which have grown out to a length of 70 μ, are cut into more compact states; or when *Tephritis* symbionts in the imaginal gut and those of *Donacia* in the Malpighian tubules fall into shorter fragments.

An unmistaken purposefulness is present wherever the changes in form strictly affect a limited number of the symbionts and also, usually, only certain areas of the sites. It is a matter, then, in most cases of origin of transmission forms and only occasionally of states used in transplantation to a different residence.

Frequently there is diffuse origin of specific infection stages. For example, the symbionts of *Icerya,* individually or in little groups or even in the total areas of a mycetocyte, are transformed here and there into more deeply staining states, that is to say, states which are more densely constructed. At the same time their forms may remain unchanged (*Icerya purchasi*), or sausage-shaped formations tightly enclosed in a membrane may be bent into spheres (*I. aegyptica*). Such forms, often still surrounded by shreds of plasma, then leave the mycetome at some point. The pseudococcines behave similarly. The infection stages of *Pseudococcus citri* have a certain similarity to those of *Icerya aegyptica.* This time the transformation, again not strictly localized, affects all the components of the symbiont balls, held together by a gelatinous substance, which are found in quantities in each mycetocyte. Sausage-shaped symbionts become shorter and thicker; the envelope surrounding each individual organism curves them more markedly into a U-shape; and the plasma becomes homogeneous. Entire balls then leave the mycetome together and remain together during egg infection. In another *Pseudococcus* type the symbionts are long thin filaments each coiled again in a membrane, and the infection stages merely appear to be miniature editions of them. Other species display no morphological differences between the forms populating the mycetome and those infecting the egg cells (*Ps. diminutus, Ps. adonidum, Ps. nipae*). With *Phenacoccus* the symbionts are visibly enlarged in the marginal areas of the mycetome, and the individual mycetocytes are disengaged from them to transmit their content immediately to the lymph. Here and there in the rectal organs one also finds that the symbionts are transformed into more deeply staining and usually distinctly larger infection forms which then inexplicably transfer at various places to the body cavity through the strongly developed musculature.

After the symbionts in both sexes of *Oryzaephilus* fall into smaller fragments, especially strongly staining stages, to be used exclusively for transmission, originate here and there side by side with unaltered states in mature females. Where irregularly developed tubes are present in psyllids, they are also divided into roundish or oval states, apparently without preference for certain areas; whereas those inmates of the syncytium which were acquired later and therefore still possess the original bacterial shape are sent in this form to the egg. The *b*-organs of the cicadas are also typical of employment of states which, though only slightly changed, are more deeply staining and shortened. In this case, as we have already seen with *Phenacoccus,* the symbionts frequently do not become smaller but are even distinctly larger in size than their fellow symbionts remaining behind in the mycetome. Departing along the whole surface of the organs, they linger first in the epithelia, thus leading one to suspect that they are undergoing a kind of maturation process there.

Also, in the ovarial ampullae of the Anoplura, rhynchophthirines, and Mallophaga, specific transmission forms are found which are distinguished by the usual stronger staining of the plasma and by special differentiations of the latter, and sometimes also by a more primitive shape, that is to say, one more like that of typical bacteria (*Haematopinus, Pediculus*). However, there is great diversity, for the changes may affect all the inmates of the ampullae (*Pediculus, Linognathus,* and others) or only those of individual mycetocytes, and in the latter may affect all inmates or only some of them. With *Turdinirmus, Columbicola,* and *Linognathus* a central granule arises in the transmission forms; with *Haematomyzus* a very striking, strongly staining ringlet attaches itself to the middle of each tube-shaped symbiont. The stronger staining sometimes occurs only with migrating, thus extracellular, stages and immediately disappears again in the oocyte (*Lipeurus*), whereas elsewhere it always starts in the mycetocytes and only in the course of embryonic development gives way to a light coloring caused by plasma which contains more fluid, thus introducing a wave of multiplication.

Constituting a special case are those *Stictococcus* species which live symbiotically with tube-shaped bacteria. Only where the diffusely scattered mycetocytes come in contact with the follicles of the small yolk-deficient eggs are the tubes transformed into large roundish forms; then they migrate to the oocytes through gaps in the follicle within a narrowly limited zone. In this case it is clearly contact with the follicle that causes the transformation!

From such strict spatial limitation in creating the transmission forms we proceed to cases where they develop in sharply separated areas of the mycetomes. In such cases, observed thus far almost exclusively with cicadas, two possibilities may be distinguished: the mycetome area may be part of the space originally assigned to the symbionts, or it may be a supplementary area composed of cell material that was originally sterile. The first description applies to the o-organs, composed of several syncytia, of a *Hysteropterum* species (issine). Here only the females separate a unilateral nest of more loosely built syncytia in which the symbionts are converted into smaller roundish forms. Originally assigned areas are also involved in the case of all *t*-organs, which play such a great role among the membracids, jassids, and euscelids. As we have seen, these *t*-organs are always uninucleate mycetocytes built more or less intimately into the *a*-organs as a medullar area. The inmates are always converted into specific transmission forms. This transformation, although it always affects the whole content of any one cell uniformly, skips about diffusely among the mycetocytes. It tends to appear first in the central areas of the organ, being easily recognizable by the rounding of the polygonal cells and their increase in size. Up to that point the symbionts have been thickly crowded, but now they are more clearly delimited. Where they were rosette-shaped, as in *Euacanthus,* they assume roundish oval forms. The staining becomes lighter, but the never-failing eosinophilic inclusions multiply and enlarge. These changes are noticed first in the middle of the organ, for the mitoses of the *t*-mycetocytes, which are still frequent even in the imago, always occur on the periphery and are so placed that they cause a marginal increase, thus making the central cells the oldest cells. Although preferring the paths marked out by the tracheae, migrating cells laden with infection stages force their way through the mycetome and its epithelium into the body cavity, where they soon disintegrate and transmit their content to the lymph. In one membracid, an unclassified polyglyptine with five symbiont types, the mycetocytes containing an accessory symbiont also become motile and expel from the body of the host their sausage-shaped inmates, which here show no change in form.

In a species of *Margarodes* described by Šulc, the origin of transmission forms is topically limited in strange mycetomes inserted in the wall of the oviduct. The mycetocytes increase in size more and more in the direction of the lumen, but the multiplying symbionts become smaller and smaller. Meanwhile a strongly staining central granule makes its appearance, as in a number of Anoplura symbionts. Thus matured, the infection stages transfer to oviduct and egg cells.

Particularly in the infection mounds of the *a*-organs it is possible to observe symbiont transformations allied with such unusual tendency to increase in number. At first tube-shaped, the symbionts begin to shorten and round off in the neighborhood of the specific cell nest, and this tendency continues after migration to the infection mound. Simultaneously they again become more deeply staining. The mature infection stages, which are now released to the body cavity, are often smaller than the states remaining in the mycetome, but just as often they are visibly more voluminous, sometimes twice their original size.

With those poiocerines where the epithelium of the *a*-mycetomes is given over to an additional supplementary symbiont and separate infection mounds are formed for both symbionts, the *a*-symbionts in the one are transformed in the ordinary fashion and the arched pale-staining filaments in the other become shortened thickened formations which are almost straight and again stain more deeply. When cicadines open their mycetomes to a second symbiont, two cell nests again provide the usual compact infection states with denser plasma. The k-organs of the poiocerines have marginal syncytia which also appear to be derived from submerged epithelial cells, originally sterile. In these syncytia only the symbionts are converted into differently shaped transmission forms, this time paler ones. However, certain conditions prevalent in the plasma of the nutrient cells in *Bladina fraterna* cause not only lively increase in numbers but also considerable growth in the n-symbionts temporarily housed there before they are transferred to egg cells.

In this connection we must refer to two completely different cases where transmission forms are produced in delimited territories: the rectal organs of the fulgoroids, and the differentiation into primary and secondary mycetocytes in *Puto* and *Macrocerococcus*. In the rectal organs, derivations of the intestinal epithelium admit the few symbionts which migrate from specific cells of the mycetome epithelium to the intestinal lumen, and only such symbionts are converted into transmission forms. With the pseudococcines mentioned, the differentiation of two types of mycetocytes occurs earlier in the unpaired mycetome of the first larval stage. Some of the mycetocytes are transformed into a few huge giant cells in which the symbionts are deformed. The other mycetrocytes do not enlarge so much, but they multiply rapidly and their symbionts are not deformed. It is these undeformed symbionts which are used later in transmission.

Where cultivation of special transmission forms is limited to such strictly defined areas—in other words the infection mounds, the migrating cells of t-symbionts, the rectal organs, and the "secondary mycetocytes" in *Puto* and *Macrocerococcus*—the devices were always created first for a single symbiont type and then at most achieved greater complexity, for instance in certain double mycetomes like the epithelial organs or the dilaminate cicadine mycetomes, when a second, well-separated re-forming zone is created for an additional guest. Furthermore, symbionts admitted later may come under the direct influence of the infection mounds or migratory cells which were originally created for older mycetome dwellers, and in this case it is not always clear whether an undisciplined inundation or an adaptation of the host animal is involved.

With *Cicadella viridis*, it is clear that an adaptation is involved. In this cicada the rod-shaped or filamentous symbionts, housed in a mycetome of their own, are never found in the a-mycetome proper but always in its infection mound, scattered among the a-symbionts in the process of transformation or joined in large uninucleate cells. It is exclusively in this location, never in their own mycetome, that the bundlets composed of short rodlets are formed. These bundlets leave the mound with the infection forms of the *a*-symbionts and migrate with them to the wedge cells and the oocytes.

In the membracid *Hypsoprora coronata* one finds a similarly regulated influencing of an accessory symbiont. Here, too, the symbionts involved are thin, rather long filaments, but they permeate the *a*-syncytia in quite undisciplined fashion. The nearer to the infection mound these bacteria come to lie, the greater is the number of bundlets of shorter filaments they form, parallel or coiled up together, such as the *Cicadella* symbionts. As such, they are taken over into the cells of the infection mound with the *a*-symbionts and are discharged into the ovary at its tip. Other membracids reveal the gradual development of the regulating process. The little bacterium which lives as the third symbiont type in a certain tragopine also invades the infection mounds, but in so doing it brings about strong disturbances of the *a*-symbionts, which nevertheless appear with them in the wedge cells. The same situation is found with the *η*-symbionts of *Bolbonota* and with the *γ*-symbionts of *Sundarion, Eualthe,* and *Hemikyphta* species.

Such a device created exclusively for *a*-symbionts is never used simultaneously with the migration cells which serve the *t*-symbionts and are also converted into transportation media for numerous accessory symbionts. Here, too, more or less strong regulation exists, and occasionally the shapes may be influenced. Certain long tubes in the migratory cells, housed chiefly in syncytia between the *t*-mycetocytes, become shorter and paler. In *Adippe* the migratory cells are all used additionally by the small accessory cocci. Such supplementary use occurs to such a degree in *Bolbonota* that these transportation media are needed in far greater numbers. In polyglyptines they are used even by yeasts living in the mycetome, and in a darniine indeed two accessory bacteria cause considerable disturbance in the *t*-organ and migrate together with the *t*-symbionts.

Thus there can be no doubt that the host organism exercises extensive control over increase and form of its guests. Indeed, the universal harmony between the needs of the host and the behavior of the symbionts is astounding. In considering how such harmony is achieved, general and specific influences must be kept apart.

It is clear that a very considerable retardation of the intensity of reproduction is generally involved in adaptation of all symbionts. The increase taking place during an individual cycle of the hosts is far less than that of free or parasitic microorganisms, and, where incontestably pure cultures of symbionts are obtained in artificial media, their propagation is naturally even greater than in the hosts. The nutritive conditions under which they live in the hosts are obviously sparse, and occasional decrease in the metabolism of the host checks propagation of the symbionts correspondingly. When embryos undergo hibernation, as with aphids, scale lice, and some cicadas (*Gargara* and *Fulgora* according to Rau and Müller, respectively), no change occurs in the symbionts at this time. With lecania, which hibernate as young larvae, the small number of symbionts does not increase (Schwartz). In the larvae of cerambycids and of *Ernobius abietis* the number of yeasts decreases sharply in winter, and, when the temperature is low in summer or when the animals are allowed to starve, the symbionts stop propagating and many are expelled into the gut. When favorable conditions return for the host animals, the propagation rate of the symbionts soon rises and the loss is recouped (Kiefer). Among blattids the egg growth may be checked when copulation fails to occur, when suitable food is lacking, or when rest periods interrupt reproduction, and in such cases the reproduction of bacteria in the ovarioles also comes to a standstill (Gier). When in the ant workers the egg growth is checked in the ovaries, the usual infecting symbionts are also checked in their reproduction (Lilienstern).

Future research will be able to explain more clearly the individual factors regulating symbiont increase. One of the questions needing explanation is the extent to which phenomena found in symbiotic cultures may exist also under natural circumstances. We know that an oversupply of certain essential vitamins may reduce the intensity of reproduction in cultures. Fink theorizes that increased synthesizing activities of the symbionts, which are regulated by the hosts, may lead to the death of the symbionts and that their reproduction may in this way be kept within certain limits.

All these regulations are connected with the healthy state of the hosts, as demonstrated by a series of observations with degenerating eggs and with aging, ill, or dead animals. Among the blattids, individual oocytes may be destroyed in the ovary with the result that the bacteria there multiply unchecked. The same observation was made in *Mastotermes*. The plasma of such eggs is so greatly inundated that it seems like a pure culture of symbionts, and when the eggs are expelled into the oviduct enormous bacteria groups are also found there (Koch). In ailing *Aleurodes* larvae, which are easily recognizable by overproduction of the orange pigment, the proliferating mycetomes are broken up into fragments and are distributed throughout the body (Buchner). The conidia of lecania usually increase by budding, but in aging animals they frequently grow out into tubes. The same thing happens in cultures, where of course the inhibiting factors are also absent (Schwartz). In

Pseudococcus mycetomes the symbionts are radically deformed after the end of egg deposit, and only degenerating symbionts are found at that period in *Oryzaephilus* mycetomes, in the modified renal vessels of the apionines, etc. In dead cerambycid larvae, chain-like sprout formations, such as do not occur in living larvae, were found 14 days after death. In such cases the sprouting was greatest the second day. Thereafter the propagation, which up to that point had been very intensive, continued to decrease until the fungi were ultimately destroyed in the putrescent larvae (Schomann). Unfortunately we know very little about the fate of the symbionts after the natural death of their hosts, but it is safe to assume, at least wherever intracellular forms are concerned, that they can no longer return to saprophytic life and are destroyed along with the hosts.

The symbiotic site is an inhibiting factor of a general kind. The specific regulation in size from case to case, along with the resistance of the other areas, necessarily check the symbiont propagation after maximum filling. During embryonic and postembryonic growth it may be observed again and again that the host organism always makes available the space necessary for the desired measure of increase according to the need of the moment. The demand may be met by cell divisions or by secondary assignment of previously sterile elements, as with the envelope cells of the *a*-organs or with the settling of certain areas of the intestinal epithelium. The demand may also be met by the creation of polyploid nuclei with corresponding growth of the cells. The *t*-organs show that symbionts may multiply even in mature animals through division of the mycetocytes, thus compensating for the shortage of infection stages leaving the organ. Where the demand for mycetocytes is particularly large, as with *Bolbonota*, where the migrating cells are infected to an unusual degree by the accessory symbionts, the number of the mitoses increases correspondingly (Rau). In the larvae of anobiids and cerambycids, where the number of symbionts may decrease sharply through cold and hunger, the increase rate corresponds to the available space, for, as soon as the deserted mycetocytes reach their normal filling again with improved conditions, it sinks to a minimum. When cockroach mycetocytes fill to a lessened degree because of infection with Microsporidia, they are resettled on all sides with reduction of the damage. In the fatty tissues and in the body cavity, on the other hand, the limitation is frequently cancelled by greater availability of space, so that extensive inundation by bacteria and yeasts is a recurrent phenomenon.

Conversely, it may often be noted that symbionts do not fill a vacuum offered to them. Thus the regulation must be complicated in character. In such cases it is chiefly a question of "lost symbionts." In the cicada

Ulopa one or another yeast cell is occasionally found in the epithelium-like envelope of the large pseudomycetomes which are formed from fat cells, but no increase occurs despite the facts that this symbiont type frequently lives in fat cells and that the mycetomes of *Ulopa* are also derived from the fatty tissue. With *Lyctus*, Koch found a larva in which, for unknown reasons, both sites were exchanged in these disymbiotic animals. The symbiont type residing in the marginal layer proper multiplied also in the central syncytia, but reproduction of the other type in the margin was checked despite abundant space. Something comparable is apparently involved in the strange "accessory cells" of the *Rhizopertha* mycetomes, which in all likelihood represent desolated sites of a one-time second symbiont. Frequently they contain several symbionts which had entered during early stages. These never increase and usually are no longer present in older larvae. Similarly, there is no increase among the symbionts which failed to transfer punctually to the oocytes and thus were left behind in the wedge cells, despite the facts that space is at their disposal and that multiplication may have been stimulated there only a short while before. In *Ixodes* the germ layer and particularly the cells of the egg socket are sparsely settled, but the symbionts multiply animatedly whenever some individual oocyte begins to develop. Where the symbionts are sent to the egg cell with the secretion of nutrient cells, they first multiply intensively in the nutrient cells, but, where the symbionts one by one reach the young oocytes directly, multiplication is retarded (snout beetles, some of the Heteroptera). In this connection we must remind the reader how little space in general is available to symbionts in the egg cells.

The formational development of symbionts depends largely on general conditions prevalent in the host metabolism. This is shown by the changes which affect male and female symbionts in the same way, more or less altering their original form, and by the other cyclical changes taking place both in male and in female host animals. On the other hand, an abundance of cases demonstrates clearly that the specific chemistry of the sexes is reflected both in intensity of multiplication and in the shape of symbionts. Three kinds of differences may be noted. The number of guests may differ, as with the symbionts of some fulgoroids and those of the frequently sharply reduced mycetomes in male coccids; or the differences may be limited to some of the symbionts of the female sex, in which case they are either distributed throughout or are localized, and thus the differences are always connected in some way with the creation of transmission forms. Thus, in view of such influences dependent upon sex, it is natural to think that the gonads would also exercise control through hormones. It is impossible to tell whether the control, which is sometimes exercised over the symbionts scattered throughout the entire body and other times causes the development of narrowly localized transmission forms, proceeds directly from the sexual glands or by way of detour over the neurosecretory elements. It seems obvious here to think of, among other things, the oenocytes, which are still very puzzling in regard to their function, and which have intimate topographical relationships to the mycetomes in a series of objects. My still unpublished studies on margarodid symbiosis have not only confirmed this anew, but also have established in certain objects a considerable growth of the oenocytes which quite early are often most intimately anchored to the surface of the mycetomes. Even more impressive is the behavior found in a species of *Eurhizococcus,* whose symbionts are limited to sparse mycetocytes scattered in the fat body. These mycetocytes are then always surrounded by either one or several oenocytes. In any case, it is the task of the future to concern itself more intensely with the relationships between oenocytes and symbiosis. The most seldom occurring phagocytosis of isolated symbionts apparently plays no essential role thereby.

For some reason it is apparently to the advantage of host animals to localize the creation of transmission forms more narrowly, for a steady progression toward increased localization can be traced among the test animals. In some cases only isolated occupants of but one cell or one syncytium are converted into transmission forms; in other cases all the occupants of a cell are affected and transformation may occur anywhere in the mycetome; in still other cases a certain degree of local delimitation may be noted, for instance, to the marginal areas of the organs; and finally there are cases where the transformation is clearly limited to narrowly defined areas. Especially striking here are the relationships between cultivation of transmission forms and the epithelium of the mycetomes. Often the forms originate only in the superficial areas bordering on the epithelium and tend to remain temporarily in the interior of the epithelial cells, where they apparently continue to be influenced in some fashion. With certain poiocerines the forms even become more distinct, for here the envelope cells sink into the interior along the whole surface of the *k*-mycetomes and only symbionts which have migrated in are turned into the states used in transmission. When the genesis of the infection mounds was being studied, the striking discovery was made that they are induced when embryonic envelope cells of the mycetome proliferate, for such proliferation sets in only where the mycetomes undergo a slight temporary fusion with the nymphal oviduct. The dominant role of female gonads or oviducts is demonstrated impressively in tubular mycetomes where the contact occurs in several places, because of the increased length, and where, consequently, many infection mounds are formed.

In the case just cited, temporary contact of oviduct cells with the larval epithelium of the mycetome causes local proliferation and separation. A much more complicated chain of influences is involved when the rectal organs serving transmission are created, yet the impulse toward multiplication and reformation proceeds from original envelope cells. Increased activity of several less directly adjacent epithelial cells take place in the females when there is close contact of the embryonic x-organ with the hindgut which is taking form. At the same time a change in the metabolism of these cells prevents further growth in size of the symbionts, which now multiply and become smaller. Thus primitive transmission forms are created; the cells containing them are dissociated from the x-organ and are united into a transitory mycetome for transmission to the midgut. They do not discharge their inmates until the rectal organ needs to be settled.

In *Puto* and *Macrocerococcus* an effect which transforms the symbionts appears to emanate from the intestine at a very early stage. In the first nymphal state, when an unpaired mycetome consisting of a few uniform mycetocytes is attached on the ventral side, the rod-shaped symbionts are reformed only in the cells bordering the gut and only in this location do the nuclei assume bizarre shapes.

In view of these interesting discoveries about the relation of sex to symbiosis, it is regrettable that these relations have not been analyzed in the laboratory, for they could doubtless be clarified by experiments on castration and transplantation which could probably be made most feasibly with the large cicada nymphs or the stately exotic fulgoroids. Meanwhile certain experiments in nature can give us a clue to the results that could be expected from such research. Among jassids and euscelids, attack by parasitic Diptera larvae (*Pipunculus*) often suppresses development of the ovaries completely or retards it. In such animals the infection mounds are lacking completely or are much less mature than the animal is, in general. The rest of the mycetome remains completely untouched (Buchner, 1925).

Histological findings also give us several valuable points of departure. Vacuoles appear quite generally in the still sterile cells of infection mounds, and one would be inclined to think that they contain the active ingredient in question and that it would be diffused into bordering mycetome areas with the gradual disappearance of changes which had become noticeable in the vicinity. If uninucleate mycetocytes are involved, the growth is checked; if syncytia are present, the inclination to fuse into synsyncytia decreases more and more with increasing closeness. At the same time the symbionts begin to shorten even before admission into the cells of the infection mound, and thus they prepare for the transformations which are starting there.

In the cells where the migratory forms of x-symbionts are to originate later, a change is found in the plasma, and it is apparently connected with this function and recognizable by stronger staining and density. An observation by H. J. Müller seems to indicate that the differentiation does not come about this time only through contact with embryonic intestinal tissue. In a female *Fulgora* embryo where this tissue had not yet quite reached the x-mycetome, he found on each side of the mycetome a very striking sterile lymphocyte, symmetrically set in the precise location where the envelope cells, which are gifted with such special capacities, are soon to be differentiated. These lymphocytes were filled with strongly eosinophilic granules, such as are usually found free in the body cavity only here and there around this time. Thus they absolutely gave the impression that they are bearers of substances contributing to the cell differentiation which now sets in.

In other cases the symbionts appear to be influenced without such complicated detours. In the ovarial ampullae of *Haematopinus*, infection stages are formed, in limited quantity, wherever an adjacent oocyte reaches a certain stage of growth. Thus it would seem that a transforming effect, emanating directly from the oocyte, is present. In a margarodine a clear progression of hormonal effects emanates from the mycetomes bordering on the lumen of the oviduct, with the result that the mycetocytes which face the oviduct become increasingly larger, while the symbionts become smaller and more numerous and gradually acquire the structure of migratory forms.

There are countless observations to prove that the release of activating substances is sharply limited as regards time and corresponds to a certain maturity of the ovaries. In aging females the supply of migratory forms is always exhausted; when infection mounds are present, new symbionts no longer migrate to them, apparently because the supply of active ingredients is gradually consumed, and thus the mounds become desolate. The substances involved cannot be very specific, for, as we have just learned in the case cited above, accessory symbionts are subjected to the same influences in infection mounds and in the migrating t-cells, and yet the form changes they undergo are radically different.

How, then, shall we judge the significance of specific migratory forms which are sometimes brought about by such complicated means? First, it can be noted that smaller coccus- or rod-shaped symbionts rarely undergo changes at all. However, longer filaments tend to be broken down into smaller fragments; rosettes are replaced by roundish or oval forms; elongated or bead-like formations are transformed into more compact shapes. All this suggests that the changes are made in the interest of convenience, to facilitate transportation of the symbionts from organ to organ and their passage through the egg follicle.

To some extent this is undoubtedly true; however, the significance of this widely used device certainly goes far beyond mere ease of transportation, for, clearly, many of the companion phenomena, such as more intensive staining, occurrence of special inclusions, and forming of bacteria bundles, are unrelated to facilitation of transport. More striking evidence is at hand wherever the transmission forms are larger (sometimes twice as large) than those in the normal mycetomes. In all probability the transformations come about chiefly because symbionts require a physiological state that renders them suitable as starting material for the new cultures which are to be established in the daughter animals. Indeed, symbionts frequently assume shapes in the mycetome which point to extensive degeneration, and the chief goal in transforming them might be to restore them to a state which is closer to the original bacterial form and would again render them capable of multiplying. The recurrent deeper staining, caused by an emission of fluid and often giving the symbionts the appearance of spores, might represent a companion phenomenon.

With this interpretation, the special lavishness entailed by the *x*-symbionts is thoroughly understandable since forms are present here which are subject to excessive transformations, and, since they also forfeit their capacity for division, they cannot be regarded as simplifications. The symbiosis can then be maintained only if a number of symbionts are transported early enough to a place where they are withdrawn from such influences. The same thing applies in all probability to the coexistence of *c*- and *d*-organs observable in *Oliarius* species, for, when male animals lack *d*-organs as well as rectal organs, they seem to reserve a less severely deformed portion of the *c*-symbionts for transmission.

Thus the simplification of unusual growth forms, at first glance the most striking phenomenon, becomes merely an occasional phenomenon accompanying changes which have a deeper purpose than providing convenient transformation forms. At the same time we can understand why the necessity for such reorganization appears in special degree with increasing age of the symbiont adaptation—we need only mention the *a*-, *x*-, and *t*-symbionts!—and conversely why such changes are rarely observable among accessory symbionts.

When the yeasts, which play such an important role in cicada symbiosis, and which also often make their appearance in scale insects and even in the aphids, transfer to the egg cells or embryos without undergoing essential changes, there is reflected not so much their later acquisition, which is now assured, as the circumstance that their constitution excludes that plasticity which is so characteristic of bacteria. Also, the fact that only in the cerambycids do spores serve in transmission may not be interpreted as a step in increased accommodation, but on the contrary it is a symptom of still incomplete adaptation.

Another impressive criterion of perfected endosymbiosis, besides control over rate of increase and symbiont shape, is absolute control over the extension of the microorganisms in the host animal. Sharply restricted zones are usually made available to the microorganisms, and, if these zones are used for other purposes, only sections of them are made available. Countless proofs of this may be found in the Specialized Section. The symbiont-containing zones of the intestinal epithelium are sometimes delimited with great exactitude against those remaining sterile, and this is true not only where such sections are detached as evaginations, but also where the intestinal tube, otherwise unchanged, is occupied for a distance. Here we may cite the hippoboscids, where the extent of the infection appears to differ from genus to genus. Where Malpighian tubules are made available in insects and ixodids, only certain sections are usually settled, and in the case of the insects only certain tubules are receptive. Thus two of the tubules are open to the symbionts among the apionines and four with the ipid *Coccotrypes*.

If the symbionts live in the fatty tissue, the symbionts are usually restricted to very definite areas and within these areas there is further restriction to well-defined subareas. Such zones are often surrounded by a margin of fat cells which are avoided by the symbionts, although the cell material there is exactly like that of the central area which is being opened to them. Where true mycetomes are formed, their well-adapted old guests are never found outside them. Spatial control over the symbionts is strikingly demonstrated again and again in processes connected with egg infection. At every turn one finds that receptivity is severely limited in certain follicle parts in the ovarioles. The cuneate cells of the Homoptera separating at the posterior pole, which are used exclusively during the passage, appear fixed in number, forming a girdle of one, two, three, or more cell circles. With other Homoptera, it is only a ring with very few cells, located between the nutrient cells and the oocytes, which gives up its defensive position. With certain cicadas and coccids, most snout beetles and the ipids, the nutrient cells or endchambers are revealed to be receptive, but in general they remain absolutely sterile. This becomes especially impressive when, as with the ants, even the young ovarioles together with all the oocytes are infected and the nutrient cells represent symbiont-free islands situated in the midst of the bacteria-laden elements. Since the associations, consisting each of one oocyte and a nutrient cell group, are derived from a mother cell, a cell immune until then is divided by mitosis into one immune and one receptive cell!

At times, sex also exercises an influence on resistance. With heteropterous bugs the male gonad rudiments are closed to symbionts, whereas the female rudiments are infected at very early stages.

Often the task of local resistance is strictly limited in time. The wedge cells are opened to the symbionts only during one period of their development and are then closed to them thereafter. The receptivity of the occytes varies considerably from case to case with respect to time and is always strictly fixed. It may exist at the earliest stages, or it may not manifest itself until the end of growth or, depending on the object, it may be connected with some phase in between. At times the oocytes are totally refractory, and only definitely fixed, early or later cleavage stages or even later stages of development are opened, as with many scale insects. With the pyrosomes a rigidly set portion of the testa cells is infected by the spores of the luminescent bacteria for purpose of transmission; furthermore, this infection extends over a rather long time and takes place in various zones of the follicle, depending on the developmental stage of the embryo.

Where several symbionts live in the same organism in a well-adjusted way, the receptivity is not generalized as long as it is a question of sites. Instead, the resistance of one certain cell type is cancelled with respect to only one symbiont type and remains in force with the others. Only in connection with transference is there any departure from this. Thus, wedge cells of the cicadas are usually equally available to all the symbionts, and the blood cells used in the disymbiotic *Tachardina silvestrii* for transferral to the ovarioles are laden with both types. During embryonic development of the disymbiotic *Lyctus* it is possible to find migrating cells which admit both types and then carry them to the place where the transitory mycetome is formed. Naturally the oocyte also gives up its resistance to all legitimate guests.

Many observations conform that the cells, organs, or organ sections which normally are placed at the disposal of symbionts represent a sanctuary from which only the symbionts serving transmission may depart unharmed. Algae of coelenterates are destroyed when they get into the false germ layer in connection with the delamination of the germ layers. The symbionts are often dislocated when *Pediculus* embryos are centrifuged, but, while this treatment does no harm to the symbionts left in the mycetome, those which as a result enter a new milieu are destroyed, and sometimes the animals are made sterile. When the posterior section of the midgut holding the symbionts in *Coptosoma* is constricted off from the anterior, the bacteria remaining behind in the latter perish there, just as do the bacteria of blattids and *Pediculus* that remain behind in the yolk-filled gut during embryonic development. The cuneate cells also represent only a temporary retreat, for symbionts which are late in transferring to the egg cell, as is so often the case, are destroyed. During blastoderm formation in curculionids the primitive sexual cells of both sexes are infected, but symbionts multiply only in females and after a while are no longer found in males. The same thing seems to be true of ixodids.

To learn how such balanced symbiosis originates one needs only observe cases where the additional symbionts display many degrees of gradual adaptation. Some groups do not show the slightest inclination to augment the supply of symbionts, for instance, blattids, aleurodids, Anoplura, mallophages; with other groups, like the hippoboscids, one finds stages ranging from undisciplined satellites to companion forms barely differentiated from older symbionts. In *Melophagus* the actual symbiont is narrowly restricted, but a bacterium, which like the symbiont is transmitted by infection of the milk glands, occurs also in the cells of the hindgut, in the renal organs, in the heart tube, and elsewhere without causing visible damage. Almost always present in *Lynchia* is a bacterium which is like the symbiont in that it is inheritable but builds its site in the musculature of the intestine, that is to say, where *Ornithomyia* develops a regular mycetome. In contrast to genuine cases of symbiosis, the degree of infection varies in *Melophagus* and *Lynchia* from case to case. A companion form, appearing almost regularly with *Hippobosca camelina* but already limited to the intestinal epithelium precisely like hippoboscid symbionts, is usually intermingled with the others, and like them is restricted to the basal sections of the cells. Rarely is it found outside the intestinal zone assigned to the symbionts. At the same time the degree of infection is already quite constant, and again the milk glands are used in transmission. Culture experiments showed that with these bacteria it is no less difficult to obtain good cultures on nutrient media which afford excellent conditions for the usual saprophytic forms than with the symbionts.

When in these cases wilder bacteria occur in the intestinal epithelium occupied by the symbionts, they settle, in the same manner as in the heteropterous bugs which have mycetomes, in the latter. With *Ischnodemus* and *Ischnorrhynchus*, little cocci and rodlets make their appearance frequently but cause no further disturbance; with *Nysius*, delicate short rodlets which are quite regularly present infecting the egg cells along with the symbionts; with the bedbug, organisms, at first not differentiable from the extremely similar symbionts, are found in a variety of tissues. With *Dacus* the actual symbiont is joined by *Bacterium luteum*, which also lives saprophytically and, although it generally multiplies within appropriate limits, increases unconstrainedly when the host approaches death or has already died.

As far as the coccids are concerned, it was formerly supposed that disymbiosis occurred but seldom. However, the as yet unpublished investigations, already mentioned several times, have shown that such a disymbiosis is widespread particularly among the margarodids. An abundance of variations gives us new insights into the different grades of an adaptation which is constantly becoming more intimate.

In the cases just discussed there was hardly any evidence of a "battle" between the two partners or of any disturbance in equilibrium worthy of mention. In two other areas, characterized by a breakthrough of the tendency to form polysymbioses, the original order is more or less extensively influenced by arrival of the newcomers and in part visibly damaged. This is the case with aphids and membracids. With aphids there is an uninterrupted sequence of species beginning with those which house the companion symbionts in envelope cells and blood cells and in the mycetome. Sometimes, as in *Schizoneura lanigera,* the symbionts invade the whole body; this seems to some extent hazardous at first glance. However, the invasion occurs in all individuals and does not impair their wellbeing. At the end of the sequence we find states in which the companion symbionts are restricted to a syncytium inserted in a particular part of the mycetome. In all instances the egg cells or embryos are always infected by a second and third symbiont type. Further details may be found in the Specialized Section of this book.

The same applies to membracids whose $a + t$-mycetomes are virtually a battleground for supplementary symbionts in all phases of adaptation. Whereas disturbances appear only here and there in aphid mycetomes, they occur in fast repetition with the membracids. Large cavities appear in the otherwise firm texture of the organs, the contours of the mycetocytes populated by t-symbionts may become blurred, and very often the cells even disintegrate completely. The sequence of pictures in Fig. 206 demonstrates more graphically than words that these double mycetomes, so well ordered up to this point, are now subject to stormy quarrels, that here and there a peace treaty is in the process of preparation, and that gradually a new order will be established. It is as though the host animal had momentarily dropped the reins of its team of horses but is about to apply a firm grip to bring order to the tangled strands again. At any rate, the virulence of even the most undisciplined guests is radically reduced, for almost always they occur only in the mycetomes and their multiplication rate is reduced to the extent that they never threaten the existence of the symbiotic organs or even the wellbeing of the hosts.

While the host organism is struggling in all these cases for a happy solution of the housing situation, it is at the same time obliged to guide the new symbionts along the regulated channels to the egg cells. We find unbalanced states and others that bear comparison with the phylogenetically older ones. With a few cicadas the yeasts populating the intestinal epithelium are not admitted to the egg cells, but, as will be explained later, these yeasts most probably do not represent an original, still imperfect state but are a sign that the displacement of symbionts which cannot come to terms with those taken up later is already far advanced. The cuneate cells are usually open to all types and thus reveal themselves to be imperfectly adapted when they are closed to the accessory symbionts, and very primitively shaped, very tiny rodlets infect the young ovarioles and oocytes instead, possibly also by infiltration of the nutrient cells. *Fulgora europaea*, which even today sends its *m*-symbionts to the eggs by way of the wedge cells, continues to produce the now sterile and unused filial mycetomes at the upper end of the ovarioles, thus reminding us that this mode of transmission is indeed in need of improvement! In general, the later-appearing symbionts share the use of devices created for their predecessors, being taken along by the migrating forms of the *t*-mycetocytes or subjected to the reshaping influences in infection mounds.

The companion bacteria, which are so frequent in aphids, use as an invasion point during embryo infection the same plasma process protruding between the follicle cells which was created for the primary symbionts. With hormaphidines which lose their primary symbionts and replace them by yeasts, the yeasts alone reap a profit from the arrangement.

As one might have predicted, the tendency toward purposeful adaptation is observed also during embryonic growth. It will be informative in this connection to compare the *m*- and *b*-symbionts in *Fulgora* and *Cixius* with the older *a*- and *x*-symbionts or with the behavior of the supplementary symbionts which are transmitted to the aphid embryos.

With all these mutual influences between host and symbiont it is all too easy to forget that reciprocal influences between the various microorganisms themselves play a significant role in the adaptation of new symbionts. Such influences result in more or less extensive damage to the older symbionts by the younger and more virulent ones, as so frequently is the case with the membracids. The surface of the primary symbionts here very often becomes irregular; in *Bolbonota corrugata*,

η-symbionts cause hypertrophy of a-symbionts; in *Hypsoprora coronata,* a-symbionts become roundish or irregular formations when the μ-symbionts are numerous, and finally fall into fragments. With a smiliine, undisciplined vibrios usually live *in* the a-symbionts; some disintegrate while others survive the attack and carry to the egg cells the organisms they contain. Something similar is found when the newcomers also use the migrating cells. For example, the t-symbionts of *Bolbonota* grow pale in these cells in the presence of σ-symbionts but become normal again after migrating to the lymph, where they are withdrawn from the harmful influence.

Other cicada symbionts are not at all compatible, for certain forms which in themselves occur very frequently are never found side by side. Therefore the phylogenetically younger yeasts and 'the so widely distributed t-symbionts are mutually exclusive. In many stictococcines the yeasts have caused the elimination of the original bacterial symbionts, while in another species an additional bacterium appears to have settled in harmony. Aside from the exceptional case of the issines, to be discussed in the following chapter, the yeasts likewise cannot live together with x-symbionts, nor the f-symbionts with a-symbionts, nor the t-symbionts with x- or f-symbionts! The accessory v-symbionts of the membracids never permit, as far as we know, the presence of other accessory symbionts next to them (*Sundarion, Eualthe, Hemikyphta* species). On the other hand, there are symbiont pairs which obviously harmonize very well together, such as $H + f$, $a + t$, $x + a$, or $H + a$.

It is obvious that such friendships and enmities among the microorganisms may become an important and even decisive factor in the founding of associations, and that the admission of one form may result in the exclusion of another form already present. Considerations to be presented in the following chapter will indeed force us to make such an assumption in the case of cicadas.

The study of these questions of progressive adaptation of the symbionts is a problem for the future. Although it is a unique field of general interest, professional bacteriologists have barely touched on it. Even Paillot, whose articles on aphids deal at length with the questions of immunity and reaction to resistance, do not take us further in this respect, since he considered the primary symbionts and the later adapted symbionts to be different states of one and the same organism.

The possibilities open to the experimenter here emerge clearly in H. J. Müller's work on *Coptosoma.* These bugs, which are reinfected on sucking out symbiont-containing capsules, will accept bacteria-containing agar clumplets added to the eggs. For example, it was possible to incorporate *Bacterium fluorescens, B. herbicola,* and *B. mycoides,* and it was found that these may be digested or they may attack the intestinal epithelial cells, and sometimes they inundate the bordering tissue, causing the death of the animals. Surprisingly, when the animals are reinfected with the legitimate symbionts which have been cultivated on agar, they are destroyed too. Thus we see clearly that in such cases of extracellular symbiosis the influence of the host upon the symbionts is not one of long duration. The intestinal crypts created for the purpose were settled approximately to the same degree as with organisms normally received only in animals to which the transmission forms, taken from corresponding intestinal sections, were added artifically.

Recently Foeckler (1961) investigated the extent to which it is possible to replace the primary symbionts of *Sitodrepa panicea* by other yeasts, among them species of *Torulopsis, Saccharomyces,* and *Candida,* which were added in both the fresh and the dried state to the diet of animals made symbiont free. The added yeasts, with the exception of *Torulopsis utilis,* remained in the intestinal lumen throughout, enabling the animals to thrive on an otherwise vitamin-deficient diet, yet the organisms never infected the blind sacs. Only when the animals were offered *Torulopsis utilis,* dried years before, was the unexpected observation made that both the blind sacs and the remainder of the midgut epithelium were infected (Fig. 359). The brush borders, lost here in colonization of the mycetocytes with the host's normal symbionts, were retained. When the sterile animals were offered both the species-specific symbionts and *Torulopsis utilis* simultaneously, only the first were taken into the blind sacs, and the other midgut cells then proved to be resistant to *Torulopsis utilis!* The great difference between such "artificial *Torulopsis*-symbiosis" and genuine symbiosis becomes clear when it is observed that with *T. utilis* the blind sacs are not reinfected during metamorphosis and the organs of symbiont transmission remain empty.

We shall report later on the successful experiments of Bewig and Schwartz and of Gumpert and Schwartz, who replaced *Triatoma* symbionts with other organisms.

Ries, using a new method to study conditions of immunity, transplanted symbiont-containing tissue both in symbiont-free animals and in animals containing symbionts of another kind. In both cases the defensive measures taken by the animals were clear-cut, although they differed in character from case to case. Bacteria-containing fatty tissue of *Blatta germanica* in the body cavity of the larvae of *Tenebrio molitor* L. is surrounded immediately by a mantle of lymphocytes and at the same time is melanized. Fatty tissue and bacteria turn an intensive brown, the plasma of the first becomes homogeneous, the nuclei clump together, but the bacterial shape surprisingly remains unchanged for months.

In grubs of the meal moth, a much stronger melanization occurs; after a few days the symbionts are mere shadows and the moth structure is disorganized and lytically altered.   In mycetomes of *Psylla buxi,* which are also encapsulated, melanized, and dissolved when transferred to meal

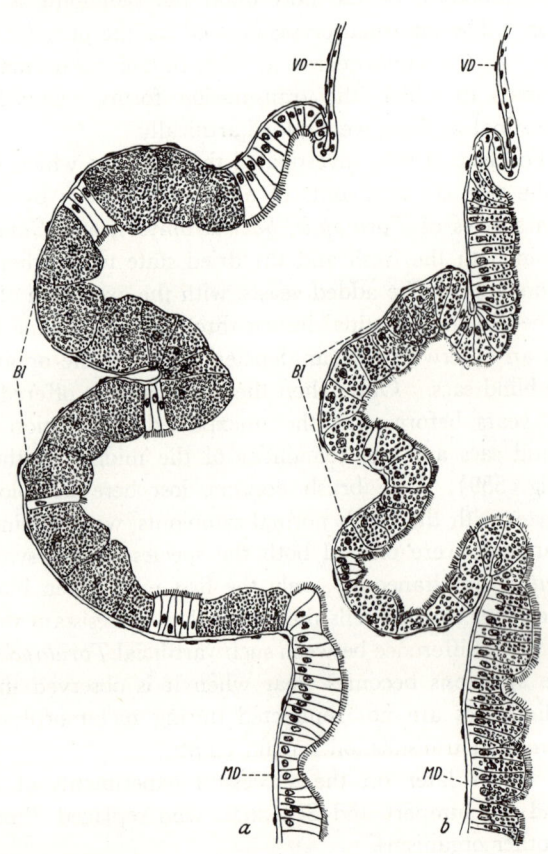

Fig. 359   *Sitodrepa panicea* L.   Larva.   The region of the ceca (blind sacs) and the midgut in an infection with (*a*) normal symbionts and (*b*) *Torulopsis utilis.*   *VD,* foregut; *Bl,* ceca; *MD,* midgut.   From Foeckler.

grubs, Ries could occasionally observe interesting form changes of the primary symbionts after 24 hours.   Swelling into large irregularly shaped structures or into tubular or clubbed branched structures, they left their uninucleate cells while the auxiliary symbionts housed in the syncytium showed no changes in form whatever.

It is no surprise that the yeast-containing midgut sections of *Sitodrepa* in *Blatta germanica* are also surrounded by lymphocytes and melanized. Yet adoption of *Blattella* symbionts in itself was enough to mobilize the defensive forces of the related host animal. Unfortunately, only tentative orientation experiments are available on this subject. If repeated with a larger amount of material with more or less closely related blattids, they would probably also show a correspondingly greater or lesser compatibility. Supporting this argument are the experiments by Pant and Fraenkel in which they successfully exchanged symbiotic yeasts in two anobiine genera without unfavorable results. This interesting experiment will occupy us more fully in our final chapter.

When Fink (1951) inoculated *Pseudococcus* symbionts in meal grubs, May beetles, and grasshoppers, he obtained results similar to those of Ries. It must remain undecided whether the melanization which occurred in all these cases was caused by the tyrosinase of the insect blood, for Fink was able to find it in the symbionts also.

Finally, we have available experiments on the interchangeability of symbiotic algae and its limits. Good results were obtained with protozoa and hydra. As one might have predicted, it was demonstrated that the ties are usually less solid here than in cases of endosymbiosis where the bacteria and fungi are more intimately adapted (Goetsch, 1924; Goetsch and Scheuring, 1926).

As shown above, the host animals have control over their symbionts with regard to their propagation rate, their development in form, the suppression of their virulence, and their strict spatial limitations. However, the control extends even to their very existence or non-existence, for in a series of cases the symbionts are completely or partially eliminated during the regular symbiotic cycle! Various routes are available. It is natural that symbionts living in the intestinal lumen or intestinal epithelium should be eliminated through the anus. Moreover, in holometabolic insects the larval sites may be fused during metamorphosis, and this is an aid to such elimination. Even in anobiids and *Dacus,* a good part of the yeasts or bacteria is expelled at this opportunity, although the symbiosis is maintained. Cerambycids and cleonids eliminate the symbionts entirely in males, and in females only keep the number needed to fill the ovipositor spaces which are used in transmission.

Where mycetomes in snout beetles are traceable to the rudiment of the stomadaeum, the bacteria are expelled by way of the intestinal lumen, but the expulsion takes place differently in the two sexes and only during imaginal life. This process has been described more fully in *Calandra* by Schneider (1956). Ten days after metamorphosis, decomposition in the imagines is in full progress. Symbionts and nuclei degenerate, and the strongly staining remains of the mycetocytes are expelled into the lumen of the intestinal tufts and from there into the gut. After 20 days the tufts are sterile, but occasionally one or another mycetocyte is left behind, though clearly in a damaged state.

In the strange dorsal organs of lagriids, which are derived from membrane invaginations, the content is presumably expelled during metamorphosis into the exuvial fluid; at any rate, the imaginal symbionts are found only in the female organs of transmission, and it is evidently only in this way that these organs can be simultaneously filled.

Other animals kill the symbionts by means of bacteriolysines. Such killing takes place on the spot and is always followed by disintegration of the organs which serve as symbiotic sites. The *Donacia* species choose this way, but here it is appendages on the gut which provide the sites, for apparently it is difficult to draw the organs into the gut because the connection has become extremely narrow. The filamentous bacteria are expelled into the lumen of the saclets and finally flow together into a homogeneous mass which is still surrounded by remains of the intestinal wall and its musculature until the organs disappear in their entirety, although no phagocytes are in evidence.

An equivalent process takes place, at least in some of the species, in the Malpighian tubules of males. In the older larvae two of the tubules are already filled with symbionts from the intestinal organs and from here are placed later at the disposal of transmission. In the process the male vessels are also infected, obviously unnecessarily, but degeneration usually sets in at once and, exactly as in the intestinal ceca, a kind of bacterial rubbish arises which is expelled into lumen of the tubes. Only a few species retain small infected sections.

The male stomach discs of *Pediculus* and *Phthirus* afford another example of such killing. When all symbionts transfer during the third molting from mycetome to ovarial ampullae in females, degeneration phenomena usually appear in males; the chamber demarcations are blurred in the organ, the microorganisms are deformed, their staining becomes irregular, and finally only a thin fibrous envelop surrounds a clump of vaguely limited bacterial remains. However, the degenerative process does not always lead to such extremes, for in some males the symbionts do not disappear completely.

With some fulgoroids there is total decomposition of certain mycetomes by lytic processes. In the young male imago of *Asiraca* the unpaired *p*-organs noticeably begin to stagnate in their development by autumn. In March the degeneration begins; in April, when mycetomes reach maximum development in females, they shrivel more and more in males; the nuclei degenerate pyknotically; the contours of the symbionts become more and more blurred; the organisms clump together and become pale. In May it is only with difficulty that one catches sight of the last remains of the mycetomes, and in still older animals they can no longer be detected with certainty. In the males of *Stenocranus* and *Kelisia,* it is the *q*-organs, much smaller from the beginning, which are reduced more and more during imaginal life and finally, together with their content, disappear completely as in *Asiraca.*

It is said of *Nosodendron* that mycetomes in male imagines become glassy and the inmates degenerate. In *Lyctus* males the two symbiont types housed in separated sections of the mycetomes do not decompose completely, but widespread degeneration and partial decomposition take place. Within the related bostrychids, the decomposition of the male mycetomes is much greater. In *Rhizopertha* and *Sinoxylon* they are completely, or almost completely, decomposed; in *Apate* species there is only partial degeneration of the mycetocytes; and in *Scobizia* they remain unchanged. The male mycetomes of *Oryzaephilus* also decompose completely, according to new observations.

On the other hand, the nycteribiid *Eucampsipoda aegyptica* proceeds toward total elimination, limited again to the male sex. In this case male and female larvae are provided in the same way with bacteria localized in the milk glands, but these bacteria reinfect the milk glands only in female larvae. In the male larval intestine they are completely digested, so that the flies issuing from them are symbiont-free.

In a number of snout beetles, lytic action and phagocytic action go hand in hand in the decomposition of the insignificant mycetome-like formations which takes place in the larval state. In *Smicronyx* the initially compact, syncytially built organ is no longer present in older larvae. It disintegrates into irregular shreds and then is approached by collections of blood cells which decompose it completely, symbiont remains being recognizable here and there in their interior. In *Ceutorrhynchus* and *Phytonomus,* the uninucleate mycetocytes, which at first formed a well-defined aggregation, are progressively loosened and take on the character of amoebocytes more and more. Elements still filled with recognizable bacteria and others looking exactly like blood cells are intermingled in confusion; the group loosens up more and more and disappears completely by pupation time. A more detailed investigation of this symbiosis type is needed to understand fully the hosts' capacity for decomposition. In all these snout beetles the symbionts settling the mycetomes are lost in both sexes.

A completely different and more radical method of symbiont elimination consists in suppressing transmission to egg cells or to embryos. At present, such an occurrence is known only in certain scale insects and aphids. Male stictococcines are short-lived, have regressed mouth parts, and are therefore unable to ingest food after birth. Since they undergo development in the maternal symbiont-containing body, they can dispense with a vitamin source of their own. Only some of the eggs are provided with symbionts, and these develop into females, while the others, which remain symbiont-free, become males. The same is true for those species whose original symbionts have been suppressed by yeasts which infect the embryos. With most species of aphids, all progeny, parthenogenetic and viviparous in origin alike, are infected with symbionts, as are also the yolk-rich winter eggs. Other species (all pemphigines and some hormaphidines) are known which in the sexupara generation suppress the infection of embryos developing into males; as adults they are greatly reduced in size and ingest no food.

It would be wrong to infer from these fairly common cases of symbiont elimination that the cooperative living is superfluous: closer examination shows that the exact opposite is true and that such living is never jeopardized thereby. Where the sites are degenerated in both sexes at the same time, transmission is ensured by infection either of organs on the ovipositor or of the Malpighian tubules (cerambycids, cleonids, lagriids, *Donacia*), or by infection of the primordial germ cells, as with most snout beetles. With the nycteribiid mentioned, in which the milk glands represent both symbiotic site and transmission organ, there is infection of the rudiments in the larvae. Where transmission is activated by transplanting symbionts to imaginal oocytes, as with the cicadas and lice mentioned, the symbionts degenerate only in the male mycetomes.

From the possibility of total elimination of symbionts, we also draw the significant conclusion, one with which we shall also be occupied in the next chapter, that these symbionts are often of great value for larval development but can sometimes be dispensed with in the imagines.

Finally, in considering the constructive achievements to which the animal body is stimulated through admission of symbionts, a question, not easy to answer at first, arises: Which of the particular cell forms observed here merely represent automatic reactions to stimuli inevitably arising from the foreign organisms, and which represent animal capacity for adaptation? In considering the cytological changes caused in animals and plants by a great variety of parasites, such as bacteria, fungi, protozoa, and worms, there is much to remind one of the construction of symbiotic sites. In many cases parasitic invasion is also linked to increased growth, indeed, gigantic growth, corresponding enlargement of nuclei, formation of bizarrely shaped nuclei, amitoses, and multinuclear complexes.

With some fungal galls the plant nuclei become larger and diversiform;

the grig galls are characterized by multinuclear giant cells; the lymphocytes of *Lumbriculus,* when attacked by microsporidia, are stimulated
to astonishing growth and enlarge their nuclei correspondingly; and something of the same nature is seen in *Potamothrix,* where, however, it
leads at the same time to amitotic nuclear divisions (Mrazek, 1910).
When attacked by *Nosema lophii,* the ganglion cells of *Lophius piscatorius*
assume enormous dimensions, although a settlement of a wide variety
of cell types by *Nosema* species may cause no growth at all, even with
insects (for example, see Schwarz, 1929).

Of special interest in this connection are cases in which an infection
by microsporidia leads to extended syncytial formation and well-defined
plasma masses which have an unmistakable similarity to some mycetomes.
Each of the spherical formations floating at times in the body cavity of
Limnodrilus was at first considered to be a parasite until it was recognized
that they are traceable to little uninucleate host cells (lymphocytes)
which develop correspondingly and form numerous amitoses after being
infected by the sporozoa *Myxocytis* multiplying briskly there. On the
exterior of such syncytia is a border of fine rigid cilia or even an
alveolar border which reminds us that a type of rodlet border is also
found on the surface of the mycetomes of *Cicada orni* (Mrázek, 1910).
The lymphocytes of *Limnodrilus* react similarly to an infection by other
microsporidia, for example, *Mrazekia* (Jirovec, 1936), and Weissenberg
(1921, 1922) showed that *Glugea* species in the stickleback and smelt
allow small uninucleate migratory cells of the connective tissue to swell
into multinucleate syncytia with diameters of 4 mm.! He also noted the
similarity of such formations, for which he proposed the term xenones,
to mycetomes. The spermatoblasts of bryozoa react in equivalent
manner to sporozoa infection (in *Alcyonella,* Korotneff, 1892). The
worm nodules, caused by an infection of nematode larvae in the fatty
tissue of *Tenebrio molitor,* represent complexes of migrating cells which
have become multinucleate. These enclose the worm bodies and are
surrounded toward the outside by a regular epitheloid envelope of connective tissue cells (Pflugfelder, 1950; and other literature cited there on
such reactions to parasitic attacks).

We can only assume that admission of symbionts to animal cells often
stimulates the cells in the manner that protozoa and other parasites
do. Here the study of accessory symbionts in membracids is especially informative. Although the adaptation process is evidently still in
progress, we find equivalent reactions, and the animal cells respond
to the often quite unregulated attack with deformation of the nuclei,
with growth, amitoses, and syncytia formation.

It therefore must be realized that the cytological processes in mycetocytes and mycetomes is the result of interplay between infection stimulus and stimulus response, and that the hosts have gradually learned to guide this influence along precise channels. The result of this arrangement differs distinctly according to the character of the infected cells. In cells already differentiated polarly, especially where intestinal epithelial cells are infected, their habitus can be preserved to a great extent. For example, in *Hippobosca,* the bacteria are still situated without exception in the basal cell-half, with the nuclei in front of them in the symbiont-free plasma, and the cells still have a well-developed brush border. In other hippoboscids with staff-shaped symbionts, they are all arranged perpendicular to the base, in conformity to the well-preserved structure of the basic plasma. Similary constituted are the mycetocytes of *Leptura,* where again only the basal region is filled as far as the nucleus by tear-shaped yeasts, all pointing toward the intestinal lumen. The settlement of cells is also extremely sharply separated in the intestinal ceca of *Cassida* and *Bromius,* yet here, as with the luminescent cells of *Physiculus japonicus,* only the distal part of the cell contains symbionts. In this connection we are reminded of the similarly situated symbiont accumulations of cuneate cells of the *Orthezia* follicle.

Other intestinal epithelial cells are infected throughout their extent but nevertheless retain their own character; the brush border, especially, is often still well developed. Thus it would seem that even their original functions are kept to a certain degree. When they are occasionally deficiently developed, as in *Melophagus,* states are prepared in which the orignal structure of the host cell is finally lost completely and non-infected and infected elements are distinguishable to a great extent. Complete loss of the brush border has been observed in the hippoboscids in *Lipoptena cervi,* among the beetles in anobiids, cerambycids, indeed, with the just-mentioned chrysomelids, where the mycetocytes moreover do not undergo such deep-seated changes; and there are other cases of this kind. Freshly and moderately infected cells still have the border at times, but increasing growth and considerable multiplication of the symbionts in these cases lead to its disintegration, cause nuclear deformations, and strip the cells more and more of their original character so that finally, having become completely apolar, they are situated between sterile epithelial cells with their slender pillars, or, if they appear in a closed group, as with *Marchalina,* they lend an entirely different character to the intestinal section.

In *Glossina* larvae, where the mycetocytes are slightly adnate and still show a symbiont-free basis, Wigglesworth found that they no longer admit the fat droplets from the milk glands; all the more, then, must we assume that in the greatly deformed intestinal epithelial cells of other symbiont bearers the functions assigned to them in accordance with their rudiments have been lost entirely. Koch was able to prove with *Sitodrepa* that abundant mitochondria are nevertheless still found there among the symbionts. Deviating development is also found in similar degree in the cells of the Malpighian tubules, which may lose their brush borders and, in certain apionids, may become fully apolar elements. The cuneate cells of the cicada follicle, in particular, lose their polarity completely at times. In their case, too, a wide range is found, beginning with cells that keep their epithelial character throughout and ending with cells that are rounded off, are filled on all sides by symbionts, and have deformed nuclei. The symbionts alone cause the irregular shapes of the nuclei here, for they assume their original shape again after migration.

In such cases of originally polar mycetocytes it is clear that the cell limits are sometimes disintegrated because they are overloaded, as with other mycetocyte types. It is, in particular, in the abundantly infected cuneate cells which flow together into syncytia that the original epithelial nature is no longer in evidence, but in rare cases the same process takes place in the intestinal epithelium also. In the membracid *Cymbomorpha*, the intestinal epithelium was filled in the usual way with an overabundance of yeast cells and was deformed into an extensive synctium with numerous, small, irregularly lodged nuclei, whereas with the only other cicada where a similar attack has been confirmed the settlement is looser by far and affects far less the character of the intestinal epithelium.

Where fatty tissue is infected, no cytological reactions whatever may occur, as with some of the lecaniines, or considerable cell growth and amitosis may be caused and fat storage reduced or eventually eliminated completely. In such cases a tendency toward cell fusion, which is rare with infected epithelial cells, is extensively manifested, often leading to syncytia of gigantic proportions and more or less organ-like appearance. The fatty tissue of *Rastrococcus iceryoides* is an isolated case. Here extensive zones of the fat lobelets are infected by enormous bacterial masses but remain completely free of nuclei. Only the marginal area is taken up by an irregular covering of little symbiont-free fat cells. Presumably this state is caused by early disintegration of fat cells.

The development of genuine mycetomes, as we have seen, is linked to the infection of indifferent embryonic cell material. Where such material is first inserted epithelially, as with blastoderm cells, or where embryonic cells first attach themselves around the symbionts in the form of an envelope, their polarity is lost immediately with the infection (*Camponotus, Formica, Cimex,* cicadas, and others). The growth may move within modest limits; this occurs particularly where mycetocytes, usually after development of transistory mycetomes, are distributed throughout or lodged in fatty tissue, as with many scale insects, with blattids and mallophages, but growth may be modest also in mycetomes, as with aleurodids, snout beetles, and others. In this case the capacity for mitotic multiplication is kept in part (coccids, aleurodids, *t*-mycetocytes, blattids, and others), and in part replaced by amitoses with subsequent cell division (*Mastotermes, Formica*). With genuine mycetomes, however, the growth more frequently leads to considerable dimensions and is always linked with loss of capacity for mitotic division.

The cells may also remain uninucleate with corresponding enlargement and contain then roundish to oval nuclei, or the nuclei may be more or less deformed, indeed, often quite branched, tattered, or punctured. More frequently the nuclei multiply amitotically, producing complexes with great differences in the number of nuclei, although at times a single amitosis may lead to dinucleate cells, as is quite regularly the case with all rectal organs or with the mycetocytes of the filial mycetome established in the ovarioles of *Bladina*. Other mycetocytes have only a few nuclei, and still others have a very large number of them.

Naturally, an increase in the number of chromosones generally goes hand in hand with this frequent progressive enlargement of mycetocytes and their nuclei. The increase may take place by fusion of daughter nuclei or by endomitotic processes; in special cases a role may be played by associations of nuclei which have not proceeded from a mother nucleus. Since we know today that endomitoses and rhythmic nuclear growth are frequent phenomena in a great variety of tissues in the insect body and usually cause the great differences in nuclear sizes, it is hardly surprising that similar phenomena occur in mycetocytes too; but there have been very few accurate accounts, and often it remains undecided whether endomitosis or nuclear fusion is involved. It is doubtless a matter of endomitosis in the complexly built mycetome of *Oryzaephilus* when the central nucleus of each chamber with symbionts becomes 128-ploid in the course of development, and the nucleus in the middle of each mycetome becomes 64-ploid. With *Periplaneta americana*, where the mycetocytes always divide mitotically shortly before each molting, the nuclei are usually octoploid, but there are also mycetocytes with giant nuclei estimated at 256- to 512-ploid. The fat cells are diploid or tetraploid. The embryonic mycetocytes of *Pediculus* are 16-ploid, like the cells of the midgut, the fat cells of the Malpighian tubules, and the yolk nuclei; the mycetocyte nuclei of *Aleurodes* are only tetraploid; and the mycetocytes of *Puto* and *Macrocerococcus*, on the other hand, have enormous numbers of chromosomes.

An exceptional case is the polyploidy of the mycetocytes which occurs in several coccids when cleavage nuclei fuse with polar bodies, in other words, with extraembryonic material.   In *Diaspis* the number originated remains at 40; in *Pseudococcus* it is increased through fusion of daughter nuclei.   *Puto* and *Macrocerococcus* are unique in that the chromosome number is doubled through fusing of an embryonic nucleus with each of the maternal mycetocytes used in transmission.   Conversely, the same animals teach us that high chromosome numbers may be considerably reduced through continuous "somatic reduction divisions."

Such fusions should not be confused with those in which the mycetocytes or syncytia, independently of each other, are directly traceable to embryonic cell material.   Occasionally only two mycetocytes flow together.   For example, the binucleate mycetocytes in the ovarial ampullae of *Haematomyzus* and *Columbicola* originate in this way.   In the depot mycetomes established in the ovarioles by some *Fulgora* species the cell limits of uninucleate mycetocytes are dissolved hesitantly, proceeding from the inside to the periphery.   Where the epithelial envelope of certain mycetomes is opened to additional symbionts, syncytia are also formed.   With the poiocerines the approximate form of the bacteria-filled zone is at least preserved, but with some membracids a complete disorganization of the original epithelium results, and the area inhabited by the *a*-symbionts is extensively fissured and deformed by the huge masses populated with yeasts (*Lycoderes, Omolon*).

Particularly with the syncytia of cicada mycetomes, there is a striking tendency to fuse together into synsyncytia.   In the *a*-syncytia this tendency is observed again and again, sometimes to the extent that only a single complex is left which, strictly considered, falls under the concept of plasmodia.

Again the differences in the constitution of the host animals are reflected when response to very similar requirements leads to great diversity. It is necessary only to recall that the intestinal epithelial cells react very differently to infection and that in wedge cells of the cicadas the degree of filling and the cytological details are typical enough to use as a criterion for systematic classification of the hosts.   Likewise, the various types of nuclear deformation in the rectal organs of fulgoroids are always characteristic of individual families and subfamilies.   As regards the diffuse, central, or marginal positions of nuclei, their special shape, and the plasma distribution, the large syncytia of fulgoroid mycetomes show a number of fine hereditary differences between the insects and occasionally there are arbitrary devices like those in the *v*-mycetomes of the megamelines or *Oryzaephilus*.

The anatomical devices used today in symbiosis result from a protracted process of adaptation, for again and again series are found with increasing degrees of improvement. The settlement of the trypetid intestine progresses step by step from primitive housing of bacteria in the intestinal lumen toward its restriction to appendages distinctly separate from the intestine; among the Heteroptera a series of forms with a few loosely placed intestinal crypts leads to others in which hundreds, arranged on the gut in four rows, follow in close succession, and in the acanthosomines they are finally separated completely. With the hippoboscids the tendency progresses from extensive infection of slightly changed epithelial cells through restriction to a short section used exclusively in symbiosis. Localization of yeasts in fatty tissue begins with widespread inundation and ends with concentration resembling that of genuine mycetomes.

The same increasing perfection applies to devices by means of which symbionts are supplied to eggs for larval infection at the moment of their deposit. Demands are made on the hosts which could be met only by complicated anatomical auxiliary devices, but the animal partner appears to be even more ingenious here than with the comparatively easier housing problem. The varied possibilities of chitinous differentiations are of great assistance in this case, as proved by the multiplicity of symbiont-filled reservoirs, jets, rockets, and crypts on the ovipositor, and the organism works just as hard to perfect them as it did to create them. Intersegmental tubules in a number of cerambycids showed an especially rapid improvement, whereas corresponding devices in anobiids, lagriids, and cleonids displayed a more gradual ascent. Investigation of trypetid ovipositors revealed simple smearing of the end gut without a specific device, reservoirs formed from longitudinal intestinal folds, and bacteria-filled crypts that are more distinctly separated. Along with these new creations we find an ever-lengthening slit-shaped connection between intestine and vagina leading to the climax represented in *Dacus*.

As is to be expected, the same demands frequently result in similar formations. Although the just-mentioned smearing apparatuses of cerambycids, anobiids, and lagriids originate independently of each other, they resemble each other extraordinarily, and similar devices are found with cleonids and even siricids. Hippoboscids and glossines, which have a similar mode of life and therefore have much in common, also deal with their symbionts in a surprisingly similar way. They frequently develop their gut appendages, even when it is a case of new formations, along similar lines; the tetralaminate construction of ovarial ampullae in Anoplura is always essentially the same, although its genesis differs from case to case, and the auxiliary devices created in cephalopods and fish by the presence of luminescent symbionts have an astonishing resemblance to each other.

More examples could easily be added. On the other hand, unusual methods may suddenly appear at some place or other that testify to the inventive ability on the part of the hosts. Species of *Bromius* and *Cassida* form for transmission vaginal tubes for which no counterpart exists; in the gut of *Coptosoma* are manufactured the complex bacteria-filled capsules to be deposited with the eggs; the acanthosomines, when separation of crypts from gut makes it necessary, establish a lubrication organ on the ovipositor which is unique in this group; and the lagriids elevate dorsal membrane invaginations into mycetome-like formations. Where occasional dryness of the wood causes the death of ambrosial fungi, the female siricids create unique hypopleural organs for forming symbiont-filled capsules. Within the capsules the oidia of the indispensable fungus survive the critical time, and after metamorphosis they serve to fill the organs of transmission.

Simple chitinous bristles are turned into excellent devices for retaining the symbionts, indeed, even into pipe-like containers, as with the acanthosomines just mentioned. In order to close off larger symbiont depots on the ovipositor, retention hairs are formed or chitinous supports are matted together, similar to those formed in some insects to tie the anterior and posterior wings more firmly together.

Where a special path is to be pointed out to symbionts, as in filling the rectal organs, net-like formations for catching them arise at the proper place and moment. Fibrous gripping processes, never observed elsewhere in insect histology, reach out for the awaited symbionts, as in the ovarial ampullae of lice or in the intestinal ceca of *Bromius* larvae; and in *Linognathus* a strange bundle of fibrous cells draws the whole mycetome into the genital rudiment. One of the most astounding inventions is found in a case of algal symbiosis, where special constructive achievements of the host are otherwise rarely seen, namely, the hyaline lens-like cell groups in the midst of the individual algal nests in *Tridacna*, which are apparently a type of illuminating apparatus for guaranteeing an increased supply of light to the deeper-lying algae.

Frequently symbiosis also causes glandular activity serving special purposes. Such activity often begins where a quick increase of freshly transplanted symbionts seems desirable. This is the case with the cephalic organ of the *Dacus* imago, which is weakly infected at first, and with the midgut evaginations of other trypetid imagines. In these two instances there are no specialized glandular cells. However, in *Lagria* regular gland cells are installed in the wall of dorsal organs; intersegmental tubules are installed in cerambycids and anobiids, and fungi jets in siricids. In cleonids and lagriids there is a secretion of the entire wall in the corresponding organs. This is also true of the open luminous organs and accessory "glands" in cephalopods. Also, the study of the manifold arrangements serving transport of ambrosia fungi in the ipids has always revealed indispensable glands brought about through symbiosis.

It is undoubtedly safe to assume that these secretions always represent a favorable culture medium for the symbionts. Moreover, the secretion in the walls of certain bacterial sites is suddenly increased when it is necessary to achieve a better gliding flotation of bacteria until then lodged thickly, for example, in the walls of terminal crypts producing specific transmission forms in the female heteropterous gut and the ampullae of the nephridia in sexually mature lumbricids. In other cases the secretion of the intersegmental organs also functions as a glue pasting the symbionts to the eggshell (*Oxymirus*). Serving a special purpose is a secretion observable frequently on cuneate cells, in which the transmission forms emerging from them are embedded. In the *Coptosoma* gut two zones can be differentiated, one furnishing secretion as a substrate for the bacteria enclosed in the capsules and the other producing the capsule envelope. The glands which open into the depots of ambrosia fungi are arrangements which prevent the drying-out of the organisms and guarantee thereby their preservation.

Where the need exists, a musculature is formed to squeeze symbionts from the ovipositor depots, as in cleonids, siricids, and others; and, where mycetomes require special provision of oxygen, the organism is provided with tracheae, the end branches sometimes penetrating deep into the organs.

Frequently such new creations are formed in advance and thus demonstrate impressively that they are deeply rooted. The intestinal evaginations of cleonids, that is to say, organs which undoubtedly arose as a direct result of symbiosis and which indeed are lacking in all other snout beetles, are formed before the larvae are infected by devouring eggshells on hatching. The same thing is true of the *Donacia* species and *Bromius,* and the latter even provides gripping fibers before hatching. The zone to be infected in the intestinal epithelium of young cerambycid larvae is likewise prepared before the yeasts are present. From the very beginning, the two club-shaped, deformed Malpighian tubules of some apionids have this shape—one that is acquired as a direct result of symbiosis. Before arrival of the bacteria, glossines and hippoboscids have a greatly swollen cell cushion on the midgut, which exists for the purpose of picking up bacteria. The intestinal epithelial cells reserved for the symbionts in *Tephritis* never form the brush border but loosen the margin instead, apparently to make infection easier. In *Oryzaephilus,* rudiments of the definitive mycetomes stand ready to receive the symbionts which are freed after dissolution of transitory mycetomes. The imaginal sites on the gut of swallow lice are already recognizable by local thickening of the musculature and by the syncytium forming in it before the bacteria penetrate the intestinal wall in clumps to take possession of these sites. The cells reserved for the blattid symbionts are separated at a time when the latter are still situated in the embryonic gut. In cephalopods and fish all luminous organs opening toward the outside, the auxiliary devices, and the accessory glands of the cephalopod, are formed under sterile conditions and stand empty until arrival of the symbionts.

Such anticipatory measures are observed again and again not only when guests are assigned to sites but also when they are transfered. Examples are the following: differentiation of certain follicle sections; construction of infection mounds; degeneration of certain zones in the ovarioles of *Bladina* to create space for future depot mycetomes; and certain preparations connected with infecting ovarial ampullae in Anoplura and Mallophaga, such as the thickening already setting in before this in *Pediculus*, the radical swelling of the section to be infected by great quantities of fluid in Mallophaga (*Sturnidoecus, Turdinirmus*), and the creation of a specific invasion port in aphid embryos.

In addition to these morphological adaptations are the specific instincts released as a direct result of symbiosis. It is obvious that they are found chiefly with ambrosia cultivators, where it becomes necessary to fill spaces serving the future inoculation and where continuous purposeful use of fungi mats is no less vital. But such adaptations are also found in two genuine cases of endosymbiosis, namely, the sucking of bacteria-filled capsules in *Coptosoma* and the instincts which cydnids develop in mother and brood after egg deposit exclusively to provide the progeny with symbionts. In this area is also the widespread phenomenon whereby hatching larvae are infected by devouring part of the eggshell or, in certain viviparous lygaeids, the maternal fecal pellet.

It is hardly necessary to emphasize how graphically such findings demonstrate the way in which the host's body develops, how intimately the symbionts are adapted to it, and how greatly the reciprocal actions of the two partners differ from those observable during parasite attack.

15

Historical problems

The whole rich array of facts unfolded in the preceding chapters is eminently historical in character. Again and again we have seen devices being added to the organization of the hosts which for the most part were in perfect equilibrium, although in some cases the hosts were obviously still struggling to achieve this harmony. As insight into the abundance of individual solutions was gained, the question arose of their phylogenetic history and today we are able to make a series of authenticated statements and to point out the direction of future work in the field.

For the most part it is possible to determine the age of the individual symbiosis types from their relations to the system of the hosts. Where similarly constituted symbionts are housed in the same manner by larger systematic units, it must be assumed that they were acquired by the primary forms and then transmitted to all later genera and species. Blattids represent a classic example. Investigation of numerous species, representing all parts of the earth and nearly all of the ten families, showed that they all had identical mycetocytes lodged in the fatty tissue which were populated by quite similar bacteria. Slight variations in arrangement and in behavior during egg infection and embryonic development must be considered secondary modifications of the same basic type resulting from the specific constitution of the various hosts. Since the blattids represent a thoroughly primitive insect order which was already fully developed in the lower Carboniferous and which early achieved the climax of its development with respect to numbers of species and individuals, one must necessarily assign to it a correspondingly old age and accustom oneself to the idea they were already in existence 300 million years ago.

Most probably those oldest blattids took over the symbiotic bacteria from their ancestors in turn. Indicative of this is the surprising and phylogenetically interesting fact that a case of bacterial symbiosis, similar to that of blattids to the last detail, has been found among the most primitive termites, namely, the mastotermitids, which today are perpetuated only in one Australian species but were still widely distributed in the Tertiary

Period (Jucci, 1930; Koch, 1938). This finding agrees nicely with a fact established beyond all doubt, namely, that extremely intimate relationships exist among Blattoidea and Isoptera. Handlirsch even believed that termites had branched off from typical Blattoidea very late, toward the end of the Jurassic Period, but according to the concept generally accepted today and first advocated by Holmgren (1910–11), the common ancestors of cockroaches and termites had already split into the two branches before the beginning of the Carboniferous. If this is true, these oldest Blattopteroidea must have lived symbiotically with the same bacteria as our present-day cockroaches and *Mastotermes*.

The phylogenetic significance of *Mastotermes* symbiosis is even more far-reaching. Among the higher families of termites (calotermitids, termopsids, hodotermitids, rhinotermitids, and termitids, which arose from the mastotermitids), such a bacterial symbiosis is lacking. There can be no doubt that it once existed and was later eliminated. The causes leading to its elimination are rather clear. *Mastotermes*, omnivorous in the manner of blattids, devours all types of cellulose-rich substances as well as wool, horn, sugar, etc. Nevertheless, living and dead wood plays such a great role in its nourishment—as it apparently also did with its extinct relatives which likewise had strongly developed head capsules and strong masticatory muscles—that acquisition of those luxurious intestinal fauna, without which, as we know, it is impossible for termites to utilize cellulose, must be presupposed. Apparently these fauna developed because the hosts had the habit of devouring humus with an abundance of organic remnants, a habit possessed also by their cockroach-like ancestors. It is clear that it was possible for such development to take place concomitantly with bacterial endosymbiosis, for the cockroach *Cryptocercus* has many termite-like traits and even shows the beginnings of a state-like organization. *Mastotermes* still unites these two forms of symbiosis, but presumably the flagellates disintegrating during digestion provided a richer source of protein and gradually made the bacteria superfluous.

Conversely, the existence of this new symbiosis depends greatly upon the hosts' mode of nutrition, for the symbiosis itself may disappear or at least recede strongly into the background as soon as these highly developed termitids devour humus or self-cultivated ambrosial fungi in place of wood or as soon as sexual forms are fed with the secretion of the salivary glands.

If the present concept of the phylogeny of the Blattopteroidea is correct, it is necessary to assume a further loss of symbiosis. Indeed, the prevailing interpretation today is that the predatory Mantoidea are highly specialized Blattopteroidea and that they branched off from the

Blattoidea during the later Carboniferous (Fig. 360). No search has ever been made to determine whether bacteria exist in the fat body of these animals, but they might quite certainly be lacking. Here, too, the reason for the loss would be a change in mode of living, as carniv-

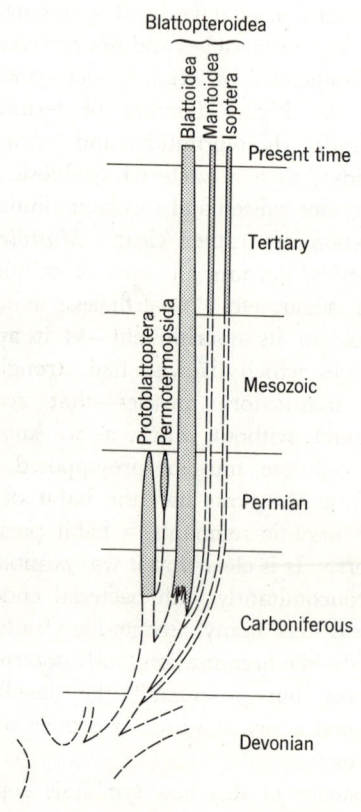

Fig. 360   Family tree of the Blattopteroidea. From Martynow and Jeannel, *Traité de Zoologie*, Vol. 9 (1949).

orous animals never live endosymbiotically. We shall never be able to tell with certainty whether bacterial symbiosis existed among the Proto-blattoptera (Handlirsch's Protoblattoidea), which became extinct in the Permian Period, but, if they actually did split off from the race of the

Blattoidea after the Isoptera originated and likewise were carniverous, one would have to conclude that the bacterial symbiosis was lost here too.

We would like to mention here that in 1959 Grassé and Noirot published a hypothetical family tree of the blattopteroids, which also takes into consideration the complex and manifold transformations within the Isoptera, i.e., their symbiotic relationships to the flagellates, to bacteria living free in the gut or in mycetocytes, to the often found and perhaps also useful amoeba, and to the establishment of fungus gardens.

Of course, the animated picture revealed here warns us against regarding well-adapted cases of symbiosis as an immutable characteristic of the hosts, for, as soon as one penetrates more deeply into the relationships between system and symbiosis, indications accumulate that elimination of symbiosis and replacement by ties of a different kind are by no means rare.

Among the higher termites there are no traces of former intracellular symbiosis, but this should be no surprise in view of the fact that the loss took place so long ago.[1] However, the traces of the former guests do not always disappear completely, still being noticeable at times in embryonic phenomena and even in the structure of fully developed animals. The first is the case with ants where, so far as we know, only the camponotines and *Formica fusca* still have bacterial symbiosis today. However, at the posterior pole in the blastoderm stage, *Formica rufa* and *F. sanguinea* continue to separate off the same cells, which in *F. fusca* develop into mycetocytes. In *F. rufa* a similarly voluminous cell cushion is constricted off and inserted as with *F. fusca;* in *F. sanguinea,* this occurs with only a few cells. In both cases these prospective mycetocytes are eventually destroyed. There can be no doubt, then, that bacterial symbiosis was once much more widespread among the formicids than it is today.

The situation is similar with *Calandra* (Fig. 361). As larvae, *Calandra oryzae* and *C. granaria* have a voluminous mycetome, usually situated ventral to the fore- and midgut, clasping the intestinal tube at the place where these two intestinal sections are joined. An Egyptian variety of *C. granaria* has a much smaller, sterile cell mass, but it still experiences the same shifts during metamorphosis as do bacteria-containing European representatives of the same species. The latter differ also in several other respects from that of *C. granaria* var. *africana* Zacher. Thus

---

[1] Nevertheless, it might after all be possible that such specific cells appear at least in the course of embryonic development, for Koch (1949) found in *Blattella germanica* that they are settled with the further development of bacteria which had migrated from the gut.

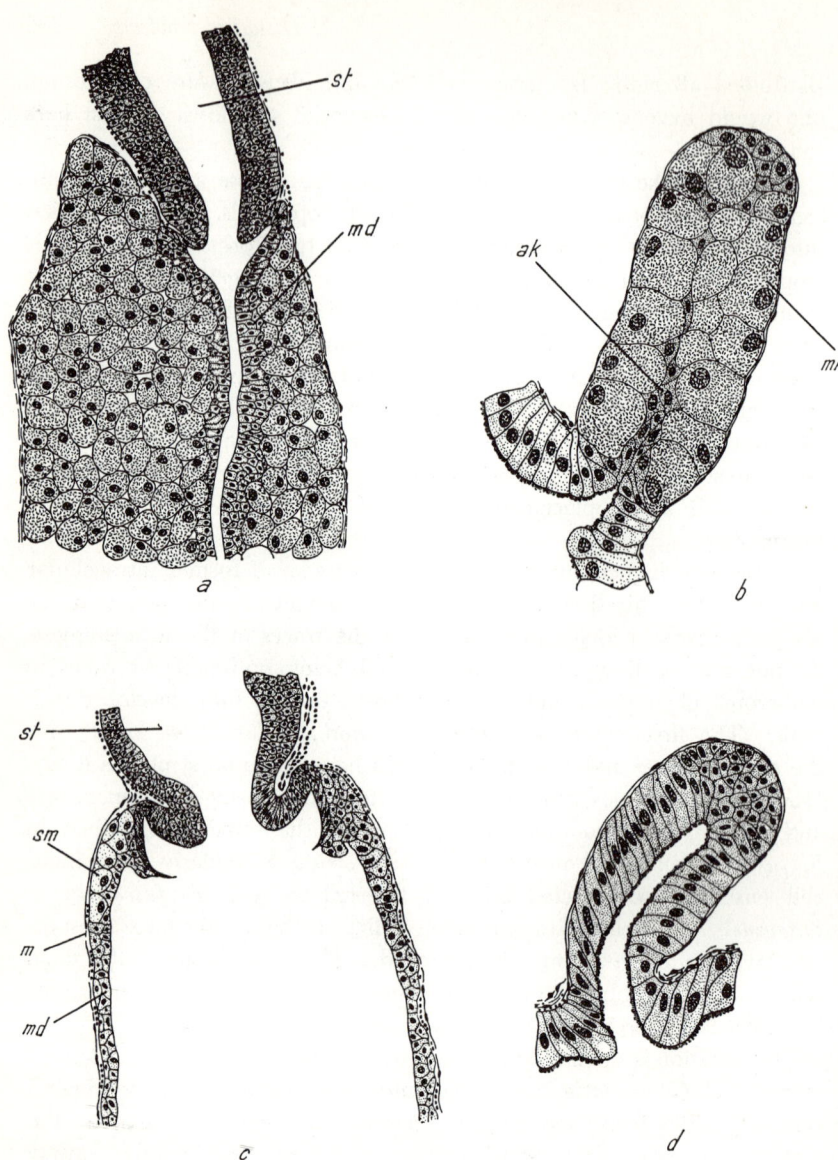

Fig. 361. (*a, b*) *Calandra oryzae* L. with symbionts. (*c, d*) *Calandra granaria* L. var. *africana* Zacher after symbiont loss. (*a*) The mycetome shifts at metamorphosis between the new midgut epithelium and the muscularis to the rear. (*b*) One of the gut villi with numerous mycetocytes. (*c*) Only a few sterile ones are transported into the area of infected cells. (*d*) A gut villus without mycetocytes, at the bottom a nest of crypt cells. *st*, stomadeum; *md*, midgut epithelium; *ak*, axis cells of the crypt; *mk*, mycetocytes of the crypt; *sm*, sterile mycetocytes; *m*, muscularis. From Mansour.

it is just as impossible to find the otherwise conspicuous mycetocytes in the crypts of the imaginal gut as in the endchambers of the ovarioles after the infection serving transmission. Although it has not been possible to account for the loss of symbionts in ants, it is known that the elimination of symbiosis in the Egyptian race of *C. granaria* is connected with the lower resistance of the symbionts to increased temperatures. It has been shown experimentally that symbiont-containing *C. granaria* also lose the bacteria after prolonged subjection to temperatures of 35°C. and produce sterile progeny, and that the symbionts of *C. oryzae* are more thermoresistant.

Certain psyllids have also undergone a similar loss of symbionts. All the psyllids investigated have two symbiont types housed in mycetomes composed of uninucleate cells and syncytia. The phylogenetically older type lives in the cells, the younger type in the syncytia. In *Strophingia ericae* and a *Troiza* species the syncytia remain empty and are accordingly forced into a narrow space by the infected mycetocytes. *Strophingia* thus reverts to the more primitive monosymbiotic state, whereas the *Troiza* species adopts a new, and obviously very young guest, this time localized in the fatty tissue, in place of the eliminated one!

Recently a clear-cut case of elimination of symbiosis was found in *Hippeococcus*, a scale insect of unusual biological interest. There are clear indications of a former mycetome. In an extensive polyploid zone of the blastoderm, which probably goes back as in *Pseudococcus* and *Diaspis* to fusion of polar bodies with cleavage nuclei, a large cell mass is formed which remains sterile but, like many other embryonic mycetomes, is pushed forward by the tip of the germ band and later glides to the ventral side, in other words, a structure which is used as a symbiotic site in other species. The animals on the various host plants reach the imaginal stage even without symbionts, but they do not develop gonads until they are fed by the ants taking care of them in their nests! Obviously, it is this symbiosis with the ants which rendered the endosymbiosis superfluous and thus led to its elimination.

In *Formica, Calandra,* the two psyllids mentioned above, and *Hippeococcus,* the continuing effect of the engrams originating through protracted cooperative living recalls a similar situation in *Fulgora europaea.* Whereas other fulgorines continue to use the individual filial mycetome erected in each ovariole for the still weakly anchored *m*-symbionts, *Fulgora europaea* transfers them by way of the cuneate cells, the same route taken by its other symbionts. Nevertheless, the old ovarial site continues to be formed, although now it remains empty. Moreover, it is only possible to explain the strange processes which occur when aphid embryos are infected—availability of a syncytial mass, behavior of the follicle, later transfer of the symbionts—by interpreting them as effects of an ovarial infection that once represented the sole transmission device. Vestigial traces of a former embryo infection are also involved where aphid males remain symbiont-free. In *Stictococcus* males, which are always sterile, the traces of the former symbiosis are even more obvious; this causes the flooding of the small ovocytes with symbionts in female embryonic development.

After such findings it is not surprising, then, that wherever symbionts were removed experimentally, as with *Sitodrepa,* cerambycids, *Pediculus,* and *Oryzaephilus,* the symbiotic sites continued to be formed and were sometimes kept sterile for generations. Details will be given in the following chapter.

In *Calandra* and the two psyllids only limited losses were involved, but comparative study of symbionts localized in the luminous organs and accessory glands of cephalopods indicates that losses occurred in several places, in some cases over a very broad front. The oldest symbiotic devices of the cuttlefish are doubtless the accessory glands, which are still quite mysterious in their significance. They are indeed the sole bacterial site in all sepiids, most loliginids, in the only genus representing the spirulids, and with the most primitive sepiolids, the sepiadariines. In several locations on the cephalopod family tree we find in place of these accessory glands luminous organs, which are also settled by bacteria and separated more or less spatially. Within the family of the sepiolids (Fig. 362), which we shall now consider in greater detail, this process has taken place with the rossiines and the sepiolines, and the latter have also transmitted them as a heritage to the heteroteuthines that sprang up at the base of the family tree. With the rossiines the new acquisition was not permanent, for the interesting *Rossia mastigophora,* in addition to the accessory glands in both sexes, has extremely rudimentary luminous organs without lenses, whereas all other litoral *Rossia* have only accessory glands and thus once again resemble the sepiadariines from which they are descended. Although no *Rossia* with well-developed light organs has ever come to light, Naef, whose arguments we are following here, suspects that they might yet be found at greater depths. No information is available on symbiotic devices in *Semirossia,* and indeed the rossiines in general have been insufficiently studied.

Within the subfamily of the sepiolids, which developed from the rossiines, there is abundant development of highly complex luminous organs and all their various auxiliary devices, such as were described in our specialized chapter for *Euprymna, Sepiola,* and others, yet at the end branches of the family tree we find likewise complete abolishment or

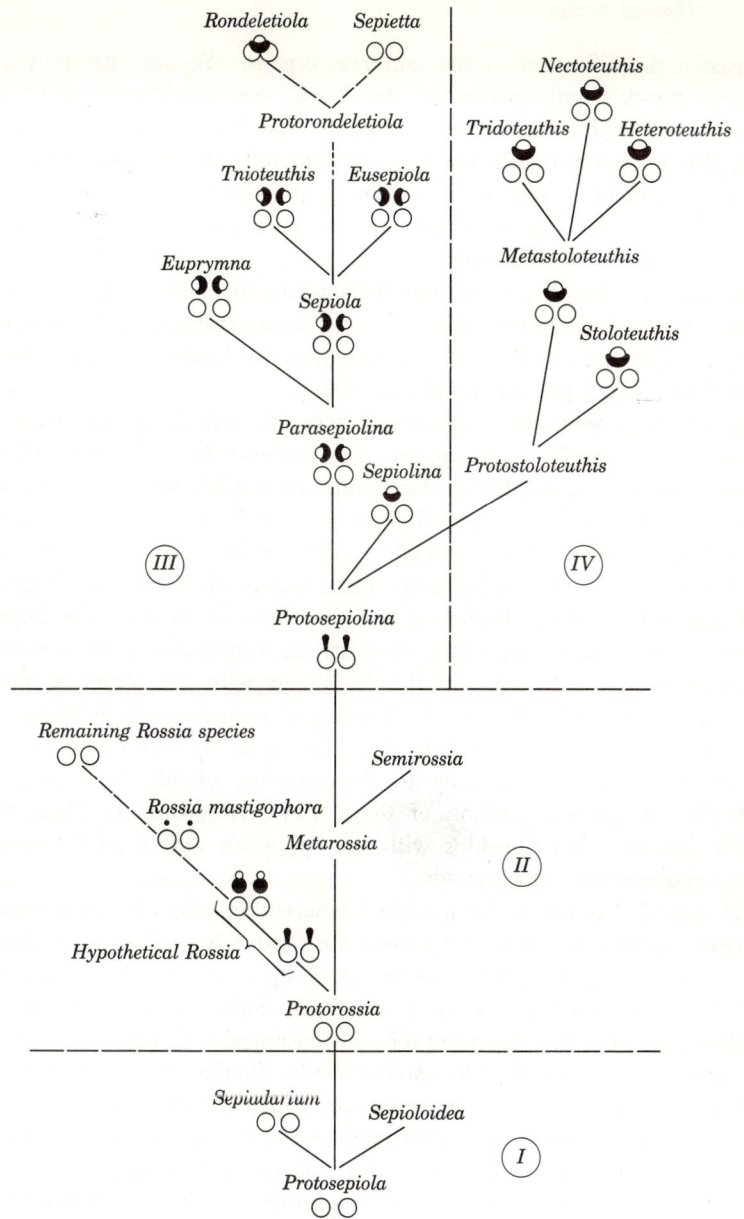

Fig. 362 Family tree of the Sepiolidae. I, Sepiadarcinae; II, Rossiinae; III, Sepiolinae; IV, Heteroteuthinae. The circles correspond to the accessory glands, which occur with and without additional luminous organs. Broken lines represent developmental lines, which lead to rudimentation of the luminous organs. With reference to Naef.

at least a simplification of the luminous organs. *Sepietta* has only accessory glands, and, according to Naef, the luminous organs of *Rondeletiola,* where the luminous portion is no longer sharply separated from the accessory glands and the lens and reflector are less perfected, do not represent a preliminary stage of the highly developed organs of other sepiolines but are organs on the path toward reduction, with *Sepietta* representing the end stage.

On the other hand, the heteroteuthids, which live pelagically at greater depths, faithfully preserved the luminous organs transmitted by the primary form of the *Protosepiolina* throughout the family tree and developed them to high perfection in both sexes.

As we have seen, the ever-present accessory glands of the sepiolids never reveal a tendency to regress, this tendency being always limited to the luminous organ, but in the family of the loliginids these glands, although somewhat developed in young animals, lag more and more behind in growth as the animals become older. At the same time the bacteria-containing tubules are more weakly developed throughout and are not so richly branched and interbraided as with the Sepia, although the organ part which contains the connective tissue becomes more prominent. But also in this family, as with the sepiolids, there occurred a splitting off of highly developed luminous organs from the accessory glands. The fact that they have some special traits, for instance, the long ducts leading to the accessory glands, is in keeping with the exceptional position of the family. According to Naef, the family has close relationships with the oegopsids and should actually not be classified with the sepioids.

Of special interest is the tendency inherent in loliginids to suppress accessory glands, for apparently even the oegopsids used to have them. The assumption, admittedly not accepted by everyone, that oegopsids of deep water developed from myopsids of shallower waters would in itself indicate that they too were formerly equipped with accessory glands. Moreover, one oegopsid, *Ctenopteryx siculus* Pfeffer, is known to have accessory glands, frequently fused into an unpaired formation! Though no information on their bacterial content exists, it is hardly to be doubted that they do contain symbionts. More or less rudimentary organs of a similar kind probably exist also among other bathyteuthids and merely have been overlooked. Naef, an expert on cephalopod organization, is convinced that all oegopsids, and thus all decapod cephalopods, were once equipped with accessory glands.

Thus it may be seen with the utmost clarity that cuttlefish progressively develop and then more or less eliminate their symbiotic devices, yet for the time being there is little that can be said about the background.

One can only surmise that among the contributing factors are changes in the mode of life, adaptation to greater depths on the one hand and transfer to littoral life on the other, and formation of floating spawn masses. It also seems plausible that the deep sea lacks the bacteria to which the shore forms are adapted and by which they are reinfected every time, thereby forcing the oegopsids to form new organs for producing a luminous secretion. Strangely, these new organs make their appearance sometimes precisely where otherwise the accessory organs are situated, and in the case of *Chiroteuthis* were actually falsely described as such.

Occasionally even males have accessory glands, but they are more weakly developed, for instance, in *Rossia mastigophora* and *Loligo forbesi*. However, the phenomenon has no phylogenetic significance, for apparently the glands are acquired by the males through the same genetic device by means of which the luminous organs acquired from the females are commonly bequeathed secondarily to the other sex.

In all probability a comparable progressive decomposition occurs within the macrurids. Among the species studied, there were some with more or less well-developed luminous organs, and others where a degenerate form of the organ could be observed only on sections. In other cases their growth decreased considerably during individual life, and in *Trachyrhynchus trachyrhynchus* they are said to be lacking completely. It is perhaps not mere chance that such signs of reduction do not occur in luminous organs of the closely related gadids, which live in shallow waters, but are numerous among the macrurids, which live in greater depths.

From all these findings it is clear that the various symbiotic devices are by no means retained by all progeny and that, when we study the relationships between system and symbiosis, we must always consider the possibility of decomposition and replacement by new guests.

It is only in Blattaria that a whole suborder of insects has the same symbiosis type throughout, for ordinarily it is only the lower systematic units which show such monotony, for instance, the family of the aleurodids among the Homoptera. Moreover, their symbiosis shows such unique and characteristic traits, for example, transmission by means of maternal mycetocytes, that one can only assume that it was acquired by the primary form. According to Handlirsch, the psyllids, which are descended from the permopsyllids and cicadopsyllids of the Permian Period, also behave in this manner, at least with respect to the older primary form of their two symbiont types. All representatives of this family found up to now have carried them in uninucleate cells of the mycetomes.

The aphids, which are also restricted to a single family, do not appear quite so uniform. Of the four subfamilies the aphidines and pemphigines (eriosomatines) are so much alike symbiotically that one may assume a common root. Their typical, ever-present, roundish primary symbionts are housed in the same way and are transmitted to ovarial eggs or embryos. The symbiosis of the adelgines (chermesines) varies somewhat from that of the subfamilies mentioned, for in one tribe, the Adelgini, tube shapes appear in place of the typical primary symbionts; and, since these tube shapes differ considerably from the primary form, the idea of independent acquisition inevitably come to mind. In the fourth subfamily, the phylloxerines, it was impossible to find symbionts. Their absence is apparently connected with the fact that this subfamily, like so many other aphids, does not suck the sieve-tube sap poor in protein and vitamins but pricks the cells of its host plants and hand in hand therewith has interrupted the connection of mid- and hindgut (Grassi, 1912). A clear-cut case of lost symbionts occurs with some of the hormaphidines, which have yeast-like symbionts in place of the primary type. As we have seen, the younger symbionts use transmission devices doubtless created for their predecessors. We would like to add that symbionts, which are distinctly different from the hereditary symbionts, are found in certain hormaphidines, and that the tendency to take up additional yeasts is very highly developed in the aphids. One would therefore have to reckon on a rather complicated evolutionary history of leaf lice.

The bacterial symbiosis of the hippoboscids must also be considered an old possession, for it has an uninterrupted existence and represents a substantial part of the adaptation complex characteristic of these animals, a phenomenon which is as much a part of their organization as their method of ingesting food, their devices of viviparity and milk glands, and the tendency to rudimentary wing formation. The variants that may be noted here in the various genera merely represent variations of the same basic type.

The nycteribiids house their symbionts in an entirely different way, and the streblids have very primitive symbiotic devices, if any, thus confirming the interpretation prevailing today that the similarities between the three groups are solely ecological in origin and that no systematic value should be ascribed to the designation Pupipara.

The glossinies resemble the hippoboscids in their symbiosis as much as they do in structure and mode of life. In all representatives of this subfamily, consisting of a single genus, the symbionts had the same requirements and were consequently localized, resettled during metamorphosis, and transmitted in exactly the same way.

Other smaller units in which symbiosis may possibly have arisen mono-phyletically are the trypetids, whose symbiosis with bacteria varies in the degree of development, the lagriines which, unlike the other four sub-families of the lagriids, showed, in principle, similarly constructed trans-mission organs in 82 of the 93 species studied, the pyrosomes and mol-gulids, among the tunicates, and the lumbricids among the oligochaetes.

If we are correct in assuming that this symbiosis originated mono-phyletically, the microorganisms of the present-day genera and species must be descendants of the original form. Although they may occasion-ally vary in shape, no contradiction is involved, for we know, on the one hand, particularly from Paillot's studies, that the same bacterium, if transplanted into various insects, may assume different shapes before our very eyes, and, on the other hand, that the symbionts, like so many parasites—for instance, the flagellates of the termite gut—may possible be transformed into new races and species during the phylogenetic devel-opment of their hosts. *Pseudococcus* illustrates strikingly the extent to which such symbionts may differ within a single genus. Here mycetomes of similar construction house symbionts which are shaped differently from species to species but nevertheless have a number of traits in com-mon and obviously are descended from the same primary form (Plates, II*e,f,* III*a,b*). Here we are clearly dealing with a matter which one day will be important for the genetic problems of bacteriology. By means of symbiotic cultures in which closely related hosts are interchanged, it will be possible to determine whether the changes are irreversible or are mere modifications entailed by milieu.

It is quite obvious in a number of cases that the symbiosis was acquired polyphyletically. Classic examples are the Heteroptera, cerambycids, cur-culionids, and ipids. From a compilation of findings with heteropterous bugs we gain a variegated picture, one that is surprising at first glance. In the family of the pentatomids all 60 species investigated from among the subfamilies of the scutellerines and pentatomines have four of the characteristic bacteria-populated crypt rows on the midgut; all acantho-somines have two rows and moreover have chitinous lubrication organs on the ovipositor which do not occur anywhere else among the Heter-optera. The fourth subfamily, that of the asopines, is without symbionts. The family of the cydnids always has two rows of crypts; in other families there are forms with and without symbionts, for instance, the coreids, where only 3 of the 14 species studied had crypts, and the lygaids, where 40 representatives had symbionts and 8 did not. In the latter, the high number of symbiotic cases may be explained chiefly by the fact that almost all of the subfamily of the aphanines form their characteristic appendages, finger-shaped or tube-shaped, instead of the shorter crypts; and furthermore, such sites were found with all three berytids investigated. In addition to lygaeid species which have such appendages, there are others with two rows of crypts and even some which develop regular paired or unpaired mycetomes. *Tropidothorax leucopterus* houses the symbionts in the gut but in a completely arbitrary manner, and its transmission methods are unique. Localization in the gut and in the mycetomes may occur in one and the same subfamily. As regards the interesting plataspids, of which only the European *Coptosoma scutellatum* had previously been investigated, I was in a position to confirm that several species from the Far East also form the unique symbiont-filled capsules. Among the phyrrhocorids, slightly organized efforts to admit symbionts were observed here and there, and no symbionts were found at all among the reduviids, nabids, anthocorids, and various families of water and shore bugs, just as in the case of the asopines.

Here it is impossible to speak of a partial later loss of symbionts, for their more or less sporadic appearance represents an original state and is clearly rooted in the animals' mode of life. Everything indicates that all the Hemiptera were carnivorous originally and only gradually turned to sucking plant juices exclusively. With the Homoptera this change, which manifestly requires symbionts, appeared everywhere; with the Heteroptera it occurred only here and there, and the pronounced predators, such as those reduviids which do not live exclusively on verte-brate blood, and asopines, therefore remained symbiont-free.

Numerous observations show that predators will suck at plants in cases of necessity. On the other hand, pentatomids, acanthosomines, and cydnids occasionally attack dead insects or carrion birds; and the symbiont-free capsids, tingids, and piesmids, as well as the corixids among water bugs, quite regularly accept animal nourishment, although they already live predominantly on a plant diet. Thus adaptation to strict plant diet is still in flux, and it is understandable that side by side in a number of subfamilies one finds species with and without symbionts in merry confusion.

Although the triatomids (*Rhodnius, Triatoma*) and cimicids live sym-biotically, they are not exceptions to the rule, as might appear at first glance. Diet is decisive here, too, in establishing symbiosis, for these animals live exclusively on vertebrate blood and these two groups thus require symbionts under all circumstances. Two possibilities, correspond-ing to the family constitution, are reflected when mycetomes are formed and the oocytes are infected, or when the intestinal epithelium is settled and the egg surface is contaminated with bacteria. Nevertheless, in each of the two groups a uniformity prevails which speaks for the monophyletic origin of their symbiosis. The same devices were found in cimicids from both the old and new worlds on one hand, and in all triatomids from South America on the other. The only exception is represented by the oldest known cimicid, which underscores its primitiveness by localizing its guests in the epithelium of the midgut instead of in mycetomes.

Yeast-like symbionts of the cerambycids also make their appearance sporadically. Among 195 species from more than 65 tribes, symbionts were found in all spondylines, asemines, and saphanines, in most lepturines, and in a single species of the 10 cerambycines tested. It is uncertain to what extent the necydalines, tillomorphines, and trichomesines are symbiont bearers, because only one or two representatives were investigated in each case and were found to be living symbiotically. Once again, the symbionts must have been acquired polyphyletically, in all probability for nutriophysiological reasons. In studies on the manner of life of the symbiont bearers, frequently unknown with exotic forms, it was shown that symbionts are never found in forms living as larvae in deciduous wood or mining in herbs, but that there is a clear relationship between symbiosis and the presence of larvae in coniferous and dead deciduous wood. The assumption that the symbionts were acquired independently is also confirmed by the fact that at least some of the symbionts are of a distinctly different type. It is true that the individual vegetative states, such as are present in the various formed conidia, may occasionally represent mere modifications connected with formation of new species, but surely this cannot be the case where variously shaped spores are used in transmission. Moreover, the cerambycid symbionts are adapted to a greatly varying extent, as we have seen. We must therefore conclude that spondylines, asemines, and saphanines acquired their symbionts at a later date, and that the most perfect and therefore presumably the oldest cases of symbiosis are found among the lepturines and necydalines.

Moreover, the siricids have a habit which was undoubtedly acquired polyphyletically, namely, that of using their oidea-filled jets to infect the wood with symbiotic food fungi. This hypothesis is supported by two facts. First, different types of fungi occur in the same host species from different localities. Second, *Xeris spectrum*, although it does not disdain fungi when they happen to be on hand, cultivates no fungi itself but lives in fresher, more nutritious wood and therefore preserves the original form of the lubricating glands on the ovipositor, whereas its relatives have converted them into organs of transmission.

No definite statement can be made about anobiids in view of the fact that recently also in one of the forms, in which at first no symbiosis was found, a symbiosis was later established. If they should be found to be incomplete, it is possible that similar differences in the character of the wood play a role, as with cerambycids, and one would surmise a polyphyletic origin.

Almost all the curculionids seem to live symbiotically, yet the devices, in contrast to those of cerambycids and anobiids, are extremely diverse and sometimes not even similar in one and the same tribe. Complete uniformity exists only in the subfamily of the cleonines, of which 55 representatives have been investigated and found to be entirely different from other snout beetles in localizing and transmitting their symbiotic bacteria. One can only assume therefore that the symbionts were acquired by their primary forms. On the other hand, completely different types of localization appear within the tribes of the calandrines, ceutorhynchines, and apionines.

The last-mentioned have very special interest for us here. A study of a large number of species of the genus *Apion,* which has approximately 1000 species distributed over a series of subgenera, reveals that even heterogeneous treatment of the symbionts may in exceptional cases be traced back to a uniform primitive type. In our section on specialized forms we described the three methods used by *Apion* to localize its symbionts: first, in larvae and imagines, two of the six Malpighian tubules are radically transformed into club-shaped, mycetome-like appendages; second, in larvae about to pupate the content of small mycetocytes clinging to the visceral lamella of the gut is sent to the Malpighian tubules, of which only four are present, where it is irregularly distributed in the renal cells and lumen; and, third, the symbiotic bacteria are localized in a lobe of the fat body which has been converted into a primitive mycetome. But even in the last case there are only four Malpighian tubules, all of them now completely symbiont-free. Since all other curculionids have six tubules, it must be assumed that the first type was originally present in all species and then was partially given up, with the consequent disappearance of the two tubules which were withdrawn entirely from their original task. In the second type, reminiscences of the former relationships to the renal organs are noticeable; in the third, these indications no longer exist (Fig. 363). The opening into the gut is usually somewhat displaced in this pair, and its lesser strength shows that the vessels which have been transformed into clubs and those which have completely disappeared are one and the same.[1]

---

[1] In Stammer's comparative study on the Malpighian tubules of the Coleoptera (1935), it is emphasized that the number four, unusual for curculionids, can be explained only by retrogression. Only *Orchestes fagi* has merely four Malpighian tubules. A symbiosis-caused reduction might possibly be involved here. However, the symbiosis has been established only through infection of the nutrient cells, and nothing is known about the localization.

To the systematicist, who from the very start assumes close relationships between curculionids and ipids, it will come as no surprise that the two groups also have great similarity with respect to endosymbiosis. In both families the symbionts are diversely localized in fatty tissue, intestinal appendages, Malpighian tubules, etc., and are transferred by end chambers of ovarioles. Like some of the curculionids, the wood-

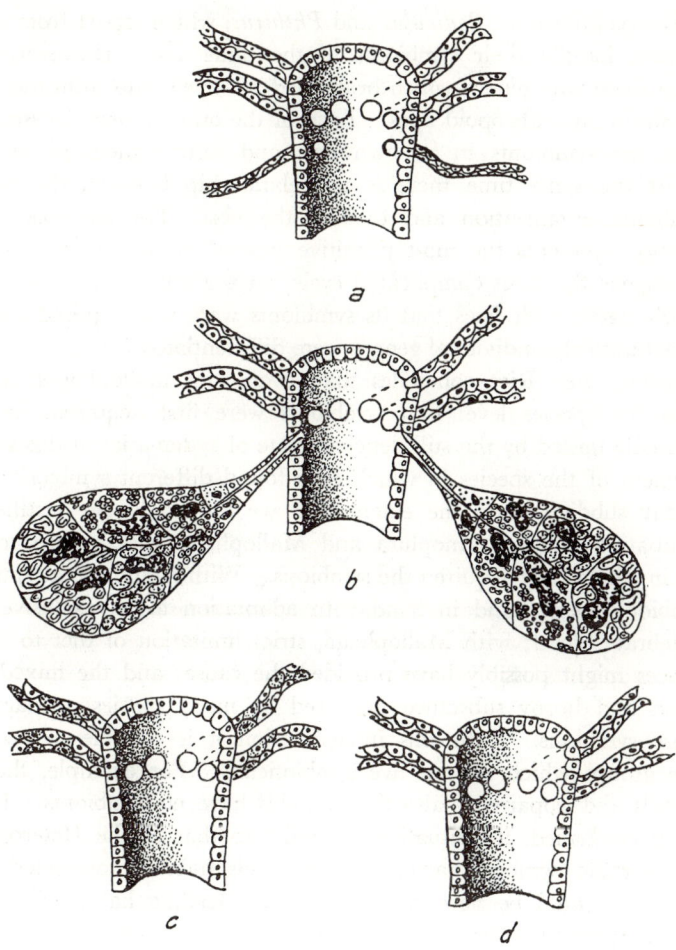

Fig. 363   Transformation and rudimentation of two Malpighian tubules in the Apioninae. (*a*) Hypothetical, symbiont-free original stage. (*b*) *Erythrapion;* two tubules have become club-shaped symbiotic sites. (*c*) *Aspidapion;* the modified tubules have disappeared, the symbiotic bacteria infect the remaining tubules during metamorphosis. (*d*) *Omphalapion;* all relationships of the bacteria to the remaining tubules have disappeared. The symbionts in (*b*) have been drawn larger than to scale.   Schematic with reference to Nolte.

breeding Xyleborini and Xyloterini have remained symbiont-free. In such cases ambrosia culture takes the place of endosymbiosis, and thus by such heterogeneity the polyphyletic character of ipid symbiosis is clearly indicated.

Especially impressive is the lack of homogeneity in Anoplura. The diversity of the devices coincides with neither families nor subfamilies; only within a genus is a characteristic device developed. Interestingly, the only exceptions are *Pediculus* and *Phthirus,* which, apart from slight differences, handle their symbionts in the same way. However, it is precisely these two genera which originated relatively late from the same parent strain on anthropoid apes. Each of the other genera investigated localizes the symbionts in its own way and infects them in its own way. At the same time there is no relationship between the degree of symbiotic organization and that of the host. For example, *Haematopinus* represents the most primitive type of Anoplura, on the one hand, yet has the most complicated cycle, on the other. Thus we must inevitably agree with Ries that its symbionts were not acquired for the most part until the individual genera were differentiated.

However, when Ries concludes that, even with mallophages, it was only on the species level that symbionts were first acquired, he was undoubtedly misled by the still deficient state of systematics in this group. The genera of the species in which Ries found different symbiosis types are today subdivided to the extent that we are no longer justified in differentiating between Anoplura and Mallophaga with regard to the period in which they acquired the symbiosis. With Anoplura, adaptation of symbionts went hand in hand with adaptation to an exclusive diet of vertebrate blood; with Mallophaga, strict limitation of diet to horny substances might possibly have provided the cause; and the mixed diet of blood and horny substance preferred by many species appears not to cause symbiosis. With this interpretation it is possible to account for the groups which do not live symbiotically. For example, the trichodectids and apparently also the liotheids have no symbionts. If the theory is confirmed, the situation is similar to that of the Heteroptera, where sporadic symbiont acquisition parallels habitual one-sided diet. However, it would be necessary to study the feeding habits of a large number of species to clarify this question.

Finally, we have even found a case with racially differentiated symbiont acquisition. In gamasids, which are parasites on reptiles, very different symbionts may appear within the same species. For example, *Ceratonyssus* of the mouse always has only one symbiont type; *Liponyssus saurarum* has no less than six, of which usually only one is present, but sometimes two and then they are cleanly divided; and animals from Istria always have forms different from those from Spain. Analogously, two vicariously appearing guests are also found in *Ophionyssus natricis*. Nevertheless, the symbiosis in each case is well-adapted and the symbionts always appear in their specific forms during egg infection and embryonic development.

The examples cited are probably sufficient to illustrate the deep-seated difference in character between monophyletic and polyphyletic symbioses. One is actually dealing with two different worlds when one compares blattid symbiosis with that of Anoplura and snout beetles, or of the coccids, which will be treated later. Homogenity on the one hand and surprising variety of devices on the other signify that symbiotic adaptations from a primary form were simply taken over almost unchanged into the further systematic development of the group. Other cases are found, however, where subfamilies, tribes, and even genera, apparently unburdened by tradition, use very different methods for localizing and transmitting the symbionts and for adapting them harmoniously to the embryonic development.

With such cases of polyphyletic symbiosis, special treatment would be required to separate the similarities rooted in the constitution of the family from the reactions characteristic of tribes, genus groups, and genera. For example, it is generally acknowledged today that a close relationship exists between Anoplura and rhynchophthirines, and mallophages, and this has been thoroughly confirmed by symbiosis research, which showed that these three groups are the only ones which form ovarial ampullae for the purpose of transmission. On the other hand, constitutional differences are involved when lice vary their methods from genus to genus in forming their complicated tetralaminate structure. The symbioses of *Cassida* and *Bromius* species obviously originated in the narrowest circle, independently of each other, yet these two genera as chrysomelids localize and transmit their guests in very similar manner. Both curculionids and ipids infect the eggshells indirectly by way of the nutrient cells, thus revealing their membership in the higher group of the Rhynchophora. The lumbricids of the Old World and the glossoscolecids of the New World house their symbionts in their nephridia, independently of one another, in exactly the same way.

Historic problems are also involved in all the numerous cases where the host animal takes up several symbionts. Here the first pressing question concerns the sequence in which they were acquired, and the age of the additional symbionts measured by the development of the group in question. Again, the principle applies that, the older the acquisition, the higher is the unit of classification which disposes of its symbionts in the same way. The psyllids stand alone in this respect, for, even if we disregard the two cases already mentioned, where the symbionts are lost later, all species here have the same symbionts, handle them always in the same way, and therefore must have acquired the second guest at early levels. This would mean that the psyllids therefore had two types of symbionts in the Upper Lias.

In general, the secondary adaptations reveal their younger age by their limitation to lower categories: subfamilies, tribes, indeed, only to groups of genera, individual genera, or even mere species and strains. With coccids the inclination to admit additional symbionts is in general moderately developed and often occurs sporadically. Among the tachardines, only the genus *Tachardiella* is disymbiotic, and only a few *Rastrococcus* species among the many pseudococcines investigated possessed two symbionts. Such disymbiotic forms are numerous only within the family of the Margarodidae and are found with Margarodini, Monophlebulini, Monophlebini, Drosichiini, Llaveiini, and Iceryini (in part also according to unpublished observations of my own). In general it is then a case of symbionts in well-balanced mycetomes. Some iceryines failed to admit a second guest at all; in others the adaptation was incompletely regulated, and in still others the adaptation reached a harmonious conclusion. A special position is occupied by *Rastrococcus,* which unfortunately has been insufficiently investigated and presumably would have to be resubdivided. One representative was found living symbiotically with rodlet-shaped bacteria, another with yeasts, and a third with yeasts and bacteria combined. In the last case the bacteria were evidently the guests acquired later, for unlike the yeasts they were unprovided with nuclei and finally filled giant anucleate areas of fatty tissue or, in another species, drifted free in the lymph in meager quantities. The only disymbiotic *Stictococcus* species has an obviously well-adapted, older type of bacteria which infects the oocytes, and a second type which finds admission only in a late stage of development but is also completely regulated.

With aphids the disymbiotic species are more frequent by far, and here and there we even find trisymbiotic types (by way of estimate, half of the numerous forms investigated have more than one symbiont), but, as among the aphidines, pemphigines, and adelgines, the adaptation is occasionally unregulated and does not occur in all representatives of the tribes. Admission of a third guest appears to be limited to the lachnines. At any rate, in all of these extensions of the symbiotic stock among aphids it is a matter of relatively late phenomena. In all these cases no disturbances were caused by acquiring differently constituted bacteria, but, when some of the hormaphidines took up "yeasts," this led to complete loss of the older symbionts.

Far more widespread, and obviously of earlier origin in some cases, are the plurisymbioses among cicadas, where species with as many as six symbionts may be found. Here the terms primary, auxiliary, companion, and accessory symbionts in themselves point to later or earlier appearance.

Up to the present, 405 cicada species have been investigated for symbiosis, a rather modest number compared to the 30,000 species, representing 3500 genera and 45 families, for which descriptions already exist according to an estimate undertaken some years ago. Thus we are informed symbiotically about only 1.3% of the species, with one-fourth of the families unrepresented. All in all, we are acquainted with about 50 cicada symbionts which are distinct in morphology and localization. Among these, only three types of primary symbionts are found, that is to say, forms which may exist alone as sole guests, although of course they rarely do occur alone. First are the *a*-symbionts, which as yet have been found without companion symbionts only with the very ancient peloridiids; second, the "yeasts," which likewise occur rarely as sole possessions, namely, with *Ledra* and *Pyrgauchenia* among the cicadoids and *Nisia* and with a derbine among the fulgoroids, and, third, the *x*-symbionts, which were found without additional forms in 11 fulgoroids.

By far the most frequent among these three primary forms are the *a*-symbionts, which are found in almost all cicadoids and in a great many fulgoroids. In second place are the *H*-symbionts which are always present, as far as is known, in a series of families such as flatids, phalaenomorphids, ledrids, and eurybrachiids, which occur frequently or rarely by turns in many other families, or which may be absent entirely, as in cixiids, tettigometrids, cercopids, and others. Whereas *a*-symbionts are more frequent with cicadoids than with fulgoroids, *H*-symbionts are much more frequent with the latter. The *x*-symbionts, finally, are limited strictly to fulgoroids, for which they are very characteristic, being found in 108 of 210 species. They are always present in cixiines and in the great majority of issines and fulgorines. They are always absent in the flatines when yeasts are present, although they are found in families without yeasts (Nogodinidae, Lophopidae).

The auxiliary symbionts designated by *f* and *t* stand but slightly behind the three primary symbionts in frequency of occurrence. Although the little *f*-symbionts, housed in typical insignificant mycetomes, are, like the *x*-symbionts, limited to the fulgoroids, they are nevertheless distributed here almost over all families in more than half the species examined. Conversely, *t*-symbionts occur only among the cicadoids, always linked with *a*; they are very typical for membracids, jassids, and euscelids, but are lacking entirely in cicadoids and cercopoids.

H. J. Müller (1962) uses the designation "essential symbiont" both for the primary symbiont, which occurs alone, and for the auxiliary symbiont, which occurs only in combination with the primary form. These two principal forms then are contrasted with the complementary symbionts, that is, the companion and accessory symbionts. The companion symbionts are characterized throughout by their limitation to certain subfamilies, tribes, or genera, though within these groups they occur more or less regularly. Often they appear as milieu-occasioned involution forms; sometimes they develop specific forms of transmission; and they are housed in well-defined organs. With fulgoroids, especially, they frequently complement *a*- and *x*-symbionts. The accessory symbionts are usually limited to individual species and display more or less clearly traits suggesting those of parasites, for instance, the variation in density and type of settlement and the infection of the habitats of other symbionts. With membracids the combination $a + t$ is extended by such guests very frequently. Quite understandably, the boundary between these two categories is sometimes obscured by transitional forms.

Table 1, which is taken from H. J. Müller's latest publication, shows the frequency of the symbiont-free forms (Typhlocybines) and the mono- to hexasymbiotic species among cicadas known at present.

TABLE 1

			Total	
	Fulgoroidea	Cicadoidea	Absolute	Per Cent
Asymbiont	0	10	10	2.5
Monosymbiont	15	7	22	5.4
Disymbiont	119	105	224	55.0
Trisymbiont	74	50	124	30.5
Tetrasymbiont	8	9	17	4.2
Pentasymbiont	1	5	6	1.5
Hexasymbiont	0	2	2	0.5
	217	188	405	

The result of such statistical observation of the temporal sequence of the individual acquisitions is confirmed by a series of other criteria. However, these were taken into consideration for the most part in the course of our general observations and will therefore only be briefly discussed at this point.

One criterion is the morphology of the symbionts. The less intimately they are adapted, the more primitive is their morphology, that is to say, the greater is their similarity to parasitic and saprophytic forms. This rule is confirmed at every turn. We need only refer to the rodlets, filamentous forms, and cocci of the additional aphid symbionts, to the "wild" accessory symbionts of membracids, to the sometimes still very unharmonious disymbioses in *Icerya* species, or to the rodlets and filaments which accompany the *a*-symbionts in *Aphrophora salicis* and by no means respect the limits of the mycetome assigned to them. To be sure, long cooperative living need not necessarily lead to the origin of deviating involutional forms, for the supplementary symbionts of psyllids have very primitive forms in many species, and among the blattids there has been no symbiont modification worthy of mention after many millions of years of intracellular living.

Criteria for determining the age of the symbioses are also furnished repeatedly by the nature of the sites. Frequently the secondary guests are housed in areas which are easily recognizable as subsequent, more or less harmonious structures in the interior of mycetomes already present. The membracid mycetomes can be arranged in an impressive sequence leading from disorder and disturbance to balanced cooperative living, and the same process is demonstrated with extreme clarity in the various disymbiotic *Icerya* species and in aphids with two and three symbionts. Some cicadas make the mycetome epithelium available to the supplementary bacteria and present thereby a kind of emergency solution; in numerous other cases the bacteria are housed in sections of fatty tissues which are more or less sharply demarcated and reveal their original function in differing degree.

Comparative study of transmission devices frequently confirms conclusions drawn from statistical observation of the occurrence and morphology of symbionts and their sites. Specific transmission forms, and thus usually a separation of their formation sites which is traceable for the most part to complicated hormone effects, are found only with widely occurring, that is, old guests (infection mounds of all *a*-symbionts, rectal organs of the *x*-symbionts, migrating cells of the *t*-symbionts). With this statement we have a firm basis for concluding that such old guests have undergone severe changes in form through long cooperative living and thus have greater need to be transformed into states more capable of reproduction than their fellows of later date.

Occasionally we saw that younger, indeed the very youngest, guests were subjected to the formative influence of effects not originally intended for them, as may be seen in a preceding compilation. It was always clear from the detailed circumstances that the resulting behavior of the symbionts, accompanied sometimes by disturbances, is always the consequence of an additional arrangement. It must also be judged as a sign of younger cooperative living when small or most minute accessory rodlets fail to transfer to older eggs with the primary and auxiliary symbionts but take their own path over the nutrient cells into the youngest oocytes, as with many membracids, or when the *m*-symbionts of *Fulgora* species settle for this purpose first in a special filial mycetome, which in the case of *F. europaea* is then given up with progressive adaptation and makes its appearance as a mere empty rudiment.

We have already mentioned that younger acquisitions are at first neglected during embryonic development. Where the embryology of cicadas with several symbionts has been investigated, dissimilarities have always been found which clearly indicate differences in the sequence of acquisition. The essential *a*-, *H*- and *t*-symbionts, which in general are particularly well adapted, are always taken into young embryonic cells shortly after the blastoderm and yolk cells first envelop the mixed symbiont ball when the germ band begins to invaginate. The supplementary *b*-, *f*-, and *p*-symbionts are also taken up by embryonic cells, but only much later, after involution of the embryo, when the transitory collecting mycetome disintegrates and the definitive sites are formed. As Ermisch showed recently with a species of *Conomelus* (delphacid), the same applies to those widespread, so-called yeasts, which are counted among the primary symbionts but undoubtedly ought to be considered supplementary guests, for they are merely housed in a loose syncytial association as development begins and only after involution of the germ band do they transfer to the embryonic fatty tissue, where they spread more or less irregularly and rarely form mycetome-like complexes.

Involved with the adaptation of symbionts acquired later are phenomena with psyllids and certain aphids. Here the nuclei which usually furnish the mycetome epithelium are given to the new partner, and, if the latter should come to be situated not peripherally but centrally, the infected elements may be strangely displaced.

The great multiplicity of cicada symbiosis and the abundant indications that the symbionts were acquired in a different sequence naturally demand phylogenetic investigations. I gained my first view of this diversity in 1925. On the basis of the knowledge available at the time, I came to the conclusion that the symbionts must have been acquired polyphyletically, for heterogeneous types exist side by side and even essentially different symbionts such as yeasts and bacteria occur vicariously. Only deeper studies of cicada symbiosis, like those of H. J. Müller and Rau, made it possible for the former to revise this interpretation and to show that the multiplicity which was so confusing at first glance is traceable to a monosymbiosis which exists at the very base of the cicada family tree. The presentation in the first edition of this book was based on Müller's publications of 1949 and 1951 and on an unpublished phylogenetic tabulation which he placed at my disposal. New data by Ermisch on the basis of delphacids and new advances in the systematics of cicadas made it possible for Müller to make substantial improvements (1962), and we follow these closely in the following discussion.

Very significant for historical consideration of cicada symbiosis are, first, the discovery that *Hemiodoecus,* the representative of the primitive peloriidids (*Coleorryncha*) which are considered to be relics from the Carboniferous, has but a single symbiont, the widespread *a*-form, and, second, the knowledge that the *H*-symbionts, though they occur very frequently, indeed sometimes as the sole guests among membracids and ledrids, are nevertheless later acquisitions that frequently caused the elimination of older inmates. Such antagonism has also been proved by new observations of *H*-symbionts among hormaphidines, pseudococcines (*Rastrococcus*), and stictococcines.

To clarify the derivation of the Fulgoridea on the one hand and that of the Jassoidea, Cercopidea, and Cicadoidea on the other from the primitive Coleorryncha, which today are represented only by the peloriidids, and to show the changes in the symbiont supply, on the basis of our present knowledge we are presenting a family tree for the cicadas viewed as from above. Although it is regrettable that only a single peloridiid could be investigated to date, the results nevertheless are of decisive importance in relation to the origin of cicada symbiosis. *Hemiodoecus fidelis* is the only Homophteron with *a*-symbionts, that is, the only symbiont form to appear again and again with great frequency in all branches of the cicada family tree. If one were dependent solely upon surmises, one would inevitably have to postulate this type for the starting form (Fig. 364).

The four families derived from the Coleorryncha are each differentiated through supplementation of a specific auxiliary symbiont: the Fulgoridea by *x*-, the Jassoidea by *t*-, the Cercopoidea by *bc*-, and the Cicadoidea by *w*-symbionts. It should be noted that these four areas are not at all equivalent, for the systematicists classify the Jassoidea, Cercopoidea, and Cicadoidea as Cicadamorpha, and the Fulgoroidea as the sole Fulgoromorpha. This is a grouping that tallies exactly with symbiotic data. With the Fulgoromorpha the symbionts are housed in separate mycetomes in the posterior half of the abdomen. Here the infection mounds are depressed and the *x*-symbionts play a significant role; *f*-symbionts are the typical auxiliary forms; the companion symbionts, also usually housed in mycetomes, are quite diverse; the accessory symbi-

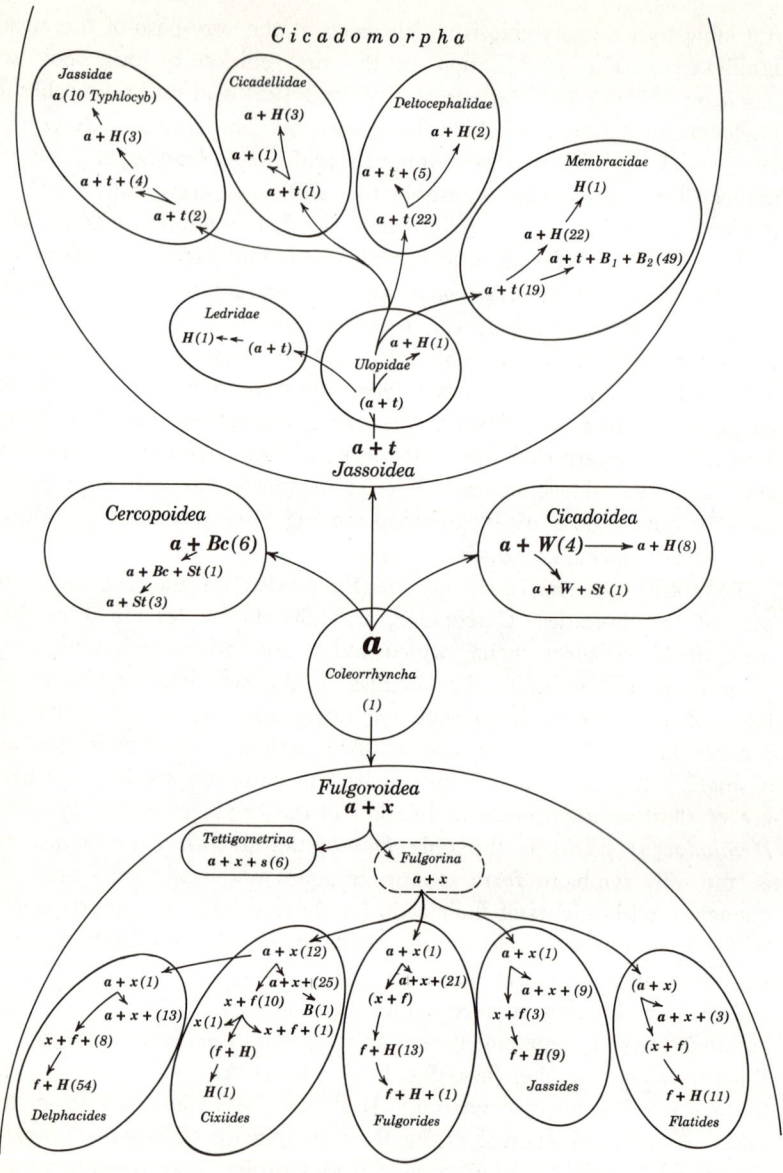

Fig. 364   Hypothetical relationships between the phylogenetic development of the cicades and their endosymbioses.   From H. J. Müller.

onts are rare; and the combination $a + H$ never appears again. The Cicadomorpha, on the contrary, characteristically unite the various symbionts more or less intimately in a single mycetome situated in the anterior half of the abdomen. When the infection mounds actually protrude, x-symbionts are never found; t-symbionts take the place of f-symbionts; companion symbionts are lacking or are few in number, in which case they are housed in the typical double mycetome $a + t$; H is now frequently combined with a.

In comparing the Fulgoromorpha families with one another, it may be noted that the fundamental combination $a + x$ occurs throughout, alone or with companions; often $x + f$ occurs, sometimes with additional symbionts. The combination $a + x + f$ is never seen, so that we are inclined to think that the acquisition of f causes a to be eliminated. In younger families capable of further development, such as the fly-like derbids, the delphacids which tend to reduce their wings, and the butterfly-like flattids, x may be eliminated by H (in 54 species of the delphacids), whereas the original combination $a + x$ is kept, particularly in the more or less primitive tettigometrids, fulgorids, and cixicids. Apparently there is a tendency for $a + x$ to arrive at $x + H$ by way of $x + f$, a tendency which, among the cixiids, in rare cases may lead to the sole possession of x or H. The occurrence of yeasts seems to favor the reduction of polysymbioses and very often renders superfluous the companion symbionts, which were found in 82% of all $a + x$ combinations.

Turning now to the development of endosymbioses within the Cicadamorpha, we soon see that its classification into the three superfamilies tallies precisely with the symbiotic findings. With the Cercopoidea and Cicadoidea the a-symbionts predominate throughout, allied only with more or less characteristic companions (bc and w), about whose possible homology nothing can be said at present but which appear to be rather loosely inserted and are not compared with the true auxiliary symbionts of the Jassoidea (t) and Fulgoroidea (x). With the singing cicadas the a-symbionts are moreover usually replaced by yeasts, and with the foam cicadas sometimes by rodlet-shaped bacteria. However, the nature of symbiosis in these two branches of the family tree is determined by the enormously developed a-organs, and thus one is reminded of the starting point of the peloriidids. In this connection Müller points to original characteristics, valid for both groups, which are morphologically and ecologically expressed, and he correctly places the two families, which apparently are particularly close to the monosymbiotic phase, at a lower level in the family tree than the Jassoidea.

The acquisition of the t-symbionts with the Jassoidea led to far greater development, just as with the x-symbionts in the fulgoroids.

The combination $a + t$ evidently facilitated the origin of polysymbioses by acquiring of companion symbionts and accessory guests, although it is impossible to determine whether this was the result or a simultaneous cause of the subdivision of this superfamily (H. J. Müller). As with the fulgoroids, the basic combination was by no means always maintained but was apparently exchanged for auxiliary and companion symbionts which presumably were more efficient. It is especially striking that an elimination of the t-symbionts by yeasts occurs just as frequently among the jassids, cicadellids, and the deltocephalids as among the more highly developed systematic units. In rare cases even the a-symbionts disappear among the membracids (*Pyrgauchenia*), and only the particularly welcome yeasts hold their ground. As already mentioned in our specialized section, new acquisitions and eliminations are still taking place with the membracids, and the adaptation of yeasts is still in progress with a number of animals. Rau had already pointed out in her membracid monograph that it was not a question of decomposition but of a new acquisition not yet fully adapted, an interpretation borne out completely by present-day research.

Although only one species with $a + H$-symbionts is known to date within the family of the ulopids, which stands at the base of the family tree, we may undoubtedly assume that more material would show the postulated $a + H$ combination. The same applies to the ledrids, of which only the yeast-carrying *Ledra aurita* has been tested, so that in this case the two primary forms were eliminated by H. Certainly indeed among the jassids all symbionts were eliminated here and there, for no symbionts at all are found among those parenchyma-sucking forms which appear in some places and which were formerly classified as typhlocybines. This exceptional position of the typhylocybines is in harmony with the concept of Evans and Wagner, who interpret them as a very young, entirely new developmental stage still in the profound process of development and splitting, and state that they are derived from the likewise tree-living idiocerines, a derivation substantially supported by Müller's finding (1951) that the hatching mechanism is the same in both groups.

Research on delphacid symbiosis by Ermisch in 1960 demonstrated the value of comparative study when it is based on extensive material and conducted with close regard to modern systematic analysis. The research was based on 76 species, nearly all genera being represented. Fortunately the same group was also studied by a systematicist (Wagner, 1961) from the aspect of taxonomy and morphology. Figure 365 shows that the family tree thus set up tallies with the symbiotic study of Ermisch. Wagner concluded that the phylogenesis of the endosymbioses coincides excellently with the system he worked out. The low-level asiracines and the more highly developed celisiines and stenocranines display the typical basic combination $a + x$, supplemented with the asiracines by p and with the celisiines and stenocranines by various bacteria presently designated as q-symbionts. But even with the slightly higher standing jassidaeines a is displaced by f, and supplementary yeasts

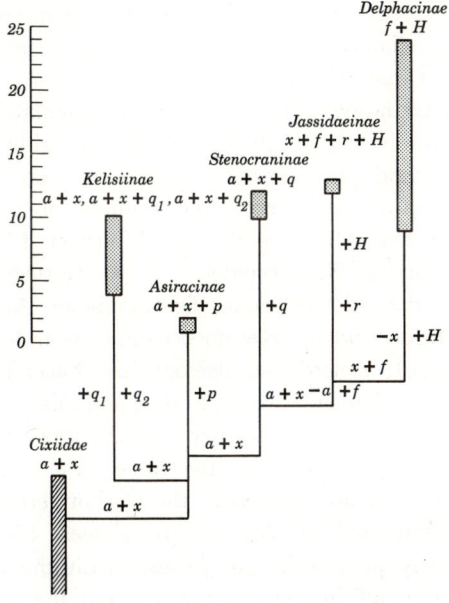

Fig. 365   Family tree of the Delphacidae and its relationships to their different combinations of symbionts.   From Wagner.

appear beside a companion symbiont r, which has also been found with Brazilian delphacids.   The highest-level delphacines then have the paired $f + H$, frequent also with other fulgoroids, which indicates that here the x-symbionts are also displaced by yeasts.

These new insights into the development of cicada symbiosis, which at first glance is so confusing, allow us to hope that research in other groups based on comprehensive material and modern systematics will further clarify phylogenetic questions.   We need hardly emphasize the usefulness of microbiological analysis of symbiont types, analysis which is almost completely lacking to date.

It is certain that the symbioses are pronouncedly monophyletic in character in psyllids, aleuroidids, and aphids, with the possible limitation made above for the aphids. Thus they are in clear contrast to coccid symbioses, which are extremely heterogeneous. Walczuch, to whom we owe a considerable increase in our knowledge of symbiotic devices in scale lice, first saw that the great differences existing between the individual families of the scale lice must have arisen polyphyletically, and my efforts to extend our knowledge of coccid symbioses are decidedly in line with her concepts. In reviewing the families investigated to date it may be seen immediately that in general far greater uniformity prevails within these families than with the cicadas.

Recently an outstanding expert on coccid systematics, Borchsenius (1958), set up a family tree for the coccids summarizing his conception of their evolution and phylogenetic correlations. Three Archaeococcoidea arise from a common root, the families of the Ortheziidae, Phenacoleachiidiae, and Margarodidae, and fourteen families summarized in the superfamily, Neococcoidea. These fourteen, divided into seven groups, are the Stictococcidae, Dactylopiidae, Apiomorphidae—Pseudococcidae—Eriococcidae, Kermococcidae—Asterolecaniinae—Coccidae (= Lecaniidae), Aclerdidae, Beesoniidae—Lacciferidae (= Tacchardiniae)—Conchaspididae, Phoenicoccidae, and Diaspididae. Exactly as with the families of the Archaeococcoidea, the root of each is placed back in the Permian, therefore at the same time when the coccid ancestors, which had already existed in the Carboniferous, were fundamentally split. According to Borchsenius, sixteen of the seventeen families were already present in the Jurassic with the Kermococcidae on oak trees splitting off in the Cretaceous from the ancestors of the present-day eriococcids, to become the youngest coccid family.

Our knowledge of coccid symbiosis has now reached the point where we have information for all families except two: the Phenacoleachiidae with its single representative in New Zealand, the beesoniids, of which only two species from China and Burma have been described; and the Phoenicococcidae, which are generally placed with the Diaspididae. However, the symbiotic devices in these families often reveal considerable uniformity: with the Ortheziidae, primitively shaped bacteria are always housed in diffusely distributed mycetocytes and characteristically infect the egg cells at the posterior pole; of the Stictococcidea, seven species of the few known to exist have been studied; in all cases the symbionts do not infect the eggs yielding males but infect the young eggs that remain yolk-free, and embryonic development, which is unusual in many aspects, always takes its course in the same way; with the Apiomorphidae both sexes remain symbiont-free; with the Eriococcidae and the Coccidae, the yeasts always live in lymph and fat cells; etc.   When the uniformity is disturbed but rarely by individual genera, it may be assumed that the primary symbiont was replaced by a yeast, as, for example, with the genus *Lecaniodiaspis* among the Asterolecaniidae and with *Lakshadia communis* among the Lacciferidae.

Moreover, the large family of the Pseudococcidae can no longer be considered to have the uniformity that was assumed earlier, for more recent studies of the *Rastrococcus, Puto,* and *Macrocerococcus* groups have yielded greatly divergent findings.   However, the greatest multiplicity has been revealed by my still unfinished investigations of the Margarodidae.   Here are found tribes which consist of only a few species, such as the Callipappini of Australia, or, as far as we know today, contain only a single species, such as the Marchalini of Turkey, but in both cases represent a type of symbiosis which is unique among the coccids.   Also the remaining tribes of this ancient coccid family are homogenous in regard to the symbiosis, with the exception of the Margarodini, among which one group, which one day will probably be eliminated by the systematists, represents a way entered upon no where else among the scale insects.   Finally, my investigations in this regard uncovered the three additional tribes, which are symbiont-free and in an interesting way also in regard to original characteristics of internal anatomy, deviate from all other coccids.

Coccid symbiosis differs accordingly in principle from that of the cicadas.   All indications which would allow, as with the latter, the exposure of a monophyletic family tree, are thoroughly lacking, and the concept, which Borchsenius proposes for the development of the coccids, harmonizes throughout with the findings of symbiosis research. There is no similarity symbiotically between the Ortheziidae and the Margarodidae.   Stictococcidae, Dactylopidae, and Apiomorphidae have nothing in common except perhaps the lack of symbionts in the two latter families.   The typical pseudococcid symbiosis has no relationship to the one previously described for three Eriococcidae but is closely related to that of *Rastrococcus spinosus*.   The Kermococcidae, which did not split off from the Eriococcidae until the Cretaceous, are symbiont-free, although the Coccidae and Aclerdidae, by way of exception, are connected by possession of yeasts.   It is no longer possible to consider Diaspididae and Conchaspididae alike, for I have discovered little paired mycetomes with filamentous symbionts in the latter (*Conchaspis lejagei* Hempel, not yet published).

We are therefore inclined to believe that it is not justifiable to assume an "archetypal homopterous symbiosis."   Instead, it should be assumed that the pronouncedly monophyletic symbiosis of aleurodids and psyllids is due to the difference in primary symbionts; that, among aphids, the aphidine and pemphigine symbioses at least possess a common root; that adelgines perhaps acquired their symbionts independently; that the lack of symbionts with phylloxerines may represent a primitive condition or one that arose through symbiont loss; and finally that coccid symbiosis represents another classic example of a symbiosis which is polyphyletic in origin.

Cicada symbiosis embodies a highly complicated type combining monophyletic and polyphyletic traits, and its symbiotic character is blurred by numerous eliminations and substitutions.

If these conceptions are correct, they involve important conclusions regarding the period of the change, for we are convinced that with all these animals carnivorous living and the habit of sucking plant juices provided the impetus for symbiosis. The locations on the family tree where the Heteroptera, coccids, psyllids, aphids, and aleurodids branch off from the common trunk of the bug ancestors must all lie within the Carboniferous, before the origin of the cicadas. In my opinion the change in feeding habit brought about by the symbionts must have set in independently in each of the groups. The change occurred very early with aphids, psyllids, and aleurodids, but for the coccids we come to the conclusion that the symbionts were acquired only after the splitting had reached essentially its present stage. On the other hand, it cannot be assumed that all these families were predatory up to that point and only then acquired the habit of puncturing the sieve-tubes, in other words, that the situation was exactly as it still exists today with the Heteroptera and as it once also existed with the Anoplura. Such an interpretation is cancelled out by all the ecological adaptations to a more or less sessile mode of living, the formation of scales, the production of galls, etc. The dilemma could possibly be resolved by the interpretation that the present feeding habit was preceded by one that required no symbiosis, namely, puncturing and sucking out the cells of the host plant. The lack of symbiotic devices, surprisingly frequent among the coccids, would then be understandable as a tendency to adhere to the old way of feeding.

Naturally it would be desirable to study in all these groups the manner of ingesting food. In contrast to the cicadas, where endosymbiosis is entirely different in structure, the coccids are only slightly inclined to build up their stock of symbionts by acquiring supplementary guests, and replacements of this kind are apparently even rarer.

The insights into the phylogeny of endosymbioses already available today will open up one of the most interesting chapters of symbiosis research. Still in its infancy, with many uncertainties, it deals with central problems that can be clarified only in cooperative work in many fields. It would be wrong to believe that the morphological bases are sufficiently known today to eliminate work in this direction; to furnish the historical foundations will require continued work, with many animals, in morphology and embryology. It will be necessary to draw on modern systematics, which in turn will benefit by confirmations and stimulations derived from these studies of ours. Moreover, historical symbiosis research must keep in constant touch with physiological research on the microorganisms and their relation to feeding habits of the hosts, and we have already indicated how desirable it would be to have the support of the microbiologists, for whom symbiosis research is opening up a world to which they have paid little attention in the past. Only with their help will it be possible to study this diversity of endosymbioses holding the answer to a variety of historical problems.

16

The significance of endosymbiosis

The objective reader of the preceding pages cannot fail to realize that cooperative living of animals with microorganisms, regulated as it is down to the last detail, must play a significant role in the economy of the host. With algae endosymbiosis, the deliberations and experiments of authors were based on this idea from the very outset, even though the relationships between the two partners are rather loose, and the exact research of today has confirmed the intuitive opinions of these older authors.

Thus it is surprising that some scholars were unwilling to accept a similar interpretation for the much more intimate cooperative living of animals with fungi and bacteria. In 1932, for example, Martini interpreted genuine mycetomes and symbiont-filled intestinal evaginations as "animal galls," equating them with the capsules in which nematodes or fly larvae are sealed off, for he had failed to take into consideration that such sites are far more deeply rooted in the organization of the animal than galls and thus in order to be formed, do not require the stimulus of foreign organisms. For Martini, symbiosis merely represented the end stage of an initial parasitism which might occasionally benefit the hosts but was unrelated to their vital needs.

On a similar basis, W. Schwartz (1935) also rejected in the beginning the idea of meaningful choice of symbionts on the part of the hosts; he interpreted them as "harmless parasites" which in some cases were converted into indispensable guests through a "reversal of parasitism." Similar ideas were expressed in various publications that appeared under the leadership of Schwartz, for example, that of W. Müller (1934), who was still interpreting anobiid and cerambycid symbiosis as "moderated hereditary parasitism" even after Koch's experiments with symbiont-free *Sitodrepa* larvae. Ripper (1931) also repudiated the idea that symbionts might be useful, interpreting the specific sites as mere reactions comparable to plant galls, although he must have been well aware of the consequences which such an interpretation in regard to the evaluation of the complicated transmission devices would involve, and Mansour (1934) was in complete agreement with him. However, since

764

the progress of research was not checked by their overcritical viewpoint, it is probably superfluous to give the complete list of those who faced the new situation with scepticism.

It is obvious that their objections would be meaningless if applied to luminous symbiosis, for here it is self-evident that the ecological considerations are important. For example, symbiotic light-producing organs on both sides of the ingestion opening in pyrosomes and those next to the mouth opening or within the interior of *Anomalops, Photoblepharon, Ceratias, Monocentris* and *Leiognathus* represent light traps to lure predatory animals; those of the cuttlefish serve for mutual identification, for pursuit of the opposite sex, and for holding the swarms together; and, where luminescent bacterial groups are expelled from these organs, which indeed are always open, the symbionts camouflage the host or provide him with a scarecrow.

Other endosymbioses must in some way be caused ecologically, for their occurrence is usually linked with unbalanced feeding, even though the idea of usefulness is understandably hard to explain in these cases. Two methods are used by experimenters: they keep the hosts symbiont-free to study deficiency symptoms and the feasibility of compensatory diets, and they obtain pure cultures of the symbionts to study the contributions of the plant partner to the symbiosis.

Over a span of years, various methods were developed to remove symbionts from animals. The easiest to make symbiont-free were forms where larvae are first infected as they emerge from the egg. In this case the egg surface can be sterilized in advance or the egg shell can be removed prematurely. With *Coptosoma*, the bacteria-filled capsules which the insects deposit along with the eggs can be removed. Where egg cells are infected in the maternal body, bactericidal substances, such as antibiotics and sulfonamides, can be fed or injected. In some cases it is sufficient to keep the insects at a temperature which they can just barely tolerate but which is lethal to the symbionts or at least to their transmission forms. In one instance it was possible to centrifuge the earlier developmental stages and thus shift the mycetome to positions where it was withdrawn from the protection of the usual preventive substances. No use has yet been made of ray-treatment.

Aschner (1932) was the first to report on methods for removing symbionts in complicated cases of endosymbiosis. Using human lice, he found it surprisingly easy to remove the stomach disc surgically. He would place on its back a louse that had sucked its fill of blood and position the perforation of a mica platelet directly over the mycetome. When he pricked this taut membrane with a fine needle, part of the myctome was thrust out by pressure within the animal. By seizing this part, it was possible then to remove the whole organ, and the punctured site would soon be closed by a clot.

Aschner published first a brief report on his interesting results and, shortly thereafter, a detailed report in cooperation with Ries (1933). When mycetomes were removed from female animals in the third nymphal stage, thus before symbiots migrate to the ovarian ampullae, the animals were unable to suck blood after a few days. Some tried again and again, sinking their mouth parts into the skin, and making the most violent efforts, but no blood was transferred to the intestine. Other animals were so apathetic and weak that they could scarcely move their limbs. Unable to cling or turn around, they perished one or two days after the appearance of these symptoms.

Symbiont deficiency also affected egg production. Normal animals produced 6 to 7 eggs daily. If eggs were laid at all by the symbiont-free animals, they were laid belatedly, with a maximum of three eggs. Most of the eggs shriveled immediately after deposit, but even those that began to develop were destroyed a few days before the normal hatching period of symbiont-containing progeny. There were always some oocytes which had advanced considerably in growth by the time the symbionts were lost. In such cases it often happened that the animals were too weak to lay the eggs which had begun to develop and these were forced to remain in the vagina.

Also, there were the same hunger symptoms as are sometimes found with symbiont-containing animals. With the latter, a single rectal blood injection enabled the animals to suck again and they regained their original vitality, but with the symbiont-free lice the hunger symptoms even appeared when an abundance of food was offered, and rectal feeding merely postponed death for a few days. That these disability symptoms were occasioned through the loss of symbionts and not by the operation itself was proved by experiments with females 2 to 3 days before their third molting, thus with animals whose symbionts had already left the mycetome and were settled in the ovarian ampullae. In such cases surgical removal of the empty stomach disc had no unfavorable effects.

The techniques described can also be used for partial removal of symbionts. As mentioned earlier, there is no increase in symbionts to compensate for the loss. In proportion to the number of symbionts present, injuries range from those that are scarcely noticeable to others that are as severe as those in animals that are completely symbiont-free, and the parallelism goes so far that the extent of the injuries can actually be used as a criterion for the quantity of symbionts.

Males in the third nymphal stage are much less affected by the total loss of symbionts, although males are generally more sensitive and short-lived than females. Whereas the females which have been operated on live an average of 4.3 days and, in the most favorable cases, 7 days, the symbiont-free males survive an average of 12 days and some-times even 3 weeks. Furthermore, no disturbances of their sexual functions were ever observed. This greater independence is undoubtedly related to the fact that also in normal males it is after the third molting that the symbionts already begin to degenerate or are sometimes completely destroyed.

It is different when the surgery is undertaken earlier. When symbionts are removed during the second developmental stage, the animals survive, male and female alike, but they all suddenly perish just before or after the third molting. When the operation is undertaken in the first larval stage, the animals survive the first molting well, but, incapable of sucking blood, they perish quite regularly on the seventh or eighth day. Partial removal of the symbionts leads here also to injuries of shorter or longer duration, but by the eighth day after surgical intervention the remaining microorganisms degenerate extensively or disappear entirely.

A study of sterile eggs during embryonic development shows the same transitory mycetome and the same sac-like abcission of the intestinal epithelium and its envelopment as in the infected embryos described in the earlier part of the book. At the same time a number of irregularities appear. The chambering of normal mycetomes may be lacking or a smaller number of them may be formed, in which case the place ordinarily taken up by the symbionts is occupied by yolk pieces and remains of the transitory mycetome, both in the process of degenerating. In the zone derived from the intestinal epithelium the nuclei often degenerate, and amitotic constrictions lead to little groups of karyomeres (Fig. 366).

In the same way the sterile lice develop the ampullae, whose only function is to infect the eggs, and these have the usual complex tri-laminate structure and the usual number of mycetocytes, approximately 1000, which vainly await infection. This failure to be infected appears to be connected with the protein granules and lumps which occur abnormally here. When symbionts are only partially removed, the ampullae are correspondingly only partially infected, but the symbionts are of gigantic size and are gnarled, and here they are unable, just as within the mycetome, to compensate for the loss by increased reproduction (Fig. 367). Corresponding to the infection of the ampullae which normally proceeds very symmetrically, partial lack of symbionts has as a consequence also partial colonization of the ampullae.

The only other histological damage occurs in the ovaries. Here the follicle epithelium shows itself to be particularly sensitive. When advanced stages of the egg growth undergo loss of symbionts, the complex egg cover with its sieve-like punctures is but slightly incomplete in

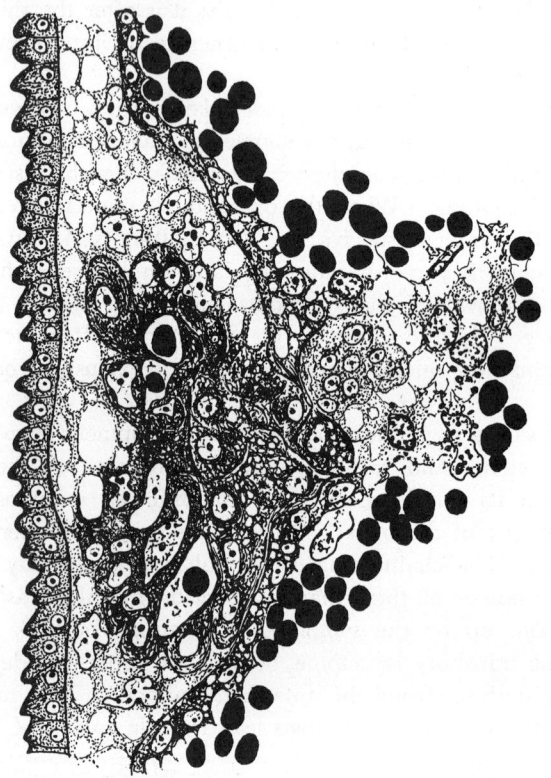

Fig. 366 *Pediculus vestimenti* Leach. Section through the stomach disc of a symbiont-free embryo. The outer syncytium appears to be established normally; in the interior of the empty chambers, which are reduced in number, there can be found yolk clumps and remainders of the transitory mycetome, which also degenerates in the yolk. 1070×. From Aschner and Ries.

its development, while the chambers which normally develop above each individual micropyle become increasingly imperfect or fail to appear at all (Fig. 368). When the loss takes place in younger oocytes, the damage is more severe. The egg then develops much more slowly and usually stops growing entirely before the shells take form. In the follicle, which ordinarily provides nutrients to the eggs, nuclear fragments are constricted off and nuclei dissolve. Disintegrating groups of follicle cells transfer to the egg, where the shell may occasionally still be indicated as a thin border, and large vacuoles in the egg plasma point to dissociation processes.

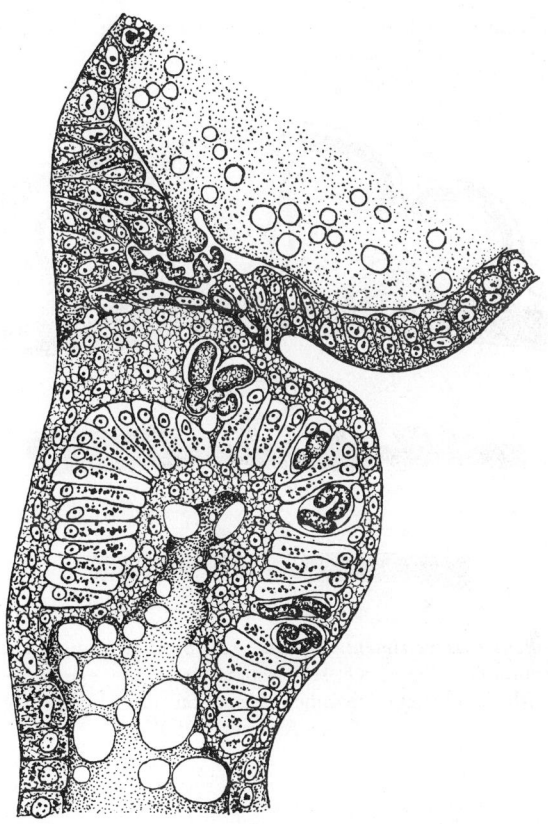

Fig. 367 *Pediculus vestimenti* Leach. Section through the ovarian ampulla of an animal which has been extensively robbed of its symbionts; the sterile, but nonetheless formed, mycetocytes contain the otherwise lacking plasma clumps; the symbionts are atypically formed. From Aschner and Ries.

Degeneration may also set in before the yolk and shell take form. If the animals surivive at all in such cases, the disintegration products of the follicle and of the young oocytes are distributed over the body cavity. In the process the nutrient cells always remain undamaged.

Aschner later (1934) removed symbionts from lice by centrifuging the embryos for about 8 hours (1500–2000 revolutions per minute) between the second and fifth day of their development, and the mycetomes were shifted in 5–10% of the larvae. The stomach disc could then be found dorsally, laterally, or near the anus, or only part of it might be displaced. The mycetomes remaining in position housed their symbionts in the usual manner, but the dislodged ones turned

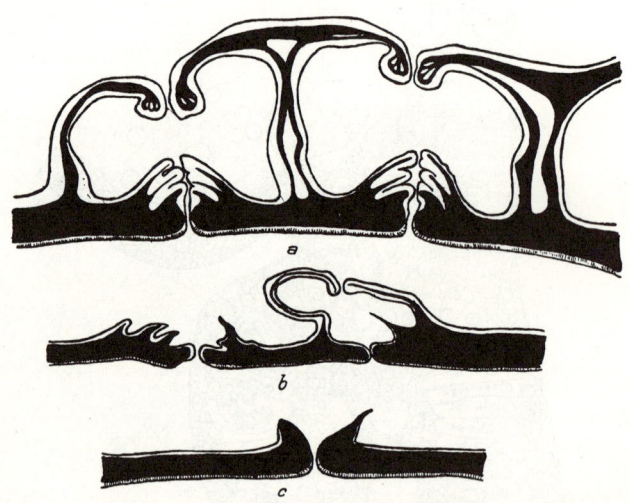

Fig. 368 *Pediculus vestimenti* Leach. (*a*) Section from a normal egg cap of a symbiont-containing louse; (*b, c*) sections from the egg caps of symbiont-free animals, with insufficient chamber formation or without the same. 1070×. From Aschner and Ries.

into massive cell masses without characteristic chambering and without bacteria, as already indicated, because only in stomach discs which are regularly constructed and situated are the symbionts protected from the lysins of the host.

The effects of centrifuging were the same as with surgical removal. Animals originating from centrifuged embryos sucked and developed as normally as the uncentrifuged ones, but those that were symbiont-free suddenly perished on the fifth or sixth day after hatching. Some died before molting, others during, but only a few thereafter. With partial sterilization the animals lived a longer time but always died earlier than normal ones.

Puchta later (1954, 1955) improved the centrifuging method to the extent that more than 50% of the animals were symbiont-free. Like Aschner, he was also able to confirm that larvae whose mycetomes had been shifted, i.e., with few symbionts or without the same, die, but he went a step further and attempted to bring them to complete development with artificial diets of differing per cent content. Avoiding the less inefficient method of rectal feeding, he fed the animals through a membrane made of specially prepared skin from very young white mice. The animals were given blood with yeast extracts (*Torula*) or defibrinized, hemolized blood containing growth factors. The latter method turned out to be the most effective, for, when the blood and yeast extracts were fed, a dangerous overdosing usually took place, with a consequent increase in mortality, making it necessary to distribute the extract needed for development over several feedings. However, the corresponding amounts of the pure vitamins could be given at the same time. It was now possible for the first time to analyze the effect of several growth substances on animals living exclusively on vertebrate blood. Omission of thiamin was almost without unfavorable results. Without riboflavin or pteroylglutamic acid, symbiont-poor animals were still able to develop. Without pyridoxin, the symbiont-free animals could achieve the third nymphal stage. Most serious was the lack of nicotinic acid, pantothenic acid, or beta-biotin; all symbiont-free or symbiont-poor offspring died during the second developmental phase at the latest. Again and again, Puchta found that the disturbance of the proportion of the various vitamins was fateful.

Baudisch (1958) further improved the centrifugation method, but he made use of it only to gain insight, in the presence of the thus disturbed embryonic development, into the normal process. Nor did Bewig and Schwartz (1956), who attempted to cultivate *Haematopinus* symbionts and studied their changes in form in non-isotonic solutions, go into the problem which is our chief interest here.

The research of Aschner and Puchta clearly proved that the significance of cooperative living, at least with *Pediculus*, rests in the availability of active substances which are indispensable to the host but are not found in the required quantity in a diet consisting exclusively of vertebrate blood. Thus the early hypotheses of Wigglesworth and Aschner in 1929 were confirmed. Apparently, the quantity of growth substances provided by the normal symbiont supply coincides precisely with the requirement of the host, for a numercial reduction of the symbionts results in noticeable deficiency. On the other hand, it is clear that the animals can store these substances, and that the stored supplies last longer with males than with females, whose requirements are greatly increased by egg production. Our present understanding of the role which symbionts play in nymphal and ovarial development also enables us to understand why *Pediculus* mycetomes are largest in the early developmental phases in proportion to body size and why the symbionts degenerate more or less extensively in growing males.

With the bug *Rhodnius,* another insect living exclusively on vertebrate blood, Brecher and Wigglesworth (1944) had results that tallied nicely with Aschner's findings. In this bug they were dealing with a form whose symbionts populate the intestinal lumen and a limited section of the intestinal epithelium and are given externally from the anus to the sterile eggs at the moment of their deposit. The surface of the eggs can easily be disinfected with a solution of gentian violet in distilled water, and nothing more is required than to keep the animals sterile after they are hatched.

Earlier morphological studies of *Rhodnius* by Wigglesworth showed that it undergoes five developmental stages and that the intervals between two moltings amount to 10 days at first but increase to 15 days up to the fourth molting. The fifth phase required 20 days for maturing to the imaginal phase. Up to the fourth phase, the loss of symbionts had little effect on the sterile animals but individual moltings were somewhat delayed at times. After the fourth molting, there was considerable change for the worse. Numerous animals failed to complete their development, and those that did required very long periods of time.

One example may serve to explain the matter. In one experiment eleven nymphs attained the penultimate stage. They sucked on March 11 but did not molt. Another ingestion of food on May 5 was also without result; indeed, one of the animals died. They sucked once more on August 5, and five specimens finally reached the imaginal phase, but only after an additional 42 days. The five remaining animals sucked blood on November 20, January 26, and on April 21 of the following year. Thus they had spent more than a year in the nymphal phase and at last completely lost the capacity to develop. In comparison, the normal period of development is slightly more than 2 months, and only a single ingestion of food is necessary between moltings.

The unfavorable effects of symbiont loss do not stop here. The few female imagines still alive at the end were unable to produce eggs and reproduce. A sterile pair mated at the end of November and sucked blood four times, but produced no eggs even 4 months after the last molting, whereas symbiont-containing females laid eggs after 10 days even without the presence of males. When the sterile male was replaced by an infected one, eggs were deposited again after several weeks, and now even the female was discovered to be a symbiont bearer, owing to the habit that these animals have of sucking the excrement of their fellows. In another case, where no eggs had been laid for 3 months although the animal had ingested food repeatedly, a section showed that no ovaries had developed.

Just as infection of sterile imagines may cause a delayed development of the ovaries, so also may the infection of younger sterile stages make possible normal progress of further moltings. Sometimes the transmission even in ordinary cultures does not function, and fifth stages are unable to molt, but even in such cases it is possible to obtain imagines capable of reproducing by adding symbiont-containing excrement to the food.

Brecher and Wigglesworth interpreted the pathological phenomena as typical symptoms of avitaminosis and thought it quite possible that it was a matter of vitamins of the B-group, which, furnished in this instance by symbiotic actinomycetes, are an indispensable requirement for producing molting hormones and for developing the ovaries. Inasmuch as development, although delayed, occurs up to the fifth phase, Brecher and Wigglesworth believe that this might be due to vitamin reserves in the egg or to small amounts of the necessary active ingredients in the blood.

Lwoff and Nicolle (1944, 1946, 1947) studied the nutrio-physiology of the triatomids, using *Triatoma infestans* as their test animal. They fed nymphs through a thin membrane, offering warmed horse serum and glucose to 100 animals, and the same mixture supplemented by ascorbic acid, nicotinamid, hemin, thiamin, riboflavin, and pantothenic acid to another 100. Only 32 animals fed with the simple mixture reached the third phase, and two the fourth, but the latter perished without further ingestion of food. Eighty-five of the animals fed with the supplemented mixture reached the third phase, although many animals died later, perhaps from bacterial infection, and four became imagines, capable of copulating and reproducing. When pantothenic acid was omitted, several animals developed into imagines but failed to produce fertile eggs. When hemin was omitted, the animals died before the last molting. Unfortunately, the problem of the utility of symbiosis was no closer to solution through these experiments. At least, it would be premature to conclude that the symbionts were incapable of rendering their hosts independent of the vitamin content of the food. The authors neglected to compare symbiont-containing and symbiont-free animals with regard to their requirements, and, on the basis of certain findings by Brecher and Wigglesworth (1944), it is even possible that Lwoff and Nicolle were working with sterile or at least insufficiently infected animals. Indeed, these authors found that cultures of *Triatoma rubrofas-*

ciata and *Eutriatoma flavida* and *E. sordida* often perished because the nymphs were no longer able to molt. Such animals were then found to be symbiont-free, because each time the eggs had been transferred to fresh, clean tubes, with the result that the animals had no opportunity to compensate for the frequently insufficient smearing of eggs by sucking the excrement of their infected fellows.

Following Brecher and Wigglesworth, and Lwoff and Nicolle, a number of other authors studied the symbiosis of triatomids (Geigy, Halff, and Kocher, 1953, 1954; Goodchild, 1955; Halff, 1956; Bewig and Schwartz, 1954, 1956; Baines, 1956; Harington, 1960; Gumpert and Schwartz, 1962, 1963; Gumpert, 1962). They all agree with their predecessors that loss of symbionts inhibits development. Halff obtained sterile animals not only by disinfecting the upper surface of the egg and safeguarding the animals against reinfection from the excrement of their fellows, but also by feeding with terramycin. However, considerable disagreement still prevails with regard to the contribution of the symbionts. Studying the performance of symbionts in pure cultures, Halff showed that the microorganism cultivated from *Triatoma infestans* is heterotrophic for riboflavin, nicotinic acid, pantothenic acid, biotin, and pyridoxin, whereas it does not require folic acid, aneurin, *p*-aminobenzoic acid, inositol, choline chloride, B_{12}, and streptogenin. Insofar as quantitative investigations were undertaken, Halff found that only folic acid is produced in excess, and he concluded that it represents the decisive metabolic product furnished by the symbiont. Again the importance of the optimum dosage was seen, for concentrations which were too high inhibited the growth.

Bewig and Schwartz, unconvinced by Halff's finding, still regard the question of vitamin provision as unclarified for the triatomids. In *Triatoma infestans* they found, besides the usual rodlet-shaped intracellular symbionts and the cocci-shaped symbionts of the intestinal lumen, only a *Streptococcus* species without significance for its host. They also found that the symbionts of *Triatoma* and *Rhodnius* could be exchanged without influencing the host. Unlike Halff and Goodchild, they consider the symbionts of the two genera to be identical. Using a new method that had certain limitations, Baines allowed bugs to suck at mice and guinea pigs that had been infected intravenously with various vitamin solutions directly before ingestion of food. In this way he came to the conclusion that *Rhodnius* is dependent upon the symbionts for pyridoxin, Ca-pantothenate, nicotinamid, thiamin, and one other undefinable factor, and that biotin and folic acid are present in sufficient quantity in the blood of the two vertebrates.

Gumpert and Schwartz also studied the question of supplying growth substances to *Rhodnius* and *Triatoma* by injecting eleven different B-vita-mins and amino acids, individually or in combination, into the hemolymph or stomach of the two species. Unlike Halff and Baines, they found that a mere watery solution of Ca-pantothenate enables sterile animals to develop to the imaginal stage and that the furnishing of pantothenic acid represents the basic principle of triatomid symbiosis. In harmony with this finding is the fact that considerable quantities of pantothenic acid could be demonstrated in filtrates of the symbiont cultures. On the other hand, the injection of folic acid, biotin, thiamin, lactoflavin, and vitamin B_{12} into sterile larvae was without effect.

As already mentioned, triatomid symbiosis has received a new, sur-prising aspect through the symbiont exchange experiments first begun by Bewig and Schwartz and then more fully developed by Gumpert and Schwartz. When symbiont-free *Triatoma* larvae were infected with 62 types of microorganisms, it was found that, besides the original symbi-onts and those of *Rhodnius*, nine additional forms were able to eliminate the disability symptoms. This is a result which not only underlines the primitiveness of the triatomid symbiosis, which is also expressed in the mode of transmission, but which also helps to explain the contra-dictory results of different investigators.

It is justifiable to draw general conclusions from these experiments with lice, *Rhodnius*, and *Triatoma* on the insufficiency of a diet of blood alone, for they are confirmed by other experiments with non-symbiotic insects kept on similar diets. By feeding larvae of *Stegomyia fasciata* (*Aedes aegypti*) nothing except sterilely removed rabbit's blood, Aschner (1931) could keep them alive more than a month at 25°C., although their development, which with normal diet takes 6 to 7 days at this temperature, was arrested. Only by adding bacteria or bacterial filtrates was it possible to cancel the retardation in growth which resulted from the diet.

The same applies to larvae of *Lucilia sericata*, normally living on meat which is broken up, usually mechanically, and absorbed in the liquid state brought about by the proteolytic enzymes contained in their excreta. Hobson (1932, 1933, 1935) could bring the animals to develop-ment on blood diets only if he added *Bacterium coli* or pure cultures of various bacteria cultivated from the gut of normally living larvae, or if he added yeast extracts; in pure sterile blood the larvae still had their initial weight, 0.05 mg., 5 days after hatching, but attained a weight of 50 mg. in the same time when yeast extracts were added. It is interesting to note here that the addition of cultures of *Rhodnius* symbionts (*Actinomyces*) also ensured the development of *Lucilia* larvae (Wigglesworth, 1936). On further analysis Hobsen found that at least three substances present in the yeast had to be added to the blood diet offered to *Lucilia:* one factor which is soluble in water, hard to dissolve in alcohol, and insoluble in ether; and two other heat-resistant factors, one of which he was inclined to identify as thiamin.

Such findings harmonize well with the circumstance that it is precisely the bloodsucking animals which contain symbionts that feed exclusively on sterile blood throughout life, in other words, the symbiont-containing hirudines, the gamasids, ixodids, Anoplura, cimicids, triatomids, glossines, hippoboscids, and streblids.   On the other hand, the search for endo-symbioses has always been futile among the forms which as larvae feed on diets rich in microorganisms and even as imagines often carry micro-organisms of many types in their intestine, that is, culicids, tabanids, Aphaniptera, stomoxyines, and phlebotomines.

Numerous experiments, especially with culicid larvae, have indeed proved that the manifold bacteria, algae, and fungi represent a vital constituent of their food and thus eliminate the necessity of symbiosis. Roubaud (1919) showed that *Stegomyia* larvae are unable to develop in a germ-free medium and that the addition of bacteria or bacterial filtrates was needed to bring the arrested development into motion again. Barber (1927, 1928) found that a diet consisting exclusively of algae, bacteria, or infusoria is sufficient for *Anopheles, Culex,* and *Aedes* larvae but that a combination of bacteria with one of the two other components is required to create optimal conditions of growth.   Hinman (1930, 1933) and Aschner (1931) likewise came to the conclusion that the bacteria must contain a factor stimulating the growth of culicid larvae.   Roze-boom (1935), studying the merit of various bacteria cultures, established that strains occurring in the natural surroundings of the larvae guarantee the best development, with other species representing more or less ac-ceptable substitutes.

Trager (1935), cultivating *Aedes* larvae in sterile liver extract with supplements of autoclaved yeasts, made a significant advance in showing that the yeasts contain an important heat-resistant factor, easily soluble in water or in alcohol that is not too strong, which, together with a substance abundant in the liver, is indispensable to the growth of the larvae.   Later, working at times with Subbarow (Trager and Sub-barow, 1938; Subbarow and Trager, 1940; Trager, 1942), he studied the dietary requirements of *Aedes* larvae much more precisely, finding that the microorganisms can be replaced by supplements of riboflavin, pantothenic acid, thiamin, pyridoxin, and nicotinic acid; this result was confirmed in its entirety by Golberg, De Meillon, and Lavoipierre (1945). The latter authors also proved that folic acid is another essential factor, dispensable up to the fourth larval stage, although pupation cannot oc-cur without it.   No other vitamins of the B-group are significant except biotin, and its influence is only stimulating at best. Lichtenstein (1948) and Trager (1948) broadened our knowledge of the requirements which growing *Aedes* larvae make on sterile dicts, and Mudrow Reichenow later (1951) confirmed their findings. Thus we are well informed in this respect.

In principle, the findings with cuclicid larvae may be transferred to tabanids, stomoxycines, Aphaniptera, etc., although very little research has been done on them. Faasch reported (1935) that it was possible to bring the larvae of some flea species to development only by giving them parental excrement and thus the necessary bacteria. Although he also reports that he regularly found short rodlets in *Hystrichopsylla talpae,* in the lumen, on the intestinal wall, and in the rectal bladder, and although he actually applied the term symbionts to them, his forms can at the very most be interpreted as a preliminary stage of such rodlets. Sharif (1937, 1948) made culture experiments with flea larvae (*Nosopsyllus* and *Xenopsylla*) and found that they also were unable to develop on a pure blood diet under aseptic conditions, but that development was possible when yeast was added.

These experiments prove beyond doubt that the deficiency of vital substances in the blood, chiefly B-vitamins, has led to symbiosis in blood-suckers, but it still remains urgently necessary to analyze in greater detail the achievements of the symbionts, for they obviously vary from case to case. It may be surprising that bloodsuckers should have a deficiency of B-vitamins, for we know that mammal and human blood contains thiamin, riboflavin, nicotinic acid, pantothenic acid, pyridoxin, folic acid, inositol, and cholin. However, exact research has shown that bloodsuckers have an insufficient supply of some of these B-vitamins. This seems evident from the mere fact that the content of thiamin, riboflavin, pantothenic acid, and pyridoxin is substantially below the amount required by symbiont-free *Tenebrio,* whereas the content of nicotinic acid substantially exceeds the amount required (cf. Table 2).

TABLE 2 *Vitamin content of ape blood and wheat, compared with the requirement in Tenebrio (according to Fraenkel and others)*

	µg/100 ml fresh blood	µg/g dry substance: blood	µg/g dry substance: wheat	µg/g dry diet requirement with *Tenebrio*
Thiamin	8.9	0.5	4	1
Riboflavin	21.2	1.1	1.1	2–8
Pantothenic acid	30	1.6	12	8
Pyridoxin	11	0.6	3.7	1
Nicotinic acid	400–700	21–37	48	16

De Meillon and his co-workers (De Meillon and Golberg, 1946, 1947; De Meillon, Thorp and Hardy, 1947), who provided the only investigations with information of this kind, confirm the hypothesis that the quantities present in the blood often fail to coincide with the needs of the parasites.

Testing the vitamin needs of *Cimex lectularius* and *Ornithodorus moubata*, these authors used a completely new method in allowing their test animals to suck at rats whose blood lacked certain B-vitamins as the result of a suitable deficiency diet.[1] Thus they tested the effect that lack of thiamin, folic acid, and riboflavin had on the bedbug primarily and on the tick secondarily. The growth rate of *Cimex* was not influenced at all by the absence of thiamin, but egg deposit seemed to be badly disturbed. The number of eggs deposited was far less that under normal conditions; a considerable percentage were sterile or at least incapable of growth, and normal larvae were hatched only now and then. On the other hand, the number of eggs capable of development rose substantially when such bug pairlets were allowed to suck on normal rats. It is possible that the damage represents the effect of toxic substances formed in the blood of thiamin-free rats, but the authors are more inclined to assume that the symbionts do not supply the vitamins in a quantity sufficient to meet the increased needs of the bugs during egg formation and that the animals are dependent in this period upon additional thiamin from the rat blood.

Ornithodorus reacts even more violently to lack of thiamin in the blood of host animals. The growth rate is reduced and the animals do not attain the size of those fed on high-quality blood. Since the experiments were broken off after the third larval stage was attained, nothing can be said at present about influence on reproductive capacity. In the case of *Cimex,* experiments with lack of folic acid in rat blood provided results similar to those with lack of thiamin. Again, the growth rate was uninfluenced, but the number of eggs was reduced.

The results are even clearer when bedbugs and ticks develop on rats which lack riboflavin. No unfavorable effect was observed on growth, egg production, and capability of the eggs to develop. Nor could injury of any kind be observed in a second generation of bedbugs developing from such eggs and likewise fed with riboflavin-free blood.

De Meillon, Thorp, and Hardy compared the riboflavin content of adult bugs which were fed rat blood without riboflavin with that of

[1] De Meillon and Golberg (1947) also tried to remove bedbug symbionts by injecting penicillin into the body cavity or administering it indirectly in the blood of guinea pigs. However, it could not be observed that either symbionts or their hosts were influenced.

animals fed normal blood, and they found that it was exactly the same in both cases, 0.020–0.024 μg. The lack of dependence of the riboflavin supply on the vitamin content of the blood is shown even more clearly by comparing the quantity sucked with the blood throughout life with the total amount of riboflavin required by an individual bug. From the first larval stage to sexual maturity, it needs an average of 20–25 mg. of blood. Since the riboflavin content of normal rat blood has been estimated at 0.2 μg. per gram of blood, 0.005 μg. of riboflavin would be the amount needed, in other words, a quantity which would be much smaller than the content found in the grown animal —0.020–0.024 μg.—even if one were inclined to believe that all the riboflavin can be stored. The same applies to human blood, in which 0.21 μg. of riboflavin per milliliter was found, again a quantity insufficient to cover the requirement of the bedbug.

Thus one must inevitably agree with the authors that in all probability the symbiotic bacteria furnish *Cimex* and *Ornithodorus* with the required quantity of the indispensable vitamins.

Finally, the hemolyzing qualities of the symbionts play a role in some bloodsuckers, usually a secondary one it is true. Roubaud (1919), Reichenow (1922), and I (1922) supported this assumption, but Wigglesworth rejected it in earlier studies (1929, 1936) and also in a later one (1952). With the Glossinae, Roubaud found that quantities of symbionts also occur free in the intestinal lumen, that the blood is thickened only before it reaches the infected section, and that hemolysis suddenly sets in when the infected section is reached. According to Wigglesworth, on the other hand, the blood is thickened in Glossinae by a loss of fluid in one of the first sections of the midgut and, in a second section preceding the zone containing the symbionts, the blood blackens on contact with the epithelial cells and hematin is deposited in quantity. Zacharias (1928) found that in *Melophagus ovinus* and *Lipoptena cervi* the blood was always undigested before the infected intestinal section, and he too was of the opinion that it was the contact with the infected section that dissolved and digested the blood. Aschner (1931) made identical observations with *Lipoptena caprina* and *Lynchia maura*. With *Hippobosca,* where the symbionts occur without sharp limits, hemolysis was not so strictly localized. With the leech *Placobdella,* Reichenow (1922) proved that the symbionts migrate from the cells of the gut appendages to the lumen and that the red blood corpuscles then begin to disintegrate in their immediate vicinity. With *Piscicola* and *Branchellion* the blood corpuscles only occasionally reaching the diverticula occupied by the symbionts disintegrate there, according to Jaschke (1933). The cultures of presumably symbiotic bacteria from the gut of *Hirudo medicinalis* (Hornbostel, 1941) clearly demonstrated hemolytic capacities on blood-agar plates, and the more penetrating studies of Büsing, Döll, and Freytag (1953) established a similar achievement on the part of the symbionts. *Pseudomonas hirudinis*, which the authors rightly consider to be the genuine symbiont of *Hirudo medicinalis*, revealed the same capacity on blood agar, and by using chloromycetin it was also possible to exclude the intestinal flora completely and thus prevent the possibility of blood feeding. The well-known fact that other germs in the *Hirudo* gut are destroyed after a while has its basis in the antibiotic capacities of the symbionts described by some authors. Moreover, cultures obtained by Faasch (1935) from Aphaniptera showed that blood-digesting bacteria appear also in the intestinal flora of insects which do not live in intimate symbiosis.

Weurman (1946) also accepted the idea that symbiotic actinomycetes in *Triatoma infestans* play a role in the blood digestion, because a verdohemochromogen appears in blood-agar cultures of the symbionts, but in spite of this new evidence Wigglesworth (1952) still maintained his skeptical attitude, pointing out that the *Triatoma* symbionts are housed in a section of the gut situated far to the front; that a much enlarged "stomach" is adjoined in which the blood remains in a thickened state for 2 to 3 weeks without showing much evidence of digestion, although it is then abundantly permeated by symbiont colonies; and that it is only in the following third midgut section that the digestion sets in rapidly. Bewig and Schwartz (1956) rejected this explanation for the Triatomidae.

No one will deny that Arthropoda are able to digest vertebrate blood without the aid of microorganisms. When Aschner fed blood under sterile conditions to freshly hatched females of *Stegomyia* and *Phlebotomus* after copulation, he found that it was normally digested and that egg deposit was not influenced in any way, although a later check showed the gut to be germ-free. Wigglesworth correctly pointed out that *Rhodnius*, when symbiont-free, digests blood as well as when it contains symbionts. The same applies to males of the nycteribiid *Eucampsipoda*, which are lacking in symbionts after metamorphosis, and to *Pediculus* males, in which symbionts finally degenerate more or less extensively. But, in spite of all this, there are enough authentic observations to indicate that in certain cases the symbionts of bloodsucking animals, localized in or on the gut, have an influence on dissolution of the blood.

We turn now to insects that feed on cellulose-rich diets, or that are descended from animals with such diets but later adapted to a diet richer in carbohydrate and therefore often acquiring importance as storage pests.   The first publications, and still the most important, studied the two representatives of the Anobiinae, *Sitodrepa panicea* and *Lasioderma serricorne*.   It seemed possible to use again superficial sterilization in this case, for these beetles transmit their symbiotic yeasts to eggs being deposited in the same way that *Rhodnius* and *Triatoma* do their actino-mycetes, except that the beetles have more complicated lubricating de-vices.   The first to carry out such experiments successfully, long before Brecher and Wigglesworth, was Koch (1933, 1934).   He had two methods.   By using a 5% solution of chloramin in 70% alcohol, he killed the yeasts in 2 minutes without influencing the development of the egg, or he safeguarded the larvae from infection by prematurely

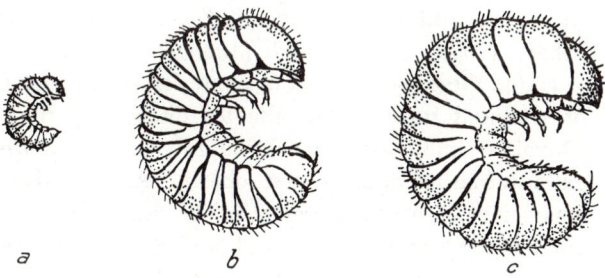

Fig. 369   Effect of symbiont removal in *Sitodrepa panicea* L.   (*a*) Ten-week-old larva without symbionts kept sterile on pea-meal; (*b*) larva of the same age without symbionts kept sterile in pea-meal and dried yeast; (*c*) symbiont-containing larva of the same age as control.   From Koch.

removing them, under a dissecting microscope, as they were about to hatch from the shell.   Just as sterile lice still developed the stomach disc, these symbiont-free larvae then developed the blind sacs that normally house the symbionts at the beginning of the midgut, yet these understandably remained small and smooth-walled (figure in Koch, 1938).

Such symbiont-free larvae of *Sitodrepa* were also greatly retarded in growth.   Although they were fed on pea-meal with many supplements, a diet which is ordinarily very acceptable to the animals, they remained small, they weakened and finally perished, although the symbiont-contain-ing control group developed normally on the same diet.   However, when beer yeasts, yeast extracts, or even wheat germ was added to the usual food, the retardation in growth was cancelled immediately and symbiont-free imagines were developed (Fig. 369).

From the very start, everything indicated that in this case too the *Sitodrepa* symbionts furnish their hosts with the same growth-promoting vitamins which are represented so numerously in the yeast and thus make it possible for the animals to exist on food in which the vitamins are lacking or are not present in sufficient quantity. Exact analysis of the growth substances furnished by the microorganisms and the amounts required by the animals had just begun when Koch was making his decisive observations; the publications of van't Hoogs (1935, 1936) on vitamins required by *Drosophila*, which appeared soon after Koch's experiments, represented the first exact attempt in a field where the data were still inexact and contradictory.

Recognizing that one of the most urgent postulates of symbiosis research, at that time in its infancy, was to analyze the vitamins especially abundant in yeasts and other factors indispensable to growth, Koch induced Offhaus (1939) and Fröbrich (1939) to work in this direction. Great merit is due these two authors not only for advancing substantially the state of symbiont research but also for discovering that *Tribolium confusum* Duval, a symbiont-free tenebrionid living like *Sitodrepa,* represents an extraordinarily valuable test object. All effective substances necessary for its development are present in the yeast. The larvae pupate at a constant temperature (31.5°C.) and 50–70% atmospheric moisture, when fed an optimal diet, after approximately 20 days, but, when even a single one of the numerous components is lacking, the larvae begin to develop more slowly or even stop developing and die. Thus a preparation of known or unknown composition can be tested for efficacy by adding it to a calorically sufficient deficiency diet.

Reichstein and his co-workers, making use of the new test, confirmed the results of Offhaus and Fröbrich in all points and even extended them in several respects. Offhaus and two other co-workers of Koch, I. Schwarz and Reitinger, continued these efforts to clarify the effective substances of the yeasts and completed them to the extent that was possible with use of the *Tribolium* test (Offhaus, 1952; Koch, 1951; Koch, Offhaus, Schwarz, and Bandier, 1951; Reitinger, 1952; Fröbrich, and Offhaus, 1952, 1953; Fröbrich, 1953, 1954). It became apparent that the factors vital to *Tribolium* which are destroyed during exhaustive extraction of the beer yeasts with boiling water—thiamin (aneurin), riboflavin (lactoflavin), pyridoxin, nicotinic acid amide, pantothenic acid, beta-biotin, folic acid (pteroylglutaminic acid), and choline chloride—are not sufficient in themselves to achieve the best possible development even when added in the optimum proportion to the deficiency diet of *Tribolium* and that, to obtain optimum development, it is always necessary in the beginning to add a certain quantity of the yeast residue to the deficiency diet. The indispensable components, according to Offhaus and Fröbrich, are a sterine (ergosterine), and a biogenic amine (carnitine). Only when these substances are added to the B-vitamins is *Tribolium* afforded a complete replacement for yeasts.

This *Tribolium* test has been used continuously for many years by I. Schwarz, co-worker of Koch, to analyze the growth-producing elements of symbionts and of a great variety of foods. In the following pages we shall often discuss her results, which, even when they were not yet published, she unselfishly made available to many of Koch's students.

For a time the war prevented Koch and his co-workers from broadening their experiments to include *Sitodrepa,* but meanwhile Blewett and Fraenkel continued them. As symbiont-free comparative test animals, they used chiefly *Tribolium* and also *Ptinus* and *Ephestia* to determine the growth substances required by the Anobiinae. Fortunately, they also used *Lasioderma serricorne,* a form which lives in tobacco supplies and several other commercial articles and belongs to a genus which includes species mining in fresh plant tissue.

When Fraenkel and Blewett (1943) first compared symbiont-containing *Sitodrepa* and *Lasioderma* with the animals mentioned above, they naturally found that the groups differed greatly in the amounts of the B-vitamins required. The nutrients indispensable for forms which were symbiont-free by nature were thiamin, riboflavin, nicotinic acid, pyridoxin, and pantothenic acid; *Lasioderma* grew well without these substances; and *Sitodrepa* needed only thiamin.

Later these two authors compared normal *Sitodrepa* and *Lasioderma* with others that had been made symbiont-free by Koch's method (Blewett and Fraenkel, 1944). First, they appropriately replaced the pea-meal preparation with wheat flour of various degrees of extraction. Although 100% flour did not completely replace the substances usually furnished by the symbionts, the growth intensity of symbiont-free and symbiont-containing animals was approximately the same on this diet. With 75% flour, the development of *Lasioderma* was substantially retarded without yeasts, and with 40% flour it was once more substantially retarded; *Sitodrepa* was even more sensitive in that symbiont-free animals did not develop at all with 40% flour, and with 75% flour only 2 of 20 larvae underwent metamorphosis, and only after a long delay. With the addition of 10% dried yeast to the diet, the findings tallied with those of Koch, for the acceleration in growth, with sterile and with infected animals, increased in proportion to the degree of extraction of the wheat flour, and the absence of symbionts had little effect.

To prove that the deficiency phenomena were actually caused by lack of B-vitamins, Blewett and Fraenkel gave their two test objects, both with and without symbionts, an artificial diet of casein, glucose, cholesterin, salt, and insoluble yeast residue, adding all B-factors to be tested or occasionally omitting one of them. Table 3, although

TABLE 3   *Growth of larvae of Lasioderma and Sitodrepa with and without symbionts, compared with that of Tribolium and Ptinus on full diet (artificial diet with eight B-vitamins) and with absence of one of the tested vitamins (from Blewett and Fraenkel, 1944, and Pant and Fraenkel, 1950, combined)*

	Lasioderma		*Sitodrepa*		*Tribolium*	*Ptinus*
	With sym-bionts	Without sym-bionts	With sym-bionts	Without sym-bionts	Without symbionts	Without sym-bionts
Full diet	++++	+++	++++	++++	++++	++++
No thiamin	++	++	−	−	+	±
No riboflavin	++++	−	+++	−	±	−
No nicotinic acid	++++	−	++++	−	−	−
No pyridoxin	++++	−	+++	−	++	+
No pantothenic acid	+++	−	++	−	+	±
No choline chloride	++	−	+++	±	+++	++
No biotin	++	++	++	−		
No folic acid	+++	+	++	−		
No mesoinositol	++++	+++	++++	++++	+++	+++
No *p*-aminobenzoic acid	++++	+++	++++	++++	+++ (+)	+++

The number of +'s corresponds to the degree of growth; ± indicates highly unfavorable growth conditions and only occasional metamorphosis of one or the other individual; + indicates high mortality and slow growth; − indicates lack of any growth possibility.

it shows a certain variability in their findings, demonstrates with great clarity that symbiont-containing animals are widely independent of the vitamin content of food but, robbed of their symbionts, are as dependent upon it as are *Tribolium* and *Ptinus*, which are symbiont-free by nature. Interestingly, it was demonstrated again that there are differences of various kinds between the two Anobiinae: *Lasioderma* can easily dispense with riboflavin, nicotinic acid, pyridoxin, and pantothenic acid and, to a certain extent, thiamin; *Sitodrepa* symbionts do not replace thiamin; and the growth of *Sitodrepa* is somewhat more influenced than that of *Lasioderma* by lack of riboflavin, pyridoxin, or pantothenic acid. The situation resembles that with biotin and folic acid, for without them *Sitodrepa* cannot develop at all, whereas it is always possible for *Lasioderma* larvae to develop at least in part (Pant and Fraenkel, 1950). On the other hand, a deficiency in choline, which makes itself felt more or less even in symbiotic animals, is tolerated somewhat better by sterile *Sitodrepa* than by sterile *Lasioderma*. Blewett and Fraenkel found that the symbionts of these two anobiines also furnish sterine, which many insects must find in their diet. When sterine is lacking in the food, the results are barely observable in normal *Lasioderma*, though slightly more so in *Sitodrepa;* when the symbionts are lost, the lack of it is far more crucial. Dispensable under all circumstances are *p*-aminobenzoic acid and mesoinositol.

When Kühlwein and Jurzitza, using cultures of *Sitodrepa* symbionts, which they place in the neighborhood of the Taphrinaceae and therefore designate as *Symbiotaphrina buchnerii* (Graebner), tested their growth, they too established that these organisms are heterotrophic for biotin and thiamin. It is interesting to note that they are partly heterotrophic for aspartic acid and glutamic acid, and in view of this the authors consider it possible that the originally saprophytic-living organisms might have lost capabilities which have become superfluous during long intracellular life and have become so dependent on the host that they cannot live outside of it and therefore their artificial culture is not possible.

Pant and Fraenkel (1950, 1954) confirmed that *Lasioderma* symbionts are superior to those of *Sitodrepa* by interchanging them; this was achieved by smearing the sterilized egg surface with the symbionts of the other genus grown in pure culture or by feeding the foreign cultures to sterile larvae. It was found that the two morphologically distinct types were immediately accepted by animals of the opposite genus and were handled in the ordinary way, without loss of their specific characteristics in the new milieu. Comparison of growth possibilities clearly indicated that lack of thiamin is tolerated less well by *Lasioderma* with *Sitodrepa* yeasts. *Sitodrepa* larvae must ordinarily find the thiamin in their food, but, when furnished with *Lasioderma* yeasts (*Symbiotaphrina kochii*), a goodly number of them achieved pupation age, even though delayed. Thus the *Lasioderma* yeasts apparently have greater synthesizing abilities than those of *Sitodrepa*. Kühlwein and Jurzitza (1961) have succeeded in replacing *Sitodrepa* symbionts with those of *Ernobius mollis*, which also were taken up only in the blind pouches. These authors have not reported whether the developmental time is favorably or unfavorably influenced by these symbionts. Recently Jurzitza also reported on symbiont exchange with *Lasioderma* and *Sitodrepa* (1964).

In view of the accomplishments of the symbionts in other closely related hosts, we must nevertheless beware of premature generalizations; this concept is supported by the experiments of Graebner (1954), who found great variation in behavior in the cultures of six different anobiid symbionts.[1]

As has already been mentioned, Foeckler (1961) obtained interesting results when he offered a great variety of yeasts to sterile *Sitodrepa* larvae which were raised on a diet of wheat grit and flour. These yeasts, added in the fresh state, supplemented the diet to such an extent that all larvae developed into adults, without becoming infected. However, after *Torulopsis utilis* (fodder-yeast) was added to the diet of the sterile larvae, there followed an infection not only of the blind-sac epithelium but also of all epithelial cells of the midgut. The mycetocytes were not so heavily infected with *T. utilis* as those of normal *Sitodrepa* with their species-specific symbionts, and in the case of the foreign infection the brush border is retained. The most interesting aspect of the infection with *T. utilis* is that the epithelial cells of the midgut itself also became infected, and the immunity of these cells, which in normal larvae never become infected with symbionts, is somehow lost. During metamorphosis, the newly formed epithelium of the midgut and blind sacs is not infected with *T. utilis*, and the transmission apparatus of the female is not supplied with these yeasts. The imago then has no intracellular symbionts. It is interesting to note that sterile *Sitodrepa* larvae, to which species-specific symbionts and dried *T. utilis* were offered in the food simultaneously, took up only their normal symbionts into the mycetocytes of the blind sacs, and neither the thin interstitial cells of the diverticula nor the epithelium of the midgut were infected with *T. utilis*. Normally infected *Sitodrepa* larvae, to which dried *T. utilis* was offered in the food, were not infected by this yeast. Thus it is evident that the presence of normal intracellular symbionts assures immunity of the midgut epithelium to invasion by *T. utilis*, although the mechanism of this phenomenon remains unknown. The addition of the other yeasts in the dry state to the food of sterile

[1] Behrenz and Technau (1959) succeeded in rendering *Anobium punctatum* symbiont-free by saturating the wood in which the animals mine with bactericidal solutions. They found that this can be done easily with a 4-aminobenzoic-sulfonamide but not with 4-methylbenzoicsulfonamide. Technologically, this method might be used to protect particularly valuable objects of wood if they are not too large. Jurzitza too (1963) tested the possibility of combating anobiids with sulfonamide, without obtaining results which could be of practical value.

larvae supplemented the diet to such an extent that all larvae developed into adults, without, however, becoming infected. The content of vitamins of the B-complex, using *Tribolium confusum* as the test object, was determined in the yeast, in their substrates both before and after inoculation, in the normal diet (wheat grit and flour), and in the yeast extract. Corresponding to the varying content of B-vitamins in the yeasts, which were added to the normal diet, different developmental times of the sterile larvae were obtained. Foeckler also showed that in the adult stage the content of pantothenic acid, riboflavin, and nicotinic acid in the diet exerts no apparent effects. These three vitamins could not be demonstrated in wheat grit and flour using the *Tribolium* test, but they were present in the yeast extract. The life span of all adults (those normally infected as well as those free of symbionts) was the same in wheat grit and flour both with and without yeast extract. These experiments of Foeckler are of considerable interest to the problems of immunology.

The possibility that the symbionts of cerambycids might also be purveyors of vitamins was suggested by the great similarity of anobiid and cerambycid symbioses. Both families have quite similar symbionts and localize and transmit them in the same way. Schomann's efforts (1937) to compare cerambycid larvae with and without symbionts by Koch's method were unsuccessful because the symbionts were difficult to cultivate. Nevertheless, he was able to prove that the symbiont-free larvae, which also develop the usual intestinal evaginations though these remain sterile, perish much sooner than those with symbionts. When Graebner (1954) succeeded in cultivating the symbionts of *Rhagium bifasciatum* and *R. inquisitor* in large quantities, it became possible to use *Tribolium* to test their dried substance for content of growth-producing factors. An experiment carried out by Reitinger seemed to indicate at first that it actually affords a complete substitute for the vitamins contained in beer yeasts (cf. Koch, 1952).

Recent data of Kühlwein and Jurzitza (1961) and Jurzitza (1959) compel us to correct this assumption. Testing a series of cerambycid symbionts, all of them *Candida* species, for ability to produce B-vitamins, the authors discovered that differences exist, as with the anobiid symbionts. When the symbionts of *Rhagium mordax, Leptura rubra,* and *Rhagium inquisitor* were compared, it was found that the first two were markedly heterotrophic for vitamins. The symbionts of *Rhagium mordax* require biotin; the symbionts of *Leptura rubra* need biotin and aneurin; and those of *Rhagium inquisitor,* although they are able to grow without these vitamins, were substantially benefited when biotin was added. Later Jurzitza, Kühlwein, and Kreger-van Rij (1960) tested six additional cerambycid symbionts, finding that five were heterotrophic and that only the *Candida* species living in *Gaurotes virginea* managed without addition of vitamins. It was also shown by means of paper chromatography that cerambycid symbionts transmit amino acids to the substrate and therefore may be of importance to the hosts as protein purveyors.

Such findings with anobiid and cerambycid symbioses make it seem probable that also the ambrosial fungi furnish growth-promoting substances to the endosymbiont-free insects which are so intimately adapted to them. This has actually been proved by I. Schwarz, with *Tribolium* tests, both for the *Hylecoetus* fungi which had been cultivated by Francke-Grosmann and which, as we have already heard, are transferred like the symbionts of anobiids and cerambycids by means of egg smearing, and also for the ambrosial fungus of the ipid *Xyleborus,* which is taken along in two grooves at the base of the wings of swarming females (reported by Koch, 1955). Table 4 compares these values with those of the beer yeasts, which contain substantially less beta-biotin but an almost similar quantity of the other growth substances.

No research has been done on the endosymbiosis of insects living primarily in wood, but studies are available on animals which, like *Sitodrepa,* are descended from wood-devouring animals but have adapted themselves to cereals and similar food, and these might quite possibly behave like *Sitodrepa.* Surprisingly it has been established that symbionts are no longer vital to the cucujid *Oryzaephilus surinamensis* and to the *Calandra* species of the curculionids. Even before Aschner eliminated the *Pediculus* symbionts, Koch (1931, 1933) found that the worldwide storage pest *Oryzaephilus,* which withstands temperatures up to 38°C., can be freed of its symbionts by increasing the temperature to 36°. In this connection it was discovered that the compact transmission forms are particularly thermosensitive, while the tube-shaped typical mycetome dwellers are somewhat more resistant. Therefore, when first-stage larvae are treated the symbionts are destroyed, but when mature females are exposed to heat only the transmission forms die, and the others merely elongate. In the descendants of animals treated in this way the mycetomes are entirely symbiont-free or have a reduced number of symbionts.

Huger (1956) found that the same thing may happen in granaries when, as a consequence of high moisture, considerable warming and decomposition set in. The mycetomes in 45% of his *Oryzaephilus* specimens were completely sterile! The same author tested the influence of lower temperatures. The mycetomes of animals kept 24 days at 4°C. and then restored to normal temperature were more or less depopu-

TABLE 4 *Vitamin content of various symbiotic microorganisms compared with that of beer yeasts, wood, and sieve-tube sap (according to Koci)*

	Beta-biotin	Panto-thenic acid	Thia-min	Riboflavin	Pyridoxin	Nicotinic acid	Folic acid	Choline chloride	Car-nitin
Hylecoetus fungus	XXXX	XXXXX	XXXXX	XXXXX	XXXXX	XXXX	XXXXX	XXXXX	+
Xyleborus fungus	XXXX	XXXXXX	XXXXX	XXXXX	XXXXX	XXXXXX	XXXXXX	XXXXXX	+
Sinobius symbiont	XXXX	XXXX	XXX	XXXXXX	XXXXX	XXXXXX	XXXX	XXXXX	+
Sitodrepa symbiont	X	XXXXXX	X	XX	XXX	—	XX	XX	+
Mesocerus symbiont	XX	XX							—
Escherichia coli	XXXXXX	XXXX	XXXXX	XXXX	XXXXXX	XXXX	XXXXXX	XXXXX	++
Beer-yeast extract	XX	XXXX	XXXX	XXXX	XXXXXX	XXXX	XXXXXX	XXXXX	+
Dead beach wood	0	0	0	0	X	0	0	XX	—
Linden wood	0	0	0	0	X	0	X	XX	—
Sieve-tube sap (*Quercus rubra*)	X	0	X	0	X	XXX	X	X	—

lated, and some of their descendants remained completely symbiont-free. When the food was saturated with Aureomycin or Terramycin, the result was even more striking, for all the mycetomes were sterile in the F_1 animals.

The appearance of mycetomes in animals exposed to a temperature of 38°C. varies according to the degree that the symbionts are injured (Koch, 1936). The bacteria may swell into clumsy vacuolized clumps or grow out into irregular tubes and bizarrely branched involutional stages until they are finally disintegrated, and these processes may involve wide areas or mere nests of varying size. Moreover, the four mycetomes of one individual may react differently. With progressive decomposition of symbionts, structural changes naturally take place in the mycetomes. The gigantic vacuoles which appeared in place of the symbionts are resorbed, lessening the size of the organs considerably. The envelope layer, highly vacuolized at first, swells, with its nuclei rounding out proportionately; the syncytia, still recognizable by their smaller nuclei, gradually lose their limitations; strangely, the especially large central nucleus which is so typical of normal mycetomes is still set off just as precisely in the completely sterile mycetome as in the infected one (Fig. 370).

These formations, symbiont-free and therefore useless, are still provided with tracheae and are always formed anew from generation to generation without substantial changes. Thus we see how firmly they took root in the heredity of *Oryzaephilus*, once the symbionts had been acquired. Koch, following them through 20 to 25 generations (1931–1936), found that normal animals and symbiont-free animals both react in the same way to the usual diet, to an extremely unbalanced diet such as potato starch, and to starvation diets.

It is not yet clear whether such data indicate that with the change in feeding habit in *Oryzaephilus* the symbionts have become superfluous but are nevertheless retained under normal environmental conditions. Koch at first adopted this view. Fraenkel and Blewett (1943), comparing the demands made by *Oryzaephilus* on the vitamin content of the food with those by *Tribolium, Ptinus, Sitodrepa,* and *Lasioderma,* believed that they had determined that the lack of riboflavin, nicotinic acid, and pantothenic acid resulted in high mortality and retarded development. Later, Pant and Fraenkel (1954) attacked the problem again, this time comparing symbiont-containing and symbiont-free *Oryzaephilus.* They found that casein used the previous time had not been vitamin-free and that sufficient quantities of thiamin and pyridoxin had been administered with the insoluble yeast residue. Nevertheless the two authors are inclined to think that a contribution is made toward the requirement of nicotinic acid, pyridoxin, and choline.

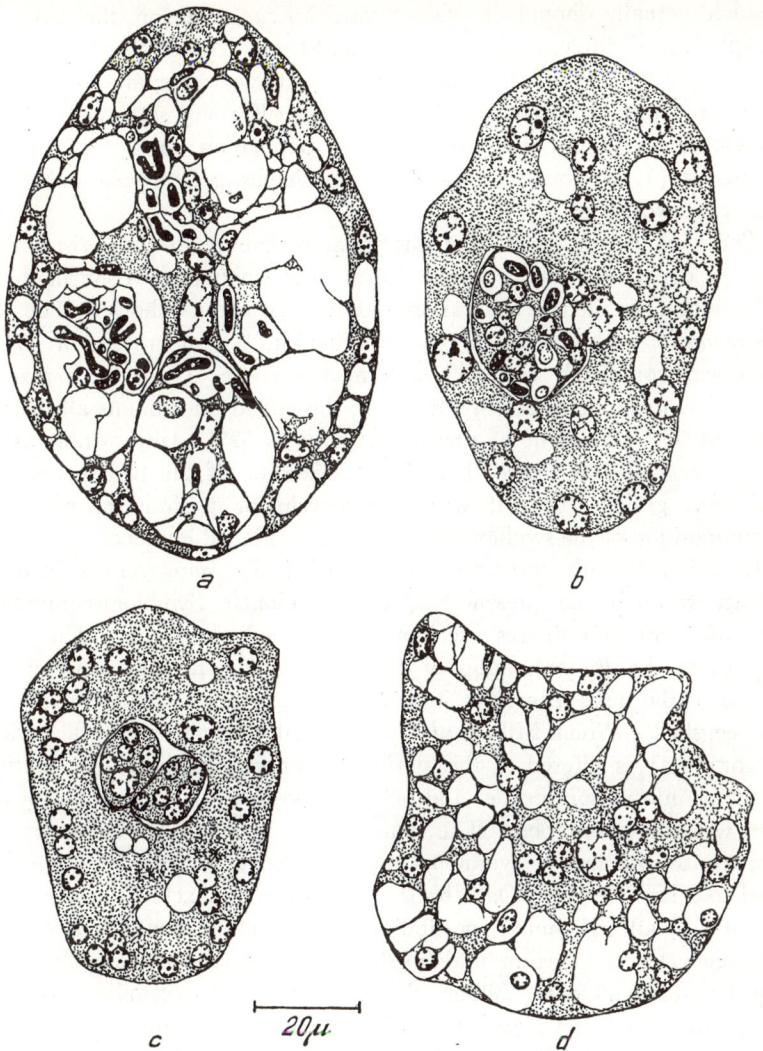

Fig. 370 *Oryzaephilus surinamensis* L. Gradual dissolution of symbionts under
the influence of increased temperature. (*a, b*) Different sized remainders of
symbionts still present; (*c, d*) symbiont-free mycetome. From Koch.

A similar situation exists with *Calandra granaria* in temperate zones
and with *Calandra oryzae* in tropical and subtropical zones. Here it
is possible at times to find animals free in nature which are symbiont-free
but decidedly viable. The first report of this kind came from Egypt,
where Mansour (1935) invariably found the indigenous *Calandra granaria*

(which actually should be called var. *africana* M. on the basis of morphological differences) to be symbiont-free. This is always a matter of secondary symbiont loss, as sterile mycetomes were always formed. The same author confirmed an extraordinarily small settlement of the mycetomes with an English *granaria* population, and recently H. Schneider (1956) found several sterile specimens of *C. oryzae* with rudimentary mycetomes.

The obvious conclusion, that this apparent liability of *Calandra* symbiosis is dependent upon temperature, was completely confirmed by Schneider's detailed investigations. His experiments showed that the *C. granaria* symbionts of imagines exposed to a temperature of 35°C. disappear completely after 33 days at the latest. *Calandra oryzae* is somewhat more thermoresistant, but in the end the symbionts all degenerate at 35° (the limit of thermosensitivity is 32°C. with symbionts of *C. granaria* and 34° with those of *C. oryzae*). If the two species are fed with grains saturated with aureomycin or terramycin, complete decomposition of the symbionts occurs with continued influence.

Experiments with sterile animals showed a clear improvement in their vitality through the presence of the symbionts. Symbiont-containing *Calandra granaria* thrives excellently in shelled oat grains. The same applies to the F_1 generation of parents treated with heat, but young larvae of the next generation invariably die directly after hatching from the eggshell. When Milo-corn (commercial designation for a variety of sorghum) is offered to the sterile F_1 generation, the F_2 generation develops into mature beetles and offers no obstacles to continued cultivation with the same food. The same applies to sterile *Calandra oryzae*. Apparently the loss of symbionts merely restricts the beetles' choice of food to a high degree. Oats, barley, and maize are excluded completely, wheat still affords limited possibilities, but sorghum contains a growth substance, for the present undefined, which compensates completely for the loss of symbionts. When the developmental possibilities of the Egyptian species of *C. granaria* (which is sterile by nature) are investigated, a great dependence on the effective-substance content of the food is also found; yet this species shows a superiority, although small, to the other two forms. Again Milo-corn and wheat are in first and second places, respectively, but oats, barley, and maize also offer at least limited possibilities for development.

When *Rhizopertha dominica*, another storage pest with relatives living principally on cellulose-rich substances, was exposed to a temperature of 38°C., the mycetomes were deserted (Huger, 1956). Again the symbionts were severely damaged by heat, but in this case some of the symbionts survived, remaining in the mycetome or migrating to the body of the host, so that the actual sites appeared completely empty. Such thermoresistant tiny states, eluding observation, must be the source of the gradual new settlement of the later generations extending over several generations. The same thing was experienced when it was possible to make *Rhizopertha* apparently symbiont-free by means of aureomycin and terramycin. Although no studies are available, one is inclined to think that here, as in *Oryzaephilus,* the symbionts have retained limited importance after the change in the host's feeding habits, and that therefore they were not eliminated.

This may be the proper place to refer to cooperative living of *Hylemyia* with bacteria. Without bacteria, this fly is not able to feed itself from the storage tissues of the potato tuber which is, of course, particularly rich in carbohydrates. Leach (1926, 1931, 1933, 1940) and Huff (1928) showed that the larvae which hatch from superficially sterilized eggs cannot grow at all on sterile potatoes, yet they grow easily when the potatoes are infected with the bacteria which normally reach the eggshell from the intestinal lumen, these bacteria being immediately inoculated into the potato tissue by the mandibles of the young larvae. We suppose that not mere decomposition fostering digestibility is involved here but bacterial supplementation of the vitamin content of the potatoes, for the latter rarely serve the insects as nourishment in the non-decomposed state, and, when sterile bean sprouts are given to germ-free larvae, growth is slow and pupation is normal.

There was no information on hand at first to indicate that also the plant-juice-sucking insects might be supplied with growth-producing substances originating from their symbionts, except the fragmentary observations by H. J. Müller (reported in Buchner, 1940). Müller had studied the behavior of the bug *Coptosoma,* an animal whose transmission method actually invites experiments with sterile cultivation and about which Schneider, the discoverer of symbiont-filled capsules that are deposited here between the eggs and sucked out by the hatching larvae, had collected the first findings of an expected nature. When the capsules are removed before the larvae hatch, the animals remain sterile, since they are not able to take up those bacteria which here and there contaminate the eggshell.

After the war H. J. Müller resumed his experiments, confirming his previous results and substantially broadening the foundation of his studies (1956). It was shown that the mortality of sterile larvae radically exceeded that of infected ones and that the growth period of the few sterile animals which did reach the imaginal phase was substantially longer. Although they finally achieved normal body size, they deposited very few eggs and the capsules which are normally filled with bacteria remained empty.   Loss of symbionts in this animal thus does not exclude possibility of growth, yet possession of them is of decisive importance. Aside from the fact that sterile animals are barely capable of reproduction, their descendants must necessarily be gradually destroyed because the larvae, as a result of their slow growth, approach winter at stages too premature for survival.   However, H. J. Müller was able to avoid, to a large extent, damage entailed in symbiont loss by cultivating the larvae in seedlings of the host plants which abound in growth substances. It was, however, not possible to reinfect them with symbiont cultures, which are obtainable without difficulty.   It is interesting to note that, when the symbionts were cultivated for a time *in vitro,* they lost the qualities which make regulated reincorporation possible.

In view of the results obtained by H. J. Müller, it is surprising that Bonnemaison (1946) arrived at different conclusions with another heteropterous bug.   He had obtained sterile larvae of the pentatomid *Eurydema ornatum* L. by removing them prematurely from the eggshells. Contrary to expectation, the sterile larvae developed as quickly in old cabbage leaves, of unknown vitamin content, as did the infected larvae; they showed no increase in mortality and yielded sterile progeny. Whereas the first generation was somewhat behind the control animals in size, this difference failed to occur with the progeny.

On the other hand, after the publication of the research by Bonnemaison, two further studies on Heteroptera symbioses confirmed the vital importance of symbionts.   L. Huber-Schneider, some of whose results we reported in 1953, cultivated the symbionts of *Mesocerus (Syromastes) marginatus* L. in pure culture and let their growth-substance content be tested with *Tribolium* (1957).   When I. Schwarz added the dried symbionts to the usual deficiency diet, the larvae had a completely normal development; in other words, the dried substance provided them with all the necessary vitamins.   However, tests showed that only thiamin was not transmitted to the agar.   As yet we know nothing about the vitamin content of the sieve-tube sap of *Rumex,* on which *Mesocerus* is found especially.   However, it may possibly be significant that, in the sieve-tube sap of *Quercus rubra* (the only sieve-tube sap tested for its vitamin content by Koch and Schwarz), whose young leaves *Mesocerus* also sucks, only nicotinic acid, which is lacking in the *Mesocerus* symbionts, is contained in larger quantity; while the other vitamins, produced by the symbionts, are lacking or are present only in small amounts (see Table 4).

Just as clear are the data compiled by Schorr (1957) on the bug *Brachypelta aterrima* Först., a member of the cydnids, which burrows in dune and steppe sand and is characterized by intensive brood care.

After hatching, the larvae remain close to the mother to suck up the fluid drops which exude from her anus on contact impulses. When prevented from doing this, they perish. By this strange instinct the descendants are ensured a supply of symbionts! As with most Heteroptera, the symbionts are housed in numerous intestinal crypts which swell enormously in the female during the breeding period but are radically reduced in the imaginal males. The droplets taken up by the young larvae represent a flotation of maternal symbionts. Only when the symbionts are provided is the sieve-tube sap of the feed plants sufficient to provide further development.

Fortunately, the positive results with Heteroptera are now augmented by several with coccids. We are referring to Fink's thorough analysis of *Pseudococcus citri* (1952). In this scale insect, too, the vital necessity of the bacteria populating its voluminous unpaired mycetome was proved by removing them. When the animals were subjected to hunger and increased temperature (39°), the symbionts suffered immediately. Their vacuoles disappeared, the tubes became thicker, degeneration spread in the form of lumps, and finally only decay products were still present, which Ries called "bacterial debris" in males of *Pediculus*. After 7 to 10 days the injuries were irreversible, the ovaries having then either completely disappeared or as in a few cases, having at least been strongly reduced. If eggs ready to hatch were still present, they could not develop; although they were infected, their symbionts degenerated. Despite no visible damage of any other nature in the animals—even the mycetomes remained fully intact—they died after 30 days at the latest.

As reported elsewhere in detail, Fink succeeded in obtaining excellent cultures of the *Pseudococcus* symbionts in which he could show that the symbionts form riboflavin, nicotinic acid, pantothenic acid, the remaining B-vitamins present in yeast, and a number of amino acids. He also tested the effect of dried symbionts on the growth of *Tribolium*, using organisms which had been cultured in vitamin-free media. With a deficiency diet containing only 5% of symbiont-containing dry substance the freshly hatched *Tribolium* larvae were not able to grow and were all dead after 40 days. By adding 5% symbiont dry substance and 5% yeast residue, he was able to obtain a delayed development of the majority of the larvae. The symbiont dry substance behaved therefore quite like the sum of the water-soluble yeast vitamins (Table 5).

Since it is known today that, in the yeast residue, ergosterin, a considerable portion of folic acid, which is difficult to extract, and carnitin—all of which are indispensable to *Tribolium* growth—are contained, it must be assumed that they are not furnished by the *Pseudococcus* sym-

TABLE 5   *Growth-accelerating effect of the dry substance of Pseudococcus citri symbionts in the Tribolium test (according to Fink and Reitinger)*

Nourishment	No. of freshly hatched larvae	Pupae	After —days	Imagines	After —days
Deficiency diet + full yeast content	12	11	23	11	29
Deficiency diet + yeast residue + symbiont dry substance	12	9 + 3 large larvae	32	9	40
Deficiency diet + symbiont dry substance without yeast residue	12	No growth, all dead after 40 days			
Deficiency diet + yeast residue without symbiont dry substance	12	No growth, all dead after 50 days			

bionts, or at least not in sufficient quantity, and that the symbionts of these conspicuous mycetomes, which have been known for such a long time, nevertheless show a vitamin composition very similar at least to that of the yeasts.

M. Köhler and W. Schwartz (1961) also studied coccids experimentally. They, too, were successful in cultivating the symbiont of *Pseudococcus citri* and, like Fink, declared it to be a *Corynebacterium*, and they also obtained cultures from *Pseudococcus maritimus* and *Orthezia insignis*. They interpret the symbiont of *P. maritimus* as a *Flavobacterium* and that of *Orthezia* as an *Achromobacter* sp. In all three it was possible to disrupt symbiosis by exposing the larvae or imagines to temperatures of 37° to 40°C., and in the case of *Pseudococcus citri* also by saturating the feed plants with penicillin solutions. Surprisingly, loss of symbionts caused death only in the two *Pseudococcus* species, whereas the *Orthezia* were cultivated, symbiont-free, for 2 years under continuous control. Later we shall discuss this surprising discovery again.

At first glance it may seem surprising that there is no uniformity in these findings on the contributions of symbionts to plant-juice-sucking animals, for this symbiosis belongs in a category which coincides particularly sharply with radically unbalanced diet, where it would be logical to presuppose uniform contributions of their symbionts. Closer examination revealed that the information available on the composition of the diet of Heteroptera and Homoptera is still quite deficient. In most cases it is the sieve tubes which are sought with astonishing perseverance by the stylets, which follow the cell interstices. Thus it would be important for the background of this significant Hemiptera symbiosis to have exact information on the vitamin content of the sieve-tube sap of all the food plants involved. Actually this branch of vitamin research is still undeveloped. As far as we know, there are only the summaries of Huber, Schmidt, and Jahnel (1937). These workers irrigated agar discs with the sieve-tube sap of a number of trees and tested their variable, usually small, content of growth substances with the *Avena* test, but here it was only a matter of auxins, that is, cell-extending hormones which are ineffective in the *Tribolium* test (Offhaus). Investigations by I. Schwarz, at the suggestion of Koch and Huber, with the sieve-tub sap *Quercus rubra* revealed striking poverty in certain growth-promoting vitamins. For example, riboflavin is lacking completely and pantothenic acid has not been demonstrated with certainty. Thiamin and beta-biotin are definitely present but only in a small quantity; the content of nicotinic acid corresponds to that of the yeasts (Table 5). On the other hand, it has been found that seed plants, roots, shoots, and pollen granules grow only when certain B-vitamins, thiamin and nicotinic acid especially, are given with the nutrient solutions (Williams, 1950). We may assume that the same situation exists among insects living on plant juice as among those living on vertebrate blood, that is, that the growth-promoting substances formed in the different plant components and at their disposal in the sieve-tube sap are not present in each case in the quantity necessary and therefore must be supplemented by the symbionts.

Another complication arose because the content of growth-promoting substances may differ considerably in the various areas of the plant. I. Schwarz tested, with the help of the valuable *Tribolium* test, various zones of dried *Coronilla varia* at the beginning of the vegetation period. It was shown that the lower-stalk parts contain no riboflavin and the upper green stalk parts no choline chloride and no beta-biotin. On the other hand, before turning green, the young shoots yielded all the B-vitamins necessary for growth in sufficient quantity. The larvae of *Coptosoma* gather and suck on the same young shoots after winter rest!

A more exact knowledge of the diets of the various plant-sap suckers is indeed necessary for symbiosis research, for in many instances it was even impossible to determine endosymbiosis. Among the scale insects, this applies to the primitive Xylococcini, Matsucoccini, and Steingeliini, which live in bark and wood, to the Apiomorphinae, which form galls on *Eucalyptus* species, and to *Kermococcus* (Buchner 1957). Among the cicadas, only the typhlocybine family is without symbionts; among the aphids, *Phylloxera* (Grassi, 1912; Maillet, 1957; and others). In all probability the reason for this lack of agreement should be sought in the fact that the forms in question do not live on sieve-tube sap, or not exclusively on it, but have learned to puncture and suck out the parenchyma cells. That this occurs with several scale lice was known as early as 1891 by Büsgen, the first to be seriously concerned with the mode of nutrition of insects sucking at plant saps. Vos, studying the course of the puncture canals of *Pseudococcus citri* in detail (1930), also found that they frequently end with the parenchyma cells, and Francke-Grosmann (1937) made the same discovery with aphids. She found that the *Dreyfusia* species (Adelginae), in contrast to *Lachnus* and *Mindarus* (Aphidinae), search out the storage parenchyma of the fir tree; that the puncture canal then branches like a tree into many canals; and that the parenchyma cells, thus strongly stimulated, become especially rich in plasma and at the same time poor in storage substances. Toward the end of egg deposit, a clear exhaustion of food tissue occurs, the amount of plasma in the punctured area decreases, starch disappears completely, and sugar can no longer be detected. Considerable differences exist here, for *Lachnus* and *Mindarus* secrete large quantities of honey dew, but *Dreyfusia*, which evidently finds much more favorably composed nourishment, shows only traces in the excrement.

We do not know whether symbiont-free Homoptera lost their symbionts in connection with a change of instinct. We do not have data to verify that they did, but an argument for the possibility is the discovery that females of *Hippeococcus* species, whose ovaries develop only after the insects have been fed with the food sap of the ants taking care of them in the nest, lose the symbionts which have thereby become superfluous but nevertheless continue to develop sterile mycetomes (Buchner, 1957). Thus it is no longer surprising that *Orthezia insignis* thrives equally well when deprived experimentally of symbionts. It is especially in the young leaves of their food plants (*Urticaria*) that we find in unusually large amounts all the important growth substances, with the exception of pantothenic acid (according to results of I. Schwarz; see Hänsel, 1960). Another orthezine, *Newsteadia floccosa* (de Geer), sucks out fungal hyphae.

The nitrogen content of the sieve-tube sap has been studied in greater detail. Indeed these investigations are of particular importance for the problem of symbiosis, for they led to a search for the role which the symbiosis of the plant-sucking animals with microorganisms plays, among others, in covering the requirement of protein. Kostytschew (1931) found in the sieve-tube sap only 1–2% protein (referring to the dry substance); Moose (1938) confirmed the presence of only 0.1% of nitrogen; whereas Lindemann (1947) in equivalent experiments with *Ribes* found 3.77% in the spring and, by the end of June, only 0.52%. Cases where substantially higher amounts were found, such as the pumpkin and the cotton plant, represent exceptions, and the high values assumed by J. D. Smith (1948) are apparently the result of a misunderstanding (see Tóth, 1950).

Although the sieve-tube sap contains only a small amount of nitrogen, relatively large quantities of protein are nevertheless found in the excrement of the aphids! Raumer (1894) found 3.5%; Michel (1942) even found that *Lachnus roboris* doubled the absorbed nitrogen (0.3 or 0.6%). On the other hand, Lindemann (1947) found 1.5% nitrogen in the honey dew of *Cryptomyzus* living on *Ribes* in the spring, and only 0.17% in June, the decrease reflecting the dwindling protein content in the sieve-tube sap.

In view of such facts, it becomes difficult to uphold the old explanation, which was advanced by Büsgen (1891) and later generally shared (Rawitscher, 1933), that the aphid compensates for the nitrogen poverty of its diet by taking up huge quantities of sieve-tube sap, giving off the enormous superfluous quantity of carbohydrates in the form of honey dew and retaining small quantities of protein within its body with the aid of the so-called filter chamber of its gut.

The source of the nitrogen assumes importance in view of the great quantity of protein required by aphids. Tóth (1940), comparing the weight of fully developed, wingless, viviparous females of *Aphis sambuci* with the total weight of the approximately 20 young aphids each bears in one day, came to the conclusion that during their most active reproductive period the mother animals must double the protein quantity of their body daily! Tóth and Wolsky (1941) provided further proof that the nitrogen supplied by the plant is insufficient to supply the amount required by the aphids. If carbohydrate consumption predominated, their respiratory quotient would amount to about 1.0, but the average value found by the two authors, 0.86, is much closer to that of the pure quotient of 0.8 which corresponds to an exclusive digestion of protein.

These results indicated that aphids and probably other Homoptera are largely independent of the host plant in regard to nitrogen and that the symbionts, which indeed are always present in aphids which produce honey dew, are apparently able to assimilate atmospheric nitrogen. Although experimental proof was lacking until Tóth and his co-workers began a series of investigations in 1942, the idea itself had made its appearance in symbiosis literature much earlier. In 1912 and 1916 the botanist Peklo had declared the aphid symbionts to be *Azotobacter* and had drawn corresponding conclusions concerning their capacities. Cleveland (1925, 1928), finding that termites thrive on a pure cellulose diet, had been forced to assume that they were able in some fashion to utilize atmospheric nitrogen, although he was unable to furnish experimental proof by the method he used. Mercier (1907) and Gropengieser (1925) stated that cultures obtained from cockroaches (which, in the light of present-day knowledge, were not identical with the symbionts) were also able to assimilate N_2. I had considered the possibility in 1921, especially in view of the generous provision of tracheae in the mycetomes, but experimental symbiosis research was at that time striking out in a different direction and for the time being I did not give attention to such a possibility.

Tóth and his co-workers employed two methods. Chiefly they used the method which Virtanen (1939) had chosen for his investigation of N_2-binding in the symbionts of legumes. Quantities of test animals were crushed; oxalic acid, which according to Virtanen is indispensable for N_2 assimilation, was added, plus glucose and NaCl; and samples of this viscid material were taken at intervals for nitrogen determination, the micro-Kjeldahl method proving to be the most suitable for the purpose.

These workers showed that a considerable amount of nitrogen accumulated during the first 24 hours but gradually disappeared after 30 hours, presumably with the death of the symbionts. In *Pterocallis*, the nitrogen content rose 110% in 24 hours, 142% in 48 hours. Without addition of oxalic acid there was a slight increase, apparently because a few traces of the acid were still present. Equivalent experiments with *Aphrophora salicis* showed an increase of 121% in 24 hours. Later Tóth, Wolsky, and Bátyka (1944) studied the extent to which other insects are able to derive nitrogen from the air. Other aphids and *Aphrophora alni* yielded the same sizable accumulation, but with the other cicadas investigated (*Philaenus, Lepyronia, Idiocercus, Cicadella,* and others) the increase in nitrogen moved between 89% and 34% in the surviving systems within 24 hours, and with *Doratura* even sank to 15%, but nevertheless increased within 48 hours 22% to 139%. In part these low amounts may be traced to the fact that fall animals with less animated metabolism were used.

When Heteroptera were finally investigated, a symbiont-containing pentatomine (*Raphigaster*) yielded 18% within 48 hours; *Pyrrhocoris*, which has at the most a very loose symbiosis, yielded 25%; *Spilostethus saxatilis* Scop. showed an accumulation of 85%. The latter, of course, is phytophagous, but it is not certain that it lives symbiotically, for Schneider found no symbionts with other species. Strangely, an increase of more than 40% appeared even with pronouncedly predatory bugs, such as *Gerris* or *Nabis* which are not truly endosymbiotic. It is no less surprising that *Trialeurodes vaporariorum*, a typical symbiont bearer, showed no capacity to assimilate N_2. It remains undecided whether a diet richer in protein is involved here or whether Weber (1931) was correct in assuming that, since the excrement of these animals completely lacks the expected carbohydrates, the hosts or symbionts may perhaps utilize the carbohydrates during protein synthesis.

Tóth and his co-workers also sought to demonstrate the nitrogen accumulation in cultures of microorganisms obtained primarily from *Aphis sambuci* and *Aphrophora salicis*. After the questionable bacteria had been placed in an optimum physiological solution and mixed with the necessary oxalic acid, the micro-Kjeldahl method actually yielded a slow increase in the nitrogen content, amounting to 6.5% in 96 hours in the aphid and 67% in the cicada. Since these values are below those obtained with cell broth, the authors are inclined to think that the substances necessary for nitrogen assimilation were still present in the surviving system, though lacking of course in the culture.

Later Tóth (1950, 1951) reported on nitrogen accumulation in cultures obtained from other aphids and cicadas, from Heteroptera (*Carpocoris* and *Pyrrhocoris*), even from the beetle *Pyrrhidium sanguineum*, of which we do not know that it actually lives symbiotically. In his reports he gave the more exact composition of the nitrogen-free culture medium, which 20 days later has its maximum N content, the maximum varying in proportion to the type of the carboxylic acid added.

Peklo and Satava (1949, 1950) seemingly provided confirmation of Tóth's findings. Peklo (1946) had already reported that in a long series of insects he had found *Azotobacter* and *Torulopsis* symbionts which safeguarded these insects from a deficiency of nitrogen, and later he extended his findings in a number of other publications. In our chapter devoted to mistaken hypotheses in symbiosis research, we have already expressed our conviction that Peklo's data were based on spherical animal-specific inclusions of the fat body and yolk clumps of the egg cells which he mistakenly considered to be symbionts. Thus it is impossible to accept the statement of the two authors that they cultivated symbionts from *Ephestia*, *Tribolium*, *Ips*, and *Sitodrepa*, that is, from animals some of which must be considered symbiont-free, and that they found varying degrees of increase in nitrogen in the cultures as well as in crushed cells.

And, even though one does not approach the Tóth publications with similar skepticism, one must at least keep in mind that they represent the first tentative explorations in new territory and that some aspects need considerable clarification. As Tóth himself explained in his publications of 1950 and 1951, the methods of crushing the animals are still subject to error and are not independent of fluctuating factors of various types, and the tested material by no means remains sterile, thus making it uncertain whether the increase in nitrogen is actually provided by the symbionts or by other microorganisms. It is also uncertain whether the "symbiont cultures" are identical with the symbionts, for the data on the manner in which they were obtained (Schmeisser, 1944) are very brief and are in need of retesting. Tóth himself admits that it is very difficult to obtain cultures and that it requires a specialist to judge the strains. Another reason for proceeding with caution in evaluating his work is that he obtained positive results in some cases with animals that are usually regarded as symbiont-free, for he even used predatory bugs. Since these bugs suffer no deficiency in protein, assimilation of atmospheric nitrogen in the body of the animals is not involved, for it is known that nitrogen-binding by microorganisms is the general rule, and it is only relatively infrequent that preliminary stages of this type develop such intimate obligatory dependence as is seen especially in aphids. Thus conclusions drawn from such experimental results about symbionts which as yet have escaped observation are just as unsatisfactory as those based on the experiments with termites, discussed earlier, in which the thoracic and abdominal sections of the body without the gut, which is so abundant in microorganisms, yielded a greater increase of nitrogen than did the gut by itself.

It is also necessary to clarify the question of the form in which the symbionts transmit protein to the host body. Of course, it is known that yeast cells continually secrete protein into their fluid surroundings, but Tóth is thinking chiefly of continuous dissolution of degenerating symbionts of the type that may be observed in aphids, according to Paillot and his own data.[1]

---

[1] This concept receives some support from observations by Graebner with the symbionts of *Ernobius* and *Rhagium*. He applied a method worked out by Strugger in which, by use of an extremely dilute solution of acridin orange, living plasma appears as a delicate green under the fluorescent microscope, and dead plasma as an intensive red. In the organs filled with the yeasts, individual cells containing dead symbionts exclusively could be detected in this way. More rarely, individual dead yeasts were found among the healthy ones. On the other hand, the yeasts which are constantly given off into the gut in *Rhagium* were found to be alive. Many cells also perish in young cultures of the anobiid symbionts and, according to Graebner, in this way transmit to the culture medium the substances necessary for growth.

We have seen that much is uncertain in Tóth's attempts to prove assimilation of nitrogen with the assistance of symbiont cultures. However, this is not true of Fink's findings (1951) with his faultlessly obtained pure cultures of *Pseudococcus* symbionts. These symbionts, subjected to all conceivable precautions whereby NH_3 in particular was excluded, yielded slow but satisfactory growth in an N_2 atmosphere on completely nitrogen-free culture medium (distilled water, NaCl, $MgSO_4$, $CaCl_2$, KH_2PO_4, galactose). By paper chromatographic methods the various amino acids could be determined, and it could be shown that leucin, valin, tyrosin, alanin, glycocol, glutaminic acid, asparagin, and histidin arise in the bacteria as products of protein synthesis and transfer from these to the fluid culture medium. The same amino acids were found again in the extract of mature *Pseudococcus* females, except that leucin and tyrosin were lacking, with isoleucin taking the place of the first-mentioned.

Proof that the slow growth of colonies was not caused by the deficiency of easily resorbable N compounds was provided by the control with culture media containing amino acids, where the multiplication of the symbionts was not much greater.

Of special interest is Fink's finding that in N-free culture solutions the *Pseudococcus* symbionts suffer form changes which cause them to resemble the symbiotic bacteria of legumes. In place of the almost completely vanishing rodlets he found Y-forms, crosses, and other irregular states which are typical of bacteroids. When these forms were inoculated on peptone or culture media containing amino acids, the aberrant shapes disappeared and gave way again to the original rodlet form.

The authentic findings of Fink are all the more valuable because the earlier data of Schanderl (1942) on atmospheric nitrogen-binding cultures of yeasts living symbiotically with *Rhagium inquisitor* are no longer valid (Jurzitza, 1959). Nevertheless, we are inclined to think that, especially with insects which are able to live in dead wood, atmospheric nitrogen can be utilized. In this connection we cite Becker's findings (1942, 1943) that the larvae of *Anobium pertinax*, *Leptura rubra*, and *Ergates faber*, in other words, of symbiont-containing species, are not dependent to the same degree on the protein content of the wood as, for instance, are the larvae of a beetle living without symbionts (*Hylotrupes bajulus*), whose development can be considerably accelerated through artificial increase of the protein content of the wood by peptones, amino acids, etc. The development of the three animals mentioned is dependent far less on additional protein, and their larvae can even grow in pure cellulose. It is among the urgent tasks of experimental symbiosis research to continue to test lignivorous insects for their capacity to assimilate nitrogen with the assistance of good cultures of symbionts. Since symbionts never appear in cerambycids that develop in fresh leaf wood and in plants, that is, in surroundings abounding in protein, we may assume without hesitation that their ability to use dead wood as a habitat is attributable to the acquisition of symbionts. Moreover, Kühlwein and Jurzitza (1959, 1961) found that cerambycid symbionts unquestionably supply the building stones of protein, for the paper chromatographic test of the nutrient solution in which the yeasts were cultivated showed clear transmission of amino acids. Conversely, they found no indication that atmospheric nitrogen was utilized.

Observations of various kinds have long indicated that symbionts may sometimes interfere with the nitrogen economy of their hosts in addition to acting as suppliers of vitamins and protein. Earlier the particularly intimate relations of many sites to the tracheal system had brought up the possibility that atmospheric nitrogen is utilized. In the same way the possibility of utilizing the end products of animal metabolism, thus of indirectly bringing them back again in part to their hosts, was suggested by localization of a number of symbionts in the area of the fat bodies which serve as storage kidneys, in some insects even in the Malpighian tubules, and, among the ixodids, in other excretory organs as well, and among cyclostomids and the annularids in the concrement gland. Moreover, in the aphids the Malpighian tubules diappear completely, and among the coccids they are greatly reduced, thus suggesting that these changes had been made possible through equivalent contributions of the symbionts (Šulc, 1910; Buchner, 1930).

Quast (1923, 1924) and Meyer (1923, 1925) actually did show that the bacteria multiplying rapidly in the storage cells of *Cyclostoma*, apparently because of a uricase deficient in the snail, were able to liquefy concrements consisting of up to 50% uric acid, xanthine, and other purine bases and could be resorbed finally by the phagocytes of the host animal. Analogously, certain stages of the fungi always present in the strange enclosed molgulid kidneys are digested by ameboid cells (Buchner, 1930).

The first equivalent data on insect symbionts were those of W. Schwartz (1924). He was able to show that his faultlessly obtained pure cultures of various *Lecanium* symbionts could utilize uric acid, guanin, xanthine, urea, and other purine bodies, but he was unable to detect any urates in the fatty tissue or in the Malpighian tubules with the aid of the murexid reaction. On the basis of his findings, it appeared likely to him that the significance of *Lecanium* symbiosis lies in the decomposition of metabolic end products of the hosts.

Schoel (1934) arrived at similar conclusions. He found that bacteria cultivated from three different monosymbiotic aphids, which he firmly believed were identical with the symbionts, produce a uricase; and that the cultures grow most rapidly with an addition of urea; that growth is somewhat less rapid with uric acid and least rapid with hippuric acid. Moreover, since Klevenhusen had found indications of a partial dissolution of the aphid symbionts, an opinion later emphatically supported by Tóth, Schoel was inclined to assume that excretions of host animals are advantageous to the symbionts.

Without referring to his predecessors, Tóth reported (1951) that in the presence of urea or uric acid his cultures of the symbionts of *Aphis brassicae* ceased assimilating atmospheric nitrogen and decomposed the two substances instead, urea again proving to be superior to uric acid. He had the same experience with bacterial strains from the gut of *Reticulitermes* and was convinced that they were able to bind atmospheric nitrogen, although he did not identify these strains more exactly. His conclusions are similar to those of Schoel. The decomposition of the urea, he believes, not only leads over the path of ammonia to the formation of amino acids and symbiont proteins but also simultaneously benefits the nitrogen metabolism of the hosts. Hellmuth (1956), using cultures of trypetid symbionts, also concluded that their significance might possibly lie in the transformation of uric acid and urea, inasmuch as syntheses of protein substances could be observed in such nutrient solutions.

The regrettable uncertainty still existing about the true nature of the strains tested by Schoel and Tóth unfortunately pertains also to the statements made by Keller (1950) on a successful pure culture of *Periplaneta* symbionts and their contributions. His cultures were able to develop, though slowly, on a nutrient medium consisting merely of uric acid, cooking salt, and agar, but they were unable to utilize urea. However, later authors have been unable to cultivate a single blattid symbiont, and we must therefore wait for confirmation of Keller's data.

It is fortunate, therefore, that other authors were able to demonstrate equivalent contributions for other insect symbionts, namely, Heteroptera, coccids, and cerambycids. Using Sörensen's method, L. Huber-Schneider (1957; provisionally reported by Koch, 1951) inoculated the bacteria originating from the midgut crypts of *Mesocerus marginatus* from beginning cultures, which were growing well on nutrient agar and glucose, onto uric acid agar or into a uric acid suspension in water with phosphate buffer; then she was able to study how the uric acid crystals dissolved, while at the same time NH_3 and CO_2 were split off as products of uric acid decomposition. Urea, on the other hand, was not attacked by the symbionts of these bugs. With respect to the possible assimilation of atmospheric nitrogen, the author determined that, with addition of gaseous NH_3 derived from the air and with suitable nutrient solutions, the symbionts are able to synthesize body-specific protein, but she assumes that the bacteria do not utilize NH_3 directly, but merely the fraction which is dissolved in the nutrient medium.

Thus the symbionts of *Mesocerus* stand in contrast to those of *Pseudococcus*, which, as recognized by Fink (1952), are excellent urea utilizers but, on the other hand, are not able to utilize uric acid and urate. With 1.0% urea as the source of nitrogen and 0.5% galactose as the C source, the symbionts thrived quite as well as with 0.5% urea without galactose. In the second case, the growth merely set in somewhat later but then rose to the same level. The *Pseudococcus* symbionts, which, as we have already seen, can assimilate atmospheric nitrogen and furnish their hosts with all the essential B-vitamins, can derive their total carbon and nitrogen requirement from urea. The NH_3 which may be expected theoretically in urea decomposition does not make its appearance, in this case because apparently it is immediately consumed again in intermediate metabolism for the aminization processes. Köhler and Schwartz, who also cultured *Pseudococcus* symbionts and thereby found that after extended cultivation they form flagella, also consider it possible that they could bind nitrogen, but they point out that neither they nor Fink has demonstrated this with isotope experiments.

Finally Jurzitza (1959) tested cerambycid symbionts for uric acid and urea and determined that they can utilize urea, and to a lesser extent uric acid, in *Harpium inquisitor* and *mordax* and in *Leptura rubra*. Experiments with *Leptura sanguinolenta* turned out negative, their symbionts in general being distinguished by slow growth and thiamin heterotrophy.

Thus there is little doubt today that symbionts, of yeasts and of bacteria, may play an important role as detoxicants of their hosts and are of considerably importance in the nitrogen-economy of plant-juice suckers with diets poor in protein.[1]

The bacterial symbiosis which is localized in all lumbricids in the ampullae of the nephridia and, interestingly, is not lacking in the glos-

---

[1] Since the fat body and the Malpighian tubules are preferred symbiont sites, it is of particular interest to prove that it is a matter here of locations where growth substances are quite generally stored and concentrated among the insects. This is shown by the investigations of Busnell and Chanoin (1940) and Busnell and Drilhorn (1943) with migratory grasshoppers and house moths, and by those of Kaudewitz (1953) with *Tenebrio molitor*. Kaudewitz found that all growth substances given with food can be detected in the fat body. Inasmuch as they are localized in the renal vessels, they are not given off during the larval period but at the time of reproduction, indicating that they are perhaps used in the construction of the gonads.

soscolecids, which in general represent a pronouncedly convergent group of the New World, apparently occupies an exceptional position among excretion symbioses but, if the ideas developed by Pandazis (1931) should be confirmed, this symbiosis would at any rate increase the nitrogen in the host animals. Pandazis found that, although the body-cavity fluid propelled into the ciliated funnels transmits its excretions to the wall of the nephridial loops, its protein content is still detectable in the ampulla. In the end vesicle, on the other hand, which is emptied slowly every 3 to 4 days, no protein is found and neither are its cleavage products. Hence Pandazis concludes that the protein in the stagnating ampullar fluid is, little by little, entirely decomposed by the bacteria and that the albumoses and the peptones forming in this way are reabsorbed by the ampullar wall. The fact that cultures of the symbiotic bacteria reveal such cleavage capacities would be an argument in favor of this assumption.

Finally we must consider numerous efforts to discover the significance of the bacterial symbiosis of blattids. Attempts to obtain sterile animals began in 1945, when Brues and Dunn reported on the effect of various antibiotics upon *Blaberus cranifer*. They found that injection of sulfonamides was without effect, but that injection of penicillin resulted in increasing damage to the symbionts according to the dosage, causing, in the case of two animals, complete or almost complete disappearance of the bacteria. The opinion of the two authors that even their limited experiments made it seem likely that the symbionts were vital to the hosts was confirmed by Glaser's experiments the next year. Glaser allowed *Periplaneta americana* to feed for 150 days on an aqueous solution of Na sulfathiazole. In males this led to degeneration of the symbionts and finally to symbiont-free animals, which however showed no injury and whose testes contained living sperm. During the course of the experiments the symbiont-free females suffered increasing injury to the ovaries, on and in which the bacteria were lacking, leading finally to maximum reduction of the gonads. When penicillin was injected into the body cavity, the results were the same. When larvae were treated, the formation of the gonads was stopped at the very beginning, whereas in animals which had been made sterile as adults, the ovaries regressed. The loss of symbionts also caused substantial delay in development, which normally takes 250 to 300 days.

Using the method of Koch and Fink in experiments with increased temperatures, Glaser found that *Periplaneta americana* dies at 40°C. and that the bacteria die at 39°. The effects of the loss were the same as when antibiotics were used.

Noland, Lilly, and Baumann (1949, 1949) studied *Blattella germanica* to determine to what extent the blattid symbionts supply vitamins. Using Blewett and Fraenkel's method, they compared symbiont-containing animals fed on a full diet containing all essential vitamins with animals in whose diet various single vitamins were omitted.   They found that absence of choline, pantothenic acid, and nicotinic acid influences growth to a great degree and that all animals die before attaining sexual maturity.   However, the same authors proved that, when betaine is used instead of choline, the effect is the same and the animals then contain almost as much choline as those fed with it.   In other words, *Blattella* can transform betaine into choline and may possibly owe this ability to its symbionts.

According to their data, lack of pyridoxin, thiamin, or riboflavin evidently checks growth, yet sooner or later the animals usually attain sexual maturity, thus making it seem likely that these three vitamins are supplied by the symbionts.   Further experiments showed this to be almost certain in the case of riboflavin.   Animals brought to maturity on a riboflavin-free diet nevertheless contain about 0.2 μ gm. of riboflavin each.   It must be admitted that the casein making up a part of the diet might also contain small quantities of this vitamin, but, to obtain such an amount of riboflavin from this source alone, the cockroaches would have to eat food amounting to 50 times their body weight.[1]

Omission of biotin, folic acid, inositol, vitamin K, and *p*-aminobenzoic acid had no effect whatever.   Since the first two are among the vitamins essential to both non-symbiotic insects and those deprived of their symbionts, it may be assumed that they are supplied here in sufficient quantity by the symbionts.   However, inositol has always proved to be superfluous to all insects, with or without symbionts.   The same might apply for vitamin K and *p*-aminobenzoic acid.

House (1949) and Hilchey (1953) studied the amino acids needed by *Blattella germanica,* arriving at results which indicate the possibility that the symbionts play a role in their formation also.

Only on the surface do the findings of Noland, House, and co-workers appear to contradict Glaser's finding that *Periplaneta americana,* when freed of symbionts, develops to the healthy adult, for Glaser fed his animals a diet of dog biscuit and sugar, in other words, a standard vitamin-rich diet, which, according to the findings of Noland and co-workers, is equivalent to their synthetic full diet.   Furthermore, the findings agree with the fact that *Blatta* and *Periplaneta,* though not adapted to an unbalanced, vitamin-poor diet, make very small demands in the way of diet and its content of growth substances.

---

[1] Metcalf and Patton (1942), however, were of the opinion that *Periplaneta americana* synthesizes little or no riboflavin.

In recent years a large number of investigators have striven to develop our knowledge of blattid symbiosis (Frank, 1954, 1956; de Haller, 1955; Brooks and Richards, 1955, 1955, 1956; Brooks, 1957, 1959, 1960, 1962, 1963; Manuta and Bernardini, 1958; Manuta, Bernardini, and Laudani, 1961; Fava and Laudani, 1960, 1960, 1961; Fava and Barbato, 1960; Selmair, 1962, 1962; Henry, 1962; Pierre, 1962, 1964; Malke, 1964). Here the question of vitamin supply by the symbionts has played a minor role; the chief interests have been the different ways of producing aposymbiotic animals and the study of the effect of symbiont loss on the mycetocytes, the ovaries, and any offspring, as well as the comparison of the two sexes. The methods used by these workers for obtaining symbiont-free animals were the same in principle as those of their predecessors. De Haller and Selmair confirmed Glaser's data on the effect of increased temperatures upon the symbionts. When the animals were kept for approximately 6 weeks at 38° or 39°C., the mycetocytes gradually washed away, the mortality increasing to 80% according to Selmair. Brooks and Richards also worked with increased temperatures but chose degrees of heat and exposition periods which were different from those used by Glaser. In their experiments the symbionts were not decomposed completely.

Of the fifteen antibiotics tested by these authors, aureomycin and sulfathiazole were the most effective. The safest and simplest method of obtaining aposymbiotic animals, in their opinion, is continuous feeding of 0.1% aureomycin with dog biscuit. On the other hand, Frank established that penicillin works most quickly, aureomycin slightly less so, and terramycin more slowly than the latter. He found that chloromycetin and streptomycin are even less efficient and that their use causes the symbionts to become resistant. The animals then deposit normal cocoons, but the larvae which arise therefrom possess only deficiently infected mycetocytes, the cause of which presumably lies in an insufficient infection of the eggs. However, aerosporin, which is effective only when gram-negative bacteria are present, has no influence at all upon the gram-positive symbionts. Selmair's data coincide approximately with those of Frank. Using chiefly 5% aureomycin, she found that the larvae became sterile after 70 to 80 days. Terramycin and penicillin had the same effect, but streptomycin and hostacycline were less effective. She recommends 5% aureomycin, with a temperature of 38°C. for 35 days, followed by normal temperatures. After 80 days with completion of the last molting, the imagines become symbiont-free.

Brooks and Richards tried in vain to reinfect the sterile animals by implantation of mycetocyte-containing fatty tissue or by injection of symbiont-containing flotations. With the first, the ovaries were never infected. With the second method, the tissue grew but the eggs were nevertheless not infected. It is hardly surprising, in view of these results, that it was impossible to transplant the symbionts of *Periplanta* and *Blatta orientalis* to aposymbiotic *Blattella germanica*.

All the authors mentioned, like Glaser before, established, in each case according to the arrangement of the experiment, retardation of development, differing in its severity, and increased mortality. According to Selmair, the cells become increasingly smaller and the symbionts more and more compact as the mycetocytes degenerate. The nuclei, which often shift toward the wall, finally disappear, or the cells grow enormously, are greatly vacuolized, and are finally destroyed. In other cases, the nuclei of wasted mycetocytes also fuse and giant cells arise. Beyond this there is also degeneration of the surrounding fat cells. But, according to Frank, larvae which undergo a completely symbiont-free embryonic development nevertheless have various mycetocytes, empty yet recognizable as such—an indication that this cell material is determined *a priori.*

The fact that the uric acid content of the fatty tissue rises considerably in the aposymbiotic animals (Selmair speaks of an inudation) is an argument that with blattids also the symbionts serve to process the excretions, but it has not yet been possible to prove it with certified pure cultures. In the meantime, with *Leucophaea maderae* as the test object, proof has been given by Pierre (1964) that the symbionts as well as the symbiont-containing and symbiont-free fat tissue contain a uricase.

Frank presents more details on the changes which occur in the form of the bacteria only a few days after the antibiotics are administered. First the bacteria always become gram-negative, as established previously by Glaser. Then elongated or locally distended forms or even cocci-like forms make their appearance and are gradually decomposed completely. Certain differences among the various antibiotics were regularly observed.

In these new investigations interesting details are given about the way the animals reacted to loss of symbionts. Here we shall follow chiefly the data of Frank, since he was the only author to separate observations of animals which were fed with penicillin, aureomycin, or terramycin from those fed with chloromycetin and streptomycin. Since he had found resistant symbionts with the latter two, it was to be expected that the differences resulting would allow us to draw important conclusions about the contributions of the symbionts. At first the animals quite generally showed some resistance to receiving food mixed with antibiotics, but this resistance was soon overcome. The consumption of food was normal at first but soon decreased, and the animals

were smaller than the control group, although their behavior in general
and their vitality in particular remained the same (Fig. 371).   The animals which received one of the first three antibiotics mentioned were not
only different in size but also underwent a change of coloring, a change
which was observed by all authors but unfortunately is not noticeable
in our illustration.   As they become sterile, the animals gradually become
lighter, and the difference is even more pronounced with the second
molting.   Only in the pro-imaginal and the imaginal states does the
coloring change to reddish brown, so that one can instantly recognize

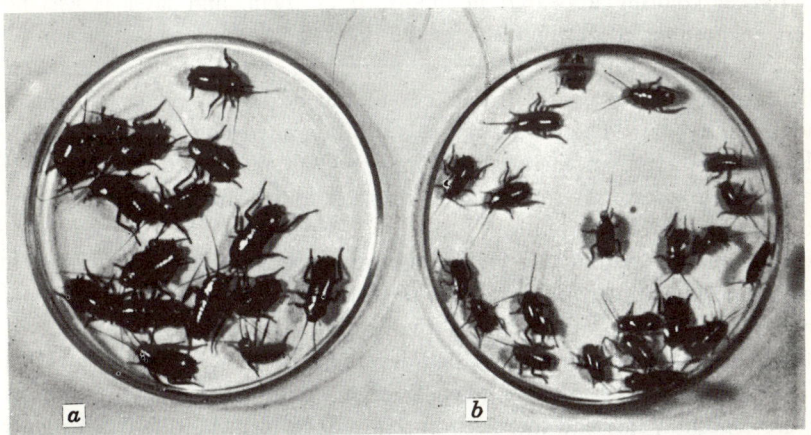

Fig. 371   Restraint of growth by removal of symbionts in *Blatta orientalis* L.
(*a*) The large dark insects possess resistant symbionts; (*b*) the small light
insects are aposymbiotic.   From Frank.

the successfully treated animals.   Tying-off experiments by Frank and
other observations indicate that this defective coloring of the cuticula
does not proceed from the cerebral endocrine system, but is directly
conditioned through the loss of symbionts.   A further result of the decrease and final elimination of the symbionts is the retarded development,
individual animals often remaining at the same stage for 2 months
or longer.   The stubby wings of the females, like the fully developed
wings of the males, are frequently spread at an angle and sometimes
incorrectly unfolded, as often occurs with *Tribolium* as the result of
vitamin deficiency.

Whereas the normal female imagines soon gain considerable weight,
the aposymbiotic females remain unchanged, a situation traceable to
the retardation in ovarial growth already noted by Glaser.   Histological

investigations show that, although the ovaries are always formed, their growth is frequently retarded in early larval states, with the result that they are barely recognizable under the dissecting microscope. If growth does occur, it is always the plasma and follicle of the oldest oocytes which exhibit clear signs of degeneration. In this experimental series of Frank, in which the animals were never fed with chick feed containing vitamins B_1, B_2, and B_{12}, coccoons were never formed.

The behavior of animals fed with chloromycetin or streptomycetin was completely different. After the first molting they all were somewhat lighter in coloring than the control group. After the second molting a darker coloring appeared in one-third of them, and later it was no longer possible to distinguish the cultures from the control group by color. Certain differences in size were also observed. The smaller, lighter cockroaches were symbiont-free; the darker, larger ones teemed with symbionts! The darker animals still deposited normal coccoons, the larvae hatched from them were somewhat lighter than the control, and not all the prospective mycetocytes formed under sterile conditions had the normal symbiont filling. From these clear informative experiments we see once more that the symbionts are indispensable for normal growth as well as for normal coloring. On the other hand, the resistant symbionts all undergo a critical gram-negative period and thereby lose some of the qualities needed for undisturbed development.

The penicillin injections given over a period of 14 days to adults were especially fast and thorough in their effect, just as in the experiments of the other authors. The bacteria were all gram-negative by the second day but as such could still be found after 14 days. During this period the females deposited a number of coccoons, but they were small and soon were shriveled up. Nevertheless, a number of young were hatched that were still light-colored and completely symbiont-free. Thus it was shown that gram-negative symbionts are unable to infect the oocytes. On the other hand, it was shown that the presence of symbionts is not necessary for embryonic development. The same was shown by the observations on centrifuged *Pediculus* eggs and by the physiological studies of Sander (1959, 1960) on eggs of the cicada *Eucelis,* in which the constricting off of polar symbiont clumps in no way disturbed development.

Frank also tested the extent to which it is possible to prevent the decreased growth of sterile larvae by administration of vitamins. A variety of vitamins in pure substance, multibionta, Baker's yeast, and *Candida* were investigated, but a substantial increase in growth was never effected. Nor was a single coccoon deposited by the 500 experimental animals. The ovaries continued to be retarded in their development, and only with feeding of polybion and folic acid did there result somewhat larger oocytes which were visible to the naked eye but never attained maturity.

In strong contradiction to these results are the experiments by Brooks and Richards with *Blattella germanica* and those by Frank with *Periplaneta orientalis*. These authors found that symbiont-free larvae that are unable to grow on the normal diet for symbiont-containing cockroaches do mature fully, though extremely slowly, after large quantities of beer yeasts are fed. In such animals reproductivity is reduced in both sexes. Crossing sterile females with normal males and vice versa, the authors found, in addition to some empty coccoons, also a few with eggs and larvae, the latter becoming infected or sterile according to whether the mother possessed symbionts or not. The second and third generations of sterile descendants showed no change in behavior and histology.

Selmair, who also worked with *Blattella germanica*, had similar results. She failed to obtain mature ovaries with animals fed on a 5% aureomycin diet at 27° or 38°C., but, when she returned them to normal conditions after 35 or 50 days and fed them 50 parts normal diet and 50 parts *Torulopsis utilis*, they formed coccoons which hatched only 10 to 15 symbiont-free larvae but these developed into imagines with well-developed ovaries.

Conclusive evidence that the intracellular symbionts of *Blattella germanica* can provide amino acids for their host was obtained through the use of radioactive isotopes. Henry and Block (1960) showed that utilization of $Na_2S^{35}O_4$ for the synthesis of cystine and methionine occurred only when the symbionts were present. Similarly the cockroach incorporated C^{14} from glucose-U-C^{14} into "essential" and "non-essential" amino acids, but only into the latter when freed of their intracellular bacteria (Henry, 1962) by treatment of the preceding generation with aureomycin. In a related study, the lighter color of aposymbiotic cockroaches was replaced by normal dark coloration when the insects were fed a diet rich in tyrosine or phenylalanine, precursors of melanine pigments (Henry). Under normal conditions, these aromatic substances, present in low concentrations in the diet, are supplemented by bacterial synthesis from glucose.

Removal of intracellular symbionts also lowers in some way the normal resistance of the cockroach to bacterial invasion. The numbers and types of gut bacteria increase significantly (Henry). However, these commensals, in contrast to the intracellular symbionts, do not affect the nutritional requirements or metabolic capabilities to any recognizably significant degree.

Although experimental analysis of blattid symbiosis has made considerable progress since the first edition of this book (1953), more study is still required. The contradictions set forth here must, of course, be resolved and the vitamin contributions of the symbionts must be studied further.[1] However, it is equally important to study more closely the effect of diet deficiency on the development of the female gonads and to investigate whether the development is inhibited indirectly by the corpora allata, whose function is inhibited by the protocerebrum and activated by the ganglion pharyngeum inferior, as Engelmann (1957) discovered with the blattid *Leucophaea maderae*. The first attempts by Springetti (1964) to transplant corpora allata and brain in *Nauphoeta* remained at first without positive results (see also Füller, 1960). Of great importance also are pure cultures of the symbionts and proof that the symbionts are able to utilize the uric acid deposited in the fat body. Furthermore, more genera should be compared, and whenever possible *Mastotermes darwiniensis* should be included, because it manifests a bacterial symbiosis identical with that of the blattids and also contains the flagellates typical of most termites. According to a personal communication of Cleveland (Brooks, 1963), *Mastotermes* is surprisingly capable of surviving an interruption of its reinfection for a shorter period of time than higher termites without bacteriocytes, when they lose their gut symbionts during molting. This two-fold symbiosis appears therefore not to be necessary for life, a fact which would make understandable the otherwise general loss of the bacterial symbiosis in termites.

The main direction of the various symbiotic contributions seems clear today. An exceptional position is occupied by a great variety of marine animals, where the light-producing capacity of certain bacteria is made to serve ecological purposes and where, in contrast to all other symbioses, the symbionts play no role in the metabolic processes of the host.

On the other hand, the supplying of growth substances, in particular the numerous vitamins of the B-group, appears to be of outstanding significance for growth and metamorphosis. Many insects provide a welcome source from which to form amino acids by pressing into service microorganisms which are able to assimilate atmospheric nitrogen. Symbionts with available uricase not only help to detoxify their hosts and thus reduce the burden of their excretory organs, but also make possible

---

[1] Manuta and her co-workers struck out in a new direction by developing a method which made it possible for them to free a considerable number of pure undamaged bacteriocytes of *Blaberus cranifer* from the fatty tissue and to undertake chromatographic investigations on this material. So far they have been able to demonstrate only the production of pantothenic acid.

a partial reutilization of their decomposition products for the synthesis of protein bodies.

In view of these important contributions, the role of furnishing digestive enzymes, which were at first frequently presumed to be present, recedes into the background. The frequent occurrence of symbionts in the lignivorous insects in particular let one think that they might possibly produce a cellulase enabling the host to use the wood as food (Buchner, 1928, 1930; Uvarov, 1929), but since then an animal-specific cellulase has been detected in a number of forms. By analysis of excrements, Falk (1930) proved a decrease in the cellulose content of the food in *Hylotrupes bajulus* and *Anobium pertinax,* and Schlottke and Becker (1942) detected cellulase in the intestinal canal of the first-mentioned. Ripper (1931), Mansour and Mansour-Bek (1933), and W. Müller (1934) were also able to confirm a greater or less capacity for cellulose digestion with other cerambycids (*Leptura, Rhagium, Cerambyx, Oxymirus, Macrotoma, Gracilia*) and other anobiines (*Anobium striatum, Xestobium*). Also, Deschamps (1951) found production of cellulase and cellobiase in cerambycids. Yet these positive results should not be generalized without further reflection. In another cerambycid (*Xystrocera*) such capacity is lacking (Mansour and Mansour-Bek), and neither the *Cossus* larvae (Ripper) nor the splint beetle *Lyctus* (Campbell, 1929) produces cellulase. With such forms, only the soluble sugars and the starch of the wood come into question as sources of carbohydrate. Accordingly, such apparently lignivorous insects tend to prefer the splintwood to the heartwood and are characterized by particularly large quantities of excrement and long periods of development frequently extending over several years.

This result is in agreement with the facts that symbiont-containing forms occur among insects producing cellulase as well as among those which do not, and that no corresponding enzymes were detectable in cultures of *Rhagium* and *Ernobius* yeasts (W. Müller).

Consequently, participation of endosymbionts in the cellulose utilization of their hosts might be confined in non-vertebrates to cases where they live in fermentation chambers and similarly extended intestinal sections, and beyond this participation would come into question, at most, with certain protozoa (polymastigines). But even here opinions differ considerably. In addition to positive results based on the intestinal flora of *Potosia, Oryctes, Osmoderma,* and *Geotrupes* larvae (Vaternahm, 1924; Werner, 1926; Wiedemann, 1930; Pochon, 1939), there are also negative results (Ripper, 1931). With reference to termites, several publications are available on cellulose-dissolving microorganisms produced from their enormously extended hindgut (Holdaway, 1933; Baldacci and

Verona, 1939, 1940, 1941; Hungate, 1946), but to all appearances they play a decisive role in instances where flagellates are lacking but wood nevertheless represents the principal food. When we consider the significant role of bacterial decomposition of cellulose in the intestine of mammals and in birds, it seems very probable that similar adaptations were made in the area of invertebrates.

Another unclarified question is the significance of bacterial symbiosis in *Pelomyxa*. Keller (1949) reported successful cultures of the ever-present inmates of these amoebae and their ability to decompose cellulose, but this ability was contradicted, on good grounds, by Leiner, Wohlfeil, and Schmidt (1951, 1954). In their opinion the *Pelomyxa* symbionts are engaged in protein metabolism. Also undecided is the role of probable bacterial symbionts in the polymastigines of the termite gut. Pierantoni considers them to be the actual cellulase producers, whereas others reject this view as unproved or do not even recognize the microorganismic nature of the inclusions in question (Grassé, 1952). On the other hand, there is agreement on the vital significance of the flagellate fauna of the termite gut in cellulose digestion, regardless of whether the termites themselves or the endosymbiotic bacteria produce the necessary cellulase.

It is important to recognize that one and the same symbiont may affect the metabolism of the host in different ways simultaneously. Thus Fink's studies with *Pseudococcus* showed that vitamins are made available, that atmospheric nitrogen is utilized, and that the urea of the hosts is used in protein synthesis. With *Mesocerus* there is detoxication of the host by decomposition of metabolic products and also production of growth substances. The blattids are indebted primarily to their symbionts for regulated growth, ability to develop mature ovaries and the dark coloring of the cuticula. With the aphids one may expect production of vitamins, assimilation of atmospheric nitrogen, and utilization of uric acid. The symbionts of certain bloodsucking animals not only furnish vitamins but also take part in the feeding and digestion of blood.

With the broadening of experimental symbiosis research, such multiple achievements of one and the same symbiont will undoubtedly become even more apparent.

Of course, it should not be forgotten that these capacities of the symbionts have often been established merely with pure cultures or experimental tests far from the body of the animal host, but nevertheless there are important grounds for believing that they play a corresponding role in the host itself. For instance, they are always significantly connected with the feeding habits of the host animals! In this respect the experimental investigations confirm *in toto* the validity of the considerations which originally led to discovering the regions of distribution. For the most part it was some insufficiency in food sources which led to the establishment of symbiosis, or, better stated, certain food sources became available to the animals only after they had symbionts at their disposal to compensate for the deficiencies. Although it has been proved that quite a number of vitamins indispensable to insects are present in vertebrate blood, they are not all present in the necessary quantity, as we have seen, and there is little doubt that only when this deficit is compensated for by symbiosis is specialized feeding on blood alone made possible. With wood destroyers, where the food is low in protein as well as vitamins, two-fold demands are made on the symbionts.

One might be misled into deciding that symbiosis is not vital to insects, for some which have very similar diets rich in carbohydrates, such as seeds of all kinds and particularly grains, flour, and flour products, in some cases possess symbionts and in other cases are able to get along without them. However, this vagueness readily disappears upon more exact consideration. The symbiont-containing forms, *Sitodrepa, Lasioderma, Oryzaephilus, Rhizopertha, Calandra,* and other snout beetles, are each and all derived from lignivorous ancestors, which, as the equivalent relatives of today show, had already acquired their symbionts. However, the symbiont-free forms, *Ephestia, Tribolium, Ptinus, Tenebrio,* all belong to groups in which no symbiont bearers have been discovered. This applies also to tenebrionids, ptinids, and Lepidoptera; indeed, among the latter even extreme lignivorous forms like *Cossus* have not learned to improve their position by acquiring symbionts. With such change in feeding habits, the symbiosis may understandably lose significance, but experiments with Anobiinae and comparison with symbiont-free forms show clearly that even in these cases the possession of symbionts permits the animals to thrive on diets impossible for *Tribolium, Ptinus,* etc. If, then, for some reason a symbiotic cereal eater such as the Egyptian variety of *Calandra granaria,* should lose its inmates, it will therewith necessarily sink to the stage of the more demanding *Tribolium.*

Similar dietary insufficiencies are the deciding factor when plant-juice suckers acquire symbionts. Here, too, the food is low both in protein and in the vitamins needed by the insect. That acquisition of symbionts and the food regime are intimately connected was demonstrated most clearly by the Heteroptera, some of which exchanged their predatory manner of life for one dependent on plant juice.

Although the reasons for the sporadically occurring Mallophaga symbiosis are at present not quite so clear-cut, there are indications that it likewise was acquired as the result of unbalanced diet, in this case, horny substance. On the other hand, the cause of the symbiosis confirmed in two forms living in tree sap, *Dasyhelea* and *Nosodendron,* remains undecided for the present.

It is understandable that the utilization of metabolic end products by symbionts is not necessarily connected with unbalanced diet. Because it can be extremely valuable for a protein-poor diet (Heteroptera, Homoptera) one is inclined to suspect that it will one day be proved even with extreme lignivorous insects; however, such utilization is also found in omnivora like the blattids. Experimental studies are needed to show in what direction to seek the significance of symbiosis with Camponotinae and Formicidae. This symbiosis is of particular interest because it is the only one known among the Hymenoptera and is apparently in the process of abolition. It already seems probable that symbionts producing B-vitamins are involved, for F. Smith (1944) found that the growth of larvae is influenced very little in *Camponotus herculaneus pennsylvanicus* when the workers feeding them are given a synthetic diet without yeast.

The foregoing indicates clearly that not only is knowledge of the respective contributions of the symbionts necessary in further developing our field, but also knowledge of the qualitative and quantitative composition of the food of the symbiont bearers. Since 1953, valuable data for symbiosis research have been obtained, and particularly by the *Tribolium* test. Pollen, leaves, wood, roots, seedlings, ambrosial fungi, cereals of various types, blood, honey, etc., have already been tested for B-vitamin content (I. Schwarz and Koch, 1954, 1961; Koch, 1955; Koch and I. Schwarz, 1955), but researchers in symbiosis are constantly confronted with questions to which there is no answer. Not only do differences exist from species to species in the composition of the respective nourishment, but even between the separate regions of the same food plant. In the case of plant-sap suckers, we need moreover to know to what extent they actually do suck out cells, and, with the insects populating galls, which so seldom have symbionts, we would like to know what the inner wall of the gall has to offer. With animals sucking vertebrate blood, the main question is whether substances indispensable for the insect are lacking in the blood or, if they are present, how the quantities of the substances taken up by the parasites in the course of development are related to their requirement. Among symbiont bearers living in "wood" there are important differences in the value of the food, according to whether they populate the splint or the heartwood, and these values vary greatly with individual trees. Little information is available on the content of protein and reserve substances in the wood serving as food, and publications and studies such as those on *Lyctus* by S. E. Wilson (1933) and Parkin (1938, 1939) or those by Becker on *Anobium* and cerambycid larvae represent commendable exceptions. Comparative studies on the vitamin content of various types of wood and wood areas appear to be lacking entirely, yet such studies are necessary, for example, to understand the fact that cerambycid symbiosis is restricted to larvae which feed on living and dead coniferous wood and dead deciduous wood and is completely absent among those which feed on fresh deciduous wood, either hard or soft, or on plants. A method exists for studying the requirements of various insects for substances in wood without withdrawing them from their natural habitat, for Becker (1942, 1943), by extracting woods or impregnating them, discovered that a concentration of watery yeast extracts promotes in the same way the development of symbiont-free cerambycids and of *Anobium punctatum.*

The complications encountered in symbiosis research have been considerable and will undoubtedly be even greater in the future in view of the increasing refinement of experimental methods. Already foreseeable is the need to guard against all generalizations regarding the contributions of the symbionts. Even the symbionts of animals so closely related as *Sitodrepa* and *Lasioderma* provide distinctly different contributions, and *Sitodrepa* in particular is at a clear disadvantage with respect to other anobiines in that its yeasts produce no thiamin. Far greater differences appear when the various *Candida* species living symbiotically with cerambycids are compared. *Cimex lectularius* is better supplied with B-vitamins by its bacteria than is *Ornithodorus,* but both are dependent upon the same blood diet. The development of *Coptosoma* merely shows considerable retardation after loss of symbionts; the larvae of the bug *Brachypelta* cannot develop without symbionts, and they perish; and *Eurydema,* another leafbug, appears to bear loss of symbionts without injury. The symbionts of *Oryzaephilus* furnish at best a fraction of the necessary B-vitamins, whereas the anobiines with very similar habits, insofar as they have been investigated, derive all or almost all their vitamins of the B-group from their yeasts. With blattids, too, the possibility must be considered that the symbiont contributions vary with the genera.

For normal development, not only must all these growth substances be present but also they must be administered in a definite dose if regulated interplay is to result. When Kreitmaier (1952), at the request of Koch, was searching for a test animal in which the effect of the individual vitamins could be tested more quickly than with *Tribolium,* he found such an object in the Paramecia. When he used them to test the vitamins of the B-complex and a number of amino acids alone and in combination with the vitamins, he found that the optimal effect of the individual substances varied greatly.   Dosages which are optimal for thiamin may have a lethal effect with beta-biotin and B_{12}. An oversupply of certain vitamins can therefore retard rather than accelerate growth.   On the other hand, it was shown that the rate of increase in the paramecia rises by leaps and bounds to far above the maximum of the partial components when vitamins and amino acids are administered together.   I. Schwarz and Koch (1961) have determined for *Tribolium* the optimal quantities, and those which represent the lowest boundary, of eight B-vitamins.   In each case it is a question of quantities that fluctuate between 0.1 and 200 mg. and yet guarantee a good effect.   The deficiency of individual vitamins also varies greatly in result. With deficiency of pantothenic acid, the larvae die very early; with subdosage of riboflavin, highly striking malformations occur, especially wings that stand off obliquely or are more or less crippled.   With still other vitamins the injuries are unspecific.   There can be no doubt that such complicated reciprocal reactions also play a role in the metabolism of the symbiont hosts and that retardation and acceleration in symbiont increase, varying with individual stages of development, are connected in each case with the desired dosage of the substances to be furnished by them.

The need for substances produced by the symbionts is apparently determined temporally, or at least attains a maximum at certain periods. This is shown, for example, by comparing symbiotic and non-symbiotic males and females of *Pediculus,* and it may also be concluded from the periodic acceleration and retardation in the reproductive rate of the symbionts.   In the bedbug, lack of thiamin has no influence on growth rate, but its presence becomes indispensable for forming a normal quantity of healthy eggs capable of development.   On the other hand, *Ornithodorus* is greatly hampered without thiamin even during the growing period.   It will also be recalled that symbiotic organs degenerate in various early stages, in a number of forms (cicadas, snout beetles), and that their inmates are consequently expelled or degenerate.   Furthermore, the two sexes may differ in this respect, apparently not merely because in the female a certain number of symbionts must be preserved for transmission, but also in connection with formation of eggs.   Such differences have been observed not only with the bedbug but also with a number of other test objects, such as *Pediculus, Rhodnius, Pseudococcus,* and blattids.

It must especially be taken into consideration that to activate pupation, or with forms lacking pupation to activate the last molting, requires special substances to release the necessary growth substances.   De Meillon,

Golberg, and Lavoipierre (1945) have shown this in a clear way with *Aedes* larvae kept under sterile conditions, for these need no folic acid up to the third larval phase, but then are incapable of pupating without them. An even more cogent illustration is presented by the larvae of certain bark-devouring beetles. The larvae of two pyrochroids, *Synchroa punctata* and *Dendroides canadensis,* are unable to pupate on sterilized oak bark but do so easily if the bark has not been sterilized or contains the mycelia of *Armillaria nigra,* a fungus which is usually present in old oak bark. However, if larvae which have lived 6 months to 6 years in sterile bark are offered non-sterilized wood or the mycelium mentioned, and therewith apparently certain vitamins, pupation sets in rapidly within 5 days (Payne, 1931).

In all previous experiments carried on to examine the significance of endosymbiosis, the best test animals were forms that live symbiotically with a single type of microorganism. Far greater complications may be expected, then, in symbioses involving two, three, or more symbiont types, and even without experimental proof the deep significance of this accumulation of symbionts emerges in the light of our present knowledge. The additional forms are usually installed with great regularity in the host organism, and cases of older acquisitions show the same complicated, reciprocal effects during embryonic development and with regard to transmission devices. However, it would be illogical, and it would destroy the unity of the phenomenon to be expected, if we were to regard the auxiliary, companion, and accessory symbionts as insignificant forms which have merged, as it were, to utilize the tolerance of the primary symbionts, simply because there are cases where the additional forms, especially the accessory symbionts, show lack of adaptation or even traits suggesting parasitism. The considerations which make it possible to construct a family tree for cicada symbiosis preclude *a priori* the assumption that one symbiont may be useful with plurisymbioses and others of no significance, for it has been shown with great cogency that forms which were originally the additional forms, for instance, the *x-* or *H*-symbionts, become the sole symbionts when older forms are excluded.

We must also consider that our earlier findings on the contributions of single symbionts easily explain such multiplicity. With additional guests the vitality of the hosts can be improved considerably. Vitamins which the primary symbiont is incapable of producing may be produced by another. Bacteria or yeasts assimilating atmospheric nitrogen may ally themselves with symbionts which furnish only vitamins. A third symbiont type may perhaps decompose the metabolic end products of the host. All these are possibilities predicated on what is known today, and we should not be intimidated by the complexity of the effects of the interrelations.

That additionally admitted symbionts apparently can bring with themselves an improvement of the vital situation of the hosts proceeds from those cases where a primary symbiont is completely pushed aside by an unquestionably younger symbiont. For instance, with some hormaphidines acquisition of the yeast causes the disappearance of the typical aphid symbionts, and with certain *Stictococcus* species it eliminates the bacteria generally distributed there. It would be of great interest in such cases to compare the contributions of the primary symbionts with those of the secondary symbionts.

Our phylogenetic discussions of plurisymbioses show that reciprocal effects take place not only between host and symbionts but also often necessarily between the individual symbiont types which are housed wall to wall. For the problem of utility it is especially significant that it must be in no way only a case of antagonistic relationships, but quite as well also harmonious cooperation and supplementation must be reckoned with, features known so well from mixed cultures. When, for example, *Mucor ramannianus* and *Torula rubra* are cultivated together in a nutrient solution which is free of such substances, by itself each form can synthesize only a single component of thiamin, but when the two are cultivated in symbiosis whole thiamin is formed (Schopfer, 1938; F. W. Müller, 1941). It is also possible that a symbiont living anaerobically would be able to utilize its capacity for assimilating atmospheric nitrogen only when aerobic symbionts make its growth possible. Influences accelerating or inhibiting growth, as frequently observed in mixed cultures, will necessarily play also a role in microorganisms associated in the insect body and may possibly have a favorable result in adapting newly acquired organisms.

To obtain understanding of these operations is a difficult though not insoluble problem of experimental symbiosis. The chief need will be to cultivate the various symbionts separately and in mixed cultures, and to analyze their contributions. Various temperature sensitivities might be utilized to eliminate one or another symbiont and then to determine the resultant deficiency phenomena. One could experiment with the chrysomelid *Bromius*, the only insect known to transfer two symbiont types simultaneously through smearing of the eggshell, by providing sterilized egg surfaces with one type each and then comparing the growth of normal animals with that of symbiont-free animals and those lacking one symbiont.

In an impressive report on enzymatic adaptations in microorganisms without change of the hereditary supply, Leiner (1958) pointed out that, by the insertion of the symbionts into the hosts' body the type and quality of very plastic, adaptive enzymes which are sometimes spontaneously equivalent to a new situation play probably, an essential role. Another field of work is thus opened to experimental symbiosis research.

The extent of the principle emerges clearly from what is known today about the significance of endosymbioses of animals with fungi and bacteria. The areas in which such symbioses occur are essentially already marked out, and great surprises are hardly to be expected. In luminescent symbiosis a new case may occur now and then in the course of time, primarily among fishes, but it is beyond doubt that the overwhelming majority of the luminous animals emit their own light. Very differently constituted types of symbiosis which play a role in the metabolism of animals are lacking, above all, in invertebrates which are predators, stir up plankton or graze on algae, are carrion feeders, or live endoparasitically, either constantly or sporadically; thus it will obviously be necessary to exclude a large numbers of the insects which are the chief domain of endosymbiosis. For instance, nothing comparable has ever been found with the Ephemeroptera, Plecoptera, Odontata, Neuroptera, or predatory Hymenoptera and Coleoptera. Also the Aculeata, which live on a mixed diet or do not use animal protein at all, are among the insects which are not inclined to establish endosymbioses. Indeed, there are only scattered cases among ants (camponotines and *Formica fusca*).

Important orders just as disinclined toward endosymbiosis are some that feed on fresh plant parts, for example, Orthoptera, Lepidoptera, and Tenthredinoidea. Endosymbiosis does not even occur with butterflies which as grubs live in wood, such as *Cossus*, and for which it would clearly be advantageous to acquire symbionts. This lack of endosymbiosis is all the more surprising because, in other insect groups where the same way of feeding plays a considerable role, symbiotic forms occur side by side with non-symbiotic representatives. This is especially true of the Coleoptera, where many curculionids and some chrysomelids which feed on fresh plant parts have symbionts, and it is true to a lesser extent of the Diptera, where the trypetids are symbiont bearers but so many other forms which mine or produce galls are not. No doubt fresh plant tissue represents a food so rich in proteins, carbohydrates, mineral salts, and vitamins that symbionts are superfluous in the overwhelming number of cases. However, as yet there is no explanation why a symbiosis was nevertheless established in the cases mentioned.

The Coleoptera and Diptera are undoubtedly orders where a good many cases of symbiosis are as yet undiscovered. One needs only to recall the many Coleoptera families, occurring for the most part only in tropical regions, which live in fresh or moldering wood, in seeds, fruit capsules, etc., such as the anthribids, aglycyderids, proterrhinids, and brenthids. All these are rhynchophores, which are closely related to the curculionids and ipids, and are thus under suspicion of symbiosis. There are also the cupedids, trichoctenids, eucnemids, monommids, cebrionids, cerophytids, and a good many others. The field might be extended by many noteworthy findings, especially if the interest of entomologists in tropical and subtropical regions could be engaged. The same situation exists with the Diptera, although here our previous findings indicate that additions might be fewer in number. Where pollen represents their principal food, no symbiotic devices can be expected, for research by Schwarz and Koch (1954, 1955) showed that almost all varieties of pollen have an extraordinarily high vitamin content which renders symbiosis superfluous.

Even in the orders where adaptation of symbionts has been pre-eminently established, much systematic, morphological investigation still remains to be done. Among the Heteroptera it is necessary to draw the line more closely between species with and without symbionts, and among the Homoptera numerous tribes are still awaiting research. Since recently a definite, though not yet thoroughly studied, symbiosis in one species of the Thysanoptera has been reported, it is also important here to determine the extent of its occurrence. The Thysanoptera are like the Heteroptera in that predatory and phytophagic forms occur side by side. Furthermore, their mouth parts strongly resemble those of the Hemiptera, and the opinion has been expressed that they have the same ancestors as do the Hemiptera. The distribution of symbioses in animals living on vertebrate blood has been almost exhaustively considered, but research on the land leech would be desirable, for a positive result might indeed be expected here. Also, the Dermaptera *Hemimerus* of hamster rats and *Arixenia* of bats ought to be tested for possible symbiosis.

Our presentation has been limited to invertebrates, but in considering the scope of the principle of endosymbiosis we must not forget that it is valid for vertebrates also. It is true that organs serving exclusively as symbiotic sites are never found among them and complicated transmission mechanisms are completely lacking, even an infection of egg cells by symbionts; nevertheless localized bacterial symbiosis plays a substantial, indeed vital, role in the intestines of birds, mammals, and man. Here we can make only brief reference to a field carried on chiefly by bacteriologists, physiologists, and physicians. (For literature see also Baumgärtel, 1940; Hungate, 1946; Stepp, 1957; Catel, 1957; Piekarski, 1965.)

Quite understandably, interest has centered chiefly on the role of bacteria in the intestine of domestic animals and man. Today there is no question that the bacteria which quite regularly line the stomach mucosa of horses contribute substantially to the decomposition of carbohydrates. Involved here are several types of lactic acid bacteria amylolytically producing the lactic acid which appears immediately after feeding and is important in the intermediate metabolism of the horse; and other fermentations caused by bacteria also play a role, as attested by concomitant production of acetic and butyric acid. The invariable presence of bacteria with a proteolytic effect explains why part of the protein content of the food is already decomposed before it comes in contact with the gastric juices.

Ruminants behave differently from herbivores with single-cavity stomachs, chiefly in regard to cellulose digestion. It is generally agreed that the enormous quantities of bacteria which populate the rumen play an essential role in decomposition of cellulose-rich food. Just as in fermentation chambers of many larvae, the bacteria disintegrate the food abundant in raw fiber by anaerobic fermentation of the cellulose and thereby facilitate later fermentative digestion in adjacent sections of the stomach. Moreover, they provide the hosts with a source of protein which should not be underestimated. Since 1 gram of the rumen content contains about 13 billion bacteria, innumerable masses of them are digested after the approximately 40 kg. of food which is processed daily moves to the abomasum and the small intestine. Furthermore, B-vitamins are formed in abundance by these bacteria. With riboflavin, the quantity exceeds that of the food by 30- to 100-fold. Thus we see that the high content of this vitamin in cow's milk is chiefly due to the bacteria of the rumen. Nowhere else among the higher animals, as Stepp wrote in 1957, is the significance of symbiosis so convincingly illustrated as with the function of the rumen. The cecum of cattle, however, is substantially smaller than that of the horse and plays a less important role in cellulose digestion.

Research with chickens shows that the very strongly developed, paired ceca serve as places of cellulose fermentation by bacteria. When they are surgically removed, the digestibility of raw fibers of various types is cancelled or at least substantially reduced. It is understandable, then, why birds of prey have slightly developed ceca in comparison to chickens, geese, swans, etc., and that wood hens, which in winter feed almost entirely on a cellulose-rich diet such as conifer needles, have enormous ceca which are sometimes as long as the intestine itself.

The term "vitamin feces" is certainly proper for the special liquid feces originating from the cecum and excreted by rats and other rodents in addition to hard fecal pellets. This liquid is easy to overlook, because it is immediately eaten. If its consumption is prevented, as happens when the rats are kept on a wide-mesh screen which allows the feces to fall through, then the animals die. Not even a diet of greens and grain can save them from perishing; only a polyvitamin preparation can replace this cecotrophy. The same is true for other rodents (Almquist, 1947; Jacquot, Armand and Rey, 1941; Williams, 1943; Kellner, 1956).

We hear also that newborn guinea pigs eat the maternal soft excrement immediately after birth and that after only 36 hours the intestine begins to be settled with the indispensable organisms, much as with the ditch bugs *Brachypelta,* the young larvae of which for days devour only the bacterial excrement of the mother.

The human newborn are already provided with the necessary intestinal flora at birth. The *Lactobacillus* so important for the infant and the so-called Döderlein bacillus are found in the maternal vagina and reaches the mouth of the child as it is born. This form of transmission also takes place with such regularity that Rimpau (1934) actually called the maternal vulva a lubricating organ equivalent to similar transmission devices of insects. Of course, many other vaginal organisms originating chiefly from the maternal intestine are transmitted along with *Bacterium bifidum* to the child's digestive tract, but it is this bacterium which controls the qualitative and quantitative settlement of the individual intestinal sections in a thoroughly regulated manner from the very first day of life.

We know of *Bacterium bifidum* that it is also of great significance for the synthesis of thiamin. It produces almost the same amount of this vitamin as yeast does, and thereby compensates for the low thiamin content of mothers' milk. Another important factor is the antibiotic effect of this bacterium on a number of pathogenic organisms (Catel, 1957; and others). The *coli* flora of the human intestine is no less important, because it can synthesize in differing amounts all vitamins necessary for growth and vitamin K (Stepp, 1953, 1956, 1957). That all these vitamins are actually absorbed by the intestinal wall is beyond all doubt.

The high degree to which normal functions of the human intestine are dependent upon a very definite composition of its flora follows from the mere fact that disturbances in digestion occur when penicillin, sulfonamides, and other medicaments with antibiotic action are administered.

In the ciliates of the ruminant stomach of mammals we have a counterpart of the symbiotic flagellates of the termite and blattid intestine, although of course the conditions are not the same in all respects. It may be regarded as certain that the great quantities of the *Diplodinium* species, in continuous and abundant multiplication in the stomach of ruminants, produce, in contrast to other infusoria living there, a cellulase by means of which they digest the cellulose in their interior, and it is also certain that they store glucose as a product of this hydrolysis (Hungate, 1943, 1946, 1950). On the other hand, in research with ruminants which were freed of ciliates it was shown that the bacteria, the only other agents capable of cellulose digestion, are sufficient to attain the same utilization of the food. At any rate, it may be assumed that both types of symbionts participate in utilization of plant fibers under normal conditions and that continuous digestion of large quantities of symbionts is a necessity for the animal host.

At the Sixteenth International Congress of Zoology in Washington, D. C., in 1963, H. Sprinz discussed many interesting immunological and physiological findings in germ-free research. He and his co-workers have made valuable contributions to this science which is so intimately related to symbiosis (Sprinz, 1962; Sprinz et al., 1961; Abrams, Bauer, and Sprinz, 1963). Sprinz and his co-workers confirmed the pathogenicity of *Vibrio cholerae* for germ-free guinea pigs, a finding first reported by Cohendy and Wollmann (1922). Sprinz showed that the natural intestinal flora of the conventional guinea pig exerts a protective action against *V. cholerae;* vibrios were eliminated from the gut within 24 hours, and no histologic sequelae were observable. A significant number of germ-free animals, however, succumbed to the disease, and a series of histologic changes in the intestinal tract accompanied the ingestion of these organisms. Three hours after oral administration of the vibrios, the mucosa of the cecum showed emptying of the goblet cells, altered shape of crypt glands, and increased cellularity of the tunica propria, and corresponding changes were found in the small intestine and colon. On this occasion Sprinz drew upon the results of Foeckler (1961) on aposymbiotic *Sitodrepa* and pointed out the relationships between vertebrates and invertebrates in regard to symbiosis and immunity.

Our knowledge of bacteria living symbiotically in the intestine of birds, mammals, and man developed independently of endosymbiotic research among invertebrates, and, because of the nature of the field, it will continue to develop chiefly within the framework of nutrio-physiology in domestic animals and of medical bacteriology. Nevertheless, it must be kept in mind that basically it is only a subdivision of the widely distributed animal symbiontology developed in the last fifty years in connection with understanding the true nature of the pseudovitellus of the Homoptera.

In the beginning, little attention was paid to this insight and its true significance was denied, but the field has developed into a scientific discipline with significance ranging far beyond the limits of zoology proper, with close relationships to the neighboring fields of comparative physiology, bacteriology, mycology, and the study of infection, and thus ultimately to medicine. Symbiosis research has provided zoological science with the knowledge of abundant devices previously unknown or misunderstood. It has discovered the multiple sites of the symbionts and the organs and devices serving transmission, and it has shown that the two partners are already intimately related during embryonic and larval development. Symbiosis research can also furnish a basis for evaluating related questions, and it already has various criteria to assist in explaining the phylogeny of the individual symbioses which sometimes have deep-reaching effects on the formation of the hosts and their mode of life. For instance, extreme adaptations to an exclusive diet of plant-juices, such as the more or less sessile and thereby considerably modified coccids or the immotile stages of the aleurodids, have doubtless been caused or at least completed by acquiring the symbionts which make such specialization possible. The same applies to hippoboscids, which are just as extensively transformed, and to some other blood-suckers; and in the last analysis the intense reproductivity of aphids is due to their symbionts, for these were acquired before the aphids became viviparous. Thus symbiosis research frequently contributes to deeper understanding of the organization of host animals.

The science of nutritional physiology of invertebrates, especially of insects, has received great stimulus from symbiosis research, and the understanding of vitamin requirements in animals is being increasingly enriched by experimental symbiosis. For instance, it has led to more penetrating analysis of the B-vitamins united in yeast and of other growth-promoting substances and has thus led indirectly to their practical utilization in man and domestic animals. This development was due primarily to Goetsch and his co-workers, who used in their experiments a yeast extract ("vitamin T Goetsch") containing all essential, vital, and growth-promoting substances (Goetsch, numerous publications from 1946 to 1957; Kaiser, 1949; Goetsch and Lang, 1949; Goetsch and Nussbaumer, 1950; Härtel, 1950; Vogel, 1954). We shall not discuss here the details of their results with hydras, tadpoles, ducks, chickens, pigs, calves, and human infants, since all these studies have only a loose connection with symbiosis research. Koch's school reached similar positive results using a purer yeast extract in feeding experiments with pigs, rats, and mice (Hedler, 1952; Bandier, see Koch, 1951; and Jobst, 1961). When Lengsfeld (1951) administered to dystrophic and atrophic infants a preparation with optimal effect upon the growth of *Tribolium,* he obtained an increase in weight, often beginning spontaneously, with more than half of his experimental subjects. All these investigations again show that the requirement for the B-vitamins of yeasts is essentially the same for both insects and vertebrates.

Symbiosis research, though it acquaints us with an abundance of regular, vital adaptations of plant microorganisms in animal bodies, is even more directly related to bacteriology, mycology, and immunology. Workers in symbiosis have discovered numerous microorganisms that were previously unknown and have demonstrated extensive changes in form that were entailed by millions of years of intracellular living. The strictly fixed cyclical changes contribute substantially to our modern understanding of the polymorphism of bacteria. The traditional theory of hygienists concerning "defense reactions" will have to be radically revised on the basis of those widespread harmonious adaptations that are "intended" by the animal organism. Professional bacteriologists will no longer be able to ignore the new findings brought to light by endosymbiosis research. In 1951 the hygienist Harmsen was expressing this idea when he wrote, "The phenomenon of intracellular symbiosis represents a new stage of knowledge for bacteriology and the theory of infection, comparable in importance with the discovery in the 70's of disease-causing capacities of bacteria and with the observations by Metschnikoff on the process of phagocytosis as a defensive function." Another happy symptom of this change was the decision of the Sixty-third Convention of the German Society for Interior Medicine to devote the entire first day of its 1957 proceedings to the relations between symbiosis research and medicine (Buchner, Koch, Stepp, Catel, Höring, and others).

May this comprehensive presentation of a field that has been casting its spell upon me for more than fifty years serve both to stimulate further work in symbiosis and to encourage the interest and cooperation of workers in related fields, so that upon the foundations which are now laid that stately edifice make take form which we already envision today.

Plate I

The symbionts of *Donacia semicuprea* Panz. From Stammer.
(a) From the organ of transmission of a middle-aged female.
(b) From the organ of transmission of a female ready to lay eggs.
(c) From the blind sacs of a 12-hour-old larva.
(d) From the blind sacs of a mature larva.
(e) From the blind sacs of a pupa-larva.
(f) Commonplace bacteria from the gut of a pupa-larva.

Plate II

(a–d) The symbionts of *Tephritis heiseri* Trfld. 1000×. From Stammer.
(a) From the blind sacs of a larva.
(b) From the gut of an imago, just arisen from the pupa.
(c) From the gut of a 1-day-old imago.
(d) From the gut of an imago ready to lay eggs.
(e–f) The symbionts of *Pseudococcus adonidum* L. 800×. From Walczuch.
(e) From an imago.
(f) From a larva.

Plate III

(a) The symbionts of a larva of a *Pseudococcus* species. 800×. From Walczuch.
(b) The symbionts of an imago of *Pseudococcus diminutus* Leonardi. 800×. From Walczuch.
(c–l) The symbionts of different hippoboscids. About 600×. From Aschner.
(c) *Melophagus ovinus* L., symbionts and rickettsia.
(d) *Hippobosca camelina* Leach.
(e) *Hippobosca camelina* Leach, symbionts and rod-shaped companion bacteria.
(f) *Hippobosca equina* L.
(g) *Hippobosca capensis* Olfers.
(h–i) *Lipoptena caprina* Austen.
(k) *Ornithomyia avicularis* L.
(l) *Lynchia maura* Bigot.

Plate IV

The symbionts of *Camponotus ligniperda* Latr. (HCl-Giemsa stain according to Piek-arski-Robinow.) About 1300×. From Kolb.
(a) From the midgut epithelium of a young pupa (worker).
(b–c) From the midgut epithelium of a 4-week-old worker (with transformation forms).

831

(*d*) Symbionts from an 8-week-old worker.
(*e*) Branching of symbionts from a 26-week-old worker.
(*f*) An extremely long thread-shaped symbiont from a 34-week-old worker.

Plate V

(*a*) Electron micrograph of symbionts from *Blatta orientalis* L.; between these are mitochondria and in the upper right are islands with staff-shaped bodies.   From Frank.
(*b–c*) Smears of symbionts from *Blatta orientalis* L.   Gram stain.   720×.   From Frank.
(*d*) Symbionts from *Pseudococcus citri* Risso; culture forms on beef bouillon-agar with peptone and glucose added (fresh mount).   From Koehler.
(*e*) Symbionts from *Pseudococcus maritimus* Ehrhorn; growth form in isolation in a hanging drop (fresh mount).   From Koehler.

Plate VI

(*a–c*) Symbionts from *Mesocerus marginatus* L. on nutrient-glucose-agar.   1600×. From Huber-Schneider.
(*a*) From the upper surface of the agar (fresh mount).
(*b*) Stained smear from a 3-day-old primary culture.
(*c*) After 20 transfers.
(*d–f*) Symbionts from *Cydnus* and *Brachypelta*.   From Schorr.
(*d*) Symbionts from the crypt gut of a young female of *Cydnus flavicornis* Fabr.
(*e*) Symbionts from the sexually mature female *Cydnus flavicornis* (transmission forms).
(*f*) Symbionts from the crypt gut of a female *Brachypelta aterrima* För st which was caring for her offspring.
(*g*) The chief symbionts from *Aphrophora salicis* De G. (Stain according to Piekarski-Robinow.)   About 1100×.
(*h*) Symbionts from *Oryzaephilus surinamensis*.   Treated with ribonuclease.   About 1240×.   From Kolb.

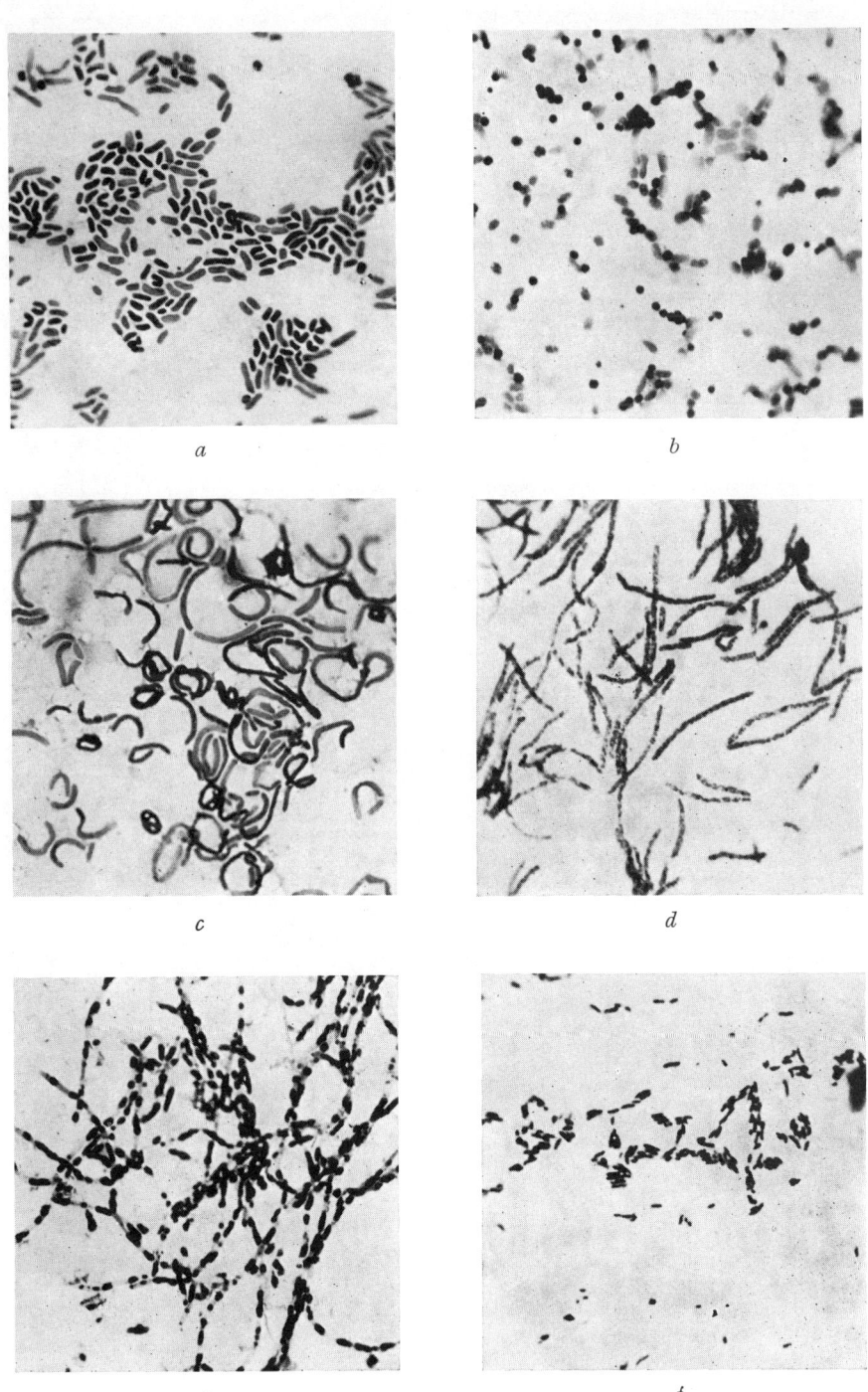

Plate I

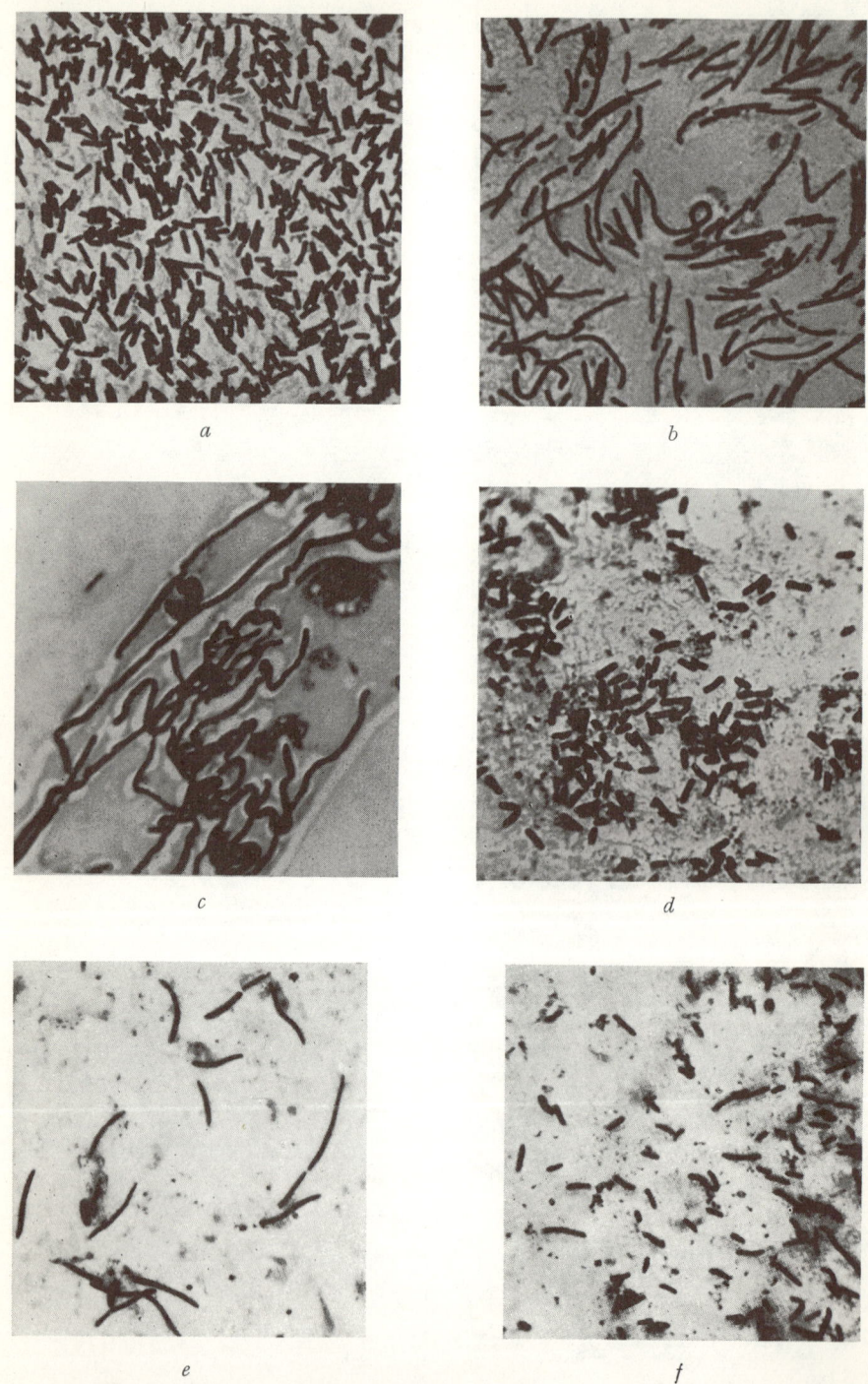

a

b

c

d

e

f

Plate II

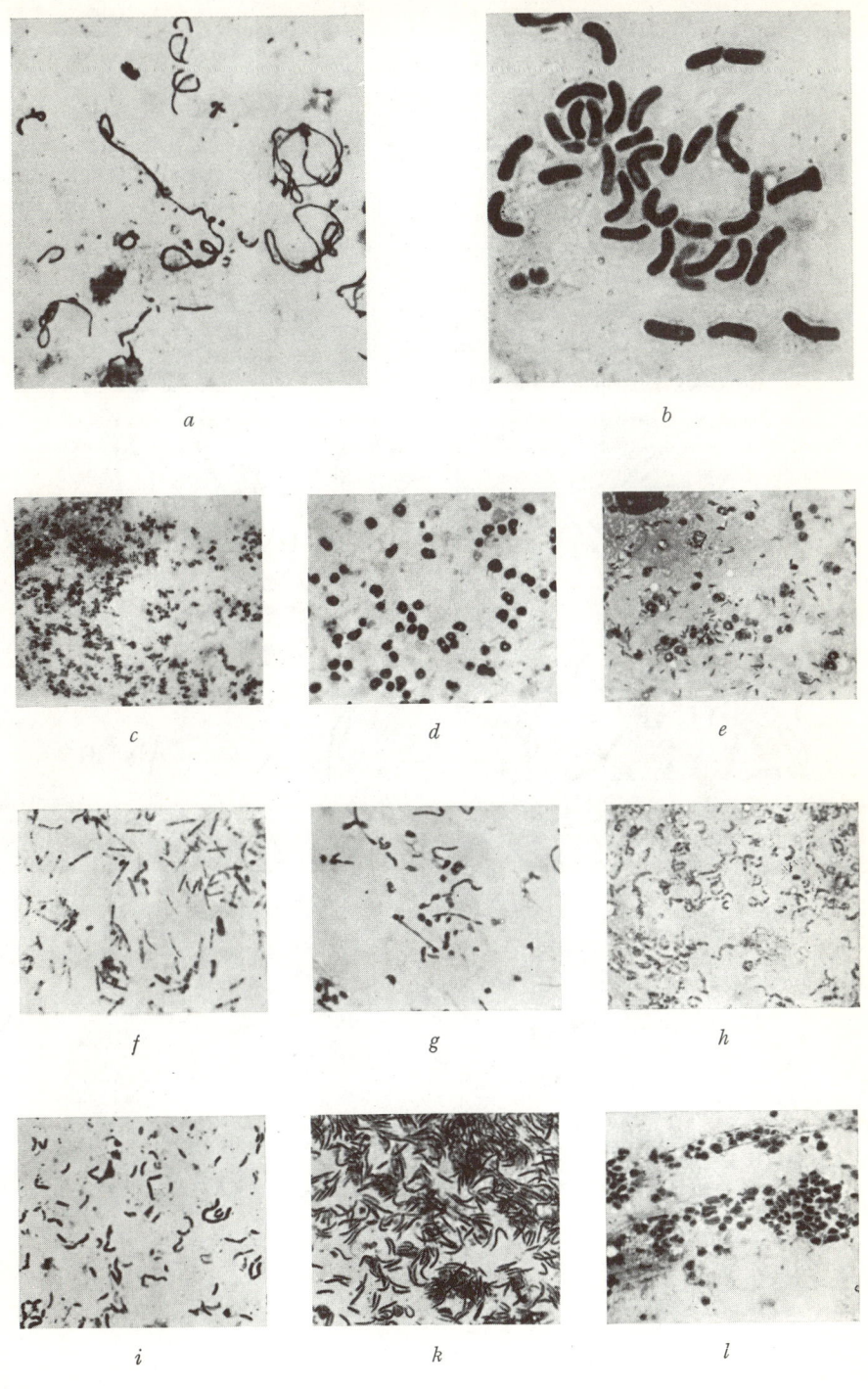

Plate III

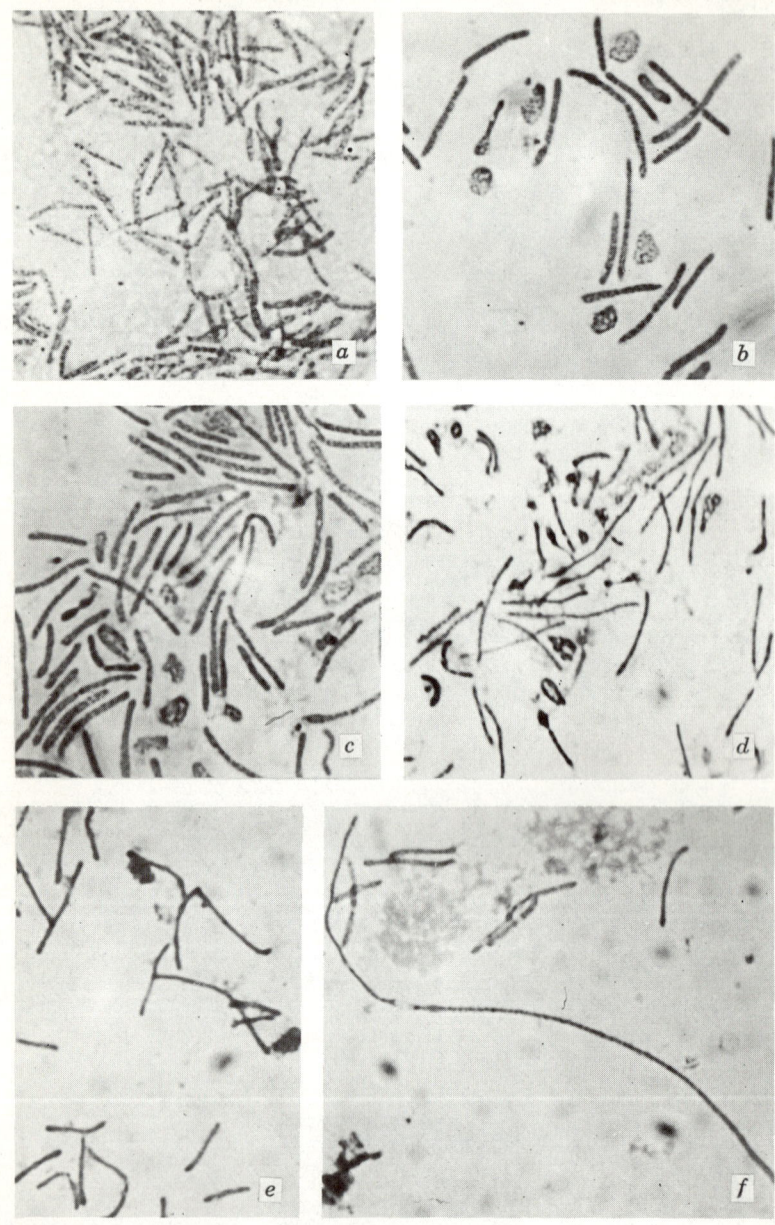

Plate IV

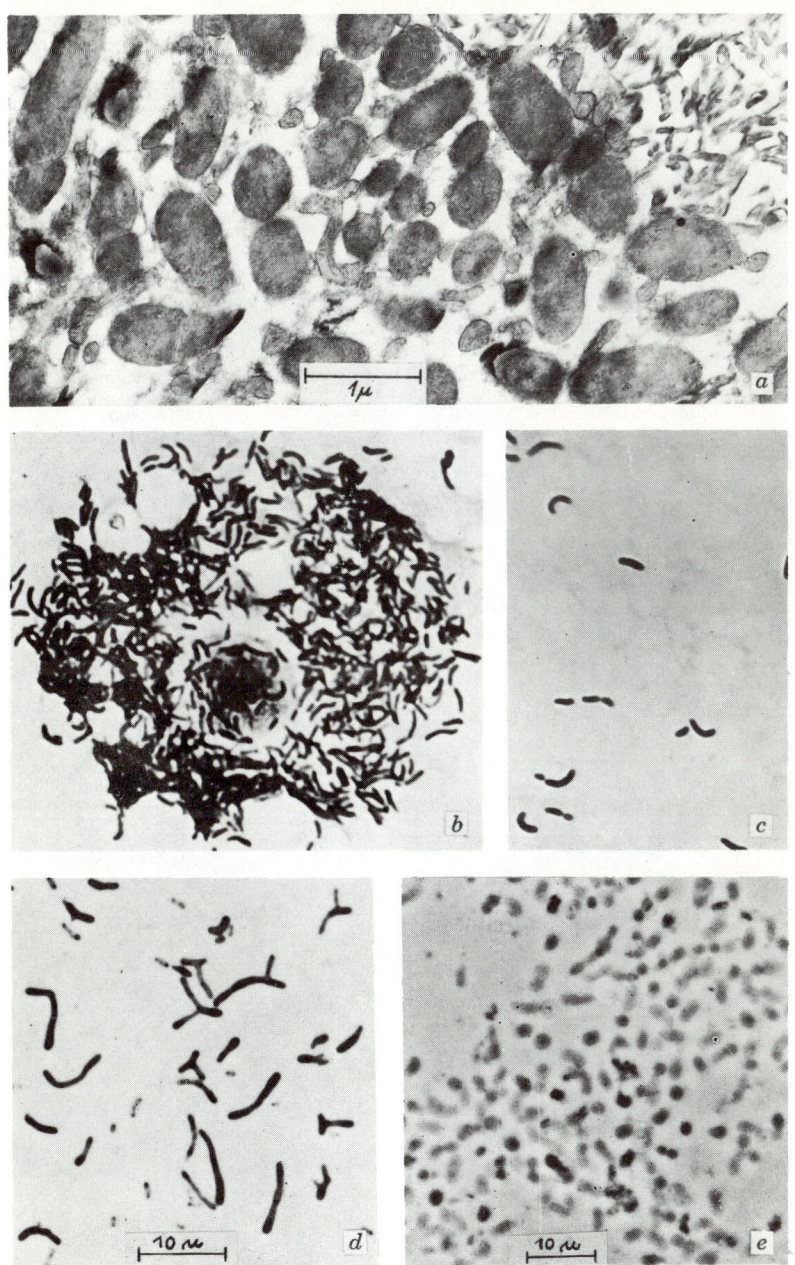

Plate V

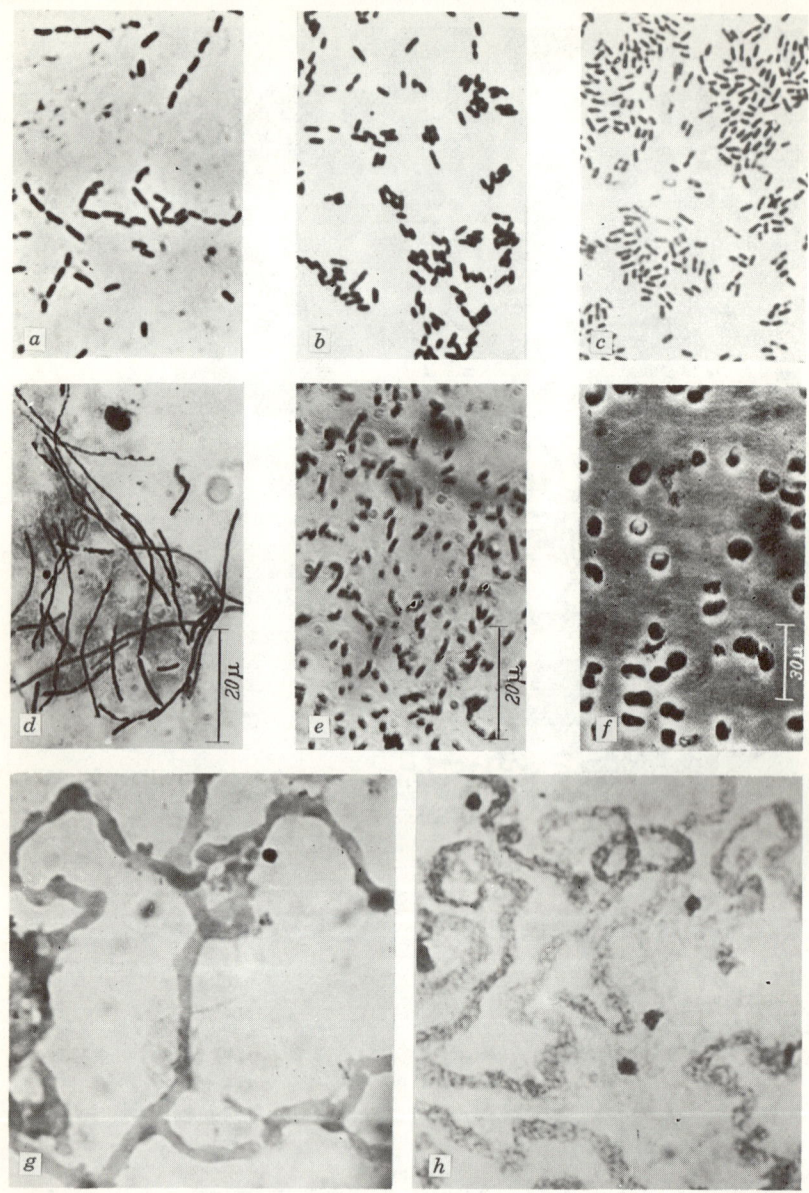

Plate VI

References

Abe, T., *A record of Anomalops katoptron Bleeker from Japan*, Annotationes Zool. Japan. *21* (1942).
——, *A record of Anomalops katoptron from Hachijo Island*, Japan. J. Ichthyol. *1* (1951).
Abrams, J. D., Bauer, H., and Sprinz, H., *Influence of the normal flora on mucosal morphology and cellular renewal in the ileum; a comparison of germ free and conventional mice.* Lab. Invest. *12* (1963).
Adlerz, G., *Om digestionssekretionen jemte några dermed sammanhängande fenomen hos insekter och myriopoder*, Bihang Kgl. Svensk. Vet. Akad. Handl. [4] *16* (1890).
Aeschlimann, A., *Développement embryonnaire d'Ornithodorus moubata (Murray) et transmission transovarienne de Borrelia duttoni*, Acta Trop. *15* (1958).
Allen, T. C., Pinckard, J. A., and Riker, A. J., *Frequent association of Phytomonas melophthora with various stages in the life cycle of the apple maggot, Rhagoletis pomonella*, Phytopathology *24* (1934).
Allen, T. C., and Riker, A. J., *A rot of apple fruit caused by Phytomonas melophthora n. sp., following invasion by the apple maggot*, Phytopathology *22* (1932).
Almquist, H. J., *Vitamin K*, Physiol. Rev. *21* (1941).
Altenburg, E., *The symbiont theory in explanation of the apparent cytoplasmatic inheritance in Paramaecium*, Am. Naturalist *80* (1946).
Altmann, R., *Die Elementarorganismen und ihre Beziehung zu den Zellen* (Veit, Leipzig 1890).
Anadón, E., *Sobre simbiosis endocelulares de ephipigerinos*, Bol. real. soc. españ. hist. nat. *41* (1943/44).
Anigstein, L., *Untersuchungen über die Morphologie und Biologie der Rickettsia melophagi Nöller*, Arch. Protistenk. *57* (1927).
Archer, W., *On some freshwater Rhizopoda*, Quart. J. Microscop. Sci. *9–11* (1869–1871).
——, *A resume of recent observations on parasitic algae*, Quart. J. Microscop. Sci. *13* (1873).
Arkwright, J. A., Atkin, E. E., and Bacot, A., *An hereditary rickettsia-like parasite of the bedbug (Cimex lectularius)*, Parasitology *13* (1921).
Arkwright, J. A., and Bacot, A., *Stages of Rickettsia in the sheep-ked, Melophagus ovinus*, Trans. Roy. Soc. Trop. Med. Hyg. *15* (1921).
Arnal, A., *Algo sobre los simbiontes de los mosquitos*, Bol. real. soc. hist. nat. *41* (1943).
Aschner, M., *Die Bakterienflora der Pupiparen (Diptera). Eine Symbiosestudie an blutsaugenden Insekten*, Z. Morphol. Ökol. Tiere *20* (1931).
——, *Experimentelle Untersuchungen über die Symbiose der Kleiderlaus*, Naturwissenschaften *1932*.
——, *Studies on the symbiosis of the body-louse, 1: Elimination of the symbionts by centrifugalisation of the eggs*, Parasitology *26* (1934).
——, *The symbiosis of Eucampsipoda aegyptica Mcq. (Diptera, Pupipara: Nycteribiidae)*, Bull. soc. Fouad I entomol. *30* (1946).

Aschner, M., and Ries, E., *Das Verhalten der Kleiderlaus beim Ausschalten der Symbionten*, Z. Morphol. Ökol. Tiere *26* (1933).

Atkin, E. E., and Bacot, A., *The relation between the hatching of the eggs and the development of the larvae of Stegomyia fasciata (Aedes calopus), and the presence of bacteria and yeasts*, Parasitology *9* (1917).

Bacot, A., *On the survival of bacteria in the alimentary canal of fleas during metamorphosis from larva to adult*, J. Hyg. *13*, Plague Suppl. III (1914).

⸻, *Reports on questions connected with the investigation of nonmalarial fevers in West Africa*, Yellow Fever Commission to West Africa *3* (1916).

Baines, S., *The role of the symbiotic bacteria in the nutrition of Rhodnius prolixus (Hemiptera)*, J. Exptl. Biol. *33* (1956).

Bakshi, B. K., *Fungi associated with ambrosia beetles in Great Britain*, Trans. Brit. Mycol. Soc. *33* (1950).

Balachowsky, A., *Les Cochenilles de France, d'Europe, du Nord de l'Afrique et du Bassin Méditerranéen*, Vol. IV (Hermann & Cie., Paris 1948).

Balbiani, M., *Mémoire sur la génération des Aphides*, Ann. sci. nat., zool. et biol. animale [5] *11* (1869); *14* (1870); *15* (1872).

⸻, *Recherches expérimentales sur la merotomie des Infusoires ciliés*, Rec. zool. suisse *5* (1889).

Baldacci, E., *Studi sulle termiti. Schizomiceti e protozoi cellulolitici nell'intestino delle termiti*, Riv. biol. coloniale (Rome) *4* (1941).

Baldacci, E., and Verona, O., *Isolamento di schizomiceti del g. Cystophaga dall'intestino delle termiti*, Boll. soc. ital. biol. sper. *14* (1939).

⸻, *Sulla presenza di schizomiceti cellulolitici nell'intestino di Reticulitermes lucifugus e Calotermes flavicollis*, Boll. soc. ital. biol. sper. *15* (1940).

Balfour, A., *Spirochaetosis of sudanese fowls*, Fourth Report of the Wellcome Tropical Research Laboratories at Gordon Memorial College Khartoum, Vol. A (1911).

Barber, M. A., *The food of anopheline larvae. Food organisms in pure culture*, Public Health Rept. *42* (1927).

⸻, *The food of culicine larvae. Food organisms in pure culture*, Public Health Rept. *43* (1928).

Barbieri, C., *Forme larvali del Cyclostoma elegans Drap.*, Zool. Anz. *32* (1907).

Batra, S. R., and Francke-Grosmann, H., *Contributions to our knowledge of ambrosia fungi*, 1: *Ascoidea hylecoeti sp. nov. (Ascomycetes)*, Am. J. Botany *48* (1961).

Baudisch, K., *Zytologische Beobachtungen an den Mycetocyten von Periplaneta americana L.*, Naturwissenschaften *43* (1956).

⸻, *Beiträge zur Zytologie und Embryologie einiger Insektensymbiosen*, Z. Morphol. Ökol. Tiere *47* (1958).

Baumgärtel, T., *Mikrobielle Symbiosen im Pflanzen- und Tierreich*, Slg. "Die Wissenschaft," Vol. 94 (Vieweg, Braunschweig 1940).

Béchamps, I., *Microzymas* (Montpellier 1875).

Becker, G., *Untersuchungen über die Ernährungsphysiologie der Hausbockkäferlarve*, Z. vergleich. Physiol. *29* (1942).

⸻, *Ökologische und physiologische Untersuchungen über die holzzerstörenden Larven von Anobium punctatum De Geer*, Z. Morphol. Ökol. Tiere *39* (1942).

⸻, *Zur Ökologie und Physiologie holzzerstörender Käfer*, Z. angew. Entomol. *30* (1943).

⸻, *Beobachtungen und experimentelle Untersuchungen zur Kenntnis des Mulmbockkäfers (Ergates faber L.)*, 2. Communication, Z. angew. Entomol. *30* (1943).

Beckwith, T. D., and Rose, E. J., *Cellulose digestion by organisms from the termite gut*, Proc. Soc. Exptl. Biol. Med. *27* (1929).

Beeson, C. F. C., *The life history of Diapus furtivus Sampson*, Indian Forest Records *6* (1917).

Begen, L., *Histologische und physiologische Untersuchungen über den Darm der Grylliden*, Dissertation, Erlangen 1948 (unpublished).

Behrenz, W., and Technau, G., *Versuche zur Bekämpfung von Anobium punctatum mit Symbionticiden (Ein Beitrag zur experimentellen Symbioseforschung)*, Z. angew. Entomol. *44* (1959).

Beijerinck, M. W., *Kulturversuche mit Zoochlorellen, Lichenogonidien und anderen niederen Algen*, Botan. Ztg. *48* (1890).

Bender, H., *Studien zum L-Cyclus der Bakterien (nach Untersuchungen an Proteus vulgaris unter Einfluss von Penicillin)*, Arch. Mikrobiol. *28* (1957).

Benedek, T., and Specht, G., *Mykologisch-bakteriologische Untersuchungen über Pilze und Bakterien als Symbionten in Kerbtieren*, Zbl. Bakteriol. [1] *130* (1933).

Bennett, Fr. D., and Brown, S. W., *Life history and sex determination in the diaspine scale, Pseudaulacaspis pentagona (Targ.) (Coccoidea)*, Can. Entomologist *90* (1958).

Berkeley, C., *Green bodies of Chaetopteridae*, Quart. J. Microscop. Sci. *73* (1930).

———, *Symbiosis of a Beroë and a flagellate*, Contrib. Can. Biol. Fish. *6* (1930).

Berlese, A., *Le cocciniglie italiane, viventi sugli agrumi*, I: *Dactylopius*, Riv. patol. vegetale *2* (1893).

———, *Sopra una nuova Mucedinea parassita del Ceroplastes rusci*, Redia *3* (1905).

Beutler, R., *Experimentelle Untersuchungen über die Verdauung bei Hydra*, Z. vergleich. Physiol. *1* (1924).

Bewig, F., and Schwartz, W., *Untersuchungen über die Symbiose dér Triatomiden Rhodnius prolixus Stål und Triatoma infestans Klug.*, Naturwissenschaften *41* (1954).

———, *Über die Symbiose von Triatomiden mit Nocardia rhodnii (Erikson)*, Naturwissenschaften *42* (1956).

———, *Untersuchungen über die Symbiose von Tieren mit Pilzen und Bakterien*, VII: *Über die Physiologie der Symbiose bei einigen blutsaugenden Insekten*, Arch. Mikrobiol. *24* (1956).

Biedermann, W., *Die Ernährung der Insekten*, in: Winterstein, *Handbuch der vergleichenden Physiologie*, Vol. 2, 1. Hälfte (G. Fischer, Jena 1911).

Bierry, H., Marchoux, E., Martin, L., and Portier, P., *Rapport de la commission nommée par la Société de Biologie*, Compt. rend. soc. biol. *83* (1920).

Bigelow, R. P., *The anatomy and development of Cassiopea xamachana*, Mem. Boston Soc. Nat. Hist. *5* (1900).

Bischoff, U., *Die postembryonale Entwicklung der Gonaden und symbiontischen Einrichtungen von Calligypona pellucida F. (Homopt. Cicadina)*, Diss. math. nat. Fak. Freie Univ. Berlin 1955 (unpublished).

Blanc, G.-R., *Les spirochètes; leur évolution chez les Ixodidae*, Thèse de Paris (Jouve et Cie. Paris 1911).

Blewett, M., and Fraenkel, G., *Intracellular symbiosis and vitamin requirements of two insects, Lasioderma serricorne and Sitodrepa panicea*, Proc. Roy. Soc. (London) *B 132* (1944).

———, see also Fraenkel, G., and Blewett, M.

Blochmann, Fr., *Über die Metamorphose der Kerne in den Ovarialeiern und über den Beginn der Blastodermbildung bei den Ameisen*, Verhandl. naturhist. med. Verein Heidelberg [N. S.] *3* (1884).

———, *Über die Reifung der Eier bei Ameisen und Wespen*, Festschr. naturhist. med. Verein Heidelberg 1886; prelim. commun. in: Verh. naturhist. med. Verein Heidelberg [N. S.] *3* (1884).

———, *Über das regelmäßige Vorkommen von bakterienähnlichen Gebilden in den Geweben und Eiern verschiedener Insekten*, Z. Biol. *24* [N. S.] *6* (1887).

Blochmann, Fr., *Über das Vorkommen von bakterienähnlichen Gebilden in den Geweben und Eiern verschiedener Insekten*, Zbl. Bakteriol. *11* (1892).

Bode, H., *Die Bakteriensymbiose bei Blattiden und das Verhalten der Blattiden bei aseptischer Aufzucht*, Arch. Mikrobiol. 7 (1936).

Bohn, G., and Drzewina, A., *Les "Convoluta." Introduction à l'étude des processus physico-chimiques chez l'être vivant*, Ann. sci. nat. zool. [10] *11* (1928).

Bonde, R., *Some conditions determining potato-seed-piece decay and blackleg induced by maggots*, Phytopathology *20* (1930).

Bonhag, Ph.F., and Wick, J. R., *The functional anatomy of the male and female reproductive systems of the milkweed bug, Oncopeltus fasciatus (Dallas) (Heteroptera, Lygaeidae)*, J. Morphol. *93* (1953).

Bonnemaison, L., *Remarques sur la symbiose chez les Pentatomidae*, Bull. soc. entomol. France *51* (1946).

Boratynski, K., *Sur l'anatomie de la femelle de Margarodes polonicus Ckll.*, Compt. rend. soc. biol. *99* (1928).

――――, *Intracelularna symbioza u czerwca polskiego*, Odbitka z Pamiętníka *14*. Zjazdu. Wrzesién (1933).

Borchsenius, H. S., *Evolution and genealogical connections of Coccoidea*, Zool. Zhur. *37* (1958).

Borghese, E., *Ricerche sul batterio simbionte della Blatella germanica L.*, Boll. soc. ital. biol. sper. *22* (1946).

――――, *Il ciclo del batterio simbionte di Blattella germanica (L.)*, Mycopathol. *4* (1947).

Boschma, H., *On the food of Madreporaria*, Proc. Acad. Sci. Amsterdam *27* (1924).

――――, *The nature of the association between Anthozoa and Zooxanthellae*, Proc. Natl. Acad. Sci. *11* (1925).

――――, *On the symbiosis of certain Bermuda Coelenterates and Zooxanthellae*, Proc. Am. Acad. Arts Sci. *60* (1925).

――――, *On the feeding reactions and digestion in the coral polyp Astrangia danae, with notes on its symbiosis with Zooxanthellae*, Biol. Bull. *49* (1925).

――――, *On the food of reef-corals*, Proc. Acad. Sci. Amsterdam *29* (1926).

――――, *On the postlarval development of the coral Maeandra areolata L.*, Carnegie Inst. Wash. Publ. No. 391 (1929).

Bournier, A., *Sur l'existence et l'evolution d'un mycétome au cours de l'embryologenèse de Caudothrips buffai Karny*, Verhandl. XI Intern. Kongr. Entomologie 1961, Vol. 1 (1961).

Boycott, A. E., *A green alga parasitic in a water snail*, Northwest. Nat. *1* (1926).

Brain, C. K., *The intracellular symbionts of South African Coccidae*, Ann. Univ. Stellenbosch, *1*, Sect. A (1923).

Brandes, C. K., *Die Ursache der Grünfärbung des Darmes von Chaetopterus*, Z. Naturwiss. *70* (1898).

Brandt, K., *Über das Zusammenleben von Tieren und Algen*, Sitzber. Ges. naturforsch. Freunde, Berlin *1881*.

――――, *Über die morphologische und physiologische Bedeutung des Chlorophylls bei Tieren*, Arch. Anat. Physiol. Abt. Physiol. *1882*.

――――, 2. Artikel: Mitt. Zool. Station Neapel *4* (1883).

――――, *Über die Symbiose von Algen und Tieren*, Arch. Anat. Physiol. *1883*.

Brauer, A., *Die Tiefseefische*, Parts 1 and 2, in: *Wissenschaftliche Ergebnisse der deutschen Tiefsee-Expedition* (G. Fischer, Jena 1906, 1908).

Braun, H., Berg, St., Kessler, S., and Mavroidi P., *Über die Anormomorphogenese bei Bakterien*, Med. Monatsschr. 7 (1954).

Brecher, G., and Wigglesworth, V. B., *The transmission of Actinomyces rhodnii Erikson in Rhodnius prolixus Stål (Hemiptera) and its influence on the growth of the host*, Parasitology *35* (1944).

Breest, F., *Zur Kenntnis der Symbiontenübertragung bei viviparen Cocciden und Psylliden*, Arch. Protistenk. *34* (1914).

Breitsprecher, E., *Beiträge zur Kenntnis der Anobiidensymbiose*, Z. Morphol. Ökol. Tiere *11* (1928).

Brien, P., *Contribution à l'étude de l'embryogenese et de la blastogenèse des Salpes*, Rec. Inst. Zool. Torley-Rousseau 2 (1928).

Brock, J., *Über die sogenannten Augen von Tridacna und das Vorkommen von Pseudochlorophyllkörpern im Gefäßsystem der Muscheln*, Z. wiss. Zool. *46* (1888).

Brooks, M. A., *Certain aspects of the histochemistry and metabolic significance of the intracellular bodies (bacteroids) of cockroaches (Blattariae)*, Thesis (unpublished) (University of Minnesota, 1954).

———, *Investigation of possible antagonism between host and symbiot*, Anat. Record *128* (1957).

———, *Some dietary factors that affect ovarial transmission of symbiots*, Proc. Helminthol. Soc., Wash. D. C. 27 (1960).

———, *The relationship between intracellular symbiots and host metabolism*, Symposia Genetica et Biologica Italica IX (Celebrazione Spallanzaniana 1959) (1962).

———, *Symbiosis and aposymbiosis in Arthropods*, Symposia Soc. General Microbiology No. XIII, Symbiotic Associations (1963).

Brooks, M. A., and Richards, A. G., *Intracellular symbiosis in cockroaches*, I: *Production of aposymbiotic cockroaches*, Biol. Bull. *109* (1955).

———, *Intracellular symbiosis in cockroaches*, II: *Mitotic division of mycetocytes*, Science *122* (1955).

———, *Intracellular symbiosis in cockroaches*, III: *Reinfection of aposymbiotic cockroaches with symbiots*, J. Exptl. Zool. *132* (1956).

Brooks, W. K., *The genus Salpa. A monograph*, Mem. Biol. Lab. Johns Hopkins Univ. *11* (1893).

Brown, S. W., *Lecanoid chromosome behavior in three more families of the Coccoidea (Homoptera)*, Chromosoma *10* (1959).

Brown, S. W., and Bennett, F. D., *On sex determination in the Diaspine scale Pseudoaulacaspis pentagone (Targ.)*, Genetics *42* (1957).

Brown, S. W., and De Lotto, G., *Cytology and sex ratios of an African species of armored scale insect (Cocciodea-Diaspididae)*, Am. Naturalist *93* (1959).

Brown, S. W., and McKenzie, Howard L., *Scale insects and their allies (Dispididae, Phoenicoccidae and Asterolecaniidae)*, Hilgardia *33* (1962).

Brues, Ch. T., and Dunn, R. C., *The effect of penicillin and certain sulfa drugs on the intracellular bacteroids of the cockroach*, Science *101* (1945).

Brues, Ch. T., and Glaser, R. W., *A symbiotic fungus occurring in the fatbody of Pulvinaria innumerabilis*, Biol. Bull. *40* (1921).

Buchner, P., *Über intrazellulare Symbionten bei zuckersaugenden Insekten und ihre Vererbung*, Sitzber. Ges. Morphol. Physiol. München *1911*.

———, *Zur Kenntnis der Aleurodes-Symbionten*, Sitzber. Ges. Morphol. Physiol. München *1912*.

———, *Studien an intrazellularen Symbionten*, 1: *Die Symbionten der Hemipteren*, Arch. Protistenk. *26* (1912).

———, *Sind die Leuchtorgane Pilzorgane?* Zool. Anz. *45* (1914).

———, *Studien an intrazellularen Symbionten*, 2: *Die Symbionten von Aleurodes, ihre Übertragung in das Ei und ihr Verhalten bei der Embryonalentwicklung*, Arch. Protistenk. *39* (1918).

———, *Vergleichende Eistudien*, 1: *Die akzessorischen Kerne des Hymenoptereneies*, Arch. Mikroskop. Anat., Abt. 2. *97* (1918) (also discusses the transmission of the symbionts).

Buchner, P., *Neue Beobachtungen an intrazellularen Symbionten*, Sitzber. Ges. Morphol. Physiol. München *1919*.

——, *Zur Kenntnis der Symbiose niederer pflanzlicher Organismen mit Pedikuliden*, Biol. Zbl. *39* (1920).

——, *Über ein neues symbiontisches Organ der Bettwanze*, Biol. Zbl. *41* (1921).

——, *Studien an intrazellularen Symbionten*, 3: *Die Symbiose der Anobiinen mit Hefepilzen*, Arch. Protistenk. *42* (1921).

——, *Tier und Pflanze in intrazellularer Symbiose* (Borntraeger, Berlin 1921), 464 pages, 103 figures, 2 plates.

——, *Rassen- und Bakteroidenbildung bei Hemipterensymbionten*, Biol. Zbl. *42* (1922).

——, *Hämophagie und Symbiose*, Naturwissenschaften *1922*.

——, *Über das "tierische Leuchten,"* Naturwissenschaften *1922*.

——, *Moderne Symbioseforskning* (from Copenhagen Lectures), Naturens Verden 7 (Copenhagen 1923).

——, *Studien an intrazellularen Symbionten*, 4: *Die Bakteriensymbiose der Bettwanze*, Arch. Protistenk. *46* (1923).

——, *System und Symbiose*, Verhandl. deut. zool. Ges. *29* (1924).

——, *Studien an intrazellularen Symbionten*, 5: *Die symbiotischen Einrichtungen der Zikaden*, Z. Morphol. Ökol. Tiere *4* (1925).

——, *Studien an intrazellularen Symbionten*, 6: *Zur Akarinen-Symbiose*, Z. Morphol. Ökol. Tiere *6* (1926).

——, *Tierisches Leuchten und Symbiose* (J. Springer, Berlin 1926).

——, *Symbiontische Einrichtungen bei blutsaugenden Tieren*, Naturwissenschaften *1927*.

——, *Holznahrung und Symbiose*, Forsch. u. Fortschr. *3* (1927).

——, *Holznahrung und Symbiose* (J. Springer, Berlin 1928).

——, *Die Grenzen des Symbioseprinzipes*, Naturwissenschaften *1928*.

——, *Ergebnisse der Symbioseforschung*, 1: *Die Übertragungseinrichtungen*, Ergebn. Biol. *4* (1928).

——, *Tier und Pflanze in Symbiose*, 2nd ed. (completely revised and enlarged edition of *Tier und Pflanze in intrazellularer Symbiose*) (Borntraeger, Berlin 1930), 900 pages, 336 figures.

——, *Symbiontengestalt und Wirtsorganismus*, Atti XI congr. intern. zool., Padova 1930, in: Arch. zool. ital. *16* (1931).

——, *Neuere Symbiosestudien*, Schles. Ges. vaterl. Cultur. *105* (1933).

——, *Studien an intrazellularen Symbionten*, 7: *Die symbiontischen Einrichtungen der Rüsselkäfer*, Z. Morphol. Ökol. Tiere *26* (1933).

——, *Symbiose der Tiere mit pflanzlichen Mikroorganismen*, Slg. Göschen No. 1128 (W. de Gruyter, Berlin 1939; 2nd ed. 1949).

——, *Symbiose und Anpassung*, Nova Acta Leopoldina [N. S.] *8* (1940).

——, *Symbiosis in Oliarius*, Nature *160* (1947).

——, *Nuove ricerche e problemi nel campo della simbiosi*, Ricerca sci. *18* (1948).

——, *Symbioseforschung und Stammesgeschichte*, Umschau *1951*.

——, *Historische Probleme der Endosymbiose bei Insekten*, Symposium sur la "Symbiose chez les Insectes," Amsterdam 1951, Tijdschr. Entomol. *95* (1952), and Union intern. Sci. Biol. Ser. B, No. 10.

——, *Endosymbiose der Tiere mit pflanzlichen Mikroorganismen*, Verlag Birkhäuser Basel/Stuttgart, 1953, *1*, 771 pages, 336 figures.

——, *Studien an intrazellularen Symbionten*, VIII: *Die symbiontischen Einrichtungen der Bostrychiden (Apatiden)*, Z. Morphol. Ökol. Tiere *42* (1954).

——, *Endosymbiosestudien an Schildläusen*, I: *Stictococcus sjoestedti*, Z. Morphol. Ökol. Tiere *43* (1954).

and *Pediculus capitis*) *with some observations on the maturation of the egg*, Quart. J. Microscop. Sci. *64* (1920).

Dorner, G., *Darstellung der Turbellarienfauna Ostpreußens*, Schrift. phys.-ökon. Ges. Königsberg *43* (1902).

Doyle, W. L., *Studies on comparative cytoplasmic cytology*, Rept. Tortugas Lab., Carnegie Inst. Year Book *33* (1934).

———, *Observations on zooxanthellae*, Rept. Tortugas Lab., Carnegie Inst. Year Book *34* (1935).

Doyle, W. L., and Doyle, M. M., *The structure of zooxanthellae*, Papers Tortugas Lab. *32* (1940).

Dreyfus, L., *Zu Krassilstschicks Mitteilungen über die vergleichende Anatomie und Systematik der Phytophthiren*, Zool. Anz. *7* (1894).

Dubois, R., *La vie et la lumière* (Alcan, Paris 1914).

Duerden, J. E., *West Indian Madreporian polypes*, Mem. Natl. Acad. Sci. U.S. *8* (1902).

———, *The coral Siderastraea radians and its postlarval development*, Carnegie Inst. Wash. Publ. *20* (1904).

———, *The rôle of mucus in corals*, Quart. J. Microscop. Sci. *49* (1906).

Duesberg, J., *Chondriosomes et bactéries dans les nodosités radicales des légumineuses*, Compt. rend. assoc. anat. *18* (1923).

Dufour, L., *Historie comparative des metamorphoses et de l'anatomie des Cetonia aurata et Dorcus parallelopipedus*, Ann. sci. nat. Paris, (2) *18* (1812).

———, *Recherches anatomiques sur les Carabigues et sur plusieurs autres insectes Coléoptères*, Ann. sci. nat. Paris, (1) *3* (1824).

———, *Recherches anatomiques et physiologiques sur les Hémiptères*, Mém. sav. etrang. acad. sci. *4* (1833).

Duncan, J. T., *On a bactericidal principle present in the alimentary canal of insects and arachnids*, Parasitology *18* (1926).

Dutton, J. E., and Todd, J. L., *A note on the morphology of Spirochaeta duttoni*, Lancet *1907*.

Eisler, M. v., *Über die Wirkung von Salzen auf Bakterien*, Zentralbl. f. Bakt. I, Orig. (1909).

Ekblom, T., *Cytological and biochemical researches into the intracellular symbiosis in the intestinal cells of Rhagium inquisitor L.*, I: Skand. Arch. Physiol. *61* (1931); II: ibid., *64* (1932).

Eliot, C., and Evans, T. J., *Doridoeides gardineri: a doriform cladohepatic nudibranch*, Quart. J. Mikroscop. Sci. *52* (1908).

Emeis, W., *Über Eientwicklung bei den Cocciden*, Zool. Jahr., Abt. Anat., *39* (1915).

Enders, H. E., *A study on the life history and habits of Chaetopterus variopedatus*, J. Morphol. *20* (1909).

Engelmann, F., *Die Steuerung der Ovarfunktion bei der ovoviviparen Schabe Leucophaea maderae (Fabr.)*, J. Insect Physiol. *1* (1957).

Engelmann, Th. W., *Neue Methode der Untersuchung der Sauerstoffausscheidung pflanzlicher und tierischer Organismen*, Botan. Ztg. *39* (1881).

———, *Über tierisches Chlorophyll*, Arch. ges. Physiol. *32* (1883).

Enriques, P., *Ricerche sui Radiolari coloniali*, I., II: R. comit. Talassogr. ital. *71* (1919/21).

Entz, G., *Értesitö a kolozsvári orvos-természettndomán yi társulat második természettndományi szaküléséröl*, Kolozsvart *1876*.

———, *Über die Natur der "Chlorophyllkörperchen" niederer Tiere*, Biol. Zentr. *1* (1881/82).

Ergene, S., *Spielen die Darmbakterien von Calotermes flavicollis bei der Assimilation des atmosphaerischen Stickstoffs eine Rolle?* Rev. fac. sci. univ. Istanbul *B14* (1949).

Erikson, D., *The pathogenic aerobic organisms of the Actinomyces group*, Spec. Rept. Ser. Med. Research Council London, No. 203 (*1935*).

Ermisch, A., *Vergleichend-anatomische Untersuchungen über die Endosymbiose der Fulgoroiden mit besonderer Berücksichtigung der Araeopiden*, Z. Morphol. Ökol. Tiere *49* (1960).

Escherich, K., *Über das regelmäßige Vorkommen von Sproßpilzen in dem Darmepithel eines Käfers*, Biol. Zbl. *20* (1900).

——, *Die Forstinsekten Mitteleuropas*, Vol. 5 (Parey, Berlin 1940–1942) (*Siriciden-Symbiose*).

Evans, J. W., *Concerning the Peloridiidae*, Australian J. Sci. *4* (1941).

Faasch, W. J., *Darmkanal und Blutverdauung bei Aphanipteren*, Z. Morphol. Ökol. Tiere *29* (1935).

Fabre, J. H., *Souvenirs entomologiques*, 1 to 5. Serie (Paris 1891–1897).

Falk, R., *Die Scheindestruktion des Koniferenholzes durch die Larven des Hausbockes, Hylotrupes bajulus*, Cellulosechemie *11* (1930).

——, *Scheindestruktion des Holzes durch die Larve von Anobium*, Cellulosechemie *11* (1930).

Famintzin, A., *Die Symbiose als Mittel der Synthese von Organismen*, Biol. Zbl. 27 (1907).

Fantham, H. B., *Some researches of the life cycle of Spirochaetes*, Ann. Trop. Med. Parasitol. *5* (1911).

Farran, G. P., *Pyrosoma spinosum*, Mem. Challenger Soc. No. 1 (London 1909).

Farris, S. H., *Ambrosia fungus storage in two species of Gnathotrichus Eichhoff (Coleoptera: Scolytidae)*, Can. Entomologist *95* (1963).

Fauré-Frémier, E.-M., *Symbiontes bactériens des Ciliés du genre Euplotes*, Compt. rend. *235* (1952).

Fava, A., and Barbato, G., *Primi risultati ottenuti nel tentativo di costruire stipiti aposimbiontici in Blattella germanica mediante somministrazione orale di sigmaamicina*, Genet. agrar. *13* (1960).

——, *Ulteriori tentativi di ottenere stipiti aposimbiotici di Blatella germanica mediante somministrazione orale di sigmaamicina*, Symposia Genetica et Biologica Italica, Pavia *1960*.

Fava, A., and Laudani, U., *Sulla risposta alla colorazione di Gram e di Neisser dei batteri simbionti di alcune specie di Blatta*, Symposia Genetica et Biologica Italica, Pavia *1960*.

——, *Sui batteriociti di Nauphoeta cinerea (Epilamprinae, Blattoidea) trattata per vie orale con sigmaamicina*, Symposia Genetica et Biologica Italica, Pavia *1960*.

——, *Sui batteriociti di Nauphoeta cinerea trattata per via orale con sigmaamicina*, Symposia Genetica et Biologica Italica, Pavia *1960*.

——, *Produzione d'stipiti aposimbiotici in Nauphoeta cinerea Burm. mediante somministrazione parenterale d'antibiotici*, Genet. agrar. *14* (1961).

Fedele, M., *Sulle strutture e funzione dei ciechi epatopancreatici nei molluschi opistobranchi*, Boll. soc. ital. biol. sper. *1* (1926).

Fekl, W., *Die Bakterienflora der Tracheen und des Blutes einiger Insekten*, Z. Morphol. Ökol. Tiere *44* (1956).

Feldman-Muhsam, R., and Havivi, Y., *A microorganism associated with sperm cells of Ornithodorus*, Nature *193* (1962).

——, *On Adlerocystis n. gen. (Phycomycetes) a symbiont of Ornithodorus ticks*, Parasitology *55* (1963).

Fernando, E. F. W., *Storage and transmission of ambrosia fungus in the adult Xyleborus fornicatus (Eich.)*, Ann. Mag. Nat. Hist., Ser. *13* [2] (1960).

Ferris, G. F., *Report upon insects collected in China (Homoptera, Coccoidea)*, Part 5, Microentomology *19* (1954).

Fink, R., *Morphologische und physiologische Untersuchungen an den intrazellularen Symbionten von Pseudococcus citri Risso*, Dissertation, Munich *1951*.

——, *Morphologische und physiologische Untersuchungen an den intrazellularen Symbionten von Pseudococcus citri Risso*, Z. Morphol. Ökol. Tiere *41* (1952).

Finnegan, R. F., *The storage of ambrosia fungus spores by the pitted ambrosia beetle, Corthylus punctatissimus Zimm.*, Can. Entomologist *95* (1963).

Fleming, A., Voureka, A., Kramer, J. R. H., and Hughes, W. H., *The morphology and motility of Proteus vulgaris and other organisms cultured in the presence of penicillin*, J. Gen. Microbiol. *4* (1950).

Flögel, J. H. L., *Monographie der Johannisbeerblattlaus (Aphis ribis L.)*, Z. wiss. Inssekten-biol. [N. S.] *1* (1905).

Florence, L., *An intracellular symbiont of the hog louse*, Am. J. Trop. Med. *4* (1924).

Florenzano, G., *Prospetto sistematico delle specie batteriche segnalate come entomosimbionti*, Redia *34* (1949).

Foeckler, Fr., *Reinfektionsversuche steriler Larven von Stegobium paniceum L. mit Fremdhefen und die Beziehungen zwischen der Entwicklungsdauer der Larven und dem B-Vitamingehalt des Futters und der Hefen*, Z. Morphol. Ökol. Tiere *50* (1961).

Forbes, S. A., *Bacterium a parasite of the chinch bug*, Am. Naturalist *16* (1882).

————*Bacteria normal to digestive organs of Hemiptera*, Bull. Ill. State Lab. Nat. Hist. *4* (1892).

Fraenkel, G., *The role of symbionts as sources of vitamins and growth factors for their insect hosts*. Symposium sur la "Symbiose chez les insectes" (Amsterdam 1951), Tijdschr. Entomol. *95* (1952), and Union Intern. Sci. Biol. Ser. B, No. 10.

————, *The nutritional value of green plants for insects*, Trans. IX Intern. Congr. Entomol. Amsterdam *1951*.

————, *The nutritional requirements of insects for known and unknown vitamins*, Trans. IX Intern. Congr. Entomol. Amsterdam *1951*.

————, *Studies on the distribution of vitamin B (carnitin)*, Biol. Bull. *104* (1953).

————, *The effect of zinc and potassium in the nutrition of Tenebrio molitor, with observations on the expression of a carnitin deficiency*, J. Nutrition *65* (1958).

————, see also Blewett, M., and Fraenkel, G.; and Pant, N. C., and Fraenkel, G.

Fraenkel, G., and Blewett, M., *Biotin, B_1, riboflavin, nicotinic acid, B_6 and pantothenic acid as growth factors for insects*, Nature *150* (1942).

————, *Vitamins of the B-group required by insects*, Nature *151* (1943).

————, *Intracellular symbionts of insects as sources of vitamins*, Nature *152* (1943).

————, *The basic food requirements of several insects*, J. Exptl. Biol. *20* (1943).

————, *The vitamin B-complex requirements of several insects*, Biochem. J. *37* (1943).

————, *The natural foods and the food requirement of several species of stored products insects*, Trans. Roy. Entomol. Soc. London *93* (1943).

————, *The dietetics of the clothes moth, Tineola bisselliella Hum.*, J. Exptl. Biol. *22* (1946).

————, *The dietetics of the caterpillars of three Ephestia species, E. kuehniella, E. elutella and E. cautella, and of a closely related species, Plodia interpunctella*, J. Exptl. Biol. *22* (1946).

————, *Linoleic acid, vitamin E and other fat-soluble substances in the nutrition of certain insects (Ephestia kuehniella, E. elutella, E. cautella and Plodia interpunctella)*, J. Exptl. Biol. *22* (1946).

————, *B_T, new vitamin of the B-complex and its relation to the folic acid group and other anti-anaemia factors*, Nature *161* (1948).

Fraenkel, G., Blewett, M., and Coles, M., *The nutrition of the mealworm, Tenebrio molitor L.*, Physiol. Zoöl. *23* (1950).

Fraenkel, G., and Leclerq, F., *Nouvelles recherches sur les besoines nutritifs de la larve du Tenebrio molitor L.*, Arch. Intern. Physiol. et Biochem. *64* (1956).

Fraenkel, G., and Chang, P.-I., *Manifestations of a vitamin B_T (Carnitin) deficiency in the larvae of the mealworm, Tenebrio molitor L.*, Physiol. Zoöl. *27* (1954).

Fränkel, H., *Die Symbionten der Blattiden im Fettgewebe und im Ei, insbesondere von Periplaneta orientalis*, Z. wiss. Zool. *119* (1921).

Francke-Grosmann, H., *Beitrage zur Kenntnis der Lebensgemeinschaft zwischen Borkenkäfern und Pilzen*, Z. Parasitenk. *3* (1931).

Francke-Grosmann, H., *Zur Kenntnis der Läuseschäden an Weisstanne (Abies pectinata)*, *Tharandt. forstl. Jahrb. 88* (1937).

———, *Beitrage zur Kenntnis der Beziehungen unserer Holzwespen zu Pilzen*, Verh. VII. Intern. Entom.-Kongr. Berlin 1939.

———, *Über das Zusammenleben von Holzwespen (Siricinae) mit Pilzen*, Z. angew. Entomol. *25* (1939).

———, *Larvenentwicklung und Generationswechsel bei Hylecoetus dermestoides*, Trans. IX Intern. Congr. Entomol. Amsterdam 1951.

———, *Über die Ambrosiazucht der beiden Kieferborkenkäfer Myelophilus minor Htg. und Ips acuminatus Gyll.*, Medd. Statens Skogsforskningsinst. *41* (1952).

———, *Grundlagen der Symbiose bei pilzzüchtenden Holzinsekten*, Verhandl. deutsch. zool. Ges. *1956*.

———, *Hautdrüsen als Träger der Pilzsymbiose bei Ambrosiakäfern*, Z. Morphol. Ökol. Tiere *45* (1956).

———, *Zur Übertragung der Nährpilze bei Ambrosiakäfern*, Naturwissenschaften *43* (1956).

———, *Über die Ambrosiazucht holzbrütender Ipiden in Hinblick auf das System*, 14. Verhandlungsbericht deutsch. Ges. angew. Entomol. *1957*.

———, *Über das Schicksal der Siricidenpilze während der Metamorphose*, Bericht 8. Wanderversammlg. deutsch. Entomol. *1957* (Berlin 1957).

———, *Some new aspects in forest entomology*. Ann. Rev. Entomol. *8* (1963).

———, *Die Übertragung der Pilzflora bei dem Borkenkafer Ips acuminatus Gyll.* Z. angew. Entomol. *52* (1963).

———, *Ein Symbioseorgan bei dem Borkenkäfer Dendroctonus frontalis Zimm*, Naturwissenschaften *52* (1965).

Francke-Grosmann, H., and Schedl, W., *Ein orales Übertragungsorgan der Nährpilze bei Xyleborus mascarensis Eich. (Scolytidae)*, Naturwissenschaften *47* (1960).

Frank, W., *Einwirkung verschiedener Antibiotica auf die Symbionten der Küchenschabe Blatta orientalis L. und die dadurch bedingten Veränderungen am Wirtstier*, Verhandl. deutsch. zool. ges. *1954*.

———, *Entfernung der intrazellulären Symbionten der Küchenschabe (Periplaneta orientalis L.) durch Einwirkung verschiedener Antibiotica, unter besonderer Berücksichtigung der Veränderungen am Wirtstier und an den Bakterien*, Z. Morphol. Ökol. Tiere *44* (1956).

Franz, V., *Die japanischen Knochenfische der Sammlungen Haberer und Doflein. Beiträge zur Naturgesch. Ostasiens*, Abhandl. bayer. Akad. wiss., Math.-Phys. Kl., Suppl. *4* (1910 to 1913).

Fraser, E. A., *Observations on the life-history and development of the hydroid Myrionema amboinense*, Sci. Rept. Great Barriere Reef Exped. *3*, No. 4 (1931).

Freytag, K., *Über die Bakterienflora der Haut und der Segmentalorgane von Hirudo officinalis und medicinalis* (Abstract, Diss. Marburg 1952) (Marburg 1952).

Fröbrich, G., *Untersuchungen über Vitaminbedarf und Wachstumsfaktoren bei Insekten*, Z. vergleich. Physiol. *27* (1939).

———, *Darstellung von Konzentraten des "Tribolium-Imago-Faktors" (TIF) und seine vermutliche chemische Natur*, Naturwissenschaften *40* (1953).

———, *Der "Tribolium-Imago-Faktor" (TIF) durch Carnitin ersetzbar*, Naturwissenschaften *40* (1953).

———, *Die Caseindosierung in syntetischen Diäten für die Aufzucht von Tribolium confusum Duval*, Naturwissenschaften *40* (1953).

———, *Neue Ergebnisse experimenteller Untersuchungen zur Ernährungsphysiologie des Reismehlkäfers Tribolium confusum Duval*, Z. Vitamin-, Hormon- u. Fermentforsch. *6* (1954).

Fröbrich, G., and Offhaus, K., *Neue Befunde zur Ernährungsphysiologie des Reismehlkäfers Tribolium confusum Duval (Vorl. Mitt.)*, Verhandl. deutsch. zool. Ges. *1952*.

Fröbrich, G., and Offhaus, K., *Ein neuer Nahrungsfaktor, der die Metamorphose von Tribolium confusum ermöglicht*, Naturwissenschaften *39* (1952).

———, *Der qualitative Vitamintest mit dem Reismehlkäfer Tribolium confusum Duv. als Testorganismus*, Z. Vitamin-, Hormon- u. Fermentforsch. *5* (1953).

Füller, H. B., *Morphologische und experimentelle Untersuchungen über die neurosekretorischen Verhältmisse im Zentralnervensystem von Blattiden und Culiciden*, Zool. Jahrb. Abt. Allgem. Zool. Physiol. *69* (1960).

Galippe, V., *Parasitisme normal et microbiose* (Masson, Paris 1917).

Gambetta, L., *Ricerche sulla simbiosi ereditaria di alcuni coleotteri silofagi*, Ric. morfol. biol. animale *1* (Napoli 1927).

Gardner, A. D., *Morphological effects of penicillin on bacteria*, Nature *1940*.

Garnault, P., *Recherches anatomiques et histologiques sur le Cyclostoma elegans*, Acta soc. Linn. Bordeaux [5] *1* (1887).

Gebhardt, A. von, *Adatok a Buprestidák bélcsövének ismeretéhez*, Folia Soc. Entomol. Hung. *2* (1929).

Geddes, P., *Sur la fonction de la chlorophylle chez les Planaires vertes*, Compt. rend. *87* (1878).

———, *Observations on the physiology and histology of Convoluta schultzii*, Proc. Roy. Soc. (London) *28* (1879).

———, *On nature and functions of the "yellow cells" of Radiolarians and Coelenterates*, Proc. Roy. Soc. Edinburgh *1882*.

———, *The yellow-cells of Radiolarians and Coelenterates*, Proc. Roy. Soc. Edinburgh *1882*.

Geigy, R., Halff, L. A., and Kocher, V., *Untersuchungen über die physiologischen Beziehungen zwischen einem Überträger der Chagas-Krankheit Triatoma infestans und dessen Darmsymbionten*, Schweiz. med. Wochschr. *83* (1953).

———, *L'acide folique comme élément important dans la symbiose intestinale de Triatoma infestans*, Acta Trop. *11* (1954).

Geigy, R., and Wagner, O., *Ovogenese und Chromosomenverhältnisse bei Ornithodorus moubata*, Acta Trop. *14* (1957).

Gelei, J. von, *Angaben zur Symbiosefrage von Chlorella*, Biol. Zbl.. *47* (1927).

Getzel, D., *I microbi della glandola nidamentale accessoria in Sepia officinalis*, Arch. zool. ital. *20* (1934).

———, *Il simbionte dell' Icerya purchasi Mask., Geotrichoides pierantonii*, Arch. zool. ital. *23* (1936).

———, *Attività vitaminiche del simbionte dell'Icerya purchasi Mask.*, Boll. zool. 7 (1936).

———, *Forme microbiche del sangue*, Boll. zool. *12* (1941).

Ghidini, G. M., *A proposito di alcune recenti ricerche sulla cellulosolisi nell'intestino delle termiti*, Boll. zool. *12* (1941).

Giard, A., *Sur les Nephromyces, genre nouveau de champignons parasites du rein des Molgulidées*, Compt. rend. *106* (1888).

Gier, H. T., *The morphology and behavior of the intracellular bacteroids of roaches*, Biol. Bull. *71* (1936).

———, *Growth of the intracellular symbionts of the cockroach, Periplaneta americana*, Anat. Record *70* (1937).

———, *Intracellular bacteroids in the cockroach (Periplaneta americana L.)*, J. Bacteriol. *53* (1947).

Glaser, R. W., *Biological studies on intracellular bacteria*, Biol. Bull. *39* (1920).

———, *On the isolation, cultivation and classification of the so-called intracellular "symbionts" or "Rickettsia" of Periplaneta americana*, J. Exptl. Med. *51* (1930).

———, *The intracellular "symbionts" and the "Rickettsiae,"* Arch. Pathol. *9* (1930).

———, *Cultivation and classification of "bacteroids," "symbionts" or "Rickettsiae" of Blattella germanica*, J. Exptl. Med. *51* (1930).

Glaser, R. W., *The "Rickettsiae" and the intracellular "symbionts,"* Arch. Pathol. *9* (1930).
——, *The intracellular bacteria of the cockroach in relation to symbiosis.* J. Parasitol. *32* (1946).
Glasgow, H., *The gastric caeca and the caecal bacteria of the Heteroptera,* Biol. Bull. *26* (1914).
Glumb, *Untersuchungen über die Übertragungsorgane der Cleoniden,* mitgeteilt in: Buchner, P., *Studien an intrazellularen Symbionten* VII, Z. Morphol. Ökol. Tiere *26* (1933).
Godoy, A., and Pinto, C., *Da presença dos symbiontes nos Ixodides,* Brasil-medico *36* (1922).
Goetsch, W., *Die Symbiose der Süßwasserhydroiden und ihre künstliche Beeinflussung,* Z. Morphol. Ökol. Tiere *1* (1924).
——, *Darmsymbionten als Eiweißquelle und Vitaminspender,* Österr. zool. Z. *1* (1946).
——, *Vitamin T—ein neuer Wirkstoff,* Österr. zool. Z. *1* (1946).
——, *Beiträge zur biologischen Analyse des Vitamin-T-Komplexes,* Z. Vitamin-, Hormon- u. Fermentforsch. *1* (1947).
——, *Der Einfluss von Vitamin T auf Gestalt und auf die Gewohnheiten von Insekten,* Österr. zool. Z. *1* (1947).
——, *Wirkstoff T,* Verhandl deutsch. zool. Ges. *1949.*
——, *Probleme der Formbildung,* in: *Neue Ergebnisse und Probleme der Zoologie,* Ergänzungsbd. Zool. Anz. *1950.*
——, *Ergebnisse und Probleme aus dem Gebiet neuer Wirkstoffe,* Österr. zool. Z. *3* (1951).
——, *Untersuchungen über Steigerung der Resistenz durch den T-Komplex,* Z. Vitamin-, Hormon-u. Fermentforsch. *6* (1954).
——, *Untersuchungen über Behebung von Alterserscheinungen,* Pubbl. staz. zool. Napoli 27 (1955).
——, *Wirkstoffprobleme,* Verhandl. deutsch. zool. Ges. *1957.*
Goetsch, W., and Nussbaumer, G., *Die Bedeutung des Wirkstoffes T für die Medizin* (T-Vitamin Goetsch), Ärztl. Praxis 2 (1950).
Goetsch, W., Offhaus, K., and Tóth, L., *Untersuchungen über Bakterien- und Flagellatensymbiosen bei Termiten,* Naturwissenschaften *1944.*
Goetsch, W., and Scheuring, L., *Parasitismus und Symbiose der Algengattung Chlorella,* Z. Morphol. Ökol. Tiere 7 (1926).
Gohar, H. A. F., *Studies on the Xeniidae of the Red Sea. Their ecology, physiology, taxonomy and phylogeny,* Publs. Marine Biol. Sta. Ghardaqa 2 (1940).
Golberg, L., and De Meillon, B., *The nutrition of the larva of Aedes aegypti L.,* 3: *Lipid requirements,* 4: *Protein and amino acid requirements,* Biochem. J. *43* (1948).
——, see also De Meillon, B., and Golberg, L.
Golberg, L., De Meillon, B., and Lavoipierre, M., *Relation of "folic acid" to the nutritional requirements of the Mosquito larva,* Nature *154* (1944).
——, *The nutrition of the larva of Aedes aegypti L.* 2: *Essential water-soluble factors from yeast,* J. Exptl. Biol. *21* (1945).
Goldschmidt, R., *Glow worms and evolution,* Rev. sci., 86ᵉ année, No. 3298 (1948).
Golgi, C., *Intorno alla struttura ed alla biologia dei considetti globuli (o piastrine) del tuorlo,* Mem. reale. ist. Lomb. sci. lett. *22/23* (1923).
Goodchild, A. J. P., *The bacteria associated with Triatoma infestans and some other species of Reduviidae,* Parasitology *45* (1955).
Gould, L. J., *Notes on the minute structure of Pelomyxa palustris Greeff,* Quart. J. Microscop. Sci. *36* (1893).
——, *A further contribution to the study of Pelomyxa palustris,* J. Linnean Soc. London 29 (1905).
Goux, L., *Notes sur une levure symbiotique de Chlamydolecanium conchioides Goux,* Compt. rend. soc. biol. *138* (1944).
Graber, V., *Anatomisch-physiologische Studien über Phthirus inguinalis,* Z. wiss. Zool. 22 (1872).

Graebner, K. E., *Vergleichend morphologische und physiologische Studien an Anobiiden- und Cerambycidensymbionten*. Z. Morphol. Ökol. Tiere *41* (1954).

Graff, L. von, *Die Organisation der Turbellaria acoela*. Appendix by G. Haberlandt: *Über den Bau und die Bedeutung der Chlorophyllzellen von Convoluta roscoffensis* (W. Engelmann, Leipzig 1891).

Graham. A., *The structure and function of the alimentary canal of aeolid molluscs, with a discussion on their nematocysts*. Trans. Roy. Soc. Edinburgh *59* (1938).

Grandori, R., *La simbiosi ereditaria del filugello*, Atti reale ist. veneto sci. lettere ed arti *79* (1919).

———, *La simbiosi creditaria nel Bombyx mori*, Ann. R. Stat. bacol. sperim. Padova *44* (1924).

———, *Microorganismi simbiotici nell'uovo di Pieris brassicae L.*, Atti reale accad. Lincei, Rend., (6) *9* (1929).

Granovsky, A. A., *Preliminary studies of the intracellular symbionts of Saissetia oleae (Bernard)*, Trans. Wisconsin Acad. Sci. *24* (1929).

Grassé, P.-P., *Isoptères* in: *Traité de Zoologie*, Vol. 9 (Masson, Paris 1949).

———, *Rôle des flagellés symbiotiques chez les blattes et les termites*, Symposium sur la "Symbiose chez les insectes," Amsterdam 1951. Tijdschr. Entomol. *95* (1952), and Union intern. sci. biol. Ser. B, No. 10.

Grassé, P.-P., and Noirot, Ch., *La transmission des flagellés symbiotiques et les aliments des termites*, Bull. biol. France Belg. *79* (1945).

———, *L'évolution de la symbiose chez les Isoptères*, Experientia *15* (1959).

Grassi, B., *Contributo alla conoscenza delle Filosserine* (Tip. naz. G. Bertero, Roma 1912).

Greeff, R., *Pelomyxa palustris, ein amoebenartiger Organismus*, Arch. mikroscop. Anat. *10* (1874).

Gresson, R. A. R., and Threadgold, L. T., *An electron microscope study of bacteria in the oocytes and follicle cells of Blatta orientalis*, Quart. J. Microscop. Sci. *101* (1960).

Grieder, F., Neipp, L., and Meier, R., *Beitrag zur Characterisierung des Wirkungstypus antibakterieller Stoffe*, Schweiz. Z. Pathol. Bakteriol. *17* (1954).

Grinbergs, J., *Untersuchungen über Vorkommen und Funktion symbiontischer Mikroorganismen bei holzfressenden Insekten Chiles*, Arch. Mikrobiol. *41/42* (1961).

Grinyer, J., and Musgrave, A. J., *Microorganisms and mitochondria in the Malpighian tubules of Sitophilus (Coleoptera)*, Can. J. Microbiol. *10* (1964).

Gropengiesser, C., *Untersuchungen über die Symbiose der Blattiden mit niederen pflanzlichen Organismen*, Zentr. Bakteriol. *64* (1925).

Gruber, A., *Über Amoeba viridis Leidy*, Zool. Jahrb. Suppl. 7 (1904).

Gubler, H. U., *Versuche über die Züchtung intrazellularer Insektensymbionten*, Dissertation, Zürich (Buchdruckerei Effingerhof, Brugg 1947); also in Schweiz. Z. Pathol. u. Bakteriol. *11* (1948).

Guilbeau, B. H., *The origin and formation of the froth in spittle-insects*, Am. Naturalist *42* (1908).

Guilliermond, A., *Mitochondries et symbiotes*, Compt. rend. soc. biol. *82* (1919).

Gumpert, J., *Die Funktion der symbiontischen Bakterien in den Triatominen*, Zentr. Bakteriol. (1), *184* (1962).

———, *Die Symbiose der Triatominen, 1: Aufzucht symbiontenhaltiger und symbiontenfreier Triatominen und Eigenschaften der bei Triatominen vorkommenden Mikroorganismen*, Z. allgem. Mikrobiol. *2* (1962).

———, *Die Symbiose der Triatominen, 2: Infektionsymbiontenfreier Triatominen mit symbiontischen und saprophytischen Mikroorganismen und gemeinsame Eigenschaften der symbiontischen Stämme*, Z. Mikrobiol. *2* (1962).

Gumpert, J., *Die Symbiose der Triatominen*, 3: *Pantothensäurelieferung als Funktion der Symbionten*, Z. Mikrobiol. *3* (1963).

Haberlandt, G., See Graff, L. von.

Haemmerling, J. *Über die Symbionten von Stentor polymorphus*, Biol. Zbl. *65* (1946).

Hänsel, G., *Über das Auftreten von Symbionten bei verschiedenen Rüsselkäfern und die Beziehungen der lurculionidensymbiose zum Wirkstoffzehalt der Nährpflanzen*, Dissertation, Munich 1960 (published privately).

Härtel, O., *Über Wirkungen des "Vitamin T" auf pflanzenphysiologische Vorgänge*, Phyton *2* (1950).

Haffner, K. von, *Untersuchungen über die Symbiose von Dalyellia viridis und Chlorohydra viridissima mit Chlorellen*, Z. wiss. Zool. *126* (1925).

Halff, L. A., *Untersuchungen über die Abhängigkeit der Entwicklung der Reduviide Triatoma infestans Klug von ihrem Darmsymbionten*, Acta Trop. *13* (1956).

Haller, G. de, *La symbiose bactérienne intracellulaire chez la Blatte, B. germanica*, Arch. sci. *8* (1955).

——, *L'isolement du symbiote intracellulaire de la Blatte (B.-germanica) (Note préliminaire)*, Rev. suisse zool. *62* (1955).

Hamann, O. *Zur Entstehung und Entwicklung der grünen Zellen bei Hydra*, Z. wiss. Zool. *37* (1882).

Handlirsch, A., *Die fossilen Insekten und die Phylogenie der recenten Formen* (W. Engelmann, Leipzig 1908).

——, *Insecta*, in: Kükenthal, *Handbuch der Zoologie*, Vol. 4 (W. de Gruyter & Co., Berlin and Leipzig 1930).

Haneda, Y., *Über den Leuchtfisch Malacocephalus laevis (Lowe)* (in Japanese), Japan. J. Med. Sci. III, Biophysics *5* (1938); also in Japan J. Physiol. *5*.

——, *Leuchtende Fische aus südlichen Meeren*, Kagaku Nanyo *1* (1938), and Zool. Mag. (Tokyo) *51*.

——, *On the luminescence of the fishes belonging to the family Leiognathidae of the tropical Pacific*, Palao Trop. Biol. Sta. Studies *2* (1940).

——, *On the photogenous organs of Anomalops katoptron, a luminous fish* (in Japanese), Kagaku Nanyo *5* (1943).

——, *Luminous organs of fish which emit light indirectly*, Pacific Sci. *4* (1950).

——, *The luminescence of some deep-sea fishes of the families Gadidae and Macruridae*, Pacific Sci. *5* (1951).

——, *Observations on luminous organisms of the Miura Peninsula, Japan*, Yokosukashi-Shi. *5* (1952).

——, *Observation on some marine luminous organisms of Hachijo Island, Japan*, Records Oceanogr. Works Japan, Vol. 12, No. 2, N. S. Vol. 1, No. 1 (1953).

——, *Luminous organisms of Japan and the Far East*. in: The Luminescence of Biological Systems. Edited by Frank H. Johnson (Washington D.C., 1955).

——, *Squid producing an abundant luminous secretion found in Suruga Bay, Japan*, Sci. Rept. Yokosuka City Museum *1* (1956).

——, *Observations on luminescence in the deep sea fish Paratrichichthys prosthemius*, Sci. Rept. Yokosuka City Museum *2* (1957).

Haneda, Y. and Johnson, F. H., *The luciferin-luciferase reaction in a fish, Parapriacanthus beryciformis, of newly discovered luminescence*, Proc. Natl. Acad. Sci. U. S. [2] *44* (1958); and Sci. Rept. Yokosuka City Museum *3* (1958).

——, *The comparative anatomy of the indirect type of photogenetic system of luminescent fishes, with special reference to Parapriacanthus beryciformis*, Sci. Rept. Yokosuka City Museum *7* (1962).

Buchner, P., *Endosymbiosestudien an Schildläusen*, II: *Stictococcus diversiseta*, Z. Morphol. Ökol. Tiere *43* (1955).

———, *Endosymbiosestudien an Schildläusen*, III: *Macrocerococcus und Puto, zwei primitive Pseudococcinen*, Z. Morphol. Ökol. Tiere *43* (1955).

———, *Neue Beiträge zur Kenntnis der Coccidensymbiosen*, Wiss. Z. Ernst-Moritz-Arndt-Universität Greifswald *5* (1955/56).

———, *Die harmonische Einbürgerung pflanzlicher Mikroorganismen in den tierischen Körper*, Verhandel. deut. Ges. inn. Med., 63. Kongr. 1957 (München 1957).

———, *Endosymbiosestudien an Schildläusen*, IV: *Hippeococcus, eine myrmekophile Pseudococcine*, Z. Morphol. Ökol. Tiere *45* (1957).

———, *Endosymbiosestudien an Schildläusen*, V: *Die Gattung Rastrococcus Ferris (Ceroputo Šulc)*, Z. Morphol. Ökol. Tiere *46* (1957).

———, *Endosymbiosestudien an Schildläusen*, VI: *Die nicht in Symbiose lebende Gattung Apiomorpha und ihre ungewöhnliche Embryonalentwicklung*, Z. Morphol. Ökol. Tiere *46* (1957).

———, *Eine neue Form der Endosymbiose bei Aphiden*, Zool. Anz. *160* (1958).

———, *Tiere als Mikrobenzüchter* (Verständliche Wiss. Vol. 75), Springer-Verlag, *1960*.

———, *Endosymbiosestudien an Ipiden*, I: *Die Gattung Coccotrypes*, Z. Morphol. Ökol. Tiere *50* (1961).

———, *Fünfzig Jahre Symbioseforschung*, Wiss. Z. Ernst-Moritz-Arndt-Universität Greifswald *10* (1961).

———, *Die geschlechtsbegrenzte Symbiose der Stictococcinen*, Proc. IV Congr. UIEIS Pavia *1961* (1963).

———, *Endosymbiosestudien an Schildläusen*, VII: *Weitere Beiträge zur Kenntnis der Stictococcinensymbiose*, Z. Morphol. Ökol. Tiere *52* (1963).

Büsgen, M., *Der Honigtau. Biologische Studien an Pflanzen und Pflanzenläusen*, Z. Naturwiss. *25* (1891).

Büsing, K.-H., *Pseudomonas hirudinis, ein bakterieller Darmsymbiont des Blutegels (Hirudo medicinalis)*, Zentr. Bakteriol., *157* (1951).

Büsing, K.-H., Döll, W., and Freytag, K., *Die Bakterienflora der medizinischen Blutegel*, Arch. Mikrobiol. *19* (1953).

Burghause, F., *Kreislauf und Herzschlag bei Pyrosoma giganteum nebst Bemerkungen zum Leuchtvermögen*, Z. wiss. Zool. *108* (1914).

Burri, R., *Weitere Beobachtungen über Formwandlungen beim Erreger der Sauerbrut der Bienen*, Beih. Schweiz. Bienenztg. *1*, H. 5 (1943).

———, *Über eine in außerordentlichem Maße zur Dissoziation neigende Bakterienart*, Mitt. naturforsch. Ges. Bern [N. S.] *2* (1944).

———, *Die Beziehungen der Bakterien zum Lebenszyklus der Honigbiene*, Schweiz. Bienen-Ztg. *1947*.

Burrill, T. J., *New species of Micrococcus*, Am. Naturalist *17* (1883).

Buscalioni, L., and Comes, S., *La digestione delle membrane vegetali per opera dei flagellati contenuti nell'intestino dei termiti e il problema della simbiosi*, Atti accad. Gioenia sci. nat. Catania [5] *3* (1910).

Bush, G. L., and Chapman, G. B., *Electron microscopy of symbiotic bacteria in developing oocytes of American cockroach, Periplaneta americana*, J. Bacteriol. *81* (1961).

Busnel, R. G., and Chanoin, R., *Dosage et repartition de la riboflavine chez le criquet pèlerin; son importance au point de vue alimentaire*, Bull. soc. zool. France *67* (1940).

Busnel, R. G., and Drilhon, A., *Presence of riboflavin in an insect, Tineola bisselliella Hum., fed a diet free of this substance*, Compt. rend. *216* (1943).

———, *Sur l'utilisation de la vitamine B$_2$ par Tenebrio molitor en présence de sulfonamide*, Compt. rend. *226* (1948).

846   *References*

Campbell, W. G., *The chemical aspect of the destruction of oakwood by powder-post and death-watch beetles: Lyctus spec. and Xestobium spec.*, Biochem. J. *23* (1929).

Carayon, J., *Les punaises des bois et leur bactéries symbiotiques*, Nature No. 3084 (Paris) *3* (1945).

——, *L'oothèque d'Hémiptères Plataspidés de l'Afrique tropicale*, Bull. soc. entomol. France *1949*.

——, *Les élements bacilliformes secrétés par les glandes génitales annexes des certains Hémiptères*, Bull. soc. zool. France *70* (1949).

——, *L'ootèque d'Hémiptères Plataspidés de l'Afrique tropicale*, Bull. soc. entomol. France *1949*.

——, *Les organes génitaux males des Hémiptères Nabidae. Absence de symbiontes dans ces organes*, Proc. Roy. Entomol. Soc. London *A26* (1951).

——, *Les organes génitaux males des Hémiptères Nabidae. Absence de symbiontes dans ces organes*, Proc. Royal Entomol. Soc. London, Ser. A. *26* (1951).

——, *Les mechanismes de transmission héréditaire des endosymbiontes chez les insectes*, Symposium sur la "Symbiose chez les insectes," Amsterdam 1951, Tijdschr. Entomol. *95* (1952), and Union intern. soc. biol. Ser. B, No. 10.

——, *Insemination par "spermalège" et cordon conducteur de spermatozoïdes chez Stricticimex brevispinosus Usinger*, Rev. zool. botan. africaines *60* (1959).

——, *La transmission héréditaire des Bactéries symbiotiques chez les Lygaeidae vivipares (Heteroptera)*, Proc. XVI Intern. Congr. of Zoology, Washington *1963*, Vol. 1.

Carrère, L., and Roux, J. *Obtention in vivo des formes L. de Salmonella typhimurium*, Compt. rend. *234* (1952).

Carter, W., *The pineapple mealy bug, Pseudococcus brevipes, and wilt of pineapples*, Phytopathology *23* (1933).

——, *The spotting of pineapple leaves caused by Pseudococcus brevipes, the pineapple mealy bug*, Phytopathology *23* (1933).

——, *The symbionts of Pseudococcus brevipes (Ckl.) in relation to a phytotoxic secretion of the insect*, Phytopathology *25* (1935).

——, *The symbionts of Pseudococcus brevipes (Ckl.)*, Ann. Entomol. Soc. Am. *28* (1935).

——, *The symbionts of Pseudococcus brevipes in relation to a phytotoxic secretion of the insect*, Phytopathology *26* (1936).

——, *The toxic dose of mealy bug wilt of pineapple*, Phytopathology *27* (1937).

——, *Injuries to plants caused by insect toxins*, Botan. Rev. Entomol. *5* (1939).

——, *The geographical distribution of mealy bug wilt with notes on some other insect pests of pineapple*, J. Econ. Entomol. *35* (1942).

Cartwright, K. St. G., *Notes on a fungus associated with Sirex cyaneus*, Ann. Appl. Biol. *16* (1929).

——, *A further note on fungus association in the Siricidae*, Ann. Appl. Biol. *25* (1938).

Casteel, D. B., *Germ cells of Argas*, J. Morphol, *28* (1916/17).

Catel, W., *Die Bedeutung der Bifidumflora für den wachsenden Organismus*, Verhandl. deut. Ges. inn. Med. *1957*.

Cerfontaine, P., *Recherches sur le système cutané et sur le système musculaire du lombric terrestre*, Arch. biol. *10* (1890).

Chamberlin, J. C., *A systematic monograph of the Tachardiinae or lac insects (Coccidae)*, Bull. Entomol. Research *14* (1923/24).

Chararas, C., *Anatomie et biologie des Coléoptères Curculionides xylophages comparées a celles des Coléoptères scolytides*, Rev. pathol. végétale et entomol. agric. France, *35* (1956).

Chatton, E., and Séguéla, J., *La continuité génétique des formations ciliaires chez les ciliés hypotriches*, Bull. biol. France et Belg. *74* (1940).

Cholodkowsky, N., *Die Entwicklung von Phyllodromia germanica*, Mém. Acad. Petersbourg [7] *38* (1891).

——, *Zur Morphologie der Pediculiden*, Zool. Anz. *27* (1914).

Chrystal, R. N., *The Sirex wood wasps and their importance in forestry*, Bull. Entomol. Research *19* (1928).

Chun, C., *Die Ctenophoren des Golfes von Neapel*, in: *Fauna und Flora des Golfes von Neapel*, Vol. 1 (R. Friedländer and Son, Berlin 1880).

——, *Die Oegopsiden der deutschen Tiefsee-Expedition*, in: *Wissenschaftliche Ergebnisse der deutschen Tiefsee-Expedition* (G. Fischer, Jena 1910).

Cienkovsky, L., *Über Schwärmerbildung bei Radiolarien*, Arch. mikroskop. Anat. *7* (1871).

Claparède, E. R., and Lachmann, K. F. I., *Etudes sur les Infusoires et les Rhizopodes*, Mém. Inst. Natl. Genève *5* (1857/58).

Clark, A. F., *The horntail borer and its fungal association*, New Zealand J. Sci. Technol. *15* (1933).

Claus, C., *Die Ephyren von Cotylorhiza und Rhizostoma*, Arb. Zool. Inst. Wien *5* (1884).

Cleveland, L. R., *The physiological and symbiotic relationships between the intestinal protozoa of termites and their host*, Biol. Bull. *46* (1924).

——, *The effects of oxygenation and starvation on the symbiosis between the termite Termopsis and its intestinal flagellates*, Biol. Bull. *48* (1925).

——, *The ability of termites to live perhaps indefinitely on a diet of pure cellulose*, Biol. Bull. *48* (1925).

——, *Further observations and experiments on the symbiosis between termites and their intestinal protozoa*, Biol. Bull. *54* (1928).

Cleveland, L. R., Hall, S. R., Sanders, E. P., and Collier, J., *The wood-feeding roach Cryptocercus, its protozoa, and the symbiosis between protozoa and roach*. Mem. Am. Acad. Arts Sci. *17* (1934).

Clusius, K., *Reaktionen mit dem Stickstoff-Isotop 15 N*, Angew. Chemie *64* (1952).

Cohendy, M., and Wollmann, E. *Anelques résultats acquise par la méthode des elevages aseptiques: I. Scorbut expérimental; II. Infection cholérique du cobaye aseptique*, Compt. rend. acad. Sci. *174* (1922).

Cohn, F., *Beiträge zur Entwicklungsgeschichte der Infusorien*, Z. wiss. Zool. *3* (1851).

Cohn, F., and Schröder, *Über parasitische Algen*, Beitr. Biol. Pflanzen *1* (1872).

Conte, A., and Faucheron, L., *Présence de levures dans le corps adipeux de divers Coccides*, Compt. rend. *145* (1907).

Convenevole, C., *La simbiosi ereditaria negli Emitteri Eterotteri (Aelia rostrata Geoffr.)*, Arch. zool. ital. *19* (1933).

Cowdry, E. V., *The distribution of Rickettsia in the tissues of insects and arachnids*, J. Exptl. Med. *37* (1923).

——, *The independence of mitochondria and the bacillus radicicola in root nodules*, Am. J. Anat. *31* (1923).

——, *The value of the study of mitochondria in cellular pathology*, Am. Naturalist *58* (1924).

——, *A group of microorganisms transmitted hereditarily in ticks and apparently unassociated with disease*, J. Exptl. Med. *41* (1925).

Cowdry, E. V., and Olitsky, P. K., *Differences between mitochondria and bacteria*, J. Exptl. Med. *36* (1922).

Crawford, R. E., McDermott, L. A., and Musgrave, A. J., *Microbial isolation from the granary weevil, Sitophilus granarius (L.)*, Can. Entomologist *92* (1960).

Csáky, T., and Tóth, L., *Enzymatic breakdown of nitrogen compounds by the nitrogen fixing bacteria of insects*, Experientia *4* (1948).

Cuénot, L., *Etudes physiologiques sur les Orthoptères*, Arch. biol. *14* (1896).

848   *References*

Cuénot, L., *Etudes physiologiques sur les Oligochetes*, Arch. biol. *15* (1898).

Cymorek, S., *Beitrag zur Kenntnis der Pockkäferarten Anobium punctatum, Anobium hederae, Anobium inexpectatum*, Entomol. Blätter *53* (1957); *55* (1960).

Dahlgren, U., *The bacterial light organ of Ceratias*, Science *68* (1928).

Dangeard, P. A., *Les zoochlorelles du Paramaecium bursaria*, Botaniste *7* (1900).

Davidoff, C., *Traité de l'embryologie comparée des Invertebrés* (Masson, Paris 1928).

Dean, R. W., *Morphology of the digestive tract of the apple maggot fly, Rhagoletis pomonella Walsh*, N. Y. State Agr. Exp. Sta. (Geneva, N. Y.) Bull. *215* (1933).

————, *Anatomy and postpupal development of the female reproductive system in the apple maggot fly, Rhagoletis pomonella Walsh*, N. Y. State Agr. Exp. Station (Geneva, N. Y.) Bull. *229* (1935).

DeHaan, W., *Mémoires sur les métamorphoses des Coléoptères*, 1: *Les Lamellicornes*, Nouv. ann. mus. nat. hist. Paris *4* (1835).

Delage, Y., *Etudes histologiques sur les Planaires rhabdocoeles acoeles (Convoluta schultzii O. Schmidt)*, Arch. zool. exptl. [2] *4* (1886).

Delage, Y., and Hérouard, E., *Traité de zoologie concrète*, Vol. 2 (Schleicher Frères, Paris 1899).

De Lerma, B., *Le recenti ricerche chimico-fisiche sulla bioluminescenza*, Riv. fis. mat. sci. nat. *11* (1937).

Delsman, H. C., *Beiträge zur Entwicklungsgeschichte von Porpita*, Treubia (Batavia) *3* (1923).

De Meillon, B., and Golberg, L., *Nutritional studies on blood-sucking Arthropods*, Nature *158* (1946).

————, *Preliminary studies on the nutritional requirements of the bedbug (Cimex lectularius L.) and the tick Ornithodorus moubata Murray*, J. Exptl. Biol. *24* (1947).

De Meillon, B., Golberg, L., and Lavoipierre, M., *The nutrition of the larva of Aedes aegypti L.*, 1, J. Exptl. Biol. *21* (1945).

De Meillon, B., Thorp, J. M., and Hardy, F., *The relationship between ectoparasite and host*, 1: *The development of Cimex lectularius and Ornithodorus moubata on riboflavin deficient rats*, S. African J. Med. Sci. *12* (1947).

Dendy, A., *On the origin, growth and arrangements of sponge spiculas: a study on symbiosis*, Quart. J. Microbiol.. Sci. *70* (1926).

Dias, E., *Estudos sobre o Schizotrypanum cruzi*, Dissertation (Rio de Janeiro 1933); and Mem. inst. Oswaldo Cruz *28* (1934).

————, *Sobre a presença de symbiontes em hemipteros hematophagos*, Mem. inst. Oswaldo Cruz *32* (1937).

Dickman, A., *Studies on the intestinal flora of termites with reference to their ability to digest cellulose*, Biol. Bull. *61* (1931).

Diddens, H. A., and Lodder, F., *Die anaskosporogenen Hefen. 2. Teil.* Amsterdam *1942*.

Dienes, L., and Weinberger, H. J., *The L-form of bacteria*, Bacteriol. Rev. *15* (1951).

DiMaria, G., *Nuove ricerche sulla attività vitaminica del simbionte d'Icerya purchasi Mask.*, Boll. zool. *9* (1938).

————, *Studii sul fabbisogno vitaminico per lo sviluppo della Sarcophaga*, Arch. zool. ital. *25* (1938).

Dittrich, H. H., *Über den Einfluss von Candida reukauffii auf die Honigbiene und den Nachweis der Pantothensäure als wirksame Substanz*, Zentr. Bakteriol. [II] *111* (1958).

Döring, W., *Über Bau und Entwicklung der weiblichen Geschlechtsorgane bei myopsiden Cephalopoden*, Z. wiss. Zool. *91* (1908).

Doflein, Fr., *Studien zur Naturgeschichte der Protozoen*, 5: *Amoebenstudien*, Arch. Protistenk. Suppl. *1* (1907).

Doncaster, L., and Cannon, H. G., *On the spermatogenesis of the louse (Pediculus corporis*

Gumpert, J., *Die Symbiose der Triatominen*, 3: *Pantothensäurelieferung als Funktion der Symbionten*, Z. Mikrobiol. *3* (1963).

Haberlandt, G., See Graff, L. von.

Haemmerling, J. *Über die Symbionten von Stentor polymorphus*, Biol. Zbl. *65* (1946).

Hänsel, G., *Über das Auftreten von Symbionten bei verschiedenen Rüsselkäfern und die Beziehungen der lurculionidensymbiose zum Wirkstoffzehalt der Nährpflanzen*, Dissertation, Munich 1960 (published privately).

Härtel, O., *Über Wirkungen des "Vitamin T" auf pflanzenphysiologische Vorgänge*, Phyton *2* (1950).

Haffner, K. von, *Untersuchungen über die Symbiose von Dalyellia viridis und Chlorohydra viridissima mit Chlorellen*, Z. wiss. Zool. *126* (1925).

Halff, L. A., *Untersuchungen über die Abhängigkeit der Entwicklung der Reduviide Triatoma infestans Klug von ihrem Darmsymbionten*, Acta Trop. *13* (1956).

Haller, G. de, *La symbiose bactérienne intracellulaire chez la Blatte, B. germanica*, Arch. sci. *8* (1955).

———, *L'isolement du symbiote intracellulaire de la Blatte (B.-germanica) (Note preliminaire)*, Rev. suisse zool. *62* (1955).

Hamann, O. *Zur Entstehung und Entwicklung der grünen Zellen bei Hydra*, Z. wiss. Zool. *37* (1882).

Handlirsch, A., *Die fossilen Insekten und die Phylogenie der recenten Formen* (W. Engelmann, Leipzig 1908).

———, *Insecta*, in: Kükenthal, *Handbuch der Zoologie*, Vol. 4 (W. de Gruyter & Co., Berlin and Leipzig 1930).

Haneda, Y., *Über den Leuchtfisch Malacocephalus laevis (Lowe)* (in Japanese), Japan. J. Med. Sci. III, Biophysics *5* (1938); also in Japan J. Physiol. *5*.

———, *Leuchtende Fische aus südlichen Meeren*, Kagaku Nanyo *1* (1938), and Zool. Mag. (Tokyo) *51*.

———, *On the luminescence of the fishes belonging to the family Leiognathidae of the tropical Pacific*, Palao Trop. Biol. Sta. Studies *2* (1940).

———, *On the photogenous organs of Anomalops katoptron, a luminous fish* (in Japanese), Kagaku Nanyo *5* (1943).

———, *Luminous organs of fish which emit light indirectly*, Pacific Sci. *4* (1950).

———, *The luminescence of some deep-sea fishes of the families Gadidae and Macruridae*, Pacific Sci. *5* (1951).

———, *Observations on luminous organisms of the Miura Peninsula, Japan*, Yokosukashi-Shi. *5* (1952).

———, *Observation on some marine luminous organisms of Hachijo Island, Japan*, Records Oceanogr. Works Japan, Vol. 12, No. 2, N. S. Vol. 1, No. 1 (1953).

———, *Luminous organisms of Japan and the Far East*. in: The Luminescence of Biological Systems. Edited by Frank H. Johnson (Washington D.C., 1955).

———, *Squid producing an abundant luminous secretion found in Suruga Bay, Japan*, Sci. Rept. Yokosuka City Museum *1* (1956).

———, *Observations on luminescence in the deep sea fish Paratrichichthys prosthemius*, Sci. Rept. Yokosuka City Museum *2* (1957).

Haneda, Y. and Johnson, F. H., *The luciferin-luciferase reaction in a fish, Parapriacanthus beryciformis, of newly discovered luminescence*, Proc. Natl. Acad. Sci. U. S. [2] *44* (1958); and Sci. Rept. Yokosuka City Museum *3* (1958).

———, *The comparative anatomy of the indirect type of photogenetic system of luminescent fishes, with special reference to Parapriacanthus beryciformis*, Sci. Rept. Yokosuka City Museum *7* (1962).

Graebner, K. E., *Vergleichend morphologische und physiologische Studien an Anobiiden- und Cerambycidensymbionten.* Z. Morphol. Ökol. Tiere *41* (1954).

Graff, L. von, *Die Organisation der Turbellaria acoela.* Appendix by G. Haberlandt: *Über den Bau und die Bedeutung der Chlorophyllzellen von Convoluta roscoffensis* (W. Engelmann, Leipzig 1891).

Graham. A., *The structure and function of the alimentary canal of aeolid molluscs, with a discussion on their nematocysts.* Trans. Roy. Soc. Edinburgh *59* (1938).

Grandori, R., *La simbiosi ereditaria del filugello,* Atti reale ist. veneto sci. lettere ed arti *79* (1919).

——, *La simbiosi creditaria nel Bombyx mori,* Ann. R. Stat. bacol. sperim. Padova *44* (1924).

——, *Microorganismi simbiotici nell'uovo di Pieris brassicae L.,* Atti reale accad. Lincei, Rend., (6) *9* (1929).

Granovsky, A. A., *Preliminary studies of the intracellular symbionts of Saissetia oleae (Bernard),* Trans. Wisconsin Acad. Sci. *24* (1929).

Grassé, P.-P., *Isoptères* in: *Traité de Zoologie,* Vol. 9 (Masson, Paris 1949).

——, *Rôle des flagellés symbiotiques chez les blattes et les termites,* Symposium sur la "Symbiose chez les insectes," Amsterdam 1951. Tijdschr. Entomol. *95* (1952), and Union intern. sci. biol. Ser. B, No. 10.

Grassé, P.-P., and Noirot, Ch., *La transmission des flagellés symbiotiques et les aliments des termites,* Bull. biol. France Belg. *79* (1945).

——, *L'évolution de la symbiose chez les Isoptères,* Experientia *15* (1959).

Grassi, B., *Contributo alla conoscenza delle Filosserine* (Tip. naz. G. Bertero, Roma 1912).

Greeff, R., *Pelomyxa palustris, ein amoebenartiger Organismus,* Arch. mikroscop. Anat. *10* (1874).

Gresson, R. A. R., and Threadgold, L. T., *An electron microscope study of bacteria in the oocytes and follicle cells of Blatta orientalis,* Quart. J. Microscop. Sci. *101* (1960).

Grieder, F., Neipp, L., and Meier, R., *Beitrag zur Characterisierung des Wirkungstypus antibakterieller Stoffe,* Schweiz. Z. Pathol. Bakteriol. *17* (1954).

Grinbergs, J., *Untersuchungen über Vorkommen und Funktion symbiontischer Mikroorganismen bei holzfressenden Insekten Chiles,* Arch. Mikrobiol. *41/42* (1961).

Grinyer, J., and Musgrave, A. J., *Microorganisms and mitochondria in the Malpighian tubules of Sitophilus (Coleoptera),* Can. J. Microbiol. *10* (1964).

Gropengiesser, C., *Untersuchungen über die Symbiose der Blattiden mit niederen pflanzlichen Organismen,* Zentr. Bakteriol. *64* (1925).

Gruber, A., *Über Amoeba viridis Leidy,* Zool. Jahrb. Suppl. *7* (1904).

Gubler, H. U., *Versuche über die Züchtung intrazellularer Insektensymbionten,* Dissertation, Zürich (Buchdruckerei Effingerhof, Brugg 1947); also in Schweiz. Z. Pathol. u. Bakteriol. *11* (1948).

Guilbeau, B. H., *The origin and formation of the froth in spittle-insects,* Am. Naturalist *42* (1908).

Guilliermond, A., *Mitochondries et symbiotes,* Compt. rend. soc. biol. *82* (1919).

Gumpert, J., *Die Funktion der symbiontischen Bakterien in den Triatominen,* Zentr. Bakteriol. (1), *184* (1962).

——, *Die Symbiose der Triatominen,* 1: *Aufzucht symbiontenhaltiger und symbiontenfreier Triatominen und Eigenschaften der bei Triatominen vorkommenden Mikroorganismen,* Z. allgem. Mikrobiol. *2* (1962).

——, *Die Symbiose der Triatominen,* 2: *Infektionsymbiontenfreier Triatominen mit symbiontischen und saprophytischen Mikroorganismen und gemeinsame Eigenschaften der symbiontischen Stämme,* Z. Mikrobiol. *2* (1962).

854   *References*

Glaser, R. W., *The "Rickettsiae" and the intracellular "symbionts,"* Arch. Pathol. *9* (1930).
————, *The intracellular bacteria of the cockroach in relation to symbiosis.* J. Parasitol. *32* (1946).
Glasgow, H., *The gastric caeca and the caecal bacteria of the Heteroptera,* Biol. Bull. *26* (1914).
Glumb, *Untersuchungen über die Übertragungsorgane der Cleoniden,* mitgeteilt in: Buchner, P., *Studien an intrazellularen Symbionten* VII, Z. Morphol. Ökol. Tiere *26* (1933).
Godoy, A., and Pinto, C., *Da presença dos symbiontes nos Ixodides,* Brasil-medico *36* (1922).
Goetsch, W., *Die Symbiose der Süßwasserhydroiden und ihre künstliche Beeinflussung,* Z. Morphol. Ökol. Tiere *1* (1924).
————, *Darmsymbionten als Eiweißquelle und Vitaminspender,* Österr. zool. Z. *1* (1946).
————, *Vitamin T—ein neuer Wirkstoff,* Österr. zool. Z. *1* (1946).
————, *Beiträge zur biologischen Analyse des Vitamin-T-Komplexes,* Z. Vitamin-, Hormon- u. Fermentforsch. *1* (1947).
————, *Der Einfluss von Vitamin T auf Gestalt und auf die Gewohnheiten von Insekten,* Österr. zool. Z. *1* (1947).
————, *Wirkstoff T,* Verhandl deutsch. zool. Ges. *1949.*
————, *Probleme der Formbildung,* in: *Neue Ergebnisse und Probleme der Zoologie,* Ergänzungsbd. Zool. Anz. *1950.*
————, *Ergebnisse und Probleme aus dem Gebiet neuer Wirkstoffe,* Österr. zool. Z. *3* (1951).
————, *Untersuchungen über Steigerung der Resistenz durch den T-Komplex,* Z. Vitamin-, Hormon-u. Fermentforsch. *6* (1954).
————, *Untersuchungen über Behebung von Alterserscheinungen,* Pubbl. staz. zool. Napoli 27 (1955).
————, *Wirkstoffprobleme,* Verhandl. deutsch. zool. Ges. *1957.*
Goetsch, W., and Nussbaumer, G., *Die Bedeutung des Wirkstoffes T für die Medizin* (T-Vitamin Goetsch), Ärztl. Praxis *2* (1950).
Goetsch, W., Offhaus, K., and Tóth, L., *Untersuchungen über Bakterien- und Flagellatensymbiosen bei Termiten,* Naturwissenschaften *1944.*
Goetsch, W., and Scheuring, L., *Parasitismus und Symbiose der Algengattung Chlorella,* Z. Morphol. Ökol. Tiere 7 (1926).
Gohar, H. A. F., *Studies on the Xeniidae of the Red Sea. Their ecology, physiology, taxonomy and phylogeny,* Publs. Marine Biol. Sta. Ghardaqa *2* (1940).
Golberg, L., and De Meillon, B., *The nutrition of the larva of Aedes aegypti L.,* 3: *Lipid requirements,* 4: *Protein and amino acid requirements,* Biochem. J. *43* (1948).
————, see also De Meillon, B., and Golberg, L.
Golberg, L., De Meillon, B., and Lavoipierre, M., *Relation of "folic acid" to the nutritional requirements of the Mosquito larva,* Nature *154* (1944).
————, *The nutrition of the larva of Aedes aegypti L.* 2: *Essential water-soluble factors from yeast,* J. Exptl. Biol. *21* (1945).
Goldschmidt, R., *Glow worms and evolution,* Rev. sci., 86e année, No. 3298 (1948).
Golgi, C., *Intorno alla struttura ed alla biologia dei considetti globuli (o piastrine) del tuorlo,* Mem. reale. ist. Lomb. sci. lett. *22/23* (1923).
Goodchild, A. J. P., *The bacteria associated with Triatoma infestans and some other species of Reduviidae,* Parasitology *45* (1955).
Gould, L. J., *Notes on the minute structure of Pelomyxa palustris Greeff,* Quart. J. Microscop. Sci. *36* (1893).
————, *A further contribution to the study of Pelomyxa palustris,* J. Linnean Soc. London *29* (1905).
Goux, L., *Notes sur une levure symbiotique de Chlamydolecanium conchioides Goux,* Compt. rend. soc. biol. *138* (1944).
Graber, V., *Anatomisch-physiologische Studien über Phthirus inguinalis,* Z. wiss. Zool. *22* (1872).

Fröbrich, G., and Offhaus, K., *Ein neuer Nahrungsfaktor, der die Metamorphose von Tribolium confusum ermöglicht*, Naturwissenschaften *39* (1952).

——, *Der qualitative Vitamintest mit dem Reismehlkäfer Tribolium confusum Duv. als Testorganismus*, Z. Vitamin-, Hormon- u. Fermentforsch. *5* (1953).

Füller, H. B., *Morphologische und experimentelle Untersuchungen über die neurosekretorischen Verhältmisse im Zentralnervensystem von Blattiden und Culiciden*, Zool. Jahrb. Abt. Allgem. Zool. Physiol. *69* (1960).

Galippe, V., *Parasitisme normal et microbiose* (Masson, Paris 1917).

Gambetta, L., *Ricerche sulla simbiosi ereditaria di alcuni coleotteri silofagi*, Ric. morfol. biol. animale *1* (Napoli 1927).

Gardner, A. D., *Morphological effects of penicillin on bacteria*, Nature *1940*.

Garnault, P., *Recherches anatomiques et histologiques sur le Cyclostoma elegans*, Acta soc. Linn. Bordeaux [5] *1* (1887).

Gebhardt, A. von, *Adatok a Buprestidák bélcsövének ismeretéhez*, Folia Soc. Entomol. Hung. *2* (1929).

Geddes, P., *Sur la fonction de la chlorophylle chez les Planaires vertes*, Compt. rend. *87* (1878).

——, *Observations on the physiology and histology of Convoluta schultzii*, Proc. Roy. Soc. (London) *28* (1879).

——, *On nature and functions of the "yellow cells" of Radiolarians and Coelenterates*, Proc. Roy. Soc. Edinburgh *1882*.

——, *The yellow-cells of Radiolarians and Coelenterates*, Proc. Roy. Soc. Edinburgh *1882*.

Geigy, R., Halff, L. A., and Kocher, V., *Untersuchungen über die physiologischen Beziehungen zwischen einem Überträger der Chagas-Krankheit Triatoma infestans und dessen Darm-symbionten*, Schweiz. med. Wochschr. *83* (1953).

——, *L'acide folique comme élément important dans la symbiose intestinale de Triatoma infestans*, Acta Trop. *11* (1954).

Geigy, R., and Wagner, O., *Ovogenese und Chromosomenverhältnisse bei Ornithodorus moubata*, Acta Trop. *14* (1957).

Gelei, J. von, *Angaben zur Symbiosefrage von Chlorella*, Biol. Zbl.. *47* (1927).

Getzel, D., *I microbi della glandola nidamentale accessoria in Sepia officinalis*, Arch. zool. ital. *20* (1934).

——, *Il simbionte dell' Icerya purchasi Mask., Geotrichoides pierantonii*, Arch. zool. ital. *23* (1936).

——, *Attività vitaminiche del simbionte dell'Icerya purchasi Mask.*, Boll. zool. 7 (1936).

——, *Forme microbiche del sangue*, Boll. zool. *12* (1941).

Ghidini, G. M., *A proposito di alcune recenti ricerche sulla cellulosolisi nell'intestino delle termiti*, Boll. zool. *12* (1941).

Giard, A., *Sur les Nephromyces, genre nouveau de champignons parasites du rein des Molgulidées*, Compt. rend. *106* (1888).

Gier, H. T., *The morphology and behavior of the intracellular bacteroids of roaches*, Biol. Bull. *71* (1936).

——, *Growth of the intracellular symbionts of the cockroach, Periplaneta americana*, Anat. Record *70* (1937).

——, *Intracellular bacteroids in the cockroach (Periplaneta americana L.)*, J. Bacteriol. *53* (1947).

Glaser, R. W., *Biological studies on intracellular bacteria*, Biol. Bull. *39* (1920).

——, *On the isolation, cultivation and classification of the so-called intracellular "symbionts" or "Rickettsia" of Periplaneta americana*, J. Exptl. Med. *51* (1930).

——, *The intracellular "symbionts" and the "Rickettsiae,"* Arch. Pathol. *9* (1930).

——, *Cultivation and classification of "bacteroids," "symbionts" or "Rickettsiae" of Blattella germanica*, J. Exptl. Med. *51* (1930).

Francke-Grosmann, H., *Zur Kenntnis der Läuseschäden an Weisstanne* (*Abies pectinata*), *Tharandt. forstl. Jahrb. 88* (1937).

————, *Beitrage zur Kenntnis der Beziehungen unserer Holzwespen zu Pilzen,* Verh. VII. Intern. Entom.-Kongr. Berlin 1939.

————, *Über das Zusammenleben von Holzwespen* (*Siricinae*) *mit Pilzen,* Z. angew. Entomol. *25* (1939).

————, *Larvenentwicklung und Generationswechsel bei Hylecoetus dermestoides,* Trans. IX Intern. Congr. Entomol. Amsterdam 1951.

————, *Über die Ambrosiazucht der beiden Kieferborkenkäfer Myelophilus minor Htg. und Ips acuminatus Gyll.,* Medd. Statens Skogsforskningsinst. *41* (1952).

————, *Grundlagen der Symbiose bei pilzzüchtenden Holzinsekten,* Verhandl. deutsch. zool. Ges. *1956.*

————, *Hautdrüsen als Träger der Pilzsymbiose bei Ambrosiakäfern,* Z. Morphol. Ökol. Tiere *45* (1956).

————, *Zur Übertragung der Nährpilze bei Ambrosiakäfern,* Naturwissenschaften *43* (1956).

————, *Über die Ambrosiazucht holzbrütender Ipiden in Hinblick auf das System,* 14. Verhandlungsbericht deutsch. Ges. angew. Entomol. *1957.*

————, *Über das Schicksal der Siricidenpilze während der Metamorphose,* Bericht 8. Wanderversammlg. deutsch. Entomol. *1957* (Berlin 1957).

————, *Some new aspects in forest entomology.* Ann. Rev. Entomol. *8* (1963).

————, *Die Übertragung der Pilzflora bei dem Borkenkafer Ips acuminatus Gyll.* Z. angew. Entomol. *52* (1963).

————, *Ein Symbioseorgan bei dem Borkenkäfer Dendroctonus frontalis Zimm,* Naturwissenschaften *52* (1965).

Francke-Grosmann, H., and Schedl, W., *Ein orales Übertragungsorgan der Nährpilze bei Xyleborus mascarensis Eich.* (*Scolytidae*), Naturwissenschaften *47* (1960).

Frank, W., *Einwirkung verschiedener Antibiotica auf die Symbionten der Küchenschabe Blatta orientalis L. und die dadurch bedingten Veränderungen am Wirtstier,* Verhandl. deutsch. zool. ges. *1954.*

————, *Entfernung der intrazellulären Symbionten der Küchenschabe* (*Periplaneta orientalis L.*) *durch Einwirkung verschiedener Antibiotica, unter besonderer Berücksichtigung der Veränderungen am Wirtstier und an den Bakterien,* Z. Morphol. Ökol. Tiere *44* (1956).

Franz, V., *Die japanischen Knochenfische der Sammlungen Haberer und Doflein. Beiträge zur Naturgesch. Ostasiens,* Abhandl. bayer. Akad. wiss., Math.-Phys. Kl., Suppl. *4* (1910 to 1913).

Fraser, E. A., *Observations on the life-history and development of the hydroid Myrionema amboinense,* Sci. Rept. Great Barriere Reef Exped. *3,* No. 4 (1931).

Freytag, K., *Über die Bakterienflora der Haut und der Segmentalorgane von Hirudo officinalis und medicinalis* (Abstract, Diss. Marburg 1952) (Marburg 1952).

Fröbrich, G., *Untersuchungen über Vitaminbedarf und Wachstumsfaktoren bei Insekten,* Z. vergleich. Physiol. *27* (1939).

————, *Darstellung von Konzentraten des "Tribolium-Imago-Faktors"* (*TIF*) *und seine vermutliche chemische Natur,* Naturwissenschaften *40* (1953).

————, *Der "Tribolium-Imago-Faktor"* (*TIF*) *durch Carnitin ersetzbar,* Naturwissenschaften *40* (1953).

————, *Die Caseindosierung in syntetischen Diäten für die Aufzucht von Tribolium confusum Duval,* Naturwissenschaften *40* (1953).

————, *Neue Ergebnisse experimenteller Untersuchungen zur Ernährungsphysiologie des Reismehlkäfers Tribolium confusum Duval,* Z. Vitamin-, Hormon- u. Fermentforsch. *6* (1954).

Fröbrich, G., and Offhaus, K., *Neue Befunde zur Ernährungsphysiologie des Reismehlkäfers Tribolium confusum Duval* (*Vorl. Mitt.*), Verhandl. deutsch. zool. Ges. *1952.*

Fleming, A., Voureka, A., Kramer, J. R. H., and Hughes, W. H., *The morphology and motility of Proteus vulgaris and other organisms cultured in the presence of penicillin*, J. Gen. Microbiol. *4* (1950).

Flögel, J. H. L., *Monographie der Johannisbeerblattlaus (Aphis ribis L.)*, Z. wiss. Inssekten-biol. [N. S.] *1* (1905).

Florence, L., *An intracellular symbiont of the hog louse*, Am. J. Trop. Med. *4* (1924).

Florenzano, G., *Prospetto sistematico delle specie batteriche segnalate come entomosimbionti*, Redia *34* (1949).

Foeckler, Fr., *Reinfektionsversuche steriler Larven von Stegobium paniceum L. mit Fremdhefen und die Beziehungen zwischen der Entwicklungsdauer der Larven und dem B-Vitamingehalt des Futters und der Hefen*, Z. Morphol. Ökol. Tiere *50* (1961).

Forbes, S. A., *Bacterium a parasite of the chinch bug*, Am. Naturalist *16* (1882).

——Bacteria normal to digestive organs of Hemiptera, Bull. Ill. State Lab. Nat. Hist. *4* (1892).

Fraenkel, G., *The role of symbionts as sources of vitamins and growth factors for their insect hosts*. Symposium sur la "Symbiose chez les insectes" (Amsterdam 1951), Tijdschr. Entomol. *95* (1952), and Union Intern. Sci. Biol. Ser. B, No. 10.

——, *The nutritional value of green plants for insects*, Trans. IX Intern. Congr. Entomol. Amsterdam *1951*.

——, *The nutritional requirements of insects for known and unknown vitamins*, Trans. IX Intern. Congr. Entomol. Amsterdam *1951*.

——, *Studies on the distribution of vitamin B (carnitin)*, Biol. Bull. *104* (1953).

——, *The effect of zinc and potassium in the nutrition of Tenebrio molitor, with observations on the expression of a carnitin deficiency*, J. Nutrition *65* (1958).

——, see also Blewett, M., and Fraenkel, G.; and Pant, N. C., and Fraenkel, G.

Fraenkel, G., and Blewett, M., *Biotin, B_1, riboflavin, nicotinic acid, B_6 and pantothenic acid as growth factors for insects*, Nature *150* (1942).

——, *Vitamins of the B-group required by insects*, Nature *151* (1943).

——, *Intracellular symbionts of insects as sources of vitamins*, Nature *152* (1943).

——, *The basic food requirements of several insects*, J. Exptl. Biol. *20* (1943).

——, *The vitamin B-complex requirements of several insects*, Biochem. J. *37* (1943).

——, *The natural foods and the food requirement of several species of stored products insects*, Trans. Roy. Entomol. Soc. London *93* (1943).

——, *The dietetics of the clothes moth, Tineola bisselliella Hum.*, J. Exptl. Biol. *22* (1946).

——, *The dietetics of the caterpillars of three Ephestia species, E. kuehniella, E. elutella and E. cautella, and of a closely related species, Plodia interpunctella*, J. Exptl. Biol. *22* (1946).

——, *Linoleic acid, vitamin E and other fat-soluble substances in the nutrition of certain insects (Ephestia kuehniella, E. elutella, E. cautella and Plodia interpunctella)*, J. Exptl. Biol. *22* (1946).

——, B_T, *new vitamin of the B-complex and its relation to the folic acid group and other anti-anaemia factors*, Nature *161* (1948).

Fraenkel, G., Blewett, M., and Coles, M., *The nutrition of the mealworm, Tenebrio molitor L.*, Physiol. Zoöl. *23* (1950).

Fraenkel, G., and Leclerq, F., *Nouvelles recherches sur les besoines nutritifs de la larve du Tenebrio molitor L.*, Arch. Intern. Physiol. et Biochem. *64* (1956).

Fraenkel, G., and Chang, P.-I., *Manifestations of a vitamin B_T (Carnitin) deficiency in the larvae of the mealworm, Tenebrio molitor L.*, Physiol. Zoöl. *27* (1954).

Fränkel, H., *Die Symbionten der Blattiden im Fettgewebe und im Ei, insbesondere von Periplaneta orientalis*, Z. wiss. Zool. *119* (1921).

Francke-Grosmann, H., *Beitrage zur Kenntnis der Lebensgemeinschaft zwischen Borkenkäfern und Pilzen*, Z. Parasitenk. *3* (1931).

Escherich, K., *Über das regelmäßige Vorkommen von Sproßpilzen in dem Darmepithel eines Käfers*, Biol. Zbl. *20* (1900).

———, *Die Forstinsekten Mitteleuropas*, Vol. 5 (Parey, Berlin 1940–1942) (*Siriciden-Symbiose*).

Evans, J. W., *Concerning the Peloridiidae*, Australian J. Sci. *4* (1941).

Faasch, W. J., *Darmkanal und Blutverdauung bei Aphanipteren*, Z. Morphol. Ökol. Tiere *29* (1935).

Fabre, J. H., *Souvenirs entomologiques*, 1 to 5. Serie (Paris 1891–1897).

Falk, R., *Die Scheindestruktion des Koniferenholzes durch die Larven des Hausbockes, Hylotrupes bajulus*, Cellulosechemie *11* (1930).

———, *Scheindestruktion des Holzes durch die Larve von Anobium*, Cellulosechemie *11* (1930).

Famintzin, A., *Die Symbiose als Mittel der Synthese von Organismen*, Biol. Zbl. *27* (1907).

Fantham, H. B., *Some researches of the life cycle of Spirochaetes*, Ann. Trop. Med. Parasitol. *5* (1911).

Farran, G. P., *Pyrosoma spinosum*, Mem. Challenger Soc. No. 1 (London 1909).

Farris, S. H., *Ambrosia fungus storage in two species of Gnathotrichus Eichhoff (Coleoptera: Scolytidae)*, Can. Entomologist *95* (1963).

Fauré-Frémier, E.-M., *Symbiontes bactériens des Ciliés du genre Euplotes*, Compt. rend. *235* (1952).

Fava, A., and Barbato, G., *Primi risultati ottenuti nel tentativo di costruire stipiti aposimbiontici in Blattella germanica mediante somministrazione orale di sigmaamicina*, Genet. agrar. *13* (1960).

———, *Ulteriori tentativi di ottenere stipiti aposimbiotici di Blatella germanica mediante somministrazione orale di sigmaamicina*, Symposia Genetica et Biologica Italica, Pavia *1960*.

Fava, A., and Laudani, U., *Sulla risposta alla colorazione di Gram e di Neisser dei batteri simbionti di alcune specie di Blatta*, Symposia Genetica et Biologica Italica, Pavia *1960*.

———, *Sui batteriociti di Nauphoeta cinerea (Epilamprinae, Blattoidea) trattata per vie orale con sigmaamicina*, Symposia Genetica et Biologica Italica, Pavia *1960*.

———, *Sui batteriociti di Nauphoeta cinerea trattata per via orale con sigmaamicina*, Symposia Genetica et Biologica Italica, Pavia *1960*.

———, *Produzione d'stipiti aposimbiotici in Nauphoeta cinerea Burm. mediante somministrazione parenterale d'antibiotici*, Genet. agrar. *14* (1961).

Fedele, M., *Sulle strutture e funzione dei ciechi epatopancreatici nei molluschi opistobranchi*, Boll. soc. ital. biol. sper. *1* (1926).

Fekl, W., *Die Bakterienflora der Tracheen und des Blutes einiger Insekten*, Z. Morphol. Ökol. Tiere *44* (1956).

Feldman-Muhsam, R., and Havivi, Y., *A microorganism associated with sperm cells of Ornithodorus*, Nature *193* (1962).

———, *On Adlerocystis n. gen. (Phycomycetes) a symbiont of Ornithodorus ticks*, Parasitology *55* (1963).

Fernando, E. F. W., *Storage and transmission of ambrosia fungus in the adult Xyleborus fornicatus (Eich.)*, Ann. Mag. Nat. Hist., Ser. *13* [2] (1960).

Ferris, G. F., *Report upon insects collected in China (Homoptera, Coccoidea)*, Part 5, Microentomology *19* (1954).

Fink, R., *Morphologische und physiologische Untersuchungen an den intrazellularen Symbionten von Pseudococcus citri Risso*, Dissertation, Munich *1951*.

———, *Morphologische und physiologische Untersuchungen an den intrazellularen Symbionten von Pseudococcus citri Risso*, Z. Morphol. Ökol. Tiere *41* (1952).

Finnegan, R. F., *The storage of ambrosia fungus spores by the pitted ambrosia beetle, Corthylus punctatissimus Zimm.*, Can. Entomologist *95* (1963).

and Pediculus capitis) *with some observations on the maturation of the egg*, Quart. J. Microscop. Sci. *64* (1920).

Dorner, G., *Darstellung der Turbellarienfauna Ostpreußens*, Schrift. phys.-ökon. Ges. Königsberg *43* (1902).

Doyle, W. L., *Studies on comparative cytoplasmic cytology*, Rept. Tortugas Lab., Carnegie Inst. Year Book *33* (1934).

————, *Observations on zooxanthellae*, Rept. Tortugas Lab., Carnegie Inst. Year Book *34* (1935).

Doyle, W. L., and Doyle, M. M., *The structure of zooxanthellae*, Papers Tortugas Lab. *32* (1940).

Dreyfus, L., *Zu Krassilstschicks Mitteilungen über die vergleichende Anatomie und Systematik der Phytophthiren*, Zool. Anz. 7 (1894).

Dubois, R., *La vie et la lumière* (Alcan, Paris 1914).

Duerden, J. E., *West Indian Madreporian polypes*, Mem. Natl. Acad. Sci. U.S. *8* (1902).

————, *The coral Siderastraea radians and its postlarval development*, Carnegie Inst. Wash. Publ. *20* (1904).

————, *The rôle of mucus in corals*, Quart. J. Microscop. Sci. *49* (1906).

Duesberg, J., *Chondriosomes et bactéries dans les nodosités radicales des légumineuses*, Compt. rend. assoc. anat. *18* (1923).

Dufour, L., *Historie comparative des metamorphoses et de l'anatomie des Cetonia aurata et Dorcus parallelopipedus*, Ann. sci. nat. Paris, (2) *18* (1812).

————, *Recherches anatomiques sur les Carabigues et sur plusieurs autres insectes Coléoptères*, Ann. sci. nat. Paris, (1) *3* (1824).

————, *Recherches anatomiques et physiologiques sur les Hémiptères*, Mém. sav. etrang. acad. sci. *4* (1833).

Duncan, J. T., *On a bactericidal principle present in the alimentary canal of insects and arachnids*, Parasitology *18* (1926).

Dutton, J. E., and Todd, J. L., *A note on the morphology of Spirochaeta duttoni*, Lancet *1907*.

Eisler, M. v., *Über die Wirkung von Salzen auf Bakterien*, Zentralbl. f. Bakt. I, Orig. (1909).

Ekblom, T., *Cytological and biochemical researches into the intracellular symbiosis in the intestinal cells of Rhagium inquisitor L.*, I: Skand. Arch. Physiol. *67* (1931); II: ibid., *64* (1932).

Eliot, C., and Evans, T. J., *Doridoeides gardineri: a doriform cladohepatic nudibranch*, Quart. J. Mikroscop. Sci. *52* (1908).

Emeis, W., *Über Eientwicklung bei den Coccinen*, Zool. Jahr., Abt. Anat., *39* (1915).

Enders, H. E., *A study on the life history and habits of Chaetopterus variopedatus*, J. Morphol. *20* (1909).

Engelmann, F., *Die Steuerung der Ovarfunktion bei der ovoviviparen Schabe Leucophaea maderae (Fabr.)*, J. Insect Physiol. *7* (1957).

Engelmann, Th. W., *Neue Methode der Untersuchung der Sauerstoffausscheidung pflanzlicher und tierischer Organismen*, Botan. Ztg. *39* (1881).

————, *Über tierisches Chlorophyll*, Arch. ges. Physiol. *32* (1883).

Enriques, P., *Ricerche sui Radiolari coloniali*, I., II: R. comit. Talassogr. ital. *71* (1919/21).

Entz, G., *Értesitö a kolozsvári orvos-természettndomán yi társulat második természettndományi szaküléséröl*, Kolozsvart *1876*.

————, *Über die Natur der "Chlorophyllkörperchen" niederer Tiere*, Biol. Zentr. *1* (1881/82).

Ergene, S., *Spielen die Darmbakterien von Calotermes flavicollis bei der Assimilation des atmosphaerischen Stickstoffs eine Rolle?* Rev. fac. sci. univ. Istanbul *B14* (1949).

Erikson, D., *The pathogenic aerobic organisms of the Actinomyces group*, Spec. Rept. Ser. Med. Research Council London, No. 203 (*1935*).

Ermisch, A., *Vergleichend-anatomische Untersuchungen über die Endosymbiose der Fulgoroiden mit besonderer Berücksichtigung der Araeopiden*, Z. Morphol. Ökol. Tiere *49* (1960).

Höring, F. O., *Parasitimus oder Symbiose? Das Infektionsproblem im Wandel der Grundlagenforschung* (J. Ebner, Ulm 1947).

———, *Zoologische Symbioseforschung und medizinische Infektionslehre*, Verhandl. deutsch. Ges. inn. Med. *1957*.

Hogg, J., *On the action of light upon the colour of the river sponges*, Mag. Nat. Hist. *4* (1840).

Holdaway, F. G., *Composition of different regions of mounds of Eutermes exitiosus Hill.* J. Council Sci. Ind. Research Australia *6* (1933).

Hollande, A. Chr., *Biologie et reproduction des Rhizopodes des genres Pelomyxa et Amoeba*, Bull. biol. France Belg. *79* (1945).

———, *L'évolution des Endosymbiontes des Termites et des Blattes*, Symposium sur la "Symbiose chez les insectes," Amsterdam 1951, Tijdschr. Entomol. *95* (1952), and Union intern. sci. biol. Ser. B, No. 10.

Hollande, A. Chr., and Favre, R., *La structure cytologique de Blattabacterium cuénoti (Mercier) n. g., symbiote du tissu adipeux chez les Blattides*, Compt. rend. soc. biol. *107* (1931).

Holmgren, N., *Über die Exkretionsorgane des Apion flavipes und Dasytes niger*, Anat. Anz. *22* (1902).

———, *Termitenstudien*, 1: *Anatomische Untersuchungen*, Kgl. Svenska Vetenskapsakad. Hdl. (N. S.) *44* (1909); 2: *Systematik der Termiten. Die Familien Mastotermitidae, Protermitidae und Mesotermitidae*, ibid *46* (1910/11).

Hood, C. J., *The zoochlorellae of Frontonia leucas*, Biol. Bull. *52* (1927).

Hooke, R., *Micrographia* (London 1665).

Hoover, Sh., *Studies on the bacteroids of Cryptocercus punctulatus*, J. Morphol. *76* (1945).

Hornbostel, H., *Über die bakteriellen Eigenschaften der Darmsymbionten beim medizinischen Blutegel (Hirudo officinalis) nebst Bemerkungen zur Symbiosefrage*, Zentr. Bakteriol. *147* (1941).

Hornell, J., *A note on the presence of symbiotic algae in the integuments of Nudibranchs of the genus Melibe*, Rept. Marine Zool. Okhamandal, London, Pt. 1 (1909).

House, H. L., *Nutritional studies with Blattella germanica L. reared under aseptic conditions*, 2: *A chemically defined diet*, 3: *Five essential amino acids*, Can. Entomologist *81* (1949).

House, H. L., and Patton, R. L., *Nutritional studies with Blatella germanica (L.) reared under aseptic conditions*, 1: *Equipment and technique*, Can. Entomologist *81* (1949).

Hovasse, R., *Bacillus cuenoti Mercier, bactéroïds de Periplaneta orientalis, a la morphologie d'une bactérie*, Arch. Zool. exptl. *70* (1913).

———, *Marchalina hellenica (Gennadius). Essai de Monographie d'une cochenille*, Bull. Biol. France Belg. *64* (1930).

Huber, B., *Die Siebröhren der Pflanzen als Nahrungsquelle fremder Organismen und als Transportbahnen von Krankheitskeimen*, Biol. Generalis. *16* (1942).

Huber, B., Schmidt, E., and Jahnel, H., *Untersuchungen über den Assimilatstrom*, 1, Tharandt. forstl. Jahr. *88* (1937).

Huber-Schneider, L., *Morphologische und physiologische Untersuchungen an der Wanze Mesocerus marginatus L. und ihren Symbionten (Heteroptera)*, Z. Morphol. Ökol. Tiere *46* (1957).

Huff, C. G., *Nutritional studies on the seed corn maggot Hylemyia cilicrura Rondani*, J. Agr. Research *36* (1928).

Huger, A., *Experimentelle Eliminierung der Symbionten aus den Mycetomen des Getreidekapuziners, Rhizopertha dominica F.*, Naturwissenschaften *41* (1954).

———, *Experimentelle Untersuchungen über die künstliche Symbiontenelimination bei Vorratsschädlingen: Rhizopertha dominica F. (Bostrychidae) und Oryzaephilus surinamensis L. (Cucujidae)*, Z. Morphol. Ökol. Tiere *44* (1956).

Hughes-Schrader, S., *Cytology of Coccids*, Advances in Genet. *2* (1948).

Hungate, R. F., *Studies on the nutrition of Zootermopsis*, 1: *The rôle of bacteria and molds in cellulose decomposition*, Z. Bakteriol. *94* (1936); 2: *The relative importance of the Termite and the Protozoa in wood digestion*, Ecology *19* (1938); 3: *The anaerobic carbohydrate dissimilation by the intestinal Protozoa*, ibid., *20* (1939).

————, *Quantitative analyses on the cellulose fermentation by Termite Protozoa*, Ann. Entomol. Soc. Am. *36* (1943).

————, *Further experiments on cellulose digestion by the Protozoa in the rumen of cattle*, Biol. Bull. *84* (1943).

————, *Studies on cellulose fermentation*, 1: *The culture and physiology of an anaerobic cellulose-digesting bacterium*, J. Bacteriol. *48* (1944).

————, *An aerobic decomposing Actinomycete, Micromonospora propionici*, J. Bacteriol. *51* (1946).

————, *The symbiotic utilization of cellulose*, J. Elisha Mitchell Sci. Soc. *62* (1946).

————, *Mutualisms in Protozoa*, Ann. Rev. Microbiol. *1950*.

————, *The anaerobic mesophilic cellulolytic Bacteria*, Bacteriol. Rev. *14* (1950).

Hurst, C. T., and Strong, J. C., *Über die Kultivierung der Mitochondrien in vitro*, Arch. Protistenk. *77* (1932).

Huxley, Th., *On the agamic reproduction and morphology of Aphids*, Trans. Linn. Soc. London *22* (1858).

Ihle, J. E., *Desmomyarier*, in: Kükenthal, *Handbuch der Zoologie*, Vol. 5, 2. Hälfte (W. de Gruyter & Co., Berlin and Leipzig 1935).

Imai, H., *Studies on the symbiotic luminous bacteria* (in Japanese), Sei-i-kwai Med. J. *61* (1942).

Iwai, T. and Asano, H., *On the luminous cardinal fish, Apogon ellioti Day.*, Sci. Rept. Yokosuka City Museum *3* (1958).

Jacquot, R., Armand, Y., and Rey, P., Bull. soc. hyg. alim. *29* (1941).

Jaschke, W., *Beitrage zur Kenntnis der symbiontischen Einrichtungen bei Hirudineen und Ixodiden*, Z. Parasitenk. *5* (1933).

Javelly, E., *Les corps bactéroides de la blatte (Periplaneta orientalis) n'ont pas encore été cultivées*, Compt. rend. soc. biol. *77* (1914).

Jírovec, O., *Notizen über parasitische Protozoen*, 2: *Symbiose von Bakterien und Trichonympha serbica (Georgevitsch)*, Zentr. Bakteriol. *123* (1932).

————, *Zur Kenntnis von in Oligochaeten parasitierenden Microsporidien aus der Familie Mrazekidae*, Arch. Protistenk. *87* (1936).

Johnson, D. E., *The relation of the cabbage maggot and other insects to the spread and development of soft rot of Cruciferae*, Phytopathology *20* (1930).

Jucci, C., *Sulla presenza di batteriociti nel tessuto adiposo dei Termitidi*, Boll. zool. *1* (1930); Atti XI congr. intern. zool., Padova 1930, in: Arch. zool. ital. *16* (1932).

————, *Symbiosis and phylogenesis in insects*, Trans. IX intern. Congr. Entomol. Amsterdam 1951.

Julin, Ch., *Les embryons de Pyrosoma sont phosphorescents: Les cellules du testa constituent les organes lumineux du cyathozoïde*, Compt. rend. soc. biol. *66* (1909).

————, *Recherches sur le développement embryonnaire de Pyrosoma giganteum Les.*, Zool. Jahr. Suppl. *15* (1912).

————, *Les charactères histologiques spécifiques des "cellules lumineuses" de Pyrosoma giganteum et de Cyclosalpa pinnata*, Compt. rend. *155* (1912).

Jungmann, P., *Untersuchungen über Schaflausrickettsien*, Deut. med. Wochschr. *44* (1918).

Jurzitza, G., *Physiologische Untersuchungen an Cerambycidensymbionten. Ein Beitrag zur Symbiose holzfressender Insektenarten*, Arch. Mikrobiol. *33* (1959).

————, *Die Symbiose der Anobiiden und Cerambyciden mit hefeartigen Pilzen; Sammelbericht*, Arch. Mikrobiol. *43* (1962).

Jurzitza, G., *Die Wirkung des Sulfanilamids auf die Symbionten einiger Anobiiden*, Z. angew, Entomol. *52* (1963).

————, *Pilze als Insektensymbionten*, Z. Pilzkunde *29* (1963).

————, *Studien an der Symbiose der Anobiiden II, Physiologische Studien an den Symbionten von Lasioderma serricorne* F., Arch. Mikrobiol. *49* (1964).

Jurzitza, G., Kühlwein, H., and Kreger-van Rij, N. J. W., *Zur Systematik einiger Cerambycidensymbionten*, Arch. Mikrobiol. *36* (1960).

Kaiser, R., *T-Vitamin*, Wien. tierärztl. Monatsschr. *6* (1949).

Kalicka-Fijalkowska, J., *Le développement embryonnaire de Margarodes polonicus Ckll.*, Compt. rend. Soc. Biol. Paris *99* (1928).

Kanz, E., *Bakterienkultivierung im Symbioseverfahren durch eine neue Form der Ammen-platte*, Arch. Hyg. u, Bakteriol *142* (1958).

Karawaiew, W., *Über Anatomie und Metamorphose des Darmkanals der Larve von Anobium paniceum*, Biol. Zentr. *19* (1899)

Kato, K., *A new type of luminous organ in fish* (in Japanese), Zool. Mag. (Tokyo) *57* (1947).

————, *On the luminous fungus gnats in Japan*, Sci. Rept. Saitama Univ. *B1* (1953).

Kaudewitz, H., *Die Wuchsstoffverteilung in larvalen Fettkörper von Tenebrio*, Z. vergleich. Physiol. *35* (1953).

Kawaguti, S., *On the physiology of reef corals I, II, III*, Palao Trop. Biol. Sta. Stud. Tokyo *2* (1937).

Keeble, F., and Gamble, F. W., *On the isolation of the infecting organism (Zoochlorella) of Convoluta roscoffensis*, Proc. Roy. Soc. (London) *77* (1905).

————, *The origin and nature of the green cells of Convoluta roscoffensis*, Quart. J. Microscop. Sci. *51* (1907).

————, *The yellow-brown cells of Convoluta paradoxa*, Quart. J. Microscop. Sci. *52* (1908).

————, *Plant-Animals. A Study in Symbiosis* (University Press, Cambridge 1910).

Keilin, D., *On the life-history of Dasyhelea obscura Winnertz, with some remarks on the parasites and hereditary bakterian symbiontes of this midge*, Ann. Mag. Nat. History [9] *8* (1921).

Keller, H., *Bakteriologische Untersuchungen an der Grillendarmflora und den intrazellulären Bakterien von Pelomyxa palustris Greeff*, Dissertation, Erlangen 1949 (unpublished).

————, *Untersuchungen über die intrazellulären Bakterien von Pelomyxa palustris Greeff*, Z. Naturforsch. *4b* (1949).

————, *Die Kultur der intrazellularen Symbionten von Periplaneta orientalis*, Z. Naturforsch. *5b* (1950).

Kellner, W., *Untersuchungen zur symbiontischen Bedeutung der Blinddärme der Nagetiere*, Z. Morphol. Ökol. Tiere *44* (1956).

Kelsey, J. M., *Symbiose and Anobium punctatum De Geer*, Proc. Roy. Entomol. Soc. London (A), *33* (1958).

Ketchel, M. M., and Williams, C. M., *Isolation of the intracellular symbiont of the roach*, Anat. Record *117* (1953).

Kiefer, H., *Der Einfluß von Kälte und Hunger auf die Symbionten der Anobiiden- und Cerambycidenlarven*, Zentr. Bakteriol. *86* (1932).

Kirby, H., *Morphology and mitosis of Dinenympha fimbriata sp. nov.*, Univ. Calif. Publs. Zool., *26* (1924).

————, *Flagellates of the genus Trichonympha in termites*, Univ. Calif. Publs. Zool., *37* (1932).

————, *The devescovinid flagellates*, Univ. Calif. Publs. Zool., *43* (1938).

Kirkaldy, G. W., *Phylogeny of Homoptera*, Can. Entomologist *42* (1910).

Kishitani, T., *Über das Leuchtorgan von Euprymna morsei Verrill*, Proc. Imp. Acad. (Tokyo) *4* (1928).

————, *Preliminary report on the luminous symbiosis in Sepiola birostrata Sasaki*, Proc. Imp. Acad. (Tokyo) *4* (1928).

Kishitani, T., *L'etude de l'organe photogène du Loligo edulis Hoyle*, Proc. Imp. Acad. (Tokyo) 4 (1928).

————, *On the luminous organs of Watasenia scintillans*, Ann. Zool. Japan *11* (1928).

————, *Studien über die Leuchtsymbiose in Physiculus japonicus Hilgendorf mit der Beilage der zwei neuen Arten der Leuchtbakterien*, Sci. Repts. Tôhoku Imp. Univ. [4th Ser. Biol.] 5 (1930).

————, *Studien über Leuchtsymbiose bei japanischen Sepien*, Folia Anat. Japon. *10* (1932).

Kitao, Z., *Notes on the anatomy of Warajicoccus corpulentus Kuwana, a scale insect noxious to various oaks*, J. Coll. Agr. Tokyo Imp. Univ. *10* (1928).

Klevenhusen, F., *Beiträge zur Kenntnis der Aphidensymbiose*, Z. Morphol. Ökol. Tiere 9 (1927).

Klieneberger-Nobel, E., *Origin development and significance of L-forms in bacterial cultures*, J. Gen. Microbiol. *3* (1949).

————, *The L-cycle: A process of regeneration in bacteria*, J. Gen. Microbiol. *5* (1951).

Knop, J. *Bakterien und Bakteroiden bei Oligochaeten*, Z. Morphol. Ökol. Tiere 6 (1926).

Koch, A., *Morphologie des Eiwachstums der Chilopoden*, Z. Zellforsch. 2 (1925).

————, *Über das Vorkommen von Mitochondrien in Mycetocyten*, Z. Morphol. Ökol. Tiere *19* (1930).

————, *Über die Symbiose von Oryzaephilus surinamensis*, Atti XI Congr. Intern. Zool., Padova 1930, in: Arch. zool. ital. *16* (1931).

————, *Die Symbiose von Oryzaephilus surinamensis L.*, Z. Morphol. Ökol. Tiere 23 (1931).

————, *Über das Verhalten symbiontenfreier Sitodrepalarven*, Biol. Zentr. 53 (1933).

————, *Über künstlich symbiontenfrei gemachte Insekten*, Verhandl. deut. Zool. Ges. *1933*.

————, *Neue Ergebnisse der Symbioseforschung, 1: Die Symbiose des Brotkäfers Sitodrepa panicea*, Prakt. Mikroskopie *13* (1934).

————, *Symbiosestudien, 1: Die Symbiose des Splintkäfers Lyctus linearis Goeze*, Z. Morphol. Ökol. Tiere 32 (1936); 2: *Experimentelle Untersuchungen an Oryzaephilus surinamensis L.*, ibid., 32 (1936); 3: *Die intrazellulare Symbiose von Mastotermes darwiniensis Froggatt*, ibid., 34 1938).

————, *Über den gegenwärtigen Stand der experimentellen Symbioseforschung*, Verhandl. intern. Kongr. Entomol. Berlin 1938, 2 (1939).

————, *Wachstumsfördernde Wirkstoffe der Hefe*, Naturwissenschaften 28 (1940).

————, *Über die vermeintliche Bakteriensymbiose von Tribolium. Ein Beitrag zur Önocyten-frage*, Z. Morphol. Ökol. Tiere 37 (1940).

————, *Wege und Ziele der experimentellen Symbioseforschung*, Naturw. Rundschau. *1948*.

————, *Die Bakteriensymbiose der Küchenschaben*, Mikrokosmos 38 (1949).

————, *Fünfzig Jahre Erforschung der Insektensymbiosen*, Naturwissenschaften 37 (1950).

————, *Untersuchungen über Wachstumsaktivatoren*, Verhandl. deut. Zool. Ges. *1951*.

————, *Biologische und medizinische Probleme der Stoffwechselphysiologie symbiontischer Mikro-organismen*, Münch. med. Wochschr. 93 (1951).

————, *Paul Buchner, Leben und Werk*. With bibliography and preface by Hans Carossa (privately published, Munich *1951*).

————, *Neuere Ergebnisse auf dem Gebiete der experimentellen Symbioseforschung*, Symposium sur la "Symbiose chez les insectes" Amsterdam 1951, Tijdschr. Entomol. 95 (1952), and Union intern. sc. biol. Ser. B, No. 10.

————, *Über die Physiologie intrazellularer Symbionten*, Zentr. Bakteriol. 158 (1952).

————, *Symbioseforschung und Wachstumsvitamine*, Landwirtsch. Forsch. 6 (1954).

————, *Das Verhältnis zwischen Symbiont und Wirt*, Verhandl. deutsch. zool. Ges. *1955*.

————, *Die experimentelle Analyse der Bedeutung der Symbionten*, Schweiz. Z. allgem. Pathol. Bakteriol. *19* (1956).

Koch, A., *The experimental elimination of symbionts and its consequences*, Exptl. Parasitol. *5* (1956).

———, *Symbiose und Ernährung*, Münch. med. Wochschr. *98* (1956).

———, *Die physiologische Bedeutung der Symbionten für den Wirtsorganismus*, Verhandl. deutsch. Ges. inn. Med. *1957*.

———, *Pourquoi la symbiose?* Scientia, 6ᵐᵉ Ser., *1959*.

———, *Intracellular symbiosis in insects*, Am. Rev. Microbiol. *14* (1960).

———, *Neuere Erkenntmisse auf dem Gebiete der Symbioseforschung*, Symposia genetica et biol. ital. *12* (1961).

———, *Neuere und neueste Untersuchungen aus dem "Paul-Buchner-Institut für experimentelle Symbioseforschung,"* Symposia Genetica et Biologica Italica IX (1959) *1962*.

———, *Grundlagen und Probleme der Symbioseforschung*, Med. Grundlagenforsch. *4* (1962).

———, *On the role of the symbionts in wood-destroying insects*, Recent Progr. Microbiol. *8* (1963).

Koch, A., Offhaus, K., Schwarz, I., and Bandier, J., *Symbioseforschung und Medizin, Ein Beitrag zur Klärung des Wirkungsmechanismus des Vitamin-B-Komplexes, nebst einer kritischen Betrachtung zum "Vitamin-T-Problem,"* Naturwissenschaften *38* (1951).

Koch, A., and Schwarz, I, *Der Wirkstoffgehalt von Blütenpollen und Waldhonigen. (Vorl. Mitt.)*, Verhandl. deutsch. Ges. angew. Entomol. *1952* (1954).

———, *Wirkstoffe der B-Gruppe in der Bienennahrung*, Communications II Congr. de l'Union Intern. pour l'Etude des Insectes Sociaux, Wurzburg *1955*. Insectes sociaux, Vol. III (1956).

Köhler, M., and Schwartz, W., *Untersuchungen über die Symbiose von Schildläusen mit pflanzlichen Mikroorganismen*, Z. allgem. Mikrobiol. *1* (1961).

———, *Reinkultur und Identifizierung der Symbionten der Cocciden Pseudococcus citri, Ps. maritimus und Orthezia insignis*, Z. allgem. Mikrobiol. 2, (1962).

———, *Über die Bezichungen zwischen Symbionten und Wirtsorganismus bei Pseudococcus citri, Ps. maritimus und Orthezia insignis*, Z. allgem. Mikrobiol. *2* (1962).

Kolb, G., *Über die Kernverhältnisse der Symbionten von Camponotus ligniperda Latr., Aphrophora salicis de G. und Pediculus vestimenti Burm.*, Naturwissenschaften *44* (1957).

———, *Untersuchungen über die Kernverhältnisse und morphologischen Eigenschaften symbiontischer Mikroorganismen bei verschiedenen Insekten*, Z. Morphol. Ökol. Tiere *48* (1959).

———, *Die Endosymbiose der Thelaxidae unter besonderer Berücksichtigung der Hormaphidinae und ihrer Embryonalentwicklung*, Z. Morphol. Ökol. Tiere *53* (1963).

Koningsberger, J. C., and Zimmermann, A., *De dierlijke vijanden der Koffiecultuur op Java*, Deel 2, Mededeel. 's Lands Plantentuin *49* (Batavia 1901).

Korotneff, A., *Myxosporidium bryozoides*, Z. wiss. Zool. *53* (1892).

———, *Zur Embryologie von Salpa cordiformis-zonaria und maculosa-punctata*, Mitt. Zool. Stat. Neapel *12* (1896).

———, *Zur Embryologie von Salpa runcinata-fusiformis*, Z. wiss. Zool. *62* (1896).

———, *Zur Embryologie von Salpa maxima-africana*, Z. wiss. Zool. *66* (1904).

Kostytschew, S., *Lehrbuch der Pflanzenphysiologie*, Vol. 2 (Springer, Berlin 1931).

Kotter, L., *Bakteriologische und mikrochemische Untersuchungen an der Magenscheibe von Pediculus vestimenti Burm.*, Arch. Mikrobiol. *23* (1955).

Kovalevsky, A., *Etude biologique de l'Haementaria costata Müller*, Mém. acad. impér. sci. St-Pétersbourg [8] *11* (1900).

Kozukue, H., *Isolation of luminous bacteria from alimentary canal of symbiotic luminous fish*, Tokyo Jikeikai Med. J. *67* (1952).

Krassilstschik, J., *Sur les bactéries biophytes*, Ann. inst. Pasteur *1889*.

———, *Zur Anatomie der Phytophthiren*, Zool. Anz. *15* (1892).

Krassilstschik, J., *Zur vergleichenden Anatomie und Systematik der Phytophthiren*, Zool. Anz. *16* (1893).

Kreitmaier, G., *B-Vitamine und Aminosäuren als Wachstumsstimulanten bei Paramaecium caudatum* (Ehrbg.), Arch. Mikrobiol. *77* (1952).

———, *Über die Beeinflussung der Vermehrungsrate von Paramaecium caudatum Ehrbg. durch B-Vitamine und Aminosäuren*, Z. Vitamin-, Hormon-, u. Fermentforsch. *4* (1952).

Kuczinski, M. H., *Die Erreger des Fleck- und Felsenfiebers* (Berlin 1927).

Kühlwein, H., und Jurzitza, G., *Zur Physiologie einiger Bockkäfersymbionten*, Naturwissenschaften *47* (1960).

———, *Studien an der Symbiose der Anobiiden, 1. Mitt. Die Kultur der Symbionten von Sitodrepa panicea L.*, Arch. Mikrobiol. *40* (1961).

Kuskop, M. *Über die Symbiose von Siphonophoren und Zooxanthellen*, Zool. Anz. *52* (1920).

———, *Bakteriensymbiosen bei Wanzen*, Arch. Protistenk. *47* (1924).

Kuwabara, S., *Some observations on the luminous organ of the fish Paratrachichthys prosthenius Jordan et Fowler*, J. Shimonoseki Coll. Fisheries *4* (1955).

Laackmann, H., *Zur Kenntnis der Alcyonariengattung Telesto Lam.*, Zool. Jahr. Suppl. *11* (1908).

Labbé, A., *Sporozoa*, in: *Das Tierreich*, Lfg. *5* (1899).

Lacaze-Duthiers, H. de, *Les Ascidies simples de côte de France*, Arch. zool. exptl. *3* (1874).

Laguesse, E., *Mitochondries et symbiotes*, Compt. rend. soc. biol. *82* (1919).

Lampel, G., *Die symbiontischen Einrichtungen im Rahmen des Generationswechsels monözischer und heterözischer Pemphiginen der Schwarz- und Pyramidenpappel*, Z. Morphol. Ökol. Tiere *47* (1958).

———, *Geschlecht und Symbiose bei den Pemphiginen*, Z. Morphol. Ökol. Tiere *48* (1959).

Landois, L., *Untersuchungen über die auf dem Menschen schmarotzenden Pediculinen*, 1: Z. wiss. Zool. *14* (1864); 2: ibid., *15* (1864).

Lanham, U. N., *Observations on the supposed intracellular symbiotic microorganisms of Aphids*, Science *115* (1952).

Lankester, E. R., *Preliminary notice on some observations with the spectroskop on animal substances*, J. Anat. Physiol. [2] *1* (1868).

Leach, J. G., *The relation of the seed-corn maggot (Phorbia fusciceps Zett.) to the spread and development of potato backleg in Minnesota*, Phytopathology *16* (1926).

———, *Potato blackleg: The survival of the pathogen in the soil and some factors influencing infection*, Phytopathology *20* (1930).

———, *Further studies on the seed-corn maggot and bacteria with special reference to potato blackleg*, Phytopathology *21* (1931).

———, *The method of survival of bacteria in the puparia of the seed-corn maggot (Hylemyia cilicrura Rond.)*, Z. angew. Entomol. *20* (1933).

———, *Insect transmission of plant diseases* (McGraw-Hill Book Co., New York and London 1940).

Lederberg, J., *Papers in Microbial Genetics, Bacteria and Bacterial Viruses, Selected by Joshua Lederberg*, University of Wisconsin Press, Madison 1951.

———, *Cell genetics and hereditary symbiosis*, Physiolog. Revs. *32* (1952).

Lehmensick, R., *Über einen neuen bakteriellen Symbionten im Darm von Hirudo officinalis L.*, Zentr. Bakteriol. *147* (1941).

———, *Weitere Untersuchungen über den bakteriellen Darmbewohner des medizinischen Blutegels*, Zentr. Bakteriol. *149* (1942).

Leidy, J., *Flora and fauna within living animals*, Smithsonian Contributions, Washington (1853).

———, *Fresh-water Rhizopods of North-America*, U. S. Geol. Survey of the Territories XII, Washington (1879).

Leiner, M., *Das Glykogen in Pelomyxa palustris Greeff mit Beiträgen zur Kenntnis des Tieres*, Arch. Protistenk. *47* (1924).

——, *Die enzymatische Anpassung bei Mikroorganismen ohne Veränderung des Erbgutes*, Ergeb. Mikrobiol., Immunitätsforsch. exptl. Therap. *31* (1958).

Leiner, M., and Wohlfeil, M., *Pelomyxa palustris Greef und ihre symbiontischen Bakterien*, Arch. Protistenk. *98* (1953).

——, *Das symbiontische Bakterium in Pelomyxa palustris Greef*, 2, Naturwissenschaften *40* (1953).

——, *Das symbiontische Bakterium in Pelomyxa palustris Greef*, 3, Z. Morphol. Ökol. Tiere *42* (1954).

Leiner, M., Wohlfeil, M., and Schmidt, D., *Das symbiontische Bakterium in Pelomyxa palustris Greef*, I, Z. Naturforsch. *6b* (1951).

——, *Pelomyxa palustris Greef*, Ann. Sci. Nat., Zool. et biol. animale, Sér. *11* (1954).

Leishman, W. B., *An adress on the mechanism of infection in tick-fever and on the hereditary transmission of Spirochaeta duttoni in the tick*, Lancet 1910.

Lemonde, A., and Bernard, R., *Aspects nutritifs des larves de Stegobium paniceu (Anobiidae) et d'Oryzaephilus surinamensis L. (Cucujidae)*, Can. Naturalist *80* (1953).

Lendenfeld, R. von, *Über Coelenteraten der Südsee*, 7: *Über die australischen rhizostomen Medusen*, Z. wiss. Zool. *47* (1888).

Lengerken, H. von, *Ekto- und Endosymbiosen zwischen phytophagen Käfern, Pilzen und Bakterien*, Biol. Generalis *16* (1942).

Lengsfeld, W., *Der Wirkstoff Bx ("Vitamin T") und seine Wirkung auf Dystrophiker und Frühgeburten*, Münch. med. Wochschr. *93* (1951).

Leonardi, G., *Monografia delle Cocciniglie italiane* (Stab. Tip. Ernesto Della Torre, Portici 1920).

Levi, G., *Chondriosomi e simbioti*, Monit. zool. ital. *33* (1922).

——, *Replica alla "Breve rettifica ecc." del Prof. Pierantoni*, Monit. zool. ital. *33* (1922).

Lewis, H. C., *The alimentary canal of Passalus*, Ohio. J. Sci. *26* (1926).

Leydig, F., *Einige Bemerkungen über die Entwicklung der Blattläuse*, Z. wiss. Zool. *2* (1850).

——, *Zur Anatomie des Coccus hesperidum*, Z. wiss. Zool. *5* (1854).

——, *Lehrbuch der Histologie des Menschen und der Tiere* (von Meidinger Sohn & Co., Frankfurt am Main 1857).

Lichtenstein, E. P., *Growth of Culex molestus under sterile conditions*, Nature *162* (1948).

Liem Soei Diong, *Onderzoekingen over Triatoma infestans als overbrenger van enkele pathogene organismen en over de complementbindingsreactie bij de ziekte van Chagas* (Dissertation, Leiden 1938).

Light, S. F., *The morphology of Eudendrium griffini*, Philippine J. Sci. Manila [Sec. D, Gen. Biol.] *8* (1913).

Lilienstern, M., *Beiträge zur Bakteriensymbiose der Ameisen*, Z. Morphol. Ökol. Tiere *26* (1932).

Limberger, A., *Über die Reinkultur der Zoochlorella aus Euspongia lacustris und Castrada viridis Volz*, Sitzber. Akad. Wiss. Wien. Math.-naturw. Kl., *1918*.

Lindemann, Chr., *Eiweißstoffwechsel bei den Blattläusen*, Naturwissenschaften *34* (1947).

Lindner, P., *Saccharomyces apiculatus parasiticus*, Zentr. Bakteriol., Abt. 2 (1895).

Lodder, J., and Kreger-van Rij, N. I. W., *The Yeasts, a Taxonomic Study*, Amsterdam *1952*.

Loefer, J. B., *Isolation and growth characteristics of the "Zoochlorellae" of Paramaecium bursaria*, Am. Naturalist *70* (1936).

Lohmann, H., *Untersuchungen zur Feststellung des vollständigen Gehaltes des Meeres an Plankton*, Wiss. Meeresunters. [N. S.] Abt. Kiel *10* (1908).

Lohmann, H., *Die Probleme der modernen Planktonforschung*, Verhandl. deut. zool. Ges. *1912.*

Lotmar, R., *Über den Einfluß der Temperatur auf den Parasiten Nosema apis*, Bei. Schweiz. Bienenztg. *1*, H. 6 (1943).

——, *Zur Wirkungssweise des Ultraschalls*, Der Ultraschall in der Medizin *5* (1952).

Lumière, A., *Le Mythe des Symbiotes* (Masson, Paris 1919).

Luther, A., *Die Eumesostominen*, 1. Teil, Z. wiss. Zool. *77* (1904).

Lwoff, A., *Nature et position systématique du bacteroïde des blattes*, Compt. rend. soc. biol. *89* (1923).

Lwoff, M., and Nicolle, P., *Thermotropisme et alimentation artificielle des Réduvidés hémophages*, Compt. rend. soc. biol. *138* (1944).

——, *Recherches sur la nutrition des Réduvidés hémophages*, 5: *Alimentation de Triatoma infestans Klug à l'aide de sérum vitaminé*, Bull. soc. pathol. exotique *39* (1946); 6: *Nécessité de l'hématine pour Triatoma infestans Klug*, ibid., *40* (1947).

——, see also Nicolle, P., and Lwoff, M.

Maassen, A., *Die teratologischen Wuchsformen (Involutionsformen) der Bakterien und ihre Bedeutung als diagnostisches Hilfsmittel*, Arb. kaiserl. Gesund. *21* (1904).

Mahdihassan, S., *Lac secretion and symbiotic fungi*, in: *Some Studies in Biochemistry by Some Students of Dr. G. J. Fowler* (The Phoenix Printing House, Bangalore 1924).

——, *Symbionts specific of wax- and pseudo lac-insects*, Arch. Protistenk. *63* (1928).

——, *Specific symbionts of a few Indian scale-insects*, Zentr. Bakteriol., Abt. 2, *78* (1929).

——, *The microorganisms of red and yellow lac-insects*, Arch. Protistenk. *68* (1929).

——, *Symbionts specific of wax- and pseudolac insects*, Zentr. Bakteriol. *78* (1929).

——, *The symbiotes of some important lac-insects*, Arch. Protistenk. *73* (1931).

——, *Sur les différents symbiotes des cochenilles productrices ou non productrices de cire*, Compt. rend. *196* (1933).

——, *Pigmentbildende Bakterien aus einer entsprechend gefärbten Cicade*, Verhandl. deut. zool. Ges. *1939.*

——, *Polyrhachis ants and bacterial symbiosis*, Current Sci. (India) *8* (1939).

——, *Insect tumours of bacterial origin*, Deccan Med. J. *1941.*

——, *The microorganisms in Melophagus ovinus*, Current Sci. (India) *15* (1946).

——, *Bacterial origin of some insect pigments*, Nature *158* (1946).

——, *Colour dimorphism in Coriococcus hibisci Green*, Current Sci. (India) *15* (1946).

——, *Two varieties of Tachardina lobata*, Current Sci. (India) *15* (1946).

——, *Two symbiotes of Psylla mali*, Nature *157* (1947).

——, *A mistaken symbiont of Oliarius cuspidatus*, Nature *159* (1947).

——, *Cicadella viridis, its symbiotes and their function*, Current Sci. (India) *16* (1947).

——, *Specificity of bacterial symbiosis in Aphrophorinae*, Proc. Indian Acad. Sci. *25* (1947).

——, *Bacterial symbiosis in Aphis rumicis*, Acta Entomol. Mus. Nat. Pragae *25* (1947).

——, *Bacterial symbiosis in a Margarodes spec.*, Current Sci. (India) *16* (1947).

——, *The role of symbiosis in the genus Coriococcus*, Coccidae. Z. angew. Entomol. *33*, (1951).

Maillet, P., *Contribution a l'étude de la biologie du Phylloxera de la vigne*, Ann. Sci. Nat., Zool. et biol. animale, Ser. 11 (1957).

Malke, H., *Wirkung von Lysozym auf die Symbionten der Blattiden*, Z. allgem. Mikrobiol. *4* (1964).

——, *Production of aposymbiotic cockroaches by means of lysozyme*, Nature *1964.*

Malouf, N. S. R., *Studies on the internal anatomy of the "stink bug" Nezara viridula L.*, Bull. soc. roy. entomol. Egypt *17* (1933).

Mangan, J., *The entry of Zooxanthellae into the ovum of Millepora and some particulars concerning the medusae*, Quart. J. Microscop. Sci. *53* (1909).

Mansour, K., *The development of the larval and adult mit-gut of Calandra oryzae (L.): The rice-weevil*, Quart. J. Microscop. Sci. *71* (1927).

————, *Preliminary studies on the bacterial cell-mass (accessory cell-mass) of Calandra oryzae (L.): The rice weevil*, Quart. J. Microscop. Sci. *73* (1930).

————, *On the so-called symbiotic relationship between Coleopterous insects and intracellular microorganisms*, Quart. J. Microscop. Sci. *77* (1934).

————, *On the intracellular microorganisms of some Bostrychid beetles*, Quart. J. Microscop. Sci. *77* (1934).

————, *Zur Frage der Holzverdauung durch Insektenlarven*, Proc. Roy. Acad. Sci. Amsterdam *36* (1933).

————, *On the microorganism-free and the infected Calandra granaria L.*, Bull. soc. roy. entomol. Egypt *1935*.

Mansour, K., and Mansour-Bek, J. J., *The digestion of wood by insects and the supposed role of micro-organisms*, Biol. Revs. *9* (1934).

————, *On the cellulase and other enzymes of the larvae of Stromatium fulvum Villers (Cerambycidae)*, Enzymologia *4* (1937).

Manuta, C., and Bernardini, P., *Saggi cromatografici preliminari su batteriociti di Blatta (Blaberus cranifer)*, Ist. Lombardo (rend. sci.) B, *92* (1958).

Manuta, C., Bernardini Mosconi, P., and Laudani, U., *Indagini fisiologiche e cromatografiche su batteriociti del corpo adiposo di Blatta (Blaberus cranifer)*, Symposia Genetica et Biologica Italica, Vol. 9 (1961).

Marchoux, E., and Couvy, L., *Argas et spirochètes. Les granules de Leishman*, Ann. inst. Pasteur *27* (1913).

Marshall, S. M., *Notes on oxygen production in coral planulae*, Sci. Rept. Great Barriere Reef Exped. *1*, No. 9 (1932).

Martini, E., *Parasitismus in der Zoologie*, Atti XI Congr. Int. Zool. Padova, Arch. zool. ital. *16* (1932).

Mathiesen-Käärik, A., *Eine Übersicht über die gewöhnlichen mit Borkenkäfern assoziierten Bläuepilze in Schweden und einige für Schweden neue Bläuepilze*, Medd. Statens Skogsforsk. Inst. *43* (1953).

Matsubara, K., *Revision of Japanese serranid fish, referable to the genus Acropoma*, Mem. Coll. Agr. Kyoto Univ. *66* (1953).

Matthai, G., *A revision of the recent colonial Astraeidae possessing distinct corallites*, Trans. Linn. Soc. London [2], Zool. *17* (1914).

Mayer, A. G., *Ecology of the Murray Island Coral Reef*, Carnegie Inst. Washington, Dep. Marine Biol., *9* (1918).

Maziarski, J., *Recherches cytologiques sur les organes segmentaires des vers de terre*, Poln. Arch. biol. med. Wiss. *2* (1905).

Meinecke, G., *Über das Vorkommen einer Mycelhefe auf hypochromen Froscherythrocyten*, Z. Hyg. *132* (1951).

————, *Genetische Zusammenhänge zwischen Amöben- und Bakterienformen*, Mikroskopie, Zentr. mikroskop. Forsch. u. Methodik, Wien *6* (1951).

————, *Veränderungen an Blutzellen in vitro*. Mikroskopie, Zentr. mikroskop. Forsch. u. Methodik, Wien *7* (1952).

Meissner, G., *Bakteriologische Untersuchungen über die symbiontischen Leuchtbakterien von Sepien aus dem Golfe von Neapel*, Biol. Zentr. *46* (1926); expanded, under the same title, in Zentr. Bakteriol. *67* (1926).

Mercier, L., *Les corps bactéroïdes de la blatte (Periplaneta orientalis): Bacillus cuenoti n. spec.*, Compt. rend. soc. biol. *61* (1906).

————, *Recherches sur les bactéroïdes des Blattides*, Arch. Protistenk. *9* (1907).

Mercier, L., *Cellules à Bacillus cuenoti dans la paroi des gaines ovariques de la Blatte*, Compt. rend. soc. biol. *62* (1907).

——, *Néoplasie du tissue adipeux chez les Blattes (Periplaneta orientalis L.) parasitées par une Microsporidie*, Arch. Protistenk. *11* (1908).

——, *Bactéries des invertebrées II, Les cellules uriques du Cyclostoma et leur bactérie symbiote*, Bull. sci. France et Belg. *15* (1911).

Mereschkovsky, C., *Über Natur und Ursprung der Chromatophoren in Pflanzenteilen*, Biol. Zentr. *25* (1905).

——, *Theorie der zwei Plasmaarten als Grundlage der Symbiogenesis, einer neuen Lehre von der Entstehung der Organismen*, Biol. Zentr. *30* (1910).

——, *La plante considérée comme un complexe symbiotique*, Bull. soc. sci. nat. Ouest France [3] *6* (1920).

Metcalf, R. L., and Patton, R. L., *A study of riboflavin metabolism in the American roach by fluorescence microscopy*, J. Cellular Comp. Physiol. *19* (1942).

Metschnikoff, E., *Untersuchungen über die Embryologie der Hemipteren*, Z. wiss. Zool. *16* (1866).

——, *Embryologische Studien an Insekten*, Z. wiss. Zool. *16* (1866).

Meves, F., *Die Plastosomentheorie der Vererbung*, Arch. mikroskop. Anat., Abt. 2, *92* (1918).

Meyer, G. F., and Frank, W., *Elektronenmikroskopische Studien zur intrazellulären Symbiose verschiedener Insekten. 1. Untersuchungen des Fettkörpers und der symbiontischen Bakterien der Küchenschabe (Blatta orientalis L.)*, Z. Zellforsch. *47* (1957).

Meyer, K. F., *Über Bakteriensymbiose bei Schnecken (Cyclostomatiden)*, Verhandl. schweiz. naturforsch. Ges. *104* (1923).

——, *On the physiological significance of the bacterial symbiosis in the concretion deposits of certain operculate land-mollusks of the family Cyclostomatidae (Annulariidae)*, Abstr. Comm. XI Int. Physiol. Congress, Edinburgh 1923.

——, *The "bacterial symbiosis" in the concretion deposits of certain operculate land mollusks of the families Cyclostomatidae and Annulariidae*, J. Infectious Diseases *36* (1925).

Michel, E., *Beiträge zur Kenntnis von Lachnus (Pterochlorus) roboris L., einer wichtigen Honigtauerzeugerin an der Eiche*, Z. angew. Entomol. *29* (1942).

Miehe, H., *Die Wärmebildung von Reinkulturen im Hinblick auf die Biologie der Selbsterhitzung pflanzlicher Stoffe*, Arch. Mikrobiol. *1*, (1930).

Migliavacca, A., *Sulla fine struttura dei globuli deutoplasmatici*, Boll. soc. med.-chir. Pavia *1927*.

Milne, D. L., *A study of the nutrition of the cigarette beetle, Lasioderma serricorne F. (Anobiidae) and a suggested new method for its control*, J. Entomol. Soc. S. Africa *26* (1963).

Milovidow, P. E., *A propos des bactéroïdes des blattes (Blattella germanica)*, Compt. rend. soc. biol. *99* (1928).

——, *Coloration différentielle des bactéries et des chrondriosomes*, Arch. anat. microscop. *24* (1928).

Minchin, E. A., *Sponges*, in: Lankester, Ray, *A Treatise on Zoology* (Adam and Charles Black, London 1900).

Mingazzini, P., *Ricerche sul canale digerente delle larve dei Lamellicorni fitofagi*, Mitt. zool. Stat. Neapel *9* (1889).

Misra, J. N., and Ranganathan, V., *Digestion of cellulose by the mound-building termite Termes (Cyclotermes) obesus (Rambur)*, Proc. Indian Acad. Sci. B. *39* (1954).

Mittler, T. E., *Amino-acids in phloem sap and their excretion by aphids*, Nature *172* (1953).

Moniez, R., *Sur un champignon parasite du Lecanium hesperidum*, Bull. soc. zool. France *12* (1887).

Montalenti, G., *Sull'elevamento dei termiti senza i protozoi dell'ampolla cecale*, Rend. reale accad. Lincei, Cl. sci. fis. mat. natur. [6] *6* (1927).

Montalenti, G., *Gli encimi digerenti e l'assorbimento delle sostanze solubili nell'intestino delle termiti*, Arch. zool. ital. *16* (1932).

Moose, C. A., *Chemical and spectroscopic analysis of phloem exudate and parenchym sap from several species of plants*, Plant Physiol. *13* (1938).

Morgan, A. F., *The water-soluble vitamins*, Ann. Rev. Biochem. *10* (1941).

Morgenthaler, O., *Das jahreszeitliche Auftreten der Bienenseuchen*, Bei. schweiz. Bienenztg. *1*, Pt. 7 (1944).

Moroff, Th., and Stiasny, G., *Über Bau und Entwicklung von Acanthometron pellucidum J. M.*, Arch. Protistenk. *16* (1909).

——, *Über vegetative und reproduktive Erscheinungen bei Thalassicola*, Festschr. R. Hertwig *1* (1910).

Morrison, H., *A classification of the higher groups and genera of the Coccid family Margarodidae*, Tech. Bull. Washington, *52* (1928).

Mortara, S., *Gli organi fotogeni di Abralia veranyi*, R. com. talassogr. ital., Mem. *95* (Venezia 1922).

——, *Sulla biofotogenesi*, Rend. reale accad. Lincei, Cl. sci. fis. mat. natur. [5a] *31* (1922).

——, *Ancora sulla biofotogenesi*, Rend. reale accad. Lincei, Cl. sci. fis. mat. natur. [5a] *31* (1922).

——, *Sulla biofotogenesi e su alcuni batteri fotogeni*, Riv. biol. *6* (1924).

Moseley, H. N., *On the structure of the Milleporidae*, Chall. Rept. Zool. *2* (1881).

——, *Pelagic life*, Nature *26* (1882).

Mouchet, S., Sta. océan. Salammbô, Notes No. 15 (1930).

Mrázek, A., *Sporozoenstudien. Zur Auffassung der Myxocystiden*, Arch. Protistenk. *18* (1910).

Mudrow, E., *Über die intrazellulären Symbionten der Zecken*, Z. Parasitenk. *5* (1932).

——, *Die keimfreie Aufzucht der Gelbfiebermücke Aedes aegypti*, Zool. Anz. *146* (1951).

Müller, F. W., *Zur Wirkstoffphysiologie des Bodenpilzes Mucor ramannianus*, Ber. schweiz. bot. Ges. *51* (1941).

Müller, H. C., *Symbiose zwischen Algen und Tieren*, Schr. phys.-ökon. Ges. Königsberg *54* (1913).

——, *Notiz über Symbionten bei Hydroiden*, Zool. Jahr. Abt. Syst., *37* (1914).

Müller, H. J., *Die intrazellulare Symbiose bei Cixius nervosus und Fulgora europaea als Beispiele polysymbiontischer Zyklen*, Verhandl. VII. intern. Kongr. Entomol. Berlin 1938 (1939).

——, *Die Symbiose der Fulgoroiden (Homoptera-Cicadina)*. Zoologica *98* (Stuttgart 1940).

——, *Formende Einflüsse des tierischen Wirtsorganismus auf symbiontische Bakterien*, Forsch. u. Fortschr. *18* (1942).

——, *Zur Systematik und Phylogenie der Zikaden-Endosymbiosen*, Biol. Zentr. *68* (1949).

——, *Über die intrazellulare Symbiose der Peloridiide Hemiodoecus fidelis Evans (Homoptera Coleorrhyncha) und ihre Stellung unter den Homopterensymbiosen*, Zool. Anz. *146* (1951).

——, *Über das Schlüpfen der Zikaden (Homoptera auchenorrhyncha) aus dem Ei*, Zoologica *103* (Stuttgart 1951).

——, *Experimentelle Studien an der Symbiose von Coptosoma scutellatum Geoffr. (Hem. Heteropt.)*, Z. Morphol. Ökol. Tiere *44* (1956).

——, *Neuere Vorstellungen über Verbreitung und Phylogenie der Endosymbiosen der Zikaden*, Z. Morph. Ökol. Tiere *51* (1962).

Müller, J., *Zur Naturgeschichte der Kleiderlaus*, Österr. Sanit.-Wesen *27* (1915).

Müller, Johannes, *Über die Thalassicollen, Polycystinen und Acanthometren des Mittelweeres*, Abhandl. Berl. Akad. 1858, Berlin 1859.

Müller, R., *Symbiose von Chromatophoren im Eiplasma*, Med. Welt *2* (1928).

——, *Virus, Bakterium, Chlorophyllkorn und Zelle*, Forsch. u. Fortschr. *21/23* (1947).

Müller, W., *Über die Pilzsymbiose holzfressender Insektenlarven*, Arch. Mikrobiol. *5* (1934).

Müller-Calé, K., and Krüger, E., *Symbiontische Algen bei Aglaophenia helleri und Sertularella polyzonias*, Mitt. zool. Stat. Neapel *21* (1913).

Murray, F. V., and Tiegs, O. W., *The metamorphosis of Calandra oryzae*, Quart. J. Microscop. Sci. [N. S.] *77* (1935).

——, see also Tiegs, O. W., and Murray, F. V.

Musgrave, A. J., Grinyer, T., and Homan, R., *Some aspects of the fine structure of the mycetomes and mycetomal microorganisms in Sitophilus (Coleoptera; Curculionidae)*, Can. J. Microbiol. *8* (1963).

Musgrave, A. J., and Homan, R., *Sitophilus sasaki (Tak.) (Coleoptera: Cucurlionidae) in Canada: Anatomy and mycetomal symbionts as valid taxonomic characters*, Can. Entomologist *94* (1962).

Musgrave, A. J., Homan, R., and Grinyer, L., *Mycetomal and other microorganisms in young and aging Sitophilus (Coleoptera, Curculionidae)*, Can. J. Microbiol. *10* (1964).

Musgrave, A. J., and Miller, J. J., *A note on some preliminary observations on the effect of the antibiotic Terramycin on insect symbiotic microorganisms*, Can. Entomologist *83* (1951).

——, *Some microorganisms associated with the weevils Sitophilus granarius (L.), and Sitophilus oryzae (L.). 1: Distribution and description of the organisms*, Can. Entomologist *85* (1953); *2: Population differences of mycetomal microorganisms in different strains of Sitophilus granarius*, ibid., *88* (1956).

——, *The possible nature and origin of the mycetomes in the Sitophilus weevils*, Rept. Entomol. Soc. Ontario *86* (1956).

——, *Studies of the association between strains and species of Sitophilus weevils and their mycetomal microorganisms*, Proc. X Intern. Congr. Entomol. *2* (1958).

Musgrave, A. J., Monro, H. A. U., and Upitis, E., *Apparent effect on the mycetomal microorganisms of repeated exposure of the host insect, Sitophilus granarius (L.) to methyl bromide fumigation*, Can. J. Microbiol. *7* (1961).

Naef, A., *Die Cephalopoden*, in: *Fauna und Flora des Golfes von Neapel* (R. Friedländer & Sohn, Berlin 1923).

Naidu, M., *Symbiosis in spittle insect Ptyelus nebulosus Fabr.* Current Sci. (India) *14* (1945).

——, *A special technique for the identification of the Membracidae species by the intracellular microorganisms of their tumours*, Current Sci. (India) *14* (1945).

Nath, V., and Piare, M., *Ovogenesis of Periplaneta americana*, J. Morphol. *48* (1929).

Naton, E., *Über die Entwicklung des schwarzbraunen Mehlkäfers, Tenebrio destructor Uyttenb., 1: Die Aufzucht von Tribolium destruktor in natürlicher Diät*, Z. angew. Entomol. *46* (1960); *2: Beitrag zur Metamorphose des Darmtraktus bei natürlicher und "synthetischer" Ernährung*, ibid., *47* (1960); *3: Die Aufzucht in "synthetischen," carnitin-unterdosierten Diäten und die darauf beruhenden Schädigungen*, ibid., *48* (1961); *4: Die Auswirkungen "carnitinfreier" Diäten auf die Entwicklungsdauer und auf den Darmtrakt.*, Zool. Beitr. [N. S.] *8* (1963); *5: Die Auswirkung "carnitinfreier" Diäten auf innersekretorische Drüsen und auf die Epidermis*, ibid., [N. S.] *8* (1963); *6: Beitrag zur Kenntnis der individuellen Empfindlichkeit gegenüber einem Carnitinmangel*, [N. S.] *8* (1963).

Naville, A., *Notes sur les Eolidiens. Un Eolidien d'eau saumâtre. Origine des nématocytes. Zooxanthelles et homochromie*, Rev. suisse zool. *33* (1926).

Neger, F. W., *Ambrosiapilze, 2: Die Ambrosia der Holzbohrkäfer*, Ber. deut. botan. Ges. *27* (1909).

Nelson-Rees, W. A., *A study of sex predetermination in the mealy bug Planococcus citri (Risso)*, J. Exptl. Zool. *144* (1960).

——, *Modification of the ovary of the mealy bug Planococcus citri (Risso) due to aging*, J. Exptl. Zool. *146* (1961).

Neukomm, A., *Action des rayons ultra-violets sur les bactéroïdes des blattes (Blattella germanica)*, Compt. rend. soc. biol. *96* (1927).

——, *Sur la structure des bactéroïdes des blattes (Blattella germanica)*, Compt. rend. soc. biol. *96* (1927).

——, *La réaction de la fixation du complément appliquée à étude des bactéroïdes des blattes (Blattella germanica)*, Compt. rend. soc. biol. *111* (1932).

Neumann, G., *Pyrosomen*, Bronns Kl. u. Ord., 3. Suppl., Abt. *2* (1909–1911).

——, *Die Pyrosomen der Deutschen Südpolarexpedition*, Deut. Südpolarexped. *14* (1913).

Nicol, J. A. C., *Observations on luminescence in pelagic animals*, J. Marine Biol. Assoc. United Kingdom *37* (1958).

Nicolle, P., and Lwoff, M., *L'acide pantothénique dans la nutrition de l'Hémiptère hémophage Triatoma infestans Klug*, Compt. rend. soc. biol. *138* (1944).

——, see also Lwoff, M., and Nicolle, P.

Nobile, M., *Contributo alla conoscenza della formazione e della struttura dei globuli del tuorlo in uova di Rana esculenta L. e Gallus gallus L.*, Reale accad. gioenia sci. nat. Catania *1927*.

Noland, J. L., *Sterol metabolism in insects: Utilization of cholesterol derivatives by the cockroach Blattella germanica L.*, Arch. Biochem. Biophys. *48* (1954).

Noland, J. L., and Baumann, C. A., *Requirement of the german cockroach for choline and related compounds*, Proc. Soc. Exptl. Biol. Med. *70* (1949).

——, *Protein requirements of the cockroach Blattella germanica (L.)*, Ann. Entomol. Soc. Am. *44* (1951).

Noland, J. L., Lilly, J. H., and Baumann, C. A., *Vitamin requirements of the cockroach Blattella germanica (L.)*, Ann. Entomol. Soc. Am. *42* (1949).

Noll, C. F., *Flußaquarien*, Zool. Garten *11* (1870).

Nöller, W., *Blut- und Insektenflagellatenzüchtung auf Platten*, Arch. Schiffs- u. Tropen-Hyg. *21* (1917).

——, *Die neuen Ergebnisse der Haemoproteus-Forschung*, Arch. Protistenk. *41* (1920).

Nolte, H. W., *Beiträge zur Kenntnis der symbiontischen Einrichtungen der Gattung Apion Herbst*, Z. Morphol. Ökol. Tiere *33* (1937).

——, *Die Legeapparate der Dorcatominen (Anobiidae) unter besonderer Berücksichtigung der symbiontischen Einrichtungen*, Verhandl. deut. zool. Ges. *1938*.

Nordenskiöld, E., *Zur Anatomie und Histologie von Ixodes reduvius*, Zool. Jahr. Abt. Anat., *25* (1908).

Nussbaum. J., *Zur Entwicklungsgeschichte der Ausführgänge der Sexualdrüsen bei den Insekten*, Zool. Anz. *5* (1882).

Nüsslin, O., *Über einige Urtiere aus dem Herrenwieser See im badischen Schwarzwald*, Z. wiss. Zool. *40* (1884).

Nutting, W. L., *Reciprocal protozoan transformation between the roach, Cryptocercus, and the termite Zootermopsis*, Biol. Bull. *110* (1956).

Öhme, B. G., *Beiträge zur Biologie und Anatomie des Baumflußkäfers Nosodendron fasciculare Olivier*, Dissertation, Erlangen 1948 (unpublished).

Offhaus, K., *Der Einfluß von wachstumsfördernden Faktoren auf die Insektenentwicklung unter besonderer Berücksichtigung der Phyto-Hormone*, Z. vergleich. Physiol. *27* (1939).

——, *Der Vitaminbedarf des Reismehlkäfers Tribolium confusum Duval. 1. Mitt. Über den für Tribolium confusum lebensnotwendigen wasserunlöslichen Hefeanteil*, Z. Vitamin-Hormon- u. Fermentforsch. *4* (1952).

Ogilvie, L., Mulligan, B. O., and Brian, P. W., *Progress report on vegetable diseases*, VI, Ann. Rept. Agr. Hort. Research Sta. Univ. of Bristol *1934* (1935).

Okada, Y. K., *On the photogenic organ of the Knight-fish (Monocentris japonicus)*, Biol. Bull. *50* (1926).

Okada, Y. K., *Contribution à l'étude des Céphalopodes lumineux, I,* Bull. inst. océanog. No. 494 (1927).

——, *Contribution à l'étude des Céphalopodes lumineux II,* Bull. inst. océanog. No. 499 (1933).

Okada, Y. K., Takagi, S., and Sugino, H., *Microchemical studies on the so-called photogenic granules of Watasenia scintillans (Berry),* Proc. Imp. Acad. (Tokyo) *10* (1933).

Oltmanns, F., *Morphologie und Biologie der Algen,* Vols. 1 and 2, 2nd ed. (G. Fischer, Jena 1904/05, 1922/23).

Omeliansky, W., *Über die Gärung der Cellulose,* Zentr. Bakteriol. Parasitenk. Abt. II, *8* (1902).

Osorio, B., *Une propriété singulière d'une bactérie phosphorescente,* Compt. rend. soc. biol. *72* (1912).

Paillot, A., *Les maladies bactériennes des insectes,* Ann. Serv. Epiphyties *8* (1921).

——, *La symbiose bactérienne et l'immunité humorale chez les Aphides,* Compt. rend. *188* (1929).

——, *Sur l'origine infectieuse des microorganismes des Aphides,* Compt. rend. *189* (1929).

——, *Sur la spécificité parasitaire des Bactéries infectant normalement les Pucerons,* Compt. rend. soc. biol. *103* (1930).

——, *Parasitisme bactérien et symbiose chez l'Aphis mali,* Compt. rend. *190* (1930).

——, *Mécanisme de la symbiose chez les Drepanosiphum platanoides,* Compt. rend. soc. biol. *103* (1930).

——, *Les réactions cellulaires et humorales d'immunité antimicrobienne dans le phénomène de la symbiose chez Macrosiphum jaceae,* Compt. rend. *190* (1930).

——, *Parasitisme et symbiose chez les Aphides,* Compt. rend. *193* (1931).

——, *Parasitisme et symbiose chez Aphis atriplicis,* Compt. rend. *193* (1931).

——, *Les variations morphologiques du bacille symbiotique de Macrosiphum tanaceti,* Compt. rend. *193* (1931).

——, *Les variations du parasitisme bactérien normale chez le Chaitophorus lyropticus Ressl.,* Compt. rend. *194* (1932).

——, *L'Infection chez les Insects. Immunité et symbiose* (G. Patissier, Trévoux 1933).

Panceri, P., *Gli organi luminosi e la luce dei Pirosomi e delle Foladi,* Atti accad. sci. fis. mat. Napoli *5* (1873).

Pandazis, G., *Zur Frage der Bakteriensymbiose bei Oligochaeten,* Zentr. Bakteriol. *120* (1931).

Pant, N. C., *Mycetomes and symbiotes in Leptocorisa varicornis Fabr. (Coreidae, Hemiptera),* Current Sci. *23* (1954).

Pant, N. C., and Fraenkel, G., *The function of the symbiotic yeasts of two insect species, Lasioderma serricorne F. and Stegobium (Sitodrepa) paniceum L.,* Science *112* (1950).

——, *Studies on the symbiotic yeasts of two insect species, Lasioderma serricorne F. and Stegobium paniceum L.,* Biol. Bull. *107* (1954).

——, *On the function of the intracellular symbionts of Oryzaephilus surinamensis L. (Cucujidae, Coleoptera),* J. Zool. Soc. India *6* (1954).

Pant, N. C., Nayar, J. K., and Gupta, P., *On the isolation and cultivation of intracellular symbiotes of Oryzaephilus surinamensis L. (Cucujidae, Col.),* Experientia *13* (1957).

——, *Intracellular bacterium-like microorganisms of Rhizopertha dominica F. (Bostrychidae, Col.),* Current Sci. *26* (1957).

——, *Physiology of intracellular symbionts of Stegobium paniceum L. with special reference to amino acid requirements of the host,* Experientia *16* (1960).

Papacostas, G., and Gaté, J., *Les associations microbiennes,* in Encyclopédie Scientifique (Gaston Doin & Cie., Paris 1928).

Parker, C. R., *Symbiosis in Paramaecium bursaria,* J. Exptl. Zool. *46* (1926).

Parkin, E. A., *A study on the food relation of the Lyctus powder-post beetles*, Ann. Appl. Biol. *23* (1936).

——, *The depletion of stark from timber in relation to attack by Lyctus-beetles*, 2–4, Forestry *12* (1938); *13* (1939).

——, *Symbiosis in larval Siricidae*, Nature *147* (1941).

——, *Symbiosis and Siricid woodwasps*, Ann. Appl. Biol. *29* (1942).

——, *Symbiosis in Ptilinus pectinicornis L.*, Nature (London) *170* (1952).

Parr, A. E., *The Macrouridae of the Western North Atlantic and Central America seas*, Bingham Oceanog. Coll. Bull. *10* (1946).

Pascher, A., *Studien über Symbionten*, 1: *Über einige Endosymbiosen von Blaualgen mit Einzellern*, Jahr. wiss. Botan. *71* (1929).

Paspaleff, G. W., *Cytologische Untersuchungen an Aphiden*, Ann. Univ. Sofia Fac. Phys.-Math., Sci. Nat. *3* (1929).

Patton, W. S., and Cragg, F. W., *A text book of medical entomology* (Christian Literature Society for India, London, Madras and Calcutta 1913).

Pax, F., *Die Aktinien*, Ergeb. u. Fortschr. Zool. *4* (1914).

Payne, N. M., *Food requirements for the pupation of two coleopterous larvae, Synchroa punctata Newm. and Dendroides canadensis Lec. (Melandryadae, Pyrochroidae)*, Entomol. News *42* (1931).

Peklo, J., *Über symbiontische Bakterien bei Aphiden*, Ber. deut. botan. Ges. *30* (1912).

——, *O mšici Krvavé [Über die Blutlaus]*. Zemědělský. arkiv 7 (1916).

——, *Symbiosis of Azotobacter with insects*, Int. Congr. Microbiol., Copenhagen 1947, Rept. Proc. 1949.

——, *Kůrovec Ips typographus L. ve světle starších i nových prací o nitrobuněcných symbion-tech u hmyzu*, Lesnická práce *25* (1946).

——, *Symbiosis of Azotobacter with insects*, Nature *158* (1946).

——, *Z histologie Drosophily* (with summary in English), Zvláštni otisk z Biologkých listů. Suppl. *2* (1951).

Peklo, J., and Satava, J., *Fixation of free nitrogen by bark beetles*, Nature *163* (1949).

——, *Fixation of free Nitrogen by insects*, Experientia 6 (1950).

Penard, E., *Sur la présence de la chlorophylle dans les animaux*, Arch. sci. phys. nat. [3] *24* (1890).

——, *Faune Rhizopodique du Bassin du Léman* (Henry Kündig, Genève 1902).

——, *Notes complementaires sur les Rhizopodes du Léman*, Rev. suisse zool. *2* (1911).

Pendergrast, J. G., *The internal anatomy of the Peloridiidae (Homopt. Coleorrhyncha)*, Trans. Roy. Entomol. Soc. London *114* (1962).

Peters, K., *Periodische Wachstumsrhythmen tierischer und pflanzlicher Zellkerne*, Z. Zellforsch. *37* (1952).

Peterson, W. H., and Peterson, M. S., *Relation of bacteria to vitamins and other growth factors*, Bacteriol. Rev. *9* (1945).

Petri, L., *Sopra la particolare localizzazione di una colonia batterica nel tubo digerente della larva della mosca olearia*, Atti reale accad. Lincei [5], Cl. sci. fis. mat. e nat. *13* (1904); additional material, *ibid.*, *14* (1905); *15* (1906); *16* (1907).

——, *Ricerche sopra i batteri intestinali della mosca olearia*, Mem. reale staz. patol. veg. Roma *1909*.

——, *Untersuchung über die Darmbakterien der Olivenfliege*, Zentr. Bakteriol. Abt. II, *26* (1910).

Pfeiffer, H., *Beiträge zu der Bakteriensymbiose der Bettwanze (Cimex lectularius) und der Schwalbenwanze (Oeciacus hirundinis)*, Zentr. Bakteriol. *123* (1931).

Pfeiffer, H., and Stammer, H. J., *Pathogenes Leuchten bei Insekten*, Z. Morphol. Ökol. Tiere *20* (1930).

Pflugfelder, O., *Zooparasiten und die Reaktionen ihrer Wirtstiere* (G. Fischer, Jena 1950).

Philiptschenko, J., *Über den Fettkörper der schwarzen Küchenschabe (Stylopyga orientalis)*, Rev. Russe Entomol. *1908*.

Piekarski, G., *Beiträge zur intrazellulären Symbiose. Entwicklungsgeschichte und Anatomie blutsaugender Gamasiden*, Z. Parasitenk. *7* (1935).

――――, *Die Zellkernäquivalente der Bakterien 2*. Colloquium deutsch. Ges. physiol. Chemie: "Mikroskop. u. chem. Organisation der Zelle." Springer-Verlag 1952.

――――, *Symbiose und Parasitismus*, Handb. d. allgem. Pathologie, Bd. XI, 2 (Springer-Verlag, Heidelberg, 1965).

Pierantoni, U., *L'origine di alcuni organi d'Icerya purchasi e la simbiosi ereditaria*, Boll. soc. nat. Napoli *23* [1909] (1910).

――――, *Ulteriori osservazioni sulla simbiosi ereditaria degli Omotteri*, Zool. Anz. *35* (1910).

――――, *Origine e struttura del corpo ovale del Dactylopius citri e del corpo verde dell'Aphis brassicae*, Boll. soc. nat. Napoli *24* [1910] (1911).

――――, *Sul corpo ovale del Dactylopius citri*, Boll. soc. nat. Napoli, *24* [1910] (1911).

――――, *Osservazioni su Aphrophora spumaria*, Boll. soc. nat. Napoli *24* [1910] (1911).

――――, *Studi sullo sviluppo d'Icerya purchasi Mask.*, 1: *Origine ed evoluzione degli elementi sessuali feminili*, Arch. zool. ital. *5* (1912); 2: *Origine ed evoluzione degli organi sessuali maschili*, ibid., *7* (1914); 3: *Osservazioni di embryologia*, ibid., *7* (1914).

――――, *Struttura ed evoluzione dell'organo simbiotico di Pseudococcus citri e ciclo biologico di Coccidomyces dactylopii Buchner*, Arch. Protistenk. *31* (1913).

――――, *La luce negli insetti luminosi e la simbiosi ereditaria*, Rend. reale accad. sci. fis. mat. Napoli *1914*.

――――, *Sulla luminosità e gli organi luminosi di Lampyris noctiluca*, Boll. soc. nat. Napoli *27* (1915).

――――, *Nuove osservazioni sulla luminosità degli animali*, Rend. reale accad. sci. fis. mat. Napoli *1917*.

――――, *Gli organi simbiotici e la luminescenza batterica dei Cephalopodi*, Pubbl. staz. zool. Napoli *2* (1918).

――――, *Le simbiosi fisiologiche e le attività dei plasmi cellulari*, Riv. biol. *1* (1919).

――――, *A proposito delle teorie sulla luminescenza batterica e sulle simbiosi fisiologiche*, Boll. soc. nat. Napoli *33* (1920).

――――, *Per una più esatta conoscenza degli organi fotogeni dei Cephalopodi abissali*, Arch. zool. ital. *9* (1920).

――――, *Gli organi luminosi simbiotici ed il loro ciclo ereditario in Pyrosoma giganteum*, Pubbl. staz. zool. Napoli *3* (1921).

――――, *Simbiosi, biofotogenesi e biocromogenesi. Stato delle conoscenze e nuove ricerche sui Pirosomi*, Arch. zool. ital. *10* (1922).

――――, *Breve rettifica alla nota critica del Prof. Levi su "Chondriosomi e simbionti,"* Monit. zool. ital. *33* (1922).

――――, *Sulla biofotogenesi simbiotica (un'ultima parola in risposta a S. Mortara)*, Boll. soc. nat. Napoli *34* (1923).

――――, *Nuove osservazioni su luminescenza e simbiosi*, 1: *La fosforescenza degli Oligocheti*, Rend. reale accad. Lincei [5], *32* (1923); 2: *La fosforescenza dei Ctenofori*, ibid., [5] *33* (1924); 3: *L'organo luminoso di Heteroteuthis dispar*, ibid., *33* (1924).

――――, *L'organo dorsale dei Pirosomi*, Pubbl. staz. zool. Napoli *4* (1923).

――――, *Le recenti ricerche sulla simbiosi fisiologica ereditaria*, Arch. sci. biol. *4* (Napoli 1923).

――――, *La fosforescenza e la simbiosi in Microscolex phosphoreus*, Boll. soc. nat. Napoli *36* (1924).

――――, *I recenti studii sulla simbiosi fisiologica ereditaria*, Atti soc. ital. progr. sci., Riun. Napoli *1924*.

Pierantoni, U., *Parasitismo, simbiosi e coltivabilità*, Riv. sci. nat. Natura *16* (Pavia 1924).

———, *I corpuscoli fotogeni di Heteroteuthis dispar*, Boll. soc. nat. Napoli *37* (1925).

———, *Microorganismi nell'economia animale*, Scientia *1925*.

———, *La vita ultramicroscopica*, Riv. fis. mat. sci. nat. [2] *1* (1926).

———, *Ancora sulla bioluminescenza da simbiosi (risposta al Prof. V. Puntoni)*, Riv. biol. *8* (1926).

———, *Nuove ricerche sugli organi luminosi simbiotici*, Ricerche morfol. biol. animale *1* (Napoli 1926).

———, *L'organo simbiotico nello sviluppo di Calandra oryzae*, Rend. reale accad. sci. fis. mat. Napoli [3], *35* (1927).

———, *Osservazioni sui cosidetti globuli del tuorlo e piastrine di Bufo viridis*, Boll. soc. nat. Napoli *39* (1927).

———, *I corpuscoli del tuorlo e la loro coltura in agar*, Mem. reale accad. Lincei Roma [6] *2* (1928).

———, *Inclusi e costituenti della sostanza vivente*, Riv. fis. mat. sci. nat. [N. S.] *3* (1929).

———, *L'organo simbiotico di Silvanus surinamensis (L.)*, Rend. reale accad. Lincei [6] *9* (1929).

———, *La trasmissione ereditaria dei simbionti fisiologici nei coleotteri*, Arch. zool. ital. *13* (1929).

———, *Origine e sviluppo degli organi simbiotici d'Oryzaephilus (Silvanus) surinamensis L.*, Atti reale accad. sci. fis. mat. Napoli [2] *18* (1930).

———, *La simbiosi ereditaria negli Eterotteri*, Atti XI Congr. Intern. Zool., Padova 1930, in Arch. zool. ital. *16* (1932).

———, *Nuove osservazioni sulla glandola nidamentale accessoria dei Cephalopodi*, Arch. zool. ital. *20* (1934).

———, *La digestione della cellulosa e del legno negli animali e la simbiosi delle Termiti*, Riv. fis. mat. sci. nat [N. S.] *9* (1934).

———, *Ancora sulla funzione della glandola accessoria dei Cefalopodi*, Boll. zool. *6* (1935).

———, *Simbiosi e digestione della cellulosa nei Termitidi e nei mammiferi*, Boll. soc. ital. biol. sper. *10* (1935).

———, *La simbiosi fisiologica nei Termitidi xilofagi e nei loro flagellati intestinali*, Arch. zool. ital. *22* (1935).

———, *Gli studii sulla endosimbiosi ereditaria nelle origini e nei più recenti sviluppi*, Attualità zool. *2* (1936).

———, *Osservazioni sulla simbiosi nei Termitidi xilofagi e nei loro flagellati intestinali*, 2: *Defaunazione per digiuno*, Arch. zool. ital. *24* (1937).

———, *Il complesso simbiotico dell'alimentazione delle Termiti*, Riv. fis. mat. Sci. nat. [N. S.] *15* (1940).

———, *Le simbiosi fisiologiche nei Vertebrati*, Riv. fis. mat. Sci. nat. [N. S.] *16* (1942).

———, *Trattato di Biologia e Zoologia Generale* (Casa ed. "Humus," Napoli 1948).

———, *Le nuove osservazioni sulla funzione e sul ciclo vitale dei globuli vitellini*, Boll. zool. *16* (1949).

———, *Die physiologische Symbiose der Termiten mit Flagellaten und Bakterien*, Naturwissenschaften *38* (1951).

———, *La simbiosi in Tropidothorax leucopterus (Heteroptera, Lygaeidae)*, Boll. zool. *18* (1951).

———, *Nuovi aspetti della convivenza fisiologica fra insetti e microorganismi*, Boll. zool. *21* (1954).

Pierre, Leon L., *Synthesis of ascorbic acid by the normal fat-body of the cockroach, Leucophaea maderae (F.), and by its symbionts*, Nature *193* (1962).

———, *Uricase activity of isolated symbionts and the aposymbiotic fat body of a cockroach*, Nature *201* (1964).

Pochon, J., *Flore bactérienne cellulolytique du tube digestive de larves xylophages*, Compt. rend. *208* (1939).

Pochon, J., De Barjac, H., and Roche, A., *Recherches sur la digestion de la cellulose chez le termite Sphaerotermes sphaerothorax*, Ann. inst. Pasteur *96* (1959).

Poisson, R., *Ordre des Heteroptères*, in: *Traité de Zoologie*, Vol. 10 (Masson & Cie., Paris 1951).

Poisson, R., and Pesson, P., *Contribution à l'étude du sang des Coccides. Le sang de Pulvinaria mesembryanthemi Vallot*, Arch. zool. exptl. gén. *81* (1937).

Polimanti, O., *Über das Leuchten von Pyrosoma elegans Les.*, Z. Biol. *55* (1911).

Ponce de Leon, S. R., *Formas microbianas de la sangre normal*, Rev. asoc. méd. arg. *49* (1935).

Porta, A., *Ricerche sull'Aphrophora spumaria*, Rend. Ist lomb. sci. lett. [2] *33* (1900).

Portier, P., *Digestion phagocytaire des chenilles xylophages des Lepidoptères. Exemple d'union symbiotique entre un insecte et un champignon*, Compt. rend. soc. biol. *70* (1911).

———, *Les Symbiotes* (Masson & Cie., Paris 1918).

Powell, W. N., *On the morphology of Pyrsonympha*, Univ. Calif. Publs. Zool. *31* (1928).

Pratt, E. M., *The assimilation and distribution of nutriment in Alcyonium digitatum*, Rept. Brit. Assoc. Sect. D. *1903*.

———, *Some Alcyonidae from Ceylon*, Herdman Rept. Ceylon Pearl Oyster Fishery, Roy. Soc. *3* (1905).

———, *The digestive organs of the Alcyonaria and their relation to the mesogloeal cell-plexus*, Quart. J. Microscop. Sci. *49* (1906).

Pratt, H. S., *Beiträge zur Kenntnis der Pupiparen*, Arch. Naturgeschichte *1893*.

———, *The anatomy of the femal genital tract of the Pupipara as observed in Melophagus ovinus*, Z. wiss. Zool. *66* (1899).

Preer, J. R., *Some properties of a genetic cytoplasmatic factor in Paramaecium*, Proc. Natl. Acad. Sci. U.S. *32* (1946).

Prenant, A., Bouin, P., and Maillard, L., *Traité d'Histologie*, Vol. 1 (Schleicher & Co., Paris 1904).

Pringsheim, E. G., *Über das Zusammenleben von Tieren und Algen*, Z. Naturwiss. (1915).

———, *Über Paramaecium bursaria, Ein Beitrag zur Symbiosefrage, Lotos 1925*.

———, *Physiologische Untersuchungen an Paramaecium bursaria*, Arch. Protistenk. *64* (1928).

Profft, J., *Beiträge zur Symbiose der Aphiden und Psylliden*, Z. Morphol. Ökol. Tiere *32* (1937).

Pucher, S., in: *Der Krebsarzt 4* (1949).

Puchta, O., *Experimentelle Untersuchungen über die Symbiose der Kleiderlaus Pediculus vestimenti Burm.*, Naturwissenschaften *41* (1954).

———, *Experimentelle Symbioseuntersuchungen an Gewebskulturen von Hühnerfibroblasten*, Arch. Mikrobiol. *21* (1955).

———, *Experimentelle Untersuchungen über die Bedeutung der Symbiose der Kleiderlaus Pediculus vestimenti Burm.*, Z. Parasitenk. *17* (1955).

———, *Züchtungsversuche an den Symbionten von Pediculus vestimenti Burm., nebst physiologischen und morphologischen Beobachtungen*, Z. Morphol. Ökol. Tiere *44* (1956).

Pulvertaft, R. J. V., *The effect of antibiotics on growing cultures of Bacterium coli*, J. Pathol. Bacteriol. *64* (1952).

Puntoni, O., *Lo stato attuale della teoria microbica della biofotogenesi*, Riv. biol. *7* (1925).

———, *Sulla biofotogenesi. Risposta ai Proff. Pierantoni e Zirpolo*, Riv. biol. *9* (1927).

Putnam, J. D., *Biological and other notes on Coccidae*, Proc. Davenport Acad. *2* (1880).

Pütter, A., *Der Stoffwechsel der Aktinien*, Z. allgem. Physiol. *12* (1911).

Quast, P., *Farbstoffinjektionsversuche bei Cyclostoma elegans Drap.*, Pflügers Arch. ges. Physiol. *200* (1923).

Quast, P., *Chemische Untersuchung des Organextraktes der Konkrementendrüse und des Nephidium von Cyclostoma elegans Drap.*, Z. Biol. *80* (1924).

——, *Der Konkrementenspeicher ("Konkrementendrüse" Claparèdes) von Cyclostoma elegans Drap.*, Z. Anat. Entwicklungsgeschichte *72* (1924).

Ramdohr, K. A., *Abhandlung über die Verdauungswerkzeuge der Insekten* (F. R. Schleicher, Halle and Leipzig 1811).

Rau, A., *Symbiose und Symbiontenerwerb bei den Membraciden (Homoptera Cicadina)*, Z. Morphol. Ökol. Tiere *39* (1943).

Raumer, von, Ed., *Über die Zusammensetzung des Honigtaus und über den Einfluss von honigtaureichem Sommer auf die Beschaffenheit des Bienenhonigs*, Z. anal. Chem. *33* (1894).

Rawtischer, F., *Wohin stechen die Pflanzenläuse?* Z. Botan. *26* (1933).

Rawlings, G. B., *The establishment of Ibalia leucosporoides in New Zealand*, For. Research Notes *1* (1951).

Regaud, C., *Mitochondries et symbiotes*, Compt. rend. soc. biol. *82* (1919).

Reichenow, E., *Haemogregarina stepanowi, die Entwicklungsgeschichte einer Gregarine*, Arch. Protistenk. *20* (1910).

——, *Los Hemococcidios de los Lacertidos*, Mus. nacl. ciencias nat. Ser. Zoologica 4, Madrid (1920).

——, *Die Haemococciden der Eidechsen*, 1. Arch. Protistenk. *42* (1921).

——, *Über intrazellulare Symbionten bei Blutsaugern*, Arch. Schiffs- u. Tropen-Hyg. *25* (1921).

——, *Intrazelluläre Symbionten bei blutsaugenden Milben und Egeln*, Arch. Protistenk. *45* (1922).

Reitinger, J., *Untersuchungen über Wirkstoffsquellen für die Entwicklung des amerikanischen Reismehlkäfers Tribolium confusum Duval*, Dissertation (Munich 1952).

Remane, E. *Intrazellulare Verdauung bei Rädertieren*, Z. vergleich. Physiol. *11* (1929).

Resühr, B., *Zur Morphologie und Protoplasmatik der bakteroiden Symbionten einiger Homopteren (Philaenus spumarius L., Cicadella viridis und Pseudococcus citri Risso)*, Arch. Mikrobiol. *9* (1938).

Reyne, A., *Hippeococcus, a new genus of Pseudococcidae from Java with peculiar habits*, Zool. Mededeel. *32* (1954).

Riccardo, S., *Un microorganismo simbionte della Ctenolepisma ciliata (Duf.)*, Boll. soc. ital. biol. sper. *20* (1945).

Richards, A. G., and Brooks, M. A., *Internal symbiosis in insects*, Ann. Rev. Entomol. *3* (1958).

Richter, G., *Untersuchungen an Homopterensymbionten*, Z. Morphol. Ökol. Tiere *10* (1928).

Ries, E., *Über die Symbionten der Läuse und Federlinge*, Zentr. Bakteriol. *117* (1930).

——, *Über ein regelmäßiges Rickettsienvorkommen bei der Hühnerlaus*, Zentr. Bakteriol. *121* (1931).

——, *Die Symbiose der Pediculiden und Mallophagen*, Arch. zool. ital. *16* (1931).

——, *Die Symbiose der Läuse und Federlinge*, Z. Morphol. Ökol. Tiere 20 (1931).

——, *Experimentelle Symbiosestudien*, 1: *Mycetomtransplantationen*, Z. Morphol. Ökol. Tiere *25* (1932).

——, *Endosymbiose und Parasitismus*, Z. Parasitenk. *6* (1933).

——, *Über den Sinn der erblichen Insektensymbiose*, Naturwissenschaften *23* (1935).

Rimpau, W., *Grundsätzliches zur pflanzlichen Endosymbiose beim Menschen*, Münch. med. Wochschr. *81* (1934).

Rippel-Baldes, A., *Grundriss der Mikrobiologie*, 2nd ed. (Springer-Verlag, Berlin-Göttingen-Heidelberg 1952).

Ripper, W., *Zur Frage des Celluloseabbaues bei der Holzverdauung xylophager Insektenlarven*, Z. vergleich. Physiol. *13* (1931).

Rizki, M. T. M., *Desoxyribose nucleic acid in the symbiotic microorganisms of the cockroach, Blattella germanica,* Science *120* (1954).

Robinson, F. A., *The Vitamin B Complex* (Chapman & Hall Ltd., London 1951).

Roesler, R., *Histologische, physiologische und serologische Untersuchungen über die Verdauung der Zeckengattung Ixodes Latr.,* Z. Morphol. Ökol. Tiere *28* (1934).

Rolle, M., and Mehnert, B., *Hefen als Symbionten der Säugetiere?* Z. Bakteriol. Parasitenkde, Infektionskrankh., Hyg. I. Orig. *168* (1957).

Rondelli, M., *Osservazioni sulla simbiosi negli ematophagi (zecche),* Atti reale accad. sci. Torino *60* (1925).

————, *Osservazioni sulla simbiosi ereditaria negli Afidi gallicoli (Erisoma),* Atti reale accad. sci. Torino *60* (1925).

————, *La simbiosi ereditaria negli Eriosomatini,* Ric. morfol. biol. animale *1* (Napoli 1928).

Rosenkranz, W., *Die Symbiose der Pentatomiden (Hemiptera heteroptera),* Z. Morphol. Ökol. Tiere *36* (1939).

Rosenthal, H., and Grob, C. A., *Über den Vitaminbedarf des amerikanischen Reismehlkäfers Tribolium confusum Duval,* 4. Mitt., Z. Vitaminforsch. *17* (1946).

Rosenthal, H., and Reichstein, T., *Vitamin requirement of the American flower beetle Tribolium confusum Duval.,* Nature *150* (1942).

————, *Der Vitaminbedarf des amerikanischen Reismehlkäfers Tribolium confusum Duval,* 2. Mitt., Z. Vitaminforsch. *15* (1945).

Roshdy, M. A., *Observations by electron microscopy and other methods on the intracellular rickettsia-like microorganisms in Argas persicus Oken (Ixodoidea, Argasidae),* J. Insect. Pathol. *3* (1961).

Roubaud, E., *Les particularités de la nutrition et la vie symbiotique chez les mouches tsetsés,* Ann. inst. Pasteur *33* (1919).

Rozeboom, L. E., *The relation of bacteria and bacterial filtrates to the development of mosquito larvae,* Am. J. Hyg. *21* (1935).

Salensky, W., *Neue Untersuchungen über die embryonale Entwicklung der Salpen,* Mitt. Zool. Stat. Neapel *4* (1883).

————, *Sur la segmentation des oeufs de Salpa maxima-africana,* Bull. acad. sci. Russ. [6] *10* (1916).

————, *Les blastomères et les calymmocytes de Salpa fusiformis,* Bull. acad. sci. Russ., [6] *10* (1916).

————, *Sur la segmentation des oeufs de Salpa fusiformis,* Bull. acad. sci. Russ. [6] *10* (1916).

————, *Sur la structure de l'appareil sexual feminin e sur la maturation de l'oeuf chez Salpa bicaudata,* Bull. acad. sci. Russ. [6], *11* (1917).

Salfi, M., *L'organo simbiotico di Troiza alacris Flor.,* Rend. unione zool. ital. Bologna *1926.*

Sander, F., *Die atypischen Bakterienformen unter besonderer Berücksichtigung des Problems bakterieller Generationswechselvorgänge,* Ergeb. Hyg. Bakteriol. *21* (1938).

Sander, K., *The early embryology of Pyrilla perpusilla Walker (Homoptera), including some observations on the later development,* Aligarh Muslim Univ. Publ. (Zool. Ser.) *1956.*

————, *Analyse des ooplasmatischen Reaktionssystems von Euscelis plebejus Fall. (Cicadina) durch Isolieren und Kombinieren von Keimteilen, 1. Mitt. Die Differenzierungsleistungen vorderer und hinterer Eiteile,* Arch. Entwicklungsmech. *151* (1959); *2. Mitt. Die Differenzierungsleistungen nach Verlagern von Hinterpolmaterial,* Ibid., *151* (1960).

Santo, E., and Rusch, H. P., *Das Gesetz von der Erhaltung der lebendigen Substanz,* Wien. med. Wochschr. *101* (1951).

Schanderl, H., *Die Bakteriensymbiose der Leguminosen und Nichtleguminosen,* Gartenbauwissenschaft *13* (1939).

————, *Über die Assimilation des elementaren Stickstoffs der Luft durch die Hefesymbionten von Rhagium inquisitor L.,* Z. Morphol. Ökol. Tiere *38* (1942).

Schanderl, H., *Ein Beitrag zur Frage der Isolierbarkeit von Mikroorganismen aus normalem pflanzlichem Gewebe und eine Kritik der sogenannten "Knöllchentheorie,"* Biol. Generalis *17* (1944).

———, *Botanische Bakteriologie und Stickstoffhaushalt der Pflanzen auf neuer Grundlage* (Eugen Ulmer, Stuttgart 1947).

———, *Über die Hefesymbiose der Cerambyciden und Aphiden,* Verhandl. deut. zool. Ges. *1949.*

———, *Über das Studium der Chondriosomen pflanzlicher Zellen intra vitam,* Züchter *20* (1950).

———, *Über die natürliche und künstliche Verwandlung von Schimmelpilzen und Hefen in Bakterien,* Mikroskopie (Wien) *6* (1951).

———, *Über die Isolierung von Bakterien aus normalem Pflanzengewebe und ihre vermutliche Herkunft,* Ber. deut. botan. Ges. *64* (1952).

———, *Methoden zur Auslösung spontaner Bakterienentwicklung in normalen Pflanzengeweben, bezw. Pflanzenzellen,* Ber. deut. botan. Ges. *66* (1953).

———, *Die spontane und künstliche Verwandlung von Sprosspilzen (Hefen) in Spaltpilze (Bakterien),* Naturwissenschaften *40* (1953).

Schanderl, H., Lauff, G., and Becker, H., *Studien über die Mycetom- und Darmsymbionten der Aphiden,* Z. Naturforsch. *4b* (1949).

Schaudinn, F., *Untersuchungen über den Generationswechsel von Trichosphaerium sieboldi,* Abhandl. Kgl. Preuss. Akad. Wiss. Berlin *1899.*

———, *Generations- und Wirtswechsel bei Trypanosoma und Spirochaeta,* Arb. kaiserl. Gesundh. *20* (1904).

Schedl, W., *Ein Beitrag zur Kenntnis der Pilzübertragungsweise bei xylomycetophagen Scolytiden (Coleoptera),* Sitzber. Österr. Akad. Wissürch., Math.- naturw. Kl. Abt. 1, Bd. *171* (1962).

———, *Biologie des gehöckerten Eichenholzbohrers, Xyleborus monographus Fab.,* Z. angew. Entomol. *53* (1964).

Scheinert, W., *Symbiose und Embryonalentwicklung bei Rüsselkäfern,* Z. Morphol. Ökol. Tiere *27* (1933).

Schewiakoff, W., *Die Acantharia des Golfes von Neapel,* in: *Fauna und Flora des Golfes von Neapel,* Vol. 37 (1926).

Schimitschek, E., *Tetropium gabrieli Weise und Tetropium fuscum F.: Ein Beitrag zu ihrer Lebensgeschichte und Lebensgemeinschaft,* Z. angew. Entomol. *15* (1929).

Schimper, W., *Untersuchungen über die Chlorophyllkörner und die homologen Gebilde,* Jahr. wiss. Botan. *16* (1885).

Schlottke, E., and Becker, G., *Verdauungsfermente im Darm der Hausbockkäferlarven,* Biol. Generalis *16* (1942).

Schmeisser, K., *Levéltetvekkel és kabócákkal együttélö mikroorganizmusok tenyésztése,* Egészégtudományi Közlemények *1944.*

Schmidt, W. J., *Untersuchungen über Bau und Lebenserscheinungen von Bursella spumosa, einem neuen Ciliaten,* Arch. mikroskop. Anat., Abt. 1, *95* (1921).

Schneider, G., *Beiträge zur Kenntnis der symbiontischen Einrichtungen der Heteropteren,* Z. Morphol. Ökol. Tiere *36* (1940).

Schneider, H., *Künstlich symbiontenfrei gemachte Kornkäfer (Calandra granaria L.),* Naturwissenschaften *41* (1954).

———, *Morphologische und experimentelle Untersuchungen über die Endosymbiose der Korn- und Reiskäfer (Calandra granaria L. und Calandra oryzae L.),* Z. Morphol. Ökol. Tiere *44* (1956).

Schneider, K. C., *Lehrbuch der vergleichenden Histologie* (G. Fischer, Jena 1902).

Schneider-Orelli, O., *Die Übertragung und Keimung des Ambrosiapilzes von Xyleborus (Anisandrus) dispar F.,* Naturw. Z. Land- u. Forstwissensch. *9* (1911).

Schneider-Orelli, O., *Untersuchungen über den pilzzüchtenden Obstbaumborkenkäfer Xyleborus (Anisandrus) dispar F. und seinen Nährpilz*, Zentr. Bakteriol. *38* (1913).

Schoel, W., *Beiträge zur Kenntnis der Aphidensymbiose*, Botan. Archiv *35* (1934).

Schoelzel, G., *Die Embryologie der Anopluren und Mallophagen*, Z. Parasitenk. *9* (1937).

Schomann, H., *Die Symbiose der Bockkäfer*, Z. Morphol. Ökol. Tiere *32* (1937).

Schopfer, W. H., *Symbiose et facteurs de croissance*, Congr. Microbiol. Paris *1938*.

Schorr, H., *Zur Verhaltungsbiologie und Symbiose von Brachypelta aterrima Först. (Cydnidae, Heteroptera)*, Z. Morphol. Ökol. Tiere *45* (1957).

Schrader, F., *Sexdetermination on the White-fly (Trialeurodes vaporariorum)*, J. Morphol. *34* (1920).

———, *The Chromosomes of Pseudococcus nipae*, Biol. Bull. *40* (1921).

———, *The sex ratio and oogenesis of Pseudococcus citri*, Z. induktive Abstammungslehre *30* (1923).

———, *The origin of the mycetocytes in Pseudococcus*, Biol. Bull. *45* (1923).

Schultz, J., St. Lawrence, P. and Newmeyer, D., *A chemical defined medium for the growth of Drosophila melanogaster*, Anat. Record *96* (1946).

Schultze, Fr. E., *Rhizopodenstudien*, Arch. mikroskop. Anat. *11* (1875).

Schultze, K. L., *Experimentelle Untersuchungen über die Chlorellensymbiose bei Ciliaten*, Biol. Generalis *19* (1951).

Schultze, M., *Beiträge zur Naturgeschichte der Turbellarien* (C. A. Koch, Greifswald 1851).

Schwartz, W., *Untersuchungen über die Pilzsymbiose der Schildläuse*, Biol. Zentr. *44* (1924).

———, *Neue Untersuchungen über die Pilzsymbiose der Schildläuse (Lecaniinen)*, Arch. Mikrobiol. *3* (1932).

———, *Der Stand unserer Kenntnisse von den physiologischen Grundlagen der Symbiose von Tieren mit Pilzen und Bakterien*, Arch. Mikrobiol. *6* (1935).

———, *Die physiologischen Grundlagen der Symbiosen von Tieren mit Pilzen und Bakterien*, VII Int. Kongr. Entomol. Berlin 1938 (1939).

Schwarz, I., *Untersuchungen an Mikrosporidien minierender Schmetterlingsraupen, den "Symbionten" Portiers*, Z. Morphol. Ökol. Tiere *13* (1929).

Schwarz, I., and Koch, A., *Vergleichende Analyse der wichtigsten Wachstumsvitamine des Blütenpollens, nebst einer Bemerkung über die Verteilung der Vitamine in Buchensämlingen*, Wiss. Z. Martin-Luther-Univ. Halle, Math. Nat., *4* (1954).

———, *Der Tribolium-Test als quantitativer Test für 8 B-Vitamine I.* Z. Vitamin-, Hormon- u. Fermentforsch. *12* (1962); *II.* (1963/64); *III* in press.

Schweizer, G., *Bacillus hirudinis, ein spezifischer Symbiont des Blutegels*, Arch. Mikrobiol. *7* (1936).

Seeliger, O., *Die Pyrosomen der Planktonexpedition*, Ergeb. Plankton-Exped. *1895*.

Sell, W., Unpublished dissertation on the embryonic development of aphids (München 1919).

Selmair, E., *Beiträge zur Wirkung wachstumsfördernder Stoffe auf die Entwicklung der Blattiden (Blattella germanica L.)*, Z. Parasitenk. *21* (1962).

———, *Über die Auslösung sog. "Grossmodifikationen" bei Insekten durch Fütterung mit "Vitamin T" und anderen wachstumsfördernden Substanzen*, Z. Ernährungswiss. *2* (1962).

———, *Über das Verhalten der Symbionten nach dem Tode des Wirtstieres und bei gleichzeitiger Infektion mit Mikrosporidien*, Arch. Mikrobiol. *43* (1962).

Sen, P., *On the occurrence of symbiotic microorganisms in the Cecidomyiidae or Gall midges (Diptera), with special reference to the larvae of Rhabdophaga saliciperda Duf.*, Arch. Protistenk. *85* (1935).

Serres, M. de, *Observations sur les usages des diverses parties du tube intestinal des insectes*, Ann. mus. nat. hist. Paris *20* (1813).

Sharif, M., *On the life history and the biology of the rat flea, Nosopsyllus fasciatus*, Parasitology *29* (1937); *38* (1948).

Shifrine, M., and Phaff, H. F., *The association of yeasts with certain bark beetles*, Mycologia *48* (1956).

Shima, G., *Über das Wesen des Leuchtens von Watasenia scintillans* (in Japanese), Tokio Izi Shinshi *1926*.

————, *Preliminary note on the nature of the luminous bodies of Watasenia scintillans (Berry)*, Proc. Imp. Acad. (Tokyo) *3* (1927).

Shinji, G. O., *Embryology of Coccids, with especial reference to the formation of the ovary, origin and differentiation of the germ cells, germ layers, rudiments of the midgut and the intracellular symbiontic organisms*, J. Morphol. *33* (1919/20).

Siebold, Th. von, *Über einzellige Pflanzen und Tiere*, Z. wiss. Zool. *1* (1849).

Siegel, *Die geschlechtliche Entwicklung von Haemoproteus stepanowi im Rüsselegel Placobdella catenigera*, Arch. Protistenk. *2* (1903).

Signoret, V., *Essai monographique sur les Aleurodes*, Ann. soc. entomol. France [4] *8* (1867).

Sikora, H., *Beiträge zur Anatomie, Physiologie und Biologie der Kleiderlaus (Pediculus vestimenti Nitzsch.)*, 1: *Anatomie des Verdauungstraktes*, Arch. Schiffs- u. Tropen-Hyg., Suppl. 1, *20* (1916).

————, *Beiträge zur Kenntnis der Rickettsien*, Arch. Schiffs- u. Tropen-Hyg., *22* (1918).

————, *Vorläufige Mitteilungen über Mycetome bei Pediculiden*, Biol. Zentr. *39* (1919).

————, *Über die Mycetome der Läuse*, Arch. Schiffs- u. Tropen-Hyg. *26* (1922).

————, *Der gegenwärtige Stand der Rickettsia-Forschung*, Klin. Wochschr. *1924*.

Silliman, W. A., *Untersuchungen über Turbellarien Nordamerikas*, Z. wiss. Zool. *41* (1885).

Skowron, S., *On the luminescence of some Cephalopods (Sepiola, Heteroteuthis)*, Riv. biol. *8* (1926).

————, *The luminous material of Microscolex phosphoreus Dug.*, Biol. Bull. *54* (1928).

Slabý, O., *O cyklické intracelulárni symbiose u hmyzu*, Acta Soc. Entomol. Čsl. Prague *43* (1946).

Smith, F., *Nutritional requirements of Camponotus ants*, Ann. Entomol. Soc. Am. *37* (1944).

Smith, H. G., *On the presence of algae in certain Ascidiacea*, Ann. Mag. Nat. Hist. [10] *15* (1935).

————, *Contribution to the anatomy and physiology of Cassiopeia frondosa*, Papers Tortugas Lab. *31* (1936).

————, *The significance of the relationship between Actinians and Zooxanthellae*, J. Exptl. Biol. *16* (1939).

Smith, J. D., *Symbiotic microorganisms of Aphids and fixation of atmospheric nitrogen*, Nature *162* (1948).

Socias, A., Gonzalez, C., and Ramirez, C., *Transformación de varias especies de Aspergillus y Penicillum en Bacillus subtilis*, Microbiol. españ. *5* (1952).

Sollas, J., *Porifera*, Cambridge Natural History *1906*.

Sorby, H. C., *On comparative vegetable chromatology*, Proc. Roy. Soc. (London) *21* (1873).

————, *On the colouring matter of Bonellia viridis*, Quart. J. Microscop. Sci. [3] *15* (1875).

Springhetti, A., *Relazioni ormonali tra insetti ed endosimbionti*, Parasitologia *6* (1964).

Sprinz, H., *Morphological response of the intestinal mucosa to enteric bacteria and its implication for sprue and Asiatic cholera*, Federation Proc. *21* (1962).

————, *The contribution of germ free research to our understanding of the cellular kinetics and responses of the intestinal mucosa*, Proc. XVI Internat. Congr. Zool. *3* (1963).

Sprinz, H., Kundel, D. W., Dammin, J. G., Horowitz, R. E., Schneider, H., and Formal, S. B., *The response of the germ free guinea pig to oral bacterial challenge with Escherichia coli and Shigella flexneri*, Am. J. Pathol. *39* (1961).

Sreenivasaya, M., and Mahdihassan, S., *A study of the symbiotic fungus from the Mysore lac insects*. J. Indian Inst. Sci. *12a* (1929).

Stammer, H. J., *Die Bakteriensymbiose der Trypetiden (Diptera)*, Z. Morphol. Ökol. Tiere *15* (1929).

———, *Die Symbiose der Lagriiden (Col.)*, Z. Morphol. Ökol. Tiere *15* (1929).

———, *Neue Symbiosen bei Coleopteren*, Verhandl. deut. zool. Ges. *1933*.

———, *Studien an Symbiosen zwischen Käfern und Mikroorganismen*, 1: *Die Symbiose der Donacien (Coleopt. Chrysomel.)*, Z. Morph. Ökol. *29* (1935); 2: *Die Symbiose des Bromius obscurus L. und der Cassida-Arten*, ibid., *31* (1936).

———, *Bau und Bedeutung der Malpighischen Gefäße der Coleopteren*, Z. Morphol. Ökol. Tiere *29* (1935).

———, *Die Verbreitung der Endosymbiose bei den Insekten*, Symposium sur la "Symbiose chez les insectes" Amsterdam 1951, Tijdschr. Entomol. *95* (1952), and Union Intern. sc. biol. Sér. B, No. 10.

Steche, O., *Die Leuchtorgane von Anomalops katoptron und Photoblepharon palpebratus*, Z. wiss. Zool. *93* (1909).

Stechow, E., *Hydroidpolypen der Japanischen Ostküste*, 1, 2 in: Doflein, *Naturgeschichte Ostasiens*, 1. Suppl. Abhandl. Bayr. Akad. Wiss. München *1909* and *1913*.

Stein, F., *Die weiblichen Geschlechtsorgane der Käfer* (Dunker & Humblot, Berlin 1847).

———, *Die Infusionstiere auf ihre Entwicklungsgeschichte untersucht* (W. Engelmann, Leipzig 1854).

———, *Der Organismus der Infusionstiere* (W. Engelmann, Leipzig 1859–1867).

Steinhaus, E. A., *A study of the bacteria associated with thirty species of insects*, J. Bacteriol. *42* (1941).

———, *Insect Microbiology* (Comstock Publishing Co., Ithaca 1946).

———, *Principles of Insect Pathology* (McGraw-Hill Book Co., New York and London 1949).

———, *Report on diagnoses of diseased insects 1944–1950*, Hilgardia *20* (1951).

———, *Observations on the symbiotes of certain Coccidae*, Hilgardia *24* (1955).

Steinhaus, E. A., Batey, M. M., and Boerke, C. L., *Bacterial symbiotes from the caeca of certain Heteroptera*, Hilgardia *24* (1956).

Stepp, W., *Einiges über die Darmbakterien als Symbionten*, Med. Monatspiegel *2* (1953).

———, *Zur Frage der Darmbakterien als Symbionten*, Verhandl. deut. Ges. inn. Med. *1957*.

Stevens, N. M., *A study on the germ cells of Aphis rosae and Aphis oenotherae*, J. Exptl. Zool. *2* (1905).

Stiasny, G., *Über die Beziehungen der sog. "gelben Zellen" zu den koloniebildenden Radiolarien*, Arch. Protistenk. *19* (1910).

———, *Zur Kenntnis der gelben Zellen der Sphaerozoen*, Biol. Zentr. *30* (1910).

Stier, A., *Beiträge zur Embryonalentwicklung der Salpa pinnata*, Z. Morphol. Ökol. Tiere *33* (1938).

Stillwell, M. A., *Decay associated with woodwasps in balsam fir weakened by insect attack*, Forest Sci. *6* (1960).

———, *The fungus associated with woodwasps occurring in beech in New Brunswick*, Can. J. Botany *42* (1964).

Stolc, A., *Beobachtungen und Versuche über die Verdauung und Bildung der Kohlehydrate bei einem amoebenartigen Organismus, Pelomyxa palustris Greeff*, Z. wiss. Zool. *68* (1900).

Strindberg, M., *Embryologische Studien an Insekten*, Z. wiss. Zool. *106* (1913).

———, *Zur Entwicklungsgeschichte der oviparen Cocciden*, Zool. Anz. *50* (1919).

Strohmeyer, H., *Die Morphologie des Chitinskelettes der Platypodiden*, Arch. Naturgeschichte [A] *84* (1918).

Strübing, H., *Über Beziehungen zwischen Oviduct, Eiablage und natürlicher Verwandtschaft einheimischer Delphaciden*, Zool. Beitr. [N. S.] *2* (1956).

Stschelkanowzew, J., *Die Entwicklung der Cunina proboscidea Metschn.*, Mitt. Zool. Stat. Neapel *17* (1906).

Stüben M., see Buchner, 1948.

Stuhlmann, F., *Beiträge zur Kenntnis der Tsetsefliege (Glossina fusca und tachinoides)*, Arb. Reichsgesundh. *26* (1907).

Subbarow, Y., and Trager, W., *The chemical nature of growth factors required by mosquito larvae, 2: Pantothenic acid and vitamin B$_6$*, J. Gen. Physiol. *23* (1940).

Sukatschow, B. W., *Beiträge zur Anatomie von Hirudineen, 1: Über den Bau von Branchellion torpedinis Sav.*, Mitt. Zool. Stat. Neapel *20* (1910–1913).

Šulc, K., *Kermincola kermesina n. gen. n. sp. und physokermina n. sp., neue Mikroendosymbiontiker der Cocciden*, Sitzber. böhm. Ges. Wiss. Prag *1906*.

———, *Symbiontische Saccharomyceten der echten Cicaden*, Sitzber. böhm. Ges. Wiss. Prag *1910*.

———, *"Pseudovitellus" und ähnliche Gewebe der Homopteren sind Wohnstätten symbiontischer Saccharomyceten*, Sitzber. böhm. Ges. Wiss. Prag *1910*.

———, *Intracellulárni hereditárni symbiosa u Margaroda (Coccidae)* (with summary in German), Publ. biol. ecole haut. etud. vétérin. *2* (Brno 1923).

———, *O biologii kvasnic a jejich symbiose s hmyzem* [*Über die Biologie der Hefepilze und ihre Symbiose mit Insekten*, Vortrag Vers. Naturw. Ges. Mährisch-Ostrau *5*. November 1909], Sbornik Přírodovědecké společnosti Mor. Ostravě *2* (1923).

———, *O intracellulárni hereditárni symbiosa u Fulgorid (Homoptera)* (with summary in German), Publ. biol. ecole haut. etud. vétérin. *3* (Brno 1924).

———, *De la symbiose intracellulaire chez les Fulgorides*, Compt. rend. soc. biol. *92* (1925).

———, *O vnitrobuěcné symbiose*, Příroda, Brno 1924/25.

———, *Symbiose*, Listy *16* (1931).

Svedelius, Nils, *Über einen Fall von Symbiose zwischen Zoochlorellen und einer marinen Hydroide*, Svensk botan. Tidskr. *1* (1907).

Swammerdam, Jan, *Algemeene Verhandeling van bloedloose Diertjens* (Utrecht 1669).

———, *Bijbel der Natuure (Biblia naturae sive historia insectorum)* (van der Aa, Leiden 1737/38).

Tacchini, J., *Ricerche sui batteri simbionti intracellulari delle Blatte*, Boll. soc. ital. biol. sper. *22* (1946).

Takagi, S., *Mitochondria in the luminous organs of Watasenia scintillans (Berry)*, Proc. Imp. Acad. (Tokyo) *9* (1933).

Tanner, V. M., *A preliminary study of the genitalia of female Coleoptera*, Trans. Am. Entomol. Soc. Philadelphia *53* (1927).

Tannreuther, G. W., *History of the germ-cells and early embryology of certain Aphids*, Zool. Jahr. Anat. Sec. *24* (1907).

Tarsia in Curia, I., *Nuove osservazioni sull'organo simbiotico di Calandra oryzae Linn.*, Arch. zool. ital. *18* (1933).

———, *La simbiosi ereditaria in Troiza alacris Fbr.*, Arch. zool. ital. *20* (1934).

———, *La simbiosi nei Coccidi*, Attualitá zool. *4* (1938).

Teodoro, G., *Ricerche sull'emolinfa dei Lecanini*, Atti accad. Veneto-Trentino-Istriana *5* (1912).

———, *Osservazioni sulla ecologia delle Cocciniglie, con speciale riguardo alla morfologia ed alla fisiologia di questi insetti*, Redia *11* (1916).

———, *Alcune osservazioni sui saccharomiceti del Lecanium persicae Fabr.*, Redia *13* (1918).

Thiel, M. E., *Die Scyphomedusen des zool. Staatsinstitutes und zool. Museums in Hamburg*, 1. Mitt. Zool. Mus. Hamburg *43* (1927).

Thiel, M. E., *Zur Frage der Ernährung der Steinkorallen und der Bedeutung ihrer Zoochlorellen,* Zool. Anz. *81* (1929).

Thienemann, A., *Das Salzwasser von Oldesloe,* Mitt. geogr. Ges. naturh. Mus. Lübeck, 2. Reihe, H. 31 (1926).

Tiegs, O. W., and Murray, F. V., *The embryonic development of Calandra oryzae,* Quart. J. Mikroscop. Sci. [N. S.] *80* (1938).

———, see also Murray, F. V., and Tiegs, O. W.

Todaro, Fr., *Studi ulteriori sullo sviluppo delle Salpe,* Mem. reale accad. Lincei [4] *1* (1886).

Tóth, L., *Über die frühembryonale Entwicklung der viviparen Aphiden,* Z. Morphol. Ökol. Tiere *27* (1933).

———, *Entwicklungszyklus und Symbiose von Pemphigus spirothecae Pass., (Aphidina),* Z. Morphol. Ökol. Tiere *33* (1937).

———, *The protein metabolism of the Aphids,* Ann. Mus. Hungar. *33* (1940).

———, *On a new category of endosymbiosis. Physiological interpretation of the endosymbiosis of plant-juice sucking insects,* Allat. Közlem. *40* (1943).

———, *Stickstoffassimilation und das symbiotische System bei Kalotermes flavicollis,* Magyar Biol. Kutatóintézet Munkái *16* (1944/45).

———, *The biological fixation of atmospheric nitrogen,* Monographs Nat. Sci. *5* (Hung. Mus. nat. Sci., Budapest 1946).

———, *Nitrogen fixing microorganisms in the alimentary canal of herbivorous farm animals,* Experientia *4* (1948).

———, *Enzymatic breakdown of nitrogen compounds by the nitrogen fixing Bacteria of insects,* Experientia *4* (1948).

———, *The biological fixation of atmospheric nitrogen by means of microorganisms living in symbiosis with animals (insects),* Proc. 6th Inter. Congr. Exptl. Cytology *1* (Stockholm 1949).

———, *Protein metabolism and nitrogen fixation by means of microorganisms living in symbiosis with insects,* Proc. 8th Intern. Congr. Entomol. Stockholm *1950.*

———, *Beiträge zur Frage des Stickstoff-Stoffwechsels der Insekten,* Ann. Agr. Coll. Sweden *17* (1950).

———, *Die Rolle der Mikroorganismen in dem Stickstoff-Stoffwechsel der Insekten,* Zool. Anz. *146* (1951).

———, *The role of nitrogen-active microorganisms in the nitrogen metabolism of insects,* Symposium sur la "Symbiose chez les insectes," Amsterdam 1951, Tijdschr. Entomol. *95* (1952), and Union intern. sc. biol. Ser. B, No. 10.

———, *Nitrogen active microorganisms living in symbiosis with animals and their role in the nitrogen metabolism of the host animal,* Arch. Mikrobiol. *18* (1953).

Tóth, L., and Wolsky, A., *Gaswechsel und respiratorischer Quotient bei den Aphiden,* Zool. Anz. *136* (1941).

Tóth, L., Wolsky, A., and Bátori, M., *Stickstoffbindung aus der Luft bei den Aphiden und bei den Homopteren,* Z. vergleich. Physiol. *30* (1942).

Tóth, L., Wolsky, A., and Bátyka, E., *Stickstoffassimilation aus der Luft bei den Rhynchoten,* Z. vergleich. Physiol. *30* (1944).

Trager, W., *A cellulase from the symbiontic flagellates of termites and of the roach, Cryptocercus punctulatus,* Biochem. J. *26* (1932).

———, *The cultivation of a cellulose digesting flagellate, Trichomonas termopsidis, and of certain other termite protozoa,* Biol. Bull. *66* (1934).

———, *The culture of mosquito larvae free from living microorganisms,* Am. J. Hyg. *22* (1935).

———, *The chemical nature of growth factors required by mosquito larvae,* Proc. 29th Ann. Meeting New Jersey Mosquito Extermination Assoc. *1942.*

Trager, W., *Insect nutrition*, Biol. Rev. *22* (1947).

——, *Biotin and fat-soluble materials with biotin activity in the nutrition of mosquito larvae*, J. Biol. Chem. *176* (1948).

Trager, W., and Lanham, U. N., *Mitochondria or microorganisms?* Science *116* (1952).

Trager, W., and Subbarow, Y., *The chemical nature of growth factors required by mosquito larvae*, Biol. Bull. *75* (1938).

Tremblay, E., *Sviluppo embrionale degli stiletti boccali dei Coccidae Diaspini (Diaspis pentagona Targ.*), Boll. lab. entomol. agr. Portici *16* (1958).

——, *Ovoviviparità, comportamento delle femmine vergini, sesso delle larve e ghiandole cefaliche larvali della Diaspis pentagona Targ.*, Boll. lab. entomol. agr. Portici *16* (1958).

——, *Osservazioni sui due tipi di uova e sulla origine dei micetociti nella Diaspis (= Pseudoaulacaspis) pentagona Targ.*, Boll. lab. entomol. agr. Portici *17* (1959).

——, *Osservazioni sulla simbiosi endocellulare di alcune Aleyrodidae (Bemisia tabaci Gennad., Aleurolobus olivinus Silv., Trialeurodes vaporariorum West.)*, Boll. lab. entomol. agr. Portici *17* (1959).

——, *Beobachtungen über das Schicksal der Richtungskörper und die Endosymbiose bei einigen Cocciden*, XI Intern. Entomologenkongress, Wien 1960.

——, *Osservazioni sul destino dei globuli polari e sulla simbiosi endocellulare di alcuni Coccidi*, Boll. lab. entomol. agr. *18* (1960).

——, *Ciclo cromosomico e simbiosi endocellulare nella Diaspis (= Pseudoaulacaspis) pentagona Targ.*, Boll. lab. entomol. agr. *19* (1960).

——, *Osservazioni sulla cariologia e sulla simbiosi endocellulare di alcuni Coccini (Sphaerolecanium prunasti Fons. ed Eulecanium coryli L.)*, Boll. lab. entomol. agr. *19* (1961).

——, *Ulteriori considerazioni sui due tipi di ovarioli della Diaspis (= Pseudoaulacaspis) pentagona Targ.*, Boll. lab. entomol. agr. *19* (1961).

Trendelenburg, W., *Versuche über den Gaswechsel bei Symbiose zwischen Algen und Tier*, Arch. Anat. Phys., Physiol. Sec. *1909*.

Tretzel, E., *Untersuchungen über die Endosymbiose der Ipiden mit Bakterien* (unpublished).

Treviranus, G. P., *Resultate einiger Untersuchungen über den inneren Bau der Insekten* (Verdauungsorgane bei Cimex rufipes), Ann. Wetterau. Ges. *1* (1809).

Trojan, E., *Bakteroiden, Mitochondrien und Chromidien. Ein Beitrag zur Entwicklung des Bindegewebes*, Arch. mikroskop. Anat., Sec. 1, *93* (1919).

——, *Die geschlossenen Leuchtorgane der Tiefseefische*, 10. Congr. Intern. Zool. Budapest 1929.

Tschirch, A., *Die Wachs-, Harz- und Farbstoffbildung bei den Cocciden. Aufbau und Abbau des Stocklackes*, Chem. Umschau Geb. Fette, Öle, Wachse u. Harze, *45/46* (1922).

——, *Handbuch der Pharmakognosie*, Vol. 3 (C. H. Tauchnitz, Leipzig 1924).

——, *Tier und Pflanze in ihren gegenseitigen Beziehungen zueinander*, Mitt. naturforsch. Ges. Bern *1924*.

Tubeuf, C. von, *Zweigtuberkulose am Ölbaum, Oleander und der Zirbelkiefer*, Naturw. Z. Forst- u. Landwirtsch. *9* (1911).

Uchida, T., *On the white markings of some Rhizostome medusae due to a cartilaginous tissue*, Japan. J. Zool. *1* (1926).

Uichanko, L. B., *Studies on the embryogeny and postnatal development of the Aphididae, with special reference to the history of the "symbiotic organ" or "mycetom,"* Philippine J. Sci. *24* (1924).

Uvarov, B. P., *Insect nutrition and metabolism*, Trans. Entomol. Soc. London *76* (1929).

Vago, C., and Laporte, M., *Microscopie electronique des symbiontes globuleux des Aphides (Hom. Aphidoidea)*, Ann. Soc. entomol. France [N. S.] *1* (1965).

van der Walt, T. P., *The mycetomsymbiont of Lasioderma serricornis*, Antonie van Leeuwenhock *27* (1961).

van t'Hoog, E. G., *Aseptic culture of insects in vitamin research*, Z. Vitaminforsch. *4* (1935); *5* (1936).

van Trigt, H., *A contribution to the physiology of the fresh water sponges (Spongillidae)*, Proc. Acad. Sci. Amsterdam *20* (1919).

——, *A contribution to the physiology of the fresh water sponges (Spongillidae)* (Dissertation, Leiden 1918); also in Biol. Zentr. *40* (1920).

Vaternahm, Th., *Zur Ernährung und Verdauung unserer einheimischen Geotrupes-Arten*, Z. wiss. Insektenbiol. *19* (1924).

Vaughan, T. W., *Corals and formation of coral-reefs*, Ann. Rept. Smithsonian Inst. Washington *1919*.

Vayssière, P., *La Symbiose des insectes avec les microorganismes*, Compt. rend. acad. agr. France *1952*.

Verona, E., and Baldacci, O., *Isolamento di schizomiceti cellulosolitici (Cytophaga), attinomiceti (Actinomyces), eumiceti dall'intestino delle termiti, e ricerche sulla attività cellulolitica degli attinomiceti*, Atti Ist. Botan. Univ. Pavia [4] *11* (1939).

Verwey, J., *Coral reef studies*, 2: *The depth of coral reefs in relation to their oxygen consumption and the penetration of light in the water*, Treubia *13* (1931).

Vinciguerra, D., *Del genere Hymenocephalus (H. italicus Gigl.)*, Ann. mus. civ. storia nat. Genova *56* (1932).

Virtanen, A. J., and Laine, T., *Investigations on the root nodule bacteria of leguminous plants*, 22: *The excretion products of root nodules. The mechanism of N-fixation*, Biochem. J. *33* (1939).

Vogel, H., *Schweinemastversuche mit dem Futterzusatz "Astoral" (Hefe-Wirkstoff-Konzentrat)*, Züchtungskunde *25* (1954).

Vogel, R., *Über die Topographie der Leuchtorgane von Phausis spendidula Leconte*, Biol. Zentr. *42* (1922).

——, *Lampyrinae*, in: *Biologie der Tiere Deutschlands* (Borntraeger, Berlin 1927).

Vollbrechtshausen, R., *Bakterien als Symbionten, Synöken und Parasiten bei Phyllodromia germanica*, Z. Parasitenk. *15* (1953).

Vonwiller, P., *Anatomische Bemerkungen über den Bau der Leuchtorgane von Lampyris splendidula*, Festschr. für Zschokke, Basel *1920*.

Vos, H., *De invloed van Pseudococcus citri (Risso) Fern. op de plant*. Dissertation Utrecht, (Baarn, 1930).

Vouk, V., *Grundriss zu einer physiologischen Auffassung der Symbiose*, Planta *2* (1926).

Wagner, W., *Dynamische Taxionomie*, Ber. G. Wandervers. deut. Entomol. Berlin *1961*.

Wagner-Jevseenko, O., *Fortpflanzung bei Ornithodorus moubata und genitale Übertragung von Borrelia duttoni*, Acta Trop. *15* (1958).

Walczuch, A., *Studien an Coccidensymbionten*, Z. Morphol. Ökol. Tiere *25* (1932).

Wallin, J. E., *Symbioticism and the Origin of Species* (Baillière, Tindall and Cox, London 1927). (Contains the publications of the author since 1922).

Weber, H., *Lebensweise und Umweltbeziehung von Trialeurodes vaporariorum Westw.*, Z. Morphol. Ökol. Tiere *23* (1931).

——, *Die postembryonale Entwicklung der Aleurodinen*, Z. Morphol. Ökol. Tiere *29* (1935).

——, *Der Bau der Imago der Aleurodinen*, Zoologica *33*, H. 89 (1935).

——, *Beiträge zur Kenntnis der Überordnung Psocoidea*, Biolog. Zentr. *59* (1939).

Weber, M. and A., *Quelques nouveaux cas de symbiose*, Zool. Ergeb. Reise Niederl. Ostindien (Leiden 1890/91).

Weber-van Bosse, A., *Sur deux nouveaux cas de symbiose entre algues et éponges*, Ann. jard. botan. Buitenzorg, 3, Suppl. *1910*.

Webster, F. M., and Phillips, W. J., *The spring grain Aphis or "greenbug,"* U. S. Dept. Agr. Bur. Entomol. Bull. *110* (Washington 1912).

Weissenberg, R., *Zur Wirtsgewebsableitung des Plasmakörpers der Glugea anomala-Cysten*, Arch. Protistenk. *42* (1921).

——, *Fremddienliche Reaktionen beim intrazellulären Parasitismus, ein Beitrag zur Kenntnis gallenähnlicher Bildungen im Tierkörper*, Verhandl. deut. zool. Ges. *1922.*

Welsh, A., *Oxygen production by Zooxanthellae in a Bermudian Turbellarian*, Biol. Bull. *70* (1936).

Weltner, W., *Spongillidenstudien*, 1: Arch. Naturgeschichte *59* (1893).

——, *Zur Biologie von Ephydatia fluviatilis und die Bedeutung der Amöbocyten für die Spongilliden*, Arch. Naturgeschichte *73* (1907).

Werner, E., *Die Ernährung der Larve von Potosia cuprea Fbr. Ein Beitrag zum Problem der Zelluloseverdauung bei Insektenlarven*, Z. Morphol. Ökol. Tiere *6* (1926).

——, *Der Erreger der Zelluloseverdauung bei der Rosenkäferlarve (Potosia cuprea Fbr.), Bacillus cellulosam fermentans n. sp.*, Zentr. Bakteriol. *67* (1926).

Wesenberg-Lund, C., *Beiträge zur Kenntnis des Lebenscyklus der Zoochlorellen*, Intern. Rev. Hydrobiol. Planktonkde. *2* (1909).

Weurman, C., *Investigations concerning the symbiosis of bacteria in Triatoma infestans (Klug)*, Antonie van Leeuwenhock *11* (1946).

Wheeler, W. M., *The embryology of Blatta germanica and Doryphora decemlineata*, J. Morphol. *3* (1889).

Wheeler, W. X., and Williams, F. H., *The luminous organ of the New-Zealand glow worm*, Psyche *22* (1915).

Wiedemann, J. F., *Die Zelluloseverdauung bei Lamellicornierlarven*, Z. Morphol. Ökol. Tiere *19* (1930).

Wigglesworth, V. B., *Digestion in the tsetse-fly: a study of structure and function*, Parasitology *21* (1929).

——, *Symbiot bacteria in a blood sucking insect, Rhodnius prolixus Stål.*, Parasitology *28* (1936).

——, *The fate of haemoglobin in Rhodinus prolixus and other blood-sucking arthropods*, Proc. Roy. Soc. (London) [B] *131* (1943).

——, *Symbiosis in blood-sucking insects*, Symposium sur la "Symbiose chez les insectes," Amsterdam *1951*, Tijdschr. Entomol. *95* (1952), and Union intern. sc. biol. Sér. B, No. 10.

——, *The Principles of Insect Physiology*, 5th ed. (Methuen and Co., London 1953).

——, *Physiologie der Insekten*, Basel-Stuttgart *1955.*

Will, L., *Entwicklungsgeschichte der viviparen Aphiden*, Zool. Jahr. Anat. Sec. *3* (1889).

Willem, V., and Minne, A., *Recherches sur l'excrétion chez quelques annélides*, Mém. cour. Mém. sav. étrang. acad. roy. belgique, C. Sc. *58* (1899).

Williams, R. J., *Water-soluble vitamins*, Ann. Rev. Biochem. *12* (1943).

——, *The Biochemistry of the B-Vitamins* (Reinhold Publishing Corporation, New York 1950).

Wilson, E. B., *The heliotropism of Hydra*, Am. Naturalist *25* (1891).

Wilson, E. E., *The olive knot disease: its inception, development and control*, Hilgardia *9* (1935).

Wilson, H. V., *On the development of Maeandrina areolata*, J. Morphol. *2* (1888).

Wilson, S. E., *Changes in the cell contents of wood (xylem-parenchyma) and their relationship to the respiration of wood and its resistance to Lyctus attack and to fungal invasion*, Ann. Appl. Biol. *20* (1933).

Windisch, S., *Beiträge zur Biologie und Systematik der Kahmhefen*, Arch. Mikrobiol. *21* (1954).

Winter, F., *Zur Kenntnis der Thalamophoren*, 1: *Untersuchung über Peneroplis pertusus Forsk.*, Arch. Protistenk. *10* (1907).

Withman, C. O., *Description of Clepsina plana*, J. Morphol. *4* (1891).

Witlaczil, E., *Zur Anatomie der Aphiden*, Arb. Zool. Inst. Wien *4* (1882).

Witlaczil, E., *Entwicklungsgeschichte der Aphiden*, Z. wiss. Zool. *40* (1884).

———, *Die Anatomie der Psylliden*, Z. wiss. Zool. *42* (1885).

Wöhler, *Über die O-Entwicklung aus dem organischen Absatz eines Solwassers*, Ann. Chem. Pharm. *15* (1843).

Wolf, J., *Contribution à la localisation des bactéroïdes dans les corps adipeux des blattes (Periplaneta orientalis)*, Compt. rend. soc. biol. *91* (1924).

———, *Contribution à la morphologie des bactéroïdes dans les Blattes (Periplaneta orientalis)*, Compt. rend. soc. biol. *91* (1924).

Wollman, E., *Observations sur une lignée aseptique de blattes (Blattella germanica) datant de cinq ans*, Compt. rend. soc. biol. *95* (1926).

Woltereck, R., *Über die Entwicklung von Velella aus einer in der Tiefe vorkommenden Larve*, Zool. Jahr. Suppl. *7* (1904).

Woodward, T. E., *The internal male reproductive organs in the genus Nabis Latreille*, Proc. Roy. Entomol. Soc. London [A] *24* (1949).

Wülker, G., *Über das Auftreten rudimentärer akzessorischer Nidamentaldrüsen bei männlichen Cephalopoden*, Zoologica *67* (1913).

Wunder, W., *Untersuchungen über Pigmentierung und Encystierung von Cercarien*, Z. Morphol. Ökol. Tiere *25* (1932).

Yasaki, Y., *On the nature of the luminescence of the Knight-fish (Monocentris japonicus Houttuyn)*, J. Exptl. Zool. *50* (1928).

Yasaki, Y., and Haneda, Y., *Über die Leuchtphänomene bei Tiefseefischen aus der Familie der Macruridae* (in Japanese). Ōyō-Dōbutsugaku-Zasshi [J. Appl. Zool.] *7* (1935).

———, *Über einen neuen Typus von Leuchtorgan in Fischen*, Proc. Imp. Acad. (Tokyo) *12* (1936).

Yonge, C. M., *The significance of the relationship between corals and zooxanthellae*, Nature *128* (1931).

———, *Origin and nature of the association between invertebrates and unicellular algae*, Nature *134* (1934).

———, *Mode of life, feeding, digestion and symbiosis with Zooxanthellae in the Tridacnidae*, Sci. Rept. Great Barriere Reef Exped. *1*, No. 11 (1936).

———, *The biology of reef-building corals*, Sci. Rept. Great Barriere Reef Exped. *1*, No. 13 (1940).

———, *Experimental analysis of the association between invertebrates and unicellular algae*, Biol. Rev. Cambridge *19* (1944).

Yonge, C. M., and Nicholas, H. M., *Structure and function of the gut and symbiosis with Zooxanthellae in Tridacnia crispata (Oerst.) Bgh.*, Papers Tortugas Lab. *32* (1940).

Yonge, C. M., and Nicholls, A. G., *The effect of starvation in light and in darkness on the relationship between corals and Zooxanthellae*, Sci. Rept. Great Barriere Reef Exped. *1*, No. 7 (1931).

———, *The structure, distribution and physiology of the Zooxanthellae*, Sci. Rept. Great Barriere Reef Exped. *1*, No. 6 (1931).

Yonge, C. M., Yonge, M. J., and Nicholls, A. G., *The relation between respiration in corals and the production of oxygen by their zooxanthellae*, Sci. Rept. Great Barriere Reef Exped. *1*, No. 8 (1932).

Yoshizawa, S., Dobutsugaku Zasshi *93* (1916).

Zacharias, A., *Untersuchungen über die intrazellulare Symbiose bei den Pupiparen*, Z. Morphol. Ökol. Tiere *10* (1928).

Zelinka, K., *Monographie der Echinoderen* (W. Engelmann, Leipzig 1928).

Zimmerman, E. C., *Insects of Hawaii*, Vol. 5, *Homoptera: Sternorhyncha*, Univ. of Hawaii Press, Honolulu 1948.

Zirpolo, G., *I batteri fotogeni degli organi luminosi di Sepiola intermedia*, Boll. soc. nat. Napoli *30* (1918).

————, *Micrococcus pierantonii—nuova specie di batterio fotogeno dell'organo luminoso di Rondeletia minor Naef*, Boll. soc. nat. Napoli *31* (1918).

————, *Studi sulla bioluminescenza batterica*, 1: Riv. biol. *2* (1920); 2: Boll. soc. nat. Napoli *32* (1920); 3: Boll. soc. nat. Napoli *33* (1920); 4: Riv. sci. nat. Pavia *72* (1921); 5: Boll soc. nat. Napoli *34* (1922); 6: Riv. sci. nat. Pavia *73* (1922); 7: Boll. soc. nat. Napoli *35* (1923); 8: Boll. soc. nat. Napoli *38* (1927); 9: Boll. soc. nat. Napoli *41* (1929); 10: Boll. soc. nat. Napoli *43* (1931); 11: Boll. soc. nat. Napoli *44* (1932); 12: Boll. soc. nat. Napoli *44* (1932); 13: Boll. zool. *9* (1938).

————, *Sulla presenza di organi simbiotici nell'Hirudo medicinalis L.*, Boll. soc. nat. Napoli *34* (1922).

————, *Ricerche sulla simbiosi fra Zooxantelle e Phyllirrhoe bucephala Per. et Less.*, Boll. soc. nat. Napoli *35* (1923).

————, *Ancora sui batteri fotogeni (Risposta a S. Mortara)*, Riv. biol. *6* (1924).

————, *Ancora sui batteri luminosi (Risposta a V. Puntoni e S. Skowron)*, Riv. biol. *8* (1926).

————, *La polemica sui batteri luminosi (Risposta al Prof. Puntoni)*, Monit. zool. ital. *38* (1927).

————, *Ricerche criobiologiche sui batteri luminosi dei Cephalopodi*, Arch. zool. ital. *18* (1933).

Author index

Abe, 56, 589
Abrams, 827
Adlerz, 23
Aeschlimann, 50, 452
Agassiz, 5
Allen, 117
Almquist, 826
Altenburg, 74
Altmann, 69
Anadón, 69
Anigstein, 50, 68
Archer, 5, 6, 19
Arkwright, 46, 50, 68, 478, 479, 480
Armand, 826
Aschner, 29, 50, 59, 60, 461, 462, 463, 465, 467, 469, 470, 625, 765, 766, 770, 771, 775, 779, 780, 788
Atkin, 46, 68, 478, 479, 480

Bacot, 46, 50, 52, 68, 478, 479, 480
Baines, 51, 774, 775
Balachowsky, 233, 254
Balbiani, 6, 24, 25, 297, 310
Baldacci, 101, 104, 815
Balfour, 44, 68
Bandier, 63, 782, 828
Barbato, 61, 516, 809
Barber, 776
Barbieri, 612
Barjac, de, 104
Batey, 37, 210, 224, 225
Bátori, 65
Batra, 40, 82
Bátyka, 65, 800
Baudisch, 42, 48, 59, 190, 421, 482, 500, 502, 517, 671, 771
Bauer, 827
Baumann, 807
Baumgärtel, 824
Béchamps, 69
Becker, 297, 301, 308, 332, 803, 815, 819

Beckwith, 104
Beeson, 41, 79
Begen, 100
Behrenz, 786
Beijerinck, 7, 577
Bender, 693
Benedek, 31, 236
Bennett, 31, 33, 240, 241
Berg, 200, 693
Berkeley, 9, 13
Berlese, 26, 235
Bernardini, 809
Beutler, 21
Bewig, 48, 51, 60, 470, 471, 472, 482, 504, 506, 696, 717, 771, 774, 775, 780
Biedermann, 97
Bierry, 70
Bigelow, 9
Blanc, 45
Blewett, 62, 783, 784, 790
Blochmann, 23, 24, 46, 516, 526, 531
Block, 61, 516, 813
Bode, 526
Boerhave, 45
Boerke, 37, 210, 224, 225
Bohn, 18
Bonde, 108
Bonhag, 37, 224
Bonnemaison, 794
Boratynski, 32, 295
Borchsenius, 760, 761
Borghese, 42, 516
Boschma, 8, 10, 20
Bournier, 41, 431, 432
Brain, 31
Brandes, 13
Brandt, 5, 7, 11, 14, 17, 19
Brauer, 605
Braun, 200, 693
Brecher, 51, 60, 224, 470, 772, 773
Breest, 29, 31, 33, 234, 239, 240, 341

Index of animal and plant names